Formulas from Geometry

Formulas for Area (A), Perimeter (P), Circumference (C), and Volume (V):

Square

$A = s^2$

$P = 4s$

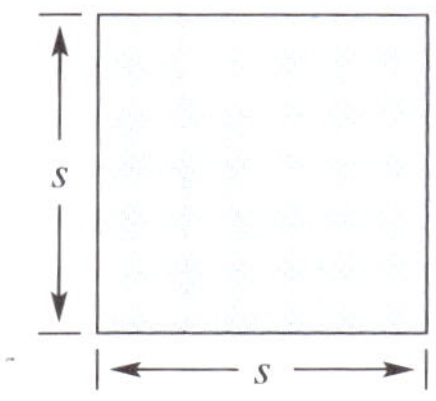

Rectangle

$A = lw$

$P = 2l + 2w$

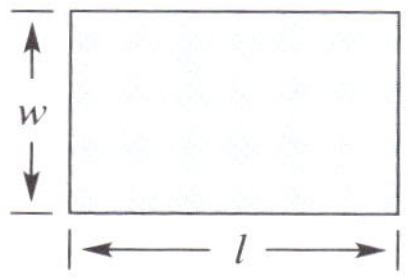

Circle

$A = \pi r^2$

$C = 2\pi r$

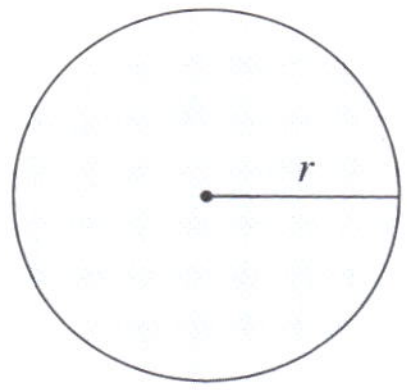

Triangle

$A = \frac{1}{2}bh$

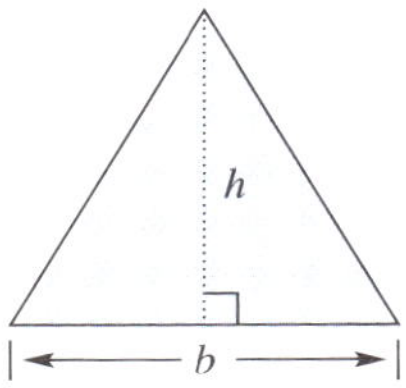

Trapezoid

$A = \frac{1}{2}h(b_1 + b_2)$

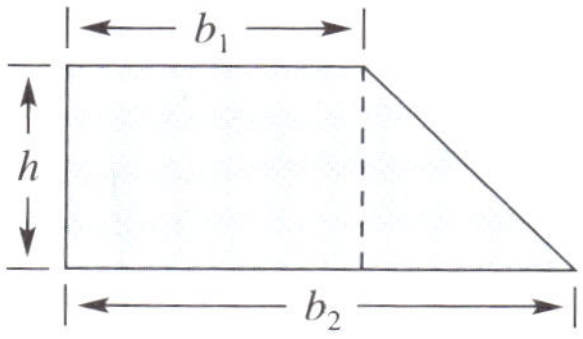

Parallelogram

$A = bh$

$P = 2a + 2b$

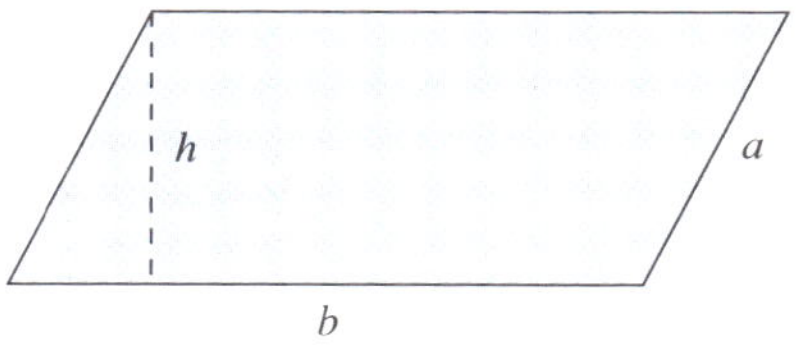

Pythagorean Theorem

$a^2 + b^2 = c^2$

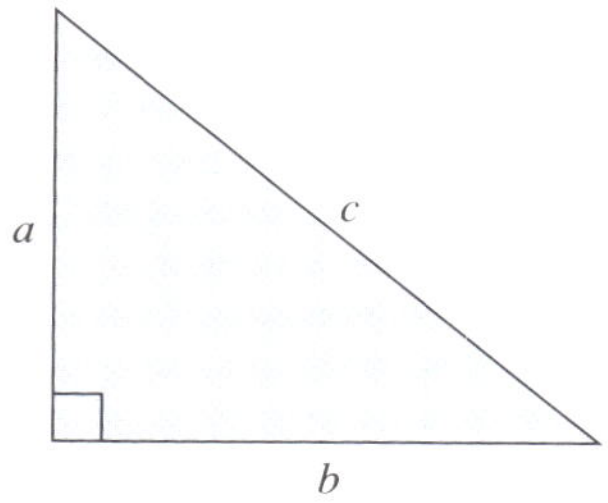

Cube

$V = s^3$

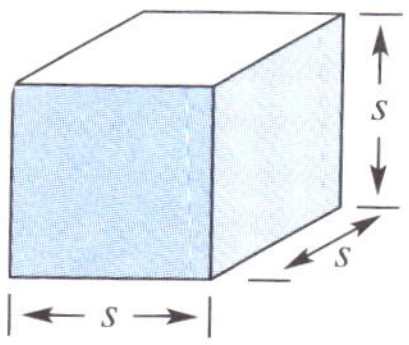

Rectangular Solid

$V = lwh$

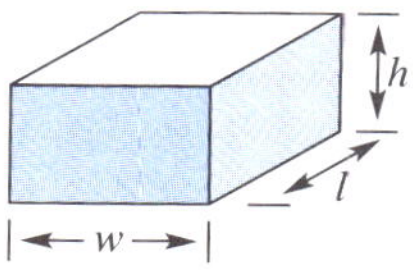

Circular Cylinder

$V = \pi r^2 h$

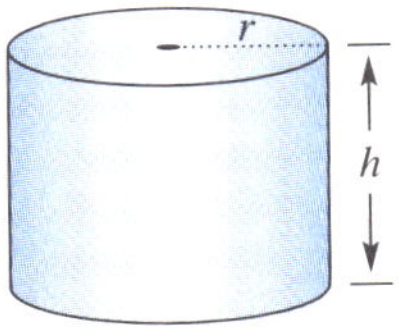

Sphere

$V = \frac{4}{3}\pi r^3$

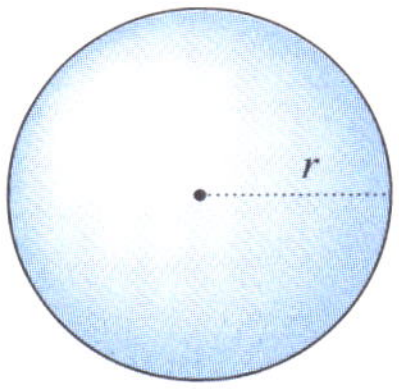

Could You Use Some Help In Your Algebra Course?

You can find it in both the ***Study Guide*** and the ***Student Solutions Guide*** accompanying this text.

The ***Study Guide*** (18769-0) includes:

- ▲ a review of key concepts for each section
- ▲ examples for review, with detailed solutions
- ▲ additional exercises to reinforce each topic
- ▲ cumulative chapter tests, with answers

In the ***Student Solutions Guide*** (18771-2), you'll find:

- ▲ detailed answers to all odd-numbered text exercises
- ▲ step-by-step solutions to Chapter Test questions from the text
- ▲ computer-generated graphics to clarify the presentation

You can get both the ***Study Guide*** and the ***Student Solutions Guide*** at your bookstore. It's an effective way to give yourself the advantage in algebra.

Many other learning aids are available with this text; you'll find them listed in the preface. If you have any questions regarding these extraordinary supplements, check with your textbook department.

Intermediate Algebra

ROLAND E. LARSON   ROBERT P. HOSTETLER
The Pennsylvania State University
The Behrend College

with the assistance of DAVID E. HEYD
The Pennsylvania State University
The Behrend College

D. C. Heath and Company
Lexington, Massachusetts   Toronto

Address editorial correspondence to:

D. C. Heath
125 Spring Street
Lexington, MA 02173

Cover and title page: Paul Klee (1879–1940). *Polyphon gefasstes Weiss, 1930.140.* Paul Klee Foundation, Museum of Fine Arts Berne. © 1991, Copyright by COSMOPRESS, Geneva, Switzerland.

Paul Klee created *Polyphonically Enclosed White* by layering different-sized blocks of color on top of one another.

Technical art: Folium, Inc.; Illustrious, Inc.

Photo credits: p. 1, p. 61 Charles Ficke; p. 135 Light Source/Paul E. Johnson; p. 196 Charles Ficke; p. 255 Tedd Benson/Benson Woodworking Co. Inc., Alstead, NH; p. 312 H. Armstrong Roberts/R. J. Bennett; p. 366 Tedd Benson/Benson Woodworking Co. Inc., Alstead, NH; p. 470 Photo Researchers/Patricia Agne; p. 508 Alan Oddie/PhotoEdit; p. 531 Timberpeg, Inc., Lebanon, NH; p. 608, p. 666 H. Armstrong Roberts.

Published simultaneously in Canada.

Printed in the United States of America.

International Standard Book Number: 0-669-18767-4

Library of Congress Catalog Card Number: 91-70669

10 9 8 7 6

PREFACE

Intermediate Algebra has two basic goals: First, to help students develop proficiency in algebra; and second, to show students how algebra can be used as a modeling language for real-life problems. To support these two functions, the text has several key pedagogical features.

- **Problem-Solving Process** One general approach to problem-solving is stressed throughout. Students are taught to form a verbal model as the first step in solving any application.
- **Applications** Numerous applications are integrated throughout every section of the text both as solved examples and as exercises. As a result, students are constantly using and reviewing their problem-solving skills. The text applications cover a wide range of relevant topics, and many use real data.
- **Discussion Problems** Appearing at the end of each section, the discussion problems offer students the opportunity to think, talk, and write about mathematics in a different way. Students are encouraged to draw new conclusions about the concepts presented and to develop a sense of how each topic studied fits into the whole concept of algebra.
- **Exercise Sets** The exercise sets contain numerous computational and applied problems dealing with a wide range of topics. Anticipating a student's needs, these problems are carefully graded to increase in difficulty as the student's problem-solving skills develop. Each pair of consecutive problems is similar, with the answer to the odd-numbered one given in the back of the text. Exercise sets appear at the end of each text section; an additional set of review exercises is given at the end of each chapter. Other chapter review features are summaries, chapter tests, and cumulative tests. The opportunity to use calculators—to show pattern, to experiment, to calculate, or to create graphic models—is available with selected topics.

These and other features of the text are described in greater detail on the following pages.

Features of the Text

The features of this text are designed to help students improve their skills and acquire an understanding of mathematical concepts. The functional use of four colors strengthens the text as a pedagogical tool.

Chapter Opener

A list of the topics to be covered and a brief chapter overview provide a survey of the contents of each chapter, showing students how the topics fit into the overall development of algebra.

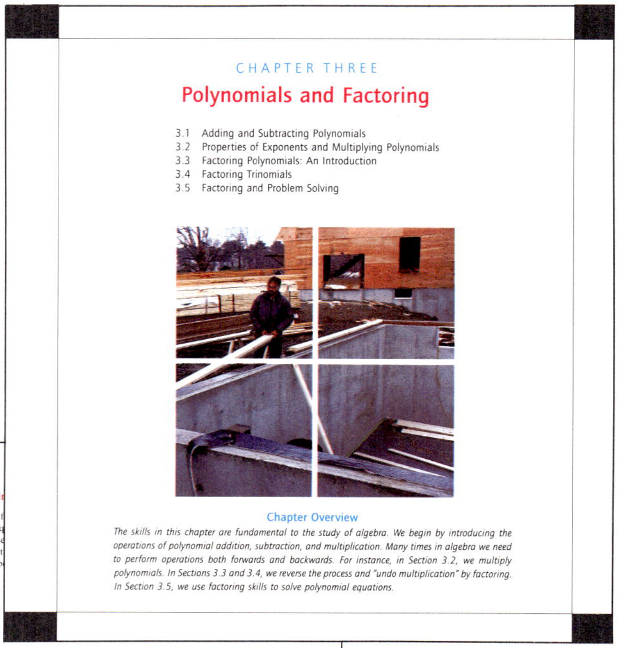
CHAPTER THREE

Polynomials and Factoring

3.1 Adding and Subtracting Polynomials
3.2 Properties of Exponents and Multiplying Polynomials
3.3 Factoring Polynomials: An Introduction
3.4 Factoring Trinomials
3.5 Factoring and Problem Solving

Chapter Overview

The skills in this chapter are fundamental to the study of algebra. We begin by introducing the operations of polynomial addition, subtraction, and multiplication. Many times in algebra we need to perform operations both forwards and backwards. For instance, in Section 3.2, we multiply polynomials. In Sections 3.3 and 3.4, we reverse the process and "undo multiplication" by factoring. In Section 3.5, we use factoring skills to solve polynomial equations.

Section Topics

Each section begins with a list of important topics that are covered in the section. These topics are also the subsection titles and can be used for easy reference and review by students.

Definitions and Rules

All of the important rules, formulas, and definitions are boxed for emphasis. Each is also titled for easy reference.

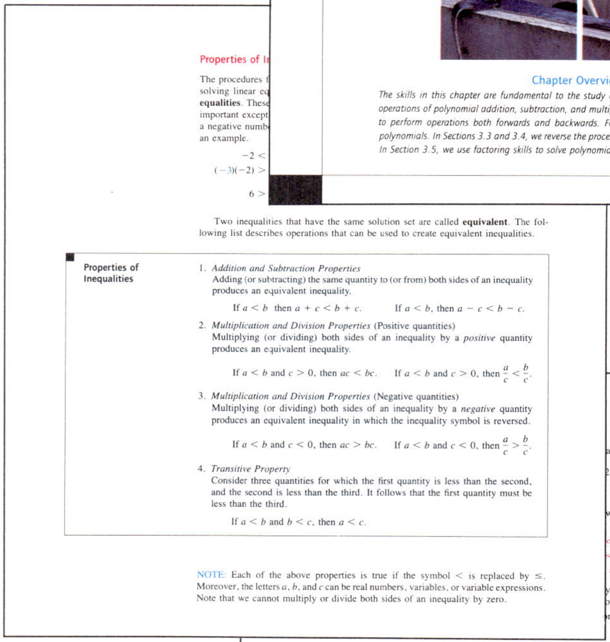
Two inequalities that have the same solution set are called **equivalent**. The following list describes operations that can be used to create equivalent inequalities.

Properties of Inequalities

1. *Addition and Subtraction Properties*
Adding (or subtracting) the same quantity to (or from) both sides of an inequality produces an equivalent inequality.
If $a < b$ then $a + c < b + c$. If $a < b$, then $a - c < b - c$.
2. *Multiplication and Division Properties* (Positive quantities)
Multiplying (or dividing) both sides of an inequality by a *positive* quantity produces an equivalent inequality.
If $a < b$ and $c > 0$, then $ac < bc$. If $a < b$ and $c > 0$, then $\frac{a}{c} < \frac{b}{c}$.
3. *Multiplication and Division Properties* (Negative quantities)
Multiplying (or dividing) both sides of an inequality by a *negative* quantity produces an equivalent inequality in which the inequality symbol is reversed.
If $a < b$ and $c < 0$, then $ac > bc$. If $a < b$ and $c < 0$, then $\frac{a}{c} > \frac{b}{c}$.
4. *Transitive Property*
Consider three quantities for which the first quantity is less than the second, and the second is less than the third. It follows that the first quantity must be less than the third.
If $a < b$ and $b < c$, then $a < c$.

NOTE: Each of the above properties is true if the symbol $<$ is replaced by $\leq$. Moreover, the letters a, b, and c can be real numbers, variables, or variable expressions. Note that we cannot multiply or divide both sides of an inequality by zero.

Notes

Notes appear after definitions and examples. Anticipating students' needs, they give additional insight, point out common errors, and describe generalizations.

Math Matters

Each chapter contains a Math Matters box that engages student interest by discussing a historical note or mathematical problem. Some of the Math Matters features pose questions—in such cases, the answers are given in the back of the text.

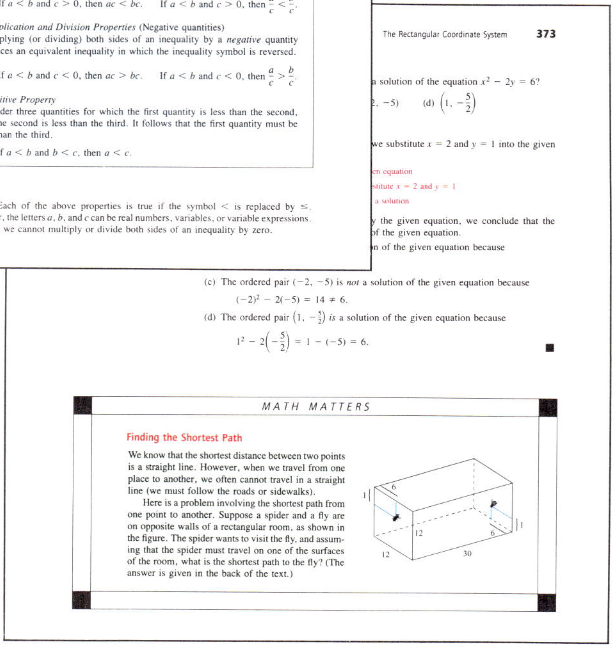
The Rectangular Coordinate System 373

(c) The ordered pair $(-2, -5)$ is *not* a solution of the given equation because $(-2)^2 - 2(-5) = 14 \neq 6$.
(d) The ordered pair $\left(1, -\frac{5}{2}\right)$ *is* a solution of the given equation because $1^2 - 2\left(-\frac{5}{2}\right) = 1 - (-5) = 6$.

MATH MATTERS

Finding the Shortest Path

We know that the shortest distance between two points is a straight line. However, when we travel from one place to another, we often cannot travel in a straight line (we must follow the roads or sidewalks).

Here is a problem involving the shortest path from one point to another. Suppose a spider and a fly are on opposite walls of a rectangular room, as shown in the figure. The spider wants to visit the fly, and assuming that the spider must travel on one of the surfaces of the room, what is the shortest path to the fly? (The answer is given in the back of the text.)

Problem-Solving Process

Students are taught the following strategies—in keeping with the spirit of NCTM standards—for solving applied problems. (1) Construct a verbal model; (2) Label variable and constant terms; (3) Construct an algebraic model; (4) Using the model, solve the problem; and (5) Check the answer in the original statement of the problem.

Examples

Each of the more than 775 text examples was carefully chosen to illustrate a particular concept or problem-solving technique and to enhance students' understanding. They are titled for easy reference.

$$\underbrace{\begin{pmatrix}\$10 \text{ per}\\ \text{pound}\end{pmatrix}\begin{pmatrix}12\\ \text{pounds}\end{pmatrix}}_{\$10 \text{ seed}} + \underbrace{\begin{pmatrix}\$15 \text{ per}\\ \text{pound}\end{pmatrix}\begin{pmatrix}8\\ \text{pounds}\end{pmatrix}}_{\$15 \text{ seed}} = \underbrace{\begin{pmatrix}\$12 \text{ per}\\ \text{pound}\end{pmatrix}\begin{pmatrix}20\\ \text{pounds}\end{pmatrix}}_{\$12 \text{ seed}}$$

$$\$120 + \$120 = \$240$$

EXAMPLE 6 ■ **A Solution Mixture Problem**

A pharmacist needs to strengthen a 20% alcohol solution so that it contains 36% alcohol. How much pure alcohol should be added to 240 milliliters of the 20% solution? (See Figure 2.1.)

FIGURE 2.1

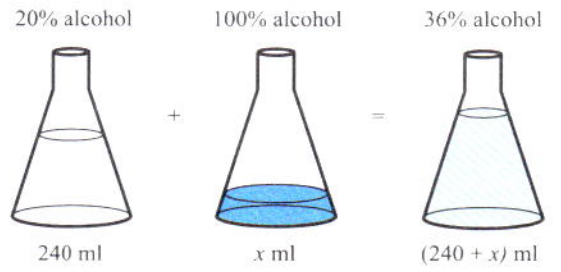

Solution

Verbal model: [Amount of alcohol in original solution] + [Amount of alcohol in pure solution] = [Amount of alcohol in final solution]

Labels: Original solution: Percent alcohol = 0.20, Amount = 0.20(240)
Pure alcohol: Percent alcohol = 1.00, Amount = 1.00(x)
Final solution: Percent alcohol = 0.36, Amount = 0.36(240 + x)

Equation:

$$0.20(240) + 1.00(x) = 0.36(240 + x)$$
$$48 + x = 86.4 + 0.36x$$
$$0.64x = 38.4$$
$$x = \frac{38.4}{0.64} = 60 \text{ ml}$$

Thus, the pharmacist should add 60 milliliters of pure alcohol to the original solution. Check this solution in the original statement of the problem. ■

Rate Problems

Time-dependent problems such as those involving distance can be classified as **rate problems**. They fit the verbal model

[Distance] = [Rate] · [Time].

EXAMPLE 6 ■ **Converting from Scientific Notat**

Convert each of the following nur

(a) 2.486×10^2 (b) $1.81 \times$

Solution

(a) For this number, the exponent
decimal notation, we move th

$2.486 \times 10^2 = 248.6$

Two places

(b) For this number, the exponent of 10 is a negative 6. Thus, to write the number in decimal notation, we move the decimal point six places, as follows.

$1.81 \times 10^{-6} = 0.00000181$

Six places ■

NOTE: To convert decimal numbers to scientific notation or conversely, we make the following associations.

Numbers ≥ 10	⇔	Positive powers of 10
Numbers < 1	⇔	Negative powers of 10
Numbers between 1 and 10	⇔	Zero power of 10

Most scientific calculators automatically switch to scientific notation when they are showing large (or small) numbers that exceed the display range. Try multiplying $86{,}500{,}000 \times 6{,}000$. If your calculator follows standard conventions, its display should be

[5.19 11] or [5.19 E 11].

This means that $c = 5.19$ and the exponent of 10 is $n = 11$, which implies that the number is 5.19×10^{11}.

To *enter* numbers in scientific notation, your calculator should have an exponential entry key labeled [EE] or [EXP]. If you were to perform the preceding multiplication using scientific notation, you could begin by writing

$$86{,}500{,}000 \times 6{,}000 = (8.65 \times 10^7)(6.0 \times 10^3)$$

and then entering the following.*

8.65 [EE] 7 [×] 6 [EE] 3 [=]   Scientific
8.65 [EE] 7 [×] 6 [EE] 3 [ENTER]   Graphing

*The graphing calculator keystrokes given in this text correspond to the *TI-81* graphing calculator from *Texas Instruments*. For other graphing calculators, the keystrokes may differ.

Calculators

Two types of calculators are discussed: scientific and graphing. Sample keystrokes are given for both. The graphing capability of graphing calculators and computer graphing software is investigated; however, coverage of this material is optional.

Discussion Problems

Discussion problems appear at the end of each section. They encourage students to think, talk, and write about mathematics, individually or in groups.

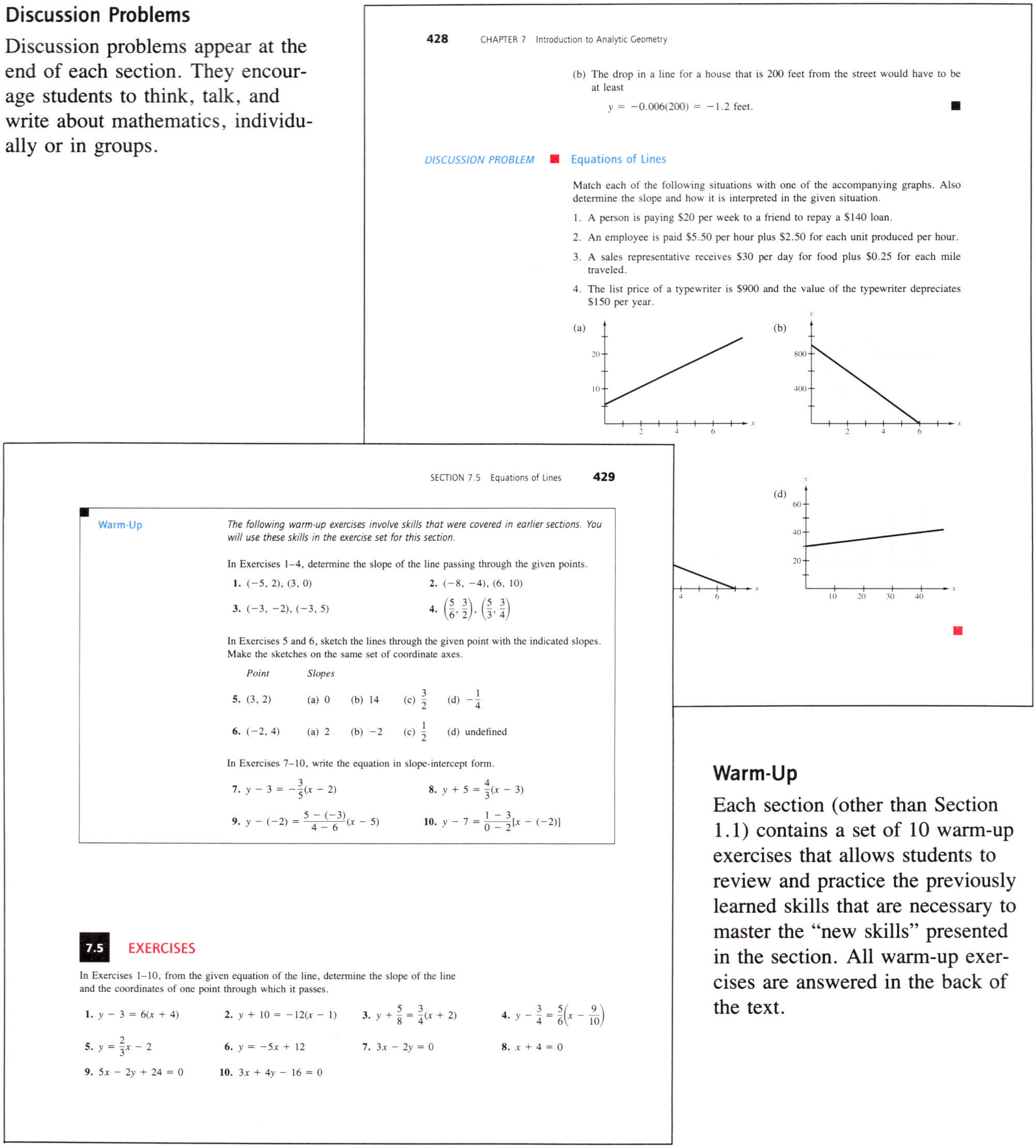

428 CHAPTER 7 Introduction to Analytic Geometry

(b) The drop in a line for a house that is 200 feet from the street would have to be at least

$$y = -0.006(200) = -1.2 \text{ feet.}$$

DISCUSSION PROBLEM ■ Equations of Lines

Match each of the following situations with one of the accompanying graphs. Also determine the slope and how it is interpreted in the given situation.

1. A person is paying $20 per week to a friend to repay a $140 loan.
2. An employee is paid $5.50 per hour plus $2.50 for each unit produced per hour.
3. A sales representative receives $30 per day for food plus $0.25 for each mile traveled.
4. The list price of a typewriter is $900 and the value of the typewriter depreciates $150 per year.

SECTION 7.5 Equations of Lines 429

Warm-Up

The following warm-up exercises involve skills that were covered in earlier sections. You will use these skills in the exercise set for this section.

In Exercises 1–4, determine the slope of the line passing through the given points.

1. $(-5, 2), (3, 0)$ **2.** $(-8, -4), (6, 10)$

3. $(-3, -2), (-3, 5)$ **4.** $\left(\frac{5}{6}, \frac{3}{2}\right), \left(\frac{5}{3}, \frac{3}{4}\right)$

In Exercises 5 and 6, sketch the lines through the given point with the indicated slopes. Make the sketches on the same set of coordinate axes.

Point	*Slopes*			
5. $(3, 2)$	(a) 0	(b) 14	(c) $\frac{3}{2}$	(d) $-\frac{1}{4}$
6. $(-2, 4)$	(a) 2	(b) -2	(c) $\frac{1}{2}$	(d) undefined

In Exercises 7–10, write the equation in slope-intercept form.

7. $y - 3 = -\frac{3}{5}(x - 2)$ **8.** $y + 5 = \frac{4}{3}(x - 3)$

9. $y - (-2) = \frac{5 - (-3)}{4 - 6}(x - 5)$ **10.** $y - 7 = \frac{1 - 3}{0 - 2}[x - (-2)]$

7.5 EXERCISES

In Exercises 1–10, from the given equation of the line, determine the slope of the line and the coordinates of one point through which it passes.

1. $y - 3 = 6(x + 4)$ **2.** $y + 10 = -12(x - 1)$ **3.** $y + \frac{5}{8} = \frac{3}{4}(x + 2)$ **4.** $y - \frac{3}{4} = \frac{5}{6}\left(x - \frac{9}{10}\right)$

5. $y = \frac{2}{3}x - 2$ **6.** $y = -5x + 12$ **7.** $3x - 2y = 0$ **8.** $x + 4 = 0$

9. $5x - 2y + 24 = 0$ **10.** $3x + 4y - 16 = 0$

Warm-Up

Each section (other than Section 1.1) contains a set of 10 warm-up exercises that allows students to review and practice the previously learned skills that are necessary to master the "new skills" presented in the section. All warm-up exercises are answered in the back of the text.

Exercises

The nearly 6,000 exercises include both computational and applied problems covering a wide range of topics. These are designed to build competence, skill, and understanding. Each exercise set is graded in difficulty to allow students to gain confidence as they progress. Answers to odd-numbered exercises are in the back of the text.

480 CHAPTER 8 Functions and Mathematical Models

6. *State and Local Taxes* Use the accompanying graph to determine the second coordinate in the ordered pair, where the first coordinate (input) is the year and the second coordinate (output) is the average number of minutes per eight-hour workday an American works to earn enough money to pay state and local taxes. (*Source: Tax Foundation Facts and Figures on Government Finance.*)

(a) (1980, ___) (b) (1984, ___)
(c) (1986, ___) (d) (1990, ___)

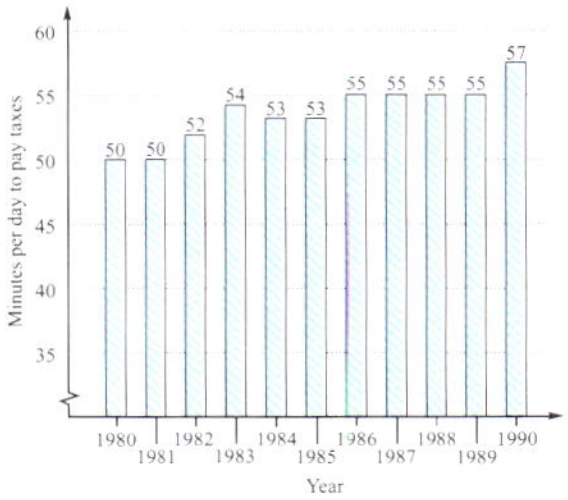

In Exercises 7 and 8, determine which sets of ordered pairs represent a function from A to B.

7. $A = \{0, 1, 2, 3\}$ and $B = \{-2, -1, 0, 1, 2\}$
(a) $\{(0, 1), (1, -2), (2, 0), (3, 2)\}$
(b) $\{(0, -1), (2, 2), (1, -2), (3, 0), (1, 1)\}$
(c) $\{(0, 0), (1, 0), (2, 0), (3, 0)\}$
(d) $\{(0, 2), (3, 0), (1, 1)\}$

8. $A = \{1, 2, 3\}$ and $B = \{9, 10, 11, 12\}$
(a) $\{(1, 10), (3, 11), (3, 12), (2, 12)\}$
(b) $\{(1, 10), (2, 11), (3, 12)\}$
(c) $\{(1, 10), (1, 9), (3, 11), (2, 12)\}$
(d) $\{(3, 9), (2, 9), (1, 12)\}$

In Exercises 9–20, determine whether the given equation represents y as a function of x.

9. $y = 10x + 12$ **10.** $y = 3 - 8x$

11. $3x + 7y - 2 = 0$ **12.** $x - 9y + 3 = 0$
13. $y = x(x - 10)$ **14.** $y = (x + 2)^2 + 3$
15. $x^2 + y^2 = 25$ **16.** $x^2 - 4y^2 = 16$
17. $y = 8 - |x|$ **18.** $|y| = x + 2$
19. $|y - 2| = x$ **20.** $y = x^3 - 2$

In Exercises 21–24, fill in the blanks using the function and the specified value of the independent variable.

21. $f(x) = 3x + 5$
(a) $f(2) = 3(\ \ ) + 5$
(b) $f(-2) = 3(\ \ ) + 5$
(c) $f(k) = 3(\ \ ) + 5$
(d) $f(k + 1) = 3(\ \ ) + 5$

22. $f(x) = 3 - x^2$
(a) $f(0) = 3 - (\ \ )^2$
(b) $f(-3) = 3 - (\ \ )^2$
(c) $f(m) = 3 - (\ \ )^2$
(d) $f(t + 2) = 3 - (\ \ )^2$

23. $f(x) = \dfrac{x}{x + 2}$
(a) $f(3) = \dfrac{(\ \ )}{(\ \ ) + 2}$
(b) $f(-4) = \dfrac{(\ \ )}{(\ \ ) + 2}$
(c) $f(s) = \dfrac{(\ \ )}{(\ \ ) + 2}$
(d) $f(s - 2) = \dfrac{(\ \ )}{(\ \ ) + 2}$

24. $f(x) = \sqrt{x + 8}$
(a) $f(1) = \sqrt{(\ \ ) + 8}$
(b) $f(-4) = \sqrt{(\ \ ) + 8}$
(c) $f(h) = \sqrt{(\ \ ) + 8}$
(d) $f(h - 8) = \sqrt{(\ \ ) + 8}$

In Exercises 25–40, evaluate the function at the specified value of the independent variable and simplify when possible.

25. $f(x) = 12x - 7$
(a) $f(3)$ (b) $f\left(\frac{3}{2}\right)$
(c) $f(a) + f(1)$ (d) $f(a + 1)$

418 CHAPTER 7 Introduction to Analytic Geometry

In Exercises 61–64, determine whether the lines L_1 and L_2 passing th of points are parallel, perpendicular, or neither.

61. L_1: (1, 3), (2, 1)
L_2: (0, 0), (4, 2)

62.

63. L_1: (−2, 0), (4, 4)
L_2: (1, −2), (4, 0)

64.

In Exercises 65–68, sketch the graphs of the pair of equations on coordinate system, and state the relationship that exists between the

65. L_1: $y = \frac{1}{2}x - 2$
L_2: $y = \frac{1}{2}x + 3$

66. L_1: $y = 3x - 2$
L_2: $y = 3x + 1$

67.

69. *Earnings per Share* The accompanying graph gives the earnings per share of common stock for Xerox Corporation for the years 1983 through 1989 (*Source:* Xerox Corporation). In what year did earnings (a) decrease most rapidly, and (b) increase most rapidly?

70. *Dividends per Share* The accompanying graph gives the declared dividends per share of common stock for Sun Company for the years 1983 through 1989 (*Source:* Sun Company). In what year did dividends (a) decrease most rapidly, and (b) increase most rapidly?

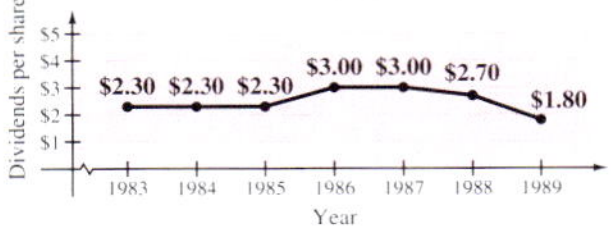

71. *Road Grade* When driving down a mountain road, you notice warning signs indicating that it is a "12% grade." This means that the slope of the road is $-\frac{12}{100}$. Determine the amount of horizontal change in your position if you note from elevation markers that you have descended 2,000 feet vertically. (See figure.)

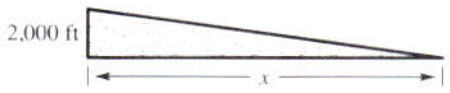

72. *Height of an Attic* The "rise to run" in determining the steepness of the roof on a house is 3 to 4. Determine the maximum height of the attic of the house if the house is 30 feet wide. (See figure.)

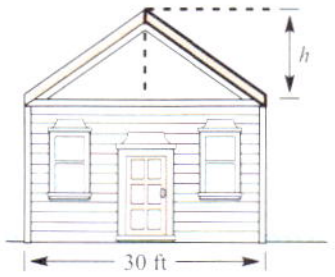

Graphics

The ability to visualize problems is a critical skill that students need in order to solve them. To encourage the development of this skill, the text has an abundance of figures, many computer-generated for accuracy.

Applications

Real-world applications are integrated throughout the text in both examples and exercises. This offers students a constant review of problem-solving skills and emphasizes the relevance of the mathematics. Many of the applications use current, real data, and all are titled for reference.

206 CHAPTER 4 Rational Expressions

Cost of Medicare In Exercises 65 and 66, use the following polynomial models, which give the projected cost of Medicare between 1990 and 1995 and the U.S. population age 65 and older (see figure).

$C = 107.1 + 12.64t + 0.54t^2$ Cost of Medicare

$P = 31.6 + 0.51t - 0.14t^2$ Population

In these models, C represents the total annual cost of Medicare (in billions of dollars), P represents the U.S. population (in millions) age 65 and older, and t represents the year, with $t = 0$ corresponding to 1990. (*Sources:* Congressional Budget Office and U.S. Bureau of Census.)

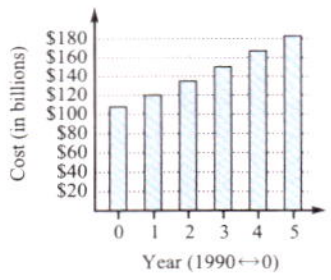

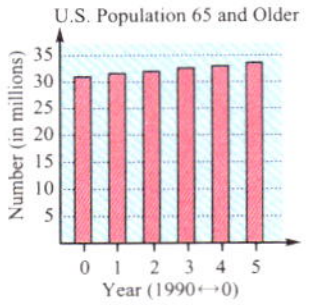

65. Find a rational model that represents the average cost of Medicare *per person age 65 or older* during the years 1990 to 1995.

66. Use the model found in Exercise 65 to complete the following table showing the average cost of Medicare per person age 65 or older.

Year	Average Cost
1990	
1991	
1992	
1993	
1994	
1995	

...tional Expressions

...ding Rational Expressions • Compound Fractions

...l Expressions

...ng rational expressions is the same as the rule for multiplying ...hat is, we *multiply numerators, multiply denominators, and* ... *in reduced form*. Here is an example.

$(x + 3)(x - 2)$ / $(x - 2)(x - 4)$ Multiply numerators and denominators

$(x + 3)(x - 2)$ / $(x - 2)(x - 4)$ Cancel common factor

$x + 3$ / $x - 4$, $x \neq 2$ Simplified form

482 CHAPTER 8 Functions and Mathematical Models

In Exercises 61–68, find an equation that represents the given function. (Note that the variables do not necessarily have to be x and y.)

61. *Perimeter of a Square* Express the perimeter P of a square as a function of the length x of a side.

62. *Surface Area of a Cube* Express the surface area S of a cube as a function of the length x of one of the edges.

63. *Volume of a Cube* Express the volume V of a cube as a function of the length x of one of the edges.

64. *Diagonal of a Square* Express the length L of the diagonal of a square as a function of the length x of a side.

65. *Volume of a Box* An open box is to be made from a square piece of material, 24 inches on a side, by cutting equal squares from each corner and turning up the sides (see figure). Write the volume V of the box as a function of x.

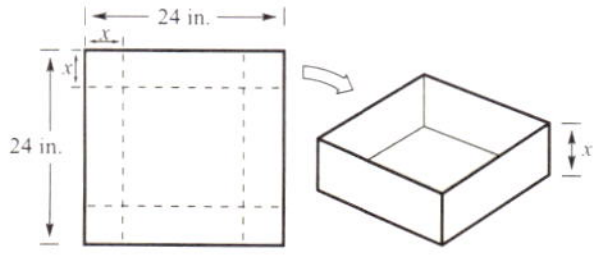

66. *Area of a Square* Strips of width x are cut from the four sides of a square that is 32 inches on a side (see figure). Write the area A of the remaining square as a function of x.

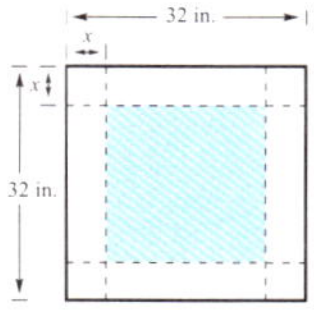

67. *Area of a Square* Strips of width x are cut from two adjacent sides of a square that is 32 inches on a side (see figure). Write the area A of the remaining square as a function of x.

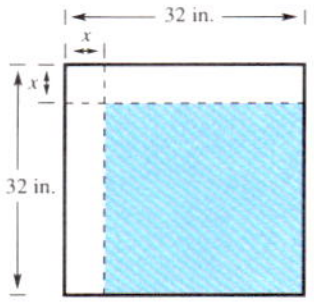

68. *Cost* The inventor of a new game believes that the variable cost for producing the game is \$1.95 per unit and the fixed costs are \$8,000. If x is the number of games produced, express the total cost C as a function of x.

In Exercises 69 and 70, determine whether the statements use the word *function* in a way that is mathematically correct.

69. (a) The sales tax on a purchased item is a function of the selling price.
(b) Your score on the next algebra exam is a function of the number of hours you study the night before the exam.

70. (a) The amount in your savings account is a function of your salary.
(b) The speed at which a free-falling baseball strikes the ground is a function of the height from which it is dropped.

Geometry

Geometric formulas and concepts are reviewed throughout the text. For reference, common formulas are given inside the front and back covers.

End-of-Chapter Study Aids

A Chapter Summary outlines all of the skills that are presented in the chapter. For review, section and exercise references are given for the major topics.

A set of Review Exercises at the end of each chapter gives students an opportunity for additional practice. Each set of review exercises includes both computational and applied problems covering a wide range of topics.

A Chapter Test allows students to assess their own level of success.

Cumulative Tests appear after Chapters 4, 7, and 11. These tests help students to judge their mastery of previously covered concepts. They also help students maintain the knowledge-base they have been building throughout the text—preparing them for other exams and future courses.

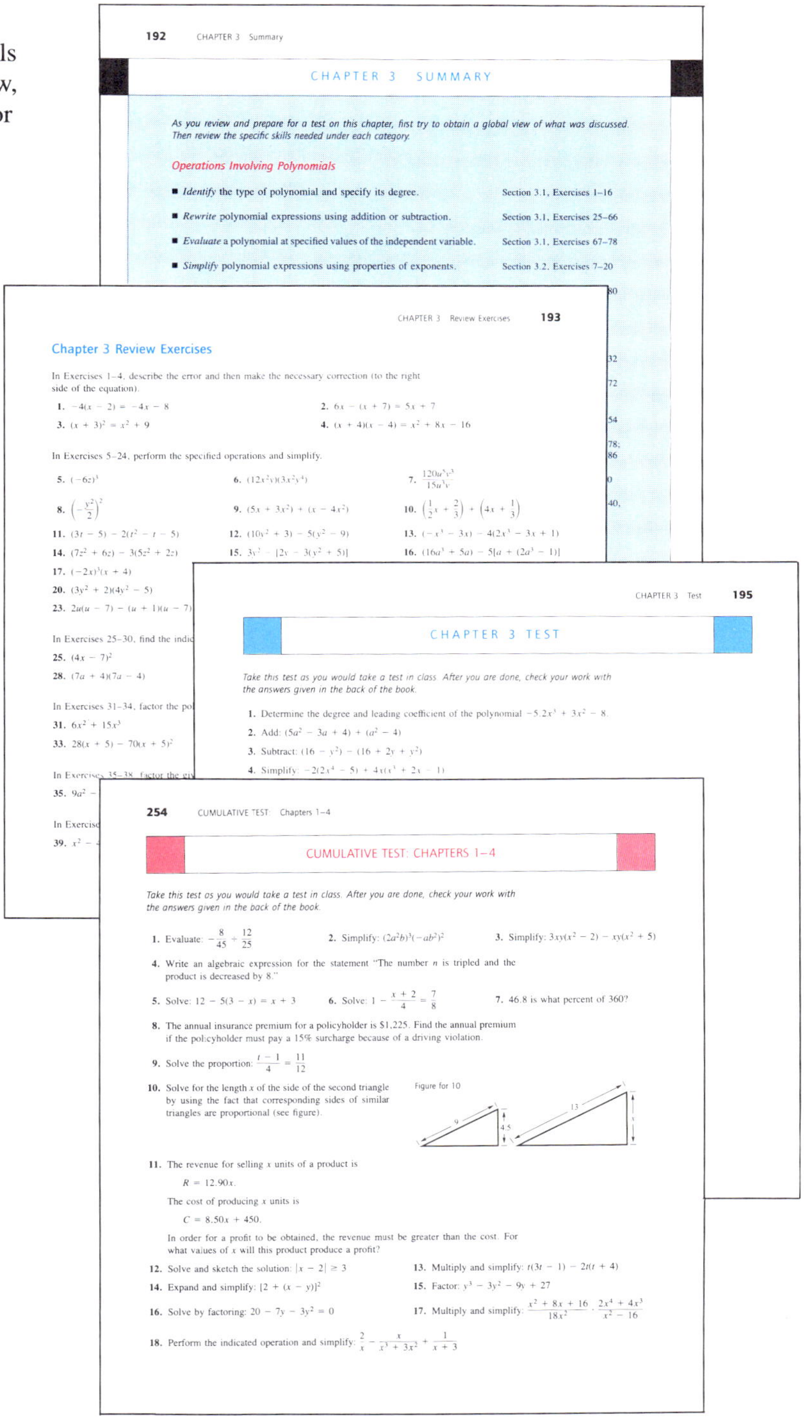

SUPPLEMENTS

Intermediate Algebra by Larson and Hostetler is accompanied by a comprehensive supplements package for maximum teaching effectiveness and efficiency.

- ***Student Solutions Guide*** by Gerry C. Vidrine, Louisiana State University
- ***Study Guide*** by Jay Wiestling, Palomar College
- ***Complete Solutions Guide*** by Gerry C. Vidrine, Louisiana State University
- ***Instructor's Guide*** by Robert P. Hostetler and Ann R. Kraus, The Pennsylvania State University
- ***Test Item File*** by Norman B. Patterson
- *Intermediate Algebra* Videotapes by Dana Mosely, Valencia Community College
- *Intermediate Algebra* TUTOR by Timothy R. Larson, Paula M. Sibeto, Kristin Winnen-Smith, and John R. Musser.
- Test-Generating Software
- Classroom Demonstration Software by Timothy R. Larson, Paula M. Sibeto, Kristin Winnen-Smith, and John R. Musser.
- Gradebook Software

This complete supplements package offers ancillary materials for students and instructors and for classroom resources. Each item is keyed directly to the textbook for ease of use. For the convenience of software users, a technical support telephone number is available with all D. C. Heath software products: (617)860-1218. The components of this comprehensive teaching and learning package are outlined in the diagram on the following page.

INTERMEDIATE ALGEBRA

FOR INSTRUCTORS

Printed Ancillaries

Complete Solutions Guide
- Solutions to all text exercises

Test Item File
- Printed test bank
- Over 3,500 test items
- Open-ended and multiple-choice test items
- Available as a computerized test bank

Instructor's Guide
- Chapter commentary with general teaching strategies
- Section topics
- Sample tests
- Transparency masters

Software

Computerized Testing Software
- Test-Generating software
- Over 3,500 test items
- Also available as a printed test bank

Classroom Demonstration Software
- Additional examples with step-by-step solutions
- Interactive, menu-driven resource bank covering major topics
- Animated color graphics

FOR STUDENTS

Printed Ancillaries

Student Solutions Guide
- Solutions to all odd-numbered text exercises

Study Guide
- Section summaries
- Additional examples with solutions
- Starter exercises with answers
- Cumulative tests

Software

TUTOR
- Interactive tutorial software follows section by section with text
- Diagnostic feedback
- Additional examples
- Chapter self-tests
- Glossary

Videotapes

Videotapes
- Comprehensive coverage
- For media/resource centers
- Additional explanation of important concepts, sample problems, and applications

CLASSROOM RESOURCES

Printed Ancillaries

Instructor's Guide
- Chapter commentary with general teaching strategies
- Section topics
- Sample tests
- Transparency masters

Software

Classroom Demonstration Software
- Additional examples with step-by-step solutions
- Interactive, menu-driven resource bank covering major topics
- Animated color graphics

Gradebook software
- Grade management system

Videotapes

Videotapes
- Comprehensive coverage
- For media/resource centers
- Additional explanation of important concepts, sample problems, and applications

ACKNOWLEDGMENTS

We would like to thank the many people who have helped us prepare the text and supplements package. Their encouragement, criticisms, and suggestions have been invaluable to us.

Reviewers: Carol J. Achs, Mesa Community College; Frank Asta, College of Du Page; Wilma M. Barlow, Liberty University; Arthur Finco, Indiana University–Purdue University at Fort Wayne; Jay Graening, University of Arkansas; Linda Padilla, Joliet Junior College; Gerald E. Rubin, Marshall University; Arthur Szylewicz, Moorpark College; and Dave Van Langeveld, Clearfield High School.

Focus Group Participants: Ruth A. Koelle, Roger Williams College; Jeri V. Love, Florida Community College at Jacksonville; Maureen A. McCarthy, Santa Rosa Junior College; Hank Martel, Broward Community College; Ellen Milosheff, Triton College; Charlotte Newsom, Tidewater Community College; and Keith Wilson, Oklahoma City Community College.

A special thanks to all of the people at D. C. Heath and Company who worked with us in the development and production of the text, especially Ann Marie Jones, Mathematics Acquisitions Editor; Cathy Cantin, Developmental Editor; Kathleen A. Savage, Production Editor; Cornelia Boynton, Designer; Carolyn Johnson, Editorial Associate; Lisa Merrill, Production Coordinator; Gary Crespo, Art Editor; and Billie L. P. Ingram, Photo Researcher.

Several other people worked on this project. David E. Heyd assisted us in writing the text and solving the exercises; Gerry C. Vidrine wrote the *Student Solutions Guide* and the *Complete Solutions Guide*; Jay Wiestling wrote the *Study Guide*; Ann R. Kraus worked on the *Instructor's Guide*; Norman B. Patterson wrote the *Test Item File* and proofed the manuscript; and Helen Medley proofed the manuscript. The following people also worked on the project: Linda L. Kifer, Linda M. Bollinger, Timothy R. Larson, Louis R. Rieger, Paula M. Sibeto, Kristin Winnen-Smith, Laurie A. Brooks, Richard J. Bambauer, Patricia S. Larson, Nancy K. Stout, and John R. Musser.

On a personal level, we are grateful to our wives, Deanna Gilbert Larson and Eloise Hostetler, for their love, patience, and support. Also, a special thanks goes to R. Scott O'Neil.

If you have suggestions for improving the text, please feel free to write to us. Over the past two decades we have received many useful comments from both instructors and students, and we value these very much.

Roland E. Larson
Robert P. Hostetler

CONTENTS

HOW TO STUDY ALGEBRA

After years of teaching and guiding students through algebra courses, we have compiled the following list of suggestions for studying algebra. These study tips may take some time and effort—but they work!

Making a Plan

Make your own course plan right now! Determine the number of hours you need to spend on algebra each week. Write your plans on your calendar or some other schedule planner, and then *stick to your plan*.

Preparing for Class

Before attending class, read the portion of the text that is to be covered. This takes a lot of self-discipline, but it pays off. By going to class prepared, you will be able to benefit much more from your instructor's presentation. Algebra, like most other technical subjects, is easier to understand the second or third time you hear it.

Attending Class

Attend every class. Arrive on time with your text, a pen or pencil, paper for notes, and your calculator.

Participating in Class

As you are reading the text before class, write down any questions that you have about the material. Then, ask your instructor during class.

Taking Notes

Take notes in class, especially on definitions, examples, concepts, and rules. Then, as soon after class as possible, read through your notes, adding any explanations that are necessary to make your notes understandable *to you*.

Doing the Homework

Learning algebra is like learning to play the piano or learning to play basketball. You cannot become skilled by just watching someone else do it. You must also do it yourself. A general guideline is to spend two to four hours of study outside of class for each hour in class. When working exercises, your ultimate goal is to be able to solve the problems accurately and quickly. When you start a new exercise set, however, understanding is much more important than speed.

Finding a Study Partner

When you get stuck on a problem, it may help to try to work with someone else. Even if you feel you are giving more help than you are getting, you will find that an excellent way to learn is by teaching others.

Building a Math Library

Start building a library of books that can help you with this and future math courses. You might consider using the *Student Solutions Guide* or the *Study Guide* that accompanies the text. Also, since you will probably be taking other math courses after you finish this course, we suggest that you keep the text. It will be a valuable reference book.

Keeping Up with the Work

Don't let yourself fall behind in the course. If you think that you are having trouble, seek help immediately. Ask your instructor, attend your school's tutoring services, talk with your study partner, use additional study aids such as videos or software tutorials—but do something. If you are having trouble with the material in one chapter of your algebra text, there is a good chance that you will also have trouble in later chapters.

Getting Stuck

Everyone who has ever taken a math course has had this experience: You are working on a problem and cannot see how to solve it, or you have solved it but your answer does not agree with the answer given in the back of the book. People have different approaches to this sort of problem. You might ask for help, take a break to clear your thoughts, sleep on it, rework the problem, or reread the section in the text. The point is, try not to get frustrated or spend too much time on a single problem.

Keeping Your Skills Sharp

Before each exercise set in the text we have included a short set of *Warm-Up Exercises*. These exercises will help you review skills that you learned in previous exercises. These sets are designed to take only a few minutes to solve. We suggest working the entire set before you start each new exercise set. (All of the Warm-Up Exercises are answered in the back of the text.)

Checking Your Work

One of the nice things about algebra is that you don't have to wonder whether your solution is correct. You can tell whether it is correct by checking it in the original statement of the problem. If, in addition to your "solving skills," you work on your "checking skills," you should find your test scores improving.

Preparing for Exams

Cramming for algebra exams seldom works. If you have kept up with the work and followed the suggestions given here, you should be almost ready for the exam. At the end of each chapter, we have included three features that should help as a final preparation. Read the *Chapter Summary*, work the *Review Exercises*, and set aside an hour to take the sample *Chapter Test*.

Taking Exams

Most instructors suggest that you do *not* study right up to the minute you are taking a test. This tends to make people anxious. The best cure for anxiousness during tests is to prepare well before taking the test. Once the test has begun, read the directions carefully, and try to work at a reasonable pace. (You might want to read the entire test first, then work the problems in the order with which you feel most comfortable.) Hurrying tends to cause people to make careless errors. If you finish early, take a few moments to clear your thoughts and then take time to go over your work.

Learning from Mistakes

When you get an exam back, be sure to go over any errors that you might have made. Don't be too quick to pass off an error as just a "dumb mistake." Take advantage of any mistakes by hunting for ways to continually improve your test-taking abilities.

WHAT IS ALGEBRA?

To some, algebra is manipulating symbols or performing mathematical operations with letters instead of numbers. To others, it is factoring, solving equations, or solving word problems. And to still others, algebra is a mathematical language that can be used to model real-world problems. In fact, algebra is all of these!

As you study this text, it is helpful to view algebra from the "big picture"—to see how the various rules, operations, and strategies fit together.

The rules of arithmetic form the foundation of algebra. These rules are generalized through the use of symbols and letters to form the basic rules of algebra, which are used to *rewrite* algebraic expressions and equations in new, more useful forms. The ability to rewrite algebraic expressions and equations is the common skill involved in the three major components of algebra—*simplifying* algebraic expressions, *solving* algebraic equations, and *graphing* algebraic functions. The following chart shows how this intermediate algebra text fits into the "big picture" of algebra.

Construct ↓ *Rewrite* ↓ *Graph*

Construct ↓ *Rewrite* ↓ *Solve*

Construct ↓ *Rewrite* ↓ *Simplify*

Rules of Arithmetic
1 Introduction to Algebra

Rules of Algebra
1 Introduction to Algebra

Algebraic Expressions
1 Introduction to Algebra
3 Polynomials and Factoring
4 Rational Expressions
5 Radicals and Complex Numbers
11 Additional Topics in Algebra

Algebraic Equations
2 Linear Equations and Inequalities
4.5 Solving Equations Involving Rational Expressions
5.5 Solving Equations Involving Radicals
6 Quadratic Equations and Inequalities
8 Functions and Mathematical Models
9 Systems of Equations and Inequalities
10 Exponential and Logarithmic Functions
11 Additional Topics in Algebra

Functions and Graphs
7 Introduction to Analytic Geometry
8 Functions and Mathematical Models
9 Systems of Equations and Inequalities
10 Exponential and Logarithmic Functions
11 Additional Topics in Algebra

CHAPTER ONE

Introduction to Algebra

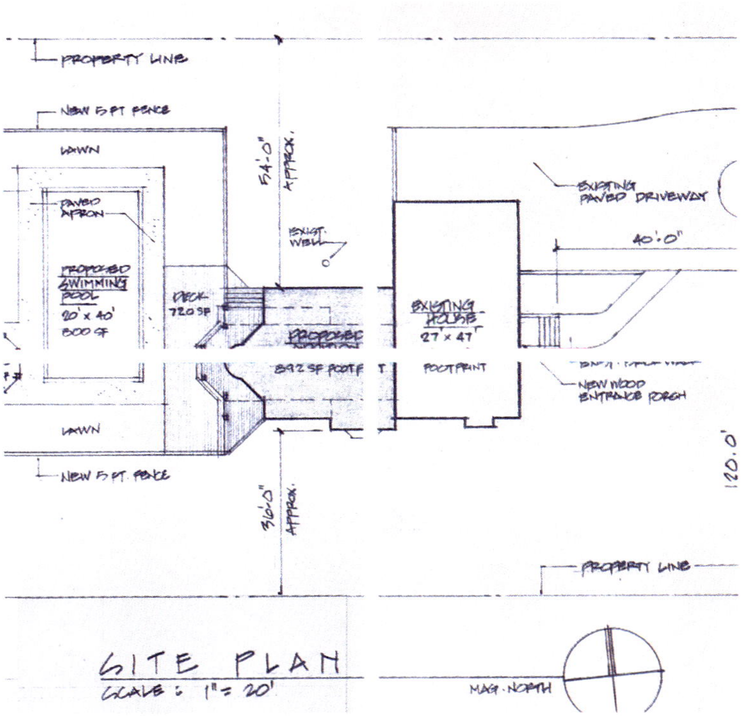

Chapter Overview

Some texts refer to the material in this chapter as "prealgebra." The reason for this term is that this material summarizes the mathematics that you will need in order to begin your formal study of algebra in Chapter 2. The first three sections of this chapter review the operations and properties of real numbers. Section 1.4 introduces the concept of an algebraic expression, and Section 1.5 describes techniques for translating verbal statements into mathematical statements.

SECTION 1.1

Real Numbers: Order and Absolute Value

Sets and Real Numbers • Subsets of Real Numbers • The Real Number Line • Ordering the Real Numbers and Distance • The Absolute Value of a Real Number

Sets and Real Numbers

The formal term that we use in mathematics to talk about a collection of objects is the word **set**.* For instance, the set

$\{1, 2, 3\}$

contains the three numbers 1, 2, and 3. Note that we use a pair of braces { } to list the members of the set. In this text a *pair* of braces will always indicate that we are describing the members of a set. Parentheses ( ) and brackets [ ] are used to represent other ideas.

The set of numbers that we use in arithmetic is called the set of **real numbers**. The term *real* distinguishes real numbers from *imaginary* or *complex numbers*—a type of number that we will study later in this text.

Subsets of Real Numbers

Mathematical problems that we face in everyday life involve different subsets of real numbers. One of the most commonly used subsets of real numbers is the set of **natural numbers** or **positive integers**

$\{1, 2, 3, 4, \ldots\}$.

Note that the three dots indicate that the pattern continues. For instance, the set also contains the numbers 5, 6, 7, and so on.

Positive integers can be used to describe many quantities that we encounter in everyday life—for instance, you might be taking four classes this term, or you might be paying \$420 a month for rent. But even in everyday life, positive integers cannot describe some concepts accurately. For instance, you could have a zero balance in your checking account, or the temperature could be $-10°$ (10 degrees below zero). To describe such quantities we need to expand the set of positive integers to include zero and the negative integers. The expanded set is called the set of **integers**, which we write as follows.

$$\{\underbrace{\ldots, -3, -2, -1}_{\text{Negative integers}}, \overbrace{0}^{\text{Zero}}, \underbrace{1, 2, 3, \ldots}_{\text{Positive integers}}\}$$

*Whenever we formally introduce a mathematical term in this text, the word will occur in boldface type. Be sure you understand the meaning of each new word—it is important that each new word becomes part of your mathematical vocabulary.

The set of integers is a **subset** of the set of real numbers, which is another way of saying that every integer is a real number.

Even using the set of integers, there are still many quantities in everyday life that we cannot describe accurately. The cost of many items is not in whole-dollar amounts, but in parts of dollars, such as \$1.19 or \$39.98. You might work $8\frac{1}{2}$ hours, or you might miss the first *half* of a movie. To describe such quantities, we expand the set of integers to include fractions. The expanded set is called the set of **rational numbers**. Formally, a real number is called **rational** if it can be written as the ratio p/q of two integers, where $q \neq 0$ (the symbol $\neq$ means **does not equal**). For instance,

$$2 = \frac{2}{1}, \quad \frac{1}{3} = 0.333\ldots, \quad \frac{1}{8} = 0.125, \quad \text{and} \quad \frac{125}{111} = 1.126126\ldots$$

are rational numbers. Real numbers that cannot be written as a ratio of two integers are called **irrational**. For instance, the numbers

$$\sqrt{2} = 1.4142135\ldots \quad \text{and} \quad \pi = 3.1415926\ldots$$

are irrational. The decimal representation of a rational number is either **terminating** or **repeating**. For instance, the decimal representation of $\frac{1}{4} = 0.25$ is terminating, and the decimal representation of $\frac{4}{11} = 0.363636\ldots = 0.\overline{36}$ is repeating. (The line over 36 indicates which digits repeat.)

The decimal representation of an irrational number neither terminates nor repeats. When we perform calculations using decimal representations for nonterminating decimals, we usually use a decimal approximation that has been **rounded*** to a certain number of decimal places. For instance, rounded to four decimal places, the decimal approximations of $\frac{2}{3}$ and π are

$$\frac{2}{3} \approx 0.6667 \quad \text{and} \quad \pi \approx 3.1416.$$

The symbol $\approx$ means **approximately equal to**.

Figure 1.1 shows several commonly used subsets of real numbers and their relationships to each other.

FIGURE 1.1

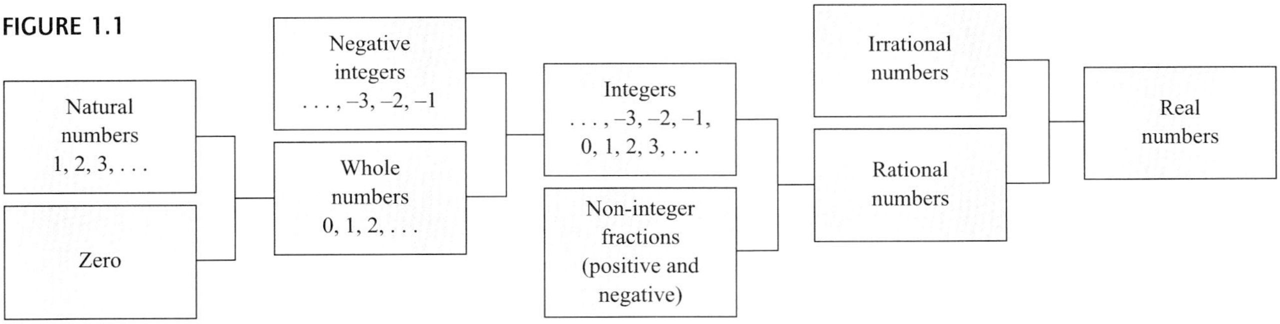

*The rounding rule we use in this text is to round *up* if the succeeding digit is 5 or more and round *down* if the succeeding digit is 4 or less. For example, if we wanted to round 7.35 to one decimal place, we would round up to 7.4. Similarly, if we wanted to round 2.364 to two decimal places, we would round down to 2.36.

EXAMPLE 1 ■ Identifying Real Numbers

Consider the following subset of real numbers.

$$\left\{-7, -\sqrt{3}, -1, \frac{1}{5}, 0, \frac{3}{4}, \sqrt{2}, \pi, 5\right\}$$

Which of the numbers in this set are (a) natural numbers, (b) integers, (c) rational numbers, and (d) irrational numbers?

Solution

(a) Natural numbers: $\{5\}$

(b) Integers: $\{-7, -1, 0, 5\}$

(c) Rational numbers: $\left\{-7, -1, \frac{1}{5}, 0, \frac{3}{4}, 5\right\}$

(d) Irrational numbers: $\{-\sqrt{3}, \sqrt{2}, \pi\}$ ■

The Real Number Line

The picture we use to represent the real numbers is called the **real number line**. It consists of a horizontal line with a point (the **origin**) labeled as 0. Numbers to the left of 0 are **negative** and numbers to the right of 0 are **positive**, as shown in Figure 1.2.

FIGURE 1.2 The Real Number Line

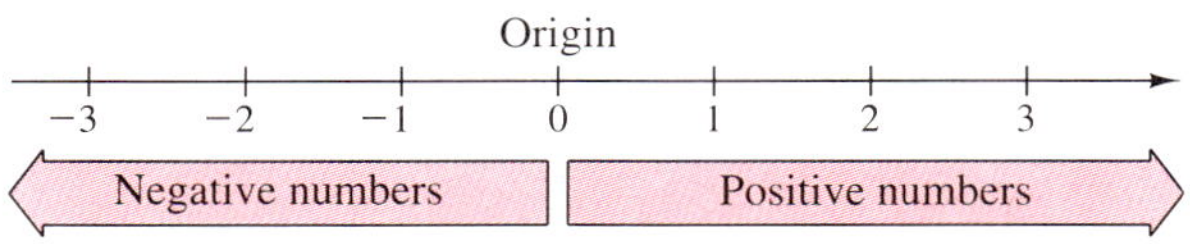

The real number zero is neither positive nor negative. Thus, when we want to talk about real numbers that might be positive or zero, we use the term **nonnegative real number**.

Each point on the real number line corresponds to exactly one real number, and each real number corresponds to exactly one point on the real number line, as shown in Figure 1.3. When we draw the point (on the real number line) that corresponds to a real number, we say we are **plotting** the real number.

FIGURE 1.3

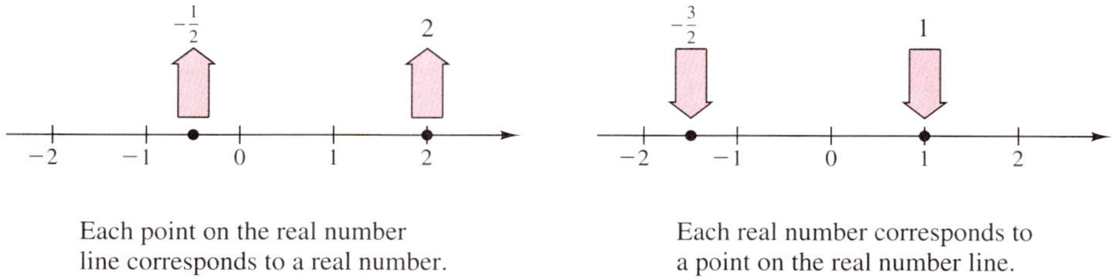

EXAMPLE 2 ■ Plotting Real Numbers on the Real Number Line

Plot the point on the real number line that represents each of the following real numbers.

(a) $-\frac{5}{3}$ (b) 2.3 (c) $\frac{9}{4}$

Solution

(a) The point representing the real number $-\frac{5}{3} = -1.666\ldots$ lies between -2 and -1 on the real number line, as shown in Figure 1.4.

(b) The point representing the real number 2.3 lies between 2 and 3 on the real number line, as shown in Figure 1.4.

(c) The point representing the real number $\frac{9}{4} = 2.25$ lies between 2 and 3 on the real number line, as shown in Figure 1.4. Note that the point representing $\frac{9}{4}$ lies slightly to the left of the point representing 2.3.

FIGURE 1.4

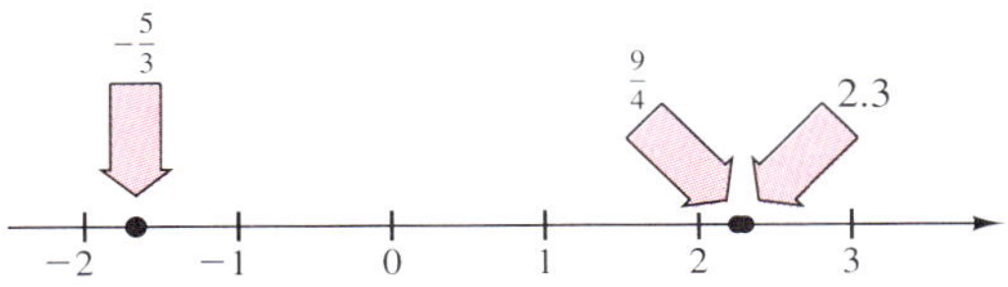

■

Ordering the Real Numbers and Distance

The real number line provides us with a way of comparing any two real numbers. For instance, if you choose any two (different) numbers on the real number line, then one of the numbers must be to the left of the other number. We say that the number to the left is **less than** the number to the right, or that the number to the right is **greater than** the number to the left, as shown in Figure 1.5.

FIGURE 1.5

$a < b$

a is to the left of b.

Definition of Order on the Real Number Line

If the real number a lies to the left of the real number b on the real number line, then we say that a is **less than** b and write

$$a < b.$$

This relationship can also be described by saying that b is **greater than** a and by writing $b > a$. The symbol $a \leq b$ means that a is **less than or equal to** b, and the symbol $b \geq a$ means that b is **greater than or equal to** a.

The symbols $<$, $>$, $\leq$, and $\geq$ are called **inequality symbols**. Figure 1.6 shows some examples of the use of inequality symbols. Note that $-3 < 2$ because -3 lies to the left of 2 on the number line. Similarly, $4 > 0$ because 4 lies to the right of 0 on the real number line.

FIGURE 1.6

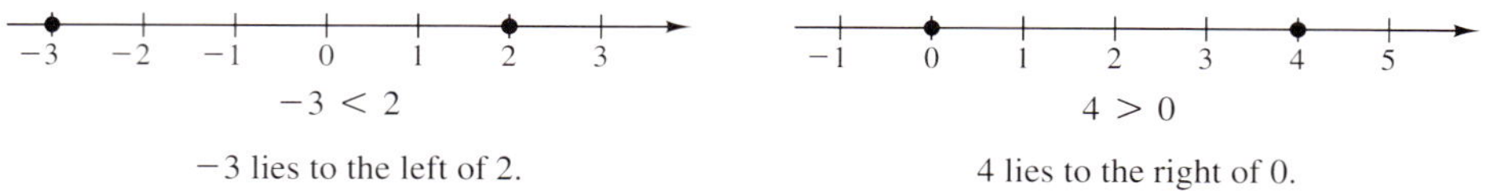

When you are asked to **order** two numbers, you are simply being asked to say which of the two numbers is greater.

EXAMPLE 3 ■ Ordering Real Numbers

Place the correct inequality symbol ($<$ or $>$) between the two numbers.

(a) -4 ▭ 0 (b) -3 ▭ -5

(c) $\frac{1}{5}$ ▭ $\frac{1}{3}$ (d) -3.09 ▭ -3.90

Solution

(a) Since -4 lies to the left of 0 on the real number line, as shown in Figure 1.7(a), we say that -4 is *less than* 0, and write

$$-4 < 0.$$

(b) Since -3 lies to the right of -5 on the real number line, as shown in Figure 1.7(b), we say that -3 is *greater than* -5, and write

$$-3 > -5.$$

(c) Since $\frac{1}{5}$ lies to the left of $\frac{1}{3}$ on the real number line, as shown in Figure 1.7(c), we say that $\frac{1}{5}$ is *less than* $\frac{1}{3}$, and write

$$\frac{1}{5} < \frac{1}{3}.$$

(d) Since -3.09 lies to the right of -3.90 on the real number line, as shown in Figure 1.7(d), we say that -3.09 is *greater than* -3.90, and write

$$-3.09 > -3.90.$$

FIGURE 1.7

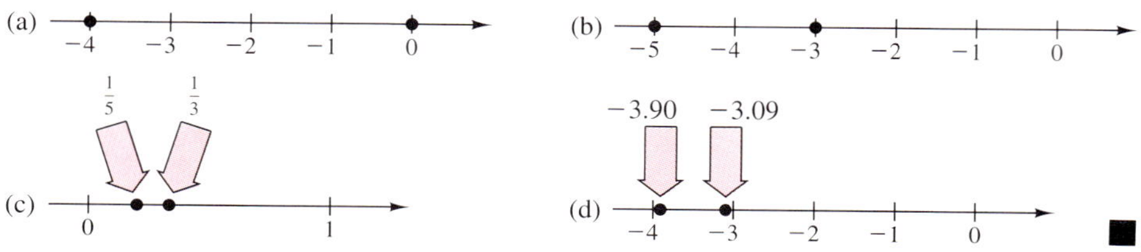

■

Once we know how to represent real numbers as points on the real number line, it is natural to talk about the **distance between two real numbers**. Specifically, if a and b are two real numbers such that $a \le b$, then we define the distance between a and b to be $b - a$.

Definition of Distance Between Two Real Numbers

If a and b are two real numbers such that $a \le b$, then the **distance between *a* and *b*** is given by

$$(\text{Distance between } a \text{ and } b) = b - a.$$

Note from this definition that if $a = b$, then the distance between a and b is zero. If $a < b$, then the distance between a and b is positive.

EXAMPLE 4 ■ Finding the Distance Between Two Real Numbers

Find the distance between the following pairs of real numbers.

(a) -2 and 3 (b) 0 and 4 (c) -4 and 0

Solution

(a) The distance between -2 and 3 is

$$3 - (-2) = 3 + 2 = 5$$

as shown in Figure 1.8(a).

(b) The distance between 0 and 4 is

$$4 - 0 = 4$$

as shown in Figure 1.8(b).

(c) The distance between -4 and 0 is

$$0 - (-4) = 0 + 4 = 4$$

as shown in Figure 1.8(c).

FIGURE 1.8

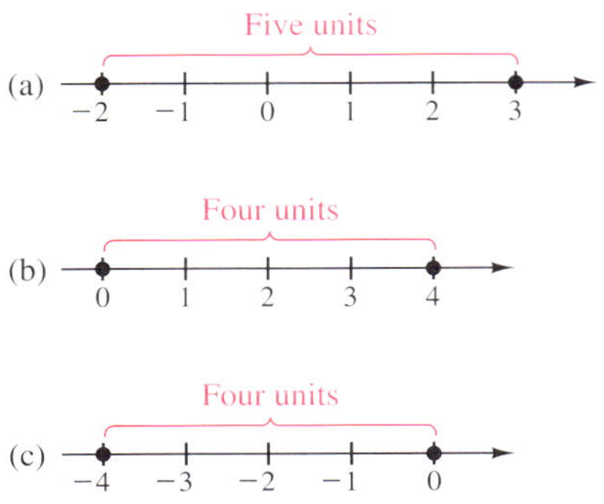

■

The Absolute Value of a Real Number

The distance between a real number a and 0 (the origin) is called the **absolute value** of a. We use double vertical bars $|\ \ |$ to denote absolute value. For example,

$$|5| = \text{distance between 5 and 0} = 5$$

and

$$|-8| = \text{distance between } -8 \text{ and } 0 = 8.$$

Definition of Absolute Value of a Real Number

The **absolute value** of a real number a is defined to be the distance between a and 0 on the real number line.

1. If $a > 0$, then $|a| = a - 0 = a$.
2. If $a = 0$, then $|a| = 0 - 0 = 0$.
3. If $a < 0$, then $|a| = 0 - a = -a$.

NOTE: The absolute value of any real number is either positive or zero. Moreover, the only real number whose absolute value is zero is 0. That is, $|0| = 0$.

Be sure you see from this definition that the absolute value of a real number is never negative. For instance, if $a = -3$, then $|a| = |-3| = -(-3) = 3 = -a$.

Two real numbers are said to be **opposites** of each other if they lie the same distance from, but on opposite sides of, zero. For example, -2 is the opposite of 2. Since opposite numbers lie the same distance from 0 on the real number line, they have the same absolute value. Thus, $|5| = 5$ and $|-5| = 5$.

The use of a negative sign to denote the *opposite* of a number gives meaning to expressions such as $-(-3)$ and $-|-3|$, as follows.

$$-(-3) = (\text{Opposite of } -3) = 3$$
$$-|-3| = \left(\text{Opposite of } |-3|\right) = -3$$

EXAMPLE 5 ■ **Finding Absolute Value**

(a) $|-10| = 10$

(b) $\left|\dfrac{3}{4}\right| = \dfrac{3}{4}$

(c) $|-3.2| = 3.2$

(d) $-|-6| = -(6) = -6$ Opposite of $|-6|$ is -6

Note that part (d) does not contradict the fact that the absolute value of a number cannot be negative. In this case, the minus sign in the answer comes from the minus sign in $-|-6|$ that is *outside* the absolute value. ■

For any two real numbers a and b, exactly one of the following must be true:

$a < b, \qquad a = b, \quad \text{or} \quad a > b.$

This property of real numbers is called the **law of trichotomy**.

EXAMPLE 6 ■ **Comparing Real Numbers**

Place the correct symbol ($<$, $>$, or $=$) between the following pairs of real numbers.

(a) $|-9|$ $\qquad$ $|9|$

(b) $|-2|$ $\qquad$ 1

(c) -4 $\qquad$ $-|-4|$

(d) $|12|$ $\qquad$ $|-15|$

Solution

(a) $|-9| = |9|$, because both are equal to 9.

(b) $|-2| > 1$, because $|-2| = 2$ and 2 is greater than 1.

(c) $-4 = -|-4|$, because both numbers are equal to -4.

(d) $|12| < |-15|$, because $|12| = 12$ and $|-15| = 15$, and 12 is less than 15. ■

When we defined the distance between a and b to be $b - a$, we specified that a was less than or equal to b. Using absolute value, we can generalize this definition as follows. If a and b are *any* two real numbers, then the distance between a and b is given by

$$(\text{Distance between } a \text{ and } b) = |a - b|.$$

For instance, the distance between -2 and 1 is given by

$$|-2 - 1| = |-3| = 3.$$

DISCUSSION PROBLEM ■ **Absolute Value of a Number**

How would you explain to someone, using everyday English, the concept of the absolute value of a real number? What is wrong with the following statement? If a is a real number, then $|a| = a$. ■

1.1 EXERCISES

1. Which of the real numbers in the following set are (a) natural numbers, (b) integers, (c) rational numbers, and (d) irrational numbers?

$$\left\{-10, -\sqrt{5}, -\frac{2}{3}, -\frac{1}{4}, 0, \frac{5}{8}, 1, \sqrt{3}, 4, 2\pi, 6\right\}$$

2. Which of the real numbers in the following set are (a) natural numbers, (b) integers, (c) rational numbers, and (d) irrational numbers?

$$\left\{-\frac{7}{2}, -\sqrt{6}, -\frac{\pi}{2}, -\frac{3}{8}, 0, \sqrt{15}, \frac{10}{3}, 8, 245\right\}$$

In Exercises 3 and 4, plot the real numbers on the real number line.

3. (a) 3 (b) $\frac{5}{2}$ (c) $-\frac{7}{2}$ (d) -5.2

4. (a) 8 (b) $\frac{4}{3}$ (c) -6.75 (d) $-\frac{9}{2}$

In Exercises 5–8, approximate the two indicated numbers on the real number line and place the correct inequality symbol ($<$ or $>$) between them.

5.

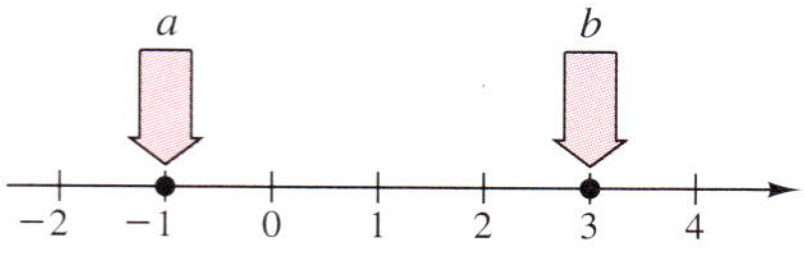

6.

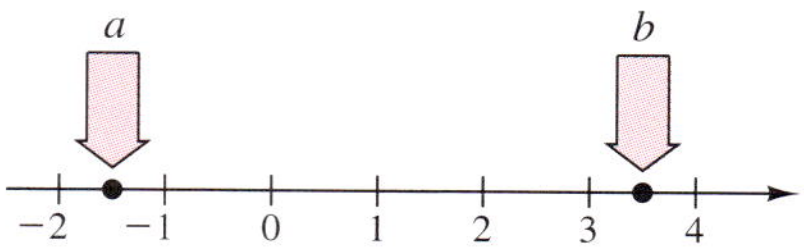

7.

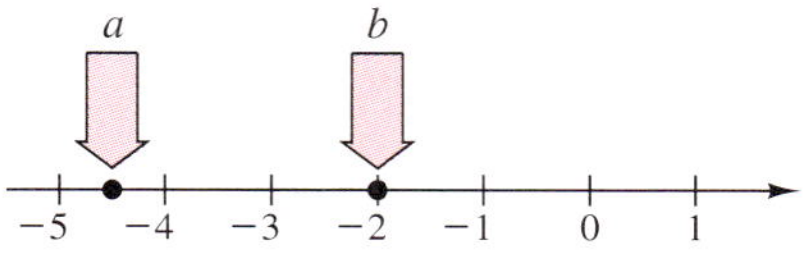

8.

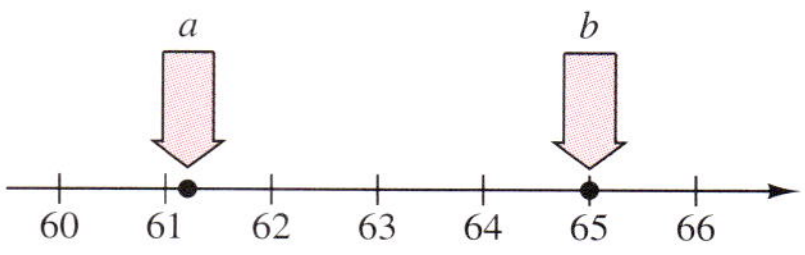

In Exercises 9–20, plot the two real numbers on the real number line and place the correct inequality symbol ($<$ or $>$) between the two numbers.

9. 2 ____ 5

10. 8 ____ 3

11. -5 ____ 2

12. -8 ____ 3

13. -7 ____ -2

14. -2 ____ -5

15. $\frac{1}{3}$ ____ $\frac{1}{4}$

16. $\frac{4}{5}$ ____ 1

17. $-\frac{5}{8}$ ____ $\frac{1}{2}$

18. $-\frac{3}{2}$ ____ $-\frac{5}{2}$

19. $-\frac{5}{3}$ ____ $-\frac{3}{2}$

20. $-\pi$ ____ -3

In Exercises 21–30, find the distance between the given pair of real numbers.

21. 4 and 10 **22.** 75 and 20 **23.** -12 and 7 **24.** 14 and -6

25. 18 and -32 **26.** -54 and 32 **27.** -8 and 0 **28.** 0 and 125

29. -6 and -45 **30.** -300 and -110

In Exercises 31–40, evaluate the given quantity.

31. $|10|$ **32.** $|62|$ **33.** $|-225|$ **34.** $|-14|$ **35.** $-|16|$ **36.** $-\left|\frac{3}{8}\right|$

37. $-|-85|$ **38.** $-|-36.5|$ **39.** $|-\pi|$ **40.** $|0|$

In Exercises 41–46, find the opposite of the given number and the absolute value of the given number.

41. 34 **42.** 225 **43.** -160 **44.** -52 **45.** $-\frac{3}{11}$ **46.** $\frac{5}{32}$

In Exercises 47–50, plot the number and its opposite on the real number line.

47. 7 **48.** $\frac{7}{4}$ **49.** $-\frac{3}{5}$ **50.** -4.25

In Exercises 51–58, write the statement using inequality notation.

51. x is negative. **52.** t is positive.

53. x is nonnegative. **54.** y is more than 25.

55. u is at least 16.

56. The Dow-Jones Average, A, will exceed 4,000 by the year 2000.

57. The price, p, of a coat will be less than \$225 during a sale.

58. The tire pressure, p, must be at least 30 pounds per square inch.

In Exercises 59–62, determine whether the statement is true or false.

59. Every integer is a rational number.

60. Every real number is either rational or irrational.

61. Every rational number is an integer.

62. The absolute value of every real number is positive.

SECTION 1.2

Operations with Real Numbers

Addition and Subtraction of Real Numbers • Multiplication of Real Numbers • Division of Real Numbers • Positive Integer Exponents • Symbols of Grouping • Calculators

Addition and Subtraction of Real Numbers

In this section we review the four basic operations of arithmetic: addition, subtraction, multiplication, and division. The first operation we consider is **addition** of two real numbers.

Addition of Two Real Numbers

1. To add two real numbers with *like signs*, add their absolute values and attach the common sign to the result.
2. To add two real numbers with *unlike signs*, subtract the smaller absolute value from the greater absolute value and attach the sign of the number with the greater absolute value.

The result of adding two real numbers is called the **sum** of the two numbers, and the two real numbers are called **terms** of the sum.

EXAMPLE 1 ■ **Adding Integers and Decimals**

(a) $-84 + 14 = -(84 - 14) = -70$

(b) $-138 + (-62) = -(138 + 62) = -200$

(c) $3.2 + (-0.4) = +(3.2 - 0.4) = 2.8$

(d) $58.06 + 24.7 = 82.76$ ■

Subtraction of one real number from another can be described as *adding the opposite* of the second number to the first number. For instance,

$$7 - 5 = 7 + (-5) \quad \text{and} \quad 10 - (-13) = 10 + 13.$$

In general, we define subtraction as follows.

Subtraction of Two Real Numbers

To **subtract** the real number b from the real number a, we add the opposite of b to a. That is,

$$a - b = a + (-b).$$

The result of this subtraction is called the **difference** of a and b.

EXAMPLE 2 ■ **Subtracting Integers and Decimals**

(a) $6 - 14 = 6 + (-14) = -(14 - 6) = -8$

(b) $-25 - (-27) = -25 + 27 = 2$

(c) $-13.8 - 7.02 = -13.8 + (-7.02) = -(13.8 + 7.02) = -20.82$ ■

To add two fractions with *like denominators*, we simply add their numerators and write the sum over the common denominator. For instance,

$$\frac{1}{5} + \frac{2}{5} = \frac{1 + 2}{5} = \frac{3}{5}.$$

To add two fractions with *unlike denominators*, we must first rewrite one (or both) of the fractions so that they have the same denominator. In choosing the common denominator, we suggest finding the smallest integer that is a multiple of both denominators. We call this integer the **least common denominator** of the two fractions. For instance, the two fractions $\frac{1}{2}$ and $\frac{1}{3}$ have a least common denominator of 6. Therefore, we can add these two fractions as follows.

$$\frac{1}{2} + \frac{1}{3} = \frac{1(3)}{2(3)} + \frac{1(2)}{3(2)} \quad \text{Common denominator is 6}$$

$$= \frac{3}{6} + \frac{2}{6} \quad \text{Fractions have like denominators}$$

$$= \frac{3 + 2}{6} \quad \text{Add numerators}$$

$$= \frac{5}{6}$$

The same principle applies to subtraction of fractions, as indicated in the following summary.

Addition and Subtraction of Fractions

1. Let a/c and b/c be two fractions with *like* denominators. The sum and difference of the two fractions is given by

$$\frac{a}{c} + \frac{b}{c} = \frac{a + b}{c} \quad \text{Sum}$$

$$\frac{a}{c} - \frac{b}{c} = \frac{a - b}{c}. \quad \text{Difference}$$

2. Let a/c and b/d be two fractions with *unlike* denominators. To find the sum or difference of the two fractions, rewrite the fractions so that they have the same denominators and apply the first rule, *or* use the following rule.

$$\frac{a}{c} + \frac{b}{d} = \frac{ad + bc}{cd} \quad \text{Sum}$$

$$\frac{a}{c} - \frac{b}{d} = \frac{ad - bc}{cd} \quad \text{Difference}$$

EXAMPLE 3 ■ **Adding Fractions**

(a) $\frac{5}{17} + \frac{9}{17} = \frac{5+9}{17}$ — Add numerators

$= \frac{14}{17}$

(b) $\frac{3}{8} + \frac{-5}{12} = \frac{3(3)}{8(3)} - \frac{5(2)}{12(2)}$ — Least common denominator is 24

$= \frac{9}{24} - \frac{10}{24}$ — Simplify

$= \frac{9-10}{24}$ — Subtract numerators

$= \frac{-1}{24}$

$= -\frac{1}{24}$ ■

NOTE: In Example 3(b) we added the fractions by the least common denominator method. Try reworking the problem using the rule

$$\frac{a}{c} + \frac{b}{d} = \frac{ad + bc}{cd}.$$

EXAMPLE 4 ■ **Adding and Subtracting Fractions or Mixed Numbers**

(a) $\frac{7}{24} - \frac{3}{20} = \frac{7(5)}{24(5)} - \frac{3(6)}{20(6)}$ — Least common denominator is 120

$= \frac{35}{120} - \frac{18}{120}$ — Simplify

$= \frac{35-18}{120}$ — Subtract numerators

$= \frac{17}{120}$

(b) $1\frac{4}{5} + \frac{11}{7} = \frac{9}{5} + \frac{11}{7}$ — Write mixed number $1\frac{4}{5}$ as $\frac{9}{5}$

$= \frac{9(7) + 11(5)}{5(7)}$ — $(a/c) + (b/d) = (ad + bc)/cd$

$= \frac{63 + 55}{35}$ — Simplify

$= \frac{118}{35}$ — Add numerators ■

Multiplication of Real Numbers

Multiplication of a real number by a positive integer can be described as repeated addition. Here are some examples.

Multiplication	*Repeated Addition*
$3(5) = 15$	$\underbrace{5 + 5 + 5}_{\text{Add 5 three times}} = 15$
$4(-2) = -8$	$\underbrace{(-2) + (-2) + (-2) + (-2)}_{\text{Add } -2 \text{ four times}} = -8$

We denote multiplication in a variety of ways. For instance,

$7 \times 3, \quad 7 \cdot 3, \quad 7(3), \quad \text{and} \quad (7)(3)$

all denote the product of 7 times 3, which we know is 21.

Multiplication of Two Real Numbers

1. To multiply two real numbers with *like signs*, find the product of their absolute values. (The final result is *positive*.)
2. To multiply two real numbers with *unlike signs*, find the product of their absolute values, and attach a minus sign. (The final result is *negative*.)

The result of multiplying two numbers is called their **product**, and each of the two numbers is called a **factor** of the product.

The product of zero and any other number is zero. For instance, if we multiply 0 and 4, we obtain $(0)(4) = 0$.

EXAMPLE 5 ■ **Multiplying Integers**

(a) $-6 \cdot 9 = -54$

(b) $(-5)(-7) = 35$

(c) To find the product of more than two numbers, we first find the product of their absolute values. If there is an *even* number of negative factors, then the product is positive. If there is an *odd* number of negative factors, then the product is negative. For instance, in the following product, there are two negative factors, so the product must be positive, and we write

$$5(-3)(-4)(7) = 420.$$ ■

To multiply two fractions, we simply multiply their numerators and their denominators. For instance, the product of $\frac{2}{3}$ and $\frac{4}{5}$ is given by

$$\left(\frac{2}{3}\right)\left(\frac{4}{5}\right) = \frac{(2)(4)}{(3)(5)} = \frac{8}{15}.$$

Multiplication of Two Fractions

The product of the two fractions a/c and b/d is given by

$$\frac{a}{c} \cdot \frac{b}{d} = \frac{ab}{cd}.$$

When operating with fractions, you should check to see if your answers can be reduced by dividing the numerator and denominator by factors that are common to the numerator and denominator. For instance, the fraction $\frac{4}{6}$ can be written in reduced form as

$$\frac{4}{6} = \frac{\cancel{2}(2)}{\cancel{2}(3)} = \frac{2}{3}.$$

EXAMPLE 6 ■ **Multiplying Fractions and Decimals**

Find the following products.

(a) $\left(-\frac{3}{8}\right)\left(\frac{11}{6}\right)$ (b) $(0.28)(3.14)$

Solution

(a) $\left(-\frac{3}{8}\right)\left(\frac{11}{6}\right) = -\frac{3(11)}{8(6)}$ Multiply numerators and denominators

$= -\frac{\cancel{3}(11)}{8(2)(\cancel{3})}$ Divide numerator and denominator by common factor

$= -\frac{11}{16}$ Reduced fraction

(b) To multiply two decimal numbers by hand, it is convenient to use a vertical arrangement, as follows.

$$\begin{array}{r} 3.14 \\ \times\ 0.28 \\ \hline 2512 \\ 628 \\ \hline 0.8792 \end{array}$$

Thus, the product of 3.14 and 0.28 is 0.8792. ■

Division of Real Numbers

The **reciprocal** of a nonzero real number a is defined to be the number that a must be multiplied by to obtain 1. For instance, the reciprocal of 3 is $\frac{1}{3}$ because

$$3\left(\frac{1}{3}\right) = 1.$$

Similarly, the reciprocal of $-\frac{4}{5}$ is $-\frac{5}{4}$ because

$$-\frac{4}{5}\left(-\frac{5}{4}\right) = 1.$$

In general, the reciprocal of a/b is b/a.

Note that the reciprocal of a positive number is positive, and the reciprocal of a negative number is negative. Also, be sure you see that zero does not have a reciprocal because there is no number that can be multiplied by zero to obtain 1.

To divide one real number by a second (nonzero) real number, we multiply the first real number by the reciprocal of the second real number.

Division of Two Real Numbers

To divide the real number a by the nonzero real number b, we multiply a by the reciprocal of b. That is,

$$a \div b = a\left(\frac{1}{b}\right).$$

The result of dividing two real numbers is called the **quotient** of the numbers. The number a is called the **dividend**, and the number b is called the **divisor**. Using the symbols

$$a/b \qquad \text{or} \qquad \frac{a}{b}$$

we call a the **numerator** and b the **denominator**.

Note that we *do not* define division by zero. For instance, the expression

$$\frac{3}{0} \qquad \text{Division by zero is undefined}$$

is undefined.

EXAMPLE 7 ■ Division of Real Numbers

(a) $-30 \div 5 = -30\left(\frac{1}{5}\right) = -\frac{30}{5} = -\frac{6(\cancel{5})}{\cancel{5}} = -6$

(b) $-\frac{9}{14} \div -\frac{1}{3} = -\frac{9}{14}\left(-\frac{3}{1}\right) = \frac{27}{14}$

(c) $\frac{5}{16} \div 2\frac{3}{4} = \frac{5}{16} \div \frac{11}{4} = \frac{5}{16}\left(\frac{4}{11}\right) = \frac{5(4)}{16(11)} = \frac{5}{44}$

(d) $\frac{-2/3}{3/5} = -\frac{2}{3} \div \frac{3}{5} = -\frac{2}{3}\left(\frac{5}{3}\right) = -\frac{10}{9}$ ■

MATH MATTERS

Creating Numbers from Fours

It is possible to create each whole number between 1 and 100 using *exactly* four 4s and the following calculator keys.

[+], [−], [×], [÷], [(], [)], [.], [x^2], [!], [=]

The repeating decimal $0.\overline{4} = 0.4444\ldots$ is also allowed. (To enter $0.4444\ldots$ into your calculator, divide 4 by 9, and store the result for later use.) Note that the symbol ! is used to represent $n!$, as in $4! = 4 \cdot 3 \cdot 2 \cdot 1$. Here is a way to represent the first 10 whole numbers. How many others can you find?

$1 = 44 \div 44$

$2 = (4 \div 4) + (4 \div 4)$

$3 = (4 + 4 + 4) \div 4$

$4 = (4! - 4 - 4) \div 4$

$5 = ((4 \times 4) + 4) \div 4$

$6 = 4 + ((4 + 4) \div 4)$

$7 = 4 + 4 - (4 \div 4)$

$8 = 4 + 4 + 4 - 4$

$9 = 4 + 4 + (4 \div 4)$

$10 = (4 \div .\overline{4}) + (4 \div 4)$

The Greeks thought that the number 4 represented harmony and justice. This made the number 4 lucky, as did the fact that the Greeks believed there are four elements: earth, air, fire, and water. The color of the number four was green and it was represented by the astrological sign of Cancer the Crab.

Positive Integer Exponents

We have seen how multiplication of a real number by a positive integer can be described as repeated addition. In a similar way, *repeated multiplication* can be written in what is called **exponential form**.

Repeated Multiplication		*Exponential Form*
$\underbrace{7 \cdot 7 \cdot 7 \cdot 7}_{\text{Four factors of 7}}$	=	7^4

Exponential Notation

Let n be a positive integer and a be a real number. Then the product of n factors of a is given by

$$a^n = \underbrace{a \cdot a \cdot a \cdots a}_{n \text{ factors}}.$$

In the exponential form a^n, a is called the **base** and n is the **exponent**. When we write the exponential form a^n, we say that we are **raising a to the nth power**.

When a number is raised to the *first* power, we usually do not write the exponent 1. For instance, we usually write 5 rather than 5^1.

EXAMPLE 8 ■ **Evaluating Exponential Expressions**

(a) $(-3)^4 = (-3)(-3)(-3)(-3) = 81$

(b) $-3^4 = -(3)(3)(3)(3) = -81$

(c) $\left(\frac{2}{5}\right)^3 = \left(\frac{2}{5}\right)\left(\frac{2}{5}\right)\left(\frac{2}{5}\right) = \frac{8}{125}$

(d) $(-2)^5 = (-2)(-2)(-2)(-2)(-2) = -32$

(e) $-2^5 = -(2)(2)(2)(2)(2) = -32$ ■

Symbols of Grouping

When adding, subtracting, multiplying, or dividing more than two numbers, it is important to use symbols of grouping (such as parentheses) to indicate the order of operation.

EXAMPLE 9 ■ **Symbols of Grouping**

Evaluate the following expressions.

(a) $7 - 3(4 - 2) = 7 - 3(2) = 7 - 6 = 1$

(b) $(4 - 5) - (3 - 6) = (-1) - (-3) = -1 + 3 = 2$

(c) $4 - 3(2)^3 = 4 - 3(8) = 4 - 24 = -20$ ■

Calculators

This text includes several examples and exercises that use a calculator. As we encounter each new calculator application, we will give general instructions for using a calculator. These instructions, however, may not agree precisely with the steps required by *your* calculator, so be sure you are familiar with the use of the keys on your own calculator.

For each of the calculator examples in the text, we will give two possible keystroke sequences: one for a standard *scientific* calculator, and one for a *graphing* calculator.*

Try the following evaluations on your own calculator to see whether the indicated keystrokes give the same display.

EXAMPLE 10 ■ **Evaluating Expressions on a Calculator**

(a) To evaluate the expression $-4 - 5$, use the following keystrokes.

Keystrokes	*Display*	
4 [+/−] [−] 5 [=]	−9	Scientific
[(−)] 4 [−] 5 [ENTER]	−9	Graphing

*The graphing calculator keystrokes in this text correspond to the *TI-81* graphing calculator from *Texas Instruments*. For other graphing calculators, the keystrokes may differ.

(b) To evaluate the expression $-3^2 + 4$, use the following keystrokes.

Keystrokes	*Display*	
3 [x²] [+/−] [+] 4 [=]	−5	Scientific
[(−)] 3 [x²] [+] 4 [ENTER]	−5	Graphing

(c) To evaluate the expression $(-3)^2 + 4$, use the following keystrokes.

Keystrokes	*Display*	
3 [+/−] [x²] [+] 4 [=]	13	Scientific
[(] [(−)] 3 [)] [x²] [+] 4 [ENTER]	13	Graphing

■

EXAMPLE 11 ■ **Evaluating an Expression on a Calculator**

(a) To evaluate the expression $7 - (5 \cdot 3)$, use the following keystrokes.

Keystrokes	*Display*	
7 [−] [(] 5 [×] 3 [)] [=]	−8	Scientific
7 [−] [(] 5 [×] 3 [)] [ENTER]	−8	Graphing

(b) To evaluate the expression $24 \div 2^3$, use the following keystrokes.

Keystrokes	*Display*	
24 [÷] 2 [yˣ] 3 [=]	3	Scientific
24 [÷] 2 [∧] 3 [ENTER]	3	Graphing

(c) To evaluate the expression $(24 \div 2)^3$, use the following keystrokes.

Keystrokes	*Display*	
[(] 24 [÷] 2 [)] [yˣ] 3 [=]	1728	Scientific
[(] 24 [÷] 2 [)] [∧] 3 [ENTER]	1728	Graphing

(d) To evaluate the expression $5/(4 + 3 \cdot 2)$, use the following keystrokes.

Keystrokes	*Display*	
5 [÷] [(] 4 [+] 3 [×] 2 [)] [=]	0.5	Scientific
5 [÷] [(] 4 [+] 3 [×] 2 [)] [ENTER]	0.5	Graphing

■

NOTE: Be sure you see the difference between the change sign key [+/−] and the subtraction key [−] on a scientific calculator, and the minus key [(−)] and the subtraction key [−] on a graphing calculator.

DISCUSSION PROBLEM ■ Adding Fractions

In Section 1.2 we described two different ways to add fractions with different denominators. (a) Rewrite the following fractions using a common denominator and then add, using the rule for adding fractions with the same denominator. (b) Use the rule

$$\frac{a}{c} + \frac{b}{d} = \frac{ad + bc}{cd}.$$

Write a short paragraph describing the advantages and disadvantages of each technique. Which of the following solutions is more efficient?

(a) $\frac{3}{5} + \frac{7}{10} = \frac{3(2)}{5(2)} + \frac{7}{10}$
$= \frac{6}{10} + \frac{7}{10}$
$= \frac{13}{10}$

(b) $\frac{3}{5} + \frac{7}{10} = \frac{3(10) + 7(5)}{5(10)}$
$= \frac{65}{50}$
$= \frac{\cancel{5}(13)}{\cancel{5}(10)}$
$= \frac{13}{10}$ ■

Warm-Up

The following warm-up exercises involve skills that were covered in earlier sections. You will use these skills in the exercise set for this section.

In Exercises 1–4, plot the two real numbers on the real number line and place the correct inequality symbol (< or >) between the two numbers.

1. $-\frac{5}{2}$ ▭ $\frac{5}{2}$

2. 6 ▭ $-\frac{1}{2}$

3. -3 ▭ $-\frac{1}{3}$

4. -4.2 ▭ -3.4

In Exercises 5–8, evaluate the given quantity.

5. $|-25|$

6. $|13|$

7. $-\left|-\frac{2}{3}\right|$

8. $-|38|$

In Exercises 9 and 10, write the following statement using inequality notation.

9. y is no more than 45.

10. x is at least 15.

1.2 EXERCISES

In Exercises 1–30, perform the indicated operations.

1. $13 + 32$
2. $16 + 84$
3. $-13 + 32$
4. $-16 + 84$
5. $13 + (-32)$
6. $16 + (-84)$
7. $-7 - 15$
8. $-22 - 6$
9. $-13 + (-8)$
10. $-5 + (-52)$
11. $4 - 16 + (-8)$
12. $-15 + (-6) + 32$
13. $5.8 - 6.2 + 1.1$
14. $46.08 - 35.1 - 16.25$
15. $\frac{3}{8} + \frac{7}{8}$
16. $\frac{5}{6} + \frac{7}{6}$
17. $\frac{3}{4} - \frac{1}{4}$
18. $\frac{5}{9} - \frac{1}{9}$
19. $\frac{3}{5} + \left(-\frac{1}{2}\right)$
20. $\frac{5}{6} - \frac{3}{4}$
21. $\frac{5}{8} + \frac{1}{4} - \frac{5}{6}$
22. $\frac{3}{11} + \frac{-5}{2}$
23. $5\frac{3}{4} + 7\frac{3}{8}$
24. $8\frac{1}{2} - 24\frac{2}{3}$
25. $85 - |-25|$
26. $-36 + |-8|$
27. $-(-11.325) + |34.625|$
28. $|-16.25| - 54.78$
29. $-|-15.667| - 12.333$
30. $-|-15\frac{2}{3}| - 12\frac{1}{3}$

In Exercises 31–34, write the given multiplication problem as a repeated addition.

31. $4(5)$
32. $5(2)$
33. $3(-4)$
34. $6(-2)$

In Exercises 35–38, write the given repeated addition as a multiplication problem.

35. $\frac{1}{4} + \frac{1}{4} + \frac{1}{4} + \frac{1}{4} + \frac{1}{4} + \frac{1}{4}$
36. $9 + 9 + 9 + 9$
37. $(-15) + (-15) + (-15) + (-15)$
38. $\left(-\frac{5}{22}\right) + \left(-\frac{5}{22}\right) + \left(-\frac{5}{22}\right)$

In Exercises 39–46, find the required product without using a calculator.

39. $5(-6)$
40. $-7(3)$
41. $6.3(5.1)$
42. $(-4.4)(-3.2)$
43. $\left(-\frac{5}{8}\right)\left(-\frac{4}{5}\right)$
44. $\left(\frac{10}{13}\right)\left(-\frac{3}{5}\right)$
45. $-\frac{9}{8}\left(\frac{16}{27}\right)\left(\frac{1}{2}\right)$
46. $\frac{2}{3}\left(-\frac{18}{5}\right)\left(-\frac{5}{6}\right)$

In Exercises 47–56, find the required quotient and round your answer to two decimal places. (A calculator may be useful.)

47. $\frac{-18}{-3}$
48. $-\frac{30}{-15}$
49. $-\frac{4}{5} \div \frac{8}{25}$
50. $\frac{8}{15} \div \frac{32}{5}$

51. $-\dfrac{-1/3}{5/6}$ **52.** $\dfrac{-11/12}{5/24}$ **53.** $5\frac{3}{4} \div 2\frac{1}{8}$ **54.** $-3\frac{5}{6} \div -2\frac{2}{3}$

55. $\dfrac{25.5}{6.325}$ **56.** $\dfrac{265.45}{25.6}$

In Exercises 57–60, write the given quantity as a repeated multiplication problem.

57. 4^3 **58.** $\left(\dfrac{2}{3}\right)^4$ **59.** $\left(-\dfrac{4}{5}\right)^6$ **60.** $(-6)^5$

In Exercises 61–64, write the given repeated multiplication problem using exponent notation.

61. $\left(\dfrac{5}{8}\right) \times \left(\dfrac{5}{8}\right) \times \left(\dfrac{5}{8}\right) \times \left(\dfrac{5}{8}\right)$ **62.** $-(7 \times 7 \times 7)$

63. $(-4)(-4)(-4)(-4)(-4)(-4)$ **64.** $(-7) \times (-7) \times (-7)$

In Exercises 65–70, evaluate the exponential expression.

65. $(-2)^4$ **66.** $(-3)^2$ **67.** $(-2)^3$ **68.** $(-3)^3$ **69.** $\left(\dfrac{4}{5}\right)^3$ **70.** $-\left(\dfrac{2}{3}\right)^4$

In Exercises 71–80, evaluate the given expression.

71. $16 - 5(6 - 10)$ **72.** $|-36| + 7(12 - 16)$ **73.** $45 + 3(16 \div 4)$ **74.** $72 - 8(6^2 \div 9)$

75. $5^3 + |-14 + 4|$ **76.** $|(-2)^5| - (25 + 7)$ **77.** $\dfrac{3^2 - 5}{12} - 3\frac{1}{6}$ **78.** $\dfrac{5^3 - 50}{-15} + 27$

79. $0.2(6 - 10)^3 + 85$ **80.** $\dfrac{3}{2} - \left(\dfrac{1}{3} \div \dfrac{5}{6}\right)$

In Exercises 81–86, use a calculator to evaluate the given expression. (Round your answer to two decimal places.)

81. $5.6[13 - 2.5(-6.3)]$ **82.** $35(1{,}032 - 4{,}650)$ **83.** $5^6 - 3(400)$

84. $300(1.09)^{10}$ **85.** $\dfrac{500}{(1.055)^{20}}$ **86.** $5(100 - 3.6^4) \div 4.1$

87. *Balance in an Account* During a given month one deposit of \$1,236.45 was made in a checking account. There were four withdrawals of \$25.62, \$455, \$125, and \$715.95. The balance at the beginning of the month was \$2,618.68. Find the balance at the end of the month. (Disregard any interest that may have been earned.)

88. *Company Profit* The midyear financial statement of a company showed a profit of \$1,415,322.62. At the close of the year the financial statement showed a profit for the year of \$916,489.26. Find the profit of the company for the second six months of the year.

89. *Savings Plan* Suppose you decide to save \$50 per month for 18 years. How much money will you set aside during the 18 years?

90. *Compound Interest* If the money in Exercise 89 is deposited in a savings account earning 9% compounded monthly, then the total amount in the fund in 18 years will be given by

$$50\left[\left(1+\frac{0.09}{12}\right)^{216}-1\right]\left(1+\frac{12}{0.09}\right).$$

Use a calculator to determine this amount.

In Exercises 91 and 92, determine the unknown fractional part of the given **pie chart**.

91.

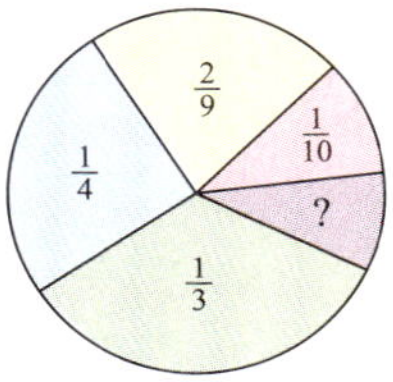

92.

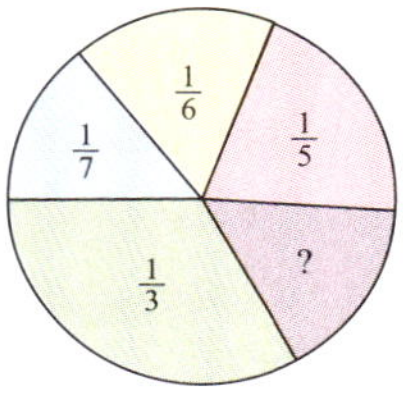

93. *Stock Values* On Monday you purchased \$500 worth of stock. The value of the stock during the remainder of the week is shown in the accompanying **bar chart**. Use the chart to complete the following table showing the daily gains and losses during the week.

Day	Daily Gain or Loss
Tuesday	
Wednesday	
Thursday	
Friday	

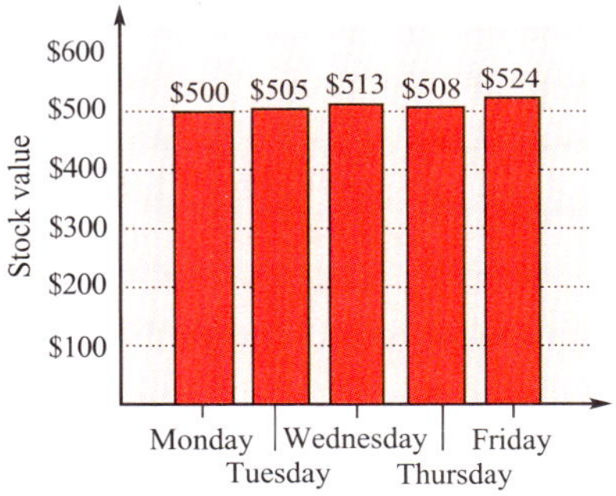

94. *Company Profits* The annual profit for a company (in millions of dollars) is shown in the accompanying bar chart. Complete the following table showing the annual profit gain or loss during the indicated years.

Year	Yearly Gain or Loss
1989	
1990	
1991	
1992	
1993	

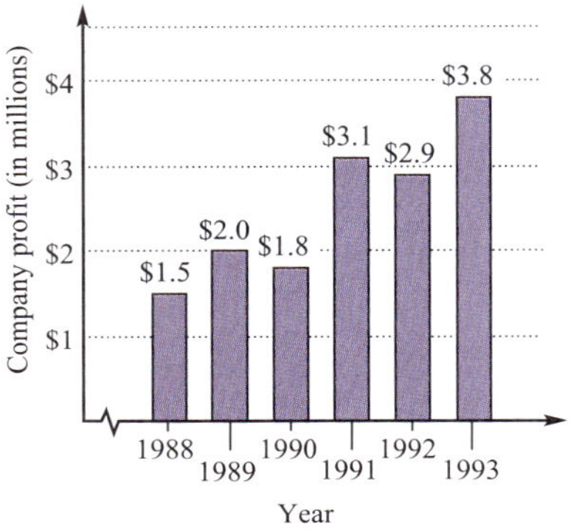

In Exercises 95 and 96, answer the questions using the following information: a bale of hay in the form of a rectangular solid has dimensions 14 inches by 18 inches by 42 inches and weighs approximately 50 pounds. (See figure.)

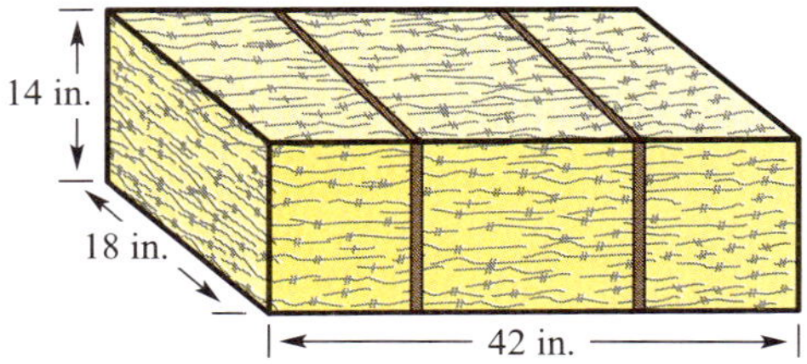

95. *Volume* Find the volume of a bale of hay in cubic feet if the volume of a rectangular solid is the product of its length, width, and height. (Use the fact that 1,728 in.^{3} = 1 ft^3.)

96. *Volume* Approximate the number of bales in a ton of hay. Then approximate the volume of a stack of baled hay that weighs 12 tons.

In Exercises 97–100, determine whether the statement is true or false.

97. The reciprocal of every nonzero integer is an integer.

98. The reciprocal of every nonzero rational number is a rational number.

99. If a negative real number is raised to the 12th power, the result will be positive.

100. If a negative real number is raised to the 11th power, the result will be positive.

SECTION 1.3

Properties of Real Numbers

Mathematical Systems • Basic Properties of Real Numbers • Additional Properties of Real Numbers

Mathematical Systems

In this section we review the properties of real numbers. These properties make up the third component of what is called a **mathematical system**. These three components are a set of numbers, operations with the set of numbers, and properties of the numbers (and operations).

Figure 1.9 shows a diagram that represents different mathematical systems. Note that the set of numbers for the system can vary. The set can consist of whole numbers, integers, rational numbers, or real numbers. (Later, we will see that the set can also consist of algebraic expressions.)

FIGURE 1.9

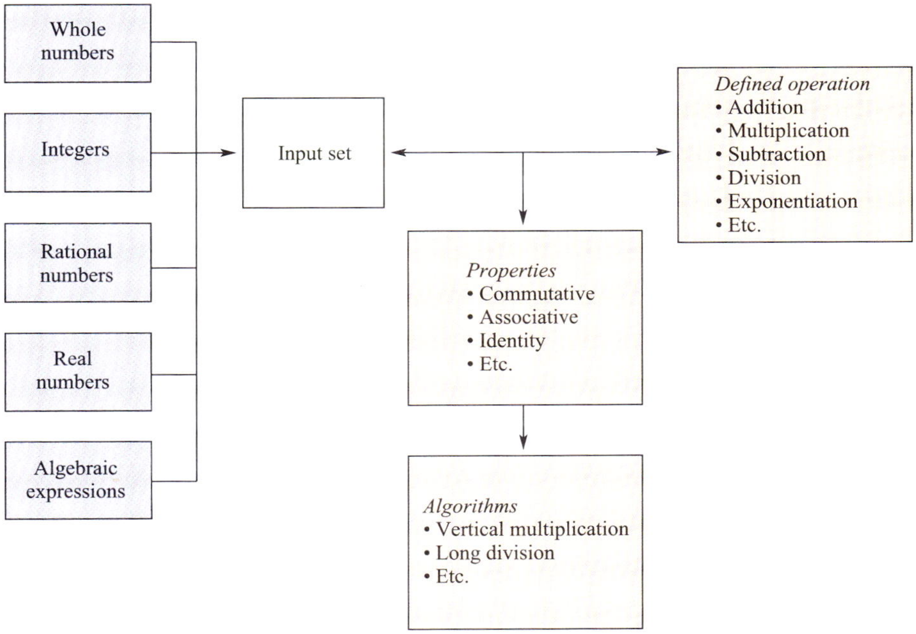

Basic Properties of Real Numbers

For the mathematical system that consists of the set of real numbers together with the operations of addition, subtraction, multiplication, and division, the resulting properties are called the **properties of real numbers**. For each property in the following list, we give a verbal description of the property, as well as an example. (In the following list, remember that the symbol $\neq$ means does not equal.)

Properties of Real Numbers

Property	*Verbal Description*
Closure Property of Addition and Multiplication	The sum and product of two real numbers are real numbers.
$a + b$ is a real number ab is a real number	Examples: $1 + 5 = 6$ is a real number $7 \cdot 3 = 21$ is a real number
Commutative Property of Addition	Two real numbers can be added in either order.
$a + b = b + a$	Example: $2 + 6 = 6 + 2$
Commutative Property of Multiplication	Two real numbers can be multiplied in either order.
$a \cdot b = b \cdot a$	Example: $3 \cdot (-5) = -5 \cdot 3$
Associative Property of Addition	When adding three real numbers, it makes no difference which two are added first.
$(a + b) + c = a + (b + c)$	Example: $(1 + 7) + 4 = 1 + (7 + 4)$
Associative Property of Multiplication	When multiplying three real numbers, it makes no difference which two are multiplied first.
$(ab)c = a(bc)$	Example: $(4 \cdot 3) \cdot 9 = 4 \cdot (3 \cdot 9)$
Distributive Property	Multiplication distributes over addition.
$a(b + c) = ab + ac$ $(b + c)a = ba + ca$	Examples: $2(3 + 4) = 2 \cdot 3 + 2 \cdot 4$ $(3 + 4)2 = 3 \cdot 2 + 4 \cdot 2$
Additive Identity Property	The sum of zero and a real number equals the number itself.
$a + 0 = 0 + a = a$	Example: $4 + 0 = 0 + 4 = 4$
Multiplicative Identity Property	The product of one and a real number equals the number itself.
$a \cdot 1 = 1 \cdot a = a$	Example: $5 \cdot 1 = 1 \cdot 5 = 5$

Additive Inverse Property	The sum of a real number and its opposite is zero.
$a + (-a) = 0$	Example: $5 + (-5) = 0$
Multiplicative Inverse Property	The product of a nonzero real number and its reciprocal is one.
$a \cdot \frac{1}{a} = 1, \quad a \neq 0$	Example: $7 \cdot \frac{1}{7} = 1$

NOTE: Why are the operations of subtraction and division not listed in the preceding collection? It is because they fail to possess many of these properties. For instance, subtraction and division are not commutative. To see this, consider $4 - 3 \neq 3 - 4$ and $15 \div 5 \neq 5 \div 15$. Similarly, the examples $8 - (6 - 2) \neq (8 - 6) - 2$ and $20 \div (4 \div 2) \neq (20 \div 4) \div 2$ illustrate the fact that subtraction and division are not associative.

EXAMPLE 1 ■ Identifying Properties of Real Numbers

Name the property of real numbers that justifies the given statement.

(a) $4(a + 3) = 4 \cdot a + 4 \cdot 3$

(b) $6 \cdot \frac{1}{6} = 1$

(c) $-3 + (2 + b) = (-3 + 2) + b$

(d) $(b + 8) + 0 = b + 8$

Solution

(a) This statement is justified by the Distributive Property.

(b) This statement is justified by the Multiplicative Inverse Property.

(c) This statement is justified by the Associative Property of Addition.

(d) This statement is justified by the Additive Identity Property. ■

EXAMPLE 2 ■ Using the Properties of Real Numbers

Complete the following statements using the specified property of real numbers.

(a) Multiplicative Identity Property
$(4a)1 =$ ____

(b) Associative Property of Addition
$(b + 8) + 3 =$ ____

(c) Additive Inverse Property
$0 = 5c +$ ____

(d) Distributive Property
$4 \cdot b + 4 \cdot 5 =$ ____

Solution

(a) By the Multiplicative Identity Property, we can write

$$(4a)1 = 4a.$$

(b) By the Associative Property of Addition, we can write

$$(b + 8) + 3 = b + (8 + 3).$$

(c) By the Additive Inverse Property, we can write

$$0 = 5c + (-5c).$$

(d) By the Distributive Property, we can write

$$4 \cdot b + 4 \cdot 5 = 4(b + 5).$$

■

Additional Properties of Real Numbers

Once we have determined the basic properties of a mathematical system (called the **axioms** of the system), we can go on to develop other properties of the system. These additional properties are often called **theorems**, and the formal arguments that justify the theorems are called **proofs**. In the following list, we summarize several additional properties of real numbers.

Additional Properties of Real Numbers

Let a, b, and c be real numbers. Then the following properties are true.

PROPERTIES OF EQUALITY

Property	*Verbal and Algebraic Description*
Addition Property of Equality	Adding a real number to both sides of a true equation produces another true equation. If $a = b$, then $a + c = b + c$.
Multiplication Property of Equality	Multiplying both sides of a true equation by a nonzero real number produces another true equation. If $a = b$, then $ac = bc$, $\quad c \neq 0$.
Cancellation Property of Addition	Subtracting a real number from both sides of a true equation produces another true equation. If $a + c = b + c$, then $a = b$.
Cancellation Property of Multiplication	Dividing both sides of a true equation by a nonzero real number produces another true equation. If $ac = bc$ and $c \neq 0$, then $a = b$.

PROPERTIES OF ZERO

Property	*Verbal and Algebraic Description*
Multiplication Property of Zero	The product of zero and any real number is zero. $0 \cdot a = 0$
Division Property of Zero	If zero is divided by any nonzero real number, the result is zero. $\frac{0}{a} = 0, \quad a \neq 0$
Division by Zero is Undefined	We do not define division by zero. $\frac{a}{0}$ is undefined.

PROPERTIES OF NEGATION

Multiplication by -1	The opposite of a real number a can be obtained by multiplying the real number by -1. $(-1)a = -a$ $(-1)(-a) = a$
Placement of Minus Signs	$(-a)(b) = -(ab) = (a)(-b)$
Product of Two Opposites	The product of the opposites of two real numbers is equal to the product of the two real numbers. $(-a)(-b) = ab$

Each of the additional properties in this list can be proved using the basic properties of real numbers. In Examples 3 and 4, we give you examples of such proofs. Proofs of some of the other properties are outlined in the exercises. (See Exercises 71 and 72.)

EXAMPLE 3 ■ Proof of the Cancellation Property of Addition

Prove that if $a + c = b + c$, then $a = b$.

Solution

$a + c = b + c$	Given equation
$(a + c) + (-c) = (b + c) + (-c)$	Addition Property of Equality
$a + (c + (-c)) = b + (c + (-c))$	Associative Property of Addition
$a + 0 = b + 0$	Additive Inverse Property
$a = b$	Additive Identity Property ■

EXAMPLE 4 ■ Proof of a Property of Negation

Prove that $(-1)a = -a$.

Solution

At first glance, it is a little difficult to see what we are being asked to prove. However, a good way to start is to carefully consider the definitions of each of the three numbers in the equation.

$$\begin{aligned} a &= \text{given real number} \\ -1 &= \text{the additive inverse of } 1 \\ -a &= \text{the additive inverse of } a \end{aligned}$$

Now, by showing that $(-1)a$ has the same properties as the additive inverse of a, we will be showing that $(-1)a$ must be the additive inverse of a.

$$\begin{aligned} (-1)a + a &= (-1)a + (1)(a) && \text{Multiplicative Identity Property} \\ &= (-1 + 1)a && \text{Distributive Property} \\ &= (0)a && \text{Additive Inverse Property} \\ &= 0 && \text{Multiplication Property of Zero} \end{aligned}$$

Since we have shown that $(-1)a + a = 0$, we now use the fact that $-a + a = 0$ to conclude that $(-1)a + a = -a + a$. From this, we complete the proof as follows.

$$\begin{aligned} (-1)a + a &= -a + a && \text{Shown in first part of proof} \\ (-1)a &= -a && \text{Cancellation Property of Addition} \end{aligned}$$ ■

The list of additional properties of real numbers forms a very important part of algebra. Knowing the names of the properties is not especially important, but knowing how to use each property is extremely important. In the next two examples, we show how several of the properties are used to solve common problems in algebra.

EXAMPLE 5 ■ Applying the Properties of Real Numbers

In the following list of equations, each equation can be justified from the previous equation using one of the properties of real numbers. Which properties of real numbers justify each step?

$$\begin{aligned} b + 2 &= 6 && \text{Given equation} \\ (b + 2) + (-2) &= 6 + (-2) \\ b + (2 + (-2)) &= 6 - 2 \\ b + 0 &= 4 \\ b &= 4 \end{aligned}$$

Solution

$b + 2 = 6$	Given equation
$(b + 2) + (-2) = 6 + (-2)$	Addition Property of Equality
$b + (2 + (-2)) = 6 - 2$	Associative Property of Addition
$b + 0 = 4$	Additive Inverse Property
$b = 4$	Additive Identity Property

EXAMPLE 6 Applying the Properties of Real Numbers

In the following list of equations, each equation can be justified from the previous equation using one of the properties of real numbers. Which properties of real numbers justify each step?

$3a = 9$	Given equation
$\left(\frac{1}{3}\right)(3a) = \left(\frac{1}{3}\right)(9)$	
$\left(\frac{1}{3} \cdot 3\right)(a) = 3$	
$(1)(a) = 3$	
$a = 3$	

Solution

$3a = 9$	Given equation
$\left(\frac{1}{3}\right)(3a) = \left(\frac{1}{3}\right)(9)$	Multiplication Property of Equality
$\left(\frac{1}{3} \cdot 3\right)(a) = 3$	Associative Property of Multiplication
$(1)(a) = 3$	Multiplicative Inverse Property
$a = 3$	Multiplicative Identity Property

DISCUSSION PROBLEM You Be the Instructor

One of the most common errors in algebra is illustrated in the following *incorrect* use of the Distributive Property.

$$-3(4 + x) = -12 + 3x$$

Suppose you are teaching an algebra class. What would you say to your class to help them avoid this type of error?

Warm-Up

The following warm-up exercises involve skills that were covered in earlier sections. You will use these skills in the exercise set for this section.

In Exercises 1–10, perform the indicated operations.

1. $625 + (-400)$

2. $-360 + 120$

3. $-2(225 - 150)$

4. $5(57 - 33)$

5. $7 \times \frac{8}{35}$

6. $-\frac{4}{15} \times \frac{15}{16}$

7. $\frac{3}{8} \div \frac{5}{16}$

8. $\frac{9}{10} \div 3$

9. $(12 - 15)^3$

10. $\left(\frac{5}{8}\right)^2$

1.3 EXERCISES

In Exercises 1–20, name the property of real numbers that justifies the given statement.

1. $3 + (-5) = -5 + 3$

2. $-5(7) = 7(-5)$

3. $(5 + 10)(8) = 8(5 + 10)$

4. $25 + 35 = 35 + 25$

5. $5(2a) = (5 \cdot 2)a$

6. $5 + 0 = 5$

7. $7 \cdot 1 = 7$

8. $3(6 + b) = 3 \cdot 6 + 3 \cdot b$

9. $3 + (12 - 9) = (3 + 12) - 9$

10. $(-4 \cdot 10) \cdot 8 = -4(10 \cdot 8)$

11. $(8 - 5)(10) = 8 \cdot 10 - 5 \cdot 10$

12. $(16 + 8) - 5 = 16 + (8 - 5)$

13. $3x + 0 = 3x$

14. $8y \cdot 1 = 8y$

15. $25 - 25 = 0$

16. $10x \cdot \frac{1}{10x} = 1$

17. $\frac{1}{y} \cdot y = 1$

18. $6(x + 3) = 6 \cdot x + 6 \cdot 3$

19. $(6 + x) - m = 6 + (x - m)$

20. $3x - 3x = 0$

In Exercises 21–30, use the given property of real numbers to fill in the missing part of the statement.

21. Associative Property of Multiplication: $3(6y) =$ ▭

22. Commutative Property of Addition: $10 + (-6) =$ ▭

23. Commutative Property of Multiplication: $15(-3) =$ ▭

24. Associative Property of Addition: $6 + (5 - y) =$ ▭

25. Distributive Property: $5(6 + z) =$ ▭

26. Distributive Property: $(8 - y)(4) =$ ▭

27. Commutative Property of Addition: $25 + (-x) =$ ▭

28. Additive Inverse: $13x - 13x =$ ▭

29. Multiplicative Identity: $(x + 8) \cdot 1 =$ ____

30. Additive Identity: $(8x) + 0 =$ ____

In Exercises 31–38, give (a) the additive inverse and (b) the multiplicative inverse of the given quantity. (Assume that the given quantity is not zero.)

31. 10

32. 18

33. -16

34. -52

35. $6z$

36. $2y$

37. $x + 1$

38. $y - 4$

In Exercises 39–44, rewrite the given expression using the Associative Property of Addition or Multiplication.

39. $(x + 5) - 3$

40. $(z - 6) + 10$

41. $3(4 \cdot 5)$

42. $(10 \cdot 8) \cdot 5$

43. $6(2y)$

44. $8(3x)$

In Exercises 45–50, rewrite the given expression using the Distributive Property.

45. $20(2 + 5)$

46. $-3(4 - 8)$

47. $5(3x - 4)$

48. $6(2x + 5)$

49. $(x + 6)(-2)$

50. $(z - 10)(12)$

In Exercises 51–54, identify the property of real numbers that justifies each equation from the previous equation.

51.

$$\begin{aligned} x + 5 &= 3 \\ (x + 5) + (-5) &= 3 + (-5) \\ x + (5 + (-5)) &= 3 - 5 \\ x + 0 &= -2 \\ x &= -2 \end{aligned}$$

52.

$$\begin{aligned} x - 8 &= 20 \\ (x - 8) + 8 &= 20 + 8 \\ x + (-8 + 8) &= 28 \\ x + 0 &= 28 \\ x &= 28 \end{aligned}$$

53.

$$\begin{aligned} 2x - 5 &= 6 \\ (2x - 5) + 5 &= 6 + 5 \\ 2x + (-5 + 5) &= 11 \\ 2x + 0 &= 11 \\ 2x &= 11 \\ \frac{1}{2}(2x) &= \frac{1}{2}(11) \\ \left(\frac{1}{2} \cdot 2\right)x &= \frac{11}{2} \\ 1 \cdot x &= \frac{11}{2} \\ x &= \frac{11}{2} \end{aligned}$$

54.

$$\begin{aligned} 3x + 4 &= 10 \\ (3x + 4) + (-4) &= 10 + (-4) \\ 3x + (4 + (-4)) &= 6 \\ 3x + 0 &= 6 \\ 3x &= 6 \\ \frac{1}{3}(3x) &= \frac{1}{3}(6) \\ \left(\frac{1}{3} \cdot 3\right)x &= 2 \\ 1 \cdot x &= 2 \\ x &= 2 \end{aligned}$$

In Exercises 55–60, the right side of the equation is *not* equal to the left side. Change the right side so that it *does* equal the left side.

55. $3(x + 5) \neq 3x + 5$

56. $4(x - 2) \neq 4x - 2$

57. $-2(x - 8) \neq -2x - 16$

58. $-9(x + 4) \neq -9x + 36$

59. $3 \cdot \left(\frac{0}{3}\right) \neq 1$

60. $6 \cdot \left(\frac{1}{6}\right) \neq 0$

In Exercises 61–66, use the Distributive Property to perform the required arithmetic mentally. For example, suppose you work in an industry where the wage is \$14 per hour with "time and a half" for overtime. Thus, your hourly wage for overtime is given by $14(1.5) = 14\left(1 + \frac{1}{2}\right) = 14 + 7 = \$21$.

61. $16(1.75)$

62. $15\left(1\frac{2}{3}\right)$

63. $7(62) = 7(60 + 2)$

64. $5(49) = 5(50 - 1)$

65. $9(6.98) = 9(7 - 0.02)$

66. $12(19.95) = 12(20 - 0.05)$

67. *Area of a Rectangle* The accompanying figure shows two adjoining rectangles. Demonstrate the Distributive Property by filling in the blanks to express the area of the rectangle in two ways.

____ (____ + ____) = ____ + ____

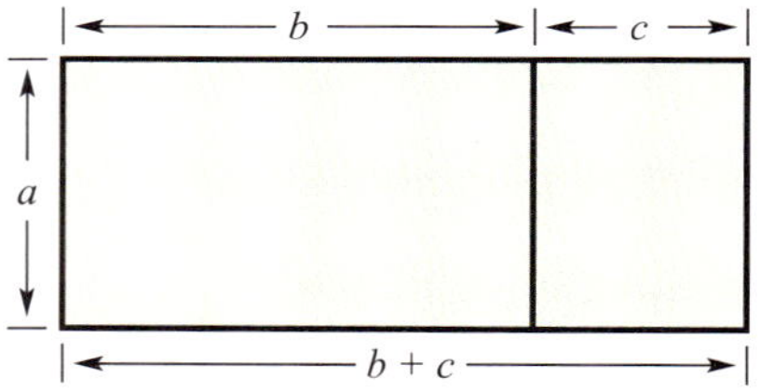

68. *Area of a Rectangle* The accompanying figure shows two adjoining rectangles. Demonstrate the "subtraction version" of the Distributive Property by filling in the blanks to express the area of the left rectangle in two ways.

____ (____ − ____) = ____ − ____

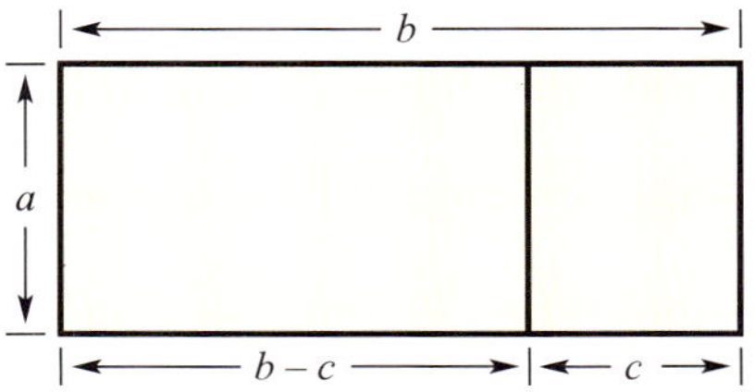

Dividends and Earnings In Exercises 69 and 70, use the following equation, which approximates the dividends paid per share for the General Electric Company.

$$\text{Dividends per share} = 0.13 + 0.367x$$

In this equation the dividends per share is measured in dollars and x represents the earnings per share in dollars (see figure). (*Source*: General Electric Company.)

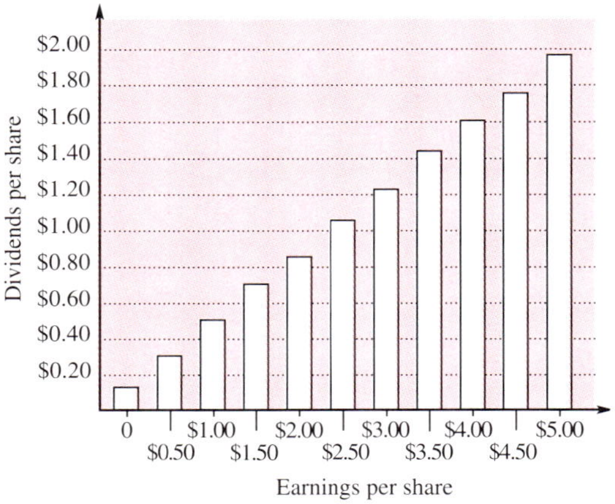

69. In 1980, General Electric's earnings per share was \$1.66. How much was the dividend per share?

70. In 1989, General Electric's earnings per share was \$4.36. How much was the dividend per share?

71. Prove that if $ac = bc$ and $c \neq 0$, then $a = b$.

72. Prove that $(-1)(-a) = a$.

SECTION **1.4**

Algebraic Expressions

Algebraic Expressions • Basic Rules of Algebra • Properties of Exponents • Simplifying Algebraic Expressions • Evaluating Algebraic Expressions

Algebraic Expressions

This section marks our transition from arithmetic to algebra. One of the basic characteristics of algebra is the use of letters (or combinations of letters) to represent numbers. The letters used to represent the numbers are called **variables**, and combinations of letters and numbers are called **algebraic expressions**. Here are a few examples.

$$3x, \qquad x + 2, \qquad \frac{x}{x^2 + 1}, \qquad 2x - 3y$$

Algebraic Expression A collection of letters (called **variables**) and real numbers (called **constants**) combined using the operations of addition, subtraction, multiplication, and division is called an **algebraic expression**.

The **terms** of an algebraic expression are those parts that are separated by addition. For example, the algebraic expression $x^2 - 3x + 6$ has three terms: x^2, $-3x$, and 6. Note that $-3x$ is a term, rather than $3x$, because

$$x^2 - 3x + 6 = x^2 + (-3x) + 6.$$

The terms x^2 and $-3x$ are called the **variable terms** of the expression, and 6 is called the **constant term** of the expression. The numerical factor of a variable term is called the **coefficient** of the variable term. For instance, the coefficient of the variable term $-3x$ is -3, and the coefficient of the variable term x^2 is 1.

EXAMPLE 1 ■ Identifying the Terms of an Algebraic Expression

Algebraic Expression	*Terms*
(a) $5x - \frac{1}{3}$	$5x, -\frac{1}{3}$
(b) $4y + 6x - 9$	$4y, 6x, -9$

■

Basic Rules of Algebra

The properties of real numbers that we discussed in Section 1.3 can be used to rewrite algebraic expressions. The following list is the same as that given in Section 1.3, except that the examples involve algebraic expressions. In other words, the properties are true for variables and algebraic expressions as well as for real numbers.

Basic Rules of Algebra

Let a, b, and c be real numbers, variables, or algebraic expressions. Then the following properties are true.

Property	*Example*
Commutative Property of Addition	
$a + b = b + a$	$5x + x^2 = x^2 + 5x$
Commutative Property of Multiplication	
$ab = ba$	$(3 + x)x^3 = x^3(3 + x)$
Associative Property of Addition	
$(a + b) + c = a + (b + c)$	$(-x + 6) + 3x^2 = -x + (6 + 3x^2)$
Associative Property of Multiplication	
$(ab)c = a(bc)$	$(5x \cdot 4y)(6) = (5x)(4y \cdot 6)$
Distributive Properties	
$a(b + c) = ab + ac$ (left)	$2x(4 + 3x) = 2x \cdot 4 + 2x \cdot 3x$
$(a + b)c = ac + bc$ (right)	$(y + 6)y = y \cdot y + 6 \cdot y$
Additive Identity Property	
$a + 0 = 0 + a = a$	$4y^2 + 0 = 4y^2$
Multiplicative Identity Property	
$a \cdot 1 = 1 \cdot a = a$	$(-5x^3)(1) = (1)(-5x^3) = -5x^3$
Additive Inverse Property	
$a + (-a) = 0$	$4x^2 + (-4x^2) = 0$
Multiplicative Inverse Property	
$a \cdot \frac{1}{a} = 1, \quad a \neq 0$	$(x^2 + 1)\left(\frac{1}{x^2 + 1}\right) = 1$

NOTE: Since subtraction is defined as "adding the opposite," the Distributive Properties are also true for subtraction. For instance, the "subtraction form" of $a(b + c) = ab + ac$ is

$$a(b - c) = a[b + (-c)] = ab + a(-c) = ab - ac.$$

In addition to these basic rules, the properties of equality, zero, and negation given in the previous section are also valid for algebraic expressions. The next example illustrates the use of a variety of the basic rules and properties.

EXAMPLE 2 ■ Identifying the Basic Rules of Algebra

Identify the rule of algebra illustrated in each of the following.

(a) $(5x^2)3 = 3(5x^2)$

(b) $(3x^2 + x) - (3x^2 + x) = 0$

(c) $(y - 6)3 + (y - 6)y = (y - 6)(3 + y)$

(d) $(5 + x^2) + 4x^2 = 5 + (x^2 + 4x^2)$

(e) $5x \cdot \dfrac{1}{5x} = 1, \quad x \neq 0$

Solution

(a) This equation illustrates the Commutative Property of Multiplication. In other words, we obtain the same result whether we multiply $5x^2$ by 3, or 3 by $5x^2$.

(b) This equation illustrates the Additive Inverse Property. In terms of subtraction, this property simply states that when any expression is subtracted from itself the result is zero.

(c) This equation illustrates the Distributive Property in reverse order.

$$ab + ac = a(b + c) \qquad \text{Distributive Property}$$
$$(y - 6)3 + (y - 6)y = (y - 6)(3 + y)$$

Note in this case that $a = y - 6$, $b = 3$, and $c = y$.

(d) This equation illustrates the Associative Property of Addition. In other words, to form the sum $5 + x^2 + 4x^2$, it doesn't matter whether 5 and x^2 are added first or x^2 and $4x^2$ are added first.

(e) This equation illustrates the Multiplicative Inverse Property. Note that it is important that x be a nonzero number. If x were allowed to be zero, we would be in trouble because the reciprocal of zero is undefined. ■

Properties of Exponents

When multiplying two exponential expressions that have the *same base*, we add exponents. To see why this is true, consider the product

$$a^3 \cdot a^2.$$

Since the first expression represents $a \cdot a \cdot a$ and the second represents $a \cdot a$, it follows that the product of the two expressions represents $a \cdot a \cdot a \cdot a \cdot a$, as follows.

$$a^3 \cdot a^2 = \underbrace{(a \cdot a \cdot a)}_{\text{Three factors}} \cdot \underbrace{(a \cdot a)}_{\text{Two factors}} = \underbrace{(a \cdot a \cdot a \cdot a \cdot a)}_{\text{Five factors}} = a^{3+2} = a^5$$

We summarize this and one other property of exponents as follows.

Properties of Exponents

Let m and n be positive integers, and let a and b be real numbers, variables, or algebraic expressions. Then the following properties are true.

1. To multiply two exponential expressions that have the *same base*, we add the exponents.

$$a^m \cdot a^n = a^{m+n}$$

2. To raise a product to a power, we raise each factor to the power and multiply the results.

$$(ab)^m = a^m b^m$$

These rules extend to three or more factors. For example,

$$a^m \cdot a^n \cdot a^k = a^{m+n+k} \quad \text{and} \quad (abc)^m = a^m b^m c^m.$$

In the next two examples, we show how these rules can be used to rewrite products involving exponential forms.

EXAMPLE 3 ■ **Applying the Properties of Exponents**

(a) $b^3 b^2 b = b^{3+2+1} = b^6$

(b) $4^2 x^2 \cdot x = (4^2)(x^2 \cdot x) = 16x^3$

(c) $(-3x^2y)(5xy)(2y^2) = (-3)(5)(2)(x^2 \cdot x)(y \cdot y \cdot y^2) = -30x^3y^4$ ■

EXAMPLE 4 ■ **Applying the Properties of Exponents**

(a) $(3x)^3 = 3^3 \cdot x^3 = 27x^3$

(b) $(-x^2)(x^2) = (-1)(x^2 \cdot x^2) = (-1)x^{2+2} = -x^4$

(c) $(-x)^2(x^2) = x^2 \cdot x^2 = x^{2+2} = x^4$ ■

Simplifying Algebraic Expressions

One of the most common uses of the basic rules of algebra is to rewrite an algebraic expression in simpler form. To **simplify** an algebraic expression generally means that we are to *remove symbols of grouping* such as parentheses or brackets and *combine like terms*.

Two or more terms of an algebraic expression can be combined only if they are *like terms*. In an algebraic expression, two terms are said to be **like terms** if they are both constant terms or if they have the same variable factor(s). For example, the terms $4x$ and $-2x$ are like terms because they have the same variable factor. Similarly, $2x^2y$, $-x^2y$, and $\frac{1}{2}(x^2y)$ are like terms. Note that $4x^2y$ and $-x^2y^2$ are not like terms because their variable factors are different.

To combine like terms in an algebraic expression, we simply add their respective coefficients and attach the common variable factor. This is actually an application of the Distributive Property, as shown in Example 5.

EXAMPLE 5 ■ Combining Like Terms

(a) $2x + 3x - 4 = (2 + 3)x - 4$ — Distributive Property

$= 5x - 4$ — Simplest form

(b) $-3 + 5 + 2y - 7y = (-3 + 5) + (2 - 7)y$ — Distributive Property

$= 2 - 5y$ — Simplest form

(c) $5x + 3y - 4x = 3y + 5x - 4x$ — Commutative Property

$= 3y + (5x - 4x)$ — Associative Property

$= 3y + (5 - 4)x$ — Distributive Property

$= 3y + x$ — Simplest form ■

As you gain experience with the rules of algebra, you may want to combine some of the steps in your work. For instance, you might feel comfortable listing only the following steps to solve Example 5(c).

$5x + 3y - 4x = 3y + (5x - 4x)$ — Group like terms

$= 3y + x$ — Combine like terms

EXAMPLE 6 ■ Combining Like Terms

(a) $7x + 7y - 4x - y = (7x - 4x) + (7y - y)$ — Group like terms

$= 3x + 6y$ — Combine like terms

(b) $2x^2 + 3x - 5x^2 - x = (2x^2 - 5x^2) + (3x - x)$ — Group like terms

$= -3x^2 + 2x$ — Combine like terms

(c) $3xy^2 - 4x^2y^2 + 2xy^2 + (xy)^2 = (3xy^2 + 2xy^2) + (-4x^2y^2 + x^2y^2)$

$= 5xy^2 - 3x^2y^2$ ■

When removing symbols of grouping, we remove the innermost symbols first and combine like terms. Then we repeat the process for any remaining symbols of grouping.

EXAMPLE 7 ■ Removing Symbols of Grouping

$3(x - 5) - (2x - 7) = 3x - 15 - 2x + 7$ — Distributive Property

$= (3x - 2x) + (-15 + 7)$ — Group like terms

$= x - 8$ — Combine like terms

■

EXAMPLE 8 ■ Removing Symbols of Grouping

(a) $5x - 2x[3 + 2(x - 7)] = 5x - 2x[3 + 2x - 14]$ Remove parentheses

$= 5x - 2x[2x - 11]$ Combine like terms

$= 5x - 4x^2 + 22x$ Remove brackets

$= -4x^2 + 27x$ Combine like terms

(b) $-3x(5x^4) + (2x)^5 = -15x^5 + (2^5)(x^5)$

$= -15x^5 + 32x^5$

$= 17x^5$ ■

NOTE: A set of parentheses preceded by a *minus* sign can be removed by changing the sign of each term inside the parentheses. For instance,

$$3x - (2x - 7) = 3x - 2x + 7.$$

A set of parentheses preceded by a *plus* sign can be removed without changing the signs of the terms inside the parentheses. For instance,

$$3x + (2x - 7) = 3x + 2x - 7.$$

Evaluating Algebraic Expressions

To **evaluate** an algebraic expression, we substitute numerical values for each of the variables in the expression. Here are some examples.

Expression	*Value of Variable*	*Substitute*	*Value of Expression*
$3x + 2$	$x = 2$	$3(2) + 2$	$6 + 2 = 8$
$4x^2 + 2x - 1$	$x = -1$	$4(-1)^2 + 2(-1) - 1$	$4 - 2 - 1 = 1$
$2x(x + 4)$	$x = -2$	$2(-2)(-2 + 4)$	$2(-2)(2) = -8$

EXAMPLE 9 ■ Evaluating Algebraic Expressions

Evaluate the following algebraic expressions when $x = -2$ and $y = 5$.

(a) $2y - 3x$ (b) $5 + x^2$ (c) $5 - x^2$

Solution

(a) When $x = -2$ and $y = 5$, the expression $2y - 3x$ has a value of

$$2(5) - 3(-2) = 10 + 6 = 16.$$

(b) When $x = -2$, the expression $5 + x^2$ has a value of

$$5 + (-2)^2 = 5 + 4 = 9.$$

(c) When $x = -2$, the expression $5 - x^2$ has a value of

$$5 - (-2)^2 = 5 - 4 = 1.$$ ■

EXAMPLE 10 ■ **Evaluating Algebraic Expressions**

Evaluate the following algebraic expressions when $x = 2$ and $y = -1$.

(a) $x - y$ (b) $|y - x|$ (c) $\dfrac{2xy}{5x + y}$

Solution

(a) When $x = 2$ and $y = -1$, the expression $x - y$ has a value of

$$2 - (-1) = 2 + 1 = 3.$$

(b) When $x = 2$ and $y = -1$, the expression $|y - x|$ has a value of

$$|-1 - 2| = |-3| = 3.$$

(c) When $x = 2$ and $y = -1$, the expression $2xy/(5x + y)$ has a value of

$$\frac{2(2)(-1)}{5(2) - 1} = \frac{-4}{10 - 1} = -\frac{4}{9}.$$ ■

DISCUSSION PROBLEM ■ **Verbalizing the Basic Rules**

In this section we listed 10 basic rules of algebra: two associative properties, two commutative properties, two distributive properties, two inverse properties, and two identity properties. Write a short verbal description of each of these properties. ■

Warm-Up

The following warm-up exercises involve skills that were covered in earlier sections. You will use these skills in the exercise set for this section.

In Exercises 1–6, evaluate the given quantity.

1. $6(3 + 5^2)$

2. $-5(6 - 10)$

3. $3(-4) + 4\left(\dfrac{3}{8}\right)$

4. $-|35 - 60| + 8(9 - 5)$

5. $\dfrac{360}{3^2 + 4^2}$

6. $\dfrac{13}{5} - \dfrac{9}{2}$

In Exercises 7–10, name the property of real numbers that justifies the given statement.

7. $8x \cdot \dfrac{1}{8x} = 1$

8. $5 + (-3 + x) = (5 - 3) + x$

9. $-4(x + 10) = -4 \cdot x + (-4)(10)$

10. $3x + 0 = 3x$

1.4 EXERCISES

In Exercises 1–8, identify the terms of the algebraic expression.

1. $10x + 5$

2. $-16t^2 + 48$

3. $-3y^2 + 2y - 8$

4. $25z^3 - 4.8z^2$

5. $4x^2 - 3y^2 - 5x + 2y$

6. $14u^2 + 25uv - 3v^2$

7. $x^2 - 2.5x - \frac{1}{x}$

8. $\frac{3}{t^2} - \frac{4}{t} + 6$

In Exercises 9–12, determine the coefficient of the term.

9. $5y^3$

10. $4x^6$

11. $-\frac{3}{4}t^2$

12. $-8.4x$

In Exercises 13–22, identify the rule of algebra that is illustrated by the given equation.

13. $4 - 3x = -3x + 4$

14. $(10 + x) - y = 10 + (x - y)$

15. $-5(2x) = (-5 \cdot 2)x$

16. $(x - 2)(3) = 3(x - 2)$

17. $(x + 5) \cdot \frac{1}{x + 5} = 1, \quad x \neq -5$

18. $(x^2 + 1) - (x^2 + 1) = 0$

19. $5(y^3 + 3) = 5y^3 + 5 \cdot 3$

20. $8x + 0 = 8x$

21. $(16t^4) \cdot 1 = 16t^4$

22. $-32(u^2 - 3u) = -32u^2 - (-32)(3u)$

In Exercises 23–28, use the given property to rewrite the expression.

23. (a) Distributive Property: $5(x + 6) =$ ____
(b) Commutative Property of Multiplication: $5(x + 6) =$ ____

24. (a) Distributive Property: $6x + 6 =$ ____
(b) Commutative Property of Addition: $6x + 6 =$ ____

25. (a) Commutative Property of Multiplication: $6(xy) =$ ____
(b) Associative Property of Multiplication: $6(xy) =$ ____

26. (a) Additive Identity Property: $3 + 0 =$ ____
(b) Commutative Property of Addition: $3 + 0 =$ ____

27. (a) Additive Inverse Property: $4 + (-4) =$ ____
(b) Commutative Property of Addition: $4 + (-4) =$ ____

28. (a) Associative Property of Addition: $(3 + 6) + (-9) =$ ____
(b) Additive Inverse Property: $9 + (-9) =$ ____

In Exercises 29–32, use the definition of exponents to write the expression as a repeated multiplication.

29. $x^3 \cdot x^4$ **30.** $z^2 \cdot z^5$ **31.** $(-2x)^3$ **32.** $-2x^4$

In Exercises 33–36, write the given expression using exponential notation.

33. $(5x)(5x)(5x)(5x)$ **34.** $(y \cdot y \cdot y)(y \cdot y \cdot y \cdot y)$

35. $(x \cdot x \cdot x)(y \cdot y \cdot y)$ **36.** $(-9t)(-9t)(-9t)(-9t)(-9t)(-9t)$

In Exercises 37–46, use the properties of exponents to rewrite the given expression.

37. $x^5 \cdot x^7$ **38.** $u^3 \cdot u^5 \cdot u$ **39.** $3^3y^4 \cdot y^2$ **40.** $6^2x^3 \cdot x^5$

41. $(-4x)^2$ **42.** $(-4x)^3$ **43.** $(2xy)(3x^2y^3)$ **44.** $(-5a^2b^3)(2ab^4)$

45. $(3uv)^2(-6u^3v)$ **46.** $(10x^2y)^3(2x^4y)$

In Exercises 47–56, simplify the expression by combining like terms.

47. $3x + 4x$ **48.** $-2x^2 + 4x^2$ **49.** $9y - 5y + 4y$

50. $8y + 7y - y$ **51.** $3x - 2y + 5x + 20y$ **52.** $-2a + \frac{1}{3}b - 7a - b$

53. $8z^2 + \frac{3}{2}z - \frac{5}{2}z^2 + 10$ **54.** $-5y^3 + 3y - 6y^2 + 8y^3 + y - 4$ **55.** $2uv + 5u^2v^2 - uv - (uv)^2$

56. $3\left(\frac{1}{x}\right) - \frac{1}{x} + 8$

In Exercises 57–64, use the Distributive Property to rewrite the expression.

57. $4(2x^2 + x - 3)$ **58.** $8(z^3 - 4z^2 + 2)$ **59.** $-3(6y^2 - y - 2)$ **60.** $-5(-x^2 + 2y + 1)$

61. $4x - 8$ **62.** $9x + 18y$ **63.** $ab + 2a$ **64.** $3bx^2 - 6bx$

In Exercises 65–74, simplify the algebraic expression by removing the symbols of grouping and combining like terms.

65. $10(x - 3) + 2x - 5$ **66.** $3(x^2 + 1) + x^2 - 6$ **67.** $-3(y^2 + 3y - 1) + 2(y - 5)$

68. $5(a + 6) - 4(a^2 - 2a - 1)$ **69.** $2[3(b - 5) - (b^2 + b + 3)]$ **70.** $4[5 - 3(x^2 + 10)]$

71. $2x(5x^2) - (x^3 + 5)$ **72.** $x(x^2 + 3) - 3(x + 4)$ **73.** $y^2(y + 1) + y(y^2 + 1)$

74. $2ab(b^2 - 3) - ab(b^2 + 2)$

In Exercises 75–86, evaluate the algebraic expression for the specified value of the variable. If not possible, state the reason.

Expression	*Value*
75. $5 - 3x$	(a) $x = \frac{2}{3}$ (b) $x = 5$
76. $\frac{3}{2}x - 2$	(a) $x = 6$ (b) $x = -3$
77. $10 - \lvert x \rvert$	(a) $x = 3$ (b) $x = -3$
78. $2x^2 + 5x - 3$	(a) $x = \frac{1}{2}$ (b) $x = -3$
79. $\frac{x}{x^2 + 1}$	(a) $x = 0$ (b) $x = 3$
80. $5 - \frac{3}{x}$	(a) $x = 0$ (b) $x = -6$
81. $3x + 2y$	(a) $x = 1,\ y = 5$ (b) $x = -6,\ y = -9$
82. $x^2 - xy + y^2$	(a) $x = 2,\ y = -1$ (b) $x = -3,\ y = -2$
83. $\frac{x}{x - y}$	(a) $x = 0,\ y = 10$ (b) $x = 4,\ y = 4$
84. $\lvert y - x \rvert$	(a) $x = 2,\ y = 5$ (b) $x = -2,\ y = -2$
85. Distance traveled: rt	(a) $r = 40,\ t = 5\frac{1}{4}$ (b) $r = 35,\ t = 4$
86. Simple interest: Prt	(a) $P = \$5{,}000,\ r = 0.085,\ t = 10$ (b) $P = \$750,\ r = 0.07,\ t = 3$

87. (a) Complete the following table by evaluating the expression $2x - 5$.

x	-1	0	1	2	3	4
$2x - 5$						

(b) From the table in part (a), determine the increase in the value of the expression for each one-unit increase in x.

(c) Use the results of parts (a) and (b) to guess the increase in the algebraic expression $\frac{3}{4}x + 5$ for each one-unit increase in x.

88. (a) Complete the following table by evaluating the expression $3x + 2$.

x	-1	0	1	2	3	4
$3x + 2$						

(b) From the table in part (a), determine the increase in the value of the expression for each one-unit increase in x.

(c) Use the results of parts (a) and (b) to guess the increase in the algebraic expression $\frac{2}{3}x - 4$ for each one-unit increase in x.

Athletic Shoes In Exercises 89 and 90, use the following equation, which approximates the annual sales of athletic and sports shoes in the United States from 1980 to 1989.

$$\text{Sales} = (41.78 - 0.40t + 0.76t^2 - 0.05t^3)^2$$

In this formula, the sales are given in millions of dollars and t represents the year, with $t = 0$ corresponding to 1980 (see figure). (*Source*: National Sporting Goods Association.)

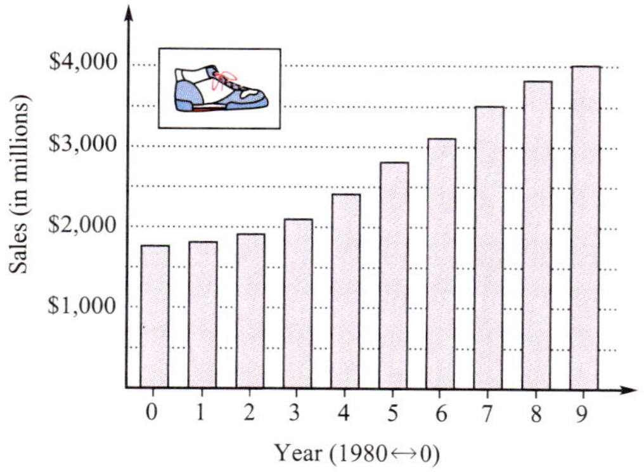

89. Use a calculator to find the sales of athletic and sports shoes in 1985.

90. Use a calculator to find the sales of athletic and sports shoes in 1989.

91. *Area of a Trapezoid* The area of the trapezoid with parallel bases of lengths b_1 and b_2 and height h in the accompanying figure is given by the following.

$$\frac{h}{2}(b_1 + b_2)$$

Use the Distributive Property to show that the area can also be expressed as

$$b_1h + \frac{1}{2}(b_2 - b_1)h.$$

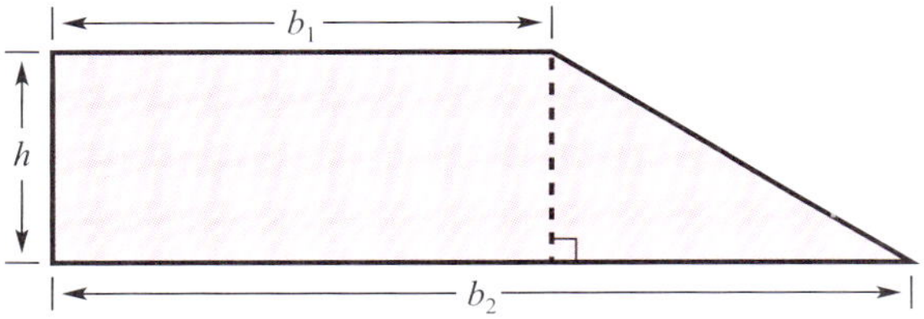

92. *Area of a Trapezoid* Use both formulas given in Exercise 91 to find the area of a trapezoid with $b_1 = 7$, $b_2 = 12$, and $h = 3$.

93. *Amount of Roofing Material* The roof shown in the accompanying figure is made up of two trapezoids and two triangles. Find the total area of the roof.

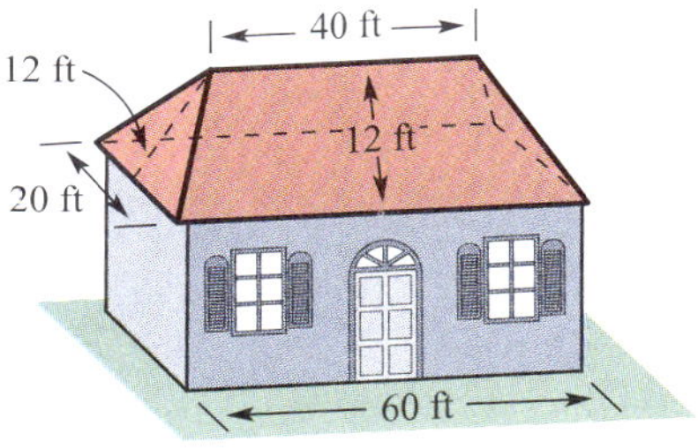

SECTION 1.5 Constructing Algebraic Expressions

Translating Phrases • Hidden Operations

Translating Phrases

In Section 1.4 we focused on rewriting and simplifying algebraic expressions. In this section we look at ways to *construct* algebraic expressions. When you translate a verbal sentence or phrase into an algebraic expression, it helps to watch for key words and phrases that indicate the four different operations of arithmetic. The following list gives several examples.

Translating Key Words and Phrases

Key Words and Phrases	*Verbal Description*	*Algebraic Expression*
Addition:		
Sum, plus, greater, increased by, more than, exceeds, total of	The sum of 5 and x.	$5 + x$
	Seven more than y.	$y + 7$

Subtraction:		
Difference, minus, less than, decreased by,	b is subtracted from 4.	$4 - b$
subtracted from, reduced by, the remainder	Three less than z.	$z - 3$
Multiplication:		
Product, multiplied by, twice, times, percent of	Two times x.	$2x$
Division:		
Quotient, divided by, ratio, per	The ratio of x to 8.	$\frac{x}{8}$

EXAMPLE 1 ■ Translating Verbal Phrases Containing a Specified Variable

Translate each of the following into an algebraic expression.

(a) Seven more than three times x

(b) Four less than the product of 6 and n

(c) y decreased by the sum of 5 and x^2

Solution

(a) *Verbal description:* Seven more than three times x
Algebraic expression: $3x + 7$ — *Think:* 7 added to what?

(b) *Verbal description:* Four less than the product of 6 and n
Algebraic expression: $6n - 4$ — *Think:* 4 subtracted from what?

(c) *Verbal description:* y decreased by the sum of 5 and x^2
Algebraic expression: $y - (5 + x^2)$ — *Think:* What is subtracted from y?

■

EXAMPLE 2 ■ Translating Verbal Phrases Containing a Specified Variable

Translate each of the following into an algebraic expression.

(a) Eight added to the product of 2 and n

(b) Four times the sum of y and 9

(c) Three times the ratio of x to 7

Solution

(a) *Verbal description:* Eight added to the product of 2 and n
Algebraic expression: $8 + 2n$ — *Think:* Add 8 to what?

(b) *Verbal description:* Four times the sum of y and 9
Algebraic expression: $4(y + 9)$ — *Think:* 4 multiplied by what?

(c) *Verbal description:* Three times the ratio of x to 7

Algebraic expression: $3\left(\frac{x}{7}\right)$ *Think:* 3 multiplied by what? ■

In most applications of algebra, the variables are not specified and it is your task to assign variables to the *appropriate* quantities. Though similar to the problems in Examples 1 and 2, the problems in the next example may seem more difficult because variables have not been assigned to the unknown quantities.

EXAMPLE 3 ■ Translating Verbal Phrases Having No Specified Variable

Translate each of the following into a variable expression.

(a) The sum of 7 and a number
(b) Four decreased by the product of 2 and a number
(c) A number increased by the sum of 8 and another number
(d) Four is subtracted from a number and the result is divided by 9.

Solution

(a) *Verbal description:* The sum of 7 and a number
Label: x = the number
Algebraic expression: $7 + x$ *Think:* 7 added to what?

(b) *Verbal description:* Four decreased by the product of 2 and a number
Label: x = the number
Algebraic expression: $4 - 2x$ *Think:* What is subtracted from 4?

(c) *Verbal description:* A number increased by the sum of 8 and another number
Labels: x = one number
y = the other number
Algebraic expression: $x + (8 + y)$ *Think:* A number added to what?

(d) *Verbal description:* Four is subtracted from a number and the result is divided by 9.
Label: x = the number
Algebraic expression: $\frac{x - 4}{9}$ *Think:* What is divided by 9? ■

A good way to learn algebra is to do it forwards and backwards. In the next two examples, we translate algebraic expressions into verbal form. Keep in mind that other key words could be used to describe the operations in each expression. In these two examples we try to use key words or phrases that keep the verbal expression clear and concise.

EXAMPLE 4 ■ Translating Algebraic Expressions into Verbal Form

Without using a variable, write a verbal description for each of the following.

(a) $5x - 10$ (b) $5(x - 10)$

Solution

(a) *Algebraic expression:* $5x - 10$
Primary operation: Subtraction
Terms: $5x$ and 10
Verbal description: 10 less than the product of 5 and a number

(b) *Algebraic expression:* $5(x - 10)$
Primary operation: Multiplication
Terms: 5 and $(x - 10)$
Verbal description: Five times 10 less than a number ■

EXAMPLE 5 ■ **Translating Algebraic Expressions into Verbal Form**

Without using a variable, write a verbal description for each of the following.

(a) $3 + \frac{x}{4}$ (b) $\frac{3 + x}{4}$

Solution

(a) *Algebraic expression:* $3 + \frac{x}{4}$
Primary operation: Addition
Terms: 3 and $\frac{x}{4}$
Verbal description: Three added to the quotient of a number and 4

(b) *Algebraic expression:* $\frac{3 + x}{4}$
Primary operation: Division
Numerator, denominator: Numerator is $3 + x$, denominator is 4
Verbal description: The sum of 3 and a number, all divided by 4 ■

Hidden Operations

Most real-life problems do not contain verbal expressions that clearly identify the arithmetic operations involved in the problem. We need to rely on common sense, past experience, and the physical nature of the problem in order to identify the operations hidden in the problem statement. For instance, products are often overlooked in verbal expressions. Watch for hidden products in the next two examples.

EXAMPLE 6 ■ **Discovering Hidden Products**

(a) A cash register contains x quarters. Write an expression for this amount of money in dollars.

(b) A cash register contains n nickels and d dimes. Write an expression for this amount of money in cents.

Solution

(a) The amount of money is a product.

Verbal model: Value of coin · Number of coins

Labels: Value of coin $= 0.25$ (in dollars)
Number of coins $= x$

Algebraic expression: $0.25x$ (in dollars)

(b) The amount of money is a sum of products.

Verbal model: [Value of nickel] · [Number of nickels] + [Value of dime] · [Number of dimes]

Labels: Value of nickel = 5 (in cents)
Number of nickels = n
Value of dime = 10 (in cents)
Number of dimes = d

Algebraic expression: $5n + 10d$ (in cents) ■

EXAMPLE 7 ■ **Discovering Hidden Products**

Write an expression showing how far a person can ride a bicycle in t hours if the person travels at a constant rate of 12 miles per hour.

Solution

For this problem, we must rely on a known *formula* for distance. This formula tells us that (distance) = (rate)(time).

Verbal model: [Rate] · [Time]

Labels: Rate = 12 miles per hour
Time = t hours

Algebraic expression: $12t$ (in miles) ■

NOTE: In Example 7 the final answer is listed in terms of miles. This makes sense in the following way.

$$12 \frac{\text{miles}}{\cancel{\text{hour}}} \cdot t\ \cancel{\text{hours}}$$

Note that the hours "cancel," leaving miles as the unit of measure in the answer. This technique is called *unit analysis* and it can be very helpful in determining the final unit of measure.

EXAMPLE 8 ■ **Discovering Hidden Operations**

A person adds k liters of a fluid containing 55% antifreeze to a car radiator. Write an expression that indicates how much antifreeze was added.

Solution

The amount of antifreeze is a product.

Verbal model: [Percentage of antifreeze] · [Number of liters]

Labels: Percentage of antifreeze $= 0.55$ (in decimal form)
Number of liters $= k$

Algebraic expression: $0.55k$ (in liters)

Note that the algebraic expression uses the decimal form of 55%. That is, we compute with 0.55 rather than 55%. ■

EXAMPLE 9 ■ Discovering Hidden Operations

(a) Suppose you received a 40% discount on an item priced at d dollars. Write an expression, in dollars, for the discount.

(b) Suppose you paid x dollars for an automobile. Your total cost included a 7% sales tax. Write an expression, in dollars, for the total cost of the automobile.

Solution

(a) The amount of the discount is a product.

Verbal model: [Percentage of discount] · [Original price]

Labels: Percentage of discount $= 0.4$ (in decimal form)
Original price $= d$ (in dollars)

Algebraic expression: $0.4d$ (in dollars)

(b) The total cost is a sum.

Verbal model: [Cost of automobile] + [Tax]

Labels: Cost of automobile $= x$ (in dollars)
Sales tax rate $= 0.07$ (in decimal form)
Sales tax $= 0.07x$ (in dollars)

Algebraic expression: $x + 0.07x = (1 + 0.07)x = 1.07x$ (in dollars) ■

Hidden operations are often involved when assigning a variable name (label) to *two* unknown quantities. For example, suppose two numbers add up to 18 and one of the numbers is assigned the variable x. What expression in x can we use to represent the second number? Let's try a specific case first, and then apply it to a general case.

Specific case: If the first number is 7, then the second number is $18 - 7 = 11$.

General case: If the first number is x, then the second number is $18 - x$.

Note the use of *subtraction* to label the second number.

The strategy of using a specific case to help determine the general case is often helpful in applications of algebra. Observe the use of this strategy in the next two examples.

EXAMPLE 10 ■ Using Hidden Operations to Label Unknown Quantities

In each of the following, use the given variable to label the unknown quantity.

(a) A person's weekly salary is d dollars. What is the annual salary for this person?

(b) A person's annual salary is y dollars. What is the monthly salary for this person?

Solution

(a) There are 52 weeks in a year.

Specific case: If the weekly salary is \$300, then the annual salary is 52(300) dollars.

General case: If the weekly salary is d dollars, then the annual salary is $52 \cdot d$ or $52d$ dollars.

(b) There are 12 months in a year.

Specific case: If the annual salary is \$24,000, then the monthly salary is $24{,}000 \div 12$ dollars.

General case: If the annual salary is y, then the monthly salary is $y \div 12$ or $y/12$ dollars. ■

EXAMPLE 11 ■ Using Hidden Operations to Label Unknown Quantities

In each of the following, use the given variable(s) to properly label the unknown quantity.

(a) One person is k inches shorter than a second person. If the first person is 66 inches tall, how tall is the second person?

(b) A consumer buys g gallons of gasoline for a total of d dollars. What is the price per gallon?

Solution

(a) Keep in mind, the first person is k inches shorter than the second person.

Specific case: If the first person is 2 inches shorter than the second person, then the second person is $66 + 2$ inches tall.

General case: If the first person is k inches shorter than the second person, then the second person is $66 + k$ inches tall.

(b) To obtain the price per gallon, we divide the total price by the total number of gallons.

Specific case: If the total price is \$12.30 and the total number of gallons is 10, then the price per gallon is $12.30 \div 10$ dollars per gallon.

General case: If the total price is d dollars and the total number of gallons is g, then the price per gallon is $d \div g$ or d/g dollars per gallon. ■

In mathematics it is useful to know how to represent certain types of integers algebraically. For instance, consider the set $\{2, 4, 6, 8, \ldots\}$ of *even* integers. What algebraic symbol could we use to denote an unknown, but even, integer? Since every even integer has 2 as a factor,

$$2 = 2 \cdot 1, \qquad 4 = 2 \cdot 2, \qquad 6 = 2 \cdot 3, \qquad 8 = 2 \cdot 4, \qquad \ldots$$

it follows that any integer n multiplied by 2 is sure to be the *even* number $2n$. Moreover, if $2n$ is even, then $2n - 1$ and $2n + 1$ are sure to be *odd* integers. For example, choose $n = 5$. Then $2n = 2 \cdot 5 = 10$ is even, whereas $2n - 1 = 10 - 1 = 9$ and $2n + 1 = 10 + 1 = 11$ are both odd.

Two integers are called **consecutive integers** if they differ by 1. Hence, for any integer n, its next two consecutive integers are $n + 1$ and $(n + 1) + 1$ or $n + 2$. Thus, we can denote three consecutive integers by n, $n + 1$, and $n + 2$. In summary, we have the following.

Labels for Integers

Let n represent an integer. Then even integers, odd integers, and consecutive integers can be represented as follows.

1. $2n$ denotes an *even* integer.
2. $2n - 1$ and $2n + 1$ denote *odd* integers.
3. $\{n, n + 1, n + 2, \ldots\}$ denotes a set of *consecutive* integers.

DISCUSSION PROBLEM

Avoiding Ambiguity

When you write a verbal model or construct an algebraic expression, watch out for statements that might be ambiguous. For instance, the statement

"The sum of 4 and x divided by 5"

could mean

$$\frac{4 + x}{5} \quad \text{or} \quad 4 + \frac{x}{5}.$$

Can you think of other statements like this that can be interpreted in two different ways?

Warm-Up

The following warm-up exercises involve skills that were covered in earlier sections. You will use these skills in the exercise set for this section.

In Exercises 1–10, use the rules of algebra to simplify the expression.

1. $2(x + 5) - 3x$

2. $3(t - 2) + 8t + 1$

3. $25 - 6(x - 3)$

4. $64 - 7(x - 1)$

5. $3(y - 4) + y(y - 3)$

6. $-2(u + 8) + u(u + 2)$

7. $3(2n + 1) - 2(2n + 3)$

8. $5(2n) + 3(2n + 2)$

9. $5\left(\frac{1}{t}\right) + 3\left(\frac{2}{t}\right)$

10. $7\left(\frac{3}{z}\right) - 2\left(\frac{4}{z}\right)$

1.5 EXERCISES

In Exercises 1–20, translate the statement into an algebraic expression.

1. The sum of 8 and a number n

2. The sum of 12 and twice a number n

3. Five more than a number n

4. Six less than a number n

5. Fifteen decreased by three times a number n

6. Six less than four times a number n

7. One-third of a number n

8. Seven-fifths of a number n

9. Thirty percent of the list price L

10. Forty percent of the cost C

11. The quotient of a number x and 6

12. The ratio of y to 3

13. The number c is quadrupled and the product is increased by 10

14. The product of a number y and 10 is decreased by 35

15. The product of 3 and the square of a number x, all decreased by 4

16. The sum of 10 and one-fourth the cube of x

17. The absolute value of the difference between a number n and 5

18. The absolute value of the quotient of y and 4

19. The sum of 3 and four times a number x, all divided by 8

20. Eight times the ratio of 3 to 5

In Exercises 21–30, write a verbal description of the algebraic expression without using the variable.

21. $3x + 2$

22. $4x - 5$

23. $8(x - 5)$

24. $-3(x + 2)$

25. $\frac{y}{8}$

26. $\frac{4x}{5}$

27. $\frac{x + 10}{3}$

28. $25 + \frac{x}{6}$

29. $x(x + 7)$

30. $x^2 + 2$

In Exercises 31–54, write an algebraic expression that represents the specified quantity in the verbal statement and simplify if possible.

31. The amount of money (in dollars) represented by n quarters

32. The amount of money (in dollars) represented by m dimes

33. The amount of money (in cents) represented by m nickels and n dimes

34. The amount of money (in cents) represented by m dimes and n quarters

35. The distance traveled in t hours at an average speed of 55 miles per hour

36. The distance traveled in 5 hours at an average speed of r miles per hour

37. The time to travel 100 miles an at average speed of r miles per hour

38. The average rate of speed when traveling 360 miles in t hours

39. The amount of antifreeze in a cooling system containing y gallons of coolant that is 45% antifreeze

40. The amount of water in q quarts of a food product that is 65% water

41. The amount of wage tax due for a taxable income of I dollars that is taxed at the rate of 1.25%

42. The amount of sales tax on a purchase valued at L dollars if the tax rate is 6%

43. The sale price of a coat that has a list price of L dollars if it is a "20% off" sale

44. The total cost for a family to stay one night at a campground if the charge is \$18 for the parents plus \$3 for each of the n children

45. The total hourly wage for an employee when the base pay is \$8.25 per hour plus 60 cents for each of q units produced per hour

46. The total hourly wage for an employee when the base pay is \$11.65 per hour plus 80 cents for each of q units produced per hour

47. The area of a square with sides of length s

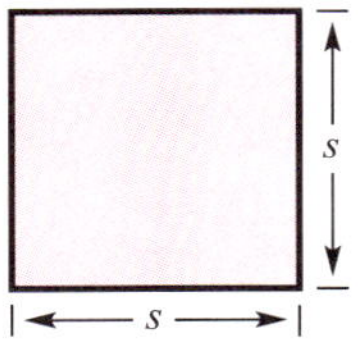

48. The perimeter of a rectangle of width w and length $2w$

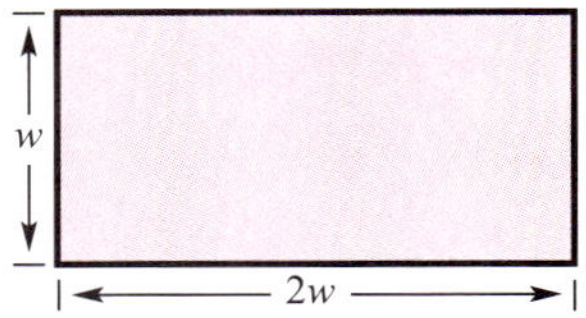

49. The perimeter of a rectangle of length l and width $0.62l$

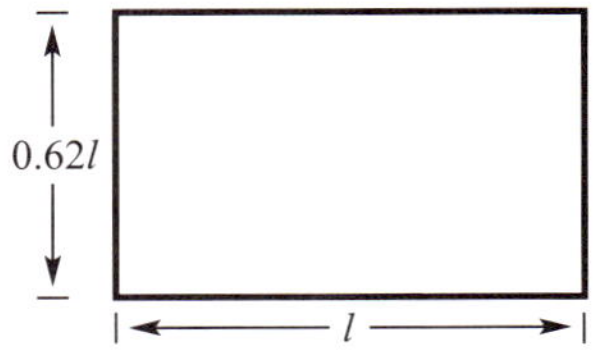

50. The area of a triangle with base b and height $0.75b$

Figure for 50

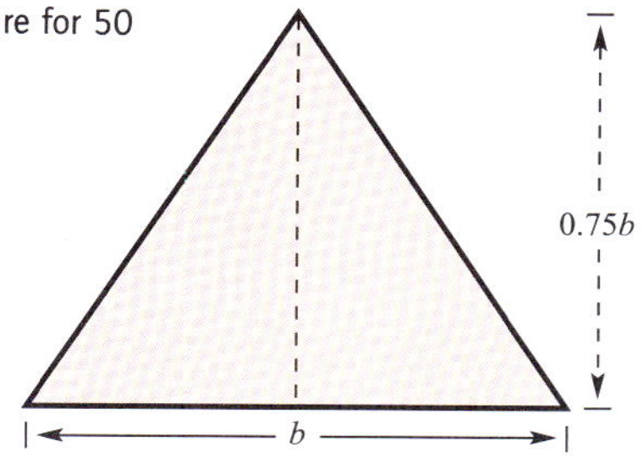

51. The sum of a number n and three times the number, all increased by two

52. The sum of two consecutive integers, the first of which is n

53. The sum of two consecutive odd integers, the first of which is $2n + 1$

54. The sum of the squares of two consecutive *even* integers, the first of which is $2n$

55. *Advertising Space* An advertising banner has a width of w and a length of $6w$, where w is measured in feet (see figure). Find an algebraic expression that represents the area of the banner. What are the units of measure for the area?

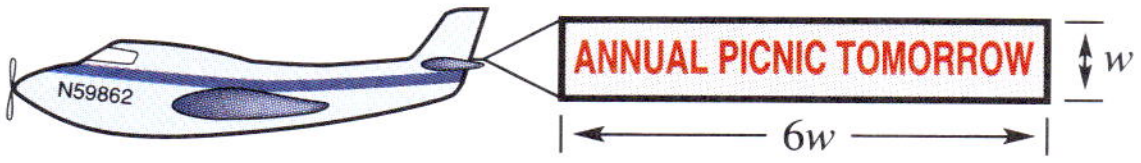

56. *Billiard Table* The top of a billiard table has a length of l and a width of $l - 6$, where l is measured in feet (see figure). Find an algebraic expression that represents the area of the top of the billiard table. What are the units of measure for the area?

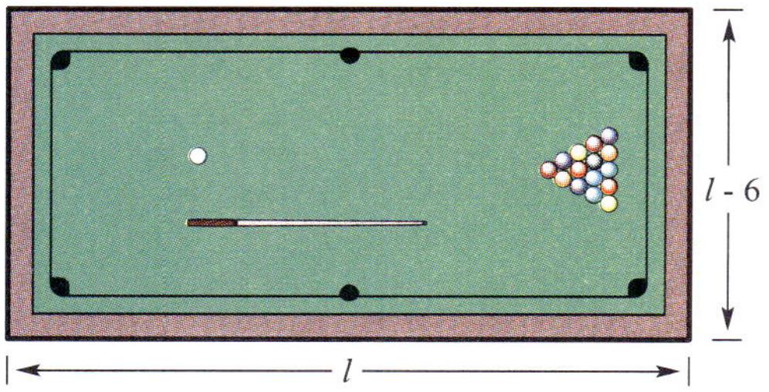

57. Complete the following table where the third row is the difference between consecutive entries of the second row.

n	0	1	2	3	4	5
$5n - 3$						
Differences						

58. Complete the following table where the third row is the difference between consecutive entries of the second row.

n	0	1	2	3	4	5
$3n + 1$						
Differences						

59. Using the results of the tables in Exercises 57 and 58, guess the third row difference that would result in a similar table if the algebraic expression were $an + b$.

60. Find a and b so that the expression $an + b$ would yield the following table.

n	0	1	2	3	4	5
$an + b$	3	7	11	15	19	23

CHAPTER 1 SUMMARY

As you review and prepare for a test on this chapter, first try to obtain a global view of what was discussed. Then review the specific skills needed under each category.

Performing Operations with Real Numbers

■ *Order* real numbers by listing two or more real numbers in increasing order using inequality symbols.	Section 1.1, Exercises 3–20
■ *Evaluate* absolute values and use them to find the distance between points on a number line.	Section 1.1, Exercises 21–46
■ *Evaluate* sums, differences, products, and quotients of real numbers.	Section 1.2, Exercises 1–56
■ *Evaluate* exponential expressions.	Section 1.2, Exercises 65–70
■ *Use a calculator* to evaluate exponential and arithmetic expressions.	Section 1.2, Exercises 47–56, Exercises 81–86
■ *Create* arithmetic expressions from verbal statements.	Section 1.1, Exercises 51–58

Using the Basic Rules of Algebra to Rewrite and Create Algebraic Expressions

■ *Identify* the rule/property demonstrated by a given statement (equation).	Section 1.3, Exercises 1–20, Exercises 51–54; Section 1.4, Exercises 13–22
■ *Rewrite* algebraic expressions according to a specified rule or property.	Section 1.3, Exercises 21–30, Exercises 39–50; Section 1.4, Exercises 23–28
■ *Rewrite and/or simplify* algebraic expressions using the basic rules of algebra.	Section 1.3, Exercises 39–50, Exercises 61–66; Section 1.4, Exercises 47–74
■ *Rewrite and/or simplify* exponential expressions using the rules of exponents.	Section 1.4, Exercises 29–46
■ *Evaluate* algebraic expressions.	Section 1.4, Exercises 75–88
■ *Create* algebraic expressions from verbal statements.	Section 1.5, Exercises 1–20, Exercises 31–54
■ *Create* a verbal statement from a given algebraic expression.	Section 1.5, Exercises 21–30

Chapter 1 Review Exercises

In Exercises 1–4, show each real number as a point on the real number line and place the correct inequality symbol (< or >) between the real numbers.

1. $3 \quad \square \quad -\frac{1}{8}$ **2.** $-2 \quad \square \quad -8$ **3.** $-\frac{8}{5} \quad \square \quad -\frac{2}{5}$ **4.** $8.4 \quad \square \quad -\pi$

In Exercises 5–8, evaluate the given expression.

5. $|-7.2|$ **6.** $|1.6|$ **7.** $-|-7.2|$ **8.** $|-3.6|$

In Exercises 9–40, evaluate the given quantity. (If not possible, state the reason.) Write all fractions in reduced form.

9. $15 + (-4)$ **10.** $-12 + 3$ **11.** $340 - 115 + 5$ **12.** $-154 + 86 - 240$

13. $|-96| - |134|$ **14.** $16(3{,}000)$ **15.** $120(-5)(7)$ **16.** $(-16)(-15)(-4)$

17. $\dfrac{-56}{-4}$ **18.** $\dfrac{85}{0}$ **19.** $\dfrac{45 - |-45|}{2}$ **20.** $\dfrac{288}{2 \cdot 3 \cdot 3}$

21. $\dfrac{4}{21} + \dfrac{7}{21}$ **22.** $\dfrac{21}{16} - \dfrac{13}{16}$ **23.** $-\dfrac{5}{6} + 1$ **24.** $\dfrac{21}{32} + \dfrac{11}{24}$

25. $8\frac{3}{4} - 6\frac{5}{8}$ **26.** $-2\frac{9}{10} + 5\frac{3}{20}$ **27.** $\dfrac{3}{8} \cdot \dfrac{-2}{15}$ **28.** $\dfrac{5}{21} \cdot \dfrac{21}{5}$

29. $-\dfrac{7}{15} \div -\dfrac{7}{30}$ **30.** $-\dfrac{2}{3} \div \dfrac{4}{15}$ **31.** $\dfrac{(3/4) - (1/2)}{5/8}$ **32.** $\dfrac{-(4/5) - (3/2)}{7/10}$

33. $-35 + 26.5 + 13.75$ **34.** $\dfrac{6.25}{1.25}$ **35.** $(-6)^3$ **36.** $-(-3)^4$

37. $\dfrac{3}{6^2}$ **38.** $-4(25 - 4^2)$ **39.** $120 - (5^2 \cdot 4)$ **40.** $45 - 45 \div 3^2$

In Exercises 41–46, name the property of real numbers that justifies the given statement.

41. $13 - 13 = 0$ **42.** $7\left(\frac{1}{7}\right) = 1$ **43.** $7(9 + 3) = 7 \cdot 9 + 7 \cdot 3$

44. $15(4) = 4(15)$ **45.** $5 + (4 - y) = (5 + 4) - y$ **46.** $6(4z) = (6 \cdot 4)z$

In Exercises 47–50, identify the rule of algebra that is illustrated by the given equation.

47. $(u - v)(2) = 2(u - v)$ **48.** $(x + y) + 0 = x + y$

49. $ab \cdot \dfrac{1}{ab} = 1$ **50.** $x(yz) = (xy)z$

In Exercises 51–56, expand the expression by using the Distributive Property.

51. $\frac{2}{3}(6s - 12t)$

52. $-5(2x - 4y)$

53. $-y(3y - 10)$

54. $x(3x + 4y)$

55. $-(-u + 3v)$

56. $(5 - 3j)(-4)$

In Exercises 57–64, simplify the algebraic expression.

57. $5(x - 4) + 10$

58. $15 - 7(z + 2)$

59. $3x - (y - 2x)$

60. $30x - (10x + 80)$

61. $-2(1 - 20u) + 2(1 - 10u)$

62. $9(11 - 2y) - 6(11 - 2y)$

63. $3[b + 5(b - a)]$

64. $-2t[8 - (6 - t)] + 5t$

In Exercises 65–70, simplify the expression by using the rules of exponents.

65. $x^2 \cdot x^3 \cdot x$

66. $y^3(-2y^2)$

67. $(xy)(-3x^2y^3)$

68. $3uv(-2uv^2)^2$

69. $(-2a^2)^3(8a)$

70. $2(a - b)^4(a - b)^2$

In Exercises 71–74, evaluate the algebraic expression for the specified value of the variable. If not possible, state the reason.

	Expression	*Value*
71.	$x^2 - 2x - 3$	(a) $x = 3$ (b) $x = 0$
72.	$y^3 - 125$	(a) $y = 5$ (b) $y = -2$
73.	$3x^2 - y(x + 1)$	(a) $x = 2,\ y = -1$ (b) $x = -1,\ y = 20$
74.	$\frac{x}{y + 2}$	(a) $x = 0,\ y = 3$ (b) $x = 5,\ y = -2$

75. Perform the indicated operations and simplify.

$$6 \cdot 10^3 + 9 \cdot 10^2 + 1 \cdot 10^1$$

76. Perform the indicated operations and simplify.

$$5 \cdot 10^3 + 4 \cdot 10^2 + 9 \cdot 10^1$$

Magazine Circulations In Exercises 77 and 78, use the accompanying figure showing the circulation for the top five magazines in the United States for the first six months of 1990. (*Source:* Audit Bureau of Circulations.)

Figure for 77 and 78

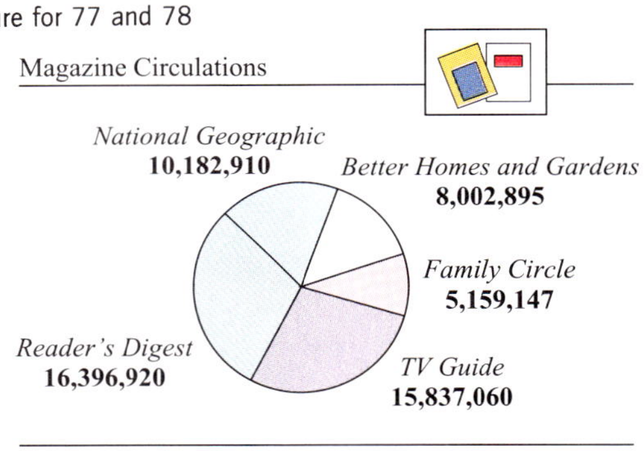

77. Determine the combined circulation for the five magazines.

78. What is the difference in circulation between *Reader's Digest* and *National Geographic*?

79. *Total Charge* Suppose you purchased a product by making a down payment of \$239 plus nine monthly payments of \$45 each. What is the total amount you paid for the product?

80. *Total Charge* Suppose you purchased a product by making a down payment of \$387 plus 12 monthly payments of \$68 each. What is the total amount you paid for the product?

81. *Calculator Experiment* Enter any number between 0 and 1 in a calculator. Take the square root of the number. Then take the square root of the result, and keep repeating this process. What number does the calculator display seem to be approaching?

82. *Calculator Experiment* Use a calculator to calculate 12^4 in two ways.*

Scientific:
(a) 12 [y^x] 4 [=]
(b) 12 [x^2] [x^2]

Graphing:
(a) 12 [^] 4 [ENTER]
(b) 12 [x^2] [x^2] [ENTER]

Why do these two methods give the same result?

In Exercises 83–86, translate the phrase into a mathematical expression. (Let n represent the arbitrary real number.)

83. Two hundred decreased by three times a number

84. One hundred increased by the product of 15 and a real number

85. The sum of the square of a real number and 49

86. The absolute value of the sum of a real number and 10

In Exercises 87–90, write a verbal description of the algebraic expression without using the variable.

87. $2y + 7$

88. $5u - 3$

89. $\frac{x - 5}{4}$

90. $-3(a - 10)$

In Exercises 91–94, write an algebraic expression that represents the quantity given by the verbal statement.

91. The amount of income tax on a taxable income of I dollars when the tax rate is 18%

92. The distance traveled when you travel eight hours at an average speed of r miles per hour

93. The area of a rectangle whose length is l inches and whose width is 5 units less than the length

94. The sum of three consecutive even integers, the first of which is $2n$

*The graphing calculator keystrokes in this text correspond to the *TI-81* graphing calculator from *Texas Instruments*. For other graphing calculators, the keystrokes may differ.

CHAPTER 1 TEST

Take this test as you would take a test in class. After you are done, check your work with the answers given in the back of the book.

1. Place the correct symbol (< or >) between the real numbers $-\frac{5}{2}$ and $-|-3|$.

2. Find the distance between the real numbers -6.2 and 5.7.

3. Evaluate: $-75 - (-18)$

4. Evaluate: $\frac{2}{3} + \left(-\frac{7}{6}\right)$

5. Evaluate: $27\left(-\frac{2}{3}\right)$

6. Evaluate: $\left(-\frac{7}{16}\right)\left(-\frac{8}{21}\right)$

7. Evaluate: $\frac{5}{18} \div \frac{15}{8}$

8. Evaluate: $-(-3)^4$

9. Evaluate: $\left(-\frac{3}{5}\right)^3$

10. Evaluate: $52 + 3(36 \div 18)$

11. Evaluate: $\frac{4^2 - 6}{5} + 13$

12. Name the property of real numbers demonstrated by $(-3 \cdot 5) \cdot 6 = -3 \cdot (5 \cdot 6)$.

13. Name the property of real numbers demonstrated by $3y \cdot \frac{1}{3y} = 1$.

14. Give the additive inverse of the quantity $5x$.

15. Rewrite the expression $5(2x - 3)$ using the Distributive Property.

16. Identify the terms of the algebraic expression $3x^2 - 5x + 6$.

17. Use properties of exponents to rewrite the expression $(-2z)^3$.

18. Simplify: $(3x^2y)(-xy)^2$

19. Simplify: $3x^2 - 2x - 5x^2 + 7x - 1$

20. Simplify: $a(5a - 4) - 2(2a^2 - 2a)$

21. Evaluate the expression $4 - (x + 1)^2$ if (a) $x = -1$ and (b) $x = 3$.

22. Translate the statement "The product of a number n and 5 is decreased by 8" into an algebraic expression.

23. Write an algebraic expression for the area of a rectangle of length l and width $0.6l$.

24. Write an algebraic expression for the sum of two consecutive even integers, the first of which is $2n$.

25. It is necessary to cut a 144-foot rope into nine pieces of equal length. What is the length of each piece?

26. A *cord* of wood is a pile 4 feet high, 4 feet wide, and 8 feet long. The volume of a rectangular solid is its length times its width times its height. Find the number of cubic feet in five cords of wood.

CHAPTER TWO

Linear Equations and Inequalities

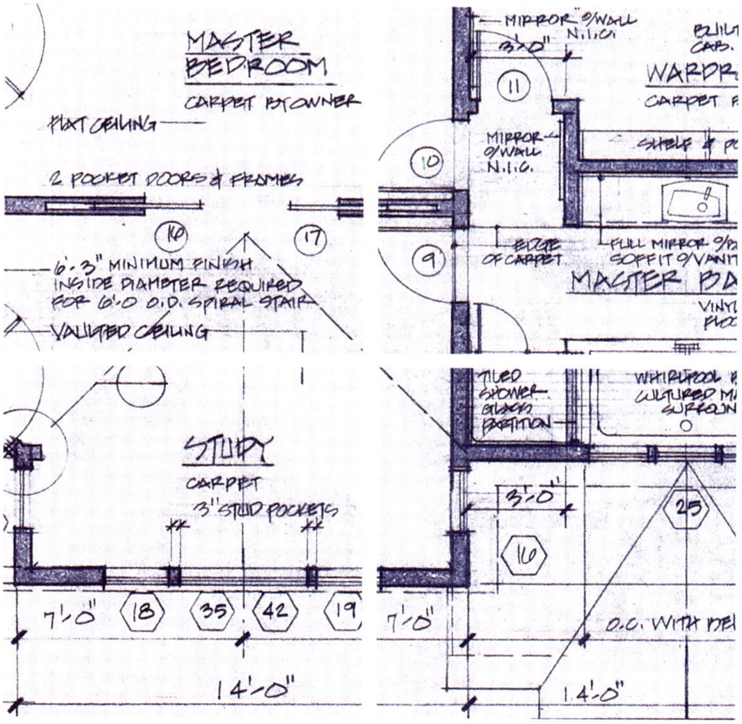

Chapter Overview

This chapter contains some of the most commonly used parts of algebra. Section 2.1 shows how to solve linear equations. Sections 2.2 and 2.3 describe a wide variety of real-life problems that can be solved using linear equations. A mastery of the material in these three sections is critical to your continued success in the course. The last two sections of the chapter describe techniques for solving linear inequalities and inequalities that involve absolute value.

SECTION 2.1

Linear Equations

Introduction to Equations • Equivalent Equations • Solving Linear Equations in Standard Form • Solving Linear Equations in Nonstandard Form

Introduction to Equations

In Chapter 1 we used the basic rules of algebra to rewrite, simplify, and construct algebraic *expressions*. In this chapter we use the same rules to rewrite, solve, and construct algebraic *equations*.

An **equation** is a statement that two mathematical expressions are equal. Some examples are

$$x = 4, \quad 4x + 3 = 15, \quad 2x - 8 = 2(x - 4), \quad \text{and} \quad x^2 - 16 = 0.$$

To **solve** an equation involving x means that we are to find all values of x for which the equation is true. Such values are called **solutions** and we say that the solutions **satisfy** the equation. For instance, 4 is a solution of $x = 4$ because $4 = 4$ is a true statement. Similarly, 3 is a solution of $4x + 3 = 15$ because $4(3) + 3 = 15$ is a true statement.

The **solution set** of an equation is the set of all real numbers that are solutions of the equation. Sometimes, an equation will have the set of all real numbers as its solution set. Such an equation is called an **identity**. For instance, the equation

$$2x - 8 = 2(x - 4) \qquad \text{Identity}$$

is an identity because the equation is true for all real values of x. Try values like 0, 1, -2, and 5 in this equation to see that each one is a solution.

An equation whose solution set is not the entire set of real numbers is called a **conditional equation**. For instance, the equation

$$x^2 - 16 = 0 \qquad \text{Conditional equation}$$

is a conditional equation because it has only two solutions, 4 and -4. In this text, we focus primarily on conditional equations.

In Examples 1 and 2, we show how to **check** whether a given value of x is a solution of an equation.

EXAMPLE 1 ■ Checking a Solution of an Equation

Determine whether -3 is a solution of the following equation.

$$-3x - 5 = 4x + 16$$

Solution

To check whether -3 is a solution of this equation, we replace the variable x by the number -3. This must be done for all occurrences of x on *both sides* of the equation. After replacing the variable, we simplify both sides of the equation. If both sides turn out to be the same number, then -3 is a solution, and we say that the solution "checks." If the two sides turn out to be different numbers, then -3 is not a solution.

$$\begin{aligned} -3x - 5 &= 4x + 16 && \text{Given equation} \\ -3(-3) - 5 &\stackrel{?}{=} 4(-3) + 16 && \text{Replace } x \text{ by } -3 \\ 9 - 5 &\stackrel{?}{=} -12 + 16 && \text{Simplify} \\ 4 &= 4 && -3 \text{ is a solution} \end{aligned}$$

Since both sides of the equation turn out to be the same number, we conclude that -3 is a solution of the given equation. ■

NOTE: When checking a solution, we like to write a question mark over the equal sign to indicate that we are uncertain whether the "equation" is true.

EXAMPLE 2 ■ A Trial Solution that Does Not Check

Determine whether -2 is a solution of the following equation.

$$-3x - 5 = 4x + 16$$

Solution

$$\begin{aligned} -3x - 5 &= 4x + 16 && \text{Given equation} \\ -3(-2) - 5 &\stackrel{?}{=} 4(-2) + 16 && \text{Replace } x \text{ by } -2 \\ 6 - 5 &\stackrel{?}{=} -8 + 16 && \text{Simplify} \\ 1 &\neq 8 && -2 \text{ is not a solution} \end{aligned}$$

Since the two sides of the equation turn out to be different after x is replaced by -2, we conclude that -2 is not a solution of the given equation. ■

Equivalent Equations

It is helpful to think of an equation as having two sides that are "in balance." Consequently, when we try to solve an equation, we must be careful to maintain that balance by performing the same operation on both sides.

Two equations that have the same set of solutions are called **equivalent**. For instance, the equations $x = 3$ and $x - 3 = 0$ are equivalent because both have only one solution—the number 3. When any one of the four techniques in the following list is applied to an equation, the resulting equation is equivalent to the original equation.

Forming Equivalent Equations

A given equation can be transformed into an *equivalent equation* by one or more of the following steps.

	Given Equation	*Equivalent Equation*
1. Remove symbols of grouping, combine like terms, or reduce fractions on one or both sides of the equation.	$3x - x = 8$	$2x = 8$
2. Add (or subtract) the same quantity to *both* sides of the equation.	$x - 3 = 5$	$x = 8$
3. Multiply (or divide) *both* sides of the equation by the same *nonzero* quantity.	$3x = 9$	$x = 3$
4. Interchange the two sides of the equation.	$7 = x$	$x = 7$

When solving an equation, the four techniques for forming equivalent equations can be used to *eliminate* terms or factors in the equation. For example, to solve the equation $x + 4 = 2$, we need to eliminate the term $+4$ on the left side. This is accomplished by adding its opposite, -4, to both sides.

$$x + 4 = 2 \qquad \text{Given equation}$$

$$x + 4 + (-4) = 2 + (-4) \qquad \text{Add } -4 \text{ to both sides}$$

$$x + 0 = -2 \qquad \text{Combine like terms}$$

$$x = -2 \qquad \text{Solution}$$

Note in this solution that we *added* -4 to both sides of the equation to eliminate 4 on the left side. We could just as easily have *subtracted* 4 from both sides. Both techniques are legitimate—which one you choose is a matter of personal preference.

Solving Linear Equations in Standard Form

The most common type of equation in one variable is a **linear equation**.

Definition of Linear Equation

A **linear equation** in one variable x is an equation that can be written in the standard form

$$ax + b = c$$

where a, b, and c are real numbers with $a \neq 0$.

A linear equation in one variable is also called a **first-degree equation** because its variable has an (implied) exponent of one. Some examples of linear equations in standard form are

$$3x = 8, \quad x + 4 = 1, \quad 3x + 2 = 0, \quad \text{and} \quad 5x - 4 = -6.$$

Remember that to *solve* an equation in x means that we are to find the values of x that satisfy the equation. For a linear equation in the standard form $ax + b = c$, the goal is to **isolate** x by rewriting the standard equation in the form

$$x = (\text{a number}).$$

Beginning with the given equation, we write a sequence of equivalent equations, each having the same solution as the original equation.

EXAMPLE 3 ■ Solving a Linear Equation in Standard Form

$4x - 2 = 10$	Given equation
$4x - 2 + 2 = 10 + 2$	Add 2 to both sides
$4x = 12$	Combine like terms
$\dfrac{4x}{4} = \dfrac{12}{4}$	Divide both sides by 4
$x = 3$	Solution

Thus, it appears that the solution of the given equation is 3. We should check this solution as follows.

Check

$4x - 2 = 10$	Given equation
$4(3) - 2 \stackrel{?}{=} 10$	Replace x by 3
$12 - 2 \stackrel{?}{=} 10$	
$10 = 10$	Solution checks ■

Be sure you see that solving an equation like the one in Example 3 has two basic steps. The first step is to *find* the solution (or solutions). The second step is to *check* that each solution you find actually satisfies the original equation. You can improve your accuracy in algebra by developing the habit of checking each solution.

We know that 3 is a solution of the equation given in Example 3. But a question you might be asking at this point is, How can we be sure that the equation does not have other solutions? The answer is that a linear equation always has *exactly one* solution. We can prove this in the following way. Let $ax + b = c$ represent an arbitrary linear equation, where $a \neq 0$.

$ax + b = c$	Given equation
$ax + b - b = c - b$	Subtract b from both sides
$ax = c - b$	Combine like terms

$$\frac{ax}{a} = \frac{c - b}{a} \qquad \text{Divide both sides by } a$$

$$x = \frac{c - b}{a} \qquad \text{Solution}$$

It is clear that the last equation has only one solution, $x = (c - b)/a$. Moreover, because the last equation is equivalent to the given equation, we can conclude that every linear equation has exactly one solution.

NOTE: Equations that are not linear may have two or more solutions. For instance, the nonlinear equation $x^2 = 4$ has two solutions: 2 and -2.

EXAMPLE 4 ■ Solving a Linear Equation in Standard Form

$$2x - 3 = -5 \qquad \text{Given equation}$$

$$2x - 3 + 3 = -5 + 3 \qquad \text{Add 3 to both sides}$$

$$2x = -2 \qquad \text{Combine like terms}$$

$$\frac{2x}{2} = \frac{-2}{2} \qquad \text{Divide both sides by 2}$$

$$x = -1 \qquad \text{Solution}$$

Thus, the solution is -1. Check this solution in the original equation. ■

As you gain experience with solving linear equations, you will probably find that you can perform some of the solution steps in your head. For instance, you might solve the equation given in Example 4 by writing only the following steps.

$$2x - 3 = -5 \qquad \text{Given equation}$$

$$2x = -2 \qquad \text{Add 3 to both sides}$$

$$x = -1 \qquad \text{Divide both sides by 2}$$

Remember, however, that you should not skip the final step—checking your solution.

Solving Linear Equations in Nonstandard Form

Linear equations often occur in nonstandard forms that contain symbols of grouping or like terms that are not combined. Here are some examples.

$$x + 2 = 2x - 6, \qquad 6(y - 1) + 4y = 3(7y + 1), \qquad \frac{x}{3} + \frac{3x}{4} = 2$$

The next three examples show how to solve these linear equations.

EXAMPLE 5 ■ Solving a Linear Equation in Nonstandard Form

$$x + 2 = 2x - 6 \qquad \text{Given equation}$$

$$-2x + x + 2 = -2x + 2x - 6 \qquad \text{Add } -2x \text{ to both sides}$$

$$-x + 2 = -6 \quad \text{Combine like terms}$$
$$-x + 2 - 2 = -6 - 2 \quad \text{Subtract 2 from both sides}$$
$$-x = -8 \quad \text{Combine like terms}$$
$$(-1)(-x) = (-1)(-8) \quad \text{Multiply both sides by } -1$$
$$x = 8 \quad \text{Solution}$$

Thus, the solution is 8. Check this solution in the original equation. ■

When solving equations that are in nonstandard form, you should be aware that it is possible for the equation to have *no solution*. For instance, the equation

$$2x - 4 = 2(x - 3) \quad \text{Given equation}$$
$$2x - 4 = 2x - 6 \quad \text{Distributive Property}$$
$$-4 = -6 \quad \text{Subtract } 2x \text{ from both sides}$$

has no solution because there is no value of x for which -4 is equal to -6.

EXAMPLE 6 ■ Solving a Linear Equation in Nonstandard Form

$$6(y - 1) + 4y = 3(7y + 1) \quad \text{Given equation}$$
$$6y - 6 + 4y = 21y + 3 \quad \text{Remove symbols of grouping}$$
$$10y - 6 = 21y + 3 \quad \text{Combine like terms}$$
$$10y - 21y - 6 = 21y - 21y + 3 \quad \text{Subtract } 21y \text{ from both sides}$$
$$-11y - 6 = 3 \quad \text{Combine like terms}$$
$$-11y - 6 + 6 = 3 + 6 \quad \text{Add 6 to both sides}$$
$$-11y = 9 \quad \text{Combine like terms}$$
$$\frac{-11y}{-11} = \frac{9}{-11} \quad \text{Divide both sides by } -11$$
$$y = -\frac{9}{11} \quad \text{Solution}$$

Thus, the solution is $-\frac{9}{11}$. Check this solution in the original equation. ■

If a linear equation contains fractions, we suggest that you first clear the equation of fractions by multiplying both sides of the equation by the least common denominator of the fractions occurring in the equation.

EXAMPLE 7 ■ Solving a Linear Equation that Contains Fractions

Solve the following linear equation.

$$\frac{x}{3} + \frac{3x}{4} = 2$$

Solution

Since the two denominators in this equation are 3 and 4, the least common denominator is 12.

$$\frac{x}{3} + \frac{3x}{4} = 2 \qquad \text{Given equation}$$

$$12\left(\frac{x}{3} + \frac{3x}{4}\right) = 12(2) \qquad \text{Multiply both sides by 12}$$

$$12 \cdot \frac{x}{3} + 12 \cdot \frac{3x}{4} = 12(2) \qquad \text{Distributive Property}$$

$$4x + 9x = 24 \qquad \text{Simplify}$$

$$13x = 24 \qquad \text{Combine like terms}$$

$$\frac{13x}{13} = \frac{24}{13} \qquad \text{Divide both sides by 13}$$

$$x = \frac{24}{13} \qquad \text{Solution}$$

Thus, the solution is $\frac{24}{13}$. Check this solution in the original equation. ■

The next example shows how to solve a linear equation involving decimals. The procedure is basically the same, but the arithmetic is somewhat messier.

EXAMPLE 8 ■ Solving a Linear Equation Involving Decimals

$$0.12x + 0.09(5{,}000 - x) = 513 \qquad \text{Given equation}$$

$$0.12x + 450 - 0.09x = 513 \qquad \text{Distributive Property}$$

$$(0.12x - 0.09x) + 450 = 513 \qquad \text{Group like terms}$$

$$0.03x + 450 - 450 = 513 - 450 \qquad \text{Subtract 450 from both sides}$$

$$0.03x = 63 \qquad \text{Combine like terms}$$

$$\frac{0.03x}{0.03} = \frac{63}{0.03} \qquad \text{Divide both sides by 0.03}$$

$$x = 2{,}100 \qquad \text{Solution}$$

Thus, the solution is 2,100. Check this solution in the original equation. ■

A different approach to solving the linear equation in Example 8 would be to begin by multiplying both sides of the equation by 100. For this particular equation, this would clear the equation of decimals and produce the equivalent equation

$$12x + 9(5{,}000 - x) = 51{,}300.$$

Try solving this equation to see that you obtain the same solution.

DISCUSSION PROBLEM ■ You Be the Instructor

Suppose you are teaching an algebra class and one of your students hands in the following problem.

$$\begin{aligned} 2x - 3(x-2) &= 5x + 12 \\ 2x - 3x - 6 &= 5x + 12 \\ -x - 6 &= 5x + 12 \\ -6x &= 6 \\ x &= -1 \end{aligned}$$

Check

$$\begin{aligned} 2(-1) - 3((-1) - 2) &\stackrel{?}{=} 5(-1) + 12 \\ -2 - 3(-3) &\stackrel{?}{=} -5 + 12 \\ -2 + 9 &\stackrel{?}{=} 7 \\ 7 &= 7 \end{aligned}$$

Is the answer correct? Are the solution steps correct? How much "partial credit" would you give the student for a correct answer that was obtained by incorrect means? ■

Warm-Up

The following warm-up exercises involve skills that were covered in earlier sections. You will use these skills in the exercise set for this section.

In Exercises 1–6, simplify the algebraic expression.

1. $-2(x-3) - 6$

2. $6(x+10) + 4$

3. $4 - 3(2x+1)$

4. $5(x+2) - 4(2x-3)$

5. $24\left(\frac{y}{3} + \frac{y}{6}\right)$

6. $0.12x + 0.05(2{,}000 - 2x)$

In Exercises 7–10, evaluate the given expression for the specified values of the variable.

Expression	*Value*	
7. $5x - 3$	(a) $x = 0$	(b) $x = 2$
8. $100 - 22x$	(a) $x = 3$	(b) $x = 6$
9. $45 - 3(x-8)$	(a) $x = 4$	(b) $x = 12$
10. $\frac{1}{2}(2x+3) - 16$	(a) $x = 8$	(b) $x = \frac{9}{2}$

2.1 EXERCISES

In Exercises 1–6, determine whether the given values of the variable are solutions of the equation.

	Equation	*Value*	
1.	$3x - 7 = 2$	(a) $x = 0$	(b) $x = 3$
2.	$5x + 9 = 4$	(a) $x = -1$	(b) $x = 2$
3.	$x + 8 = 3x$	(a) $x = 4$	(b) $x = -4$
4.	$10x - 3 = 7x$	(a) $x = 0$	(b) $x = -1$
5.	$3x + 3 = 2(x - 4)$	(a) $x = -11$	(b) $x = 5$
6.	$7x - 1 = 5(x + 5)$	(a) $x = 2$	(b) $x = 13$

In Exercises 7–10, state the rule of algebra used in each step of the solution process. Also verify that the solution of the given equation is a solution of *each* equation in the solution process.

7.
$$\begin{aligned} 3x + 15 &= 0 \\ 3x + 15 - 15 &= 0 - 15 \\ 3x &= -15 \\ \frac{3x}{3} &= \frac{-15}{3} \\ x &= -5 \end{aligned}$$

8.
$$\begin{aligned} 7x - 21 &= 0 \\ 7x - 21 + 21 &= 0 + 21 \\ 7x &= 21 \\ \frac{7x}{7} &= \frac{21}{7} \\ x &= 3 \end{aligned}$$

9.
$$\begin{aligned} -2x + 5 &= 12 \\ -2x + 5 - 5 &= 12 - 5 \\ -2x &= 7 \\ \frac{-2x}{-2} &= \frac{7}{-2} \\ x &= -\frac{7}{2} \end{aligned}$$

10.
$$\begin{aligned} 25 - 3x &= 10 \\ 25 - 3x + 3x &= 10 + 3x \\ 25 &= 10 + 3x \\ 25 - 10 &= 10 + 3x - 10 \\ 15 &= 3x \\ \frac{15}{3} &= \frac{3x}{3} \\ 5 &= x \end{aligned}$$

In Exercises 11–54, solve the equation and check the result. (If not possible, state the reason.)

11. $3x = 12$ **12.** $-14x = 28$ **13.** $6x + 4 = 0$ **14.** $8z - 10 = 0$

15. $23x - 4 = 42$ **16.** $15x - 18 = 27$ **17.** $4y - 3 = 3y$ **18.** $24 - 2x = x$

19. $7 - 8x = 13x$ **20.** $2s - 16 = 34s$ **21.** $8 - 5t = 20 + t$ **22.** $3y + 14 = y + 20$

23. $15t = 0$

24. $6a + 2 = 6a$

25. $-8t + 7 = -8t$

26. $4x = -12x$

27. $8(x - 8) = 24$

28. $6(x + 2) = 30$

29. $5 - (2y - 4) = 15$

30. $26 - (3x - 10) = 6$

31. $-4(t + 2) = 0$

32. $8(z - 8) = 0$

33. $12(x + 3) = 7(x + 3)$

34. $-25(x - 100) = 16(x - 100)$

35. $8x - 3(x - 2) = 12$

36. $12 = 6(y + 1) - 8(1 - y)$

37. $2(x + 7) - 9 = 5(x - 4)$

38. $6[x - (5x - 7)] = 4 - 5x$

39. $\frac{u}{5} = 10$

40. $-\frac{z}{2} = 7$

41. $t - \frac{2}{5} = \frac{3}{2}$

42. $z + \frac{1}{15} = -\frac{3}{10}$

43. $\frac{1}{3}x + 1 = \frac{1}{12}x - 4$

44. $\frac{1}{9}x + \frac{1}{3} = \frac{11}{18}$

45. $\frac{t + 4}{14} = \frac{2}{7}$

46. $\frac{11x}{6} + \frac{1}{3} = 0$

47. $\frac{t}{5} - \frac{t}{2} = 1$

48. $\frac{t}{6} + \frac{t}{8} = 1$

49. $\frac{25 - 4u}{3} = \frac{5u + 12}{4} + 6$

50. $\frac{8 - 3x}{4} - 4 = \frac{x}{6}$

51. $0.3x + 1.5 = 8.4$

52. $16.3 - 0.2x = 7.1$

53. $1.2(x - 3) = 10.8$

54. $6.5(1 - 2x) = 13$

In Exercises 55–58, solve the given equation and round your answer to two decimal places. (A calculator may be helpful.)

55. $1.234x + 3 = 7.805$

56. $325x - 4{,}125 = 612$

57. $\frac{x}{10.625} = 2.850$

58. $2x + \frac{1}{8.6} = \frac{3}{2}$

59. *Dimensions of a Rectangle* The length of a certain rectangle is t times its width. Thus, the perimeter P is given by $P = 2w + 2(tw)$, where w is the width of the rectangle. The perimeter of the rectangle is 1,200 meters. Complete the following table giving the length, width, and area of the rectangle for the specified values of t.

t	1	1.5	2	3	4	5
Width						
Length						
Area						

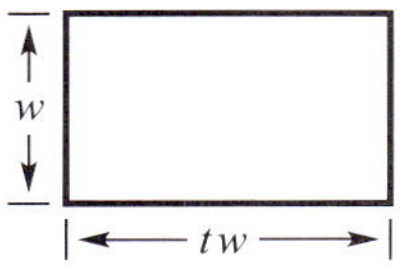

60. *Dimensions of a Rectangle* Use the table in Exercise 59 to write a short paragraph describing the relationship between the width, length, and area of a rectangle that has a *fixed* perimeter.

61. *Maximum Height of a Fountain* Consider the water fountain shown in the accompanying figure. The initial velocity of the stream of water is 48 feet per second. The velocity v of the water at any time t (in seconds) is then given by $v = 48 - 32t$. Find the time when the maximum height of the fountain is obtained. (*Hint:* The maximum height is reached when $v = 0$.)

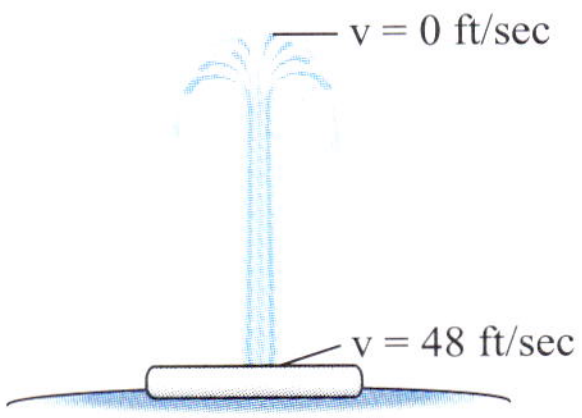

62. *Maximum Height of an Object* The velocity v of an object projected vertically with an initial velocity of 64 feet per second is given by $v = 64 - 32t$ where t is the time in seconds and air resistance is neglected. Find the time when the maximum height of the object is obtained.

63. *Work Rate Problem* Two people can complete a task in t hours where t must satisfy the equation

$$\frac{t}{10} + \frac{t}{15} = 1.$$

Find the required time t.

64. *Work Rate Problem* Two people can complete a task in t hours where t must satisfy the equation

$$\frac{t}{12} + \frac{t}{20} = 1.$$

Find the required time t.

65. *Tuition and Fees* The total amount of tuition and fees collected by colleges and universities in the United States from 1980 to 1985 can be approximated by the linear equation

$$y = 13{,}916.9 + 1{,}862.5t$$

where y represents the amount of tuition and fees (in millions of dollars) and where t represents the year with $t = 0$ corresponding to 1980 (see figure). According to this model, during which year did the total amount of tuition and fees reach $23,229.4 million? (*Source:* U.S. National Center for Education Statistics.)

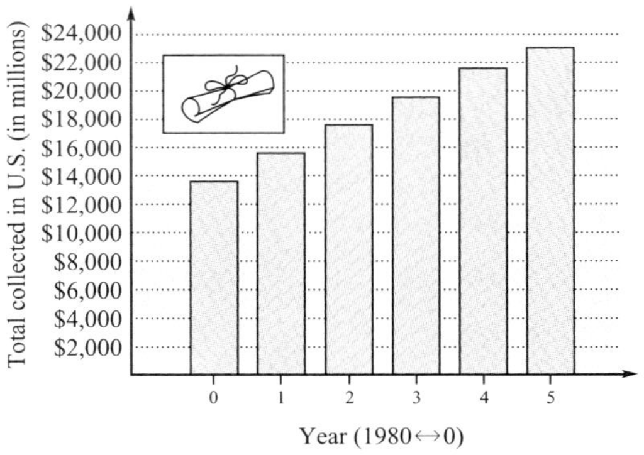

66. *Private School Expenditures* The total amount spent by primary and secondary private schools in the United States from 1975 to 1989 can be approximated by the linear equation

$$y = 7{,}717.4 + 894.6t$$

where y represents the amount spent in millions of dollars and t represents the year with $t = 0$ corresponding to 1980 (see figure). According to this model, during which year did the total amount spent reach $13,085 million? (*Source:* U.S. National Center for Education Statistics.)

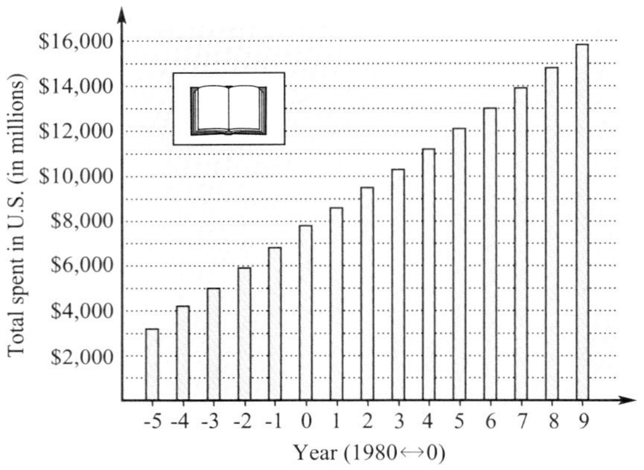

Cable Television Subscribers In Exercises 67 and 68, use the following *two-part* model, which approximates the number of cable television subscribers in the United States between 1970 and 1988.

$$y = 4{,}193 + 1{,}099.4t, \quad 0 \le t \le 9$$
$$y = -21{,}786 + 3{,}657t, \quad 10 \le t \le 18$$

In this model, y represents the number of subscribers (in thousands) and t represents the year with $t = 0$ corresponding to 1970 (see figure on next page). (*Source:* Corporation for Public Broadcasting.)

Figure for 67 and 68

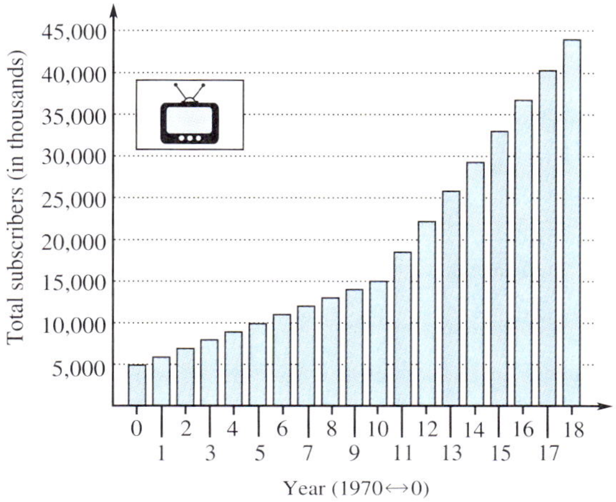

67. During which year were there 10,789.4 (thousand) cable television subscribers?

68. During which year were there 40,383 (thousand) cable television subscribers?

SECTION **2.2**

Applications: Percent, Ratio, and Proportion

Introduction to Problem Solving • Applications Involving Percents • Ratios • Proportions

Introduction to Problem Solving

In this section we will see how algebra can be used to solve problems that occur in real-life situations. The first step in the procedure is *to construct* one or more equations that represent the real-life situation. We call this procedure **mathematical modeling**.

A good approach to mathematical modeling is to use two stages. The first stage is to use the given verbal description of the problem to form a *verbal model*. Then, after assigning labels to the unknown quantities in the verbal model, we form a *mathematical model* or *algebraic equation*.

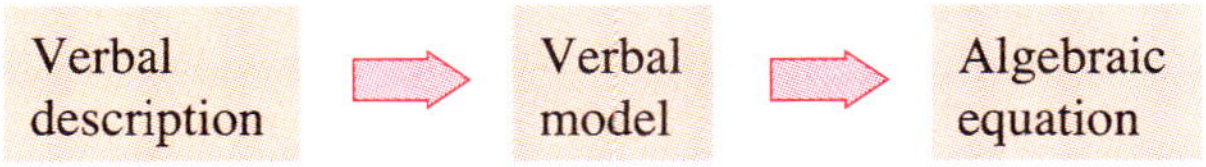

When you are trying to construct a verbal model, it is helpful to look for a hidden equality—a statement that two algebraic expressions are equal. These two expressions might be stated explicitly as being equal, or they might be known to be equal (based on prior knowledge or experience).

EXAMPLE 1 ■ Constructing a Verbal Model

Construct a verbal model for the following.

The sale price of a shirt is 18 dollars. If the sale price is 7 dollars less than the original price, what is the original price?

Solution

In this problem the hidden equality is that "the sale price is *equal* to 7 dollars less than the original price." This hidden equality can be translated into a verbal model as follows.

Verbal model: [Sale price] = [Original price] − [Discount]

[18 dollars] = [Original price] − [7 dollars]

Now, by letting x represent the original price, we could rewrite the verbal model as $18 = x - 7$. Then, solving for x would show that the original price is \$25. ■

EXAMPLE 2 ■ Using a Verbal Model to Construct an Algebraic Equation

Write an algebraic equation that represents the following problem.

> You have accepted a job at an annual salary of \$27,630. This salary includes a year-end bonus of \$750. If you are paid twice a month, what will your gross pay be for each paycheck?

Solution

Since there are 12 months in a year and you will be paid twice a month, it follows that you will receive 24 paychecks during the year. We can construct an algebraic equation for this problem as follows. Note that we begin with a verbal model, then assign labels, and finally form an algebraic equation.

Verbal model: [Income for year] = [24 paychecks] + [Bonus]

Labels: Income for year = \$27,630
Amount of each paycheck = x (in dollars)
Bonus = \$750

Algebraic equation: $27{,}630 = 24x + 750$

The algebraic equation for this problem is a *linear equation* in the variable x. Using the techniques discussed in Section 2.1, you can solve this equation for x. If you do that, you will find that the solution is $x = \$1{,}120$. ■

You could solve the problems in Examples 1 and 2 *without* algebra. For instance, in Example 2 you could solve the problem by subtracting the bonus of \$750 from the annual salary of \$27,630, and dividing the result by 24 pay periods. The reason for listing these examples is that we want you to practice writing algebraic versions of the problem-solving skills that you already possess. Your goal in this section is to be able to solve problems by commonsense reasoning *and* to use this reasoning to write out an algebraic version of the problem. Later, we will encounter more complicated problems in which algebra is a necessary part of the solution.

Applications Involving Percents

Numbers that describe rates, increases, decreases, and discounts are often given as percents, denoted by the symbol %. **Percent** means *per hundred*, or *parts of 100*. (The Latin word for 100 is *centum*.) For example, 40% means 40 parts of 100, which is equivalent to the fraction $\frac{40}{100}$ or $\frac{4}{10}$.

In applications involving percents, we convert the percent number to decimal (or fraction) form before performing any arithmetic operations. Here are some examples.

$$28\% = \frac{28}{100} = 0.28$$ Convert percent to decimal

$$62.5\% = \frac{62.5}{100} = \frac{625}{1{,}000} = \frac{5}{8}$$ Convert percent to fraction

Some percents occur so commonly that we tend to have them memorized. Table 2.1 shows the decimal and fraction conversions for several common percents.

TABLE 2.1 Common Percent Conversions

Percent	10%	$12\frac{1}{2}\%$	20%	25%	$33\frac{1}{3}\%$	50%	$66\frac{2}{3}\%$	75%
Decimal	0.1	0.125	0.2	0.25	0.33 . . .	0.5	0.66 . . .	0.75
Fraction	$\frac{1}{10}$	$\frac{1}{8}$	$\frac{1}{5}$	$\frac{1}{4}$	$\frac{1}{3}$	$\frac{1}{2}$	$\frac{2}{3}$	$\frac{3}{4}$

The primary use of percents is to compare two numbers. For example, we say 3 is 50% of 6. In this statement, note that we are comparing 3 to 6. Hence, we call 6 the **base number**, and 3 is the number being compared to the base number. The following model, which we call the **percent equation**, is often helpful.

Verbal model: Compared number = Percent (decimal form) · Base number

Labels:
a = number being compared to b
b = base number
p = percent (in decimal form)

Percent equation: $a = p \cdot b$

Remember to convert p to a decimal value before multiplying by b.

EXAMPLE 3 ■ Solving a Percent Equation

What is 20% of 50?

Solution

Verbal model: [Compared number] = [Percent (decimal form)] · [Base number]

Labels:
a = compared number
b = base number = 50
p = 20% = 0.2

Percent equation:
$$a = p \cdot b$$
$$a = (0.2)(50)$$
$$a = 10$$

Therefore, 10 is 20% of 50. ■

EXAMPLE 4 ■ Solving a Percent Equation

The number 15.6 is 26% of what number?

Solution

Verbal model: [Compared number] = [Percent (decimal form)] · [Base number]

Labels:
a = compared number = 15.6
b = base number
p = 26% = 0.26

Percent equation:
$$a = p \cdot b$$
$$15.6 = 0.26b$$
$$\frac{15.6}{0.26} = b$$
$$60 = b$$

Therefore, 15.6 is 26% of 60. ■

EXAMPLE 5 ■ Solving a Percent Equation

The number 28 is what percent of 80?

Solution

Verbal model: [Compared number] = [Percent (decimal form)] · [Base number]

Labels: a = compared number = 28
b = 80
p = percent (in decimal form)

Percent equation:

$$a = p \cdot b$$
$$28 = p(80)$$
$$\frac{28}{80} = p$$
$$0.35 = p$$

Therefore, 28 is 35% of 80. ■

In most real-life applications, the base number b and the number a are much more disguised than they were in Examples 3, 4, and 5. It sometimes helps to think of a as a "new" amount and b as the "original" amount.

EXAMPLE 6 ■ **A Percent Application**

A real estate agency receives a commission of $8,092.50 for the sale of a $124,500 house. What percent commission is this?

Solution

Verbal model: [Commission] = [Percent (decimal form)] · [Sale price]

Labels: a = commission = $8,092.50
p = percent (in decimal form)
b = sale price = $124,500

Percent equation:

$$a = p \cdot b$$
$$8{,}092.50 = p(124{,}500)$$
$$\frac{8{,}092.50}{124{,}500} = p$$
$$0.065 = p$$

Therefore, the real estate agency receives a commission of 6.5%. Check this solution in the original statement of the problem. ■

EXAMPLE 7 ■ **A Percent Application**

Suppose you missed getting an A in your chemistry course by only two points. If your point total for the course was 493, how many points were possible in the course? (Assume that you needed 90% of the course total for an A.)

Solution

Verbal model: [Your points] + [Two points] = [90%] · [Total points]

Labels:

$a = 493 + 2 = 495$
$p = 0.9$
$b = \text{total points}$

Percent equation:

$$a = p \cdot b$$
$$495 = 0.9b$$
$$\frac{495}{0.9} = b$$
$$550 = b$$

Therefore, there were 550 total points for the course. Check this solution in the original statement of the problem. ■

EXAMPLE 8 ■ **A Percent Application**

The seating capacity of the university football stadium was increased from 82,000 to 98,400. The old capacity is what percent of the new capacity?

Solution

Verbal model: [Old capacity] = [Percent (decimal form)] · [New capacity]

Labels:

$a = \text{old capacity} = 82{,}000$
$p = \text{percent (in decimal form)}$
$b = \text{new capacity} = 98{,}400$

Percent equation:

$$a = p \cdot b$$
$$82{,}000 = p(98{,}400)$$
$$\frac{82{,}000}{98{,}400} = p$$
$$0.833 \approx p$$

Therefore, the old capacity is approximately $83\frac{1}{3}\%$ of the new capacity. Check this solution in the original statement of the problem. ■

Ratios

A comparison of one number to another by division is called a **ratio**. For example, if a car rental agency has 58 compact cars out of a total fleet of 241 cars, then the ratio of compact cars to the total number of cars is

$$\frac{\text{Number of compact cars}}{\text{Total number of cars}} = \frac{58}{241}.$$

Note the *order* implied by a ratio. The ratio of a to b means a/b, whereas the ratio of b to a means b/a.

Definition of Ratio

The **ratio** of the real number a to the real number b is given by

$$\frac{a}{b}.$$

The ratio of a to b is sometimes written as $a : b$.

There are many applications of ratios in real life. They are used to describe opinion surveys (for/against), populations (male/female, unemployed/employed, predator/prey), mixtures (oil/gasoline, water/alcohol, active/inert), unit prices (dollars/ounce), and so on.

When comparing two *measurements* by a ratio, we prefer to use the same unit of measurement in both the numerator and the denominator. For example, to find the ratio of 4 feet to 8 inches, we could convert 4 feet to 48 inches (by multiplying by 12) to obtain

$$\frac{4 \text{ feet}}{8 \text{ inches}} = \frac{48 \cancel{\text{ inches}}}{8 \cancel{\text{ inches}}} = \frac{48}{8} = \frac{6}{1}.$$

Or, we could convert 8 inches to $\frac{8}{12}$ feet (by dividing by 12) to obtain

$$\frac{4 \text{ feet}}{8 \text{ inches}} = \frac{4 \cancel{\text{ feet}}}{\frac{8}{12} \cancel{\text{ feet}}} = 4 \cdot \frac{12}{8} = \frac{6}{1}.$$

The point is that if you use different units of measurement in the numerator and denominator, then you *must* include the units in the ratio. If you use the same units of measurement in the numerator and denominator, then it is not necessary to write the units.

EXAMPLE 9 ■ Comparing Measurements

Find a ratio to compare the relative sizes of the following. Use the same unit of measurement in the numerator and denominator.

(a) 6 quarts to 14 quarts

(b) 2 meters to 55 centimeters

Solution

(a) Since the units of measurement are the same, the ratio is $\frac{6}{14} = \frac{3}{7}$.

(b) Since the units of measurement are different, we begin by converting meters to centimeters *or* centimeters to meters. Here, it is easier to convert meters to centimeters by multiplying by 100. (Note that 1 meter is equivalent to 100 centimeters.) Thus, the ratio is

$$\frac{2 \text{ meters}}{55 \text{ centimeters}} = \frac{2(100) \cancel{\text{ centimeters}}}{55 \cancel{\text{ centimeters}}} = \frac{200}{55} = \frac{40}{11}.$$

Therefore, the ratio is 40 to 11. ■

EXAMPLE 10 ■ Comparing Measurements

Find a ratio to compare the relative sizes of the following. Use the same unit of measurement in the numerator and denominator.

(a) 200 cents to 5 dollars (b) 28 months to $1\frac{1}{2}$ years

Solution

(a) Since 200 cents is the same as 2 dollars, the ratio is

$$\frac{200 \text{ cents}}{5 \text{ dollars}} = \frac{2 \cancel{\text{dollars}}}{5 \cancel{\text{dollars}}} = \frac{200}{500} = \frac{2}{5}.$$

Note that if we had converted 5 dollars to 500 cents, we would have obtained the same ratio since

$$\frac{200 \text{ cents}}{5 \text{ dollars}} = \frac{200 \cancel{\text{cents}}}{500 \cancel{\text{cents}}} = \frac{200}{500} = \frac{2}{5}.$$

(b) Since $1\frac{1}{2}$ years = 18 months, the ratio is

$$\frac{28 \text{ months}}{1\frac{1}{2} \text{ years}} = \frac{28 \cancel{\text{months}}}{18 \cancel{\text{months}}} = \frac{28}{18} = \frac{14}{9}.$$ ■

The **unit price** of an item is given by the ratio of the total price to the total units. That is,

$$\text{Unit price} = \frac{\text{Total price}}{\text{Total units}}.$$

To state unit prices, we usually use the word *per*. For instance, the unit price for a particular brand of coffee might be 4.79 dollars *per* pound. This unit price could also be listed as \$4.79 per pound.

EXAMPLE 11 ■ Comparing Unit Prices

Which is the better buy: a 12-ounce box of breakfast cereal for \$2.69 or a 16-ounce box of the same cereal for \$3.49?

Solution

The unit price for the smaller box is

$$\text{Unit price} = \frac{\text{Total price}}{\text{Total units}} = \frac{\$2.69}{12 \text{ ounces}} \approx \$0.224 \text{ per ounce}.$$

The unit price for the larger box is

$$\text{Unit price} = \frac{\text{Total price}}{\text{Total units}} = \frac{\$3.49}{16 \text{ ounces}} \approx \$0.218 \text{ per ounce}.$$

Thus, the larger box has a slightly smaller unit price and is the better buy. ■

Proportions

A **proportion** is a statement that equates two ratios. For example, if the ratio of a to b is the same as the ratio of c to d, we can write the proportion as

$$\frac{a}{b} = \frac{c}{d}.$$

In typical applications we know values for three of the letters (quantities) and we are required to find the value of the fourth. The quantities a and d are called the **extremes** of the proportion, whereas b and c are called the **means** of the proportion.

EXAMPLE 12 ■ Solving a Proportion

The ratio of a certain number to 8 is the same as the ratio of 2 to 5. What is the number?

Solution

We let x represent the unknown number and write the following proportion.

$$\frac{x}{8} = \frac{2}{5}$$

We can solve this equation for x by multiplying both sides of the equation by 8 to obtain

$$x = 8 \cdot \frac{2}{5} = \frac{16}{5}.$$

Thus, the number is $\frac{16}{5}$. Check this solution in the original statement of the problem. ■

EXAMPLE 13 ■ An Application of Proportion

Suppose you are driving from Arizona to New York, a trip of 2,750 miles. You begin the trip with a full tank of gas, and after traveling 424 miles you refill the tank for \$22. How much should you plan to spend on gasoline for the entire trip? (Assume that the driving conditions are similar throughout the trip.)

Solution

Verbal model: $\dfrac{\text{Dollars for trip}}{\text{Miles for trip}} = \dfrac{\text{Dollars for tank}}{\text{Miles for tank}}$

Labels:	Entire trip:	Cost of gasoline $= x$ (dollars),	Miles $= 2{,}750$
	First part of trip:	Cost of gasoline $=$ \$22,	Miles $= 424$

Proportion: $\dfrac{x}{2{,}750} = \dfrac{22}{424}$

$$x = (2{,}750)\left(\frac{22}{424}\right)$$

$$x \approx 142.69$$

Thus, you should plan to spend approximately \$142.69 for gasoline on the trip. Check this solution in the original statement of the problem. ■

EXAMPLE 14 ■ Using a Proportion to Calculate Taxes

You have just moved into a new house valued at \$120,000. If your next-door neighbor pays \$1,670 in real estate taxes each year on a house valued at \$93,000, how much should you expect to pay in real estate taxes each year? (Assume that the rate is the same.)

Solution

Verbal model: $\dfrac{\text{Your taxes}}{\text{House value}} = \dfrac{\text{Neighbor's taxes}}{\text{House value}}$

Your house:	Taxes = x (dollars),	Value = \$120,000
Neighbor's house:	Taxes = \$1,670,	Value = \$93,000

Proportion: $\dfrac{x}{120{,}000} = \dfrac{1{,}670}{93{,}000}$

$$x = (120{,}000)\left(\frac{1{,}670}{93{,}000}\right)$$

$$x = 2{,}154.84$$

Thus, you should expect your real estate taxes to be about \$2,155. Check this solution in the original statement of the problem. ■

As a summary of what we have been doing and for future reference in work with applied problems, we recommend the following approach to solving word problems.

Strategy for Solving Word Problems

To solve a word problem, we suggest the following steps.

1. *Search* for the hidden equality—two expressions said to be equal or known to be equal. A sketch may be helpful.
2. *Write* a verbal model that equates these two expressions.
3. *Assign* numbers to the known quantities and letters (or algebraic expressions) to the variable quantities.

4. *Rewrite* the verbal model as an algebraic equation using the assigned labels.
5. *Solve* the resulting algebraic equation.
6. *Check* to see that the answer satisfies the word problem as stated.

DISCUSSION PROBLEM ■ Checking the Sensibility of an Answer

When solving problems related to real-life situations, you should always ask yourself whether the answer makes sense. Write a short paragraph explaining why the following answers are suspicious.

(a) A problem asked you to find the sales tax on a car, and your answer is \$1,948.56.

(b) A problem asked you to find the number of square feet in a kitchen, and your answer is 736 square feet.

(c) A problem asked you to find the interest on a \$2,000 deposit that is left in a savings account for one month, and your answer is \$105.75.

(d) A problem asked you to find the time it takes for a passenger plane to travel from New York to London, and your answer is $1\frac{1}{2}$ hours. ■

Warm-Up

The following warm-up exercises involve skills that were covered in earlier sections. You will use these skills in the exercise set for this section.

In Exercises 1–10, solve the given equation. (If not possible, state the reason.)

1. $4 - \frac{x}{2} = 6$

2. $\frac{x}{3} + 1 = 10$

3. $8(x - 14) = 0$

4. $3(2x + 4) - (6x - 5) = 0$

5. $-(2x + 8) + \frac{1}{3}(6x + 5) = 0$

6. $12(3 - x) = 5 - 7(2x + 1)$

7. $x + \frac{x}{2} = 4$

8. $0.75x = 235$

9. $543r = 434.4$

10. $(1 + r)500 = 550$

2.2 EXERCISES

In Exercises 1–6, construct a verbal model and write an algebraic equation that represents the problem. Do not solve the equation.

1. *Number Problem* Find a number such that the sum of that number and 30 is 82.
2. *Markup* The selling price of a jacket in a department store is \$85. The cost of the jacket to the store is \$62.95. What is the markup?
3. *Dimensions of a Triangle* The Slow Moving Vehicle (SMV) sign (see figure) is an equilateral triangle. Find the length of the sides of the triangle if its perimeter is 129 centimeters.

4. *Dimensions of a Rectangle* The length of a rectangle is three times its width. The perimeter of the rectangle is 64 inches. Find the dimensions of the rectangle.
5. *Repair Time* The bill for the repair of an automobile is \$380. Included in this bill is a charge of \$275 for parts. If the remainder of the bill is for labor, how many hours were spent repairing the car? (The charge for labor is \$35 per hour.)
6. *Units Produced* Suppose you have a job on an assembly line for which you are paid \$10 per hour plus \$0.75 per unit assembled. Find the number of units produced in an eight-hour day if your earnings for the day are \$146.

In Exercises 7–12, determine the equivalent forms of a percent, the number of parts out of 100, a decimal, and a fraction (in reduced form).

	Percent	*Parts out of 100*	*Decimal*	*Fraction*
7.	30%			
8.	75%			
9.			0.075	
10.				$\frac{2}{3}$
11.				$\frac{1}{8}$
12.		100		

In Exercises 13 and 14, what percent of the entire figure is shaded? (Assume that each rectangular portion of the figure has the same area.)

13.

14.

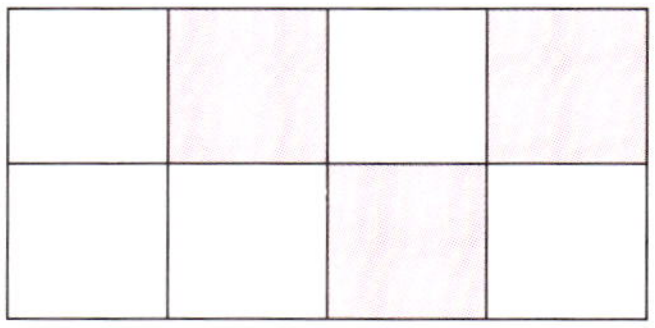

In Exercises 15–30, solve the given percent problem.

15. What is 35% of 250?
16. What is 68% of 800?
17. What is 8.5% of 816?
18. What is $33\frac{1}{3}\%$ of 816?
19. What is 0.4% of 150,000?
20. What is 300% of 16?
21. 84 is 24% of what number?
22. 416 is 65% of what number?
23. 42 is $10\frac{1}{2}\%$ of what number?

24. 168 is 350% of what number?

25. 96 is 0.8% of what number?

26. 496 is what percent of 800?

27. 1,650 is what percent of 5,000?

28. 2.4 is what percent of 480?

29. 2,100 is what percent of 1,200?

30. 900 is what percent of 500?

31. *Monthly Rent* If you spend 15% of your monthly income of $2,800 for rent, what is your monthly rent payment?

32. *College Enrollment* Thirty-eight percent of a college's enrollment are freshmen. Determine the number of freshmen if the enrollment of the college is 3,000.

33. *Company Layoff* Because of slumping sales, a small company laid off 25 of its 160 employees. What percentage of the work force was laid off?

34. *Number of Eligible Voters* The news media reported that 7,387 votes were cast in the last election. This represented 63% of the eligible voters of a district. Assuming this to be true, how many eligible voters are in the district?

35. *Defective Parts* A quality control engineer reported that 1.5% of a sample of parts were defective. Find the size of the sample if the engineer detected three defective parts.

36. *Price Inflation* The price of a new van is approximately 115% of what it was three years ago. What was the approximate price three years ago if the current price is $25,750?

37. *Course Grade* Suppose you missed getting a B in your mathematics course by eight points. If your point total for the course was 372, how many points were possible in the course? (Assume that you needed 80% of the course total for a B.)

38. *Room Dimensions* The floor of a room 10 feet by 12 feet is partially covered by a circular rug with a radius of 4 feet, as shown in the accompanying figure. What percentage of the floor is covered by the rug? (The area of a circle is $A = \pi r^2$, where r is the radius of the circle.)

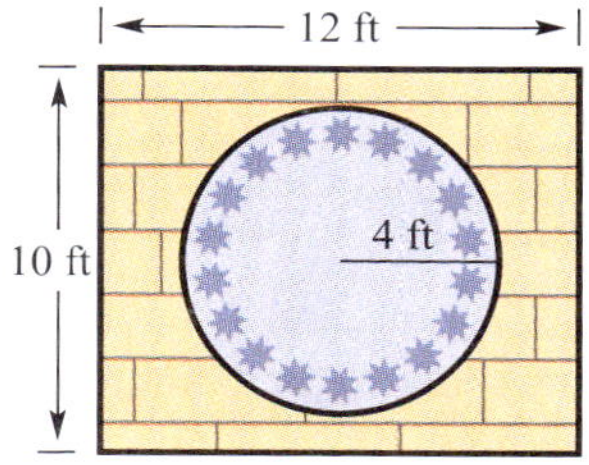

39. *Floor Space* Suppose you are planning to build a tool shed, but are undecided about the size. The two sizes you are considering are 12 feet by 15 feet and 16 feet by 20 feet. The floor space of the larger shed is what percent of the floor space of the smaller? The floor space of the smaller shed is what percent of the floor space of the larger?

40. *Real Estate Commission* A real estate agency receives a commission of $9,100 for the sale of a $130,000 house. What percent commission is this?

In Exercises 41–44, use the given double bar graph to answer the questions. The graph shows the percentage of respondents of faculty and staff at The Pennsylvania State University who are subject to common health-related risk factors. There were 3,789 female and 3,083 male respondents in the study. (*Source:* The Pennsylvania State University.)

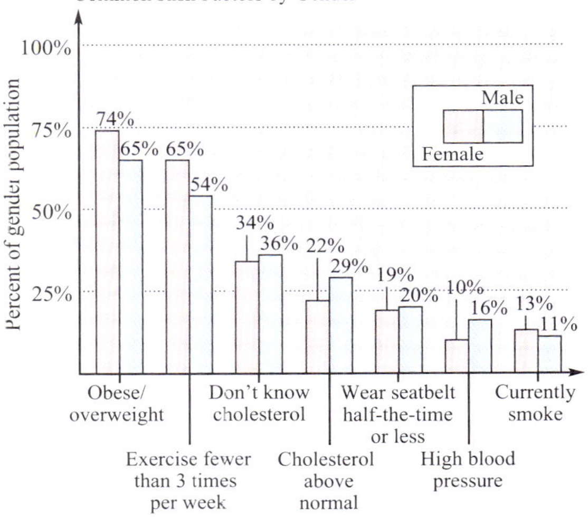

41. *Health Risks* Estimate the number of males in the sample who are overweight.

42. *Health Risks* Estimate the number of females in the sample who have high blood pressure.

43. *Survey Response* Estimate the number of female employees at the university if 51.8% returned their survey forms.

44. *Survey Response* Approximate the percentage of respondents who are nonsmokers.

In Exercises 45–50, express the ratio as a fraction. Use the same units in both the numerator and denominator, and write your answer in reduced form.

45. 36 inches to 48 inches **46.** 5 pounds to 24 ounces

47. 1 pint to 1 gallon

48. 125 centimeters to 2 meters

49. 40 milliliters to 1 liter **50.** 45 minutes to 2 hours

51. *State Income Tax* You have \$12.50 in state tax withheld from your paycheck each week when your gross pay is \$625. Find the ratio of tax to gross pay.

52. *Price-Earnings Ratio* The ratio of the price of a stock to its earnings is called the **price-earnings ratio**. Find the price-earnings ratio of a stock that sells for \$56.25 per share and earns \$6.25 per share.

53. *Compression Ratio* The **compression ratio** of a cylinder is the ratio of its expanded volume to its compressed volume (see figure). The expanded volume of one cylinder of a small diesel engine is 425 cubic centimeters, and its compressed volume is 20 cubic centimeters. Find the compression ratio of this engine.

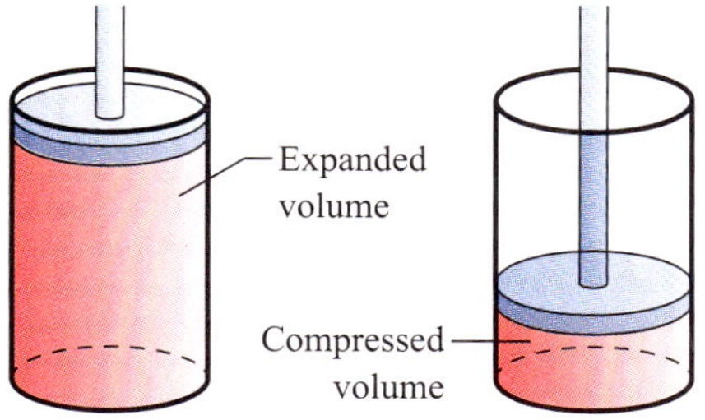

54. *Gear Ratio* The **gear ratio** of two gears is the number of teeth in one gear to the number of teeth in a second gear (see figure). If two gears in a gearbox have 60 teeth and 40 teeth, find the ratio of the larger gear to the smaller.

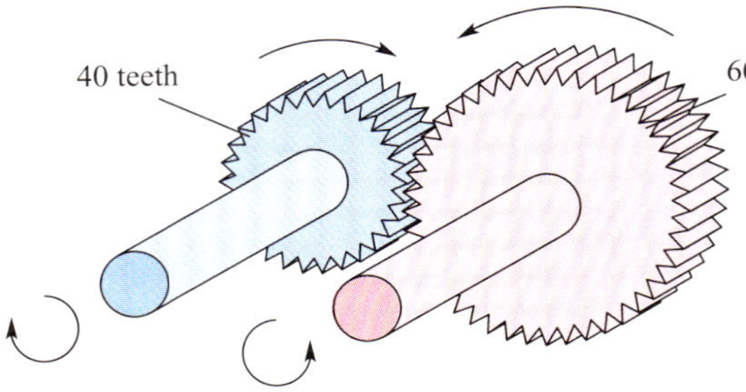

In Exercises 55–58, find the unit price (in dollars per ounce) of the product.

55. A 20-ounce can of pineapple for 95¢

56. A 32-ounce bottle of cola for \$1.69

57. A 1-pound, 4-ounce loaf of bread for \$1.39

58. An 18-ounce box of cereal for \$2.49

In Exercises 59–62, use unit prices to determine the better buy.

59. A $14\frac{1}{2}$-ounce bag of chips for \$2.32 or a 32-ounce bag for \$4.85

60. A $10\frac{1}{2}$-ounce package of coconut cookies for \$1.79 or a 16-ounce package for \$2.39

61. An 8-ounce tube of toothpaste for \$1.69 or a 12-ounce tube for \$2.39

62. A 2-pound package of hamburger for \$3.49 or a 3-pound package for \$5.29

In Exercises 63–68, solve the given proportion.

63. $\frac{x}{6} = \frac{2}{3}$ **64.** $\frac{y}{36} = \frac{6}{7}$

65. $\frac{4}{5} = \frac{6}{t}$ **66.** $\frac{8}{7} = \frac{2}{x}$

67. $\frac{y + 5}{6} = \frac{y - 2}{4}$ **68.** $\frac{z - 3}{3} = \frac{z + 8}{12}$

69. *Fuel Usage* A tractor uses five gallons of diesel fuel to plow for 105 minutes. Assuming conditions remain the same, determine the number of gallons of fuel used in six hours.

70. *Spring Length* A force of 32 pounds stretches a spring six inches. Determine the number of pounds of force required to stretch it nine inches.

71. *Fuel Mixture* The gasoline-to-oil ratio of a two-cycle engine is 40 to 1. Determine the amount of gasoline required to produce a mixture that has one-half pint of oil.

72. *Recipe Proportions* Three cups of flour are required to make one batch of cookies. How many cups are required to make $3\frac{1}{2}$ batches?

73. *Property Tax* The taxes on property with an assessed value of $75,000 are $1,125. Find the taxes on property with an assessed value of $120,000.

74. *Public Opinion Poll* In a public opinion poll, 870 people from a sample of 1,500 indicated they would vote for a specific candidate. Assuming this poll to be a correct indicator of the electorate, how many votes can the candidate expect to receive from 80,000 votes cast?

75. *Defective Units* A quality control engineer for a manufacturer found one defective unit in a sample of 75. At that rate, what is the expected number of defective units in a shipment of 200,000?

76. *Map Scale* Use the map scale in the accompanying figure to approximate the straight line distance from Los Angeles to San Francisco.

77. *Pumping Time* A pump can fill a 500-gallon tank in 45 minutes. Determine the time required to fill a 1,250-gallon tank with this pump.

78. *Weight of Sand* The ratio of cement to sand in an 80-pound bag of dry mix is 1 to 5. Find the number of pounds of sand in the bag.

79. *Monthly Expenses* The expenses for a small company for January are shown in the accompanying pie chart. What percent is each budget item of the total monthly expenses?

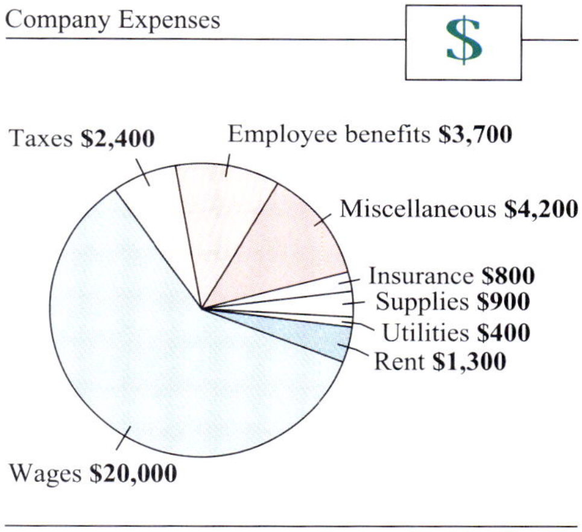

80. *Population* The populations of six counties are shown in the accompanying pie chart. What percent is each county population of the total population of the six counties?

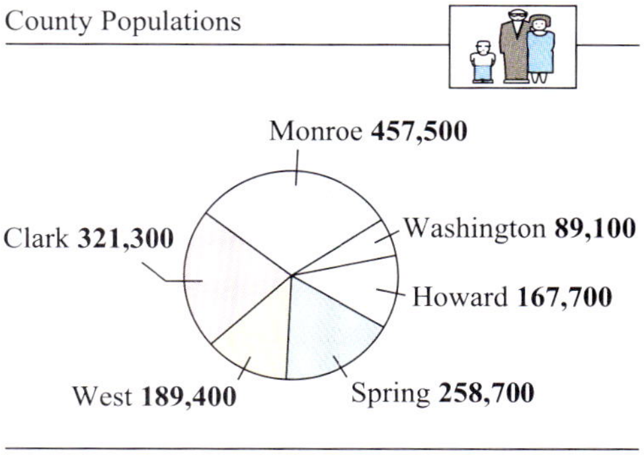

Similar Triangles In Exercises 81–84, solve for the length x of the side of a triangle by using the fact that corresponding sides of similar triangles are proportional.

81.

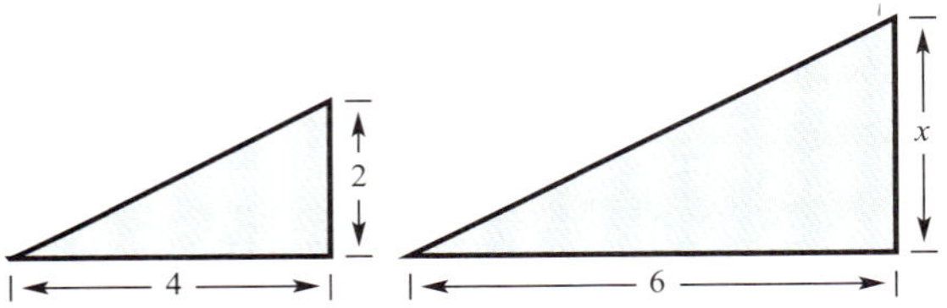

82.

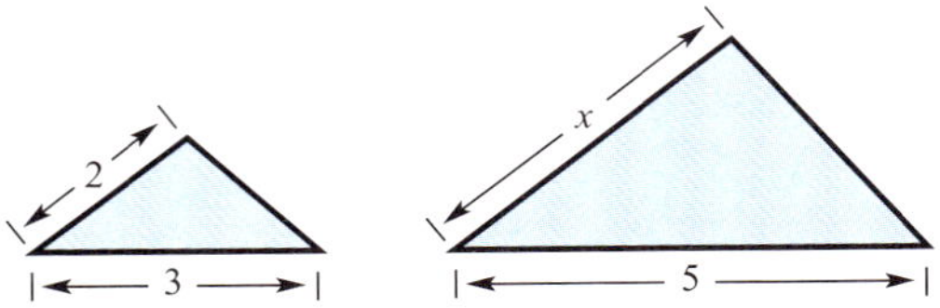

83.

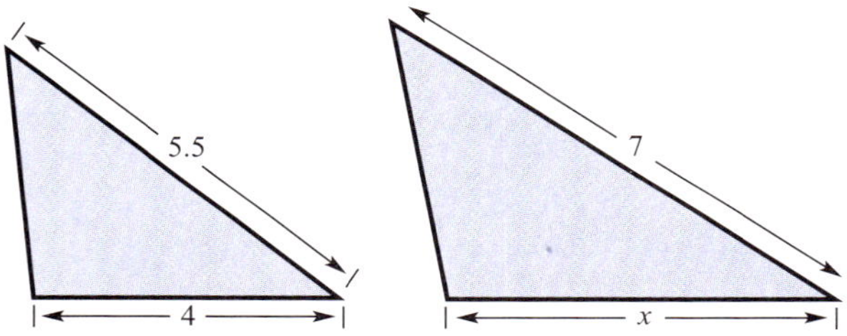

84.

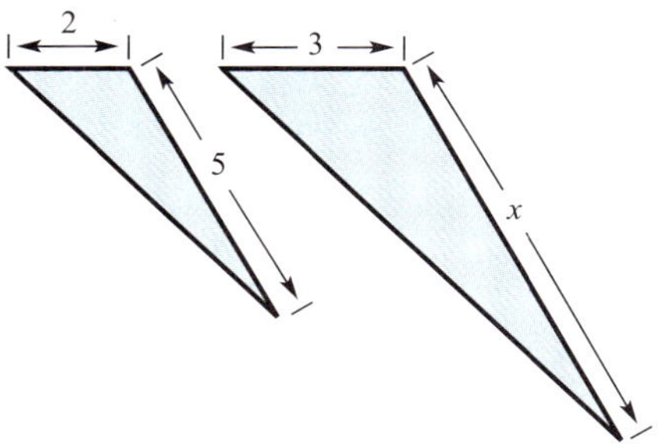

85. *Shadow Length* Find the length of the shadow of a man who is 6 feet tall and is standing 15 feet from a streetlight that is 20 feet high. (See the accompanying figure.)

Figure for 85

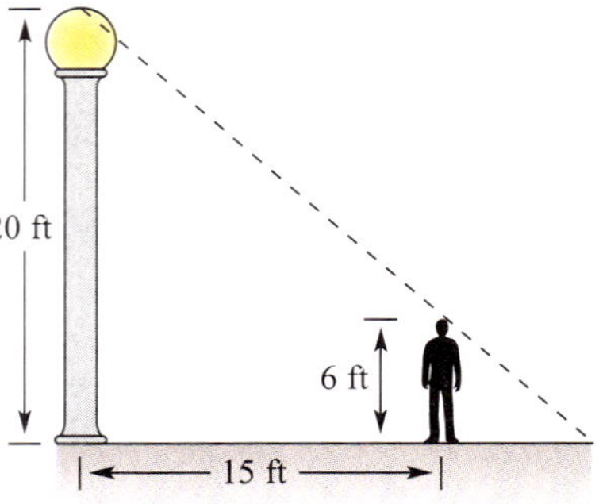

86. *Tree Height* A man who is 6 feet tall walks directly toward the tip of the shadow of a tree. When the man is 75 feet from the tree, he starts forming his own shadow beyond the shadow of the tree. Find the height of the tree, if the length of the shadow of the tree beyond this point is 11 feet.

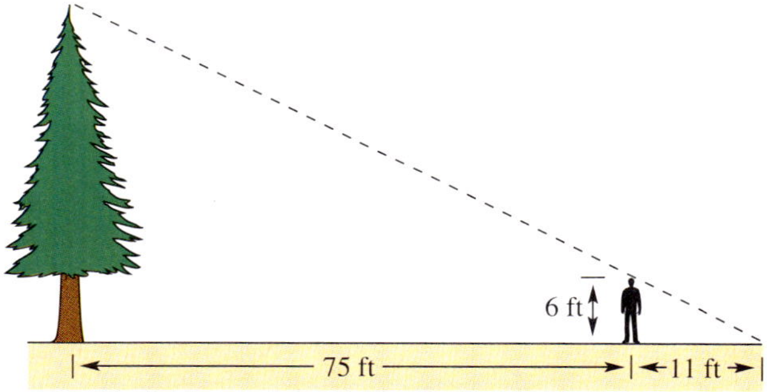

87. *Area of a Circle* Find the ratio of the areas of the two circles in the accompanying figure. (*Note:* The area of a circle is $A = \pi r^2$.)

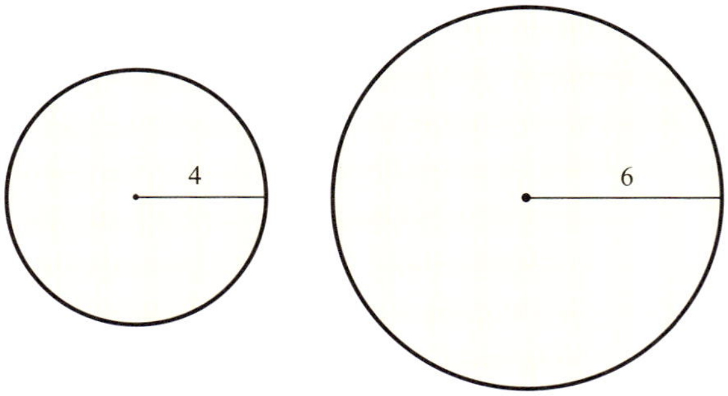

SECTION 2.3

Applications: Consumer and Scientific Problems

Consumer Problems • Mixture Problems • Rate Problems • Formulas

Consumer Problems

Many consumer problems have a mathematical model that involves the sum of a fixed term and a variable term. The variable term is often a hidden product in which one of the factors is a percent or some other type of rate. Watch for these occurrences in the discussion and examples that follow.

The **markup** on a consumer item is the difference between the amount (**cost**) a retailer pays for an item and the **price** at which the retailer sells the item. A verbal model for this relationship is as follows.

Selling price = Cost + Markup

In any such problem, the markup may be known or it may be expressed as a *percent of the cost*. This percent is called the **markup rate**.

Markup = Markup rate · Cost

EXAMPLE 1 ■ Finding the Markup Rate

A clothing store sells a pair of jeans for \$42. If the cost of the jeans is \$16.80, what is the markup rate?

Solution

Verbal model: Selling price = Cost + Markup

Labels: Selling price = \$42
Cost = \$16.80
p = markup rate
Markup = $p(16.80)$ (in dollars)

Equation:

$$\begin{aligned} 42 &= 16.8 + p(16.8) \\ 42 - 16.8 &= p(16.8) \\ 25.2 &= p(16.8) \\ \frac{25.2}{16.8} &= p \\ 1.5 &= p \end{aligned}$$

Thus, since $p = 1.5$, it follows that the markup rate is 150%. (Remember that when we convert the decimal form $p = 1.5$ to percent form, we obtain 150%.) Check this solution in the original statement of the problem. ■

As consumers, we all like a bargain! We often delay buying an item until it goes on sale at a discounted price. The model for this situation is

$$\boxed{\text{Sale price}} = \boxed{\text{List price}} - \boxed{\text{Discount}}$$

where the **discount** is given in dollars. The **discount rate** is the percent that the discount is of the list price. That is,

$$\boxed{\text{Discount}} = \boxed{\text{Discount rate}} \cdot \boxed{\text{List price}}.$$

Keep this in mind as you study the following example.

EXAMPLE 2 ■ Finding the Discount and the Discount Rate

A compact disc player is marked down from its list price of \$820 to a sale price of \$574. What is the discount rate?

Solution

Verbal model: $\boxed{\text{Discount}} = \boxed{\text{Discount rate}} \cdot \boxed{\text{List price}}$

Labels: Discount $= 820 - 574 = \$246$
List price $= \$820$
$p =$ discount rate

Equation:

$$246 = p(820)$$

$$\frac{246}{820} = p$$

$$0.30 = p$$

Thus, the discount rate is 30%. Check this solution in the original statement of the problem. ■

As consumers, we pay a wide variety of fees, bills, and charges that include descriptive statements such as "parts plus labor," "daily rate plus mileage," "food plus tips," or "down payment plus so much a month." The next two examples show how algebra can be used to solve these types of problems.

EXAMPLE 3 ■ Finding the Hours of Labor

An auto repair bill of \$338 lists \$170 for parts and the rest for labor. If the labor is \$28 per hour, how many hours did it take to repair the auto?

Solution

Verbal model: [Total bill] = [Price of parts] + [Price of labor]

Labels:
Total bill = \$338
Price of parts = \$170
x = number of hours of labor
Hourly rate for labor = \$28
Price of labor = $28x$ (in dollars)

Equation:

$$338 = 170 + 28x$$
$$168 = 28x$$
$$\frac{168}{28} = x$$
$$6 = x$$

Thus, it took six hours (of labor) to repair the auto. Check this solution in the original statement of the problem. ■

EXAMPLE 4 ■ Finding Hours of Overtime

An employee's regular hourly rate is \$10.50, and the overtime hourly rate is \$15.75 (for each hour over 40 hours in a week). During a given week the employee earned \$561.75. How many hours of overtime did the employee work?

Solution

Verbal model: [Total pay] = [Regular pay] + [Overtime pay]

Labels:
Total pay = \$561.75
Regular pay = (10.50)(40) (in dollars)
x = number of hours of overtime
Overtime pay = $15.75x$ (in dollars)

Equation:

$$561.75 = (10.50)(40) + 15.75x$$
$$561.75 - 420 = 15.75x$$
$$141.75 = 15.75x$$
$$\frac{141.75}{15.75} = x$$
$$9 = x$$

Therefore, the employee worked nine hours of overtime during the week. Check this solution in the original statement of the problem. ■

Mixture Problems

Many real-life problems involve combinations of two or more quantities that make up a new or different quantity. We call such problems **mixture problems**. They are usually composed of the sum of two or more "hidden products" that fit the following verbal model.

First component + Second component = Final mixture

First rate · Amount + Second rate · Amount = Final rate · Final amount

EXAMPLE 5 ■ **A Mixture Problem**

A nursery wants to mix two types of lawn seed. One type sells for \$10 per pound and the other type sells for \$15 per pound. To obtain 20 pounds of a mixture whose price is \$12 per pound, how many pounds of each type of seed are needed?

Solution

Verbal model: Total cost of \$10 seed + Total cost of \$15 seed = Total cost of \$12 seed

Labels:
\$10 seed: Cost per pound = \$10, Number of pounds = x
\$15 seed: Cost per pound = \$15, Number of pounds = $20 - x$
\$12 seed: Cost per pound = \$12, Number of pounds = 20

Equation:

$$10x + 15(20 - x) = 12(20)$$
$$10x + 300 - 15x = 240$$
$$-5x = -60$$
$$x = 12$$

Therefore, the mixture should contain 12 pounds of the \$10 seed and $20 - 12 = 8$ pounds of the \$15 seed. ■

There are a couple of additional guidelines that you should remember when working mixture problems. First, when you set up your verbal model, be sure to check that you are working with the *same type of units* in each part of the model. For instance, in Example 5, note that each of the three parts of the verbal model are measuring cost. (If two parts were measuring cost and the other part pounds, you would know that the model was incorrect.)

Second, when you have found the solution, go back to the original statement of the problem and check to see that the solution makes sense—both algebraically and from a common sense point of view. For instance, in Example 5, you might perform a check as follows.

$$\underbrace{\begin{pmatrix}\$10 \text{ per} \\ \text{pound}\end{pmatrix}\begin{pmatrix}12 \\ \text{pounds}\end{pmatrix}}_{\$10 \text{ seed}} + \underbrace{\begin{pmatrix}\$15 \text{ per} \\ \text{pound}\end{pmatrix}\begin{pmatrix}8 \\ \text{pounds}\end{pmatrix}}_{\$15 \text{ seed}} = \underbrace{\begin{pmatrix}\$12 \text{ per} \\ \text{pound}\end{pmatrix}\begin{pmatrix}20 \\ \text{pounds}\end{pmatrix}}_{\$12 \text{ seed}}$$

$$\$120 + \$120 = \$240$$

EXAMPLE 6 ■ A Solution Mixture Problem

A pharmacist needs to strengthen a 20% alcohol solution so that it contains 36% alcohol. How much pure alcohol should be added to 240 milliliters of the 20% solution? (See Figure 2.1.)

FIGURE 2.1

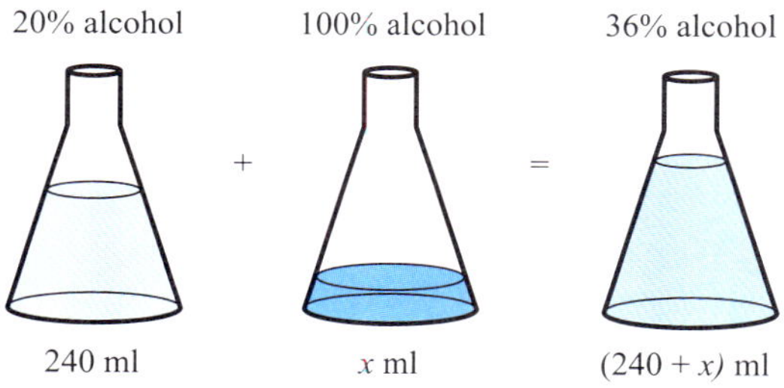

Solution

Verbal model: [Amount of alcohol in original solution] + [Amount of alcohol in pure solution] = [Amount of alcohol in final solution]

Labels: Original solution: Percent alcohol = 0.20, Amount = 0.20(240)
Pure alcohol: Percent alcohol = 1.00, Amount = 1.00(x)
Final solution: Percent alcohol = 0.36, Amount = 0.36(240 + x)

Equation:

$$0.20(240) + 1.00(x) = 0.36(240 + x)$$
$$48 + x = 86.4 + 0.36x$$
$$0.64x = 38.4$$
$$x = \frac{38.4}{0.64} = 60 \text{ ml}$$

Thus, the pharmacist should add 60 milliliters of pure alcohol to the original solution. Check this solution in the original statement of the problem. ■

Rate Problems

Time-dependent problems such as those involving distance can be classified as **rate problems**. They fit the verbal model

[Distance] = [Rate] · [Time] .

For instance, if you are traveling at a constant (or average) rate of 55 miles per hour for 45 minutes, then the total distance traveled is given by

$$\left(55 \frac{\text{miles}}{\text{hour}}\right)\left(\frac{45}{60} \text{ hour}\right) = 41.25 \text{ miles}.$$

As with all problems involving applications, be sure to check that the units in the verbal model make sense. For instance, in this problem, the rate is given in *miles per hour*. Therefore, in order for the solution to be given in *miles*, we must convert the time (from minutes) to *hours*. In the model, we think of the two "hours" as canceling, as follows.

$$\left(55 \frac{\text{miles}}{\cancel{\text{hour}}}\right)\left(\frac{45}{60} \cancel{\text{hour}}\right) = 41.25 \text{ miles}$$

EXAMPLE 7 ■ **Distance-Rate-Time Problem**

Suppose you jogged at an average rate of 9 kilometers per hour. How long will it take you to jog 15 kilometers?

Solution

Verbal model: Distance = Rate · Time

Labels: Distance = 15 kilometers
Rate = 9 kilometers per hour
t = time (in hours)

Equation:

$$15 = 9(t)$$

$$\frac{15}{9} = t$$

$$\frac{5}{3} = t$$

Thus, it would take you $1\frac{2}{3}$ hours (or one hour and 40 minutes). Check this solution in the original statement of the problem. ■

In work problems, the **rate of work** is the *reciprocal* of the time needed to do the entire job. For instance, if it takes 5 hours to complete a job, then the per hour work rate is

$$\frac{1}{5} \text{ job per hour.}$$

Similarly, if it takes $3\frac{1}{2}$ minutes to complete a job, then the per minute rate is

$$\frac{1}{3\frac{1}{2}} = \frac{1}{\frac{7}{2}} = \frac{2}{7} \text{ job per minute.}$$

EXAMPLE 8 ■ **Work Rate Problem**

Consider two machines in a paper manufacturing plant. Machine 1 can produce 2,000 pounds of paper in 4 hours. Machine 2 is newer and can produce 2,000 pounds of paper in $2\frac{1}{2}$ hours. How long will it take the two machines working together to produce 2,000 pounds of paper?

Solution

Verbal model: [Work done] = [Portion done by Machine 1] + [Portion done by Machine 2]

Labels:

Both machines:	Work done = 1 complete job,	Time = t
Machine 1:	Rate = $\frac{1}{4}$ job per hour,	Time = t
Machine 2:	Rate = $\frac{2}{5}$ job per hour,	Time = t

Equation:

$$1 = \left(\frac{1}{4}\right)(t) + \left(\frac{2}{5}\right)(t)$$

$$1 = \left(\frac{1}{4} + \frac{2}{5}\right)(t)$$

$$1 = \left(\frac{13}{20}\right)(t)$$

$$\frac{1}{13/20} = t$$

$$\frac{20}{13} = t$$

Thus, it would take $\frac{20}{13}$ hours (or about 1.54 hours) for both machines to complete the job. Check this solution in the original statement of the problem. ■

Note in Example 8 that the "2,000 pounds" of paper was unnecessary information. We simply represented the 2,000 pounds as "one complete job." This type of unnecessary information in an applied problem is sometimes called a *red herring*.

Formulas

Many common types of geometric, scientific, and investment problems use ready-made equations, called **formulas**. Knowing formulas like those in the following list will help you translate and solve a wide variety of real-life problems involving perimeter, area, volume, temperature, interest, and principal.

Common Formulas for Area and Perimeter

The following formulas give the area A and perimeter P (or circumference C) of the indicated figures.

Square

$A = s^2$

$P = 4s$

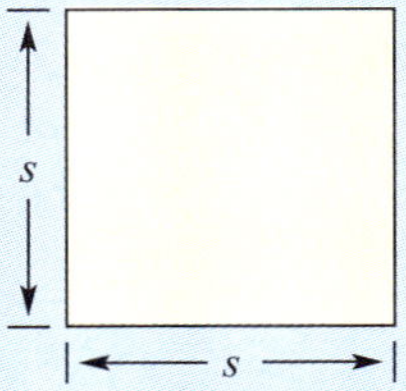

Rectangle

$A = lw$

$P = 2l + 2w$

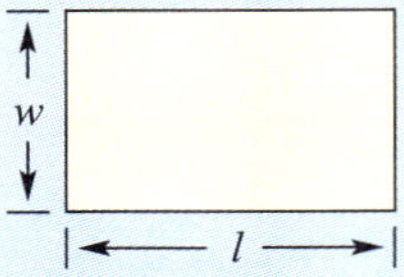

Circle

$A = \pi r^2$

$C = 2\pi r$

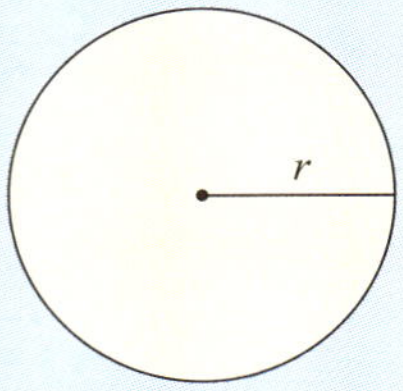

Triangle

$A = \frac{1}{2}bh$

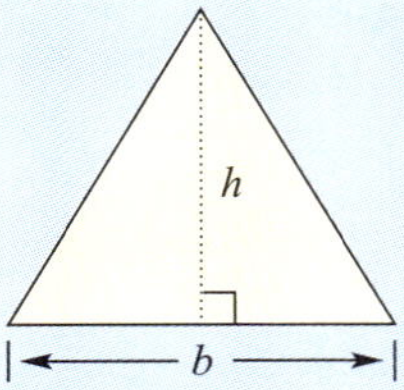

Common Formulas for Volume

The following formulas give the volume V of the indicated figures.

Cube

$V = s^3$

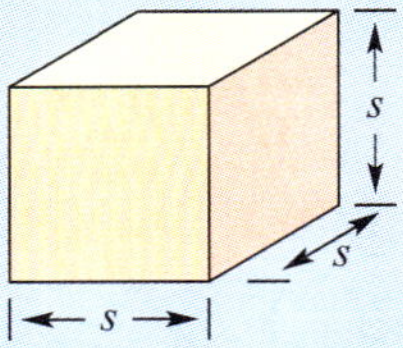

Rectangular Solid

$V = lwh$

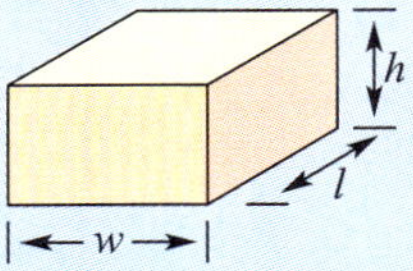

Circular Cylinder

$V = \pi r^2 h$

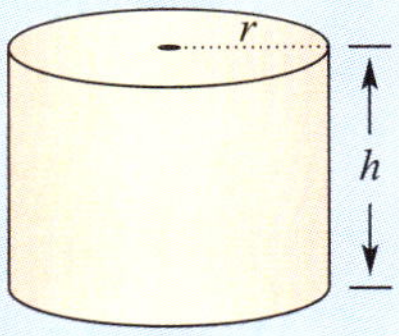

Sphere

$V = \frac{4}{3}\pi r^3$

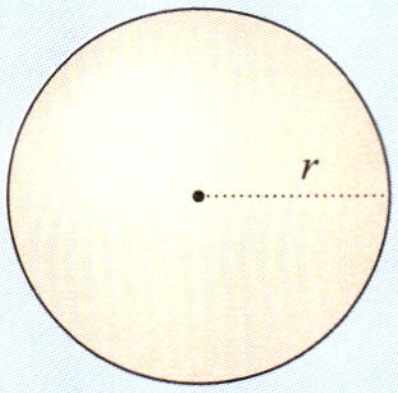

Miscellaneous Common Formulas

Temperature $F = \frac{9}{5}C + 32$

F = degrees Fahrenheit C = degrees Celsius

Simple Interest $I = Prt$

I = interest P = principal

r = annual interest rate t = time in years

Compound Interest	$A = P\left(1 + \frac{r}{n}\right)^{nt}$	
	A = balance	P = principal
	r = annual interest rate	n = compoundings per year
	t = time in years	
Distance	$d = rt$	
	d = distance traveled	r = rate
	t = time	

When working with applied problems you often need to rewrite one of the common formulas. For instance, the formula $P = 2l + 2w$ (the perimeter of a rectangle) can be rewritten to solve for w in the following manner.

$$P = 2l + 2w \quad \text{Given formula}$$

$$P - 2l = 2w$$

$$\frac{P - 2l}{2} = w \quad \text{Solve for } w$$

EXAMPLE 9 ■ **Using a Geometric Formula**

A rectangular plot of land has an area of 120,000 square feet. The plot is 300 feet wide. How long is it?

Solution

In a problem like this, we find it helpful to begin by drawing a picture, as shown in Figure 2.2. In this picture, we label the width of the rectangle as $w = 300$ feet, and label the unknown length as l.

FIGURE 2.2

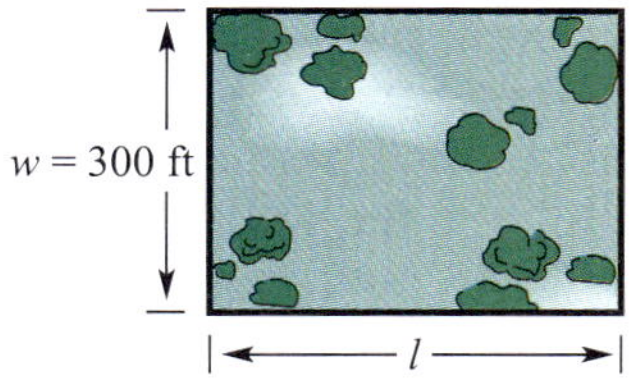

Now, to solve for the unknown length, we use the following steps.

Common formula: $A = lw$

Labels: A = 120,000 square feet
w = 300 feet
l = length (in feet)

Equation:

$$A = lw$$

$$120{,}000 = l(300)$$

$$\frac{120{,}000}{300} = l$$

$$400 = l$$

Thus, the length of the rectangular plot is 400 feet. (Note that we can check this by multiplying 300 feet by 400 feet to obtain an area of 120,000 square feet.) ■

EXAMPLE 10 ■ Simple Interest

A deposit of \$8,000 was left in an account paying simple interest. After six months, the account had earned \$300 in interest. What is the annual interest rate for this account?

Solution

Common formula: $I = Prt$

Labels:
$I =$ interest $=$ \$300
$P =$ principal $=$ \$8,000
$t =$ time $= \frac{1}{2}$ year
$r =$ annual interest rate

Equation:

$$I = Prt$$

$$300 = 8{,}000(r)\left(\frac{1}{2}\right)$$

$$\frac{2(300)}{8{,}000} = r$$

$$0.075 = r$$

Therefore, the annual interest rate is $r = 0.075$ (or 7.5%). Check this solution in the original statement of the problem. ■

DISCUSSION PROBLEM ■ Red Herrings

Most applied problems in textbooks give precisely the right amount of information that is necessary to solve a given problem. In real life, however, we often must sort through the given information and discard the portion that is irrelevant to the problem. Such irrelevant information is called a **red herring**. Find the red herrings in the following problems.

(a) Suppose you are hired for a job that pays \$8 per hour. After a 90-day review, your salary will be increased to \$8.50 per hour. During your first two weeks on the job, you work 80 hours. How much will you be paid for your first two weeks of work?

(b) A person leaves home at noon and drives 50 miles before stopping to fill the car with gas. The person then continues driving until 3 P.M. At the beginning of the trip the odometer reading for the car is 45,768, and at the end of the trip, the reading is 45,930. Find the average speed of the car during the entire trip. ■

Warm-Up

The following warm-up exercises involve skills that were covered in earlier sections. You will use these skills in the exercise set for this section.

In Exercises 1–4, solve the equation.

1. $44 - 16x = 0$

2. $-4(x - 5) = 0$

3. $3[4 + 5(x - 1)] = 6x + 2$

4. $\frac{3x}{8} + \frac{3}{4} = 2$

In Exercises 5–10, solve the given "percent problem."

5. What is $7\frac{1}{2}\%$ of 25?

6. What is 150% of 6,000?

7. 225 is what percent of 150?

8. What percent of 240 is 160?

9. 400 is 0.5% of what number?

10. 132 is 48% of what number?

2.3 EXERCISES

1. *Energy Use* The accompanying figure shows how 17.3 million barrels of oil are consumed each day in the United States. (a) How many barrels are used in the transportation sector? (b) We use approximately how many times as much oil in transportation as in heating and cooking? (*Source:* American Petroleum Institute, Department of Energy.)

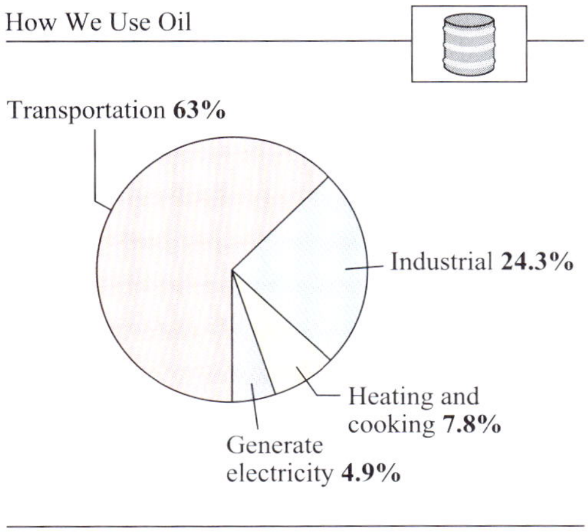

2. *Generation of Electricity* Use the accompanying figure to approximate the percentage of electricity generated in the United States by hydroelectric plants. (*Source:* Energy Information Administration, *Annual Energy Review*.)

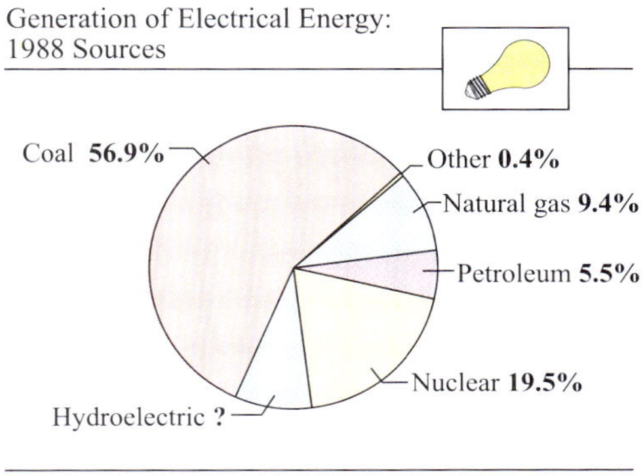

In Exercises 3–10, find the missing quantities. (Assume the markup rate is a percentage based on the cost.)

	Merchandise	*Cost*	*Selling Price*	*Markup*	*Markup Rate*
3.	Wristwatch	\$45.95	\$64.33		
4.	Bicycle	\$84.20	\$113.67		
5.	Sleeping bag		\$250.80	\$98.80	
6.	VCR		\$603.72	\$184.47	
7.	Van		\$26,922.50	\$4,672.50	
8.	Tractor		\$16,440.50	\$3,890.50	
9.	Camera	\$225.00			85.2%
10.	Refrigerator	\$732.00			$33\frac{1}{3}\%$

In Exercises 11–18, find the missing quantities. (Assume the discount rate is a percentage based on the list price.)

	Merchandise	*List Price*	*Sale Price*	*Discount*	*Discount Rate*
11.	Wheel alignment	\$49.50	\$25.74		
12.	Earrings	\$119.00	\$79.73		
13.	Coffee service	\$300.00		\$189.00	
14.	Luggage	\$345.00		\$134.55	
15.	Athletic shoes	\$95.00			65%
16.	Gallon of paint		\$15.92		20%
17.	Camcorder		\$893.10	\$251.90	
18.	Typewriter		\$257.32	\$202.18	

19. *Tire Cost* An auto store gives the list price of a tire as \$79.42. During a promotional sale, the store is selling four tires for the price of three. The store needs a markup on cost of 10% during the sale. What is the cost to the store of each tire?

20. *Price per Pound* The produce manager of a supermarket pays \$22.60 for a 100-pound box of bananas. From past experience, the manager estimates that 10% of the bananas will spoil before they are sold. At what price per pound should the bananas be sold to give the supermarket an average markup rate on cost of 30%?

21. *Comparison Shopping* A department store is offering a discount of 20% on a sewing machine with a list price of \$279.95. A mail-order catalog has the same machine for \$228.95 plus \$4.32 for shipping. Which is the better buy?

22. *Labor Charges* An appliance repair store charges \$35 for the first one-half hour of a service call. For each additional half hour of labor there is a charge of \$18. Find the length of a service call for which the charge is \$89.

23. *Tip Rate* A customer left a total of \$10 for a meal that cost \$8.45. Determine the tip rate.

24. *Tip Rate* A customer left a total of \$40 for a meal that cost \$34.73. Determine the tip rate.

25. *Long Distance Rates* The weekday rate for a telephone call is \$0.75 for the first minute plus \$0.55 for each additional minute. Determine the length of a call that cost \$5.15. What would have been the cost of the call if it had been made during the weekend when there is a 60% discount?

26. *Insurance Premiums* The annual insurance premium for a policyholder is \$862. Find the annual premium if the policyholder must pay a 20% surcharge because of an accident.

27. *Amount Financed* A customer bought a lawn tractor that cost \$4,450 plus 6% sales tax. Find the amount of the sales tax and the total bill. Find the amount financed if a down payment of \$1,000 was made.

28. *Weekly Pay* The weekly salary of an employee is \$250 plus a 6% commission on the employee's total sales. Find the weekly pay for a week in which the sales are \$5,500.

29. *Commission Rate* Determine the commission rate for an employee who earned \$450 in commissions for sales of \$5,000.

30. *Overtime Hours* Last week you earned \$740. If you are paid \$14.50 per hour for the first 40 hours and \$20 for each hour over 40, how many hours of overtime did you work?

31. *Number of Stamps* You have a set of 70 stamps with a total value of \$16.50. If this set includes 15¢ stamps and 30¢ stamps, find the number of each type.

32. *Coin Problem* A person has 50 coins in dimes and quarters with a combined value of \$10.25. Determine the number of coins of each type.

33. *Opinion Poll* Fourteen hundred people were surveyed in an opinion poll. Political candidates A and B received approximately the same preference, but Candidate C was preferred by twice the number of people as either A or B. Determine the number in the sample that preferred Candidate C.

34. *Ticket Sales* Ticket sales for a play total \$2,200. There are three times as many adult tickets sold, and the prices of the tickets for adults and children are \$6 and \$4, respectively. Find the number of children's tickets sold.

35. *Feed Mixture* A rancher must purchase 500 bushels of a feed mixture for cattle. She is considering oats and corn, which cost \$1.70 per bushel and \$3.00 per bushel, respectively. Complete the following table where x is the number of bushels of oats in the mixture.

Oats x	Corn $500 - x$	Price/bushel of the mixture
0		
100		
200		
300		
400		
500		

(a) How does the increase in the number of bushels of oats affect the number of bushels of corn in the mixture?
(b) How does the increase in the number of bushels of oats affect the price per bushel of the mixture?
(c) If there were equal numbers of bushels of oats and corn in the mixture, how would the price of the mixture be related to the price of each component?

36. *Nut Mixture* A grocer mixes two kinds of nuts costing \$3.88 per pound and \$4.88 per pound, to make 100 pounds of a mixture costing \$4.13 per pound. How many pounds of each kind of nut was put into the mixture?

Mixture Problem In Exercises 37–40, determine the number of units of Solution 1 and Solution 2 needed to obtain the desired amount and concentration of the final solution.

	Concentration			*Amount of Final Solution*
	Sol. 1	*Sol. 2*	*Final*	
37.	20%	60%	40%	100 gal
38.	50%	75%	60%	10 L

	Concentration			*Amount of Final Solution*
	Sol. 1	*Sol. 2*	*Final*	
39.	15%	60%	45%	24 qt
40.	60%	80%	75%	55 gal

41. *Antifreeze Coolant* The cooling system on a truck contains five gallons of coolant that is 40% antifreeze. How much must be withdrawn and replaced with 100% antifreeze to bring the coolant in the system to 50% antifreeze?

42. *Fuel Mixture* Suppose you mix gasoline and oil to obtain $2\frac{1}{2}$ gallons of mixture for an engine. The mixture is 40 parts gasoline and 1 part two-cycle oil. How much gasoline must be added to bring the mixture to 50 parts gasoline and 1 part oil?

Distance Problem In Exercises 43–48, determine the unknown distance, rate, or time.

	Distance, d	*Rate, r*	*Time, t*
43.		650 mi/hr	$3\frac{1}{2}$ hrs
44.		45 ft/sec	10 sec
45.	1,000 km	110 km/hr	
46.	250 ft	32 ft/sec	
47.	1,000 ft		$\frac{3}{2}$ sec
48.	385 mi		7 hr

49. *Average Speed* Determine the time for the space shuttle to travel a distance of 5,000 miles when its average speed is 17,000 miles per hour. (See figure.)

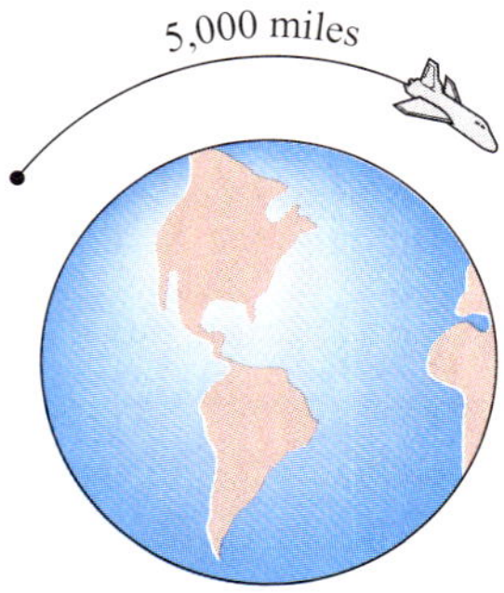

50. *Speed of Light* Determine the time for light to travel from the Sun to the Earth. The distance between the Sun and the Earth is approximately 93,000,000 miles, and the speed of light is approximately 186,000 miles per second. (See figure.)

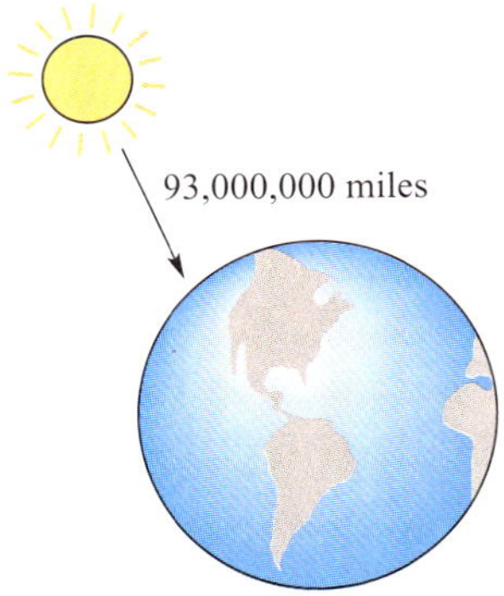

51. *Flying Distance* Two planes leave an airport at approximately the same time and fly in opposite directions. How far apart are the planes after $1\frac{1}{3}$ hours if their speeds are 480 miles per hour and 600 miles per hour?

52. *Average Speed* An Olympic runner completes a 5,000-meter race in 13 minutes and 20 seconds. What is the average speed of the runner?

53. *Travel Time* Two cars start at a given point and travel in the same direction at average speeds of 30 miles per hour and 45 miles per hour. (See figure.) How much time must elapse before the two cars are five miles apart?

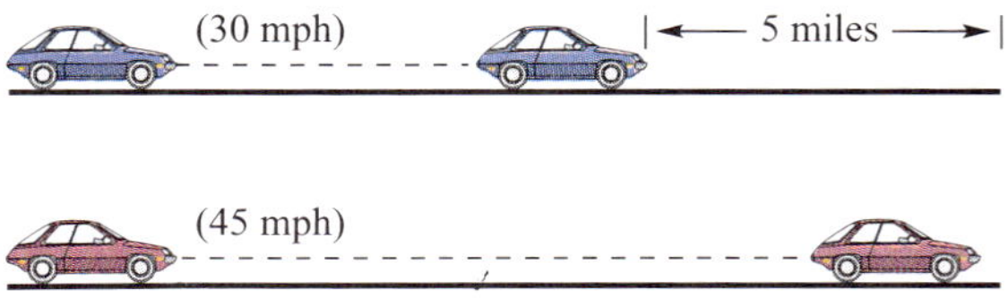

54. *Travel Time* On the first part of a 317-mile trip a sales representative averaged 58 miles per hour. The sales representative averaged only 52 miles per hour on the last part of the trip because of the increased volume of traffic. (See figure.) Find the amount of driving time at each speed if the trip took 5 hours and 45 minutes.

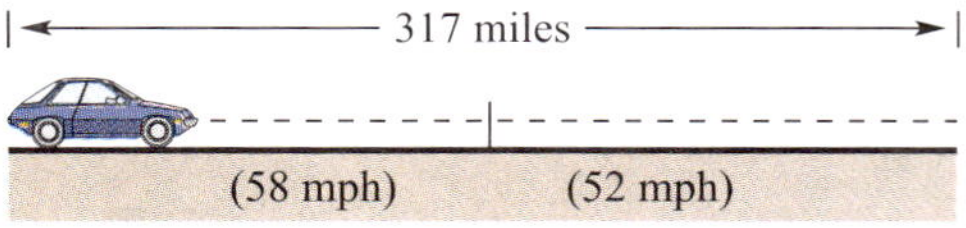

55. *Average Speed* A truck driver traveled at an average speed of 48 miles per hour on a 100-mile trip to pick up a load of freight. On the return trip with the truck fully loaded, the average speed was 40 miles per hour. Find the average speed for the round trip.

56. *Running Time* A jogger leaves a given point on a fitness trail, running at a rate of four miles per hour. Ten minutes later a second jogger leaves from the same location, running at five miles per hour. How long will it take the second jogger to overtake the first and how far will they have run?

57. *Work Rate Problem* Suppose you can mow the lawn in three hours and your friend can mow it in four hours. What fractional part of the lawn can each of you mow in one hour? How long will it take both of you to mow the lawn together?

58. *Work Rate Problem* Suppose you can complete a typing project in five hours, and a friend estimates that it would take him eight hours. What fractional part of the task can be accomplished by each typist in one hour? If you work on the project together, in how many hours could it be completed?

59. *Age Differences* Your age is one and one-half times the age of one of your cousins. What is the age of your cousin if the sum of your two ages is 45?

60. *Age Differences* The difference in ages of a father and his daughter is 28 years. Determine the age of the father when his age is twice that of his daughter.

61. *Number Problem* Find three consecutive numbers whose sum is 966.

62. *Number Problem* Find two consecutive odd numbers whose sum is 228.

63. *Number Problem* One number is five times a second. Find the numbers if their difference is 336. (Assume both numbers are positive.)

64. *Number Problem* One number is one-fourth of the second. Find the numbers if their difference is 108. (Assume both numbers are positive.)

In Exercises 65–70, solve for the specified variable.

65. Solve for R.
Ohm's Law: $E = IR$

66. Solve for C.
Markup: $S = C + rC$

67. Solve for r.
Simple Interest: $A = P + Prt$

68. Solve for P.
Compound Interest: $A = P\left(1 + \frac{r}{n}\right)^{nt}$

69. Solve for b.
Area of a Trapezoid: $A = \frac{1}{2}(a + b)h$

70. Solve for α.
Thermal Expansion: $L = L_0[1 + \alpha(\Delta t)]$

Dimensions of a Box In Exercises 71–73, use the closed rectangular box shown in the accompanying figure to answer the question.

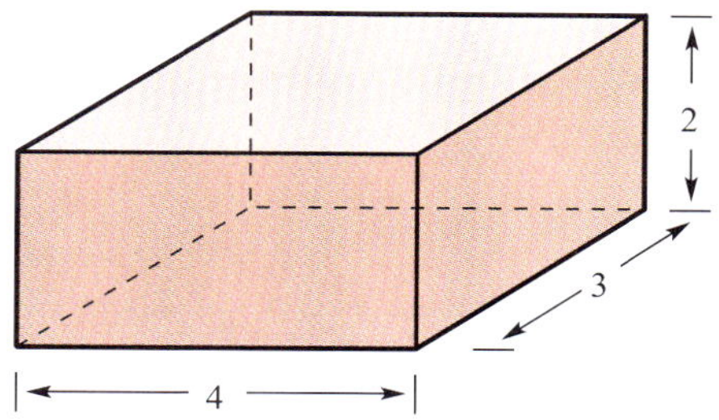

71. Find the perimeter of the base.

72. Find the volume of the box.

73. Find the surface area of the box. (*Note:* This is the combined area of the six surfaces.)

74. *Area of a Square* The accompanying figure shows three squares. The perimeter of square I is 12 inches, and the area of square II is 36 square inches. Find the area of square III.

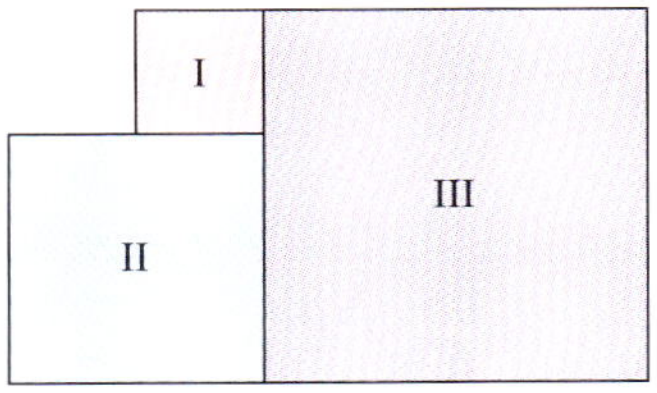

75. *Simple Interest* Find the interest on a \$5,000 bond that pays an annual rate of $9\frac{1}{2}\%$ for six years.

76. *Simple Interest* Find the annual interest rate on a certificate of deposit that accumulated \$400 interest in two years on a principal of \$2,500.

77. *Simple Interest* Find the principal required to earn \$500 in interest in two years if the annual interest rate is 7%.

78. *Simple Interest* Suppose you borrow \$36,000 for six months. You agree to pay back the principal and the interest (finance charges) in one lump sum. What will be the amount of the payment if the interest rate is 13%?

79. *Simple Interest* An inheritance of \$40,000 is divided into two investments earning 8.5% and 10% simple interest. How much is in each investment if the total interest for one year is \$3,550?

80. *Simple Interest* Four thousand dollars is divided into two investments earning $6\frac{1}{2}\%$ and 9% simple interest. How much is in each investment if the total interest for one year is \$335?

81. *Average Salary* The average salary for bus drivers at public schools in the United States between 1975 and 1989 can be approximated by the linear model

$$y = 5.45 + 0.366t$$

where y represents the salary (in dollars per hour) and t represents the year, with $t = 0$ corresponding to 1980 (see figure). During which year was the average salary \$7.28? What was the average annual raise for bus drivers during this 15-year period? (*Source:* U.S. Equal Employment Opportunities Commission.)

82. *Average Salary* The average salary for cafeteria workers at public schools in the United States between 1975 and 1989 can be approximated by the linear model

$$y = 3.96 + 0.286t$$

where y represents the salary (in dollars per hour) and t represents the year, with $t = 0$ corresponding to 1980 (see figure). During which year was the average salary \$5.39? What was the average annual raise for cafeteria workers during this 15-year period? (*Source:* U.S. Equal Employment Opportunities Commission.)

Figure for 81

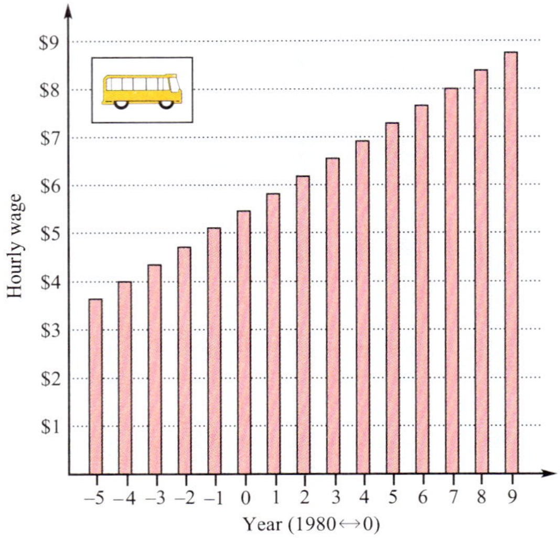

Figure for 82

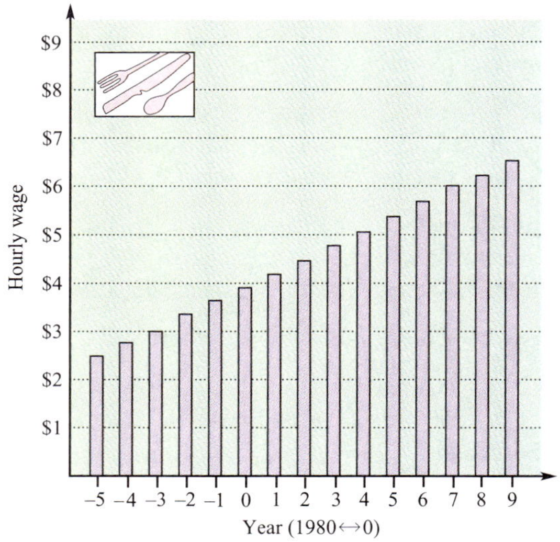

83. *Comparing Salary Increases* Use the information given in Exercises 81 and 82 to determine which of the two groups' average salaries was increasing at a greater annual rate during the 15-year period from 1975 to 1989.

SECTION 2.4 Linear Inequalities and Applications

Inequalities and Intervals on the Real Number Line • Properties of Inequalities • Solving a Linear Inequality • Solving a Double Linear Inequality • Applications of Linear Inequalities

Inequalities and Intervals on the Real Number Line

Simple inequalities were introduced in Section 1.1 to *order* the real numbers. There we compared two real numbers using the inequality symbols $<$, $\leq$, $>$, and $\geq$. For instance, because -5 lies to the left of 3 on the number line, we say that -5 is less than 3 and write $-5 < 3$. In this section we study **algebraic inequalities**, which are inequalities that contain one or more variable terms. Some examples are

$$x \leq 4, \qquad z \geq -3, \qquad y + 2 < 7, \quad \text{and} \quad 4x - 6 < 3x + 8.$$

As with an equation, we **solve an inequality** in the variable x by finding all values of x for which the inequality is true. Such values are called **solutions**, and we say that the solutions **satisfy** the inequality. The set of all real numbers that are solutions of an inequality is called the **solution set** of the inequality.

The set of all points on the real number line that represents the solution set is called the **graph** of the inequality. Graphs of many types of inequalities consist of intervals on the real number line. We summarize the four different types of **bounded** intervals on the real number line as follows. (Note that we use a parenthesis to denote the inequality symbols $<$ and $>$, and a bracket to denote the inequality symbols $\leq$ and $\geq$.)

Bounded Intervals on the Real Number Line

Let a and b be real numbers such that $a < b$. The following intervals on the real number line are called **bounded intervals**. The numbers a and b are called the **endpoints** of each interval.

Notation	*Interval Type*	*Inequality*	*Graph*
$[a, b]$	Closed	$a \leq x \leq b$	
(a, b)	Open	$a < x < b$	
$[a, b)$	Half-open	$a \leq x < b$	
$(a, b]$	Half-open	$a < x \leq b$	

Note that a closed interval contains both of its endpoints, a half-open interval contains only one of its endpoints, and an open interval does not contain either of its endpoints.

EXAMPLE 1 ■ Sketching the Graph of a Bounded Inequality

Sketch the graphs of the following inequalities, and write the interval notation for each.

(a) $-3 < x \le 1$ (b) $0 < x < 2$ (c) $1 \le x \le 5$

Solution

(a) The graph of this inequality is the set of all numbers on the real line that are greater than -3 and less than or equal to 1, as shown in Figure 2.3(a). The interval notation for this graph is

$(-3, 1]$. Half-open interval

This interval does not contain the left endpoint, -3, but it does contain the right endpoint, 1.

(b) The graph of this inequality is the set of all numbers on the real line that are greater than 0 and less than 2, as shown in Figure 2.3(b). The interval notation for this graph is

$(0, 2)$. Open interval

This interval does not contain the left endpoint, 0, or the right endpoint, 2.

(c) The graph of this inequality is the set of all numbers on the real line that are greater than or equal to 1 and less than or equal to 5, as shown in Figure 2.3(c). The interval notation for this graph is

$[1, 5]$. Closed interval

This interval contains its left endpoint, 1, and its right endpoint, 5.

FIGURE 2.3

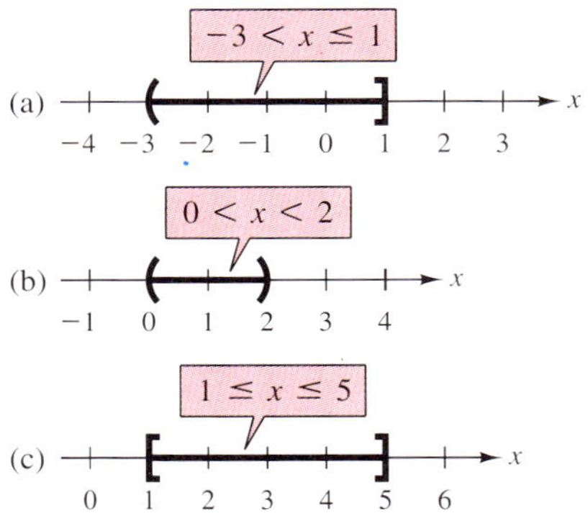

■

The **length** of the interval $[a, b]$ is simply the distance between its endpoints: $b - a$. The lengths of (a, b), $(a, b]$, and $[a, b)$ are the same. The reason that these four types of intervals are called **bounded** is that each has a finite length. An interval that *does not* have a finite length is called **unbounded** (or **infinite**). The five basic types of unbounded intervals are summarized in the following list.

Unbounded Intervals on the Real Number Line

Let a and b be real numbers. The following intervals on the real number line are called **unbounded intervals**.

Notation	*Interval Type*	*Inequality*	*Graph*
$[a, \infty)$	Half-open	$a \le x$	
(a, ∞)	Open	$a < x$	
$(-\infty, b]$	Half-open	$x \le b$	
$(-\infty, b)$	Open	$x < b$	
$(-\infty, \infty)$	Entire real line		

The symbols ∞ (**positive infinity**) and $-\infty$ (**negative infinity**) do not represent real numbers. They are simply convenient symbols used to describe the unboundedness of an interval such as $(1, \infty)$.

EXAMPLE 2 ■ Sketching the Graph of an Unbounded Interval

Sketch the graphs of the following inequalities, and write the interval notation for each.

(a) $-2 < x$ (b) $x \le 2$

Solution

(a) The graph of this inequality is the set of all numbers on the real line that are greater than -2, as shown in Figure 2.4(a). The interval notation for this graph is

$$(-2, \infty).$$ Open, unbounded interval

This interval does not contain its left endpoint, -2. Moreover, it has no right endpoint (because it is unbounded to the right).

(b) The graph of this inequality is the set of all numbers on the real line that are less than or equal to 2, as shown in Figure 2.4(b). The interval notation for this graph is

$(-\infty, 2]$. Half-open, unbounded interval

This interval has no left endpoint, but does have a right endpoint, 2, which is included in the interval.

FIGURE 2.4

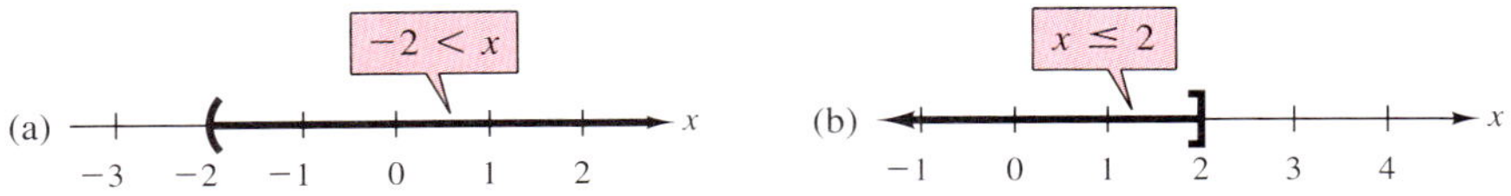

MATH MATTERS

Miles or Kilometers?

To convert from miles per hour to kilometers per hour, multiply miles per hour by 1.609344.

Miles per hour × 1.609344 = Kilometers per hour

To convert from kilometers per hour to miles per hour, multiply kilometers per hour by 0.621371.

Kilometers per hour × 0.621371 = Miles per hour

For a quick approximation that you can perform in your head, use the following rules.

1. *Quick Approximation* To convert miles per hour to kilometers per hour, multiply by 8 and divide by 5. For instance, 50 miles per hour is approximately

$(50 \times 8) \div 5 = 400 \div 5 = 80$ kilometers per hour.

2. *Quick Approximation* To convert kilometers per hour to miles per hour, multiply by 5 and divide by 8. For instance, 88 kilometers per hour is approximately

$(88 \times 5) \div 8 = 440 \div 8 = 55$ miles per hour.

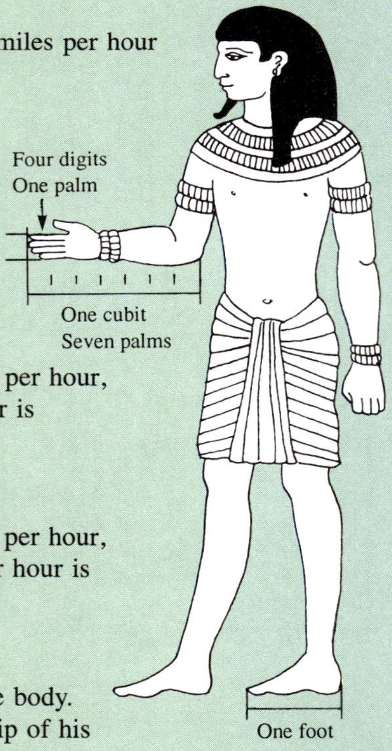

In ancient Egypt, the units of measure were related to parts of the body. The *cubit* was the distance from the tip of a man's elbow to the tip of his middle finger. The *digit* was the width of one finger. These units fit together like this: four digits are equal to one *palm*, and seven palms are equal to one cubit.

Properties of Inequalities

The procedures for solving linear inequalities in one variable are much like those for solving linear equations. To isolate the variable, we make use of **properties of inequalities**. These properties are similar to the properties of equality, but there are two important exceptions. When both sides of an inequality are multiplied or divided by a negative number, the direction of the inequality symbol must be reversed. Here is an example.

$$-2 < 5 \qquad \text{Given inequality}$$

$$(-3)(-2) > (-3)(5) \qquad \text{Multiply both sides by } -3 \text{ and reverse the inequality}$$

$$6 > -15$$

Two inequalities that have the same solution set are called **equivalent**. The following list describes operations that can be used to create equivalent inequalities.

Properties of Inequalities

1. *Addition and Subtraction Properties*
   Adding (or subtracting) the same quantity to (or from) both sides of an inequality produces an equivalent inequality.

   If $a < b$, then $a + c < b + c$. If $a < b$, then $a - c < b - c$.

2. *Multiplication and Division Properties* (Positive quantities)
   Multiplying (or dividing) both sides of an inequality by a *positive* quantity produces an equivalent inequality.

   If $a < b$ and $c > 0$, then $ac < bc$. If $a < b$ and $c > 0$, then $\frac{a}{c} < \frac{b}{c}$.

3. *Multiplication and Division Properties* (Negative quantities)
   Multiplying (or dividing) both sides of an inequality by a *negative* quantity produces an equivalent inequality in which the inequality symbol is reversed.

   If $a < b$ and $c < 0$, then $ac > bc$. If $a < b$ and $c < 0$, then $\frac{a}{c} > \frac{b}{c}$.

4. *Transitive Property*
   Consider three quantities for which the first quantity is less than the second, and the second is less than the third. It follows that the first quantity must be less than the third.

   If $a < b$ and $b < c$, then $a < c$.

NOTE: Each of the above properties is true if the symbol $<$ is replaced by $\leq$. Moreover, the letters a, b, and c can be real numbers, variables, or variable expressions. Note that we cannot multiply or divide both sides of an inequality by zero.

Solving a Linear Inequality

In the next several examples we apply the properties of inequalities to solve inequalities. The only type of inequality we will work with in this section is a **linear inequality** in a single variable. For instance, the inequalities

$$2x + 3 > 4, \qquad 3x - 4 \leq 2x + 5, \quad \text{and} \quad 3 \geq -5x + 4$$

are each linear inequalities in the variable x because the (implied) exponent of x in each inequality is 1. (The inequality $x^2 < 4$ is *not* linear in the variable x because x has an exponent of 2.)

As you read through each of the following examples, pay special attention to the steps in which the inequality symbol is reversed. Remember that when you multiply or divide an inequality by a negative number, you must reverse the inequality symbol.

EXAMPLE 3 ■ Solving a Linear Inequality

$$x + 6 < 9 \qquad \text{Given inequality}$$

$$x + 6 - 6 < 9 - 6 \qquad \text{Subtract 6 from both sides}$$

$$x < 3 \qquad \text{Solution set}$$

Thus, the solution set consists of all real numbers that are less than 3. The interval notation for this solution set is

$$(-\infty, 3). \qquad \text{Solution set}$$

The graph of this solution set is shown in Figure 2.5.

FIGURE 2.5

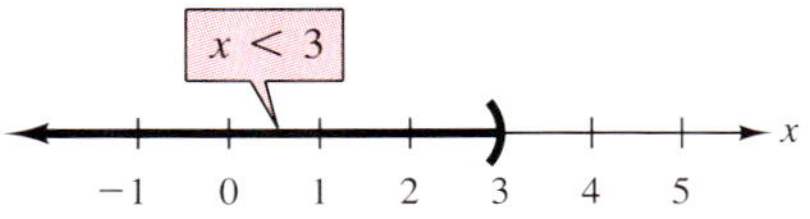

■

Checking the solution set of an inequality is not as simple as checking the solutions of an equation. (There are usually too many x-values to substitute back into the original inequality.) We can, however, get an indication of the validity of a solution set by substituting a few convenient values of x. For instance, in Example 3, we found the solution set of $x + 6 < 9$ to be $x < 3$. Try checking that $x = 0$ satisfies the original inequality, whereas $x = 4$ does not.

EXAMPLE 4 ■ Solving a Linear Inequality

$$2y - 3 \leq -9 \qquad \text{Given inequality}$$

$$2y - 3 + 3 \leq -9 + 3 \qquad \text{Add 3 to both sides}$$

$$2y \leq -6 \qquad \text{Combine like terms}$$

$$\frac{2y}{2} \leq \frac{-6}{2} \qquad \text{Divide both sides by (positive) 2}$$

$$y \leq -3 \qquad \text{Solution set}$$

Thus, the solution set consists of all real numbers that are less than or equal to -3. The interval notation for this solution set is

$(-\infty, -3]$. Solution set

The graph of this solution set is shown in Figure 2.6.

FIGURE 2.6

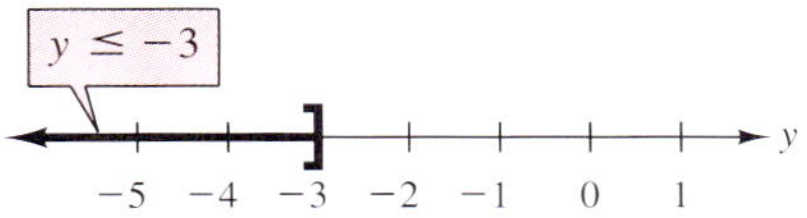

■

EXAMPLE 5 ■ Solving a Linear Inequality

$8 - 3x \leq 20$	Given inequality
$8 - 8 - 3x \leq 20 - 8$	Subtract 8 from both sides
$-3x \leq 12$	Combine like terms
$\dfrac{-3x}{-3} \geq \dfrac{12}{-3}$	Divide both sides by -3 and reverse inequality symbol
$x \geq -4$	Solution set

Thus, the solution set consists of all real numbers that are greater than or equal to -4. The interval notation for this solution set is

$[-4, \infty)$. Solution set

The graph of this solution set is shown in Figure 2.7.

FIGURE 2.7

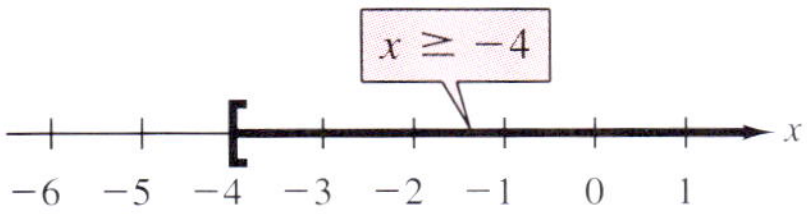

■

In the next example, we solve a linear inequality in which the variable occurs more than once.

EXAMPLE 6 ■ Solving a Linear Inequality

$9x - 4 > 5x + 2$	Given inequality
$9x - 4 + 4 > 5x + 2 + 4$	Add 4 to both sides
$9x > 5x + 6$	Simplify
$9x - 5x > 6$	Subtract $5x$ from both sides
$4x > 6$	Combine like terms
$x > \dfrac{3}{2}$	Divide both sides by 4

Thus, the solution set consists of all real numbers that are greater than $\frac{3}{2}$. The interval notation for this solution set is

$$\left(\tfrac{3}{2}, \infty\right). \qquad \text{Solution set}$$

The graph of this solution set is shown in Figure 2.8.

FIGURE 2.8

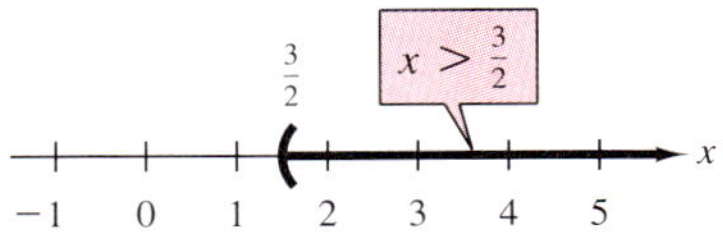

When one (or more) of the terms in a linear inequality involve constant denominators, we suggest multiplying both sides of the equation by the least common denominator. This clears the inequality of fractions. For instance, as a first step in solving the linear inequality

$$10 - \frac{3x}{2} \geq x - 5$$

we would multiply both sides of the inequality by 2. (Try solving this inequality—you should find that the solution set is $x \leq 6$.) This procedure is further demonstrated in Example 7.

EXAMPLE 7 ■ Solving a Linear Inequality

$$\frac{2x}{3} + 12 < \frac{x}{6} + 18 \qquad \text{Given inequality}$$

$$4x + 72 < x + 108 \qquad \text{Multiply both sides by 6}$$

$$4x - x < 108 - 72 \qquad \text{Subtract } x \text{ and 72 from both sides}$$

$$3x < 36 \qquad \text{Combine like terms}$$

$$x < 12 \qquad \text{Divide both sides by 3}$$

Thus, the solution set consists of all real numbers that are less than 12. The interval notation for this solution set is

$$(-\infty, 12). \qquad \text{Solution set}$$

The graph of this solution set is shown in Figure 2.9.

FIGURE 2.9

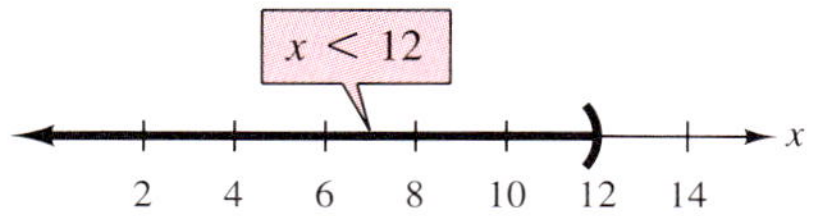

■

Solving a Double Linear Inequality

Sometimes it is convenient to write two inequalities as a **double inequality**. For instance, we can write the two inequalities $-4 \leq 5x - 2$ and $5x - 2 < 7$ more simply as

$$-4 \leq 5x - 2 < 7.$$

This form allows us to solve the two given inequalities together, as demonstrated in Example 8.

EXAMPLE 8 ■ Solving a Double Inequality

$-7 \leq 5x - 2 < 8$	Given inequality
$-7 + 2 \leq 5x - 2 + 2 < 8 + 2$	Add 2 to all three parts
$-5 \leq 5x < 10$	Combine like terms
$\dfrac{-5}{5} \leq \dfrac{5x}{5} < \dfrac{10}{5}$	Divide each part by 5
$-1 \leq x < 2$	Solution set

Thus, the solution set consists of all real numbers that are greater than or equal to -1 and less than 2. The interval notation for this solution set is

$[-1, 2)$. Solution set

The graph of this solution set is shown in Figure 2.10.

FIGURE 2.10

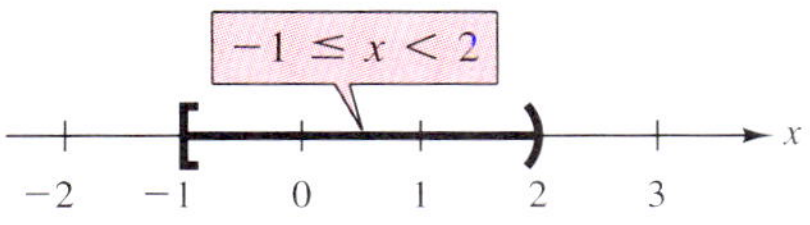

■

The double inequality in Example 8 could have been solved in two parts as follows.

$-7 \leq 5x - 2$	and	$5x - 2 < 8$
$-5 \leq 5x$		$5x < 10$
$-1 \leq x$		$x < 2$

The solution set consists of all real numbers that satisfy *both* inequalities. In other words, the solution set is the set of all values of x for which $-1 \leq x < 2$.

Applications of Linear Inequalities

Linear inequalities in real-life problems arise from statements that involve phrases such as "at least," "no more than," "minimum value," and so on. In the last two examples in this section, we look at applications involving linear inequalities.

Before looking at some applications, we give one example of translating verbal statements into inequalities.

EXAMPLE 9 ■ **Translating Verbal Statements**

	Verbal Statement	*Inequality*
(a)	x is at most 3.	$x \leq 3$
(b)	x is no more than 3.	$x \leq 3$
(c)	x is at least 3.	$x \geq 3$
(d)	x is more than 3.	$x > 3$
(e)	x is less than 3.	$x < 3$

■

EXAMPLE 10 ■ **An Application of an Inequality**

A subcompact car can be rented from Company A for \$240 per week with no extra charge for mileage. A similar car can be rented from Company B for \$100 per week, plus 25 cents for each mile driven. How many miles must a person drive in a week to make the rental fee for Company A less than that for Company B?

Solution

Verbal model: [Weekly cost for Company B] > [Weekly cost for Company A]

Labels: m = Number of miles driven in one week
Weekly cost for Company A = \$240
Weekly cost for Company B = 100 + 0.25m (in dollars)

Inequality:

$$100 + 0.25m > 240$$

$$0.25m > 140$$

$$m > 560 \text{ miles}$$

Thus, the car from Company A is cheaper if the person plans to drive more than 560 miles in a week. ■

EXAMPLE 11 ■ An Application of an Inequality

An overnight delivery service will not accept packages whose girth exceeds 132 inches. (The girth of a rectangular package is the sum of its length and the perimeter of its cross section.) Suppose you are sending a rectangular package that has square cross sections. If the length of the package is 68 inches, what is the maximum width of one of the sides?

Solution

We begin by making a sketch. In Figure 2.11 notice that the length of the package is 68 inches, and each side is x inches long.

FIGURE 2.11

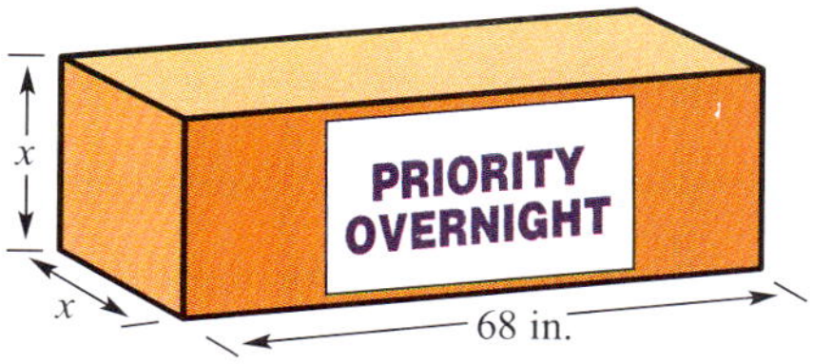

Verbal model: Girth $\leq$ 132 inches

Labels:
x = Width of a side (in inches)
Length = 68 inches
Girth = $4x + 68$ (in inches)

Inequality:
$$4x + 68 \leq 132$$
$$4x \leq 64$$
$$x \leq 16 \text{ inches}$$

Thus, the width of each side of the package must be less than or equal to 16 inches. ■

DISCUSSION PROBLEM ■ You Be the Instructor

Suppose you are teaching an algebra class and one of your students asks the following question.

> "Why is it that when we add a *negative* number to both sides of an equation we don't have to reverse the inequality, but when we multiply or divide both sides of an inequality by a *negative* number, we do have to reverse the inequality?"

How would you answer this question? ■

Warm-Up

The following warm-up exercises involve skills that were covered in earlier sections. You will use these skills in the exercise set for this section.

In Exercises 1–4, place the correct inequality symbol (< or >) between the two real numbers.

1. $-\frac{3}{4}$ ____ -5

2. $-\frac{1}{5}$ ____ $-\frac{1}{3}$

3. π ____ -3

4. 6 ____ $\frac{13}{2}$

In Exercises 5–10, solve the equation.

5. $-2n + 12 = 5$

6. $16 + 7z = 64$

7. $\frac{12 + x}{4} = 13$

8. $20 - \frac{x}{9} = 3$

9. $55 - 4(3 - y) = -5$

10. $8(t - 24) = 0$

2.4 EXERCISES

In Exercises 1–4, determine whether the given value of x satisfies the inequality.

Inequality	*Value*			
1. $7x - 10 > 0$	(a) $x = 3$	(b) $x = -2$	(c) $x = \frac{5}{2}$	(d) $x = \frac{1}{2}$
2. $3x + 2 < \frac{7x}{5}$	(a) $x = 0$	(b) $x = 4$	(c) $x = -4$	(d) $x = -1$
3. $0 < \frac{x + 5}{6} < 2$	(a) $x = 10$	(b) $x = 4$	(c) $x = 0$	(d) $x = -6$
4. $-2 < \frac{3 - x}{2} \le 2$	(a) $x = 0$	(b) $x = 3$	(c) $x = 9$	(d) $x = -12$

In Exercises 5–10, match the given inequality with its graph. [The graphs are labeled (a), (b), (c), (d), (e), and (f).]

5. $-1 < x \le 2$

6. $-1 < x \le 1$

7. $-1 \le x \le 1$

8. $-1 < x < 2$

9. $-2 \le x < 1$

10. $-2 < x < 1$

(a)
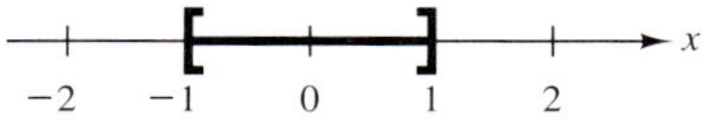

(b)
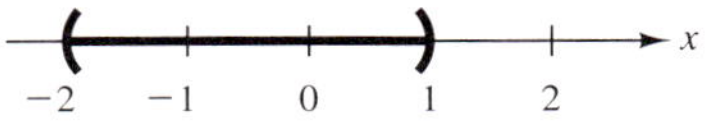

(c)
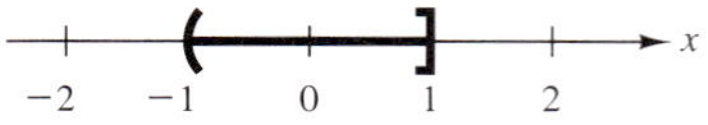

(d)
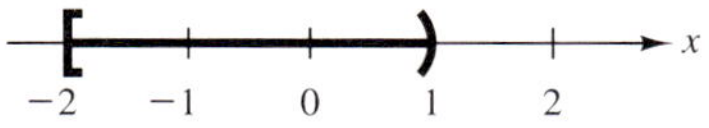

(e)
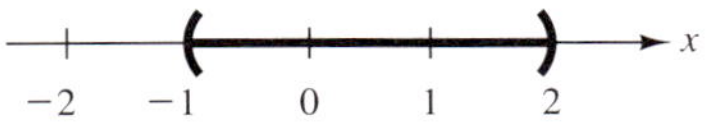

(f)
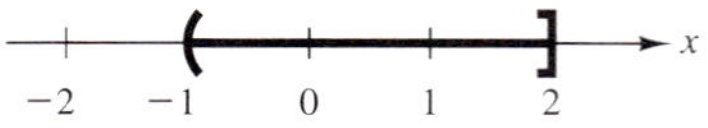

In Exercises 11–16, sketch the graph of the inequality and write the interval notation for the inequality.

11. $-5 < x \leq 3$

12. $4 > x \geq 1$

13. $\frac{3}{2} \geq x > 0$

14. $-7 < x \leq -3$

15. $-\frac{15}{4} < x < -\frac{5}{2}$

16. $-\pi > x > -5$

In Exercises 17–46, solve the inequality and sketch the solution on the real number line.

17. $x + 7 \leq 9$

18. $z - 4 > 0$

19. $4x < 22$

20. $2x > 5$

21. $-9x \geq 36$

22. $-6x \leq 24$

23. $-\frac{3}{4}x < -6$

24. $-\frac{1}{5}x > -2$

25. $2x - 5 > 9$

26. $3x + 4 \leq 22$

27. $5 - x \leq -2$

28. $1 - y \geq -5$

29. $5 - 3x < 7$

30. $12 - 5x > 5$

31. $3x - 11 > -x + 7$

32. $21x - 11 \leq 6x + 19$

33. $16 < 4(y + 2) - 5(2 - y)$

34. $4[z - 2(z + 1)] < 2 - 7z$

35. $-3(y + 10) \geq 4(y + 10)$

36. $2(4 - z) \geq 8(1 + z)$

37. $\frac{x}{6} - \frac{x}{4} \leq 1$

38. $\frac{x + 3}{6} + \frac{x}{8} \geq 1$

39. $0 < 2x - 5 < 9$

40. $-4 \leq 2 - 3(x + 2) < 11$

41. $-3 < \frac{2x - 3}{2} < 3$

42. $0 \leq \frac{x - 5}{2} < 4$

43. $1 > \frac{x - 4}{-3} > -2$

44. $-\frac{2}{3} < \frac{x - 4}{-6} \leq \frac{1}{3}$

45. $\frac{2}{5} < x + 1 < \frac{4}{5}$

46. $-1 < -\frac{x}{6} < 1$

In Exercises 47–50, use inequality notation to denote each of the given statements.

47. x is nonnegative.

48. y is more than -2.

49. z is at least 2.

50. x is at least 450 but no more than 500.

In Exercises 51–54, write a statement describing the set of real numbers that satisfy the inequality.

51. $x \geq \frac{5}{2}$

52. $t < 4$

53. $0 < z \leq \pi$

54. $-4 \leq t \leq 4$

55. *Travel Budget* A student group has \$4,500 budgeted for a field trip. The cost of transportation for the trip is \$1,900. To stay within the budget, all other costs C must be no more than what amount?

56. *Monthly Budget* Suppose you budgeted \$1,200 a month for your total expenses. Rent is \$400 per month, and the food costs \$275. To stay within your budget, all other costs C must be no more than what amount?

57. *Comparing Average Temperatures* The average temperature in Miami is greater than the average temperature in Washington, D.C., and the average temperature in Washington, D.C. is greater than the average temperature in New York City. How does the average temperature in Miami compare with the average temperature in New York City?

58. *Comparing Elevations* The elevation (above sea level) of San Francisco is less than the elevation of Dallas, and the elevation of Dallas is less than the elevation of Denver. How does the elevation of San Francisco compare with the elevation of Denver?

59. *Operating Costs* A utility company has a fleet of vans. The annual operating cost per van is

$$C = 0.28m + 2{,}900$$

where m is the number of miles traveled by a van in a year. What number of miles will yield an annual operating cost that is less than $10,000?

60. *Operating Costs* A fuel company has a fleet of trucks. The annual operating cost per truck is

$$C = 0.58m + 7{,}800$$

where m is the number of miles traveled by a truck in a year. What number of miles will yield an annual operating cost that is less than $25,000?

61. *Profit* The revenue for selling x units of a product is

$$R = 89.95x.$$

The cost of producing x units is

$$C = 61x + 875.$$

In order to obtain a profit, the revenue must be greater than the cost. For what values of x will this product produce a profit?

62. *Profit* The revenue for selling x units of a product is

$$R = 105.45x.$$

The cost of producing x units is

$$C = 78x + 25{,}850.$$

In order to obtain a profit, the revenue must be greater than the cost. For what values of x will this product produce a profit?

63. *Long Distance Charges* The cost for a long-distance telephone call is $0.96 for the first minute and $0.75 for each additional minute. If the total cost of the call cannot exceed $5, find the interval of time that is available for the call.

64. *Long Distance Charges* The cost for a long-distance telephone call is $1.45 for the first minute and $0.95 for each additional minute. If the total cost of the call cannot exceed $15, find the interval of time that is available for the call.

65. *Distance* If you live five miles from school and your friend lives three miles from you, then the distance d that your friend lives from school is in what interval? (See figure.)

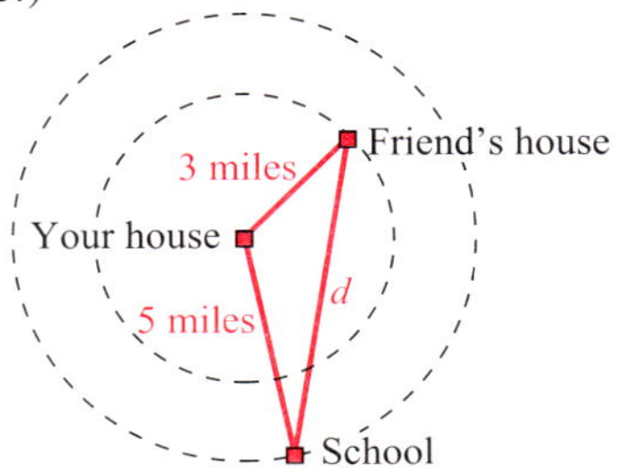

66. *Distance* If you live 10 miles from your work and your work is 5 miles from school, then the distance d that you live from school is in what interval?

67. If $-5 \le t < 8$, then $-t$ must be in what interval?

68. If $-3 \le x \le 10$, then $-x$ must be in what interval?

69. Four times a number n must be at least 12 and no more than 30. What interval contains this number?

70. Five times a number n must be at least 15 and no more than 45. What interval contains this number?

71. Determine all real numbers n such that one-third n must be more than seven.

72. Determine all real numbers n such that one-fourth n must be more than eight.

Air Pollutant Emissions In Exercises 73 and 74, use the following equation, which gives the approximate amount of air pollutant emissions of lead in the continental United States from 1970 to 1986 (see figure).

$$y = 199.85 - 11.88t$$

In this model y represents the amount of pollutant in thousands of metric tons (a metric ton is 2,204.6 pounds) and t represents the year with $t = 0$ corresponding to 1970. (*Source:* U.S. Environmental Protection Agency.)

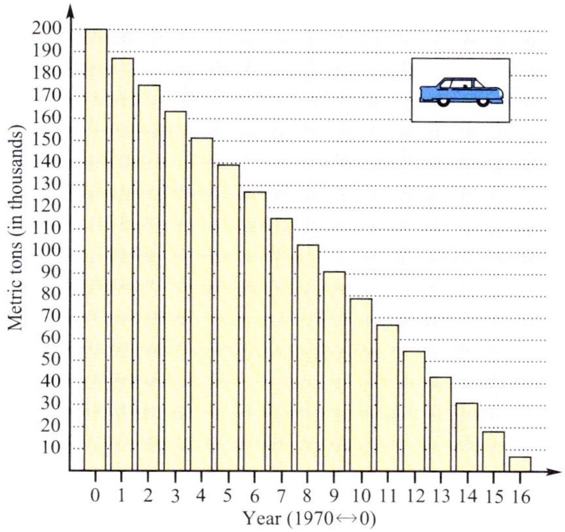

73. During which years between 1970 and 1986 was the air pollutant emission of lead greater than 140.45 thousand metric tons? (This corresponds to an emission of 309,636 pounds per year.)

74. During which years between 1970 and 1986 was the air pollutant emission of lead less than 33.53 thousand metric tons? (This corresponds to an emission of 73,920 pounds per year.)

75. If a, b, and c are the lengths of the sides of the triangle in the accompanying figure, then $a + b$ must be greater than what value?

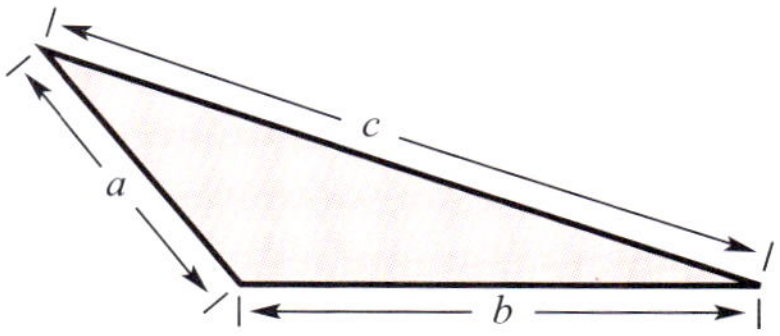

SECTION **2.5**

Equations and Inequalities Involving Absolute Value

Solving Equations Involving Absolute Value • *Solving Inequalities Involving Absolute Value*

Solving Equations Involving Absolute Value

We know from Section 1.1 that the absolute value of a real number can be described geometrically as the *distance* between the number and zero. By placing a variable inside the absolute value sign and specifying a distance, we can form an **absolute value equation**. For instance, using x as the variable and 3 as the distance, we obtain the absolute value equation

$$|x| = 3.$$

The only solutions of this equation are -3 and 3, because these are the only two real numbers whose distance from zero is 3. (See Figure 2.12.) In other words, the absolute value equation $|x| = 3$ has exactly two solutions: $x = -3$ and $x = 3$.

FIGURE 2.12

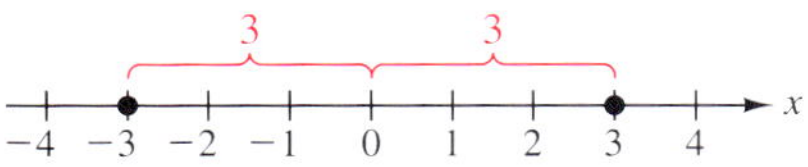

Solving an Absolute Value Equation

Let x be a variable or a variable expression and let a be a real number such that $a \geq 0$. The solutions of the equation $|x| = a$ are given by $x = -a$ or $x = a$. That is,

$$|x| = a \quad \Rightarrow \quad x = -a \quad \text{or} \quad x = a.$$

NOTE: The strategy for solving absolute value equations is to *rewrite* the equation in *equivalent forms* that can be solved by previously learned methods. This is a common strategy in mathematics. That is, when we encounter a new type of problem, we try to rewrite the problem so that it can be solved by previously learned techniques.

EXAMPLE 1 ■ Solving Absolute Value Equations

Solve the following absolute value equations.

(a) $|x| = 10$ (b) $|x| = 0$ (c) $|y| = -1$

Solution

(a) This absolute value equation is equivalent to the following two linear equations.

$$x = -10 \quad \text{or} \quad x = 10$$

Thus, the given equation has two solutions: -10 and 10.

(b) This absolute value equation is equivalent to the equations

$$x = 0 \quad \text{or} \quad x = 0.$$

Since both of these equations are the same, we conclude that the given equation has only one solution: 0.

(c) This absolute value equation has *no solution* because it is not possible for the absolute value of a real number to be negative. ■

EXAMPLE 2 ■ Solving Absolute Value Equations

Solve the following absolute value equation.

$$|x - 2| = 4$$

Solution

$\lvert x - 2\rvert = 4$			Given equation
$x - 2 = -4$	or	$x - 2 = 4$	Equivalent equations
$x - 2 + 2 = -4 + 2$		$x - 2 + 2 = 4 + 2$	Add 2 to both sides
$x = -2$		$x = 6$	Solutions

Thus, the given equation has two solutions: -2 and 6. We can check the solutions as follows.

Check

$$\begin{array}{rclcrcll}
|x-2| &=& 4 & \text{or} & |x-2| &=& 4 & \text{Given equation} \\
|-2-2| &\stackrel{?}{=}& 4 & & |6-2| &\stackrel{?}{=}& 4 & \text{Replace } x \text{ by } -2 \text{ and } 6 \\
|-4| &\stackrel{?}{=}& 4 & & |4| &\stackrel{?}{=}& 4 & \text{Simplify} \\
4 &=& 4 & & 4 &=& 4 & \text{Both solutions check}
\end{array}$$

■

EXAMPLE 3 ■ Solving Absolute Value Equations

Solve the following absolute value equation.

$$|3x + 4| = 10$$

Solution

$$\begin{array}{rclcrcll}
|3x+4| &=& 10 & & & & & \text{Given equation} \\
3x+4 &=& -10 & \text{or} & 3x+4 &=& 10 & \text{Equivalent equations} \\
3x+4-4 &=& -10-4 & & 3x+4-4 &=& 10-4 & \text{Subtract 4 from both sides} \\
3x &=& -14 & & 3x &=& 6 & \text{Combine like terms} \\
x &=& -\dfrac{14}{3} & & x &=& 2 & \text{Solutions}
\end{array}$$

Thus, the given equation has two solutions: $-\frac{14}{3}$ and 2. Check these solutions in the original equation. ■

EXAMPLE 4 ■ Solving Absolute Value Equations

Solve the following absolute value equation.

$$|x + 1| = -2$$

Solution

This equation is tricky. If we were to use the technique shown in the first three examples, we would solve the two equations $x + 1 = -(-2)$ and $x + 1 = -2$. The resulting solutions would be $x = 1$ and $x = -3$. After checking, however, we would discover that neither of these numbers is a solution! Can you see what went wrong? Notice in the original equation that the right-hand side is negative. This type of equation

$$|ax + b| = \text{(negative number)}$$

never has a solution because the absolute value of a real number cannot be negative. Thus, the given equation has no solution. ■

The equation in the next example is not given in the **standard form**

$$|ax + b| = c, \quad c \geq 0.$$

Notice that the first step in solving such an equation is to write it in standard form.

EXAMPLE 5 ■ An Absolute Value Equation in Nonstandard Form

Solve the following absolute value equation.

$$|2x - 1| + 3 = 8$$

Solution

$\|2x - 1\| + 3 = 8$		Given equation
$\|2x - 1\| = 5$		Standard form
$2x - 1 = -5$	or $2x - 1 = 5$	Equivalent equations
$2x = -4$	$2x = 6$	Add 1 to both sides
$x = -2$	$x = 3$	Solutions

Thus, the given equation has two solutions: -2 and 3. Check these solutions in the original equation. ■

If two algebraic expressions are equal in absolute value, then they must either be *equal* to each other, or they must be the *opposites* of each other. Thus, to solve equations of the form

$$|ax + b| = |cx + d|$$

we form the two linear equations

$$\underbrace{ax + b = -(cx + d)}_{\text{Expressions opposite}} \quad \text{and} \quad \underbrace{ax + b = cx + d}_{\text{Expressions equal}}.$$

We use this technique in the next two examples.

EXAMPLE 6 ■ Solving an Equation Involving Absolute Values

Solve the following equation.

$$|3x - 4| = |7x - 16|$$

Solution

$\|3x - 4\| = \|7x - 16\|$		Given equation
$3x - 4 = -(7x - 16)$	or $3x - 4 = 7x - 16$	Equivalent equations
$3x - 4 = -7x + 16$	$3x = 7x - 12$	
$10x = 20$	$-4x = -12$	
$x = 2$	$x = 3$	Solutions

Thus, the given equation has two solutions: 2 and 3. Check these solutions in the original equation. ■

When solving equations of the form $|ax + b| = |cx + d|$, it is possible that one of the following equations will not have a solution.

$$ax + b = -(cx + d) \quad \text{and} \quad ax + b = cx + d$$

This is the case in the next example.

EXAMPLE 7 ■ Solving an Equation Involving Absolute Values

Solve the following equation.

$$|x + 5| = |x + 11|$$

Solution

By equating the expression $(x + 5)$ to the opposite of $(x + 11)$, we obtain

$$\begin{aligned} x + 5 &= -(x + 11) \\ x + 5 &= -x - 11 \\ x &= -x - 16 \\ 2x &= -16 \\ x &= -8. \end{aligned}$$

Thus, $x = -8$ is one solution. However, by setting the two expressions equal to each other, we obtain

$$\begin{aligned} x + 5 &= x + 11 \\ x &= x + 6 \\ 0 &= 6 \end{aligned}$$

which makes no sense. Therefore, the equation has only one solution: -8. Check this solution in the original equation. ■

Solving Inequalities Involving Absolute Value

To see how to solve inequalities involving absolute values, consider the following comparisons.

Equation or Inequality	*Geometric Solution*	*Graph*
$\lvert x \rvert = 2$	Values of x that lie two units from 0. $x = -2$ and $x = 2$	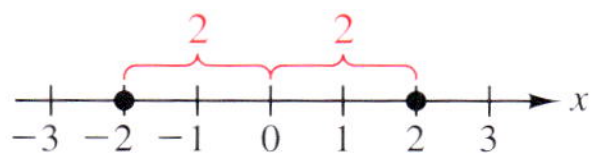

$|x| < 2$ — Values of x that lie *less than* two units from 0.

$$-2 < x < 2$$

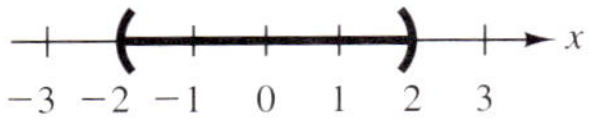

$|x| > 2$ — Values of x that lie *more than* two units from 0.

$$x < -2 \text{ or } x > 2$$

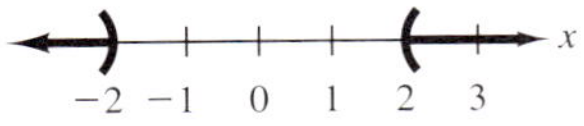

These comparisons suggest the following rule for solving inequalities involving absolute values.

Solving an Absolute Value Inequality

Let x be a variable or an algebraic expression and let a be a real number such that $a > 0$.

1. The solutions of $|x| < a$ are all values of x that lie between $-a$ and a. That is,

$$|x| < a \quad \text{if and only if} \quad -a < x < a.$$

2. The solutions of $|x| > a$ are all values of x that are less than $-a$ or greater than a. That is,

$$|x| > a \quad \text{if and only if} \quad x < -a \quad \text{or} \quad x > a.$$

These two rules are also valid if $<$ is replaced by $\leq$ and $>$ is replaced by $\geq$.

EXAMPLE 8 ■ Solving an Absolute Value Inequality

Solve the following inequality and sketch the graph of its solution set.

$$|x - 5| < 2$$

Solution

$\|x - 5\| < 2$	Given inequality
$-2 < x - 5 < 2$	Equivalent inequalities
$-2 + 5 < x - 5 + 5 < 2 + 5$	Add 5 to all three parts
$3 < x < 7$	Solution set

Thus, the solution set consists of all real numbers that are greater than 3 and less than 7. The interval notation for this solution set is

$(3, 7)$. Solution set

The graph of this solution set is shown in Figure 2.13.

FIGURE 2.13

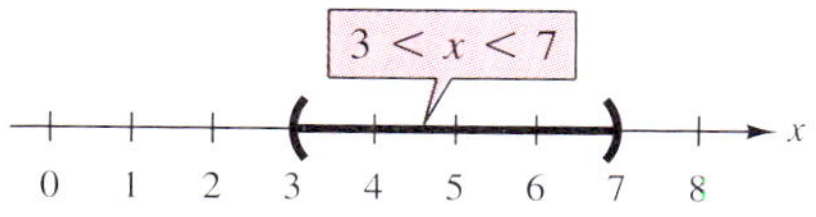

■

In Example 8 note that absolute value inequalities of the form $|x| < a$ can be solved using a "double inequality" (see Example 8 in Section 2.4). Inequalities of the form $|x| > a$ cannot be solved using a double inequality. Instead, we must solve two separate inequalities, as demonstrated in the next example.

EXAMPLE 9 ■ **Solving an Absolute Value Inequality**

Solve the following inequality and sketch the graph of its solution set.

$$|3x - 4| \geq 5$$

Solution

$|3x - 4| \geq 5$			Given inequality
$3x - 4 \leq -5$	or	$3x - 4 \geq 5$	Equivalent inequalities
$3x - 4 + 4 \leq -5 + 4$		$3x - 4 + 4 \geq 5 + 4$	Add 4 to both sides
$3x \leq -1$		$3x \geq 9$	Combine like terms
$\frac{3x}{3} \leq \frac{-1}{3}$		$\frac{3x}{3} \geq \frac{9}{3}$	Divide both sides by 3
$x \leq -\frac{1}{3}$		$x \geq 3$	Solution set

Thus, the solution set consists of all real numbers that are less than or equal to $-\frac{1}{3}$ or greater than or equal to 3. The interval notation for this solution set is

$$\left(-\infty, -\frac{1}{3}\right] \cup \left[3, \infty\right).$$ Solution set

The symbol $\cup$ is called a **union** symbol, and it is used to denote the combining of two sets. The graph of this solution set is shown in Figure 2.14.

FIGURE 2.14

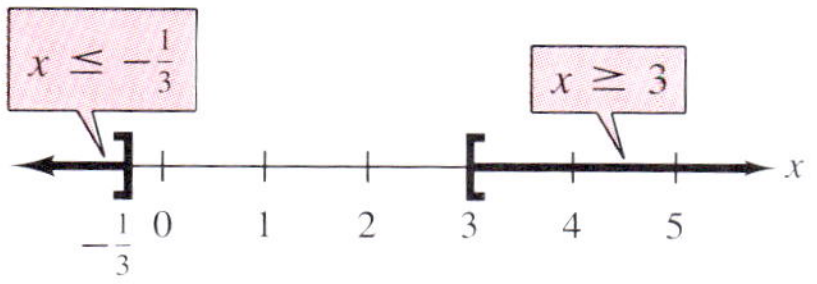

■

EXAMPLE 10 ■ **Solving an Absolute Value Inequality**

Solve the following inequality and sketch the graph of its solution set.

$$\left|2 - \frac{x}{3}\right| \leq 0.01$$

Solution

$$\left|2 - \frac{x}{3}\right| \le 0.01$$ Given inequality

$$-0.01 \le 2 - \frac{x}{3} \le 0.01$$ Equivalent inequalities

$$-2.01 \le -\frac{x}{3} \le -1.99$$ Subtract 2 from all three parts

$$6.03 \ge x \ge 5.97$$ Multiply all three parts by -3 and reverse both inequality symbols

$$5.97 \le x \le 6.03$$ Solution set in standard form

Thus, the solution set consists of all real numbers that are greater than or equal to 5.97 and less than or equal to 6.03. The interval notation for this solution set is

$$[5.97, 6.03].$$ Solution set

The graph of this solution set is shown in Figure 2.15.

FIGURE 2.15

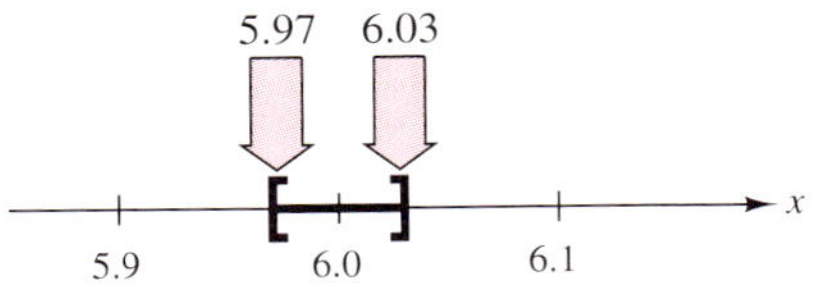

DISCUSSION PROBLEM ■ A Challenging Problem

The following inequality is not covered by the standard solution techniques in this section. Try to find the solution set of this inequality and then write a short paragraph describing the procedure that you used.

$$|x - 4| \le \frac{x}{2}$$

Warm-Up

The following warm-up exercises involve skills that were covered in earlier sections. You will use these skills in the exercise set for this section.

In Exercises 1–6, evaluate the given expression.

1. $|-15|$ **2.** $|-3.2|$ **3.** $-|72|$

4. $-|-4|$ **5.** $|14 - 32|$ **6.** $-|46.8 - 27|$

In Exercises 7–10, solve the inequality.

7. $-4 < 10x + 1 < 6$ **8.** $-2 \le 1 - 2x \le 2$

9. $-3 \le \frac{x}{2} \le 3$ **10.** $-5 < x - 25 < 5$

2.5 EXERCISES

In Exercises 1–22, solve the given equation. (Some of the equations have no solution.)

1. $|t| = 45$ **2.** $|s| = 16$ **3.** $|h| = 0$ **4.** $|x| = -82$

5. $|x - 16| = 5$ **6.** $|z - 100| = 100$ **7.** $|2s + 3| = 25$ **8.** $|7a + 6| = 8$

9. $|32 - 3y| = 16$ **10.** $|3 - 5x| = 13$ **11.** $|3x + 4| = -16$ **12.** $|20 - 5t| = 50$

13. $|5x - 3| + 8 = 22$ **14.** $4|x + 5| = 9$ **15.** $\left|\frac{2}{3}x + 4\right| = 9$ **16.** $\left|\frac{3}{2} - \frac{4}{5}x\right| = 1$

17. $|0.32x - 2| = 4$ **18.** $|3.2 - 1.054x| = 2$ **19.** $|x + 8| = |2x + 1|$ **20.** $|10 - 3x| = |x + 7|$

21. $|45 - 4x| = |32 - 3x|$ **22.** $|5x + 4| = |3x + 25|$

In Exercises 23–30, determine whether the given values of the variable are solutions of the inequality.

	Inequality	*Value*			*Inequality*	*Value*	
23.	$\|x\| < 3$	(a) $x = 2$	(b) $x = -4$	**24.**	$\|x\| \leq 5$	(a) $x = -7$	(b) $x = -4$
		(c) $x = 4$	(d) $x = -1$			(c) $x = 4$	(d) $x = 9$
25.	$\|x\| \geq 3$	(a) $x = 2$	(b) $x = -4$	**26.**	$\|x\| > 5$	(a) $x = -7$	(b) $x = -4$
		(c) $x = 4$	(d) $x = -1$			(c) $x = 4$	(d) $x = 9$
27.	$\|x - 7\| < 3$	(a) $x = 9$	(b) $x = -4$	**28.**	$\|x - 3\| \leq 5$	(a) $x = 16$	(b) $x = 3$
		(c) $x = 11$	(d) $x = 6$			(c) $x = -2$	(d) $x = -3$
29.	$\|x - 7\| \geq 3$	(a) $x = 9$	(b) $x = -4$	**30.**	$\|x - 3\| > 5$	(a) $x = 16$	(b) $x = 3$
		(c) $x = 11$	(d) $x = 6$			(c) $x = -2$	(d) $x = -3$

In Exercises 31–34, sketch a graph that shows the real numbers that satisfy the given statement.

31. All real numbers greater than -2 *and* less than 5

32. All real numbers greater than or equal to 3 *and* less than 10

33. All real numbers less than or equal to 4 *or* greater than 7

34. All real numbers less than -6 *or* greater than or equal to 6

In Exercises 35–54, solve the inequality and sketch the graph of the solution.

35. $|y| < 4$ **36.** $|x| < 6$

37. $|y - 2| \leq 4$ **38.** $|x - 3| \leq 6$

39. $|y + 2| < 4$ **40.** $|x + 3| < 6$

41. $|y| \geq 4$ **42.** $|x| \geq 6$

43. $|y - 2| > 4$ **44.** $|x - 3| > 6$

45. $|y + 2| \geq 4$ **46.** $|x + 3| \geq 6$

47. $|2x| < 14$ **48.** $|4z| \leq 9$

49. $\left|\frac{y}{3}\right| \leq 3$ **50.** $\left|\frac{t}{2}\right| < 4$

51. $|3x + 2| > 4$ **52.** $|2x - 1| \geq 3$

53. $|x - 5| + 3 > 5$ **54.** $|a + 1| - 4 < 0$

In Exercises 55–64, solve the inequality.

55. $|0.2x - 3| < 4$

56. $|1.5t - 8| \leq 16$

57. $\dfrac{|x + 2|}{10} \leq 8$

58. $\dfrac{|y - 16|}{4} < 30$

59. $|6t + 15| \geq 30$

60. $|3t + 1| > 5$

61. $\dfrac{|s - 3|}{5} > 4$

62. $\dfrac{|a + 6|}{2} \geq 16$

63. $\left|\dfrac{z}{10} - 3\right| > 8$

64. $\left|\dfrac{x}{8} + 1\right| < 0$

65. *Temperature* The operating temperature for an electronic device must satisfy the inequality

$$|t - 72| < 10$$

where t is given in degrees Fahrenheit. Sketch the graph of the solution of the inequality.

66. *Time Study* A time study was conducted to determine the length of time required to perform a particular task in a manufacturing process. The time required by approximately two-thirds of the workers in the study satisfied the inequality

$$\left|\frac{t - 15.6}{1.9}\right| < 1$$

where t is the time in minutes. Sketch the graph of the solution of the inequality.

In Exercises 67–72, write an inequality involving absolute values to represent the given interval.

67.

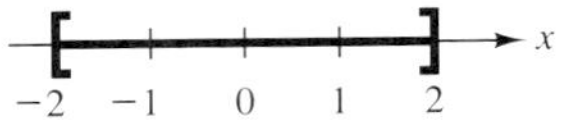

68.

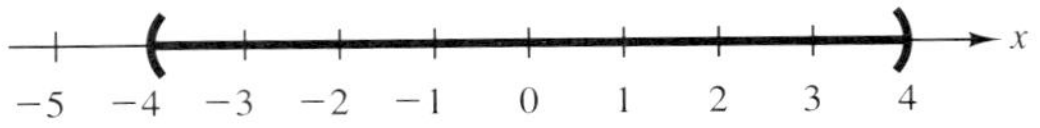

69.

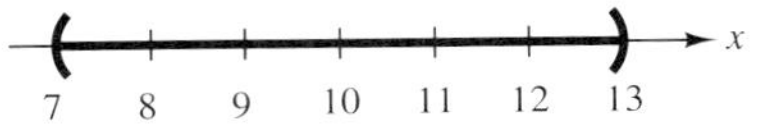

70.

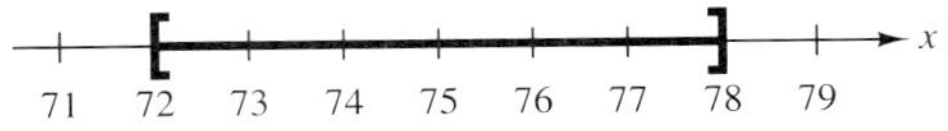

71.

72.

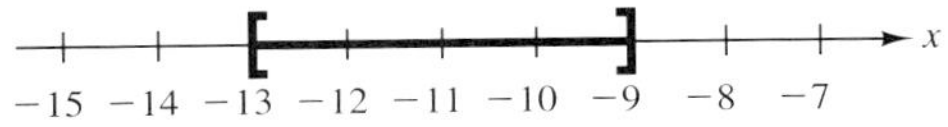

In Exercises 73–76, write an inequality involving absolute values to represent the verbal statement.

73. The set of all real numbers x whose distance from 0 is less than 3

74. The set of all real numbers x whose distance from 0 is more than 2

75. The set of all real numbers x whose distance from 5 is more than 6

76. The set of all real numbers x whose distance from 16 is less than 5

CHAPTER 2 SUMMARY

As you review and prepare for a test on this chapter, first try to obtain a global view of what was discussed. Then review the specific skills needed under each category.

Creating and Solving Linear Equations

- *Determine* if a specified number is a solution of a linear equation. — Section 2.1, Exercises 1–10
- *Solve and check* linear equations. — Section 2.1, Exercises 11–58
- *Create* a linear equation from a verbal statement. — Section 2.2, Exercises 1–6
- *Create and/or solve* linear equations that involve percents. — Section 2.2, Exercises 7–40
- *Create* ratios from verbal statements. — Section 2.2, Exercises 45–54
- *Create and solve* linear equations that involve unit prices and proportions. — Section 2.2, Exercises 55–75, Exercises 81–84
- *Create and solve* linear equations that involve price, cost, markup, or discount. — Section 2.3, Exercises 1–22
- *Create and solve* linear equations that involve rates or mixtures. — Section 2.3, Exercises 23–48
- *Create and solve* linear equations that involve formulas. — Section 2.3, Exercises 65–80

Creating and Solving Linear Inequalities

- *Determine* if a specified number satisfies a given inequality. — Section 2.4, Exercises 1–4
- *Identify and sketch* the graph of a given inequality. — Section 2.4, Exercises 5–16
- *Solve* linear inequalities and sketch their solution sets. — Section 2.4, Exercises 17–46
- *Create* linear inequalities from verbal statements and conversely. — Section 2.4, Exercises 47–54
- *Create and solve* linear inequalities from verbal statements. — Section 2.4, Exercises 55–72

Solving Equations and Inequalities that Involve Absolute Value

- *Solve* equations involving the absolute value of a linear expression. — Section 2.5, Exercises 1–22
- *Determine* if specified numbers are solutions of absolute value inequalities. — Section 2.5, Exercises 23–30
- *Create* an absolute value inequality for a given interval. — Section 2.5, Exercises 67–72
- *Solve and sketch* the solution sets of absolute value inequalities. — Section 2.5, Exercises 35–54
- *Create* an absolute value inequality from a verbal statement. — Section 2.5, Exercises 73–76

Chapter 2 Review Exercises

In Exercises 1–4, determine whether the values of the variable are solutions of the equation.

Equation	*Value*	
1. $45 - 7x = 3$	(a) $x = 3$	(b) $x = 6$
2. $3(5 - x) = 6 - x$	(a) $x = \frac{9}{2}$	(b) $x = -\frac{2}{3}$
3. $\frac{x}{7} + \frac{x}{5} = 1$	(a) $x = \frac{35}{12}$	(b) $x = -\frac{2}{35}$
4. $\frac{t + 2}{6} = \frac{7}{2}$	(a) $t = -12$	(b) $t = 19$

In Exercises 5–16, solve the equation and check the result. (Some of the equations have no solution.)

5. $17 - 7x = 3$

6. $3 + 6x = 51$

7. $4y - 6(y - 5) = 2$

8. $7x + 2(7 - x) = 8$

9. $1.4t + 2.1 = 0.9t - 2$

10. $8(x - 2) = 3(x - 2)$

11. $\frac{3x}{4} = 4$

12. $-\frac{5x}{14} = \frac{1}{2}$

13. $\frac{4}{5}x - \frac{1}{10} = \frac{3}{2}$

14. $\frac{1}{4}s + \frac{3}{8} = \frac{5}{2}$

15. $\frac{v - 20}{-8} = 2v$

16. $x + \frac{2x}{5} = 1$

In Exercises 17–20, solve the equation and round your answer to two decimal places. (A calculator may be helpful.)

17. $382x - 575 = 715$

18. $3.625x + 3.5 = 22.125$

19. $\frac{x}{2.33} = 14.302$

20. $\frac{7x}{3} + 2.5 = 8.125$

In Exercises 21 and 22, complete the table showing the equivalent forms of a percent, the number of parts out of 100, a decimal, and a fraction (in reduced form).

21.

Percent	Parts out of 100	Decimal	Fraction
87%			

22.

Percent	Parts out of 100	Decimal	Fraction
			$\frac{1}{6}$

In Exercises 23–28, solve the given percent problem.

23. What is 130% of 50?

24. What is 0.4% of 7,350?

25. 645 is $21\frac{1}{2}\%$ of what number?

26. 498 is 83% of what number?

27. 250 is what percent of 200?

28. 162.5 is what percent of 6,500?

29. *Instant Replays* Complete the accompanying table, which shows National Football League data on instant replays. (*Source:* National Football League.)

Year	Reviewed	Reversed	Percent Reversed
1986	374	38	
1987	490	57	
1988	537	53	
1989	492	65	

30. *Census Costs* Use the information in the accompanying figure to determine the percentage increase in the per capita cost from the 1980 census to the 1990 census. If the population of the United States is 250 million, determine the total cost for conducting the 1990 census. (*Source:* U.S. Bureau of Census.)

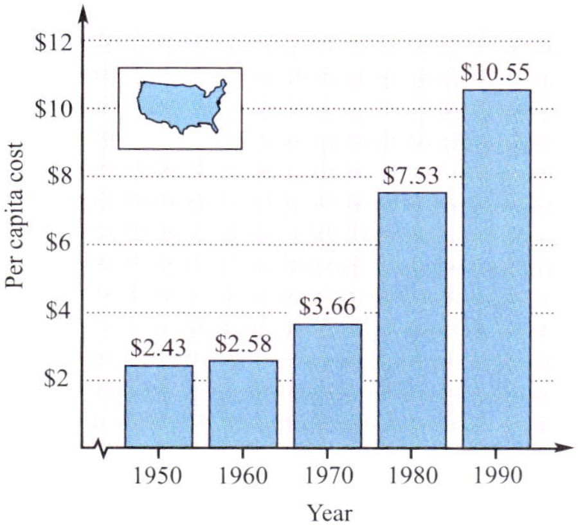

31. *Revenue Increase* The revenues for a corporation in millions of dollars in the years 1988 and 1989 were \$4,521.4 and \$4,679.0, respectively. Determine the percentage increase in revenue from 1988 to 1989.

32. *Price Increase* The manufacturer's suggested retail price for a certain truck model is \$25,750. Estimate the price of a comparably equipped truck for the next model year if it is projected that truck prices will increase by $5\frac{1}{2}\%$.

33. *Retail Price* A camera that costs a retailer \$259.95 is marked up by 35%. Find the price to the consumer.

34. *Markup Rate* A calculator with a cost to the consumer of \$175 costs the retailer \$95. Find the markup rate.

35. *Sale Price* The list price of a coat is \$259. Find the sale price of the coat if it is reduced by 25%.

36. *Comparison Shopping* A mail-order catalog has an attaché case with a list price of \$99.97 plus \$4.50 for shipping and handling. A local department store has the same attaché case for \$125.95. If the department store has a special 20%-off sale on the case, which is the better buy?

37. *Tip Rate* A customer left a total of \$25 for a meal that cost \$21.65. Determine the tip rate.

38. *Sales Goal* The weekly salary of an employee is \$150 plus a 6% commission on total sales. The employee needs a minimum salary of \$650 per week. How much must be sold to produce this salary?

39. *Coin Problem* You have 60 coins in dimes and quarters with a combined value of \$11.25. Determine the number of coins of each type.

40. *Opinion Poll* Twelve hundred people were surveyed in an opinion poll. Political candidates A and B received the same preference, but Candidate C was preferred by one and one-third times the number of people as either A or B. Determine the number in the sample that preferred Candidate C.

41. *Travel Time* Determine the time for a bus to travel 330 miles if its average speed is 52 miles per hour.

42. *Average Speed* An Olympic cross-country skier completed the 15 kilometer event in 41 minutes and 20 seconds. What was the average speed for this skier?

43. *Distance* Determine the distance an air force jet can travel in $2\frac{1}{3}$ hours if its average speed is 1,200 miles per hour.

44. *Average Speed* For two hours of a 400-mile trip, your average speed is only 40 miles per hour. Determine the average speed that must be maintained for the remainder of the trip if you want the average speed for the entire trip to be 50 miles per hour.

45. *Work Rate Problem* Find the time it takes two people working together to complete a task if working individually it takes them 4.5 hours and 6 hours.

46. *Dimensions of a Rectangle* The width of a rectangle is six inches less than its length. Find its dimensions if the perimeter of the rectangle is 42.

47. *Simple Interest* Find the total simple interest you will earn on a $1,000 corporate bond that matures in four years and has an 8.5% interest rate.

48. *Simple Interest* Find the annual simple interest rate on a certificate of deposit that pays $37.50 per year in interest on a principal of $500.

49. *Simple Interest* Find the principal required to have an annual interest income of $20,000 if the annual simple interest rate on the principal is 9.5%.

50. *Simple Interest* A corporation borrows $3\frac{1}{4}$ million dollars for two years to modernize one of its manufacturing facilities. If it pays an annual simple interest rate of 12%, what will be the total principal and interest that must be repaid?

51. *Simple Interest* An inheritance of $50,000 is divided between two investments earning 8.5% and 10% simple interest. How much is in each investment if the total interest for one year is $4,700?

52. *Simple Interest* You invest $1,000 in a certificate of deposit that has an annual simple interest rate of 7%. After six months the interest is computed and added to the principal. During the second six months the interest is computed using the original investment plus the interest earned during the first six months. What is the total interest earned during the first year of the investment?

53. *Dimensions of a Rectangle* The area of the shaded region in the accompanying figure is 48 square inches. Find the dimensions of the rectangle.

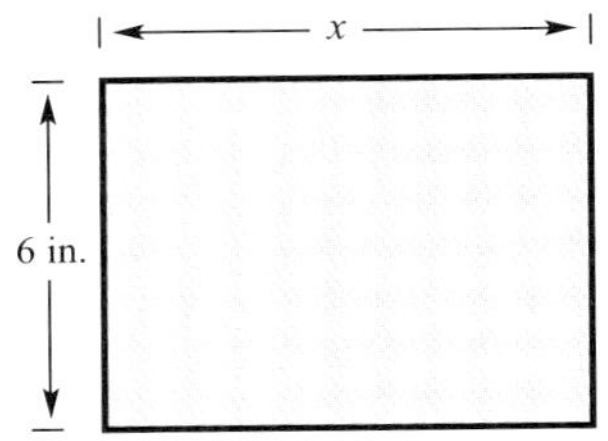

54. *Dimensions of a Rectangle* The perimeter of the rectangle in the accompanying figure is 110 feet. Find the dimensions of the rectangle.

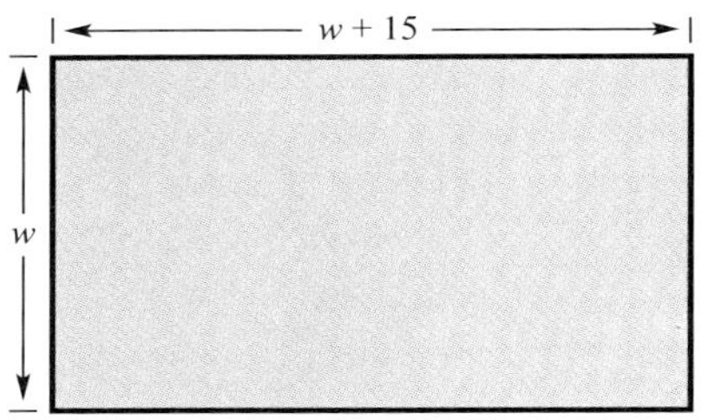

In Exercises 55–58, express the ratio as a fraction in reduced form. (Use the same units for the numerator and denominator.)

55. 16 feet to 4 yards

56. 3 quarts to 5 pints

57. 45 seconds to 5 minutes

58. 3 meters to 150 centimeters

In Exercises 59–62, solve the proportion.

59. $\frac{7}{8} = \frac{y}{4}$

60. $\frac{x}{16} = \frac{5}{12}$

61. $\frac{b}{15} = \frac{5}{6}$

62. $\frac{x + 1}{3} = \frac{x - 1}{2}$

63. *Property Tax* The tax on property with an assessed value of \$80,000 is \$1,350. Find the tax on property with an assessed value of \$110,000. (Assume that the properties are taxed at the same rate.)

64. *Recipe Enlargement* One and a half cups of milk are required to make one batch of pudding. How much milk is required to make $2\frac{1}{2}$ batches?

65. *Map Scale* One-third inch represents 50 miles on a map. Approximate the distance between two cities that are $3\frac{1}{4}$ inches apart on the map.

66. *Fuel Mixture* The gasoline-to-oil ratio is fifty-to-one for a lawn mower engine. Determine the amount of gasoline required if one-half pint of oil is used in producing the mixture.

Similar Triangles In Exercises 67 and 68, solve for the length x of the side of a triangle by using the fact that corresponding sides of similar triangles are proportional. (*Note:* In each pair, assume that the triangles are similar.)

67.

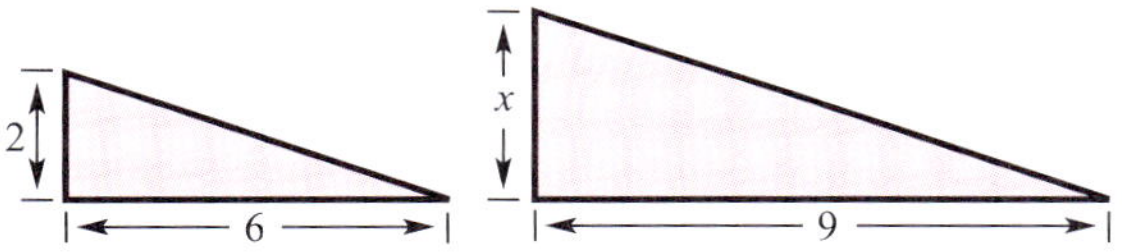

68.

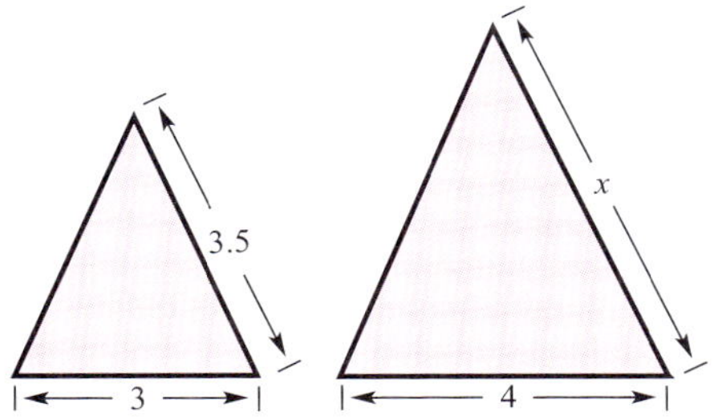

In Exercises 69–82, solve the inequality and sketch the solution on the real number line.

69. $5x + 3 > 18$

70. $-11x \geq 44$

71. $\frac{1}{3} - \frac{1}{2}y < 12$

72. $3(2 - y) \geq 2(1 + y)$

73. $-4 < \frac{x}{5} \leq 4$

74. $-13 \leq 3 - 4x < 13$

75. $5 > \frac{x + 1}{-3} > 0$

76. $12 \geq \frac{x - 3}{2} > 1$

77. $|2x - 7| < 15$

78. $|5x - 1| < 9$

79. $|x - 4| > 3$

80. $|t + 3| > 2$

81. $|b + 2| - 6 > 1$

82. $\left|\frac{t}{3}\right| < 1$

In Exercises 83–86, write an inequality for the given statement.

83. z is no more than 10.

84. x is nonnegative.

85. y is at least 7 but less than 14.

86. The volume V is less than 27 cubic feet.

CHAPTER 2 TEST

Take this test as you would take a test in class. After you are done, check your work with the answers given in the back of the book.

1. Determine whether the given value of x is a solution of the equation $3(5 - 2x) - (3x - 1) = -2$.
(a) $x = -4$
(b) $x = 2$

2. Solve: $6x - 5 = 19$

3. Solve: $15 - 7(1 - x) = 3(x + 8)$

4. Solve: $\frac{2x}{3} = \frac{x}{2} + 4$

5. Solve: $\frac{t - 5}{12} = \frac{3}{8}$

6. Solve using a calculator and round the solution to two decimal places: $3.2x + \frac{x}{5.8} = 50$.

7. When the sum of a number and 10 is divided by 5, the result is 7. Find the number.

8. What is 27% of 3,200?

9. 1,200 is what percent of 800?

10. A store is offering a 20% discount on all items in its inventory. Find the list price of a tractor that has a sale price of $6,400.

11. Use unit prices to determine which is the better buy: a 12-ounce package of candy for $2.49 or a 15-ounce package for $2.99.

12. The tax on property with an assessed value of $90,000 is $1,200. Find the tax on property with an assessed value of $110,000. (Assume the same tax rate.)

13. The bill (including parts and labor) for the repair of a home appliance was $165. The cost for parts was $85. How many hours were spent to repair the appliance if the cost of labor was $16 per half hour?

14. A person has 30 coins in dimes and quarters with a combined value of $5.70. Determine the number of coins of each type.

15. Two solutions—one of 10% concentration and one of 40% concentration—are mixed to create 100 liters of a 30% solution. Determine the number of liters of the 10% solution and the 40% solution that are required.

16. Two cars start at a given time and travel in the same direction at average speeds of 40 miles per hour and 55 miles per hour. How much time must elapse before the two cars are 10 miles apart?

17. One number is four times a second. Find the numbers if their difference is 39.

18. Find the principal required to earn $300 in simple interest in two years if the annual interest rate is 7.5%.

19. Solve and sketch the solution: $1 + 2x > 7 - x$

20. Solve and sketch the solution: $0 \leq \frac{1 - x}{4} < 2$

21. Solve and sketch the solution: $|x - 3| \leq 2$

22. Solve and sketch the solution: $|x + 4| > 1$

23. Use inequality notation to denote the phrase "t is at least 8."

CHAPTER THREE

Polynomials and Factoring

Chapter Overview

The skills in this chapter are fundamental to the study of algebra. We begin by introducing the operations of polynomial addition, subtraction, and multiplication. Many times in algebra we need to perform operations both forwards and backwards. For instance, in Section 3.2, we multiply polynomials. In Sections 3.3 and 3.4, we reverse the process and "undo multiplication" by factoring. In Section 3.5, we use factoring skills to solve polynomial equations.

SECTION 3.1

Adding and Subtracting Polynomials

Basic Definitions • Adding and Subtracting Polynomials • Evaluating Polynomials • Applications

Basic Definitions

An algebraic expression whose terms are all of the form ax^k, where a is any real number and k is a nonnegative integer, is called a **polynomial in one variable** or simply a **polynomial**. Here are some examples of polynomials in one variable.

$$3x - 8, \qquad x^4 + 3x^3 - x^2 - 8x + 1, \quad \text{and} \quad x^3 - 8$$

In the term ax^k, a is called the **coefficient** and k the **degree** of the term. Note that the degree of the term ax is one, and the degree of the constant term is zero. Since a polynomial is an algebraic *sum*, the coefficients take on the signs between the terms. For instance,

$$x^3 - 4x^2 + 3 = (1)x^3 + (-4)x^2 + (0)x + 3$$

has coefficients 1, -4, 0, and 3.

We usually write polynomials in order of descending powers of the variable. This is called **standard form**. For example, the standard form of $3x^2 - 5 - x^3 + 2x$ is

$$-x^3 + 3x^2 + 2x - 5. \qquad \text{Standard form}$$

The **degree of a polynomial** is defined as the degree of the term with the highest power, and the coefficient of this term is called the **leading coefficient** of the polynomial. For instance, the polynomial

$$-3x^4 + 4x^2 + x + 7$$

is of fourth degree and its leading coefficient is -3.

Definition of a Polynomial in x

Let $a_n, \ldots, a_2, a_1, a_0$ be real numbers and let n be a *nonnegative integer*. A **polynomial in x** is an expression of the form

$$a_nx^n + a_{n-1}x^{n-1} + \cdots + a_2x^2 + a_1x + a_0$$

where $a_n \neq 0$. The polynomial is of **degree** n, and the number a_n is called the **leading coefficient**. The number a_0 is called the **constant term**.

NOTE: The following are *not* polynomials, for the reasons stated.

$2x^{-1} + 5$	Exponent in $2x^{-1}$ is *not* nonnegative.
$x^3 + 3x^{1/2}$	Exponent in $3x^{1/2}$ is *not* an integer.

EXAMPLE 1 ■ Identifying Leading Coefficients and Degrees of Polynomials

	Polynomial	*Standard Form*	*Degree*	*Leading Coefficient*
(a)	$5x^2 - 2x^7 + 4 - 2x$	$-2x^7 + 5x^2 - 2x + 4$	7	-2
(b)	$16 - 8x^3$	$-8x^3 + 16$	3	-8
(c)	10	10	0	10
(d)	$5 + x^4 - 6x^3$	$x^4 - 6x^3 + 5$	4	1 ■

A polynomial with only one term is called a **monomial**. Polynomials with two *unlike* terms are called **binomials**, and those with three *unlike* terms are called **trinomials**. Here are some examples.

$5x^3$ Monomial

$-4x + 3$ Binomial

$2x^2 + 3x - 7$ Trinomial

NOTE: The prefix *poly* means *many*. For instance, a polygon is a many-sided figure. Similarly, the prefix *mono* means one, the prefix *bi* means two, and the prefix *tri* means three.

Adding and Subtracting Polynomials

To add two polynomials, we simply combine like terms. This can be done in either a horizontal or vertical format, as shown in Examples 2 and 3.

EXAMPLE 2 ■ Adding Polynomials Horizontally

(a) $(2x^3 + x^2 - 5) + (x^2 + x + 6)$ — Given polynomials

$= (2x^3) + (x^2 + x^2) + (x) + (-5 + 6)$ — Group like terms

$= 2x^3 + 2x^2 + x + 1$ — Standard form

(b) $(3x^2 + 2x + 4) + (3x^2 - 6x + 3) + (-x^2 + 2x - 4)$

$= (3x^2 + 3x^2 - x^2) + (2x - 6x + 2x) + (4 + 3 - 4)$

$= 5x^2 - 2x + 3$ ■

EXAMPLE 3 ■ Using a Vertical Format to Add Polynomials

Use a vertical format to perform the following operations.

$$(5x^3 + 2x^2 - x + 7) + (3x^2 - 4x + 7) + (-x^3 + 4x^2 - 8)$$

Solution

To use a vertical format to add polynomials, we align the terms of the polynomials by their degrees, as follows.

$$\begin{array}{r} 5x^3 + 2x^2 - x + 7 \\ 3x^2 - 4x + 7 \\ -x^3 + 4x^2 - 8 \\ \hline 4x^3 + 9x^2 - 5x + 6 \end{array}$$

■

To subtract one polynomial from another, we must change the sign of each of the terms of the polynomial that is being subtracted and then add the resulting like terms. For instance, we can subtract the polynomial $(2x^2 + 3x - 4)$ from the polynomial $(x^2 + x - 1)$ as follows.

$$\begin{aligned} (x^2 + x - 1) - (2x^2 + 3x - 4) &= (x^2 + x - 1) + (-2x^2 - 3x + 4) \\ &= (x^2 - 2x^2) + (x - 3x) + (-1 + 4) \\ &= -x^2 - 2x + 3 \end{aligned}$$

EXAMPLE 4 ■ Subtracting Polynomials Horizontally

$$\begin{aligned} &(3x^3 - 5x^2 + 3) - (x^3 + 2x^2 - x - 4) && \text{Given polynomials} \\ &\quad = (3x^3 - 5x^2 + 3) + (-x^3 - 2x^2 + x + 4) && \text{Change signs and add} \\ &\quad = (3x^3 - x^3) + (-5x^2 - 2x^2) + (x) + (3 + 4) && \text{Group like terms} \\ &\quad = 2x^3 - 7x^2 + x + 7 && \text{Standard form} \end{aligned}$$

■

Be especially careful to get the correct signs when you are subtracting one polynomial from another. One of the most common mistakes in algebra is to forget to change signs correctly when subtracting one expression from another. Here is an example.

Wrong sign ↓

$$(x^2 - 2x + 3) - (x^2 + 2x - 2) \neq x^2 - 2x + 3 - x^2 + 2x - 2 \qquad \text{Common error}$$

↑ Wrong sign

Note that the error was forgetting to change two of the signs in the polynomial that was being subtracted. Here is the correct way to perform the subtraction.

Correct sign ↓

$$(x^2 - 2x + 3) - (x^2 + 2x - 2) = x^2 - 2x + 3 - x^2 - 2x + 2 \qquad \text{Correct}$$

↑ Correct sign

EXAMPLE 5 ■ Using a Vertical Format to Subtract Polynomials

Use a vertical format to perform the following operations.

$$(4x^4 - 2x^3 + 5x^2 - x + 8) - (3x^4 - 2x^3 + 3x - 4)$$

Solution

$$\begin{array}{rcr} (4x^4 - 2x^3 + 5x^2 - x + 8) & \Rightarrow & 4x^4 - 2x^3 + 5x^2 - x + 8 \\ \underline{-(3x^4 - 2x^3 + 3x - 4)} & \Rightarrow & \underline{-3x^4 + 2x^3 - 3x + 4} \\ & & x^4 + 5x^2 - 4x + 12 \end{array}$$ ■

EXAMPLE 6 ■ Combining Polynomials

$$\begin{aligned} &(2x^2 - 2x + 1) - [(x^2 + x - 3) + (-2x^2 - 4x + 5)] \\ &\quad = (2x^2 - 2x + 1) - [(x^2 - 2x^2) + (x - 4x) + (-3 + 5)] \\ &\quad = (2x^2 - 2x + 1) - [-x^2 - 3x + 2] \\ &\quad = (2x^2 - 2x + 1) + [x^2 + 3x - 2] \\ &\quad = (2x^2 + x^2) + (-2x + 3x) + (1 - 2) \\ &\quad = 3x^2 + x - 1 \end{aligned}$$ ■

EXAMPLE 7 ■ Combining Polynomials

(a) $(2x^2 - 7x + 2) - (4x^2 + 5x - 1) + (-x^2 + 4x + 4)$

$$\begin{aligned} &= 2x^2 - 7x + 2 - 4x^2 - 5x + 1 - x^2 + 4x + 4 \\ &= (2x^2 - 4x^2 - x^2) + (-7x - 5x + 4x) + (2 + 1 + 4) \\ &= -3x^2 - 8x + 7 \end{aligned}$$

(b) $(-x^2 + 4x - 3) - [(4x^2 - 3x + 8) - (-x^2 + x + 7)]$

$$\begin{aligned} &= (-x^2 + 4x - 3) - [4x^2 - 3x + 8 + x^2 - x - 7] \\ &= (-x^2 + 4x - 3) - [(4x^2 + x^2) + (-3x - x) + (8 - 7)] \\ &= (-x^2 + 4x - 3) - [5x^2 - 4x + 1] \\ &= -x^2 + 4x - 3 - 5x^2 + 4x - 1 \\ &= (-x^2 - 5x^2) + (4x + 4x) + (-3 - 1) \\ &= -6x^2 + 8x - 4 \end{aligned}$$ ■

Evaluating Polynomials

The next two examples show how to **evaluate** a polynomial in x for a given value of x.

EXAMPLE 8 ■ Evaluating a Polynomial

Find the value of the following polynomial when $x = 4$.

$$x^3 - 5x^2 + 6x - 3$$

Solution

When $x = 4$, we have the following.

$\text{Value} = x^3 - 5x^2 + 6x - 3$	Given polynomial
$= (4)^3 - 5(4)^2 + 6(4) - 3$	Replace x by 4
$= 64 - 80 + 24 - 3$	Evaluate terms
$= 5$	Simplify ■

EXAMPLE 9 ■ Evaluating Polynomials

Find the value of the polynomial $2x^3 - 9x^2 + 4x + 12$ for the following values of x.

(a) $x = 1$ (b) $x = -1$

Solution

(a) When $x = 1$, we have the following.

$\text{Value} = 2x^3 - 9x^2 + 4x + 12$	Given polynomial
$= 2(1)^3 - 9(1)^2 + 4(1) + 12$	Replace x by 1
$= 2 - 9 + 4 + 12$	Evaluate terms
$= 9$	Simplify

(b) When $x = -1$, we have the following.

$\text{Value} = 2x^3 - 9x^2 + 4x + 12$	Given polynomial
$= 2(-1)^3 - 9(-1)^2 + 4(-1) + 12$	Replace x by -1
$= -2 - 9 - 4 + 12$	Evaluate terms
$= -3$	Simplify ■

Applications

There are many applications that require the evaluation of polynomials. One commonly used second-degree polynomial is called a **position polynomial**. This polynomial has the form

$$-16t^2 + v_0 t + s_0 \qquad \text{Position polynomial}$$

where t is the time, measured in seconds. The value of this polynomial gives the height (in feet) of a free-falling object above the ground. The coefficient of t, v_0, is called the **initial velocity** of the object, and the constant term, s_0, is called the **initial height** of the object. If the initial velocity is positive, then the object was projected upward (at $t = 0$), and if the initial velocity is negative, then the object was projected downward.

EXAMPLE 10 ■ Finding the Height of a Free-Falling Object

An object is thrown downward from the top of a 200-foot building. The initial velocity is -10 feet per second. Use the position polynomial

$$-16t^2 - 10t + 200$$

to find the height of the object when $t = 1$, $t = 2$, and $t = 3$. (See Figure 3.1.)

FIGURE 3.1

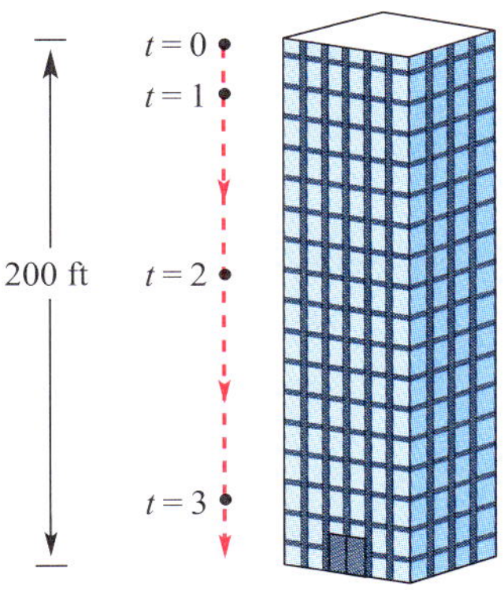

Solution

When $t = 1$, the height of the object is

$$\begin{aligned}\text{Height} &= -16(1)^2 - 10(1) + 200\\ &= -16 - 10 + 200\\ &= 174 \text{ feet.}\end{aligned}$$

When $t = 2$, the height of the object is

$$\begin{aligned}\text{Height} &= -16(2)^2 - 10(2) + 200\\ &= -64 - 20 + 200\\ &= 116 \text{ feet.}\end{aligned}$$

When $t = 3$, the height of the object is

$$\begin{aligned}\text{Height} &= -16(3)^2 - 10(3) + 200\\ &= -144 - 30 + 200\\ &= 26 \text{ feet.}\end{aligned}$$

■

A second commonly used mathematical model that often involves polynomials is called a **profit equation**. It has the following form.

$$\text{Profit} = \text{Revenue} - \text{Cost}$$

$$P = R - C$$

The **revenue** is the total amount of money received for selling x units of an item, and the **cost** is the total amount of money spent to produce x units of the item.

EXAMPLE 11 ■ Using the Profit Equation

A small manufacturing company can produce and sell x flashlights per week. The total cost (in dollars) for producing x flashlights is given by

$$C = 2x + 900$$

and the total revenue for selling x flashlights is given by

$$R = 6x.$$

Find the profit obtained by selling 700 flashlights per week.

Solution

Verbal model: Profit = Revenue − Cost

Labels: P = Weekly profit (in dollars)
R = Weekly revenue (in dollars) = $6x$
C = Weekly cost (in dollars) = $2x + 900$
x = Number of flashlights = 700

Equation:

$$P = R - C$$
$$P = 6x - (2x + 900)$$
$$P = 4x - 900$$
$$P = 4(700) - 900 = 2{,}800 - 900 = \$1{,}900$$

Therefore, the total weekly profit for selling 700 flashlights is \$1,900.

DISCUSSION PROBLEM

The Position Polynomial

Consider the following polynomials in the variable t to represent the height above ground of an object moving vertically up or down through the air.

$$-16t^2 + 80t + 30, \quad -16t^2 - 80t + 30, \quad -16t^2 + 30, \quad -16t^2 + 30t$$

Write a short paragraph to support your answer to each of the following questions.

(a) Which polynomial represents the height of an object thrown *downward*?

(b) Which polynomial represents the height of an object that is *dropped* from a height of 30 feet?

(c) Which polynomial represents the height of an object thrown *upward* from ground level?

Warm-Up

The following warm-up exercises involve skills that were covered in earlier sections. You will use these skills in the exercise set for this section.

In Exercises 1–4, expand the given expression.

1. $23(y - 3)$ **2.** $6(2 - 5z)$ **3.** $-\frac{2}{3}(15 - 6x)$ **4.** $-5(5x - 4)$

In Exercises 5 and 6, list the terms of the algebraic expression.

5. $32x - 10y - 7$ **6.** $-3s + 4t + 1$

In Exercises 7–10, simplify the expression by combining like terms.

7. $6y - 3x + 3x - 10y$ **8.** $\frac{3}{8}x - \frac{1}{12}x + 8$

9. $3x^2 - 5x + 3 + 28x - 33x^2$ **10.** $4x^3 - 3x^2y + 4xy^2 + 15x^2y + y^3$

3.1 EXERCISES

In Exercises 1–10, write the given polynomial in standard form and find its degree and leading coefficient.

1. $10x - 4$

2. $3x^2 + 8$

3. $5 - 3y^4$

4. $-3x^3 - 2x^2 - 3$

5. $6t + 4t^5 - t^2 + 3$

6. $8z - 16z^2$

7. -4

8. 28

9. $v_0 t - 16t^2$ (v_0 is a constant)

10. $48 - \frac{1}{2}at^2$ (a is a constant)

In Exercises 11–16, determine whether the polynomial is a monomial, binomial, or trinomial.

11. $12 - 5y^2$

12. $-6y + 3$

13. $x^3 + 2x^2 - 4$

14. t^3

15. 5

16. $25 - 2u^2$

In Exercises 17–20, state why the algebraic expression is not a polynomial.

17. $\frac{8}{x}$

18. $\frac{2}{x - 4}$

19. $y^{-3} - 2$

20. $x^3 + 4x^{1/3}$

In Exercises 21–24, give an example of a polynomial in one variable satisfying the given conditions. (*Note:* Each problem has many correct answers.)

21. A monomial of degree 3

22. A trinomial of degree 4 and leading coefficient -2

23. A binomial of degree 2 and leading coefficient 8

24. A monomial of degree 0

In Exercises 25–34, perform the addition using a horizontal format.

25. $(2x^2 - 3) + (5x^2 + 6)$

26. $(3x^3 - 2x + 8) + (3x - 5)$

27. $(8 - t^4) + (5 + t^4)$

28. $(z^3 + 6z - 2) + (3z^2 - 6z)$

29. $(2 - 8y) + (-2y^4 + 3y + 2)$

30. $(3a^2 + 5a) + (7 - a^2 - 5a) + (2a^2 + 8)$

31. $(x^2 - 3x + 8) + (2x^2 - 4x) + 3x^2$

32. $(5y + 6) + (4y^2 - 6y - 3)$

33. Add $3x^2 + 8$ to $7 - 5x^2$.

34. Add $20s - 12s^2 - 32$ to $15s^2 + 6s$.

In Exercises 35–40, perform the addition using a vertical format.

35. $(5x^2 - 3x + 4) + (-3x^2 - 4)$

36. $(4x^3 - 2x^2 + 8x) + (4x^2 + x - 6)$

37. $(5p^2 - 4p + 2) + (-3p^2 + 2p - 7)$

38. $(16 - 32t) + (64 + 48t - 16t^2)$

39. $(2b - 3) + (b^2 - 2b) + (7 - b^2)$

40. $(v^2 + v - 3) + (4v + 1) + (2v^2 - 3v)$

In Exercises 41–50, perform the subtraction using a horizontal format.

41. $(3x^2 - 2x + 1) - (2x^2 + x - 1)$

42. $(5y^4 - 2) - (3y^4 + 2)$

43. $(4 - y^3) - (4 + y^3)$

44. $(6t^3 - 12) - (-t^3 + t - 2)$

45. $(5q^2 - 3q + 5) - (4q^2 - 3q - 10)$

46. $(-10s^2 - 5) - (2s^2 + 6s)$

47. $(8x^3 - 4x^2 + 3x) - [(x^3 - 4x^2 + 5) + (x - 5)]$

48. $(5y^2 - 2y) - [(y^2 + y) - (3y^2 - 6y + 2)]$

49. Subtract $6x^3 - x + 11$ from $10x^3 + 15$.

50. Subtract $y^4 - y^2$ from $y^2 + 3y^4$.

In Exercises 51–56, perform the subtraction using a vertical format.

51. $(x^2 - x + 3) - (x - 2)$

52. $(z^2 - 5) - (z^2 - 5)$

53. $(25 - 15x - 2x^3) - (12 - 13x + 2x^3)$

54. $(0.2t^4 - 0.5t^2) - (0.3t^2 - t^4 - 1.4)$

55. $(4x^2 + 5x - 6) - (2x^2 - 4x + 5)$

56. $(13x^3 - 9x^2 + 4x - 5) - (5x^3 + 7x + 3)$

In Exercises 57–66, perform the indicated operations.

57. $-(2x^3 - 3) + (4x^3 - 2x)$

58. $(8z^4 - 7z^2) + 3(z^4 + 4)$

59. $(10x^2 - 11) - (-7x^3 - 12x^2 - 15)$

60. $(15y^4 - 18y - 18) - (-11y^4 - 8y - 8)$

61. $5s - [6s - (30s + 8)]$

62. $(p^3 + 4) - [(p^2 + 4) + (3p - 9)]$

63. $2(t^2 + 12) - 5(t^2 + 5) + 6(t^2 + 5)$

64. $-10(v + 2) + 8(v - 1) - 3(v - 9)$

65. $15v - 3(3v - v^2) + 9(8v + 3)$

66. $9(7x^2 - 3x + 3) - 4(15x + 2) - (3x^2 - 7x)$

In Exercises 67–70, find the value of the polynomial at the given values of the variable.

	Polynomial	*Value*
67.	$x^3 - 12x$	(a) $x = -2$ (b) $x = 0$ (c) $x = 2$ (d) $x = 4$
68.	$\frac{1}{4}x^4 - 2x^2$	(a) $x = -2$ (b) $x = 0$ (c) $x = 2$ (d) $x = 3$
69.	$x^4 - 4x^3 + 16x - 16$	(a) $x = -1$ (b) $x = 0$ (c) $x = 2$ (d) $x = \frac{5}{2}$
70.	$3t^4 + 4t^3$	(a) $t = -1$ (b) $t = -\frac{2}{3}$ (c) $t = 0$ (d) $t = 1$

Free-Falling Object In Exercises 71–74, find the height of a free-falling object at the specified times using the given position polynomial.

Position Polynomial	*Time*	
71. $-16t^2 + 32$	(a) $t = 0$	(b) $t = \frac{5}{4}$
72. $-16t^2 + 128$	(a) $t = 1$	(b) $t = 2$
73. $-16t^2 + 80t + 50$	(a) $t = 0$	(b) $t = 2$
	(c) $t = 4$	(d) $t = 5$
74. $-16t^2 + 96t$	(a) $t = 0$	(b) $t = 2$
	(c) $t = 3$	(d) $t = 6$

Free-Falling Object In Exercises 75–78, use the given position polynomial to determine whether the free-falling object was dropped, thrown upward, or thrown downward. Also determine the height of the object at time $t = 0$.

75. $-16t^2 + 100$

76. $-16t^2 + 50t$

77. $-16t^2 - 24t + 50$

78. $-16t^2 + 32t + 300$

79. *Free-Falling Object* An object is thrown upward from the top of a 200-foot building (see figure). The initial velocity is 40 feet per second. Use the position polynomial

$$\text{Height} = -16t^2 + 40t + 200$$

to find the height of the object when $t = 1$, $t = 2$, and $t = 3$.

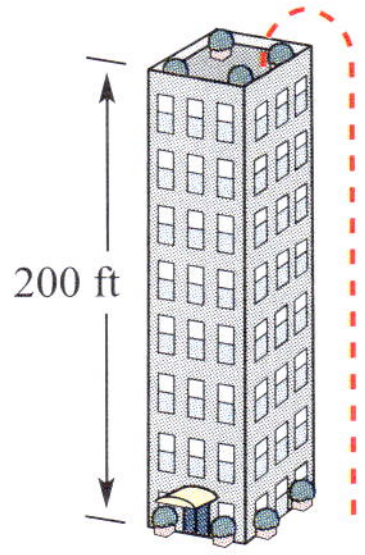

80. *Free-Falling Object* An object is dropped from a hot-air balloon that is 200 feet above the ground (see figure). Use the position polynomial

$$\text{Height} = -16t^2 + 200$$

to find the height of the object when $t = 1$, $t = 2$, and $t = 3$.

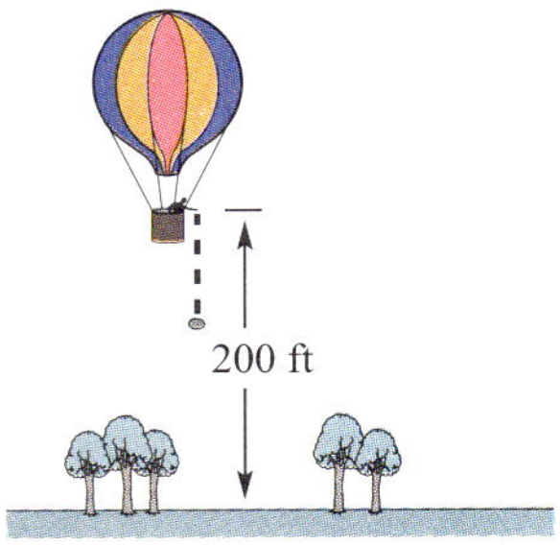

81. *Profit* A manufacturer can produce and sell x radios per week. The total cost (in dollars) for producing the radios is given by

$$C = 8x + 15{,}000$$

and the total revenue is given by

$$R = 14x.$$

Find the profit obtained by selling 5,000 radios per week.

82. *Profit* A manufacturer can produce and sell x golf clubs per week. The total cost (in dollars) for producing the golf clubs is given by

$$C = 12x + 8{,}000$$

and the total revenue is given by

$$R = 17x.$$

Find the profit obtained by selling 10,000 golf clubs per week.

83. *Horner's Rule* Use the Distributive Property to verify the equivalence of

$$2x^3 - 3x^2 + 10x - 12 \text{ and}$$
$$[(2x - 3)x + 10]x - 12.$$

The second form of the polynomial is said to be rewritten using **Horner's Rule**, which enables a person to evaluate the polynomial using only multiplication, addition, and subtraction.

84. *Horner's Rule* Evaluate both forms of the polynomial given in Exercise 83 when $x = 4$ and $x = 8$. Note that the second form requires only the operations of addition, subtraction, and multiplication.

Horner's Rule In Exercises 85–88, rewrite the polynomial in the form shown in Exercise 83.

85. $x^3 - 4x^2 + 2x + 10$

86. $x^3 + 10x^2 - 3x - 1$

87. $4x^3 - 3x^2 - 2x + 25$

88. $x^4 - 2x^2 - 5x + 2$

89. *Area of a Rectangle* Find the area of the region shown in the accompanying figure.

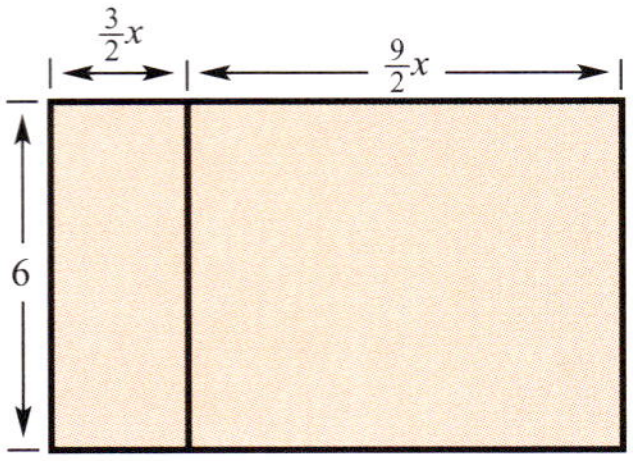

90. *Area of a Rectangle* Find the area of the region shown in the accompanying figure.

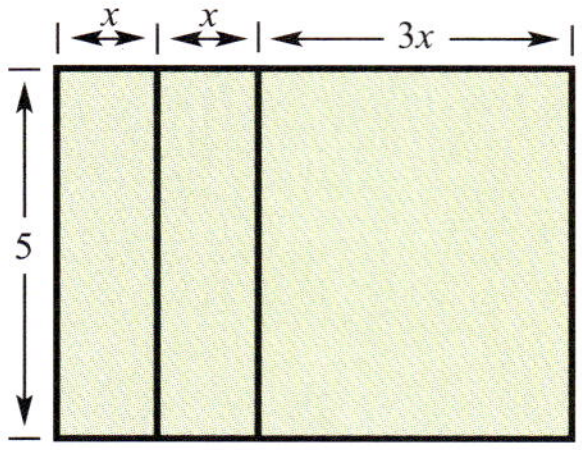

91. *Consumption of Milk* The per capita consumption (average consumption per person) of whole milk and lowfat milk in the United States between 1970 and 1987 can be approximated by the following two polynomial models.

$$y = 25.4 - 0.96t + 0.013t^2 \qquad \text{Whole milk}$$
$$y = 6.03 + 0.41t \qquad \text{Lowfat milk}$$

In these models, y represents the average consumption in gallons per person and t represents the year, with $t = 0$ corresponding to 1970. (*Source:* U.S. Department of Agriculture.)

(a) Find a polynomial model that represents the per capita consumption of milk (of either type) during this time period. Use this model to find the per capita consumption of milk in 1975 and in 1985.

(b) Between 1970 and 1985, the per capita consumption of whole milk was decreasing and the per capita consumption of lowfat milk was increasing (see figure). Was the per capita consumption of milk (of either type) increasing or decreasing?

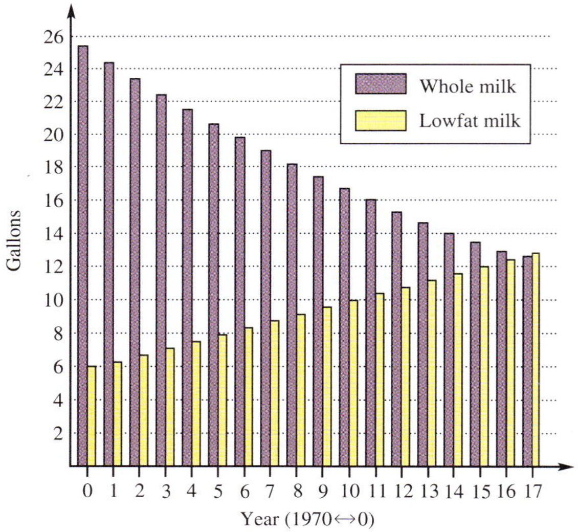

SECTION 3.2

Properties of Exponents and Multiplying Polynomials

Properties of Exponents • Multiplying Polynomials • Special Products • Applications

Properties of Exponents

In Section 1.4 we introduced the following two properties of exponents:

$$a^m \cdot a^n = a^{m+n} \quad \text{and} \quad (ab)^m = a^m \cdot b^m.$$

In this section we introduce three more properties. All five properties are summarized in the following list.

Properties of Exponents

Let m and n be positive integers, and let a and b represent real numbers, variables, or algebraic expressions.

Property	*Example*
1. $a^m \cdot a^n = a^{m+n}$	$x^3(x^2) = x^{3+2} = x^5$
2. $(ab)^m = a^m \cdot b^m$	$(2x)^3 = 2^3(x^3) = 8x^3$
3. $(a^m)^n = a^{mn}$	$(x^2)^3 = x^{2\cdot 3} = x^6$
4. $\dfrac{a^m}{a^n} = a^{m-n}, \quad m > n, a \neq 0$	$\dfrac{x^4}{x^2} = x^{4-2} = x^2$
5. $\left(\dfrac{a}{b}\right)^m = \dfrac{a^m}{b^m}, \quad b \neq 0$	$\left(\dfrac{x}{2}\right)^3 = \dfrac{x^3}{2^3} = \dfrac{x^3}{8}$

The validity of each of these five properties of exponents is illustrated in Examples 1 and 2.

EXAMPLE 1 ■ Illustrating the Properties of Exponents

(a) To multiply two exponential expressions that have the *same base*, we add exponents.

$$x^2 \cdot x^3 = \underbrace{x \cdot x}_{2 \text{ factors}} \cdot \underbrace{x \cdot x \cdot x}_{3 \text{ factors}} = \underbrace{x \cdot x \cdot x \cdot x \cdot x}_{5 \text{ factors}} = x^{2+3} = x^5$$

(b) To raise the product of two factors to the *same power*, we raise each factor to the power and multiply the results.

$$(3x)^3 = \underbrace{3x \cdot 3x \cdot 3x}_{3 \text{ factors}} = \underbrace{3 \cdot 3 \cdot 3}_{3 \text{ factors}} \cdot \underbrace{x \cdot x \cdot x}_{3 \text{ factors}} = 3^3 \cdot x^3 = 27x^3$$

(c) To raise an exponential expression to a power, we multiply the powers.

$$(x^3)^2 = \underbrace{(x \cdot x \cdot x)}_{\text{3 factors}} \cdot \underbrace{(x \cdot x \cdot x)}_{\text{3 factors}} = \underbrace{(x \cdot x \cdot x \cdot x \cdot x \cdot x)}_{\text{6 factors}} = x^{3 \cdot 2} = x^6$$

EXAMPLE 2 ■ **Illustrating the Properties of Exponents**

(a) To divide two exponential expressions that have the *same base*, we subtract exponents.

$$\frac{x^4}{x^2} = \frac{\overbrace{x \cdot x \cdot x \cdot x}^{\text{4 factors}}}{\underbrace{x \cdot x}_{\text{2 factors}}} = x^{4-2} = x^2$$

(b) To raise the quotient of two expressions to the *same power*, we raise each expression to the power and divide the results.

$$\left(\frac{x}{3}\right)^3 = \frac{x}{3} \cdot \frac{x}{3} \cdot \frac{x}{3} = \frac{\overbrace{x \cdot x \cdot x}^{\text{3 factors}}}{\underbrace{3 \cdot 3 \cdot 3}_{\text{3 factors}}} = \frac{x^3}{3^3} = \frac{x^3}{27}$$

EXAMPLE 3 ■ **Applying Properties of Exponents**

(a) $(x^2y^4)(3x) = 3(x^2 \cdot x)(y^4) = 3x^3y^4$

(b) $(-2y^2)^3 = (-2)^3(y^2)^3 = -8y^6$

(c) $-2(y^2)^3 = (-2)(y^{2(3)}) = -2y^6$

EXAMPLE 4 ■ **Applying Properties of Exponents**

(a) $\dfrac{14a^5b^3}{7a^2b^2} = 2(a^{5-2})(b^{3-2}) = 2a^3b$

(b) $\left(\dfrac{x^2}{2y}\right)^3 = \dfrac{(x^2)^3}{(2y)^3} = \dfrac{x^6}{2^3y^3} = \dfrac{x^6}{8y^3}$

(c) $\dfrac{x^{n+2}y^{3n}}{x^2y^n} = x^{(n+2)-2}y^{3n-n} = x^ny^{2n}$

Multiplying Polynomials

The simplest type of polynomial multiplication involves a monomial multiplier. The product is obtained by direct application of the Distributive Property. For instance, to multiply the monomial $3x$ by the polynomial $(2x^2 - 5x + 3)$, we multiply *each* term of the polynomial by $3x$.

$$\begin{aligned}(3x)(2x^2 - 5x + 3) &= (3x)(2x^2) - (3x)(5x) + (3x)(3) \\ &= 6x^3 - 15x^2 + 9x\end{aligned}$$

EXAMPLE 5 ■ Finding Products with Monomial Multipliers

(a) $(2x - 7)(-3x) = 2x(-3x) - 7(-3x)$ Distributive Property

$= -6x^2 + 21x$ Standard form

(b) $4x^2(3x - 2x^3 + 1) = 4x^2(3x) - 4x^2(2x^3) + 4x^2(1)$ Distributive Property

$= 12x^3 - 8x^5 + 4x^2$ Properties of exponents

$= -8x^5 + 12x^3 + 4x^2$ Standard form

(c) $(-x)(5x^2 - x) = (-x)(5x^2) - (-x)(x)$ Distributive Property

$= -5x^3 + x^2$ Standard form ■

To multiply two binomials, we make use of both (left and right) forms of the Distributive Property. For example, if we treat the binomial $(2x + 7)$ as a single quantity, we can multiply $(3x - 2)$ by $(2x + 7)$ as follows.

$$\begin{aligned}(3x - 2)(2x + 7) &= 3x(2x + 7) - 2(2x + 7)\\ &= (3x)(2x) + (3x)(7) - (2)(2x) - 2(7)\\ &= 6x^2 + 21x - 4x - 14\end{aligned}$$

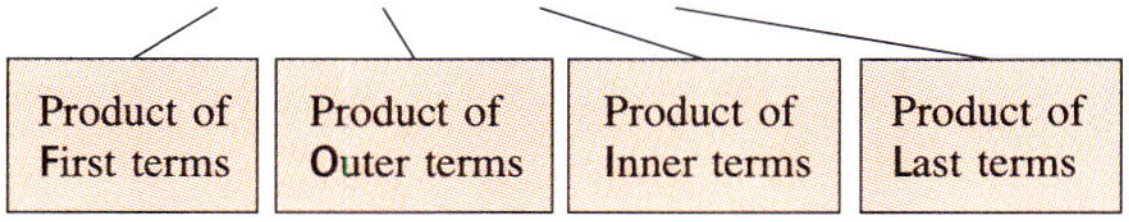

$$= 6x^2 + 17x - 14$$

With practice, you should be able to multiply two binomials without writing out all of the above steps. In fact, the four products in the boxes above suggest we can put the product of two binomials in the FOIL form in just one step. This is called the **FOIL Method**. Note that the words *first*, *outer*, *inner*, and *last* refer to the position of the terms in the original product.

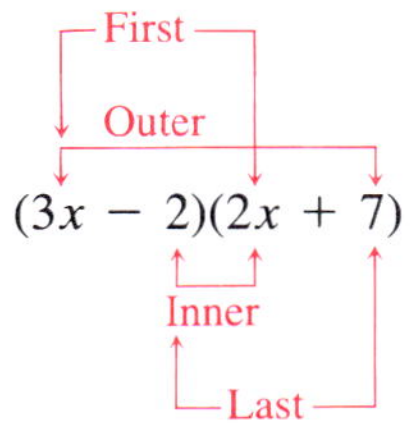

EXAMPLE 6 ■ Multiplying Binomials (Distributive Property)

$(x + 2)(x - 3) = (x + 2)(x) - (x + 2)(3)$ Left Distributive Property

$= (x^2 + 2x) - (3x + 6)$ Right Distributive Property

$= x^2 + (2x - 3x) - 6$ Collect like terms

$= x^2 - x - 6$ Combine like terms ■

EXAMPLE 7 ■ **Multiplying Binomials (FOIL Method)**

Use the FOIL Method to multiply the following.

(a) $(3x + 4)(2x + 1)$ (b) $(x - 3)(x + 3)$

Solution

(a) Applying the FOIL Method produces the following.

$$(3x + 4)(2x + 1) = \overbrace{6x^2}^{F} + \overbrace{3x}^{O} + \overbrace{8x}^{I} + \overbrace{4}^{L}$$

$$= 6x^2 + 11x + 4 \qquad \text{Combine like terms}$$

(b) Note how the outer and inner products add up to zero in this case.

$$(x - 3)(x + 3) = \overbrace{x^2}^{F} + \overbrace{3x}^{O} - \overbrace{3x}^{I} - \overbrace{9}^{L}$$

$$= x^2 - 9 \qquad \text{Combine like terms}$$ ■

EXAMPLE 8 ■ **Simplifying Polynomial Expressions**

Simplify the following expression and write the result in standard form.

$$6x(x^2 - 3x + 7) - (x^3 + 11x)$$

Solution

$$6x(x^2 - 3x + 7) - (x^3 + 11x) = 6x^3 - 18x^2 + 42x - x^3 - 11x$$

$$= (6x^3 - x^3) - 18x^2 + (42x - 11x)$$

$$= 5x^3 - 18x^2 + 31x$$ ■

When multiplying two polynomials that have three or more terms, we use the same basic principle as we do when multiplying monomials and binomials. That is, *each term of one polynomial must be multiplied by each term of the other polynomial*. This can be done using either a horizontal or vertical format, as shown in the next two examples.

EXAMPLE 9 ■ **Multiplying Polynomials (Horizontal Format)**

Use a horizontal format of factors to find the product $(4x^2 - 3x - 1)(2x - 5)$.

Solution

To help assure that each term of one polynomial gets multiplied by each term of the other, we use the Distributive Property in a horizontal format.

$$(4x^2 - 3x - 1)(2x - 5)$$

$$= (4x^2 - 3x - 1)(2x) - (4x^2 - 3x - 1)(5) \qquad \text{Distributive Property}$$

$$= 8x^3 - 6x^2 - 2x - (20x^2 - 15x - 5) \quad \text{Distributive Property}$$
$$= 8x^3 - 6x^2 - 2x - 20x^2 + 15x + 5$$
$$= 8x^3 - 26x^2 + 13x + 5 \quad \text{Combine like terms}$$

■

When multiplying two polynomials, it is best to write each in standard form before using either the horizontal or vertical formats. This is illustrated in the next example.

EXAMPLE 10 ■ Multiplying Polynomials (Vertical Format)

Multiply the following.

$$(x + 4x^2 - 2)(5 + 3x - x^2)$$

Solution

With the vertical format, we line up like terms in the same vertical columns, much as we align digits in whole-number multiplication.

$$\begin{array}{rrrrrl}
 & & 4x^2 & + x & - 2 & \text{Standard form} \\
\times & & -x^2 & + 3x & + 5 & \text{Standard form} \\
\hline
 & & 20x^2 & + 5x & - 10 & \Leftarrow 5(4x^2 + x - 2) \\
 & 12x^3 & + 3x^2 & - 6x & & \Leftarrow 3x(4x^2 + x - 2) \\
-4x^4 & - x^3 & + 2x^2 & & & \Leftarrow -x^2(4x^2 + x - 2) \\
\hline
-4x^4 & + 11x^3 & + 25x^2 & - x & - 10 &
\end{array}$$

■

EXAMPLE 11 ■ Multiplying Polynomials

Multiply $(x - 4)^3$.

Solution

We can raise $(x - 4)$ to the third power in two steps. First, since $(x - 4)^3 = (x - 4)^2(x - 4)$, we find the square of $(x - 4)$ as follows.

$$(x - 4)(x - 4) = x^2 - 4x - 4x + 16 \quad \text{Find } (x - 4)^2$$
$$= x^2 - 8x + 16 \quad \text{Combine like terms}$$

Now, using a vertical format, we find $(x - 4)^3$ as follows.

$$\begin{array}{rrrr}
 & x^2 & - 8x & + 16 \\
\times & & x & - 4 \\
\hline
 & -4x^2 & + 32x & - 64 \\
x^3 & - 8x^2 & + 16x & \\
\hline
x^3 & - 12x^2 & + 48x & - 64
\end{array}$$

■

Special Products

Some binomial products, like those in Example 7(b), have special forms that occur frequently in algebra. Let's look at this product again. We call $(x + 3)(x - 3)$ the **product of the sum and difference of two terms**. With such products, the two middle terms cancel as follows.

$$(x + 3)(x - 3) = x^2 - 3x + 3x - 9$$ Sum and difference of two terms

$$= x^2 - 9$$ Product has no middle term

Another common type of product is the **square of a binomial**. With this type of product, the middle term is always twice the product of the terms in the binomial.

$$(2x + 5)^2 = (2x + 5)(2x + 5)$$ Square of a binomial

$$= 4x^2 + 10x + 10x + 25$$

$$= 4x^2 + 20x + 25$$ Middle term is twice the product of the terms of the binomial

You should learn to recognize the patterns of these two special products. We give the general form of these special products in the following statements.

Special Products

Let u and v be real numbers, variables, or algebraic expressions. Then the following formulas are true.

Special Product	*Example*
Sum and Difference of Two Terms	
$(u + v)(u - v) = u^2 - v^2$	$(3x - 4)(3x + 4) = 9x^2 - 16$
Square of a Binomial	
$(u + v)^2 = u^2 + 2uv + v^2$	$(2x + 5)^2 = 4x^2 + 2(2x)(5) + 25$ $= 4x^2 + 20x + 25$
$(u - v)^2 = u^2 - 2uv + v^2$	$(x - 6)^2 = x^2 - 2(x)(6) + 36$ $= x^2 - 12x + 36$

NOTE: When squaring a binomial, the resulting middle term is always *twice* the product of the two terms.

$$\underbrace{(x + y)}_{\text{Terms}}{}^2 = x^2 + \underbrace{2(xy)}_{\text{Twice the product of the terms}} + y^2$$

EXAMPLE 12 ■ Finding the Product of the Sum and Difference of Two Terms

Find the following products.

(a) $(3x - 2)(3x + 2)$ (b) $(6 + 5x)(6 - 5x)$

Solution

(a) This special product represents the sum and difference of two terms. Therefore, the product has the form $(a - b)(a + b) = a^2 - b^2$.

$$\begin{aligned}(3x - 2)(3x + 2) &= (3x)^2 - 2^2 \\ &= 9x^2 - 4\end{aligned}$$

(b) This special product also represents the sum and difference of two terms. Therefore, the product has the form $(a + b)(a - b) = a^2 - b^2$.

$$\begin{aligned}(6 + 5x)(6 - 5x) &= 6^2 - (5x)^2 \\ &= 36 - 25x^2\end{aligned}$$

■

EXAMPLE 13 ■ Squaring a Binomial

Find the product $(2x - 7)^2$.

Solution

Note that the *middle* term of the product is twice the product of the two terms of the binomial.

$$\begin{aligned}(2x - 7)^2 &= (2x)^2 - 2(2x)(7) + 7^2 \\ &= 4x^2 - 28x + 49\end{aligned}$$

■

Applications

A common application involving the product of polynomials is the calculation of area and volume. We illustrate this type of application in Example 14.

EXAMPLE 14 ■ Finding Area and Volume

The closed box shown in Figure 3.2 has sides whose measurements (in inches) are consecutive integers. Find a polynomial that describes the volume of the box. What is the volume if the length of the shortest side is 4 inches?

FIGURE 3.2

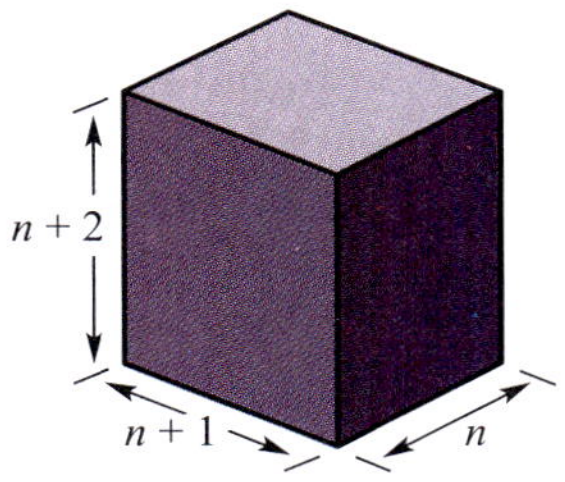

Solution

Verbal model: Volume = Length · Width · Height

Labels:
n = Length (in inches)
$n + 1$ = Width (in inches)
$n + 2$ = Height (in inches)
V = Volume (in cubic inches)

Equation:

$$\begin{aligned} V &= n(n + 1)(n + 2) \\ &= n(n^2 + 3n + 2) \\ &= n^3 + 3n^2 + 2n \end{aligned}$$

If $n = 4$, then the volume of the box is

$$V = 4^3 + 3(4^2) + 2(4) = 64 + 48 + 8 = 120 \text{ cubic inches.}$$

DISCUSSION PROBLEM — The Revenue Polynomial

Suppose the flashlight company (see Example 11 in Section 3.1) determined that it can sell x flashlights per week according to the equation

$$x = 700 - 8p$$

where p is the price (in dollars) of each flashlight sold.

(a) Write a *sentence* that describes the total weekly revenue obtained by selling x flashlights.

(b) Make up a table showing the revenue R obtained when the flashlights are priced at $p = \$4$, $p = \$5$, $p = \$7$, and $p = \$10$.

(c) Write a polynomial in the variable p that gives the weekly revenue R.

(d) Write a polynomial in the variable x that gives the weekly revenue R.

Warm-Up

The following warm-up exercises involve skills that were covered in earlier sections. You will use these skills in the exercise set for this section.

In Exercises 1 and 2, rewrite each quantity in exponential form.

1. $(-5)(-5)(-5)$

2. $y \cdot y \cdot y \cdot y$

In Exercises 3–6, rewrite the quantity as a repeated multiplication.

3. $\left(\frac{2}{3}\right)^4$

4. $(2.1)^5$

5. $(2x)^3$

6. $(-y)^2$

In Exercises 7–10, simplify the algebraic expression.

7. $3(x - 5) + 4x$

8. $5(4 - y) + 8(y + 2)$

9. $-3(u - 6) - (u + 18)$

10. $(v - 3) - 3(2v + 5)$

3.2 EXERCISES

In Exercises 1–6, use the given expression to illustrate one of the properties of exponents. (See Example 1.)

1. $t^3 \cdot t^4$ **2.** $(2y)^3$ **3.** $(-5x)^5$ **4.** $(x^2)^3$ **5.** $\left(\frac{y}{5}\right)^4$ **6.** $\left(\frac{3}{t}\right)^5$

In Exercises 7–10, use the given expression to illustrate one of the properties of exponents. (See Example 2.)

7. $\frac{x^6}{x^4}$ **8.** $\frac{2^4y^5}{2^2y^3}$ **9.** $\frac{3^7x^5}{3^3x^3}$ **10.** $\frac{4^3(x+2)^4}{4(x+2)^2}$

In Exercises 11–20, simplify the given expression.

11. (a) $-3x^3 \cdot x^5$ (b) $(-3x)^2 \cdot x^5$

12. (a) $5^2y^4 \cdot y^2$ (b) $(5y)^2 \cdot y^4$

13. (a) $(-5z)^3$ (b) $(-5z)^2$

14. (a) $-5z^3$ (b) $-z^2 \cdot z^2$

15. (a) $(-5z)^4$ (b) $-5z^4 \cdot (-5z)^4$

16. (a) $(3y)^3(2y^2)$ (b) $3y^3 \cdot 2y^2$

17. (a) $\frac{28x^2y^3}{21xy^2}$ (b) $\frac{21xy^2}{28x^2y^3}$

18. (a) $\frac{(-3xy)^3}{9xy^2}$ (b) $\frac{(-3xy)^2}{-3(xy)^2}$

19. (a) $\frac{[2(x+3)]^4}{6(x+3)^2}$ (b) $\frac{2(x+3)^4}{[6(x+3)]^2}$

20. (a) $\left(\frac{2a}{3y}\right)^5$ (b) $-\left(\frac{2a}{3y}\right)^2$

In Exercises 21–40, perform the indicated multiplication and simplify.

21. $(-2a^2)(-8a)$ **22.** $(-6n)(3n^2)$ **23.** $5x(2x)^2$

24. $(3y)^3\left(\frac{y}{2}\right)^2$ **25.** $2y(5-y)$ **26.** $5z(2z-7)$

27. $4x(2x^2-3x+5)$ **28.** $3y^2(-3y^2+7y-3)$ **29.** $-2x(5+3x^2-7x^3)$

30. $-3a^2(11a-3)$ **31.** $-3x(-5x)(5x+2)$ **32.** $4t(-3t)(t^2-1)$

33. $\left(4y-\frac{1}{3}\right)(12y+9)$ **34.** $\left(5t-\frac{3}{4}\right)(2t-16)$ **35.** $(2x+y)(3x+2y)$

36. $(2x-y)(3x-2y)$ **37.** $5a(a+2)-3a(2a-3)$ **38.** $(2t-1)(t+1)+3(2t-5)$

39. $(s-3t)(s+t)-(s-3t)(s-t)$ **40.** $2x(x-5y)+3y(x+5y)$

In Exercises 41–46, use a horizontal format to perform the indicated multiplication.

41. $(x^3-3x+2)(x-2)$ **42.** $(t+3)(t^2-5t+1)$ **43.** $(u+5)(2u^2+3u-4)$

44. $(x^2+4)(x^2-2x-4)$ **45.** $(x-1)(x^2-4x+6)$ **46.** $(2x^2-3)(2x^2-2x+3)$

In Exercises 47–56, use a vertical format to perform the indicated multiplication.

47. $(x + 3)(7x^2 - 14x + 9)$

48. $(2x^2 + 3)(4x^4 - 6x^2 + 9)$

49. $(t^2 + t - 2)(t^2 - t + 2)$

50. $(y^2 + 3y + 5)(2y^2 - 3y - 1)$

51. $(x^3 + x)(x^2 - 4)$

52. $(z - 2)(z^2 + z + 1)$

53. $(2x + 1)(-3x^2 + 2x - 1)$

54. $(3s - 4)(2s^2 - 5s + 6)$

55. $(u - 2)(2u^2 + 5u + 3)$

56. $(2x + 5)(5x^2 - 13x + 6)$

In Exercises 57–74, perform the indicated multiplication.

57. $(x + 2)(x - 2)$

58. $(4 + 3z)(4 - 3z)$

59. $(2 + 7y)(2 - 7y)$

60. $(5u + 12v)(5u - 12v)$

61. $\left(2x - \frac{1}{4}\right)\left(2x + \frac{1}{4}\right)$

62. $\left(\frac{2}{3}x + 7\right)\left(\frac{2}{3}x - 7\right)$

63. $(0.2t + 0.5)(0.2t - 0.5)$

64. $(4a - 0.1b)(4a + 0.1b)$

65. $(x + 5)^2$

66. $(x + 9)^2$

67. $(5x - 2)^2$

68. $(3x - 8)^2$

69. $(2x - 7y)^2$

70. $(5 - 3z)^2$

71. $[(x + 2) + y]^2$

72. $[(x - 4) - y]^2$

73. $[u - (v - 3)]^2$

74. $[2z + (y + 1)]^2$

In Exercises 75 and 76, perform the indicated operations and simplify.

75. $(t + 3)^2 - (t - 3)^2$

76. $(a + 6)^2 + (a - 6)^2$

In Exercises 77–80, perform the indicated multiplication. (See Example 11.)

77. $(x + 3)^3$

78. $(y - 2)^3$

79. $(u + v)^3$

80. $(u - v)^3$

In Exercises 81 and 82, find the area of the shaded portion of the figure.

81.

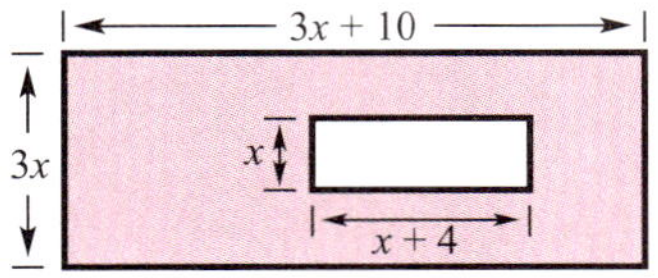

82.

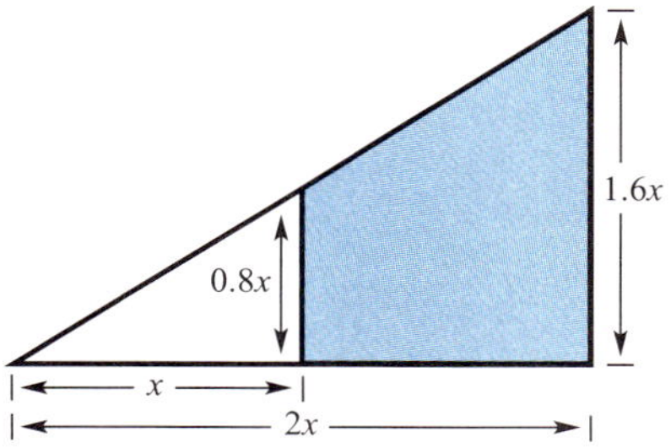

83. *Dimensions of a Rectangle* The length of a rectangle is one and one-half times its width w. Find (a) the perimeter, and (b) the area of the rectangle.

84. *Area of a Triangle* The base of a triangle is $3x$ and its height is $x + 5$. Find the area A of the triangle.

85. *Compound Interest* After two years, an investment of \$1,000 compounded annually at an interest rate of r will yield an amount of $1{,}000(1 + r)^2$. Find this product.

86. *Compound Interest* After two years, an investment of \$1,000 compounded annually at an interest rate of 9.5% will yield an amount of $1{,}000(1 + 0.095)^2$. Find this product.

87. *Area of a Rectangle* Find the area of the rectangle in two ways (see figure) and state the formula that the two areas represent.

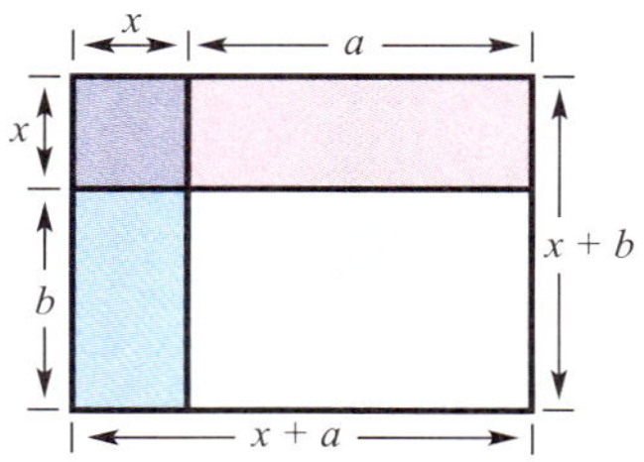

88. *Area of a Square* Find the area of the square in two ways (see figure) and state the special product that the area represents.

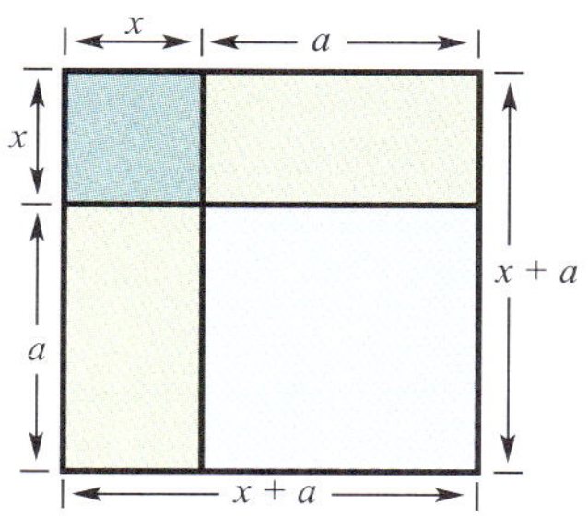

89. *Finding a Pattern* Perform the following multiplications.
(a) $(x - 1)(x + 1)$
(b) $(x - 1)(x^2 + x + 1)$
(c) $(x - 1)(x^3 + x^2 + x + 1)$
From the pattern formed by these products, can you predict the result of $(x - 1)(x^4 + x^3 + x^2 + x + 1)$?

90. *Evaluating Expressions* Verify that $(x + y)^2$ is not equal to $x^2 + y^2$ by letting $x = 3$, $y = 4$, and evaluating both expressions.

SECTION **3.3**

Factoring Polynomials: An Introduction

Introduction • Factoring Polynomials with Common Factors • Factoring by Grouping • Factoring the Difference of Two Squares • Factoring the Sum or Difference of Two Cubes • Factoring Completely

Introduction

We now switch from multiplying polynomials to the *reverse* process—**factoring polynomials**. In this and the next two sections, we will deal only with polynomials that have integer coefficients. Remember that in Section 3.2 we used the Distributive Property to *multiply* and *remove* parentheses as follows.

Multiply →

$$3x(4 - 5x) = 12x - 15x^2$$ Distributive Property

In this and the next two sections, we use the Distributive Property in the reverse direction to *factor* and *create* parentheses.

Factor →

$$12x - 15x^2 = 3x(4 - 5x)$$ Distributive Property

Factoring an expression (by the Distributive Property) changes a *sum of terms* into a *product of factors*. Later we will see that this is an important strategy for solving equations and for simplifying algebraic expressions.

Factoring Polynomials with Common Factors

To be efficient in factoring an expression, you need to understand the concept of the **greatest common factor** of two (or more) integers or terms. Recall from arithmetic that every integer can be factored into a product of prime numbers. The **greatest common factor** of two or more integers is the greatest integer that is a factor of each number. For example, the greatest common factor of 18 and 42 is 6.

$$\left.\begin{aligned}18 &= 2 \cdot 3 \cdot 3\\ 42 &= 2 \cdot 3 \cdot 7\end{aligned}\right\} \text{ } 2 \cdot 3 = 6 \text{ is common to both integers}$$

In a similar way, an algebraic term can be factored into a product of prime factors, as shown in Example 1.

EXAMPLE 1 ■ Finding the Greatest Common Factor

Find the greatest common factor of the following.

$6x^5, \quad 30x^4, \quad \text{and} \quad 12x^3$

Solution

From the factorizations

$$\begin{aligned}6x^5 &= 2 \cdot 3 \cdot x \cdot x \cdot x \cdot x \cdot x = (6x^3)(x^2)\\ 30x^4 &= 2 \cdot 3 \cdot 5 \cdot x \cdot x \cdot x \cdot x = (6x^3)(5x)\\ 12x^3 &= 2 \cdot 2 \cdot 3 \cdot x \cdot x \cdot x = (6x^3)(2)\end{aligned}$$

we conclude that the greatest common factor is $6x^3$. ■

Suppose we consider the three terms given in Example 1 as terms of the polynomial

$6x^5 + 30x^4 + 12x^3$.

The greatest common factor, $6x^3$, of these terms is called the **greatest common monomial factor** of the polynomial. When we use the Distributive Property to remove this factor from each term of the polynomial, we say that we are **factoring out** the greatest common monomial factor.

$$\begin{aligned}6x^5 + 30x^4 + 12x^3 &= 6x^3(x^2) + 6x^3(5x) + 6x^3(2) && \text{Factor each term}\\ &= 6x^3(x^2 + 5x + 2) && \text{Factor out common monomial factor}\end{aligned}$$

EXAMPLE 2 ■ Factoring Out the Greatest Common Monomial Factor

Factor out the greatest common monomial factor from the following polynomial.

$24x^3 - 32x^2$

Solution

The greatest common factor of $24x^3$ and $32x^2$ is $8x^2$. Thus, we can factor the given polynomial as follows.

$$24x^3 - 32x^2 = (8x^2)(3x) - (8x^2)(4) = 8x^2(3x - 4)$$ ■

NOTE: If a polynomial in x (with integer coefficients) has a greatest common monomial factor of the form ax^n, then the following statements must be true. (1) The coefficient, a, of the greatest common monomial factor must be the greatest integer that *divides* each of the coefficients in the polynomial. (2) The variable factor, x^n, of the greatest common monomial factor has the *least* power of x of all terms of the polynomial.

We usually consider the greatest common monomial factor of a polynomial to have a positive coefficient. However, sometimes it is convenient to factor a negative number out of a polynomial. You can see how this is done in the next example.

EXAMPLE 3 ■ A Negative Common Monomial Factor

Factor the following polynomial in two ways.

(a) Factor out 3. (b) Factor out -3.

$$-3x^2 + 12x - 18$$

Solution

(a) To factor out the common monomial factor of 3, we write the following.

$$-3x^2 + 12x - 18 = 3(-x^2) + 3(4x) + 3(-6)$$ Factor 3 out of each term

$$= 3(-x^2 + 4x - 6)$$ Factor 3 out of the polynomial

(b) To factor -3 out of the polynomial, we write the following.

$$-3x^2 + 12x - 18 = -3(x^2) + (-3)(-4x) + (-3)(6)$$ Factor -3 out of each term

$$= -3(x^2 - 4x + 6)$$ Factor -3 out of the polynomial ■

Factoring by Grouping

There are occasions when a common factor of a polynomial is not simply a monomial. For instance, the polynomial

$$x^2(2x - 3) + 4(2x - 3)$$

has the common *binomial* factor $(2x - 3)$. Factoring out this common factor produces

$$x^2(2x - 3) + 4(2x - 3) = (2x - 3)(x^2 + 4).$$

This type of factoring is part of a more general procedure called **factoring by grouping**.

EXAMPLE 4 ■ **Common Binomial Factors**

Factor the following.

$$5x^2(6x - 5) - 2(6x - 5)$$

Solution

Each of the terms of this polynomial has a binomial factor of $(6x - 5)$. Factoring this binomial out of each term produces the following.

$$5x^2(6x - 5) - 2(6x - 5) = (6x - 5)(5x^2 - 2)$$ ■

In Example 4 the given polynomial was already grouped so that it was easy to determine the common binomial factor. In practice, you will have to do the grouping as well as the factoring. To see how this works, let's consider the expression

$$x^3 - 3x^2 - 5x + 15$$

and try to *factor* it. We note first that there is no common monomial factor to take out of all four terms. But suppose we *group* the first two terms together and the last two terms together. Then we have the following.

$$x^3 - 3x^2 - 5x + 15 = (x^3 - 3x^2) - (5x - 15)$$ Group terms

$$= x^2(x - 3) - 5(x - 3)$$ Take out common factor in each group

$$= (x - 3)(x^2 - 5)$$ Take out common binomial factor

EXAMPLE 5 ■ **Factoring by Grouping**

Factor the following polynomial.

$$x^3 - 5x^2 + x - 5$$

Solution

By grouping the first two terms together, and the third and fourth terms together, we obtain the following.

$$\begin{aligned} x^3 - 5x^2 + x - 5 &= (x^3 - 5x^2) + (x - 5) \\ &= x^2(x - 5) + (x - 5) \\ &= (x - 5)(x^2 + 1) \end{aligned}$$ ■

NOTE: You can check to see that you have factored an expression correctly by multiplying out the factors and comparing the result to the original expression.

MATH MATTERS

The Rolling Coin Problem

The history of mathematics contains many paradoxes—problems that appear to be contradictory or to have absurd solutions. Here is one such problem that concerns two coins.

Suppose one coin is fixed to the edge of a table. A second coin is placed to the left of the fixed coin, with the head upright, as shown in the figure. If the second coin is rolled along the fixed coin with *no slippage*, will it arrive on the right side of the fixed coin with its head upright or turned down?

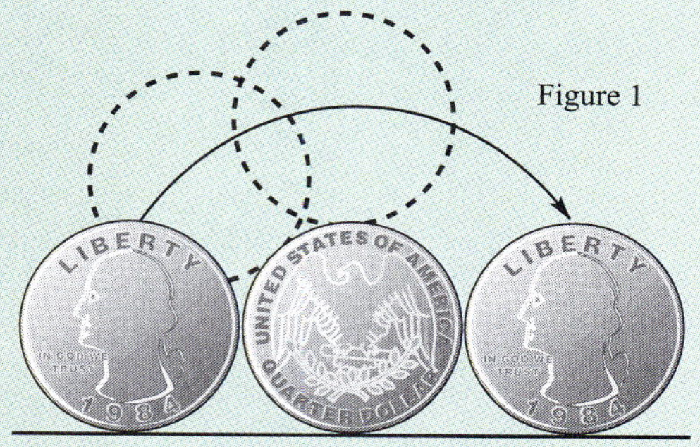

Figure 1

To many people, the correct answer appears to be the second possibility because, they reason, the coin rolled around half of its circumference, which should result in the rotation of 180° degrees, and should therefore have turned the head of the coin down. Which of these two possibilities seems most likely to you? Try the experiment using two quarters. What is the result? (The answer is given in the back of the text.)

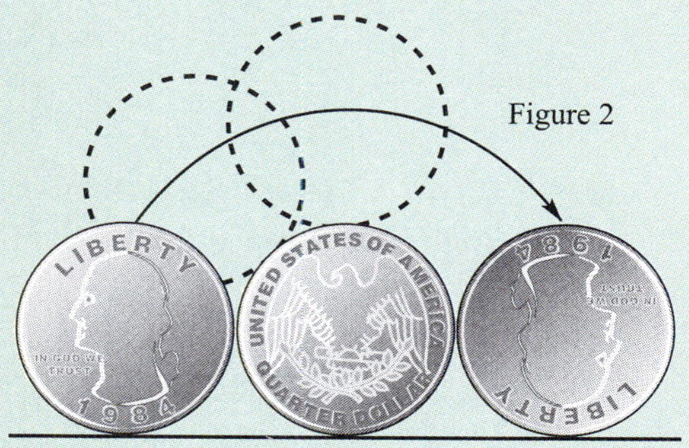

Figure 2

Factoring the Difference of Two Squares

Some polynomials have special forms that you should learn to recognize so that they can be factored easily. Here are some examples of forms that you should be able to recognize by the time you have completed this section.

$x^2 - 9 = (x + 3)(x - 3)$ Difference of two squares

$x^3 + 8 = (x + 2)(x^2 - 2x + 4)$ Sum of two cubes

$x^3 - 1 = (x - 1)(x^2 + x + 1)$ Difference of two cubes

One of the easiest special polynomial forms to recognize and to factor is the form $u^2 - v^2$, called a **difference of two squares**. It factors according to the following pattern.

Difference of Two Squares

Let u and v be real numbers, variables, or algebraic expressions. Then the expression $u^2 - v^2$ can be factored as follows.

$$u^2 - v^2 = (u + v)(u - v)$$

Difference (pointing to the minus sign in $u^2 - v^2$) — Opposite signs (pointing to the signs in $(u + v)(u - v)$)

To recognize perfect squares, look for coefficients that are squares of integers and for variables raised to *even* powers. Here are some examples.

Given Polynomial		*Think: Difference of Squares*		*Factored Form*
$x^2 - 4$	⇨	$x^2 - 2^2$	⇨	$(x + 2)(x - 2)$
$4x^2 - 25$	⇨	$(2x)^2 - 5^2$	⇨	$(2x + 5)(2x - 5)$
$25 - 49x^2$	⇨	$5^2 - (7x)^2$	⇨	$(5 + 7x)(5 - 7x)$

EXAMPLE 6 ■ **Factoring the Difference of Two Squares**

Factor the following polynomial.

$$x^2 - 64$$

Solution

Because x^2 and 64 are both perfect squares, we recognize this polynomial to be the difference of two squares. Therefore, the polynomial factors as follows.

$x^2 - 64 = x^2 - 8^2$ — Write as difference of two squares

$= (x + 8)(x - 8)$ — Factored form ■

EXAMPLE 7 ■ **Factoring the Difference of Two Squares**

Factor the following polynomial.

$$49x^2 - 81$$

Solution

Because $49x^2$ and 81 are both perfect squares, we recognize this polynomial to be the difference of two squares. Therefore, the polynomial factors as follows.

$49x^2 - 81 = (7x)^2 - 9^2$ — Write as difference of two squares

$= (7x + 9)(7x - 9)$ — Factored form ■

Remember that the rule $u^2 - v^2 = (u + v)(u - v)$ applies to polynomials or expressions in which u and v are themselves expressions. The next two examples illustrate this possibility.

EXAMPLE 8 ■ **Factoring the Difference of Two Squares**

$$\begin{aligned}(x + 2)^2 - 9 &= (x + 2)^2 - 3^2 && \text{Write as difference of two squares}\\ &= [(x + 2) + 3][(x + 2) - 3] && \text{Factored form}\\ &= (x + 5)(x - 1) && \text{Simplify}\end{aligned}$$ ■

Factoring the Sum or Difference of Two Cubes

The last type of special factoring that we will study in this section is the sum or difference of two cubes. The formulas for these two special forms are summarized as follows. In these formulas pay particular attention to the signs of the terms.

Sum or Difference of Two Cubes

Let u and v be real numbers, variables, or algebraic expressions. Then the expressions $u^3 + v^3$ and $u^3 - v^3$ can be factored as follows.

Like signs

1. $u^3 + v^3 = (u + v)(u^2 - uv + v^2)$

Unlike signs

Like signs

2. $u^3 - v^3 = (u - v)(u^2 + uv + v^2)$

Unlike signs

EXAMPLE 9 ■ **Factoring Sums and Differences of Cubes**

Factor the following polynomials.

(a) $x^3 - 125$ (b) $8y^3 + 1$

Solution

(a) This polynomial is the difference of two cubes because x^3 is the cube of x and 125 is the cube of 5. Therefore, we can factor the polynomial as follows.

$$\begin{aligned}x^3 - 125 &= x^3 - 5^3 && \text{Write as difference of two cubes}\\ &= (x - 5)(x^2 + 5x + 5^2) && \text{Factored form}\\ &= (x - 5)(x^2 + 5x + 25) && \text{Simplify}\end{aligned}$$

(b) This polynomial is the sum of two cubes because $8y^3$ is the cube of $2y$ and 1 is its own cube. Therefore, we can factor the polynomial as follows.

$8y^3 + 1 = (2y)^3 + 1^3$	Write as sum of two cubes
$= (2y + 1)((2y)^2 - (2y)(1) + 1^2)$	Factored form
$= (2y + 1)(4y^2 - 2y + 1)$	Simplify ■

It is easy to make arithmetic errors when applying the formulas for factoring the sum or difference of two cubes. When you use these formulas, we suggest that you check your work by multiplying the two factors. For instance, we can check the factors given in Example 9 as follows.

$$\begin{array}{rrrr} & x^2 + & 5x + & 25 \\ \times & & x - & 5 \\ \hline & -5x^2 & - 25x & - 125 \\ x^3 + & 5x^2 & + 25x & \\ \hline x^3 & & & - 125 \end{array} \qquad \begin{array}{rrrr} & 4y^2 & - 2y & + 1 \\ \times & & 2y & + 1 \\ \hline & 4y^2 & - 2y & + 1 \\ 8y^3 - & 4y^2 & + 2y & \\ \hline 8y^3 & & & + 1 \end{array}$$

Factoring Completely

Sometimes the difference of two squares can be hidden by the presence of a common monomial factor. Remember that with *all* factoring techniques, we should first remove any common monomial factors. This is demonstrated in Example 10.

EXAMPLE 10 ■ **Removing a Common Monomial Factor First**

Factor the following polynomial.

$$125x^2 - 80$$

Solution

The polynomial $125x^2 - 80$ has a common monomial factor of 5. After removing this factor, we see that the remaining polynomial is the difference of two squares.

$125x^2 - 80 = 5(25x^2 - 16)$	Remove common monomial factor
$= 5((5x)^2 - 4^2)$	Write as difference of two squares
$= 5(5x + 4)(5x - 4)$	Factored form ■

The polynomial in Example 10 is said to be **completely factored** because none of its factors can be factored further using integer coefficients.

EXAMPLE 11 ■ Removing a Common Monomial Factor First

Factor the following polynomial completely.

$$3x^3 - 81$$

Solution

$3x^3 - 81 = 3(x^3 - 27)$	Remove common monomial factor
$= 3(x^3 - 3^3)$	Write as difference of two cubes
$= 3(x - 3)(x^2 + 3x + 9)$	Factored form ■

To completely factor a polynomial, you should always check to see whether the factors obtained might themselves be factorable. That is, can any of the factors be factored? For instance, after factoring the polynomial $x^4 - 16$ once as the difference of two squares

$$x^4 - 16 = (x^2)^2 - 4^2 = (x^2 + 4)(x^2 - 4)$$

we see that the second factor is itself the difference of two squares. Thus, to factor the polynomial *completely*, we must continue the factoring process as follows.

$$x^4 - 16 = (x^2 + 4)(x^2 - 4) = (x^2 + 4)(x + 2)(x - 2)$$

Other instances of "repeated factoring" are given in the next example.

EXAMPLE 12 ■ Factoring Completely

Factor the following polynomials completely.

(a) $x^4 - y^4$ (b) $81m^4 - 1$

Solution

(a) Recognizing $x^4 - y^4$ as a difference of two squares, we write

$$x^4 - y^4 = (x^2)^2 - (y^2)^2 = (x^2 + y^2)(x^2 - y^2).$$

Note that the second factor $(x^2 - y^2)$ is itself a difference of two squares and we therefore obtain

$$x^4 - y^4 = (x^2 + y^2)(x^2 - y^2) = (x^2 + y^2)(x + y)(x - y).$$

(b) Recognizing $81m^4 - 1$ as a difference of two squares, we write

$$81m^4 - 1 = (9m^2)^2 - (1)^2 = (9m^2 + 1)(9m^2 - 1).$$

Now, since the second factor $(9m^2 - 1)$ is itself a difference of two squares, we continue the factoring process as follows.

$$81m^4 - 1 = (9m^2 + 1)(9m^2 - 1) = (9m^2 + 1)(3m + 1)(3m - 1)$$ ■

Note in Example 12 that we did not attempt to factor further the *sum of two squares*. The reason for this is that a second-degree polynomial that is the sum of two squares is *not factorable*. For instance, the second-degree polynomials $x^2 + 4$ and $4x^2 + 9$ are not factorable.* Some other polynomials that are not factorable are $x^2 - 2$ and $2x^2 - 9$.

DISCUSSION PROBLEM ■ Factoring Higher-Powered Polynomials

Find a formula for factoring the polynomial $u^6 - v^6$. Then use your result to factor the following polynomials.

(a) $x^6 - 1$ (b) $x^6 - 64$ ■

Warm-Up

The following warm-up exercises involve skills that were covered in earlier sections. You will use these skills in the exercise set for this section.

In Exercises 1–10, perform the indicated multiplication.

1. $12x(2x - 3)$ **2.** $7y(4 - 3y)$

3. $-6x(10 - 7x)$ **4.** $-2y(y + 1)$

5. $t(t^2 + 1) - t(t^2 - 1)$ **6.** $2z(z + 5) - 7(z + 5)$

7. $(11 - x)(11 + x)$ **8.** $(6r + 5s)(6r - 5s)$

9. $(x - 2)(x^2 + 2x + 4)$ **10.** $(x + 1)(x^2 - x + 1)$

3.3 EXERCISES

In Exercises 1–10, find the greatest common factor of the given expressions.

1. 48, 90 **2.** 36, 150, 100 **3.** $3x^2$, $12x$

4. $27x^4$, $18x^3$ **5.** $30z^2$, $-12z^3$ **6.** $-45y$, $150y^3$

7. $28b^2$, $14b^3$, $42b^5$ **8.** $16x^2y$, $84xy^2$, $36x^2y^2$ **9.** $42(x + 8)^2$, $63(x + 8)^3$

10. $66(3 - y)$, $44(3 - y)^2$

*In this text when we say that a polynomial is not factorable, we mean that it is not factorable *using integer coefficients*.

In Exercises 11–26, factor by removing the greatest common monomial factor. (Some of the polynomials have no common monomial factor other than 1 and -1.)

11. $8z - 8$ **12.** $5x + 5$ **13.** $4u + 10$ **14.** $-15t - 10$

15. $24x^2 - 18$ **16.** $14z^3 + 21$ **17.** $2x^2 + x$ **18.** $-a^3 - 4a$

19. $21u^2 - 14u$ **20.** $36y^4 + 24y^2$ **21.** $3x^2y^2 - 15y$ **22.** $4uv + 6u^2v^2$

23. $28x^2 + 16x - 8$ **24.** $9 - 27y - 15y^2$ **25.** $14x^4 + 21x^3 + 9x^2$ **26.** $17x^5y^3 - xy^2 + 34y^2$

In Exercises 27–32, factor the given polynomial by factoring out the common binomial factor.

27. $2y(y - 3) + 5(y - 3)$ **28.** $7t(s + 9) - 6(s + 9)$

29. $a(a + 6) - a^2(a + 6)$ **30.** $y^4(y - 12)^2 + y^2(y - 12)^3$

31. $(10 + b)(c + 7) - (10 + b)(c + 7)^2$ **32.** $(5x + y)(x - y) - 5x(x - y)$

In Exercises 33–38, factor a negative real number from the polynomial and then write the polynomial factor in standard form.

33. $10 - x$ **34.** $32 - x^4$ **35.** $7 - 14x$

36. $15 - 5x$ **37.** $5 + 4x - x^2$ **38.** $12x - 6x^2 - 18$

In Exercises 39–44, factor the expression as indicated.

39. $\frac{3}{2}x + \frac{5}{4} = \frac{1}{4}(\quad)$ **40.** $\frac{1}{3}x - \frac{5}{6} = \frac{1}{6}(\quad)$ **41.** $2y - \frac{3}{5} = \frac{1}{5}(\quad)$

42. $3z + \frac{3}{8} = \frac{1}{8}(\quad)$ **43.** $\frac{5}{8}x + \frac{5}{16}y = \frac{5}{16}(\quad)$ **44.** $\frac{7}{12}u - \frac{21}{8}v = \frac{7}{24}(\quad)$

In Exercises 45–54, factor the expression by grouping.

45. $x^2 + 25x + x + 25$ **46.** $x^2 - 7x + x - 7$ **47.** $y^2 - 6y + 2y - 12$

48. $y^2 + 3y + 4y + 12$ **49.** $x^3 + 2x^2 + x + 2$ **50.** $t^3 - 11t^2 + t - 11$

51. $a^3 - 4a^2 + 2a - 8$ **52.** $3s^3 + 6s^2 + 5s + 10$ **53.** $z^4 + 3z^3 - 2z - 6$

54. $4u^4 - 2u^3 - 6u + 3$

In Exercises 55–64, factor the given difference of two squares.

55. $x^2 - 64$ **56.** $y^2 - 144$ **57.** $16y^2 - 9$

58. $9z^2 - 25$ **59.** $225 - 9y^2$ **60.** $625 - 49x^2$

61. $u^2 - \frac{1}{16}$ **62.** $v^2 - \frac{9}{25}$ **63.** $81 - (z + 5)^2$

64. $(x - 3)^2 - 4$

In Exercises 65–72, factor the given sum or difference of cubes.

65. $x^3 - 8$

66. $t^3 - 27$

67. $y^3 + 64$

68. $z^3 + 125$

69. $8t^3 - 27$

70. $27s^3 + 64$

71. $27u^3 + 1$

72. $64v^3 - 125$

In Exercises 73–78, factor the algebraic expression completely.

73. $8 - 50x^2$

74. $a^3 - 16a$

75. $y^4 - 81$

76. $u^4 - 256$

77. $2x^3 - 54$

78. $5y^3 - 625$

In Exercises 79 and 80, evaluate the given quantity mentally using the following example as a model.

$$48 \cdot 52 = (50 - 2)(50 + 2) = 50^2 - 2^2 = 2{,}496$$

79. $79 \cdot 81$

80. $18 \cdot 22$

In Exercises 81 and 82, factor the given expression. (Assume $n > 0$.)

81. $4x^{2n} - 25$

82. $81 - 16y^{4n}$

In Exercises 83 and 84, show all of the different groupings that can be used to completely factor the given polynomial. Carry out the various factorizations to show that they yield the same result.

83. $3x^3 + 4x^2 - 3x - 4$

84. $6x^3 - 8x^2 + 9x - 12$

85. *Revenue and Price* The revenue for selling x units of a product at a price of p dollars per unit is given by xp. For a particular commodity, the revenue is given by

$$R = 800x - 0.25x^2.$$

Factor the expression for the revenue and determine an expression that gives the price in terms of x.

86. *Revenue and Price* The revenue for selling x units of a product at a price of p dollars per unit is given by xp. For a particular commodity, the revenue is given by

$$R = 1{,}000x - 0.4x^2.$$

Factor the expression for the revenue and determine an expression that gives the price in terms of x.

87. *Width of a Rectangle* The area of a rectangle of length l is given by $45l - l^2$. Factor this expression to determine the width of the rectangle.

88. *Length of a Rectangle* The area of a rectangle of width w is given by $32w - w^2$. Factor this expression to determine the length of the rectangle.

89. *Simple Interest* The total amount of money from a principal of P invested at r percent simple interest for t years is given by $P + Prt$. Factor this expression.

90. *Chemical Reaction* The rate of change of a chemical reaction is given by

$$kQx - kx^2$$

where Q is the amount of the original substance, x is the amount of the substance formed, and k is a constant of proportionality. Factor the expression for this rate of change.

91. *Product Design* A washer on the drive train of a car has an inside radius of r centimeters and an outside radius of R centimeters (see figure). Find the area of one of the flat surfaces of the washer and express the area in factored form.

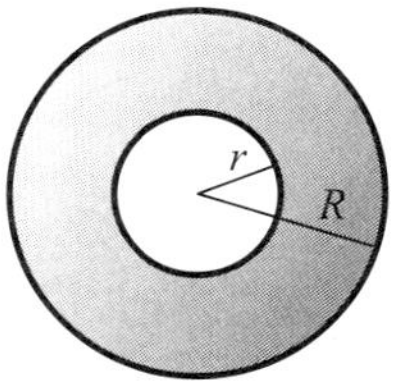

SECTION 3.4

Factoring Trinomials

Perfect Square Trinomials • *Factoring Trinomials of the Form $x^2 + bx + c$* • *Factoring Trinomials of the Form $ax^2 + bx + c$* • *Factoring Trinomials by Grouping (Optional)* • *Summary of Factoring*

Perfect Square Trinomials

A **perfect square trinomial** is the square of a binomial. For instance,

$$x^2 + 6x + 9 = (x + 3)^2$$

is the square of the binomial $(x + 3)$, and

$$4x^2 - 20x + 25 = (2x - 5)^2$$

is the square of the binomial $(2x - 5)$. Perfect square trinomials come in two forms—one in which the middle term is positive and the other in which the middle term is negative.

Perfect Square Trinomials

Let u and v represent real numbers, variables, or algebraic expressions. Then the following perfect square trinomials can be factored as indicated.

1. $u^2 + 2uv + v^2 = (u + v)^2$ (Same sign)
2. $u^2 - 2uv + v^2 = (u - v)^2$ (Same sign)

To recognize a perfect square trinomial, remember that the first and last terms must be perfect squares and positive, and the middle term must be twice the product of u and v. (Note that the middle term can be positive or negative.)

EXAMPLE 1 ■ Factoring Perfect Square Trinomials

(a) $x^2 - 4x + 4 = x^2 - 2(2x) + 2^2 = (x - 2)^2$

(b) $16y^2 + 24y + 9 = (4y)^2 + 2(4y)(3) + 3^2 = (4y + 3)^2$

(c) $9x^2 - 30xy + 25y^2 = (3x)^2 - 2(3x)(5y) + (5y)^2 = (3x - 5y)^2$ ■

EXAMPLE 2 ■ Removing a Common Monomial Factor First

Factor the following trinomials.

(a) $3x^2 - 30x + 75$ (b) $16y^3 + 80y^2 + 100y$

Solution

(a) This polynomial has a common monomial factor of 3. After removing this factor we find that the remaining polynomial is a perfect square trinomial.

$$\begin{aligned} 3x^2 - 30x + 75 &= 3(x^2 - 10x + 25) && \text{Remove common monomial factor} \\ &= 3(x - 5)^2 && \text{Factor as perfect square trinomial} \end{aligned}$$

(b) This polynomial has a common monomial factor of $4y$. After removing this factor we find that the remaining polynomial is a perfect square trinomial.

$$\begin{aligned} 16y^3 + 80y^2 + 100y &= 4y(4y^2 + 20y + 25) && \text{Remove common monomial factor} \\ &= 4y(2y + 5)^2 && \text{Factor as perfect square trinomial} \end{aligned}$$

■

Factoring Trinomials of the Form $x^2 + bx + c$

Try covering the factored forms in the left-hand column below. Can you determine the factored forms from the trinomial forms?

Factored Form	F O I L	*Trinomial Form*
$(x - 1)(x + 4) =$	$x^2 + 4x - x - 4$	$= x^2 + 3x - 4$
$(x - 3)(x - 2) =$	$x^2 - 2x - 3x + 6$	$= x^2 - 5x + 6$
$(3x + 5)(x + 1) =$	$3x^2 + 3x + 5x + 5$	$= 3x^2 + 8x + 5$

Our goal here is to factor trinomials of the form $x^2 + bx + c$. To begin, let's consider the following factorization.

$$x^2 + bx + c = (x + m)(x + n)$$

By multiplying the right-hand side, we obtain the following.

$$(x + m)(x + n) = x^2 + nx + mx + mn = x^2 + \underbrace{(m + n)}_{\text{Sum of terms}}x + \underbrace{mn}_{\text{Product of terms}}$$

$$= x^2 + b\,x + c$$

Thus, to *factor* a trinomial of the form $x^2 + bx + c$ into a product of two binomials we must find factors of c whose sum is b. Note how this strategy is used in the following examples.

EXAMPLE 3 ■ **Factoring Trinomials**

Factor the following trinomials.

(a) $x^2 - 2x - 8$ (b) $x^2 - 5x + 6$

Solution

(a) For this trinomial, we have $x^2 + bx + c = x^2 - 2x - 8$. Thus, $b = -2$ and $c = -8$, and we need to find two numbers whose product is -8 and whose sum is -2. After considering the various possibilities, we see that the numbers 2 and -4 work. Therefore, the factorization is as follows.

$$x^2 - 2x - 8 = (x + \quad)(x + \quad)$$

Think: We need factors of -8 whose sum is -2

$$= (x + 2)(x - 4)$$

$(2)(-4) = -8, \quad 2 - 4 = -2$

(b) For this trinomial, we have $x^2 + bx + c = x^2 - 5x + 6$. Thus, $b = -5$ and $c = 6$, and we need to find two numbers whose product is 6 and whose sum is -5. After considering the various possibilities, we see that the numbers -2 and -3 work. Therefore, the factorization is as follows.

$$x^2 - 5x + 6 = (x + \quad)(x + \quad)$$

Think: We need factors of 6 whose sum is -5

$$= (x - 2)(x - 3)$$

$(-2)(-3) = 6, \quad -2 - 3 = -5$

■

NOTE: When the constant term of the trinomial is positive, its factors must have *like* signs; otherwise, its factors have *unlike* signs.

When factoring a trinomial of the form $x^2 + bx + c$, if you have trouble finding two factors of c whose sum is b, it may be helpful to make a list of all the distinct pairs of factors and then choose the appropriate one from the list. For instance, consider the trinomial

$$x^2 - 5x - 24.$$

For this trinomial, we have $c = -24$ and $b = -5$. Thus, we need to find two factors of -24 whose sum is -5, as follows.

Factors of -24	*Sum of Factors*	
$(1)(-24)$	$1 - 24 = -23$	
$(-1)(24)$	$-1 + 24 = 23$	
$(2)(-12)$	$2 - 12 = -10$	
$(-2)(12)$	$-2 + 12 = 10$	
$(3)(-8)$	$3 - 8 = -5$	⇦ Correct factorization
$(-3)(8)$	$-3 + 8 = 5$	
$(4)(-6)$	$4 - 6 = -2$	
$(-4)(6)$	$-4 + 6 = 2$	

With experience, you will be able to *mentally* narrow this list down to only two or three possibilities whose sums could then be tested to determine the correct factorization.

EXAMPLE 4 ■ Factoring a Trinomial

Factor the following trinomial.

$$x^2 - 7x - 18$$

Solution

To factor this trinomial, we need to find two factors of -18 whose sum is -7. After testing some possibilities, we see that 2 and -9 work. Thus, the factorization is as follows.

$$x^2 - 7x - 18 = (x + \square)(x + \square)$$ *Think:* We need factors of -18 whose sum is -7

$$= (x + 2)(x - 9)$$ $(2)(-9) = -18, \quad 2 - 9 = -7$ ■

Applications of algebra sometimes involve trinomials that have a common monomial factor. To completely factor such trinomials, we first factor out the common monomial factor. Then we try to factor the resulting trinomial by the methods in this section. For instance, the trinomial

$$5x^2 - 5x - 30 = 5(x^2 - x - 6) = 5(x - 3)(x + 2)$$

is factored completely.

EXAMPLE 5 ■ Factoring Completely

Factor the following trinomial completely.

$$4x^3 - 8x^2 - 60x$$

Solution

This trinomial has a common monomial factor of $4x$. Thus, we start the factoring process by factoring $4x$ out of each term.

$$4x^3 - 8x^2 - 60x = 4x(x^2 - 2x - 15)$$ Take out common factor

$$= 4x(x + \square)(x + \square)$$ *Think:* Factors of -15 whose sum is -2

$$= 4x(x + 3)(x - 5)$$ $(3)(-5) = -15, \quad 3 - 5 = -2$ ■

Factoring Trinomials of the Form $ax^2 + bx + c$

To factor a trinomial whose leading coefficient is not 1, we use the following pattern.

Factors of a

$$ax^2 + bx + c = (\square\, x + \square)(\square\, x + \square)$$

Factors of c

The goal is to find a combination of factors of a and c so that the outer and inner products add up to the middle term bx. In the trinomial $6x^2 + 17x + 5$, for instance, we have $a = 6$, $c = 5$, and $b = 17$. After some experimentation, we find that the factorization is

$$(2x + 5)(3x + 1) = \overset{\text{F}}{6x^2} + \overset{\text{O}}{2x} + \overset{\text{I}}{15x} + \overset{\text{L}}{5} = 6x^2 + 17x + 5.$$

Note that the outer (O) and inner (I) products add up to $17x$.

EXAMPLE 6 ■ Factoring a Trinomial of the Form $ax^2 + bx + c$

Factor the following trinomial.

$$6x^2 + 5x - 4$$

Solution

For this trinomial, we have $ax^2 + bx + c = 6x^2 + 5x - 4$, which implies that $a = 6$, $c = -4$, and $b = 5$. The possible factors of 6 are (1)(6) and (2)(3), and the possible factors of -4 are $(-1)(4)$, $(1)(-4)$, and $(2)(-2)$. By trying the *many* different combinations of these factors, we obtain the following list.

$$(x + 1)(6x - 4) = 6x^2 + 2x - 4$$
$$(x - 1)(6x + 4) = 6x^2 - 2x - 4$$
$$(x + 4)(6x - 1) = 6x^2 + 23x - 4$$
$$(x - 4)(6x + 1) = 6x^2 - 23x - 4$$
$$(x + 2)(6x - 2) = 6x^2 + 10x - 4$$
$$(x - 2)(6x + 2) = 6x^2 - 10x - 4$$
$$(2x + 1)(3x - 4) = 6x^2 - 5x - 4$$
$$(2x - 1)(3x + 4) = 6x^2 + 5x - 4$$ ⇐ Correct factorization
$$(2x + 4)(3x - 1) = 6x^2 + 10x - 4$$
$$(2x - 4)(3x + 1) = 6x^2 - 10x - 4$$
$$(2x + 2)(3x - 2) = 6x^2 + 2x - 4$$
$$(2x - 2)(3x + 2) = 6x^2 - 2x - 4$$

Thus, we conclude that the correct factorization is

$$6x^2 + 5x - 4 = (2x - 1)(3x + 4).$$ ■

To help shorten the list of *possible* factorizations of a trinomial, we suggest the following guidelines.

Guidelines for Limiting Possible Factorizations of a Trinomial

When factoring a trinomial of the form $ax^2 + bx + c$, consider the following guidelines.

1. If the given trinomial has a *common monomial factor*, then you should remove the monomial factor before trying to find binomial factors. For instance, the trinomial $12x^2 + 10x - 8$ has a common factor of 2. By removing this common factor, we obtain $12x^2 + 10x - 8 = 2(6x^2 + 5x - 4)$.
2. Do not switch the signs of the factors of c unless the middle term is correct except in sign. In Example 6, after determining that $(x + 4)(6x - 1)$ is not the correct factorization, for instance, it is unnecessary to test $(x - 4)(6x + 1)$.
3. Do not use binomial factors that have a common monomial factor. Such a factor cannot be correct because the trinomial has no common monomial factor. (Any common monomial factor was already removed in Step 1.) For instance, in Example 6, it is unnecessary to test $(x + 1)(6x - 4) = 6x^2 + 2x - 4$ because the factor $(6x - 4)$ has a common factor of 2.

With these suggestions, we could shorten the list given in Example 6 to the following three possible factorizations.

$$(x + 4)(6x - 1) = 6x^2 + 23x - 4$$
$$(2x + 1)(3x - 4) = 6x^2 - 5x - 4$$
$$(2x - 1)(3x + 4) = 6x^2 + 5x - 4 \quad \text{Correct factorization}$$

Do you see why we can cut the list from 12 possible factorizations to only 3?

EXAMPLE 7 ■ **Factoring a Trinomial of the Form $ax^2 + bx + c$**

Factor the trinomial $2x^2 - x - 21$.

Solution

For this trinomial we have $a = 2$, which factors as (1)(2), and $c = -21$, which factors as (1)(−21), (−1)(21), (3)(−7), or (−7)(3). We test the possible factors as follows.

$$(2x + 1)(x - 21) = 2x^2 - 41x - 21$$
$$(2x + 21)(x - 1) = 2x^2 + 19x - 21$$
$$(2x + 3)(x - 7) = 2x^2 - 11x - 21$$
$$(2x + 7)(x - 3) = 2x^2 + x - 21 \quad \text{Middle term has incorrect sign}$$
$$(2x - 7)(x + 3) = 2x^2 - x - 21 \quad \text{Correct factorization}$$

Therefore, the correct factorization is

$$2x^2 - x - 21 = (2x - 7)(x + 3).$$

■

NOTE: Notice in Example 7 that if the middle term is correct except in sign, we need only change the signs of the factors of c.

EXAMPLE 8 ■ Factoring a Trinomial of the Form $ax^2 + bx + c$

Factor the following trinomial.

$$3x^2 + 11x + 10$$

Solution

For this trinomial we have $a = 3$, which factors as (1)(3), and $c = 10$, which factors as (1)(10) or (2)(5). We test the possible factors as follows.

$$(x + 10)(3x + 1) = 3x^2 + 31x + 10$$
$$(x + 1)(3x + 10) = 3x^2 + 13x + 10$$
$$(x + 5)(3x + 2) = 3x^2 + 17x + 10$$
$$(x + 2)(3x + 5) = 3x^2 + 11x + 10$$ ⇐ Correct factorization

Therefore, the correct factorization is

$$3x^2 + 11x + 10 = (x + 2)(3x + 5).$$ ■

We already mentioned that if a trinomial has a common monomial factor, then the common monomial factor should be removed first. The next two examples illustrate this technique.

EXAMPLE 9 ■ Factoring Completely

Factor the following trinomial completely.

$$8x^3 - 60x^2 + 28x$$

Solution

This trinomial has a common monomial factor of $4x$, so we remove this factor first.

$$8x^3 - 60x^2 + 28x = 4x(2x^2 - 15x + 7)$$

Now, for the new trinomial $2x^2 - 15x + 7$, we have $a = 2$ and $c = 7$. The possible factorizations of this trinomial are as follows.

$$(2x - 7)(x - 1) = 2x^2 - 9x + 7$$
$$(2x - 1)(x - 7) = 2x^2 - 15x + 7$$ ⇐ Correct factorization

Therefore, the complete factorization of the original trinomial is

$$8x^3 - 60x^2 + 28x = 4x(2x^2 - 15x + 7) = 4x(2x - 1)(x - 7).$$ ■

When factoring a trinomial with a negative leading coefficient, we suggest that you first factor -1 out of the trinomial. This technique is demonstrated in Example 10.

EXAMPLE 10 ■ Factoring a Trinomial with a Negative Leading Coefficient

Factor the following trinomial.

$$-3x^2 + 16x + 35$$

Solution

This trinomial has a negative leading coefficient, so we begin by factoring this out.

$$-3x^2 + 16x + 35 = (-1)(3x^2 - 16x - 35)$$

Now, for the new trinomial $3x^2 - 16x - 35$, we have $a = 3$ and $c = -35$. The possible factorizations of this trinomial are as follows.

$$\begin{aligned}
(3x - 1)(x + 35) &= 3x^2 + 104x - 35 \\
(3x - 35)(x + 1) &= 3x^2 - 32x - 35 \\
(3x - 7)(x + 5) &= 3x^2 + 8x - 35 \\
(3x - 5)(x + 7) &= 3x^2 + 16x - 35 && \text{Middle term has incorrect sign} \\
(3x + 5)(x - 7) &= 3x^2 - 16x - 35 && \text{Correct factorization}
\end{aligned}$$

Thus, the correct factorization is

$$-3x^2 + 16x + 35 = (-1)(3x + 5)(x - 7) = (3x + 5)(-x + 7).$$ ■

Not all trinomials are factorable using only integers. For instance, to factor $x^2 + 3x + 5$, we need factors of 5 that add up to 3. This is not possible since the only integer factors of 5 are 1 and 5, and their sum is not 3. We say that such a trinomial is **not factorable**. Try factoring $2x^2 - 3x + 2$ to see that it is also not factorable. Watch for other trinomials that are not factorable in the exercises for this section.

Factoring Trinomials by Grouping (Optional)

In this section you have seen that factoring a trinomial can involve quite a bit of trial and error. An alternative technique that some people like to use is factoring by grouping. For instance, suppose we rewrite the trinomial $2x^2 + x - 15$ as

$$2x^2 + x - 15 = 2x^2 + 6x - 5x - 15.$$

Then, by grouping the first two terms and the third and fourth terms, we could factor the trinomial as follows.

$$\begin{aligned}
2x^2 + x - 15 &= 2x^2 + (6x - 5x) - 15 && \text{Rewrite middle term} \\
&= (2x^2 + 6x) - (5x + 15) && \text{Group terms} \\
&= 2x(x + 3) - 5(x + 3) && \text{Factor groups} \\
&= (2x - 5)(x + 3) && \text{Distributive Property}
\end{aligned}$$

The key to this method of factoring is knowing how to rewrite the middle term. In general, *to factor a trinomial of the form $ax^2 + bx + c$ by grouping, we choose factors of the product ac that add up to b and use these factors to rewrite the middle term.* This technique is illustrated in Example 11.

EXAMPLE 11 ■ **Factoring a Trinomial by Grouping**

Use factoring by grouping to factor the following trinomial.

$$2x^2 + 5x - 3$$

Solution

In the trinomial $2x^2 + 5x - 3$, we have $a = 2$ and $c = -3$, which implies that the product ac is -6. Now, since -6 factors as $(6)(-1)$, and $6 - 1 = 5 = b$, we rewrite the middle term as $5x = 6x - x$. This produces the following.

$$\begin{aligned} 2x^2 + 5x - 3 &= 2x^2 + 6x - x - 3 && \text{Rewrite middle term} \\ &= (2x^2 + 6x) - (x + 3) && \text{Group terms} \\ &= 2x(x + 3) - (x + 3) && \text{Factor groups} \\ &= (x + 3)(2x - 1) && \text{Distributive Property} \end{aligned}$$

Therefore, the given trinomial factors as

$$2x^2 + 5x - 3 = (x + 3)(2x - 1).$$ ■

What do you think of this optional technique? Some people think that it is more efficient than the trial and error process, especially when the coefficients a and c have many factors.

Summary of Factoring

We have discussed the basic factoring techniques one at a time, but from this point on you must decide which technique to apply to any given problem situation. To assist you in this selection process, we provide the following guidelines.

Guidelines for Factoring Polynomials

1. Factor out any common factors.
2. Factor according to one of the special polynomial forms: difference of squares, sum or difference of cubes, or perfect square trinomials.
3. Factor trinomials, $ax^2 + bx + c$, using the methods for $a = 1$ or $a \neq 1$.
4. Factor by grouping—for polynomials with four terms.
5. Check to see if the factors themselves can be factored further.
6. Check the results by multiplying the factors.

EXAMPLE 12 ■ **Factoring Polynomials**

Factor the following polynomials.

(a) $3x^2 - 108$ (b) $4x^3 - 32x^2 + 64x$

(c) $6x^3 + 27x^2 - 15x$ (d) $x^3 - 3x^2 - 4x + 12$

Solution

(a) $3x^2 - 108 = 3(x^2 - 36)$	Remove common factor
$= 3(x + 6)(x - 6)$	Difference of two squares
(b) $4x^3 - 32x^2 + 64x = 4x(x^2 - 8x + 16)$	Remove common factor
$= 4x(x - 4)^2$	Perfect square trinomial
(c) $6x^3 + 27x^2 - 15x = 3x(2x^2 + 9x - 5)$	Remove common factor
$= 3x(2x - 1)(x + 5)$	Factor trinomial
(d) $x^3 - 3x^2 - 4x + 12 = (x^3 - 3x^2) + (-4x + 12)$	Group terms
$= x^2(x - 3) - 4(x - 3)$	Remove common factors
$= (x - 3)(x^2 - 4)$	Distributive Property
$= (x - 3)(x + 2)(x - 2)$	Difference of two squares

DISCUSSION PROBLEM You Be the Instructor

Suppose you are making up a test on this chapter for your algebra class. The test has 20 problems and you decide that 8 of the problems will come from Sections 3.3 and 3.4 (on factoring). Create 8 factoring problems that you think would represent a fair test of your class's factoring skills.

Warm-Up

The following warm-up exercises involve skills that were covered in earlier sections. You will use these skills in the exercise set for this section.

In Exercises 1–6, perform the indicated multiplication.

1. $(x + 1)^2$

2. $(x - 5)^2$

3. $(2 - y)^2$

4. $(2 - y)(3 + 2y)$

5. $(5z + 3)(2z - 7)$

6. $t(t - 4)(2t + 3)$

In Exercises 7–10, factor the expression completely.

7. $3x^2 - 21x$

8. $-x^3 + 3x^2 - x + 3$

9. $4t^2 - 169$

10. $y^3 - 64$

3.4 EXERCISES

In Exercises 1–10, factor the perfect square trinomial.

1. $x^2 + 4x + 4$
2. $z^2 + 6z + 9$
3. $a^2 - 12a + 36$
4. $y^2 - 14y + 49$
5. $25y^2 - 10y + 1$
6. $4z^2 + 28z + 49$
7. $9b^2 + 12b + 4$
8. $4x^2 - 4x + 1$
9. $u^2 + 8uv + 16v^2$
10. $4y^2 + 20yz + 25z^2$

In Exercises 11–14, find two real numbers b such that the algebraic expression is a perfect square trinomial.

11. $x^2 + bx + 81$
12. $x^2 + bx + \frac{9}{16}$
13. $4x^2 + bx + 9$
14. $16x^2 + bxy + 25y^2$

In Exercises 15–18, find a real number c such that the algebraic expression is a perfect square trinomial.

15. $x^2 + 8x + c$
16. $x^2 + 12x + c$
17. $y^2 - 6y + c$
18. $z^2 - 20z + c$

In Exercises 19 and 20, fill in the missing number.

19. $x^2 + 12x + 50 = (x + 6)^2 + \square$
20. $x^2 + 10x + 22 = (x + 5)^2 + \square$

In Exercises 21–26, find the missing factor.

21. $x^2 + 5x + 4 = (x + 4)(\quad)$
22. $a^2 + 2a - 8 = (a + 4)(\quad)$
23. $y^2 - y - 20 = (y + 4)(\quad)$
24. $y^2 + 6y + 8 = (y + 4)(\quad)$
25. $z^2 - 6z + 8 = (z - 4)(\quad)$
26. $z^2 + 2z - 24 = (z - 4)(\quad)$

In Exercises 27–40, factor the expression.

27. $x^2 + 4x + 3$
28. $x^2 + 7x + 10$
29. $x^2 - 5x + 6$
30. $x^2 - 10x + 24$
31. $y^2 + 7y - 30$
32. $m^2 - 3m - 10$
33. $t^2 - 4t - 21$
34. $x^2 + 4x - 12$
35. $x^2 - 20x + 96$
36. $y^2 - 35y + 300$
37. $x^2 + 30x + 216$
38. $u^2 + 5uv + 6v^2$
39. $x^2 - 2xy - 35y^2$
40. $a^2 - 21ab + 110b^2$

In Exercises 41–46, find all integers b such that the trinomial can be factored.

41. $x^2 + bx + 18$
42. $x^2 + bx + 14$
43. $x^2 + bx - 21$
44. $x^2 + bx - 7$
45. $x^2 + bx + 35$
46. $x^2 + bx - 38$

In Exercises 47–50, find *two* integers c such that the trinomial can be factored.

47. $x^2 + 6x + c$
48. $x^2 + 9x + c$
49. $x^2 - 3x + c$
50. $x^2 - 12x + c$

In Exercises 51–56, find the missing factor.

51. $5x^2 + 18x + 9 = (x + 3)(\quad)$
52. $5x^2 + 19x + 12 = (x + 3)(\quad)$
53. $5a^2 + 12a - 9 = (a + 3)(\quad)$
54. $5c^2 + 11c - 12 = (c + 3)(\quad)$
55. $2y^2 - 3y - 27 = (y + 3)(\quad)$
56. $3y^2 - y - 30 = (y + 3)(\quad)$

In Exercises 57–66, factor the algebraic expression, if possible. Some of the expressions are not factorable using integer coefficients.

57. $3x^2 + 4x + 1$
58. $5x^2 + 7x + 2$
59. $2x^2 - 9x + 9$
60. $2t^2 - 13t + 20$
61. $6b^2 + 19b - 7$
62. $10x^2 - 24x - 18$
63. $24x^2 - 14xy - 3y^2$
64. $20x^2 + x - 12$
65. $3y^2 + 4y + 12$
66. $10x^2 + 9xy - 9y^2$

In Exercises 67–70, factor the given trinomial.

67. $-2x^2 - x + 6$
68. $-6x^2 + 5x - 6$
69. $1 - 11x - 60x^2$
70. $2 + 5x - 12x^2$

In Exercises 71–76, factor the trinomial by grouping. (See Example 11.)

71. $3x^2 + 10x + 8$
72. $2x^2 + 9x + 9$
73. $6x^2 + x - 2$
74. $6x^2 - x - 15$
75. $15x^2 - 11x + 2$
76. $12x^2 - 14x + 1$

In Exercises 77–86, factor the algebraic expression completely.

77. $3x^4 - 12x^3$
78. $20y^2 - 45$
79. $10t^3 + 2t^2 - 36t$
80. $16z^2 - 56z + 49$
81. $4x(3x - 2) + (3x - 2)^2$
82. $6 - x - 6x^2 + x^3$
83. $36 - (z + 3)^2$
84. $3t^3 - 24$
85. $54x^3 - 2$
86. $v^3 + 3v^2 + 5v$

In Exercises 87 and 88, evaluate the given quantity mentally using the following as a model.

$$29^2 = (30 - 1)^2 = 30^2 - 2(30)(1) + 1^2 = 900 - 60 + 1 = 841$$

87. 52^2
88. 39^2

In Exercises 89 and 90, factor the trinomial and show the result using the concept of area as illustrated in the accompanying figure.

$x^2 + 3x + 2 = (x + 1)(x + 2)$

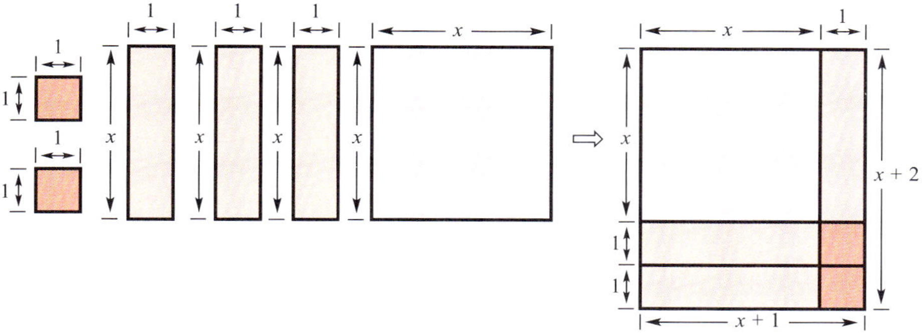

89. $x^2 + 4x + 3$

90. $x^2 + 5x + 4$

SECTION **3.5**

Factoring and Problem Solving

The Zero-Factor Property • Solving Equations by Factoring • Applications

The Zero-Factor Property

We have spent the first part of this chapter developing skills for *rewriting* (simplifying and factoring) polynomials. Now we are ready to combine these skills with the following **Zero-Factor Property** to *solve* equations. The Zero-Factor Property indicates that if the product of two factors is zero, then one (or both) of the factors must also be zero.

Zero-Factor Property

Let u and v represent real numbers, variables, or algebraic expressions. If u and v are factors such that

$$uv = 0$$

then $u = 0$ or $v = 0$. This property applies to three or more factors as well.

NOTE: The Zero-Factor Property is just another way of saying that the only way the product of two (or more) real numbers can be zero is if one (or more) of the real numbers is zero.

The Zero-Factor Property is the primary property we use for solving equations in algebra. For instance, to solve the equation

$$(x - 2)(x + 3) = 0$$

we use the Zero-Factor Property to conclude that either $(x - 2)$ or $(x + 3)$ must be zero. Setting the first factor equal to zero implies that $x = 2$ is a solution. That is,

$$x - 2 = 0 \implies x = 2.$$

Similarly, setting the second factor equal to zero implies that $x = -3$ is a solution. That is,

$$x + 3 = 0 \implies x = -3.$$

Thus, the equation $(x - 2)(x + 3) = 0$ has exactly two solutions: 2 and -3.

Solving Equations by Factoring

In each of the examples in this section, note how we combine our factoring skills with the Zero-Factor Property to solve equations.

EXAMPLE 1 ■ Using Factoring to Solve an Equation

Solve the following polynomial equation.

$$x^2 - x - 12 = 0$$

Solution

The strategy here is to first check to see that the right side of the equation is zero. Next, we factor the left side of the equation. Finally, we apply the Zero-Factor Property to find the solutions.

$x^2 - x - 12 = 0$	Given equation in standard form
$(x + 3)(x - 4) = 0$	Factor left side of equation
$x + 3 = 0 \implies x = -3$	Set first factor equal to 0
$x - 4 = 0 \implies x = 4$	Set second factor equal to 0

Thus, the given equation has two solutions: -3 and 4.

Check First Solution

$x^2 - x - 12 = 0$	Given equation
$(-3)^2 - (-3) - 12 \stackrel{?}{=} 0$	Replace x by -3
$9 + 3 - 12 \stackrel{?}{=} 0$	
$0 = 0$	Solution checks

Check Second Solution

$x^2 - x - 12 = 0$	Given equation
$4^2 - 4 - 12 \stackrel{?}{=} 0$	Replace x by 4
$16 - 4 - 12 \stackrel{?}{=} 0$	
$0 = 0$	Solution checks ■

NOTE: Factoring and the Zero-Factor Property allow us to solve a polynomial equation of degree 2 by converting it into two *linear* equations that we already know how to solve. This is a common strategy of algebra—to break down a given problem into simpler parts, each solved by previously learned methods.

To use the Zero-Factor Property, a polynomial equation *must* be written in the **standard form**. That is, the polynomial must be on the left side of the equation and zero must be the only term on the right side of the equation. For instance, to write the equation $x^2 - 3x = 18$ in standard form, we must subtract 18 from both sides of the equation, as follows.

$x^2 - 3x = 18$	Given equation (nonstandard form)
$x^2 - 3x - 18 = 18 - 18$	Subtract 18 from both sides
$x^2 - 3x - 18 = 0$	Standard form

EXAMPLE 2 ■ Solving a Polynomial Equation

Solve the following polynomial equation.

$$3x^2 + 5x = 12$$

Solution

$3x^2 + 5x = 12$	Given equation
$3x^2 + 5x - 12 = 0$	Write in standard form
$(3x - 4)(x + 3) = 0$	Factor left side of equation
$3x - 4 = 0 \Rightarrow x = \frac{4}{3}$	Set first factor equal to 0
$x + 3 = 0 \Rightarrow x = -3$	Set second factor equal to 0

Therefore, the solutions are $\frac{4}{3}$ and -3. Check these solutions in the original equation. ■

In Examples 1 and 2, the given equations each involved a second-degree polynomial, and each had *two different* solutions. Sometimes we encounter a second-degree polynomial equation that has only one (repeated) solution. This occurs when the left side of the equation is a perfect square trinomial, as shown in Example 3.

EXAMPLE 3 ■ A Polynomial Equation with a Repeated Solution

Solve the following polynomial equation.

$$x^2 - 6x + 11 = 2$$

Solution

$x^2 - 6x + 11 = 2$	Given equation
$x^2 - 6x + 9 = 0$	Write in standard form
$(x - 3)^2 = 0$	Factor
$x - 3 = 0$	Set factor equal to 0
$x = 3$	Solution is 3

Note that even though the left side of this equation has two factors, the two factors are the same. Thus, we conclude that the only solution of the equation is 3. Check this solution in the original equation. ■

In the next example, we look at an equation that has three different solutions. It is no coincidence that the equation in the example is of degree 3. In general, a polynomial equation can have *at most* as many solutions as its degree. For instance, a second-degree equation can have zero, one, or two real solutions, but it cannot have three or more solutions.

EXAMPLE 4 ■ Solving a Polynomial Equation with Three Factors

Solve the following polynomial equation.

$$3x^3 = 15x^2 + 18x$$

Solution

$3x^3 = 15x^2 + 18x$		Given equation
$3x^3 - 15x^2 - 18x = 0$		Standard form
$3x(x^2 - 5x - 6) = 0$		Remove common monomial factor
$3x(x - 6)(x + 1) = 0$		Factor completely
$x = 0$ ⇨	$x = 0$	Set first factor equal to 0
$x - 6 = 0$ ⇨	$x = 6$	Set second factor equal to 0
$x + 1 = 0$ ⇨	$x = -1$	Set third factor equal to 0

Note that we set only the **variable** factors equal to zero. (Setting the constant factor 3 equal to zero would yield no solution.) Therefore, the given equation has three solutions: 0, 6, and -1. Check these solutions in the original equation. ■

Be sure you see that the Zero-Factor Property can only be applied to a product that is equal to *zero*. For instance, we cannot conclude from the equation

$$x(x - 1) = 6$$

that $x = 6$ and $x - 1 = 6$ yield solutions. Instead, we must first write the equation in standard form and then factor the left side, as follows.

$$x^2 - x - 6 = 0 \quad \Rightarrow \quad (x - 3)(x + 2) = 0$$

Now, from the factored form, we see that the solutions are 3 and -2.

EXAMPLE 5 ■ **Solving a Polynomial Equation**

Solve the following polynomial equation.

$$(x + 2)(x + 4) = 3$$

Solution

To write this equation in standard form, we must first multiply the factors on the left side.

$(x + 2)(x + 4) = 3$		Given equation
$x^2 + 6x + 8 = 3$		Multiply factors
$x^2 + 6x + 5 = 0$		Standard form
$(x + 1)(x + 5) = 0$		Factor left side of equation
$x + 1 = 0$	$\Rightarrow$ $x = -1$	Set first factor equal to 0
$x + 5 = 0$	$\Rightarrow$ $x = -5$	Set second factor equal to 0

Therefore, the equation has two solutions: -1 and -5. Check these solutions in the original equation. ■

Some polynomial equations do not have solutions that are real numbers. For instance, there is no real number that satisfies the equation $x^2 + 4 = 0$. To see why, we can rewrite the equation as $x^2 = -4$. In this form, it is clear that there is no real number that can be multiplied by itself to produce -4. We will say more about this type of equation in Chapter 5.

Applications

In the last part of this section, we look at applications that involve polynomial equations. The general approach to such problems is similar to that studied in Chapter 2.

EXAMPLE 6 ■ An Application of a Polynomial Equation

A rectangular room has an area of 192 square feet. The length of the room is 4 feet more than its width. Find the dimensions of the room.

Solution

FIGURE 3.3

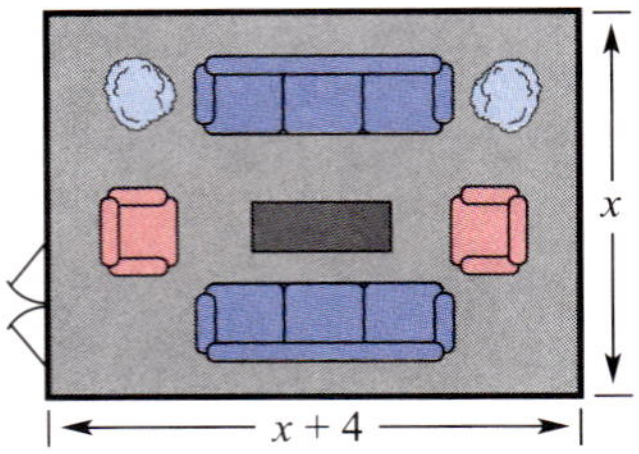

To begin, we make a sketch of the room, as shown in Figure 3.3. Note that we have labeled the width of the room as x, and the length of the room as $x + 4$, because the length is 4 feet more than the width.

Verbal model: Length · Width = Area

Labels: Width $= x$ (in feet)
Length $= x + 4$ (in feet)
Area $= 192$ square feet

Equation:

$$x(x + 4) = 192$$
$$x^2 + 4x - 192 = 0$$
$$(x + 16)(x - 12) = 0$$
$$x = -16 \text{ or } 12$$

Note that, mathematically, the equation has two solutions: -16 and 12. However, for this application, the negative solution makes no sense, so we discard it and keep the positive solution. Thus, we conclude that the width of the room is 12 feet, and the length of the room is 16 feet.

Check

To check this solution, go back to the original statement of the problem. Note that a length of 16 feet is 4 feet more than a width of 12 feet. Moreover, a rectangular room with dimensions 16 feet by 12 feet has an area of 192 square feet. Thus, the solution checks. ■

EXAMPLE 7 ■ An Application of a Polynomial Equation

The product of two consecutive positive integers is 132. What are the integers?

Solution

Verbal model: First integer · Second integer = 132

Labels: First integer $= n$
Second integer $= n + 1$

Equation:

$$n(n+1) = 132$$
$$n^2 + n - 132 = 0$$
$$(n+12)(n-11) = 0$$
$$n = -12 \text{ or } 11$$

Since the original statement of the problem required that the integers be positive, we discard -12 as a solution and choose $n = 11$. Thus, the two integers are $n = 11$ and $n + 1 = 12$. Check this solution in the original statement of the problem. ■

EXAMPLE 8 ■ **An Application of a Polynomial Equation**

A roadside sign in the shape of a triangle must have a base that is 4 feet less than twice its height. A local zoning ordinance restricts the area of signs to a maximum of 24 square feet. Find the base and height of the largest triangular sign that meets the zoning ordinance.

Solution

Since the base is to be 4 feet less than twice the height of the triangle, we let h be the height (in feet) and label the base as $2h - 4$, as shown in Figure 3.4.

FIGURE 3.4

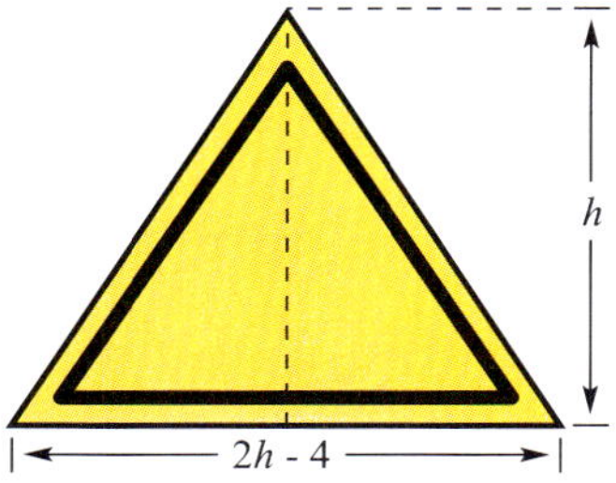

Verbal model: $\frac{1}{2}$ · Base · Height = Area

Labels: Height $= h$ (in feet)
Base $= 2h - 4$ (in feet)
Area $= 24$ square feet

Equation:

$$\frac{1}{2}(2h-4)(h) = 24$$
$$h^2 - 2h = 24$$
$$h^2 - 2h - 24 = 0$$
$$(h+4)(h-6) = 0$$
$$h = -4 \text{ or } 6$$

Since the height must be positive, we discard -4 as a solution and choose $h = 6$. Thus, the height of the sign is 6 feet, and the base is $2(6) - 4 = 8$ feet. Check this solution in the original statement of the problem. ■

EXAMPLE 9 An Application of a Polynomial Equation

A rock is dropped into a well that is 64 feet deep, as shown in Figure 3.5. The height of the rock (above the water level) is given by the equation

$$\text{Height} = -16t^2 + 64$$

where the height is measured in feet and the time t is measured in seconds. How long will it take the rock to hit the water at the bottom of the well?

FIGURE 3.5

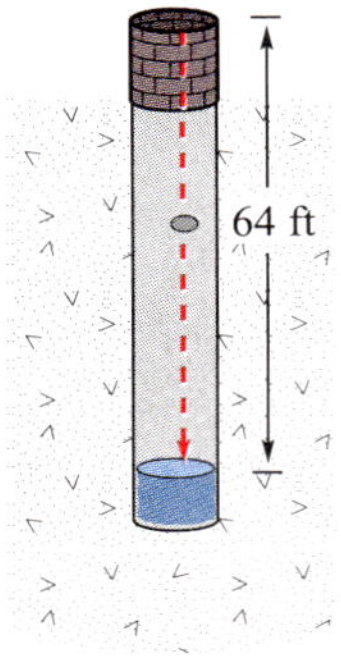

Solution

From Figure 3.5, note that the water level in the well corresponds to a height of 0 feet. Thus, we substitute a height of 0 into the given equation and solve for t.

$$0 = -16t^2 + 64 \quad \text{Set height equal to 0}$$
$$16t^2 - 64 = 0 \quad \text{Standard form}$$
$$16(t^2 - 4) = 0 \quad \text{Remove common monomial factor}$$
$$16(t + 2)(t - 2) = 0 \quad \text{Factor left side of equation}$$
$$t = -2 \text{ or } 2 \quad \text{Solutions}$$

Since a time of -2 seconds doesn't make sense in this problem, we choose the positive solution and conclude that the rock hits the bottom of the well 2 seconds after it is dropped. Check this solution in the original statement of the problem. ■

DISCUSSION PROBLEM A Calculator Experiment

The following equation cannot be solved by factoring (using any of the techniques described in this text).

$$x^3 - x - 3 = 0$$

Use your calculator to approximate the real solution of this equation. For instance, by filling in the following table, you can find an approximation that is correct to *one* decimal place.

x	1.0	1.1	1.2	1.3	1.4	1.5	1.6	1.7	1.8	1.9	2.0
$x^3 - x - 3$	-3										3

Keep trying various real numbers until your approximation is correct to *two* decimal places. ■

Warm-Up

The following warm-up exercises involve skills that were covered in earlier sections. You will use these skills in the exercise set for this section.

In Exercises 1–4, solve the given equation.

1. $5x - 22 = 0$

2. $3x + 42 = 0$

3. $3t - 2(t + 5) = 2t - 15$

4. $2x + \frac{x}{2} = \frac{10}{3}$

In Exercises 5–10, factor the algebraic expression.

5. $4v^2 + 18v$

6. $16x^2 - 25$

7. $36 - (x - 11)^2$

8. $x^2 + 6x - 27$

9. $-8x^2 + 10x - 3$

10. $6x^2 - 73x + 12$

3.5 EXERCISES

In Exercises 1–10, solve the given equation.

1. $2x(x - 8) = 0$

2. $z(z + 6) = 0$

3. $(y - 3)(y + 10) = 0$

4. $(s - 16)(s + 15) = 0$

5. $25(a + 4)(a - 2) = 0$

6. $17(t - 3)(t + 8) = 0$

7. $(2t + 5)(3t + 1) = 0$

8. $(5x - 3)(x - 8) = 0$

9. $4x(2x - 3)(2x + 25) = 0$

10. $(y - 39)(2y + 7)(y + 12) = 0$

In Exercises 11–50, solve the equation by factoring.

11. $x^2 - 25 = 0$

12. $x^2 - 121 = 0$

13. $3y^2 - 48 = 0$

14. $25z^2 - 100 = 0$

15. $(t - 2)^2 - 16 = 0$

16. $(s + 4)^2 - 49 = 0$

17. $9x^2 + 15x = 0$

18. $4x^2 - 6x = 0$

19. $x(x + 10) - 2(x + 10) = 0$

20. $u(u - 3) + 3(u - 3) = 0$

21. $x^2 - 3x - 10 = 0$

22. $x^2 - x - 12 = 0$

23. $m^2 - 8m + 16 = 0$

24. $a^2 + 4a + 4 = 0$

25. $x^2 + 16x + 64 = 0$

26. $x^2 - 12x + 36 = 0$

27. $4z^2 - 12z + 9 = 0$

28. $16t^2 + 48t + 36 = 0$

29. $7 + 13x - 2x^2 = 0$

30. $11 + 32y - 3y^2 = 0$

31. $4x^2 + 15x - 25 = 0$

32. $14x^2 + 9x + 1 = 0$

33. $x(x - 3) = 10$

34. $x(x - 1) = 12$

35. $y(y + 6) = 72$

36. $s(s + 4) = 96$

37. $t(2t - 3) = 35$

38. $3u(3u + 1) = 20$

39. $(a + 2)(a + 5) = 10$

40. $(x - 8)(x - 7) = 20$

41. $(x - 4)(x + 5) = 10$

42. $(5 - u)(7 + u) = 11$

43. $x^3 - 19x^2 + 84x = 0$

44. $x^3 + 18x^2 + 45x = 0$

45. $6t^3 - t^2 - t = 0$

46. $3u^3 - 5u^2 - 2u = 0$

47. $x^2(x - 25) - 16(x - 25) = 0$

48. $y^2(y + 250) - (y + 250) = 0$

49. $a^3 + 2a^2 - 9a - 18 = 0$

50. $x^3 - 2x^2 - 4x + 8 = 0$

51. *Dimensions of a Rectangle* The rectangular floor of a storage shed has an area of 330 square feet. The length of the floor is 7 feet more than its width (see figure). Find the dimensions of the floor.

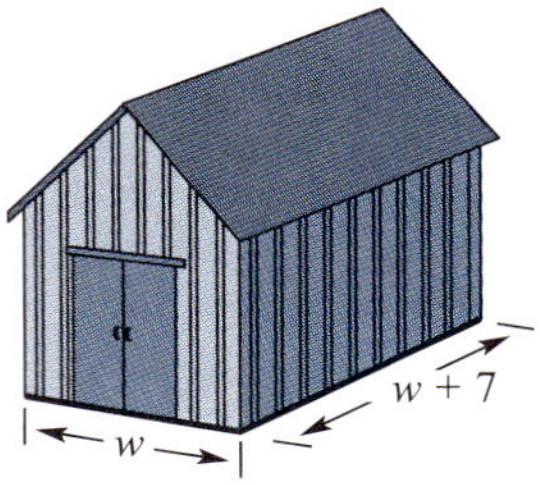

52. *Dimensions of a Rectangle* The length of a rectangle is two and one-fourth times the width. Find the dimensions of the rectangle if its area is 900 square inches.

53. *Dimensions of a Triangle* The triangular cross section of a machine part must have an area of 48 square inches (see figure). Find the base and height of the triangle if the height is one and one-half times the base.

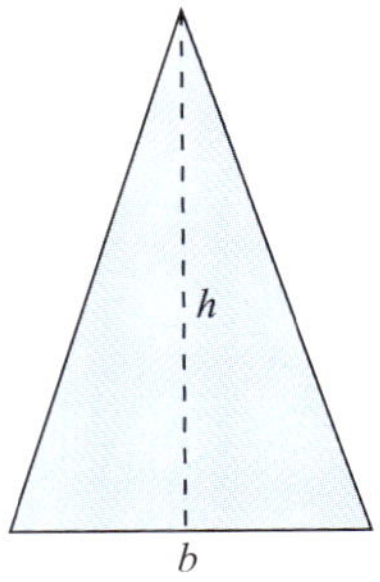

54. *Dimensions of a Triangle* The height of a triangle is 4 inches less than its base. Find the base and height of the triangle if its area is 70 square inches.

55. *Number Problem* Find two consecutive positive integers whose product is 156.

56. *Number Problem* Find two consecutive positive even integers whose product is 168.

57. *Free-Falling Object* An object is dropped from a weather balloon 6,400 feet above the ground. Find the time t in seconds for the object to reach the ground by solving the following equation.

$$-16t^2 + 6{,}400 = 0$$

58. *Free-Falling Object* An object is thrown upward from a height of 64 feet with an initial velocity of 48 feet per second. Find the time t in seconds for the object to reach the ground by solving the following equation.

$$-16t^2 + 48t + 64 = 0$$

59. *Dimensions of an Open Box* An open box is to be made from a square piece of material by cutting 5-inch squares from each corner and turning up the sides (see figure). The volume V of a rectangular solid is the product of its length, width, and height, and x is the length of an edge of the square base.
(a) Show that the volume is given by $V = 5x^2$.
(b) Complete the following table.

x	2	4	6	8
V				

(c) Find the size of the original piece of material if $V = 720$ cubic inches.

Figure for 59

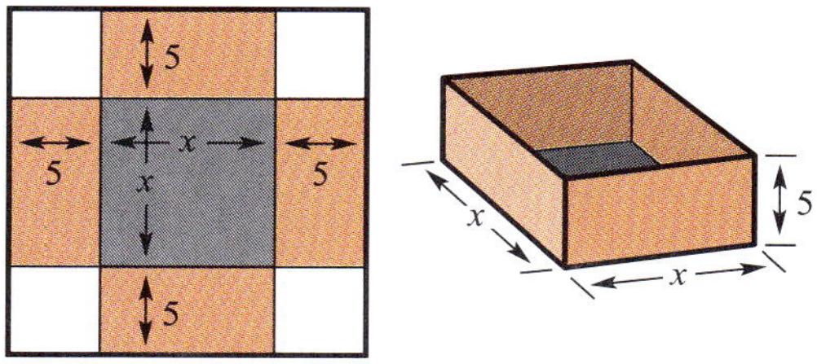

60. *Dimensions of an Open Box* An open box with a square base is to be constructed from 880 square inches of material (see figure). What should be the dimensions of the base if the height of the box is to be 6 inches? (*Hint:* The surface area is given by $S = x^2 + 4xh$.)

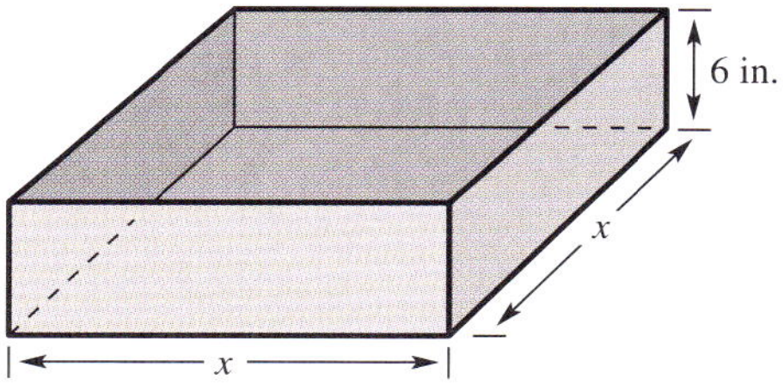

61. Let a and b be real numbers such that $a \neq 0$. Find the solutions of $ax^2 + bx = 0$.

62. Let a be a nonzero real number. Find the solutions of $ax^2 - ax = 0$.

63. (a) Complete the following table finding the product $P = (x + 5)(x - 4)$ for each value of x.

x	3	4	5	6	7	8
P						

(b) Use the table to determine how P changes for each one-unit increase in x.

(c) Use the values of P in the table to solve the equation $(x + 5)(x - 4) = 70$.

64. Solve the equation $2(x + 3)^2 + (x + 3) - 15 = 0$ in two ways.

(a) Let $u = x + 3$, and solve the resulting equation. Then find the corresponding values of x, which will be solutions to the given equation.

(b) Expand and collect like terms in the given equation, and solve the resulting equation for x.

65. Solve each of the following equations using either or both methods described in Exercise 64.

(a) $3(x + 6)^2 - 10(x + 6) - 8 = 0$

(b) $8(x + 2)^2 - 18(x + 2) + 9 = 0$

CHAPTER 3 SUMMARY

As you review and prepare for a test on this chapter, first try to obtain a global view of what was discussed. Then review the specific skills needed under each category.

Operations Involving Polynomials

- *Identify* the type of polynomial and specify its degree. — Section 3.1, Exercises 1–16
- *Rewrite* polynomial expressions using addition or subtraction. — Section 3.1, Exercises 25–66
- *Evaluate* a polynomial at specified values of the independent variable. — Section 3.1, Exercises 67–78
- *Simplify* polynomial expressions using properties of exponents. — Section 3.2, Exercises 7–20
- *Rewrite* polynomial expressions using multiplication or special product formulas. — Section 3.2, Exercises 21–80

Factoring Polynomials

- *Factor* a polynomial by removing the greatest common factor. — Section 3.3, Exercises 11–32
- *Factor* the special polynomial forms: difference of two squares, the sum or difference of two cubes. — Section 3.3, Exercises 55–72
- *Factor* polynomial expressions by grouping. — Section 3.3, Exercises 45–54
- *Factor* polynomial expressions completely. — Section 3.3, Exercises 73–78; Section 3.4, Exercises 77–86
- *Factor* perfect square trinomials. — Section 3.4, Exercises 1–10
- *Factor* trinomials. — Section 3.4, Exercises 21–40, Exercises 51–70

Using Factoring to Solve Polynomial Equations

- *Solve* polynomial equations by factoring. — Section 3.5, Exercises 11–50
- *Create and solve* polynomial equations from verbal applications. — Section 3.5, Exercises 51–58

Chapter 3 Review Exercises

In Exercises 1–4, describe the error and then make the necessary correction (to the right side of the equation).

1. $-4(x - 2) = -4x - 8$

2. $6x - (x + 7) = 5x + 7$

3. $(x + 3)^2 = x^2 + 9$

4. $(x + 4)(x - 4) = x^2 + 8x - 16$

In Exercises 5–24, perform the specified operations and simplify.

5. $(-6z)^3$

6. $(12x^2y)(3x^2y^4)$

7. $\dfrac{120u^5v^3}{15u^3v}$

8. $\left(-\dfrac{y^2}{2}\right)^2$

9. $(5x + 3x^2) + (x - 4x^2)$

10. $\left(\dfrac{1}{2}x + \dfrac{2}{3}\right) + \left(4x + \dfrac{1}{3}\right)$

11. $(3t - 5) - 2(t^2 - t - 5)$

12. $(10y^2 + 3) - 5(y^2 - 9)$

13. $(-x^3 - 3x) - 4(2x^3 - 3x + 1)$

14. $(7z^2 + 6z) - 3(5z^2 + 2z)$

15. $3y^2 - [2y - 3(y^2 + 5)]$

16. $(16a^3 + 5a) - 5[a + (2a^3 - 1)]$

17. $(-2x)^3(x + 4)$

18. $3y(-4y)(y - 2)$

19. $(5x + 3)(3x - 4)$

20. $(3y^2 + 2)(4y^2 - 5)$

21. $(2x^2 - 3x + 2)(2x + 3)$

22. $(5s^3 + 4s - 3)(4s - 5)$

23. $2u(u - 7) - (u + 1)(u - 7)$

24. $(3v + 2)(-5v) + 5v(3v + 2)$

In Exercises 25–30, find the indicated product by using the special product formulas.

25. $(4x - 7)^2$

26. $(8 - 3x)^2$

27. $(5u - 8)(5u + 8)$

28. $(7a + 4)(7a - 4)$

29. $[(u - 3) + v][(u - 3) - v]$

30. $[(m - 5) + n]^2$

In Exercises 31–34, factor the polynomial by removing the common factor.

31. $6x^2 + 15x^3$

32. $8y - 12y^4$

33. $28(x + 5) - 70(x + 5)^2$

34. $(u - 9v)(u - v) + v(u - 9v)$

In Exercises 35–38, factor the given difference of two squares.

35. $9a^2 - 100$

36. $b^2 - 900$

37. $(u + 6)^2 - 81$

38. $(y - 3)^2 - 16$

In Exercises 39–42, factor the perfect square trinomial.

39. $x^2 - 40x + 400$

40. $y^2 + 26y + 169$

41. $4s^2 + 40s + 100$

42. $u^2 - 10uv + 25v^2$

In Exercises 43–54, factor the expression completely, if possible.

43. $4a - 64a^3$

44. $3b + 27b^3$

45. $8x(2x - 3) - 4(2x - 3)$

46. $x^3 + 3x^2 - 4x - 12$

47. $8x^3 + 1$

48. $t^3 - 216$

49. $4u^2 - 28u + 49$

50. $x^2 - \frac{2}{3}x + \frac{1}{9}$

51. $\frac{1}{4}x^2 + x - 1$

52. $3x^2 + 23x - 8$

53. $4t^2 - t + 13$

54. $6h^2 - h - \frac{1}{3}$

In Exercises 55–58, find all integers b such that the trinomial is factorable.

55. $x^2 + bx + 5$

56. $x^2 + bx + 6$

57. $2x^2 + bx - 15$

58. $3x^2 + bx - 21$

In Exercises 59 and 60, find *two* integers c such that the trinomial is factorable.

59. $x^2 + 7x + c$

60. $3x^2 - 5x + c$

In Exercises 61–70, solve the given equation.

61. $10x(x - 3) = 0$

62. $3x(4x + 7) = 0$

63. $v^2 - 100 = 0$

64. $(x + 3)^2 - 25 = 0$

65. $x^2 - 25x = -150$

66. $4t^2 - 12t = -9$

67. $3s^2 - 2s - 8 = 0$

68. $2y^3 + 2y^2 - 24y = 0$

69. $z(5 - z) + 36 = 0$

70. $b^3 - 6b^2 - b + 6 = 0$

71. *Probability* The probability of three successes in five trials of an experiment is given by $10p^3(1 - p)^2$. Find this product.

72. *Metal Expansion* When the temperature of an iron steam pipe 200 feet long is increased by 75°C, the length of the pipe will increase by the following amount.

$$75(200)(10 \times 10^{-6})$$

Find this amount of increase in length.

73. *Dimensions of a Rectangle* The length of a rectangle is l units, and its width is $l - 5$ units. Find (a) the perimeter and (b) the area of the rectangle.

74. The accompanying figure shows two squares with sides of length x and y.

(a) Remove the small square with side of length y from the larger square. What is the area of the remaining figure?

(b) After removing the small square, slide and rotate the remaining top rectangle so that it fits against the right side of the figure. What are the dimensions of the resulting rectangle and what special product formula have you demonstrated geometrically?

Figure for 74

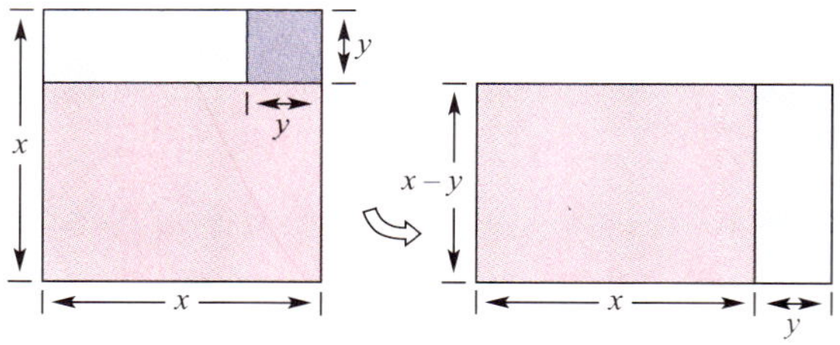

75. *Free-Falling Object* An object is thrown upward from a height of 48 feet with an initial velocity of 32 feet per second. Find the time t in seconds for the object to reach the ground by solving the following equation.

$$-16t^2 + 32t + 48 = 0$$

76. *Number Problem* The product of two consecutive positive odd integers is 195. Find the two integers.

77. *Dimensions of a Rectangle* The width of a rectangle is three-fourths of its length. Find the dimensions of the rectangle if its area is 432 square inches.

CHAPTER 3 TEST

Take this test as you would take a test in class. After you are done, check your work with the answers given in the back of the book.

1. Determine the degree and leading coefficient of the polynomial $-5.2x^3 + 3x^2 - 8$.
2. Add: $(5a^2 - 3a + 4) + (a^2 - 4)$
3. Subtract: $(16 - y^2) - (16 + 2y + y^2)$
4. Simplify: $-2(2x^4 - 5) + 4x(x^3 + 2x - 1)$
5. Simplify: $4t - [3t - (10t + 7)]$
6. Multiply and simplify: $2y\left(\frac{y}{4}\right)^2$
7. Multiply and simplify: $(2x - 3y)(x + 5y)$
8. Multiply and simplify: $(x - 1)[2x + (x - 3)]$
9. Multiply and simplify: $(2s - 3)(3s^2 - 4s + 7)$
10. Multiply and simplify: $(4x - 3)^2$
11. Multiply and simplify: $[4 - (a + b)][4 + (a + b)]$
12. Factor: $18y^2 - 12y$
13. Factor: $v^2 - \frac{16}{9}$
14. Factor: $x^3 - 3x^2 - 4x + 12$
15. Factor: $9u^2 - 6u + 1$
16. Factor: $6x^2 - 26x - 20$
17. Solve: $(y + 2)^2 - 9 = 0$
18. Solve: $12 + 5y - 3y^2 = 0$
19. Find the area of the shaded region in the accompanying figure.

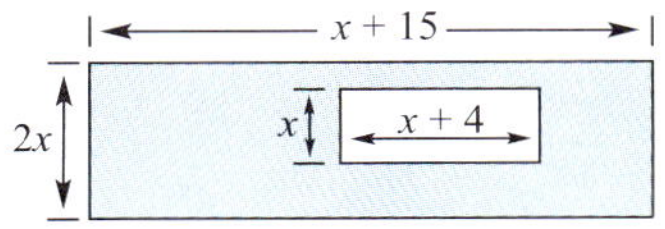

20. The length of a rectangle is one and one-half times its width. Find the dimensions of the rectangle if its area is 54 square centimeters.

CHAPTER FOUR

Rational Expressions

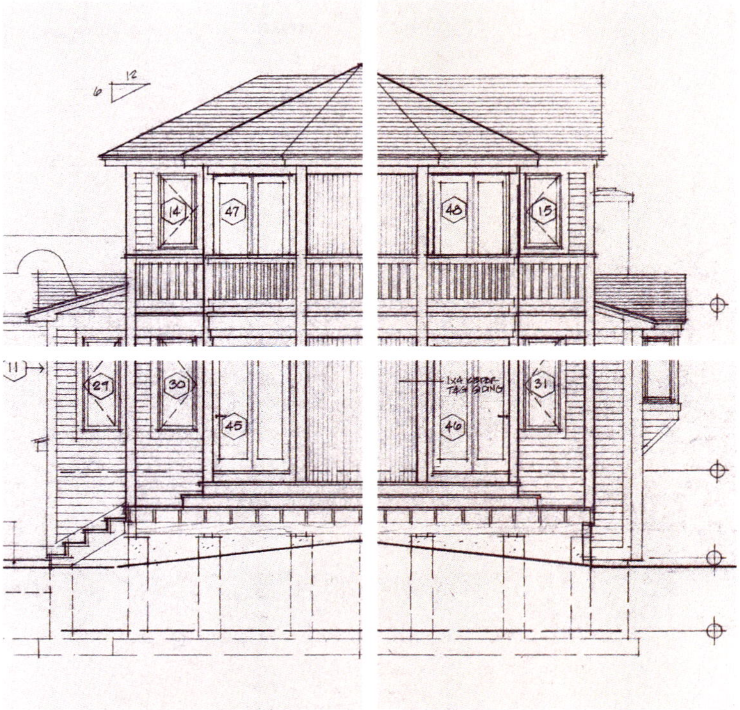

Chapter Overview

We have now studied two different algebraic systems. In Chapter 2 we studied the algebra of linear expressions, and in Chapter 3 we studied the algebra of polynomials. In this chapter we study the algebra of rational expressions. A rational expression is a fraction whose numerator and denominator are polynomials. We will look at techniques for simplifying, adding, subtracting, multiplying, and dividing rational expressions, and solving equations involving rational expressions.

SECTION 4.1

Simplifying Rational Expressions

The Domain of a Rational Expression • *Simplifying Rational Expressions*

The Domain of a Rational Expression

A fraction whose numerator and denominator are polynomials is called a **rational expression**. Some examples are

$$\frac{3}{x+4}, \qquad \frac{2x}{x^2-4x+4}, \qquad \text{and} \qquad \frac{x^2-5x}{x^2+2x-3}.$$

In our work with rational expressions, we will assume that all real number values of the variable that make the denominator zero are excluded. For the three fractions above, we exclude $x = -4$ from the first fraction, $x = 2$ from the second, and both $x = 1$ and $x = -3$ from the third fraction. We refer to the *usable* values of the variable as the **domain** of the rational expression.

Definition of a Rational Expression

Let u and v be polynomials. The algebraic expression

$$\frac{u}{v}$$

is called a **rational expression**. The **domain** of this rational expression is the set of all real numbers for which $v \neq 0$.

EXAMPLE 1 ■ **Finding the Domain of a Rational Expression**

Find the domain of each of the following rational expressions.

(a) $\dfrac{4}{x-2}$ (b) $\dfrac{2x+5}{8}$ (c) $3x^2 + 2x - 5$

Solution

(a) The denominator is zero when $x - 2 = 0$ or $x = 2$. Therefore, the domain is all real values of x such that $x \neq 2$. In interval notation, we can write the domain as

$$\text{Domain} = (-\infty, 2) \cup (2, \infty).$$

(b) The denominator, 8, is never zero; hence, the domain is the set of *all* real numbers. In interval notation, we can write the domain as

$$\text{Domain} = (-\infty, \infty).$$

(c) Note that any polynomial is also a rational expression, because we can consider its denominator to be 1. For instance, in this case we can write

$$3x^2 + 2x - 5 = \frac{3x^2 + 2x - 5}{1}.$$

Since the denominator of this expression is not zero for any value of x, it follows that the domain is the set of all real numbers. ■

NOTE: The domain of any polynomial is the set of all real numbers.

EXAMPLE 2 ■ Finding the Domain of a Rational Expression

Find the domain of each of the following rational expressions.

(a) $\dfrac{3x}{x^2 - 25}$ (b) $\dfrac{x^2 + 3x}{x^2 + 5x - 6}$

Solution

(a) The denominator of this rational expression is

$$x^2 - 25 = (x + 5)(x - 5).$$

Since this denominator is zero when $x = -5$ or when $x = 5$, the domain of the fraction is all real values of x such that $x \neq -5$ and $x \neq 5$.

(b) The denominator of this rational expression is

$$x^2 + 5x - 6 = (x + 6)(x - 1).$$

Since this denominator is zero when $x = -6$ or when $x = 1$, the domain of the fraction is all real values of x such that $x \neq -6$ and $x \neq 1$. ■

In applications involving rational expressions, it is often necessary to restrict the domain further. To indicate such a restriction, we write the domain to the right of the fraction. For instance, the domain of the rational expression

$$\frac{x^2 + 20}{x + 4}, \quad x > 0$$

is the set of all positive real numbers, indicated by the inequality $x > 0$.

EXAMPLE 3 ■ An Application Involving a Restricted Domain

Suppose you have started a small manufacturing business. The initial investment for the business is \$120,000. The cost of each unit that you manufacture is \$15. Thus, your total cost of producing x units is

$$\text{Total cost} = 15x + 120{,}000.$$

Your average cost per unit depends on the number of units produced. If you produce 100 units, your average cost per unit is $(15(100) + 120{,}000)/100 = \$1{,}215$. The average cost per unit decreases as the number of units increases. If you produce 1,000 units, then your average cost per unit is (15(1,000) + 120,000)/1,000 = $135. In general, if you produce x units, then your average cost per unit is

$$\text{Average cost per unit} = \frac{15x + 120{,}000}{x}.$$

What is the domain of this rational expression?

Solution

If we were simply considering the expression $(15x + 120{,}000)/x$ as a mathematical quantity, we would say that the domain is all real values of x such that $x \neq 0$. However, because this fraction is a mathematical model representing a real-life situation, we must consider which values of x make sense in real life. For this model, the variable x represents the number of units that you produce. Assuming that you cannot produce a fractional number of units, we conclude that the domain is the set of positive integers. That is,

$$\text{Domain} = \{1, 2, 3, 4, \ldots\}.$$

Simplifying Rational Expressions

The rules for operating with rational expressions are like those for numerical fractions (see the properties of fractions in Section 1.2). As with numerical fractions, we say that a rational expression is **simplified** or **in reduced form** if its numerator and denominator have no factors in common (other than ± 1). To reduce fractions, we apply the following law.

Reduction Law for Fractions

Let u, v, and w represent numbers, variables, or algebraic expressions such that $v \neq 0$ and $w \neq 0$. Then the following reduction law is valid.

$$\frac{u\cancel{w}}{v\cancel{w}} = \frac{u}{v}$$

Be sure you see that this reduction law allows us only to divide out *factors, not terms*. For instance, consider the following.

$$\frac{\cancel{2} \cdot x}{\cancel{2}(x + 5)}$$ We CAN divide out common *factor* 2

$$\frac{3 + x}{3 + 2x}$$ We CANNOT divide out common *term* 3

$$\frac{\cancel{x}(x^2 + 6)}{\cancel{x} \cdot x}$$ We CAN divide out common *factor* x

$$\frac{x + 4}{x}$$ We CANNOT divide out common *term* x

Using the reduction law to simplify a rational expression requires two steps. First, we must completely factor the numerator and denominator. Then we apply the reduction law to divide the numerator and denominator by any *factors* that are common to both. Thus, your success in simplifying rational expressions actually lies in your ability to completely factor the polynomials in both the numerator and denominator.

EXAMPLE 4 ■ **Simplifying a Rational Expression**

Simplify the following rational expression. $\dfrac{2x^3 - 6x}{6x^2}$

Solution

To begin, we completely factor both the numerator and denominator. If the numerator and denominator have any common factors, then we divide out the common factors to obtain the simplified form. If the numerator and denominator do not have any common factors, then the rational expression is already in simplified form.

$$\frac{2x^3 - 6x}{6x^2} = \frac{2x(x^2 - 3)}{6x^2} \qquad \text{Factor numerator and denominator}$$

$$= \frac{\cancel{2x}(x^2 - 3)}{3(\cancel{2x})(x)} \qquad \text{Divide out common factor } 2x$$

$$= \frac{x^2 - 3}{3x} \qquad \text{Simplified form}$$

■

EXAMPLE 5 ■ **Adjusting the Domain After Simplifying**

$$\frac{x^2 + 2x - 15}{3x - 9} = \frac{(x + 5)(x - 3)}{3(x - 3)} \qquad \text{Factor numerator and denominator}$$

$$= \frac{(x + 5)\cancel{(x - 3)}}{3\cancel{(x - 3)}} \qquad \text{Divide out common factor } (x - 3)$$

$$= \frac{x + 5}{3}, \quad x \neq 3 \qquad \text{Simplified form}$$

The simplification process used in this example has a tricky technicality. This technicality arises from the fact that dividing out common factors from the numerator and denominator of a rational expression can change the domain of the expression. For instance, in this example, the domain of the original expression is all real values of x such that $x \neq 3$. Thus, in order for us to equate the original expression to the simplified expression, we must restrict the domain of the simplified expression to exclude 3. ■

In this text, when simplifying a rational expression, we follow the convention of listing *by the simplified expression* all values of x that must be specifically excluded from the domain in order to make the domains of the simplified and original expressions agree. For instance, in the next example the restriction $x \neq 4$ must be listed with the simplified expression in order to make the two domains agree. (Note that the value of -2 is excluded from both domains, so it is not necessary to list that value.)

EXAMPLE 6 ■ **Simplifying a Rational Expression**

$$\frac{x^3 - 16x}{x^2 - 2x - 8} = \frac{x(x^2 - 16)}{(x + 2)(x - 4)} \qquad \text{Partially factor}$$

$$= \frac{x(x + 4)(x - 4)}{(x + 2)(x - 4)} \qquad \text{Factor completely}$$

$$= \frac{x(x + 4)\cancel{(x - 4)}}{(x + 2)\cancel{(x - 4)}} \qquad \text{Divide out common factor } (x - 4)$$

$$= \frac{x(x + 4)}{x + 2}, \quad x \neq 4 \qquad \text{Simplified form}$$

■

Be sure to *completely* factor the numerator and denominator of an algebraic fraction before concluding that there is no common factor. This may involve a change in sign to see if further reduction is possible. Watch for this in the next example.

EXAMPLE 7 ■ **Simplification Involving a Change of Sign**

$$\frac{2x^2 - 9x + 4}{12 + x - x^2} = \frac{(2x - 1)(x - 4)}{(4 - x)(3 + x)} \qquad \text{Factor completely}$$

$$= \frac{(2x - 1)(x - 4)}{-(x - 4)(3 + x)} \qquad (4 - x) = -(x - 4)$$

$$= \frac{(2x - 1)\cancel{(x - 4)}}{-\cancel{(x - 4)}(3 + x)} \qquad \text{Divide out common factor } (x - 4)$$

$$= -\frac{2x - 1}{3 + x}, \quad x \neq 4 \qquad \text{Simplified form}$$

Be sure you see that when we divided the numerator and denominator by the common factor, $(x - 4)$, we kept the negative sign. In the simplified form of the fraction, we usually like to write the negative sign in front of the fraction. However, this is merely a personal preference. Any of the following forms are legitimate.

$$-\frac{2x - 1}{3 + x} = \frac{-(2x - 1)}{3 + x} = \frac{2x - 1}{-(3 + x)} = \frac{2x - 1}{-3 - x} = \frac{1 - 2x}{3 + x}$$

■

In the next two examples, we apply the reduction law to simplify rational expressions that involve more than one variable.

EXAMPLE 8 ■ **Simplifying a Rational Expression Involving Two Variables**

$$\frac{3xy + y^2}{2y} = \frac{y(3x + y)}{2y} \qquad \text{Factor numerator and denominator}$$

$$= \frac{\cancel{y}(3x + y)}{2\cancel{y}} \qquad \text{Divide out common factor } y$$

$$= \frac{3x + y}{2}, \quad y \neq 0 \qquad \text{Simplified form}$$

■

EXAMPLE 9 ■ Simplifying a Rational Expression Involving Two Variables

Simplify the following rational expression.

$$\frac{2x^2 + 2xy - 4y^2}{5x^3 - 5xy^2}$$

Solution

$$\frac{2x^2 + 2xy - 4y^2}{5x^3 - 5xy^2} = \frac{2(x - y)(x + 2y)}{5x(x - y)(x + y)} \quad \text{Factor numerator and denominator}$$

$$= \frac{2\cancel{(x - y)}(x + 2y)}{5x\cancel{(x - y)}(x + y)} \quad \text{Divide out common factor } (x - y)$$

$$= \frac{2(x + 2y)}{5x(x + y)}, \quad x \neq y \quad \text{Simplified form}$$

■

REMINDER: As you study the examples and work the exercises in this and the following three sections, keep in mind that you are *rewriting expressions in simpler forms*. You are not solving equations. We introduce equal signs in the steps of the simplification process only to indicate that the new form of the expression is *equivalent* to the previous one.

DISCUSSION PROBLEM ■ Equivalent Expressions

(a) Evaluate the two given expressions for the x-values listed in the following table.

x-Values	-2	-1	0	1	2	3	4
$\frac{x^2 - x - 6}{x - 3}$							
$x + 2$							

(b) Write a paragraph describing the equivalence (or nonequivalence) of these two expressions. Support your argument with appropriate algebra from this section *and* a discussion of the domains of the two expressions. ■

Warm-Up

The following warm-up exercises involve skills that were covered in earlier sections. You will use these skills in the exercise set for this section.

In Exercises 1–4, find the missing real number so that the two fractions are equivalent.

1. $-\frac{5}{6} = \frac{(\quad)}{24}$

2. $\frac{3}{7} = \frac{30}{(\quad)}$

3. $\frac{4}{3} = \frac{-28}{(\quad)}$

4. $-\frac{9}{16} = \frac{(\quad)}{48}$

In Exercises 5–8, write the fraction in reduced form.

5. $\frac{45}{78}$

6. $\frac{55}{115}$

7. $\frac{160}{256}$

8. $\frac{189}{324}$

In Exercises 9 and 10, factor the algebraic expression.

9. $20ax^2 + 60ax + 80a$

10. $2x^2y^2 - 9xy^2 - 56y^2$

4.1 EXERCISES

In Exercises 1–10, find the domain of the rational expression.

1. $\frac{5}{x - 8}$

2. $\frac{9}{x - 13}$

3. $\frac{7x}{x + 4}$

4. $x^4 + 2x^2 - 5$

5. $\frac{4}{x^2 + 9}$

6. $\frac{y^2 - 3}{7}$

7. $\frac{5t}{t^2 - 16}$

8. $\frac{z + 2}{z(z - 4)}$

9. $\frac{u^2}{u^2 - 4u - 5}$

10. $\frac{y + 5}{4y^2 - 5y - 6}$

In Exercises 11 and 12, determine the restricted domain of the mathematical model.

11. *Inventory Cost* The inventory cost when x units of a product are ordered from a supplier is given by

$$\frac{0.25x + 2,000}{x}.$$

12. *Average Cost* The average cost for a manufacturer to produce x units of a product is given by

$$\frac{1.35x + 4,570}{x}.$$

In Exercises 13–20, find the missing algebraic expression so that the two expressions are equivalent.

13. $\frac{5}{6} = \frac{5(\quad)}{6(x + 3)}, \quad x \neq -3$

14. $\frac{7}{15} = \frac{7(\quad)}{45(x - 10)^2}, \quad x \neq 10$

15. $\dfrac{x}{2} = \dfrac{3x(x + 16)^2}{2(\boxed{})}, \quad x \neq -16$

16. $\dfrac{5x}{12} = \dfrac{25x^2(x - 10)}{12(\boxed{})}, \quad x \neq 10$

17. $\dfrac{x + 5}{3x} = \dfrac{(x + 5)(\boxed{})}{3x^2(x - 2)}, \quad x \neq 2$

18. $\dfrac{3y - 7}{y + 2} = \dfrac{(3y - 7)(\boxed{})}{y^2 - 4}, \quad y \neq 2$

19. $\dfrac{8x}{x - 5} = \dfrac{8x(\boxed{})}{x^2 - 3x - 10}, \quad x \neq -2$

20. $\dfrac{3 - z}{z^2} = \dfrac{(3 - z)(\boxed{})}{z^3 + 2z^2}, \quad z \neq -2$

In Exercises 21–50, write the expression in reduced form.

21. $\dfrac{5x}{25}$

22. $\dfrac{32y}{24}$

23. $\dfrac{12y^2}{2y}$

24. $\dfrac{15z^3}{15z^3}$

25. $\dfrac{18x^2y}{15xy^4}$

26. $\dfrac{16y^2z^2}{60y^5z}$

27. $\dfrac{x^2(x - 8)}{x(x - 8)}$

28. $\dfrac{a^2b(b - 3)}{b^3(b - 3)^2}$

29. $\dfrac{2x - 3}{4x - 6}$

30. $\dfrac{y^2 - 81}{2y - 18}$

31. $\dfrac{5 - x}{3x - 15}$

32. $\dfrac{x^2 - 36}{6 - x}$

33. $\dfrac{3xy^2}{xy^2 + x}$

34. $\dfrac{x + 3x^2y}{3xy + 1}$

35. $\dfrac{y^2 - 64}{5(3y + 24)}$

36. $\dfrac{x^2 - 25z^2}{x + 5z}$

37. $\dfrac{a + 3}{a^2 + 6a + 9}$

38. $\dfrac{u^2 - 12u + 36}{u - 6}$

39. $\dfrac{x^2 - 7x}{x^2 - 14x + 49}$

40. $\dfrac{z^2 + 22z + 121}{3z + 33}$

41. $\dfrac{y^3 - 4y}{y^2 + 4y - 12}$

42. $\dfrac{x^2 - 7x}{x^2 - 4x - 21}$

43. $\dfrac{3 - x}{2x^2 - 3x - 9}$

44. $\dfrac{2y^2 + 13y + 20}{2y^2 + 17y + 30}$

45. $\dfrac{15x^2 + 7x - 4}{15x^2 + x - 2}$

46. $\dfrac{56z^2 - 3z - 20}{49z^2 - 16}$

47. $\dfrac{5xy + 3x^2y^2}{xy^3}$

48. $\dfrac{4u^2v - 12uv^2}{18uv}$

49. $\dfrac{3m^2 - 12n^2}{m^2 + 4mn + 4n^2}$

50. $\dfrac{x^2 + xy - 2y^2}{x^2 + 3xy + 2y^2}$

In Exercises 51–54, demonstrate that the two quantities are *not* equivalent by evaluating each side when $x = 10$.

51. $\dfrac{x - 4}{4} \neq x - 1$

52. $\dfrac{x - 4}{x} \neq -4$

53. $\dfrac{3x + 2}{4x + 2} \neq \dfrac{3}{4}$

54. $\dfrac{1 - x}{2 - x} \neq \dfrac{1}{2}$

In Exercises 55 and 56, write the fraction in reduced form. (Assume n is a positive integer.)

55. $\dfrac{x^{2n} - 4}{x^n + 2}$

56. $\dfrac{x^{2n} + x^n - 12}{x^{n+1} + 4x}$

In Exercises 57 and 58, evaluate the two given expressions for the x-values listed in the table. Comment on the domain and equivalence of the two expressions.

57.

x	$\dfrac{x^2 - x - 2}{x - 2}$	$x + 1$
-2		
-1		
0		
1		
2		
3		
4		

58.

x	$\dfrac{x^2 + 5x}{x}$	$x + 5$
-2		
-1		
0		
1		
2		
3		
4		

In Exercises 59 and 60, assume that a poll was taken and 70% of the sample preferred a certain product. The fraction of the sample that preferred the product can be given by

$$\frac{0.7x}{x}$$

where x is the number of people sampled.

59. *Market Survey* If 900 people were included in the sample, how many preferred the product?

60. *Market Survey* If 350 people preferred the product, how large was the sample?

61. *Average Cost* A machine shop has a setup cost of \$2,500 for the production of a new product. The cost for labor and material used to produce each unit is \$9.25.
(a) Write an algebraic fraction that gives the average cost per unit when x units are produced.
(b) Determine the domain of the fraction in part (a).
(c) Find the average cost per unit when $x = 100$ units are produced.

62. *Pollution Removal* The cost in dollars of removing p percent of the air pollutants from the stack emission of a utility company is given by the rational expression

$$\frac{80{,}000p}{100 - p}.$$

Determine the domain of the rational expression.

63. *Distance Traveled* A van started on a trip and traveled at an average speed of 45 miles per hour. Three hours later a car started on the same trip and traveled at an average speed of 60 miles per hour (see figure).
(a) Find the distance each vehicle has traveled when the car has been on the road for t hours.
(b) Use the result of part (a) to determine the ratio of the distance the car has traveled to the distance the van has traveled.

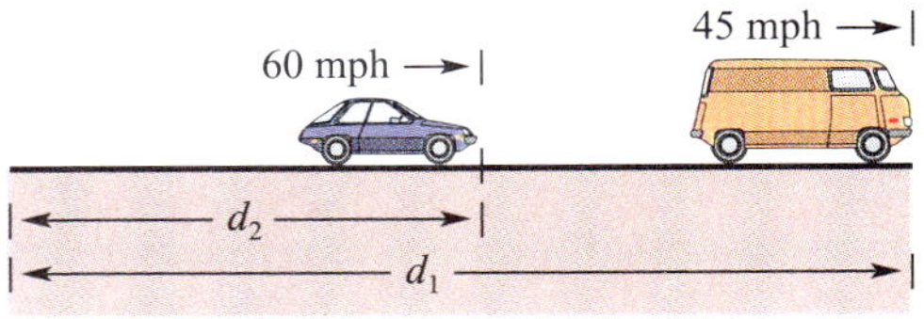

64. Find the ratio of the shaded portion of the accompanying figure to the total area of the figure.

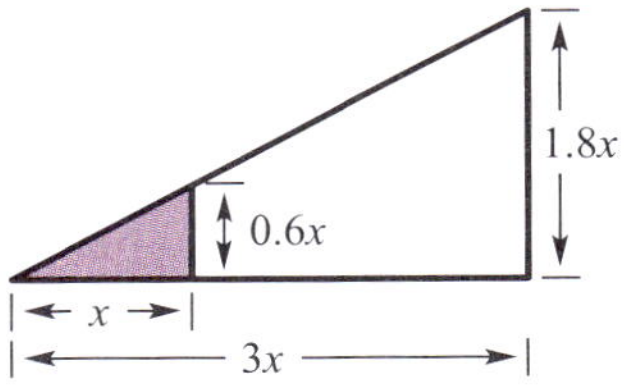

Cost of Medicare In Exercises 65 and 66, use the following polynomial models, which give the projected cost of Medicare between 1990 and 1995 and the U.S. population age 65 and older (see figure).

$$C = 107.1 + 12.64t + 0.54t^2 \qquad \text{Cost of Medicare}$$

$$P = 31.6 + 0.51t - 0.14t^2 \qquad \text{Population}$$

In these models, C represents the total annual cost of Medicare (in billions of dollars), P represents the U.S. population (in millions) age 65 and older, and t represents the year, with $t = 0$ corresponding to 1990. (*Sources:* Congressional Budget Office and U.S. Bureau of Census.)

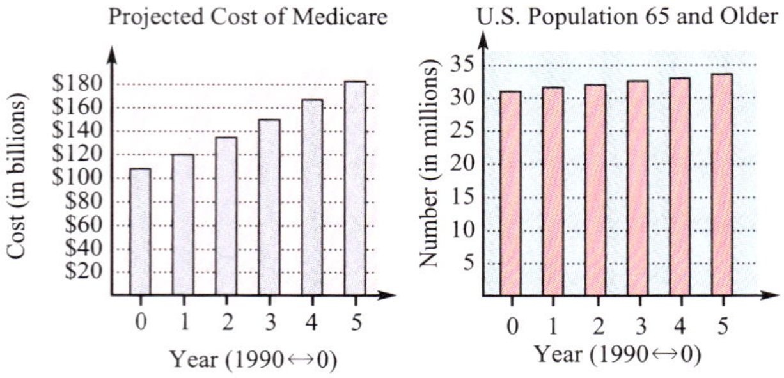

65. Find a rational model that represents the average cost of Medicare *per person age 65 or older* during the years 1990 to 1995.

66. Use the model found in Exercise 65 to complete the following table showing the average cost of Medicare per person age 65 or older.

Year	Average Cost
1990	
1991	
1992	
1993	
1994	
1995	

SECTION 4.2

Multiplying and Dividing Rational Expressions

Multiplying Rational Expressions • Dividing Rational Expressions • Compound Fractions

Multiplying Rational Expressions

The rule for multiplying rational expressions is the same as the rule for multiplying numerical fractions. That is, we *multiply numerators, multiply denominators, and write the new fraction in reduced form*. Here is an example.

$$\frac{x+3}{x-2} \cdot \frac{x-2}{x-4} = \frac{(x+3)(x-2)}{(x-2)(x-4)} \qquad \text{Multiply numerators and denominators}$$

$$= \frac{(x+3)\cancel{(x-2)}}{\cancel{(x-2)}(x-4)} \qquad \text{Divide out common factor}$$

$$= \frac{x+3}{x-4}, \quad x \neq 2 \qquad \text{Simplified form}$$

Multiplying Rational Expressions

Let u, v, w, and z be real numbers, variables, or algebraic expressions such that $v \neq 0$ and $z \neq 0$. Then the product of u/v and w/z is given by

$$\frac{u}{v} \cdot \frac{w}{z} = \frac{uw}{vz}.$$

In order to recognize common factors in the product of fractions, you should write the numerators and denominators in completely factored form, as demonstrated in the following examples.

EXAMPLE 1 ■ Multiplying Rational Expressions

Multiply the following rational expressions.

$$\frac{4x^3y}{3xy^4} \cdot \frac{-6x^2y^2}{10x^4}$$

Solution

$$\frac{4x^3y}{3xy^4} \cdot \frac{-6x^2y^2}{10x^4} = \frac{-24x^5y^3}{30x^5y^4}$$ Multiply numerators and denominators

$$= \frac{-4(6)(x^5)(y^3)}{5(6)(x^5)(y^3)(y)}$$ Divide out common factors

$$= -\frac{4}{5y}, \quad x \neq 0$$ Simplified form ■

EXAMPLE 2 ■ Multiplying Rational Expressions

Multiply the following rational expressions.

$$\frac{x}{5x^2 - 20x} \cdot \frac{x - 4}{2x^2 + x - 3}$$

Solution

$$\frac{x}{5x^2 - 20x} \cdot \frac{x - 4}{2x^2 + x - 3}$$

$$= \frac{x(x - 4)}{5x(x - 4)(x - 1)(2x + 3)}$$ Factor and multiply

$$= \frac{x(x - 4)(1)}{5x(x - 4)(x - 1)(2x + 3)}$$ Divide out common factors

$$= \frac{1}{5(x - 1)(2x + 3)}, \quad x \neq 0, x \neq 4$$ Simplified form ■

EXAMPLE 3 ■ Multiplying Rational Expressions

$$\begin{aligned}
\frac{4x^2 - 4x}{x^2 + 2x - 3} \cdot \frac{x^2 + x - 6}{4x} &= \frac{4x(x - 1)(x + 3)(x - 2)}{(x - 1)(x + 3)(4x)} && \text{Factor and multiply} \\
&= \frac{\cancel{4x}\cancel{(x - 1)}\cancel{(x + 3)}(x - 2)}{\cancel{(x - 1)}\cancel{(x + 3)}\cancel{(4x)}} && \text{Divide out common factors} \\
&= x - 2, \quad x \neq 0,\ x \neq 1,\ x \neq -3 && \text{Simplified form}
\end{aligned}$$

■

The rule for multiplying fractions can be extended to cover products involving expressions that are not in fractional form. To do this, we rewrite the nonfractional expression as a fraction with a denominator of 1. Here is a simple example.

$$\frac{x + 3}{x - 2} \cdot (5x) = \frac{x + 3}{x - 2} \cdot \frac{5x}{1} = \frac{(x + 3)(5x)}{x - 2} = \frac{5x(x + 3)}{x - 2}$$

EXAMPLE 4 ■ Multiplying Rational Expressions

$$\begin{aligned}
\frac{3x}{2x^2 - 9x + 10} \cdot (2x - 5) &= \frac{3x}{2x^2 - 9x + 10} \cdot \frac{2x - 5}{1} && \text{Write in fractional form} \\
&= \frac{3x(2x - 5)}{(2x - 5)(x - 2)} && \text{Factor and multiply} \\
&= \frac{3x\cancel{(2x - 5)}}{\cancel{(2x - 5)}(x - 2)} && \text{Divide out common factors} \\
&= \frac{3x}{x - 2}, \quad x \neq \frac{5}{2} && \text{Simplified form}
\end{aligned}$$

■

In the next two examples, note how to divide out factors that differ only in sign.

EXAMPLE 5 ■ Multiplying Rational Expressions

$$\begin{aligned}
&\frac{x^3 - 125}{x^2 + 5x + 25} \cdot \frac{x + 4}{10 - 2x} \\
&= \frac{(x - 5)(x^2 + 5x + 25)}{x^2 + 5x + 25} \cdot \frac{x + 4}{2(5 - x)} && \text{Factor completely} \\
&= \frac{(x - 5)(x^2 + 5x + 25)}{x^2 + 5x + 25} \cdot \frac{x + 4}{-2(x - 5)} && 2(5 - x) = -2(x - 5) \\
&= \frac{(x - 5)(x^2 + 5x + 25)(x + 4)}{(x^2 + 5x + 25)(-2)(x - 5)} && \text{Multiply numerators and denominators} \\
&= \frac{\cancel{(x - 5)}\cancel{(x^2 + 5x + 25)}(x + 4)}{\cancel{(x^2 + 5x + 25)}(-2)\cancel{(x - 5)}} && \text{Divide out common factors} \\
&= -\frac{x + 4}{2}, \quad x \neq 5 && \text{Simplified form}
\end{aligned}$$

■

EXAMPLE 6 ■ Multiplying Rational Expressions

$$\frac{x-y}{y^2-x^2}\cdot\frac{x^2-xy-2y^2}{3x-6y}$$

$$=\frac{x-y}{(y+x)(y-x)}\cdot\frac{(x-2y)(x+y)}{3(x-2y)} \qquad \text{Factor completely}$$

$$=\frac{x-y}{(y+x)(-1)(x-y)}\cdot\frac{(x-2y)(x+y)}{3(x-2y)} \qquad (y-x)=(-1)(x-y)$$

$$=\frac{(x-y)(x-2y)(x+y)}{(y+x)(-1)(x-y)(3)(x-2y)} \qquad \text{Multiply numerators and denominators}$$

$$=\frac{\cancel{(x-y)}\cancel{(x-2y)}\cancel{(x+y)}}{\cancel{(x+y)}(-1)\cancel{(x-y)}(3)\cancel{(x-2y)}} \qquad \text{Divide out common factors}$$

$$=-\frac{1}{3},\quad x\neq y,\ x\neq -y,\ x\neq 2y \qquad \text{Simplified form}$$

■

The rule for multiplying rational expressions can be extended to cover the product of three or more fractions. This procedure is illustrated in Example 7.

EXAMPLE 7 ■ Multiplying Three Rational Expressions

$$\frac{x^2-3x+2}{x+2}\cdot\frac{3x}{x-2}\cdot\frac{2x+4}{x^2-5x}$$

$$=\frac{(x-1)(x-2)(3)(x)(2)(x+2)}{(x+2)(x-2)(x)(x-5)} \qquad \text{Factor and multiply}$$

$$=\frac{(x-1)\cancel{(x-2)}(3)\cancel{(x)}(2)\cancel{(x+2)}}{\cancel{(x+2)}\cancel{(x-2)}\cancel{(x)}(x-5)} \qquad \text{Divide out common factors}$$

$$=\frac{6(x-1)}{x-5},\quad x\neq 0,\ x\neq 2,\ x\neq -2 \qquad \text{Simplified form}$$

■

Dividing Rational Expressions

To divide two rational expressions, we simply *invert the divisor and multiply*. That is, we multiply the first fraction by the *reciprocal* of the second. For instance, to perform the following division

$$\frac{x}{x+3}\div\frac{4}{x-1}$$

we invert the fraction $4/(x-1)$ and multiply as follows.

$$\frac{x}{x+3}\div\frac{4}{x-1}=\frac{x}{x+3}\cdot\frac{x-1}{4}=\frac{x(x-1)}{(x+3)(4)}=\frac{x(x-1)}{4(x+3)},\quad x\neq 1$$

Dividing Rational Expressions

Let u, v, w, and z be real numbers, variables, or algebraic expressions such that $v \neq 0$, $w \neq 0$, and $z \neq 0$. Then the quotient of u/v and w/z is given by

$$\frac{u}{v} \div \frac{w}{z} = \frac{u}{v} \cdot \frac{z}{w} = \frac{uz}{vw}.$$

EXAMPLE 8 ■ Dividing Rational Expressions

Perform the following division.

$$\frac{3x}{x+2} \div \frac{x-6}{x+2}$$

Solution

$$\frac{3x}{x+2} \div \frac{x-6}{x+2} = \frac{3x}{x+2} \cdot \frac{x+2}{x-6} \qquad \text{Invert and multiply}$$

$$= \frac{(3x)(x+2)}{(x+2)(x-6)} \qquad \text{Multiply numerators and denominators}$$

$$= \frac{(3x)\cancel{(x+2)}}{\cancel{(x+2)}(x-6)} \qquad \text{Divide out common factors}$$

$$= \frac{3x}{x-6}, \quad x \neq -2 \qquad \text{Simplified form}$$

■

EXAMPLE 9 ■ Dividing Rational Expressions

Perform the following division.

$$\frac{2x}{3x-12} \div \frac{x^2-2x}{x^2-6x+8}$$

Solution

$$\frac{2x}{3x-12} \div \frac{x^2-2x}{x^2-6x+8} = \frac{2x}{3x-12} \cdot \frac{x^2-6x+8}{x^2-2x} \qquad \text{Invert and multiply}$$

$$= \frac{(2x)(x-2)(x-4)}{(3)(x-4)(x)(x-2)} \qquad \text{Factor and multiply}$$

$$= \frac{(2\cancel{x})\cancel{(x-2)}\cancel{(x-4)}}{(3)\cancel{(x-4)}\cancel{(x)}\cancel{(x-2)}} \qquad \text{Divide out common factors}$$

$$= \frac{2}{3}, \quad x \neq 0, x \neq 2, x \neq 4 \qquad \text{Simplified form}$$

■

Compound Fractions

Problems involving the division of two rational expressions are sometimes written as **compound fractions**. The rules for dividing fractions still apply in such cases. For instance, consider the following compound fraction.

$$\frac{\left(\dfrac{x-1}{5}\right)}{\left(\dfrac{x+3}{x}\right)} \quad \begin{array}{l} \text{Numerator fraction} \\ \rightarrow \text{Main fraction line} \\ \text{Denominator fraction} \end{array}$$

(Note that for compound fractions, we make the main fraction line slightly longer than the fraction lines in the numerator and denominator.) To perform the division implied by this compound fraction, we invert the denominator and multiply, as follows.

$$\frac{\left(\dfrac{x-1}{5}\right)}{\left(\dfrac{x+3}{x}\right)} = \frac{x-1}{5} \cdot \frac{x}{x+3} = \frac{x(x-1)}{5(x+3)}, \quad x \neq 0$$

EXAMPLE 10 ■ Simplifying a Compound Fraction

Simplify the following compound fraction.

$$\frac{\left(\dfrac{x^2+2x-3}{x-3}\right)}{4x+12}$$

Solution

In this case the denominator is not written in fractional form, so we begin by converting it to fractional form.

$$\frac{\left(\dfrac{x^2+2x-3}{x-3}\right)}{4x+12} = \frac{\left(\dfrac{x^2+2x-3}{x-3}\right)}{\left(\dfrac{4x+12}{1}\right)} \qquad \text{Rewrite denominator}$$

$$= \frac{x^2+2x-3}{x-3} \cdot \frac{1}{4x+12} \qquad \text{Invert and multiply}$$

$$= \frac{(x-1)(x+3)}{(x-3)(4)(x+3)} \qquad \text{Factor and multiply}$$

$$= \frac{(x-1)\cancel{(x+3)}}{(x-3)(4)\cancel{(x+3)}} \qquad \text{Divide out common factors}$$

$$= \frac{x-1}{4(x-3)}, \quad x \neq -3 \qquad \text{Simplified form}$$

■

DISCUSSION PROBLEM ■ You Be the Instructor

Suppose you are the instructor in an algebra course. One of your students turned in the following solution. Is the solution correct? If not, find the errors and write a correct solution.

$$\frac{9 - 4x^2}{6x^2 + 7x - 3} \cdot \frac{6x - 2}{2x - 3} \div \frac{3x + 1}{3x - 1}$$
$$= \frac{(3 + 2x)(3 - 2x)}{(3x + 1)(2x + 3)} \cdot \frac{2(3x - 1)}{2x - 3} \cdot \frac{3x + 1}{3x - 1}$$
$$= \frac{(2x + 3)(-1)(2x - 3)}{(3x + 1)(2x + 3)} \cdot \frac{2(3x - 1)}{(2x - 3)} \cdot \frac{3x + 1}{3x - 1}$$
$$= \frac{(2x + 3)(-1)(2x - 3)(2)(3x - 1)(3x + 1)}{(3x + 1)(2x + 3)(2x - 3)(3x - 1)}$$
$$= -2$$

■

Warm-Up

The following warm-up exercises involve skills that were covered in earlier sections. You will use these skills in the exercise set for this section.

In Exercises 1–6, evaluate the given quantity and write the answer in reduced form.

1. $\frac{3}{16} \cdot \frac{8}{5}$　　**2.** $\frac{5}{12} \cdot \frac{9}{75}$　　**3.** $-\frac{13}{35} \cdot \frac{-25}{104}$

4. $\frac{-225}{-448} \div \frac{-105}{28}$　　**5.** $\frac{14}{3} \div \frac{42}{45}$　　**6.** $\frac{-8}{105} \div \frac{2}{3}$

In Exercises 7–10, factor the algebraic expression.

7. $x^2 - 3x + 2$　　**8.** $2x^2 + 5x - 7$

9. $11x^2 + 6x - 5$　　**10.** $4x^2 - 28x + 49$

4.2 EXERCISES

In Exercises 1–6, find the missing algebraic expression so that the two fractions are equivalent.

1. $\frac{7}{3y} = \frac{7x^2}{3y(\quad)}, \quad x \neq 0$　　**2.** $\frac{2x}{x - 3} = \frac{14x(x - 3)^2}{(x - 3)(\quad)}$

3. $\frac{3u}{7v} = \frac{3u(\quad)}{7v(u + 1)}, \quad u \neq -1$　　**4.** $\frac{3t + 5}{t} = \frac{(3t + 5)(\quad)}{5t^2(3t - 5)}, \quad t \neq \frac{5}{3}$

5. $\frac{13x}{x - 2} = \frac{13x(\quad)}{4 - x^2}, \quad x \neq -2$　　**6.** $\frac{x^2}{10 - x} = \frac{x^2(\quad)}{x^2 - 10x}, \quad x \neq 0$

In Exercises 7–50, perform the indicated operation.

7. $\dfrac{7x^2}{3} \cdot \dfrac{9}{14x}$

8. $\dfrac{6}{5a} \cdot (25a)$

9. $\dfrac{8s^3}{9s} \cdot \dfrac{6s^2}{32s}$

10. $\dfrac{3x^4}{7x} \cdot \dfrac{8x^2}{9}$

11. $16u^4 \cdot \dfrac{12}{8u^2}$

12. $\dfrac{25}{8x} \cdot \dfrac{8x}{35}$

13. $\dfrac{8}{3 + 4x} \cdot (9 + 12x)$

14. $\dfrac{1 - 3x}{4} \cdot \dfrac{46}{15 - 45x}$

15. $\dfrac{8u^2v}{3u + v} \cdot \dfrac{u + v}{12u}$

16. $\dfrac{x + 25}{8} \cdot \dfrac{8}{x + 25}$

17. $\dfrac{12 - r}{3} \cdot \dfrac{3}{r - 12}$

18. $\dfrac{8 - z}{8 + z} \cdot \dfrac{z + 8}{z - 8}$

19. $\dfrac{(2x - 3)(x + 8)}{x^3} \cdot \dfrac{x}{3 - 2x}$

20. $\dfrac{x + 14}{x^3(10 - x)} \cdot \dfrac{x(x - 10)}{5}$

21. $\dfrac{6r}{r - 2} \cdot \dfrac{r^2 - 4}{33r^2}$

22. $\dfrac{5y - 20}{5y + 15} \cdot \dfrac{2y + 6}{y - 4}$

23. $\dfrac{2t^2 - t - 15}{t + 2} \cdot \dfrac{t^2 - t - 6}{t^2 - 6t + 9}$

24. $\dfrac{y^2 - 16}{2y^3} \cdot \dfrac{4y}{y^2 - 6y + 8}$

25. $(x^2 - 4y^2) \cdot \dfrac{xy}{(x - 2y)^2}$

26. $(u - 2v)^2 \cdot \dfrac{u + 2v}{u - 2v}$

27. $\dfrac{x^2 + 2xy - 3y^2}{(x + y)^2} \cdot \dfrac{x^2 - y^2}{x + 3y}$

28. $\dfrac{x - 2y}{x + 2y} \cdot \dfrac{x^2 + 4y^2}{x^2 - 4y^2}$

29. $\dfrac{xu - yu + xv - yv}{xu + yu - xv - yv} \cdot \dfrac{xu + yu + xv + yv}{xu - yu - xv + yv}$

30. $\dfrac{x^3 + 3x^2 - 4x - 12}{x^3 - 3x^2 - 4x + 12} \cdot \dfrac{x^2 - 9}{x}$

31. $\dfrac{x + 5}{x - 5} \cdot \dfrac{2x^2 - 9x - 5}{3x^2 + x - 2} \cdot \dfrac{x^2 - 1}{x^2 + 7x + 10}$

32. $\dfrac{t^2 + 4t + 3}{2t^2 - t - 10} \cdot \dfrac{t}{t^2 + 3t + 2} \cdot \dfrac{2t^2 + 4t^3}{t^2 + 3t}$

33. $\dfrac{7xy^2}{10u^2v} \div \dfrac{21x^3}{45uv}$

34. $\dfrac{25x^2y}{60x^3y^2} \div \dfrac{5x^4y^3}{16x^2y}$

35. $\dfrac{3(a + b)}{4} \div \dfrac{(a + b)^2}{2}$

36. $\dfrac{x^2 + 9}{5(x + 2)} \div \dfrac{x + 3}{5(x^2 - 4)}$

37. $\dfrac{(x^3y)^2}{(x + 2y)^2} \div \dfrac{x^2y}{(x + 2y)^3}$

38. $\dfrac{x^2 - y^2}{2x^2 - 8x} \div \dfrac{(x - y)^2}{2xy}$

39. $\dfrac{\left(\dfrac{x^2}{12}\right)}{\left(\dfrac{5x}{18}\right)}$

40. $\dfrac{\left(\dfrac{(3u^2v)^2}{6v^3}\right)}{\left(\dfrac{(uv^3)^2}{3uv}\right)}$

41. $\dfrac{\left(\dfrac{25x^2}{x - 5}\right)}{\left(\dfrac{10x}{5 - x}\right)}$

42. $\dfrac{\left(\dfrac{5x}{x + 7}\right)}{\left(\dfrac{10}{x^2 + 8x + 7}\right)}$

43. $\dfrac{16x^2 + 8x + 1}{3x^2 + 8x - 3} \div \dfrac{4x^2 - 3x - 1}{x^2 + 6x + 9}$

44. $\dfrac{x^2 - 25}{x} \div \dfrac{x^3 - 5x^2}{x^2 + x}$

45. $\dfrac{x(x + 3) - 2(x + 3)}{x^2 - 4} \div \dfrac{x}{x^2 + 4x + 4}$

46. $\dfrac{t^3 + t^2 - 9t - 9}{t^2 - 5t + 6} \div \dfrac{t^2 + 6t + 9}{t - 2}$

47. $\dfrac{2x^2 + 5x - 25}{3x^2 + 5x + 2} \cdot \dfrac{3x^2 + 2x}{x + 5} \div \left(\dfrac{x}{x + 1}\right)^2$

48. $\dfrac{t^2 - 100}{4t^2} \cdot \dfrac{t^3 - 5t^2 - 50t}{t^4 + 10t^3} \div \dfrac{(t - 10)^2}{5t}$

49. $x^3 \cdot \dfrac{x^{2n} - 9}{x^{2n} + 4x^n + 3} \div \dfrac{x^{2n} - 2x^n - 3}{x}$, n is a positive integer

50. $\dfrac{x^{n+1} - 8x}{x^{2n} + 2x^n + 1} \cdot \dfrac{x^{2n} - 4x^n - 5}{x} \div x^n$, n is a positive integer

In Exercises 51 and 52, evaluate the expression at each value of x. (If not possible, state the reason.)

	Expression	*Value*	
51.	$\dfrac{x - 10}{4x}$	(a) $x = 10$	(b) $x = 0$
		(c) $x = -2$	(d) $x = 12$
52.	$\dfrac{x^2 - 4x}{x^2 - 9}$	(a) $x = 0$	(b) $x = 4$
		(c) $x = 3$	(d) $x = -3$

53. *Photocopy Rate* A photocopier produces copies at a rate of 20 pages per minute.
(a) Determine the time required to copy one page.
(b) Determine the time required to copy x pages.
(c) Determine the time required to copy 35 pages.

54. *Pumping Rate* The rated volume for a pump is 15 gallons per minute.
(a) Determine the time required to pump one gallon.
(b) Determine the time required to pump x gallons.
(c) Determine the time required to pump 130 gallons.

55. Determine the area of the shaded region in the accompanying figure.

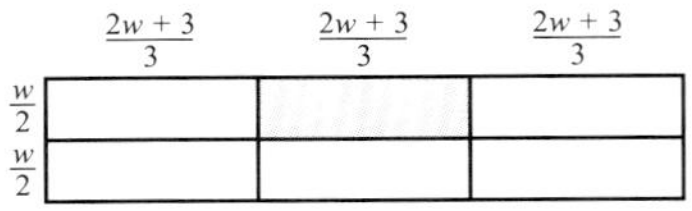

56. Determine the area of the shaded region in the accompanying figure.

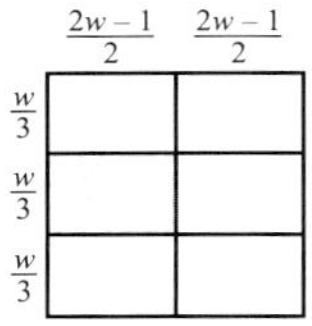

SECTION 4.3

Adding and Subtracting Rational Expressions

Combining Rational Expressions with Like Denominators • The Least Common Multiple of Polynomials • Combining Rational Expressions with Unlike Denominators • Compound Fractions

Combining Rational Expressions with Like Denominators

As with numerical fractions, the procedure used to add (or subtract) two rational expressions depends upon whether the expressions have *like* or *unlike* denominators. To add (or subtract) two rational expressions with *like* denominators, we simply combine their numerators and place the result over the common denominator. Here are some examples.

$$\frac{2}{7} + \frac{4}{7} = \frac{2 + 4}{7} = \frac{6}{7}$$

$$\frac{4}{3x} - \frac{5}{3x} = \frac{4 - 5}{3x} = -\frac{1}{3x}$$

$$\frac{3x + 2}{3} - \frac{x}{3} = \frac{(3x + 2) - (x)}{3} = \frac{2x + 2}{3}$$

$$\frac{2x}{x + 3} - \frac{4}{x + 3} = \frac{2x - 4}{x + 3}$$

In each of these cases, we used one of the following rules.

Adding or Subtracting Rational Expressions with Like Denominators

If u, v, and w are real numbers, variables, or variable expressions, and $w \neq 0$, then the following rules are valid.

1. $\dfrac{u}{w} + \dfrac{v}{w} = \dfrac{u + v}{w}$ — Add fractions with like denominators
2. $\dfrac{u}{w} - \dfrac{v}{w} = \dfrac{u - v}{w}$ — Subtract fractions with like denominators

EXAMPLE 1 ■ Adding Rational Expressions with Like Denominators

Add the rational expressions $\dfrac{x}{4} + \dfrac{5 - x}{4}$.

Solution

For these two rational expressions, the denominators are the same. Therefore, we can add the rational expressions by simply adding their numerators, as follows.

$$\frac{x}{4} + \frac{5 - x}{4} = \frac{x + (5 - x)}{4} = \frac{5}{4}$$

■

EXAMPLE 2 ■ Subtracting Rational Expressions with Like Denominators

Subtract the following rational expressions.

$$\frac{7}{2x - 3} - \frac{3x}{2x - 3}$$

Solution

Since the denominators are the same, we use the rule for subtracting rational expressions with like denominators, as follows.

$$\frac{7}{2x - 3} - \frac{3x}{2x - 3} = \frac{7 - 3x}{2x - 3}$$ ■

After adding or subtracting two (or more) rational expressions, you should check the resulting fraction to see if it can be simplified. This procedure is illustrated in the next example.

EXAMPLE 3 ■ Subtracting Rational Expressions and Simplifying

$$\frac{x}{x^2 - 2x - 3} - \frac{3}{x^2 - 2x - 3} = \frac{x - 3}{x^2 - 2x - 3}$$ Add numerators

$$= \frac{x - 3}{(x - 3)(x + 1)}$$ Factor completely

$$= \frac{\cancel{(x - 3)}(1)}{\cancel{(x - 3)}(x + 1)}$$ Divide out common factor

$$= \frac{1}{x + 1}, \quad x \neq 3$$ Simplified form ■

The rules for adding and subtracting rational expressions with like denominators can be extended to cover sums and differences involving three or more rational expressions. Example 4 demonstrates this procedure.

EXAMPLE 4 ■ Combining Three Rational Expressions with Like Denominators

$$\frac{x^2 - 26}{x - 5} - \frac{2x + 4}{x - 5} + \frac{10 + x}{x - 5} = \frac{(x^2 - 26) - (2x + 4) + (10 + x)}{x - 5}$$

$$= \frac{x^2 - 26 - 2x - 4 + 10 + x}{x - 5}$$

$$= \frac{x^2 - x - 20}{x - 5}$$

$$= \frac{\cancel{(x - 5)}(x + 4)}{\cancel{x - 5}}$$

$$= x + 4, \quad x \neq 5$$ ■

MATH MATTERS

Roman Numerals

The standard numerals 0, 1, 2, 3, 4, 5, 6, 7, 8, and 9 that we use to represent numbers are called *Arabic numerals*. Although almost all modern representation of numbers is done with Arabic numerals, we still occasionally use *Roman numerals* to represent numbers. For instance, chapter numbers in books, numbers in an outline, and numbers on watches are sometimes written with Roman numerals. The list on the right compares the two numeral systems.

To represent a number using Roman numerals, use the following two rules.

- If a larger numeral is in front of a smaller numeral, the smaller is added to the larger. For example, XVI = 10 + 5 + 1 = 16.
- If a smaller numeral is in front of a larger numeral, the smaller is subtracted from the larger. For example, XIV = 10 + (5 − 1) = 14.

The following Roman numerals represent famous dates in American history. Write the Arabic numerals for these dates. (The answers are given in the back of the text.)

(a) MDCCLXXVI (b) MCDXCII (c) MDCXX

Arabic	*Roman*
1	I
2	II
3	III
4	IV
5	V
6	VI
7	VII
8	VIII
9	IX
10	X
50	L
100	C
500	D
1,000	M

Modern	1	2	3	4	5	6	7	8	9	10	100
Egyptian	|	||	|||	||||	||| ||	||| |||	|||| |||	|||| ||||	||| ||| |||	𓎆	𓍢
Babylonian	▼	▼▼	▼▼▼	▼▼▼▼	▼▼▼ ▼▼	▼▼▼ ▼▼▼	▼▼▼▼ ▼▼▼	▼▼▼▼ ▼▼▼▼	▼▼▼▼▼ ▼▼▼▼	◀	▼◀◀
Early Roman	I	II	III	IIII	V	VI	VII	VIII	IX	X	C
Chinese	一	二	三	四	五	六	七	八	九	十	百
Hindu	१	२	३	४	५	६	७	८	९	१०	१००
Mayan	•	••	•••	••••	—	• —	•• —	••• —	•••• —	— —	— 𝋠

The Least Common Multiple of Polynomials

To add or subtract rational expressions with *unlike* denominators, we must first rewrite each expression using the **least common multiple** of the denominators of the individual expressions. The least common multiple of two (or more) polynomials is the simplest polynomial that is a multiple of each of the original polynomials. Here are some examples.

Polynomials	*Least Common Multiple*
27 and 18	$\rlap{\overbrace{}^{27}} 3 \cdot \underbrace{3 \cdot 3 \cdot 2}_{18} = 3^3 \cdot 2 = 54$
x and $x - 4$	$x \cdot (x - 4) = x(x - 4)$
x^2 and $x(x - 3)$	$\rlap{\overbrace{}^{x^2}} x \cdot \underbrace{x \cdot (x - 3)}_{x(x-3)} = x^2(x - 3)$
$x^2 - 4$ and $x + 2$	$\overbrace{\underbrace{(x + 2)}_{x+2} \cdot (x - 2)}^{x^2 - 4} = x^2 - 4$

These examples illustrate the following guidelines for finding the least common multiple of two or more polynomials.

Guidelines for Finding the Least Common Multiple of Polynomials

1. Factor each polynomial completely.
2. List all the *different* factors of the polynomials.
3. For each factor, determine the highest power (of that factor) that occurs in the polynomials.
4. The least common multiple is the product of the different factors, each raised to the appropriate power.

EXAMPLE 5 ■ **Finding Least Common Multiples**

Find the least common multiple of each of the following sets of polynomials.

(a) $6x,\ 2x^2$ (b) $x^2 - x,\ 2x - 2$ (c) $3x^2 + 6x,\ x^2 + 4x + 4$

Solution

(a) These two polynomials factor as follows.

$$6x = 2 \cdot 3 \cdot x \quad \text{and} \quad 2x^2 = 2 \cdot x^2$$

The different factors are 2, 3, and x. Using the highest powers of these factors, we conclude that the least common multiple is $6x^2$.

(b) These two polynomials factor as follows.

$$x^2 - x = x(x - 1) \quad \text{and} \quad 2x - 2 = 2(x - 1)$$

The different factors are 2, x, and $x - 1$. Using the highest powers of these factors, we conclude that the least common multiple is $2x(x - 1)$.

(c) These two polynomials factor as follows.

$$3x^2 + 6x = 3x(x + 2) \quad \text{and} \quad x^2 + 4x + 4 = (x + 2)^2$$

The different factors are 3, x, and $x + 2$. Using the highest powers of these factors, we conclude that the least common multiple is $3x(x + 2)^2$. ■

Combining Rational Expressions with Unlike Denominators

To add or subtract rational expressions with *unlike* denominators, we must first rewrite the rational expressions so that they have *like* denominators. The (like) denominator that we use is the least common multiple of the original denominators, and it is called the **least common denominator** of the original rational expressions. Once the rational expressions have been written with like denominators, we simply add or subtract the rational expressions using the rules given at the beginning of this section. This procedure is demonstrated in the next four examples.

EXAMPLE 6 ■ Adding Rational Expressions with Unlike Denominators

Add the following rational expressions.

$$\frac{7}{6x} + \frac{5}{8x}$$

Solution

By factoring the denominators, $6x = 2 \cdot 3 \cdot x$ and $8x = 2^3 \cdot x$, we conclude that the least common denominator is $24x$.

$$\frac{7}{6x} + \frac{5}{8x} = \frac{7(4)}{6x(4)} + \frac{5(3)}{8x(3)} \quad \text{Rewrite fractions using least common denominator}$$

$$= \frac{28}{24x} + \frac{15}{24x} \quad \text{Like denominators}$$

$$= \frac{28 + 15}{24x} \quad \text{Add fractions}$$

$$= \frac{43}{24x} \quad \text{Simplified form}$$

■

EXAMPLE 7 ■ Subtracting Rational Expressions with Unlike Denominators

Subtract the rational expressions $\dfrac{3}{x-3} - \dfrac{5}{x+2}$.

Solution

The only factors of the denominators are $(x - 3)$ and $(x + 2)$. Therefore, the least common denominator is $(x - 3)(x + 2)$.

$$\frac{3}{x-3} - \frac{5}{x+2} = \frac{3(x+2)}{(x-3)(x+2)} - \frac{5(x-3)}{(x-3)(x+2)} \qquad \text{Rewrite fractions using least common denominator}$$

$$= \frac{3x+6}{(x-3)(x+2)} - \frac{5x-15}{(x-3)(x+2)} \qquad \text{Like denominators}$$

$$= \frac{(3x+6)-(5x-15)}{(x-3)(x+2)} \qquad \text{Subtract fractions}$$

$$= \frac{3x+6-5x+15}{(x-3)(x+2)} \qquad \text{Remove parentheses}$$

$$= \frac{-2x+21}{(x-3)(x+2)} \qquad \text{Simplified form}$$

■

EXAMPLE 8 ■ Adding Rational Expressions with Unlike Denominators

Add the rational expressions $\dfrac{6x}{x^2-4} + \dfrac{3}{2-x}$.

Solution

The factors in the denominators are $x^2 - 4 = (x + 2)(x - 2)$ and $2 - x$. Since $(2 - x) = (-1)(x - 2)$, we can rewrite the original addition problem as a subtraction problem, as follows.

$$\frac{6x}{x^2-4} + \frac{3}{(-1)(x-2)} = \frac{6x}{(x+2)(x-2)} - \frac{3}{x-2}$$

Now, using the least common denominator of $(x + 2)(x - 2)$, we subtract as follows.

$$\frac{6x}{(x+2)(x-2)} - \frac{3}{x-2}$$

$$= \frac{6x}{(x+2)(x-2)} - \frac{3(x+2)}{(x+2)(x-2)} \qquad \text{Rewrite fractions using least common denominator}$$

$$= \frac{6x}{(x+2)(x-2)} - \frac{3x+6}{(x+2)(x-2)} \qquad \text{Like denominators}$$

$$= \frac{6x-(3x+6)}{(x+2)(x-2)} \qquad \text{Subtract fractions}$$

$$= \frac{6x - 3x - 6}{(x + 2)(x - 2)} \qquad \text{Remove parentheses}$$

$$= \frac{3x - 6}{(x + 2)(x - 2)}$$

$$= \frac{3\cancel{(x - 2)}}{(x + 2)\cancel{(x - 2)}} \qquad \text{Divide out common factors}$$

$$= \frac{3}{x + 2}, \quad x \neq 2 \qquad \text{Simplified form}$$

■

In the next example, we show how a least common denominator can be used to combine three rational expressions.

EXAMPLE 9 ■ Combining Rational Expressions with Unlike Denominators

Combine the following rational expressions.

$$\frac{2x - 5}{6x + 9} - \frac{4}{2x^2 + 3x} + \frac{1}{x}$$

Solution

The denominators factor as $6x + 9 = 3(2x + 3)$, $2x^2 + 3x = x(2x + 3)$, and x. Therefore, we conclude that the least common denominator is $3x(2x + 3)$.

$$\frac{2x - 5}{6x + 9} - \frac{4}{2x^2 + 3x} + \frac{1}{x}$$

$$= \frac{(2x - 5)(x)}{3(2x + 3)(x)} - \frac{(4)(3)}{x(2x + 3)(3)} + \frac{3(2x + 3)}{(x)(3)(2x + 3)} \qquad \text{Rewrite fractions using least common denominator}$$

$$= \frac{2x^2 - 5x}{3x(2x + 3)} - \frac{12}{3x(2x + 3)} + \frac{6x + 9}{3x(2x + 3)} \qquad \text{Like denominators}$$

$$= \frac{2x^2 - 5x - 12 + 6x + 9}{3x(2x + 3)} \qquad \text{Add fractions}$$

$$= \frac{2x^2 + x - 3}{3x(2x + 3)} \qquad \text{Combine like terms}$$

$$= \frac{(x - 1)(2x + 3)}{3x(2x + 3)} \qquad \text{Factor numerator}$$

$$= \frac{(x - 1)\cancel{(2x + 3)}}{3x\cancel{(2x + 3)}} \qquad \text{Divide out common factors}$$

$$= \frac{x - 1}{3x}, \quad x \neq -\frac{3}{2} \qquad \text{Simplified form}$$

■

Compound Fractions

Compound fractions can have numerators or denominators that are the sum or difference of fractions. Here are a couple of examples.

$$\frac{\left(\frac{4}{x}+5\right)}{\left(2-\frac{3}{x}\right)} \quad \text{and} \quad \frac{\left(\frac{1}{x+4}-\frac{2}{x}\right)}{\left(\frac{3}{x}+\frac{3}{2x^2}\right)}$$

To simplify a compound fraction, we first combine its numerator into a single fraction and its denominator into a single fraction. Then we divide as demonstrated in the previous section (by inverting the denominator and multiplying).

EXAMPLE 10 ■ Simplifying a Compound Fraction

$$\frac{\left(\frac{x}{4}+\frac{3}{2}\right)}{\left(2-\frac{3}{x}\right)} = \frac{\left(\frac{x}{4}+\frac{6}{4}\right)}{\left(\frac{2x}{x}-\frac{3}{x}\right)}$$ Find least common denominators

$$= \frac{\left(\frac{x+6}{4}\right)}{\left(\frac{2x-3}{x}\right)}$$ Add fractions in numerator and in denominator

$$= \frac{x+6}{4} \cdot \frac{x}{2x-3}$$ Invert and multiply

$$= \frac{x(x+6)}{4(2x-3)}, \quad x \neq 0$$ Simplified form ■

Another way to simplify the compound fraction given in Example 10 is to multiply the numerator and denominator by the least common denominator of *every* fraction in the numerator and denominator. For this fraction, notice what happens when we multiply the numerator and denominator by $4x$.

$$\frac{\left(\frac{x}{4}+\frac{3}{2}\right)}{\left(2-\frac{3}{x}\right)} = \frac{\left(\frac{x}{4}+\frac{3}{2}\right)}{\left(2-\frac{3}{x}\right)} \cdot \frac{4x}{4x} = \frac{\frac{x}{4}(4x)+\frac{3}{2}(4x)}{2(4x)-\frac{3}{x}(4x)} = \frac{x^2+6x}{8x-12}, \quad x \neq 0$$

EXAMPLE 11 ■ **Compound Fractions**

Simplify the following compound fraction.

$$\frac{\left(\dfrac{2}{x+2}\right)}{\left(\dfrac{1}{x+2}+\dfrac{2}{x}\right)}$$

Solution

To simplify this compound fraction, we multiply the numerator and denominator by the least common denominator of all fractions occurring in the numerator and denominator, which in this case is $x(x+2)$.

$$\frac{\left(\dfrac{2}{x+2}\right)}{\left(\dfrac{1}{x+2}+\dfrac{2}{x}\right)} = \frac{\left(\dfrac{2}{x+2}\right)(x)(x+2)}{\dfrac{1}{x+2}(x)(x+2)+\dfrac{2}{x}(x)(x+2)}$$

Multiply numerator and denominator by $x(x+2)$

$$= \frac{2x}{x+2(x+2)}$$

$$= \frac{2x}{3x+4}, \quad x \neq -2, x \neq 0$$

Simplified form ■

DISCUSSION PROBLEM ■ **You Be the Instructor**

Suppose you are the instructor in an algebra course. One of your students turned in the following solution. Is the solution correct? If not, find the error(s) and write a correct solution.

$$\frac{4}{x-1} - \frac{4}{x} + \frac{x+3}{x^2-1} = \frac{4(x+1)}{x(x^2-1)} - \frac{4(x^2-1)}{x(x^2-1)} + \frac{x(x+3)}{x(x^2-1)}$$

$$= \frac{4x+4}{x(x^2-1)} - \frac{4x^2-4}{x(x^2-1)} + \frac{x^2+3x}{x(x^2-1)}$$

$$= \frac{4x+4-4x^2-4+x^2+3x}{x(x^2-1)}$$

$$= \frac{-3x^2+7x}{x(x^2-1)}$$

$$= \frac{x(-3x+7)}{x(x^2-1)}$$

$$= \frac{-3x+7}{x^2-1}, \quad x \neq 0$$

■

Warm-Up

The following warm-up exercises involve skills that were covered in earlier sections. You will use these skills in the exercise set for this section.

In Exercises 1–10, perform the indicated operations.

1. $\frac{4}{9} + \frac{3}{9}$ **2.** $\frac{3}{20} + \frac{7}{20}$ **3.** $\frac{11}{32} - \frac{5}{32}$ **4.** $\frac{7}{10} - \frac{13}{10}$

5. $\frac{3}{4} + \frac{5}{18}$ **6.** $\frac{11}{24} + \frac{5}{16}$ **7.** $-\frac{8}{9} + \frac{18}{75}$ **8.** $\frac{27}{3} - \frac{8}{21}$

9. $\frac{28}{5} \cdot \frac{65}{14}$ **10.** $\frac{15}{16} \div 12$

4.3 EXERCISES

In Exercises 1–14, perform the indicated operation and write your answer in reduced form.

1. $\frac{x}{9} - \frac{x+2}{9}$ **2.** $\frac{z^2}{3} + \frac{z^2-2}{3}$ **3.** $\frac{4-y}{4} + \frac{3y}{4}$

4. $\frac{10x^2+1}{3} - \frac{10x^2}{3}$ **5.** $\frac{2}{3a} - \frac{11}{3a}$ **6.** $\frac{16+z}{5z} - \frac{11-z}{5z}$

7. $\frac{2x+5}{3} + \frac{1-x}{3}$ **8.** $\frac{6x}{13} - \frac{7x}{13}$ **9.** $\frac{3y}{3} - \frac{3y-3}{3} - \frac{7}{3}$

10. $\frac{-16u}{9} - \frac{27-16u}{9} + \frac{2}{9}$ **11.** $\frac{2x-1}{x(x-3)} + \frac{1-x}{x(x-3)}$ **12.** $\frac{5x-1}{x+4} + \frac{5-4x}{x+4}$

13. $\frac{3y-22}{y-6} - \frac{2y-16}{y-6}$ **14.** $\frac{7s-5}{2s+5} + \frac{3(s+10)}{2s+5}$

In Exercises 15–20, find the least common multiple of the pair of polynomials.

15. $5x^2, 20x^3$ **16.** $14t^2, 42t^5$ **17.** $15x^2, 3(x+5)$

18. $18y^3, 27y(y-3)^2$ **19.** $6(x^2-4), 2x(x+2)$ **20.** $t^3+3t^2+9t, 2t^2(t^2-9)$

In Exercises 21–26, find the least common denominator of the two fractions and rewrite each fraction using the least common denominator.

21. $\frac{n+8}{3n-12}, \frac{10}{6n^2}$ **22.** $\frac{8s}{(s+2)^2}, \frac{3}{s^3+s^2-2s}$ **23.** $\frac{2}{x^2(x-3)}, \frac{5}{x(x+3)}$

24. $\frac{5t}{2t(t-3)^2}, \frac{4}{t(t-3)}$ **25.** $\frac{x-8}{x^2-25}, \frac{9x}{x^2-10x+25}$ **26.** $\frac{3y}{y^2-y-12}, \frac{y-4}{y^2+3y}$

In Exercises 27–58, perform the specified operations and simplify.

27. $\frac{5}{4x} - \frac{3}{5}$

28. $\frac{10}{b} + \frac{1}{10b^2}$

29. $\frac{7}{a} + \frac{14}{a^2}$

30. $\frac{1}{6u^2} - \frac{2}{9u}$

31. $\frac{20}{x-4} + \frac{20}{4-x}$

32. $\frac{15}{2-t} - \frac{7}{t-2}$

33. $\frac{3x}{x-8} - \frac{6}{8-x}$

34. $\frac{1}{y-6} + \frac{y}{6-y}$

35. $25 + \frac{10}{x+4}$

36. $\frac{100}{x-10} - 8$

37. $-\frac{1}{6x} + \frac{1}{6(x-3)}$

38. $\frac{1}{x} - \frac{1}{x+2}$

39. $\frac{x}{x+3} - \frac{5}{x-2}$

40. $\frac{3}{t(t+1)} + \frac{4}{t}$

41. $\frac{3}{x+1} - \frac{2}{x}$

42. $\frac{5}{x-4} - \frac{3}{x}$

43. $\frac{3}{x-5} + \frac{2}{x+5}$

44. $\frac{7}{2x-3} + \frac{3}{2x+3}$

45. $\frac{4}{x^2} - \frac{4}{x^2+1}$

46. $\frac{2}{y} + \frac{1}{2y^2}$

47. $\frac{x}{x^2-9} + \frac{3}{x(x-3)}$

48. $\frac{x}{x^2-x-30} - \frac{1}{x+5}$

49. $\frac{4}{x-4} + \frac{16}{(x-4)^2}$

50. $\frac{3}{x-2} - \frac{1}{(x-2)^2}$

51. $\frac{4}{x} - \frac{2}{x^2} + \frac{4}{x+3}$

52. $\frac{5}{2(x+1)} - \frac{1}{2x} - \frac{3}{2(x+1)^2}$

53. $\frac{x+2}{x-1} - \frac{2}{x+6} - \frac{14}{x^2+5x-6}$

54. $\frac{x}{x^2+15x+50} + \frac{7}{2(x+10)} - \frac{3}{2(x+5)}$

55. $\frac{y}{x^2+xy} - \frac{x}{xy+y^2}$

56. (a) $\frac{5}{x+y} + \frac{5}{x-y}$
(b) $\frac{3}{x^2+5x+6} + \frac{2}{x^2+3x+2}$

57. (a) $\frac{3u}{u^2-2uv+v^2} + \frac{2}{u-v}$
(b) $\frac{1}{x^2-1} - \frac{1}{x^2+x-2}$

58. (a) $\frac{1}{x} - \frac{3}{y} + \frac{3x-y}{xy}$
(b) $\frac{2}{x^2+x-6} - \frac{2}{x^2+4x+3}$

In Exercises 59–68, simplify the compound fraction.

59. $\dfrac{\frac{1}{2}}{\left(3+\frac{1}{x}\right)}$

60. $\dfrac{\left(\frac{1}{t}-1\right)}{\left(\frac{1}{t}+1\right)}$

61. $\dfrac{\left(16x-\frac{1}{x}\right)}{\left(\frac{1}{x}-4\right)}$

62. $\dfrac{\left(\frac{36}{y}-y\right)}{6+y}$

63. $\dfrac{\left(3+\frac{9}{x-3}\right)}{\left(4+\frac{12}{x-3}\right)}$

64. $\dfrac{\left(x+\frac{2}{x-3}\right)}{\left(x+\frac{6}{x-3}\right)}$

65. $\dfrac{\left(1-\frac{1}{y^2}\right)}{\left(1-\frac{4}{y}+\frac{3}{y^2}\right)}$

66. $\dfrac{\left(\frac{x+1}{x+2}-\frac{1}{x}\right)}{\left(\frac{2}{x+2}\right)}$

67. $\dfrac{\left(\frac{y}{x}-\frac{x}{y}\right)}{\left(\frac{x+y}{xy}\right)}$

68. $\dfrac{\left(x-\frac{2y^2}{x-y}\right)}{x-2y}$

69. *Work Rate* After two people work together for t hours on a common task, the fractional part of the job done by each of the two workers is $t/4$ and $t/6$. What fractional part of the task has been completed?

70. *Work Rate* After two people work together for t hours on a common task, the fractional part of the job done by each of the two workers is $t/3$ and $t/5$. What fractional part of the task has been completed?

71. *Average of Two Numbers* Determine the average of the two real numbers $x/4$ and $x/6$.

72. *Average of Two Numbers* Determine the average of the two real numbers $x/3$ and $x/5$.

73. *Equal Parts* Find three real numbers that divide the real number line between $x/6$ and $x/2$ into four equal parts.

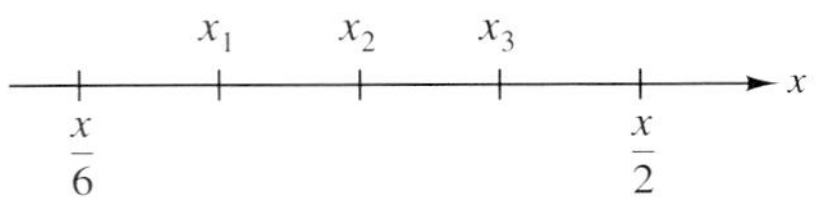

74. *Equal Parts* Find two real numbers that divide the real number line between $x/5$ and $x/3$ into three equal parts.

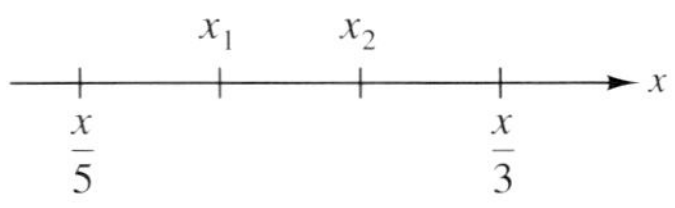

75. *Parallel Resistance* When two resistors are connected in parallel (see figure), the total resistance is given by

$$\frac{1}{\left(\frac{1}{R_1} + \frac{1}{R_2}\right)}.$$

Simplify this compound fraction.

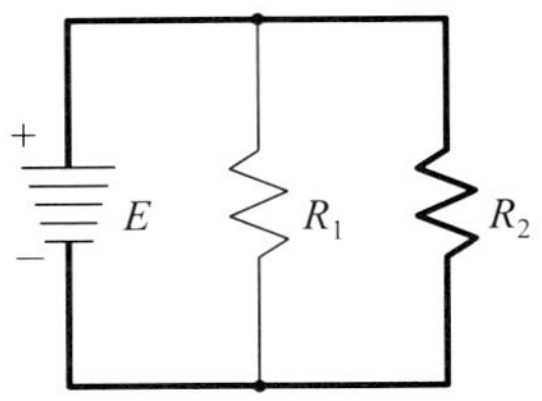

76. Determine whether the following is correct. If it is not, find and correct any errors.

$$\frac{2}{x} - \frac{3}{x+1} + \frac{x+1}{x^2} = \frac{2x(x+1) - 3x^2 + (x+1)^2}{x^2(x+1)}$$

$$= \frac{2x^2 + x - 3x^2 + x^2 + 1}{x^2(x+1)}$$

$$= \frac{x+1}{x^2(x+1)}$$

$$= \frac{1}{x^2}$$

SECTION 4.4

Dividing Polynomials

Dividing a Polynomial by a Monomial • *Long Division* • *Synthetic Division* • *Factoring and Division*

Dividing a Polynomial by a Monomial

To divide a polynomial by a monomial, we reverse the procedure used to add (or subtract) two rational expressions. For example, we can add the expressions 2 and $1/x$ as follows.

$$2 + \frac{1}{x} = \frac{2x}{x} + \frac{1}{x} = \frac{2x + 1}{x}$$

By reversing this process, we can divide the polynomial $2x + 1$ by the monomial x. That is,

$$\frac{2x + 1}{x} = \frac{2x}{x} + \frac{1}{x} = 2 + \frac{1}{x}.$$

Dividing a Polynomial by a Monomial

Let u, v, and w be real numbers, variables, or algebraic expressions such that $w \neq 0$. Then the following rules are valid.

1. $\dfrac{u + v}{w} = \dfrac{u}{w} + \dfrac{v}{w}$ 2. $\dfrac{u - v}{w} = \dfrac{u}{w} - \dfrac{v}{w}$

Note that when dividing a polynomial by a monomial, we usually reduce the resulting expressions to simplest form.

EXAMPLE 1 ■ **Dividing a Polynomial by a Monomial**

Perform the following division and simplify.

$$\frac{12x^2 - 20x + 8}{4x}$$

Solution

$$\frac{12x^2 - 20x + 8}{4x} = \frac{12x^2}{4x} - \frac{20x}{4x} + \frac{8}{4x}$$

$$= \frac{3(\cancel{4x})(x)}{\cancel{4x}} - \frac{5(\cancel{4x})}{\cancel{4x}} + \frac{2(\cancel{4})}{\cancel{4}x}$$

$$= 3x - 5 + \frac{2}{x}$$

■

EXAMPLE 2 ■ **Dividing a Polynomial by a Monomial**

Perform the following division and simplify.

$$\frac{5y^3 + 8y^2 - 10y}{5y^2}$$

Solution

$$\frac{5y^3 + 8y^2 - 10y}{5y^2} = \frac{5y^3}{5y^2} + \frac{8y^2}{5y^2} - \frac{10y}{5y^2}$$

$$= \frac{\cancel{5y^2}(y)}{\cancel{5y^2}} + \frac{8\cancel{y^2}}{5\cancel{y^2}} - \frac{2(\cancel{5y})}{\cancel{5y}(y)}$$

$$= y + \frac{8}{5} - \frac{2}{y}$$ ■

EXAMPLE 3 ■ **Dividing a Polynomial by a Monomial**

Perform the following division and simplify. $\dfrac{-12x^3y^2 + 6x^2y + 9y^3}{3xy^2}$

Solution

$$\frac{-12x^3y^2 + 6x^2y + 9y^3}{3xy^2} = -\frac{12x^3y^2}{3xy^2} + \frac{6x^2y}{3xy^2} + \frac{9y^3}{3xy^2}$$

$$= -\frac{4(\cancel{3x})(x^2)(\cancel{y^2})}{(\cancel{3x})(\cancel{y^2})} + \frac{2(\cancel{3x})(x)(\cancel{y})}{\cancel{3x}(\cancel{y})(y)} + \frac{3y(\cancel{3y^2})}{x(\cancel{3y^2})}$$

$$= -4x^2 + \frac{2x}{y} + \frac{3y}{x}$$ ■

Long Division

In Section 4.2 we saw how to divide one polynomial by another by factoring and canceling common factors. For instance, we can divide $(x^2 - 2x - 3)$ by $(x - 3)$ as follows.

$$(x^2 - 2x - 3) \div (x - 3) = \frac{x^2 - 2x - 3}{x - 3} = \frac{(x + 1)(x - 3)}{x - 3} = x + 1, \quad x \neq 3$$

This procedure works well for polynomials that are easily factored. We will now look at a more *general* procedure, which follows a "long division algorithm" that is similar to the algorithm used for dividing positive integers. We review that procedure in the following example.

EXAMPLE 4 ■ **Long Division Algorithm for Positive Integers**

Use the long division algorithm to divide 6,584 by 28.

Solution

	Think $\frac{65}{28} \approx 2$
	Think $\frac{98}{28} \approx 3$
	Think $\frac{144}{28} \approx 5$
235	
$28\overline{)6584}$	
56	Multiply $2 \cdot 28$
98	Subtract and bring down 8
84	Multiply $3 \cdot 28$
144	Subtract and bring down 4
140	Multiply $5 \cdot 28$
4	Remainder

Thus, we have

$$6{,}584 \div 28 = 235 + \frac{4}{28} = 235 + \frac{1}{7}.$$

■

In Example 4 the number 6,584 is called the **dividend**, 28 is called the **divisor**, 235 is called the **quotient**, and 4 is called the **remainder**. In the next several examples, we show how the long division algorithm can be extended to cover division of one polynomial by another.

EXAMPLE 5 ■ Long Division Algorithm for Polynomials

Use the long division algorithm to perform the following division.

$$(x^2 + 2x + 4) \div (x - 1)$$

Solution

	Think $\frac{x^2}{x} = x$
	Think $\frac{3x}{x} = 3$
$x + 3$	
$x - 1\overline{)x^2 + 2x + 4}$	
$x^2 - x$	Multiply $x(x - 1)$
$3x + 4$	Subtract and bring down 4
$3x - 3$	Multiply $3(x - 1)$
7	Subtract

Considering the remainder as a fractional part of the divisor, we can write the result as

$$\overbrace{\frac{x^2 + 2x + 4}{\underbrace{x - 1}_{\text{Divisor}}}}^{\text{Dividend}} = \overbrace{x + 3}^{\text{Quotient}} + \frac{\overbrace{7}^{\text{Remainder}}}{\underbrace{x - 1}_{\text{Divisor}}}.$$

■

You can check a long division problem by multiplying. For instance, we can check the results of Example 5 as follows.

$$\frac{x^2 + 2x + 4}{x - 1} \stackrel{?}{=} x + 3 + \frac{7}{x - 1}$$ Result to be checked

$$x^2 + 2x + 4 \stackrel{?}{=} (x + 3)(x - 1) + 7$$ Multiply both sides by $(x - 1)$

$$x^2 + 2x + 4 \stackrel{?}{=} (x^2 + 2x - 3) + 7$$

$$x^2 + 2x + 4 = x^2 + 2x + 4$$ Result checks

EXAMPLE 6 ■ **Long Division Algorithm for Polynomials**

Use the long division algorithm to perform the following division.

$$(3x^3 + 10x^2 + 6x - 4) \div (x + 2)$$

Solution

Think $\frac{3x^3}{x} = 3x^2$

Think $\frac{4x^2}{x} = 4x$

Think $-\frac{2x}{x} = -2$

$$\begin{array}{r} 3x^2 + 4x - 2 \\ x + 2\overline{)3x^3 + 10x^2 + 6x - 4} \\ \underline{3x^3 + 6x^2} \\ 4x^2 + 6x \\ \underline{4x^2 + 8x} \\ -2x - 4 \\ \underline{-2x - 4} \\ 0 \end{array}$$

Multiply $3x^2(x + 2)$

Subtract and bring down $6x$

Multiply $4x(x + 2)$

Subtract and bring down -4

Multiply $-2(x + 2)$

Remainder

Thus, we have

$$\frac{3x^3 + 10x^2 + 6x - 4}{x + 2} = 3x^2 + 4x - 2, \quad x \neq -2.$$

Note that the remainder in this problem is zero. In such cases we say that the denominator (or divisor) **divides evenly** into the numerator (or dividend).

Check

We can check this result by multiplying, as follows.

$$\begin{array}{rrrrr} & & 3x^2 & +\ 4x & -\ 2 \\ \times & & & x & +\ 2 \\ \hline & & 6x^2 & +\ 8x & -\ 4 \\ 3x^3 & + & 4x^2 & -\ 2x & \\ \hline 3x^3 & + & 10x^2 & +\ 6x & -\ 4 \end{array}$$

■

When using the long division algorithm for polynomials, be sure that both the divisor and the dividend are written in standard form before beginning the division process. This procedure is illustrated in Example 7.

EXAMPLE 7 ■ Writing in Standard Form Before Dividing

Use the long division algorithm to perform the following division.

$$(-13x^3 + 10x^4 + 8x - 7x^2 + 4) \div (3 - 2x)$$

Solution

We first write the divisor and dividend in *standard polynomial form* (with decreasing powers of x).

$$\begin{array}{r} -5x^3 - x^2 + 2x - 1 \\ -2x + 3 \overline{) 10x^4 - 13x^3 - 7x^2 + 8x + 4} \\ \underline{10x^4 - 15x^3} \qquad\qquad\qquad \\ 2x^3 - 7x^2 \qquad\qquad \\ \underline{2x^3 - 3x^2} \qquad\qquad \\ -4x^2 + 8x \qquad \\ \underline{-4x^2 + 6x} \qquad \\ 2x + 4 \\ \underline{2x - 3} \\ 7 \end{array}$$

Thus, we have

$$\frac{10x^4 - 13x^3 - 7x^2 + 8x + 4}{-2x + 3} = -5x^3 - x^2 + 2x - 1 + \frac{7}{-2x + 3}.$$

We leave the checking of this result to you. ■

When the dividend is missing some powers of x, the long division algorithm requires that you account for the missing powers, as shown in Example 8.

EXAMPLE 8 ■ Accounting for Missing Powers of x

Divide $(x^3 - 2)$ by $(x - 1)$.

Solution

Because there are no x^2-terms or x-terms in the dividend, we line up the subtractions by using *zero* coefficients (or by leaving a space) for these missing terms.

$$\begin{array}{r}
x^2 + x + 1 \\
x - 1\overline{)x^3 + 0x^2 + 0x - 2} \\
\underline{x^3 - x^2} \\
x^2 + 0x \\
\underline{x^2 - x} \\
x - 2 \\
\underline{x - 1} \\
-1
\end{array}$$

Thus, we can write

$$\frac{x^3 - 2}{x - 1} = x^2 + x + 1 - \frac{1}{x - 1}.$$ ■

In each of the long division examples so far, the divisor has been a first-degree polynomial. The long division algorithm works just as well with polynomial divisors of degree 2 or more, as shown in Example 9.

EXAMPLE 9 ■ A Second-Degree Divisor

Use the long division algorithm to perform the following division.

$$\frac{x^4 + 6x^3 + 6x^2 - 10x - 3}{x^2 + 2x - 3}$$

Solution

$$\begin{array}{r}
x^2 + 4x + 1 \\
x^2 + 2x - 3\overline{)x^4 + 6x^3 + 6x^2 - 10x - 3} \\
\underline{x^4 + 2x^3 - 3x^2} \\
4x^3 + 9x^2 - 10x \\
\underline{4x^3 + 8x^2 - 12x} \\
x^2 + 2x - 3 \\
\underline{x^2 + 2x - 3} \\
0
\end{array}$$

Thus, $(x^2 + 2x - 3)$ divides evenly into $(x^4 + 6x^3 + 6x^2 - 10x - 3)$, and we have

$$\frac{x^4 + 6x^3 + 6x^2 - 10x - 3}{x^2 + 2x - 3} = x^2 + 4x + 1, \quad x \neq 1, x \neq -3.$$ ■

Synthetic Division

There is a nice shortcut for division by polynomials of the form $(x - k)$. This shortcut is called **synthetic division**, and it is outlined for a third-degree polynomial as follows.

Synthetic Division for a Third-Degree Polynomial

To use synthetic division to divide $(ax^3 + bx^2 + cx + d)$ by $(x - k)$, we use the following pattern.

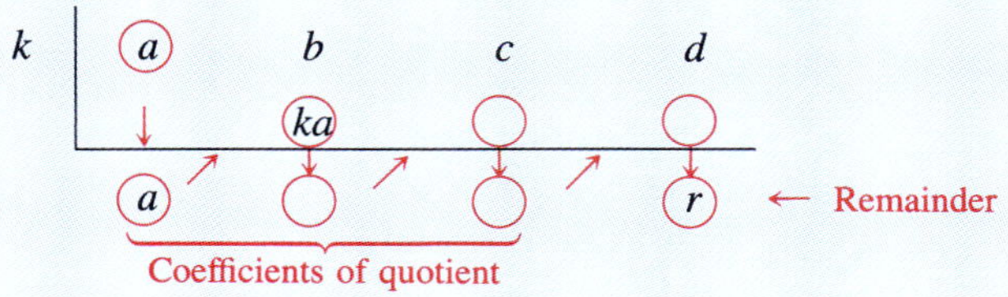

Vertical pattern: Add terms.

Diagonal pattern: Multiply by k.

Be sure you see that synthetic division works *only* for divisors of the form $(x - k)$. Moreover, the degree of the quotient is always one less than the degree of the dividend. We demonstrate the process in the next two examples.

EXAMPLE 10 ■ Using Synthetic Division

Use synthetic division to divide $(x^3 + 3x^2 - 4x - 10)$ by $(x - 2)$.

Solution

The coefficients of the dividend form the top row of the synthetic division table, and since we are dividing by $(x - 2)$, we write 2 at the top left of the table. To begin the algorithm, we bring down the first coefficient. Then we multiply this coefficient by 2, write the result in the second row of the table, and add the two numbers in the second column, as follows.

2	1	3	−4	−10
		2		
	1	5		

By continuing this pattern, we obtain the following completed synthetic division table.

2	1	3	−4	−10
		2	10	12
	1	5	6	(2) ← Remainder

Finally, from the bottom row of the table, we can determine that the quotient is $(1)x^2 + (5)x + (6)$ and the remainder is 2. Thus, the result of the division problem is

$$\frac{x^3 + 3x^2 - 4x - 10}{x - 2} = x^2 + 5x + 6 + \frac{2}{x - 2}.$$

■

If the remainder in a synthetic division problem turns out to be zero, then we can conclude that the divisor divides *evenly* into the dividend, as shown in Example 11.

EXAMPLE 11 ■ **Using Synthetic Division**

Use synthetic division to divide $(x^4 - 10x^2 - 2x + 3)$ by $(x + 3)$.

Solution

The coefficients of the dividend form the top row of the synthetic division table, and since we are dividing by $(x + 3)$, we write -3 at the top left of the table. The completed synthetic division table is as follows. (Note that we include a zero for the missing term in the dividend.)

$$\begin{array}{r|rrrrr} -3 & 1 & 0 & -10 & -2 & 3 \\ & & -3 & 9 & 3 & -3 \\ \hline & 1 & -3 & -1 & 1 & \textcircled{0} \end{array} \quad \leftarrow \text{ Remainder}$$

Since the remainder is zero, we can conclude that

$$\frac{x^4 - 10x^2 - 2x + 3}{x + 3} = x^3 - 3x^2 - x - 1, \quad x \neq -3.$$

■

Factoring and Division

One of the uses of synthetic division (or long division) is to factor polynomials. This is demonstrated in Example 12.

EXAMPLE 12 ■ **Factoring a Polynomial**

Factor the following polynomial completely.

$$x^3 - 7x + 6$$

Use the fact that $(x - 1)$ is one of the factors.

Solution

Since we are given one of the factors, we divide this factor into the given polynomial. (In this case we use synthetic division, but long division could have been used also.)

$$\begin{array}{r|rrrr} 1 & 1 & 0 & -7 & 6 \\ & & 1 & 1 & -6 \\ \hline & 1 & 1 & -6 & \textcircled{0} \end{array} \quad \leftarrow \text{ Remainder}$$

Thus, we can conclude that

$$\frac{x^3 - 7x + 6}{x - 1} = x^2 + x - 6$$

which implies that

$$x^3 - 7x + 6 = (x - 1)(x^2 + x - 6), \quad x \neq 1.$$

Thus, we have used division to factor the polynomial $(x^3 - 7x + 6)$ into the product of a first-degree polynomial and a second-degree polynomial. To complete the factorization, we factor the second-degree polynomial as follows.

$$\begin{aligned} x^3 - 7x + 6 &= (x - 1)(x^2 + x - 6) \\ &= (x - 1)(x + 3)(x - 2) \end{aligned}$$

■

DISCUSSION PROBLEM ■ Factoring and Problem Solving

Consider the following polynomial.

$$2x^4 + 7x^3 - 4x^2 - 27x - 18 = (x - 2)(x + 3)(2x + 3)(x + 1)$$

(a) Use the given complete factorization to find all of the solutions to the following polynomial equation.

$$2x^4 + 7x^3 - 4x^2 - 27x - 18 = 0$$

(b) Evaluate the given polynomial at the x-values listed in the following table.

x-Values	-3	-2	$-\frac{3}{2}$	-1	0	2	3
Polynomial Values							

(c) Use synthetic division to find the remainders indicated in the following table.

x-Values	-3	-2	$-\frac{3}{2}$	-1	0	2	3
Divisors	$x + 3$	$x + 2$	$x + \frac{3}{2}$	$x + 1$	x	$x - 2$	$x - 3$
Remainders							

(d) Compare the polynomial values in the first table with the remainders in the second table. What conclusion can you draw? ■

Warm-Up

The following warm-up exercises involve skills that were covered in earlier sections. You will use these skills in the exercise set for this section.

In Exercises 1–6, rewrite the fraction in reduced form.

1. $\frac{36}{144}$ **2.** $\frac{58}{180}$ **3.** $\frac{-16x^2}{12x}$

4. $\frac{5t^4}{45t^2}$ **5.** $\frac{6u^2v}{27uv^3}$ **6.** $\frac{14r^4s^2}{-98rs^2}$

In Exercises 7 and 8, find the indicated product.

7. $-3x^3(5x - 4)$ **8.** $10y(7y^2 - 2)$

In Exercises 9 and 10, perform the indicated subtraction.

9. $(3x^3 + 5x^2 - 2x - 1) - (x^3 - 2x^2 + x + 3)$

10. $(7x^2 + 6) - (-x^2 - 10x + 3)$

4.4 EXERCISES

In Exercises 1–50, perform the specified division.

1. $\frac{6z + 10}{2}$ **2.** $\frac{9x + 12}{3}$ **3.** $\frac{10z^2 + 4z - 12}{4}$ **4.** $\frac{4u^2 + 8u - 24}{16}$

5. $(7x^3 - 2x^2) \div x$ **6.** $(6a^2 + 7a) \div a$ **7.** $\frac{50z^3 + 30z}{-5z}$ **8.** $\frac{18c^4 - 24c^2}{-6c}$

9. $\frac{8z^3 + 3z^2 - 2z}{2z}$ **10.** $\frac{6x^4 + 8x^3 - 18x^2}{3x^2}$ **11.** $\frac{m^4 + 2m^2 - 7}{m}$ **12.** $\frac{l^2 - 8}{-l}$

13. $(5x^2y - 8xy + 7xy^2) \div 2xy$ **14.** $(-14s^4t^2 + 7s^2t^2 - 18t) \div 2s^2t$ **15.** $\frac{4(x + 5)^2 + 8(x + 5)}{x + 5}$

16. $\frac{12(r - 9)^3 - 18(r - 9)^2}{4(r - 9)^2}$ **17.** $\frac{x^2 - 8x + 15}{x - 3}$ **18.** $\frac{t^2 - 18t + 72}{t - 6}$

19. $(x^2 + 15x + 50) \div (x + 5)$ **20.** $(y^2 - 6y - 16) \div (y + 2)$ **21.** Divide $21 - 4x - x^2$ by $3 - x$.

22. Divide $5 + 4x - x^2$ by $1 + x$. **23.** Divide $2y^2 + 7y + 3$ by $2y + 1$.

24. Divide $10t^2 - 7t - 12$ by $2t - 3$. **25.** $(12t^2 - 40t + 25) \div (2t - 5)$

26. $(15 - 14u - 8u^2) \div (5 + 2u)$

27. $\dfrac{16x^2 - 1}{4x + 1}$

28. $\dfrac{81y^2 - 25}{9y - 5}$

29. $\dfrac{x^3 + 125}{x + 5}$

30. $\dfrac{x^3 - 27}{x - 3}$

31. $\dfrac{x^3 - 2x^2 + 4x - 8}{x - 2}$

32. $\dfrac{x^3 - 28x - 48}{x + 4}$

33. $(2x + 9) \div (x + 2)$

34. $(12x - 5) \div (2x + 3)$

35. $\dfrac{x^2 + 16}{x + 4}$

36. $\dfrac{y^2 + 8}{y + 2}$

37. $\dfrac{5x^2 + 2x + 3}{x + 2}$

38. $\dfrac{2x^2 + 5x + 2}{x + 4}$

39. $\dfrac{12x^2 - 17x - 5}{3x + 2}$

40. $\dfrac{8x^2 + 2x + 3}{4x - 1}$

41. $\dfrac{6z^2 + 7z}{5z - 1}$

42. $\dfrac{8y^2 - 2y}{3y + 5}$

43. $\dfrac{2x^3 - 5x^2 + x - 6}{x - 3}$

44. $\dfrac{5x^3 + 3x^2 + 12x + 20}{x + 1}$

45. Divide $x^6 - 1$ by $x - 1$.

46. Divide x^3 by $x - 1$.

47. $(x^3 + 4x^2 + 7x + 6) \div (x^2 + 2x + 3)$

48. $(2x^3 + 2x^2 - 2x - 15) \div (2x^2 + 4x + 5)$

49. $\dfrac{x^{3n} + 3x^{2n} + 6x^n + 8}{x^n + 2}$, n is a positive integer

50. $\dfrac{x^{3n} - x^{2n} + 5x^n - 5}{x^n - 1}$, n is a positive integer

In Exercises 51–60, use synthetic division to perform the indicated division. Use the result to write the dividend in factored form.

51. $\dfrac{x^2 - 15x + 56}{x - 8}$

52. $\dfrac{x^2 + 3x - 154}{x + 14}$

53. $\dfrac{2a^2 + 13a - 45}{a + 9}$

54. $\dfrac{24 + 13t - 2t^2}{8 - t}$

55. $\dfrac{15x^2 - 2x - 8}{x - \frac{4}{5}}$

56. $\dfrac{18x^2 - 9x - 20}{x + \frac{5}{6}}$

57. $\dfrac{2t^3 + 15t^2 + 19t - 30}{t + 5}$

58. $\dfrac{5t^3 - 27t^2 - 14t - 24}{t - 6}$

59. $\dfrac{x^4 - x^3 - 3x^2 + 4x - 1}{x - 1}$

60. $\dfrac{x^4 - 16}{x - 2}$

In Exercises 61–68, use synthetic division to perform the indicated division.

61. $\dfrac{x^3 + 3x^2 - 1}{x + 4}$

62. $\dfrac{x^4}{x + 2}$

63. $\dfrac{x^4 - 4x^3 + x + 10}{x - 2}$

64. $\dfrac{2x^5 - 3x^3 + x}{x - 3}$

65. $\dfrac{5x^3 + 12}{x + 5}$

66. $\dfrac{8x + 35}{x - 10}$

67. $\dfrac{0.1x^2 + 0.8x + 1}{x - 0.2}$

68. $\dfrac{x^3 - 0.8x + 2.4}{x + 1}$

In Exercises 69 and 70, find the constant c such that the denominator will divide evenly into the numerator.

69. $\dfrac{x^3 + 2x^2 - 4x + c}{x - 2}$ **70.** $\dfrac{x^4 - 3x^2 + c}{x + 6}$

In Exercises 71–74, simplify the algebraic expression.

71. $\dfrac{4x^4}{x^3} - 2x$ **72.** $\dfrac{15x^3y}{10x^2} + \dfrac{3xy^2}{2y}$

73. $\dfrac{8u^2v}{2u} + \dfrac{3(uv)^2}{uv}$

74. $\dfrac{x^2 + 2x - 3}{x - 1} - (3x - 4)$

In Exercises 75–78, determine whether the cancellation shown is valid or invalid. If it is invalid, state what is wrong.

75. $\dfrac{5 + 12}{5} = \dfrac{\cancel{5} + 12}{\cancel{5}} = 12$

76. $\dfrac{6 - 3}{6 + 11} = \dfrac{\cancel{6} - 3}{\cancel{6} + 11} = \dfrac{-3}{11}$

77. $\dfrac{9 \cdot 12}{19 \cdot 9} = \dfrac{\cancel{9} \cdot 12}{19 \cdot \cancel{9}} = \dfrac{12}{19}$

78. $\dfrac{28}{83} = \dfrac{2\cancel{8}}{\cancel{8}3} = \dfrac{2}{3}$

79. When a polynomial is divided by $x - 6$, the quotient is $x^2 + x + 1$ and the remainder is -4. Find the polynomial.

80. When a polynomial is divided by $x + 3$, the quotient is $x^3 + x^2 - 4$ and the remainder is 8. Find the polynomial.

81. *Dimensions of a Rectangle* The area of a rectangle is $2x^3 + 3x^2 - 6x - 9$ (see figure). Find its length if its width is $2x + 3$.

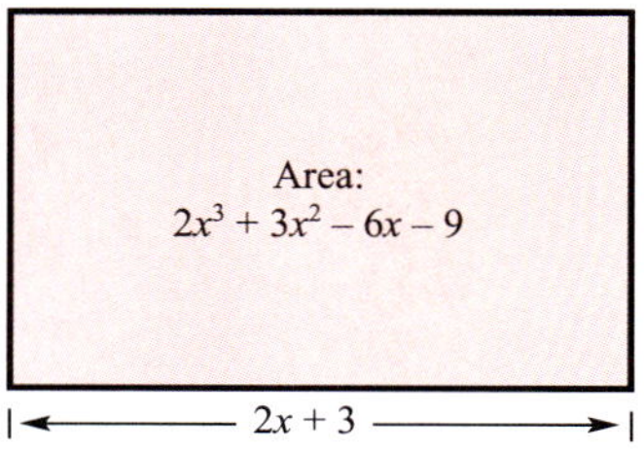

82. *Dimensions of a Box* The volume of a rectangular box is $h^4 + 3h^3 + 2h^2$ (see figure). Find its length if its height and width are h and $h + 1$, respectively.

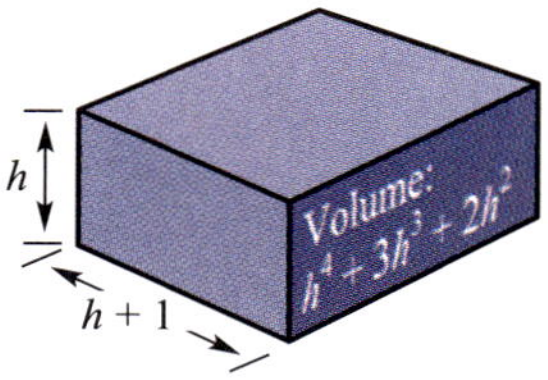

83. Consider the polynomial $x^3 + 4x^2 - 15x - 2$.

(a) Evaluate the polynomial at the x-values listed in the table.

x-Values	Polynomial Values
-2	
$-\frac{1}{2}$	
1	
3	
5	

(b) Use synthetic division to find the remainders indicated in the following table.

x-Values	Divisors	Remainders
-2	$x + 2$	
$-\frac{1}{2}$	$x + \frac{1}{2}$	
1	$x - 1$	
3	$x - 3$	
5	$x - 5$	

(c) What conclusion can you draw as you compare the polynomial values in the first table with the remainders in the second? Use your conclusion and synthetic division to evaluate the polynomial when $x = 4$.

SECTION **4.5**

Solving Equations Involving Rational Expressions

Introduction • Equations Containing Constant Denominators • Equations Containing Variable Denominators • Applications

Introduction

In this section we switch from *rewriting* rational expressions to *solving* equations that involve rational expressions. The least common denominator continues to play a key role in the solution process.

Equations that involve fractions are common in real-life experiences. Consider the following example.

> A manufacturing plant can produce x units of a certain item for \$26 per unit *plus* an initial investment of \$80,000. How many units must be produced in order for the average cost per unit to be \$30?

The solution to this problem requires that we solve the fractional equation

$$30 = \frac{26x + 80{,}000}{x}.$$

In Example 8 we will discuss this application further.

Equations Containing Constant Denominators

In Section 2.1 we showed how to solve equations containing fractions with *constant* denominators. We review that procedure here because it is the basis for solving more general equations involving fractions. Recall from Section 2.1 that we can "clear an equation of fractions" by multiplying both sides of the equation by the least common denominator of the fractions in the equation. Note how this is done in Example 1.

EXAMPLE 1 ■ An Equation Containing Constant Denominators

Solve the equation $\frac{3}{5} = \frac{x}{2}$.

Solution

For this equation, the least common denominator is 10. Therefore, we begin by multiplying both sides of the equation by 10.

$$\frac{3}{5} = \frac{x}{2} \qquad \text{Given equation}$$

$$10\left(\frac{3}{5}\right) = 10\left(\frac{x}{2}\right) \qquad \text{Multiply both sides by 10}$$

$$6 = 5x \qquad \text{Simplify fractions}$$

$$\frac{6}{5} = x \qquad \text{Solution}$$

Check

$\frac{3}{5} = \frac{x}{2}$	Given equation
$\frac{3}{5} \stackrel{?}{=} \frac{6/5}{2}$	Replace x by $\frac{6}{5}$
$\frac{3}{5} \stackrel{?}{=} \left(\frac{6}{5}\right)\left(\frac{1}{2}\right)$	Invert and multiply
$\frac{3}{5} = \frac{3}{5}$	Solution checks

Note in Example 1 that the given equation involves only two fractions—one on each side of the equation. To solve this type of equation, we can use an alternative technique called **cross multiplication**. For instance, in Example 1, we could have solved the equation as follows.

$\frac{3}{5} = \frac{x}{2}$	Given equation
$3(2) = 5x$	Cross multiply
$6 = 5x$	
$\frac{6}{5} = x$	Solution

Although this cross-multiplication technique can be a little quicker than multiplying by the least common denominator, remember that it can *only be used with equations that have a single fraction on each side* of the equation.

EXAMPLE 2 An Equation Containing Constant Denominators

Solve the equation $\frac{x}{6} = 7 - \frac{x}{12}$.

Solution

For this equation, the least common denominator is 12. Therefore, we begin by multiplying both sides of the equation by 12.

$\frac{x}{6} = 7 - \frac{x}{12}$	Given equation
$12\left(\frac{x}{6}\right) = 12\left(7 - \frac{x}{12}\right)$	Multiply both sides by 12
$2x = 84 - x$	Distribute and simplify
$3x = 84$	Add x to both sides
$x = 28$	Solution

Therefore, the solution of the equation is 28. Check this solution in the original equation.

EXAMPLE 3 ■ An Equation Containing Constant Denominators

Solve the equation $\frac{x+2}{6} - \frac{x-4}{8} = \frac{2}{3}$.

Solution

For this equation, the least common denominator is 24. Therefore, we begin by multiplying both sides of the equation by 24.

$\frac{x+2}{6} - \frac{x-4}{8} = \frac{2}{3}$	Given equation
$24\left(\frac{x+2}{6} - \frac{x-4}{8}\right) = 24\left(\frac{2}{3}\right)$	Multiply both sides by 24
$4(x+2) - 3(x-4) = 8(2)$	Distribute and simplify
$4x + 8 - 3x + 12 = 16$	Distributive Property
$x = -4$	Solution

Therefore, the solution of the equation is -4. Check this solution in the original equation. ■

Equations Containing Variable Denominators

As stated in Section 4.1, we always *exclude* those values of a variable that make the denominator of an algebraic fraction zero. This is especially critical for solving equations that contain variable denominators. You will see why in the examples that follow.

EXAMPLE 4 ■ An Equation Containing Variable Denominators

Solve the following equation.

$$\frac{2}{x} + \frac{5}{3} = \frac{7}{x}$$

Solution

For this equation, the least common denominator is $3x$. Therefore, we begin by multiplying both sides of the equation by $3x$.

$\frac{2}{x} + \frac{5}{3} = \frac{7}{x}$	Given equation
$3x\left(\frac{2}{x} + \frac{5}{3}\right) = 3x\left(\frac{7}{x}\right)$	Multiply both sides by $3x$
$\frac{6x}{x} + \frac{15x}{3} = \frac{21x}{x}$	Distributive Property
$6 + 5x = 21, \quad x \neq 0$	Simplify fractions
$5x = 15$	Combine like terms
$x = 3$	Solution

Therefore, the solution appears to be 3. We check this solution in the original equation as follows.

Check

$$\frac{2}{x} + \frac{5}{3} = \frac{7}{x} \qquad \text{Given equation}$$

$$\frac{2}{3} + \frac{5}{3} \stackrel{?}{=} \frac{7}{3} \qquad \text{Replace } x \text{ by } 3$$

$$\frac{7}{3} = \frac{7}{3} \qquad \text{Solution checks}$$

After checking, we conclude that the solution of the equation is 3. ■

Throughout the text, we have emphasized the importance of checking solutions. Up to this point, the main reason for checking has been to make sure that we don't make errors in the solution process. In the next example we will see that there is another reason for checking solutions in the *original* equation. That is, even with no mistakes in the solution process, it can happen that a "trial solution" does not satisfy the original equation. This type of "solution" is called **extraneous**. (An extraneous solution of an equation does not, by definition, satisfy its original equation, and must therefore *not* be listed as an actual solution.)

EXAMPLE 5 ■ An Equation with No Solution

Solve the following equation.

$$\frac{5x}{x-2} = 7 + \frac{10}{x-2}$$

Solution

The least common denominator for this equation is $x - 2$. Therefore, we begin by multiplying both sides of the equation by $x - 2$.

$$\frac{5x}{x-2} = 7 + \frac{10}{x-2} \qquad \text{Given equation}$$

$$(x-2)\left(\frac{5x}{x-2}\right) = (x-2)\left(7 + \frac{10}{x-2}\right) \qquad \text{Multiply both sides by } x - 2$$

$$5x = 7(x-2) + 10, \quad x \neq 2 \qquad \text{Distribute and simplify}$$

$$5x = 7x - 4 \qquad \text{Group like terms}$$

$$-2x = -4 \qquad \text{Subtract } 7x \text{ from both sides}$$

$$x = 2 \qquad \text{Trial solution}$$

At this point, the solution appears to be 2. However, by performing the following check, we see that this "trial solution" is extraneous.

Check

$$\frac{5x}{x-2} = 7 + \frac{10}{x-2} \qquad \text{Given equation}$$

$$\frac{5(2)}{2-2} \stackrel{?}{=} 7 + \frac{10}{2-2} \qquad \text{Replace } x \text{ by } 2$$

$$\frac{10}{0} \stackrel{?}{=} 7 + \frac{10}{0} \qquad \text{Division by zero is undefined}$$

Since the check resulted in *division by zero*, we conclude that 2 is extraneous. Therefore, the given equation has no solution. ■

NOTE: In Example 5 can you see why $x = 2$ is extraneous? By looking back at the original equation we can see that 2 is excluded from the domain of both of the fractions that occur in the equation.

EXAMPLE 6 ■ An Equation Containing Variable Denominators

Solve the following equation.

$$\frac{4}{x-2} + \frac{3x}{x+1} = 3$$

Solution

The least common denominator for this equation is $(x - 2)(x + 1)$. Therefore, we begin by multiplying both sides of the equation by $(x - 2)(x + 1)$.

$$\frac{4}{x-2} + \frac{3x}{x+1} = 3 \qquad \text{Given equation}$$

$$(x-2)(x+1)\left(\frac{4}{x-2} + \frac{3x}{x+1}\right) = 3(x-2)(x+1) \qquad \text{Multiply both sides by } (x-2)(x+1)$$

$$4(x+1) + 3x(x-2) = 3(x^2 - x - 2), \quad x \neq 2, x \neq -1 \qquad \text{Distribute and simplify}$$

$$4x + 4 + 3x^2 - 6x = 3x^2 - 3x - 6 \qquad \text{Distributive Property}$$

$$x = -10 \qquad \text{Solution}$$

After checking, we conclude that the solution is -10. Check this solution in the original equation. ■

So far in this section each of the equations we have solved has had one solution (or no solution). In the next example we look at an equation that has two solutions.

EXAMPLE 7 ■ An Equation that Has Two Solutions

Solve the following equation.

$$\frac{3x}{x+1} = \frac{12}{x^2-1} + 2$$

Solution

The least common denominator for this equation is $(x + 1)(x - 1) = x^2 - 1$. Therefore, we begin by multiplying both sides of the equation by $x^2 - 1$.

$\dfrac{3x}{x+1} = \dfrac{12}{x^2 - 1} + 2$	Given equation
$(x^2 - 1)\left(\dfrac{3x}{x+1}\right) = (x^2 - 1)\left(\dfrac{12}{x^2 - 1} + 2\right)$	Multiply both sides by $x^2 - 1$
$(x - 1)(3x) = 12 + 2(x^2 - 1), \quad x \neq \pm 1$	Distribute and simplify
$3x^2 - 3x = 12 + 2x^2 - 2$	Distributive Property
$x^2 - 3x - 10 = 0$	Standard form
$(x + 2)(x - 5) = 0$	Factor
$x + 2 = 0 \Rightarrow x = -2$	Set first factor equal to 0
$x - 5 = 0 \Rightarrow x = 5$	Set second factor equal to 0

After checking *each* solution, we conclude that the equation has two solutions: -2 and 5. Check these solutions in the original equation. ■

Applications

At the beginning of this section we looked at an application involving total cost and average cost per unit. Now let's see how to solve this problem using an equation that involves fractions.

EXAMPLE 8 ■ **An Application Involving Fractions**

A manufacturing plant can produce x units of a certain item for \$26 per unit *plus* an initial investment of \$80,000. How many units must be produced in order for the average cost per unit to be \$30?

Solution

We will follow the problem-solving strategy discussed in Chapter 2.

Verbal model: [Average cost per unit] = [Total cost] ÷ [Number of units]

Labels:
x = Number of units
Total cost = $26x + 80{,}000$ (in dollars)
Average cost per unit = \$30

Equation:

$$30 = \frac{26x + 80{,}000}{x}$$

$$30x = 26x + 80{,}000, \quad x \neq 0$$

$$4x = 80{,}000$$

$$x = 20{,}000$$

Therefore, in order for the average cost per unit to be \$30, the plant must manufacture 20,000 units. Check this solution in the original statement of the problem. ■

EXAMPLE 9 ■ **An Application Involving Fractions**

With only the cold water valve open, it takes eight minutes to fill the tub of an automatic washer. With both the hot and cold water valves fully open, it takes only five minutes to fill the tub. How long will it take to fill the tub with only the hot water valve open?

Solution

Verbal model: [Rate for cold water] + [Rate for hot water] = [Rate for warm water]

Labels:

Warm water: Time = 5 minutes, Rate = $\frac{1}{5}$ tub per minute

Cold water: Time = 8 minutes, Rate = $\frac{1}{8}$ tub per minute

Hot water: Time = t minutes, Rate = $\frac{1}{t}$ tub per minute

Equation:

$$\frac{1}{8} + \frac{1}{t} = \frac{1}{5}$$

$$40t\left(\frac{1}{8} + \frac{1}{t}\right) = 40t\left(\frac{1}{5}\right)$$

$$5t + 40 = 8t$$

$$40 = 3t$$

$$\frac{40}{3} = t$$

Thus, it will take about $13\frac{1}{3}$ minutes to fill the tub with hot water alone. Check this solution in the original statement of the problem. ■

EXAMPLE 10 ■ **An Application Involving Fractions**

In this year's playing season, a baseball player has been up to bat 140 times and has hit the ball successfully 35 times. Thus, the batting average for the player is $35/140 = 0.250$. How many consecutive times must the player successfully hit the ball in order to obtain a batting average of 0.300?

Solution

Verbal model: [Batting average] = [Total hits] ÷ [Total times at bat]

Labels: Current times at bat = 140
Current hits = 35
Additional consecutive hits = x

Equation:

$$0.300 = \frac{x + 35}{x + 140}$$

$$0.300(x + 140) = x + 35$$

$$0.3x + 42 = x + 35$$

$$7 = 0.7x$$

$$10 = x$$

Thus, the player must successfully hit the ball for the next 10 times at bat. After that, the player's batting average will be $45/150 = 0.300$. Check this solution in the original statement of the problem. ■

DISCUSSION PROBLEM ■ **Extraneous Solutions**

Consider the steps in the following solution.

$$\frac{x^2 + 2x + 2}{x - 1} = \frac{2x + 3}{x - 1}$$ Given equation

$$x^2 + 2x + 2 = (2x + 3), \quad x \neq 1$$ Multiply both sides by $(x - 1)$

$$x^2 - 1 = 0$$

$$(x - 1)(x + 1) = 0$$ Factor

$$x + 1 = 0 \Rightarrow x = -1$$ Set first factor equal to zero

$$x - 1 = 0 \Rightarrow x = 1$$ Set second factor equal to zero

Which of these two possible solutions is extraneous? In the second step, when we multiplied both sides of the equation by $(x - 1)$, we noted that the new equation was not valid for $x = 1$. Rework Example 5 in this section. At what stage in the solution can you determine that $x = 2$ could not be a solution? ■

Warm-Up

The following warm-up exercises involve skills that were covered in earlier sections. You will use these skills in the exercise set for this section.

In Exercises 1–10, solve the given equation.

1. $3x = 16$ **2.** $-5x = 95$ **3.** $15x + 3 = 48$

4. $125 - 50x = 0$ **5.** $(3x - 2)(x + 8) = 0$

6. $(2x + 3)(x - 16) = 0$ **7.** $x(2x - 21) = 0$

8. $x(10 - x) = 25$ **9.** $x^2 + x - 42 = 0$ **10.** $t^2 - 8t = 0$

4.5 EXERCISES

In Exercises 1–4, determine whether the given value of x is a solution to the equation.

Equation	*Value*			
1. $\frac{x}{3} - \frac{x}{5} = \frac{4}{3}$	(a) $x = 0$	(b) $x = -1$	(c) $x = \frac{1}{8}$	(d) $x = 10$
2. $x = 4 + \frac{21}{x}$	(a) $x = 0$	(b) $x = -3$	(c) $x = 7$	(d) $x = -1$
3. $\frac{x}{4} + \frac{3}{4x} = 1$	(a) $x = -1$	(b) $x = 1$	(c) $x = 3$	(d) $x = 2$
4. $5 - \frac{1}{x - 3} = 2$	(a) $x = \frac{10}{3}$	(b) $x = -\frac{1}{3}$	(c) $x = 0$	(d) $x = 1$

In Exercises 5–14, solve the given equation.

5. $\frac{x}{4} = \frac{3}{8}$

6. $\frac{x}{10} = \frac{12}{5}$

7. $\frac{t}{2} = \frac{1}{8}$

8. $\frac{y}{5} = \frac{3}{2}$

9. $\frac{z + 2}{3} = \frac{z}{12}$

10. $5 + \frac{y}{3} = y + 2$

11. $\frac{4t}{3} = 15 - \frac{t}{6}$

12. $\frac{x}{3} + \frac{x}{6} = 10$

13. $\frac{h + 2}{5} - \frac{h - 1}{9} = \frac{2}{3}$

14. $\frac{u - 2}{6} + \frac{2u + 5}{15} = 3$

In Exercises 15–50, solve the given equation. (Be sure to check each solution.)

15. $\frac{7}{x} = 21$

16. $\frac{9}{t} = -\frac{4}{3}$

17. $\frac{9}{25 - y} = -\frac{1}{4}$

18. $\frac{2}{u + 4} = \frac{5}{8}$

19. $5 - \frac{12}{a} = \frac{5}{3}$

20. $\frac{6}{b} + 22 = 24$

21. $\frac{12}{y + 5} + \frac{1}{2} = 2$

22. $\frac{7}{8} - \frac{16}{t - 2} = \frac{3}{4}$

23. $\frac{5}{x} = \frac{25}{3(x + 2)}$

24. $\frac{10}{x + 4} = \frac{15}{4(x + 1)}$

25. $\frac{8}{3x + 5} = \frac{1}{x + 2}$

26. $\frac{500}{3x + 5} = \frac{50}{x - 3}$

27. $\frac{3}{x + 2} - \frac{1}{x} = \frac{1}{5x}$

28. $\frac{12}{x + 5} + \frac{5}{x} = \frac{20}{x}$

29. $\frac{4}{2x + 3} + \frac{17}{5(2x + 3)} = 3$

30. $\frac{2}{6q + 5} - \frac{3}{4(6q + 5)} = \frac{1}{28}$

31. $\frac{10}{x(x - 2)} + \frac{4}{x} = \frac{5}{x - 2}$

32. $3\left(\frac{1}{x} + 4\right) = 2 + \frac{4}{3x}$

33. $\frac{x}{x + 4} + \frac{4}{x + 4} + 2 = 0$

34. $\frac{2}{x^2 - 6x + 8} = \frac{1}{x - 4} + \frac{2}{x - 2}$

35. $\frac{1}{x - 5} + \frac{1}{x + 5} = \frac{x + 3}{x^2 - 25}$

36. $\frac{2}{x - 10} - \frac{3}{x - 2} = \frac{6}{x^2 - 12x + 20}$

37. $\frac{1}{2} = \frac{18}{x^2}$

38. $\frac{1}{6} = \frac{150}{z^2}$

39. $\frac{32}{t} = 2t$

40. $\frac{45}{u} = \frac{u}{5}$

41. $x + 1 = \frac{72}{x}$

42. $t + 4\left(\frac{2t + 5}{t + 4}\right) = 0$

43. $1 = \frac{16}{y} - \frac{39}{y^2}$

44. $\frac{15}{x^2} + \frac{7}{x} = 2$

45. $\frac{1}{x - 1} + \frac{3}{x + 1} = 2$

46. $\frac{x + 42}{x} = x$

47. $x - \frac{24}{x} = 5$

48. $\frac{3x}{2} + \frac{4}{x} = 5$

49. $\frac{x}{2} = \frac{2 - \frac{3}{x}}{1 - \frac{1}{x}}$

50. $\frac{2x}{3} = \frac{1 + \frac{2}{x}}{1 + \frac{1}{x}}$

51. *Number Problem* Find a number such that the sum of the number and its reciprocal is $\frac{65}{8}$.

52. *Number Problem* Find a number such that the sum of two times the number and three times its reciprocal is $\frac{97}{4}$.

53. *Wind Speed* A plane has a speed of 300 miles per hour in still air. Find the speed of the wind if the plane traveled a distance of 680 miles with a tail wind in the same amount of time it took to travel 520 miles into a head wind.

54. *Average Speed* During the first 145 miles of a 320-mile trip you travel at an average speed of r miles per hour. Because of the difference in speed limits among states, you increased your average speed by 10 miles per hour during the last part of the trip. Find the slower average speed if the total time for the trip was six hours.

55. *Partnership Costs* A group of outdoor enthusiasts plan to buy a piece of property for $78,000 by sharing the cost equally. To ease the financial burden they look for three additional partners so that the cost per person will be reduced by $1,300. Determine the number of partners they would like to have in the group.

56. *Average Cost* The average cost for producing x units of a product is given by

$$\text{Average cost} = 1.50 + \frac{4{,}200}{x}.$$

Determine the number of units that must be produced in order to have an average cost of $2.90.

57. *Pollution Removal* The cost C in dollars of removing p percent of the air pollution in the stack emission of a utility company that burns coal to generate electricity is given by

$$C = \frac{120{,}000p}{100 - p}.$$

Determine the percentage of the stack emission that can be removed for $680,000.

58. *Population Growth* A biologist introduces 100 insects into a culture. The population P of the culture is approximated by the model

$$P = \frac{500(1 + 3t)}{5 + t}$$

where t is the time in hours. Find the time required for the population to increase to 1,000 insects.

Work Rate In Exercises 59 and 60, complete the table by finding the time it takes for two individuals working together to complete a task. The first two columns in the table give the time required for two individuals working alone to complete the task. (Assume that when they work together their individual rates are not changed.)

59.

Person #1	Person #2	Together
6 hours	6 hours	
3 minutes	5 minutes	
5 hours	$2\frac{1}{2}$ hours	

60.

Person #1	Person #2	Together
4 days	4 days	
$5\frac{1}{2}$ hours	3 hours	
a days	b days	

61. *Work Rate* One landscaper works one and one-half times as fast as a second landscaper. Find their individual times if, on a certain job, it takes them nine hours working together.

62. *Work Rate* Assume that the slower of the two landscapers in Exercise 61 is given another job after 4 hours. The faster of the two must work an additional 10 hours to complete the task. Find the individual times of the two people.

63. *Swimming Pool* The flow rate for one pipe is one and one-fourth that of a second pipe. If a swimming pool can be filled in five hours using both pipes, find the time required to fill the pool using only the pipe with the smaller flow rate.

64. *Swimming Pool* Assume that the pipe with the faster flow rate in Exercise 63 is shut off after 1 hour and that it takes an additional 10 hours to fill the pool. Find the individual time of each pipe.

Car Rental Revenues In Exercises 65 and 66, use the following model, which approximates the total revenue for the car rental industry in the United States between 1984 and 1990 (see figure).

$$y = \frac{1}{2.39 - 0.326t + 0.015t^2}$$

In this model y represents the total revenue in billions of dollars and t represents the year, with $t = 0$ corresponding to 1980. (*Source:* Alex Brown and Sons.)

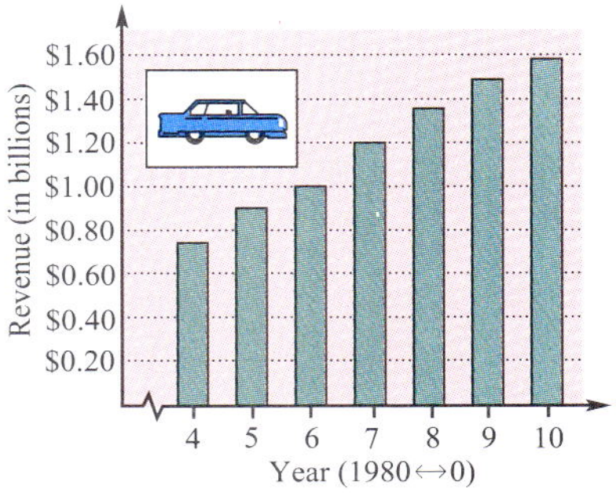

65. Use the bar graph to determine the year the total revenue approximated $1.03 billion. Then use the model to verify the reasonableness of your answer.

66. Use the bar graph to determine the year the total revenue approximated $1.35 billion. Then use the model to verify the reasonableness of your answer.

CHAPTER 4 SUMMARY

As you review and prepare for a test on this chapter, first try to obtain a global view of what was discussed. Then review the specific skills needed under each category.

Simplifying Rational Expressions

- *Determine* the domain of rational expressions. — Section 4.1, Exercises 1–12
- *Rewrite* rational expressions in reduced form. — Section 4.1, Exercises 21–50

Operations Involving Rational Expressions

- *Multiply* rational expressions. — Section 4.2, Exercises 7–32
- *Divide* rational expressions. — Section 4.2, Exercises 33–49
- *Add and subtract* rational expressions with like denominators. — Section 4.3, Exercises 1–14
- *Add and subtract* rational expressions with unlike denominators. — Section 4.3, Exercises 21–58
- *Simplify* compound fractions. — Section 4.3, Exercises 59–68
- *Create* rational expressions from verbal statements. — Section 4.3, Exercises 69–74
- *Divide* a polynomial by a monomial. — Section 4.4, Exercises 1–14
- *Divide* a polynomial by a binomial, using long division. — Section 4.4, Exercises 21–46
- *Divide* a polynomial by a binomial, using synthetic division. — Section 4.4, Exercises 51–68

Solving Equations that Involve Rational Expressions

- *Solve* equations containing fractions with constant denominators. — Section 4.5, Exercises 5–14
- *Solve* equations containing fractions with variable denominators. — Section 4.5, Exercises 15–50
- *Create and solve* fractional equations from verbal statements. — Section 4.5, Exercises 51–64

Chapter 4 Review Exercises

In Exercises 1–4, find the domain of the rational expression.

1. $\dfrac{3y}{y-8}$

2. $\dfrac{t+4}{t+12}$

3. $\dfrac{u}{u^2-7u+6}$

4. $\dfrac{x-12}{x(x^2-16)}$

In Exercises 5–10, rewrite the fraction in reduced form.

5. $\dfrac{6x^4y^2}{15xy^2}$

6. $\dfrac{2(y^3z)^2}{28(yz^2)^2}$

7. $\dfrac{5b-15}{30b-120}$

8. $\dfrac{4a}{10a^2+26a}$

9. $\dfrac{9x-9y}{y-x}$

10. $\dfrac{x+3}{x^2-x-12}$

In Exercises 11–40, perform the indicated operations and simplify your answer.

11. $\dfrac{7}{8}\cdot\dfrac{2x}{y}\cdot\dfrac{y^2}{14x^2}$

12. $\dfrac{15(x^2y)^3}{3y^3}\cdot\dfrac{12y}{x}$

13. $\dfrac{60z}{z+6}\cdot\dfrac{z^2-36}{5}$

14. $\dfrac{1}{6}(x^2-16)\cdot\dfrac{3}{x^2-8x+16}$

15. $\dfrac{u}{u-3}\cdot\dfrac{3u-u^2}{4u^2}$

16. $x^2\cdot\dfrac{x+1}{x^2-x}\cdot\dfrac{(5x-5)^2}{x^2+6x+5}$

17. $\dfrac{\left(\dfrac{6}{x}\right)}{\left(\dfrac{2}{x^3}\right)}$

18. $\dfrac{0}{\left(\dfrac{5x^2}{2y}\right)}$

19. $25y^2\div\dfrac{xy}{5}$

20. $\dfrac{6}{z^2}\div 4z^2$

21. $\dfrac{x^2-7x}{x+1}\div\dfrac{x^2-14x+49}{x^2-1}$

22. $\left(\dfrac{6x}{y^2}\right)^2\div\left(\dfrac{3x}{y}\right)^3$

23. $\dfrac{4}{9}-\dfrac{11}{9}$

24. $\dfrac{2(3y+4)}{2y+1}+\dfrac{3-y}{2y+1}$

25. $\dfrac{15}{16}-\dfrac{5}{24}-1$

26. $-\dfrac{3}{8}+\dfrac{7}{6}-\dfrac{1}{12}$

27. $\dfrac{1}{x+5}+\dfrac{3}{x-12}$

28. $\dfrac{2}{x-10}+\dfrac{3}{4-x}$

29. $5x+\dfrac{2}{x-3}-\dfrac{3}{x+2}$

30. $4-\dfrac{4x}{x+6}+\dfrac{7}{x-5}$

31. $\dfrac{6}{x}-\dfrac{6x-1}{x^2+4}$

32. $\dfrac{5}{x+2}+\dfrac{25-x}{x^2-3x-10}$

33. $\dfrac{5}{x+3}-\dfrac{4x}{(x+3)^2}-\dfrac{1}{x-3}$

34. $\dfrac{8}{y}-\dfrac{3}{y+5}+\dfrac{4}{y-2}$

35. $\dfrac{\left(\dfrac{6x^2}{x^2+2x-35}\right)}{\left(\dfrac{x^3}{x^2-25}\right)}$

36. $\dfrac{\left(\dfrac{24-18x}{(2-x)^2}\right)}{\left(\dfrac{60-45x}{x^2-4x+4}\right)}$

37. $\dfrac{3t}{\left(5-\dfrac{2}{t}\right)}$

38. $\dfrac{\left(x-3+\dfrac{2}{x}\right)}{\left(1-\dfrac{2}{x}\right)}$

39. $\dfrac{\left(\dfrac{1}{a^2-16}-\dfrac{1}{a}\right)}{\left(\dfrac{1}{a^2+4a}+4\right)}$

40. $\dfrac{\left(\dfrac{1}{x^2}-\dfrac{1}{y^2}\right)}{\left(\dfrac{1}{x}+\dfrac{1}{y}\right)}$

In Exercises 41–46, perform the indicated division.

41. $(4x^3 - x) \div 2x$

42. $(10x + 15) \div (5x - 2)$

43. $\dfrac{6x^3 + 2x^2 - 4x + 2}{3x - 1}$

44. $\dfrac{4x^4 - x^3 - 7x^2 + 18x}{x - 2}$

45. $\dfrac{x^4 - 3x^2 + 2}{x^2 - 1}$

46. $\dfrac{3x^6}{x^2 - 1}$

In Exercises 47–50, use synthetic division to perform the indicated division.

47. $\dfrac{x^3 + 7x^2 + 3x - 14}{x + 2}$

48. $\dfrac{x^4 - 2x^3 - 15x^2 - 2x + 10}{x - 5}$

49. $(x^4 - 3x^2 - 25) \div (x - 3)$

50. $(2x^3 + 5x - 2) \div \left(x + \dfrac{1}{2}\right)$

In Exercises 51–60, solve the equation.

51. $\dfrac{3x}{8} = -15$

52. $\dfrac{t + 1}{8} = \dfrac{1}{2}$

53. $3\left(8 - \dfrac{12}{t}\right) = 0$

54. $\dfrac{1}{3y - 4} = \dfrac{6}{4(y + 1)}$

55. $\dfrac{2}{y} - \dfrac{1}{3y} = \dfrac{1}{3}$

56. $8\left(\dfrac{6}{x} - \dfrac{1}{x + 5}\right) = 15$

57. $r = 2 + \dfrac{24}{r}$

58. $\dfrac{3}{y + 1} - \dfrac{8}{y} = 1$

59. $\dfrac{2}{x} - \dfrac{x}{6} = \dfrac{2}{3}$

60. $\dfrac{2x}{x + 3} - \dfrac{3}{x} = 0$

61. *Average Speed* Suppose you drove 56 miles (one way) on a service call for your company. On the return trip, which took 10 minutes less than the original trip, your average speed was 8 miles per hour faster. What was your average speed on the return trip?

62. *Batting Average* In this year's playing season, a baseball player has been up to bat 150 times and has hit the ball successfully 45 times. Thus, the player's batting average is $45/150 = 0.300$. How many consecutive times must the player successfully hit the ball in order to obtain a batting average of 0.400?

63. *Forming a Partnership* A group of people agree to share equally in the cost of a \$60,000 piece of machinery. If they could find two more people to join the group, each person's share of the cost would decrease by \$5,000. How many people are presently in the group?

64. *Forming a Partnership* An individual is planning to start a small business that will require \$28,000 before any income can be generated. Because it is difficult to borrow money for new ventures, the individual wants a group of friends to divide the cost equally and, in return, receive a share of the future profits. The person has found some investors, but three more are needed so that the price per person will be \$1,200 less. How many investors are currently in the group?

65. *Work Rate* Suppose that in 12 minutes your supervisor can do a task that you require 15 minutes to complete. Determine the time required to complete the task if you work together.

CHAPTER 4 TEST

Take this test as you would take a test in class. After you are done, check your work with the answers given in the back of the book.

1. Find the domain of the rational expression: $\dfrac{3y}{y^2 - 25}$

2. Find the missing algebraic expression so that the two expressions are equivalent.

$$\frac{3x}{x-3} = \frac{3x(\quad)}{x^2 - x - 6}, \quad x \neq -2$$

3. Simplify: $\dfrac{2-x}{3x-6}$

4. Simplify: $\dfrac{2a^2 - 5a - 12}{5a - 20}$

5. Multiply: $\dfrac{4z^3}{5} \cdot \dfrac{25}{12z^2}$

6. Multiply: $\dfrac{y^2 + 8y + 16}{2(y-2)} \cdot \dfrac{8y - 16}{(y+4)^3}$

7. Multiply: $(4x^2 - 9) \cdot \dfrac{2x+3}{2x^2 - x - 3}$

8. Divide: $\dfrac{(2xy^2)^3}{15} \div \dfrac{12x^3}{21}$

9. Divide: $\dfrac{\left(\dfrac{3x}{x+2}\right)}{\left(\dfrac{12}{x^3 + 2x^2}\right)}$

10. Find the lowest common denominator of the fractions.

$$\frac{3}{x^2}, \quad \frac{x}{x-3}, \quad \frac{2x}{x^3(x+3)}, \quad \frac{10}{x^2 + 6x + 9}$$

11. Add: $2x + \dfrac{1 - 4x^2}{x+1}$

12. Subtract: $\dfrac{5x}{x+2} - \dfrac{2}{x^2 - x - 6}$

13. Perform the specified operations and simplify.

$$\frac{3}{x} - \frac{5}{x^2} + \frac{2x}{x^2 + 2x + 1}$$

14. Simplify: $\dfrac{\left(9x - \dfrac{1}{x}\right)}{\left(\dfrac{1}{x} - 3\right)}$

15. Divide (use long division): $\dfrac{t^4 + t^2 - 6t}{t^2 - 2}$

16. Divide (use synthetic division): $\dfrac{5x^3 + 6x^2 + 3x - 10}{x+5}$

17. Divide (use synthetic division): $\dfrac{2x^4 - 15x^2 - 7}{x-3}$

18. Solve (if possible): $\dfrac{3}{h+2} = \dfrac{1}{8}$

19. Solve (if possible): $\dfrac{2}{x+5} - \dfrac{3}{x+3} = \dfrac{1}{x}$

20. Solve (if possible): $\dfrac{1}{x+1} + \dfrac{1}{x-1} = \dfrac{2}{x^2 - 1}$

21. One painter works one and one-half times as fast as a second painter. Find their individual times for painting a room if, working together, they can complete the job in four hours.

CUMULATIVE TEST: CHAPTERS 1–4

Take this test as you would take a test in class. After you are done, check your work with the answers given in the back of the book.

1. Evaluate: $-\frac{8}{45} \div \frac{12}{25}$

2. Simplify: $(2a^2b)^3(-ab^2)^2$

3. Simplify: $3xy(x^2 - 2) - xy(x^2 + 5)$

4. Write an algebraic expression for the statement "The number n is tripled and the product is decreased by 8."

5. Solve: $12 - 5(3 - x) = x + 3$

6. Solve: $1 - \frac{x + 2}{4} = \frac{7}{8}$

7. 46.8 is what percent of 360?

8. The annual insurance premium for a policyholder is $1,225. Find the annual premium if the policyholder must pay a 15% surcharge because of a driving violation.

9. Solve the proportion: $\frac{t - 1}{4} = \frac{11}{12}$

10. Solve for the length x of the side of the second triangle by using the fact that corresponding sides of similar triangles are proportional (see figure).

Figure for 10

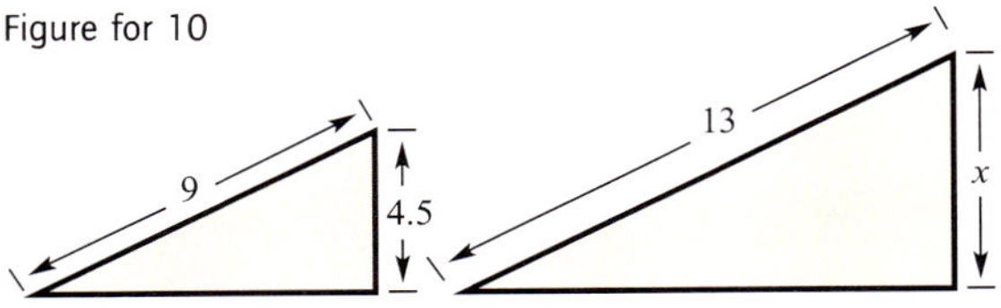

11. The revenue for selling x units of a product is

$$R = 12.90x.$$

The cost of producing x units is

$$C = 8.50x + 450.$$

In order for a profit to be obtained, the revenue must be greater than the cost. For what values of x will this product produce a profit?

12. Solve and sketch the solution: $|x - 2| \geq 3$

13. Multiply and simplify: $t(3t - 1) - 2t(t + 4)$

14. Expand and simplify: $[2 + (x - y)]^2$

15. Factor: $y^3 - 3y^2 - 9y + 27$

16. Solve by factoring: $20 - 7y - 3y^2 = 0$

17. Multiply and simplify: $\frac{x^2 + 8x + 16}{18x^2} \cdot \frac{2x^4 + 4x^3}{x^2 - 16}$

18. Perform the indicated operation and simplify: $\frac{2}{x} - \frac{x}{x^3 + 3x^2} + \frac{1}{x + 3}$

19. Simplify: $\dfrac{\left(\frac{x}{y} - \frac{y}{x}\right)}{\left(\frac{x - y}{xy}\right)}$

20. Divide (use synthetic division): $\frac{x^3 + 27}{x + 3}$

21. Solve: $x + \frac{4}{x} = 4$

CHAPTER FIVE

Radicals and Complex Numbers

Chapter Overview

In preparation for studying the algebra of radical expressions, we review negative exponents in Section 5.1. Then, in Sections 5.2 through 5.5, we study techniques for simplifying radical expressions, operations with radical expressions, and techniques for solving equations involving radical expressions. In the last section, we introduce the concept of complex numbers.

SECTION 5.1

Integer Exponents and Scientific Notation

Integer Exponents • *Scientific Notation*

Integer Exponents

Our study of algebra thus far has involved only positive integer exponents. In this section we extend the concept of an exponent to include zero and negative integers.

If a is a real number such that $a \neq 0$, then we define a^0 to be 1. Moreover, if m is an integer, then we define a^{-m} to be the reciprocal of a^m.

Definition of Zero Exponents and Negative Exponents

Let a be a real number such that $a \neq 0$, and let m be an integer.

1. $a^0 = 1, \quad a \neq 0$ 2. $a^{-m} = \dfrac{1}{a^m}, \quad a \neq 0$

These definitions of zero and negative exponents are consistent with the properties for exponents given in Section 3.2. For instance, consider the following.

$$x^0 \cdot x^m = x^{0+m} = x^m = 1 \cdot x^m$$

(x^0 is the same as 1)

$$x^{-m} \cdot x^m = x^{-m+m} = x^0 = 1 = \frac{1}{x^m} \cdot x^m$$

(x^{-m} is the same as $1/x^m$)

EXAMPLE 1 Zero Exponents and Negative Exponents

Rewrite the following numbers without using zero exponents or negative exponents.

(a) 3^0 (b) 0^0 (c) 2^{-1} (d) 3^{-2}

Solution

(a) $3^0 = 1$

(b) 0^0 is undefined. Remember that when raising a number to the zero power, you must be sure the number is not zero.

(c) $2^{-1} = \dfrac{1}{2^1} = \dfrac{1}{2}$

(d) $3^{-2} = \dfrac{1}{3^2} = \dfrac{1}{9}$

The following properties of exponents are valid for all integer exponents, including integer exponents that are zero or negative.

Properties of Exponents Let m and n be integers, and let a and b represent real numbers, variables, or algebraic expressions.

Property	*Example*
1. $a^m \cdot a^n = a^{m+n}$	$x^4(x^3) = x^{4+3} = x^7$
2. $(ab)^m = a^m \cdot b^m$	$(3x)^2 = 3^2(x^2) = 9x^2$
3. $(a^m)^n = a^{mn}$	$(x^3)^3 = x^{3 \cdot 3} = x^9$
4. $\dfrac{a^m}{a^n} = a^{m-n}, \quad a \neq 0$	$\dfrac{x^3}{x} = x^{3-1} = x^2, \quad x \neq 0$
5. $\left(\dfrac{a}{b}\right)^m = \dfrac{a^m}{b^m}, \quad b \neq 0$	$\left(\dfrac{x}{3}\right)^2 = \dfrac{x^2}{3^2} = \dfrac{x^2}{9}$
6. $\left(\dfrac{a}{b}\right)^{-m} = \left(\dfrac{b}{a}\right)^m, \quad a \neq 0, b \neq 0$	$\left(\dfrac{x}{3}\right)^{-2} = \left(\dfrac{3}{x}\right)^2 = \dfrac{3^2}{x^2} = \dfrac{9}{x^2}$
7. $a^0 = 1, \quad a \neq 0$	$(x^2 + 1)^0 = 1$
8. $a^{-m} = \dfrac{1}{a^m}, \quad a \neq 0$	$x^{-2} = \dfrac{1}{x^2}, \quad x \neq 0$

EXAMPLE 2 ■ **Using Properties of Exponents**

Rewrite the following expressions using only positive exponents. (For each expression, assume that $x \neq 0$.)

(a) $2x^{-1}$ (b) $(2x)^{-1}$ (c) $\dfrac{3}{x^{-2}}$ (d) $\dfrac{1}{(3x)^{-2}}$

Solution

(a) $2x^{-1} = 2(x^{-1}) = 2\left(\dfrac{1}{x}\right) = \dfrac{2}{x}$

(b) $(2x)^{-1} = \dfrac{1}{(2x)^1} = \dfrac{1}{2x}$

(c) $\dfrac{3}{x^{-2}} = \dfrac{3}{\left(\dfrac{1}{x^2}\right)} = 3\left(\dfrac{x^2}{1}\right) = 3x^2$

(d) $\dfrac{1}{(3x)^{-2}} = \dfrac{1}{\left(\dfrac{1}{(3x)^2}\right)} = \dfrac{1}{\left(\dfrac{1}{3^2x^2}\right)} = \dfrac{1}{\left(\dfrac{1}{9x^2}\right)} = (1)\left(\dfrac{9x^2}{1}\right) = 9x^2$ ■

As you become accustomed to working with negative exponents, you will probably not write as many steps as are shown in Example 2. To rewrite a fraction involving exponents, we like to use the following simplified rule: *To move a factor from the numerator to the denominator or vice versa, change the sign of its exponent.* For instance, in the expression

$$\frac{3}{x^{-2}}$$

we can eliminate the negative exponent by "moving" the factor x^{-2} to the numerator and changing the exponent to 2. That is,

$$\frac{3}{x^{-2}} = 3x^2.$$

EXAMPLE 3 ■ **Using Properties of Exponents**

Rewrite the following expressions using only positive exponents. (For each expression, assume that $x \neq 0$ and $y \neq 0$.)

(a) $(-5x^{-3})^2$

(b) $-\left(\frac{7x}{y^2}\right)^{-2}$

Solution

(a) $(-5x^{-3})^2 = (-5)^2(x^{-3})^2 = 25x^{-6} = \dfrac{25}{x^6}$

(b) $-\left(\frac{7x}{y^2}\right)^{-2} = -\left(\frac{y^2}{7x}\right)^2 = -\dfrac{(y^2)^2}{(7x)^2} = -\dfrac{y^4}{49x^2}$ ■

EXAMPLE 4 ■ **Using Properties of Exponents**

Rewrite the following expressions using only positive exponents. (For each expression, assume that $x \neq 0$ and $y \neq 0$.)

(a) $\left(\frac{8x^{-1}y^4}{4x^3y^2}\right)^{-3}$ (b) $\dfrac{3xy^0}{x^2(5y)^0}$

Solution

(a) $\left(\frac{8x^{-1}y^4}{4x^3y^2}\right)^{-3} = \left(\frac{2y^2}{x^4}\right)^{-3} = \left(\frac{x^4}{2y^2}\right)^3 = \dfrac{x^{12}}{2^3y^6} = \dfrac{x^{12}}{8y^6}$

(b) $\dfrac{3xy^0}{x^2(5y)^0} = \dfrac{3x(1)}{x^2(1)} = \dfrac{3}{x}$ ■

Scientific Notation

Exponents provide an efficient way of writing and computing with very large (or very small) numbers. For instance, a drop of water contains more than 33 billion billion molecules—that is 33 followed by 18 zeros.

$$33{,}000{,}000{,}000{,}000{,}000{,}000$$

It is convenient to write such numbers in **scientific notation**. This notation has the form $c \times 10^n$, where $1 \le c < 10$ and n is an integer. Thus, the number of molecules in a drop of water can be written in scientific notation as

$$3.3 \times 10{,}000{,}000{,}000{,}000{,}000{,}000 = 3.3 \times 10^{19}.$$

The *positive* exponent 19 indicates that the number is large (10 or more) and that the decimal point has been moved 19 places. A *negative* exponent in scientific notation indicates that the number is *small* (less than 1). For instance, the mass (in grams) of one electron is approximately

$$9.0 \times 10^{-28} = 0.0000000000000000000000000009.$$

28 decimal places

EXAMPLE 5 ■ **Converting from Decimal Notation to Scientific Notation**

Write each of the following real numbers in scientific notation.

(a) 0.0000684

(b) 937,200,000

Solution

(a) This decimal number is less than 1. Thus, to write the number in scientific notation, we move the decimal point as follows.

$$0.0000684 = 6.84 \times 10^{-5}$$

Five places

(b) This number is larger than 10. Thus, to write the number in scientific notation, we move the decimal point as follows.

$$937{,}200{,}000.0 = 9.372 \times 10^{8}$$

Eight places

■

EXAMPLE 6 ■ **Converting from Scientific Notation to Decimal Notation**

Convert each of the following numbers from scientific notation to decimal notation.

(a) 2.486×10^2 (b) 1.81×10^{-6}

Solution

(a) For this number, the exponent of 10 is a positive 2. Thus, to write the number in decimal notation, we move the decimal point two places, as follows.

$$2.486 \times 10^2 = 248.6$$

Two places

(b) For this number, the exponent of 10 is a negative 6. Thus, to write the number in decimal notation, we move the decimal point six places, as follows.

$$1.81 \times 10^{-6} = 0.00000181$$

Six places ■

NOTE: To convert decimal numbers to scientific notation or conversely, we make the following associations.

Numbers ≥ 10	⟺	Positive powers of 10
Numbers < 1	⟺	Negative powers of 10
Numbers between 1 and 10	⟺	Zero power of 10

Most scientific calculators automatically switch to scientific notation when they are showing large (or small) numbers that exceed the display range. Try multiplying $86{,}500{,}000 \times 6{,}000$. If your calculator follows standard conventions, its display should be

[5.19 11] or [5.19 E 11].

This means that $c = 5.19$ and the exponent of 10 is $n = 11$, which implies that the number is 5.19×10^{11}.

To *enter* numbers in scientific notation, your calculator should have an exponential entry key labeled [EE] or [EXP]. If you were to perform the preceding multiplication using scientific notation, you could begin by writing

$$86{,}500{,}000 \times 6{,}000 = (8.65 \times 10^7)(6.0 \times 10^3)$$

and then entering the following.*

8.65 [EXP] 7 [×] 6 [EXP] 3 [=]   Scientific

8.65 [EE] 7 [×] 6 [EE] 3 [ENTER]   Graphing

*The graphing calculator keystrokes given in this text correspond to the *TI-81* graphing calculator from *Texas Instruments*. For other graphing calculators, the keystrokes may differ.

EXAMPLE 7 ■ **Using Scientific Notation with a Calculator**

Use a scientific calculator to find the following.

(a) $65{,}000 \times 3{,}400{,}000{,}000$ (b) $0.000000348 \div 870$

Solution

(a) Since $65{,}000 = 6.5 \times 10^4$ and $3{,}400{,}000{,}000 = 3.4 \times 10^9$, we can multiply the two numbers using the following calculator steps.

6.5 [EXP] 4 [×] 3.4 [EXP] 9 [=] Scientific

6.5 [EE] 4 [×] 3.4 [EE] 9 [ENTER] Graphing

After entering these keystrokes, the calculator display should read [2.21 14]. Therefore, the product of the two numbers is

$$(6.5 \times 10^4)(3.4 \times 10^9) = 2.21 \times 10^{14} = 221{,}000{,}000{,}000{,}000.$$

(b) Since $0.000000348 = 3.48 \times 10^{-7}$ and $870 = 8.7 \times 10^2$, we can divide the two numbers as follows.

3.48 [EXP] 7 [+/−] [÷] 8.7 [EXP] 2 [=] Scientific

3.48 [EE] [(−)] 7 [÷] 8.7 [EE] 2 [ENTER] Graphing

After entering these keystrokes, the calculator display should read [4.0 −10]. Therefore, the quotient of the two numbers is

$$\frac{3.48 \times 10^{-7}}{8.7 \times 10^2} = 4.0 \times 10^{-10} = 0.0000000004.$$ ■

EXAMPLE 8 ■ **Using Scientific Notation and Properties of Exponents**

Use scientific notation and the properties of exponents to simplify the following expression. Then evaluate the expression.

$$\frac{(2{,}400{,}000{,}000)(0.00000345)}{(0.00007)(3{,}800)}$$

Solution

We begin by rewriting each number in scientific notation and simplifying.

$$\begin{aligned}\frac{(2{,}400{,}000{,}000)(0.00000345)}{(0.00007)(3{,}800)} &= \frac{(2.4 \times 10^9)(3.45 \times 10^{-6})}{(7.0 \times 10^{-5})(3.8 \times 10^3)}\\ &= \frac{(2.4)(3.45)(10^3)}{(7)(3.8)(10^{-2})}\\ &= \frac{(8.28)(10^5)}{26.6}\\ &\approx 0.3112782(10^5)\\ &= 31{,}127.82\end{aligned}$$ ■

DISCUSSION PROBLEM ■ Limiting Value of an Expression

(a) Use a calculator to complete the following tables.

x	10^0	10^1	10^2	10^3	10^4	10^5	10^6
x^{-1}							

x	10^0	10^{-1}	10^{-2}	10^{-3}	10^{-4}	10^{-5}	10^{-6}
x^{-1}							

(b) Write a short paragraph describing the value of the expression x^{-1} as the value of x gets closer and closer to zero (through positive values).

(c) Write a short paragraph describing the value of the expression x^{-1} as the value of x grows larger and larger. ■

Warm-Up

The following warm-up exercises involve skills that were covered in earlier sections. You will use these skills in the exercise set for this section.

In Exercises 1–10, simplify the expression.

1. $a^4 \cdot a^6$

2. $x^2 \cdot y^4 \cdot x^3$

3. $\dfrac{81y^5}{3^2y^2}, \quad y \neq 0$

4. $\dfrac{64r^2s^4}{16rs^2}, \quad r \neq 0, s \neq 0$

5. $(-t^2)^3$

6. $(4z)^2$

7. $(-3x^2y^3)^2 \cdot (4xy^2)$

8. $(-4uv^2)^3 \cdot (2u^2v)$

9. $\left(\dfrac{2a^2}{3b}\right)^4, \quad b \neq 0$

10. $\left(\dfrac{3x}{4y^3}\right)^2, \quad y \neq 0$

5.1 EXERCISES

In Exercises 1–30, evaluate the given expression.

1. 5^{-2}

2. 2^{-4}

3. -10^{-3}

4. -20^{-2}

5. $(-3)^{-5}$

6. 25^0

7. $\dfrac{1}{4^{-3}}$

8. $\dfrac{1}{-8^{-2}}$

9. $\dfrac{1}{(-2)^{-5}}$

10. $-\dfrac{1}{6^2}$

11. $\left(\dfrac{2}{3}\right)^{-1}$

12. $\left(\dfrac{4}{5}\right)^{-3}$

Mission College

Math Tutoring

Do you want help on a regular basis all semester?

INDIVIDUAL TUTORING

* Fill out an application in the Math Lab
* Meet with your tutor every week
  (you arrange the time)
* Make a commitment to meet with your tutor
  for at least 4 weeks

Do you need help only occasionally?

DROP-IN TUTORING

* Fill out an application in the Math Lab
* Come to the Drop-In tables in the Math Lab
  (no appointment needed)
* Study in the Math Lab
  (ask questions when you want to)

or more information: Come to the Math Lab (S2-301) or call 748-2763

TH LAB HOURS:

M-TH 9AM-7:30PM
F 9AM-2:00PM
Sat 10AM-2:00PM

13. $\left(\frac{3}{16}\right)^0$ **14.** $\left(-\frac{5}{8}\right)^{-2}$ **15.** $27 \cdot 3^{-3}$ **16.** $4^2 \cdot 4^{-3}$ **17.** $\frac{3^4}{3^{-2}}$ **18.** $\frac{5^{-1}}{5^2}$

19. $\frac{10^3}{10^{-2}}$ **20.** $\frac{10^{-5}}{10^{-6}}$ **21.** $(2^{-3})^2$ **22.** $(-4^{-1})^{-2}$ **23.** $(4^2 \cdot 4^{-1})^{-2}$ **24.** $(5^3 \cdot 5^{-4})^{-3}$

25. $2^{-3} + 2^{-4}$ **26.** $4 - 3^{-2}$ **27.** $\left(\frac{3}{4} + \frac{5}{8}\right)^{-2}$

28. $\left(\frac{1}{2} - \frac{2}{3}\right)^{-1}$ **29.** $(5^0 - 4^{-2})^{-1}$ **30.** $(32 + 4^{-3})^0$

In Exercises 31–54, rewrite the expression using only positive exponents and simplify. (Assume any variables in the expression are nonzero.)

31. $y^4 \cdot y^{-2}$ **32.** $x^{-2} \cdot x^{-5}$ **33.** $\frac{1}{x^{-6}}$ **34.** $\frac{x^{-3}}{y^{-1}}$

35. $\frac{x^4}{x^{-2}}$ **36.** $\frac{s^{-3}}{s^3}$ **37.** $\frac{a^{-6}}{a^{-7}}$ **38.** $\frac{6u^{-2}}{15u^{-1}}$

39. $\frac{(4t)^0}{t^{-2}}$ **40.** $\frac{(5u)^4}{(5u)^4}$ **41.** $(-3x^{-3}y^2)(4x^2y^{-5})$ **42.** $(5s^5t^{-5})\left(\frac{3s^{-2}}{50t^{-1}}\right)$

43. $(3x^2y^{-2})^{-2}$ **44.** $(-4y^{-3}z)^{-3}$ **45.** $\frac{6^2x^3y^{-3}}{12x^{-2}y}$ **46.** $\frac{2^{-4}y^{-1}z^{-3}}{4^{-2}yz^{-3}}$

47. $\left(\frac{3u^2v^{-1}}{3^3u^{-1}v^3}\right)^{-2}$ **48.** $\left(\frac{5^2x^3y^{-3}}{125xy}\right)\left(\frac{5x}{3y}\right)^{-1}$ **49.** $(u + v^{-2})^{-1}$ **50.** $x^{-2}(x^2 + y^2)$

51. $\frac{a + b}{ba^{-1} - ab^{-1}}$ **52.** $\frac{u^{-1} - v^{-1}}{u^{-1} + v^{-1}}$ **53.** $[(2x^{-3}y^{-2})^2]^{-2}$ **54.** $\left[\left(\frac{2x^2}{4y}\right)^{-3}\right]^2$

In Exercises 55–60, write the real number in scientific notation.

55. 3,600,000 **56.** 98,100,000 **57.** 68,430,000

58. 0.000012 **59.** 0.0000000381 **60.** 0.0007384

In Exercises 61–66, write the real number in decimal notation.

61. 6×10^7 **62.** 5.05×10^{12} **63.** 1.359×10^{-7}

64. 8.6×10^{-9} **65.** 9.3×10^{-3} **66.** 2.5×10^4

In Exercises 67–72, evaluate the quantity without the use of a calculator.

67. $(2 \times 10^9)(3.4 \times 10^{-4})$ **68.** $(5 \times 10^4)^3$ **69.** $\frac{3.6 \times 10^9}{9 \times 10^5}$

70. $\frac{2.5 \times 10^{-3}}{5 \times 10^2}$ **71.** $(4{,}500{,}000)(2{,}000{,}000{,}000)$ **72.** $\frac{64{,}000{,}000}{0.00004}$

In Exercises 73–76, evaluate the quantity using a calculator. (Write your answer in scientific notation.)

73. $\dfrac{1.357 \times 10^{12}}{(4.2 \times 10^{2})(6.87 \times 10^{-3})}$

74. $(8.67 \times 10^{4})^{7}$

75. $\dfrac{(0.0000565)(2{,}850{,}000{,}000{,}000)}{0.00465}$

76. $\dfrac{(5{,}000{,}000)^{3}(0.000037)^{2}}{(0.005)^{4}}$

77. *Distance to the Sun* The distance from the Earth to the Sun is approximately 93 million miles. Write this distance in scientific notation.

78. *Copper Electrons* A cube of copper that measures 1 centimeter on an edge has approximately 8×10^{22} free electrons. Write this real number in decimal form.

79. *Light-Year* One light-year (the distance light can travel in one year) is approximately 9.45×10^{15} meters. Approximate the time for light to travel from the Sun to the Earth if that distance is approximately 1.49×10^{11} meters.

80. *Distance to a Star* Determine the distance (in meters) to the star Alpha Andromedae if it is 90 light-years away from the Earth (see figure). See Exercise 79 for the definition of a light-year.

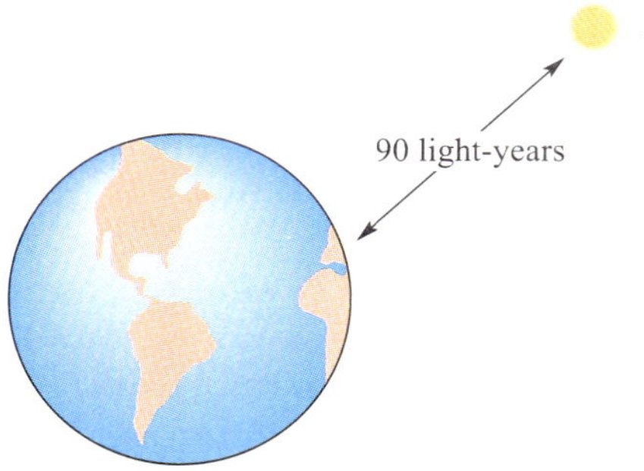

81. *Mass of Earth and Sun* The masses of the Earth and the Sun are about 5.975×10^{24} kg and 1.99×10^{30} kg, respectively. The mass of the Sun is approximately how many times that of the Earth?

82. *Federal Debt* From the reference section of a library find the population of the United States and the magnitude of the debt of the federal government. Use these two numbers to determine the amount each person would have to pay to remove the debt.

SECTION 5.2 Radicals and Rational Exponents

Roots and Radicals • *Rational Exponents* • *Radicals and Calculators*

Roots and Radicals

We already know how to square a number—we raise it to the second power by using the number *twice* as a factor. For instance, $5 \cdot 5 = 5^2 = 25$, and we say that 5 squared is 25. Conversely, we define a **square root of a number** as one of its two equal factors. For example, 5 is a square root of 25 because 5 is one of the two equal factors of 25. In a similar way, a **cube root** of a number is one of its three equal factors. Consider the following examples.

Number	*Equal Factors*	*Root*	
$9 = 3^2$	$3 \cdot 3$	3	(square root)
$25 = 5^2$	$5 \cdot 5$	5	(square root)
$25 = (-5)^2$	$(-5)(-5)$	-5	(square root)
$-27 = (-3)^3$	$(-3)(-3)(-3)$	-3	(cube root)
$16 = 2^4$	$2 \cdot 2 \cdot 2 \cdot 2$	2	(fourth root)

In general, we define the ***n*****th root** of a number as follows.

Definition of *n*th Root of a Number

Let a and b be real numbers and let n be a positive integer such that $n \geq 2$. If

$$a = b^n$$

then we say that b is an ***n*****th root of *a***. If $n = 2$, we call the root a **square root**, and if $n = 3$, we call the root a **cube root**.

By applying this definition, we see that some numbers have more than one nth root. For example, both 5 and -5 are square roots of 25 because $25 = 5^2$ and $25 = (-5)^2$. Similarly, both 2 and -2 are fourth roots of 16 because $16 = 2^4$ and $16 = (-2)^4$. To avoid ambiguity about which root of a number we are talking about, we define the **principal *n*th root** of the number in terms of the radical symbol $\sqrt[n]{}$

Principal *n*th Root of a Number

Let a be a real number that has at least one (real number) nth root. The **principal *n*th root of *a*** is the nth root that has the same sign as a, and it is denoted by the **radical symbol**

$$\sqrt[n]{a}. \qquad \text{Principal } n\text{th root}$$

The positive integer n is called the **index** of the radical, and the number a is called the **radicand**. If $n = 2$, we omit the index and write $\sqrt{a}$ rather than $\sqrt[2]{a}$.

NOTE: "Having the same sign" means that the principal nth root of a is positive if a is positive and negative if a is negative. For example, $\sqrt{4} = 2$ and $\sqrt[3]{-8} = -2$.

EXAMPLE 1 ■ Finding the Principal Square Root of a Number

Find the following principal square roots.

(a) $\sqrt{36}$ (b) $\sqrt{0}$ (c) $\sqrt{-4}$

Solution

(a) The number $36 = 6^2 = (-6)^2$ has two square roots, 6 and -6. Because the principal square root is the root that has the same sign as 36, we can write

$$\sqrt{36} = 6.$$

(b) The number $0 = 0^2$ has only one square root, which is itself. Therefore, the principal square root of 0 is given by

$$\sqrt{0} = 0.$$

(c) The number -4 has no square root because there is no real number that can be squared to produce -4. Therefore, we say that $\sqrt{-4}$ is not a real number. The same comment can be made of any negative number. That is, if a is negative, then $\sqrt{a}$ is not a real number. ■

EXAMPLE 2 ■ **Finding the Principal *n*th Root of a Number**

Find the following principal nth roots.

(a) $\sqrt[3]{8}$ (b) $\sqrt[3]{-27}$ (c) $\sqrt[4]{256}$

Solution

(a) The number $8 = 2^3$ has only one real cube root, 2. Because it has the same sign as 8, we can write

$$\sqrt[3]{8} = 2.$$

(b) The number $-27 = (-3)^3$ has only one real cube root, -3. Because it has the same sign as -27, we can write

$$\sqrt[3]{-27} = -3.$$

(c) The number $256 = 4^4 = (-4)^4$ has two fourth roots, 4 and -4. Because the principal fourth root is the root that has the same sign as 256, we can write

$$\sqrt[4]{256} = 4.$$ ■

From Examples 1 and 2, we make the following generalizations about nth roots of a real number. (Remember that for nth roots n is an integer that is greater than or equal to 2.)

1. If a is a positive real number and n is *even*, then a has exactly two (real) nth roots. We denote these roots by

$$\sqrt[n]{a} \quad \text{and} \quad -\sqrt[n]{a}.$$

2. If a is any real number and n is *odd*, then a has only one (real) nth root, which is denoted by $\sqrt[n]{a}$.
3. If a is a negative real number and n is *even*, then a has no (real) nth root.

Integers such as 1, 4, 9, 16, 49, and 81 are called **perfect squares** because they have integer square roots. Similarly, integers such as 1, 8, 27, 64, and 125 are called **perfect cubes** because they have integer cube roots.

EXAMPLE 3 ■ Evaluating Expressions Involving Radicals

Evaluate the following radical expressions.

(a) $\sqrt{144}$ (b) $-\sqrt{81}$ (c) $-\sqrt[3]{-64}$ (d) $\sqrt[4]{1}$

Solution

(a) $\sqrt{144} = 12$

(b) $-\sqrt{81} = -9$

(c) $-\sqrt[3]{-64} = -(-4) = 4$

(d) $\sqrt[4]{1} = 1$ ■

Raising a number to the nth power and taking the principal nth root of a number can be thought of as *inverse* operations. For instance,

$$\left(\sqrt{4}\right)^2 = 2^2 = 4 \quad \text{and} \quad \sqrt{2^2} = \sqrt{4} = 2.$$

Similarly,

$$\left(\sqrt[3]{27}\right)^3 = 3^3 = 27 \quad \text{and} \quad \sqrt[3]{3^3} = \sqrt[3]{27} = 3.$$

For *even* roots involving *negative* numbers, this pattern breaks down a little. However, using absolute value signs, we can make the following generalizations.

Inverse Properties of nth Powers and nth Roots

Let a be a real number, and let n be an integer such that $n \geq 2$.

1. If a has a principal nth root, then
$$\left(\sqrt[n]{a}\right)^n = a.$$
2. If n is *odd*, then
$$\sqrt[n]{a^n} = a.$$
If n is *even*, then
$$\sqrt[n]{a^n} = |a|.$$

EXAMPLE 4 ■ Evaluating Radical Expressions

Evaluate the following radical expressions.

(a) $\sqrt[3]{5^3}$ (b) $\sqrt[3]{(-2)^3}$ (c) $\left(\sqrt{7}\right)^2$ (d) $\sqrt{(-3)^2}$

Solution

(a) Since the index of the radical is odd, we can write

$$\sqrt[3]{5^3} = 5.$$

(b) The index of this radical is also odd, and we can write

$$\sqrt[3]{(-2)^3} = -2.$$

(c) Using the inverse property of powers and roots, we can write

$$\left(\sqrt{7}\right)^2 = 7.$$

(d) Since the index of this radical is even, we must include absolute value signs, and write

$$\sqrt{(-3)^2} = |-3| = 3.$$

Rational Exponents

We are now ready to give meaning to rational exponents such as $4^{1/2}$ and $8^{1/3}$. We do this with the following definition.

Definition of Rational Exponents

Let a be a real number, and let n be an integer such that $n \geq 2$. If the principal nth root of a exists, then we define $a^{1/n}$ to be

$$a^{1/n} = \sqrt[n]{a}.$$

Moreover, if m is a positive integer that has no common factor with n, then

$$a^{m/n} = (a^{1/n})^m = \left(\sqrt[n]{a}\right)^m \quad \text{and} \quad a^{m/n} = (a^m)^{1/n} = \sqrt[n]{a^m}.$$

The numerator of a rational exponent denotes the *power* to which the base is raised, and the denominator denotes the *root* to be taken. Moreover, it doesn't matter in which order the two operations are performed, provided the nth root exists. Here is an example.

$$8^{2/3} = \left(\sqrt[3]{8}\right)^2 = 2^2 = 4 \qquad \text{Cube root, then second power}$$
$$8^{2/3} = \sqrt[3]{8^2} = \sqrt[3]{64} = 4 \qquad \text{Second power, then cube root}$$

The properties of exponents that we listed in Section 5.1 also apply to rational exponents (provided the roots indicated by the denominators exist). We relist those properties here, with different examples.

Properties of Exponents

Let r and s be rational numbers, and let a and b be real numbers, variables, or algebraic expressions. If the roots indicated by the rational exponents exist, then the following properties are true.

Property	*Example*
1. $a^r \cdot a^s = a^{r+s}$	$4^{1/2}(4^{1/3}) = 4^{5/6}$
2. $(ab)^r = a^r \cdot b^r$	$(2x)^{1/2} = 2^{1/2}(x^{1/2})$
3. $(a^r)^s = a^{rs}$	$(x^3)^{1/3} = x$
4. $\dfrac{a^r}{a^s} = a^{r-s}, \quad a \neq 0$	$\dfrac{x^2}{x^{1/2}} = x^{2-(1/2)} = x^{3/2}$
5. $\left(\dfrac{a}{b}\right)^r = \dfrac{a^r}{b^r}, \quad b \neq 0$	$\left(\dfrac{x}{3}\right)^{1/3} = \dfrac{x^{1/3}}{3^{1/3}}$
6. $\left(\dfrac{a}{b}\right)^{-r} = \left(\dfrac{b}{a}\right)^r, \quad a \neq 0, b \neq 0$	$\left(\dfrac{x}{4}\right)^{-1/2} = \left(\dfrac{4}{x}\right)^{1/2} = \dfrac{2}{x^{1/2}}$
7. $a^{-r} = \dfrac{1}{a^r}, \quad a \neq 0$	$4^{-1/2} = \dfrac{1}{4^{1/2}} = \dfrac{1}{2}$

EXAMPLE 5 ■ **Evaluating Expressions with Rational Exponents**

Evaluate the following expressions.

(a) $8^{4/3}$ (b) $(4^2)^{3/2}$ (c) $25^{-3/2}$

Solution

(a) $8^{4/3} = \left(\sqrt[3]{8}\right)^4 = 2^4 = 16$

(b) $(4^2)^{3/2} = \left(\sqrt{4^2}\right)^3 = 4^3 = 64$

(c) $25^{-3/2} = \dfrac{1}{25^{3/2}} = \dfrac{1}{\left(\sqrt{25}\right)^3} = \dfrac{1}{5^3} = \dfrac{1}{125}$ ■

EXAMPLE 6 ■ **Evaluating Expressions with Rational Exponents**

Evaluate the following expressions.

(a) $\left(\dfrac{64}{125}\right)^{2/3}$ (b) $-9^{1/2}$ (c) $(-9)^{1/2}$

Solution

(a) $\left(\dfrac{64}{125}\right)^{2/3} = \left(\sqrt[3]{\dfrac{64}{125}}\right)^2 = \left(\dfrac{\sqrt[3]{64}}{\sqrt[3]{125}}\right)^2 = \left(\dfrac{4}{5}\right)^2 = \dfrac{16}{25}$

(b) $-9^{1/2} = -\sqrt{9} = -3$

(c) $(-9)^{1/2} = \sqrt{-9}$ is not a real number. ■

EXAMPLE 7 ■ Using Properties of Exponents

Rewrite the expressions using rational exponents.

(a) $x\sqrt[4]{x^3}$ (b) $\dfrac{\sqrt[3]{x^2}}{\sqrt{x^3}}$

Solution

(a) $x\sqrt[4]{x^3} = x(x^{3/4}) = x^{1+(3/4)} = x^{7/4}$

(b) $\dfrac{\sqrt[3]{x^2}}{\sqrt{x^3}} = \dfrac{x^{2/3}}{x^{3/2}} = x^{(2/3)-(3/2)} = x^{-5/6}$ ■

EXAMPLE 8 ■ Using Properties of Exponents

Use the properties of exponents to simplify the following expressions.

(a) $\sqrt{\sqrt[3]{x}}$ (b) $\dfrac{(2x-1)^{4/3}}{\sqrt[3]{2x-1}}$

Solution

(a) $\sqrt{\sqrt[3]{x}} = \sqrt{x^{1/3}} = (x^{1/3})^{1/2} = x^{1/6} = \sqrt[6]{x}$

(b) $\dfrac{(2x-1)^{4/3}}{\sqrt[3]{2x-1}} = \dfrac{(2x-1)^{4/3}}{(2x-1)^{1/3}} = (2x-1)^{(4/3)-(1/3)} = 2x-1$ ■

Radicals and Calculators

There are two methods of evaluating radicals on most calculators. For square roots, you can use the *square root key* [√]. For other roots, you should first convert the radical to exponential form and then use the *exponential key* [y^x] or [∧].

NOTE: For cube roots, check to see whether your calculator has a cube root key. For instance, the *T1-81* has such a key as part of its "MATH" menu.

EXAMPLE 9 ■ Evaluating a Square Root with a Calculator

Use a calculator to evaluate $\sqrt{5}$. Round your answer to three decimal places.

Solution

Using the following calculator steps*

5 [√]	Scientific
[√] 5 [ENTER]	Graphing

*The graphing calculator keystrokes given in this text correspond to the *TI-81* graphing calculator from *Texas Instruments*. For other graphing calculators, the keystrokes may differ.

we obtain a calculator display of 2.236068. Rounded to three decimal places, we obtain

$$\sqrt{5} \approx 2.236.$$

Try entering $5^{1/2}$ on your calculator to see that it yields the same results. ■

EXAMPLE 10 ■ Evaluating a Fifth Root with a Calculator

Use a calculator to evaluate $\sqrt[5]{25}$. Round your answer to three decimal places.

Solution

First we write $\sqrt[5]{25}$ in exponential form.

$$\sqrt[5]{25} = 25^{1/5}$$

Now use one of the following keystroke sequences.

25 [y^x] [(] 1 [÷] 5 [)] [=]	Scientific
25 [∧] [(] 1 [÷] 5 [)] [ENTER]	Graphing

For either of these keystroke sequences, the calculator display should read 1.9036539. Thus, we have

$$\sqrt[5]{25} \approx 1.904.$$ ■

EXAMPLE 11 ■ Evaluating Radicals with a Calculator

Use a calculator to evaluate the following expressions. (Round your answers to three decimal places.)

(a) $\sqrt[3]{-4}$ (b) $(1.4)^{-2/5}$

Solution

(a) Most scientific calculators do not have a cube root key. Thus, to find the cube root of a negative number with a scientific calculator, we use the fact that

$$\sqrt[3]{-4} = \sqrt[3]{(-1)(4)} = \sqrt[3]{-1}\sqrt[3]{4} = -\sqrt[3]{4}$$

and attach the negative sign of the radicand at the end of the keystroke sequence.

4 [y^x] [(] 1 [÷] 3 [)] [=] [+/−]	Scientific
[$\sqrt[3]{\ }$] [(−)] 4 [ENTER]	Graphing

The calculator display is −1.5874011, which implies that

$$\sqrt[3]{-4} \approx -1.587.$$

NOTE: To enter [$\sqrt[3]{\ }$] on a *TI-81*, press [MATH], cursor to $\sqrt[3]{\ }$, and press [ENTER].

(b) Using the following keystroke sequence

1.4 [y^x] [(] 2 [÷] 5 [+/−] [)] [=] — Scientific

1.4 [∧] [(] [(−)] 2 [÷] 5 [)] [ENTER] — Graphing

we obtain a calculator display of 0.8740751. Thus, we have

$$(1.4)^{-2/5} \approx 0.874.$$

DISCUSSION PROBLEM ■ The Domain of a Radical Expression

Describe the domain of each of the following radical expressions. Write a short paragraph justifying each of your answers.

(a) $\sqrt{x}$ (b) $\dfrac{1}{\sqrt{x}}$ (c) $\sqrt[3]{x}$ (d) $\dfrac{1}{\sqrt[3]{x}}$

Warm-Up

The following warm-up exercises involve skills that were covered in earlier sections. You will use these skills in the exercise set for this section.

In Exercises 1–6, evaluate the given expression.

1. $(-13)^2$ **2.** -13^2 **3.** $\left(\dfrac{2}{3}\right)^3$

4. $\dfrac{5}{6}\left(\dfrac{3}{5}\right)^2$ **5.** $\dfrac{3}{4^{-2}}$ **6.** $\left(\dfrac{2}{3}\right)^{-3}$

In Exercises 7–10, simplify the given expression.

7. $(x^3 \cdot x^{-2})^{-3}$ **8.** $(5x^{-4}y^5)(-3x^2y^{-1})$

9. $\left(\dfrac{2x}{3y}\right)^{-2}$ **10.** $\left(\dfrac{7u^{-4}}{3v^{-2}}\right)\left(\dfrac{14u}{6v^2}\right)^{-1}$

5.2 EXERCISES

In Exercises 1–6, fill in each blank with the appropriate real number or word.

1. Since $7^2 = 49$, ______ is a square root of 49.

2. Since $24.5^2 = 600.25$, ______ is a square root of 600.25.

3. Since $4.2^3 = 74.088$, ______ is a cube root of 74.088.

4. Since $6^4 = 1{,}296$, ______ is a fourth root of 1,296.

5. Since $45^2 = 2{,}025$, 45 is called the ______ of 2,025 and is denoted by ______

6. Since $12^3 = 1{,}728$, 12 is called the ______ of 1,728 and is denoted by ______

In Exercises 7–10, find the indicated square root of the real number without using a calculator. (If not possible, state the reason.)

7. $\sqrt{81}$ **8.** $\sqrt{-144}$ **9.** $\sqrt{\frac{9}{16}}$ **10.** $\sqrt{0.09}$

In Exercises 11–30, evaluate the indicated *n*th root of the real number without using a calculator. (If not possible, state the reason.)

11. $\sqrt{64}$ **12.** $-\sqrt{100}$ **13.** $\sqrt{-100}$ **14.** $\sqrt{144}$

15. $-\sqrt{\frac{4}{9}}$ **16.** $\sqrt{-\frac{9}{25}}$ **17.** $\sqrt{0.16}$ **18.** $-\sqrt{0.0009}$

19. $\sqrt{49 - 4(2)(-15)}$ **20.** $\sqrt{\frac{75}{3}}$ **21.** $\sqrt[3]{125}$ **22.** $\sqrt[3]{-8}$

23. $\sqrt[3]{-\frac{1}{64}}$ **24.** $-\sqrt[3]{0.008}$ **25.** $\sqrt[4]{81}$ **26.** $\sqrt[5]{32}$

27. $-\sqrt[4]{-625}$ **28.** $-\sqrt[4]{\frac{1}{625}}$ **29.** $\sqrt[5]{-0.00243}$ **30.** $\sqrt[6]{64}$

In Exercises 31–34, determine whether the square root is a rational or an irrational number.

31. $\sqrt{6}$ **32.** $\sqrt{\frac{9}{16}}$ **33.** $\sqrt{900}$ **34.** $\sqrt{72}$

In Exercises 35–40, fill in the missing description.

	Radical Form	*Rational Exponent Form*		*Radical Form*	*Rational Exponent Form*
35.	$\sqrt{16} = 4$	______	**36.**	$\sqrt[4]{81} = 3$	______
37.	$\sqrt[3]{27^2} = 9$	______	**38.**	______	$125^{1/3} = 5$
39.	______	$256^{3/4} = 64$	**40.**	______	$32^{3/5} = 8$

In Exercises 41–50, evaluate the quantity without the use of a calculator.

41. $25^{1/2}$ **42.** $-121^{1/2}$ **43.** $16^{3/4}$ **44.** $81^{-3/4}$ **45.** $32^{-2/5}$ **46.** $243^{-3/5}$

47. $\left(\frac{8}{27}\right)^{2/3}$ **48.** $\left(\frac{256}{625}\right)^{1/4}$ **49.** $\left(\frac{121}{9}\right)^{-1/2}$ **50.** $\left(\frac{27}{1{,}000}\right)^{-4/3}$

In Exercises 51–60, use a calculator to approximate the given quantity accurate to four decimal places. (If not possible, state the reason.)

51. $\sqrt{73}$ **52.** $\sqrt{-532}$ **53.** $\dfrac{8-\sqrt{35}}{2}$ **54.** $\dfrac{-5+\sqrt{3{,}215}}{10}$

55. $1{,}698^{-3/4}$ **56.** $962^{2/3}$ **57.** $\sqrt[4]{342}$ **58.** $\sqrt[3]{159}$ **59.** $\sqrt[3]{545^2}$ **60.** $\sqrt[5]{-35^3}$

In Exercises 61–76, simplify the given expression.

61. $\sqrt{t^2}$ **62.** $\sqrt[3]{z^3}$ **63.** $\sqrt[3]{y^9}$ **64.** $\sqrt[4]{a^8}$

65. $x^{2/3} \cdot x^{7/3}$ **66.** $z^{3/5} \cdot z^{-2/5}$ **67.** $(3x^{-1/3}y^{3/4})^2$ **68.** $(-2u^{3/5}v^{-1/5})^3$

69. $\dfrac{18y^{4/3}z^{-1/3}}{24y^{-2/3}z}$ **70.** $\dfrac{a^{3/4} \cdot a^{1/2}}{a^{5/2}}$ **71.** $\left(\dfrac{x^{1/4}}{x^{1/6}}\right)^3$ **72.** $\left(\dfrac{3m^{1/6}n^{1/3}}{4n^{-2/3}}\right)^2$

73. $(c^{3/2})^{1/3}$ **74.** $(k^{-1/3})^{3/2}$ **75.** $\sqrt{\sqrt[4]{y}}$ **76.** $\sqrt[3]{\sqrt{2x}}$

In Exercises 77–80, perform the indicated multiplication.

77. $x^{1/2}(2x - 3)$ **78.** $x^{4/3}(3x^2 - 4x + 5)$

79. $y^{-1/3}(y^{1/3} + 5y^{4/3})$ **80.** $(x^{1/2} - 3)(x^{1/2} + 3)$

81. Use a calculator to evaluate each of the following.

(a) $\left(\sqrt{3.4}\right)^2$ (b) $\left(\sqrt[3]{142}\right)^3$

(c) $\left(\sqrt[4]{22}\right)^4$ (d) $\left(\sqrt{1{,}623}\right)^2$

82. *Perfect Squares* Find all possible "last digits" of perfect squares. (For instance, the last digit of 81 is 1 and the last digit of 64 is 4.) Is it possible that 4,322,788,986 is a perfect square?

83. *Microwave Oven* The usable space in a particular microwave oven is in the form of a cube (see figure). The sales brochure indicates that the interior space of the oven is 2,197 cubic inches. Find the inside dimensions of the oven.

Figure for 83

84. *Dimensions of a Room* Find the dimensions of a piece of carpet for a classroom with 529 square feet of floor space, assuming the floor is square.

85. *Velocity of a Stream* A stream of water moving at the rate of v feet per second can carry particles of diameter $0.03\sqrt{v}$ inches. Find the particle size that can be carried by a stream flowing at the rate of $\frac{3}{4}$ feet per second.

SECTION 5.3

Simplifying and Combining Radicals

Domain of a Radical • Properties of Radicals • Simplifying Radicals • Combining Radicals

Domain of a Radical

The **domain** of the radical $\sqrt[n]{x}$ is the set of all real numbers such that x has a principal nth root.

Domain of a Radical

Let n be an integer that is greater than or equal to 2.

1. If n is odd, then the domain of $\sqrt[n]{x}$ is the set of all real numbers.
2. If n is even, then the domain of $\sqrt[n]{x}$ is the set of all nonnegative real numbers.

EXAMPLE 1 ■ Finding the Domain of a Radical

(a) The domain of $\sqrt{x}$ is the set of all nonnegative real numbers. For instance, 2 is in the domain of $\sqrt{x}$, but -2 is not, because $\sqrt{-2}$ is not a real number.

(b) The domain of $\sqrt{x^2}$ is the set of all real numbers because for any real number x, the expression x^2 is a nonnegative real number.

(c) Because x^3 is nonnegative when x is nonnegative, and x^3 is negative when x is negative, we can conclude that the domain of $\sqrt{x^3}$ is the set of all nonnegative real numbers.

(d) The domain of $\sqrt[3]{x}$ is the set of all real numbers. ■

Properties of Radicals

To simplify radical expressions, we use the following two properties of radicals.

Properties of Radicals

Let u and v be real numbers, variables, or algebraic expressions. If the nth roots of u and v are real, then the following properties are true.

1. $\sqrt[n]{uv} = \sqrt[n]{u}\sqrt[n]{v}$ — Multiplication Property
2. $\sqrt[n]{\dfrac{u}{v}} = \dfrac{\sqrt[n]{u}}{\sqrt[n]{v}}, \quad v \neq 0$ — Division Property

We can use these properties of radicals to simplify radical expressions as follows.

$$\sqrt{48} = \sqrt{(16)(3)} = \sqrt{16}\sqrt{3} = 4\sqrt{3}$$

This simplification process is called **removing perfect square factors from the radical**.

EXAMPLE 2 ■ Simplifying Radicals with Constant Radicands

For each of the following radical expressions, factor the radicand and simplify the radical by removing as many factors as possible.

(a) $\sqrt{75}$ (b) $\sqrt{72}$ (c) $\sqrt{162}$

Solution

(a) $\sqrt{75} = \sqrt{25(3)} = \sqrt{5^2(3)} = \sqrt{5^2}\sqrt{3} = 5\sqrt{3}$

(b) $\sqrt{72} = \sqrt{36(2)} = \sqrt{6^2(2)} = \sqrt{6^2}\sqrt{2} = 6\sqrt{2}$

(c) $\sqrt{162} = \sqrt{81(2)} = \sqrt{9^2(2)} = \sqrt{9^2}\sqrt{2} = 9\sqrt{2}$ ■

Removing *variable* factors from a square root radical is a little tricky. For instance, it is not valid to say that $\sqrt{x^2}$ is equal to x unless we happen to know that x is nonnegative. Without knowing anything about x, the only way we can simplify $\sqrt{x^2}$ is to include absolute value signs when we remove x from the radical. Thus, we have

$\sqrt{x^2} = |x|$. Restricted by absolute value signs

When simplifying the expression $\sqrt{x^3}$, it is not necessary to include absolute value signs because the domain of this expression does not include negative numbers. Thus, we can write

$\sqrt{x^3} = \sqrt{x^2(x)} = x\sqrt{x}$. Restricted by domain of radical

EXAMPLE 3 ■ Simplifying Radicals with Variable Radicands

For each of the following radical expressions, factor the radicand and simplify the radical by removing as many factors as possible.

(a) $\sqrt{25x^2}$ (b) $\sqrt{12x^3}$, $x \geq 0$ (c) $\sqrt{144x^4}$

Solution

(a) $\sqrt{25x^2} = \sqrt{5^2x^2} = \sqrt{5^2}\sqrt{x^2} = 5|x|$ $\sqrt{x^2} = |x|$

(b) $\sqrt{12x^3} = \sqrt{2^2x^2(3x)} = 2x\sqrt{3x}$ $\sqrt{2^2}\sqrt{x^2} = 2x$, $x \geq 0$

(c) $\sqrt{144x^4} = \sqrt{12^2(x^2)^2} = 12x^2$ $\sqrt{12^2}\sqrt{(x^2)^2} = 12|x^2| = 12x^2$

■

In the same way that perfect squares can be removed from square root radicals, perfect nth powers can be removed from nth root radicals. This is demonstrated in Example 4.

EXAMPLE 4 ■ **Simplifying Radicals**

For each of the following radical expressions, factor the radicand and simplify the radical by removing as many perfect nth powers as possible.

(a) $\sqrt[3]{40}$ (b) $\sqrt[4]{x^5}$, $x \geq 0$ (c) $\sqrt[3]{54x^3y^5}$

Solution

(a) $\sqrt[3]{40} = \sqrt[3]{8(5)} = \sqrt[3]{2^3(5)} = 2\sqrt[3]{5}$ $\sqrt[3]{2^3} = 2$

(b) $\sqrt[4]{x^5} = \sqrt[4]{x^4(x)} = x\sqrt[4]{x}$ $\sqrt[4]{x^4} = x,\quad x \geq 0$

(c) $\sqrt[3]{54x^3y^5} = \sqrt[3]{27x^3y^3(2y^2)}$ $\sqrt[3]{3^3}\sqrt[3]{x^3}\sqrt[3]{y^3} = 3xy$

$= \sqrt[3]{3^3x^3y^3(2y^2)}$

$= 3xy\sqrt[3]{2y^2}$ ■

MATH MATTERS

Standard Prefixes Used with the Metric System

In the metric system, a *kilo*gram is equal to 1,000 grams, a *centi*meter is equal to $\frac{1}{100}$ of a meter, and a *milli*liter is equal to $\frac{1}{1,000}$ of a liter. "Kilo," "centi," and "milli" are the most well-known prefixes in the metric system. However, there are several other prefixes that can be used to denote the powers of 10 in the metric system.

How many of the prefixes listed have you heard used? Use the list of prefixes to answer the following questions. (The answers are given in the back of the text.)

How many nanoseconds are in a second?
How many millimeters are in a meter?
How many watts are in a kilowatt?

Factor by Which Unit Is Multiplied	*Prefix*
10^{12}	tera
10^{9}	giga
10^{6}	mega
10^{3}	kilo
10^{2}	hecto
10^{1}	deca
10^{-1}	deci
10^{-2}	centi
10^{-3}	milli
10^{-6}	micro
10^{-9}	nano
10^{-12}	pico
10^{-15}	femto
10^{-18}	atto

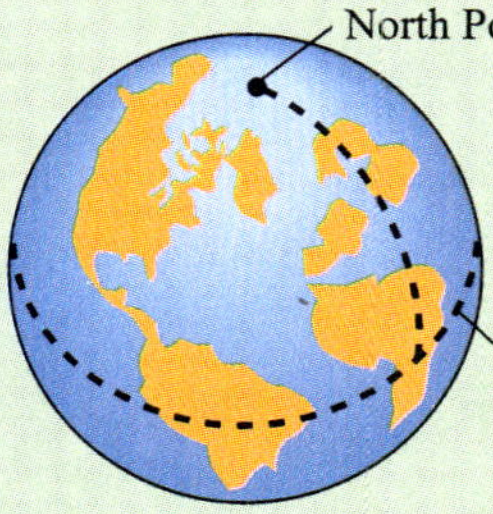

In 1793 the French government adopted the metric system, in which the basic unit of length was defined to be one ten-millionth of the distance between the North Pole and the equator. For working standards, a platinum bar was marked with lines a meter apart. Later, when refinements in measuring the earth were discovered, the meter was redefined in terms of the platinum bar that had been constructed.

Simplifying Radicals

Removing factors from a radical is only one of three techniques that we use to simplify radicals. We summarize all three techniques as follows.

Simplifying Radical Expressions

A radical expression is said to be in simplest form if all three of the following are true.

1. All possible factors have been removed from the radical.
2. No radical contains a fraction.
3. No denominator of a fraction contains a radical.

To meet the last two conditions, we use a technique called **rationalizing the denominator**. This technique involves multiplying both the numerator and the denominator by a factor that creates a perfect nth power in the denominator. For square roots, we want a perfect square in the denominator; for cube roots, we want a perfect cube; and so on.

EXAMPLE 5 ■ Rationalizing the Denominator

Simplify the following expressions so that their denominators are free of radicals.

(a) $\sqrt{\dfrac{3}{5}}$ (b) $\dfrac{4}{\sqrt[3]{9}}$ (c) $\dfrac{8}{3\sqrt{18}}$

Solution

(a) $\sqrt{\dfrac{3}{5}} = \dfrac{\sqrt{3}}{\sqrt{5}} = \dfrac{\sqrt{3}}{\sqrt{5}} \cdot \dfrac{\sqrt{5}}{\sqrt{5}} = \dfrac{\sqrt{15}}{\sqrt{5^2}} = \dfrac{\sqrt{15}}{5}$ Multiply by $\sqrt{5}$ to create a perfect square in the denominator

(b) $\dfrac{4}{\sqrt[3]{9}} = \dfrac{4}{\sqrt[3]{9}} \cdot \dfrac{\sqrt[3]{3}}{\sqrt[3]{3}} = \dfrac{4\sqrt[3]{3}}{\sqrt[3]{3^3}} = \dfrac{4\sqrt[3]{3}}{3}$ Multiply by $\sqrt[3]{3}$ to create a perfect cube in the denominator

(c) $\dfrac{8}{3\sqrt{18}} = \dfrac{8}{3\sqrt{18}} \cdot \dfrac{\sqrt{2}}{\sqrt{2}} = \dfrac{8\sqrt{2}}{3\sqrt{36}} = \dfrac{8\sqrt{2}}{3(6)} = \dfrac{4\sqrt{2}}{9}$ Multiply by $\sqrt{2}$ to create a perfect square in the denominator

■

EXAMPLE 6 ■ Rationalizing the Denominator

Simplify the following expressions so that their denominators are free of radicals.

(a) $\sqrt{\dfrac{8x}{12y^5}}$ (b) $\sqrt[3]{\dfrac{54x^6y^3}{5z^2}}$

Solution

(a) $\sqrt{\dfrac{8x}{12y^5}} = \sqrt{\dfrac{2x}{3y^5}} = \dfrac{\sqrt{2x}}{\sqrt{3y^5}} \cdot \dfrac{\sqrt{3y}}{\sqrt{3y}} = \dfrac{\sqrt{6xy}}{\sqrt{9y^6}} = \dfrac{\sqrt{6xy}}{3y^3}$

(b) $\sqrt[3]{\dfrac{54x^6y^3}{5z^2}} = \dfrac{\sqrt[3]{(3^3)(2)(x^6)(y^3)}}{\sqrt[3]{5z^2}} \cdot \dfrac{\sqrt[3]{25z}}{\sqrt[3]{25z}} = \dfrac{3x^2y\sqrt[3]{50z}}{\sqrt[3]{5^3z^3}} = \dfrac{3x^2y\sqrt[3]{50z}}{5z}$ ■

Combining Radicals

Two or more radical expressions are *alike* if they have the same radicands and the same indices. For instance, $\sqrt{2}$ and $3\sqrt{2}$ are alike, but $\sqrt{3}$ and $\sqrt[3]{3}$ are not alike. Two radical expressions that are alike can be added or subtracted by simply adding or subtracting their coefficients. For instance, the sum of $\sqrt{2}$ and $3\sqrt{2}$ is

$$\sqrt{2} + 3\sqrt{2} = (1 + 3)\sqrt{2} = 4\sqrt{2}.$$

EXAMPLE 7 ■ **Combining Radicals**

Simplify the following radical expressions by combining radicals that are alike.

(a) $\sqrt{7} + 5\sqrt{7} - 2\sqrt{7}$

(b) $6\sqrt{x} - \sqrt[3]{4} - 5\sqrt{x} + 2\sqrt[3]{4}$

(c) $3\sqrt[3]{x} + \sqrt[3]{8x}$

Solution

(a) $\sqrt{7} + 5\sqrt{7} - 2\sqrt{7} = (1 + 5 - 2)\sqrt{7} = 4\sqrt{7}$

(b) $6\sqrt{x} - \sqrt[3]{4} - 5\sqrt{x} + 2\sqrt[3]{4} = 6\sqrt{x} - 5\sqrt{x} - \sqrt[3]{4} + 2\sqrt[3]{4}$
$= (6 - 5)\sqrt{x} + (-1 + 2)\sqrt[3]{4}$
$= \sqrt{x} + \sqrt[3]{4}$

(c) As written, the radical expressions $3\sqrt[3]{x}$ and $\sqrt[3]{8x}$ are not alike. However, by simplifying the second expression, they can be made to be alike.

$$3\sqrt[3]{x} + \sqrt[3]{8x} = 3\sqrt[3]{x} + 2\sqrt[3]{x} = (3 + 2)\sqrt[3]{x} = 5\sqrt[3]{x}$$ ■

Notice in Example 7(c) that *before* concluding that two radicals cannot be combined, we should check to see that they are written in simplest form.

EXAMPLE 8 ■ **Simplifying Radical Expressions**

Simplify the following radical expressions.

(a) $\sqrt{45x} + 3\sqrt{20x}$ (b) $5\sqrt{x^3} - x\sqrt{4x}$ (c) $\sqrt[3]{54y^5} + 4\sqrt[3]{2y^2}$

Solution

(a) $\sqrt{45x} + 3\sqrt{20x} = 3\sqrt{5x} + 6\sqrt{5x} = 9\sqrt{5x}$

(b) $5\sqrt{x^3} - x\sqrt{4x} = 5x\sqrt{x} - 2x\sqrt{x} = 3x\sqrt{x}$

(c) $\sqrt[3]{54y^5} + 4\sqrt[3]{2y^2} = 3y\sqrt[3]{2y^2} + 4\sqrt[3]{2y^2} = (3y + 4)\sqrt[3]{2y^2}$ ■

In some instances, it may be necessary to rationalize denominators before combining radicals. This is demonstrated in Examples 9 and 10.

EXAMPLE 9 ■ **Rationalizing Denominators Before Simplifying**

Simplify the following radical expressions.

(a) $\sqrt{7} - \dfrac{5}{\sqrt{7}}$ (b) $\dfrac{\sqrt{12}}{6} + \sqrt{\dfrac{1}{3}} - \dfrac{2\sqrt{3}}{3}$

Solution

(a)
$$\begin{aligned}\sqrt{7} - \frac{5}{\sqrt{7}} &= \sqrt{7} - \left(\frac{5}{\sqrt{7}} \cdot \frac{\sqrt{7}}{\sqrt{7}}\right)\\ &= \sqrt{7} - \frac{5\sqrt{7}}{7}\\ &= \left(1 - \frac{5}{7}\right)\sqrt{7}\\ &= \frac{2}{7}\sqrt{7}\end{aligned}$$

(b)
$$\begin{aligned}\frac{\sqrt{12}}{6} + \sqrt{\frac{1}{3}} - \frac{2\sqrt{3}}{3} &= \frac{2\sqrt{3}}{6} + \left(\frac{1}{\sqrt{3}} \cdot \frac{\sqrt{3}}{\sqrt{3}}\right) - \frac{2\sqrt{3}}{3}\\ &= \frac{\sqrt{3}}{3} + \frac{\sqrt{3}}{3} - \frac{2\sqrt{3}}{3}\\ &= \left(\frac{1}{3} + \frac{1}{3} - \frac{2}{3}\right)\sqrt{3}\\ &= 0\end{aligned}$$
■

EXAMPLE 10 ■ **Rationalizing Denominators Before Simplifying**

Simplify the following radical expression.

$$\frac{1}{\sqrt{2x}} + \frac{\sqrt{8x}}{x}$$

Solution

$$\begin{aligned}\frac{1}{\sqrt{2x}} + \frac{\sqrt{8x}}{x} &= \left(\frac{1}{\sqrt{2x}} \cdot \frac{\sqrt{2x}}{\sqrt{2x}}\right) + \frac{2\sqrt{2x}}{x}\\ &= \frac{\sqrt{2x}}{2x} + \frac{2\sqrt{2x}}{x}\end{aligned}$$

$$= \left(\frac{1}{2x} + \frac{2}{x}\right)\sqrt{2x}$$
$$= \left(\frac{1}{2x} + \frac{4}{2x}\right)\sqrt{2x}$$
$$= \frac{5}{2x}\sqrt{2x}$$

DISCUSSION PROBLEM Properties of Radicals

In general, $\sqrt{a+b}$ is *not* equal to $\sqrt{a} + \sqrt{b}$. One nice example that demonstrates this is to let $a = 9$ and $b = 16$. Then

$$\sqrt{9+16} = \sqrt{25} = 5 \quad \text{whereas} \quad \sqrt{9} + \sqrt{16} = 3 + 4 = 7.$$

Can you find another example in which a, b, and $a + b$ are each perfect squares?

Warm-Up

The following warm-up exercises involve skills that were covered in earlier sections. You will use these skills in the exercise set for this section.

In Exercises 1–4, find the indicated nth root without the use of a calculator.

1. $\sqrt{10{,}000}$ **2.** $-\sqrt[3]{1{,}000}$ **3.** $\sqrt[3]{-\frac{27}{125}}$ **4.** $\sqrt[4]{\frac{16}{625}}$

In Exercises 5–10, simplify the given expression.

5. $x^{5/3} \cdot x^{1/6}$ **6.** $\frac{a^{7/4}}{a^{3/2}}$ **7.** $(2x)^{3/2}(2x)^{5/2}$

8. $(-4x^{1/3}y^{4/3})^6$ **9.** $\frac{32x^{1/4}y^{3/4}}{2x^{1/2}y}$ **10.** $\frac{8y^{3/2}z^{-2}}{y^{-1/2}z^{1/2}}$

5.3 EXERCISES

In Exercises 1–6, find the domain of the given expression.

1. $\sqrt{2x}$ **2.** $\sqrt{4x^3}$ **3.** $\sqrt{x^4}$ **4.** $\sqrt[3]{x^3}$ **5.** $\sqrt[4]{x}$ **6.** $\sqrt[5]{x}$

In Exercises 7–14, write the given expression as a single radical.

7. $\sqrt{3} \cdot \sqrt{10}$ **8.** $\sqrt[5]{9} \cdot \sqrt[5]{19}$ **9.** $\sqrt[3]{11} \cdot \sqrt[3]{10}$ **10.** $\sqrt[4]{35} \cdot \sqrt[4]{3}$

11. $\dfrac{\sqrt{15}}{\sqrt{31}}$ **12.** $\dfrac{\sqrt[3]{85}}{\sqrt[3]{9}}$ **13.** $\dfrac{\sqrt[5]{152}}{\sqrt[5]{3}}$ **14.** $\dfrac{\sqrt[4]{633}}{\sqrt[4]{5}}$

In Exercises 15–22, simplify the given expression.

15. $\sqrt{9 \cdot 35}$ **16.** $\sqrt[3]{27 \cdot 4}$ **17.** $\sqrt[4]{81 \cdot 11}$ **18.** $\sqrt[5]{100{,}000 \cdot 3}$

19. $\sqrt{\dfrac{35}{9}}$ **20.** $\sqrt[4]{\dfrac{165}{16}}$ **21.** $\sqrt[3]{\dfrac{1{,}000}{11}}$ **22.** $\sqrt[5]{\dfrac{243}{2}}$

In Exercises 23–36, simplify the given radical.

23. $\sqrt{20}$ **24.** $\sqrt{50}$ **25.** $\sqrt{27}$ **26.** $\sqrt{125}$ **27.** $\sqrt[3]{24}$ **28.** $\sqrt[3]{54}$

29. $\sqrt[4]{30{,}000}$ **30.** $\sqrt[5]{96}$ **31.** $\sqrt{\dfrac{15}{4}}$ **32.** $\sqrt{\dfrac{5}{36}}$ **33.** $\sqrt[3]{\dfrac{35}{64}}$ **34.** $\sqrt[4]{\dfrac{5}{16}}$

35. $\sqrt[5]{\dfrac{15}{243}}$ **36.** $\sqrt[3]{\dfrac{1}{1{,}000}}$

In Exercises 37–40, use the multiplication property of radicals to simplify the given expression.

37. $\sqrt{4 \times 10^{-4}}$ **38.** $\sqrt{8.5 \times 10^{3}}$ **39.** $\sqrt[3]{2.4 \times 10^{6}}$ **40.** $\sqrt[4]{4.4 \times 10^{-4}}$

In Exercises 41–52, simplify the given expression.

41. $\sqrt{9x^5}$ **42.** $\sqrt{64x^3}$ **43.** $\sqrt[3]{x^4y^3}$ **44.** $\sqrt[3]{a^5b^6}$

45. $\sqrt[5]{32x^5y^6}$ **46.** $\sqrt[4]{128u^4v^7}$ **47.** $\sqrt{\dfrac{13}{25}}$ **48.** $\sqrt{\dfrac{15}{36}}$

49. $\sqrt[5]{\dfrac{32x^2}{y^5}}$ **50.** $\sqrt[3]{\dfrac{16z^3}{y^6}}$ **51.** $\sqrt[3]{\dfrac{54a^4}{b^9}}$ **52.** $\sqrt[4]{\dfrac{3u^2}{16v^8}}$

In Exercises 53–66, rationalize the denominator and simplify further, if possible.

53. $\sqrt{\dfrac{1}{3}}$ **54.** $\sqrt{\dfrac{1}{5}}$ **55.** $\dfrac{12}{\sqrt{3}}$ **56.** $\dfrac{5}{\sqrt{10}}$ **57.** $\sqrt[4]{\dfrac{5}{4}}$ **58.** $\sqrt[3]{\dfrac{9}{25}}$

59. $\dfrac{6}{\sqrt[3]{32}}$ **60.** $\dfrac{10}{\sqrt[5]{16}}$ **61.** $\dfrac{1}{\sqrt{y}}$ **62.** $\sqrt{\dfrac{5}{c}}$ **63.** $\sqrt[3]{\dfrac{2x}{3y}}$ **64.** $\sqrt[3]{\dfrac{20x^2}{9y^2}}$

65. $\dfrac{a^3}{\sqrt[3]{ab^2}}$ **66.** $\dfrac{3u^2}{\sqrt[4]{8u^3}}$

In Exercises 67–80, combine the radical expressions, if possible.

67. $3\sqrt{2} - \sqrt{2}$

68. $\frac{2}{5}\sqrt{5} - \frac{6}{5}\sqrt{5}$

69. $\sqrt[4]{3} - 5\sqrt[4]{7} - 12\sqrt[4]{3}$

70. $9\sqrt[3]{17} + 7\sqrt[3]{2} - 4\sqrt[3]{17} + \sqrt{2}$

71. $12\sqrt{8} - 3\sqrt[3]{8}$

72. $4\sqrt{32} + 7\sqrt{32}$

73. $2\sqrt[3]{54} + 12\sqrt[3]{16}$

74. $4\sqrt[4]{48} - \sqrt[4]{243}$

75. $5\sqrt{x} - 3\sqrt[4]{x}$

76. $3\sqrt{x+1} + 10\sqrt{x+1}$

77. $\sqrt{25y} + \sqrt{64y}$

78. $\sqrt[3]{16t^4} - \sqrt[3]{54t^4}$

79. $10\sqrt[3]{z} - \sqrt[3]{z^4}$

80. $5\sqrt[3]{24u^2} + 2\sqrt[3]{81u^5}$

In Exercises 81–84, perform the indicated addition or subtraction and simplify your answer.

81. $\sqrt{5} - \frac{3}{\sqrt{5}}$

82. $\sqrt{10} + \frac{5}{\sqrt{10}}$

83. $\sqrt{20} - \sqrt{\frac{1}{5}}$

84. $\frac{x}{\sqrt{3x}} + \sqrt{27x}$

85. *Vibrating String* The frequency f in cycles per second of a vibrating string is given by

$$f = \frac{1}{100}\sqrt{\frac{400 \times 10^6}{5}}$$

Use a calculator to approximate this number. (Round your answer to two decimal places.)

86. *Period of a Pendulum* The period T in seconds of a pendulum (see figure) is given by

$$T = 2\pi\sqrt{\frac{L}{32}}$$

where L is the length of the pendulum in feet. Find the period of a pendulum whose length is 4 feet. (Round your answer to two decimal places.)

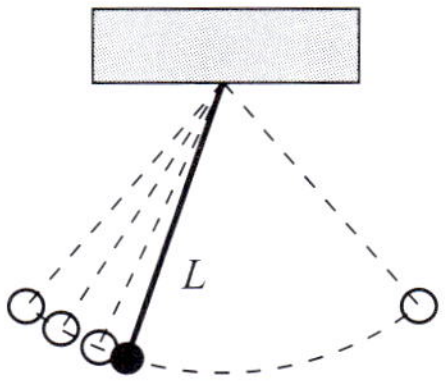

In Exercises 87–90, place the correct inequality symbol (<, >, or =) between the given numbers.

87. $\sqrt{7} + \sqrt{18}$ ▭ $\sqrt{7 + 18}$

88. $\sqrt{10} - \sqrt{6}$ ▭ $\sqrt{10 - 6}$

89. 5 ▭ $\sqrt{3^2 + 2^2}$

90. 5 ▭ $\sqrt{3^2 + 4^2}$

91. *Calculator Experiment* Enter any positive real number in your calculator and find its square root. Then repeatedly take the square root of the result.

$$\sqrt{x},\ \sqrt{\sqrt{x}},\ \sqrt{\sqrt{\sqrt{x}}},\ \ldots$$

What real number does the display appear to be approaching?

92. *Number Experiment* Square the real number $5/\sqrt{3}$ and note that the radical is eliminated from the denominator. Is this equivalent to rationalizing the denominator? Why or why not?

SECTION 5.4

Multiplying and Dividing Radical Expressions

Multiplying Radical Expressions • Dividing Radical Expressions

Multiplying Radical Expressions

We can multiply radicals by using the Distributive Property and the FOIL method. In both procedures, we also make use of the multiplication property of radicals. Recall from Section 5.3 that the product of two radicals is given by

$$\sqrt[n]{a}\sqrt[n]{b} = \sqrt[n]{ab}$$

where a and b are real numbers whose nth roots are also real numbers.

EXAMPLE 1 ■ Multiplying Radicals

$\sqrt{3}(2 + \sqrt{5}) = 2\sqrt{3} + \sqrt{3}\sqrt{5}$ — Distributive Property

$= 2\sqrt{3} + \sqrt{15}$ — Multiplication property of radicals ■

EXAMPLE 2 ■ Multiplying Radicals

(a) $\sqrt{2}(4 - \sqrt{8}) = 4\sqrt{2} - \sqrt{2}\sqrt{8}$ — Distributive Property

$= 4\sqrt{2} - \sqrt{16}$ — Multiplication property of radicals

$= 4\sqrt{2} - 4$ — Simplify

(b) $\sqrt{6}(\sqrt{12} - \sqrt{3}) = \sqrt{6}\sqrt{12} - \sqrt{6}\sqrt{3}$ — Distributive Property

$= \sqrt{72} - \sqrt{18}$ — Multiplication property of radicals

$= 6\sqrt{2} - 3\sqrt{2}$ — Find perfect square factors

$= 3\sqrt{2}$ — Simplify ■

EXAMPLE 3 ■ Multiplying Radicals

(a) $(2\sqrt{7} - 4)(\sqrt{7} + 1) = \underbrace{2(\sqrt{7})^2}_{F} + \underbrace{2\sqrt{7}}_{O} - \underbrace{4\sqrt{7}}_{I} - \underbrace{4}_{L}$ — FOIL Method

$= 2(7) + (2 - 4)\sqrt{7} - 4$ — Combine like radicals

$= 10 - 2\sqrt{7}$ — Combine like terms

(b) $(3 - \sqrt{x})(1 + \sqrt{x}) = 3 + 3\sqrt{x} - \sqrt{x} - (\sqrt{x})^2$ — FOIL Method

$= 3 + 2\sqrt{x} - x, \quad x \geq 0$ ■

EXAMPLE 4 ■ **Multiplying Radicals**

(a) $(2 - \sqrt{5})(2 + \sqrt{5}) = 2^2 - (\sqrt{5})^2$ Special product formula

$= 4 - 5$

$= -1$

(b) $(\sqrt{3} + \sqrt{x})(\sqrt{3} - \sqrt{x}) = (\sqrt{3})^2 - (\sqrt{x})^2$ Special product formula

$= 3 - x, \quad x \geq 0$ ■

In Example 4 the expressions $(2 - \sqrt{5})$ and $(2 + \sqrt{5})$ are called **conjugates** of each other. The product of two conjugates is the difference of two squares, which is given by the special product formula $(a + b)(a - b) = a^2 - b^2$. Here are some other examples.

Expression	*Conjugate*	*Product*
$(1 - \sqrt{3})$	$(1 + \sqrt{3})$	$(1)^2 - (\sqrt{3})^2 = 1 - 3 = -2$
$(\sqrt{5} + \sqrt{2})$	$(\sqrt{5} - \sqrt{2})$	$(\sqrt{5})^2 - (\sqrt{2})^2 = 5 - 2 = 3$
$(\sqrt{10} - 3)$	$(\sqrt{10} + 3)$	$(\sqrt{10})^2 - (3)^2 = 10 - 9 = 1$
$\sqrt{x} + 2$	$\sqrt{x} - 2$	$(\sqrt{x})^2 - (2)^2 = x - 4, \quad x \geq 0$

Dividing Radical Expressions

To simplify a *quotient* involving radicals, we rationalize the denominator. For single-term denominators, we use the rationalizing process described in Section 5.3. For instance, to divide $\sqrt{3}$ by $\sqrt{5}$, we multiply both the numerator and denominator by $\sqrt{5}$, as follows.

$$\frac{\sqrt{3}}{\sqrt{5}} = \frac{\sqrt{3}}{\sqrt{5}} \cdot \frac{\sqrt{5}}{\sqrt{5}} = \frac{\sqrt{15}}{5}$$

We can rationalize a denominator involving two terms by multiplying both the numerator and denominator by the *conjugate of the denominator*. Here is an example.

$$\frac{\sqrt{3}}{1 - \sqrt{5}} = \frac{\sqrt{3}}{1 - \sqrt{5}} \cdot \frac{1 + \sqrt{5}}{1 + \sqrt{5}}$$
$$= \frac{\sqrt{3}(1 + \sqrt{5})}{(1 - \sqrt{5})(1 + \sqrt{5})}$$
$$= \frac{\sqrt{3} + \sqrt{15}}{1 - 5}$$
$$= \frac{\sqrt{3} + \sqrt{15}}{-4}$$

EXAMPLE 5 ■ **Simplifying Quotients Involving Radicals**

Simplify the expression $\frac{4}{2-\sqrt{3}}$ by rationalizing its denominator.

Solution

$$\frac{4}{2-\sqrt{3}} = \frac{4}{2-\sqrt{3}} \cdot \frac{2+\sqrt{3}}{2+\sqrt{3}} \qquad \text{Multiply by conjugate}$$

$$= \frac{4(2+\sqrt{3})}{2^2-(\sqrt{3})^2} \qquad \text{Special product formula}$$

$$= \frac{8+4\sqrt{3}}{4-3}$$

$$= 8+4\sqrt{3} \qquad \text{Simplest form}$$

■

EXAMPLE 6 ■ **Simplifying Quotients Involving Radicals**

(a) $$\frac{5\sqrt{2}}{\sqrt{7}+\sqrt{2}} = \frac{5\sqrt{2}}{\sqrt{7}+\sqrt{2}} \cdot \frac{\sqrt{7}-\sqrt{2}}{\sqrt{7}-\sqrt{2}} \qquad \text{Multiply by conjugate}$$

$$= \frac{5(\sqrt{14}-\sqrt{4})}{(\sqrt{7})^2-(\sqrt{2})^2}$$

$$= \frac{5(\sqrt{14}-2)}{7-2}$$

$$= \frac{\cancel{5}(\sqrt{14}-2)}{\cancel{5}} \qquad \text{Divide out common factor}$$

$$= \sqrt{14}-2$$

(b) $$\frac{6}{\sqrt{x}-2} = \frac{6}{\sqrt{x}-2} \cdot \frac{\sqrt{x}+2}{\sqrt{x}+2} \qquad \text{Multiply by conjugate}$$

$$= \frac{6(\sqrt{x}+2)}{(\sqrt{x})^2-(2)^2}$$

$$= \frac{6\sqrt{x}+12}{x-2}, \quad x \ge 0$$

■

EXAMPLE 7 ■ **Dividing Radical Expressions**

Perform the indicated division and simplify your answer.

$$(2-\sqrt{3}) \div (\sqrt{6}+\sqrt{2})$$

Solution

$$\frac{2-\sqrt{3}}{\sqrt{6}+\sqrt{2}} = \frac{2-\sqrt{3}}{\sqrt{6}+\sqrt{2}} \cdot \frac{\sqrt{6}-\sqrt{2}}{\sqrt{6}-\sqrt{2}} \qquad \text{Multiply by conjugate}$$

$$= \frac{2\sqrt{6}-2\sqrt{2}-\sqrt{18}+\sqrt{6}}{(\sqrt{6})^2-(\sqrt{2})^2}$$

$$= \frac{3\sqrt{6} - 2\sqrt{2} - 3\sqrt{2}}{6 - 2}$$

$$= \frac{3\sqrt{6} - 5\sqrt{2}}{4} \qquad \text{Simplest form}$$

DISCUSSION PROBLEM ■ The Golden Section

The ratio of the width of the Temple of Hephaestus to its height (see Figure 5.1) is approximately

$$\frac{w}{h} \approx \frac{2}{\sqrt{5} - 1}.$$

This number is called the **golden section**. Early Greeks believed that the most aesthetically pleasing rectangles were those with sides of this ratio. Rationalize the denominator for this number. Approximate your answer, rounded to two decimal places.

FIGURE 5.1

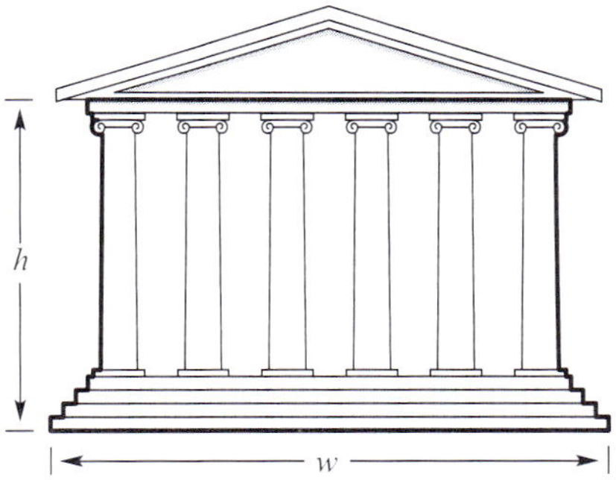

Warm-Up

The following warm-up exercises involve skills that were covered in earlier sections. You will use these skills in the exercise set for this section.

In Exercises 1–6, perform the indicated operations and simplify.

1. $19 - 5(x - 3)$ **2.** $32 + 6(3x - 10)$ **3.** $\left(t + \frac{1}{3}\right)^2$

4. $(2x - 5)^2$ **5.** $(5x - 4)(5x + 4)$ **6.** $\left(\frac{1}{2} - 3x\right)(4 + x)$

In Exercises 7–10, simplify the given expression. (Assume all variables are positive.)

7. $\sqrt[3]{16x^4y^3}$ **8.** $\sqrt{72x^2y^3}$

9. $\sqrt{3x^2} - 2\sqrt{12}$ **10.** $\frac{14x}{\sqrt{7}}$

5.4 EXERCISES

In Exercises 1–30, perform the indicated multiplication and simplify the result. (Assume all variables are positive.)

1. $\sqrt{2} \cdot \sqrt{8}$
2. $\sqrt{6} \cdot \sqrt{18}$
3. $\sqrt{5}(2 - \sqrt{3})$
4. $\sqrt{11}(\sqrt{5} - 3)$
5. $\sqrt{2}(\sqrt{20} + 8)$
6. $\sqrt{7}(\sqrt{14} + 3)$
7. $(\sqrt{3} + 2)(\sqrt{3} - 2)$
8. $(\sqrt{15} + 3)(\sqrt{15} - 3)$
9. $(2\sqrt{2} + \sqrt{4})(2\sqrt{2} - \sqrt{4})$
10. $(4\sqrt{3} + \sqrt{2})(4\sqrt{3} - \sqrt{2})$
11. $(\sqrt{20} + 2)^2$
12. $(4 - \sqrt{20})^2$
13. $(\sqrt{5} + 3)(\sqrt{3} - 5)$
14. $(\sqrt{30} + 6)(\sqrt{2} + 6)$
15. $\sqrt{y}(\sqrt{y} + 4)$
16. $\sqrt{x}(5 - \sqrt{x})$
17. $(3\sqrt{x} - 5)(3\sqrt{x} + 5)$
18. $(7 - 3\sqrt{3t})(7 + 3\sqrt{3t})$
19. $(10 + \sqrt{2x})^2$
20. $(5 - \sqrt{3v})^2$
21. $(9\sqrt{x} + 2)(5\sqrt{x} - 3)$
22. $(16\sqrt{u} - 3)(\sqrt{u} - 1)$
23. $(\sqrt{x} + \sqrt{y})(\sqrt{x} - \sqrt{y})$
24. $(3\sqrt{u} + \sqrt{3v})(3\sqrt{u} - \sqrt{3v})$
25. $\sqrt[3]{4}(\sqrt[3]{2} - 7)$
26. $(\sqrt[3]{9} + 5)(\sqrt[3]{5} - 5)$
27. $(\sqrt[3]{2x} + 5)^2$
28. $(\sqrt[3]{y} + 2)(\sqrt[3]{y^2} - 5)$
29. $(\sqrt[3]{2y} + 10)(\sqrt[3]{4y^2} - 10)$
30. $(\sqrt[3]{t} + 1)(\sqrt[3]{t^2} + 4\sqrt[3]{t} - 3)$

In Exercises 31–36, insert the missing factor.

31. $5\sqrt{3} + 15\sqrt{3} = 5(\quad)$
32. $x\sqrt{7} - x^2\sqrt{7} = x(\quad)$
33. $4\sqrt{12} - 2x\sqrt{27} = 2\sqrt{3}(\quad)$
34. $5\sqrt{50} + 10y\sqrt{8} = 5\sqrt{2}(\quad)$
35. $6u^2 + \sqrt{18u^3} = 3u(\quad)$
36. $12s^3 - \sqrt{32s^4} = 4s^2(\quad)$

In Exercises 37–40, write the fraction in reduced form.

37. $\dfrac{4 - 8\sqrt{x}}{12}$
38. $\dfrac{-3 + 27\sqrt{2y}}{18}$
39. $\dfrac{-2y + \sqrt{12y^3}}{8y}$
40. $\dfrac{-t^2 - \sqrt{2t^3}}{3t}$

In Exercises 41–48, determine the conjugate of the expression and then find the product of the expression and its conjugate. (Assume all variables are positive.)

41. $2 + \sqrt{5}$
42. $\sqrt{2} - 9$
43. $\sqrt{11} - \sqrt{3}$
44. $\sqrt{10} + \sqrt{7}$
45. $\sqrt{x} - 3$
46. $\sqrt{t} + 7$
47. $\sqrt{2u} - \sqrt{3}$
48. $\sqrt{5a} + \sqrt{2}$

In Exercises 49–60, rationalize the denominator of the given expression. (Assume all variables are positive.)

49. $\dfrac{6}{\sqrt{22} - 2}$
50. $\dfrac{3}{2\sqrt{10} - 5}$
51. $\dfrac{8}{\sqrt{7} + 3}$

52. $\dfrac{10}{\sqrt{9}+\sqrt{5}}$

53. $(\sqrt{7}+2) \div (\sqrt{7}-2)$

54. $(5-3\sqrt{3}) \div (3+\sqrt{3})$

55. $\dfrac{3x}{\sqrt{15}-\sqrt{3}}$

56. $\dfrac{6(y+1)}{y^2+\sqrt{y}}$

57. $\dfrac{2t^2}{\sqrt{5t}-\sqrt{t}}$

58. $\dfrac{5x}{\sqrt{x}-\sqrt{2}}$

59. $(\sqrt{x}-5) \div (2\sqrt{x}-1)$

60. $(2\sqrt{t}+1) \div (2\sqrt{t}-1)$

61. *Strength of a Wooden Beam* The rectangular cross section of a wooden beam cut from a log of diameter 24 inches will have maximum strength if its width w and height h are given by

$$w = 8\sqrt{3} \quad \text{and} \quad h = \sqrt{24^2 - (8\sqrt{3})^2}$$

Find the area of the rectangular cross section and express the area in simplest form.

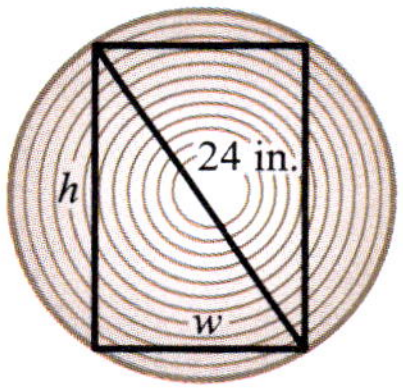

62. *Force to Move a Block* The force required to slide a steel block weighing 500 pounds across a milling machine is given by

$$F = \frac{500k}{\dfrac{1}{\sqrt{k^2+1}} + \dfrac{k^2}{\sqrt{k^2+1}}}$$

where k is a constant determined by the amount of friction (see figure). Simplify this expression for force.

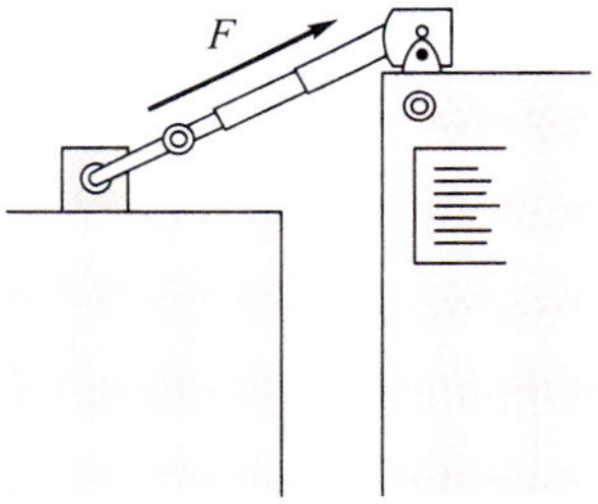

SECTION **5.5**

Solving Equations Involving Radicals

Solving Equations Involving Radicals • Applications of Radicals

Solving Equations Involving Radicals

Solving equations involving radicals is somewhat like solving equations that contain fractions—we try to get rid of the radicals and obtain a polynomial equation. Then we solve the polynomial equation using the standard procedures. The following property plays a key role.

Raising Both Sides of an Equation to the nth Power

Let u and v be real numbers, variables, or algebraic expressions, and let n be a positive integer. If $u = v$, then it follows that $u^n = v^n$. We call this operation **raising both sides of an equation to the nth power**.

You will see in this section that raising both sides of an equation to the nth power often introduces *extraneous* solutions. So when you use this procedure, it is critical that you check each solution in the *original* equation.

To use this property to solve an equation, we first try to isolate one of the radicals on one side of the equation. Watch how this is done in the examples that follow.

EXAMPLE 1 ■ Solving an Equation Having One Radical

$$\sqrt{x} - 8 = 0 \qquad \text{Given equation}$$
$$\sqrt{x} = 8 \qquad \text{Isolate radical}$$
$$\left(\sqrt{x}\right)^2 = 8^2 \qquad \text{Square both sides}$$
$$x = 64 \qquad \text{Solution}$$

Check

$$\sqrt{x} - 8 = 0 \qquad \text{Given equation}$$
$$\sqrt{64} - 8 \stackrel{?}{=} 0 \qquad \text{Replace } x \text{ by } 64$$
$$8 - 8 = 0 \qquad \text{Solution checks}$$

Therefore, the equation has one solution: $x = 64$. ■

EXAMPLE 2 ■ Solving Equations Having One Radical

$$\sqrt[3]{2x + 1} - 2 = 3 \qquad \text{Given equation}$$
$$\sqrt[3]{2x + 1} = 5 \qquad \text{Isolate radical}$$
$$\left(\sqrt[3]{2x + 1}\right)^3 = 5^3 \qquad \text{Cube both sides}$$
$$2x + 1 = 125 \qquad \text{Simplify}$$
$$2x = 124$$
$$x = 62 \qquad \text{Solution}$$

Check

$$\sqrt[3]{2x + 1} - 2 = 3 \qquad \text{Given equation}$$
$$\sqrt[3]{2(62) + 1} - 2 \stackrel{?}{=} 3 \qquad \text{Replace } x \text{ by } 62$$
$$\sqrt[3]{125} - 2 \stackrel{?}{=} 3$$
$$5 - 2 = 3 \qquad \text{Solution checks}$$

Therefore, the equation has one solution: $x = 62$. ■

EXAMPLE 3 ■ Solving Equations Having One Radical

$\sqrt{3x} + 6 = 0$	Given equation
$\sqrt{3x} = -6$	Isolate radical
$(\sqrt{3x})^2 = (-6)^2$	Square both sides
$3x = 36$	
$x = 12$	Trial solution

Check

$\sqrt{3x} + 6 = 0$	Given equation
$\sqrt{3(12)} + 6 \stackrel{?}{=} 0$	Replace x by 12
$6 + 6 \neq 0$	Solution does not check

Therefore, the equation has no solution. ■

EXAMPLE 4 ■ Solving an Equation Having Two Radicals

$\sqrt{5x + 3} = \sqrt{x + 11}$	Given equation
$(\sqrt{5x + 3})^2 = (\sqrt{x + 11})^2$	Square both sides
$5x + 3 = x + 11$	
$4x = 8$	
$x = 2$	Solution

Check

$\sqrt{5x + 3} = \sqrt{x + 11}$	Given equation
$\sqrt{5(2) + 3} \stackrel{?}{=} \sqrt{2 + 11}$	Replace x by 2
$\sqrt{13} = \sqrt{13}$	Solution checks

Therefore, the equation has one solution: $x = 2$. ■

EXAMPLE 5 ■ Solving an Equation Having Two Radicals

$\sqrt[3]{6x - 4} - \sqrt[3]{2 - x} = 0$	Given equation
$\sqrt[3]{6x - 4} = \sqrt[3]{2 - x}$	Isolate radicals
$6x - 4 = 2 - x$	Cube both sides
$7x = 6$	
$x = \frac{6}{7}$	Solution

Checking this result is a good test of your skill with fractions. You should end up with the true statement, $\sqrt[3]{\frac{8}{7}} - \sqrt[3]{\frac{8}{7}} = 0$. ■

EXAMPLE 6 ■ **Solving an Equation Having Two Radicals**

$$\sqrt[4]{3x} + \sqrt[4]{2x-5} = 0 \quad \text{Given equation}$$
$$\sqrt[4]{3x} = -\sqrt[4]{2x-5} \quad \text{Isolate radicals}$$
$$\left(\sqrt[4]{3x}\right)^4 = \left(-\sqrt[4]{2x-5}\right)^4 \quad \text{Raise both sides to fourth power}$$
$$3x = 2x - 5$$
$$x = -5 \quad \text{Trial solution}$$

A check will show that -5 is not a solution because it yields fourth roots of negative radicands. Therefore, this equation has no solution. ■

In the next two examples, we show how squaring both sides of an equation can yield a *nonlinear* equation.

EXAMPLE 7 ■ **A Radical Equation that Converts to a Nonlinear Equation**

$$\sqrt{x} + 2 = x \quad \text{Given equation}$$
$$\sqrt{x} = x - 2 \quad \text{Isolate radical}$$
$$\left(\sqrt{x}\right)^2 = (x-2)^2 \quad \text{Square both sides}$$
$$x = x^2 - 4x + 4$$
$$-x^2 + 5x - 4 = 0 \quad \text{Standard form}$$
$$(-1)(x-4)(x-1) = 0 \quad \text{Factor}$$
$$x - 4 = 0 \Rightarrow x = 4 \quad \text{Set first factor equal to 0}$$
$$x - 1 = 0 \Rightarrow x = 1 \quad \text{Set second factor equal to 0}$$

Try checking each of these solutions. When you do, you will find that $x = 4$ is a valid solution, but that $x = 1$ is extraneous. Therefore, the equation has only one solution: $x = 4$. ■

When an equation contains two radicals, it may not be possible to isolate both. In such cases, we may have to raise both sides of the equation to a power at *two* different stages in the solution. We demonstrate this procedure in the next two examples.

EXAMPLE 8 ■ **Repeatedly Squaring Both Sides of an Equation**

$$\sqrt{3t+1} = 2 - \sqrt{3t} \quad \text{Given equation}$$
$$\left(\sqrt{3t+1}\right)^2 = \left(2 - \sqrt{3t}\right)^2 \quad \text{Square both sides (first time)}$$
$$3t + 1 = 4 - 4\sqrt{3t} + 3t$$
$$-3 = -4\sqrt{3t} \quad \text{Isolate radical}$$

$$(-3)^2 = \left(-4\sqrt{3t}\right)^2 \qquad \text{Square both sides (second time)}$$
$$9 = 16(3t)$$
$$\frac{9}{48} = t$$
$$\frac{3}{16} = t$$

A check will show that $t = \frac{3}{16}$ is a valid solution. Therefore, this equation has only one solution: $t = \frac{3}{16}$. ■

EXAMPLE 9 ■ **Repeatedly Squaring Both Sides of an Equation**

$$\sqrt{x^2 + 11} - \sqrt{x^2 - 9} = 2 \qquad \text{Given equation}$$
$$\sqrt{x^2 + 11} = 2 + \sqrt{x^2 - 9} \qquad \text{Isolate one of the radicals}$$
$$\left(\sqrt{x^2 + 11}\right)^2 = \left(2 + \sqrt{x^2 - 9}\right)^2 \qquad \text{Square both sides (first time)}$$
$$x^2 + 11 = 4 + 4\sqrt{x^2 - 9} + (x^2 - 9)$$
$$16 = 4\sqrt{x^2 - 9}$$
$$4 = \sqrt{x^2 - 9} \qquad \text{Isolate radical}$$
$$4^2 = \left(\sqrt{x^2 - 9}\right)^2 \qquad \text{Square both sides (second time)}$$
$$16 = x^2 - 9$$
$$0 = x^2 - 25$$
$$0 = (x + 5)(x - 5)$$
$$x + 5 = 0 \Rightarrow x = -5 \qquad \text{Set first factor equal to 0}$$
$$x - 5 = 0 \Rightarrow x = 5 \qquad \text{Set second factor equal to 0}$$

A check will show that both 5 and -5 are solutions. Therefore, this equation has two solutions: 5 and -5. ■

Applications of Radicals

FIGURE 5.2

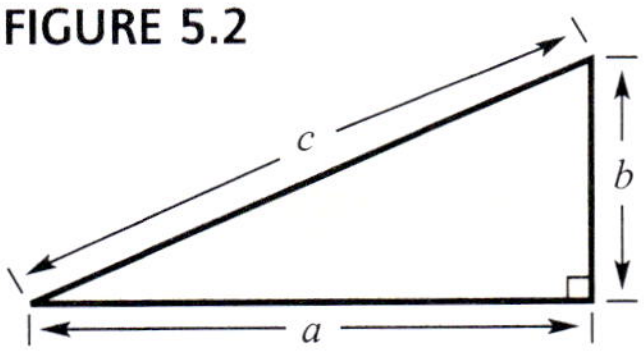

A common use of radicals occurs in applications involving right triangles. Recall that a right triangle is one that contains a right (or 90°) angle, as shown in Figure 5.2. The relationship among the three sides of a right triangle is described by the **Pythagorean Theorem**, which says that if a and b are the lengths of the legs and c is the length of the hypotenuse, then

$$c^2 = a^2 + b^2 \qquad \text{Pythagorean Theorem}$$
$$c = \sqrt{a^2 + b^2}.$$

EXAMPLE 10 ■ An Application Involving a Right Triangle

A softball diamond has the shape of a square with 60-foot sides (see Figure 5.3). The catcher is 5 feet behind home plate. How far does the catcher have to throw the ball to reach second base?

Solution

In Figure 5.3 let x be the hypotenuse of a right triangle with 60-foot legs. Thus, by the Pythagorean Theorem, we have the following equation.

$$x = \sqrt{60^2 + 60^2} \qquad \text{Pythagorean Theorem}$$
$$x = \sqrt{7{,}200}$$
$$x \approx 84.9 \text{ feet}$$

Thus, the distance from home plate to second base is approximately 84.9 feet. Since the catcher is 5 feet behind home plate, the catcher must make a throw of

$$x + 5 \approx 84.9 + 5 = 89.9 \text{ feet.}$$

Check this solution in the original equation.

FIGURE 5.3

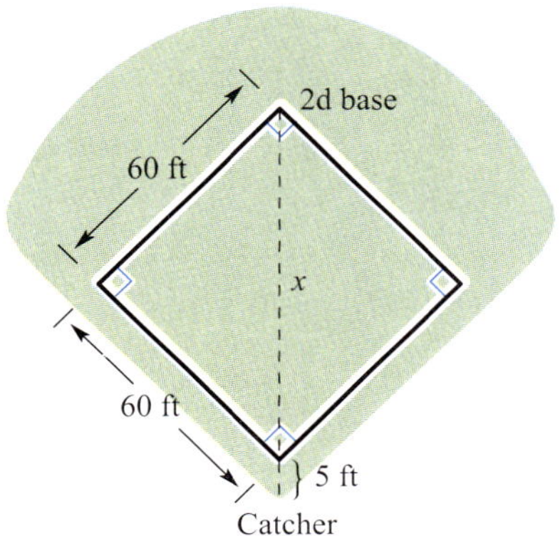

■

EXAMPLE 11 ■ An Application Involving the Use of Electricity

The amount of power consumed by an electrical appliance is given by the equation

$$I = \sqrt{\frac{P}{R}}$$

where I is the current measured in amps, R is the resistance measured in ohms, and P is the power measured in watts. Find the power used by an electric heater, for which $I = 10$ amps and $R = 16$ ohms.

Solution

$$I = \sqrt{\frac{P}{R}} \qquad \text{Given equation}$$

$$10 = \sqrt{\frac{P}{16}} \qquad \text{Substitute known values}$$

$$(10)^2 = \left(\sqrt{\frac{P}{16}}\right)^2 \qquad \text{Square both sides}$$

$$100 = \frac{P}{16}$$

$$1{,}600 = P \qquad \text{Solve for } P$$

Therefore, the electric heater uses 1,600 watts of power. Check this solution in the original equation. ■

EXAMPLE 12 ■ An Application Involving the Velocity of a Falling Object

The velocity of a free-falling object can be determined from the equation

$$v = \sqrt{2gh}$$

where v is the velocity measured in feet per second, $g = 32$ feet per second, and h is the distance (in feet) the object has fallen. Find the height from which a rock has been dropped if it strikes the ground with a velocity of 50 feet per second.

Solution

$$v = \sqrt{2gh} \qquad \text{Given equation}$$

$$50 = \sqrt{2(32)h} \qquad \text{Substitute known values}$$

$$(50)^2 = \left(\sqrt{64h}\right)^2 \qquad \text{Square both sides}$$

$$2{,}500 = 64h$$

$$39 \approx h \qquad \text{Solve for } h$$

Thus, the rock has fallen approximately 39 feet when it hits the ground, as shown in Figure 5.4. Check this solution in the original equation.

FIGURE 5.4

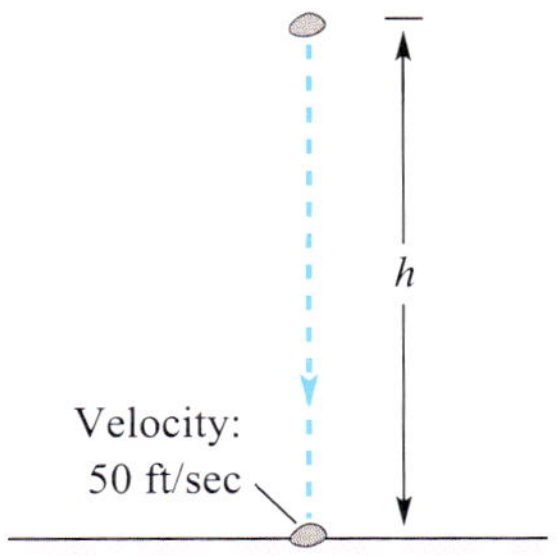

■

DISCUSSION PROBLEM ■ Constructing Lengths to Represent Radicals

FIGURE 5.5

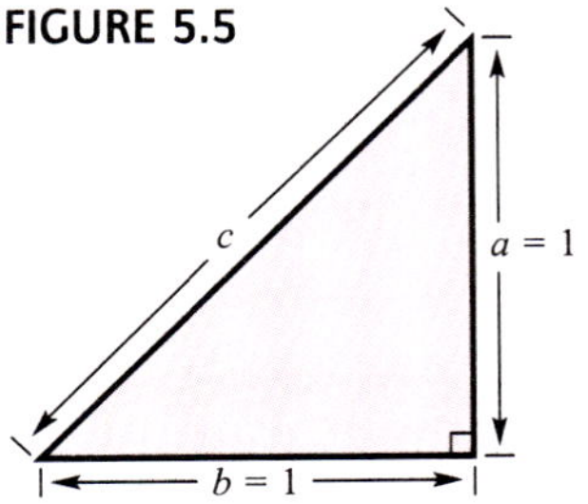

Consider a right triangle that has two sides of length 1, as shown in Figure 5.5. From the Pythagorean Theorem, we can conclude that the length of the hypotenuse is

$$c = \sqrt{a^2 + b^2} = \sqrt{1^2 + 1^2} = \sqrt{2}.$$

Can you construct a right triangle whose three sides have lengths of 1, 2, and $\sqrt{3}$? (*Note:* The **Pythagorean Theorem** states that the sides of a *right triangle* satisfy the equation $a^2 + b^2 = c^2$, where a and b are the lengths of the legs and c is the length of the hypotenuse.) ■

Warm-Up

The following warm-up exercises involve skills that were covered in earlier sections. You will use these skills in the exercise set for this section.

In Exercises 1–6, solve the given equation.

1. $13x + 6 = 32$ **2.** $7x = 18 + 3x$

3. $17 - 5(x - 2) = 0$ **4.** $32 + 3(2x - 11) = 0$

5. $3(x - 2)(x + 25) = 0$ **6.** $-2(12 - x)(114 + x) = 0$

In Exercises 7–10, perform the indicated operations and simplify. (Assume all variables are positive.)

7. $(\sqrt{x} + 3)(\sqrt{x} - 3)$ **8.** $\sqrt{u}(\sqrt{20} - \sqrt{5})$

9. $(2\sqrt{t} + 3)^2$ **10.** $\dfrac{50x}{\sqrt{2}}$

5.5 EXERCISES

In Exercises 1–30, solve the given equation. (Some of the equations have no solution.)

1. $\sqrt{x} = 20$ **2.** $\sqrt{x} = 5$ **3.** $\sqrt{y} - 7 = 0$

4. $\sqrt{t} - 13 = 0$ **5.** $\sqrt{u} + 13 = 0$ **6.** $\sqrt{y} + 15 = 0$

7. $\sqrt{a + 100} = 25$ **8.** $\sqrt{b + 12} = 13$ **9.** $\sqrt{10x} = 30$

10. $\sqrt{8x} = 6$ **11.** $\sqrt{3y + 5} - 3 = 4$ **12.** $\sqrt{5z - 2} + 7 = 10$

13. $5\sqrt{x + 2} = 8$ **14.** $2\sqrt{x + 4} = 7$ **15.** $\sqrt{x^2 + 5} = x + 3$

16. $\sqrt{x^2 - 4} = x - 2$

17. $\sqrt{x + 3} = \sqrt{2x - 1}$

18. $\sqrt{3t + 1} = \sqrt{t + 15}$

19. $\sqrt{3y - 5} - 3\sqrt{y} = 0$

20. $\sqrt{2u + 10} - 2\sqrt{u} = 0$

21. $\sqrt[3]{3x - 4} = \sqrt[3]{x + 10}$

22. $2\sqrt[3]{10 - 3x} = \sqrt[3]{2 - x}$

23. $\sqrt[3]{2x + 15} - \sqrt[3]{x} = 0$

24. $\sqrt[4]{2x} + \sqrt[4]{x + 3} = 0$

25. $\sqrt{2x} = x - 4$

26. $\sqrt{x} = x - 6$

27. $\sqrt{8x + 1} = x + 2$

28. $\sqrt{3x + 7} = x + 3$

29. $\sqrt{2t + 3} = 3 - \sqrt{2t}$

30. $\sqrt{x + 3} - \sqrt{x - 1} = 1$

In Exercises 31–36, find the length x of the unknown side of the right triangle. (Round your answer to two decimal places.)

31.

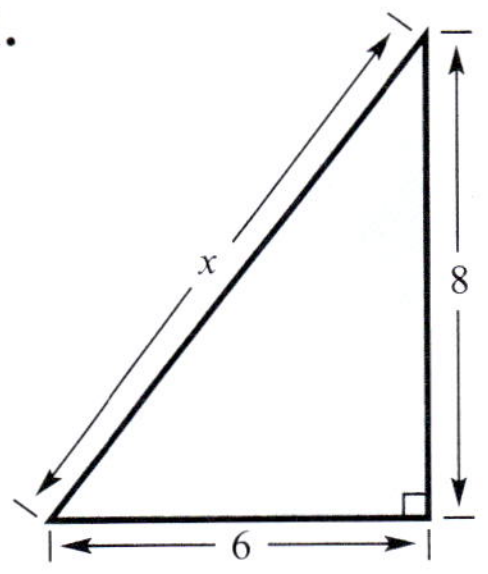

32.

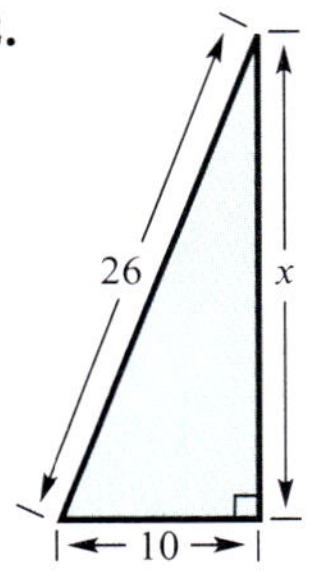

33.

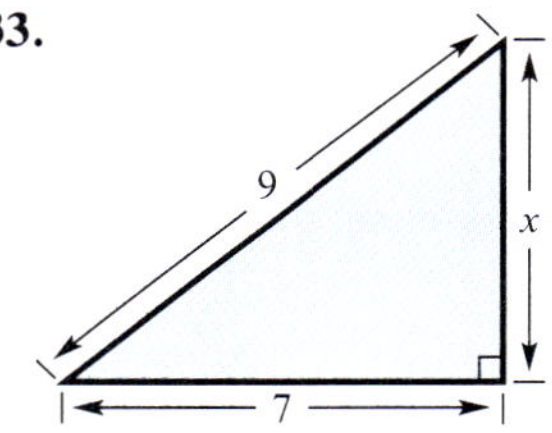

34.

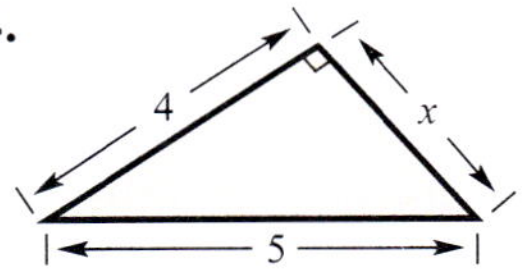

35.

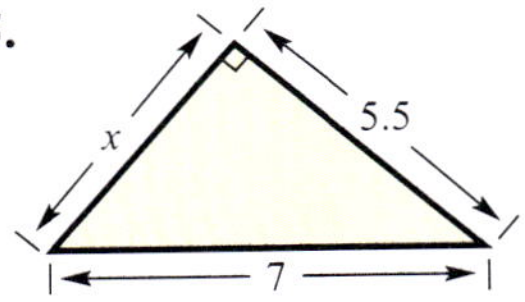

36.

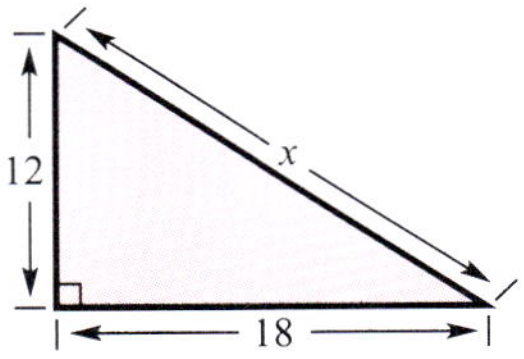

37. *Height of a Ladder* A ladder is 20 feet long, and the bottom of the ladder is 4 feet from the wall of a house (see figure). Determine the height at which the top of the ladder rests against the wall.

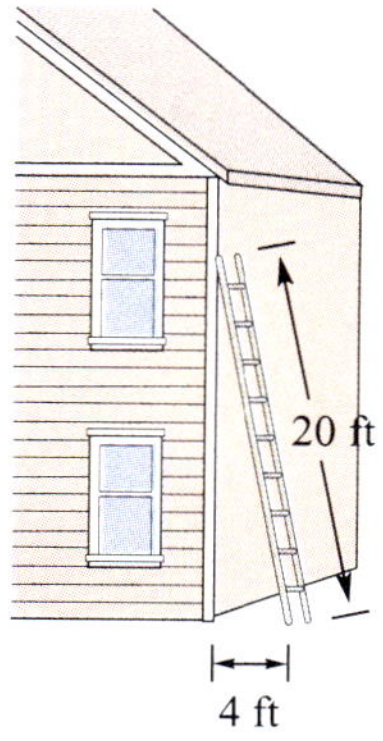

38. *Length of a Wire* A guy wire on a radio tower is attached to the top of the tower and to an anchor 50 feet from the base of the tower (see figure). Determine the length of the guy wire if the tower is 100 feet high.

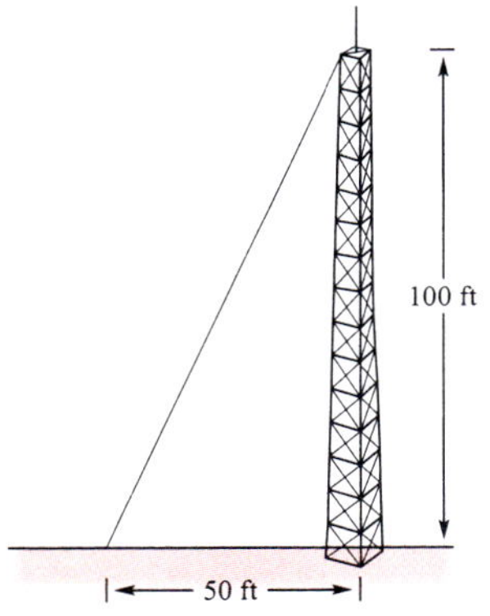

Free-Falling Object In Exercises 39–42, use the equation for the velocity of a free-falling object ($v = \sqrt{2gh}$) as described in Example 12.

39. An object is dropped from a height of 50 feet. Find the velocity of the object when it strikes the ground.

40. An object is dropped from a height of 200 feet. Find the velocity of the object when it strikes the ground.

41. An object that was dropped strikes the ground with a velocity of 60 feet per second. Find the height from which the object was dropped.

42. An object that was dropped strikes the ground with a velocity of 120 feet per second. Find the height from which the object was dropped.

Height of an Object The time t in seconds for a free-falling object to fall d feet is given by

$$t = \sqrt{\frac{d}{16}}.$$

43. A construction worker drops a nail from the top of a building and observes it strike a water puddle after approximately 2 seconds. Estimate the height of the worker.

44. A construction worker drops a nail from the top of a building and observes it strike a water puddle after approximately 3 seconds. Estimate the height of the worker.

Length of a Pendulum The time t in seconds for a pendulum of length L feet to go through one complete cycle (its period) is given by

$$t = 2\pi\sqrt{\frac{L}{32}}.$$

45. How long is the pendulum of a grandfather clock with a period of 1.5 seconds (see figure)?

46. How long is the pendulum of a mantle clock with a period of 0.75 seconds?

47. *Demand for a Product* The demand equation for a certain product is given by

$$p = 50 - \sqrt{0.8(x - 1)}$$

where x is the number of units demanded per day and p is the price per unit. Find the demand if the price is \$30.02.

48. *Airline Passengers* An airline offers daily flights between Denver and Seattle. The total monthly cost of the flights is given by

$$C = \frac{1}{10}\sqrt{0.1(x + 15)}$$

where C is measured in millions of dollars and x is measured in thousands of passengers. The total cost of the flights for a certain month was 1.75 million dollars. Approximately how many passengers flew that month?

49. *Surface Area of a Cone* The surface area of a cone is given by

$$S = \pi r\sqrt{r^2 + h^2}$$

as shown in the accompanying figure. Solve this equation for h.

50. *Dimensions of a Rectangle* Determine the length and width of a rectangle with a perimeter of 68 inches and a diagonal of length 26 inches.

Figure for 45

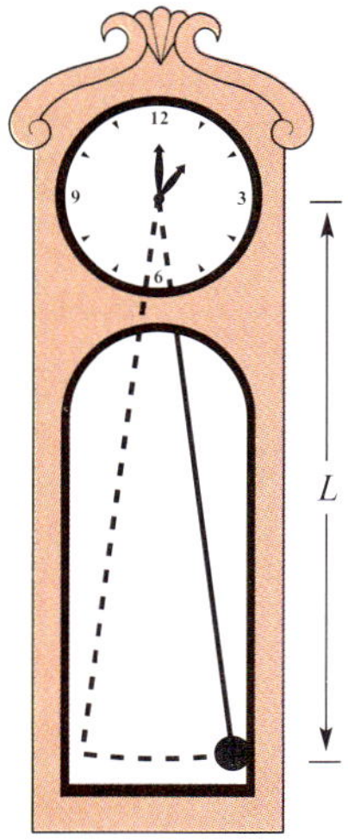

Figure for 49

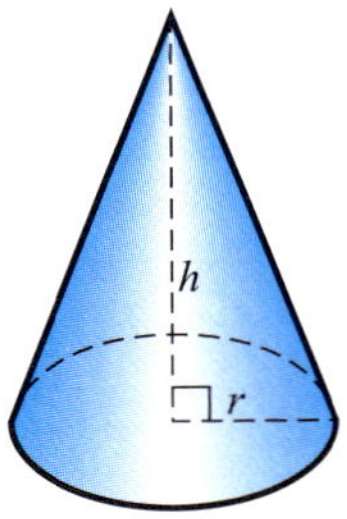

SECTION **5.6**

Complex Numbers

The Imaginary Unit i • *Complex Numbers* • *Operations with Complex Numbers* • *Complex Conjugates*

The Imaginary Unit i

In Section 5.2 we noted that a negative number has no *real* square root. For instance, we say that $\sqrt{-1}$ is not real because there is no real number x such that $x^2 = -1$. Thus, as long as we are dealing only with real numbers, we say that the equation

$$x^2 = -1$$

has no solution. To overcome this deficiency, mathematicians have expanded the set of real numbers, using the **imaginary unit** $\boldsymbol{i}$, which we define as

$$i = \sqrt{-1}. \qquad \text{Imaginary unit}$$

This number has the property that $i^2 = -1$. Thus, the imaginary unit i is a solution of the equation $x^2 = -1$.

The Square Root of a Negative Number

Let c be a positive real number. Then the square root of $-c$ is given by

$$\sqrt{-c} = \sqrt{c(-1)} = \sqrt{c}\sqrt{-1} = \sqrt{c}i.$$

When we write $\sqrt{-c}$ in the form $\sqrt{c}i$, we say that we are writing the number in ***i*-form**.

EXAMPLE 1 ■ **Writing Square Roots of Negative Numbers in *i*-Form**

(a) $\sqrt{-36} = \sqrt{36(-1)} = \sqrt{36}\sqrt{-1} = 6i$

(b) $\sqrt{-\frac{16}{25}} = \sqrt{\frac{16}{25}(-1)} = \sqrt{\frac{16}{25}}\sqrt{-1} = \frac{4}{5}i$

(c) $\sqrt{-5} = \sqrt{5(-1)} = \sqrt{5}\sqrt{-1} = \sqrt{5}i$

(d) $\sqrt{-54} = \sqrt{54(-1)} = \sqrt{54}\sqrt{-1} = 3\sqrt{6}i$ ■

To perform operations with square roots of negative numbers, we must first write the numbers in i-form. Once the numbers are written in i-form, we add, subtract, and multiply as follows.

$$ai + bi = (a + b)i \qquad \text{Addition}$$

$$ai - bi = (a - b)i \qquad \text{Subtraction}$$

$$(ai)(bi) = ab(i^2) = ab(-1) = -ab \qquad \text{Multiplication}$$

EXAMPLE 2 ■ Adding Square Roots of Negative Numbers

Perform the following addition.

$$\sqrt{-9} + \sqrt{-49}$$

Solution

$$\begin{aligned}\sqrt{-9} + \sqrt{-49} &= \sqrt{9}\sqrt{-1} + \sqrt{49}\sqrt{-1} \\ &= 3i + 7i && \text{Write in } i\text{-form} \\ &= (3 + 7)i && \text{Distributive Property} \\ &= 10i && \text{Simplified form}\end{aligned}$$ ■

When performing operations with numbers in i-form, we sometimes need to be able to evaluate powers of the imaginary unit i. The first several powers of i are as follows.

$$\begin{aligned} i^1 &= i \\ i^2 &= -1 \\ i^3 &= i(i^2) = i(-1) = -i \\ i^4 &= (i^2)(i^2) = (-1)(-1) = 1 \\ i^5 &= i(i^4) = i(1) = i \\ i^6 &= (i^2)(i^4) = (-1)(1) = -1 \\ i^7 &= (i^3)(i^4) = (-i)(1) = -i \\ i^8 &= (i^4)(i^4) = (1)(1) = 1 \end{aligned}$$

Note how the pattern of values i, -1, $-i$, and 1 repeats for powers greater than 4.

EXAMPLE 3 ■ Operations with Square Roots of Negative Numbers

(a)
$$\begin{aligned}\sqrt{-15}\sqrt{-15} &= (\sqrt{15}i)(\sqrt{15}i) && \text{Write in standard form} \\ &= (\sqrt{15})^2 i^2 \\ &= 15(-1) \\ &= -15 && \text{Simplified form}\end{aligned}$$

(b)
$$\begin{aligned}\sqrt{-5}(\sqrt{-45} - \sqrt{-4}) &= \sqrt{5}i(3\sqrt{5}i - 2i) && \text{Write in standard form} \\ &= (\sqrt{5}i)(3\sqrt{5}i) - (\sqrt{5}i)(2i) && \text{Distributive Property} \\ &= 3(5)(-1) - 2\sqrt{5}(-1) \\ &= -15 + 2\sqrt{5} && \text{Simplified form}\end{aligned}$$ ■

When multiplying square roots of negative numbers, be sure to write them in i-form *before multiplying*. If you don't, you may obtain incorrect answers. For instance, in Example 3(a), be sure you see the following.

$$\sqrt{-15}\sqrt{-15} \neq \sqrt{(-15)(-15)} = \sqrt{225} = 15$$

Complex Numbers

A number of the form $a + bi$, where a and b are real numbers, is called a **complex number**.

Definition of a Complex Number

If a and b are real numbers, then the number

$$a + bi$$

is called a **complex number**, written in **standard form**. If $b = 0$, then the number $a + bi = a$ is a real number. If $b \neq 0$, then the number $a + bi$ is called an **imaginary number**.

A number cannot be both real and imaginary. For instance, the numbers -2, 0, 1, $\frac{1}{2}$, and $\sqrt{2}$ are real numbers (but they are *not* imaginary numbers), and the numbers $-3i$, $2 + 4i$, and $-1 + i$ are imaginary numbers (but they are *not* real numbers). The diagram shown in Figure 5.6 further illustrates the relationships among real, complex, and imaginary numbers.

FIGURE 5.6

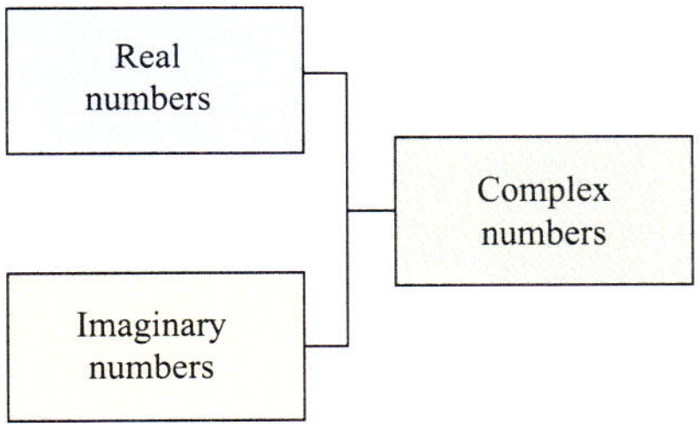

Two complex numbers $a + bi$ and $c + di$, in standard form, are equal if and only if $a = c$ and $b = d$.

EXAMPLE 4 ■ **Equality of Two Complex Numbers**

(a) Determine whether the complex numbers $\sqrt{9} + \sqrt{-48}$ and $3 - 4\sqrt{3}i$ are equal.

(b) Find x and y so that the following equation is valid.

$$3x - \sqrt{-25} = -6 + 3yi$$

Solution

(a) Writing the first number in standard form produces the following.

$$\sqrt{9} + \sqrt{-48} = \sqrt{3^2} + \sqrt{4^2(3)(-1)} = 3 + 4\sqrt{3}i$$

This complex number is not equal to $3 - 4\sqrt{3}i$, because the two numbers have imaginary parts that differ in sign.

(b) By writing the left side of the equation in standard form, we have the following.

$3x - \sqrt{-25} = -6 + 3yi$	Given equation
$3x - 5i = -6 + 3yi$	Both sides in standard form

For these two numbers to be equal, their real parts must be equal to each other and their imaginary parts must be equal to each other. Thus, we can write the following.

$$3x = -6 \quad \text{and} \quad -5 = 3y$$

$$x = -2 \qquad\qquad y = -\frac{5}{3}$$

■

Operations with Complex Numbers

The real number a is called the **real part** of the complex number $a + bi$, and the real number b is called the **imaginary part** of the complex number. To add or subtract two complex numbers, we add (or subtract) the real and imaginary parts separately. This is similar to combining like terms of a polynomial.

$(a + bi) + (c + di) = (a + c) + (b + d)i$	Addition of complex numbers
$(a + bi) - (c + di) = (a - c) + (b - d)i$	Subtraction of complex numbers

EXAMPLE 5 ■ Adding and Subtracting Complex Numbers

(a) $(3 - i) + (-2 + 4i) = (3 - 2) + (-1 + 4)i = 1 + 3i$

(b) $3i + (5 - 3i) = 5 + (3 - 3)i = 5$

(c) $4 - (-1 + 5i) + (7 + 2i) = (4 - (-1) + 7) + (-5 + 2)i = 12 - 3i$

Note in part (b) that the sum of two complex numbers can be a real number. ■

The Commutative, Associative, and Distributive Properties of real numbers are also valid for complex numbers. Note how we use these properties to find the product of two complex numbers.

EXAMPLE 6 ■ Multiplying Complex Numbers

Find the following products and write your answer in standard form.

(a) $(1 - i)(\sqrt{-9})$ (b) $(2 - i)(4 + 3i)$

Solution

(a) $(1 - i)(\sqrt{-9}) = (1 - i)(3i)$	Write in standard form
$= (1)(3i) - (i)(3i)$	Distributive Property
$= 3i - 3(i^2)$	
$= 3i - 3(-1)$	
$= 3 + 3i$	Standard form

$$\text{(b)}\ (2 - i)(4 + 3i) = \overbrace{8}^{F} + \overbrace{6i}^{O} - \overbrace{4i}^{I} - \overbrace{3i^2}^{L} \qquad \text{FOIL Method}$$
$$= 8 - 3(-1) + 6i - 4i \qquad \text{Collect like terms}$$
$$= 11 + 2i \qquad \text{Standard form}$$

EXAMPLE 7 ■ **Multiplying Complex Numbers**

Find the following products and write your answer in standard form.

(a) $(3 + 2i)(3 - 2i)$ (b) $(1 + 4i)^2$

Solution

$$\text{(a)}\ (3 + 2i)(3 - 2i) = 3^2 - (2i)^2 \qquad \text{Special product}$$
$$= 9 - 2^2i^2$$
$$= 9 - 4(-1)$$
$$= 9 + 4$$
$$= 13 \qquad \text{Standard form}$$
$$\text{(b)}\ (1 + 4i)^2 = 1 + 2(4i) + (4i)^2 \qquad \text{Special product}$$
$$= 1 + 8i + 4^2i^2$$
$$= 1 + 8i + 16(-1)$$
$$= -15 + 8i \qquad \text{Standard form}$$

Complex Conjugates

In Example 7(a), note that the product of two complex numbers can be a real number. This occurs with pairs of complex numbers of the form $a + bi$ and $a - bi$, called **complex conjugates**. In general, the product of complex conjugates has the following form.

$$(a + bi)(a - bi) = a^2 - (bi)^2$$
$$= a^2 - b^2i^2$$
$$= a^2 - b^2(-1)$$
$$= a^2 + b^2$$

Here are some examples.

Complex Number	*Complex Conjugate*	*Product*
$4 - 5i$	$4 + 5i$	$4^2 + 5^2 = 41$
$3 + 2i$	$3 - 2i$	$3^2 + 2^2 = 13$
$-2 = -2 + 0i$	$-2 = -2 - 0i$	$(-2)^2 + 0^2 = 4$
$i = 0 + i$	$-i = 0 - i$	$0^2 + 1^2 = 1$

Complex conjugates are used to divide one complex number by another. To do this, we multiply the numerator and denominator by the complex conjugate of the denominator, as shown in Example 8.

EXAMPLE 8 ■ **Division of Complex Numbers**

Perform the following division and write your answer in standard form.

$5 \div (3 - 2i)$

Solution

$$\frac{5}{3-2i} = \frac{5}{3-2i} \cdot \frac{3+2i}{3+2i}$$ Multiply by complex conjugate

$$= \frac{5(3+2i)}{(3-2i)(3+2i)}$$

$$= \frac{5(3+2i)}{3^2+2^2}$$

$$= \frac{15+10i}{13}$$

$$= \frac{15}{13} + \frac{10}{13}i$$ Standard form ■

EXAMPLE 9 ■ **Division of Complex Numbers**

$$\frac{1}{1+i} = \frac{1}{1+i} \cdot \frac{1-i}{1-i}$$ Multiply by conjugate

$$= \frac{1-i}{1^2+1^2}$$

$$= \frac{1-i}{2}$$

$$= \frac{1}{2} - \frac{1}{2}i$$ Standard form ■

EXAMPLE 10 ■ **Division of Complex Numbers**

(a) $$\frac{2+3i}{4-2i} = \frac{2+3i}{4-2i} \cdot \frac{4+2i}{4+2i}$$ Multiply by conjugate

$$= \frac{8+16i+6i^2}{4^2+2^2}$$

$$= \frac{8+16i+6(-1)}{20}$$

$$= \frac{2+16i}{20}$$

$$= \frac{1}{10} + \frac{4}{5}i$$ Standard form

(b) $$\frac{7-i}{5i} = \frac{7-i}{5i} \cdot \frac{-5i}{-5i}$$ Multiply by conjugate

$$= \frac{(7-i)(-5i)}{(5i)(-5i)}$$

$$= \frac{-35i + 5i^2}{-25i^2}$$

$$= \frac{-35i + 5(-1)}{-25(-1)}$$

$$= \frac{-35i - 5}{25}$$

$$= -\frac{1}{5} - \frac{7}{5}i$$ Standard form

DISCUSSION PROBLEM ■ Factoring the Sum of Two Squares

We know from Section 3.3 that the *difference* of two squares, such as $x^2 - 9$, factors as $(x + 3)(x - 3)$. There we said that the *sum* of two squares, such as $x^2 + 9$, is irreducible over the real numbers.

(a) Using complex numbers, it *is* possible to factor the polynomial $x^2 + 9$. Write a short paragraph showing how this can be done.

(b) As long as we were dealing only with real numbers, we said that equations such as $x^2 + 9 = 0$ have no solutions. Using complex numbers, does this equation have solutions? If so, what are the solutions?

Warm-Up

The following warm-up exercises involve skills that were covered in earlier sections. You will use these skills in the exercise set for this section.

In Exercises 1–6, simplify the given expression.

1. $\sqrt{48}$ **2.** $\sqrt{108}$ **3.** $\sqrt{128} + 3\sqrt{50}$

4. $3\sqrt{5}\sqrt{500}$ **5.** $\frac{8}{\sqrt{10}}$ **6.** $\frac{5}{\sqrt{12} - 2}$

In Exercises 7–10, solve the given equation.

7. $x(2x - 35) = 0$ **8.** $(x + 4)\left(x - \frac{5}{4}\right) = 0$

9. $x^2 - 7x - 18 = 0$ **10.** $2x^2 - x - 3 = 0$

5.6 EXERCISES

In Exercises 1–12, write the number in i-form.

1. $\sqrt{-4}$
2. $\sqrt{-9}$
3. $-\sqrt{-144}$
4. $\sqrt{-49}$
5. $\sqrt{-\dfrac{4}{25}}$
6. $-\sqrt{-\dfrac{36}{121}}$
7. $\sqrt{-0.09}$
8. $\sqrt{-0.0004}$
9. $\sqrt{-8}$
10. $\sqrt{-75}$
11. $\sqrt{-7}$
12. $\sqrt{-15}$

In Exercises 13–16, perform the indicated addition or subtraction and write your answer in standard form.

13. $\sqrt{-16} + 6i$
14. $\sqrt{-\dfrac{1}{4}} - \dfrac{3}{2}i$
15. $\sqrt{-50} - \sqrt{-8}$
16. $\sqrt{-500} + \sqrt{-45}$

In Exercises 17–24, perform the indicated operations and write your answer in standard form.

17. $\sqrt{-8}\sqrt{-2}$
18. $\sqrt{-25}\sqrt{-6}$
19. $\sqrt{-18}\sqrt{-3}$
20. $\sqrt{-0.16}\sqrt{-1.21}$
21. $\sqrt{-3}(\sqrt{-3} + \sqrt{-4})$
22. $\sqrt{-12}(\sqrt{-3} - \sqrt{-12})$
23. $\sqrt{-5}(\sqrt{-16} - \sqrt{-10})$
24. $\sqrt{-24}\left(\sqrt{-\dfrac{4}{9}} + \sqrt{-\dfrac{1}{4}}\right)$

In Exercises 25–30, determine a and b.

25. $3 - 4i = a + bi$
26. $-8 + 6i = a + bi$
27. $-4 - \sqrt{-8} = a + bi$
28. $\sqrt{-36} - 3 = a + bi$
29. $(a + 5) + (b - 1)i = 7 - 3i$
30. $(2a + 1) + (2b + 3)i = 5 + 12i$

In Exercises 31–60, perform the specified operations and write your answer in standard form.

31. $(4 - 3i) + (6 + 7i)$
32. $(-10 + 2i) + (4 - 7i)$
33. $(-4 - 7i) + (-10 - 33i)$
34. $(4 + 6i) + (15 + 24i) - (1 - i)$
35. $(15 + 10i) - (2 + 10i)$
36. $(-21 - 50i) + (21 - 20i)$
37. $13i - (14 - 7i)$
38. $22 + (-5 + 8i) + 10i$
39. $\left(\dfrac{4}{3} + \dfrac{1}{3}i\right) + \left(\dfrac{5}{6} + \dfrac{7}{6}i\right)$
40. $(0.05 + 2.50i) - (6.2 + 11.8i)$
41. $15i - (3 - 25i) + \sqrt{-81}$
42. $(-1 + i) - \sqrt{2} - \sqrt{-2}$
43. $(3i)(12i)$
44. $(-5i)(4i)$
45. $(-6i)(-i)(6i)$
46. $\dfrac{1}{2}(10i)(12i)(-3i)$

47. $(-3i)^3$

48. $(8i)^2$

49. $-5(13 + 2i)$

50. $10(8 - 6i)$

51. $4i(-3 - 5i)$

52. $-3i(10 - 15i)$

53. $(4 + 3i)(-7 + 4i)$

54. $(3 + 5i)(2 + 15i)$

55. $(3 - 4i)^2$

56. $(7 + i)^2$

57. $(-3 - \sqrt{-12})(4 - \sqrt{-12})$

58. $\sqrt{-9}(1 + \sqrt{-16})$

59. $(2 + i)^3$

60. $(3 - 2i)^3$

In Exercises 61–70, find the conjugate of the given complex number. Then find the product of the number and its conjugate.

61. $2 + i$

62. $3 + 2i$

63. $5 - \sqrt{6}i$

64. $10 - 3i$

65. $-2 - 8i$

66. $-4 + \sqrt{2}i$

67. $10i$

68. 20

69. $1 + \sqrt{-3}$

70. $-3 - \sqrt{-5}$

In Exercises 71–80, perform the indicated division and write your answer in standard form.

71. $\dfrac{4}{1 - i}$

72. $\dfrac{20}{3 + i}$

73. $\dfrac{-12}{2 + 7i}$

74. $\dfrac{15}{2(1 - i)}$

75. $\dfrac{4i}{1 - 3i}$

76. $\dfrac{17i}{5 + 3i}$

77. $\dfrac{20}{2i}$

78. $\dfrac{1 + i}{3i}$

79. $\dfrac{2 + 3i}{1 + 2i}$

80. $\dfrac{4 - 5i}{4 + 5i}$

In Exercises 81–84, find the indicated sum or difference and write your answer in standard form.

81. $\dfrac{1}{1 - 2i} + \dfrac{4}{1 + 2i}$

82. $\dfrac{3i}{1 + i} + \dfrac{2}{2 + 3i}$

83. $\dfrac{i}{4 - 3i} - \dfrac{5}{2 + i}$

84. $\dfrac{1 + i}{i} - \dfrac{3}{5 - 2i}$

In Exercises 85–88, perform the indicated operations on the complex number $a + bi$ and its conjugate $a - bi$.

85. $(a + bi) + (a - bi)$

86. $(a + bi)(a - bi)$

87. $(a + bi) - (a - bi)$

88. $(a + bi)^2 + (a - bi)^2$

89. Show that $-1 + 2i$ is a solution of the equation $x^2 + 2x + 5 = 0$.

90. Show that $2 - 3i$ is a solution of the equation $x^2 - 4x + 13 = 0$.

CHAPTER 5 SUMMARY

As you review and prepare for a test on this chapter, first try to obtain a global view of what was discussed. Then review the specific skills needed under each category.

Simplifying Expressions Involving Integer Exponents, Rational Exponents, or Radicals

■ *Evaluate* numerical expressions involving integer exponents, rational exponents, or radicals.	Section 5.1, Exercises 1–30; Section 5.2, Exercises 7–30, Exercises 41–50
■ *Evaluate* numerical expressions involving rational exponents or radicals by using a calculator.	Section 5.2, Exercises 51–60
■ *Rewrite or simplify* expressions, using only positive exponents.	Section 5.1, Exercises 31–54; Section 5.2, Exercises 61–76
■ *Simplify* expressions containing radicals or rational exponents.	Section 5.2, Exercises 61–80
■ *Rewrite* decimal numbers in scientific notation and conversely.	Section 5.1, Exercises 55–66

Operations Involving Radical Expressions

■ *Rewrite* expressions involving radicals.	Section 5.3, Exercises 7–52
■ *Simplify* expressions involving radicals by rationalizing the denominator.	Section 5.3, Exercises 53–66
■ *Combine* radical expressions.	Section 5.3, Exercises 67–80
■ *Multiply* expressions involving radicals.	Section 5.4, Exercises 1–30
■ *Divide* expressions involving radicals by rationalizing the denominator.	Section 5.4, Exercises 49–60

Solving Equations Involving Radicals

■ *Solve* equations involving radicals.	Section 5.5, Exercises 1–30
■ *Create and solve* equations involving radicals.	Section 5.5, Exercises 31–50

Complex Numbers

■ *Rewrite* complex numbers in standard form.	Section 5.6, Exercises 1–16
■ *Add, subtract, and multiply* complex numbers.	Section 5.6, Exercises 17–60
■ *Divide* complex numbers by using the conjugate.	Section 5.6, Exercises 71–80

Chapter 5 Review Exercises

In Exercises 1–6, evaluate the given expression. (Do not use a calculator.)

1. $(2^3 \cdot 3^2)^{-1}$

2. $(2^{-2} \cdot 5^2)^{-2}$

3. $\left(\frac{2}{5}\right)^{-3}$

4. $\left(\frac{1}{3^{-2}}\right)^2$

5. $(6 \times 10^3)^2$

6. $(3 \times 10^{-3})(8 \times 10^7)$

In Exercises 7–12, write the given expression without negative or zero exponents and simplify. (Assume that none of the variables are zero.)

7. $(x^3y^{-4})^2$

8. $5yx^0$

9. $\frac{t^{-5}}{t^{-2}}$

10. $\frac{a^5 \cdot a^{-3}}{a^{-2}}$

11. $\left(\frac{y}{3}\right)^{-3}$

12. $(2x^2y^4)^4(2x^2y^4)^{-4}$

In Exercises 13–18, evaluate the given expression. (Do not use a calculator.)

13. $\sqrt{1.44}$

14. $\sqrt{0.16}$

15. $\sqrt{\frac{25}{36}}$

16. $-\sqrt{\frac{64}{225}}$

17. $\sqrt{169 - 25}$

18. $\sqrt{16 + 9}$

In Exercises 19 and 20, use a calculator to approximate the given expression. (Round your answer to two decimal places.)

19. $\sqrt{13^2 - 4(2)(7)}$

20. $\frac{-3.7 + \sqrt{15.8}}{2(2.3)}$

In Exercises 21 and 22, evaluate the given expression. (Do not use a calculator.)

21. $27^{4/3}$

22. $243^{-2/5}$

In Exercises 23 and 24, use a calculator to approximate the given expression. (Round your answer to two decimal places.)

23. $75^{-3/4}$

24. $510^{5/3}$

In Exercises 25–28, simplify the given expression. (Assume all variables are positive.)

25. $x^{3/4} \cdot x^{-1/6}$

26. $(2y^2)^{3/2}(2y^{-4})^{1/2}$

27. $\frac{15x^{1/4}y^{3/5}}{5x^{1/2}y}$

28. $\frac{48a^2b^{5/2}}{14a^{-3}b^{-1/2}}$

In Exercises 29–34, simplify the given expression. (Assume all variables are positive.)

29. $\sqrt{360}$

30. $\sqrt{\frac{50}{9}}$

31. $\sqrt{0.25x^4y}$

32. $\sqrt{0.16s^6t^3}$

33. $\sqrt[3]{48a^3b^4}$

34. $\sqrt[4]{32u^4v^5}$

In Exercises 35–40, rationalize the denominator and simplify further when possible. (Assume all variables are positive.)

35. $\sqrt{\frac{5}{6}}$

36. $\sqrt{\frac{3}{20}}$

37. $\frac{3}{\sqrt{12x}}$

38. $\frac{4y}{\sqrt{10z}}$

39. $\frac{2}{\sqrt[3]{2x}}$

40. $\sqrt[3]{\frac{16t}{s^2}}$

In Exercises 41–50, perform the indicated operations and simplify. (Assume all variables are positive.)

41. $3\sqrt{40} - 10\sqrt{90}$

42. $9\sqrt{50} - 5\sqrt{8} + \sqrt{48}$

43. $10\sqrt[4]{y+3} - 3\sqrt[4]{y+3}$

44. $\sqrt{25x} + \sqrt{49x} - \sqrt[3]{8x}$

45. $(\sqrt{5} + 6)^2$

46. $(\sqrt{3} - \sqrt{x})(\sqrt{3} + \sqrt{x})$

47. $(2\sqrt{3} + 7)(\sqrt{6} - 2)$

48. $\frac{15}{\sqrt{x} + 3}$

49. $(\sqrt{x} + 10) \div (\sqrt{x} - 10)$

50. $(3\sqrt{s} + 4) \div (\sqrt{s} + 2)$

In Exercises 51–56, solve the given equation.

51. $\sqrt{y} = 15$

52. $\sqrt{3x} + 9 = 0$

53. $\sqrt{2(a - 7)} = 14$

54. $\sqrt{3(2x + 3)} = \sqrt{x + 15}$

55. $\sqrt{2(x + 5)} = x + 5$

56. $\sqrt{5t} = 1 + \sqrt{5(t - 1)}$

In Exercises 57–60, write the complex number in standard form.

57. $\sqrt{-48}$

58. $3 + 2\sqrt{-500}$

59. $\frac{3}{4} - 5\sqrt{-\frac{3}{25}}$

60. $-0.5 + 3\sqrt{-1.21}$

In Exercises 61–70, perform the indicated operations and write the answer in standard form.

61. $(-4 + 5i) - (-12 + 8i)$

62. $5(3 - 8i) + (5 + 12i)$

63. $(-2)(15i)(-3i)$

64. $-10i(4 - 7i)$

65. $(4 - 3i)(4 + 3i)$

66. $(6 - 5i)^2$

67. $(12 - 5i)(2 + 7i)$

68. $\frac{4}{5i}$

69. $\frac{5i}{2 + 9i}$

70. $\frac{2 + i}{1 - 9i}$

71. *Length of a Wire* A guy wire on a radio tower is attached to the top of the tower and to an anchor 80 feet from the base of the tower (see figure). Determine the length of the guy wire if the tower is 200 feet high.

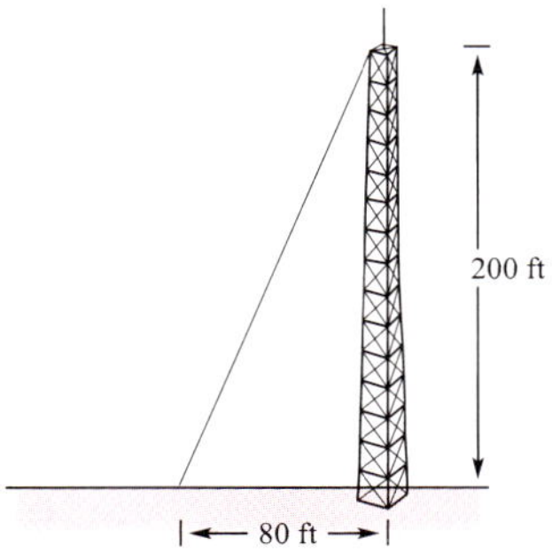

72. *Height of a Bridge* The time t in seconds for a free-falling object to fall d feet is given by

$$t = \sqrt{\frac{d}{16}}.$$

A child drops a pebble from a bridge and observes it strike the water after approximately 4 seconds (see figure). Estimate the height of the bridge.

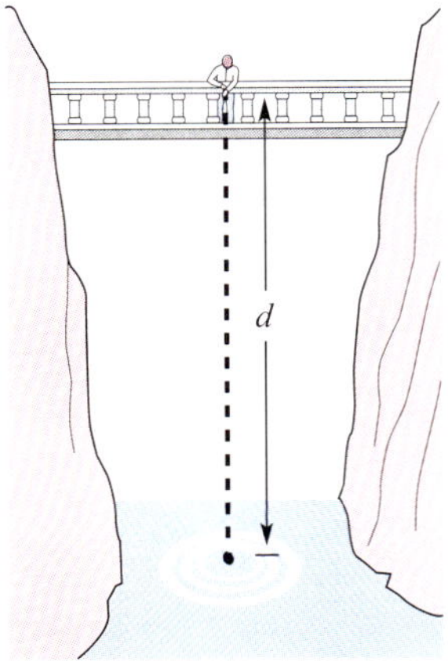

73. *Length of a Pendulum* The time t in seconds for a pendulum of length L in feet to go through one complete cycle (its period) is given by

$$t = 2\pi\sqrt{\frac{L}{32}}.$$

How long is the pendulum of a grandfather clock with a period of 1.3 seconds?

74. *Dimensions of a Rectangle* Determine the length and width of a rectangle with a perimeter of 84 inches and a diagonal of 30 inches.

CHAPTER 5 TEST

Take this test as you would take a test in class. After you are done, check your work with the answers given in the back of the book.

1. Evaluate: $2^{-2} + 2^{-3}$

2. Simplify: $\dfrac{12t^{-2}}{20t^{-1}}$

3. Simplify: $(x + y^{-2})^{-1}$

4. Write the real number 0.000032 in scientific notation.

5. Write the real number 3.04×10^7 in decimal notation.

6. Evaluate (do not use a calculator): $\dfrac{6.3 \times 10^{-3}}{2.1 \times 10^2}$

7. Evaluate (do not use a calculator): $27^{-2/3}$

8. Simplify: $\left(\dfrac{x^{1/2}}{x^{1/3}}\right)^2$

9. Multiply: $\sqrt[3]{5} \cdot \sqrt[3]{6}$

10. Simplify: $\sqrt{\dfrac{32}{9}}$

11. Simplify: $\sqrt[3]{24}$

12. Rationalize the denominator: $\dfrac{3}{\sqrt{6}}$

13. Rationalize the denominator: $\sqrt[3]{\dfrac{1}{4}}$

14. Combine: $5\sqrt{3x} - 3\sqrt{75x}$

15. Multiply and simplify (assume x is positive): $\sqrt{5}\left(\sqrt{15x} + 3\right)$

16. Expand (assume x is positive): $\left(4 - \sqrt{2x}\right)^2$

17. Factor: $7\sqrt{27} + 14y\sqrt{12} = 7\sqrt{3}(\quad\quad)$

18. Solve: $\sqrt{x^2 - 1} = x - 2$

19. Solve: $\sqrt{x} - x + 6 = 0$

20. Subtract: $(2 + 3i) - \sqrt{-25}$

21. Multiply: $(3 - 2i)(1 + 5i)$

22. Multiply: $\sqrt{-16}\left(1 + \sqrt{-4}\right)$

23. Divide: $\dfrac{2i}{5 - 4i}$

24. A ladder is 12 feet long, and the bottom of the ladder is 3 feet from the wall of a house. Determine the height at which the top of the ladder rests against the wall of the house.

CHAPTER SIX

Quadratic Equations and Inequalities

6.1 The Factoring and Square Root Methods
6.2 Completing the Square
6.3 The Quadratic Formula and the Discriminant
6.4 Applications of Quadratic Equations
6.5 Quadratic and Rational Inequalities

Chapter Overview

In this chapter we look at several ways to solve quadratic equations. In Section 6.1 we review the factoring method and introduce a method called extracting square roots. In Section 6.2 we see how a quadratic equation can be solved by completing the square, and in Section 6.3 we introduce the Quadratic Formula. Quadratic equations have many applications in real life, and we survey several of these in Section 6.4. Section 6.5 discusses quadratic and rational inequalities.

SECTION **6.1**

The Factoring and Square Root Methods

Solving Quadratic Equations by Factoring • Solving Quadratic Equations by Extracting Square Roots • Quadratic Equations with Complex Solutions • Equations of Quadratic Type

Solving Quadratic Equations by Factoring

In this chapter we look at second-degree polynomial equations, such as

$$x^2 - 3x - 4 = 0, \qquad 2x^2 + x - 3 = 0, \quad \text{and} \quad x^2 + 8x = 0.$$

We call this type of equation a **quadratic equation** in x.

Definition of a Quadratic Equation in x

A **quadratic equation** in x is an equation that can be written in the standard form

$$ax^2 + bx + c = 0$$

where a, b, and c are real numbers with $a \neq 0$.

In Section 3.5 we looked at one method for solving a quadratic equation—factoring. In this chapter we review that method and look at other ways of solving quadratic equations.

Remember from Section 3.5 that the first step in solving a quadratic equation by factoring is to write the equation in standard form (with the polynomial on the left side of the equation and zero on the right side). Next, we attempt to factor the left side. Finally, we set each factor equal to zero and solve for x.

EXAMPLE 1 ■ Solving Quadratic Equations by Factoring

Solve the following quadratic equations.

(a) $x^2 + 5x = 24$ (b) $3x^2 = 4 - 11x$

Solution

(a)

$x^2 + 5x = 24$		Given equation
$x^2 + 5x - 24 = 0$		Standard form
$(x + 8)(x - 3) = 0$		Factor
$x + 8 = 0$	⇨ $x = -8$	Set first factor equal to 0
$x - 3 = 0$	⇨ $x = 3$	Set second factor equal to 0

Thus, this quadratic equation has two solutions: -8 and 3. Check these solutions in the original equation.

(b)
$$3x^2 = 4 - 11x \quad \text{Given equation}$$
$$3x^2 + 11x - 4 = 0 \quad \text{Standard form}$$
$$(3x - 1)(x + 4) = 0 \quad \text{Factor}$$
$$3x - 1 = 0 \Rightarrow x = \frac{1}{3} \quad \text{Set first factor equal to 0}$$
$$x + 4 = 0 \Rightarrow x = -4 \quad \text{Set second factor equal to 0}$$

Thus, this quadratic equation has two solutions: $\frac{1}{3}$ and -4. Check these solutions in the original equation. ■

If the two factors of a quadratic equation in standard form are identical, then the corresponding solution is called a **double** or **repeated solution**. This occurs in the next example.

EXAMPLE 2 ■ A Quadratic Equation with a Repeated Solution

$$9x^2 + 12 = 3 + 12x + 5x^2 \quad \text{Given equation}$$
$$4x^2 - 12x + 9 = 0 \quad \text{Standard form}$$
$$(2x - 3)(2x - 3) = 0 \quad \text{Factor}$$
$$2x - 3 = 0 \quad \text{Set factor equal to 0}$$
$$x = \frac{3}{2} \quad \text{Repeated solution}$$

Thus, this quadratic equation has only one (repeated) solution: $\frac{3}{2}$. Check this solution in the original equation. ■

EXAMPLE 3 ■ Solving Quadratic Equations by Factoring

Solve the following equations by factoring.

(a) $(x + 1)(x - 2) = 18$ (b) $5x^2 = 45$

Solution

(a) Don't be tricked by this quadratic equation. Even though the left side is factored, the right side is not zero. Thus, we must first multiply the left side, then rewrite the equation in standard form, and finally refactor, as follows.

$$(x + 1)(x - 2) = 18 \quad \text{Given equation}$$
$$x^2 - x - 2 = 18 \quad \text{Multiply}$$
$$x^2 - x - 20 = 0 \quad \text{Standard form}$$
$$(x - 5)(x + 4) = 0 \quad \text{Factor}$$
$$x - 5 = 0 \Rightarrow x = 5 \quad \text{Set first factor equal to 0}$$
$$x + 4 = 0 \Rightarrow x = -4 \quad \text{Set second factor equal to 0}$$

Thus, this equation has two solutions: 5 and -4. Check these solutions in the original equation.

(b)

$5x^2 = 45$		Given equation
$5x^2 - 45 = 0$		Standard form
$5(x^2 - 9) = 0$		Common monomial factor
$5(x + 3)(x - 3) = 0$		Factor as difference of squares
$x + 3 = 0$	⇨ $x = -3$	Set first factor equal to 0
$x - 3 = 0$	⇨ $x = 3$	Set second factor equal to 0

Thus, this equation has two solutions: -3 and 3. Check these solutions in the original equation. ■

In Example 3(b) note that it is not necessary to set the *constant* factor 5 equal to zero (because it would produce no solution). Because of this, when an equation in standard form has a constant factor, some people like to divide both sides of the equation by the factor. This procedure is valid because the equations

$$5(x^2 - 9) = 0 \quad \text{and} \quad x^2 - 9 = 0$$

are *equivalent*, which means that both equations have the same solutions. If you use this technique, be sure that you apply it *only* to constant factors. We cannot divide both sides of an equation by a variable factor without risking the loss of one or more solutions.

Solving Quadratic Equations by Extracting Square Roots

There is a nice shortcut for solving quadratic equations of the form

$$u^2 = d$$

where $d > 0$ and u is an algebraic expression. By factoring, we can see that this equation has two solutions.

$u^2 = d$		Given equation
$u^2 - d = 0$		Standard form
$(u + \sqrt{d})(u - \sqrt{d}) = 0$		Factor
$u + \sqrt{d} = 0$	⇨ $u = -\sqrt{d}$	Set first factor equal to 0
$u - \sqrt{d} = 0$	⇨ $u = \sqrt{d}$	Set second factor equal to 0

Thus, the equation $u^2 = d$ (where $d > 0$) has two solutions. Since the solutions differ only in sign, we sometimes write both of the solutions together, using a "plus or minus sign," as follows.

$$u = \pm\sqrt{d}$$

This form of the solution is read as "u is equal to plus or minus the square root of d." When we solve an equation of the form $u^2 = d$ without going through the steps of factoring, we say that we are **extracting square roots**.

Extracting Square Roots

The equation $u^2 = d$, where $d > 0$, has exactly two solutions:

$$u = \sqrt{d} \quad \text{and} \quad u = -\sqrt{d}.$$

These solutions can also be written as

$$u = \pm\sqrt{d}.$$

EXAMPLE 4 ■ Solving a Quadratic Equation by Extracting Square Roots

$$\begin{aligned} 3x^2 &= 15 && \text{Given equation} \\ x^2 &= 5 && \text{Divide both sides by 3} \\ x &= \pm\sqrt{5} && \text{Extract square roots} \end{aligned}$$

Thus, the equation has two solutions: $\sqrt{5}$ and $-\sqrt{5}$. Note that $x^2 - 5 = 0$ factors as $(x + \sqrt{5})(x - \sqrt{5}) = 0$, which gives the same two solutions. Check these solutions in the original equation. ■

EXAMPLE 5 ■ Solving a Quadratic Equation by Extracting Square Roots

Solve the following quadratic equation by extracting square roots.

$$(x - 2)^2 = 10$$

Solution

In this case an extra step is needed after extracting square roots.

$$\begin{aligned} (x - 2)^2 &= 10 && \text{Given equation} \\ x - 2 &= \pm\sqrt{10} && \text{Extract square roots} \\ x &= 2 \pm \sqrt{10} && \text{Add 2 to both sides} \end{aligned}$$

Thus, the equation has two solutions: $2 + \sqrt{10}$ and $2 - \sqrt{10}$. Check these solutions in the original equation. ■

EXAMPLE 6 ■ Solving a Quadratic Equation by Extracting Square Roots

Solve the following quadratic equation by extracting square roots.

$$3(5x + 4)^2 - 81 = 0$$

Solution

In this case some preliminary steps are needed before taking square roots.

$$\begin{aligned} 3(5x + 4)^2 - 81 &= 0 && \text{Given equation} \\ 3(5x + 4)^2 &= 81 && \\ (5x + 4)^2 &= 27 && \text{Divide both sides by 3} \end{aligned}$$

$$5x + 4 = \pm\sqrt{27} \qquad \text{Extract square roots}$$
$$5x = -4 \pm 3\sqrt{3} \qquad \text{Subtract 4 from both sides}$$
$$x = \frac{-4 \pm 3\sqrt{3}}{5} \qquad \text{Divide both sides by 5}$$

Thus, the solutions are

$$\frac{-4 + 3\sqrt{3}}{5} \quad \text{and} \quad \frac{-4 - 3\sqrt{3}}{5}$$

Check

$$3(5x + 4)^2 - 81 = 0 \qquad \text{Given equation}$$
$$3\left(5\left[\frac{-4 + 3\sqrt{3}}{5}\right] + 4\right)^2 \stackrel{?}{=} 81 \qquad \text{Replace } x \text{ by } (-4 + 3\sqrt{3})/5$$
$$3\left(\left[-4 + 3\sqrt{3}\right] + 4\right)^2 \stackrel{?}{=} 81$$
$$3\left(3\sqrt{3}\right)^2 \stackrel{?}{=} 81$$
$$3(3^2)\left(\sqrt{3}\right)^2 \stackrel{?}{=} 81$$
$$81 = 81 \qquad \text{Solution checks}$$

Check the other solution in the original equation. ■

Quadratic Equations with Complex Solutions

So far in the text, the only solutions we have been finding have been real numbers. But now, having introduced complex numbers (in Section 5.6), we can talk about other types of solutions. For instance, although the quadratic equation $x^2 + 1 = 0$ has no solutions that are real numbers, it does have two solutions that are complex numbers: i and $-i$. To check this, we replace x by i and $-i$, as follows.

$$i^2 + 1 = (-1) + 1 = 0 \quad \Rightarrow \quad i \text{ is a solution}$$
$$(-i)^2 + 1 = (-1) + 1 = 0 \quad \Rightarrow \quad -i \text{ is a solution}$$

One way to find complex solutions of a quadratic equation is to extend the *extraction of square roots* technique to cover the case where d is a negative number.

Extracting Complex Square Roots

The equation $u^2 = d$, where $d < 0$, has exactly two solutions:

$$u = \sqrt{d}i \quad \text{and} \quad u = -\sqrt{d}i.$$

These solutions can also be written as

$$u = \pm\sqrt{d}i.$$

EXAMPLE 7 ■ **Extracting Complex Square Roots**

$x^2 + 8 = 0$	Given equation
$x^2 = -8$	Write in standard form
$x = \pm\sqrt{8}i$	Extract complex square roots
$x = \pm 2\sqrt{2}i$	Simplify

Thus, the given equation has two solutions: $2\sqrt{2}i$ and $-2\sqrt{2}i$. Check these solutions in the original equation. ■

EXAMPLE 8 ■ **Extracting Complex Square Roots**

$2(3x - 5)^2 + 32 = 0$	Given equation
$2(3x - 5)^2 = -32$	Write in standard form
$(3x - 5)^2 = -16$	Divide both sides by 2
$3x - 5 = \pm 4i$	Extract complex square roots
$3x = 5 \pm 4i$	Add 5 to both sides
$x = \dfrac{5 \pm 4i}{3}$	Divide both sides by 3

Thus, the given equation has two solutions: $(5 + 4i)/3$ and $(5 - 4i)/3$. Check these solutions in the original equation. ■

Equations of Quadratic Type

Both the factoring method and the extraction of square roots method can be applied to nonquadratic equations that are of **quadratic type**. An equation is said to be of quadratic type if it has the form

$$au^2 + bu + c = 0$$

where $a \neq 0$ and u is an algebraic expression. Here are some examples.

Equation	*Written in Quadratic Form*	*Substitution*
$x^4 + 5x^2 + 4 = 0$	$(x^2)^2 + 5(x^2) + 4 = 0$	$u = x^2$
$2x^{2/3} + 3x^{1/3} - 9 = 0$	$2(x^{1/3})^2 + 3(x^{1/3}) - 9 = 0$	$u = x^{1/3}$
$x - 5\sqrt{x} + 6 = 0$	$(\sqrt{x})^2 - 5\sqrt{x} + 6 = 0$	$u = \sqrt{x}$

To solve an equation of quadratic type, we rewrite the equation in terms of u to obtain $au^2 + bu + c = 0$, and solve for u. Then we replace u by its original value and solve for the original variable. This procedure is demonstrated in Examples 9, 10, and 11.

EXAMPLE 9 ■ **Solving an Equation of Quadratic Type**

Solve the following equation for x.

$$x^4 - 13x^2 + 36 = 0$$

Solution

This equation is of quadratic type because it can be written in the form

$$(x^2)^2 - 13(x^2) + 36 = 0.$$

Thus, we let $u = x^2$, and solve the equation as follows.

$x^4 - 13x^2 + 36 = 0$	Given equation
$(x^2)^2 - 13(x^2) + 36 = 0$	Write in quadratic form
$u^2 - 13u + 36 = 0$	Replace x^2 by u
$(u - 4)(u - 9) = 0$	Factor
$u - 4 = 0 \Rightarrow u = 4$	Set first factor equal to 0
$u - 9 = 0 \Rightarrow u = 9$	Set second factor equal to 0

At this point, we have found the "u-solutions." To find the "x-solutions" we replace u by x^2, as follows.

$$u = 4 \Rightarrow x^2 = 4 \Rightarrow x = \pm 2$$
$$u = 9 \Rightarrow x^2 = 9 \Rightarrow x = \pm 3$$

Thus, the original equation has *four* solutions: 2, −2, 3, and −3. Check these solutions in the original equation. ■

Be sure you see in Example 9 that the u-solutions of 4 and 9 represent only a temporary step in the solution. They are not (in general) solutions of the original equation.

EXAMPLE 10 ■ **Solving an Equation of Quadratic Type**

Solve the equation $x - 5\sqrt{x} + 6 = 0$ for x.

Solution

This equation is of quadratic type with $u = \sqrt{x}$.

$x - 5\sqrt{x} + 6 = 0$	Given equation
$(\sqrt{x})^2 - 5(\sqrt{x}) + 6 = 0$	Write in quadratic form
$u^2 - 5u + 6 = 0$	Replace $\sqrt{x}$ by u
$(u - 2)(u - 3) = 0$	Factor
$u - 2 = 0 \Rightarrow u = 2$	Set first factor equal to 0
$u - 3 = 0 \Rightarrow u = 3$	Set second factor equal to 0

Now, using the u-solutions of 2 and 3, we obtain the following x-solutions.

$$u = 2 \Rightarrow \sqrt{x} = 2 \Rightarrow x = 4$$
$$u = 3 \Rightarrow \sqrt{x} = 3 \Rightarrow x = 9$$

Thus, the original equation has two solutions: 4 and 9. Check these solutions in the original equation. ■

EXAMPLE 11 ■ **Solving an Equation of Quadratic Type**

Solve the following equation for x.

$$(x^2 - 4)^2 - 25 = 0$$

Solution

This equation is of quadratic type with $u = x^2 - 4$.

$(x^2 - 4)^2 - 25 = 0$	Given equation
$u^2 - 25 = 0$	Replace $x^2 - 4$ by u
$(u + 5)(u - 5) = 0$	Factor
$u + 5 = 0 \Rightarrow u = -5$	Set first factor equal to 0
$u - 5 = 0 \Rightarrow u = 5$	Set second factor equal to 0

Now, using the u-solutions of -5 and 5, we obtain the following x-solutions.

$u = -5$	and	$u = 5$
$x^2 - 4 = -5$		$x^2 - 4 = 5$
$x^2 = -1$		$x^2 = 9$
$x = \pm i$		$x = \pm 3$

Thus, the original equation has four solutions: i, $-i$, 3, and -3. Note that two of these solutions are real numbers and two are complex numbers. Check these solutions in the original equation. ■

DISCUSSION PROBLEM ■ **Constructing a Quadratic Equation from Its Solutions**

Suppose you are tutoring a student in algebra and want to make up several practice equations for the student to solve by factoring. You might want to start by writing several simple equations with easy solutions such as 1 and 2. Then you might want to put in some tougher problems that have solutions such as 13 and $-\frac{1}{2}$.

(a) Find a quadratic equation that has $x = 1$ and $x = 2$ as solutions.

(b) Find a quadratic equation that has $x = 13$ and $x = -\frac{1}{2}$ as solutions.

(c) Write a short paragraph describing a general procedure for creating a quadratic equation that has two given numbers as solutions. ■

Warm-Up

The following warm-up exercises involve skills that were covered in earlier sections. You will use these skills in the exercise set for this section.

In Exercises 1–6, completely factor the expression.

1. $16x^2 - 121$

2. $9t^2 - 24t + 16$

3. $5x^2 - 13x - 6$

4. $x(x - 10) - 4(x - 10)$

5. $4y^3 + 4y^2 - 3y$

6. $4x^3 - 12x^2 + 16x$

In Exercises 7–10, solve the given equation.

7. $5y + 6 = 0$

8. $2s - 7 = 0$

9. $(x + 9)(2x - 15) = 0$

10. $(5x - 8)(x - 12) = 0$

6.1 EXERCISES

In Exercises 1–20, solve the equation by factoring.

1. $4x^2 - 12x = 0$

2. $25y^2 - 75y = 0$

3. $u(u - 9) - 12(u - 9) = 0$

4. $16x(x - 8) - 12(x - 8) = 0$

5. $4x^2 - 25 = 0$

6. $16y^2 - 121 = 0$

7. $x^2 - 12x + 36 = 0$

8. $9x^2 + 24x + 16 = 0$

9. $x^2 + 60x + 900 = 0$

10. $8x^2 - 10x + 3 = 0$

11. $(y - 4)(y - 3) = 6$

12. $(6 + u)(1 - u) = 10$

13. $2x(3x + 2) = 5 - 6x^2$

14. $(2z + 1)(2z - 1) = -4z^2 - 5z + 2$

15. $3x(x - 6) - 5(x - 6) = 0$

16. $3(4 - x) - 2x(4 - x) = 0$

17. $6x^2 = 54$

18. $5t^2 = 125$

19. $\frac{y^2}{2} = 32$

20. $\frac{x^2}{6} = 24$

In Exercises 21–36, solve the quadratic equation by extracting square roots.

21. $x^2 = 64$

22. $z^2 = 169$

23. $25x^2 = 16$

24. $9z^2 = 121$

25. $4u^2 - 225 = 0$

26. $16x^2 - 1 = 0$

27. $(x + 4)^2 = 169$

28. $(y - 20)^2 = 625$

29. $(x - 3)^2 = 0.25$

30. $(x + 2)^2 = 0.81$

31. $(x - 2)^2 = 7$

32. $(x + 8)^2 = 28$

33. $(2x + 1)^2 = 50$

34. $(3x - 5)^2 = 48$

35. $(4x - 3)^2 - 98 = 0$

36. $(5x + 11)^2 - 300 = 0$

In Exercises 37–50, solve the equation by extracting complex square roots.

37. $x^2 + 4 = 0$

38. $x^2 = -9$

39. $(t - 3)^2 = -25$

40. $(x + 5)^2 + 81 = 0$

41. $(3z + 4)^2 + 144 = 0$

42. $(2y - 3)^2 + 25 = 0$

43. $9(x + 6)^2 = -121$

44. $4(x - 4)^2 = -169$

45. $\left(c - \frac{2}{3}\right)^2 + \frac{1}{9} = 0$

46. $\left(u + \frac{5}{8}\right)^2 + \frac{49}{16} = 0$

47. $(x - 1)^2 = -27$

48. $(2x + 3)^2 = -54$

49. $\left(x + \frac{7}{3}\right)^2 = -\frac{38}{9}$

50. $\left(y - \frac{5}{6}\right)^2 = -\frac{4}{5}$

In Exercises 51–56, find all the real solutions of the given equation.

51. $x^4 - 20x^2 + 64 = 0$

52. $x^6 + 7x^3 - 8 = 0$

53. $3x^{2/3} + 8x^{1/3} - 3 = 0$

54. $2x - 9\sqrt{x} + 10 = 0$

55. $(x^2 - 2)^2 - 36 = 0$

56. $(u^2 + 4)^2 - 64 = 0$

In Exercises 57–66, solve the equation. (Find all the real *and* complex solutions.)

57. $x^2 - 100 = 0$

58. $y^2 - 400 = 0$

59. $(x - 5)^2 - 100 = 0$

60. $(y + 12)^2 - 400 = 0$

61. $(x - 5)^2 + 100 = 0$

62. $(y + 12)^2 + 400 = 0$

63. $(x^2 - 5)^2 - 100 = 0$

64. $(y^2 + 12)^2 - 400 = 0$

65. $x^4 - 16 = 0$

66. $y^4 - 81 = 0$

In Exercises 67–70, find a quadratic equation having the given solutions. (*Note:* An equation with solutions r_1 and r_2 is $(x - r_1)(x - r_2) = 0$.)

67. $5, -2$

68. $-2, \frac{1}{3}$

69. $1 + \sqrt{2}, 1 - \sqrt{2}$

70. $1 + \sqrt{2}i, 1 - \sqrt{2}i$

Figure for 71

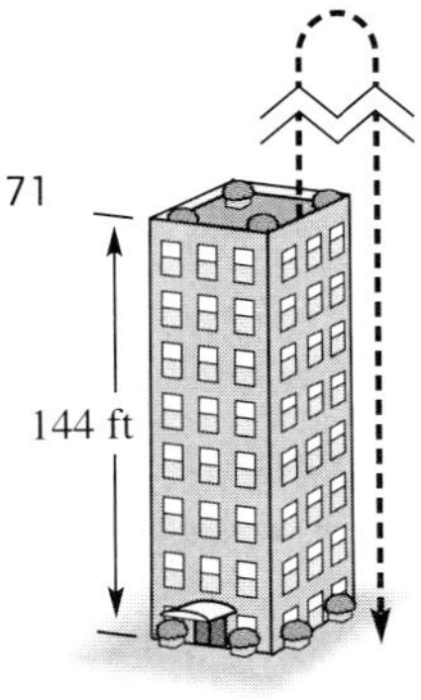

71. *Free-Falling Object* The height h (in feet) of an object propelled upward from a building 144 feet high is given by

$$h = 144 + 128t - 16t^2$$

where t measures the time in seconds from when the object is released. How long does it take for the object to reach the ground?

72. *Revenue* The revenue R (in dollars) when x units of a product are sold is given by

$$R = x\left(120 - \frac{1}{2}x\right).$$

Determine the number of units that must be sold to produce a revenue of $7,000.

Fire Loss In Exercises 73 and 74, use the following model, which gives the approximate amount of fire loss to private property in the United States between 1970 and 1988.

$$y = 1007.775 + 10.675t^2$$

In this model, y represents the value of private property lost to fire (in millions of dollars) and t represents the year, with $t = 0$ corresponding to 1970 (see figure). (*Source:* Insurance Information Institute.)

73. Which year had a private property loss of approximately \$2,075 million due to fire?

74. Which year had a private property loss of approximately \$3,740 million due to fire?

Figure for 73 and 74

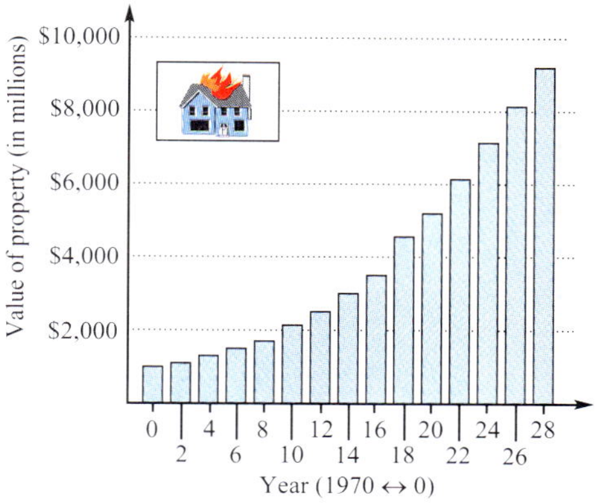

SECTION 6.2

Completing the Square

Constructing Perfect Square Trinomials • Solving Quadratic Equations by Completing the Square

Constructing Perfect Square Trinomials

Consider the quadratic equation

$$(x - 2)^2 = 10. \qquad \text{Completed square form}$$

We know from Example 5 of the previous section that this equation has two solutions: $2 + \sqrt{10}$ and $2 - \sqrt{10}$. Now, let's suppose that you were given the equation in its standard form

$$x^2 - 4x - 6 = 0. \qquad \text{Standard form}$$

How would you solve this equation if you were given only the standard form? We could try factoring, but after attempting to do that we would find that the left side of the equation is not factorable (using integer coefficients).

In this section we look at a technique for rewriting an equation so that the equation is in completed square form. We call this process **completing the square**. To complete the square, we must realize that all perfect square trinomials have a similar form. For instance, consider the following perfect square trinomials.

$$x^2 + 6x + 9 = (x + 3)^2$$
$$x^2 - 12x + 36 = (x - 6)^2$$
$$x^2 + 5x + \frac{25}{4} = \left(x + \frac{5}{2}\right)^2$$

In each case note that the constant term of the perfect square trinomial is the square of half the coefficient of the x-term. That is, a perfect square trinomial has the following form.

Perfect Square Trinomial		*Square of Binomial*
$x^2 + bx + \left(\frac{b}{2}\right)^2$	$=$	$\left(x + \frac{b}{2}\right)^2$

(half)^2

Thus, to complete the square for an expression of the form $x^2 + bx$, we must add $(b/2)^2$ to the expression.

Completing the Square To **complete the square** for the expression

$$x^2 + bx$$

we add $(b/2)^2$, which is the square of half the coefficient of x. Consequently,

$$x^2 + bx + \left(\frac{b}{2}\right)^2 = \left(x + \frac{b}{2}\right)^2.$$

EXAMPLE 1 ■ Constructing a Perfect Square Trinomial

What term should be added to the expression $x^2 - 8x$ so that it becomes a perfect square trinomial?

Solution

For this expression, the coefficient of the x-term is -8. By taking half of this and squaring the result, we see that $(-4)^2 = 16$ should be added to the expression to make it a perfect square trinomial. That is,

$$x^2 - 8x + 16 = x^2 - 2(4)x + (-4)^2 = (x - 4)^2.$$ ■

EXAMPLE 2 ■ Constructing Perfect Square Trinomials

What terms should be added to the following expressions so that each becomes a perfect square trinomial?

(a) $x^2 + 14x$ (b) $x^2 - 9x$

Solution

(a) For this expression, the coefficient of the x-term is 14. By taking half of this and squaring the result, we see that $7^2 = 49$ should be added to the expression to make it a perfect square trinomial. That is,

$$x^2 + 14x + 49 = x^2 + 2(7)x + 7^2 = (x + 7)^2.$$

(b) For this expression, the coefficient of the x-term is -9. By taking half of this and squaring the result, we see that $\left(-\frac{9}{2}\right)^2 = \frac{81}{4}$ should be added to the expression to make it a perfect square trinomial. That is,

$$x^2 - 9x + \frac{81}{4} = x^2 - 2\left(\frac{9}{2}\right)x + \left(-\frac{9}{2}\right)^2 = \left(x - \frac{9}{2}\right)^2.$$

■

Solving Quadratic Equations by Completing the Square

We now look at how completing the square can be used to solve a quadratic equation. When using this procedure with an equation, remember that it is essential to *preserve the equality*. Thus, when you add a constant term to one side of the equation, you must be sure to add the same constant to the other side of the equation.

EXAMPLE 3 ■ **Completing the Square: Leading Coefficient is 1**

$x^2 + 12x = 0$ — Given equation

$x^2 + 12x + (6)^2 = 36$ — Add $\left(\frac{12}{2}\right)^2 = 36$ to *both* sides

(half)^2

$(x + 6)^2 = 36$ — Binomial squared

$x + 6 = \pm\sqrt{36}$ — Extract square roots

$x = -6 \pm 6$ — Solve for x

$x = 0$ or $x = -12$ — Solutions

Thus, this equation has two solutions: 0 and -12. Check these solutions in the original equation. ■

In Example 3 we used completing the square to solve the quadratic equation simply for the sake of illustration. This particular equation would be easier to solve by factoring. Try reworking the problem by factoring to see that you obtain the same two solutions. In the next example, we look at an equation that cannot be solved by factoring (using integer coefficients).

EXAMPLE 4 ■ **Completing the Square: Leading Coefficient is 1**

$x^2 - 6x + 7 = 0$ — Given equation

$x^2 - 6x = -7$ — Subtract 7 from both sides

$x^2 - 6x + (-3)^2 = -7 + 9$ — Add $(-3)^2 = 9$ to *both* sides

(half)^2

$$(x - 3)^2 = 2 \qquad \text{Binomial squared}$$

$$x - 3 = \pm\sqrt{2} \qquad \text{Extract square roots}$$

$$x = 3 \pm \sqrt{2} \qquad \text{Solve for } x$$

Thus, this equation has two solutions: $3 + \sqrt{2}$ and $3 - \sqrt{2}$. Check these solutions in the original equation. ■

If the leading coefficient of a quadratic expression is not 1, we must divide both sides of the equation by this coefficient *before* completing the square, as shown in the following example.

EXAMPLE 5 ■ Completing the Square: Leading Coefficient is not 1

Solve the following equation by completing the square.

$$3x^2 + 5x = 2$$

Solution

$$3x^2 + 5x = 2 \qquad \text{Given equation}$$

$$x^2 + \frac{5}{3}x = \frac{2}{3} \qquad \text{Divide both sides by 3}$$

$$x^2 + \frac{5}{3}x + \left(\frac{5}{6}\right)^2 = \frac{2}{3} + \frac{25}{36} \qquad \text{Add } \left(\tfrac{5}{6}\right)^2 = \tfrac{25}{36} \text{ to } \textit{both} \text{ sides}$$

(half)^2

$$\left(x + \frac{5}{6}\right)^2 = \frac{49}{36} \qquad \text{Binomial squared}$$

$$x + \frac{5}{6} = \pm\frac{7}{6} \qquad \text{Extract square roots}$$

$$x = -\frac{5}{6} \pm \frac{7}{6} \qquad \text{Subtract } \tfrac{5}{6} \text{ from both sides}$$

$$x = \frac{1}{3} \text{ or } x = -2 \qquad \text{Solutions}$$

Thus, this equation has two solutions: $\frac{1}{3}$ and -2. Check these solutions in the original equation. ■

NOTE: If you solve a quadratic equation by completing the square and obtain solutions that do not involve radicals, then you could have solved the equation by factoring. For instance, in Example 5 verify that the equation could have been factored as

$$(3x - 1)(x + 2) = 0.$$

In the next example, we use completing the square to solve a quadratic equation that cannot be solved by factoring (using integer coefficients).

EXAMPLE 6 ■ **Completing the Square: Leading Coefficient is not 1**

$$2x^2 - x - 2 = 0 \qquad \text{Given equation}$$

$$2x^2 - x = 2 \qquad \text{Add 2 to both sides}$$

$$x^2 - \frac{1}{2}x = 1 \qquad \text{Divide both sides by 2}$$

$$x^2 - \frac{1}{2}x + \left(-\frac{1}{4}\right)^2 = 1 + \frac{1}{16} \qquad \text{Add } \left(-\tfrac{1}{4}\right)^2 = \tfrac{1}{16} \text{ to } \textit{both} \text{ sides}$$

(half)^2

$$\left(x - \frac{1}{4}\right)^2 = \frac{17}{16} \qquad \text{Binomial squared}$$

$$x - \frac{1}{4} = \pm\frac{\sqrt{17}}{4} \qquad \text{Extract square roots}$$

$$x = \frac{1}{4} \pm \frac{\sqrt{17}}{4} \qquad \text{Solve for } x$$

Thus, this equation has two solutions:

$$x = \frac{1}{4} + \frac{\sqrt{17}}{4} = \frac{1 + \sqrt{17}}{4}$$

and

$$x = \frac{1}{4} - \frac{\sqrt{17}}{4} = \frac{1 - \sqrt{17}}{4}.$$

Check these solutions in the original equation. ■

EXAMPLE 7 ■ **Completing the Square**

Solve the following quadratic equation and use a calculator to estimate the solutions to two decimal places.

$$3x^2 - 5x - 4 = 0$$

Solution

$$3x^2 - 5x - 4 = 0 \qquad \text{Given equation}$$

$$3x^2 - 5x = 4 \qquad \text{Add 4 to both sides}$$

$$x^2 - \frac{5}{3}x = \frac{4}{3} \qquad \text{Divide both sides by 3}$$

$$x^2 - \frac{5}{3}x + \left(-\frac{5}{6}\right)^2 = \frac{4}{3} + \frac{25}{36} \qquad \text{Add } \left(-\tfrac{5}{6}\right)^2 = \tfrac{25}{36} \text{ to } \textit{both} \text{ sides}$$

(half)^2

$$\left(x - \frac{5}{6}\right)^2 = \frac{73}{36} \qquad \text{Binomial squared}$$

$$x - \frac{5}{6} = \pm\frac{\sqrt{73}}{6} \qquad \text{Extract square roots}$$

$$x = \frac{5}{6} \pm \frac{\sqrt{73}}{6} \qquad \text{Solve for } x$$

Thus, this equation has two solutions:

$$x = \frac{5}{6} + \frac{\sqrt{73}}{6} = \frac{5 + \sqrt{73}}{6} \approx 2.26$$

and

$$x = \frac{5}{6} - \frac{\sqrt{73}}{6} = \frac{5 - \sqrt{73}}{6} \approx -0.59.$$

Check these solutions in the original equation. ■

The method of completing the square can be used to solve *any* quadratic equation, including equations that have complex solutions. We demonstrate this procedure in the next example.

EXAMPLE 8 ■ A Quadratic Equation with Complex Solutions

Use completing the square to solve the equation $x^2 - 4x + 8 = 0$.

Solution

We begin by writing the equation in completed square form, as follows.

$$x^2 - 4x + 8 = 0 \qquad \text{Given equation}$$

$$x^2 - 4x = -8 \qquad \text{Subtract 8 from both sides}$$

$$x^2 - 4x + (-2)^2 = -8 + 4 \qquad \text{Add } (-2)^2 = 4 \text{ to } \textit{both} \text{ sides}$$

$(\text{half})^2$

$$(x - 2)^2 = -4 \qquad \text{Binomial squared}$$

$$x - 2 = \pm 2i \qquad \text{Extract complex square roots}$$

$$x = 2 \pm 2i \qquad \text{Solve for } x$$

Thus, the given equation has two solutions: $2 + 2i$ and $2 - 2i$.

Check

$$x^2 - 4x + 8 = 0 \qquad \text{Given equation}$$

$$(2 + 2i)^2 - 4(2 + 2i) + 8 \stackrel{?}{=} 0 \qquad \text{Replace } x \text{ by } 2 + 2i$$

$$4 + 8i - 4 - 8 - 8i + 8 \stackrel{?}{=} 0$$

$$0 = 0 \qquad \text{Solution checks}$$

Check the other solution in the original equation. ■

DISCUSSION PROBLEM ■ Solving a Quadratic Equation in Completed Square Form

Consider the following quadratic equation.

$$(x - 1)^2 = d$$

(a) What value or values of d will produce a quadratic equation that has exactly one (repeated) solution?

(b) Describe the values of d that will produce two different solutions, both of which are *rational* numbers.

(c) Describe the values of d that will produce two different solutions, both of which are *irrational* numbers.

(d) Describe the values of d that will produce two different solutions, both of which are *complex* (nonreal) numbers. ■

Warm-Up

The following warm-up exercises involve skills that were covered in earlier sections. You will use these skills in the exercise set for this section.

In Exercises 1–4, expand and simplify the expression.

1. $(x + 3)^2 - 1$

2. $(x + 10)^2 + 50$

3. $\left(u - \frac{1}{2}\right)^2$

4. $(v - 12)^2 - 100$

In Exercises 5–10, solve the equation. (Find all real *and* complex solutions.)

5. $x^2 = \frac{4}{49}$

6. $y^2 = -\frac{9}{16}$

7. $(y - 4)^2 = -36$

8. $(z + 2)^2 = 20$

9. $(3x - 2)^2 = 5$

10. $3(12 - x) - 8x(12 - x) = 0$

6.2 EXERCISES

In Exercises 1–10, determine the constant that must be added to the expression so that it becomes a perfect square trinomial.

1. $x^2 + 8x + \square$

2. $x^2 + 12x + \square$

3. $y^2 - 20y + \square$

4. $y^2 - 2y + \square$

5. $t^2 + 5t + \square$

6. $u^2 + 7u + \square$

7. $x^2 - \frac{6}{5}x + \square$

8. $y^2 + \frac{4}{3}y + \square$

9. $y^2 - \frac{3}{5}y + \square$

10. $a^2 - \frac{1}{3}a + \square$

In Exercises 11–20, solve the quadratic equation by (a) completing the square, and (b) factoring.

11. $x^2 - 25x = 0$

12. $x^2 + 32x = 0$

13. $t^2 - 8t + 7 = 0$

14. $y^2 - 8y + 12 = 0$

15. $x^2 + 2x - 24 = 0$

16. $x^2 + 12x + 27 = 0$

17. $x^2 + 7x + 12 = 0$

18. $z^2 + 3z - 10 = 0$

19. $x^2 - 3x - 18 = 0$

20. $t^2 - 5t - 36 = 0$

In Exercises 21–50, solve the quadratic equation by the method of completing the square. Give the solutions in exact form and in decimal form rounded to two decimal places. (The solutions to some equations are complex numbers.)

21. $x^2 - 4x - 3 = 0$

22. $x^2 - 6x + 7 = 0$

23. $x^2 + 4x - 3 = 0$

24. $x^2 + 6x + 7 = 0$

25. $u^2 - 4u + 1 = 0$

26. $a^2 - 10a - 15 = 0$

27. $x^2 + 2x + 3 = 0$

28. $x^2 - 6x + 12 = 0$

29. $x^2 - 10x - 2 = 0$

30. $x^2 + 8x - 4 = 0$

31. $y^2 + 20y + 10 = 0$

32. $y^2 + 6y - 24 = 0$

33. $x^2 - \frac{2}{3}x - 3 = 0$

34. $x^2 + \frac{4}{5}x - 1 = 0$

35. $t^2 + 5t + 3 = 0$

36. $u^2 - 9u - 1 = 0$

37. $v^2 + 3v - 2 = 0$

38. $z^2 - 7z + 9 = 0$

39. $-x^2 + x - 1 = 0$

40. $1 - x - x^2 = 0$

41. $2x^2 + 8x + 3 = 0$

42. $3x^2 - 24x - 5 = 0$

43. $3x^2 + 9x + 5 = 0$

44. $5x^2 - 15x + 7 = 0$

45. $4y^2 + 4y - 9 = 0$

46. $4z^2 - 3z + 2 = 0$

47. $0.1x^2 + 0.5x + 0.2 = 0$

48. $0.02x^2 + 0.10x - 0.05 = 0$

49. $x(x - 7) = 2$

50. $2x\left(x + \frac{4}{3}\right) = 5$

In Exercises 51–54, solve the equation and check your results.

51. $\frac{x}{2} - \frac{1}{x} = 1$

52. $\frac{x}{2} + \frac{5}{x} = 4$

53. $\sqrt{2x + 1} = x - 3$

54. $\sqrt{3x - 2} = x - 2$

55. *Completing the Square*

(a) Find the area of the two adjoining rectangles and the square in the accompanying figure.

(b) Find the area of the small square region that is missing in the lower right-hand corner of the figure and add it to the area found in part (a).

(c) Find the dimensions and the area of the entire figure after adjoining the small square in the lower right-hand corner of the given figure. Note that you have demonstrated how to complete the square geometrically.

Figure for 55

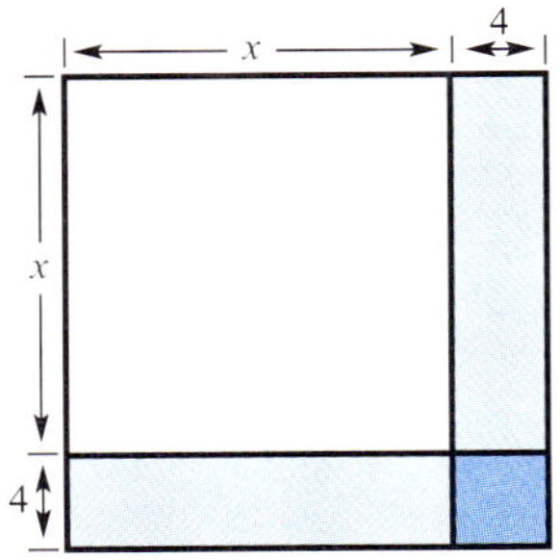

56. *Completing the Square* Use the model given in Exercise 55 to geometrically represent completing the square for $x^2 + 6x$.

57. *Revenue* The revenue R for selling x units of a certain product is given by

$$R = x\left(50 - \frac{1}{2}x\right).$$

Find the number of units that must be sold to produce a revenue of \$1,218.

58. *Revenue* The revenue R for selling x units of a certain product is given by

$$R = x\left(100 - \frac{1}{10}x\right).$$

Find the number of units that must be sold to produce a revenue of \$12,000.

SECTION 6.3 The Quadratic Formula and the Discriminant

The Quadratic Formula • Solving Quadratic Equations by the Quadratic Formula • The Discriminant

The Quadratic Formula

In Section 6.2 we solved quadratic equations by completing the square for *each problem separately*. Perhaps you have thought of the possibility of completing the square *once* for the general quadratic equation $ax^2 + bx + c = 0$, and producing a solution model for solving any quadratic equation. This is precisely what is done in this section to develop what is called the **Quadratic Formula**.

$$ax^2 + bx + c = 0 \qquad \text{Standard form, } a \neq 0$$

$$ax^2 + bx = -c \qquad \text{Subtract } c \text{ from both sides}$$

$$x^2 + \frac{b}{a}x = -\frac{c}{a} \qquad \text{Divide both sides by } a$$

$$x^2 + \frac{b}{a}x + \left(\frac{b}{2a}\right)^2 = -\frac{c}{a} + \left(\frac{b}{2a}\right)^2 \qquad \text{Complete the square}$$

(half)^2

$$\left(x + \frac{b}{2a}\right)^2 = \frac{b^2 - 4ac}{4a^2} \qquad \text{Simplify}$$

$$x + \frac{b}{2a} = \pm\sqrt{\frac{b^2 - 4ac}{4a^2}} \qquad \text{Extract square roots}$$

$$x = -\frac{b}{2a} \pm \frac{\sqrt{b^2 - 4ac}}{2|a|}$$

$$x = \frac{-b \pm \sqrt{b^2 - 4ac}}{2a} \qquad \text{Solutions}$$

Note that since $\pm 2|a|$ represents the same numbers as $\pm 2a$, we can omit the absolute value sign.

The Quadratic Formula The solutions of a quadratic equation in the standard form

$$ax^2 + bx + c = 0, \quad a \neq 0$$

are given by the **Quadratic Formula**

$$x = \frac{-b \pm \sqrt{b^2 - 4ac}}{2a}.$$

The Quadratic Formula is one of the most important formulas in algebra, and you should memorize it. We have found that it helps to try to memorize a verbal statement of the rule. For instance, you might try to remember the following verbal statement of the Quadratic Formula: "Minus b, plus or minus the square root of b squared minus $4ac$, all divided by $2a$."

Solving Quadratic Equations by the Quadratic Formula

When using the Quadratic Formula, remember that *before* the formula can be applied, you must first write the quadratic equation in standard form.

EXAMPLE 1 ■ **Using the Quadratic Formula: Two Distinct Solutions**

Use the Quadratic Formula to solve the equation $x^2 + 6x = 16$.

Solution

To begin, we write the equation in standard form, $ax^2 + bx + c = 0$. Then we determine the values of a, b, and c. Finally, we substitute these values into the Quadratic Formula to obtain the solutions.

$$x^2 + 6x = 16 \quad \text{Given equation}$$

$$x^2 + 6x - 16 = 0 \quad \text{Standard form with } a = 1,\ b = 6,\ c = -16$$

$$x = \frac{-b \pm \sqrt{b^2 - 4ac}}{2a} \quad \text{Quadratic Formula}$$

$$x = \frac{-6 \pm \sqrt{6^2 - 4(1)(-16)}}{2(1)} \quad \text{Substitute}$$

$$x = \frac{-6 \pm \sqrt{36 + 64}}{2}$$

$$x = \frac{-6 \pm \sqrt{100}}{2}$$

$$x = \frac{-6 \pm 10}{2}$$

$$x = 2 \quad \text{or} \quad x = -8 \quad \text{Solutions}$$

Therefore, the equation has two solutions: 2 and -8. Check these solutions in the original equation. (Because the simplified versions of the solutions have no radicals, we know that the equation could have been solved by factoring. Try solving the equation that way.) ■

EXAMPLE 2 ■ **Using the Quadratic Formula: Two Distinct Solutions**

Use the Quadratic Formula to solve the following equation.

$$-x^2 - 4x + 8 = 0$$

Solution

This equation is already written in standard form, $ax^2 + bx + c = 0$. However, the leading coefficient is negative. We find that it is cumbersome to have negative leading coefficients, and we suggest multiplying both sides of the equation by -1. (This will produce a positive leading coefficient.)

$-x^2 - 4x + 8 = 0$ Leading coefficient is negative

$x^2 + 4x - 8 = 0$ Standard form with $a = 1$, $b = 4$, $c = -8$

$$x = \frac{-b \pm \sqrt{b^2 - 4ac}}{2a} \quad \text{Quadratic Formula}$$

$$x = \frac{-4 \pm \sqrt{4^2 - 4(1)(-8)}}{2(1)} \quad \text{Substitute}$$

$$x = \frac{-4 \pm \sqrt{16 + 32}}{2}$$

$$x = \frac{-4 \pm \sqrt{48}}{2}$$

$$x = \frac{-4 \pm 4\sqrt{3}}{2} \quad \text{Simplify radical}$$

$$x = \frac{\cancel{2}(-2 \pm 2\sqrt{3})}{\cancel{2}} \quad \text{Cancel common factor}$$

$$x = -2 \pm 2\sqrt{3} \quad \text{Solutions}$$

Therefore, the equation has two solutions: $-2 + 2\sqrt{3}$ and $-2 - 2\sqrt{3}$. Check these solutions in the original equation. ■

EXAMPLE 3 ■ **Using the Quadratic Formula: One Repeated Solution**

Use the Quadratic Formula to solve the following equation.

$$18x^2 - 24x + 8 = 0$$

Solution

We note that this equation has a common factor of 2. To simplify things, we first divide both sides of the equation by 2.

$18x^2 - 24x + 8 = 0$ Common factor of 2

$9x^2 - 12x + 4 = 0$ Standard form with $a = 9$, $b = -12$, $c = 4$

$$x = \frac{-b \pm \sqrt{b^2 - 4ac}}{2a} \quad \text{Quadratic Formula}$$

$$x = \frac{-(-12) \pm \sqrt{(-12)^2 - 4(9)(4)}}{2(9)} \quad \text{Substitute}$$

$$x = \frac{12 \pm \sqrt{144 - 144}}{18}$$

$$x = \frac{12 \pm \sqrt{0}}{18} = \frac{2}{3} \quad \text{Repeated solution}$$

Therefore, this quadratic equation has only one solution: $\frac{2}{3}$. Check this solution in the original equation. ■

Note in the next example how the Quadratic Formula can be used to solve a quadratic equation that has complex solutions.

EXAMPLE 4 ■ Using the Quadratic Formula: Complex Solutions

Use the Quadratic Formula to solve the following equation.

$$2x^2 - 4x + 5 = 0$$

Solution

$2x^2 - 4x + 5 = 0$ Standard form with $a = 2$, $b = -4$, $c = 5$

$$x = \frac{-b \pm \sqrt{b^2 - 4ac}}{2a} \quad \text{Quadratic formula}$$

$$x = \frac{-(-4) \pm \sqrt{(-4)^2 - 4(2)(5)}}{2(2)} \quad \text{Substitute}$$

$$x = \frac{4 \pm \sqrt{16 - 40}}{4}$$

$$x = \frac{4 \pm \sqrt{-24}}{4}$$

$$x = \frac{4 \pm 2\sqrt{6}i}{4}$$

$$x = \frac{\cancel{2}(2 \pm \sqrt{6}i)}{\cancel{2} \cdot 2} \quad \text{Cancel common factor of 2}$$

$$x = \frac{2 \pm \sqrt{6}i}{2}$$ Solutions

Thus, the given equation has two complex solutions: $(2 + \sqrt{6}i)/2$ and $(2 - \sqrt{6}i)/2$. Check these solutions in the original equation. ■

MATH MATTERS

Olympic Games

As this book is being written, the twenty-fifth modern Summer Olympic Games are scheduled to be held in Barcelona, Spain, in the summer of 1992. The original Olympic Games were held once every four years in ancient Greece. The earliest recorded games took place in 776 B.C., and the games were abandoned in A.D. 393.

In 1896 the games were revived. Since then, they have taken place every four years, except during the two World Wars. Note that the three games that were not held had already been planned, and therefore kept their planned numbers.

The number of participating countries has tended to increase each year. Do you know why the number of participants was down in 1976 and 1980?

No.	Year	City	Participants
1	1896	Athens, Greece	13
2	1900	Paris, France	22
3	1904	St. Louis, USA	13
4	1908	London, England	22
5	1912	Stockholm, Sweden	28
6	1916	Berlin, Germany	—
7	1920	Antwerp, Belgium	29
8	1924	Paris, France	44
9	1928	Amsterdam, Holland	46
10	1932	Los Angeles, USA	37
11	1936	Berlin, Germany	49
12	1940	Tokyo, Japan	—
13	1944	London, England	—
14	1948	London, England	59
15	1952	Helsinki, Finland	69
16	1956	Melbourne, Australia	67
17	1960	Rome, Italy	83
18	1964	Tokyo, Japan	93
19	1968	Mexico City, Mexico	112
20	1972	Munich, West Germany	122
21	1976	Montreal, Canada	92
22	1980	Moscow, USSR	81
23	1984	Los Angeles, USA	140
24	1988	Seoul, South Korea	160
25	1992	Barcelona, Spain	

The Discriminant

By looking back at the radicand, $b^2 - 4ac$, in the solutions of the equations in Examples 1–4, we can see that its value determines whether the solutions are rational, irrational, repeated, or complex. For this reason, the radicand $b^2 - 4ac$ in the Quadratic Formula is called the **discriminant**. We summarize the use of the discriminant as follows.

Using the Discriminant

Let a, b, and c be rational numbers such that $a \neq 0$. The **discriminant** of the quadratic equation $ax^2 + bx + c = 0$ is given by $b^2 - 4ac$, and can be used to classify the solutions of the equation as follows.

Discriminant	*Solution Types*
1. $b^2 - 4ac =$ perfect square	Two distinct rational solutions (Example 1)
2. $b^2 - 4ac =$ positive nonperfect square	Two distinct irrational solutions (Example 2)
3. $b^2 - 4ac = 0$	One rational repeated solution (Example 3)
4. $b^2 - 4ac =$ negative number	Two distinct complex solutions (Example 4)

By reexamining Examples 1–4, we can see that the equations with rational or repeated solutions could have been solved by *factoring*. In general, quadratic equations (with integer coefficients) for which the discriminant is either zero or a perfect square are always factorable (using integer coefficients). Consequently, a quick test of the discriminant can be used to help us decide which solution method to use to solve a quadratic equation.

EXAMPLE 5 ■ Using the Discriminant

Use the discriminant to determine the type of solutions for each of the following. Which of the equations can be solved by factoring using integer coefficients?

(a) $x^2 - x + 2 = 0$ (b) $2x^2 - 3x - 2 = 0$

(c) $x^2 - 2x + 1 = 0$ (d) $x^2 - 2x - 1 = 9$

Solution

(a) For this equation, we have $a = 1$, $b = -1$, and $c = 2$, which yields a discriminant of

$$b^2 - 4ac = (-1)^2 - 4(1)(2) = 1 - 8 = -7.$$

Therefore, the equation has two distinct complex solutions. The quadratic cannot be factored using integer coefficients.

(b) For this equation, we have $a = 2$, $b = -3$, and $c = -2$, which yields a discriminant of

$$b^2 - 4ac = (-3)^2 - 4(2)(-2) = 9 + 16 = 25.$$

Therefore, the equation has two distinct rational solutions. The quadratic can be factored as $(x - 2)(2x + 1)$.

(c) For this equation, we have $a = 1$, $b = -2$, and $c = 1$, which yields a discriminant of

$$b^2 - 4ac = (-2)^2 - 4(1)(1) = 4 - 4 = 0.$$

Therefore, the equation has one repeated rational solution. The quadratic can be factored as $(x - 1)^2$.

(d) For this equation, we have $a = 1$, $b = -2$, and $c = -1$, which yields a discriminant of

$$b^2 - 4ac = (-2)^2 - 4(1)(-1) = 4 + 4 = 8.$$

Therefore, the equation has two distinct irrational solutions. The quadratic cannot be factored using integer coefficients. ■

We have now looked at four ways to solve quadratic equations.

1. Factoring
2. Extracting square roots
3. Completing the square
4. The Quadratic Formula

Of these four methods, we suggest that you first check to see whether the equation is in a form in which you can extract square roots. For instance, it is easy to see that the solutions of $x^2 = 10$ are $x = \pm\sqrt{10}$. Next, we suggest that you try factoring. For instance, by rewriting the equation $2x^2 - x - 1$ in the factored form $(2x + 1)(x - 1) = 0$, we can see that it has two solutions: $-\frac{1}{2}$ and 1. If neither of these two methods work, we suggest using the Quadratic Formula (or completing the square); they will work for any quadratic equation.

One type of quadratic equation for which the Quadratic Formula is useful is an equation whose coefficients are decimals. We illustrate the use of the Quadratic Formula for this type of equation in Example 6.

EXAMPLE 6 ■ Using a Calculator with the Quadratic Formula

Use the Quadratic Formula and a calculator to solve the following equation.

$$1.2x^2 - 17.8x + 8.05 = 0$$

Solution

This quadratic equation is already in standard form. From this form, we see that $a = 1.2$, $b = -17.8$, and $c = 8.05$. Therefore, we can use the Quadratic Formula to solve the equation as follows.

$$1.2x^2 - 17.8x + 8.05 = 0$$

$$x = \frac{-b \pm \sqrt{b^2 - 4ac}}{2a}$$

$$x = \frac{-(-17.8) \pm \sqrt{(-17.8)^2 - 4(1.2)(8.05)}}{2(1.2)}$$

To evaluate these solutions, we begin by calculating the square root, which we *store* for later use.*

17.8 [+/−] [x^2] [−] 4 [×] 1.2 [×] 8.05 [=] [√] — Scientific

[√] [(] [(] [(−)] 17.8 [)] [x^2] [−] 4 [×] 1.2 [×] 8.05 [)] [ENTER] — Graphing

The display for either of these keystroke sequences should be 16.67932852. Storing this result and using the recall key, we find the following two solutions.

$$x \approx \frac{17.8 + 16.67932852}{2.4} \approx 14.366 \quad \text{Add stored value}$$

$$x \approx \frac{17.8 - 16.67932852}{2.4} \approx 0.467 \quad \text{Subtract stored value}$$

DISCUSSION PROBLEM Solutions of a Quadratic Equation

Solve the given quadratic equations and complete the following table.

Equation	*Solutions:* x_1, x_2	*Sum:* $x_1 + x_2$	*Product:* x_1x_2
(a) $x^2 - x - 6 = 0$			
(b) $2x^2 + 5x - 3 = 0$			
(c) $4x^2 - 9 = 0$			
(d) $x^2 - 10x + 34 = 0$			

Consider a general quadratic equation $ax^2 + bx + c = 0$ whose solutions are x_1 and x_2. Can you determine a relationship among the coefficients a, b, and c, and the sum, $x_1 + x_2$, and the product, x_1x_2, of the solutions?

*The graphing calculator keystrokes given in this text correspond to the *TI-81* graphing calculator from *Texas Instruments*. For other graphing calculators, the keystrokes may differ.

Warm-Up

The following warm-up exercises involve skills that were covered in earlier sections. You will use these skills in the exercise set for this section.

In Exercises 1–4, solve the quadratic equation by factoring.

1. $x^2 - 10x + 21 = 0$

2. $x^2 + 14x - 32 = 0$

3. $3x^2 + x - 30 = 0$

4. $5x^2 - 18x + 9 = 0$

In Exercises 5 and 6, solve the quadratic equation by completing the square. (Find all real *and* complex solutions.)

5. $x^2 + 3x - 2 = 0$

6. $3x^2 + 6x + 4 = 0$

In Exercises 7–10, simplify the radical.

7. $\sqrt{25 - 4(2)(2)}$

8. $\sqrt{9 - 4(1)(5)}$

9. $\sqrt{64 - 4(2)(-4)}$

10. $\sqrt{81 - 4(3)(-6)}$

6.3 EXERCISES

In Exercises 1–4, write the quadratic equation in standard form.

1. $2x^2 = 7 - 2x$

2. $7x^2 + 15x = 5$

3. $x(10 - x) = 5$

4. $x(3x + 8) = 15$

In Exercises 5–14, solve the quadratic equation by (a) using the Quadratic Formula, and (b) factoring.

5. $x^2 - 11x + 28 = 0$

6. $x^2 - 12x + 27 = 0$

7. $x^2 + 6x + 8 = 0$

8. $x^2 + 9x + 14 = 0$

9. $x^2 - \frac{4}{3}x + \frac{4}{9} = 0$

10. $x^2 + x + \frac{1}{4} = 0$

11. $4x^2 + 12x + 9 = 0$

12. $9x^2 - 30x + 25 = 0$

13. $6x^2 - x - 2 = 0$

14. $10x^2 - 11x + 3 = 0$

In Exercises 15–20, use the discriminant to determine the type of solutions to the quadratic equation.

15. $2x^2 - 5x - 4 = 0$

16. $10x^2 + 5x + 1 = 0$

17. $x^2 + 7x + 15 = 0$

18. $3x^2 - 2x - 5 = 0$

19. $4x^2 - 12x + 9 = 0$

20. $2x^2 + 10x + 6 = 0$

In Exercises 21–40, solve the quadratic equation by using the Quadratic Formula. (Find all real *and* complex solutions.)

21. $x^2 - 2x - 4 = 0$
22. $x^2 - 2x - 6 = 0$
23. $t^2 + 4t + 1 = 0$
24. $y^2 + 6y + 4 = 0$
25. $x^2 + 6x - 3 = 0$
26. $x^2 + 8x - 4 = 0$
27. $x^2 - 10x + 23 = 0$
28. $u^2 - 12u + 29 = 0$
29. $x^2 + 3x + 3 = 0$
30. $2x^2 - x + 1 = 0$
31. $2v^2 - 2v - 1 = 0$
32. $4x^2 + 6x + 1 = 0$
33. $2x^2 + 4x - 3 = 0$
34. $2x^2 + 3x + 3 = 0$
35. $9z^2 + 6z - 4 = 0$
36. $8a^2 - 8a - 1 = 0$
37. $x^2 - 0.4x - 0.16 = 0$
38. $x^2 + 0.6x - 0.41 = 0$
39. $2.5x^2 + x - 0.9 = 0$
40. $0.09x^2 - 0.12x - 0.26 = 0$

In Exercises 41–50, solve the quadratic equation by the most convenient method. (Find all real *and* complex solutions.)

41. $z^2 - 169 = 0$
42. $t^2 = 150$
43. $y^2 + 15y = 0$
44. $2y(y - 18) + 3(y - 18) = 0$
45. $25(x - 3)^2 - 36 = 0$
46. $4u^2 + 49 = 0$
47. $x^2 - 24x + 128 = 0$
48. $1.2x^2 - 0.8x - 5.5 = 0$
49. $x^2 + 8x + 25 = 0$
50. $2x^2 + 8x + 4.5 = 0$

In Exercises 51–54, use a calculator to solve the equation. (Round your answer to three decimal places.)

51. $5x^2 - 18x + 6 = 0$
52. $15x^2 + 3x - 105 = 0$
53. $-0.04x^2 + 4x - 0.8 = 0$
54. $3.7x^2 - 10.2x + 3.2 = 0$

In Exercises 55–58, solve the given equation. (Be sure to check your solutions.)

55. $\frac{2x^2}{5} - \frac{x}{2} = 1$
56. $\frac{x}{3} + \frac{1}{x} = 4$
57. $\sqrt{x + 3} = x - 1$
58. $\sqrt{2x - 3} = x - 2$

59. *Free-Falling Object* A ball is thrown upward at a velocity of 40 feet per second from a bridge that is 50 feet above the level of the water (see figure). The height h (in feet) of the ball at time t seconds after it is thrown is given by

$$h = -16t^2 + 40t + 50.$$

(a) Find the time when the ball is again at the 50-foot level above the water.
(b) Find the time when the ball strikes the water.

Figure for 59

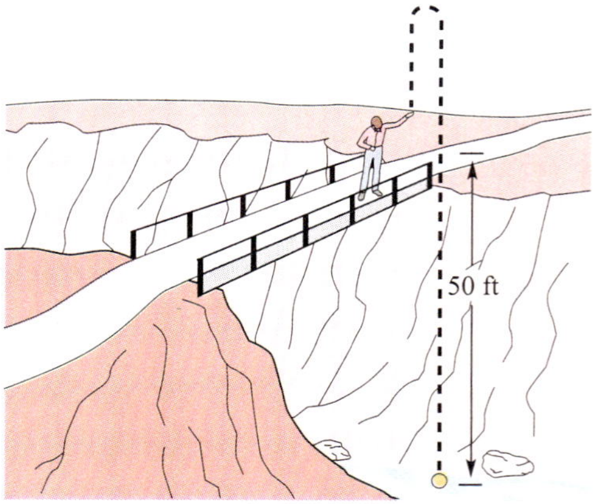

60. *Number Problem* Find two consecutive positive even integers whose product is 288.

Practicing Lawyers In Exercises 61 and 62, use the following equation, which gives the approximate number of practicing lawyers in the United States between 1960 and 1985 (see figure).

$$y = 273{,}063 + 2{,}225t + 471t^2$$

In this model, y represents the number of practicing lawyers and t represents the year, with $t = 0$ corresponding to 1960. (*Source:* American Bar Foundation.)

61. Find the year in which there were approximately 321,000 practicing lawyers in the United States.

62. Find the year in which there were approximately 527,500 practicing lawyers in the United States.

Figure for 61 and 62

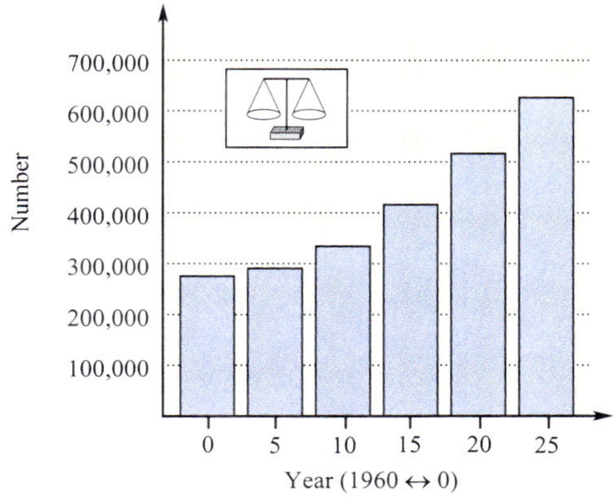

SECTION 6.4

Applications of Quadratic Equations

Review of Problem-Solving Strategy • Applications of Quadratic Equations

Review of Problem-Solving Strategy

We have completed our study of the technical parts of the algebra of quadratic equations. In this section we look at some practical problems involving quadratic equations. To construct these equations from verbal problems, we again rely on the problem-solving strategy developed in Chapter 2. For convenience, we review the steps involved in this strategy, as follows.

Strategy for Solving Word Problems

To solve a word problem, we suggest the following steps.

1. *Search* for the hidden equality—two expressions said to be equal or known to be equal. A sketch may be helpful.
2. *Write* a verbal model that equates these two expressions.
3. *Assign* numbers to the known quantities, and letters (or algebraic expressions) to the variable quantities.
4. *Rewrite* the verbal model as an algebraic equation using the assigned labels.
5. *Solve* the resulting algebraic equation.
6. *Check* to see that the answer satisfies the word problem as stated.

For word problems that involve quadratic equations, the verbal model is often a *product* of two variable quantities. On other occasions, the model may be a previously known formula that describes the situation. Watch for these variations in the examples that follow.

Applications of Quadratic Equations

EXAMPLE 1 ■ An Application Involving Area

A picture is 4 inches taller than it is wide and has an area of 192 square inches, as shown in Figure 6.1. What are the outer dimensions of the frame?

FIGURE 6.1

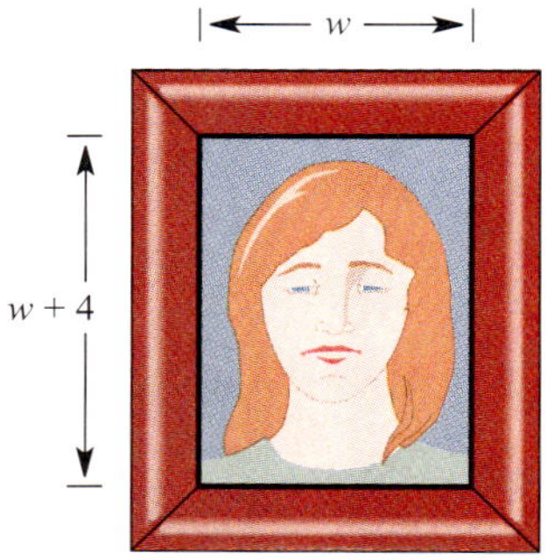

Solution

Verbal model: Area of picture = Width · Height

Labels: Picture width = w (in inches)
Picture height = $w + 4$ (in inches)
Area = 192 (in square inches)

Equation:

$$192 = w(w + 4)$$
$$0 = w^2 + 4w - 192$$
$$0 = (w + 16)(w - 12)$$
$$w + 16 = 0 \Rightarrow w = -16$$
$$w - 12 = 0 \Rightarrow w = 12$$

Of the two possible solutions, we choose the positive value of w and conclude that the width of the picture is $w = 12$ inches. Finally, since the picture is 4 inches taller than it is wide, we see that the height of the picture is $12 + 4 = 16$ inches. Try checking these dimensions in the original statement of the problem. ■

EXAMPLE 2 ■ An Interest Problem

The formula

$$A = P(1 + r)^2$$

represents the amount, A, of money in an account in which P dollars was deposited for two years, at an annual percentage rate of r (in decimal form). Find the interest rate if a deposit of \$6,000 increased to an amount of \$6,933.75 over a period of two years.

Solution

$$A = P(1 + r)^2 \quad \text{Given formula}$$
$$6{,}933.75 = 6{,}000(1 + r)^2 \quad \text{Substitute known values}$$

$$1.155625 = (1 + r)^2 \qquad \text{Divide both sides by 6,000}$$

$$\pm 1.075 = 1 + r \qquad \text{Extract square roots}$$

$$0.075 = r \qquad \text{Choose positive solution}$$

Thus, the annual percentage rate is $r = 0.075 = 7.5\%$. Check this solution in the original equation. ■

EXAMPLE 3 ■ **An Investment Problem**

A car dealer bought a fleet of cars from a car rental agency for a total of \$90,000. By the time the dealer had sold all but six of the cars, at an average profit of \$2,500 each, the original investment of \$90,000 had been regained. How many cars did the dealer sell, and what was the average price per car?

Solution

Although this problem is stated in terms of average prices and average profits per car, we can use a model that assumes that each car sold for the same price.

Verbal model: [Selling price per car] = [Cost per car] + [Profit per car]

Labels:
x = Number of cars sold
$x + 6$ = Number of cars bought
Selling price per car $= \dfrac{90{,}000}{x}$ (in dollars)
Cost per car $= \dfrac{90{,}000}{x + 6}$ (in dollars)
Profit per car = \$2,500

Equation:

$$\frac{90{,}000}{x} = \frac{90{,}000}{x + 6} + 2{,}500$$

$$90{,}000(x + 6) = 90{,}000x + 2{,}500x(x + 6), \quad x \neq 0,\ x \neq -6$$

$$90{,}000x + 540{,}000 = 90{,}000x + 2{,}500x^2 + 15{,}000x$$

$$0 = 2{,}500x^2 + 15{,}000x - 540{,}000$$

$$0 = x^2 + 6x - 216$$

$$0 = (x - 12)(x + 18)$$

$$x - 12 = 0 \Rightarrow x = 12$$

$$x + 18 = 0 \Rightarrow x = -18$$

Choosing the positive value, we conclude that the dealer sold 12 cars at an average price of

$$\frac{90{,}000}{12} = \$7{,}500 \text{ per car.}$$

Check this solution in the original equation. ■

EXAMPLE 4 ■ Reduced Rates

A ski club chartered a bus for a ski trip at a cost of \$520. In an attempt to lower the bus fare per skier, the club invited nonmembers to go along. After five nonmembers joined the trip, the fare per skier decreased by \$5.20. How many club members are going on the bus?

Solution

Verbal model: [Cost per skier] · [Number of skiers] = [\$520]

Labels: x = Number of ski club members

$x + 5$ = Number of skiers

Original cost per skier $= \dfrac{520}{x}$ (in dollars)

New cost per skier $= \dfrac{520}{x} - 5.20$ (in dollars)

Equation:

$$\left(\frac{520}{x} - 5.20\right)(x + 5) = 520$$

$$\left(\frac{520 - 5.2x}{x}\right)(x + 5) = 520$$

$$(520 - 5.2x)(x + 5) = 520x, \quad x \neq 0$$

$$520x - 5.2x^2 - 26x + 2{,}600 = 520x$$

$$-5.2x^2 - 26x + 2{,}600 = 0$$

$$x^2 + 5x - 500 = 0$$

$$(x + 25)(x - 20) = 0$$

$$x + 25 = 0 \quad \Rightarrow \quad x = -25$$

$$x - 20 = 0 \quad \Rightarrow \quad x = 20$$

Choosing the positive value of x, we have 20 ski club members. Check this solution in the original equation. ■

EXAMPLE 5 ■ An Application Involving the Pythagorean Theorem

An L-shaped sidewalk from Building A to Building B on a college campus is 200 meters long, as shown in Figure 6.2. By cutting diagonally across the grass, students shorten the walking distance to 150 meters. What are the lengths of the two legs of the existing sidewalk?

Solution

Common formula: $a^2 + b^2 = c^2$ Pythagorean Theorem

Labels:

a = Length of one leg = x (in meters)
b = Length of other leg = $200 - x$ (in meters)
c = Length of diagonal = 150 (in meters)

Equation:

$$x^2 + (200 - x)^2 = (150)^2$$
$$2x^2 - 400x + 40{,}000 = 22{,}500$$
$$2x^2 - 400x + 17{,}500 = 0$$
$$x^2 - 200x + 8{,}750 = 0$$

By the Quadratic Formula, we find the solutions of this equation as follows.

$$x = \frac{200 \pm \sqrt{(-200)^2 - 4(1)(8{,}750)}}{2(1)}$$
$$= \frac{200 \pm \sqrt{5{,}000}}{2}$$
$$= \frac{200 \pm 50\sqrt{2}}{2}$$
$$= 100 \pm 25\sqrt{2}$$

Both solutions are positive, and it does not matter which one we choose. If we let

$$x = 100 + 25\sqrt{2} \approx 135.4 \text{ meters}$$

then the length of the other leg is

$$200 - x \approx 200 - 135.4 \approx 64.6 \text{ meters.}$$

Try choosing the other value of x to see that the same two lengths result.

FIGURE 6.2

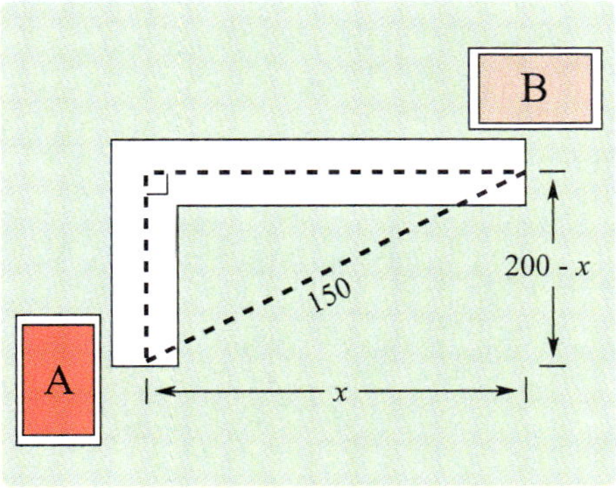

■

EXAMPLE 6 ■ Work Problem

An office contains two copy machines. Machine B is known to take 12 minutes longer than Machine A to copy the company's monthly report. The two machines working together take 8 minutes to reproduce the report. How long would it take each machine alone to reproduce the report?

Solution

Verbal model: [Work done by Machine A] + [Work done by Machine B] = [1 complete job]

([Rate for A] · [Time for both]) + ([Rate for B] · [Time for both]) = [1]

Labels: Both machines: Time = 8 minutes, Rate = $\frac{1}{8}$ job per minute

Machine A: Time = t minutes, Rate = $\frac{1}{t}$ job per minute

Machine B: Time = $t + 12$ minutes, Rate = $\frac{1}{t + 12}$ job per minute

Equation:

$$\frac{1}{t}(8) + \frac{1}{t + 12}(8) = 1$$

$$8\left(\frac{1}{t} + \frac{1}{t + 12}\right) = 1$$

$$8\left(\frac{t + 12 + t}{t(t + 12)}\right) = 1$$

$$8t(t + 12)\left(\frac{2t + 12}{t(t + 12)}\right) = t(t + 12)$$

$$8(2t + 12) = t^2 + 12t$$

$$16t + 96 = t^2 + 12t$$

$$0 = t^2 - 4t - 96$$

$$0 = (t - 12)(t + 8)$$

$$t - 12 = 0 \quad \Rightarrow \quad t = 12$$

$$t + 8 = 0 \quad \Rightarrow \quad t = -8$$

We choose the positive value for t and find that

Time for Machine A = t = 12 minutes

Time for Machine B = $t + 12$ = 24 minutes.

Check these solutions in the original equation. ■

EXAMPLE 7 ■ **The Height of a Model Rocket**

A model rocket is projected upward from ground level according to the height equation

$$h = -16t^2 + 192t, \quad t \geq 0$$

where h is the height in feet and t is the time in seconds.

(a) After how many seconds will the height be 432 feet?

(b) When will the rocket hit the ground?

Solution

(a)

$h = -16t^2 + 192t$		Given equation
$432 = -16t^2 + 192t$		Replace h by 432 feet
$16t^2 - 192t + 432 = 0$		Standard form
$t^2 - 12t + 27 = 0$		Divide both sides by 16
$(t - 3)(t - 9) = 0$		Factor
$t - 3 = 0$	⇨ $t = 3$	Set first factor equal to 0
$t - 9 = 0$	⇨ $t = 9$	Set second factor equal to 0

Thus, the rocket will have obtained a height of 432 feet at two different times—once (going up), after 3 seconds, and again (coming down), after 9 seconds. (See Figure 6.3.)

FIGURE 6.3

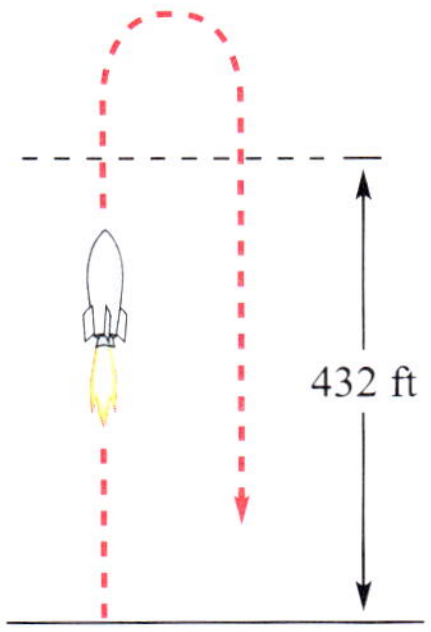

(b) To find the time it takes for the rocket to hit the ground, we let the height be 0.

$$0 = -16t^2 + 192t$$
$$0 = t^2 - 12t$$
$$0 = t(t - 12)$$
$$t = 0 \quad \text{or} \quad t = 12$$

Thus, the rocket will hit the ground after 12 seconds. (Note that a time of $t = 0$ seconds corresponds to the time of lift-off.) ■

DISCUSSION PROBLEM ■ The Rocket Problem Revisited

At what time will the rocket described in Example 7 reach its maximum height? What will the maximum height be? Write a short paragraph describing the procedure you used to determine your answers. ■

Warm-Up

The following warm-up exercises involve skills that were covered in earlier sections. You will use these skills in the exercise set for this section.

In Exercises 1–10, solve the equation. (Find all real *and* complex solutions.)

1. $3(x + 7) = 0$

2. $3x(15x - 2) = 0$

3. $2n(n - 2) + (n - 2) = 3$

4. $2n(n + 2) = 240$

5. $3(x + 8)^2 = 243$

6. $t + 3\sqrt{t} - 4 = 0$

7. $x^2 + 3x + 5 = 0$

8. $u^2 + 3u - 5 = 0$

9. $t - \dfrac{6}{t} = 5$

10. $\sqrt{2s + 3} = s$

6.4 EXERCISES

Number Problems In Exercises 1–6, find two positive integers that satisfy the given requirement. (*Hint:* Remember that even and odd integers can be represented by $2n$ and $2n + 1$, respectively.)

1. The product of two consecutive integers is 240.
2. The product of two consecutive integers is 1,122.
3. The product of two consecutive *even* integers is 224.
4. The product of two consecutive *odd* integers is 255.
5. The sum of the squares of two consecutive integers is 313.
6. The sum of the squares of two consecutive integers is 421.

Dimensions of a Rectangle In Exercises 7–16, find the dimensions of the rectangle and complete the table where l and w represent the length and width of a rectangle, respectively.

	Width	Length	Perimeter	Area
7.	$0.75l$	l	42 in.	
8.	w	$1.5w$	40 m	
9.	w	$2.5w$		250 ft^2
10.	w	$1.5w$		216 cm^2
11.	$\frac{1}{3}l$	l		192 in.^{2}
12.	$\frac{3}{4}l$	l		2,700 in.^{2}
13.	w	$w + 3$	54 km	
14.	$l - 6$	l	108 ft	
15.	$l - 20$	l		12,000 m^2
16.	w	$w + 5$		500 ft^2

17. *Lumber Storage Area* A retail lumberyard stores lumber in a rectangular region adjoining the sales office (see figure). The region will be fenced on three sides, and the fourth side will be bounded by the existing building of the business. Find the dimensions of the region if 350 feet of fencing is available and the area of the region is 12,500 square feet.

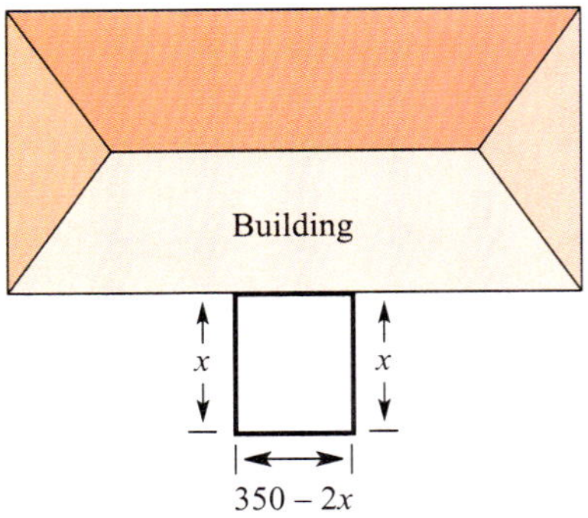

18. *Open Conduit* An open-topped rectangular conduit for carrying water in a manufacturing process is made by folding up the edges of a sheet of aluminum 48 inches wide (see figure). A cross section of the conduit must have an area of 288 square inches. Find the width and height of the conduit.

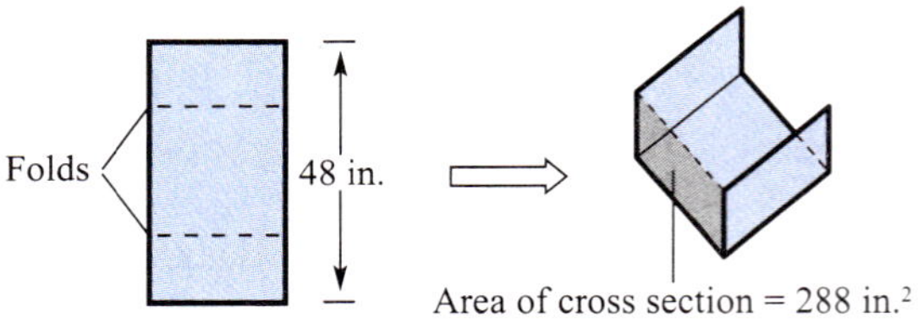

19. *Dimensions of a Triangle* The height of a triangle is two-thirds of its base. The area of the triangle is 192 square inches. Find the dimensions of the triangle.

20. *Dimensions of a Triangle* The height of a triangle is twice its base. The area of the triangle is 625 square inches. Find the dimensions of the triangle.

In Exercises 21 and 22, use the compound interest formula $A = P(1 + r)^2$ where A represents the amount after two years when a principal of P dollars is invested at an annual percentage rate r (in decimal form). Find the interest rate needed to obtain the amount A when P dollars are invested for two years.

21. $P = \$3{,}000,\ A = \$3{,}499.20$
22. $P = \$10{,}000,\ A = \$11{,}990.25$

23. *Selling Price* A store owner bought a case of grade A large eggs for \$21.60. By the time all but six dozen of the eggs had been sold at a profit of \$0.30 per dozen, the original investment of \$21.60 had been regained. How many dozen eggs did the owner sell, and what was the selling price per dozen?

24. *Selling Price* A manager of a computer store bought some computers of the same model for \$27,000. When all but three of the computers had been sold at a profit of \$750 per computer, the original investment of \$27,000 had been regained. How many computers were sold, and what was the selling price of each computer?

25. *Reduced Ticket Price* A service organization obtained a block of tickets for a ball game for \$240. The block contained eight more tickets than the organization needed for its members. By inviting eight more people to attend (and share the cost), the organization lowered the price per ticket by \$1. How many people are going to the game?

26. *Reduced Fare* A science club charters a bus to attend a science fair at a cost of \$480. In an attempt to lower the bus fare per person, the club invites nonmembers to go along. After two nonmembers join the trip, the fare per person is decreased by \$1. How many people are going on the excursion?

27. *Delivery Route* Suppose you are asked to deliver pizza to Offices B and C in your city (see figure), and you are required to keep a log of the mileages between stops. You forget to look at the odometer at stop B, but after getting to stop C you record the total distance traveled from the pizza shop as 18 miles. The return distance from C to A is 16 miles. The route forms a right triangle, and the distance from A to B is greater than the distance from B to C. Find the distance from A to B.

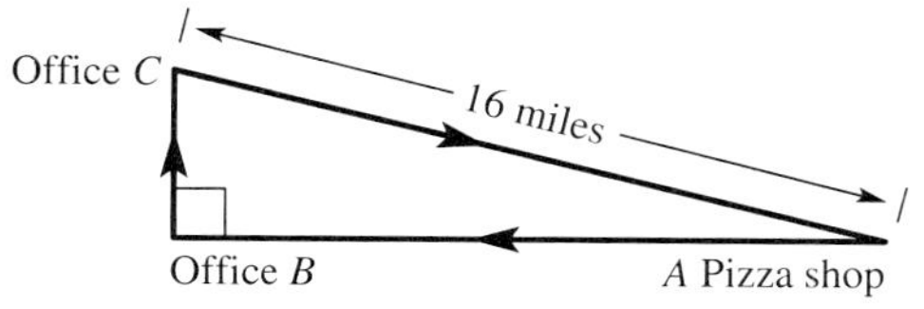

28. *Dimensions of a Rectangle* The perimeter of a rectangle is 102 inches and the length of the diagonal is 39 inches. Find the dimensions of the rectangle.

29. *Air Speed* An airline runs a commuter flight between two cities that are 720 miles apart. If the average speed of the planes could be increased by 40 miles per hour, the travel time would be decreased by 12 minutes. What air speed is required to obtain this decrease in travel time?

30. *Average Speed* A truck traveled the first 100 miles of a trip at one speed and the last 135 miles at an average speed of 5 miles per hour less. If the entire trip took five hours, what was the average speed for the first part of the trip?

31. *Work Rate Problem* Working together, two people can complete a task in five hours. Working alone, how long would it take each to do the task if one person takes two hours longer than the other?

32. *Work Rate Problem* An office contains two printers. Machine B is known to take three minutes longer than Machine A to produce the company's monthly financial report. With both machines working together, it takes six minutes to produce the report. How long would it take each machine alone to produce the report?

Free-Falling Object In Exercises 33–36, find the time necessary for an object to fall to ground level from an initial height of h_0 feet if its height h at any time t (in seconds) is given by

$$h = h_0 - 16t^2.$$

33. $h_0 = 144$

34. $h_0 = 625$

35. $h_0 = 1{,}454$ (Height of Sears Tower)

36. $h_0 = 984$ (Height of Eiffel Tower)

37. *Height of a Baseball* The height h in feet of a baseball that is hit 3 feet above the ground is given by

$$h = 3 + 75t - 16t^2$$

where t is the time in seconds. Find the time when the ball hits the ground in the outfield.

SECTION 6.5

Quadratic and Rational Inequalities

Introduction • Quadratic Inequalities • Rational Inequalities • Applications

Introduction

When working with inequalities involving polynomials, it is important to realize that the value of a polynomial can change signs only at its **zeros**. That is, a polynomial can change signs only at the x-values that make the polynomial zero. For instance, the first-degree polynomial $x + 2$ has a zero at -2, and it changes sign at that zero.

$x + 2 < 0$ for $x < -2$ — Polynomial is negative

$x + 2 = 0$ for $x = -2$ — Zero of polynomial

$x + 2 > 0$ for $x > -2$ — Polynomial is positive

We can picture this result on the real number line, as shown in Figure 6.4.

FIGURE 6.4

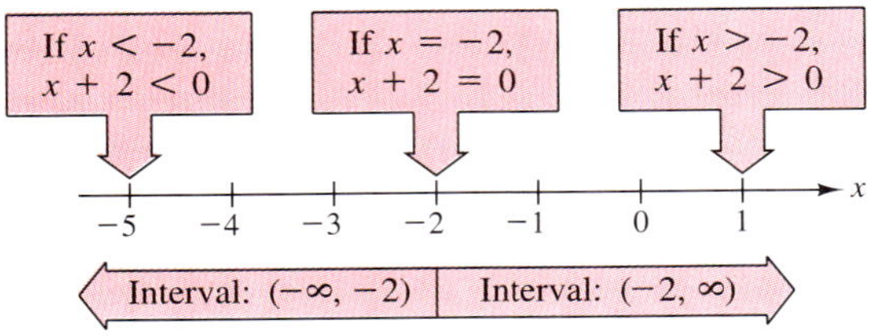

Note in Figure 6.4 that the polynomial $x + 2$ has one zero (given by $x = -2$), and that this zero partitions the real number line into two **test intervals**. The value of this polynomial is negative for *every* x-value in the first test interval $(-\infty, -2)$, and the value of the polynomial is positive for *every* x-value in the second test interval $(-2, \infty)$.

We use the same basic approach to determine the test intervals for any polynomial.

Finding Test Intervals for a Polynomial

To determine the intervals on which the values of a polynomial are entirely negative or entirely positive, use the following steps.

1. Find all real zeros of the polynomial, and arrange the zeros in increasing order (from smallest to largest). We call the zeros of a polynomial its **critical numbers**.
2. Use the critical numbers of the polynomial to determine its **test intervals**.
3. Choose one representative x-value in each test interval and evaluate the polynomial at that value. If the value of the polynomial is negative, then the polynomial will have negative values for *every* x-value in the interval. If the value of the polynomial is positive, then the polynomial will have positive values for *every* x-value in the interval.

EXAMPLE 1 ■ Finding Test Intervals for a Quadratic Polynomial

Determine the intervals on which the following polynomial is entirely negative or entirely positive.

$$x^2 - x - 6$$

Solution

By factoring the given quadratic as

$$x^2 - x - 6 = 0$$
$$(x + 2)(x - 3) = 0$$

we see that the critical numbers occur at $x = -2$ and $x = 3$. Therefore, the test intervals for the quadratic are

$$(-\infty, -2), \quad (-2, 3), \quad \text{and} \quad (3, \infty). \qquad \text{Test intervals}$$

In each test interval, we choose a representative x-value and evaluate the polynomial, as shown in Table 6.1.

TABLE 6.1

Test Interval	Representative x-Value	Value of Polynomial	Conclusion
$(-\infty, -2)$	$x = -3$	$(-3)^2 - (-3) - 6 = 6$	Polynomial is positive
$(-2, 3)$	$x = 0$	$(0)^2 - (0) - 6 = -6$	Polynomial is negative
$(3, \infty)$	$x = 4$	$(4)^2 - (4) - 6 = 4$	Polynomial is positive

Therefore, the polynomial has positive values for every x in the intervals $(-\infty, -2)$ and $(3, \infty)$, and negative values for every x in the interval $(-2, 3)$. This result is shown graphically in Figure 6.5.

FIGURE 6.5

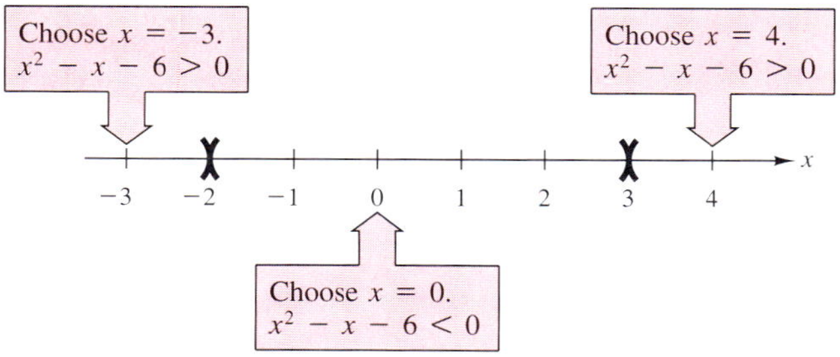

■

Quadratic Inequalities

In Section 2.4 we looked at methods for solving linear inequalities of the form

$$ax + b < 0, \quad ax + b \le 0, \quad ax + b > 0, \quad \text{or} \quad ax + b \ge 0.$$

We now look at methods for solving **quadratic inequalities** of the form

$$ax^2 + bx + c < 0, \qquad ax^2 + bx + c \le 0, \qquad ax^2 + bx + c > 0,$$
$$\text{or} \quad ax^2 + bx + c \ge 0.$$

To solve such an inequality, we use the guidelines for finding the intervals on which a polynomial is entirely negative or entirely positive. This procedure is demonstrated in Example 2.

EXAMPLE 2 ■ Solving a Quadratic Inequality

Solve the following quadratic inequality.

$$x^2 - 5x < 0$$

Solution

$$x^2 - 5x < 0 \qquad \text{Given inequality}$$
$$x(x - 5) < 0 \qquad \text{Factor}$$

Critical numbers: $x = 0,\ x = 5$

Test intervals: $(-\infty, 0),\ (0, 5),\ (5, \infty)$

Test: Is $x(x - 5) < 0$?

To test an interval, we first choose a convenient number in the interval and then compute the sign of $x(x - 5)$. The results are shown in Figure 6.6.

FIGURE 6.6

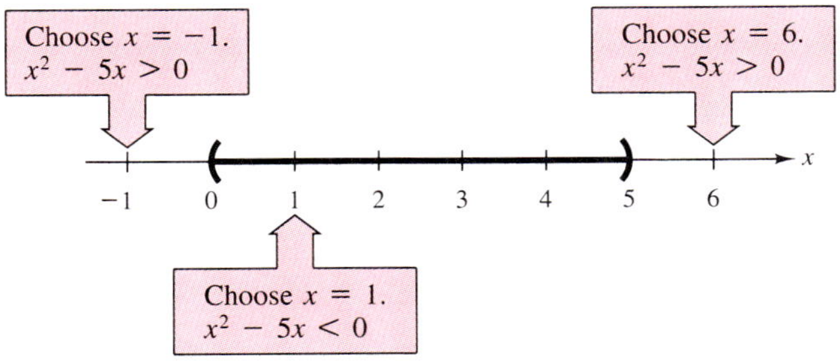

Since the inequality $x^2 - 5x < 0$ is satisfied only by the middle test interval, we conclude that the solution set of the inequality is the interval

$$(0, 5). \qquad \text{Solution set}$$ ■

In Example 2 note that we would have used the same basic procedure if the inequality had been written using the symbols $\le$, $>$, or $\ge$. For instance, from Figure

6.6 we can see that the solution set of the inequality

$$x^2 - 5x \geq 0$$

consists of the union of the half-open intervals $(-\infty, 0]$ and $[5, \infty)$, which we write as

$$(-\infty, 0] \cup [5, \infty). \qquad \text{Solution set}$$

Just as with solving quadratic *equations*, the first step in solving a quadratic *inequality* is to write the inequality in **standard form**, with the polynomial on the left and zero on the right, as demonstrated in Example 3.

EXAMPLE 3 ■ Solving a Quadratic Inequality

Solve the following quadratic inequality.

$$2x^2 + 5x > 12$$

Solution

$$2x^2 + 5x > 12 \qquad \text{Given inequality}$$
$$2x^2 + 5x - 12 > 0 \qquad \text{Write in standard form}$$
$$(x + 4)(2x - 3) > 0 \qquad \text{Factor}$$

Critical numbers: $x = -4$, $x = \frac{3}{2}$

Test intervals: $(-\infty, -4)$, $\left(-4, \frac{3}{2}\right)$, $\left(\frac{3}{2}, \infty\right)$

Test: Is $(x + 4)(2x - 3) > 0$?

After testing these intervals, as shown in Figure 6.7, we see that the polynomial $2x^2 + 5x - 12$ is positive in the open intervals $(-\infty, -4)$ and $\left(\frac{3}{2}, \infty\right)$. Therefore, the solution set of the inequality is

$$(-\infty, -4) \cup \left(\tfrac{3}{2}, \infty\right). \qquad \text{Solution set}$$

FIGURE 6.7

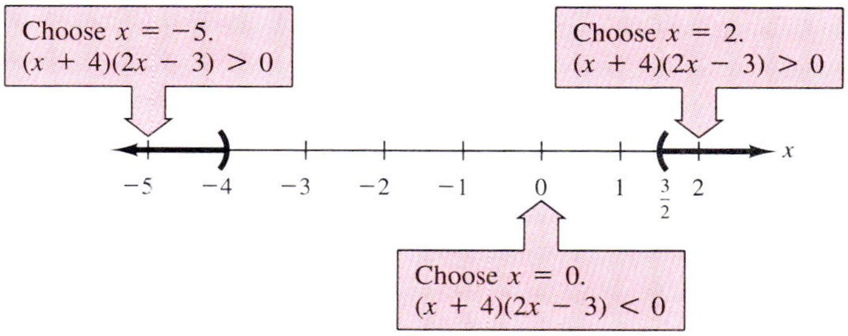

■

Example 4 shows how the Quadratic Formula can be used to solve a quadratic inequality.

EXAMPLE 4 ■ **Using the Quadratic Formula**

Solve the following quadratic inequality.

$$x^2 - 2x - 1 \le 0$$

Solution

Since the quadratic $x^2 - 2x - 1$ does not factor (using integer coefficients), we use the Quadratic Formula to find its zeros. That is, the solutions of $x^2 - 2x - 1 = 0$ are given by

$$x = \frac{-(-2) \pm \sqrt{(-2)^2 - 4(1)(-1)}}{2} = \frac{2 \pm \sqrt{8}}{2} = \frac{2 \pm 2\sqrt{2}}{2} = 1 \pm \sqrt{2}.$$

Now, having found the zeros of the polynomial $x^2 - 2x - 1$, we proceed as usual.

Critical numbers: $x = 1 - \sqrt{2}$, $x = 1 + \sqrt{2}$

Test intervals: $(-\infty, 1 - \sqrt{2})$, $(1 - \sqrt{2}, 1 + \sqrt{2})$, $(1 + \sqrt{2}, \infty)$

Test: Is $x^2 - 2x - 1 \le 0$?

After testing these intervals, as shown in Figure 6.8, we see that the polynomial $x^2 - 2x - 1$ is less than or equal to zero in the *closed* interval $[1 - \sqrt{2}, 1 + \sqrt{2}]$. Therefore, the solution set of the inequality is

$$[1 - \sqrt{2}, 1 + \sqrt{2}].$$ Solution set

Note that $1 - \sqrt{2} \approx -0.414$ and $1 + \sqrt{2} \approx 2.414$.

FIGURE 6.8

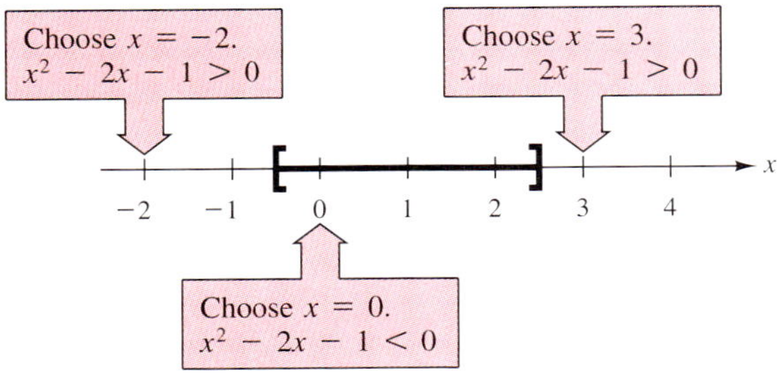

■

When solving a quadratic inequality, be sure you have accounted for the particular type of inequality symbol given in the inequality. For instance, in Example 4 note that the solution is a closed interval because the original inequality contained a "*less than or equal to*" symbol. If the original inequality had been $x^2 - 2x - 1 < 0$, the solution would have been the *open* interval $(1 - \sqrt{2}, 1 + \sqrt{2})$.

Each of the quadratic inequalities we have looked at so far have had solution sets that consisted of a single interval or the union of two intervals. When solving the exercises for this section, you should be on the watch for some unusual solution sets, as illustrated in Example 5.

EXAMPLE 5 ■ Unusual Solution Sets

Solve each of the following inequalities to verify that the given solution set is correct.

(a) The solution set of the quadratic inequality

$$x^2 + 2x + 4 > 0$$

consists of the entire set of real numbers, $(-\infty, \infty)$. In other words, the quadratic $x^2 + 2x + 4$ is positive for every real value of x. (Note that this quadratic inequality has *no* critical numbers. In such a case, there is only one test interval—the entire real line.)

(b) The solution set of the quadratic inequality

$$x^2 + 2x + 1 \le 0$$

consists of the single real number $\{-1\}$.

(c) The solution set of the quadratic inequality

$$x^2 + 3x + 5 < 0$$

is empty. In other words, the quadratic $x^2 + 3x + 5$ is not less than zero for any value of x.

(d) The solution set of the quadratic inequality

$$x^2 - 4x + 4 > 0$$

consists of all real numbers *except* the number 2. In interval notation, this solution set can be written as $(-\infty, 2) \cup (2, \infty)$. ■

Rational Inequalities

The concepts of critical numbers and test intervals can be extended to inequalities involving rational expressions. To do this, we use the fact that the value of a rational expression can change sign only at its *zeros* (the x-values for which its numerator is zero) and its *undefined values* (the x-values for which its denominator is zero). These two types of numbers make up the **critical numbers** of a rational inequality. For instance, the critical numbers of the inequality

$$\frac{x - 2}{(x - 1)(x + 3)} < 0$$

are $x = 2$ (the numerator is zero), $x = 1$, and $x = -3$ (the denominator is zero). From these three critical numbers we see that the given inequality has *four* test intervals:

$$(-\infty, -3), \qquad (-3, 1), \qquad (1, 2), \quad \text{and} \quad (2, \infty).$$

EXAMPLE 6 ■ Solving a Rational Inequality

Solve the following rational inequality.

$$\frac{x}{x-2} > 0$$

Solution

Critical numbers: $x = 0$, $x = 2$

Test intervals: $(-\infty, 0)$, $(0, 2)$, $(2, \infty)$

Test: Is $\dfrac{x}{x-2} > 0$?

After testing these intervals, as shown in Figure 6.9, we see that the rational expression $x/(x-2)$ is positive in the open intervals $(-\infty, 0)$ and $(2, \infty)$. Therefore, the solution set of the inequality is

$$(-\infty, 0) \cup (2, \infty). \qquad \text{Solution set}$$

FIGURE 6.9

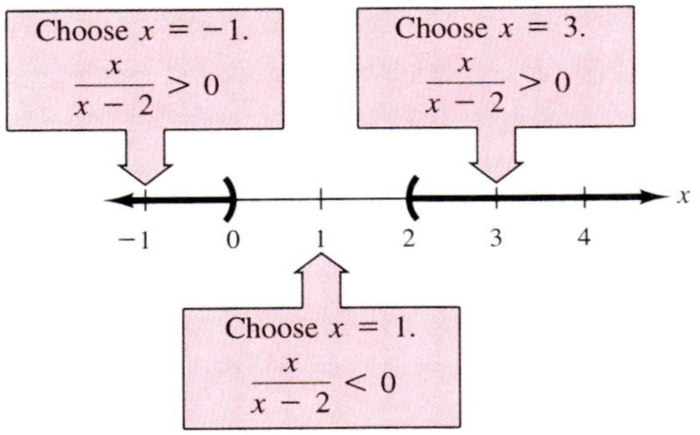

■

In the next example, note that the original inequality is not written in standard form, with the rational expression on the left (written as a *single* fraction) and zero on the right. The first step in solving such an inequality is to write it in standard form.

EXAMPLE 7 ■ Writing in Standard Form First

Solve the rational inequality $\dfrac{2x}{x+3} < 4$.

Solution

$$\frac{2x}{x+3} < 4 \qquad \text{Given inequality}$$

$$\frac{2x}{x+3} - 4 < 0 \qquad \text{Subtract 4 from both sides}$$

$$\frac{2x - 4(x+3)}{x+3} < 0 \qquad \text{Subtract fractions}$$

$$\frac{-2x - 12}{x + 3} < 0 \qquad \text{Standard form}$$

$$\frac{-2(x + 6)}{x + 3} < 0 \qquad \text{Factor}$$

Critical numbers: $x = -6,\ x = -3$

Test intervals: $(-\infty, -6),\ (-6, -3),\ (-3, \infty)$

Test: Is $\dfrac{-2(x + 6)}{x + 3} < 0$?

After testing these intervals, as shown in Figure 6.10, we see that the rational expression $-2(x + 6)/(x + 3)$ is negative in the open intervals $(-\infty, -6)$ and $(-3, \infty)$. Therefore, the solution set of the inequality is

$$(-\infty, -6) \cup (-3, \infty). \qquad \text{Solution set}$$

FIGURE 6.10

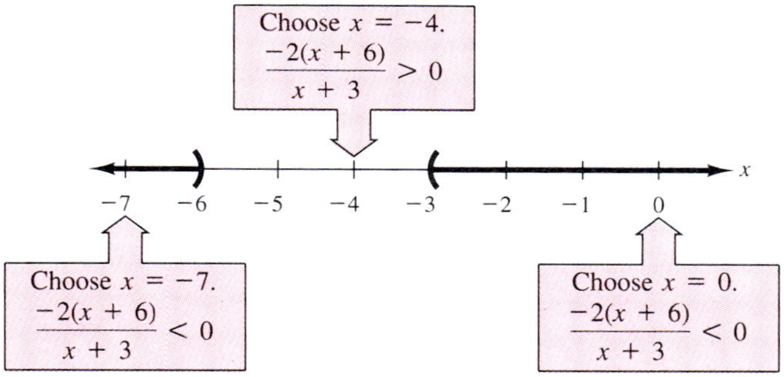

■

Applications

In Examples 8 and 9, we give some applications of quadratic inequalities.

EXAMPLE 8 ■ **The Height of a Projectile**

A projectile is fired straight up from ground level with an initial velocity of 256 feet per second, so that its height at any time t is given by

$$h = -16t^2 + 256t$$

where the height h is measured in feet and the time t is measured in seconds. During what interval of time will the height of the projectile exceed 960 feet?

Solution

To solve this problem, we must solve the inequality given by

$$\text{Height} > 960$$

$$-16t^2 + 256t > 960.$$

Using the techniques described in this section, we solve this inequality as follows.

$-16t^2 + 256t > 960$	Given inequality
$-16t^2 + 256t - 960 > 0$	Subtract 960 from both sides
$t^2 - 16t + 60 < 0$	Divide both sides by -16 and reverse inequality
$(t - 6)(t - 10) < 0$	Factor

Thus, the critical numbers are $t = 6$ and $t = 10$. By testing the intervals $(-\infty, 6)$, $(6, 10)$, and $(10, \infty)$, we find that the solution interval is $(6, 10)$. Therefore, the height of the object will exceed 960 feet during the time interval given by

$$6 \text{ seconds} < t < 10 \text{ seconds}.$$

(See Figure 6.11.)

FIGURE 6.11

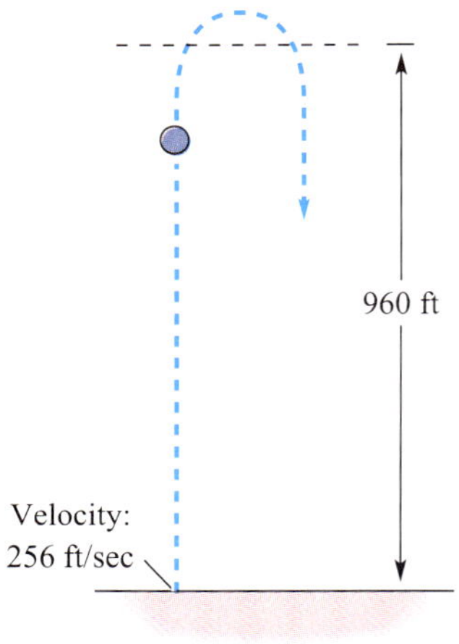

■

EXAMPLE 9 ■ Maintaining a Profit Level

A bakery sells decorated cakes for a profit of

$$P = 5x - \frac{1}{100}x^2$$

where P is the profit measured in dollars and x is the number of cakes sold. If the bakery wants its weekly profit for cakes to exceed \$600, how many cakes must it sell each week?

Solution

$600 < P$	Profit must exceed \$600
$600 < 5x - \frac{1}{100}x^2$	Profit equation

$$\frac{1}{100}x^2 - 5x + 600 < 0 \qquad \text{Standard form}$$

$$x^2 - 500x + 60{,}000 < 0 \qquad \text{Multiply both sides by 100}$$

$$(x - 200)(x - 300) < 0 \qquad \text{Factor}$$

The critical numbers are $x = 200$ and $x = 300$. By testing the intervals $(-\infty, 200)$, $(200, 300)$, and $(300, \infty)$, we find that the solution interval is $(200, 300)$. Therefore, the profit will exceed \$600 provided the production level of cakes falls in the interval

$$200 \text{ cakes} < x < 300 \text{ cakes}.$$

■

The solution of Example 9 might seem a little unusual, because you might think that the more cakes the bakery sells the more profit it makes. However, in this example the profit actually falls below \$600 when the bakery sells more than 300 cakes. If you take a course in economics, you will see that this type of situation often occurs in business. The reason is that as a business begins to saturate the market, the only way it can increase sales volume is to lower its price per unit.

DISCUSSION PROBLEM ■ A Graphing Calculator Experiment

In the next chapter, we will look at techniques for sketching the graph of an equation involving *two* variables. We can use this technique as an alternative way of solving inequalities. For instance, to solve the inequality in Example 2

$$x^2 - 5x < 0$$

we can sketch the graph of $y = x^2 - 5x$. Using a graphing calculator or a computer with graphing capabilities, we can see that the graph of this equation is a "U-shaped" curve called a parabola, as shown in Figure 6.12. From this figure we see that the only part of the curve that lies below the x-axis is the portion for which $0 < x < 5$. Thus, the solution of $x^2 - 5x < 0$ is $0 < x < 5$.

Try using this graphing approach to solve some of the other examples in this section.

FIGURE 6.12

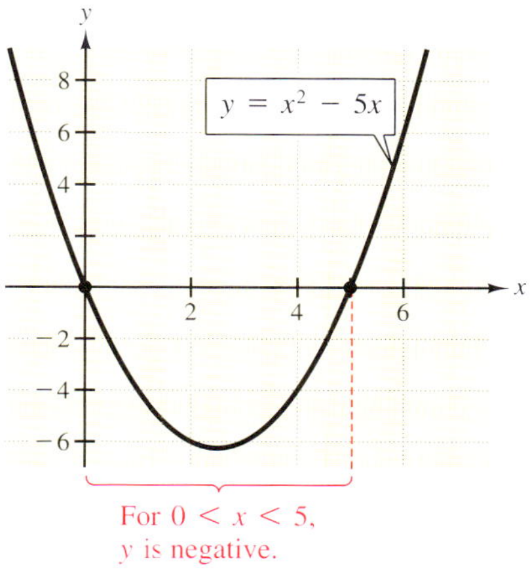

■

Warm-Up

The following warm-up exercises involve skills that were covered in earlier sections. You will use these skills in the exercise set for this section.

In Exercises 1–6, solve the inequality and sketch the graph of the solution on the real number line.

1. $3x - 15 \le 0$ **2.** $3x + 5 \ge 0$ **3.** $7 - 3x > 4 - x$

4. $2(x + 6) - 20 < 2$ **5.** $|x - 3| < 2$ **6.** $|x - 5| > 3$

In Exercises 7–10, determine whether the product is zero, positive, or negative. (It is not necessary to actually find the product.)

7. $(35 - 10)(13 - 25)$ **8.** $(254 - 254)(625 - 500)$

9. $(64 - 82)(12 - 15)$ **10.** $\left(\frac{3}{4} - \frac{3}{5}\right)\left(\frac{1}{6} - \frac{2}{3}\right)$

6.5 EXERCISES

In Exercises 1–6, find the critical numbers of the polynomial and locate them on the real number line.

1. $2x - 6$ **2.** $4x + 6$ **3.** $x(2x - 5)$

4. $5x(x - 3)$ **5.** $x^2 - 4x + 3$ **6.** $3x^2 - 2x - 8$

In Exercises 7–12, determine the intervals on which the polynomial is entirely negative and entirely positive. Show the intervals graphically on the real number line.

7. $x - 4$ **8.** $3 - x$ **9.** $2x(x - 4)$

10. $7x(3 - x)$ **11.** $x^2 - 4x - 5$ **12.** $2x^2 - 4x - 3$

In Exercises 13–36, solve the inequality and sketch the graph of the solution on the real number line. (Some of the inequalities have no solution.)

13. $2x + 6 \ge 0$ **14.** $5x - 20 < 0$ **15.** $3x(x - 2) < 0$

16. $2x(x - 6) > 0$ **17.** $3x(2 - x) < 0$ **18.** $2x(6 - x) > 0$

19. $x^2 - 4 \ge 0$ **20.** $z^2 \le 9$ **21.** $x^2 + 3x \le 10$

22. $t^2 - 15t + 50 < 0$ **23.** $-2u^2 + 7u + 4 < 0$ **24.** $-3x^2 - 4x + 4 \le 0$

25. $x^2 + 4x + 5 < 0$ **26.** $x^2 + 6x + 10 > 0$ **27.** $x^2 + 2x + 1 \ge 0$

28. $25 \geq (x - 3)^2$

29. $x^2 - 4x + 2 > 0$

30. $-x^2 + 8x - 11 \leq 0$

31. $(x - 5)^2 < 0$

32. $(y + 3)^2 \geq 0$

33. $6 - (x - 5)^2 < 0$

34. $(y + 3)^2 - 6 \geq 0$

35. $x(x - 2)(x + 2) > 0$

36. $x^2(x - 2) \leq 0$

In Exercises 37–40, determine the critical numbers of the rational expression and locate them on the real number line.

37. $\dfrac{5}{x - 3}$

38. $\dfrac{-6}{x + 2}$

39. $\dfrac{2x}{x + 5}$

40. $\dfrac{x - 2}{x - 10}$

In Exercises 41–52, solve the rational inequality and sketch the graph of the solution on the real number line.

41. $\dfrac{5}{x - 3} > 0$

42. $\dfrac{3}{4 - x} > 0$

43. $\dfrac{-5}{x - 3} > 0$

44. $\dfrac{-3}{4 - x} > 0$

45. $\dfrac{x}{x - 4} \leq 0$

46. $\dfrac{z - 1}{z + 3} < 0$

47. $\dfrac{y - 3}{2y - 11} \geq 0$

48. $\dfrac{x + 5}{3x + 2} \geq 0$

49. $\dfrac{6}{x - 4} > 2$

50. $\dfrac{1}{x + 2} > -3$

51. $\dfrac{x}{x - 3} \leq 2$

52. $\dfrac{x + 4}{x - 5} \geq 10$

53. *Height of a Projectile* A projectile is fired straight up from ground level with an initial velocity of 128 feet per second, so that its height at any time t is given by

$$h = -16t^2 + 128t$$

where the height h is measured in feet and the time t is measured in seconds. During what interval of time will the height of the projectile exceed 240 feet?

54. *Height of a Projectile* A projectile is fired straight up from ground level with an initial velocity of 88 feet per second, so that its height at any time t is given by

$$h = -16t^2 + 88t$$

where the height h is measured in feet and the time t is measured in seconds. During what interval of time will the height of the projectile exceed 50 feet?

55. *Length of Fencing* Suppose you have 64 feet of fencing to enclose a rectangular region. Determine the interval for the length such that the area will exceed 240 square feet. (*Hint:* The area is given by $A = l(32 - l)$ where l is the length of the rectangle.)

56. *Annual Interest Rate* You plan to invest \$1,000 in a certificate of deposit for two years, and you want the interest for that time period to exceed \$150. Find the annual interest rate that must be exceeded if the interest is compounded annually. (*Hint:* You must solve the inequality $1{,}000(1 + r)^2 > 1{,}150$.)

CHAPTER 6 SUMMARY

As you review and prepare for a test on this chapter, first try to obtain a global view of what was discussed. Then review the specific skills needed under each category.

Solving Quadratic Equations

- *Solve* quadratic equations by factoring. — Section 6.1, Exercises 1–20; Section 6.2, Exercises 11–20; Section 6.3, Exercises 5–14
- *Solve* quadratic equations by extracting square roots. — Section 6.1, Exercises 21–50, Exercises 57–66
- *Solve* equations of quadratic type. — Section 6.1, Exercises 51–56
- *Solve* quadratic equations by completing the square. — Section 6.2, Exercises 11–20, Exercises 21–50
- *Solve* quadratic equations by using the Quadratic Formula. — Section 6.3, Exercises 5–14, Exercises 21–40
- *Determine* the number and type of solutions of a quadratic equation by use of the discriminant. — Section 6.3, Exercises 15–20
- *Solve* quadratic equations by the most convenient method. — Section 6.3, Exercises 41–50

Applications of Quadratic Equations

- *Create and solve* quadratic equations that arise from interest and consumer problems. — Section 6.4, Exercises 21–25
- *Create and solve* quadratic equations that arise from geometric problems. — Section 6.4, Exercises 7–20

Solving Quadratic and Rational Inequalities

- *Solve and sketch* the graph of a quadratic inequality. — Section 6.5, Exercises 13–36
- *Solve and sketch* the graph of a rational inequality. — Section 6.5, Exercises 41–52

Chapter 6 Review Exercises

In Exercises 1–10, solve the quadratic equation by factoring.

1. $x^2 + 12x = 0$

2. $u^2 - 18u = 0$

3. $3z(z + 10) - 8(z + 10) = 0$

4. $7x(2x - 9) + 4(2x - 9) = 0$

5. $4y^2 - 1 = 0$

6. $2z^2 - 72 = 0$

7. $4y^2 + 20y + 25 = 0$

8. $x^2 + \frac{8}{3}x + \frac{16}{9} = 0$

9. $2x^2 - 2x - 180 = 0$

10. $15x^2 - 30x - 45 = 0$

In Exercises 11–16, solve the quadratic equation by extracting square roots.

11. $x^2 = 10{,}000$

12. $x^2 = 98$

13. $y^2 - 2.25 = 0$

14. $y^2 - 8 = 0$

15. $(x - 16)^2 = 400$

16. $(x + 3)^2 = 0.04$

In Exercises 17–22, solve the quadratic equation by completing the square. (Find all real *and* complex solutions.)

17. $x^2 - 6x - 3 = 0$

18. $x^2 + 12x + 6 = 0$

19. $x^2 - 3x + 3 = 0$

20. $t^2 + \frac{1}{2}t - 1 = 0$

21. $2y^2 + 10y + 3 = 0$

22. $3x^2 - 2x + 2 = 0$

In Exercises 23–28, solve the quadratic equation by using the Quadratic Formula. (Find all real *and* complex solutions.)

23. $y^2 + y - 30 = 0$

24. $x^2 - x - 72 = 0$

25. $2y^2 + y - 21 = 0$

26. $2x^2 - 3x - 20 = 0$

27. $0.3t^2 - 2t + 5 = 0$

28. $-u^2 + 2.5u + 3 = 0$

In Exercises 29–40, solve the equation by the method of your choice. (Find all real *and* complex solutions.)

29. $(v - 3)^2 = 250$

30. $x^2 - 36x = 0$

31. $-x^2 + 5x + 84 = 0$

32. $9x^2 + 6x + 1 = 0$

33. $(x - 9)^2 - 121 = 0$

34. $60 - (x - 6)^2 = 0$

35. $z^2 - 6z + 10 = 0$

36. $z^2 - 14z + 5 = 0$

37. $2y^2 + 3y + 1 = 0$

38. $0.25y^2 + 0.35y - 0.50 = 0$

39. $\frac{1}{x} + \frac{1}{x + 1} = \frac{1}{2}$

40. $x - 5 = \sqrt{x - 2}$

In Exercises 41–50, solve the inequality and sketch the graph of the solution on the real number line. (Some of the inequalities have no solution.)

41. $4x - 12 < 0$

42. $3(x + 2) > 0$

43. $5x(7 - x) > 0$

44. $-2x(x - 10) \le 0$

45. $16 - (x - 2)^2 \leq 0$

46. $(x - 5)^2 - 36 > 0$

47. $2x^2 + 3x - 20 < 0$

48. $3x^2 - 2x - 8 > 0$

49. $\dfrac{x}{2x - 7} \geq 0$

50. $\dfrac{2x - 9}{x - 1} \leq 0$

51. *Number Problem* Find two consecutive positive integers such that the sum of their squares is 265.

52. *Number Problem* Find two consecutive positive integers whose product is 156.

53. *Falling Time* The height h in feet of an object above the ground is given by

$$h = 200 - 16t^2, \quad t > 0$$

where t is the time in seconds. Find the time when the object strikes the ground.

54. *Falling Time* The height h in feet of an object above the ground is given by

$$h = -16t^2 + 64t + 192, \quad t > 0$$

where t is the time in seconds. Find the time when the object strikes the ground.

55. *Area of a Rectangle* The perimeter of a rectangle of length l and width w is 48 feet (see figure).

(a) Show that $w = 24 - l$.

(b) Show that the area A is given by $A = lw = l(24 - l)$.

(c) Complete the following table by using the equation in part (b).

l	2	4	6	8	10	12	14	16	18
A									

Figure for 55

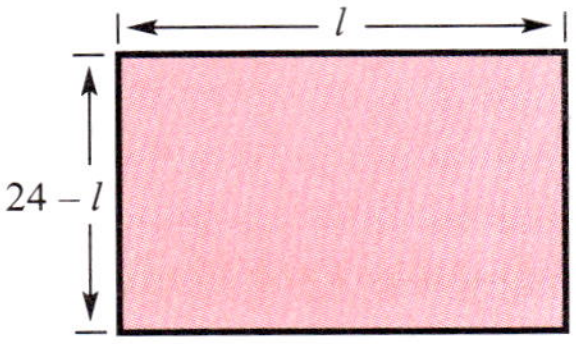

56. *Dimensions of a Triangle* Find the dimensions of a triangle if its height is one and two-thirds times its base and its area is 3,000 square inches.

57. *Decreased Price* A Little League baseball team obtained a block of tickets for a ball game for \$96. The block contained three more tickets than the team needed for its members. By inviting three more people to attend (and share the cost), the team lowered the price per ticket by \$1.60. How many people are going to the game?

58. *Average Speed* A train travels the first 220 miles of a trip at one speed and the last 180 miles at an average speed of 5 miles per hour faster. If the entire trip took seven hours, find the faster of the two speeds.

59. *Work Rate Problem* Working together, two people can complete a task in six hours. Working alone, how long would it take each to do the task if one person takes three hours longer than the other?

60. Find a quadratic equation with the given solutions.

(a) $x = -3, \ x = 3$

(b) $x = -\frac{1}{2}, \ x = 2$

(c) $x = -5, \ x = -1$

(d) $x = -10, \ x = 8$

CHAPTER 6 TEST

Take this test as you would take a test in class. After you are done, check your work with the answers given in the back of the book.

1. Solve by factoring: $x(x + 5) - 10(x + 5) = 0$

2. Solve by factoring: $8x^2 - 21x - 9 = 0$

3. Solve by extracting square roots: $(x - 2)^2 = 0.09$

4. Solve by extracting complex square roots: $(x + 3)^2 + 81 = 0$

5. Find the real number c such that $x^2 - 3x + c$ is a perfect square trinomial.

6. Solve by completing the square: $2x^2 - 6x + 3 = 0$

7. Find the discriminant and use it to determine the type of solutions for the quadratic equation: $5x^2 - 12x + 10 = 0$

8. Solve by using the Quadratic Formula: $3x^2 - 8x + 3 = 0$

9. Solve: $\sqrt{3x + 7} - \sqrt{x + 12} = 1$

10. Solve and sketch the solution: $2x(x - 3) < 0$

11. Solve and sketch the solution: $16 \le (x - 2)^2$

12. Solve and sketch the solution: $\dfrac{3}{x - 2} > 4$

13. Find a quadratic equation having the solutions -4 and 5.

14. Find two consecutive positive integers whose product is 210.

15. The width of a rectangle is 8 feet less than the length. The area of the rectangle is 240 square feet. Find the dimensions of the rectangle.

16. A train traveled the first 125 miles of a trip at one speed and the last 180 miles at an average speed of 5 miles per hour less. If the total time for the trip was $6\frac{1}{2}$ hours, what was the average speed for the first part of the trip?

17. The height of an object dropped from an initial height of 75 feet is given by

$$\text{Height} = -16t^2 + 75$$

where the height is measured in feet and the time is measured in seconds. How long will it take the object to fall to a height of 35 feet?

CHAPTER SEVEN

Introduction to Analytic Geometry

Chapter Overview

We have been studying algebra from a purely algebraic point of view. In this chapter, we take a geometric look at algebra. This combination of algebra and geometry is called analytic geometry. To begin our study, we introduce the rectangular coordinate system in Section 7.1. Then, in Sections 7.2 and 7.3, we discuss the graph of an equation. Finally, in Sections 7.4 through 7.7, we discuss special types of graphs including lines, parabolas, circles, ellipses, and hyperbolas.

SECTION 7.1

The Rectangular Coordinate System

The Rectangular Coordinate System • Ordered Pairs as Solutions of Equations • The Distance Between Two Points in the Plane

The Rectangular Coordinate System

Just as we can represent real numbers by points on the real number line, we can represent ordered pairs of real numbers by points in a plane. This plane is called a **rectangular coordinate system** or the **Cartesian plane**, after the French mathematician René Descartes (1596–1650).

A rectangular coordinate system is formed by two real lines intersecting at a right angle, as shown in Figure 7.1. The horizontal number line is usually called the ***x*-axis**, and the vertical number line is usually called the ***y*-axis**. (The plural of axis is *axes*.) The point of intersection of the two axes is called the **origin**, and the axes separate the plane into four regions called **quadrants**.

FIGURE 7.1

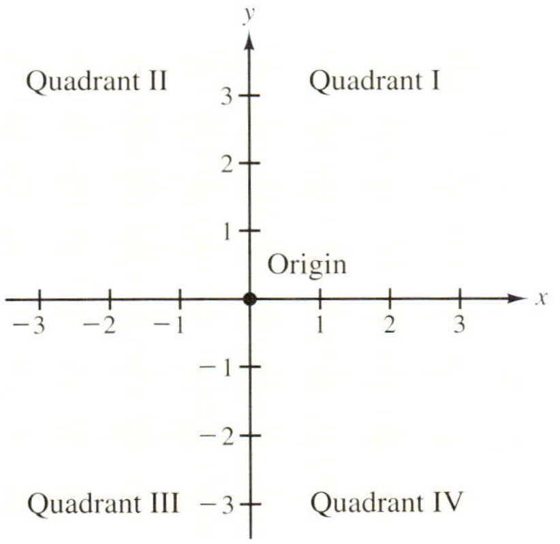

Each point in the plane corresponds to an **ordered pair** (x, y) of real numbers x and y, called the **coordinates** of the point. The first number (or ***x*-coordinate**) tells how far to the left or right the point is from the vertical axis, and the second number (or ***y*-coordinate**) tells how far up or down the point is from the horizontal axis, as shown in Figure 7.2.

FIGURE 7.2

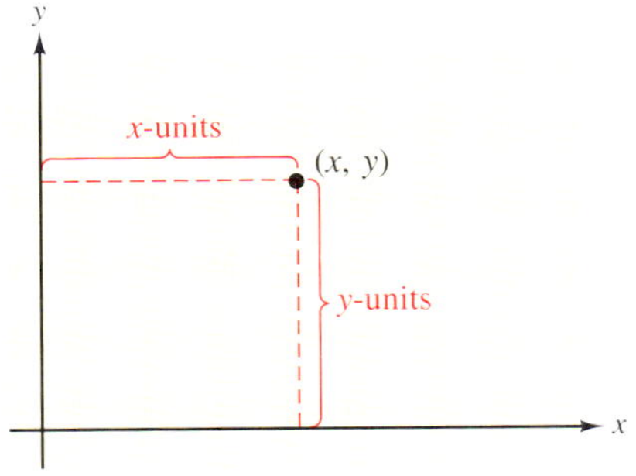

NOTE: A positive x-coordinate implies that the point lies to the right of the vertical axis; a negative x-coordinate implies that the point lies to the left of the vertical axis; and an x-coordinate of zero implies that the point lies on the vertical axis. Similarly, a positive y-coordinate implies that the point lies above the horizontal axis; a negative y-coordinate implies that the point lies below the horizontal axis; and a y-coordinate of zero implies that the point lies on the horizontal axis.

Locating a given point in the plane is called **plotting** the point. We show how this is done in Example 1.

EXAMPLE 1 ■ Plotting Points in a Rectangular Coordinate System

Plot the points $(-2, 1)$, $(4, 0)$, $(3, -1)$, $(4, 3)$, $(0, 0)$, and $(-1, -3)$ in a rectangular coordinate system.

Solution

The point $(-2, 1)$ is two units to the *left* of the vertical axis and one unit *above* the horizontal axis. Similarly, the point $(4, 0)$ is four units to the *right* of the vertical axis and *on* the horizontal axis. (It is on the horizontal axis because its y-coordinate is zero.) The other four points can be plotted in a similar way, as shown in Figure 7.3.

FIGURE 7.3

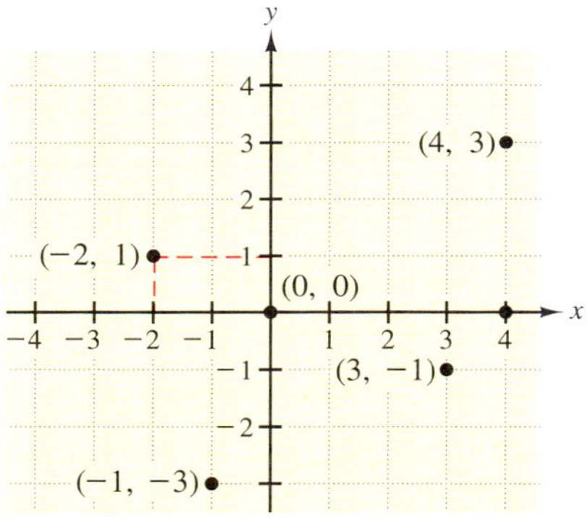

■

In Example 1 we were given the coordinates of several points and asked to plot the points in a rectangular coordinate system. In Example 2 we look at the reverse problem. That is, we are given points in a rectangular coordinate system and asked to determine their coordinates.

EXAMPLE 2 ■ Finding Coordinates of Points

Determine the coordinates for each of the points shown in Figure 7.4.

Solution

Point A lies two units to the *right* of the vertical axis and one unit *below* the horizontal axis. Therefore, point A must be given by the ordered pair $(2, -1)$. The coordinates of the other four points can be determined in a similar way, and we summarize the results as follows.

Point	*Position*	*Coordinates*
A	2 units *right*, 1 unit *down*	$(2, -1)$
B	1 unit *left*, 5 units *up*	$(-1, 5)$
C	4 units *right*, 0 units *up*	$(4, 0)$
D	3 units *left*, 2 units *down*	$(-3, -2)$
E	2 units *right*, 4 units *up*	$(2, 4)$

FIGURE 7.4

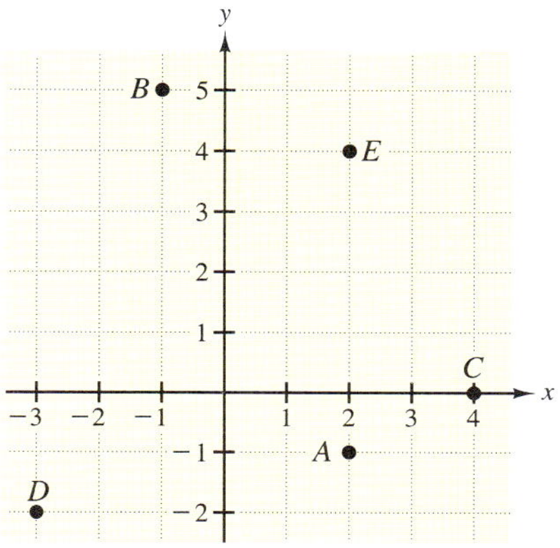

■

The primary value of a rectangular coordinate system is that it allows us to visualize relationships between two variables. Today, Descartes's ideas are in common use in virtually every scientific and business-related field. Ordered pairs can give us visual pictures of how total sales are related to the price of an item, how temperature is related to time of day, how interest rates are related to months of a year, and so on.

EXAMPLE 3 ■ An Application of the Rectangular Coordinate System

The population of the state of California from 1976 to 1989 is given in Table 7.1. Plot these points in a rectangular coordinate system. (Note that the population numbers given in the table are in thousands. Thus, the population in 1976 was 21,522 thousand, or 21,522,000.)

TABLE 7.1

Year	1976	1977	1978	1979	1980	1981	1982
Population	21,522	21,900	22,314	23,255	23,668	24,265	24,783

Year	1983	1984	1985	1986	1987	1988	1989
Population	25,308	26,358	26,981	26,997	27,653	28,314	28,607

Solution

For these points, we choose to plot the years on the x-axis and the population on the y-axis. The points are shown in Figure 7.5. Note that the break in the x-axis indicates that we have omitted the numbers between 0 and 1976.

FIGURE 7.5

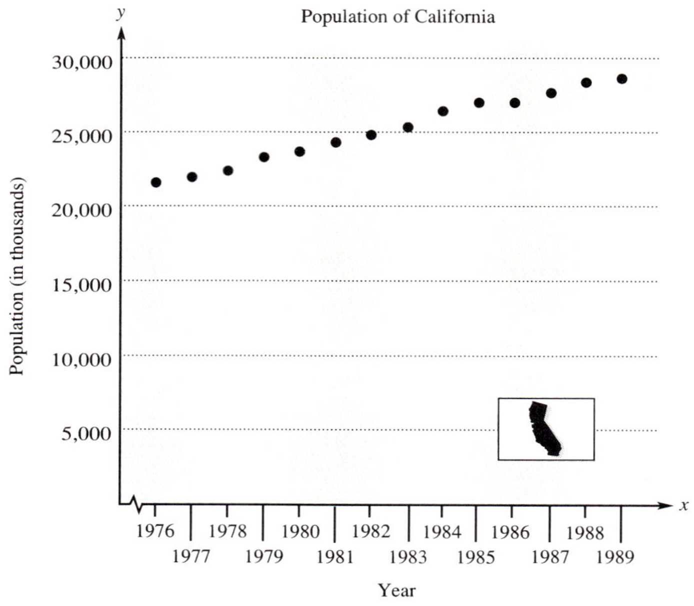

■

Ordered Pairs as Solutions of Equations

In Example 3 the relationship between the year and the population was given by a **table of values**. In mathematics the relationship between the variables x and y is often given by an equation. From the equation, we can then construct our own table of values. For instance, consider the equation

$$y = 3x + 2.$$

To construct a table of values for this equation, we choose several x-values and then calculate the corresponding y-values. For example, if we choose $x = 1$, then the y-value is

$$y = 3(1) + 2 = 5.$$

The corresponding ordered pair $(x, y) = (1, 5)$ is called a **solution point** (or simply a **solution**) of the equation. Table 7.2 shows a table of values (and their corresponding solution points) using x-values of -3, -2, -1, 0, 1, 2, and 3.

TABLE 7.2

Choose x	Calculate y from $y = 3x + 1$	Solution Points
$x = -3$	$y = 3(-3) + 2 = -7$	$(-3, -7)$
$x = -2$	$y = 3(-2) + 2 = -4$	$(-2, -4)$
$x = -1$	$y = 3(-1) + 2 = -1$	$(-1, -1)$
$x = 0$	$y = 3(0) + 2 = 2$	$(0, 2)$
$x = 1$	$y = 3(1) + 2 = 5$	$(1, 5)$
$x = 2$	$y = 3(2) + 2 = 8$	$(2, 8)$
$x = 3$	$y = 3(3) + 2 = 11$	$(3, 11)$

Once we have constructed a table of values, we can get a visual idea of the relationship between the variables x and y by plotting the solution points in a rectangular coordinate system. For instance, the solution points shown in Table 7.2 are plotted in Figure 7.6.

FIGURE 7.6

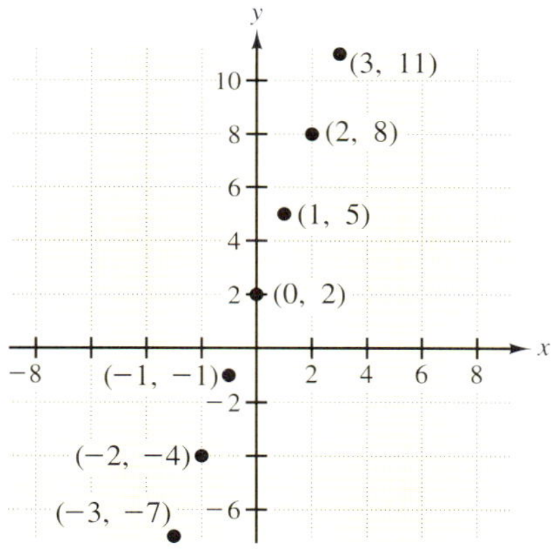

When making up a table of values for an equation, it is helpful to first solve the equation for y. For instance, the equation $5x + 3y = 4$ can be solved for y as follows.

$$5x + 3y = 4 \qquad \text{Given equation}$$

$$3y = -5x + 4 \qquad \text{Subtract } 5x \text{ from both sides}$$

$$y = -\frac{5}{3}x + \frac{4}{3} \qquad \text{Solve for } y$$

This procedure is demonstrated further in Example 4.

EXAMPLE 4 ■ Making Up a Table of Values

Make up a table of values showing five solution points for the equation

$$2x^2 - 3y = 5.$$

Then plot the solution points in the rectangular coordinate system. (Choose x-values of -2, -1, 0, 1, and 2.)

Solution

To begin we solve the given equation for y.

$$\begin{aligned} 2x^2 - 3y &= 5 \\ -3y &= -2x^2 + 5 \\ y &= \frac{2}{3}x^2 - \frac{5}{3} \end{aligned}$$

Now, using the equation $y = \frac{2}{3}x^2 - \frac{5}{3}$, we construct the table of values, as shown in Table 7.3.

TABLE 7.3

x	-2	-1	0	1	2
$y = \frac{2}{3}x^2 - \frac{5}{3}$	1	-1	$-\frac{5}{3}$	-1	1
Solution Points	$(-2, 1)$	$(-1, -1)$	$\left(0, -\frac{5}{3}\right)$	$(1, -1)$	$(2, 1)$

Finally, from Table 7.3, we plot the five solution points in the rectangular coordinate system, as shown in Figure 7.7.

FIGURE 7.7

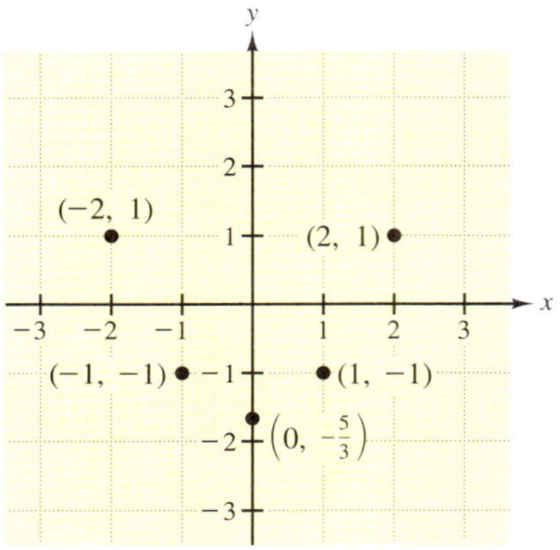

In the next example, you are given several ordered pairs and asked to determine whether they are solutions of the given equation. To do this you need to substitute the values of x and y into the equation. If the substitution produces a true equation, then the ordered pair (x, y) is a solution. Otherwise, the ordered pair is not a solution.

NOTE: When a substitution of (x, y) produces a true equation, we say that the ordered pair (x, y) **satisfies** the equation.

EXAMPLE 5 ■ Verifying Solutions of an Equation

Which of the following ordered pairs is a solution of the equation $x^2 - 2y = 6$?

(a) $(2, 1)$ (b) $(0, -3)$ (c) $(-2, -5)$ (d) $\left(1, -\frac{5}{2}\right)$

Solution

(a) For the ordered pair $(x, y) = (2, 1)$, we substitute $x = 2$ and $y = 1$ into the given equation.

$x^2 - 2y = 6$	Given equation
$2^2 - 2(1) \stackrel{?}{=} 6$	Substitute $x = 2$ and $y = 1$
$2 \neq 6$	Not a solution

Since the substitution did not satisfy the given equation, we conclude that the ordered pair $(2, 1)$ is *not* a solution of the given equation.

(b) The ordered pair $(0, -3)$ *is* a solution of the given equation because

$$0^2 - 2(-3) = 0 + 6 = 6.$$

(c) The ordered pair $(-2, -5)$ is *not* a solution of the given equation because

$$(-2)^2 - 2(-5) = 14 \neq 6.$$

(d) The ordered pair $\left(1, -\frac{5}{2}\right)$ *is* a solution of the given equation because

$$1^2 - 2\left(-\frac{5}{2}\right) = 1 - (-5) = 6.$$

■

MATH MATTERS

Finding the Shortest Path

We know that the shortest distance between two points is a straight line. However, when we travel from one place to another, we often cannot travel in a straight line (we must follow the roads or sidewalks).

Here is a problem involving the shortest path from one point to another. Suppose a spider and a fly are on opposite walls of a rectangular room, as shown in the figure. The spider wants to visit the fly, and assuming that the spider must travel on one of the surfaces of the room, what is the shortest path to the fly? (The answer is given in the back of the text.)

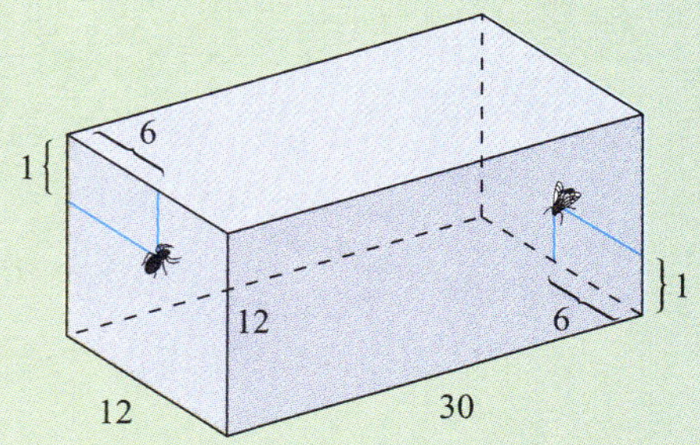

The Distance Between Two Points in the Plane

We know from Section 1.1 that the distance d between two points a and b on the real number line is simply

$$d = |b - a|.$$

The same "absolute value rule" is used to find the distance between two points that lie on the same *vertical* or *horizontal* line, as shown in Example 6.

EXAMPLE 6 ■ Finding Horizontal and Vertical Distances

(a) Find the distance between the points $(2, -2)$ and $(2, 4)$.

(b) Find the distance between the points $(-3, -2)$ and $(2, -2)$.

Solution

(a) Because the x-coordinates are equal, we visualize a vertical line through the points $(2, -2)$ and $(2, 4)$, as shown in Figure 7.8. The distance between these two points is given by the absolute value of the difference of their y-coordinates. That is,

$$\text{Vertical distance} = |4 - (-2)| = 6. \qquad \text{Subtract } y\text{-coordinates}$$

(b) Because the y-coordinates are equal, we visualize a horizontal line through the points $(-3, -2)$ and $(2, -2)$, as shown in Figure 7.8. The distance between these two points is given by the absolute value of the difference of their x-coordinates. That is,

$$\text{Horizontal distance} = |2 - (-3)| = 5. \qquad \text{Subtract } x\text{-coordinates}$$

FIGURE 7.8

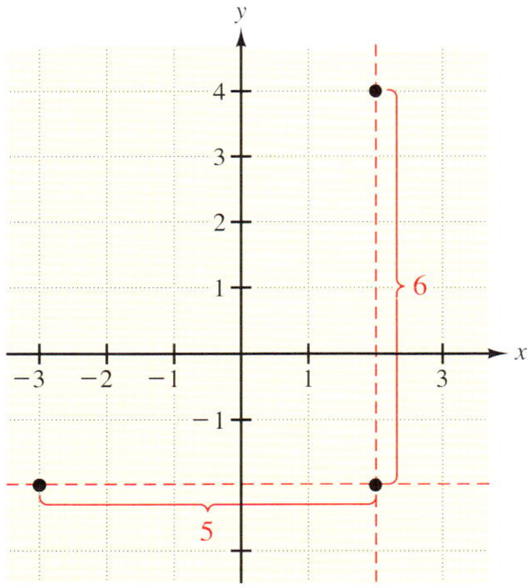

■

The technique used in Example 6 can be used to develop a general formula for finding the distance between two points in the plane. This general formula will work for any two points, even if they do not lie on the same vertical or horizontal line. To develop the formula, we use the Pythagorean Theorem, which states that for a right triangle, the hypotenuse c and sides a and b are related by the formula $a^2 + b^2 = c^2$, as shown in Figure 7.9. (The converse is also true. That is, if $a^2 + b^2 = c^2$, then the triangle is a right triangle.)

FIGURE 7.9 Pythagorean Theorem

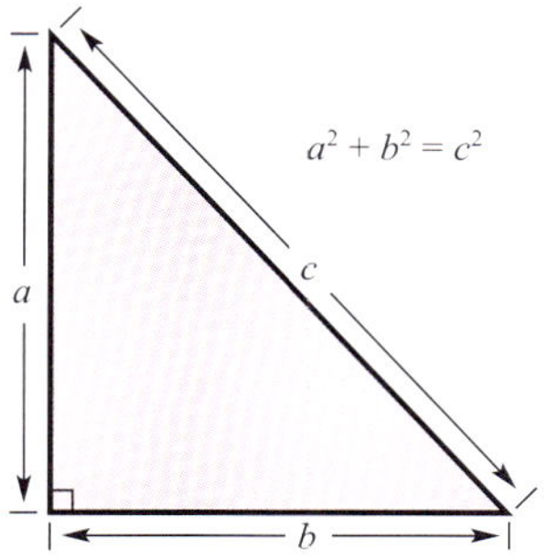

FIGURE 7.10 Distance Between Two Points

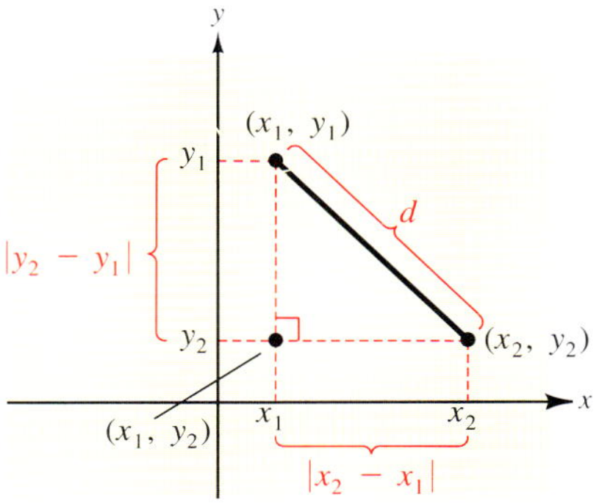

To develop a general formula for the distance between two points, we let (x_1, y_1) and (x_2, y_2) represent two points in the plane (that do not lie on the same horizontal or vertical line). With these two points, a right triangle can be formed, as shown above in Figure 7.10. Note that the third vertex of the triangle is (x_1, y_2). Since (x_1, y_1) and (x_1, y_2) lie on the same vertical line, the length of the vertical side of the triangle is $|y_2 - y_1|$. Similarly, the length of the horizontal side is $|x_2 - x_1|$. Thus, by the Pythagorean Theorem, the distance between (x_1, y_1) and (x_2, y_2) is given by

$$d^2 = |x_2 - x_1|^2 + |y_2 - y_1|^2.$$

Since the distance d must be positive, we choose the positive square root and write

$$d = \sqrt{|x_2 - x_1|^2 + |y_2 - y_1|^2}.$$

Finally, replacing $|x_2 - x_1|^2$ and $|y_2 - y_1|^2$ by the equivalent expressions $(x_2 - x_1)^2$ and $(y_2 - y_1)^2$ gives us the following formula for the distance between two points in a rectangular coordinate plane.

The Distance Formula The distance d between the two points (x_1, y_1) and (x_2, y_2) in a coordinate plane is

$$d = \sqrt{(x_2 - x_1)^2 + (y_2 - y_1)^2}.$$

Note that for the special case in which the two points lie on the same vertical or horizontal line, the Distance Formula still works. For instance, applying the Distance Formula to the points $(2, -2)$ and $(2, 4)$ produces

$$d = \sqrt{(2 - 2)^2 + [4 - (-2)]^2} = \sqrt{6^2} = 6$$

which is the same result we obtained in Example 6.

EXAMPLE 7 ■ **Finding the Distance Between Two Points**

Find the distance between the points $(-1, 2)$ and $(2, 4)$.

Solution

Letting $(x_1, y_1) = (-1, 2)$ and $(x_2, y_2) = (2, 4)$, we apply the Distance Formula to obtain

$$\begin{aligned} d &= \sqrt{[2 - (-1)]^2 + (4 - 2)^2} \\ &= \sqrt{3^2 + 2^2} \\ &= \sqrt{13} \\ &\approx 3.61. \end{aligned}$$

(See Figure 7.11.)

FIGURE 7.11

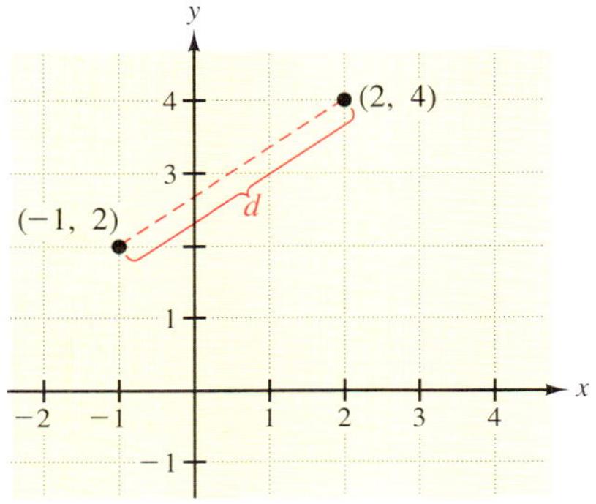

■

In Example 7 the figure provided was not essential to the solution of the problem. Nevertheless, we recommend that you include graphs with your solutions.

EXAMPLE 8 ■ **An Application of the Distance Formula**

Show that the points $(1, 2)$, $(3, 1)$, and $(4, 3)$ are vertices of a right triangle.

Solution

FIGURE 7.12

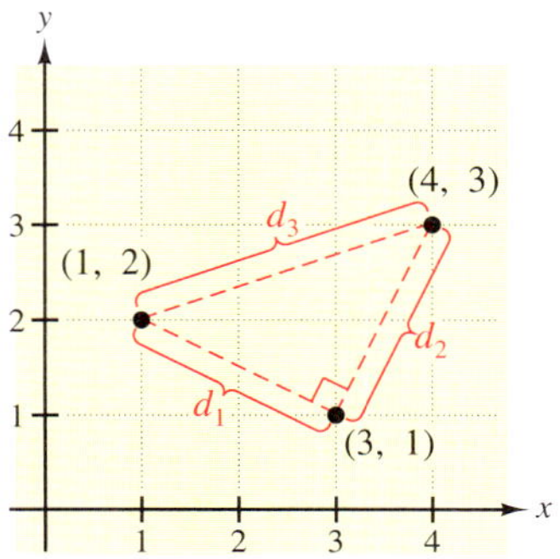

The three points are plotted in Figure 7.12. Using the Distance Formula, we find the lengths of the three sides of the triangle.

$$d_1 = \sqrt{(3 - 1)^2 + (1 - 2)^2} = \sqrt{4 + 1} = \sqrt{5}$$
$$d_2 = \sqrt{(4 - 3)^2 + (3 - 1)^2} = \sqrt{1 + 4} = \sqrt{5}$$
$$d_3 = \sqrt{(4 - 1)^2 + (3 - 2)^2} = \sqrt{9 + 1} = \sqrt{10}$$

Since $d_1{}^2 + d_2{}^2 = 5 + 5 = 10 = d_3{}^2$, we can conclude from the Pythagorean Theorem that the triangle is a right triangle.

■

DISCUSSION PROBLEM ■ Finding the Midpoint of a Line Segment

Plot the points $A = (-1, -3)$ and $B = (5, 2)$ in a rectangular coordinate system, and draw the line segment connecting the two points.

(a) Try to find the coordinates of the point C that is the midpoint of the line segment.

(b) Give a verbal explanation of how you arrived at your choice of coordinates for the point C.

(c) Write a paragraph describing how you could verify that the point C is the midpoint of the line segment. (Why is it not sufficient to show that the distance between A and C is the same as the distance between C and B?)

(d) Try to find a general formula for the coordinates of the midpoint of the line segment connecting the points (x_1, y_1) and (x_2, y_2). ■

Warm-Up

The following warm-up exercises involve skills that were covered in earlier sections. You will use these skills in the exercise set for this section.

In Exercises 1–6, plot each real number on the real number line and place the correct inequality symbol (< or >) between the real numbers.

1. $-6 \;\square\; 2$ **2.** $\frac{11}{3} \;\square\; 2$ **3.** $-3 \;\square\; -7$

4. $-5 \;\square\; 0$ **5.** $\frac{13}{3} \;\square\; -\frac{3}{2}$ **6.** $\frac{3}{5} \;\square\; \frac{13}{16}$

In Exercises 7–10, evaluate the specified quantity.

7. $5 - |-2|$ **8.** $-15 - 3(4 - 18)$

9. $\dfrac{3 - 2(5 - 20)}{4}$ **10.** $\dfrac{|3 - 12|}{3}$

7.1 EXERCISES

In Exercises 1–6, plot the points in a rectangular coordinate system.

1. $(4, 3), (-5, 3), (3, -5)$ **2.** $(-2, 5), (-2, -5), (3, 5)$ **3.** $(-8, -2), (6, -2), (6, 5)$

4. $(0, 4), (0, 0), (3, 0)$ **5.** $\left(\frac{5}{2}, -2\right), \left(-2, \frac{1}{4}\right), \left(\frac{3}{2}, -\frac{7}{2}\right)$ **6.** $\left(-\frac{2}{3}, 3\right), \left(\frac{1}{4}, -\frac{5}{4}\right), \left(-5, -\frac{7}{4}\right)$

In Exercises 7 and 8, without plotting the points, determine the quadrant in which each point is located.

7. (a) $\left(3, -\frac{5}{8}\right)$ (b) $(-6.2, 8.05)$

8. (a) $(200, 1365.6)$ (b) $\left(-\frac{5}{11}, -\frac{3}{8}\right)$

In Exercises 9 and 10, approximate the coordinates of the plotted points.

9.

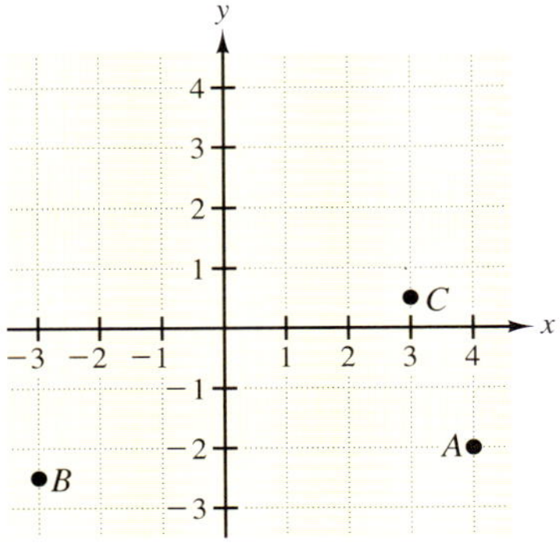

10.

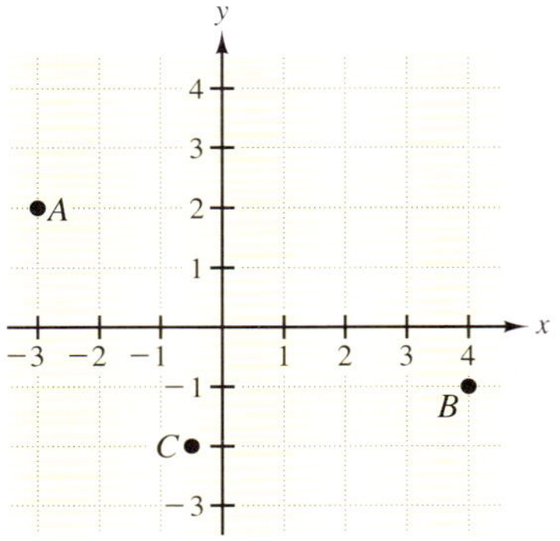

In Exercises 11–16, plot the given points and connect them with line segments in such a way as to form the specified figure. (*Note:* A **rhombus** is a parallelogram whose sides are all the same length.)

11. Triangle: $(-1, 2)$, $(2, 0)$, $(3, 5)$

12. Triangle: $(1, 4)$, $(0, -1)$, $(5, 9)$

13. Square: $(0, 6)$, $(3, 3)$, $(0, 0)$, $(-3, 3)$

14. Rectangle: $(7, 0)$, $(9, 1)$, $(4, 6)$, $(6, 7)$

15. Parallelogram: $(4, 0)$, $(6, -2)$, $(0, -4)$, $(-2, -2)$

16. Rhombus: $(-3, -3)$, $(-2, -1)$, $(-1, -2)$, $(0, 0)$

In Exercises 17–20, determine the quadrant or quadrants in which the point must be located in order to satisfy the given requirement.

17. $(-3, y)$

18. $(x, 5)$

19. (x, y), $xy > 0$

20. (x, y), $x > 0$, $y < 0$

In Exercises 21–24, find the coordinates of the indicated point.

21. The point is located three units to the right of the y-axis and four units below the x-axis.

22. The point is located two units to the left of the y-axis and five units above the x-axis.

23. The coordinates of the point are equal, and it is located in the third quadrant 10 units to the left of the y-axis.

24. The coordinates of the point are equal in magnitude and opposite in sign, and the point is located seven units to the right of the y-axis.

In Exercises 25–28, plot the points whose coordinates are given in the table.

25. *Exam Score* The accompanying table gives the time x in hours invested in concentrated study for five different algebra exams and the resulting score y.

x	5	2	3	6.5	4
y	81	71	88	92	86

26. *Fuel Efficiency* The accompanying table gives the speed x of a car in miles per hour and the approximate fuel efficiency y in miles per gallon.

x	50	55	60	65	70
y	28	26.4	24.8	23.4	22

27. *Average Temperature* The normal temperature y (Fahrenheit) for Duluth, Minnesota, for each month of the year is given in the following table. (*Source*: *PC USA*.) The months are numbered 1 through 12, with 1 corresponding to January.

x	1	2	3	4	5	6
y	6.3	12.0	22.9	38.3	50.3	59.4

x	7	8	9	10	11	12
y	65.3	63.2	54.0	44.2	28.2	13.8

28. *Price of Stock* The book value y per common share of stock in the Chrysler Corporation for the years 1983 through 1989 is given in the accompanying table. The time in years is given by x. (*Source*: Chrysler Corporation.)

x	y
1983	$ 4.17
1984	12.10
1985	18.50
1986	24.67
1987	29.40
1988	32.58
1989	32.42

In Exercises 29–34, determine whether the ordered pairs are solution points to the given equation.

29. $y = 3x + 8$
(a) $(3, 17)$ (b) $(-1, 10)$
(c) $(0, 0)$ (d) $(-2, 2)$

30. $y = 10x - 7$
(a) $(2, 10)$ (b) $(-2, -27)$
(c) $(5, 43)$ (d) $(1, 5)$

31. $4y - 2x + 1 = 0$
(a) $(0, 0)$ (b) $\left(\frac{1}{2}, 0\right)$
(c) $\left(-3, -\frac{7}{4}\right)$ (d) $\left(1, -\frac{3}{4}\right)$

32. $5x - 2y + 50 = 0$
(a) $(-10, 0)$ (b) $(-5, 5)$
(c) $(0, 25)$ (d) $(20, -2)$

33. $y = \frac{7}{8}x$
(a) $\left(\frac{8}{7}, 1\right)$ (b) $\left(4, \frac{7}{2}\right)$
(c) $(0, 0)$ (d) $(-16, 14)$

34. $y = \frac{5}{8}x - 2$
(a) $(0, 0)$ (b) $(2, 2)$
(c) $(-4, -7)$ (d) $(32, 18)$

In Exercises 35–38, complete the following table by finding the indicated value of y.

35.

x	-2	0	2	4	6
$y = 5x - 1$					

36.

x	-2	0	2	4	6
$y = \frac{3}{4}x + 2$					

37.

x	-4	$\frac{2}{5}$	4	8	12
$y = -\frac{5}{2}x + 4$					

38.

x	-6	-3	0	$\frac{3}{4}$	10
$y = \frac{4}{3}x - \frac{1}{3}$					

39. *Cost* The cost y for producing x units is given by $y = 28x + 3{,}000$. Complete the following table to determine the cost for producing the specified number of units.

x	100	150	200	250	300
$y = 28x + 3{,}000$					

40. *Hourly Wages* When an employee produces x units per hour, the wage y paid per hour is given by $y = 0.75x + 8$. Complete the following table to determine the hourly wages for producing the specified number of units.

x	2	4	8	10	20
$y = 0.75x + 8$					

In Exercises 41–46, plot the points and find the distance between them.

41. $(3, -2), (3, 5)$

42. $(-2, 8), (-2, 1)$

43. $(3, 2), (10, 2)$

44. $(-120, -2), (130, -2)$

45. $\left(\frac{3}{4}, 1\right), \left(\frac{3}{4}, -10\right)$

46. $\left(\frac{1}{2}, \frac{7}{8}\right), \left(\frac{11}{2}, \frac{7}{8}\right)$

In Exercises 47–50, find (a) the missing coordinates of the one vertex of the right triangle, (b) the lengths of the vertical and horizontal sides of the triangle, (c) the length of the hypotenuse, and (d) the distance between the two given points.

47. $(10, 8), (2, 2)$

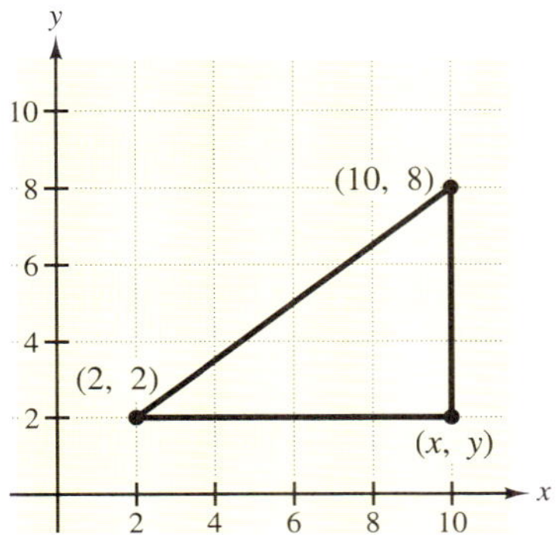

48. $(-2, 1), (10, 6)$

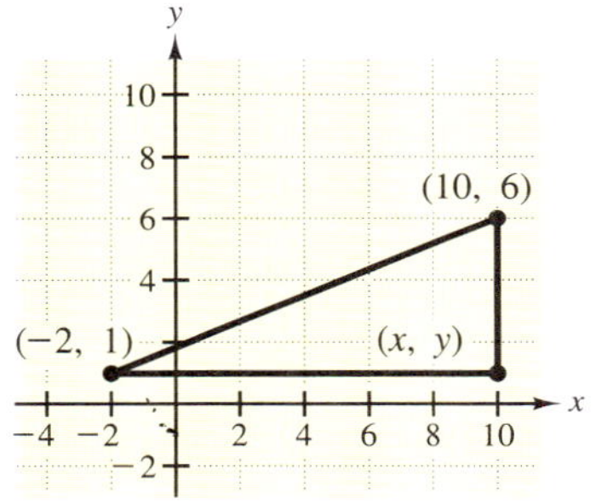

49. $(-3, -4), (4, 4)$

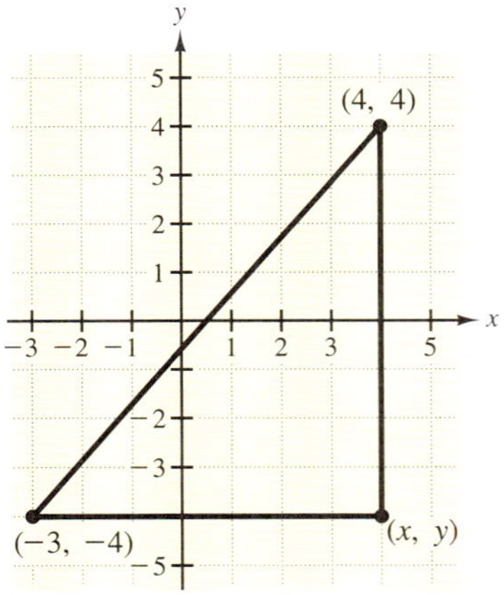

50. (0, 6), (6, −2)

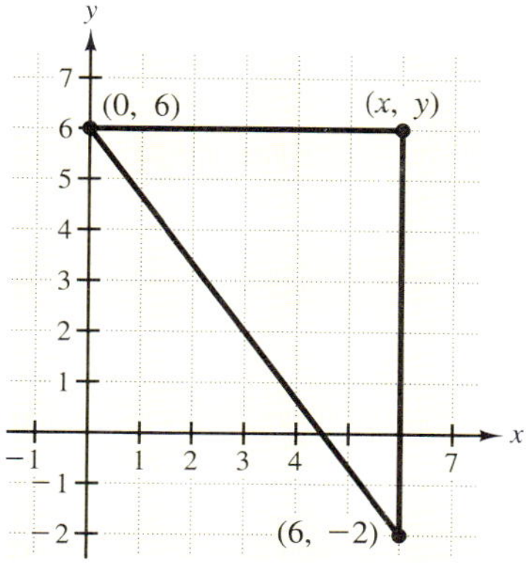

In Exercises 51–58, find the distance between the two given points.

51. (1, 3), (5, 6)

52. (3, 10), (15, 5)

53. (0, 0), (8, −8)

54. (−5, 0), (0, 5)

55. (−2, −3), (4, 1)

56. (−5, 4), (10, −1)

57. $\left(\frac{1}{3}, 3\right), \left(3, \frac{2}{3}\right)$

58. (0.3, 1), (1.5, −0.5)

In Exercises 59–62, use the Distance Formula to determine whether the three points are collinear.

59. (2, 1), (4, 2), (6, 3)

60. (0, 2), (1, 0), (2, −2)

61. (−1, 3), (2, 1), (4, 0)

62. (−5, 1), (0, 3), (5, 4)

In Exercises 63 and 64, find the perimeter of the triangle having the given vertices.

63. (−2, 0), (0, 5), (1, 0)

64. (−5, −2), (−1, 4), (3, −1)

In Exercises 65–68, plot the given points and the midpoint of the line segment joining the given points. The coordinates of the midpoint of the line segment joining the points (x_1, y_1) and (x_2, y_2) are given by the following.

$$\text{Midpoint} = \left(\frac{x_1 + x_2}{2}, \frac{y_1 + y_2}{2}\right)$$

65. (−2, 0), (4, 8)

66. (−3, −2), (7, 2)

67. (1, 6), (6, 3)

68. (2, 7), (9, −1)

69. Plot the points (2, 1), (−3, 5), and (7, −3) in a rectangular coordinate system. Then change the sign of the x-coordinate of each point and plot the three new points in the same rectangular coordinate system. What inference can you make about the location of a point when the sign of the x-coordinate is changed?

70. Plot the points (2, 1), (−3, 5), and (7, −3) in a rectangular coordinate system. Then change the sign of the y-coordinate of each point and plot the three new points in the same rectangular coordinate system. What inference can you make about the location of a point when the sign of the y-coordinate is changed?

71. *Housing Construction* A house is 30 feet wide and the ridge of the roof is 6 feet above the top of the walls (see figure). Find the length of the rafters if they overhang the edge of the walls by 2 feet.

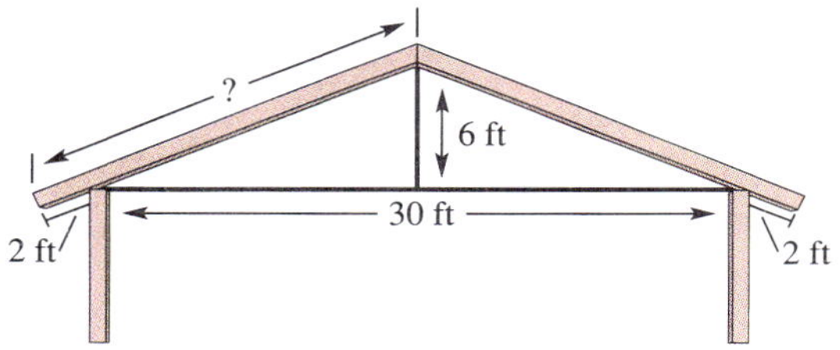

72. *Housing Construction* Determine the length of the handrail over the stairs shown in the accompanying figure.

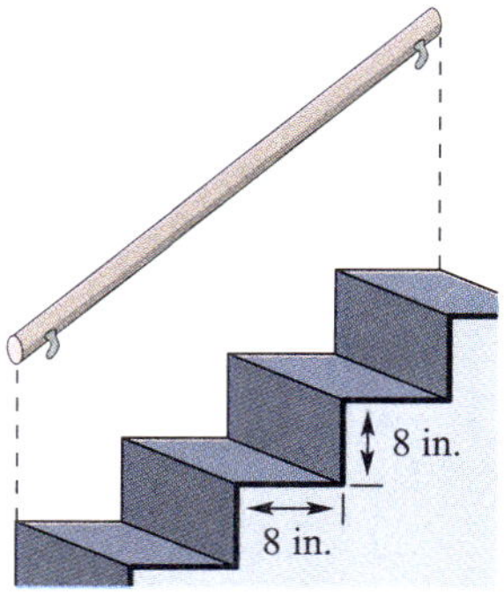

SECTION 7.2

Graphs of Equations

The Graph of an Equation • *Intercepts: An Aid to Sketching Graphs* • *Applications*

The Graph of an Equation

We have already seen that the solutions of an equation involving two variables can be represented by points in a rectangular coordinate system. The set of all solutions of an equation is called its **graph**. In Section 7.1 we took the first step in sketching the graph of an equation. That is, we constructed a table of values and plotted the corresponding solution points in a rectangular coordinate system. In this section we take the process one step further by connecting the solution points with a smooth curve or line.

To see how this works, let's make a sketch of the graph of the equation

$$y = -2x + 3.$$

To begin, we construct a table of values, as shown in Table 7.4.

Next, from the table of values, we plot the solution points in a rectangular coordinate system, as shown in Figure 7.13(a). The seven points plotted in this figure represent only a few of the points on the graph of the equation. To complete our sketch, we try to determine whether the representative points form a pattern. In this case all seven points appear to lie on a line, so we complete the sketch by drawing a straight line through the points, as shown in Figure 7.13(b).

TABLE 7.4

x	$y = -2x + 3$	Solution Points
-3	9	$(-3, 9)$
-2	7	$(-2, 7)$
-1	5	$(-1, 5)$
0	3	$(0, 3)$
1	1	$(1, 1)$
2	-1	$(2, -1)$
3	-3	$(3, -3)$

FIGURE 7.13

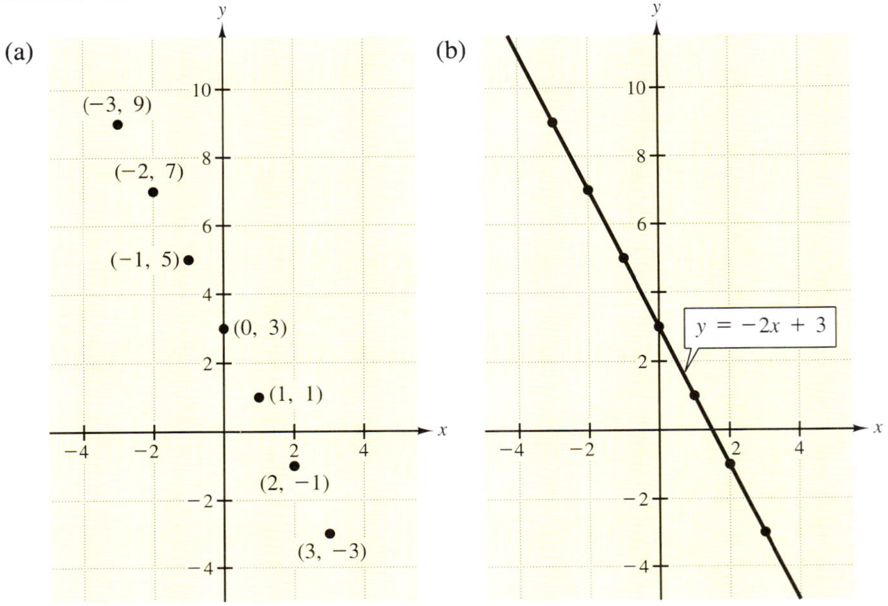

This method for sketching the graph of an equation is called the **point-plotting method**. We summarize the basic steps of the method as follows.

The Point-Plotting Method for Sketching a Graph

To sketch the graph of an equation by point-plotting, use the following steps.

1. If possible, rewrite the equation by isolating one of the variables.
2. Make up a table of values showing several solution points.
3. Plot these points in a rectangular coordinate system.
4. Connect the points with a smooth curve or line.

EXAMPLE 1 ■ Sketching the Graph of an Equation

Sketch the graph of the equation $3x - y = 2$.

Solution

To begin, we rewrite the equation so that y is isolated on the left.

$$y = 3x - 2$$

Next, we make up a table of values, as shown in Table 7.5. (The choice of x-values to use in the table is somewhat arbitrary. However, the more x-values we choose, the easier it is to recognize a pattern.)

TABLE 7.5

x	-2	-1	0	1	2	3
$y = 3x - 2$	-8	-5	-2	1	4	7
Solution Points	$(-2, -8)$	$(-1, -5)$	$(0, -2)$	$(1, 1)$	$(2, 4)$	$(3, 7)$

Now, we plot the six solution points, as shown in Figure 7.14(a). It appears that all six points lie on a line, so we complete the sketch by drawing a line through these points, as shown in Figure 7.14(b).

FIGURE 7.14

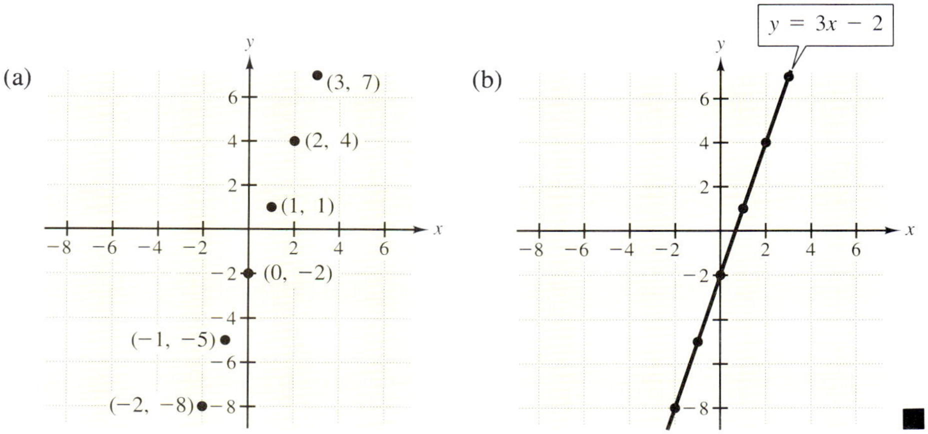

■

The equation in Example 1 is an example of a **linear equation** in two variables—it is of first degree in both variables and its graph is a straight line. (We will study this type of equation in Sections 7.4 and 7.5.) The graph of a nonlinear equation is not a straight line. The next two examples show how we can use the point-plotting method to sketch the graph of a nonlinear equation.

EXAMPLE 2 ■ Sketching the Graph of a Nonlinear Equation

Sketch the graph of the following equation.

$$-x^2 + 2x + y = 0$$

Solution

To begin, we rewrite the equation so that y is isolated on the left.

$$y = x^2 - 2x$$

Next, we make up a table of values, as shown in Table 7.6. Watch out for the signs of the numbers when making up such a table. For instance, when $x = -2$, the value of y is

$$y = (-2)^2 - 2(-2) = 4 + 4 = 8.$$

TABLE 7.6

x	$y = x^2 - 2x$	Solution Points
-2	8	$(-2, 8)$
-1	3	$(-1, 3)$
0	0	$(0, 0)$
1	-1	$(1, -1)$
2	0	$(2, 0)$
3	3	$(3, 3)$
4	8	$(4, 8)$

Now, we plot the seven solution points, as shown in Figure 7.15(a). Finally, we connect the points with a smooth curve, as shown in Figure 7.15(b).

FIGURE 7.15

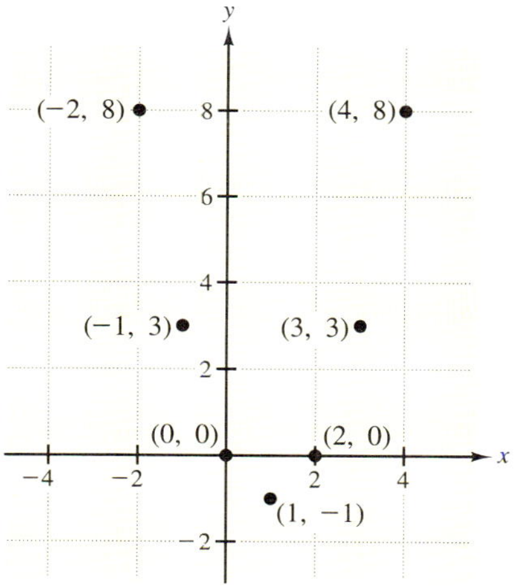

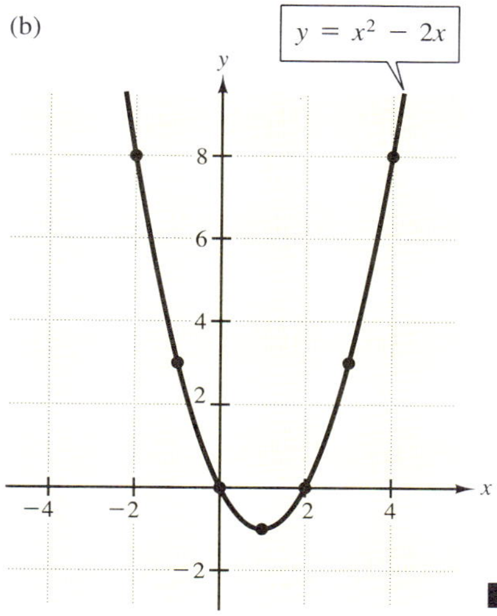

The graph of the equation given in Example 2 is called a **parabola**. We will study this type of graph in detail in Section 7.6.

In Example 3 we look at the graph of an equation that involves an absolute value. Remember from Chapter 1 that to find the absolute value of a number, we disregard the sign of the number. For instance, $|-5| = 5$, $|2| = 2$, and $|0| = 0$.

EXAMPLE 3 ■ The Graph of an Equation Involving an Absolute Value

Sketch a graph of the following equation.

$$y = |x - 2|$$

Solution

This equation is already written in a form with y isolated on the left. So we begin by making up a table of values, as shown in Table 7.7. Be sure that you understand how the absolute value is evaluated. For instance, when $x = -2$, the value of y is

$$y = |-2 - 2| = |-4| = 4.$$

Similarly, when $x = 3$, the value of y is

$$y = |3 - 2| = |1| = 1.$$

TABLE 7.7

x	-2	-1	0	1	2	3	4	5
$y = \|x - 2\|$	4	3	2	1	0	1	2	3
Solution Points	$(-2, 4)$	$(-1, 3)$	$(0, 2)$	$(1, 1)$	$(2, 0)$	$(3, 1)$	$(4, 2)$	$(5, 3)$

Next, we plot the eight solution points, as shown in Figure 7.16(a). It appears that the points lie in a "V-shaped" pattern, with the point (2, 0) lying at the bottom of the "V." Following this pattern, we connect the points to form the graph shown in Figure 7.16(b).

FIGURE 7.16

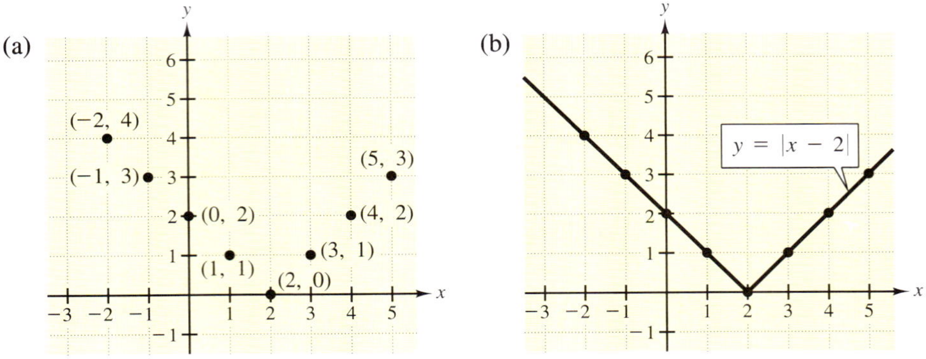

Intercepts: An Aid to Sketching Graphs

Two types of solution points that are especially useful are those having zero as either the x- or y-coordinate. Such points are called **intercepts** because they are the points at which the graph intersects the x- or y-axis.

Definition of Intercepts

The point $(a, 0)$ is called an ***x*-intercept** of the graph of an equation if it is a solution point of the equation. To find the x-intercepts, let y be zero and solve the equation for x.

The point $(0, b)$ is called a ***y*-intercept** of the graph of an equation if it is a solution point of the equation. To find the y-intercepts, let x be zero and solve the equation for y.

EXAMPLE 4 ■ **Finding the Intercepts of a Graph**

Find the intercepts and sketch a graph of the following equation.

$$y = 2x - 3$$

Solution

To determine whether the graph has any x-intercepts, we let y be zero and solve the resulting equation for x.

$y = 2x - 3$ Given equation

$0 = 2x - 3$ Let $y = 0$

$\frac{3}{2} = x$ Solve equation for x

Therefore, the graph has one x-intercept, which occurs at the point $\left(\frac{3}{2}, 0\right)$.

To determine whether the graph has any y-intercepts, we let x be zero and solve the resulting equation for y.

$y = 2x - 3$ Given equation

$y = 2(0) - 3$ Let $x = 0$

$y = -3$ Solve equation for y

Therefore, the graph has one y-intercept, which occurs at the point $(0, -3)$.

Now, to sketch the graph of the equation, we make up a table of values, as shown in Table 7.8. (Notice that the two intercepts are incorporated into the table.) Finally, using the solution points given in the table, we sketch the graph of the equation, as shown in Figure 7.17.

TABLE 7.8

x	-1	0	1	$\frac{3}{2}$	2	3	4
$y = 2x - 3$	-5	-3	-1	0	1	3	5
Solution Points	$(-1, -5)$	$(0, -3)$	$(1, -1)$	$\left(\frac{3}{2}, 0\right)$	$(2, 1)$	$(3, 3)$	$(4, 5)$

FIGURE 7.17

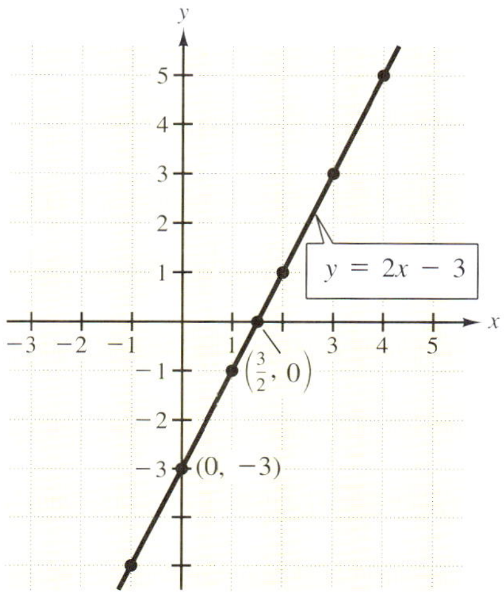

■

It is possible for a graph to have no intercepts or several intercepts. For instance, consider the three graphs in Figure 7.18.

FIGURE 7.18

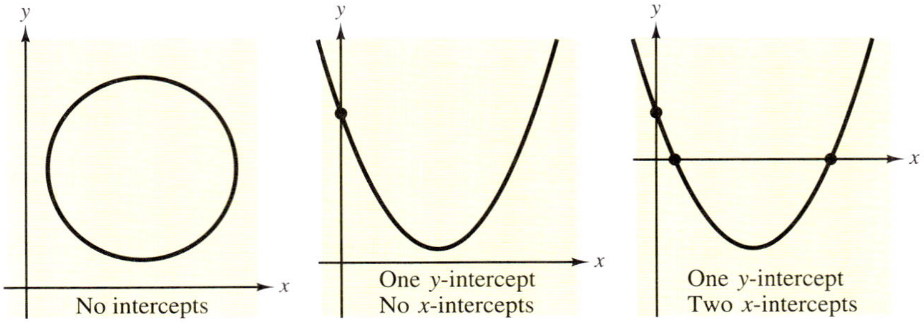

EXAMPLE 5 ■ A Graph that Has Two x-Intercepts

Find the intercepts and sketch a graph of the following equation.

$$y = x^2 - 5x + 4$$

Solution

To determine whether the graph has any x-intercepts, we let y be zero and solve the resulting equation for x.

$y = x^2 - 5x + 4$	Given equation
$0 = x^2 - 5x + 4$	Let $y = 0$
$0 = (x - 1)(x - 4)$	Factor

By setting each of these factors equal to zero, we conclude that the equation $0 = x^2 - 5x + 4$ has two solutions: 1 and 4. Therefore, it follows that the graph has two x-intercepts, the points (1, 0) and (4, 0).

To determine whether the graph has any y-intercepts, we let x be zero and solve the resulting equation for y.

$y = x^2 - 5x + 4$	Given equation
$y = 0^2 - 5(0) + 4$	Let $x = 0$
$y = 4$	Solve equation for y

Therefore, the graph has one y-intercept, the point (0, 4).

Now, to sketch the graph of the equation, we make up a table of values, as shown in Table 7.9. Finally, using the solution points given in the table, we sketch the graph of the equation, as shown in Figure 7.19.

TABLE 7.9

x	-1	0	1	2	3	4	5
$y = x^2 - 5x + 4$	10	4	0	-2	-2	0	4
Solution Points	$(-1, 10)$	$(0, 4)$	$(1, 0)$	$(2, -2)$	$(3, -2)$	$(4, 0)$	$(5, 4)$

FIGURE 7.19

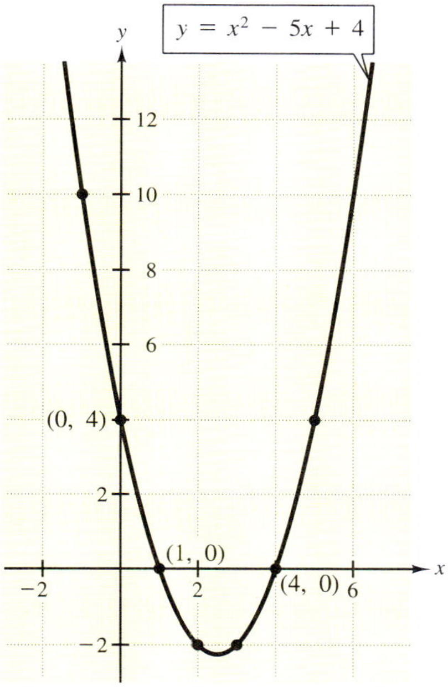

Applications

Newspapers and news magazines frequently use graphs to show time-of-year comparisons with rate of inflation, housing costs, wholesale prices, or the unemployment rate. Industrial firms and businesses use graphs to report monthly or yearly production and sales statistics. Such graphs provide a simple geometric picture of the way one quantity changes with respect to another. In the following example, we show how a graph can help us visualize the concept of straight-line depreciation.

EXAMPLE 6 ■ **An Application**

A small business buys a new printing press for \$65,000. For income tax purposes, the owner of the business decides to depreciate the printing press over a 10-year period. At the end of the 10 years, the salvage value of the printing press is expected to be \$5,000. Find an equation that relates the value of the printing press to the number of years. Then sketch the graph of the equation. (Assume that the depreciation is the same each year. This type of depreciation is called **straight-line depreciation**.)

Solution

The total depreciation over the 10-year period is $65,000 − $5,000 = $60,000. Since the same amount is depreciated each year, it follows that the annual depreciation is $60,000 ÷ 10 = $6,000. Thus, after one year, the value of the printing press is

Value after one year = $65,000 − $6,000 = $59,000.

By similar reasoning, we can see that the value after two years is

Value after two years = $65,000 − $6,000(2) = $53,000.

Now, by letting y represent the value of the printing press after t years, we follow the pattern given by the first two years and conclude that the equation is

$$y = 65{,}000 - 6{,}000t.$$

A sketch of the graph of this equation is shown in Figure 7.20.

FIGURE 7.20

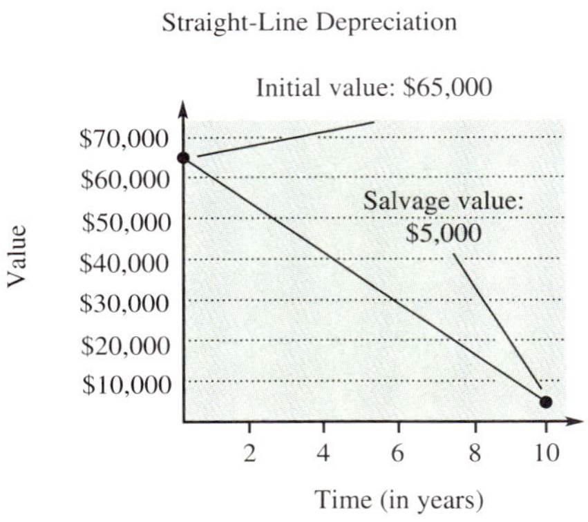

DISCUSSION PROBLEM ■ Vertical Shifts of Graphs

Sketch the graphs of the following equations in the same rectangular coordinate system.

$$y = x^2, \qquad y = x^2 + 1, \qquad y = x^2 + 2, \qquad y = x^2 + 3$$

Write a short paragraph describing the effect of the constant term a in the graph of the equation $y = x^2 + a$.

Warm-Up

The following warm-up exercises involve skills that were covered in earlier sections. You will use these skills in the exercise set for this section.

In Exercises 1–4, plot the given points in a rectangular coordinate system.

1. $(-3, 2), (5, -4)$

2. $(2, 8), (7, -3)$

3. $\left(\frac{5}{2}, \frac{7}{2}\right), \left(\frac{7}{3}, -2\right)$

4. $\left(-\frac{9}{4}, -\frac{1}{4}\right), \left(-3, \frac{9}{2}\right)$

In Exercises 5 and 6, solve the given equation.

5. $\frac{5}{8}x - 6 = 0$

6. $14 - \frac{7}{3}x = 0$

In Exercises 7–10, find the missing coordinate of the solution point.

7. $y = \frac{3}{5}x + 4, \quad (15, \quad)$

8. $y = 3 - \frac{5}{9}x, \quad (12, \quad)$

9. $y = 5.5 - 0.95x, \quad (\quad , -1)$

10. $y = 3 + 0.2x, \quad (\quad , 4.4)$

7.2 EXERCISES

In Exercises 1–4, complete the table and use the resulting solution points to sketch the graph of the equation.

1. $2x + y = 3$

x	-4			2	4
y		7	3		
(x, y)					

2. $2x - 3y = 6$

x	-3	0		3	
y			$-\frac{2}{3}$		5
(x, y)					

3. $y = 4 - x^2$

x		-1		2	
y	0		4		-5
(x, y)					

4. $y = \frac{1}{2}x^3 - 4$

x	-1		1		3
y		-4		0	
(x, y)					

In Exercises 5–10, sketch the graph of the equation.

5. $y = 4 - x$ **6.** $y = \frac{1}{2}x - 2$ **7.** $y = x^2 - 4$ **8.** $y = x^3$

9. $y = |x|$ **10.** $y = |x + 3|$

In Exercises 11–20, find the x- and y-intercepts (if any) of the graph of the given equation.

11. $x + 2y = 10$ **12.** $3x - 2y + 12 = 0$ **13.** $y = \frac{3}{4}x + 15$ **14.** $y = 12 - \frac{2}{5}x$

15. $y = x^2 - 25$ **16.** $y = x^2 + 2x + 1$ **17.** $y = x^2 + 3$ **18.** $x = 3$

19. $y = 3x^2 + x - 2$ **20.** $y = x(x - 4) - 3(x - 4)$

In Exercises 21–40, sketch the graph of the equation and show the coordinates of three solution points (including intercepts).

21. $y = 3 - x$ **22.** $y = x - 3$ **23.** $y = 2x - 3$ **24.** $y = -4x + 8$

25. $y = 4$ **26.** $x = -6$ **27.** $2x - 3y = 6$ **28.** $3x + 4y = 12$

29. $x + 5y = 10$ **30.** $5x - y = 10$ **31.** $4x + y = 3$ **32.** $y - 2x = -4$

33. $x - 1 = 0$ **34.** $y + 3 = 0$ **35.** $y = x^2 - 9$ **36.** $y = 9 - x^2$

37. $y = x^2 - 2x$ **38.** $y = -x^2 - 4x$ **39.** $y = |x| - 3$ **40.** $y = |x - 3|$

41. *Straight-Line Depreciation* A manufacturing plant purchases a new molding machine for \$225,000. The depreciated value y after t years is given by

$$y = 225{,}000 - 20{,}000t, \quad 0 \le t \le 8.$$

Sketch the graph of the equation over the given interval for t.

42. *Maximum Area* A rectangle of length l and width w has a perimeter of 12 meters.
(a) Show that the width of the rectangle is $w = 6 - l$ and its area is $A = l(6 - l)$.
(b) Sketch the graph of the equation for the area.
(c) From the graph of part (b), estimate the dimensions of the rectangle that yield a maximum area.

In Exercises 43 and 44, sketch the graph of each equation on the same set of coordinate axes. What conclusions can you draw by comparing the graphs?

43. (a) $y = x^2$
(b) $y = (x - 2)^2$
(c) $y = (x + 2)^2$
(d) $y = (x + 4)^2$

44. (a) $y = x^2$
(b) $y = x^2 - 2$
(c) $y = x^2 - 4$
(d) $y = x^2 + 4$

SECTION **7.3**

Graphing Calculators

Introduction • Basic Graphing • Special Features

Introduction

In Section 7.2 we introduced the point-plotting method for sketching the graph of an equation. One of the disadvantages of the point-plotting method is that in order to get a good idea about the shape of a graph we need to plot *many* points. With only a few points, we could badly misrepresent the graph. For instance, consider the equation

$$y = \frac{1}{30}x(39 - 10x^2 + x^4).$$

Suppose we plotted only five points, $(-3, -3)$, $(-1, -1)$, $(0, 0)$, $(1, 1)$, and $(3, 3)$, as shown in Figure 7.21(a). From these five points, a person might assume that the graph of the equation is a straight line. That, however, is not correct. By plotting several more points we can see that the actual graph is not straight at all! [See Figure 7.21(b).]

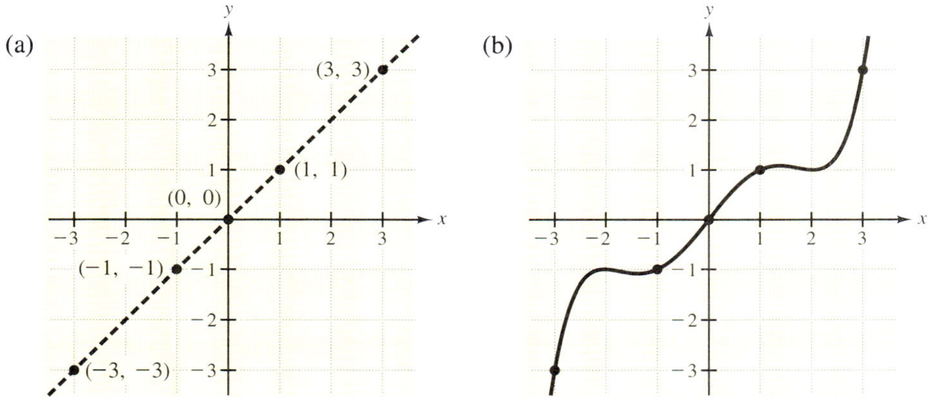

FIGURE 7.21

Thus, the point-plotting method leaves us with a dilemma. On the one hand, the method can be very inaccurate if only a few points are plotted. On the other hand, it is very time-consuming to plot a dozen (or more) points. Technology can help us solve this dilemma. Plotting several (even several hundred) points in a rectangular coordinate system is something that a computer or calculator can do easily.

The point-plotting method is the method used by *all* graphing packages for computers and *all* graphing calculators. Each computer or calculator screen is made up of a grid of hundreds or thousands of small areas called **pixels**. Screens that have many pixels per inch are said to have a higher **resolution** than screens that have fewer pixels. For instance, the screen shown in Figure 7.22(a) has a higher resolution than the screen shown in Figure 7.22(b). Note that the "graph" of the line on the first screen looks more like a line than the "graph" on the second screen.

FIGURE 7.22

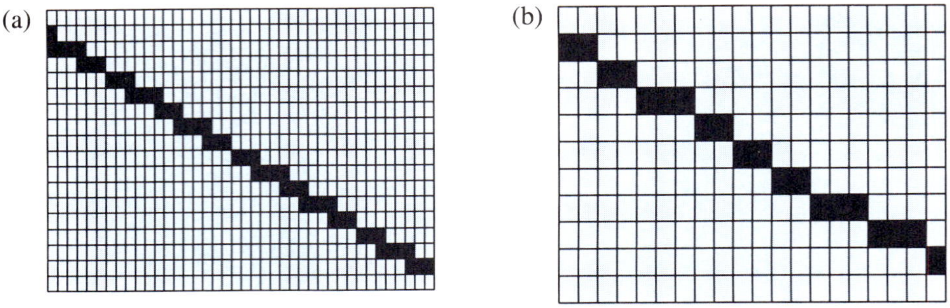

Screens on most graphing calculators have 48 pixels per inch, whereas screens on computer monitors typically have between 32 and 100 pixels per inch.

EXAMPLE 1 ■ Using Pixels to Sketch a Graph

Use the grid shown in Figure 7.23 to sketch a graph of $y = \frac{1}{2}x^2$. Each pixel on the grid must be either on (shaded black) or off (unshaded).

Solution

To shade the grid we use the following rule. If a pixel contains a plotted point of the graph, then it will be on; otherwise the pixel will be off. Using this rule, the graph of the curve looks like that shown in Figure 7.24.

FIGURE 7.23

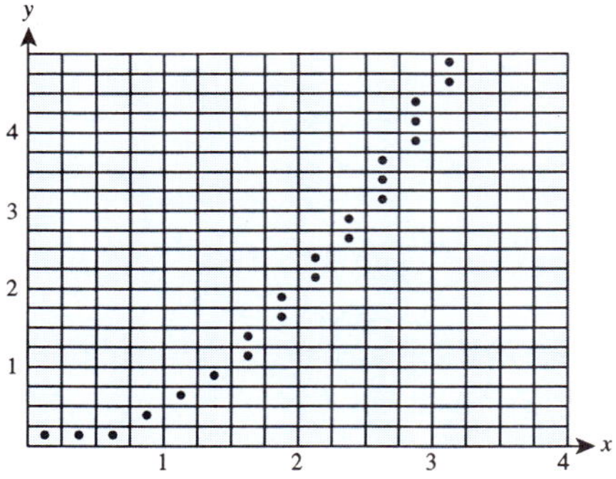

FIGURE 7.24

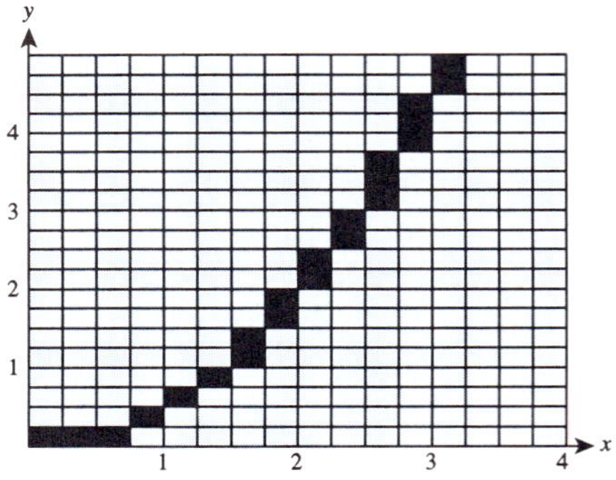

Basic Graphing

There are many different graphing packages you can buy for computers and graphing calculators. In this section we describe only one—the *TI-81* from *Texas Instruments*. The keys for the *TI-81* are shown in Figure 7.25. The keys on the top row are used for graphing. The steps used to sketch a simple graph are summarized in the following list.

Basic Graphing Steps for the *TI-81* Graphing Calculator

To sketch the graph of an equation involving x and y on the *TI-81*, use the following steps. (Before performing these steps, you should set your calculator so that all of the standard defaults are active. Press [ZOOM], cursor to "Standard," and press [ENTER].)

1. Rewrite the equation so that y is isolated on the left side of the equation.
2. Press the [Y=] key. Then enter the right side of the equation on the first line of the display. (The first line is labeled as $Y_1=$.)
3. Press the [GRAPH] key.

FIGURE 7.25

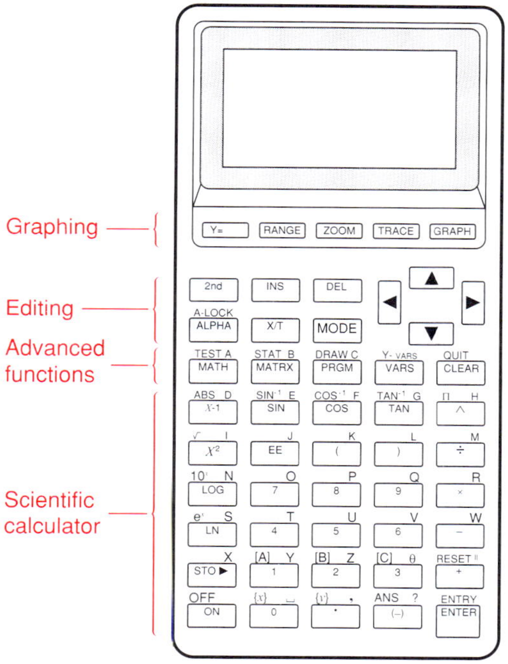

EXAMPLE 2 ■ Sketching the Graph of a Linear Equation

Sketch the graph of $x + 2y = 6$.

Solution

To begin, we solve the given equation for y in terms of x.

$x + 2y = 6$	Given equation
$2y = -x + 6$	Subtract x from both sides
$y = -\frac{1}{2}x + 3$	Divide both sides by 2

Next, after pressing the [Y=] key, we enter the following keystrokes.

[(−)] [X|T] [÷] 2 [+] 3

The top row of the display should now be as follows.

Y_1=X/2+3

Now, press the [GRAPH] key, and the screen should look like the one shown in Figure 7.26.

FIGURE 7.26 $y = -\frac{x}{2} + 3$

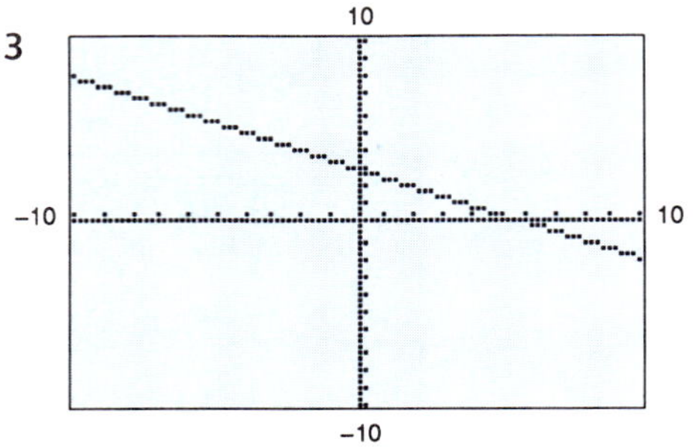

■

In Figure 7.26, notice that the calculator screen does not label the tic marks on the x-axis or the y-axis. To see what the tic marks represent, you can press [RANGE]. If you set your calculator to the standard graphing defaults before solving Example 2, the screen should be as follows.

RANGE

Xmin=−10	The minimum x-value is −10.
Xmax=10	The maximum x-value is 10.
Xscl=1	The x-scale is one unit per tic mark.
Ymin=−10	The minimum y-value is −10.
Ymax=10	The maximum y-value is 10.
Yscl=1	The y-scale is one unit per tic mark.
Xres=1	The x-resolution is one plotted point per *one* pixel.

These settings are summarized visually in Figure 7.27. We suggest leaving the x-resolution as 1. It can be set at any integer between 1 and 8. For instance, if you set the x-resolution as 2, then you will obtain one plotted point per *two* pixels. By doing this, you will have increased the speed at which the graph is sketched, but you will have decreased the accuracy of the graph.

FIGURE 7.27

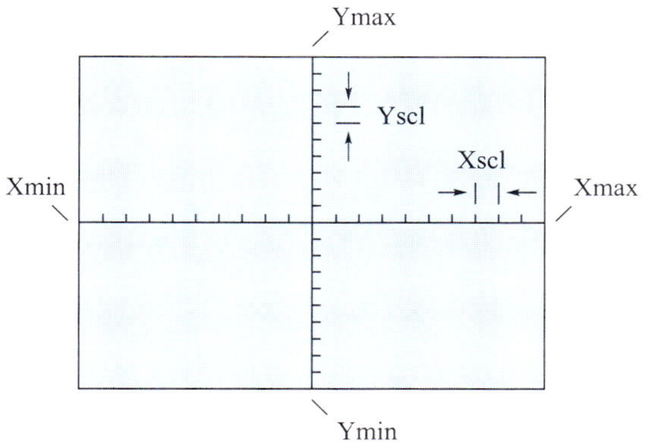

EXAMPLE 3 ■ **Sketching the Graph of an Equation Involving Absolute Value**

Sketch the graph of $y = |x - 3|$.

Solution

This equation is already written so that y is isolated on the left side of the equation. Thus, after pressing the [Y=] key, we enter the following keystrokes.

[ABS] [(] [X|T] [−] 3 [)]

The top row of the display should now be as follows.

Y_1=abs (X−3)

Now, press the [GRAPH] key, and the screen should look like the one shown in Figure 7.28. (To enter [ABS] on the *TI-81*, press [2nd] and [x^{-1}].)

FIGURE 7.28 $y = |x - 3|$

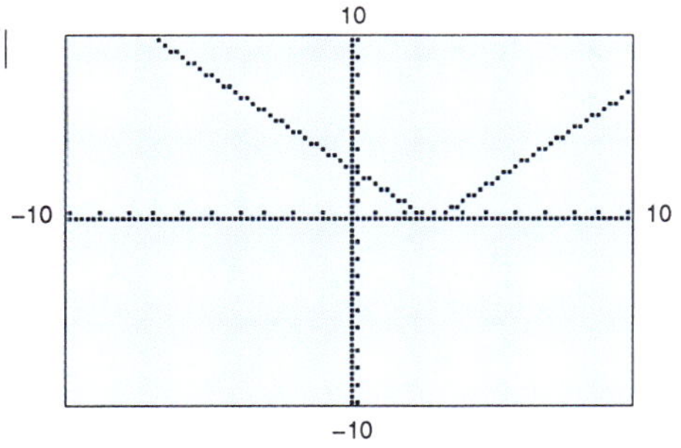

■

Special Features

In order to be able to use your graphing calculator to its best advantage, you must be able to use the [RANGE] key and the [ZOOM] key. The next two examples show how these keys can be used.

EXAMPLE 4 ■ **Resetting the Scales on an Axis**

Sketch the graph of $y = x^2 + 12$.

Solution

We begin as usual. Press [Y=] and enter $y = x^2 + 12$ on the first line.

[X|T] [x^2] [+] 12

Press the [GRAPH] key, and you will notice that nothing appears on the screen (provided your calculator is set to standard defaults). The reason for this is that the lowest point on the graph of $y = x^2 + 12$ occurs at the point (0, 12). With *standard* range settings, we obtain a screen whose largest y-value is 10. In other words, none of the graph is visible on a screen whose y-values range between -10 and 10, as shown in Figure 7.29(a).

To change these settings, press [RANGE] and change Ymax=10 to Ymax=30. Then, change Yscl=1 to Yscl=5. Now press [GRAPH], and you will obtain the graph shown in Figure 7.29(b). On this graph, note that each tic mark on the y-axis represents five units because we changed the y-scale to 5. Also note that the highest point on the y-axis is now 30 because we changed the maximum value of y to 30.

FIGURE 7.29 $y = x^2 + 12$

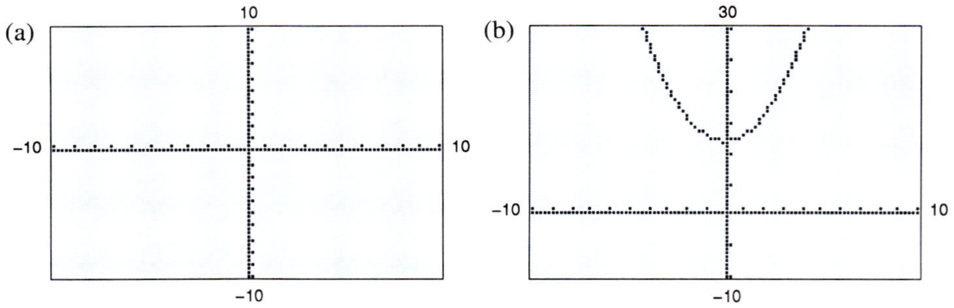

■

NOTE: If you changed the y-maximum and y-scale on your calculator as indicated in Example 4, you should return to the standard settings before working Example 5. To do this, press [ZOOM], cursor to "Standard," and press [ENTER].

EXAMPLE 5 ■ Using the Zoom Key

Sketch the graph of $y = x^2 + 2x - \frac{1}{2}$. Use the sketch to estimate the x-intercepts of the graph.

Solution

We begin as usual. Press [Y=] and enter $y = x^2 + 2x - \frac{1}{2}$ on the first line. Press the [GRAPH] key, and you will obtain the graph shown in Figure 7.30. From Figure 7.30, it is clear that the graph has two intercepts, but it is difficult to estimate the x-coordinates of the x-intercepts. To obtain a better view of the graph at its intercepts, we press [ZOOM], cursor to "Zoom-In," and press [ENTER] *twice*. The redrawn screen is shown in Figure 7.31. From this screen, we can tell that the graph has x-intercepts whose x-coordinates are approximately -2.2 and 0.2. [Using the Quadratic Formula, we can verify that these x-coordinates occur at $(-2-\sqrt{6})/2 \approx -2.225$ and $(-2+\sqrt{6})/2 \approx 0.225$.]

FIGURE 7.30 $y = x^2 + 2x - \frac{1}{2}$

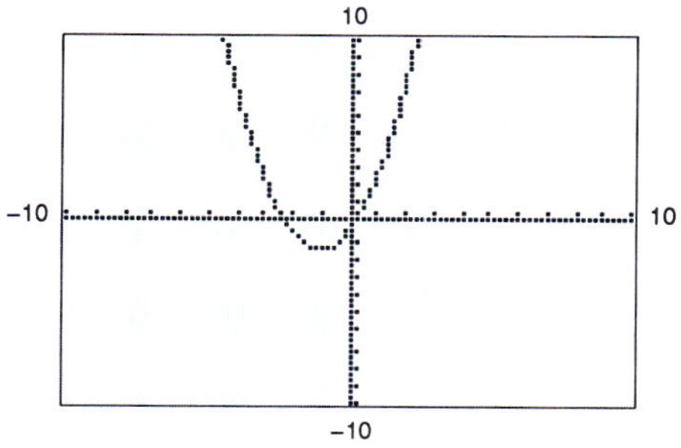

FIGURE 7.31 $y = x^2 + 2x - \frac{1}{2}$

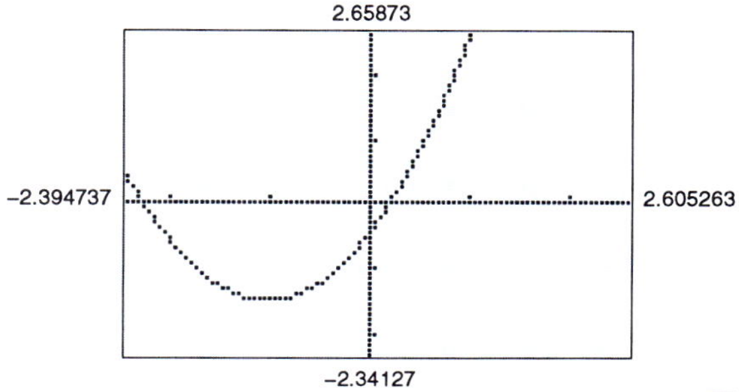

■

EXAMPLE 6 ■ Sketching More than One Graph on the Same Screen

Sketch the graphs of $y = x^2$ and $y = (x-3)^2$ on the same screen.

Solution

To begin, press [Y=] and enter the two equations on the first two lines. The display should now be as follows.

$Y_1 = X^2$

$Y_2 = (X-3)^2$

$Y_3 =$

$Y_4 =$

Press the [GRAPH] key, and you will obtain the graph shown in Figure 7.32.

FIGURE 7.32 $y_1 = x^2, \quad y_2 = (x-3)^2$

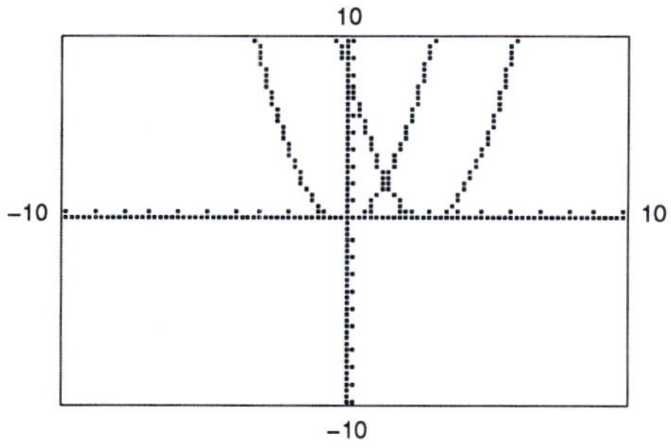

■

DISCUSSION PROBLEM ■ A Misleading Graph

Set your calculator in its standard setting and sketch the graph of $y = x^2 - 12x$. The graph appears to be a straight line, as shown in Figure 7.33. However, this is misleading because the screen doesn't show an important portion of the graph. Can you find a range setting that reveals a better view of this graph?

FIGURE 7.33 $y = x^2 - 12x$

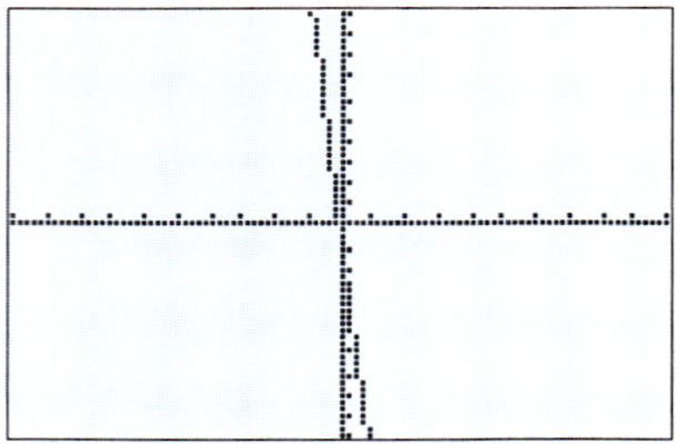

■

Warm-Up

The following warm-up exercises involve skills that were covered in earlier sections. You will use these skills in the exercise set for this section.

In Exercises 1–10, solve for y in terms of x.

1. $3x + y = 4$ **2.** $x - y = 0$ **3.** $2x + 3y = 2$

4. $4x - 5y = -2$ **5.** $3x + 4y - 5 = 0$ **6.** $-2x - 3y + 6 = 0$

7. $x^2 + y - 4 = 0$ **8.** $-2x^2 + 3y + 2 = 0$

9. $x^2 + 3y = 4$ **10.** $x^2 - 2y = 9$

7.3 EXERCISES

*In Exercises 1–20, use a graphing calculator to sketch the graph of the equation. Use the standard setting of the calculator.

RANGE
Xmin=−10
Xmax=10
Xscl=1
Ymin=−10
Ymax=10
Yscl=1
Xres=1

1. $y = 2x - 6$ **2.** $y = x - 2$ **3.** $y = \frac{2}{3}x + 1$

4. $y = \frac{1}{4}x - 2$ **5.** $y = -\frac{3}{4}x + 4$ **6.** $y = -2x + 5$

7. $y = x^2 - 3$ **8.** $y = 6 - x^2$ **9.** $y = -x^2 + 4x - 1$

10. $y = \frac{1}{4}x^2 - x$ **11.** $y = x^3 - 2$ **12.** $y = 1 - x^3$

*The symbol indicates problems that are designed to be solved with a graphing calculator or some other graphing facility.

13. $y = \sqrt{x + 4}$

14. $y = 2\sqrt{2x + 3}$

15. $y = 2(1 - \sqrt{x})$

16. $y = x\sqrt{x + 3}$

17. $y = |x| - 4$

18. $y = 4 - |x|$

19. $y = |x - 4|$

20. $y = \frac{1}{2}|x|$

In Exercises 21–26, use a graphing calculator to sketch the graph of the equation. Begin by using the standard setting of the calculator and then sketch the graph a second time using the specified setting.

21. $y = \frac{1}{2}x - 10$

RANGE
Xmin=−1
Xmax=20
Xscl=1
Ymin=−15
Ymax=5
Yscl=1
Xres=1

22. $y = -2x + 25$

RANGE
Xmin=−1
Xmax=20
Xscl=1
Ymin=−2
Ymax=28
Yscl=1
Xres=1

23. $y = 20 - \frac{1}{4}x^2$

RANGE
Xmin=−10
Xmax=10
Xscl=1
Ymin=−2
Ymax=22
Yscl=1
Xres=1

24. $y = x^2 - 12x + 36$

RANGE
Xmin=−5
Xmax=15
Xscl=1
Ymin=−2
Ymax=20
Yscl=1
Xres=1

25. $y = 2x^3 - 5x^2$

RANGE
Xmin=−5
Xmax=5
Xscl=1
Ymin=−5
Ymax=5
Yscl=1
Xres=1

26. $y = 6\sqrt{x + 4}$

RANGE
Xmin=−5
Xmax=5
Xscl=1
Ymin=−1
Ymax=20
Yscl=2
Xres=1

In Exercises 27–32, find a setting on a graphing calculator so that the graph of the equation agrees with the given graph.

27. $y = -3x + 15$

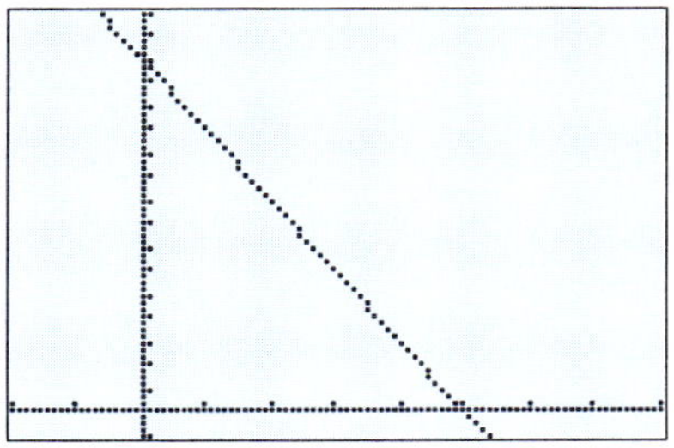

28. $y = \frac{2}{3}x - 8$

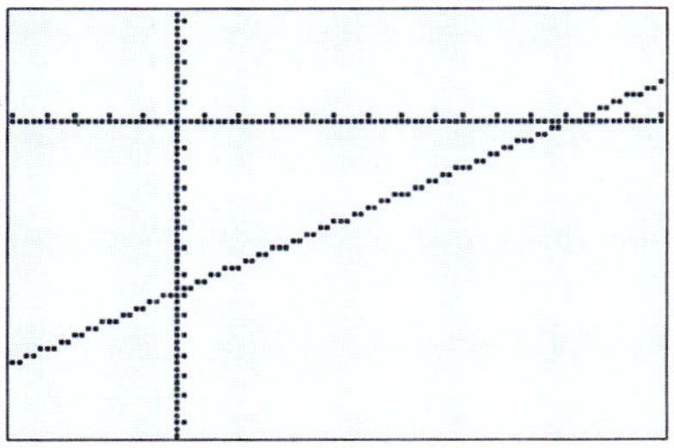

29. $y = -\frac{1}{8}x^2 + x$

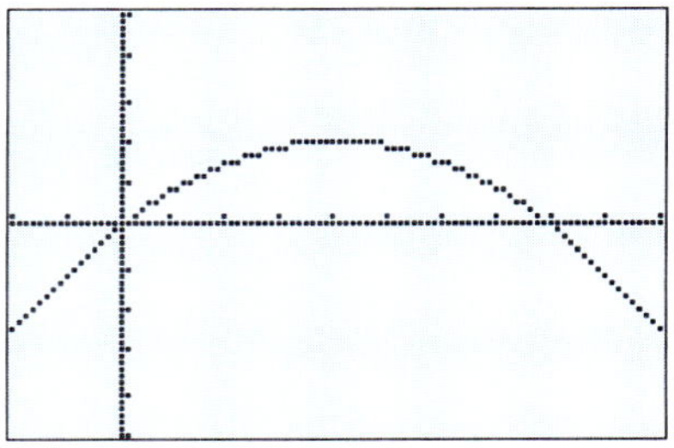

30. $y = 2x^2 - 16x + 26$

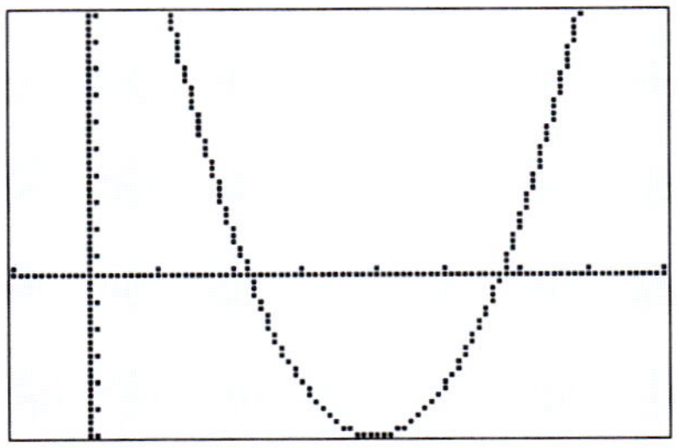

31. $y = x^3 - 3x + 2$

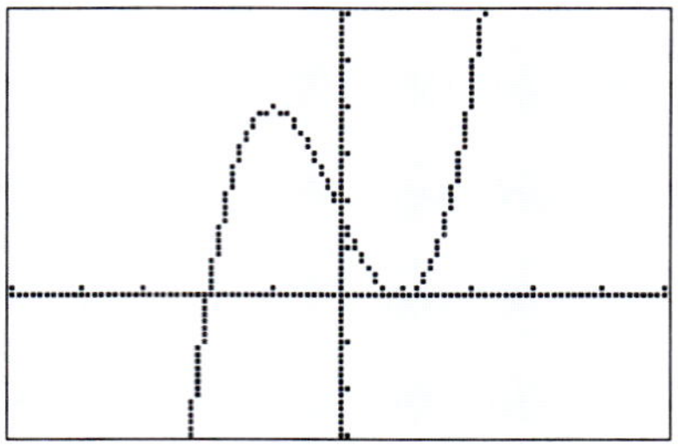

32. $y = (x^2 - 4)^2$

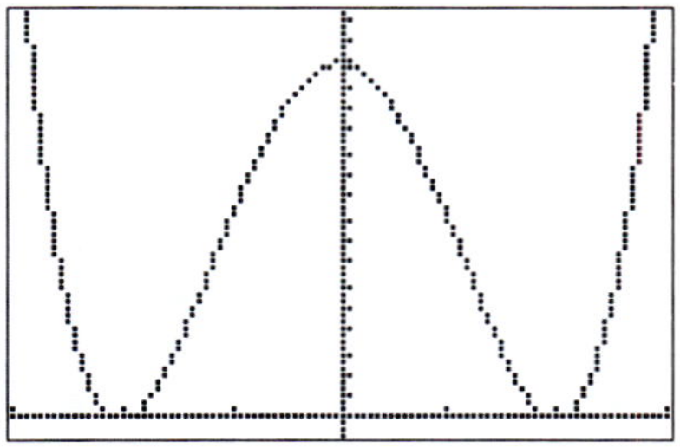

In Exercises 33–36, use the standard setting of a graphing calculator to sketch the graph of the equation. Then "zoom in" on the graph using the procedure shown in Example 5. (After performing this procedure, press the RANGE key and observe the changes from the standard setting of the calculator.)

33. $y = \frac{1}{4}(x^2 - 1)$

34. $y = -\frac{1}{3}x^3 + x$

35. $y = x^3(3x + 4)$

36. $y = x\sqrt{x + 1}$

In Exercises 37–40, use a graphing calculator to sketch the graphs of both equations on the same screen.

37. $y = x^2$
$y = 4x - x^2$

38. $y = x + 2$
$y = 2 + 4x - x^2$

39. $y = x$
$y = 3\sqrt{x}$

40. $y = 2x^2$
$y = x^4 - 2x^2$

In Exercises 41–44, use a graphing calculator to find the number of x-intercepts of the given equation.

41. $y = x^2 - 4x - 3$

42. $y = -2x^2 - x + 8$

43. $y = -x^2 + 12x - 37$

44. $y = x^2 + 14x + 49$

In Exercises 45–48, use a graphing calculator to approximate the x-intercepts of the graph of the equation. (After pressing the TRACE key, the cursor will move from one plotted point to the next as you press the right and left cursor keys. The coordinates of each point are displayed at the bottom of the viewing window.)

45. $y = 0.4x^2 - 0.2x - 3$

46. $y = 14 + 3x - 2x^2$

47. $y = x^3 - 3x^2 + 2$

48. $y = x^3 + x - 1$

Stopping Distance In Exercises 49 and 50, consider the stopping distance y (in feet) of an automobile approximated by

$$y = 30.00 + 0.08x^2, \quad 25 \le x \le 75$$

where x is the speed in miles per hour.

49. Use a graphing calculator to sketch the graph of the equation using the following range settings.

RANGE
Xmin=20
Xmax=80
Xscl=5
Ymin=20
Ymax=500
Yscl=50
Xres=1

50. Use the TRACE key to approximate the stopping distance when the car's speed is (a) $x = 35$ miles per hour, and (b) $x = 55$ miles per hour.

Strength of Copper In Exercises 51 and 52, use the following model, which relates the percentage strength of copper in terms of its Celsius temperature. The percentage strength is relative to 100% at 20° C. The approximate model is

$$y = \sqrt{10{,}700 - 17.6x}, \quad 100 \le x \le 500$$

where y is the percentage strength and x is the temperature.

51. Use a graphing calculator to sketch the graph of the equation using the following range settings.

RANGE
Xmin=0
Xmax=500
Xscl=50
Ymin=30
Ymax=110
Yscl=10
Xres=1

52. Use the TRACE key to approximate the percentage strength when the temperature is (a) $x = 150°$ C, and (b) $x = 400°$ C.

SECTION 7.4

Slope of a Line, Parallel and Perpendicular Lines

The Slope of a Line • Slope: An Aid to Sketching Lines • Parallel and Perpendicular Lines

The Slope of a Line

The **slope** of a nonvertical line represents the number of units a line rises or falls vertically for each unit of horizontal change from left to right. For example, the line in Figure 7.34 rises two units for each unit of horizontal change from left to right.

FIGURE 7.34

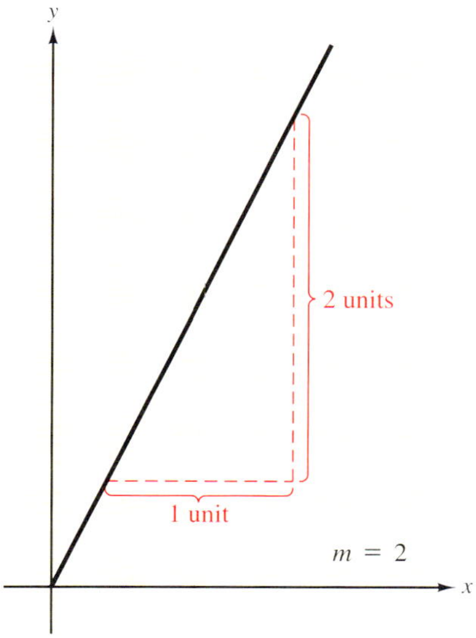

Definition of the Slope of a Line

The **slope** m of the nonvertical line passing through the points (x_1, y_1) and (x_2, y_2) is

$$m = \frac{y_2 - y_1}{x_2 - x_1} = \frac{\text{Change in } y}{\text{Change in } x}$$

where $x_1 \neq x_2$. (See Figure 7.35.)

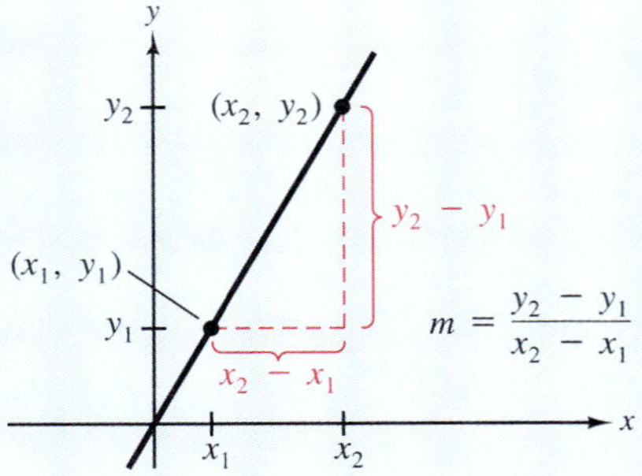

FIGURE 7.35

When this formula is used, the *order of subtraction* is important. Given two points on a line, you are free to label either one of them as (x_1, y_1), and the other as (x_2, y_2). However, once this is done, you must form the numerator and denominator using the same order of subtraction.

$$\underbrace{m = \frac{y_2 - y_1}{x_2 - x_1}}_{\text{Correct}} \qquad \underbrace{m = \frac{y_1 - y_2}{x_1 - x_2}}_{\text{Correct}} \qquad \underbrace{m = \frac{y_2 - y_1}{x_1 - x_2}}_{\text{Incorrect}}$$

EXAMPLE 1 ■ **Finding the Slope of a Line Passing Through Two Points**

Find the slope of the line passing through each of the following pairs of points.

(a) $(2, 4)$ and $(0, -2)$ (b) $(-2, 4)$ and $(3, 4)$ (c) $(2, -3)$ and $(0, 1)$

Solution

(a) The slope of the line through $(x_1, y_1) = (2, 4)$ and $(x_2, y_2) = (0, -2)$ is

$$\begin{aligned} m &= \frac{y_2 - y_1}{x_2 - x_1} && \text{Difference in } y\text{-values} \; / \; \text{Difference in } x\text{-values} \\ &= \frac{-2 - 4}{0 - 2} \\ &= \frac{-6}{-2} \\ &= 3. \end{aligned}$$

(b) The slope of the line through $(-2, 4)$ and $(3, 4)$ is

$$m = \frac{4 - 4}{3 - (-2)} = \frac{0}{5} = 0.$$

(c) The slope of the line through $(2, -3)$ and $(0, 1)$ is

$$m = \frac{1 - (-3)}{0 - 2} = \frac{4}{-2} = -2.$$

The graphs of the three lines are shown in Figure 7.36.

FIGURE 7.36

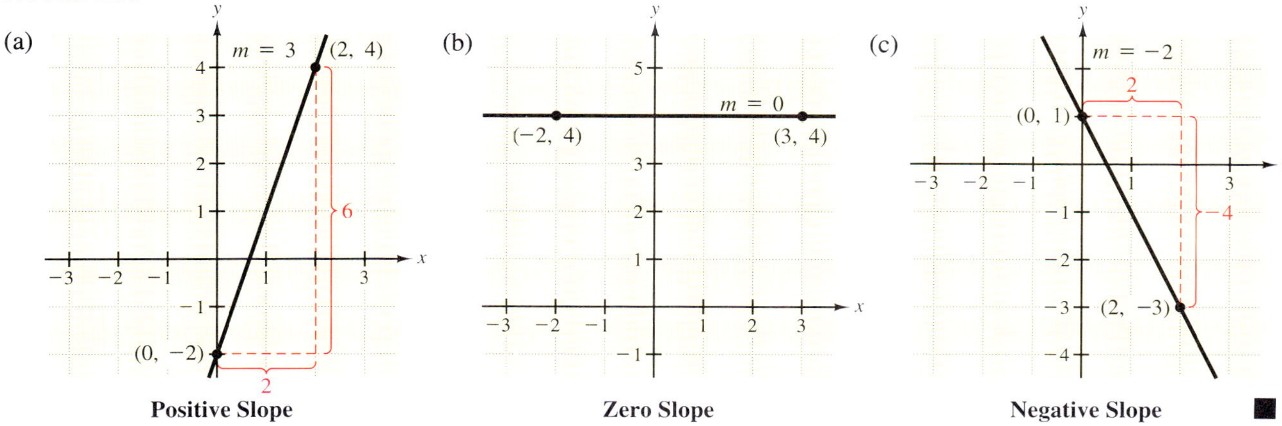

Note that the definition of slope does not apply to vertical lines. For instance, consider the points (2, 3) and (2, −1) on the vertical line shown in Figure 7.37. Applying the formula for slope, we have

$$\frac{-1 - 3}{2 - 2}. \qquad \text{Undefined division by zero}$$

Because division by zero is not defined, we do not define the slope of a vertical line.

FIGURE 7.37

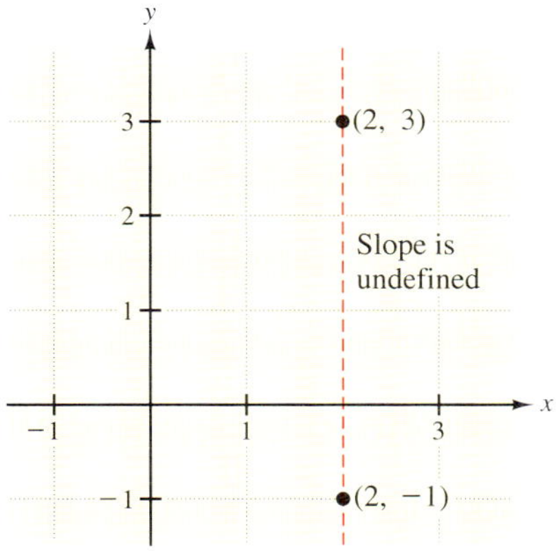

From the slopes of the lines shown in Figures 7.36 and 7.37, we make the following generalizations about the slope of a line.

1. A line with positive slope ($m > 0$) *rises* from left to right.
2. A line with negative slope ($m < 0$) *falls* from left to right.
3. A line with zero slope ($m = 0$) is *horizontal*.
4. A line with undefined slope is *vertical*.

EXAMPLE 2 ■ Using Slope to Describe Lines

Use slope to determine whether the line through each of the following pairs rises, falls, is horizontal, or is vertical.

(a) $(3, 3), (4, 1)$ (b) $(-2, 3), (-2, -1)$

(c) $(-1, -3), (2, 2)$ (d) $(-1, 2), (2, 2)$

Solution

(a) Since

$$m = \frac{1 - 3}{4 - 3} = \frac{-2}{1} = -2 < 0 \qquad \text{Negative slope}$$

the line falls (from left to right).

(b) Since

$$m = \frac{-1 - 3}{-2 - (-2)} = \frac{-4}{0} \qquad \text{Undefined slope}$$

is undefined, the line is vertical.

(c) Since

$$m = \frac{2 - (-3)}{2 - (-1)} = \frac{5}{3} > 0 \qquad \text{Positive slope}$$

the line rises (from left to right).

(d) Since

$$m = \frac{2 - 2}{2 - (-1)} = \frac{0}{3} = 0 \qquad \text{Zero slope}$$

the line is horizontal.

The graphs of these four lines are shown in Figure 7.38.

FIGURE 7.38

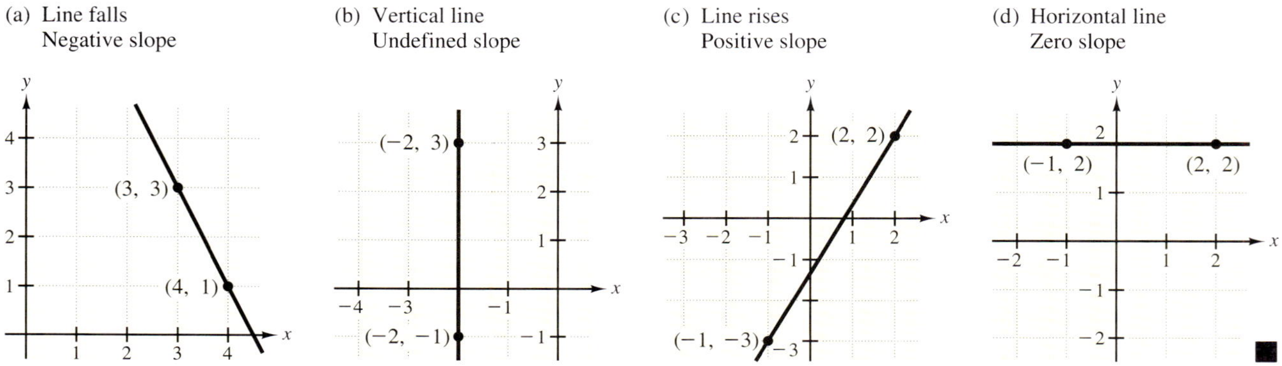

■

Any two points on a line can be used to calculate its slope. This is demonstrated in the next example.

EXAMPLE 3 ■ Finding the Slope of a Line

Sketch the line given by the equation

$$2x - 3y = 6.$$

Then find the slope of the line. (Choose two different pairs of points on the line and show that the same slope is obtained from either pair.)

Solution

We begin by solving the given equation for y.

$$y = \frac{2}{3}x - 2$$

Then we construct the table of values shown in Table 7.10.

TABLE 7.10

x	-3	0	3	6
$y = \frac{2}{3}x - 2$	-4	-2	0	2
Solution Points	$(-3, -4)$	$(0, -2)$	$(3, 0)$	$(6, 2)$

From the solution points in the table, we sketch the graph of the line, as shown in Figure 7.39.

FIGURE 7.39

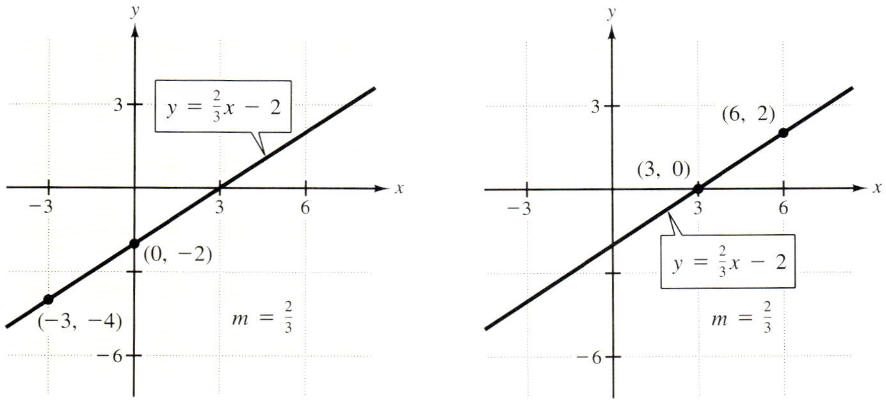

To calculate the slope of the line using two different sets of points, we first use the points $(-3, -4)$ and $(0, -2)$, and obtain a slope of

$$m = \frac{-2 - (-4)}{0 - (-3)} = \frac{2}{3}.$$

Next, we use the points $(3, 0)$ and $(6, 2)$ to obtain a slope of

$$m = \frac{2 - 0}{6 - 3} = \frac{2}{3}.$$

Note that the same slope is obtained using either set of points. (Try some other pairs of points on the line to see that you obtain a slope of $m = \frac{2}{3}$ regardless of which two points you use.) ■

Slope: An Aid to Sketching Lines

As we learned earlier, before making up a table of values for an equation, it is helpful to first solve the equation for y. When we do this for a linear equation, we obtain some very useful information. Consider the results of Example 3.

$$2x - 3y = 6 \qquad \text{Given equation}$$
$$-3y = -2x + 6$$
$$y = \frac{2}{3}x - 2 \qquad \text{Solve for } y$$

Observe that the coefficient of x is the slope of the graph for this equation (see Example 3). Moreover, the constant term, -2, gives the y-intercept of the graph.

$$y = \boxed{\tfrac{2}{3}}\, x + \boxed{-2}$$

↑ Slope ↑ y-intercept $(0, -2)$

We call this the **slope-intercept** form of the equation of the line.

Slope-Intercept Form of the Equation of a Line

The graph of the equation

$$y = mx + b$$

is a line whose slope is m and whose y-intercept is $(0, b)$.

So far, we have been plotting several points to graph the equation of a line. However, now that we can recognize equations of lines, we don't have to plot as many points—two points are enough. (You might remember from geometry that *two points are all that are necessary to determine a line*.) We demonstrate this technique in the next two examples.

EXAMPLE 4 ■ Using the Slope and y-Intercept to Graph a Line

Use the slope and y-intercept to sketch the graph of the following equation.

$$y = \frac{1}{2}x + 1$$

Solution

The equation is already in slope-intercept form.

$$y = mx + b$$

$$y = \frac{1}{2}x + 1$$

Thus, we see that the slope of the line is $m = \frac{1}{2}$ and the y-intercept is $(0, b) = (0, 1)$. Knowing this information, we can sketch the graph of the line as follows. First, we plot the y-intercept, as shown in Figure 7.40(a). Then, using a slope of $\frac{1}{2}$

$$m = \frac{1}{2} = \frac{\text{Change in } y}{\text{Change in } x}$$

we locate a second point on the line by moving two units to the right and one unit up, as shown in Figure 7.40(b). Finally, we obtain the graph by drawing the line that passes through the two points.

FIGURE 7.40

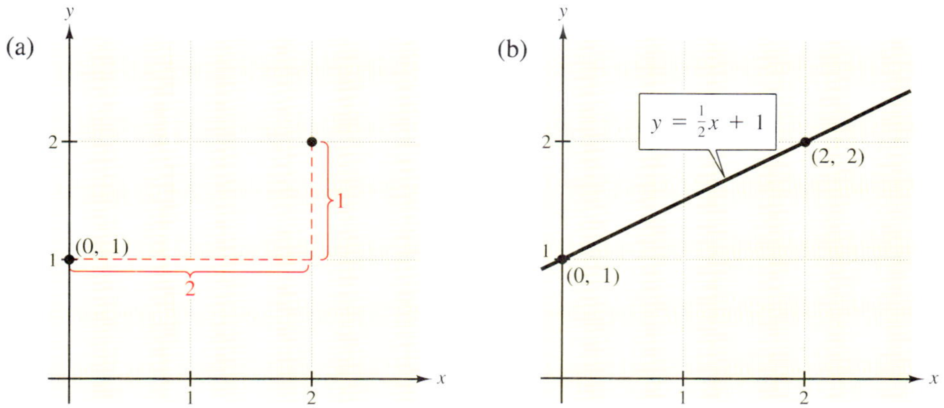

■

EXAMPLE 5 ■ Using the Slope and y-Intercept to Graph a Line

Use the slope and y-intercept to sketch the graph of the following equation.

$$4x + 3y - 2 = 0$$

Solution

We begin by writing the equation in slope-intercept form.

$$4x + 3y - 2 = 0 \qquad \text{Given equation}$$
$$3y = -4x + 2$$
$$y = -\frac{4}{3}x + \frac{2}{3} \qquad \text{Slope-intercept form}$$

From this form, we see that the slope is $m = -\frac{4}{3}$ and the y-intercept is $(0, b) = \left(0, \frac{2}{3}\right)$. Thus, to sketch the line, we first plot the y-intercept, as shown in Figure 7.41(a). Then, using a slope of $-\frac{4}{3}$

$$m = \frac{-4}{3} = \frac{\text{Change in } y}{\text{Change in } x}$$

we locate a second point on the line by moving three units to the right and four units down, as shown in Figure 7.41(b). (Note that we moved down because the slope is negative.)

FIGURE 7.41

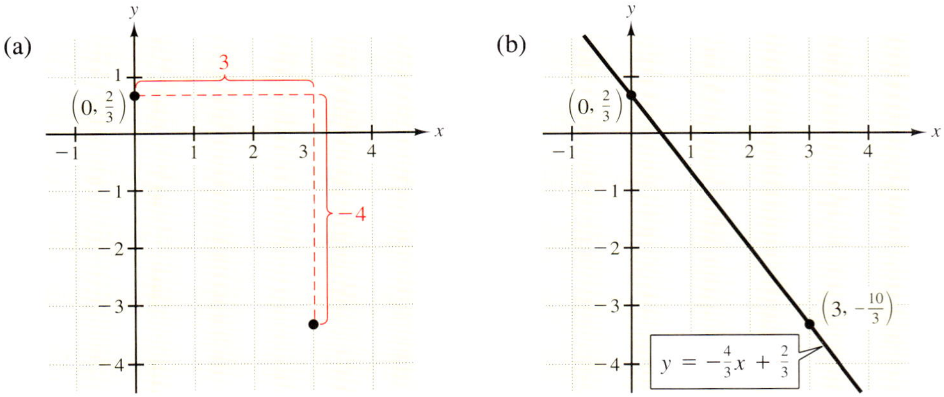

■

Parallel and Perpendicular Lines

We know from geometry that two lines in a plane are *parallel* if they do not intersect. What this means in terms of their slopes is suggested by the following example.

EXAMPLE 6 ■ Lines that Have the Same Slope

In the same rectangular coordinate system, sketch the two lines given by the following equations.

$$y = -2x \quad \text{and} \quad y = -2x + 3$$

Solution

For the line given by

$$y = -2x$$

the slope is $m = -2$ and the y-intercept is $(0, 0)$. For the line

$$y = -2x + 3$$

the slope is also $m = -2$ and the y-intercept is $(0, 3)$. The graphs of these two lines are shown in Figure 7.42.

FIGURE 7.42

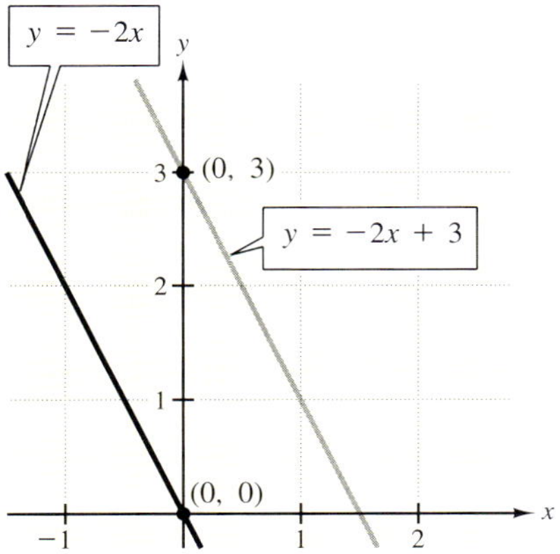

In Example 6 notice that the two lines have the same slope *and* that the two lines appear to be parallel. The following rule tells us that this is always the case—that is, two (nonvertical) lines are parallel if and only if they have the same slope.

Parallel Lines Two distinct nonvertical lines are parallel if and only if they have the same slope.

NOTE: The phrase "if and only if" in this rule is used in mathematics as a way to write two statements in one. The first statement says that *if two distinct nonvertical lines have the same slope, then they must be parallel.* The second statement says that *if two distinct nonvertical lines are parallel, then they must have the same slope.*

From geometry, we know that two lines in a plane are *perpendicular* if they intersect at right (90°) angles. Two *nonvertical* lines are perpendicular if their slopes are negative reciprocals of each other. For instance, the line given by

$$y = 2x$$

has a slope of $m_1 = 2$, and the line given by

$$y = -\frac{1}{2}x + 1$$

has a slope of $m_2 = -\frac{1}{2}$. Note that the slope of the second line is the negative reciprocal of the slope of the first line. That is,

$$m_2 = -\frac{1}{2} = -\frac{1}{m_1}.$$

From Figure 7.43, it appears that the two lines are perpendicular. We generalize this result in the following rule.

FIGURE 7.43

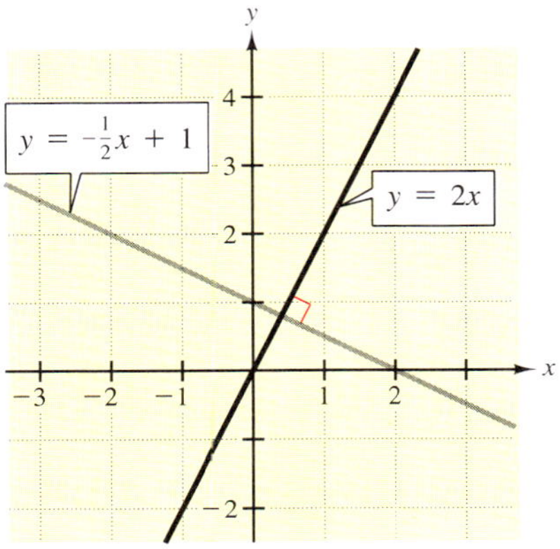

Perpendicular Lines

Consider two nonvertical lines whose slopes are m_1 and m_2. The two lines are perpendicular if and only if their slopes are *negative reciprocals* of each other. That is,

$$m_1 = -\frac{1}{m_2}.$$

EXAMPLE 7 ■ **Determining Whether Lines are Parallel or Perpendicular**

Determine whether the following pairs of lines are parallel, perpendicular, or neither.

(a) Line 1: $y = -2x - 5$ Line 2: $y = \frac{1}{2}x + 1$

(b) Line 1: $y = \frac{1}{2}x + 3$ Line 2: $y = \frac{1}{2}x - 1$

Solution

(a) For this pair of lines, the first line has a slope of $m_1 = -2$ and the second line has a slope of $m_2 = \frac{1}{2}$. Since these slopes are negative reciprocals of each other, the two lines must be perpendicular, as shown in Figure 7.44.

(b) Each of these two lines has a slope of $m = \frac{1}{2}$. Therefore, the two lines must be parallel, as shown in Figure 7.45.

FIGURE 7.44

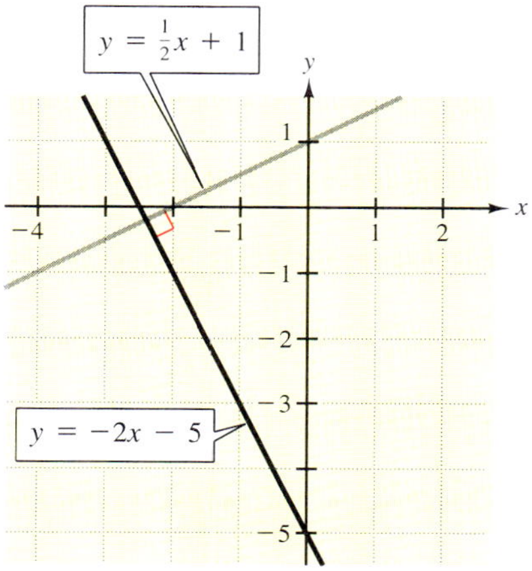

FIGURE 7.45

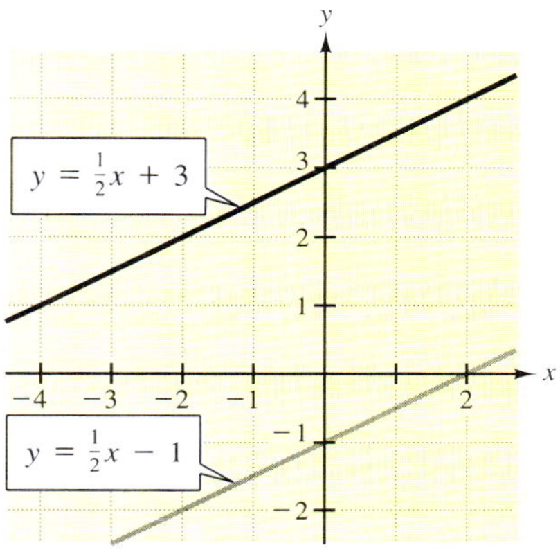

■

EXAMPLE 8 ■ A Graphing Calculator Problem

Use a graphing calculator* to sketch the graphs of the lines given by

$$y = x + 1 \quad \text{and} \quad y = -x + 3.$$

Sketch *both* graphs on the same display screen. The lines are supposed to be perpendicular (they have slopes of $m_1 = 1$ and $m_2 = -1$). Do they appear to be perpendicular on the calculator display?

Solution

If the zoom feature of your calculator is set on *standard*, then the tic marks on both the x-axis and the y-axis will vary between -10 and 10, as shown in Figure 7.46. However, because the display screen of the calculator is not square, the lines will *not* appear to be perpendicular. In other words, the graphs of two perpendicular lines will only appear to be perpendicular if the tic marks on the x-axis have the same spacing as the tic marks on the y-axis. To correct this problem, set the zoom feature on your calculator to *square*, press ENTER, and draw the graph again. You should obtain the graph shown in Figure 7.47.

*The graphing calculator keystrokes in this text correspond to the *TI-81* graphing calculator from *Texas Instruments*. However, you can also use other graphing calculators and computer graphing software for these examples and exercises.

FIGURE 7.46 $y_1 = x + 1$
$y_2 = -x + 3$

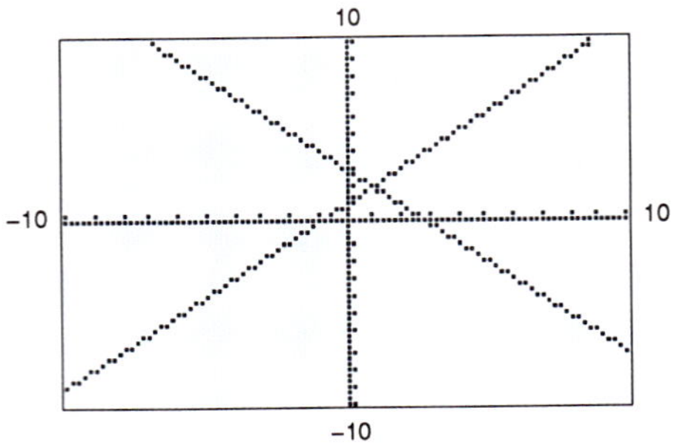

FIGURE 7.47 $y_1 = x + 1$
$y_2 = -x + 3$

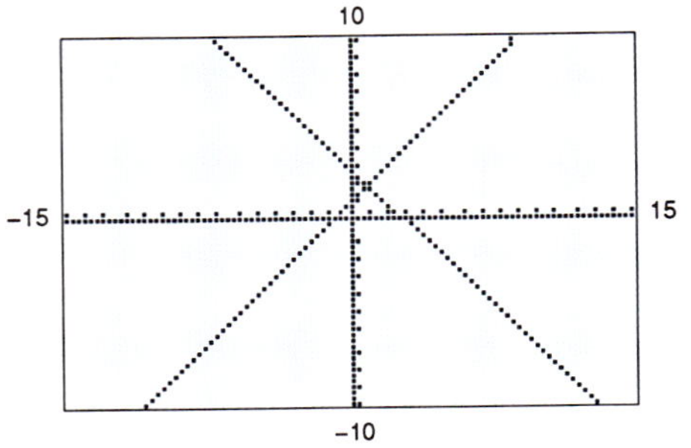

DISCUSSION PROBLEM ■ An Application of Slope

In 1980, a college had an enrollment of 4,000 students. By 1990, the enrollment had increased to 6,000 students.

(a) What was the average annual change in enrollment from 1980 to 1990?

(b) Use the average annual change in enrollment to estimate the enrollment in 1984, 1988, and 1992.

Year	1980	1984	1988	1990	1992
Enrollment	4,000			6,000	

(c) Sketch the line represented by the data given in the table in part (b). What is the slope of this line?

(d) Write a short paragraph that compares the concepts of *slope* and *average rate of change*.

Warm-Up

The following warm-up exercises involve skills that were covered in earlier sections. You will use these skills in the exercise set for this section.

In Exercises 1–4, evaluate the given quantity.

1. $\dfrac{8 - 3}{12 - 2}$

2. $\dfrac{-3 - (-6)}{11 - 5}$

3. $\dfrac{14 - (-15)}{-13 - (-11)}$

4. $\dfrac{-3 - 7}{0 - (-20)}$

In Exercises 5 and 6, find $-1/m$ for the given value of m.

5. $m = -\dfrac{5}{16}$

6. $m = \dfrac{3}{7}$

In Exercises 7–10, solve for y in terms of x.

7. $5x - 8y = 5$

8. $9x + 2y = 0$

9. $y - (-7) = 2[x - (-3)]$

10. $y - 14 = \frac{3}{4}(x - 12)$

7.4 EXERCISES

In Exercises 1–6, estimate the slope of the given line from its graph.

1.

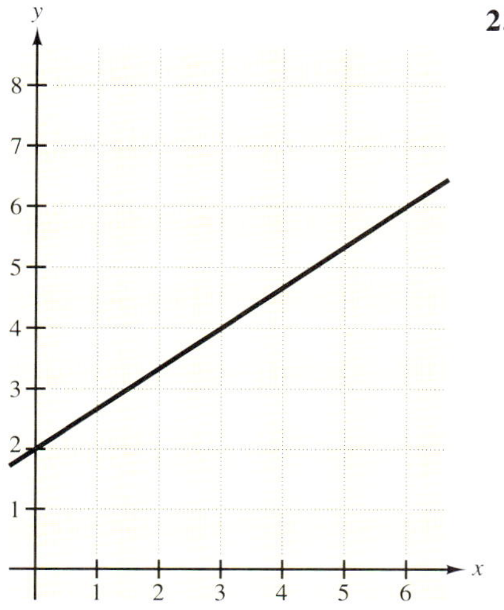

2.

3.

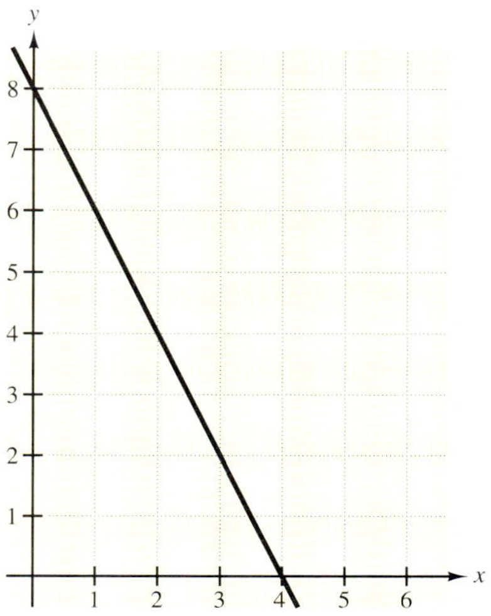

4.

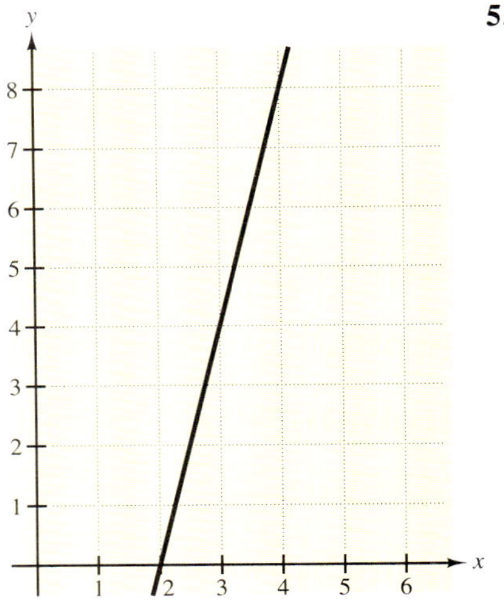

5.

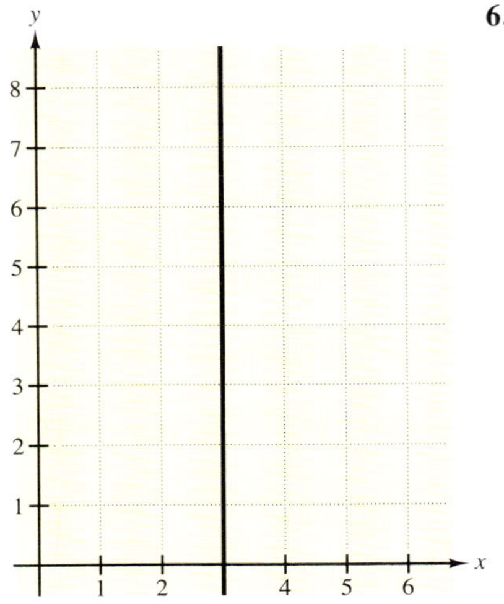

6.

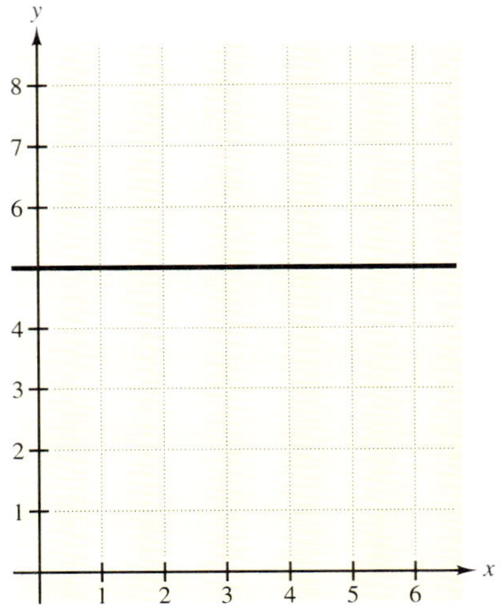

In Exercises 7 and 8, identify the line in the accompanying figure that has the specified slope m.

7. (a) $m = \frac{3}{4}$

(b) $m = 0$

(c) $m = -3$

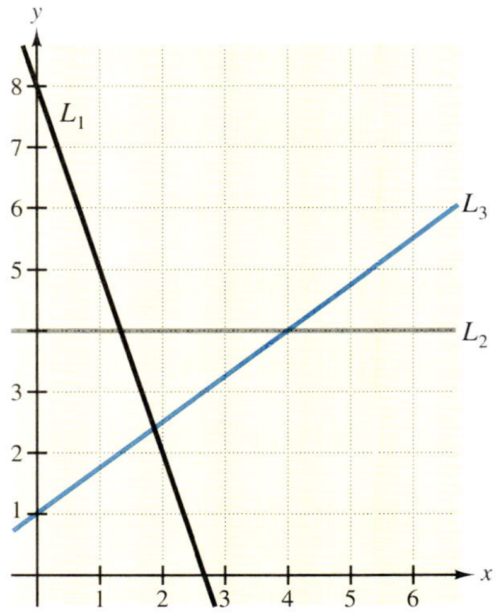

8. (a) $m = -\frac{5}{2}$

(b) m is undefined

(c) $m = 2$

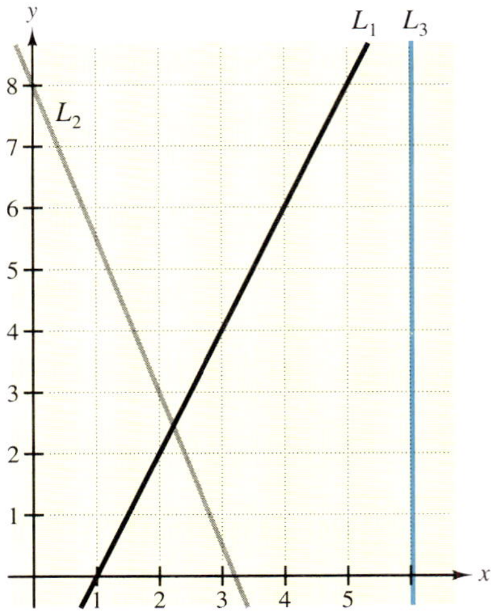

In Exercises 9–26, plot the points and find the slope (if possible) of the line passing through each pair of points. State whether the line through the pair of points rises, falls, is horizontal, or is vertical.

9. (0, 0), (7, 5)

10. (0, 0), (−3, −4)

11. (−2, −3), (6, 1)

12. (0, −4), (6, 0)

13. (0, 12), (8, 0)

14. (3, 6), (5, −2)

15. $(-5, -3), (-5, 4)$

16. $(0, -8), (-5, 0)$

17. $(2, -5), (7, -5)$

18. $(-2, 1), (-4, -3)$

19. $\left(\frac{3}{4}, 2\right), \left(5, -\frac{5}{2}\right)$

20. $\left(-\frac{3}{2}, -\frac{1}{2}\right), \left(\frac{5}{8}, \frac{1}{2}\right)$

21. $(4.2, -1), (-4.2, 6)$

22. $(3.4, 0), (3.4, 1)$

23. $(2.5, -2), (4.75, 5.25)$

24. $(0, 4.5), (3, 4.5)$

25. $(-3, -5), (3, 5)$

26. $(-5, 0), (0, 6)$

In Exercises 27–30, find the unknown coordinate of the indicated point so that the slope through the two points will have the specified magnitude.

27. $(4, 5), (x, 7), m = -\frac{2}{3}$

28. $(x, -2), (5, 0), m = \frac{3}{4}$

29. $(-3, y), (9, 3), m = \frac{3}{2}$

30. $(-3, 20), (2, y), m = -6$

In Exercises 31–40, a point on a line and the slope of the line are given. Find two additional points on the line. (There are many correct answers to each problem.)

31. $(5, 2), \quad m = 0$

32. $(-4, 3), \quad m$ is undefined

33. $(3, -4), \quad m = 3$

34. $(-1, -5), \quad m = 2$

35. $(0, 3), \quad m = -1$

36. $(-2, 6), \quad m = -3$

37. $(-5, 0), \quad m = \frac{4}{3}$

38. $(-1, 1), \quad m = -\frac{3}{4}$

39. $(4, 2), \quad m$ is undefined

40. $(-2, -2), \quad m = 0$

In Exercises 41–46, sketch the graph of a line through the point (0, 1) having the given slope.

41. $m = 2$

42. $m = 0$

43. m is undefined

44. $m = -1$

45. $m = -\frac{4}{3}$

46. $m = \frac{2}{3}$

In Exercises 47–50, plot the x- and y-intercepts and sketch the graph of the line.

47. $2x - y + 4 = 0$

48. $3x + 5y + 15 = 0$

49. $-5x + 2y - 20 = 0$

50. $3x - 5y - 15 = 0$

In Exercises 51–60, determine the slope and y-intercept of the given line. Then sketch the line.

51. $3x - y - 2 = 0$

52. $x - y - 5 = 0$

53. $x + y = 0$

54. $x - y = 0$

55. $3x + 2y - 2 = 0$

56. $x - 2y - 2 = 0$

57. $x - 4y + 2 = 0$

58. $8x + 6y - 3 = 0$

59. $y - 2 = 0$

60. $y + 4 = 0$

In Exercises 61–64, determine whether the lines L_1 and L_2 passing through the given pairs of points are parallel, perpendicular, or neither.

61. L_1: (1, 3), (2, 1)
L_2: (0, 0), (4, 2)

62. L_1: (−3, −3), (1, 7)
L_2: (0, 4), (5, −2)

63. L_1: (−2, 0), (4, 4)
L_2: (1, −2), (4, 0)

64. L_1: (−5, 3), (3, 0)
L_2: $(1, 2), \left(3, \frac{22}{3}\right)$

In Exercises 65–68, sketch the graphs of the pair of equations on the same rectangular coordinate system, and state the relationship that exists between the resulting lines.

65. L_1: $y = \frac{1}{2}x - 2$
L_2: $y = \frac{1}{2}x + 3$

66. L_1: $y = 3x - 2$
L_2: $y = 3x + 1$

67. L_1: $y = \frac{3}{4}x - 3$
L_2: $y = -\frac{4}{3}x + 1$

68. L_1: $y = -\frac{2}{3}x - 5$
L_2: $y = \frac{3}{2}x + 1$

69. *Earnings per Share* The accompanying graph gives the earnings per share of common stock for Xerox Corporation for the years 1983 through 1989 (*Source*: Xerox Corporation). In what year did earnings (a) decrease most rapidly, and (b) increase most rapidly?

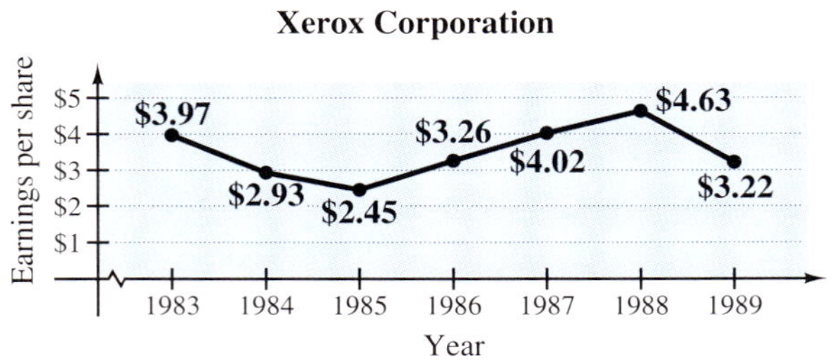

70. *Dividends per Share* The accompanying graph gives the declared dividends per share of common stock for Sun Company for the years 1983 through 1989 (*Source*: Sun Company). In what year did dividends (a) decrease most rapidly, and (b) increase most rapidly?

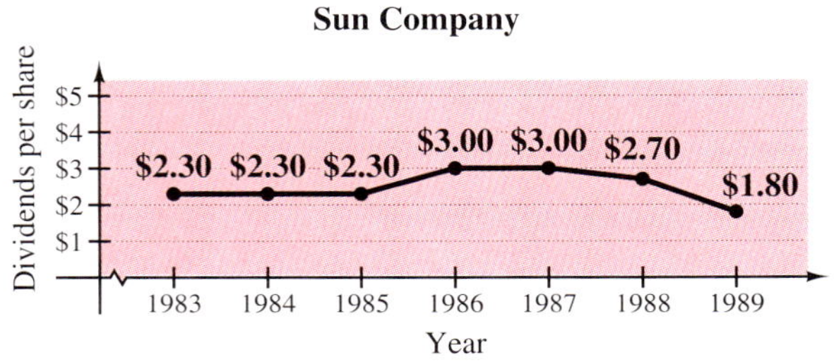

71. *Road Grade* When driving down a mountain road, you notice warning signs indicating that it is a "12% grade." This means that the slope of the road is $-\frac{12}{100}$. Determine the amount of horizontal change in your position if you note from elevation markers that you have descended 2,000 feet vertically. (See figure.)

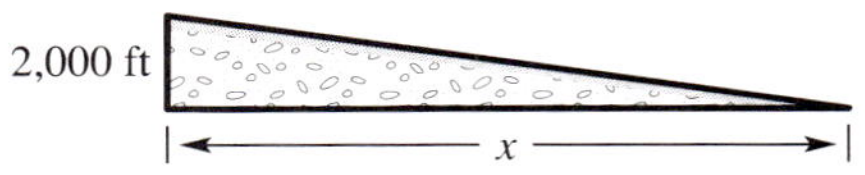

72. *Height of an Attic* The "rise to run" in determining the steepness of the roof on a house is 3 to 4. Determine the maximum height of the attic of the house if the house is 30 feet wide. (See figure.)

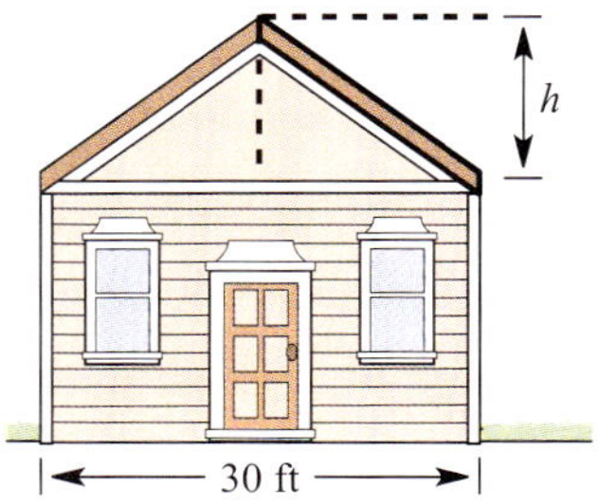

SECTION **7.5**

Equations of Lines

Introduction • *The Point-Slope Equation of a Line* • *Summary of Equations of Lines* • *Applications*

Introduction

There are two basic types of problems in coordinate geometry.

1. Given an algebraic equation, sketch its graph.
2. Given some geometric conditions about a graph, construct its algebraic equation.

The first type of problem can be thought of as moving from algebra to geometry, whereas the second type can be thought of as moving the other way—from geometry to algebra. So far in this chapter we have been working primarily with the first type of problem. In this section we look at the second problem. Solving this second type of problem is much like solving a word problem, as we did in Chapter 2. That is, we are given certain facts about a straight line, and are asked to write an equation for the line.

The Point-Slope Equation of a Line

If you know the slope of a line *and* the coordinates of one point on the line, then you can find an equation for the line. Before giving a general formula for doing this, let's look at an example.

EXAMPLE 1 ■ **Finding an Equation of a Line Given Its Slope and a Point on the Line**

A line has a slope of $\frac{4}{3}$ and passes through the point $(-2, 1)$. Find an equation for this line.

FIGURE 7.48

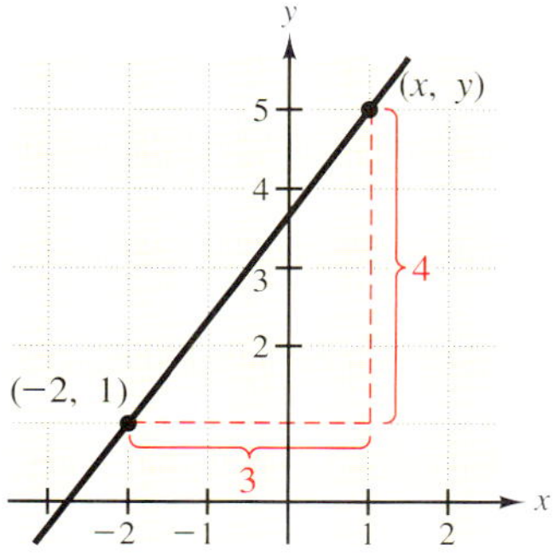

Solution

Using methods from the previous section, we sketch the graph of the line, as shown in Figure 7.48. We know that the slope of a line is the same through any two points on the line. Thus, to find an equation of the line, we let (x, y) represent *any* point on the line. Now, using the representative point (x, y) and the given point $(-2, 1)$, it follows that the slope of the line is given by

$$m = \frac{y - 1}{x - (-2)}.$$

Difference in y-values (numerator)

Difference in x-values (denominator)

Since the slope of the line is known to be $m = \frac{4}{3}$, we can rewrite this equation as follows.

$$\frac{4}{3} = \frac{y - 1}{x + 2} \qquad \text{Slope formula}$$

$$4(x + 2) = 3(y - 1) \qquad \text{Cross multiply}$$

$$4x + 8 = 3y - 3$$

$$4x - 3y = -11 \qquad \text{Equation of line}$$

Therefore, we have found an equation for the line. It is $4x - 3y = -11$. ■

The procedure in Example 1 can be used to derive a *formula* for the equation of a line, given its slope and a point on the line. In Figure 7.49, let (x_1, y_1) be a given point on the line whose slope is m. If (x, y) is any *other* point on the line, then it follows that

$$\frac{y - y_1}{x - x_1} = m.$$

This equation in variables x and y can be rewritten in the form

$$y - y_1 = m(x - x_1)$$

which is called the **point-slope form** of the equation of a line.

FIGURE 7.49

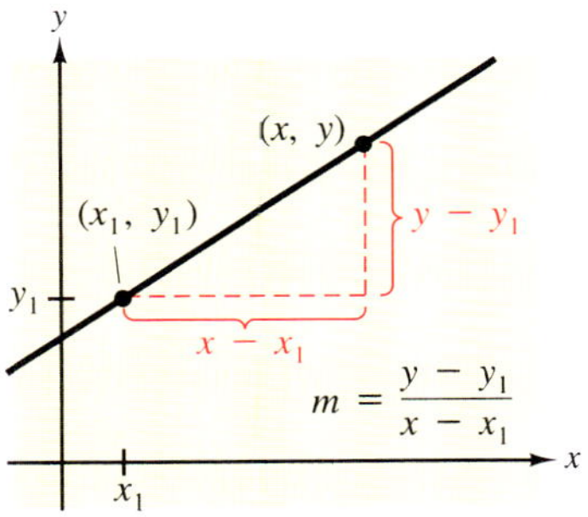

Point-Slope Form of the Equation of a Line

The **point-slope** form of the equation of the line that passes through the point (x_1, y_1) and has a slope of m is

$$y - y_1 = m(x - x_1).$$

EXAMPLE 2 ■ The Point-Slope Form of the Equation of a Line

Find an equation of the line that passes through the point $(2, -3)$ and has a slope of -2.

Solution

Using the point-slope form with $(x_1, y_1) = (2, -3)$ and $m = -2$, we have the following.

$$y - y_1 = m(x - x_1) \quad \text{Point-slope form}$$
$$y - (-3) = -2(x - 2) \quad \text{Substitute } y_1 = -3,\ x_1 = 2, \text{ and } m = -2$$
$$y + 3 = -2x + 4$$
$$y = -2x + 1 \quad \text{Equation of line}$$

The graph of this line is shown in Figure 7.50.

FIGURE 7.50

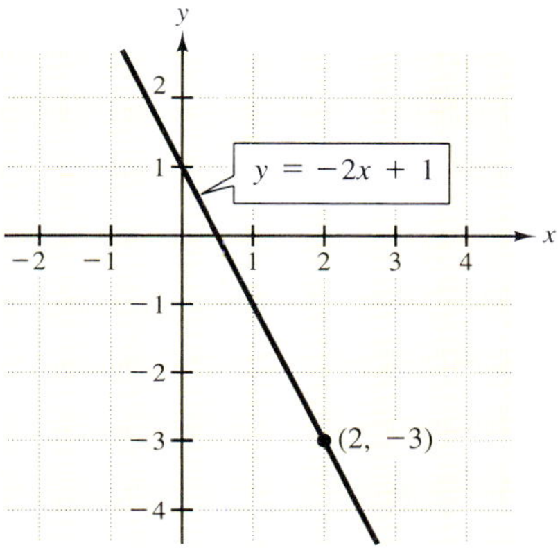

The point-slope form can be used to find the equation of a line passing through two points (x_1, y_1) and (x_2, y_2). First, we use the formula for the slope of the line passing through two points.

$$m = \frac{y_2 - y_1}{x_2 - x_1}$$

Then, once we know the slope, we use the point-slope form to obtain the equation

$$y - y_1 = \frac{y_2 - y_1}{x_2 - x_1}(x - x_1).$$

This is sometimes called the **two-point form** of the equation of a line.

EXAMPLE 3 ■ Finding an Equation of a Line Passing Through Two Points

Find an equation of the line that passes through the points (4, 2) and (−2, 3).

Solution

If we let $(x_1, y_1) = (4, 2)$ and $(x_2, y_2) = (-2, 3)$, then we can apply the formula for the slope of a line passing through two points as follows.

$$m = \frac{y_2 - y_1}{x_2 - x_1} = \frac{3 - 2}{-2 - 4} = -\frac{1}{6}$$

Now, using the point-slope form, we can find the equation of the line.

$$y - y_1 = m(x - x_1)$$

$$y - 2 = -\frac{1}{6}(x - 4)$$

$$y - 2 = -\frac{1}{6}x + \frac{2}{3}$$

$$y = -\frac{1}{6}x + \frac{8}{3}$$

The graph of this line is shown in Figure 7.51.

FIGURE 7.51

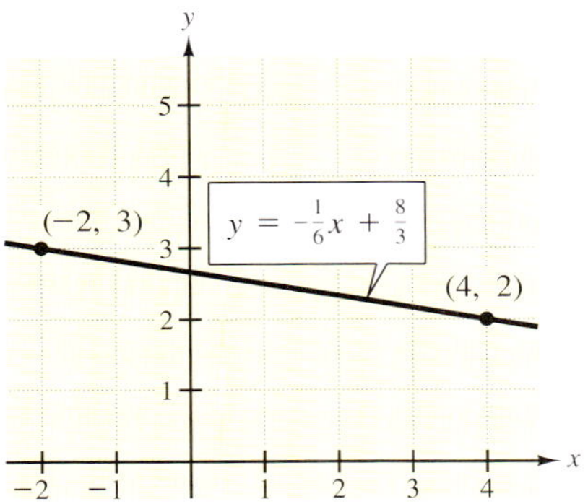

■

In Example 3 it did not matter which of the two points we labeled as (x_1, y_1) and (x_2, y_2). Try switching these labels to $(x_1, y_1) = (-2, 3)$ and $(x_2, y_2) = (4, 2)$ and reworking the problem to see that you obtain the same equation.

Summary of Equations of Lines

From the slope-intercept form of the equation of a line, we see that a horizontal line $(m = 0)$ has an equation of the form

$$y = (0)x + b \quad \text{or} \quad y = b. \qquad \text{Horizontal line}$$

This is consistent with the fact that each point on a horizontal line through $(0, b)$ has a y-coordinate of b, as shown in Figure 7.52.

FIGURE 7.52

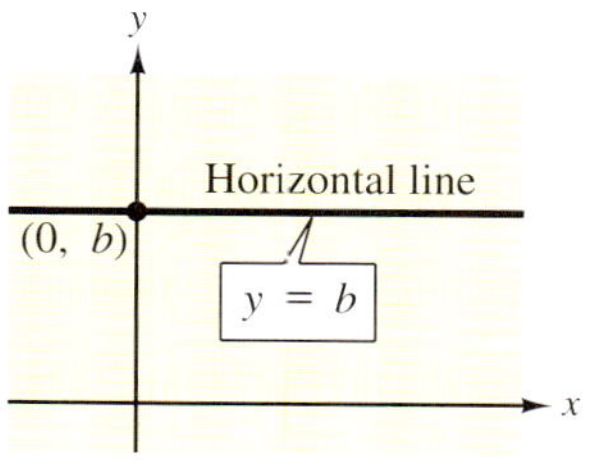

Similarly, each point on a vertical line through $(a, 0)$ has an x-coordinate of a, as shown in Figure 7.53. Hence, a vertical line has an equation of the form

$x = a$. Vertical line

FIGURE 7.53

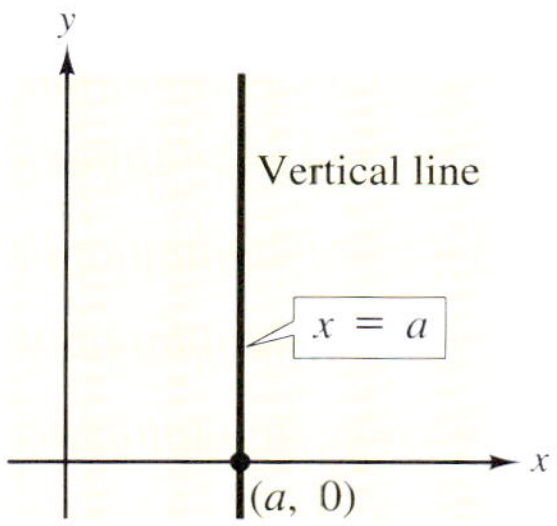

EXAMPLE 4 ■ **Writing Equations of Horizontal and Vertical Lines**

Write an equation for each of the following lines.

(a) Vertical line through $(-2, 4)$

(b) Line passing through $(-2, 3)$ and $(3, 3)$

(c) Line passing through $(-1, 2)$ and $(-1, 3)$

Solution

(a) Since the line is vertical and passes through the point $(-2, 4)$, we know that every point on the line has an x-coordinate of -2. Therefore, the equation of the line is

$$x = -2.$$

(b) The line through $(-2, 3)$ and $(3, 3)$ is horizontal. Thus, its equation is

$$y = 3.$$

(c) The line through $(-1, 2)$ and $(-1, 3)$ is vertical. Thus, its equation is

$$x = -1.$$ ■

The equation of a vertical line cannot be written in the slope-intercept form because the slope of a vertical line is undefined. However, *every* line has an equation that can be written in the **general form**

$ax + by + c = 0$ General form

where a and b are not *both* zero. If $a = 0$ (and $b \neq 0$), the equation can be reduced to the form $y = -c/b$, which represents a horizontal line. If $b = 0$ (and $a \neq 0$), the general equation can be reduced to the form $x = -c/a$, which represents a vertical line.

For convenience, we summarize our discussion of lines in the following list.

Summary of Equations of Lines

1. Slope of line through (x_1, y_1) and (x_2, y_2): $m = \dfrac{y_2 - y_1}{x_2 - x_1}$
2. General form of equation of line: $ax + by + c = 0$
3. Equation of vertical line: $x = a$
4. Equation of horizontal line: $y = b$
5. Slope-intercept form of equation of line: $y = mx + b$
6. Point-slope form of equation of line: $y - y_1 = m(x - x_1)$
7. Parallel lines have *equal* slopes: $m_1 = m_2$
8. Perpendicular lines have *negative reciprocal* slopes: $m_2 = -\dfrac{1}{m_1}$

EXAMPLE 5 ■ Equations of Parallel Lines

Find an equation of the line that passes through the point $(3, -2)$ and is parallel to the line $x - 4y = 6$, as shown in Figure 7.54.

FIGURE 7.54

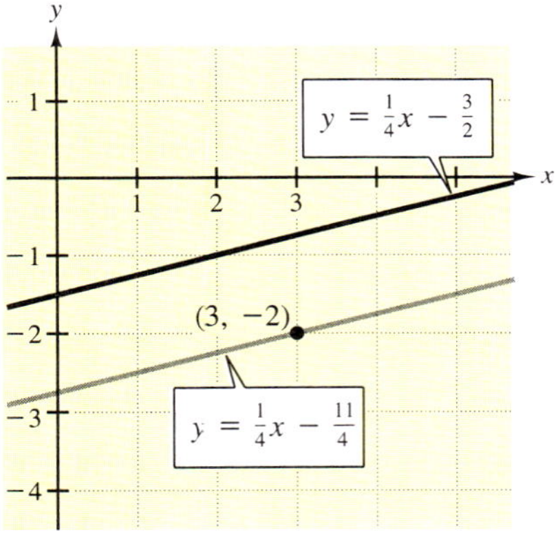

Solution

To begin, we write the given equation in slope-intercept form.

$$x - 4y = 6 \qquad \text{Given equation}$$
$$-4y = -x + 6$$
$$4y = x - 6$$
$$y = \frac{1}{4}x - \frac{3}{2} \qquad \text{Slope-intercept form}$$

Therefore, the given line has a slope of $m = \frac{1}{4}$. Since any line parallel to the given line must also have a slope of $\frac{1}{4}$, the required line through $(3, -2)$ has the following equation.

$$y - y_1 = m(x - x_1)$$

$$y - (-2) = \frac{1}{4}(x - 3)$$

$$y = \frac{1}{4}x - \frac{11}{4}$$

■

EXAMPLE 6 ■ Equations of Perpendicular Lines

Find an equation of the line that passes through the point $(-2, 3)$ and is perpendicular to the line $5x - 2y = 7$, as shown in Figure 7.55.

FIGURE 7.55

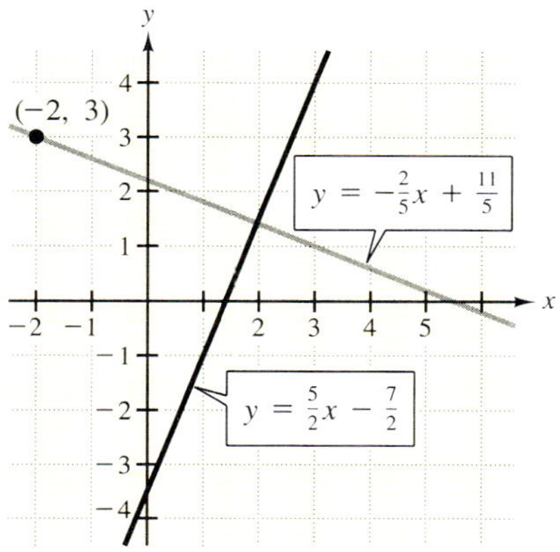

Solution

To begin, we write the given equation in slope-intercept form.

$$5x - 2y = 7 \qquad \text{Given equation}$$

$$-2y = -5x + 7$$

$$2y = 5x - 7$$

$$y = \frac{5}{2}x - \frac{7}{2} \qquad \text{Slope-intercept form}$$

Therefore, the given line has a slope of $m = \frac{5}{2}$. Hence, any line perpendicular to this line must have a slope of $-\frac{2}{5}$. (Note that $-\frac{2}{5}$ is the negative reciprocal of $\frac{5}{2}$.) Therefore, the equation of the required line through $(-2, 3)$ has the following form.

$$y - y_1 = m(x - x_1)$$
$$y - 3 = -\frac{2}{5}(x - (-2))$$
$$y - 3 = -\frac{2}{5}x - \frac{4}{5}$$
$$y = -\frac{2}{5}x + \frac{11}{5}$$

Applications

Linear equations are used frequently as mathematical models in business. The next example gives you one idea of how useful such a model can be.

EXAMPLE 7 ■ **An Application**

The total sales of a new computer software company were \$500,000 for the second year and \$1,000,000 for the fourth year. Using only this information, what would you estimate the total sales to be during the fifth year?

Solution

FIGURE 7.56

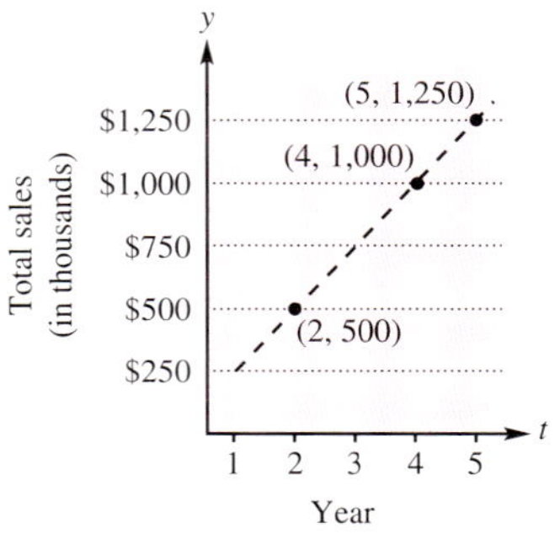

To solve this problem, we will use a *linear model*, with y representing the total sales in thousands and t representing the year. That is, in Figure 7.56, we let (2, 500) and (4, 1,000) be two points on the line representing the total sales for the company. The slope of the line passing through these points is

$$m = \frac{1{,}000 - 500}{4 - 2} = 250.$$

Now, using the point-slope form, we find the equation of the line as follows.

$$y - y_1 = m(t - t_1)$$
$$y - 500 = 250(t - 2)$$
$$y = 250t$$

Finally, we estimate the total sales during the fifth year ($t = 5$) to be

$$y = 250(5) = \$1{,}250 \text{ thousand} = \$1{,}250{,}000.$$

The estimation method illustrated in Example 7 is called **linear extrapolation**. Note in Figure 7.57 that for linear extrapolation, the estimated point lies to the right of the given points. When the estimated point lies *between* two given points, we call the procedure **linear interpolation**.

FIGURE 7.57

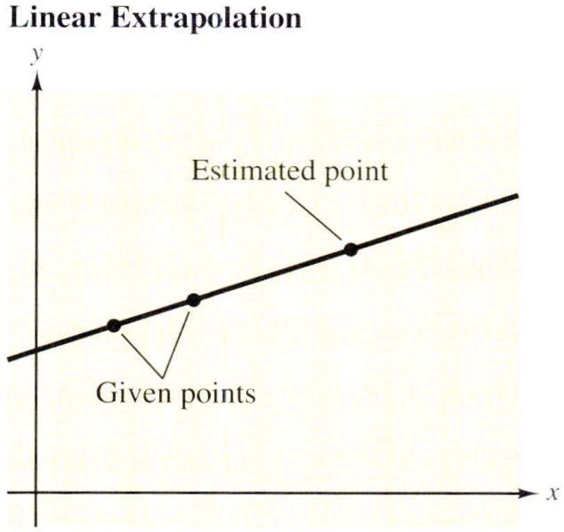

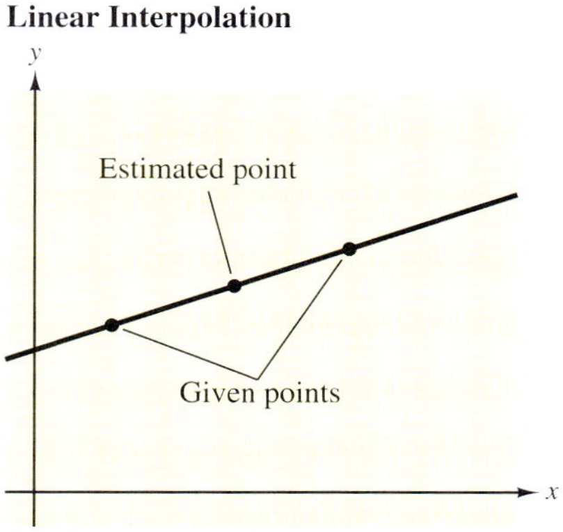

EXAMPLE 8 ■ An Application

A city ordinance requires a minimum drop of 0.6% in a home sewer line to the main sewer line (along the street). This means that for each foot from the house to the street, the sewer line must drop 0.006 feet, as shown in Figure 7.58.

(a) Write an equation that expresses the minimum vertical drop y in terms of the horizontal length x of the pipe.

(b) How much drop is needed for a house that is 200 feet from the main line?

FIGURE 7.58

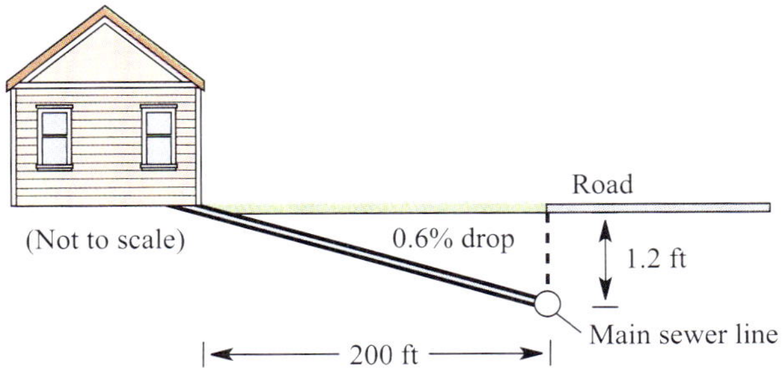

Solution

(a) In this case the slope is

$$m = -0.006$$

because the line *drops* 0.006 feet vertically for each foot of horizontal length. Since a pipe of zero length would have a zero drop, it follows that $(x_1, y_1) = (0, 0)$ is a point that satisfies the equation relating x and y. Therefore, using the point-slope form, we can write the following equation.

$$y - y_1 = m(x - x_1)$$
$$y - 0 = -0.006(x - 0)$$
$$y = -0.006x$$

(b) The drop in a line for a house that is 200 feet from the street would have to be at least

$$y = -0.006(200) = -1.2 \text{ feet.}$$

■

DISCUSSION PROBLEM ■ Equations of Lines

Match each of the following situations with one of the accompanying graphs. Also determine the slope and how it is interpreted in the given situation.

1. A person is paying \$20 per week to a friend to repay a \$140 loan.
2. An employee is paid \$5.50 per hour plus \$2.50 for each unit produced per hour.
3. A sales representative receives \$30 per day for food plus \$0.25 for each mile traveled.
4. The list price of a typewriter is \$900 and the value of the typewriter depreciates \$150 per year.

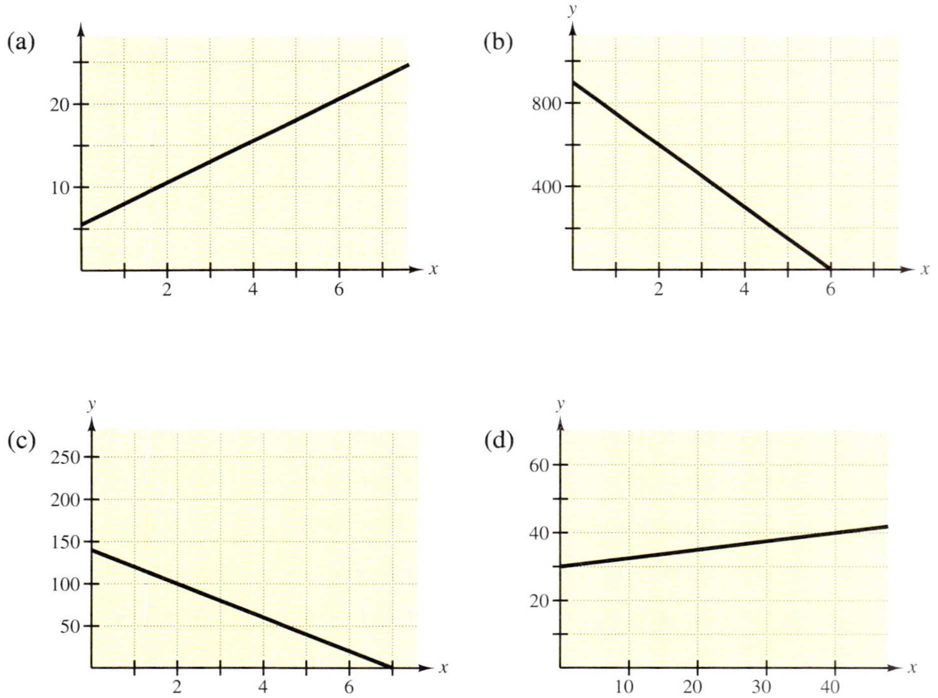

■

Warm-Up

The following warm-up exercises involve skills that were covered in earlier sections. You will use these skills in the exercise set for this section.

In Exercises 1–4, determine the slope of the line passing through the given points.

1. $(-5, 2), (3, 0)$

2. $(-8, -4), (6, 10)$

3. $(-3, -2), (-3, 5)$

4. $\left(\frac{5}{6}, \frac{3}{2}\right), \left(\frac{5}{3}, \frac{3}{4}\right)$

In Exercises 5 and 6, sketch the lines through the given point with the indicated slopes. Make the sketches on the same set of coordinate axes.

Point	*Slopes*			
5. $(3, 2)$	(a) 0	(b) 14	(c) $\frac{3}{2}$	(d) $-\frac{1}{4}$
6. $(-2, 4)$	(a) 2	(b) -2	(c) $\frac{1}{2}$	(d) undefined

In Exercises 7–10, write the equation in slope-intercept form.

7. $y - 3 = -\frac{3}{5}(x - 2)$

8. $y + 5 = \frac{4}{3}(x - 3)$

9. $y - (-2) = \frac{5 - (-3)}{4 - 6}(x - 5)$

10. $y - 7 = \frac{1 - 3}{0 - 2}[x - (-2)]$

7.5 EXERCISES

In Exercises 1–10, from the given equation of the line, determine the slope of the line and the coordinates of one point through which it passes.

1. $y - 3 = 6(x + 4)$

2. $y + 10 = -12(x - 1)$

3. $y + \frac{5}{8} = \frac{3}{4}(x + 2)$

4. $y - \frac{3}{4} = \frac{5}{6}\left(x - \frac{9}{10}\right)$

5. $y = \frac{2}{3}x - 2$

6. $y = -5x + 12$

7. $3x - 2y = 0$

8. $x + 4 = 0$

9. $5x - 2y + 24 = 0$

10. $3x + 4y - 16 = 0$

In Exercises 11–16, find the point-slope form of the equation of the line that passes through the given point and has the specified slope. Sketch the graph of the line.

11. $(0, 0), \quad m = -\frac{1}{2}$

12. $(0, 0), \quad m = \frac{2}{3}$

13. $(0, -4), \quad m = 3$

14. $(0, 6), \quad m = -\frac{3}{4}$

15. $\left(-2, \frac{7}{2}\right), \quad m = -4$

16. $\left(1, -\frac{3}{2}\right), \quad m = \frac{3}{4}$

In Exercises 17–26, find the general form of the equation of the line through the given points and sketch a graph of the line.

17. $(0, 0), (2, 3)$

18. $(0, 0), (3, -5)$

19. $(-2, 3), (5, 0)$

20. $(-6, -3), (4, 3)$

21. $(-2, 2), (4, 5)$

22. $(0, 10), (5, 0)$

23. $(-2, 3), (4, 3)$

24. $(1, -2), (1, 8)$

25. $\left(\frac{3}{2}, 3\right), \left(\frac{9}{2}, -4\right)$

26. $\left(4, \frac{7}{3}\right), \left(-1, \frac{2}{3}\right)$

In Exercises 27–36, find the general form of the equation of the line passing through the given point with the specified slope.

27. $(0, -6), \quad m = \frac{1}{2}$

28. $(0, 9), \quad m = -\frac{1}{3}$

29. $(-2, 8), \quad m = -2$

30. $(0, 0), \quad m = 3$

31. $(5, 0), \quad m = -\frac{2}{3}$

32. $(-3, -7), \quad m = \frac{5}{4}$

33. $(2, -1), \quad m$ is undefined

34. $(-8, 5), \quad m = 0$

35. $\left(\frac{3}{4}, \frac{5}{2}\right), \quad m = \frac{4}{3}$

36. $\left(-\frac{3}{2}, \frac{1}{2}\right), \quad m = -3$

In Exercises 37–44, find the general form of the equation of the line passing through the two given points.

37. $(-5, 2), (5, -2)$

38. $(5, 4), (-3, 5)$

39. $\left(10, \frac{1}{2}\right), \left(\frac{3}{2}, \frac{7}{4}\right)$

40. $(-2, 12), (6, 12)$

41. $(7.5, 2), (7.5, 9)$

42. $(2, -8), \left(6, \frac{8}{3}\right)$

43. $(2, 0.6), (8, -4.2)$

44. $(-5, 0.6), (3, -3.4)$

In Exercises 45–50, write an equation of the line through the given point (a) parallel to the given line, and (b) perpendicular to the given line.

45. $(2, 1), \quad 6x - 2y = 3$

46. $(-3, 4), \quad x + 6y = 12$

47. $(-5, 4), \quad 5x + 4y = 24$

48. $\left(\frac{5}{8}, \frac{9}{4}\right), \quad -5x + 4y = 0$

49. $(-1, 2), \quad y + 5 = 0$

50. $(3, -4), \quad x - 10 = 0$

In Exercises 51–54, find an equation of the line with intercepts $(a, 0)$ and $(0, b)$ where the equation is given by the following.

$$\frac{x}{a} + \frac{y}{b} = 1, \quad a \neq 0, b \neq 0$$

51. x-intercept: $(3, 0)$
y-intercept: $(0, 2)$

52. x-intercept: $(-6, 0)$
y-intercept: $(0, 2)$

53. x-intercept: $\left(-\frac{5}{6}, 0\right)$
y-intercept: $\left(0, -\frac{7}{3}\right)$

54. x-intercept: $\left(-\frac{8}{3}, 0\right)$
y-intercept: $(0, -4)$

55. *Cost* The cost in dollars of producing x units of a certain product is given by $C = 20x + 5{,}000$. Use this equation to complete the following table.

x	0		100		1,000
C		6,000		15,000	

56. *Temperature Conversion* The linear relationship between the Fahrenheit and Celsius temperature scales is given by $C = \frac{5}{9}F - \frac{160}{9}$. Use this equation to complete the following table.

F	C
$-20°$	
	$-17.8°$
$20°$	
	$0°$
$100°$	
	$100°$

57. *Sales Commission* The salary for a sales representative is \$1,500 per month plus a 3% commission on total monthly sales. Write a linear equation giving the salary S in terms of the monthly sales M.

58. *Reimbursed Expenses* A sales representative is reimbursed \$125 per day for lodging and meals plus \$0.22 per mile driven. Write a linear equation giving the daily cost C to the company in terms of x, the number of miles driven.

59. *Discount Price* A store is offering a 30% discount on all items in its inventory. Write a linear equation giving the sale price S for an item in terms of its list price L.

60. *Straight-Line Depreciation* A small business purchases a photocopier for \$7,400. It is estimated that after four years its depreciated value will be \$1,500. Assuming straight-line depreciation, write a linear equation giving the value V of the copier in terms of time t.

61. *Study Time* Suppose you can get a score of 72 on an exam with no additional study. For each additional hour of studying you can increase the score by five points. Write a linear equation giving your estimated score y in terms of the number of hours t of additional study.

62. *College Enrollment* A small college had an enrollment of 1,500 students in 1980. During the next 10 years, the enrollment increased by approximately 60 students per year. Write a linear equation giving the enrollment N in terms of the year t. (Let $t = 0$ correspond to the year 1980.) If this constant rate of growth continues, predict the enrollment in the year 2000.

63. *Rental Occupancy* A real estate office handles an apartment complex with 50 units. When the rent per unit is \$450 per month, all 50 units are occupied. However, when the rent is \$525 per month, the average number of occupied units drops to 45. Assume the relationship between the monthly rent p and the demand x is linear.
(a) Write the equation of the line giving the demand x in terms of the rent p.
(b) (*Linear Extrapolation*) Use this equation to predict the number of units occupied if the rent is \$570.
(c) (*Linear Interpolation*) Use this equation to predict the number of units occupied if the rent is \$480.

64. *Depth Markers* A swimming pool is 40 feet long, 20 feet wide, 4 feet deep at the shallow end, and 9 feet deep at the deep end. Position the side of the pool on the rectangular coordinate system as shown in the figure that follows and find an equation of the line representing the edge of the inclined bottom of the pool. Use this equation to determine the distance from the deep end where the markers must be placed to indicate that the depth of the pool has changed by 1 foot.

Figure for 64

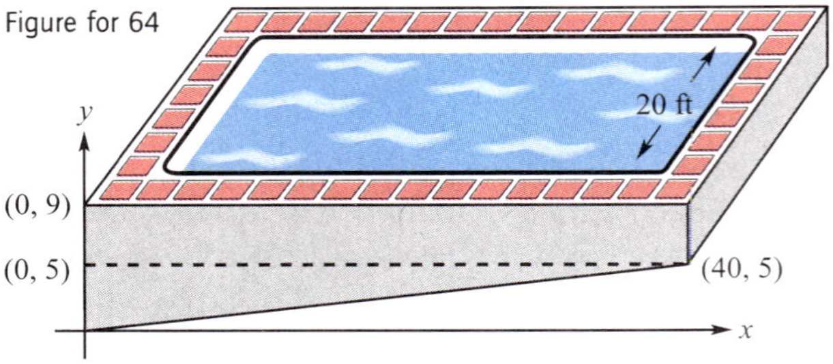

65. *Best-Fitting Line* An instructor gives 20-point quizzes and 100-point exams in a mathematics course. The average quiz and test scores for six students are given as ordered pairs (x, y) where x is the average quiz score and y is the average test score. The ordered pairs are (18, 87), (10, 55), (19, 96), (16, 79), (13, 76), and (15, 82).
(a) Plot the points.
(b) Use a ruler to sketch the best-fitting line through the points.
(c) Find an equation for the line sketched in part (b).
(d) Use the equation in part (c) to estimate the average test score for a person with an average quiz score of 17.

In Exercises 66–69, use a graphing calculator to sketch the graph of the line.

66. $y = -\frac{3}{4}x + 5$

67. $y = 2x - 1$

68. $6x - 9y + 18 = 0$

69. $0.5x + 0.33y - 1 = 0$

In Exercises 70 and 71, use a graphing calculator to sketch the graph of both equations on the same screen and determine if the lines are parallel or perpendicular. (Press ZOOM and cursor to "Square" so that the tic marks are equally spaced on each axis.)

70. $x + 2y - 3 = 0$
$-2x - 4y + 1 = 0$

71. $3x - 4y - 1 = 0$
$4x + 3y + 2 = 0$

SECTION 7.6

Parabolas

Equations of Parabolas • *Sketching a Parabola* • *Finding the Equation of a Parabola* • *Applications*

Equations of Parabolas

In this section we study graphs of second-degree equations of the form

$$y = ax^2 + bx + c.$$

Figure 7.59 shows the graph of a simple second-degree equation, $y = x^2$. This graph is an example of a **parabola**. The lowest point on this graph, (0, 0), is called the **vertex** of the parabola, and the vertical line that passes through the vertex (the y-axis, in this case) is called the **axis** of the parabola. Every parabola is *symmetric* about its axis, which means that if folded along its axis, the two parts of the parabola would match. In general, the graph of an equation of the form $y = ax^2 + bx + c$, $a \neq 0$ is a parabola.

FIGURE 7.59

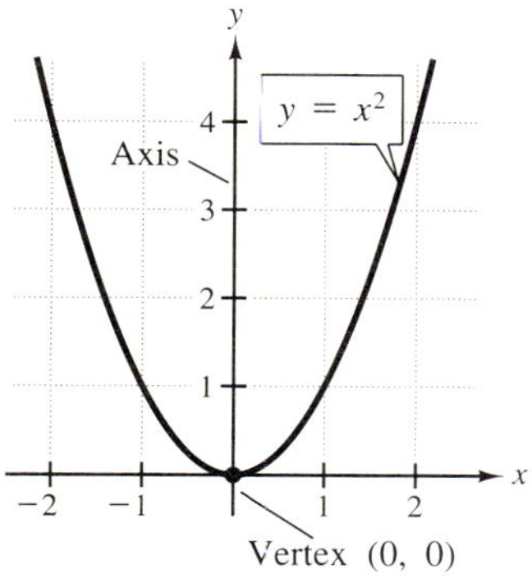

Equations of Parabolas The graph of $y = ax^2 + bx + c$, $a \neq 0$ is a **parabola**. The completed-square form of the equation

$$y = a(x - h)^2 + k \qquad \text{Standard form}$$

is called the **standard form** of the equation. The **vertex** of the parabola occurs at the point (h, k), and the vertical line passing through the vertex is called the **axis** of the parabola.

Parabolas that correspond to equations of the form $y = ax^2 + bx + c$ open *upward* if the leading coefficient a is positive, and open *downward* if the leading coefficient is negative, as shown in Figure 7.60.

FIGURE 7.60

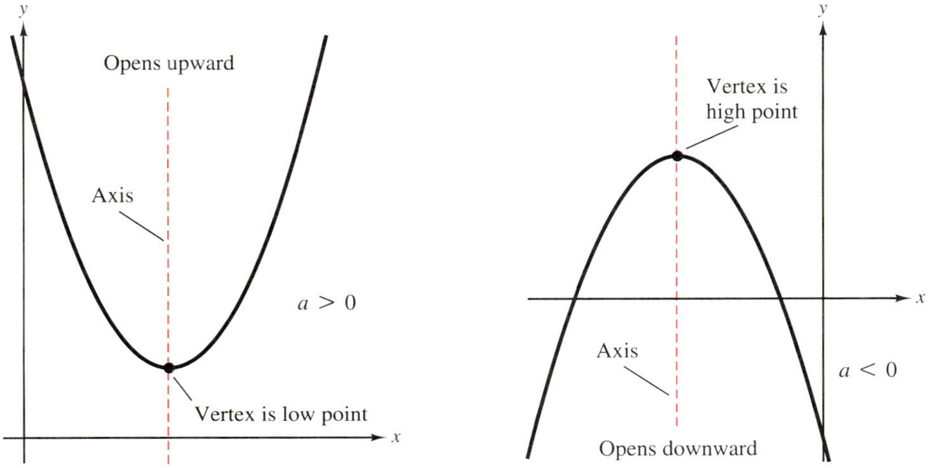

EXAMPLE 1 ■ Finding the Vertex of a Parabola

Find the vertex of the parabola given by the equation $y = x^2 - 6x + 5$.

Solution

We begin by writing the equation in standard form.

$$y = x^2 - 6x + 5 \qquad \text{Given equation}$$

$$y = x^2 - 6x + (-3)^2 - (-3)^2 + 5 \qquad \text{Add and subtract } (-3)^2$$

half of -6

$$y = (x^2 - 6x + 3^2) - 9 + 5 \qquad \text{Regroup terms}$$

$$y = (x - 3)^2 - 4 \qquad \text{Standard form}$$

Now, from the standard form, we can see that the vertex of the parabola occurs at the point $(3, -4)$, as shown in Figure 7.61.

FIGURE 7.61

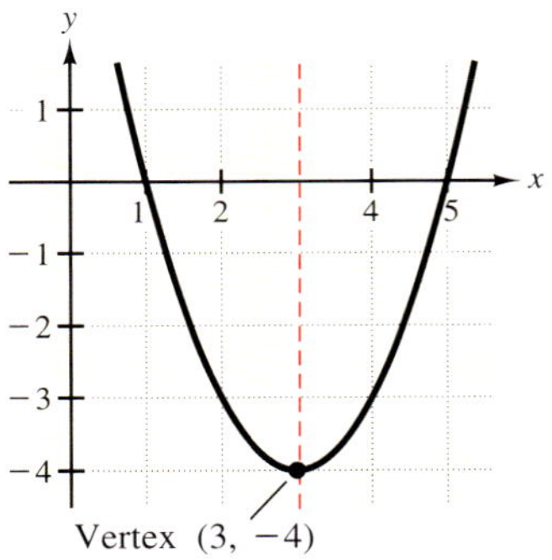

■

EXAMPLE 2 ■ Finding the Vertex of a Parabola

Find the vertex of the parabola given by the following equation.

$$y = x^2 + x$$

Solution

We begin by writing the equation in standard form.

$$y = x^2 + x \qquad \text{Given equation}$$

$$y = x^2 + x + \left(\frac{1}{2}\right)^2 - \left(\frac{1}{2}\right)^2 \qquad \text{Add and subtract } \left(\tfrac{1}{2}\right)^2$$

half of 1

$$y = \left(x^2 + x + \frac{1}{4}\right) - \frac{1}{4} \qquad \text{Regroup terms}$$

$$y = \left(x + \frac{1}{2}\right)^2 - \frac{1}{4} \qquad \text{Standard form}$$

Now, from the standard form, we can see that the vertex of the parabola occurs at the point $\left(-\frac{1}{2}, -\frac{1}{4}\right)$, as shown in Figure 7.62. Note that the vertex is $\left(-\frac{1}{2}, -\frac{1}{4}\right)$ because the standard form is

$k = -\frac{1}{4}$

$$y = a(x - h)^2 + k = 1\left(x - \left[-\frac{1}{2}\right]\right)^2 + \left(-\frac{1}{4}\right).$$

$h = -\frac{1}{2}$

■

FIGURE 7.62

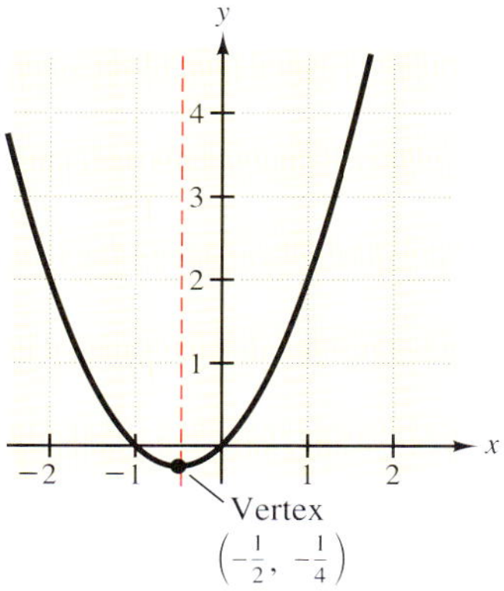

EXAMPLE 3 ■ Finding the Vertex of a Parabola

Find the vertex of the parabola given by the following equation.

$$y = 2x^2 - 4x$$

Solution

We begin by writing the equation in standard form.

$$y = 2x^2 - 4x \qquad \text{Given equation}$$

$$y = 2(x^2 - 2x) \qquad \text{Factor out leading coefficient}$$

$$y = 2(x^2 - 2x + [-1]^2 - [-1]^2) \qquad \text{Complete the square}$$

half of -2

$$y = 2(x - 2x + 1^2) - 2(1^2)$$

$$y = 2(x - 1)^2 - 2 \qquad \text{Standard form}$$

Now, from the standard form, we can see that the vertex of the parabola occurs at the point $(1, -2)$, as shown in the Figure 7.63.

FIGURE 7.63

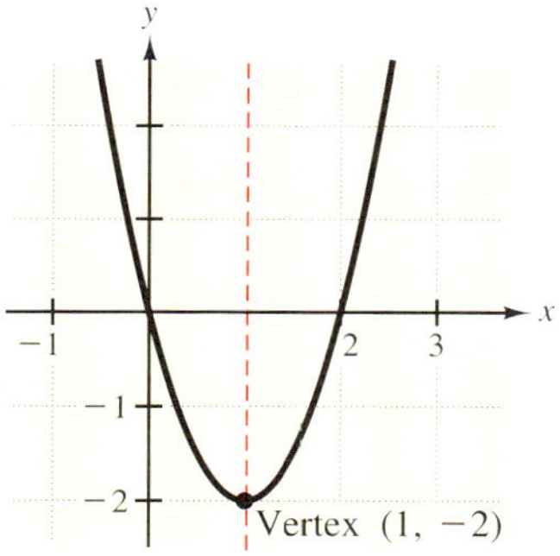

■

EXAMPLE 4 ■ **Finding the Vertex of a Parabola**

Find the vertex of the parabola given by the following equation.

$$y = -x^2 - 4x - 3$$

Solution

We begin by writing the equation in standard form.

$$y = -x^2 - 4x - 3 \qquad \text{Given equation}$$

$$y = -1(x^2 + 4x) - 3 \qquad \text{Factor out leading coefficient}$$

$$y = -(x^2 + 4x + \underbrace{2^2 - 2^2}_{\text{half of 4}}) - 3 \qquad \text{Complete the square}$$

$$y = -(x^2 + 4x + 2^2) + 2^2 - 3$$

$$y = -(x + 2)^2 + 1 \qquad \text{Standard form}$$

Now, from the standard form, we can see that the vertex of the parabola occurs at the point $(-2, 1)$, as shown in Figure 7.64.

FIGURE 7.64

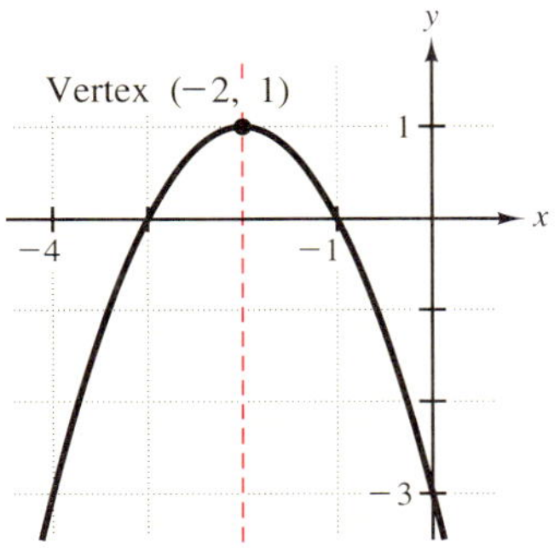

■

In Examples 1–4 we found the vertex of the given parabola by completing the square *for each* equation. Another technique we could use would be to complete the square once for the general equation, as follows.

$$y = ax^2 + bx + c$$

$$y = a\left(x^2 + \frac{b}{a}x\right) + c$$

$$y = a\left(x + \frac{b}{2a}\right)^2 + c - \frac{b^2}{4a}$$

From this form, we see that the vertex occurs when $x = -b/2a$. Try using this technique to find the vertex of each of the parabolas in Examples 1–4.

Sketching a Parabola

To sketch a parabola, we suggest the following steps.

Sketching a Parabola

To sketch the graph of the parabola given by $y = ax^2 + bx + c$, use the following steps.

1. Write the given equation in standard form and determine the vertex and axis of the parabola.
2. Plot the vertex, axis, and a few additional points on the parabola. (Using the symmetry about the axis can reduce the number of points you need to plot.)
3. Use the fact that the parabola opens *upward* if $a > 0$ and opens *downward* if $a < 0$ to complete the sketch.

EXAMPLE 5 ■ Sketching a Parabola

Sketch the graph of the parabola given by the following second-degree equation.

$$x^2 - y + 6x + 8 = 0$$

Solution

We begin by writing the equation in standard form.

$$x^2 - y + 6x + 8 = 0 \qquad \text{Given equation}$$

$$-y = -x^2 - 6x - 8$$

$$y = x^2 + 6x + 8$$

$$y = (x^2 + 6x + 3^2 - 3^2) + 8$$

half of 6

$$y = (x^2 + 6x + 3^2) - 9 + 8$$

$$y = (x + 3)^2 - 1 \qquad \text{Standard form}$$

Therefore, the vertex occurs at the point $(-3, -1)$ and the axis is given by the line $x = -3$. After plotting this information, we calculate a few additional points on the parabola, as shown in Table 7.11.

TABLE 7.11

x-Value	-5	-4	-3	-2	-1
y-Value	3	0	-1	0	3
Solution Points	$(-5, 3)$	$(-4, 0)$	$(-3, -1)$	$(-2, 0)$	$(-1, 3)$

The graph of the parabola is shown in Figure 7.65. Note that the parabola opens upward because the leading coefficient (in the standard form) is positive.

FIGURE 7.65

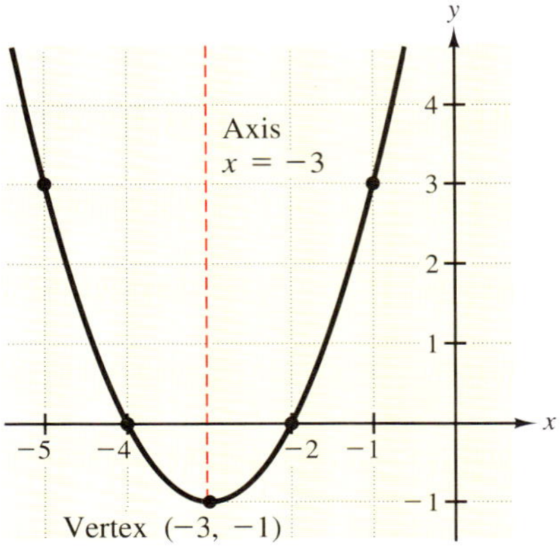

EXAMPLE 6 ■ Sketching a Parabola

Sketch the graph of the parabola given by the following second-degree equation.

$$x^2 + 2y - 4x + 8 = 0$$

Solution

We begin by writing the equation in standard form.

$$x^2 + 2y - 4x + 8 = 0 \qquad \text{Given equation}$$

$$2y = -x^2 + 4x - 8$$

$$y = -\frac{1}{2}x^2 + 2x - 4$$

$$y = -\frac{1}{2}(x^2 - 4x) - 4$$

$$y = -\frac{1}{2}(x^2 - 4x + [-2]^2 - [-2]^2) - 4$$

half of -4

$$y = -\frac{1}{2}(x^2 - 4x + 2^2) + 2 - 4$$

$$y = -\frac{1}{2}(x - 2)^2 - 2 \qquad \text{Standard form}$$

Therefore, the vertex occurs at the point $(2, -2)$ and the axis is given by the line $x = 2$. After plotting this information, we calculate a few additional points on the parabola, as shown in Table 7.12.

TABLE 7.12

x-Value	0	1	2	3	4
y-Value	-4	$-\frac{5}{2}$	-2	$-\frac{5}{2}$	-4
Solution Points	$(0, -4)$	$\left(1, -\frac{5}{2}\right)$	$(2, -2)$	$\left(3, -\frac{5}{2}\right)$	$(4, -4)$

The graph of the parabola is shown in Figure 7.66. Note that the parabola opens downward because the leading coefficient (in the standard form) is negative.

FIGURE 7.66

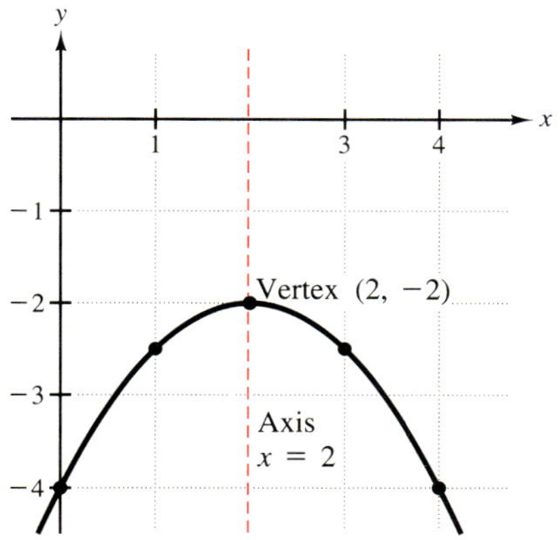

■

Finding the Equation of a Parabola

To write the equation of a parabola with a vertical axis, we use the fact that its standard equation has the form

$$y = a(x - h)^2 + k$$

where (h, k) is the vertex.

EXAMPLE 7 ■ **Finding the Equation of a Parabola**

Find the equation of the parabola whose vertex is $(-2, 1)$ and whose y-intercept is $(0, -3)$, as shown in Figure 7.67.

FIGURE 7.67

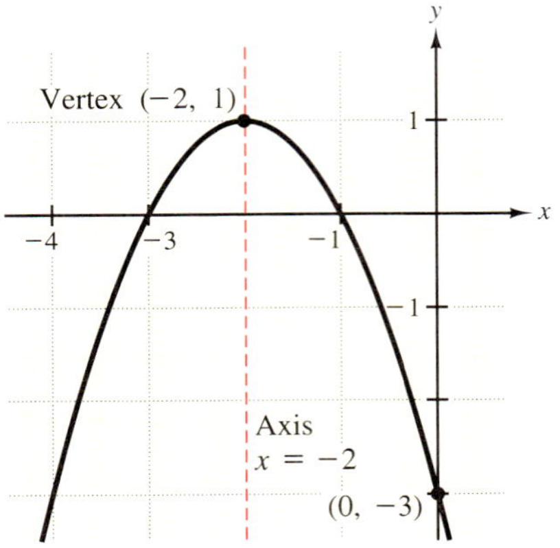

Solution

Since the vertex occurs at $(h, k) = (-2, 1)$, we can write the following.

$$y = a(x - h)^2 + k \qquad \text{Standard form}$$
$$y = a(x - (-2))^2 + 1$$
$$y = a(x + 2)^2 + 1$$

Now, to find the value of a, we use the fact that the y-intercept is $(0, -3)$.

$$y = a(x + 2)^2 + 1 \qquad \text{Standard form}$$
$$-3 = a(0 + 2)^2 + 1 \qquad \text{Substitute } x = 0 \text{ and } y = -3$$
$$-3 = 4a + 1$$
$$-4 = 4a$$
$$-1 = a \qquad \text{Solve for } a$$

Finally, we can conclude that the standard form of the equation of the parabola is

$$y = -(x + 2)^2 + 1.$$ ■

EXAMPLE 8 ■ Finding the Equation of a Parabola

The vertex of a parabola is $(0, -4)$ and the parabola passes through the point $(3, 1)$, as shown in Figure 7.68. Find the equation of the parabola.

Solution

Since the vertex occurs at $(h, k) = (0, -4)$, we can write the following.

$$y = a(x - h)^2 + k \qquad \text{Standard form}$$
$$y = a(x - 0)^2 - 4$$
$$y = ax^2 - 4$$

FIGURE 7.68

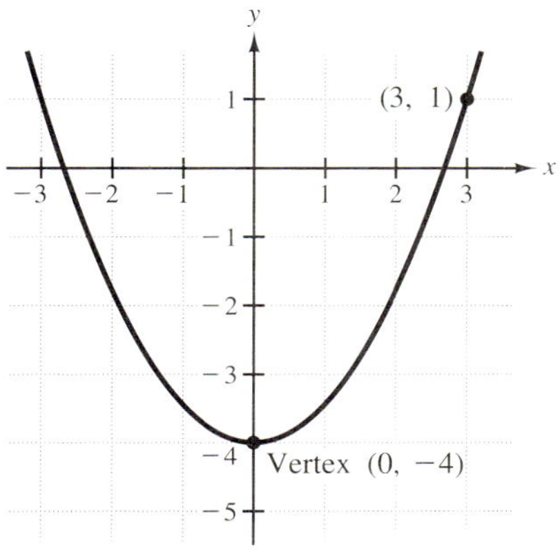

Now, to find the value of a, we use the fact that the parabola passes through the point (3, 1).

$$y = ax^2 - 4 \qquad \text{Standard form}$$
$$1 = a(3)^2 - 4 \qquad \text{Substitute } x = 3 \text{ and } y = 1$$
$$1 = 9a - 4$$
$$\frac{5}{9} = a \qquad \text{Solve for } a$$

Finally, we can conclude that the standard form of the equation of the parabola is

$$y = \frac{5}{9}x^2 - 4.$$

Applications

EXAMPLE 9 ■ An Application Involving a Parabola

A suspension bridge is 100 feet long, as shown in Figure 7.69(a). The bridge is supported by cables attached to the top of the tower at each end of the bridge. Each cable hangs in the shape of a parabola given by $y = 0.01x^2 - x + 35$, where x and y are both measured in feet.

(a) Find the distance between the lowest point of the cable and the roadbed of the bridge.

(b) How tall are the towers?

FIGURE 7.69

(a)

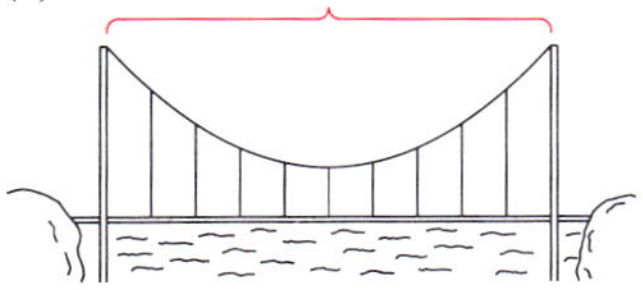

(b)

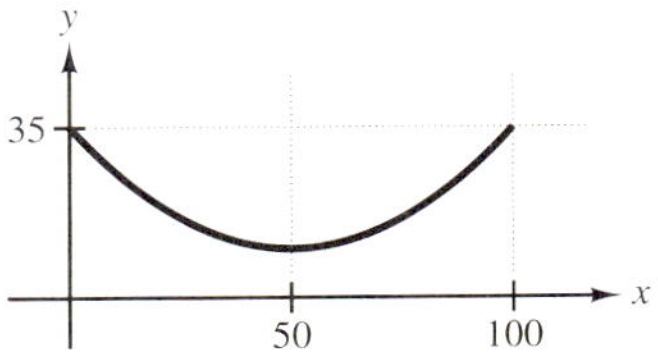

Solution

(a) By writing the equation of the parabola in standard form

$$y = 0.01x^2 - x + 35$$
$$y = 0.01(x^2 - 100x) + 35$$
$$y = 0.01(x^2 - 100x + 50^2 - 50^2) + 35$$
$$y = 0.01(x^2 - 100x + 50^2) - 25 + 35$$
$$y = 0.01(x - 50)^2 + 10$$

we see that the vertex occurs at the point (50, 10). Therefore, the minimum distance between the cable and the roadbed is 10 feet. [See Figure 7.69(b).]

(b) Since the vertex of the parabola occurs at the midpoint of the bridge, the two towers are located at the points where $x = 0$ and $x = 100$. Substituting an x-value of 0, we find that the corresponding y-value is

$$y = 0.01(0^2) - 0 + 35 = 35 \text{ feet.}$$

Therefore, the towers are each 35 feet high. (Try substituting $x = 100$ in the equation to see that you obtain the same y-value.) ■

DISCUSSION PROBLEM ■ **Graphing Calculator Experiment**

Figure 7.70 shows the display screen for a graphing calculator.* Can you find an equation for each of the four parabolas shown in the figure? If you have access to a graphing calculator, try entering each of the four equations to obtain a display that looks like the one in Figure 7.70.

FIGURE 7.70

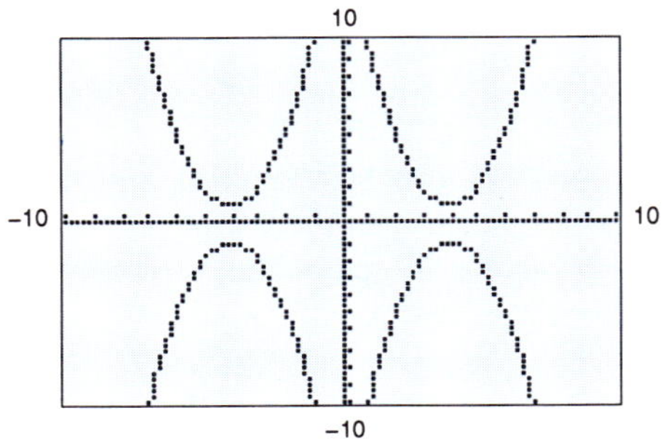

■

*The graphing calculator keystrokes given in this text correspond to the *TI-81* graphing calculator from *Texas Instruments*. However, you can also use other graphing calculators and computer graphing software for these examples and exercises.

Warm-Up

The following warm-up exercises involve skills that were covered in earlier sections. You will use these skills in the exercise set for this section.

In Exercises 1–4, sketch the graph of the given equation.

1. $y = 3x - 2$

2. $y = \frac{1}{2}x + 2$

3. $y = -3x + 5$

4. $y = -\frac{1}{2}x + 2$

In Exercises 5–8, solve the given equation.

5. $10(x - 2) = 25$

6. $30 - 3x = 70 - 5x$

7. $10x(x + 8) = 0$

8. $3x(2x + 1) = 15$

In Exercises 9 and 10, expand and simplify the given expression.

9. $2(x + 5)^2 - 30$

10. $-3(x - 2)^2 - 7$

7.6 EXERCISES

In Exercises 1–6, match the equation with the correct graph. [The graphs are labeled (a), (b), (c), (d), (e), and (f).]

1. $y = 4 - 2x$

2. $y = \frac{1}{2}x - 4$

3. $y = x^2 - 3$

4. $y = -x^2 + 3$

5. $y = (x - 2)^2$

6. $y = 2 - (x - 2)^2$

(a)

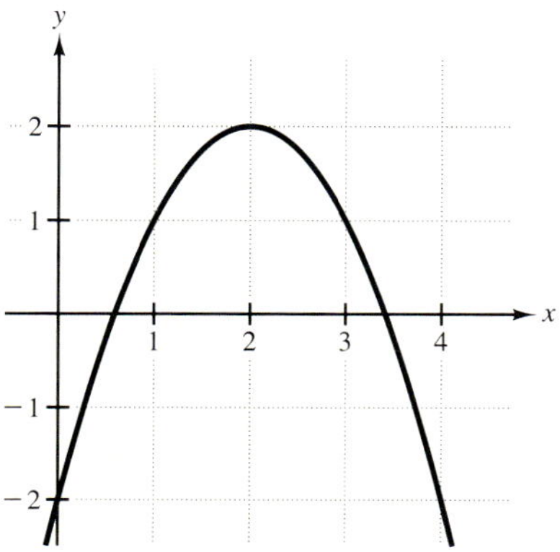

(b)

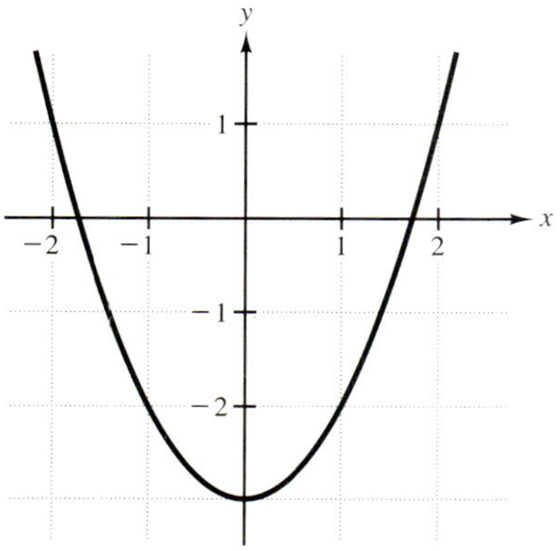

(c)

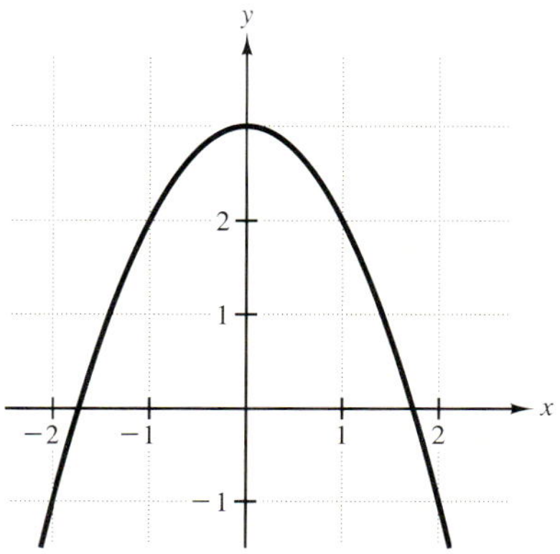

(d)

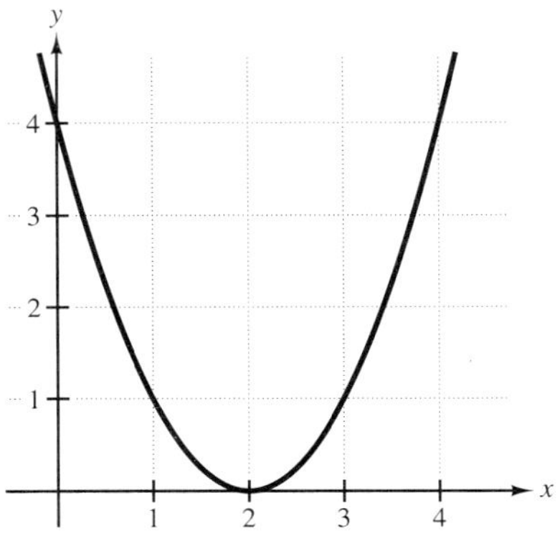

(e)

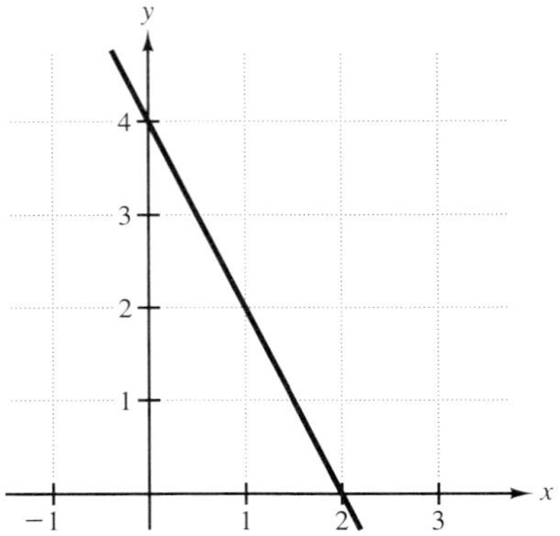

(f)

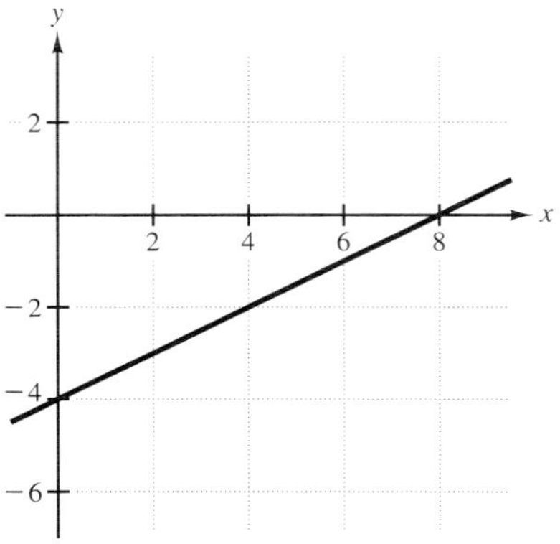

In Exercises 7–12, determine whether the graph of the equation is a parabola that opens up or opens down. Then find the vertex of the parabola.

7. $y = 2(x - 0)^2 + 2$ **8.** $y = -3(x + 5)^2 - 3$ **9.** $y = 4 - (x - 10)^2$ **10.** $y = 2(x - 12)^2 + 3$

11. $y = x^2 - 6$ **12.** $y = -(x + 1)^2$

In Exercises 13–20, find any x-intercepts of the given parabola.

13. $y = 25 - x^2$ **14.** $y = x^2 - 49$ **15.** $y = x^2 - 9x$ **16.** $y = x^2 + 4x$

17. $y = 4x^2 - 12x + 9$ **18.** $y = 10 - x - 2x^2$ **19.** $y = x^2 - 3x + 3$ **20.** $y = x^2 - 3x - 10$

In Exercises 21–26, write the given equation in standard form and find the vertex of the parabola.

21. $y = x^2 - 4x + 7$ **22.** $y = x^2 + 6x - 5$ **23.** $y = -x^2 + 2x - 7$ **24.** $y = -x^2 - 10x + 10$

25. $y = 2x^2 + 6x + 2$ **26.** $y = 3x^2 - 3x - 9$

In Exercises 27–50, sketch the graph of the given quadratic equation. Identify the vertex and any x-intercepts.

27. $y = x^2 - 4$ **28.** $y = x^2 - 9$ **29.** $y = -x^2 + 4$ **30.** $y = -x^2 + 9$

31. $y = x^2 - 3x$ **32.** $y = x^2 - 4x$ **33.** $y = -x^2 + 3x$ **34.** $y = -x^2 + 4x$

35. $y = (x - 4)^2$ **36.** $y = -(x + 4)^2$ **37.** $y = x^2 - 8x + 15$ **38.** $y = x^2 + 4x + 7$

39. $y = -(x^2 + 6x + 5)$ **40.** $y = -x^2 + 2x + 8$ **41.** $y = -x^2 + 6x - 7$ **42.** $y = x^2 + 4x + 2$

43. $y = 2(x^2 + 6x + 8)$ **44.** $y = 3x^2 - 6x + 4$ **45.** $y = \frac{1}{2}(x^2 - 2x - 3)$ **46.** $y = -\frac{1}{2}(x^2 - 6x + 7)$

47. $y = 5 - \frac{x^2}{3}$ **48.** $y = \frac{x^2}{3} - 2$ **49.** $y = \frac{1}{5}(3x^2 - 24x + 38)$ **50.** $y = \frac{1}{5}(2x^2 - 4x + 7)$

In Exercises 51–54, sketch the graph of the given equation and determine the values of x (if any exist) for which the graph is at the specified height y.

51. $y = -x^2 + 6, \quad y = 2$

52. $y = x^2 - 6x + 8, \quad y = 3$

53. $y = \frac{1}{2}x^2 - 3x + \frac{13}{2}, \quad y = 3$

54. $y = -2x^2 - 4x, \quad y = 1$

In Exercises 55–66, find the general form, $y = ax^2 + bx + c$, of the equation of the parabola satisfying the given criteria.

55. Vertex: $(2, 1)$, $a = 1$

56. Vertex: $(-3, -3)$, $a = 1$

57. Vertex: $(-3, 4)$, $a = -1$

58. Vertex: $(3, -2)$, $a = -1$

59. Vertex: $(2, -4)$, Passes through the point $(0, 0)$

60. Vertex: $(-2, -4)$, Passes through the point $(0, 0)$

61. Vertex: $(3, 2)$, Passes through the point $(1, 4)$

62. Vertex: $(-1, -1)$, Passes through the point $(0, 4)$

63. Vertex: $(-1, 5)$, Passes through the point $(0, 1)$

64. Vertex: $(5, 10)$, Passes through the point $(6, 6)$

65. Vertex: $(5, 2)$, Passes through the point $(10, 3)$

66. Vertex: $(0, 20)$, Passes through the point $(10, 15)$

67. *Path of a Ball* The height y (in feet) of a ball thrown by a child is given by

$$y = -\frac{1}{12}x^2 + 2x + 4$$

where x is the horizontal distance (in feet) from where the ball was thrown.

(a) Sketch the path of the ball.
(b) How high was the ball when it left the child's hand?
(c) How high was the ball when it was at its maximum height?
(d) How far from the child did the ball strike the ground?

68. *Path of a Ball* The height y (in feet) of a ball thrown by a child is given by

$$y = -\frac{1}{16}x^2 + 2x + 5$$

where x is the horizontal distance (in feet) from where the ball was thrown.

(a) Sketch the path of the ball.
(b) How high was the ball when it left the child's hand?
(c) How high was the ball when it was at its maximum height?
(d) How far from the child did the ball strike the ground?

69. *Maximum Profit* A company manufactures radios that cost (the company) \$60 each. For buyers who purchase 100 or less radios, the purchase price is \$90 per radio. In order to encourage large orders, the company will reduce the price *per radio* for orders over 100, as follows. If 101 radios are purchased, the price is \$89.95 per unit. If 102 radios are purchased, the price is \$89.70 per unit. If $(100 + x)$ radios are purchased, the price is $[90 - x(0.15)]$ dollars per unit.

(a) Show that the profit P is given by

$$P = (100 + x)[90 - x(0.15)] - (100 + x)60$$
$$= 3{,}000 + 15x - \frac{3}{20}x^2$$

where x is the number over 100 in the order.

(b) Find the vertex of the profit curve and determine the order size for maximum profit.
(c) Suppose you work for this company and are in charge of setting the price for this model of radio. Would you recommend this pricing scheme with no restrictions? Would you recommend this pricing scheme if price reductions were restricted to orders between 100 and 150 units?

70. *Maximum Profit* The profit P (in thousands of dollars) for a certain company is given by

$$P = 230 + 20s - \frac{1}{2}s^2$$

where s is the amount (in hundreds of dollars) spent on advertising. Sketch the graph of the profit equation and determine the amount of advertising that yields a maximum profit.

71. *Bridge Design* A bridge is to be constructed over a gorge with the main supporting arch being a parabola (see figure). If the equation of the parabola is

$$y = 4\left(100 - \frac{x^2}{2{,}500}\right)$$

where x and y are measured in feet, find the following.

(a) The length of the road across the gorge.

(b) The height of the parabolic arch at the center of the span.

(c) The length of the vertical girders at intervals of 100 feet from the center of the bridge.

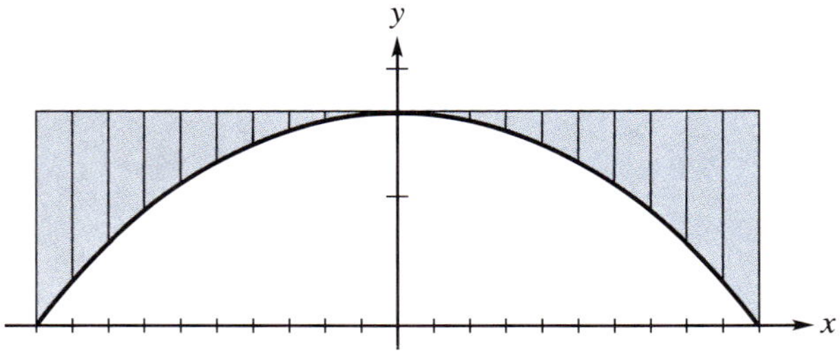

72. *Highway Design* An engineer must design a parabolic arc to create a turn in a freeway around a city. The vertex of the parabola is placed at the origin and must connect with roads represented by the equations $y = -0.4x - 100$ when $x < -500$, and $y = 0.4x - 100$ when $x > 500$ (see figure). Find an equation for the parabolic arc.

Figure for 72

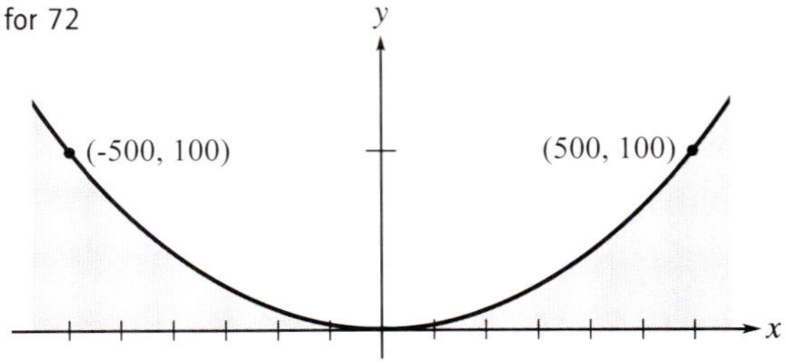

73. *Area* The area of a rectangle is given by the equation

$$A = \frac{2}{\pi}(100x - x^2), \quad 0 < x < 100$$

where x is the length of its base in feet. Use a graphing calculator to sketch the graph of this equation and use the TRACE key to approximate the value of x when A is maximum.* Use the following range settings.

RANGE
Xmin=10
Xmax=90
Xscl=10
Ymin=0
Ymax=1600
Yscl=100
Xres=1

74. *Cost* The cost for producing x units of a product is given by

$$C = 800 - 10x + \frac{1}{4}x^2, \quad 0 < x < 40.$$

Use a graphing calculator to sketch the graph of this equation and use the TRACE key to approximate the value of x when C is minimum. Use the following range settings.

RANGE
Xmin=0
Xmax=40
Xscl=5
Ymin=600
Ymax=1000
Yscl=40
Xres=1

*The graphing calculator keystrokes given in this text correspond to the *TI-81* graphing calculator from *Texas Instruments*. However, you can also use other graphing calculators and computer graphing software for these examples and exercises.

SECTION **7.7**

Circles, Ellipses, and Hyperbolas

Introduction • Circles • Ellipses • Hyperbolas • Applications

Introduction

In Section 7.6 we saw that the graph of $y = ax^2 + bx + c$, a second-degree equation, is a parabola. A parabola is one of four types of curves that are called **conics** or **conic sections**. The other three types are circles, ellipses, and hyperbolas. As indicated in Figure 7.71, the name "conic" relates to the fact that each of these figures can be obtained by intersecting a plane with a double-napped cone.

FIGURE 7.71

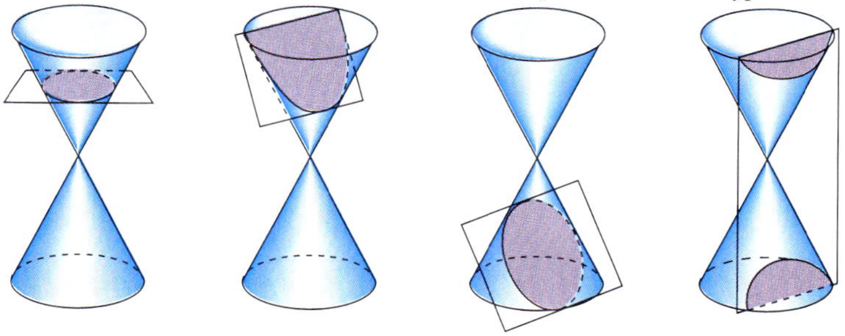

Conic sections occur in many practical applications. Reflective surfaces in satellite dishes, flashlights, and telescopes often have a parabolic shape. The orbits of planets are elliptical, and the orbits of comets are usually either elliptical or hyperbolic. Ellipses and parabolas are also used in the construction of archways and bridges.

Circles

A **circle** in the rectangular coordinate plane consists of all points (x, y) that are a given positive distance r from a fixed point, called the **center** of the circle. The positive distance r is called the **radius** of the circle. If the center of the circle is the origin, as shown in Figure 7.72, then the relationship between the coordinates of any point (x, y) on the circle and the radius r is given by

$$\text{Radius} = r = \sqrt{(x-0)^2 + (y-0)^2} \qquad \text{Distance Formula}$$
$$= \sqrt{x^2 + y^2}.$$

By squaring both sides of this equation, we obtain the following equation, which we call the **standard form of the equation of a circle**.

FIGURE 7.72

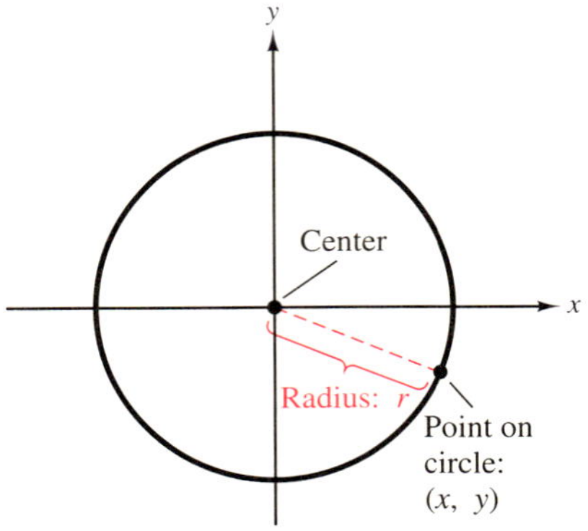

Standard Form of the Equation of a Circle (Center at Origin)

The **standard form of the equation of a circle** whose center is the origin is

$$x^2 + y^2 = r^2.$$

The positive number r is called the **radius** of the circle.

EXAMPLE 1 ■ **Finding an Equation of a Circle**

Find an equation of the circle whose center is (0, 0) and whose radius is 2.

Solution

Using the standard form of the equation of a circle (with center at the origin)

$$x^2 + y^2 = r^2$$

and $r = 2$, we obtain

$$x^2 + y^2 = 2^2$$
$$x^2 + y^2 = 4.$$

The circle given by this equation is shown in Figure 7.73.

FIGURE 7.73

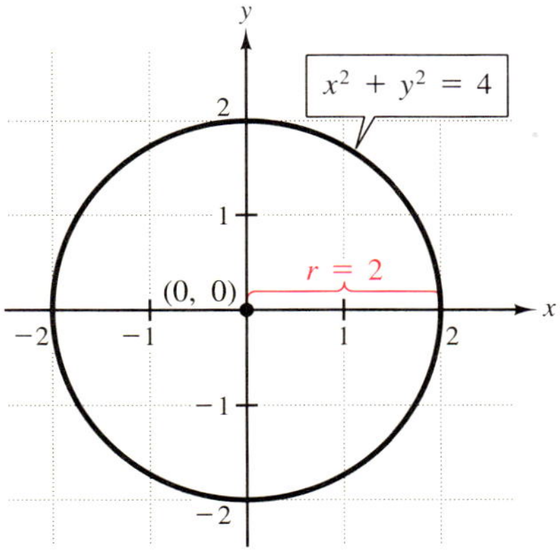

■

To sketch the circle for a given equation, we first write the equation in standard form. Then, from the standard form, we identify the center and radius and sketch the circle.

EXAMPLE 2 ■ **Sketching a Circle**

Identify the center and radius of the circle given by the following equation, and sketch the circle.

$$4x^2 + 4y^2 - 25 = 0$$

Solution

We begin by writing the given equation in standard form.

$$4x^2 + 4y^2 - 25 = 0 \quad \text{Given equation}$$
$$4x^2 + 4y^2 = 25 \quad \text{Add 25 to both sides}$$
$$x^2 + y^2 = \frac{25}{4} \quad \text{Divide both sides by 4}$$
$$x^2 + y^2 = \left(\frac{5}{2}\right)^2 \quad \text{Standard form}$$

Now, from this standard form, we see that the graph of the equation is a circle whose center is at the origin and whose radius is $\frac{5}{2}$, as shown in Figure 7.74.

FIGURE 7.74

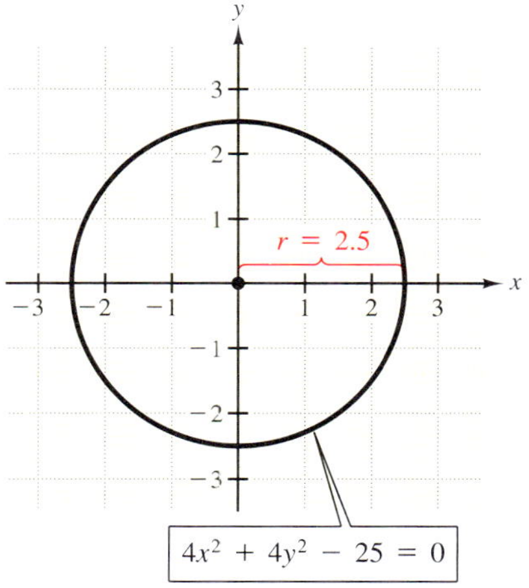

■

Ellipses

Another type of conic is called an **ellipse**, and it is defined as follows. An **ellipse** in the rectangular coordinate system consists of all points (x, y) such that the sum of the distances between (x, y) and two distinct fixed points is a constant, as shown in Figure 7.75. Each of the two fixed points is called a **focus** of the ellipse. (The plural of focus is **foci**.) In this text we will restrict our study of ellipses to those whose center is at the origin and whose foci lie on the same horizontal or vertical line.

FIGURE 7.75

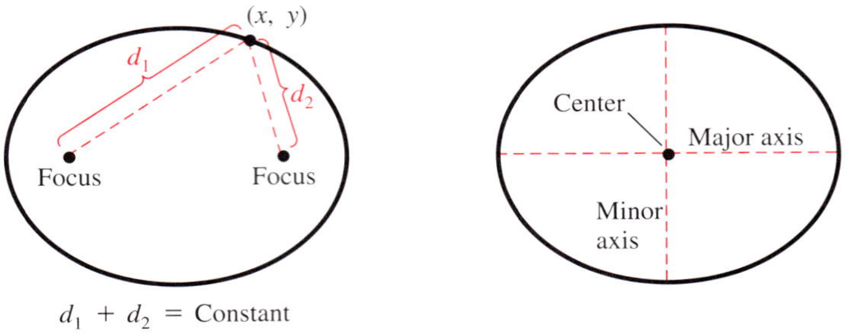

The line through the foci intersects the ellipse at two points, called the **vertices**. The line segment joining the vertices is called the **major axis**, and its midpoint is called the **center** of the ellipse. The line segment perpendicular to the major axis at the center is called the **minor axis** of the ellipse, and the points at which the minor axis intersect the ellipse are called **co-vertices**.

You can visualize the definition of an ellipse by imagining two thumbtacks placed at the foci, as shown in Figure 7.76. If the ends of a fixed length of string are fastened to the thumbtacks and the string is drawn taut with a pencil, the path traced by the pencil will be an ellipse.

FIGURE 7.76

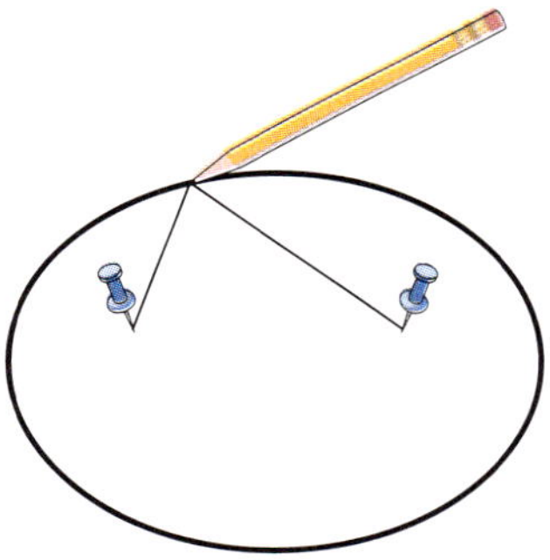

The standard form of the equation of an ellipse takes one of two forms, depending on whether the major axis is horizontal or vertical.

Standard Form of the Equation of an Ellipse (Center at Origin)

The **standard form of the equation of an ellipse** with center at the origin and major and minor axes of lengths $2a$ and $2b$, respectively, is

$$\frac{x^2}{a^2} + \frac{y^2}{b^2} = 1 \quad \text{or} \quad \frac{x^2}{b^2} + \frac{y^2}{a^2} = 1, \quad 0 < b < a.$$

The vertices lie on the major axis, a units from the center, and the co-vertices lie on the minor axis, b units from the center, as shown in Figure 7.77.

FIGURE 7.77

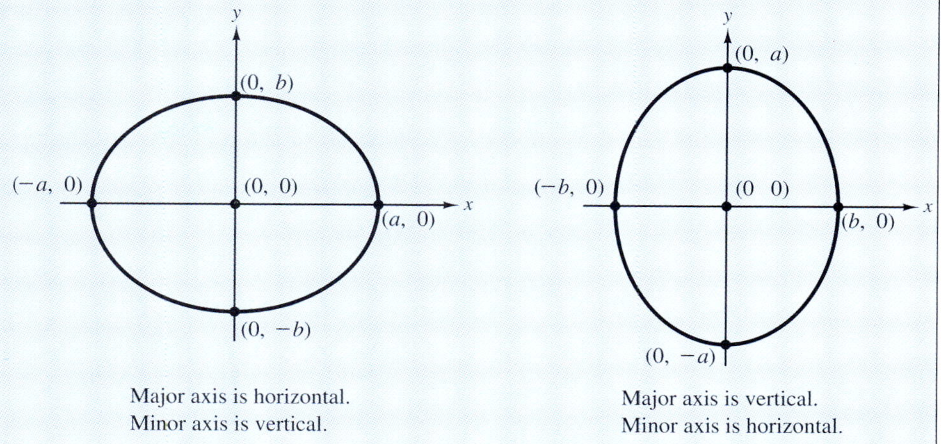

EXAMPLE 3 ■ Finding an Equation of an Ellipse

Find an equation of the ellipse whose vertices are $(-3, 0)$ and $(3, 0)$ and whose co-vertices are $(0, -2)$ and $(0, 2)$.

Solution

We begin by plotting the two vertices and two co-vertices, as shown in Figure 7.78. The center of the ellipse is $(0, 0)$ because it is the point that lies halfway between the vertices (or halfway between the co-vertices). From the figure, we see that the major axis is horizontal. Thus, the equation of the ellipse has the form

$$\frac{x^2}{a^2} + \frac{y^2}{b^2} = 1.$$

In this equation a is the distance between the center and either vertex, which implies that $a = 3$. Similarly, b is the distance between the center and either co-vertex, which implies that $b = 2$. Thus, the standard equation of the ellipse is

$$\frac{x^2}{3^2} + \frac{y^2}{2^2} = 1.$$

FIGURE 7.78

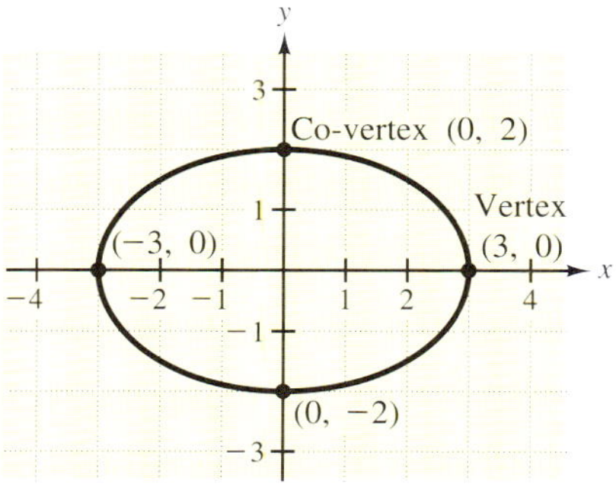

■

EXAMPLE 4 ■ **Sketching an Ellipse**

Sketch the ellipse given by $4x^2 + y^2 = 36$, and identify the vertices and co-vertices.

Solution

We begin by writing the equation in standard form.

$$4x^2 + y^2 = 36 \qquad \text{Given equation}$$

$$\frac{4x^2}{36} + \frac{y^2}{36} = \frac{36}{36} \qquad \text{Divide both sides by 36}$$

$$\frac{x^2}{9} + \frac{y^2}{36} = 1 \qquad \text{Simplify}$$

$$\frac{x^2}{3^2} + \frac{y^2}{6^2} = 1 \qquad \text{Standard form}$$

Since the denominator of the y^2 term is larger than the denominator of the x^2 term, we conclude that the major axis is vertical. Moreover, since $a = 6$, the vertices are $(0, -6)$ and $(0, 6)$. Finally, since $b = 3$, the co-vertices are $(-3, 0)$ and $(3, 0)$, as shown in Figure 7.79.

FIGURE 7.79

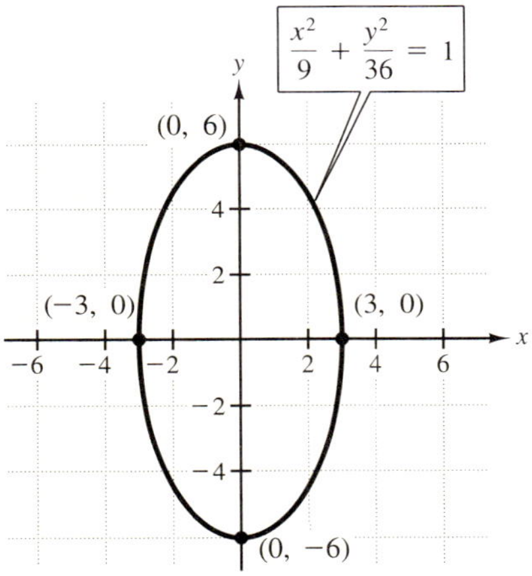

■

In Example 4 note that from the standard form of the equation we can sketch the ellipse by locating the vertices and co-vertices. Since 3^2 is the denominator of the x^2 term, we move three units to the *right* and *left* of the center to locate the endpoints of the horizontal axis. Similarly, since 6^2 is the denominator of the y^2 term, we move six units *up* and *down* from the center to locate the endpoints of the vertical axis.

Hyperbolas

The fourth basic type of conic is called a **hyperbola**. A hyperbola in the rectangular coordinate system consists of all points (x, y) such that the *difference* between (x, y) and two fixed points is a constant, as shown in Figure 7.80. The two fixed points are called the **foci** of the hyperbola. As with ellipses, we will consider only equations of hyperbolas whose foci lie on the x-axis or on the y-axis. The line on which the foci lie is called the **transverse axis** of the hyperbola.

FIGURE 7.80

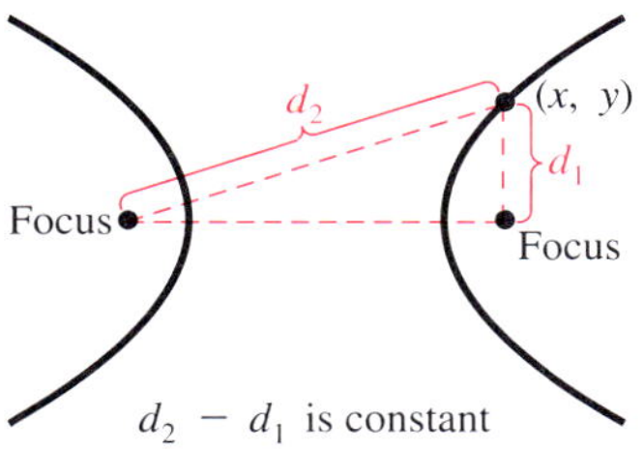

Standard Form of the Equation of a Hyperbola (Center at Origin)

The **standard form of the equation of a hyperbola** whose center is at the origin is given by

$$\frac{x^2}{a^2} - \frac{y^2}{b^2} = 1 \qquad \text{Transverse axis is horizontal}$$

or

$$\frac{y^2}{a^2} - \frac{x^2}{b^2} = 1 \qquad \text{Transverse axis is vertical}$$

where a and b are positive real numbers. The **vertices** of the hyperbola lie on the transverse axis, a units from the center, as shown in Figure 7.81.

FIGURE 7.81

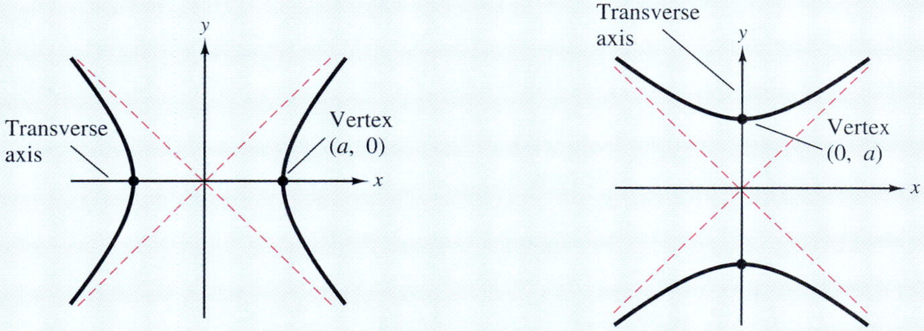

A hyperbola has two disconnected parts, each of which is called a **branch** of the hyperbola. The two branches approach a pair of intersecting straight lines called **asymptotes** of the hyperbola. The two asymptotes intersect at the center of the hyperbola. To sketch a hyperbola, we form a **central rectangle** whose center is the origin and whose width and height are $2a$ and $2b$. Note in Figure 7.82 that the asymptotes pass through the corners of the central rectangle and the vertices of the hyperbola lie at the centers of opposite sides of the central rectangle.

FIGURE 7.82

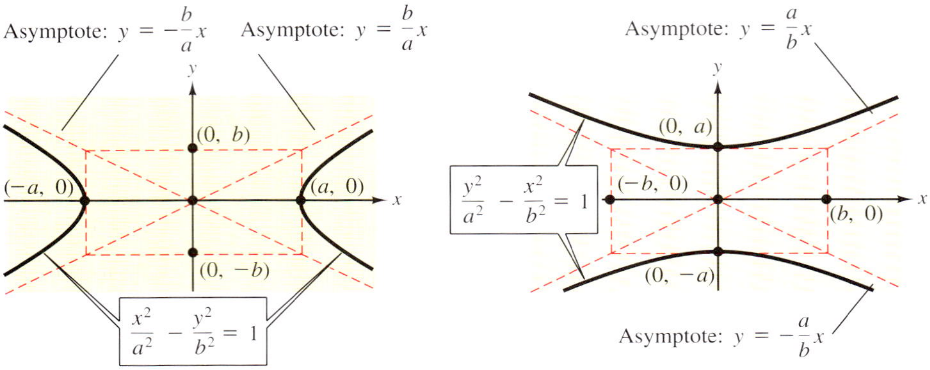

Transverse axis is horizontal. **Transverse axis is vertical.**

EXAMPLE 5 ■ Sketching a Hyperbola

Identify the vertices of the hyperbola given by the following equation, and sketch the hyperbola.

$$\frac{x^2}{36} - \frac{y^2}{16} = 1$$

Solution

From the standard form of the equation

$$\frac{x^2}{6^2} - \frac{y^2}{4^2} = 1$$

we see that the center of the hyperbola is the origin and the transverse axis is horizontal. Therefore, the vertices lie six units to the left and right of the center at the points $(-6, 0)$ and $(6, 0)$. Since $a = 6$ and $b = 4$, we can sketch the hyperbola by first drawing a central rectangle whose width is $2a = 12$ and whose height is $2b = 8$, as shown in Figure 7.83(a). Next we draw the asymptotes of the hyperbola through the corners of the central rectangle and plot the vertices. Finally, we draw the hyperbola, as shown in Figure 7.83(b).

FIGURE 7.83

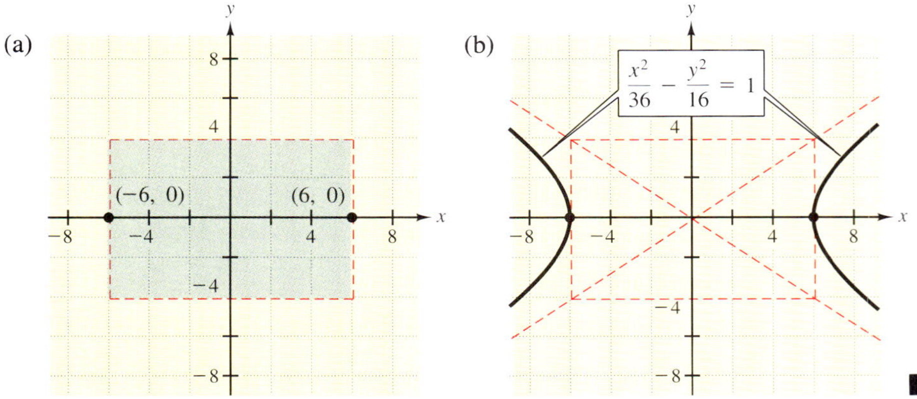

■

EXAMPLE 6 ■ Sketching a Hyperbola

Identify the vertices of the hyperbola given by the following equation, and sketch the hyperbola.

$$4y^2 - 9x^2 = 36$$

Solution

To begin, we write the given equation in standard form.

$$4y^2 - 9x^2 = 36 \qquad \text{Given equation}$$

$$\frac{4y^2}{36} - \frac{9x^2}{36} = \frac{36}{36} \qquad \text{Divide both sides by 36}$$

$$\frac{y^2}{9} - \frac{x^2}{4} = 1 \qquad \text{Simplify}$$

$$\frac{y^2}{3^2} - \frac{x^2}{2^2} = 1 \qquad \text{Standard form}$$

Now, from the standard form, we see that the center of the hyperbola is the origin, and that the transverse axis is vertical. Thus, the vertices lie three units above and below the center at $(0, 3)$ and $(0, -3)$. Since $a = 3$ and $b = 2$, we can sketch the hyperbola by first drawing a central rectangle whose height is $2a = 6$ and whose width is $2b = 4$, as shown in Figure 7.84(a). Next, we draw the asymptotes of the hyperbola through the corners of the central rectangle and plot the vertices. Finally, we draw the hyperbola, as shown in Figure 7.84(b).

FIGURE 7.84

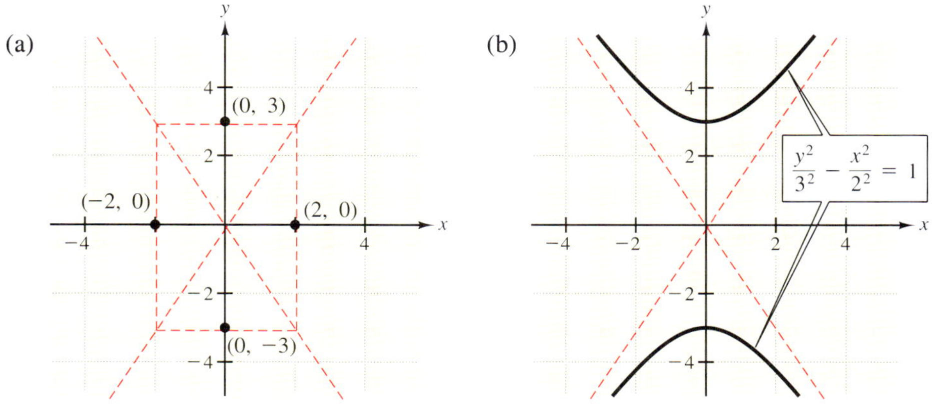

■

Finding the equation of a hyperbola is a little more difficult than finding the equation of one of the other three types of conics. However, if we know the vertices and the asymptotes, we can find the values of a and b, which enables us to write the equation. Notice in Example 7 that the key to this procedure is knowing that the central rectangle has a width of $2a$ and a height of $2b$.

EXAMPLE 7 ■ Finding the Equation of a Hyperbola

Find an equation of the hyperbola with a vertical transverse axis whose vertices are (0, 3) and (0, −3). The asymptotes of the hyperbola are given by $y = \frac{3}{5}x$ and $y = -\frac{3}{5}x$.

Solution

To begin, we sketch the lines that represent the asymptotes, as shown in Figure 7.85(a). Note that these two lines intersect at the origin, which implies that the center of the hyperbola is (0, 0). Next, we plot the two vertices at the points (0, 3) and (0, −3). Since we know where the vertices are located, we can sketch the central rectangle of the hyperbola, as shown in Figure 7.85(a). Note that the corners of the central rectangle occur at the points (−5, 3), (5, 3), (−5, −3), and (5, −3). Since the width of the central rectangle is $2b = 10$, it follows that $b = 5$. Similarly, since the height of the central rectangle is $2a = 6$, it follows that $a = 3$. Thus, we can conclude that the standard form of the equation of the hyperbola is

$$\frac{y^2}{3^2} - \frac{x^2}{5^2} = 1.$$

The graph is shown in Figure 7.85(b).

FIGURE 7.85

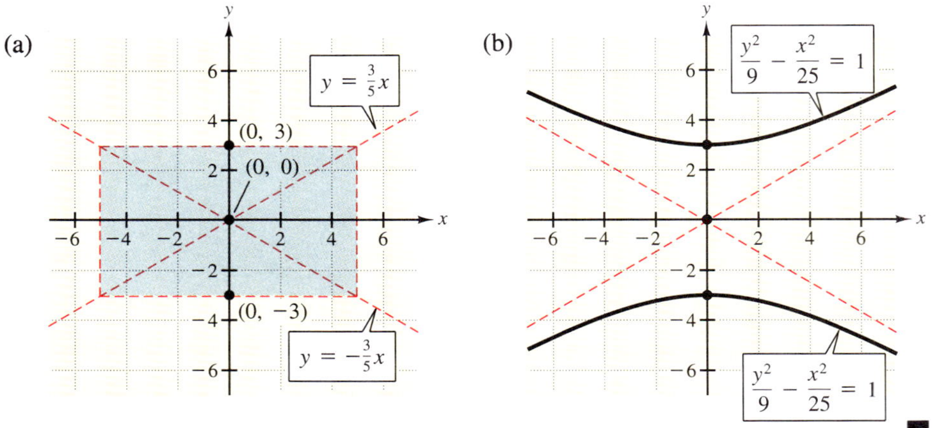

■

Applications

EXAMPLE 8 ■ An Application Involving an Ellipse

Suppose you are responsible for designing a semi-elliptical archway, as shown in Figure 7.86. The height of the archway is 10 feet and its width is 30 feet. Find an equation of the ellipse and use the equation to sketch an accurate drawing of the archway.

Solution

To make the equation simple, let's place the origin at the center of the ellipse. This means that the standard form of the equation is

$$\frac{x^2}{a^2} + \frac{y^2}{b^2} = 1.$$

Since the major axis is horizontal, we see that $a = 15$ and $b = 10$, which implies that the equation is

$$\frac{x^2}{15^2} + \frac{y^2}{10^2} = 1.$$

FIGURE 7.86

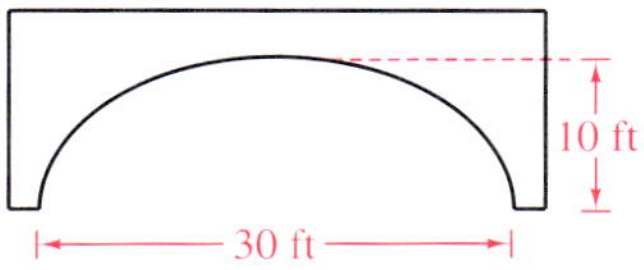

In order to make an accurate sketch of the ellipse, it is helpful to solve this equation for y as follows.

$$\begin{aligned}
\frac{x^2}{15^2} + \frac{y^2}{10^2} &= 1 \\
\frac{x^2}{225} + \frac{y^2}{100} &= 1 \\
\frac{y^2}{100} &= 1 - \frac{x^2}{225} \\
y^2 &= 100\left(1 - \frac{x^2}{225}\right) \\
y &= 10\sqrt{1 - \frac{x^2}{225}}
\end{aligned}$$

(Note that we chose the positive square root because we know that the y-values must be positive.) Finally, we calculate several y-values for the archway, as shown in Table 7.13.

TABLE 7.13

x-Value	± 15	± 12.5	± 10	± 7.5	± 5	± 2.5	0
y-Value	0	5.53	7.45	8.66	9.43	9.86	10

Using the values shown in the table, we can make a sketch of the archway, as shown in Figure 7.87.

FIGURE 7.87

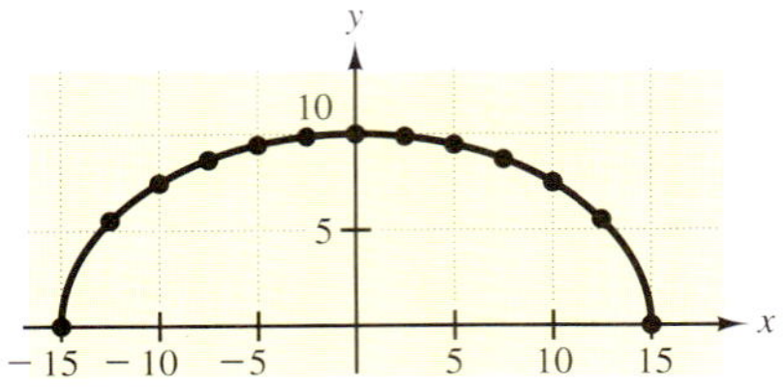

DISCUSSION PROBLEM ■ A Graphing Calculator Experiment

Graphing calculators are designed to sketch the graphs of equations in which y is isolated on the left side of the equation. Thus, if we want to sketch the graph of the circle given by

$$x^2 + y^2 = 36$$

we can solve the equation for y and obtain the following *two* equations.

$$y = \sqrt{36 - x^2} \quad \text{and} \quad y = -\sqrt{36 - x^2}$$

The first of these two equations represents the upper half of the circle and the second represents the lower half of the circle. To sketch the graph of both the upper and lower halves of the circle on the same screen, we press [Y=] and enter the two equations on the first two lines.* This display should now be as follows.

$Y_1 = \sqrt{\ }(36 - X^2)$	[√] [(] 36 [−] [X|T] [x^2] [)]
$Y_2 = -\sqrt{\ }(36 - X^2)$	[(−)] [√] [(] 36 [−] [X|T] [x^2] [)]
$Y_3 =$	
$Y_4 =$	

By pressing [GRAPH], we obtain the graph shown in Figure 7.88(a). The reason the graph doesn't look like a circle is that with the standard settings, the tic marks on the x-axis are further apart than the tic marks on the y-axis. To correct this, press [ZOOM], cursor to "Square," and press [ENTER]. The redrawn screen is shown in Figure 7.88(b). Notice that in this screen the graph appears to be more circular.

*The graphing calculator keystrokes given in this text correspond to the *TI-81* graphing calculator from *Texas Instruments*. However, you can also use other graphing calculators and computer graphing software for these examples and exercises.

FIGURE 7.88

(a) $y_1 = \sqrt{(36 - x^2)}$
$y_2 = -\sqrt{(36 - x^2)}$

(b) $y_1 = \sqrt{(36 - x^2)}$
$y_2 = -\sqrt{(36 - x^2)}$

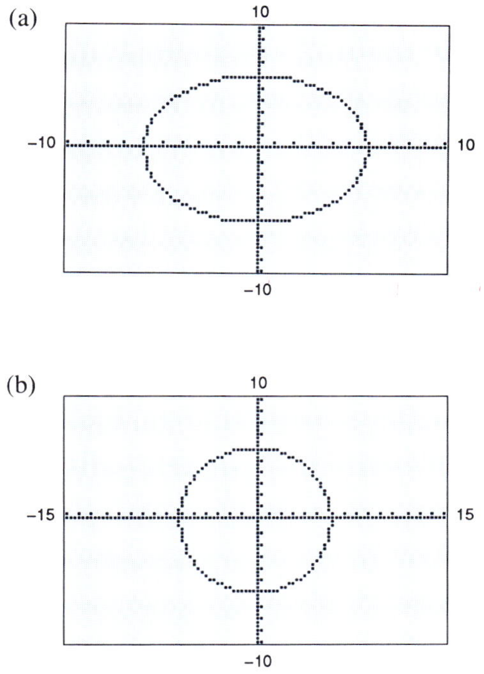

Warm-Up

The following warm-up exercises involve skills that were covered in earlier sections. You will use these skills in the exercise set for this section.

In Exercises 1–4, solve the equation for y.

1. $2x + 3y - 9 = 0$

2. $x^2 + y^2 = 36$

3. $2x^2 + 3y^2 - 6 = 0$

4. $2x^2 - y^2 - 3 = 0$

In Exercises 5 and 6, simplify the radical.

5. $\sqrt{450}$

6. $\sqrt{1,350}$

In Exercises 7–10, sketch the graph of the given equation.

7. $y = 2x - 3$

8. $y = -\frac{3}{4}x + 2$

9. $y = x^2 - 4x + 4$

10. $y = 9 - x^2$

7.7 EXERCISES

In Exercises 1–6, match the equation with the correct graph. [The graphs are labeled (a), (b), (c), (d), (e), and (f).]

1. $x^2 + y^2 = 9$

2. $4x^2 + 4y^2 = 9$

3. $\frac{x^2}{4} + \frac{y^2}{9} = 1$

4. $\frac{x^2}{9} + \frac{y^2}{4} = 1$

5. $x^2 - y^2 = 4$

6. $x^2 - y^2 = -4$

(a)

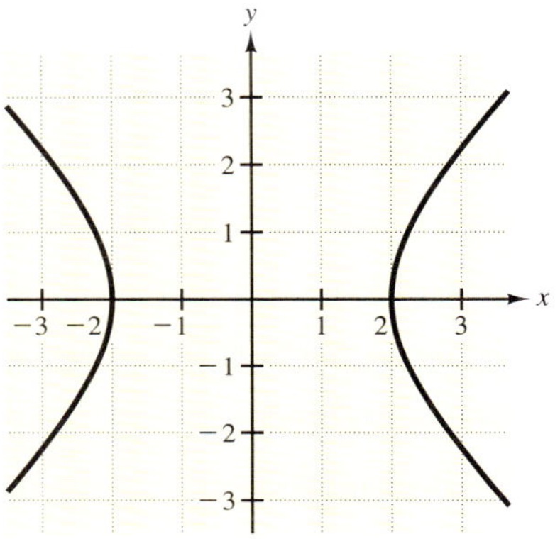

(b)

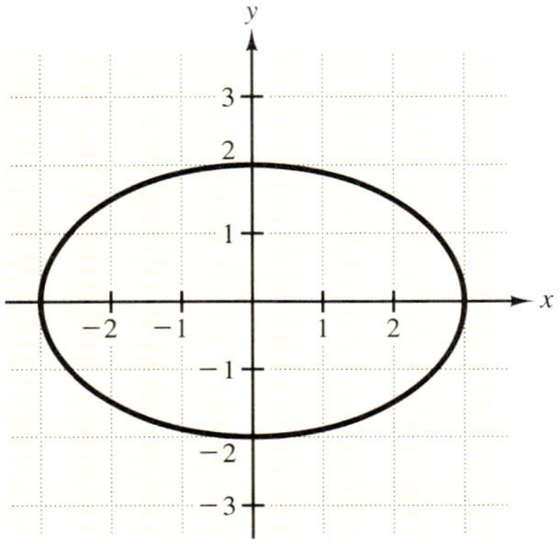

(c)

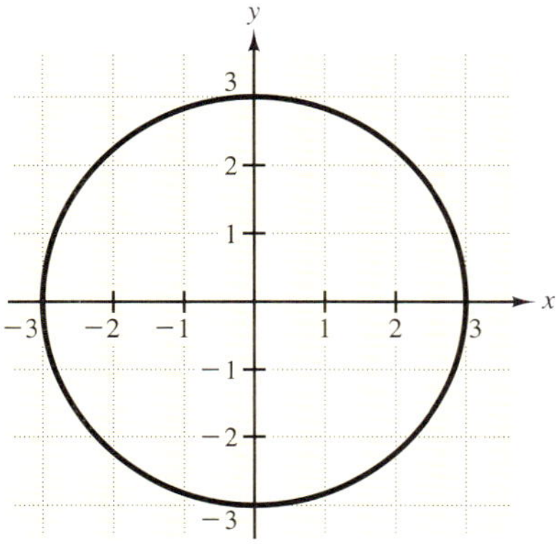

(d)

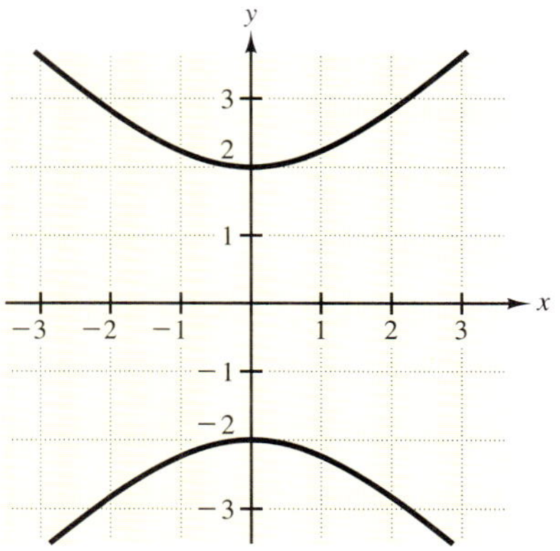

(e)

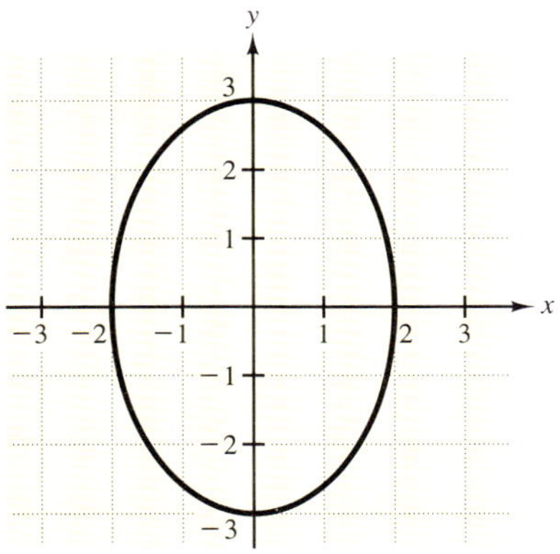

(f)

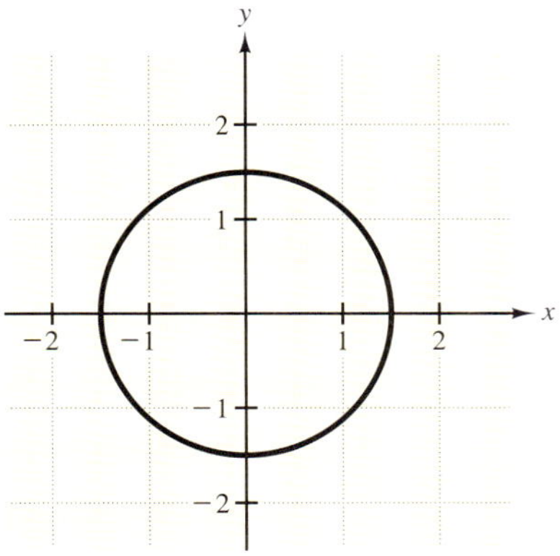

In Exercises 7–14, find an equation of the circle whose center is (0, 0) and satisfies the given criteria.

7. Radius: 5

8. Radius: 7

9. Radius: $\frac{2}{3}$

10. Radius: $\frac{5}{2}$

11. Passes through the point $(0, 8)$

12. Passes through the point $(-2, 0)$

13. Passes through the point $(5, 2)$

14. Passes through the point $(-1, -4)$

In Exercises 15–20, identify the radius of the circle centered at $(0, 0)$ and sketch its graph.

15. $x^2 + y^2 = 16$

16. $x^2 + y^2 = 25$

17. $4x^2 + 4y^2 = 1$

18. $9x^2 + 9y^2 = 64$

19. $25x^2 + 25y^2 - 144 = 0$

20. $\frac{x^2}{4} + \frac{y^2}{4} - 1 = 0$

In Exercises 21–26, find the standard form of the equation of the ellipse that is centered at the origin and that satisfies the given criteria.

	Vertices	*Co-vertices*
21.	$(-4, 0), (4, 0)$	$(0, -3), (0, 3)$
22.	$(-4, 0), (4, 0)$	$(0, -1), (0, 1)$
23.	$(0, -4), (0, 4)$	$(-3, 0), (3, 0)$
24.	$(0, -5), (0, 5)$	$(-1, 0), (1, 0)$

25. The major axis is horizontal with length 20 and the minor axis has length 12.

26. The minor axis is horizontal with length 30 and the major axis has length 50.

In Exercises 27–34, identify the vertices and co-vertices of the ellipse and sketch the graph of the ellipse.

27. $\frac{x^2}{16} + \frac{y^2}{4} = 1$

28. $\frac{x^2}{25} + \frac{y^2}{9} = 1$

29. $\frac{x^2}{4} + \frac{y^2}{16} = 1$

30. $\frac{x^2}{9} + \frac{y^2}{25} = 1$

31. $\frac{x^2}{25/9} + \frac{y^2}{16/9} = 1$

32. $\frac{x^2}{1} + \frac{y^2}{1/4} = 1$

33. $4x^2 + y^2 - 4 = 0$

34. $4x^2 + 9y^2 - 36 = 0$

In Exercises 35–44, identify the vertices of the hyperbola and sketch its graph. Show the asymptotes and write an equation for each.

35. $x^2 - y^2 = 9$

36. $x^2 - y^2 = 1$

37. $y^2 - x^2 = 9$

38. $y^2 - x^2 = 1$

39. $\frac{x^2}{9} - \frac{y^2}{25} = 1$

40. $\frac{x^2}{4} - \frac{y^2}{9} = 1$

41. $\frac{y^2}{9} - \frac{x^2}{25} = 1$

42. $\frac{y^2}{4} - \frac{x^2}{9} = 1$

43. $4y^2 - x^2 + 16 = 0$

44. $4y^2 - 9x^2 - 36 = 0$

In Exercises 45–50, find an equation of the hyperbola that is centered at the origin and that satisfies the given criteria.

	Vertices	*Asymptotes*
45.	$(-4, 0), (4, 0)$	$y = 2x, \quad y = -2x$
46.	$(-2, 0), (2, 0)$	$y = \frac{1}{3}x, \quad y = -\frac{1}{3}x$

47. $(0, -4), (0, 4)$ $\quad y = \frac{1}{2}x, \quad y = -\frac{1}{2}x$

48. $(0, -2), (0, 2)$ $\quad y = 3x, \quad y = -3x$

49. $(-9, 0), (9, 0)$ $\quad y = \frac{2}{3}x, \quad y = -\frac{2}{3}x$

50. $(0, -5), (0, 5)$ $\quad y = x, \quad y = -x$

In Exercises 51–60, identify the graph of the equation as a line, circle, parabola, ellipse, or hyperbola.

51. $y = 2x^2 - 8x + 2$

52. $y = 10 - \frac{3}{2}x$

53. $4x^2 + 9y^2 - 36 = 0$

54. $4x^2 + 4y^2 - 36 = 0$

55. $4x^2 - 9y^2 - 36 = 0$

56. $x^2 - 4y + 2x = 0$

57. $x^2 + y^2 - 1 = 0$

58. $2x^2 + 2y^2 - 9 = 0$

59. $3x + 2 = 0$

60. $y^2 = x^2 + 2$

In Exercises 61–66, use a graphing calculator to sketch the graph of the equation.

61. $x^2 + y^2 = 30$

62. $4x^2 + 4y^2 = 45$

63. $3x^2 + y^2 - 12 = 0$

64. $x^2 + 4y^2 = 18$

65. $6x^2 - y^2 = 40$

66. $5x^2 - 2y^2 + 10 = 0$

67. *Satellite Orbit* Find an equation of the circular orbit of a satellite 500 miles above the surface of the Earth. Place the origin of the rectangular coordinate system at the center of the Earth and assume the radius of the Earth is 4,000 miles.

68. *Architecture* The top portion of a stained glass window is in the form of a pointed Gothic arch (see figure). Each side of the arch is an arc of a circle that has a radius of 12 feet and a center at the base of the opposite arch. Find an equation of one of the circles and use it to determine the height of the point of the arch above the rectangular portion of the window.

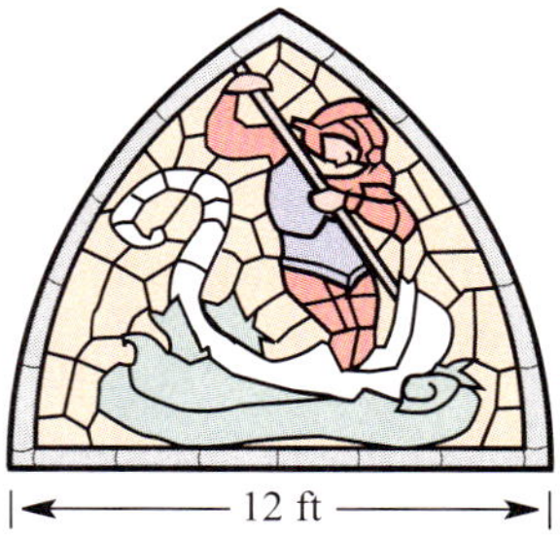

69. A rectangle centered at the origin with sides parallel to the coordinate axes is placed in a circle of radius 25 inches centered at the origin (see figure). If the length of the rectangle is $2x$ inches, show that its width and area are given by $2\sqrt{625 - x^2}$ and $4x\sqrt{625 - x^2}$, respectively.

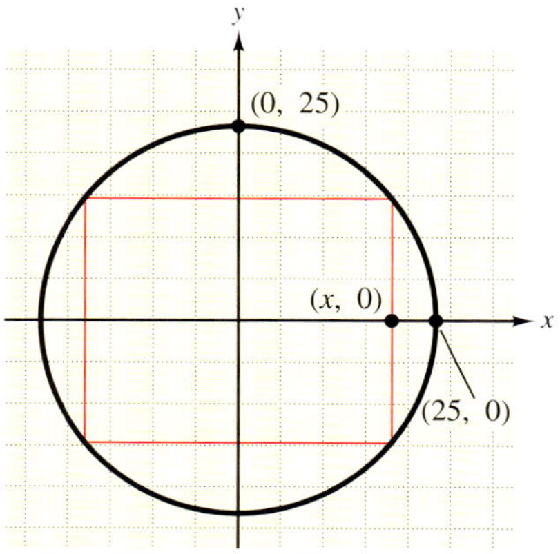

70. *Height of an Arch* A *semicircular* arch for a tunnel under a river has a diameter of 100 feet (see figure). Determine the height of the arch 5 feet from the edge of the tunnel.

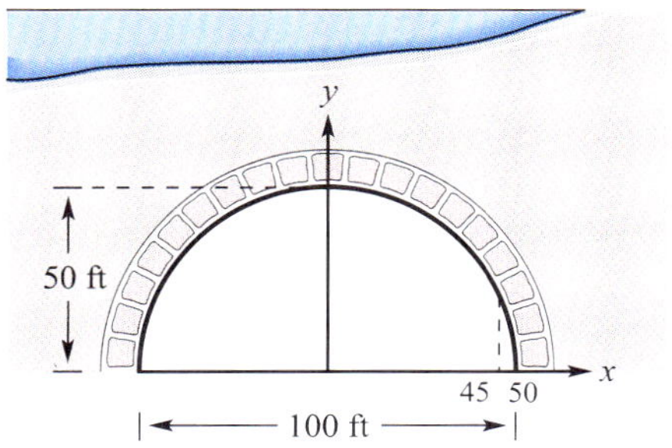

71. *Height of an Arch* A *semi-elliptical* arch for a tunnel under a river has a width of 100 feet and a height of 40 feet (see figure). Determine the height of the arch 5 feet from the edge of the tunnel.

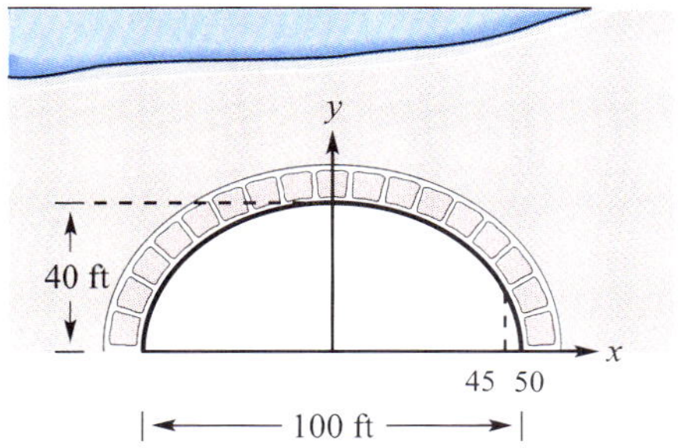

72. Sketch a graph of the ellipse that consists of all points (x, y) such that the sum of the distances between (x, y) and two fixed points is 15 units and the foci are located at the centers of the two sets of concentric circles in the accompanying figure.

73. Sketch a graph of the hyperbola that consists of all points (x, y) such that the difference in the distances between (x, y) and two fixed points is eight units and the foci are located at the centers of the two sets of concentric circles in the accompanying figure.

Figure for 72 and 73

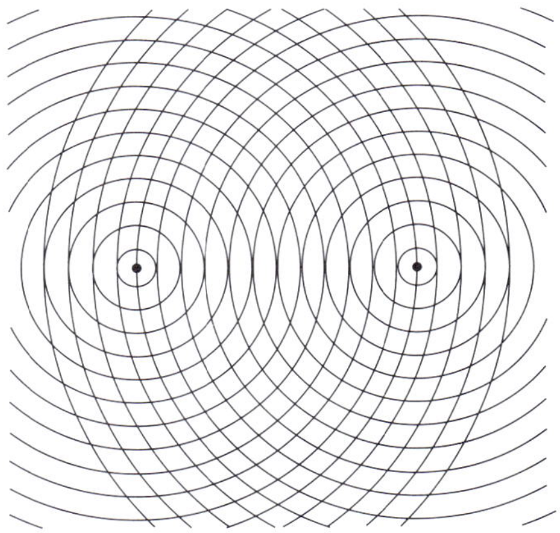

CHAPTER 7 SUMMARY

As you review and prepare for a test on this chapter, first try to obtain a global view of what was discussed. Then review the specific skills needed under each category.

Graphing Equations

■ *Plot* specified points in a rectangular coordinate system.	Section 7.1, Exercises 1–16
■ *Determine and plot* solution points of equations.	Section 7.1, Exercises 29–40
■ *Find* the distance between two points in a plane by using the Distance Formula.	Section 7.1, Exercises 41–64
■ *Construct* a table of values and *sketch* the graph of an equation.	Section 7.2, Exercises 1–10, Exercises 21–40
■ *Sketch* the graph of an equation using a graphing calculator.	Section 7.3, Exercises 1–26, Exercises 33–40

Slope and Equations of Lines

■ *Calculate* the slope of a line that passes through a given pair of points.	Section 7.4, Exercises 9–26
■ *Determine* an additional point on the line that passes through a given point and has a specified slope.	Section 7.4, Exercises 31–46
■ *Sketch* the graph of a line using intercepts and slope as an aid.	Section 7.4, Exercises 47–60
■ *Determine* whether specified lines are parallel, perpendicular, or neither, by using their slopes.	Section 7.4, Exercises 61–68
■ *Determine* the slope of a line from its equation.	Section 7.5, Exercises 1–10
■ *Write* the equation of a line given its slope and a point on the line, or given two points on the line.	Section 7.5, Exercises 11–44
■ *Write* the equation of a line, through a given point, that is parallel or perpendicular to a given line.	Section 7.5, Exercises 45–50
■ *Create* the equation of a line described by a verbal statement.	Section 7.5, Exercises 55–65

Conics and Their Graphs

■ *Find* the vertex and the x-intercepts of a parabola.	Section 7.6, Exercises 7–26
■ *Sketch* the graph of a quadratic equation and locate the vertex and the x-intercepts.	Section 7.6, Exercises 27–50
■ *Determine* the equation of a parabola, given its vertex and a point on the parabola.	Section 7.6, Exercises 55–66
■ *Solve* verbal statements about parabolas by using the vertex and the x-intercepts.	Section 7.6, Exercises 67–72
■ *Sketch* the graph of a circle, ellipse, or hyperbola, given its equation.	Section 7.7, Exercises 15–20, Exercises 27–44
■ *Write* the standard equation for a circle, ellipse, or hyperbola, given appropriate parts of the conic.	Section 7.7, Exercises 7–14, Exercises 21–26, Exercises 45–50

Chapter 7 Review Exercises

In Exercises 1 and 2, plot the points in a rectangular coordinate system.

1. $(0, -3), \left(\frac{5}{2}, 5\right), (-2, -4)$

2. $\left(1, -\frac{3}{2}\right), \left(-2, 2\frac{3}{4}\right), (5, 10)$

In Exercises 3–6, plot the points and find the distance between them.

3. $(1, 3), (5, 6)$

4. $(2, -5), (6, -5)$

5. $(-5, -1), (1, 2)$

6. $(-2, 10), (3, -2)$

In Exercises 7–12, sketch the graph of the equation. Label the x- and y-intercepts.

7. $y = 6 - \frac{1}{3}x$

8. $y = \frac{3}{4}x - 2$

9. $3y - 2x - 3 = 0$

10. $3x + 4y + 12 = 0$

11. $x = |y - 3|$

12. $y = 1 - x^3$

In Exercises 13–16, find the slope of the line through the given points.

13. $(-1, 1), (6, 3)$

14. $(-2, 5), (3, -8)$

15. $(-1, 3), (4, 3)$

16. $(7, 2), (7, 8)$

In Exercises 17–20, find t so that the three points are collinear. (*Note*: Collinear means that the points lie on the same straight line.)

17. $(-3, -3), (0, t), (1, 3)$

18. $(2, 1), (1, t), (8, 3)$

19. $(-1, 4), (t, 3), (4, 0)$

20. $(-5, 3), (t, 0), (5, -2)$

In Exercises 21 and 22, plot the points and verify that the points form the vertices of the specified polygon.

21. Right triangle: $(1, 1), (12, 9), (4, 20)$

22. Parallelogram: $(0, 0), (7, 1), (8, 4), (1, 3)$

In Exercises 23–26, a point on a line and the slope of the line are given. Find two additional points on the line. (Each problem has many correct solutions.)

23. $(2, -4), \quad m = -3$

24. $\left(-4, \frac{1}{2}\right), \quad m = 2$

25. $(3, 1), \quad m = \frac{5}{4}$

26. $\left(-3, -\frac{3}{2}\right), \quad m = -\frac{1}{3}$

In Exercises 27–34, find the general form of the equation of the line passing through the given point with the specified slope.

27. $(1, -4), \quad m = 2$

28. $(-5, -5), \quad m = 3$

29. $(-1, 4), \quad m = -4$

30. $(5, -2), \quad m = -2$

31. $\left(\frac{5}{2}, 4\right)$, $m = -\frac{2}{3}$

32. $\left(-2, -\frac{4}{3}\right)$, $m = \frac{3}{2}$

33. $(7, 8)$, m is undefined

34. $(-6, 5)$, $m = 0$

In Exercises 35–40, find the general equation of the line passing through the two given points.

35. $(-6, 0), (0, -3)$

36. $(-2, -3), (4, 6)$

37. $(0, 10), (6, 10)$

38. $(-10, 2), (4, -7)$

39. $\left(\frac{4}{3}, \frac{1}{6}\right), \left(4, \frac{7}{6}\right)$

40. $\left(\frac{5}{2}, 0\right), \left(\frac{5}{2}, 5\right)$

In Exercises 41–44, find an equation of the line through the indicated point (a) parallel to the given line, and (b) perpendicular to the given line.

41. $(-1, 5)$, $2x + 4y = 1$

42. $\left(\frac{3}{5}, -\frac{4}{5}\right)$, $3x + y = 2$

43. $\left(\frac{3}{8}, 3\right)$, $4x - 3y = 12$

44. $(12, 1)$, $5x = 3$

In Exercises 45–54, identify and sketch the graph of the conic represented by the given equation.

45. $x^2 - 2y = 0$

46. $x^2 + y^2 = 64$

47. $x^2 - y^2 = 64$

48. $x^2 + 4y^2 = 64$

49. $y = x(x - 6)$

50. $y = 9 - (x - 3)^2$

51. $\frac{x^2}{25} + \frac{y^2}{4} = 1$

52. $\frac{x^2}{25} - \frac{y^2}{4} = -1$

53. $4x^2 + 4y^2 - 9 = 0$

54. $x^2 + 9y^2 - 9 = 0$

In Exercises 55–60, find the general form of the equation for the conic meeting the given criteria.

55. Parabola: Vertex $(5, 0)$
Passes through the point $(1, 1)$

56. Parabola: Vertex $(-2, 5)$
Passes through the point $(0, 1)$

57. Ellipse: Vertices $(0, -5), (0, 5)$
Co-vertices $(-2, 0), (2, 0)$

58. Ellipse: Vertices $(-10, 0), (10, 0)$
Co-vertices $(0, -6), (0, 6)$

59. Circle: Center $(0, 0)$
Radius 20

60. Hyperbola: Vertices $(0, -4), (0, 4)$
Asymptotes $y = 2x$, $y = -2x$

61. *Rocker Arm Construction* Find the distance d between the centers of the two small bolt holes in the rocker arm shown in the accompanying figure.

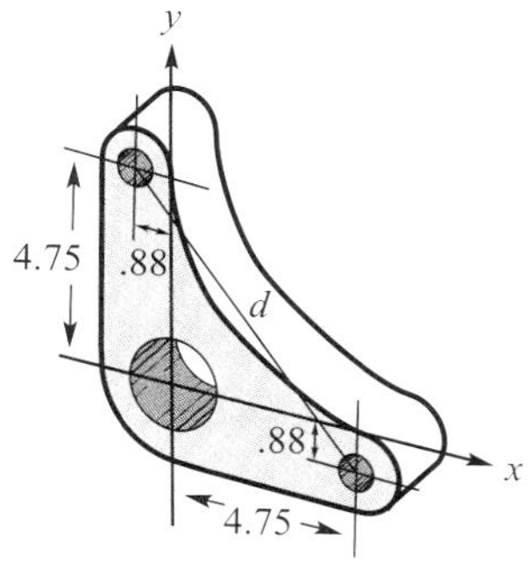

62. *Velocity of a Ball* The velocity of a ball thrown upward from ground level is given by $v = -32t + 80$ where t is the time in seconds and v is the velocity in feet per second.
(a) Find the velocity when $t = 2$.
(b) Find the time when the ball reaches its maximum height. (*Hint*: Find the time when $v = 0$.)
(c) Find the velocity when $t = 3$.

63. *Cost and Profit* A company produces a product for which the variable cost is \$8.55 per unit and the fixed costs are \$25,000. The product is sold for \$12.60, and the company can sell all it produces.
(a) Write the cost C as a linear equation in terms of x, the number of units produced.
(b) Write the profit P as a linear equation in terms of x, the number of units sold.

64. *Path of a Projectile* The height y (in feet) of a projectile is given by

$$y = -\frac{1}{16}x^2 + 5x$$

where x is the horizontal distance (in feet) from where the projectile was launched.
(a) Sketch the path of the projectile.
(b) How high is the projectile when it is at its maximum height?
(c) How far from the launch point will the projectile strike the ground?

65. *Satellite Orbit* Find an equation of the circular orbit of a satellite 1,000 miles above the surface of the Earth. Place the origin of the rectangular coordinate system at the center of the Earth and assume the radius of the Earth is 4,000 miles.

66. *Satellite Orbit* Find an equation of the elliptic orbit of a satellite that varies between 500 miles and 1,000 miles above the surface of the Earth. Place the origin of the rectangular coordinate system at the center of the Earth and assume the radius of the Earth is 4,000 miles.

CHAPTER 7 TEST

Take this test as you would take a test in class. After you are done, check your work with the answers given in the back of the book.

1. Determine the quadrants in which the point (x, y) lies if $xy < 0$.

2. Find the distance between the points $(0, 5)$ and $(3, 1)$.

3. Find the x- and y-intercepts of the graph of the equation $y = x(x + 1) - 3(x + 1)$.

4. Sketch the graph of the equation $y = |x - 2|$.

5. Find the slope (if possible) of the line passing through each pair of points.

(a) $(-4, 7), (2, 3)$ (b) $(3, -2), (3, 6)$

6. Sketch the line passing through the point $(0, -6)$ with slope $m = \frac{4}{3}$.

7. Plot the x- and y-intercepts and sketch the graph of the line given by the equation $2x + 5y - 10 = 0$.

8. Find the slope-intercept form of the equation of the line passing through the point $(0, 4)$ with slope $m = -\frac{3}{2}$.

9. Find the general form of the equation of the line passing through points $(25, -15)$ and $(75, 10)$.

10. Find the slope of a line perpendicular to the line given by $5x + 3y - 9 = 0$.

11. Sketch the graph of the parabola given by the equation $y = -2(x - 2)^2 + 8$. Identify the coordinates of the vertex and any intercepts of the graph.

12. Find the general form of the equation of the parabola with vertex $(3, -2)$ and passing through point $(0, 4)$.

13. Find an equation of the circle with center at the origin and passing through the point $(-3, 4)$.

14. Find the standard form of the equation of the ellipse centered at the origin with vertices $(0, -10)$ and $(0, 10)$ and co-vertices $(-3, 0)$ and $(3, 0)$.

15. Find the standard form of the equation of the hyperbola centered at the origin with vertices $(-3, 0)$, and $(3, 0)$ and asymptotes $y = \frac{1}{2}x$ and $y = -\frac{1}{2}x$.

16. Sketch the graph of each of the following equations.

(a) $x^2 + y^2 = 9$ (b) $\frac{x^2}{9} + \frac{y^2}{16} = 1$

(c) $\frac{x^2}{9} - \frac{y^2}{16} = 1$ (d) $\frac{x}{3} - \frac{y}{4} = 1$

17. Use a graphing calculator to sketch the graph of the equation $y = -0.3x^2 + 2x + 5$. Use the TRACE key to approximate the coordinates of the vertex and the x-intercepts.*

18. A small business purchases a car for \$16,000. It is estimated that after four years its depreciated value will be \$7,000. Assuming straight-line depreciation, write a linear equation giving the value V of the car in terms of t. Use the equation to determine the time t when the value of the car is \$10,000.

19. The revenue R for a charter bus company is given by

$$-\frac{1}{20}(n^2 - 240n), \quad 80 \leq n \leq 160$$

where n is the number of passengers that charter buses for a particular trip. Find the number of passengers that will give the bus company maximum revenue.

*The graphing calculator keystrokes given in this text correspond to the *TI-81* graphing calculator from *Texas Instruments*. However, you can also use other graphing calculators and computer graphing software for these examples and exercises.

CUMULATIVE TEST: CHAPTERS 5–7

Take this test as you would take a test in class. After you are done, check your work with the answers given in the back of the book.

1. Simplify: $\dfrac{-4x^{-3}y^4}{6xy^{-2}}$

2. Simplify: $\left(\dfrac{t^{1/2}}{t^{1/4}}\right)^2$

3. Evaluate without the aid of a calculator: $(4 \times 10^3)^2$

4. Combine the radicals: $10\sqrt{20x} + 3\sqrt{125x}$

5. Multiply: $\left(\sqrt{2x} - 3\right)^2$

6. Rationalize the denominator and simplify: $\dfrac{6}{\sqrt{10} - 2}$

7. Perform the indicated operations and simplify: $\sqrt{-2}\left(\sqrt{-8} + 3\right)$

8. Solve by extracting complex square roots: $(x - 5)^2 + 50 = 0$

9. Solve by completing the square: $3x^2 + 6x + 2 = 0$

10. Solve: $\sqrt{x + 10} = x - 2$

11. Solve: $\dfrac{4}{x} - x = 2$

12. Working together, two people can complete a task in eight hours. Working alone, how long would it take each to do the task if one person takes two hours longer than the other?

13. Given the two points $(-4, 0)$ and $(4, 6)$, find (a) the distance between the points, and (b) the equation of the line passing through the points.

14. Sketch the graph of each of the following equations.

(a) $4x + 3y - 12 = 0$ (b) $x^2 + y^2 = 8$

(c) $\dfrac{x^2}{1} + \dfrac{y^2}{4} = 1$ (d) $\dfrac{x^2}{1} - \dfrac{y^2}{4} = 1$

15. Sketch the graph of the parabola given by the equation $y = \frac{1}{2}(x^2 - 2x - 3)$. Identify the vertex and any intercepts of the graph.

16. Find an equation of the ellipse centered at the origin with vertices $(-20, 0)$ and $(20, 0)$ and co-vertices $(0, -10)$ and $(0, 10)$.

17. The profit for a small business in 1990 was \$250,000. The owners forecast that profits will increase by \$25,000 per year from 1990 to 2000. Write a linear equation giving the estimated profit P in terms of the year t. (Let $t = 0$ correspond to the year 1990.) Assuming this constant growth rate is correct, predict the profit in the year 2000.

18. A *semicircular* arch is positioned over the roadway onto the grounds of an estate. The roadway is 10 feet wide and the arch is sitting on pillars that are 8 feet tall. Find the maximum height of a truck that can be driven onto the estate if the truck is 8 feet wide.

CHAPTER EIGHT

Functions and Mathematical Models

Chapter Overview

This chapter introduces one of the most important concepts of modern mathematics — the concept of a function. The chapter begins by defining a function. Then, in Sections 8.2 through 8.4 we study the graph of a function, operations with functions, and the inverse of a function. Finally, in Section 8.5 we discuss variation and mathematical modeling.

SECTION 8.1

Functions and Function Notation

Definition of a Function • *Function Notation* • *Finding the Domain and Range of a Function* • *Applications*

Definition of a Function

Many everyday occurrences involve two quantities that are related to each other by some rule of correspondence. Here are some examples.

1. The simple interest I earned on \$1,000 for one year is related to the annual percentage rate r by the formula $I = 1{,}000r$.
2. The distance d traveled on a bicycle in two hours is related to the speed s of the bicycle by the formula $d = 2s$.
3. The area A of a circle is related to its radius by the formula $A = \pi r^2$.

Not all correspondences between two quantities have simple mathematical formulas. For instance, we commonly match up quantities such as NFL starting quarterbacks with touchdown passes, days of the year with the Dow-Jones Industrial Average, and hours of the day with temperature. In each of these cases, however, there is some rule of correspondence that matches each item in one set with exactly one item in another set. Such a rule of correspondence is called a **function**.

Definition of a Function

A **function** f from a set A to a set B is a rule of correspondence that assigns to each element x in the set A exactly one element y in the set B.

The set A is called the **domain** (or set of inputs) of the function f, and the set B contains the **range** (or set of outputs) of the function.

The rule of correspondence for a function establishes a set of "input-output" ordered pairs of the form (x, y) where x is an input and y is the corresponding output. In some cases, the rule may generate only a finite set of ordered pairs, while in other cases the rule may generate an infinite set of ordered pairs. Example 1 lists some examples of "input-output" ordered pairs that represent functions.

EXAMPLE 1 ■ Input-Output Ordered Pairs for Functions

Rule of Correspondence	*Set of Ordered Pairs*
(a) Winners of the Super Bowl from 1986 to 1991	Pairs: (year, team) Set: {(1986, Bears), (1987, Giants), (1988, Redskins), (1989, 49ers), (1990, 49ers), (1991, Giants)}
(b) $y = x - 2$	Pairs: (x, y) Set: {All points on the line given by $y = x - 2$}

Rule of Correspondence	*Set of Ordered Pairs*
(c) The squares of all positive integers that are less than 7	Pairs: (n, n^2) Set: $\{(1, 1), (2, 4), (3, 9), (4, 16), (5, 25), (6, 36)\}$
(d) The squares of all positive real numbers that are less than 7	Pairs: (x, x^2) Set: {All points (x, x^2) such that $0 < x < 7$}

Note that the sets in parts (a) and (c) have only a finite number of ordered pairs, whereas the sets in parts (b) and (d) have an infinite number of ordered pairs. ■

EXAMPLE 2 ■ Representing a Function by a Set of Ordered Pairs

Find the set of ordered pairs that represents the function shown in Figure 8.1.

FIGURE 8.1

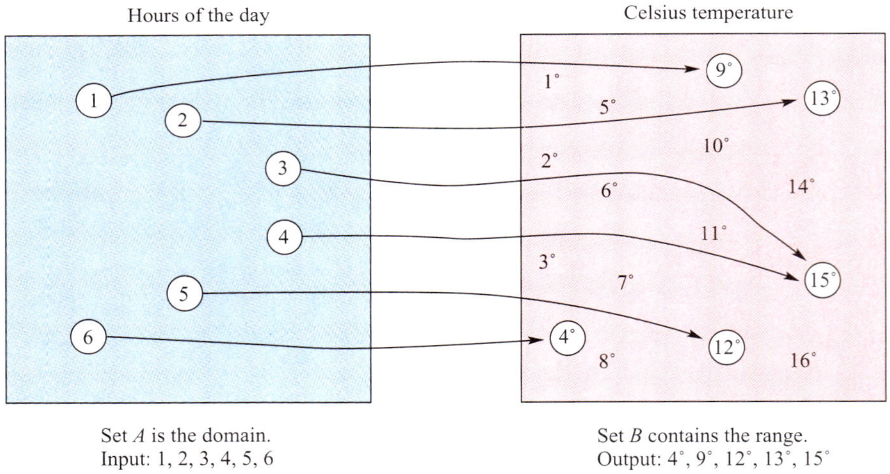

Function from Set A to Set B

Solution

From the figure, we let the first coordinate of each ordered pair represent an input and the second coordinate represent the corresponding output. This produces the following set.

$$\{(1, 9°), (2, 13°), (3, 15°), (4, 15°), (5, 12°), (6, 4°)\}$$ ■

From Examples 1 and 2, we can make up the following list of characteristics of a function. Study this list carefully.

Characteristics of a Function

1. *Each* element in the domain A must be matched with an element in the range, which is contained in the set B.
2. Some elements in the set B may not be matched with any element in the domain A.
3. Two or more elements of the domain may be matched with the *same* element in the range.
4. *No* element of the domain is matched with two different elements in the range.

Any set of ordered pairs is a **relation**. Some relations are functions and some are not. For instance, the relation $\{(0, 1), (1, 1), (2, 4)\}$ is a function, but the relation $\{(0, 1), (0, 2), (2, 4)\}$ is not a function. Other examples are given in Example 3.

EXAMPLE 3 ■ **Testing for Functions Represented by Ordered Pairs**

Let $A = \{a, b, c\}$ and let $B = \{1, 2, 3, 4, 5\}$. Determine which of the following sets of ordered pairs or figures represents a function from set A to set B.

(a) $\{(a, 2), (b, 3), (c, 4)\}$ (b) $\{(a, 4), (b, 5)\}$

(c)

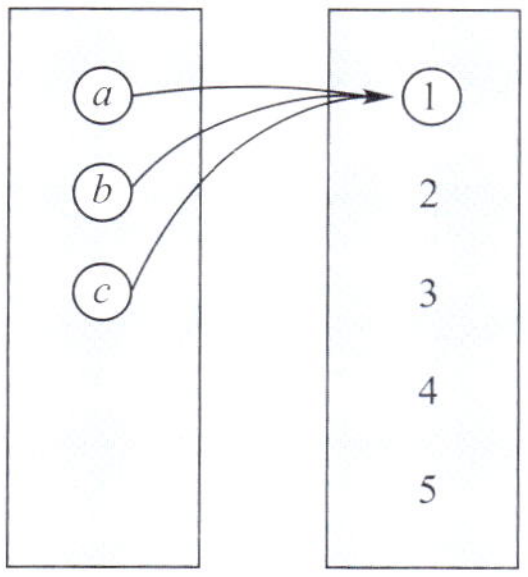

(d)

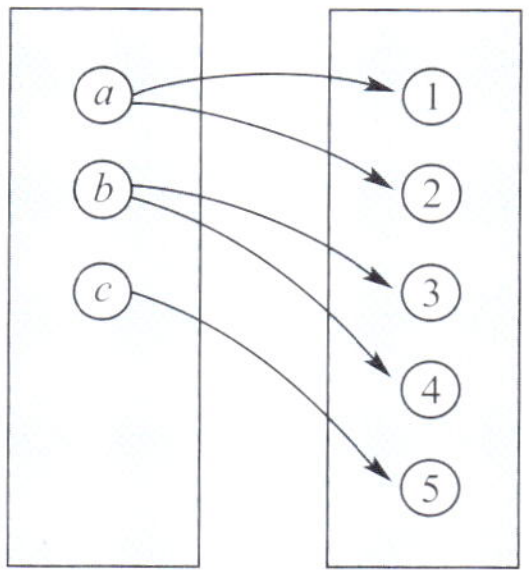

Solution

(a) This collection of ordered pairs *does* represent a function from A to B. Each element of A is matched with exactly one element of B.

(b) This collection of ordered pairs *does not* represent a function from A to B. Not all elements of A are matched with an element of B.

(c) This figure *does* represent a function from A to B. It does not matter that each element of A is matched with the same element in B.

(d) This figure *does not* represent a function from A to B. The element a in A is matched with *two* elements, 1 and 2, of B. This is also true of the element b. ■

Representing functions by sets of ordered pairs is a common practice in the study of *discrete mathematics*, which deals mainly with finite sets of data or with finite subsets of the set of real numbers. In algebra, however, it is more common to represent functions by equations or formulas involving two variables. For instance, the equation

$y = x^2$ Squaring function

represents the variable y as a function of the variable x. We call x the **independent variable** and y the **dependent variable**. In this context the domain of the function is the set of all *allowable* real values for the independent variable x, and the range of the function is the *resulting* set of all values taken on by the dependent variable y.

EXAMPLE 4 ■ **Testing for Functions Represented by Equations**

Determine which of the following equations represents y as a function of x.

(a) $y = x^2 + 1$ (b) $y^2 = x$

Solution

(a) From the equation

$$y = x^2 + 1$$

we see that for each value of x there corresponds just one value of y. For instance, when $x = 1$, y is determined to be $1^2 + 1 = 2$. Therefore, y *is* a function of x.

(b) From the equation

$$y^2 = x$$

we see that y is *not* a function of x because we can find two different y-values that correspond to the same x-value. For instance, when $x = 4$, y can be 2 or -2. ■

Function Notation

When using an equation to represent a function, it is convenient to name the function so that it can be easily referenced. For example, we know from Example 4(a) that the equation $y = x^2 + 1$ describes y as a function of x. Suppose we give this function the name "f." Then we can use the following **function notation**.

Input	*Output*	*Equation*
x	$f(x)$	$f(x) = x^2 + 1$

The symbol $f(x)$ is read as the **value of f at x** or simply "**f of x**." Because $f(x)$ represents the y-value for a given x, we can write $y = f(x)$. Keep in mind that f is the *name* of the function, while $f(x)$ is the *value* of the function at x. For instance, the function given by

$$f(x) = 4 - 3x$$

has *function values* denoted by $f(-1)$, $f(0)$, $f(1)$, and so on. To find these values, we substitute the specified input values into the given equation, as follows.

x-Value	*Function Value*
$x = -1$	$f(-1) = 4 - 3(-1) = 4 + 3 = 7$
$x = 0$	$f(0) = 4 - 3(0) = 4 - 0 = 4$
$x = 1$	$f(1) = 4 - 3(1) = 4 - 3 = 1$

The process of finding the value of $f(x)$ for a given value of x is called **evaluating a function**.

Although we often use f as a convenient function name, and x as the independent variable, we can use other letters. For instance, $f(x) = 2x^2 - 3x + 5$, $f(t) = 2t^2 - 3t + 5$, and $g(s) = 2s^2 - 3s + 5$ all define the same function. In fact, the role of the independent variable in a function is simply that of a "placeholder." Consequently, the function given by $f(x) = 2x^2 - 3x + 5$ could be described by the form

$$f(\quad) = 2(\quad)^2 - 3(\quad) + 5$$

where the parentheses are used in place of a letter. Therefore, to evaluate $f(-2)$, we simply place -2 in each set of parentheses, as follows.

$$\begin{aligned} f(-2) &= 2(-2)^2 - 3(-2) + 5 \\ &= 8 + 6 + 5 \\ &= 19 \end{aligned}$$

When evaluating a function, we aren't restricted to substituting only numerical values into the parentheses. For instance, the value of $f(3x)$ is

$$\begin{aligned} f(3x) &= 2(3x)^2 - 3(3x) + 5 \\ &= 18x^2 - 9x + 5. \end{aligned}$$

EXAMPLE 5 ■ Evaluating a Function

Let $g(x) = 3x - x^2$ and find the following.

(a) $g(1)$ (b) $g(x + 1)$ (c) $g(x) + g(1)$

Solution

(a) Replacing x by 1 produces

$$g(1) = 3(1) - 1^2 = 3 - 1 = 2.$$

(b) Replacing x with $x + 1$ produces

$$\begin{aligned} g(x + 1) &= 3(x + 1) - (x + 1)^2 \\ &= 3x + 3 - (x^2 + 2x + 1) \\ &= -x^2 + (3x - 2x) + (3 - 1) \\ &= -x^2 + x + 2. \end{aligned}$$

(c) Using the result of part (a), we have

$$g(x) + g(1) = (3x - x^2) + 2 = -x^2 + 3x + 2.$$

Note that $g(x + 1) \neq g(x) + g(1)$. In general, $g(a + b)$ is not equal to $g(a) + g(b)$. ■

Sometimes a function is defined by more than one equation. An illustration of this is given in Example 6.

EXAMPLE 6 ■ **A Function Defined by Two Equations**

Evaluate the function given by

$$f(x) = \begin{cases} x^2 + 1, & \text{if } x < 0 \\ x - 2, & \text{if } x \geq 0 \end{cases}$$

at (a) $x = -1$, (b) $x = 0$, and (c) $x = 1$.

Solution

(a) Since $x = -1 < 0$, we use $f(x) = x^2 + 1$ to obtain

$$f(-1) = (-1)^2 + 1 = 2.$$

(b) Since $x = 0 \geq 0$, we use $f(x) = x - 2$ to obtain

$$f(0) = 0 - 2 = -2.$$

(c) Since $x = 1 \geq 0$, we use $f(x) = x - 2$ to obtain

$$f(1) = 1 - 2 = -1.$$

■

Finding the Domain and Range of a Function

The domain of a function may be explicitly described along with the function, or it may be *implied* by the expression used to define the function. The **implied domain** is the set of all real numbers for which we can find real number values for the function. For instance, the function given by

$$f(x) = \frac{1}{x^2 - 9}$$

has an implied domain that consists of all real values of x other than $x = \pm 3$. These two values are excluded from the domain because division by zero is undefined. Another common type of implied domain is that used to avoid even roots of negative numbers. For instance, the function given by

$$f(x) = \sqrt{x}$$

is defined only for $x \geq 0$. Therefore, its implied domain is the interval $[0, \infty)$.

EXAMPLE 7 ■ **Finding the Domain and Range of a Function**

Find the domain and range of each of the following functions.

(a) f: $\{(-3, 0), (-1, 2), (0, 4), (2, 4), (4, -1)\}$ (b) Area of a circle: $A = \pi r^2$

Solution

(a) The domain of f consists of all first coordinates in the set of ordered pairs, and the range consists of all second coordinates in the set of ordered pairs.

Thus, the domain is

$$\text{Domain} = \{-3, -1, 0, 2, 4\}$$

and the range is

$$\text{Range} = \{0, 2, 4, -1\}.$$

(b) For the area of a circle, we must choose positive values for the radius, r. Thus, the domain is the set of all real numbers r such that $r > 0$. The range is therefore the set of all real numbers A such that $A > 0$. ■

Note in Example 7(b) that the domain of a function can be implied by a physical context. For instance, from the equation $A = \pi r^2$, we would have no reason to restrict r to positive values. However, because we know this function represents the area of a circle, we conclude that the radius must be positive.

EXAMPLE 8 ■ Finding the Domain of a Function

Find the domain of each of the following functions.

(a) $f(x) = \dfrac{1}{x + 4}$ (b) $g(x) = \sqrt{1 - x^2}$

Solution

(a) For this function, we exclude all values of x for which the denominator is 0. Thus, the domain of f consists of all real values of x such that $x \neq -4$.

(b) For this function, we must exclude all values of x for which the radicand $1 - x^2$ is negative. Thus, the domain consists of all x such that $1 - x^2 \geq 0$. Using the techniques described in Section 6.5, we find that the domain is the closed interval $[-1, 1]$. ■

Applications

To find mathematical models to represent functions, we use the same guidelines that we outlined in Chapter 2.

EXAMPLE 9 ■ Finding an Equation to Represent a Function

Is the area of a square a *function* of the length of one of its sides? If so, find an equation that represents this function.

FIGURE 8.2

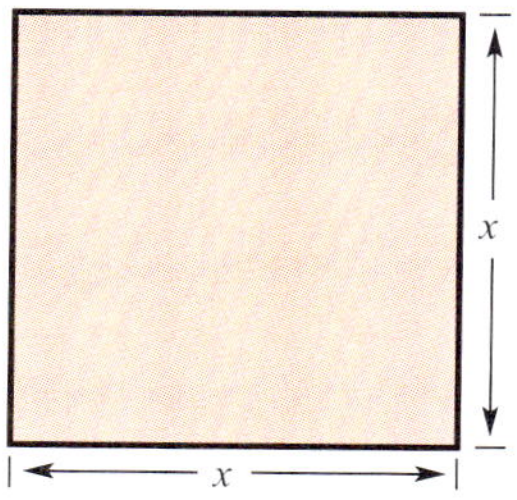

Solution

Figure 8.2 shows a typical square. For this square, we let the variable A represent the area of the square, and we let the variable x represent the length of any one of its sides. (Note that, by definition, all sides of the square have the same length.) Because the area of a square is completely determined from the length of its side, we can see that A *is* a function of x. The equation that represents the function is

$$A = x^2.$$ ■

In Example 9 we saw that the area of a square is a function of the length of one of its sides. In the next example, we list some other relationships between two variables. Some of the relationships are functions and some are not.

EXAMPLE 10 ■ Determining Relationships that Are Functions

Which of the following statements use the word *function* in a way that is *mathematically* correct?

(a) Your happiness is a function of the grade you receive in this course.

(b) Your federal income tax is a function of your "adjusted gross income."

(c) The average number of miles per gallon that you obtain is a function of the type of car that you drive.

(d) Your total calorie usage during a day is a function of the total amount of exercise you perform in the day.

(e) The sales commission that you pay a real estate agent for selling your house is a function of the sale price of the house.

Solution

(a) This is *not* a correct mathematical use of the word function. The word "happiness" is ambiguous.

(b) This *is* a correct mathematical use of the word function. Once you have determined your adjusted gross income, then your income tax can be determined.

(c) This is *not* a correct mathematical use of the word function. While the average number of miles per gallon that you obtain is related to the type of car that you drive, it is also affected by the way you drive, the condition of the road, and the type of fuel you use.

(d) This is *not* a correct mathematical use of the word function. While calorie usage tends to be related to activity level, individual body chemistry has been shown to also affect the number of calories a person uses.

(e) This *is* a correct mathematical use of the word function. Under normal circumstances, a real estate agent will charge a commission that is a percentage of the sale price of the house. ■

DISCUSSION PROBLEM ■ Determining Relationships that Are Functions

Follow the model given in Example 10 to write two statements describing relationships in everyday life that *are* functions and two that are *not* functions. ■

Warm-Up

The following warm-up exercises involve skills that were covered in earlier sections. You will use these skills in the exercise set for this section.

In Exercises 1–6, simplify the given expression.

1. $2(x + h) - 3 - (2x - 3)$

2. $3(y + t) + 5 - (3y + 5)$

3. $2(x - 4)^2 - 5$

4. $-3(x + 2)^2 + 10$

5. $4 - 2[3 + 4(x + 1)]$

6. $5x + x[3 - 2(x - 3)]$

In Exercises 7–10, evaluate the expression at the given values of the variable. (If not possible, state the reason.)

7. $3x^2 - 4$ (a) $x = 0$ (b) $x = -2$ (c) $x = 2$ (d) $x = 5$

8. $9 - (x - 3)^2$ (a) $x = 0$ (b) $x = 3$ (c) $x = 6$ (d) $x = 2.5$

9. $\dfrac{x + 4}{x - 3}$ (a) $x = -4$ (b) $x = 3$ (c) $x = 4$ (d) $x = 2$

10. $x + \dfrac{1}{x}$ (a) $x = 1$ (b) $x = -1$ (c) $x = \dfrac{1}{2}$ (d) $x = 10$

8.1 EXERCISES

In Exercises 1–4, determine the set of ordered pairs for the given rule of correspondence.

1. In a given week, a salesperson travels a distance d in t hours at an average speed of 50 mph. The travel times for each day are three hours, two hours, eight hours, six hours, and one-half hour.

2. The cubes of all positive integers less than or equal to 6

3. The winners of the World Series from 1986 to 1990

4. The men inaugurated as President of the United States in 1969, 1973, 1977, 1981, 1985, and 1989

5. *Beef Consumption* Use the accompanying graph to determine the second coordinate in the ordered pair, where the first coordinate (input) is the year and the second coordinate (output) is annual U.S. per capita beef consumption. (*Source:* U.S. Department of Agriculture.)

(a) (1974, ____) (b) (1976, ____)
(c) (1985, ____) (d) (1990, ____)

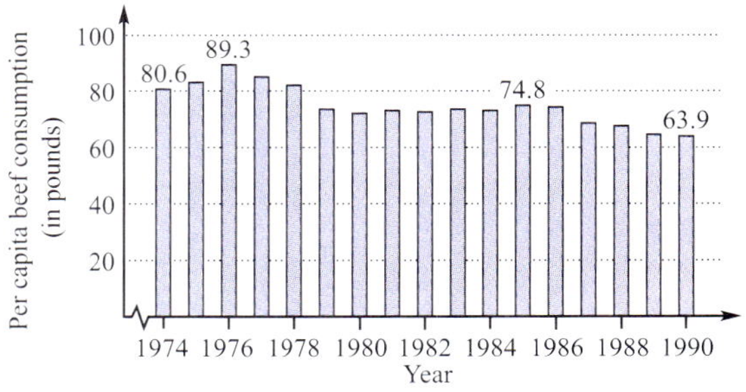

6. *State and Local Taxes* Use the accompanying graph to determine the second coordinate in the ordered pair, where the first coordinate (input) is the year and the second coordinate (output) is the average number of minutes per eight-hour workday an American works to earn enough money to pay state and local taxes. (*Source: Tax Foundation Facts and Figures on Government Finance.*)

(a) $(1980, \underline{\quad\quad})$ (b) $(1984, \underline{\quad\quad})$
(c) $(1986, \underline{\quad\quad})$ (d) $(1990, \underline{\quad\quad})$

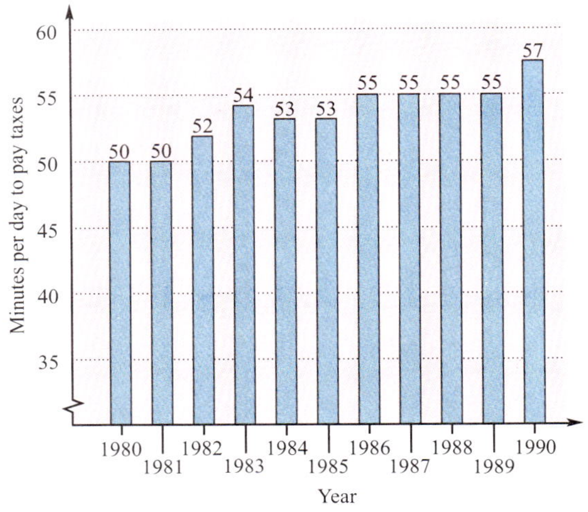

In Exercises 7 and 8, determine which relation (set of ordered pairs) represents a function from A to B.

7. $A = \{0, 1, 2, 3\}$ and $B = \{-2, -1, 0, 1, 2\}$
(a) $\{(0, 1), (1, -2), (2, 0), (3, 2)\}$
(b) $\{(0, -1), (2, 2), (1, -2), (3, 0), (1, 1)\}$
(c) $\{(0, 0), (1, 0), (2, 0), (3, 0)\}$
(d) $\{(0, 2), (3, 0), (1, 1)\}$

8. $A = \{1, 2, 3\}$ and $B = \{9, 10, 11, 12\}$
(a) $\{(1, 10), (3, 11), (3, 12), (2, 12)\}$
(b) $\{(1, 10), (2, 11), (3, 12)\}$
(c) $\{(1, 10), (1, 9), (3, 11), (2, 12)\}$
(d) $\{(3, 9), (2, 9), (1, 12)\}$

In Exercises 9–20, determine whether the given equation represents y as a function of x.

9. $y = 10x + 12$ **10.** $y = 3 - 8x$

11. $3x + 7y - 2 = 0$ **12.** $x - 9y + 3 = 0$

13. $y = x(x - 10)$ **14.** $y = (x + 2)^2 + 3$

15. $x^2 + y^2 = 25$ **16.** $x^2 - 4y^2 = 16$

17. $y = 8 - |x|$ **18.** $|y| = x + 2$

19. $|y - 2| = x$ **20.** $y = x^3 - 2$

In Exercises 21–24, fill in the blanks using the function and the specified value of the independent variable.

21. $f(x) = 3x + 5$
(a) $f(2) = 3(\underline{\quad\quad}) + 5$
(b) $f(-2) = 3(\underline{\quad\quad}) + 5$
(c) $f(k) = 3(\underline{\quad\quad}) + 5$
(d) $f(k + 1) = 3(\underline{\quad\quad}) + 5$

22. $f(x) = 3 - x^2$
(a) $f(0) = 3 - (\underline{\quad\quad})^2$
(b) $f(-3) = 3 - (\underline{\quad\quad})^2$
(c) $f(m) = 3 - (\underline{\quad\quad})^2$
(d) $f(t + 2) = 3 - (\underline{\quad\quad})^2$

23. $f(x) = \dfrac{x}{x + 2}$

(a) $f(3) = \dfrac{(\underline{\quad\quad})}{(\underline{\quad\quad}) + 2}$

(b) $f(-4) = \dfrac{(\underline{\quad\quad})}{(\underline{\quad\quad}) + 2}$

(c) $f(s) = \dfrac{(\underline{\quad\quad})}{(\underline{\quad\quad}) + 2}$

(d) $f(s - 2) = \dfrac{(\underline{\quad\quad})}{(\underline{\quad\quad}) + 2}$

24. $f(x) = \sqrt{x + 8}$
(a) $f(1) = \sqrt{(\underline{\quad\quad}) + 8}$
(b) $f(-4) = \sqrt{(\underline{\quad\quad}) + 8}$
(c) $f(h) = \sqrt{(\underline{\quad\quad}) + 8}$
(d) $f(h - 8) = \sqrt{(\underline{\quad\quad}) + 8}$

In Exercises 25–40, evaluate the function at the specified value of the independent variable and simplify when possible.

25. $f(x) = 12x - 7$
(a) $f(3)$ (b) $f\left(\dfrac{3}{2}\right)$
(c) $f(a) + f(1)$ (d) $f(a + 1)$

26. $f(x) = 3 - 7x$

(a) $f(-1)$ (b) $f\left(\frac{1}{2}\right)$

(c) $f(t) + f(-2)$ (d) $f(w)$

27. $g(x) = \frac{1}{2}x^2$

(a) $g(4)$ (b) $g\left(\frac{2}{3}\right)$

(c) $g(2y)$ (d) $g(4) + g(6)$

28. $h(x) = x(x - 2)$

(a) $h(2)$ (b) $h(0)$

(c) $h(1)$ (d) $h(t + 2)$

29. $f(x) = \sqrt{x + 5}$

(a) $f(-1)$ (b) $f(4)$

(c) $f\left(\frac{16}{3}\right)$ (d) $f(5z)$

30. $g(x) = 8 - |x - 4|$

(a) $g(0)$ (b) $g(8)$

(c) $g(16) - g(-1)$ (d) $g(x - 2)$

31. $f(x) = \dfrac{3x}{x - 5}$

(a) $f(0)$ (b) $f\left(\frac{5}{3}\right)$

(c) $f(2) - f(-1)$ (d) $f(x + 4)$

32. $g(x) = \dfrac{|x + 1|}{x + 1}$

(a) $g(2)$ (b) $g\left(-\frac{1}{3}\right)$

(c) $g(-4)$ (d) $g(3) + g(-5)$

33. $f(x) = \begin{cases} x + 8 & \text{if } x < 0 \\ 10 - 2x, & \text{if } x \geq 0 \end{cases}$

(a) $f(4)$ (b) $f(-10)$

(c) $f(0)$ (d) $f(6) - f(-2)$

34. $f(x) = \begin{cases} -x, & \text{if } x \leq 0 \\ x^2 - 3x, & \text{if } x > 0 \end{cases}$

(a) $f(0)$ (b) $f\left(-\frac{3}{2}\right)$

(c) $f(4)$ (d) $f(-2) + f(25)$

35. $h(x) = \begin{cases} 4 - x^2, & \text{if } x \leq 2 \\ x - 2, & \text{if } x > 2 \end{cases}$

(a) $h(2)$ (b) $h\left(-\frac{3}{2}\right)$

(c) $h(5)$ (d) $h(-3) + h(7)$

36. $f(x) = \begin{cases} x^2, & \text{if } x < 1 \\ x^2 - 3x + 2, & \text{if } x \geq 1 \end{cases}$

(a) $f(1)$ (b) $f(-1)$

(c) $f(2)$ (d) $f(-3) + f(3)$

37. $f(x) = 2x + 5$

(a) $\dfrac{f(x + 2) - f(2)}{x}$ (b) $\dfrac{f(x - 3) - f(3)}{x}$

38. $f(x) = x^2 + 4$

(a) $\dfrac{f(x + 1) - f(1)}{x}$ (b) $\dfrac{f(x - 5) - f(5)}{x}$

39. $g(x) = 1 - x^2$

(a) $g(x + h)$ (b) $\dfrac{g(x + h) - g(x)}{h}$

40. $f(x) = x^3$

(a) $f(x + h)$ (b) $\dfrac{f(x + h) - f(x)}{h}$

In Exercises 41–44, find the domain and range of the function.

41. f: $\{(0, 0), (2, 1), (4, 8), (6, 27)\}$

42. f: $\left\{\left(-3, -\frac{17}{2}\right), \left(-1, -\frac{5}{2}\right), (4, 2), (10, 11)\right\}$

43. Circumference of a circle: $C = 2\pi r$

44. Area of a square: $A = s^2$

In Exercises 45–60, find the domain of the function.

45. $f(x) = \dfrac{2x}{x - 3}$

46. $g(x) = \dfrac{x + 5}{x + 4}$

47. $h(x) = \dfrac{9}{x^2 + 1}$

48. $h(x) = 4x - 3$

49. $f(t) = \dfrac{t + 3}{t(t + 2)}$

50. $g(s) = \dfrac{s - 2}{(s - 6)(s - 10)}$

51. $g(x) = \dfrac{5x}{x^2 - 3x + 2}$

52. $f(x) = \dfrac{4}{x^2 - 4}$

53. $f(x) = \sqrt{2x - 1}$

54. $f(x) = \sqrt{2 - x}$

55. $H(x) = \sqrt{25 - x^2}$

56. $G(x) = \sqrt{x^2 - 25}$

57. $f(x) = \sqrt[3]{x + 1}$

58. $h(x) = \sqrt[4]{x}$

59. $f(t) = t^2 + 4t - 1$

60. $f(x) = |x + 3|$

In Exercises 61–68, find an equation that represents the given function. (Note that the variables do not necessarily have to be x and y.)

61. *Perimeter of a Square* Express the perimeter P of a square as a function of the length x of a side.

62. *Surface Area of a Cube* Express the surface area S of a cube as a function of the length x of one of the edges.

63. *Volume of a Cube* Express the volume V of a cube as a function of the length x of one of the edges.

64. *Diagonal of a Square* Express the length L of the diagonal of a square as a function of the length x of a side.

65. *Volume of a Box* An open box is to be made from a square piece of material, 24 inches on a side, by cutting equal squares from each corner and turning up the sides (see figure). Write the volume V of the box as a function of x.

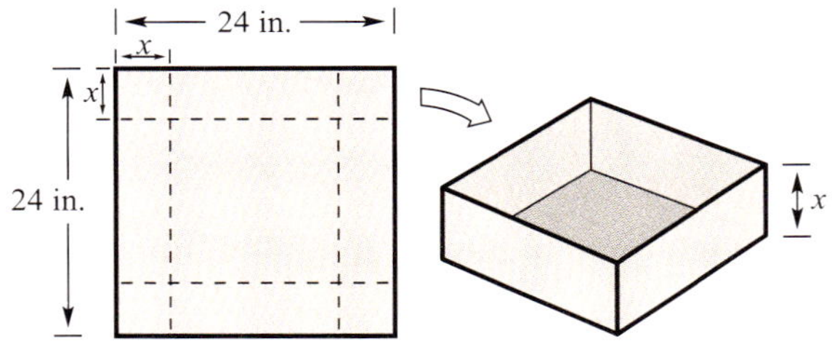

66. *Area of a Square* Strips of width x are cut from the four sides of a square that is 32 inches on a side (see figure). Write the area A of the remaining square as a function of x.

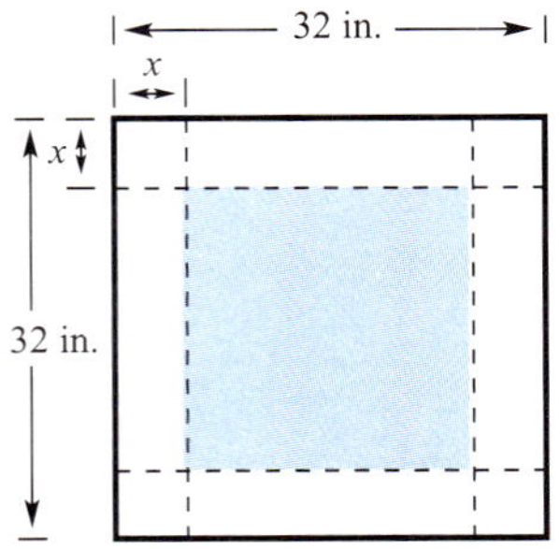

67. *Area of a Square* Strips of width x are cut from two adjacent sides of a square that is 32 inches on a side (see figure). Write the area A of the remaining square as a function of x.

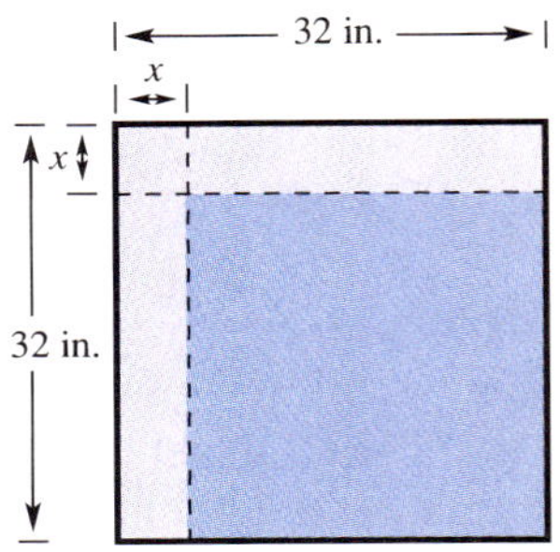

68. *Cost* The inventor of a new game believes that the variable cost for producing the game is \$1.95 per unit and the fixed costs are \$8,000. If x is the number of games produced, express the total cost C as a function of x.

In Exercises 69 and 70, determine whether the statements use the word *function* in a way that is mathematically correct.

69. (a) The sales tax on a purchased item is a function of the selling price.
(b) Your score on the next algebra exam is a function of the number of hours you study the night before the exam.

70. (a) The amount in your savings account is a function of your salary.
(b) The speed at which a free-falling baseball strikes the ground is a function of the height from which it is dropped.

SECTION **8.2**

The Graph of a Function

The Graph of a Function • Properties of Graphs of Functions • Graphs of Basic Functions • Transformations of Graphs of Functions

The Graph of a Function

In Section 8.1 we discussed many different types of functions. In this section we limit our discussion of functions to those for which the independent variable and the dependent variable represent real numbers. The **graph** of such a function f is the set of ordered pairs $(x, f(x))$, where x is in the domain of f.

$$x = x\text{-coordinate of the ordered pair}$$
$$f(x) = y\text{-coordinate of the ordered pair}$$

Figure 8.3 shows a typical graph of a function. To sketch the graph of a function, we use the same procedures discussed in Section 7.2 for sketching the graph of an equation.

FIGURE 8.3

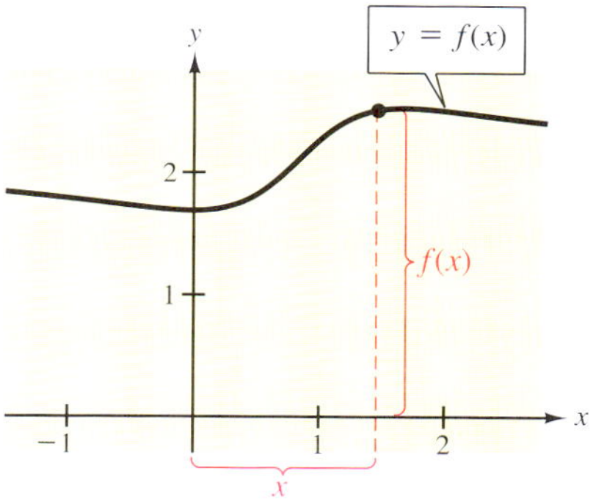

EXAMPLE 1 ■ Sketching the Graph of a Function

Sketch the graph of the following function.

$$f(x) = 2x - 1$$

Solution

Another way to write this equation is

$$y = 2x - 1.$$

Moreover, from our work in Chapter 7, we know that the graph of $y = 2x - 1$ is a line whose slope is 2 and whose y-intercept is -1. Thus, we can sketch the graph of the function as shown in Figure 8.4.

FIGURE 8.4

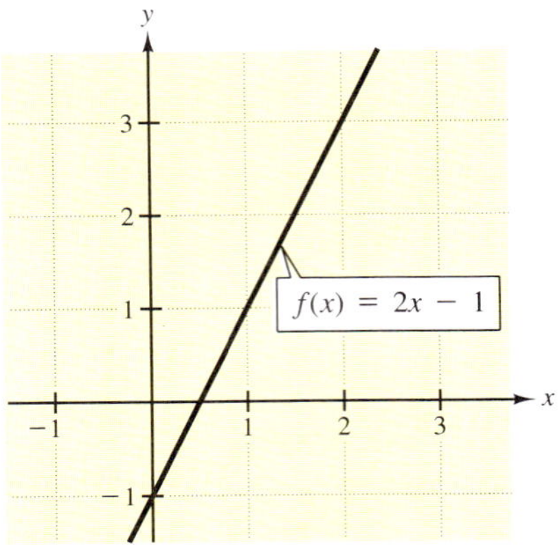

In Example 1 the (implied) domain of the function is the set of all real numbers. When writing the equation of a function, we often choose to restrict its domain by writing a condition to the right of the equation. For instance, the domain of the function

$$f(x) = 4x + 5, \quad x \geq 0$$

is the set of all nonnegative real numbers. When sketching the graph of a function, be sure you look for any such restrictions.

EXAMPLE 2 ■ A Function with a Restricted Domain

Sketch the graph of the function $f(x) = x^2, \quad -2 < x \leq 2$.

Solution

From Section 7.5, we know that the graph of $y = x^2$ is a parabola whose vertex is at the origin, as shown in Figure 8.5(a). The graph of the given function, however, is not the entire parabola, because its domain has been restricted. Using only x-values in the half-open interval $(-2, 2]$, we see that the graph of the given function is the bottom part of the parabola, as shown in Figure 8.5(b). Note in this figure that the point

FIGURE 8.5

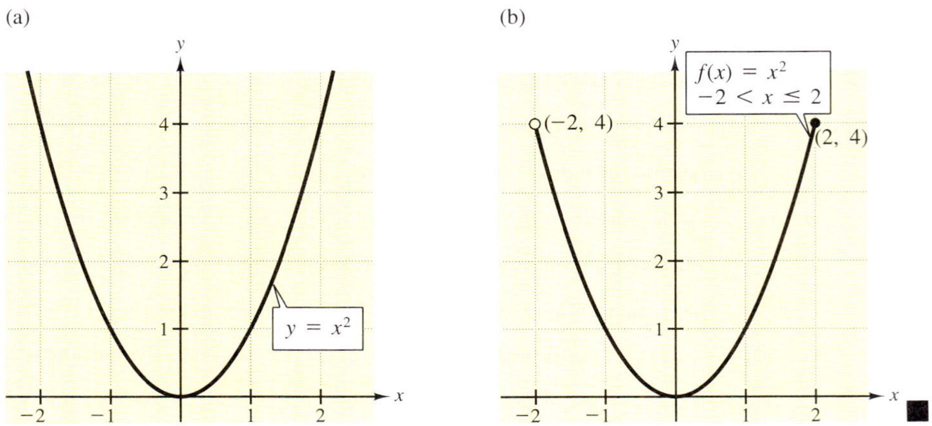

$(-2, 4)$ is represented by an *open* dot, because this point is not part of the graph. The point $(2, 4)$, on the other hand, is represented by a *solid* dot, because this point is part of the graph.

Determining the range of a function is often more difficult than determining its domain. Once we have sketched the graph of the function, however, we can use the graph to help determine the range. From Figure 8.5(b), for instance, we can see that the range of the function

$$f(x) = x^2, \quad -2 < x \leq 2$$

is given by $0 \leq y \leq 4$. Remember that the domain of a function is the set of all allowable x-values, and the range is the resulting set of y-values.

Properties of Graphs of Functions

By the definition of a function, at most one y-value corresponds to a given x-value. It follows, then, that a vertical line can intersect the graph of a function at most once. This important observation provides us with a convenient visual test for a function. We call this test the **vertical line test for functions**.

Vertical Line Test for Functions

A set of points in a rectangular coordinate system is the graph of y as a function of x if and only if no vertical line intersects the graph at more than one point.

The vertical line test provides us with an easy way to determine whether an equation represents y as a function of x. If the graph of an equation has the property that no vertical line intersects the graph at two (or more) points, then the equation represents y as a function of x [see Figure 8.6(a)]. On the other hand, if you can find a vertical line that intersects the graph at two (or more) points, then the equation does not represent y as a function of x [see Figure 8.6(b)].

FIGURE 8.6

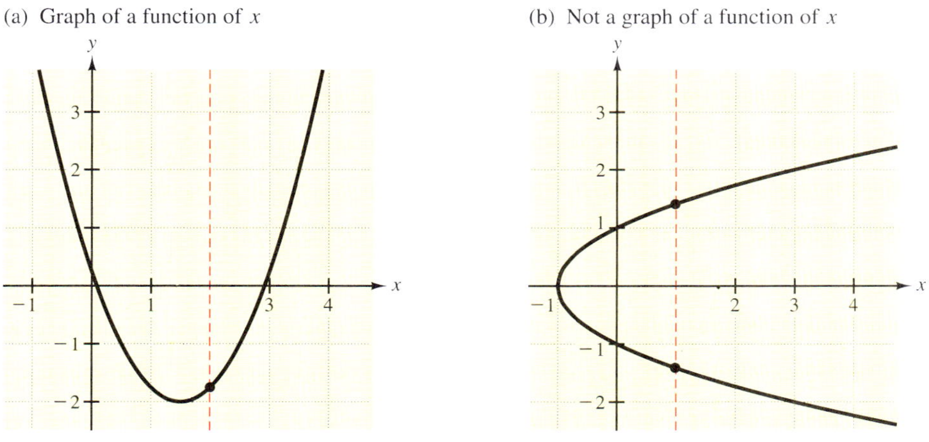

Vertical line intersects once. Vertical line intersects twice.

EXAMPLE 3 ■ Using the Vertical Line Test for Functions

The graphs of the following equations are shown in Figure 8.7. Use the vertical line test to determine whether the equations represent y as a function of x.

(a) $y = x - 1$ (b) $y^2 = x - 1$

Solution

Notice that for the first equation there is no vertical line that intersects the graph twice. Therefore, the equation $y = x - 1$ *does* represent y as a function of x. However, for the second equation, there are some vertical lines that intersect the graph twice. Therefore, the equation $y^2 = x - 1$ *does not* represent y as a function of x.

FIGURE 8.7

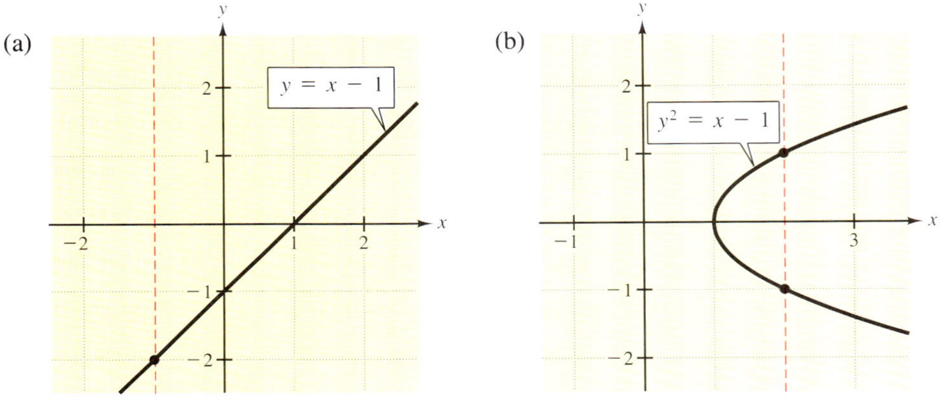

■

EXAMPLE 4 ■ Testing for *y* as a Function of *x*

Which of the following equations represent y as a function of x?

(a) $x^2 = y - 5$ (b) $y^2 = x - 5$ (c) $y = 2$

Solution

(a) To begin, we rewrite the equation in the form

$$y = x^2 + 5.$$

Since we were able to isolate y on the left side of the equation, it appears that the equation does represent y as a function of x. We can confirm this conclusion by the vertical line test. Notice in Figure 8.8 that no vertical line intersects the graph more than once. Therefore, this equation *does* represent y as a function of x.

(b) The graph of this equation is shown in Figure 8.9. Note that it is possible for a vertical line to intersect the graph at more than one point. Therefore, this equation *does not* represent y as a function of x. Remember that if an equation represents y as a function of x, then no two different solution points of the equation can have the same x-value. For this equation, we see that the points $(6, 1)$ and $(6, -1)$ are both solution points of the equation. The fact that these two points have the same x-value is another way we can tell that the equation

$$y^2 = x - 5$$

does not represent y as a function of x.

(c) At first glance, you might be tempted to say that this equation does not represent y as a function of x, because x does not appear in the equation. However, in Figure 8.10, we see that the graph of this equation is simply a horizontal line (two units above the x-axis). Since this graph passes the vertical line test, we conclude that the equation

$$y = 2$$

does represent y as a function of x. We call this type of function a **constant function**.

FIGURE 8.8

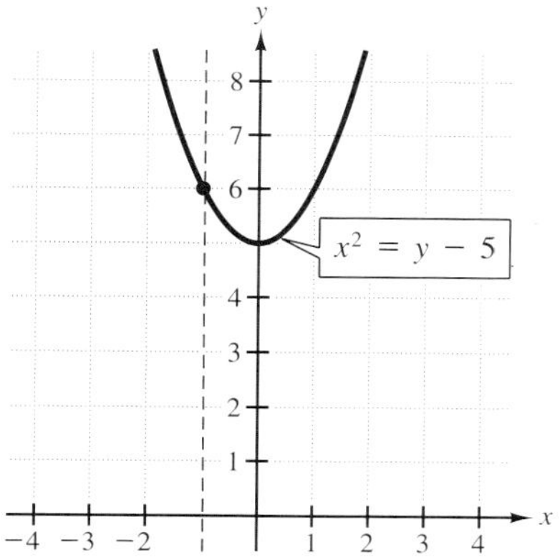

FIGURE 8.9

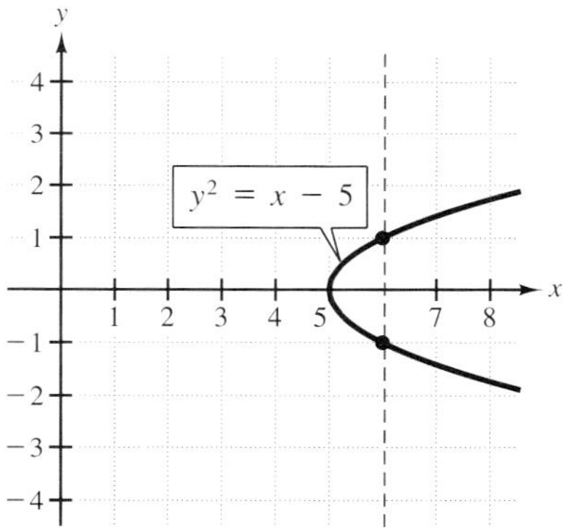

FIGURE 8.10

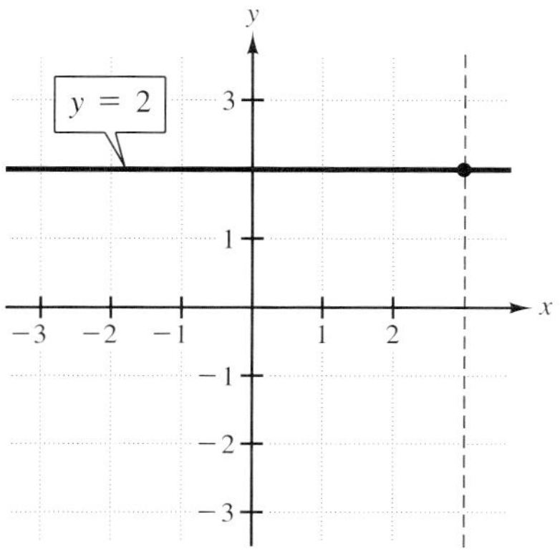

■

Graphs of Basic Functions

To become good at sketching the graphs of functions, it helps to be familiar with the graphs of some basic functions. The functions shown in Figure 8.11, and variations of them, occur frequently in applications.

FIGURE 8.11

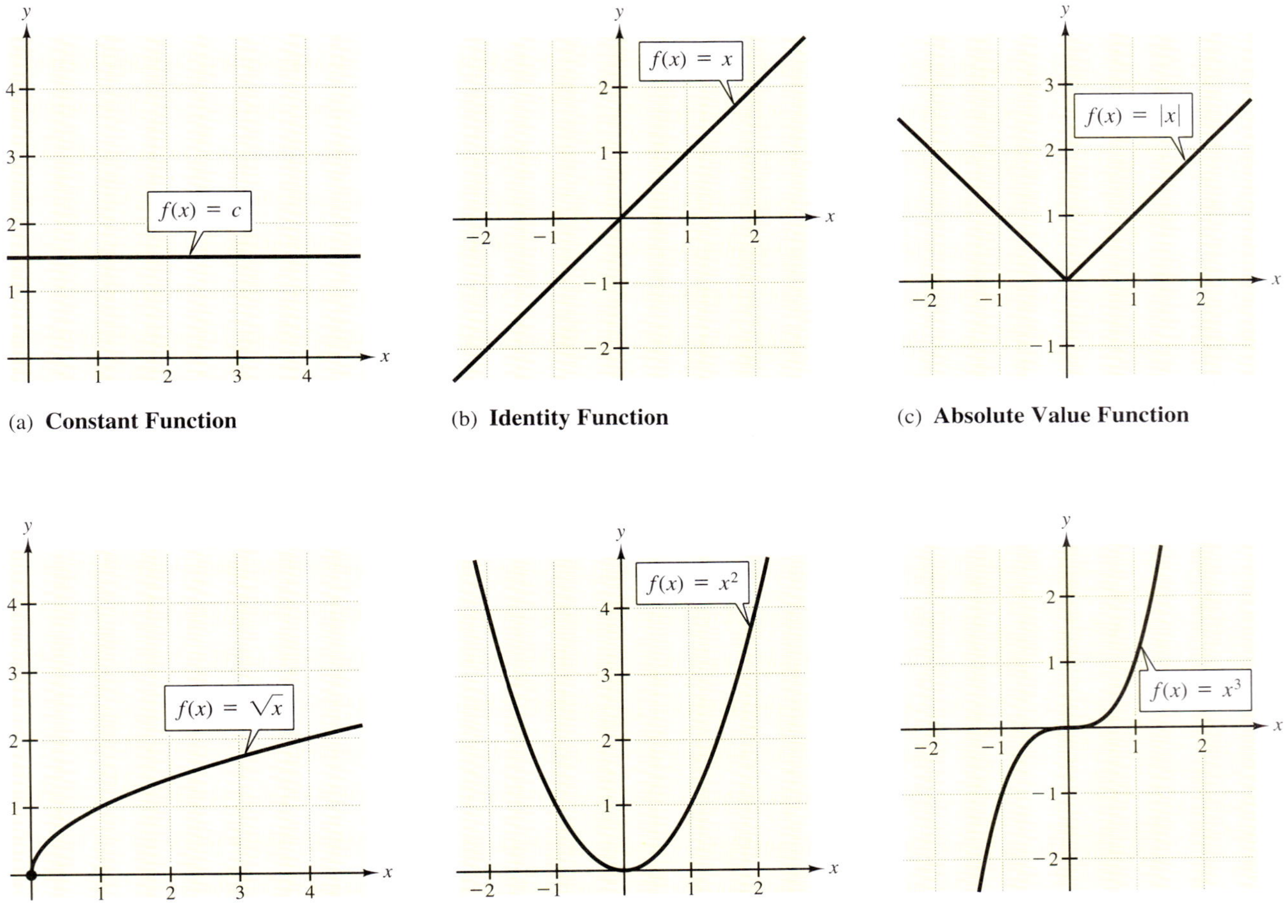

(a) **Constant Function** (b) **Identity Function** (c) **Absolute Value Function**

(d) **Square Root Function** (e) **Squaring Function** (f) **Cubing Function**

Transformations of Graphs of Functions

Many functions have graphs that are simple transformations of the basic graphs shown in Figure 8.11. For example, we can obtain the graph of $h(x) = x^2 + 1$ by shifting the graph of $f(x) = x^2$ *upward* one unit, as shown in Figure 8.12. In function notation,

h and f are related as follows.

$$h(x) = x^2 + 1 = f(x) + 1 \qquad \text{Upward shift of 1}$$

Similarly, we can obtain the graph of $g(x) = (x - 1)^2$ by shifting the graph of $f(x) = x^2$ to the *right* one unit, as shown in Figure 8.13. In this case, the functions g and f have the following relationship.

$$g(x) = (x - 1)^2 = f(x - 1) \qquad \text{Right shift of 1}$$

FIGURE 8.12

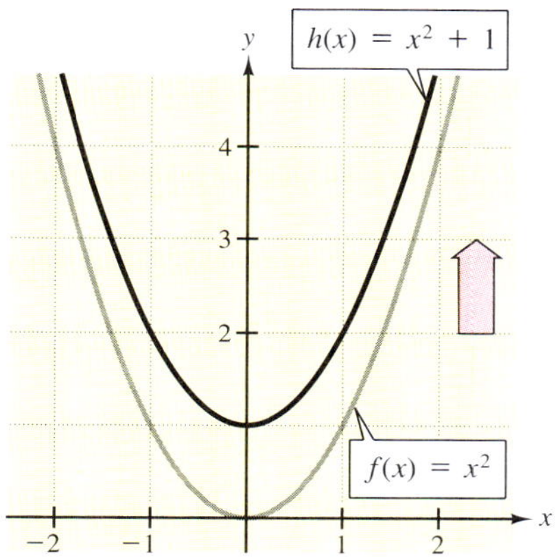

Vertical Shift: One Unit Up

FIGURE 8.13

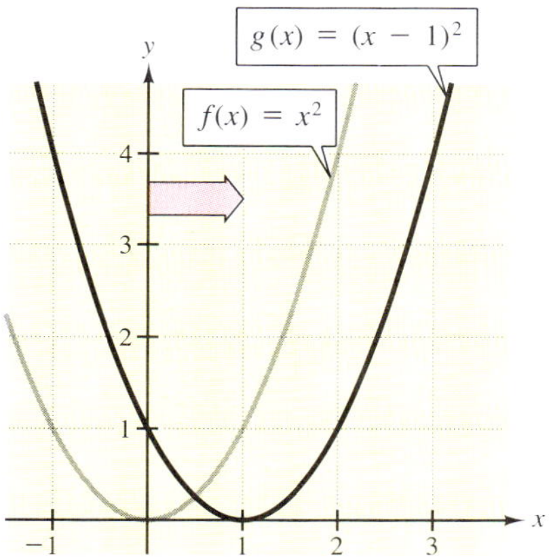

Horizontal Shift: One Unit Right

We summarize the various types of **vertical** and **horizontal shifts** of the graphs of functions as follows.

Vertical and Horizontal Shifts

Let c be a positive real number. **Vertical** and **horizontal shifts** of the graph of the function $y = f(x)$ are represented as follows.

1. Vertical shift c units **upward**: $h(x) = f(x) + c$
2. Vertical shift c units **downward**: $h(x) = f(x) - c$
3. Horizontal shift c units to the **right**: $h(x) = f(x - c)$
4. Horizontal shift c units to the **left**: $h(x) = f(x + c)$

EXAMPLE 5 ■ Shifts of the Graphs of Functions

Use the graph of $f(x) = x^2$ to sketch the graph of each of the following functions.

(a) $g(x) = x^2 - 2$ (b) $h(x) = (x + 3)^2$

Solution

(a) Relative to the graph of $f(x) = x^2$, the graph of $g(x) = x^2 - 2$ represents a *downward shift* of two units, as shown in Figure 8.14.

(b) Relative to the graph of $f(x) = x^2$, the graph of $h(x) = (x + 3)^2$ represents a *left shift* of three units, as shown in Figure 8.15.

FIGURE 8.14

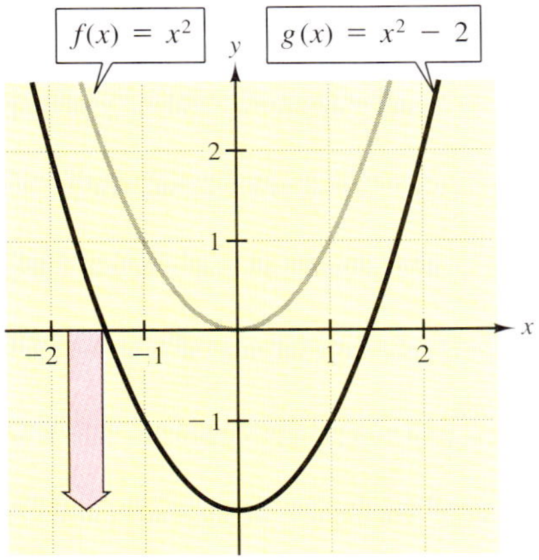

Vertical Shift: Two Units Down

FIGURE 8.15

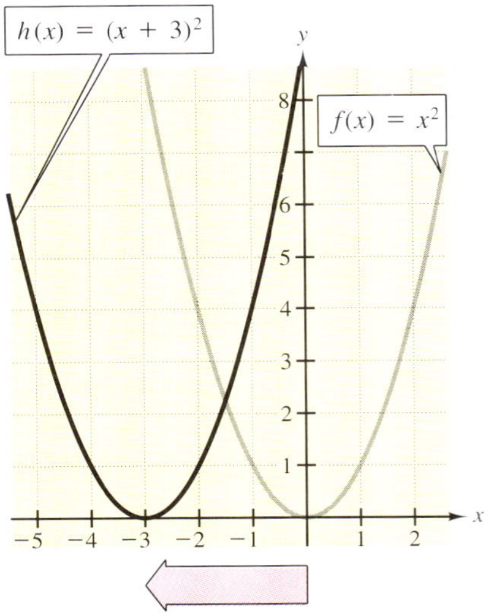

Horizontal Shift: Three Units Left ■

Some graphs can be obtained from a *combination* of vertical and horizontal shifts, as shown in part (b) of the next example.

EXAMPLE 6 ■ Shifts of the Graphs of Functions

Use the graph of $f(x) = x^3$ to sketch the graph of each of the following functions.

(a) $g(x) = x^3 + 2$ (b) $h(x) = (x - 1)^3 + 2$

Solution

(a) Relative to the graph of $f(x) = x^3$, the graph of $g(x) = x^3 + 2$ represents an *upward shift* of two units, as shown in Figure 8.16.

(b) Relative to the graph of $f(x) = x^3$, the graph of $h(x) = (x - 1)^3 + 2$ represents a *right shift* of one unit, followed by an *upward shift* of two units, as shown in Figure 8.17.

FIGURE 8.16

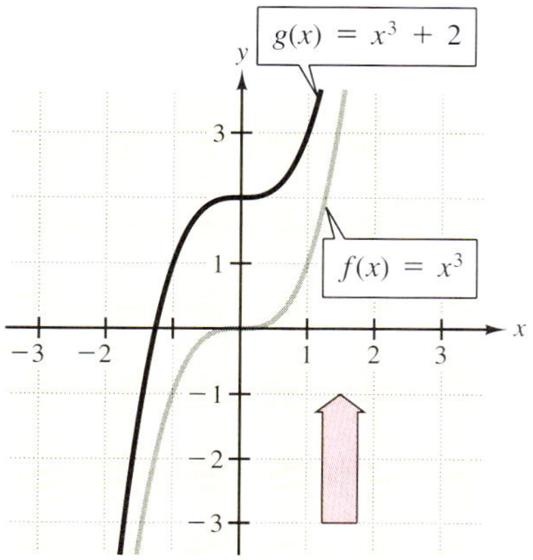

Vertical Shift: Two Units Up

FIGURE 8.17

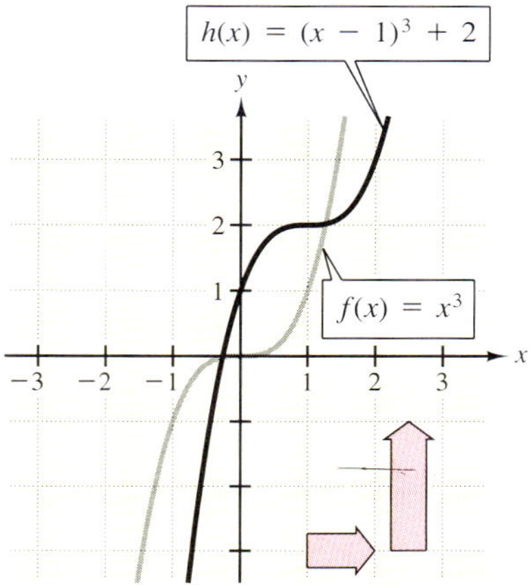

Horizontal Shift: One Unit Right
Vertical Shift: Two Units Up ■

The second basic type of transformation is called a **reflection**. For instance, if we imagine that the x-axis represents a mirror, then the graph of $h(x) = -x^2$ is the mirror image (or reflection) of the graph of $f(x) = x^2$, as shown in Figure 8.18.

FIGURE 8.18

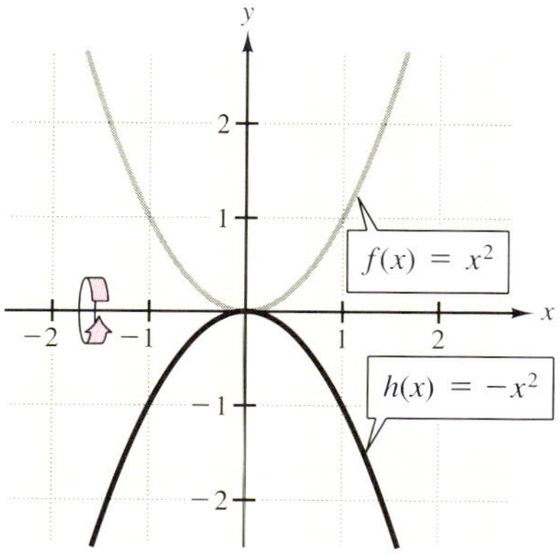

Reflection

Reflections in the Coordinate Axes

Reflections, in the coordinate axes, of the graph of $y = f(x)$ are represented as follows.

1. **Reflection in the x-axis:** $h(x) = -f(x)$
2. **Reflection in the y-axis:** $h(x) = f(-x)$

EXAMPLE 7 ■ Reflections of the Graphs of Functions

Use the graph of $f(x) = \sqrt{x}$ to sketch the graph of each of the following functions.

(a) $g(x) = -\sqrt{x}$ (b) $h(x) = \sqrt{-x}$

Solution

(a) Relative to the graph of $f(x) = \sqrt{x}$, the graph of $g(x) = -\sqrt{x}$ represents a *reflection in the x-axis*, as shown in Figure 8.19.

(b) Relative to the graph of $f(x) = \sqrt{x}$, the graph of $h(x) = \sqrt{-x}$ represents a *reflection in the y-axis*, as shown in Figure 8.20.

FIGURE 8.19

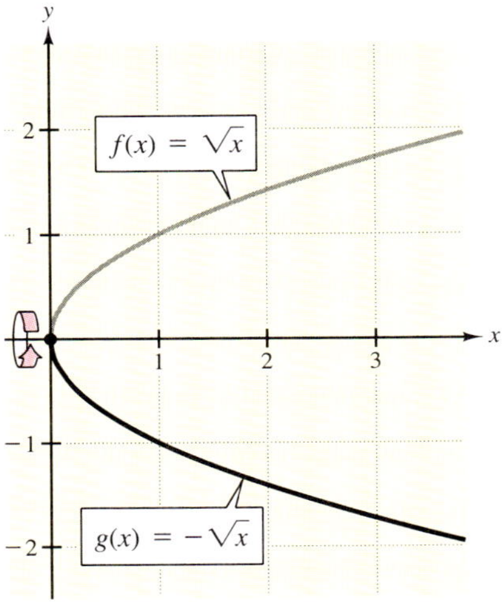

Reflection in *x*-axis

FIGURE 8.20

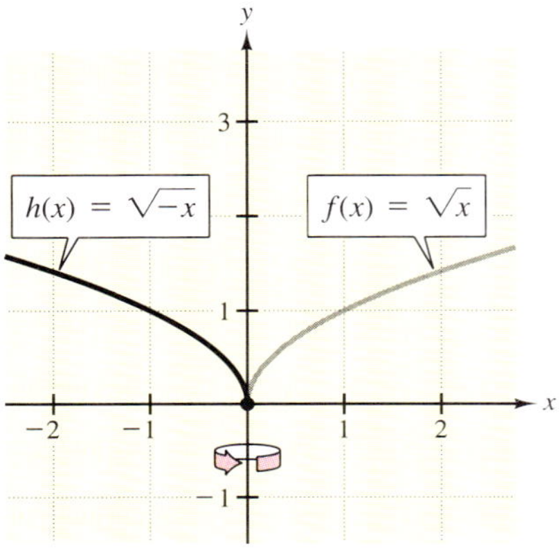

Reflection in *y*-axis

■

DISCUSSION PROBLEM ■ Representing an Equation by More than One Function

We know that the graph of the equation

$$x^2 + y^2 = 1$$

is a circle whose center is the origin and whose radius is 1, as shown in Figure 8.21. From the graph, we see that the equation does not represent y as a function of x, because the graph does not pass the vertical line test for functions. Can you think of a way to write *two* different functions to represent this graph? Try finding one function to represent the upper semicircle and another function to represent the lower semicircle.

FIGURE 8.21

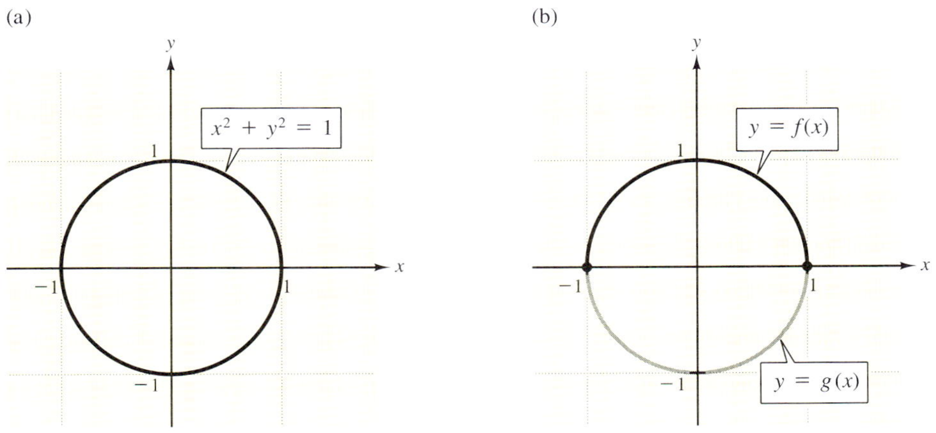

Warm-Up

The following warm-up exercises involve skills that were covered in earlier sections. You will use these skills in the exercise set for this section.

In Exercises 1–6, sketch the graph of the given equation.

1. $y = 3 - 2x$ **2.** $y = \frac{1}{2}x - 1$ **3.** $y = (x - 2)^2 - 4$

4. $y = 9 - (x + 1)^2$ **5.** $x - y^2 = 0$ **6.** $y = |x| + 1$

In Exercises 7–10, evaluate the given function at the indicated values.

Function	*Function Value*	
7. $f(x) = \frac{1}{3}x^2$	(a) $f(6)$	(b) $f\left(\frac{3}{4}\right)$
8. $f(x) = 3 - 2x$	(a) $f(5)$	(b) $f(x + 3) - f(3)$
9. $f(x) = \frac{x}{x + 10}$	(a) $f(5)$	(b) $f(c - 6)$
10. $f(x) = \sqrt{x - 4}$	(a) $f(16)$	(b) $f(t + 3)$

8.2 EXERCISES

In Exercises 1–20, sketch the graph of the given function. Determine the domain and range of each function.

1. $f(x) = 2x - 7$

2. $f(x) = 3 - 2x$

3. $g(x) = \frac{1}{2}x^2$

4. $h(x) = \frac{1}{4}x^2 - 1$

5. $h(x) = x^2 - 6x + 8$

6. $f(x) = -x^2 - 2x + 1$

7. $f(t) = \sqrt{t - 2}$

8. $h(x) = \sqrt{4 - x}$

9. $g(s) = \frac{1}{2}s^3$

10. $f(x) = x^3 - 4$

11. $f(x) = |x + 3|$

12. $g(x) = 2 - |x - 1|$

13. $f(x) = 6 - 3x, \quad 0 \le x \le 2$

14. $f(x) = \frac{1}{3}x - 2, \quad 6 \le x \le 12$

15. $h(x) = x^3, \quad -2 \le x \le 2$

16. $h(x) = x(6 - x), \quad 0 \le x \le 6$

17. $f(x) = \begin{cases} x + 6, & \text{if } x < 0 \\ 6 - 2x, & \text{if } x \ge 0 \end{cases}$

18. $f(x) = \begin{cases} -x, & \text{if } x \le 0 \\ x^2 - 4x, & \text{if } x > 0 \end{cases}$

19. $h(x) = \begin{cases} 4 - x^2, & \text{if } x \le 2 \\ x - 2, & \text{if } x > 2 \end{cases}$

20. $f(x) = \begin{cases} x^2, & \text{if } x < 1 \\ x^2 - 3x + 2, & \text{if } x \ge 1 \end{cases}$

In Exercises 21–24, determine whether y is a function of x.

21. $y = (x + 2)^2$

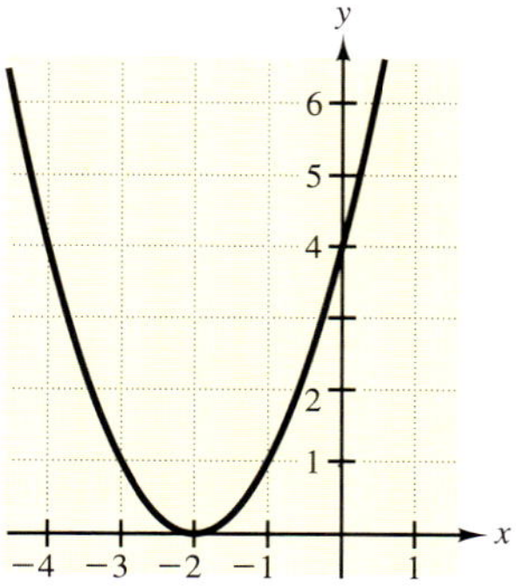

22. $x - 2y^2 = 0$

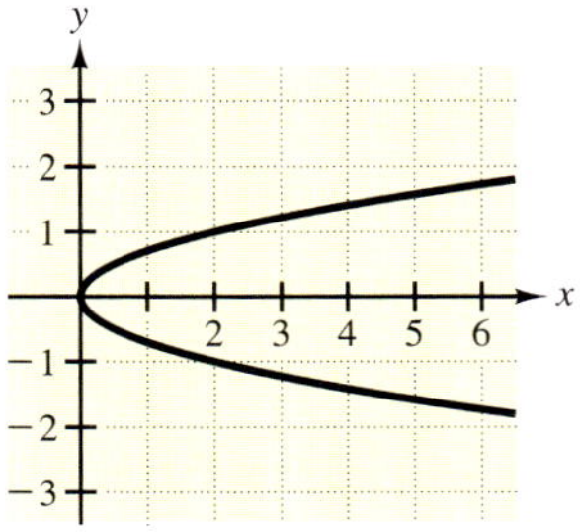

23. $x^2 + y^2 = 16$

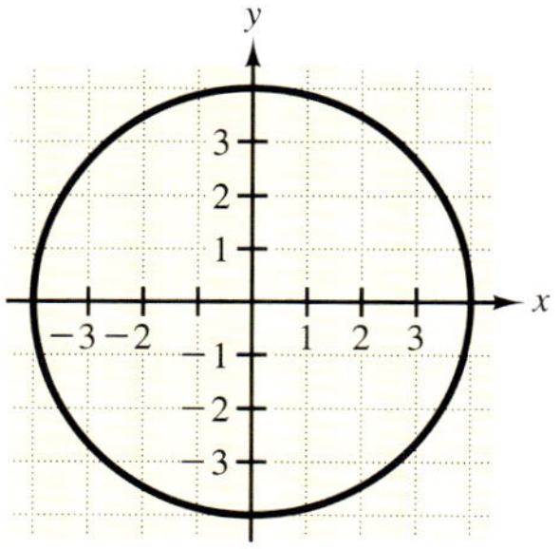

24. $y = x^3 - 3$

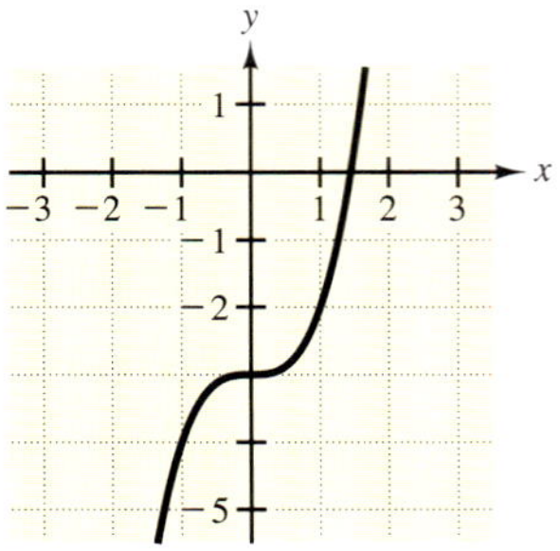

In Exercises 25–30, sketch the graph of the given equation. Use the vertical line test to determine whether y is a function of x.

25. $3x - 5y = 15$

26. $y = x^2 + 2$

27. $y^2 = x + 1$

28. $2|y| = x$

29. $y = 1 - x^3$

30. $x = y^4$

In Exercises 31–36, identify the transformation of the graph of $f(x) = x^2$ and sketch a graph of h.

31. $h(x) = x^2 + 2$ **32.** $h(x) = x^2 - 4$ **33.** $h(x) = (x + 2)^2$ **34.** $h(x) = (x - 4)^2$

35. $h(x) = -x^2$ **36.** $h(x) = -x^2 + 4$

In Exercises 37–44, identify the transformation of the graph of $f(x) = x^3$ and sketch a graph of h.

37. $h(x) = x^3 + 3$ **38.** $h(x) = x^3 - 5$ **39.** $h(x) = (x - 3)^3$ **40.** $h(x) = (x + 2)^3$

41. $h(x) = (-x)^3$ **42.** $h(x) = -x^3$ **43.** $h(x) = 2 - (x - 1)^3$ **44.** $h(x) = (x + 2)^3 - 3$

In Exercises 45–50, use the graph of $f(x) = \sqrt{x}$ to write an equation to represent the transformation of the given graph.

45.

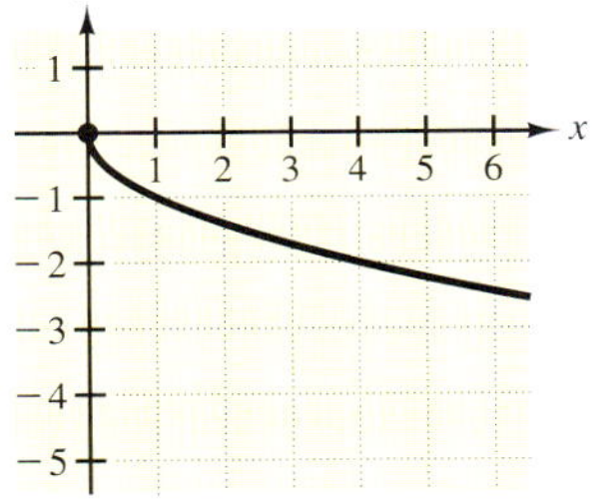

46.

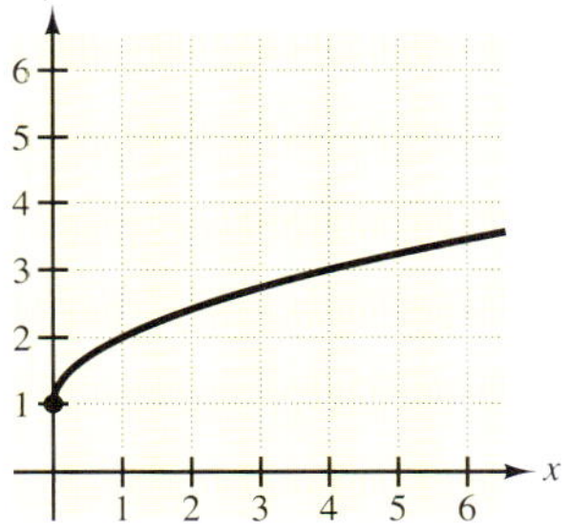

47.

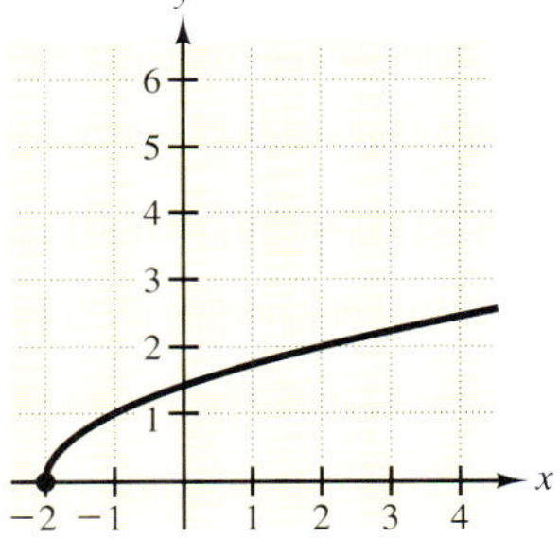

48.

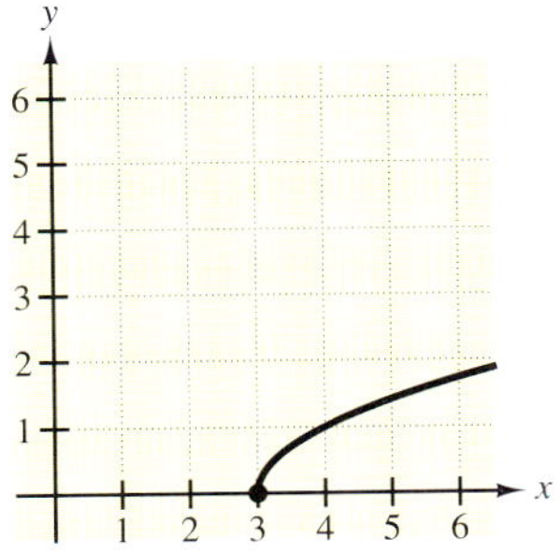

49.

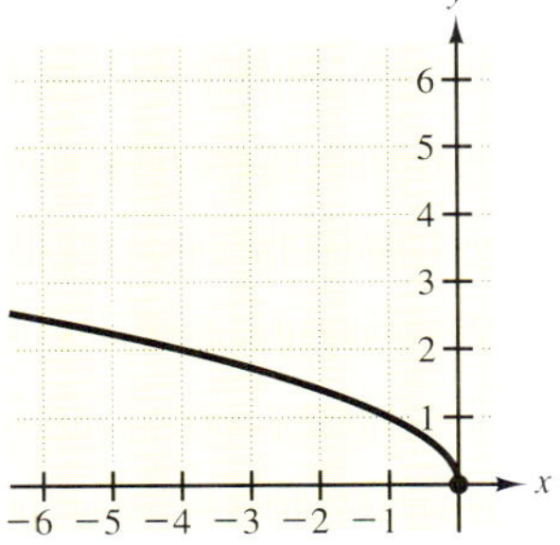

50.

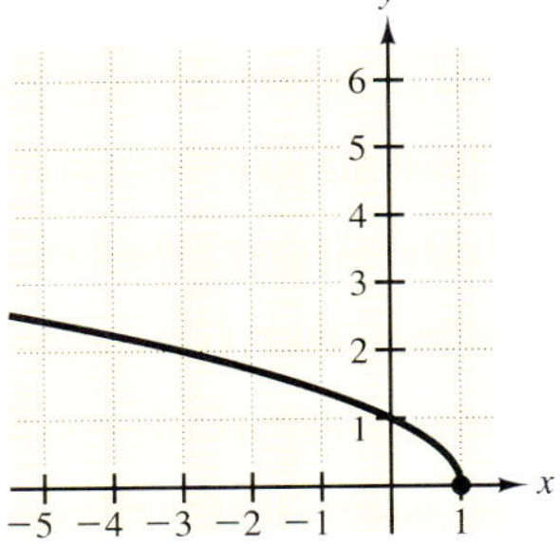

In Exercises 51–54, solve for y in terms of x. Let f and g be those functions where f represents the positive square root and g the negative square root. Sketch the graph of f and g on different sets of coordinate axes.

51. $x^2 + y^2 = 4$ **52.** $x^2 - y^2 = 4$ **53.** $x^2 + 4y^2 = 4$ **54.** $x - y^2 = 0$

In Exercises 55–60, identify the transformation of the graph of $f(x) = |x|$ and use a graphing calculator to sketch the graph of h.

55. $h(x) = |x - 5|$ **56.** $h(x) = |x + 3|$ **57.** $h(x) = |x| - 5$ **58.** $h(x) = |-x|$

59. $h(x) = -|x|$ **60.** $h(x) = 5 - |x|$

SECTION 8.3

Combinations of Functions

Arithmetic Combinations of Functions • Composition of Functions

Arithmetic Combinations of Functions

Just as two real numbers can be combined by the operations of addition, subtraction, multiplication, and division to form other real numbers, two functions can be combined to form new functions. For example, if

$$f(x) = 3x - 1 \quad \text{and} \quad g(x) = x^2 - 4$$

we can form the sum, difference, product, and quotient of f and g as follows.

$$f(x) + g(x) = (3x - 1) + (x^2 - 4) = x^2 + 3x - 5 \qquad \text{Sum}$$

$$f(x) - g(x) = (3x - 1) - (x^2 - 4) = -x^2 + 3x + 3 \qquad \text{Difference}$$

$$f(x)g(x) = (3x - 1)(x^2 - 4) = 3x^3 - x^2 - 12x + 4 \qquad \text{Product}$$

$$\frac{f(x)}{g(x)} = \frac{3x - 1}{x^2 - 4}, \quad x \neq \pm 2 \qquad \text{Quotient}$$

The domain of an arithmetic combination of functions f and g consists of all real numbers that are common to the domains of f and g. In the case of the quotient $f(x)/g(x)$, there is the further restriction that $g(x) \neq 0$.

Definition of Sum, Difference, Product, and Quotient of Functions

Let f and g be two functions with overlapping domains. Then, for all x common to both domains, the sum, difference, product, and quotient of f and g are defined as follows.

1. **Sum:** $(f + g)(x) = f(x) + g(x)$
2. **Difference:** $(f - g)(x) = f(x) - g(x)$
3. **Product:** $(fg)(x) = f(x) \cdot g(x)$
4. **Quotient:** $\left(\frac{f}{g}\right)(x) = \frac{f(x)}{g(x)}, \quad g(x) \neq 0$

EXAMPLE 1 ■ **Finding the Sum and Difference of Two Functions**

Given $f(x) = 4x - 3$ and $g(x) = 2x^2 + x$, find the following.

(a) $(f + g)(x)$ (b) $(f - g)(x)$

Then evaluate each of these combinations when $x = 3$.

Solution

(a) The sum of the functions f and g is given by

$$\begin{aligned}(f + g)(x) &= f(x) + g(x)\\ &= (4x - 3) + (2x^2 + x)\\ &= 2x^2 + 5x - 3.\end{aligned}$$

When $x = 3$, the value of this sum is

$$(f + g)(3) = 2(3^2) + 5(3) - 3 = 30.$$

(b) The difference of the functions f and g is given by

$$\begin{aligned}(f - g)(x) &= f(x) - g(x)\\ &= (4x - 3) - (2x^2 + x)\\ &= -2x^2 + 3x - 3.\end{aligned}$$

When $x = 3$, the value of this difference is

$$(f - g)(3) = -2(3^2) + 3(3) - 3 = -12.$$ ■

In Example 1 both f and g have domains that consist of all real numbers. Thus, the domain of both $(f + g)$ and $(f - g)$ is also the set of all real numbers. In the next example, the domain of the sum of f and g is smaller than the domains of f and g.

EXAMPLE 2 ■ **Finding the Domain of the Sum of Two Functions**

Given the functions $f(x) = \sqrt{x + 1}$ and $g(x) = \sqrt{1 - x}$, find the sum of f and g. Then find the domain of $(f + g)$.

Solution

The sum of f and g is given by

$$\begin{aligned}(f + g)(x) &= f(x) + g(x)\\ &= \sqrt{x + 1} + \sqrt{1 - x}.\end{aligned}$$

Now, since f is defined for $x + 1 \geq 0$ and g is defined for $1 - x \geq 0$, it follows that the domains of f and g are as follows.

$$(\text{Domain of } f) = [-1, \infty)$$

$$(\text{Domain of } g) = (-\infty, 1]$$

The set that is common to both of these domains is the closed interval $[-1, 1]$, as shown in Figure 8.22. Therefore, the domain of $(f + g)$ is

$$(\text{Domain of } f + g) = [-1, 1].$$

FIGURE 8.22

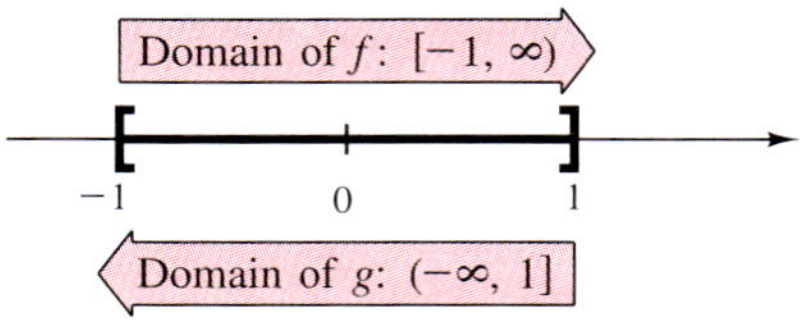

EXAMPLE 3 ■ Finding the Product of Two Functions

Given $f(x) = x^2$ and $g(x) = x - 3$, find the product of f and g. Then evaluate the product when $x = 4$.

Solution

The product of f and g is given by

$$\begin{aligned}(fg)(x) &= f(x)g(x)\\ &= (x^2)(x - 3)\\ &= x^3 - 3x^2.\end{aligned}$$

The value of this product when $x = 4$ is

$$(fg)(4) = 4^3 - 3(4^2) = 16.$$

EXAMPLE 4 ■ Finding the Quotient of Two Functions

Given $f(x) = x^2$ and $g(x) = x - 3$, find the quotient of f and g. Then evaluate the quotient when $x = 5$.

Solution

The quotient of f and g is given by

$$\begin{aligned}\left(\frac{f}{g}\right)(x) &= \frac{f(x)}{g(x)}\\ &= \frac{x^2}{x - 3}, \quad x \neq 3.\end{aligned}$$

(Note that the domain of the quotient has to be restricted to exclude the number 3.) The value of this quotient when $x = 5$ is

$$\left(\frac{f}{g}\right)(5) = \frac{5^2}{5 - 3} = \frac{25}{2}.$$

Composition of Functions

Another way of combining two functions is to form the **composition** of one with the other. For instance, if $f(x) = 2x^2$ and $g(x) = x - 1$, then the composition of f with g is given by

$$f(g(x)) = f(x - 1) = 2(x - 1)^2.$$

We denote this composition as $\boldsymbol{f \circ g}$.

Definition of Composition of Two Functions

The **composition** of the functions f and g is given by

$$(f \circ g)(x) = f(g(x)).$$

The domain of $(f \circ g)$ is the set of all x in the domain of g such that $g(x)$ is in the domain of f. (See Figure 8.23.)

FIGURE 8.23

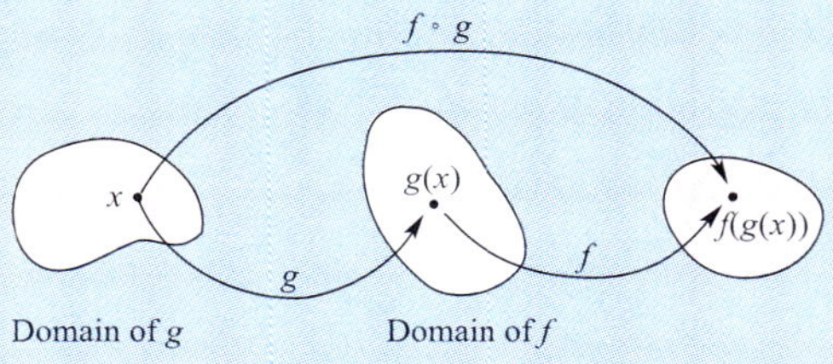

EXAMPLE 5 ■ Forming the Composition of Two Functions

Given $f(x) = 2x + 4$ and $g(x) = 3x - 1$, find the composition of f with g. Then find the value of the composition when $x = 1$ in two ways.

(a) $(f \circ g)(1)$ (b) $f(g(1))$

Solution

The composition of f with g is given by

$$\begin{aligned}(f \circ g)(x) &= f(g(x)) \\ &= f(3x - 1) \\ &= 2(3x - 1) + 4 \\ &= 6x - 2 + 4 \\ &= 6x + 2.\end{aligned}$$

To evaluate this composition when $x = 1$, we can use either of the following two methods. Note that we obtain the same value with either method.

(a) $(f \circ g)(1) = 6(1) + 2 = 8$

(b) $f(g(1)) = f(3(1) - 1) = f(2) = 2(2) + 4 = 8$ ■

The composition of f with g is generally *not* the same as the composition of g with f. This is illustrated in Example 6.

EXAMPLE 6 ■ **Comparing the Compositions of Functions**

Given $f(x) = 2x - 3$ and $g(x) = x^2 + 1$, find the following.

(a) $(f \circ g)(x)$ (b) $(g \circ f)(x)$

Solution

(a) The composition of f with g is as follows.

$$\begin{aligned} (f \circ g)(x) &= f(g(x)) && \text{Definition of } f \circ g \\ &= f(x^2 + 1) && \text{Definition of } g(x) \\ &= 2(x^2 + 1) - 3 && \text{Definition of } f(x) \\ &= 2x^2 + 2 - 3 \\ &= 2x^2 - 1 \end{aligned}$$

(b) The composition of g with f is as follows.

$$\begin{aligned} (g \circ f)(x) &= g(f(x)) && \text{Definition of } g \circ f \\ &= g(2x - 3) && \text{Definition of } f(x) \\ &= (2x - 3)^2 + 1 && \text{Definition of } g(x) \\ &= 4x^2 - 12x + 9 + 1 \\ &= 4x^2 - 12x + 10 \end{aligned}$$

Note that $(f \circ g)(x) \neq (g \circ f)(x)$. ■

EXAMPLE 7 ■ **Finding the Domain of the Composition of Two Functions**

Given $f(x) = x^2$ and $g(x) = \sqrt{x}$, find the composition of f with g, and find the domain of the composition.

Solution

The composition of f with g is given by

$$\begin{aligned} (f \circ g)(x) &= f(g(x)) \\ &= f\left(\sqrt{x}\right) \\ &= \left(\sqrt{x}\right)^2 \\ &= x, \quad x \geq 0. \end{aligned}$$

To find the domain of $(f \circ g)$, we note that the domain of g consists of all positive real numbers, $[0, \infty)$. Thus, we must restrict the composition of f with g to the same set, and conclude that

$$(\text{Domain of } f \circ g) = [0, \infty).$$ ■

DISCUSSION PROBLEM ■ Comparing the Compositions of Functions

We have seen that the composition of f with g is generally not the same as the composition of g with f. There are, however, special cases for which these two compositions are the same. For instance, let $f(x) = 2x$ and $g(x) = \frac{1}{2}x$. Show that for these two functions, the compositions $(f \circ g)$ and $(g \circ f)$ are the same. Can you find other examples of functions f and g such that the compositions $(f \circ g)$ and $(g \circ f)$ are the same? ■

Warm-Up

The following warm-up exercises involve skills that were covered in earlier sections. You will use these skills in the exercise set for this section.

In Exercises 1–4, find the domain of the given function.

1. $f(x) = x^3 + 3x^2 - 2$

2. $A(x) = \sqrt{10 - x}$

3. $H(x) = \dfrac{12}{x - 5}$

4. $g(x) = \dfrac{25}{x^2 - 25}$

In Exercises 5–10, sketch the graph of the given function.

5. $f(x) = \dfrac{1}{2}x$

6. $f(x) = -2x + 3$

7. $f(x) = x^2 - 4$

8. $f(x) = 2x^2 - x$

9. $f(x) = |x| + 1$

10. $f(x) = x^3 + 3$

8.3 EXERCISES

In Exercises 1–8, use the given functions f and g to find the combinations $(f + g)(x)$, $(f - g)(x)$, $(fg)(x)$, and $(f/g)(x)$. Find the domain of each combination.

1. $f(x) = 2x$
$g(x) = x^2$

2. $f(x) = x + 1$
$g(x) = \sqrt{x}$

3. $f(x) = 4x - 3$
$g(x) = x^2 - 9$

4. $f(x) = x^3$
$g(x) = x^2 + 1$

5. $f(x) = \dfrac{1}{x}$
$g(x) = 5x$

6. $f(x) = \dfrac{1}{x + 2}$
$g(x) = \dfrac{1}{x - 2}$

7. $f(x) = \sqrt{x + 4}$
$g(x) = \sqrt{4 - x}$

8. $f(x) = \sqrt{x}$
$g(x) = \sqrt{x^2 - 16}$

In Exercises 9–16, evaluate the indicated combination of the functions f and g at the given values of the independent variable. (If not possible, state the reason.)

9. $f(x) = x^2$, $g(x) = 2x + 3$
(a) $(f + g)(2)$ (b) $(f - g)(3)$

10. $f(x) = x^3$, $g(x) = x - 5$
(a) $(f + g)(-2)$ (b) $(f - g)(5)$

11. $f(x) = \sqrt{x}$, $g(x) = x^2 + 1$
(a) $(f + g)(4)$ (b) $(f - g)(9)$

12. $f(x) = \sqrt{x - 3}$, $g(x) = x^2 - 4$
(a) $(f + g)(7)$ (b) $(f - g)(1)$

13. $f(x) = |x|$, $g(x) = 5$
(a) $(fg)(-2)$ (b) $\left(\dfrac{f}{g}\right)(-2)$

14. $f(x) = |x - 3|$, $g(x) = (x - 2)^3$
(a) $(fg)(3)$ (b) $\left(\dfrac{f}{g}\right)\left(\dfrac{5}{2}\right)$

15. $f(x) = \dfrac{1}{x - 4}$, $g(x) = x$
(a) $(fg)(-2)$ (b) $\left(\dfrac{f}{g}\right)(4)$

16. $f(x) = \dfrac{1}{x^2}$, $g(x) = \dfrac{1}{x + 1}$
(a) $(fg)\left(\dfrac{1}{2}\right)$ (b) $\left(\dfrac{f}{g}\right)(0)$

In Exercises 17–24, find the indicated composite function.

17. $f(x) = x - 3$, $g(x) = x^2$ (a) $(f \circ g)(4)$ (b) $(g \circ f)(7)$ (c) $(f \circ g)(x)$ (d) $(g \circ f)(x)$

18. $f(x) = x + 5$, $g(x) = x^3$ (a) $(f \circ g)(2)$ (b) $(g \circ f)(-3)$ (c) $(f \circ g)(x)$ (d) $(g \circ f)(x)$

19. $f(x) = |x - 3|$, $g(x) = 3x$ (a) $(f \circ g)(1)$ (b) $(g \circ f)(2)$ (c) $(f \circ g)(x)$ (d) $(g \circ f)(x)$

20. $f(x) = |x|$, $g(x) = 2x + 5$ (a) $(f \circ g)(-2)$ (b) $(g \circ f)(-4)$ (c) $(f \circ g)(x)$ (d) $(g \circ f)(x)$

21. $f(x) = \sqrt{x}$, $g(x) = x + 5$ (a) $(f \circ g)(4)$ (b) $(g \circ f)(9)$ (c) $(f \circ g)(x)$ (d) $(g \circ f)(x)$

22. $f(x) = \sqrt{x + 6}$, $g(x) = 2x - 3$ (a) $(f \circ g)(3)$ (b) $(g \circ f)(-2)$ (c) $(f \circ g)(x)$ (d) $(g \circ f)(x)$

23. $f(x) = \dfrac{1}{x - 3}$, $g(x) = \sqrt{x}$ (a) $(f \circ g)(49)$ (b) $(g \circ f)(12)$ (c) $(f \circ g)(x)$ (d) $(g \circ f)(x)$

24. $f(x) = \dfrac{4}{x^2 - 4}$, $g(x) = \dfrac{1}{x}$ (a) $(f \circ g)(-2)$ (b) $(g \circ f)(1)$ (c) $(f \circ g)(x)$ (d) $(g \circ f)(x)$

In Exercises 25–30, find the domain of the composition (a) $f \circ g$, and (b) $g \circ f$.

25. $f(x) = x^2 + 1$
$g(x) = 2x$

26. $f(x) = 2 - 3x$
$g(x) = 5x + 3$

27. $f(x) = \sqrt{x}$
$g(x) = x - 2$

28. $f(x) = \sqrt{x - 5}$
$g(x) = x^2$

29. $f(x) = \dfrac{9}{x + 9}$
$g(x) = x^2$

30. $f(x) = \dfrac{x}{x - 4}$
$g(x) = \sqrt{x}$

In Exercises 31–40, find the indicated combination of the functions $f(x) = x^2 - 3x$ and $g(x) = 5x + 3$.

31. $f(t)$

32. $g(x + 2)$

33. $(f + g)(t - 2)$

34. $(fg)(z)$

35. $\dfrac{g(x + h) - g(x)}{h}$

36. $\dfrac{f(x + h) - f(x)}{h}$

37. $(f \circ g)(-1)$

38. $(g \circ f)(3)$

39. $(g \circ f)(y)$

40. $(f \circ g)(z)$

In Exercises 41–46, find the indicated combination of the following functions.

$$f = \{(-2, 3), (-1, 1), (0, 0), (1, -1), (2, -3)\}$$
$$g = \{(-3, 1), (-1, -2), (0, 2), (2, 2), (3, 1)\}$$

41. $(f - g)(0)$

42. $(f + g)(-1)$

43. $\left(\dfrac{f}{g}\right)(0)$

44. $(fg)(2)$

45. $(f \circ g)(-3)$

46. $(g \circ f)(1)$

In Exercises 47–50, sketch the graph of the functions f, g, and $f + g$ on the same set of coordinate axes.

47. $f(x) = \frac{1}{2}x$
$g(x) = x - 1$

48. $f(x) = \frac{1}{3}x$
$g(x) = -x + 4$

49. $f(x) = x^2$
$g(x) = -2x$

50. $f(x) = 4 - x^2$
$g(x) = x$

51. *Stopping Time* While traveling in a car at x miles per hour, you are required to make a panic stop in order to avoid an accident. The distance (in feet) the car travels during your reaction time is given by $R(x) = \frac{3}{4}x$. The distance traveled while braking is given by $B(x) = \frac{1}{15}x^2$. Find the function giving the total stopping distance T (in feet). Graph the functions R, B, and T on the same set of coordinate axes for $0 \le x \le 60$.

In Exercises 52 and 53, use a graphing calculator to sketch the graphs of the functions f, g, and $f + g$ on the same set of coordinate axes.* Observe from the graphs which of the two functions is most significant in determining the magnitude of the sum for different values of x. Use the given range settings.

52. $f(x) = x^3$
$g(x) = 3(1 - x^2)$

RANGE
Xmin=−2
Xmax=3
Xscl=1
Ymin=−3
Ymax=3
Yscl=1
Xres=1

53. $f(x) = x^3$
$g(x) = -3x$

RANGE
Xmin=−3
Xmax=3
Xscl=1
Ymin=−4
Ymax=4
Yscl=1
Xres=1

SECTION **8.4**

Inverse Functions

The Inverse of a Function • *Finding the Inverse of a Function* • *The Graph of the Inverse of a Function*

The Inverse of a Function

When we introduced the concept of a function in Section 8.1, we noted that one way to represent a function is by a set of ordered pairs. For instance, the function $f(x) = x + 2$ from the set $A = \{1, 2, 3, 4\}$ to the set $B = \{3, 4, 5, 6\}$ can be written as follows.

$$f(x) = x + 2: \{(1, 3), (2, 4), (3, 5), (4, 6)\}$$

By interchanging the first and second coordinates of each of these ordered pairs, we can form another function that we call the **inverse function** of f. We denote this function by f^{-1}. It is a function from the set B to the set A, and can be written as follows.

$$f^{-1}(x) = x - 2: \{(3, 1), (4, 2), (5, 3), (6, 4)\}$$

*The graphing calculator keystrokes given in this text correspond to the *TI-81* graphing calculator from *Texas Instruments*. However, you can also use other graphing calculators and computer graphing software for these examples and exercises.

FIGURE 8.24

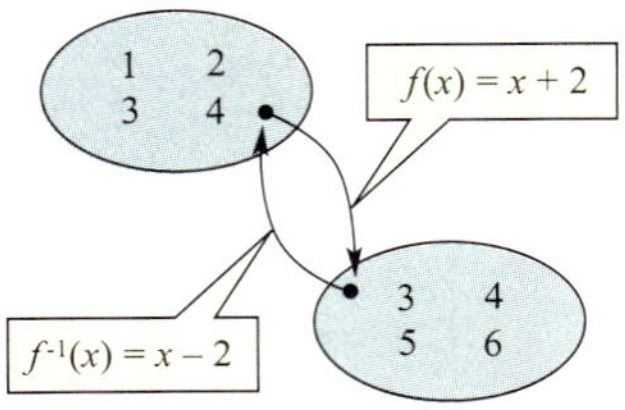

Note that the domain of f is equal to the range of f^{-1}, and vice versa, as shown in Figure 8.24. Also note that the functions f and f^{-1} have the effect of "undoing" each other. In other words, when we form the composition of f with f^{-1}, or the composition of f^{-1} with f, we obtain the identity function, as follows.

$$f(f^{-1}(x)) = f(x - 2) = (x - 2) + 2 = x$$
$$f^{-1}(f(x)) = f^{-1}(x + 2) = (x + 2) - 2 = x$$

NOTE: Don't be confused by the use of -1 to denote the inverse function f^{-1}. Whenever we write f^{-1}, we will *always* be referring to the inverse of the function f and *not* to the reciprocal of $f(x)$.

EXAMPLE 1 ■ Finding Inverse Functions Informally

Find the inverse of the function $f(x) = 3x$ and verify that both $f(f^{-1}(x))$ and $f^{-1}(f(x))$ are equal to the identity function.

Solution

The given function is a function that *multiplies* each input by 3. To "undo" this function, we need to *divide* each input by 3. Thus, the inverse function of $f(x) = 3x$ is given by

$$f^{-1}(x) = \frac{x}{3}.$$

We can verify that both $f(f^{-1}(x))$ and $f^{-1}(f(x))$ are equal to the identity function, as follows.

$$f(f^{-1}(x)) = f\left(\frac{x}{3}\right) = 3\left(\frac{x}{3}\right) = x$$

$$f^{-1}(f(x)) = f^{-1}(3x) = \frac{3x}{3} = x$$ ■

EXAMPLE 2 ■ Finding Inverse Functions Informally

Find the inverse of the function $f(x) = x - 4$ and verify that both $f(f^{-1}(x))$ and $f^{-1}(f(x))$ are equal to the identity function.

Solution

The given function is a function that *subtracts* 4 from each input. To "undo" this function, we need to *add* 4 to each input. Thus, the inverse function of $f(x) = x - 4$ is given by

$$f^{-1}(x) = x + 4.$$

We can verify that both $f(f^{-1}(x))$ and $f^{-1}(f(x))$ are equal to the identity function, as follows.

$$f(f^{-1}(x)) = f(x + 4) = (x + 4) - 4 = x$$
$$f^{-1}(f(x)) = f^{-1}(x - 4) = (x - 4) + 4 = x$$

The formal definition of the inverse of a function is given as follows.

Definition of the Inverse of a Function

Let f and g be two functions such that

$$f(g(x)) = x \qquad \text{for every } x \text{ in the domain of } g$$

and

$$g(f(x)) = x \qquad \text{for every } x \text{ in the domain of } f.$$

Then we say that the function g is the **inverse** of the function f, and we denote g by f^{-1} (read "f-inverse"). Thus, $f(f^{-1}(x)) = x$ and $f^{-1}(f(x)) = x$. The domain of f must be equal to the range of f^{-1}, and vice versa.

Note from this definition that if the function g is the inverse of the function f, then it must also be true that the function f is the inverse of the function g. For this reason, we will sometimes refer to the functions f and g as being *inverses of each other*.

EXAMPLE 3 **Verifying Inverse Functions**

Show that the functions $f(x) = x^3 + 1$ and $g(x) = \sqrt[3]{x - 1}$ are inverses of each other.

Solution

We begin by noting that the domain and range of both functions is the entire set of real numbers. To show that f and g are inverses of each other, we need to show that $f(g(x)) = x$ and $g(f(x)) = x$, as follows.

$$\begin{aligned} f(g(x)) &= f\left(\sqrt[3]{x - 1}\right) \\ &= \left(\sqrt[3]{x - 1}\right)^3 + 1 \\ &= (x - 1) + 1 \\ &= x \end{aligned}$$

$$\begin{aligned} g(f(x)) &= g(x^3 + 1) \\ &= \sqrt[3]{(x^3 + 1) - 1} \\ &= \sqrt[3]{x^3} \\ &= x \end{aligned}$$

Note that you can see that the two functions f and g "undo" each other in the following verbal sense. The function f first cubes the input x and then adds 1, whereas the function g first subtracts 1, and then takes the cube root of the result.

EXAMPLE 4 ■ Verifying Inverse Functions

Which of the following functions is the inverse of the function $f(x) = 1/(5x)$?

$$g(x) = \frac{5}{x} \quad \text{or} \quad h(x) = \frac{1}{5x}$$

Solution

By forming the composition of f with g, we find that

$$\begin{aligned} f(g(x)) &= f\left(\frac{5}{x}\right) \\ &= \frac{1}{5(5/x)} \\ &= \frac{1}{25/x} \\ &= \frac{x}{25}. \end{aligned}$$

Because this composition does not yield x, we conclude that g is *not* the inverse of f. By forming the composition of f with h, we find that

$$\begin{aligned} f(h(x)) &= f\left(\frac{1}{5x}\right) \\ &= \frac{1}{5(1/5x)} \\ &= \frac{1}{1/x} \\ &= x. \end{aligned}$$

Thus, it appears that h is the inverse of f. We can confirm this by showing that the composition of h with f is also equal to the identity function. (Try doing that.) ■

Finding the Inverse of a Function

For simple functions (like the ones in Examples 1 and 2), we can find inverse functions by inspection. For instance, the inverse of $f(x) = 10x$ is equal to $f^{-1}(x) = x/10$. For more complicated functions, however, it is best to use the following steps for finding the inverse of a function. The key step in these guidelines is switching the roles of x and y. This step corresponds to the fact that inverse functions have ordered pairs with the coordinates reversed.

Finding the Inverse of a Function

To find the inverse of a function f, use the following steps.

1. In the equation for $f(x)$, replace $f(x)$ by y.
2. Interchange the roles of x and y.
3. If the new equation does not represent y as a function of x, then the function f does not have an inverse function. If the new equation does represent y as a function of x, then solve the new equation for y.
4. Replace y by $f^{-1}(x)$.
5. Verify that f and f^{-1} are inverses of each other by showing that $f(f^{-1}(x)) = x = f^{-1}(f(x))$.

EXAMPLE 5 ■ Finding the Inverse of a Function

Determine whether the following function has an inverse. If it does, find its inverse.

$$f(x) = 2x + 3$$

Solution

$f(x) = 2x + 3$	Given function
$y = 2x + 3$	Replace $f(x)$ by y
$x = 2y + 3$	Interchange x and y
$y = \dfrac{x - 3}{2}$	Solve for y
$f^{-1}(x) = \dfrac{x - 3}{2}$	Replace y by $f^{-1}(x)$

Thus, the inverse of $f(x) = 2x + 3$ is

$$f^{-1}(x) = \frac{x - 3}{2}.$$

We leave the verification that $f(f^{-1}(x)) = x$ and $f^{-1}(f(x)) = x$ to you. ■

Note in Step 3 of the guidelines for finding the inverse of a function that it is possible that a function has no inverse. We illustrate this possibility in Example 6.

EXAMPLE 6 ■ A Function that Has No Inverse

Determine whether the following function has an inverse. If it does, find its inverse.

$$f(x) = x^2$$

Solution

$f(x) = x^2$	Given function
$y = x^2$	Replace $f(x)$ by y
$x = y^2$	Interchange x and y

Because the equation $x = y^2$ does not represent y as a function of x, we conclude that the original function f does not have an inverse. ■

MATH MATTERS

Annual Salary Versus Hourly Wage

Here is a quick method you can use to approximate the annual salary that corresponds to a given hourly wage.

$$\text{Hourly wage} \times 2 \times 1{,}000 \approx \text{Annual salary}$$

For example, \$5 an hour is approximately \$10,000 a year, and \$12 an hour is approximately \$24,000. This approximation assumes that there are only fifty 40-hour weeks in a year. To obtain the exact conversion from an hourly wage to an annual salary, you should multiply by 40 (hours per week) and then multiply by 52 (weeks per year).

To approximate the hourly wage that corresponds to a given annual salary, use the following rule.

$$\text{Annual salary} \div 1{,}000 \div 2 \approx \text{Hourly wage}$$

For example, \$12,000 a year is approximately \$6 an hour, and \$20,000 a year is approximately \$10 an hour. To obtain the exact conversion from an annual salary to an hourly wage, you should divide by 52 (weeks per year) and then divide the result by 40 (hours per week).

Use the quick approximations technique to determine which of the following wages is greater.

(a) \$7.50 an hour or \$17,500 a year

(b) \$12.50 an hour or \$22,500 a year

The Graph of the Inverse of a Function

The graphs of a function f and its inverse f^{-1} are related to each other in the following way. If the point (a, b) lies on the graph of f, then the point (b, a) must lie on the graph of f^{-1}, and vice versa. This means that the graph of f^{-1} is a reflection of the graph of f in the line $y = x$, as shown in Figure 8.25.

FIGURE 8.25

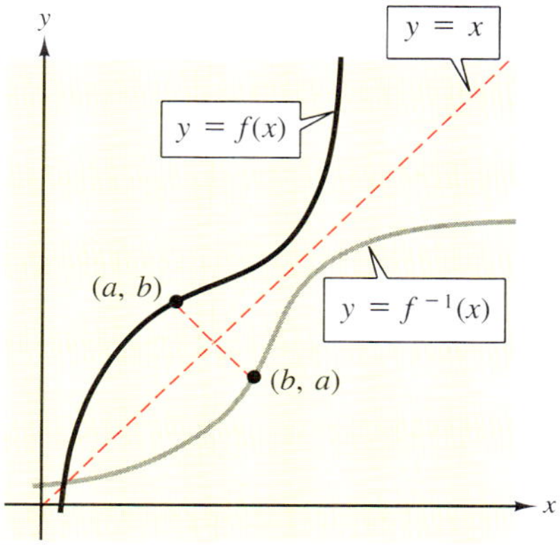

The graph of f^{-1} is a reflection of the graph of f in the line $y = x$

We illustrate this "reflective property" of the graphs of f and f^{-1} in Examples 7 and 8.

EXAMPLE 7 ■ The Graphs of f and f^{-1}

Sketch the graphs of the inverse functions $f(x) = 2x - 3$ and $f^{-1}(x) = \dfrac{x + 3}{2}$ in the same rectangular coordinate system, and show that the graphs are reflections of each other in the line $y = x$.

Solution

The graphs of f and f^{-1} are shown in Figure 8.26. Visually, it appears that the graphs are reflections of each other in the line $y = x$. We can further verify this reflective property by testing a few points on each graph. Note in the following list that if the point (a, b) is on the graph of f, then the point (b, a) is on the graph of f^{-1}.

$f(x) = 2x - 3$	$f^{-1}(x) = \frac{1}{2}(x + 3)$
$(-1, -5)$	$(-5, -1)$
$(0, -3)$	$(-3, 0)$
$(1, -1)$	$(-1, 1)$
$(2, 1)$	$(1, 2)$
$(3, 3)$	$(3, 3)$

FIGURE 8.26

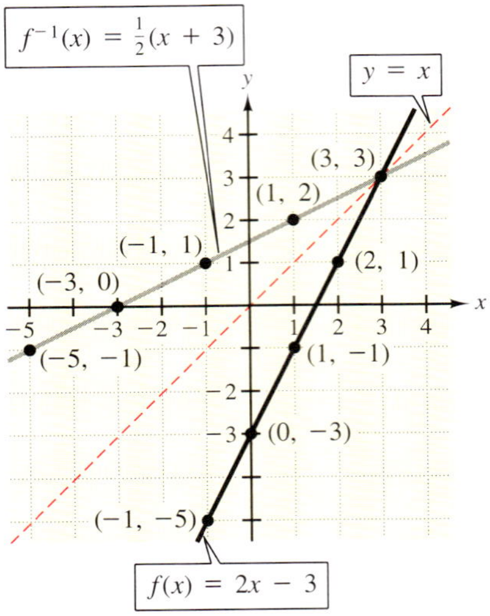

■

In Example 6 we noted that the function $f(x) = x^2$ has no inverse. What we really mean when we say this is that *assuming the domain of f is the entire real line*, the function $f(x) = x^2$ has no inverse. If, however, we restrict the domain of f to the positive real numbers, then f does have an inverse, as demonstrated in Example 8.

EXAMPLE 8 ■ The Graphs of f and f^{-1}

Sketch the graphs of the inverse functions

$$f(x) = x^2, \ x \geq 0, \quad \text{and} \quad f^{-1}(x) = \sqrt{x}$$

in the same rectangular coordinate system, and show that the graphs are reflections of each other in the line $y = x$.

Solution

The graphs of f and f^{-1} are shown in Figure 8.27. Visually, it appears that the graphs are reflections of each other in the line $y = x$. We can further verify this reflective property by testing a few points on each graph. Note in the following list that if the point (a, b) is on the graph of f, then the point (b, a) is on the graph of f^{-1}.

$f(x) = x^2, \ x \geq 0$	$f^{-1}(x) = \sqrt{x}$
(0, 0)	(0, 0)
(1, 1)	(1, 1)
(2, 4)	(4, 2)
(3, 9)	(9, 3)

FIGURE 8.27

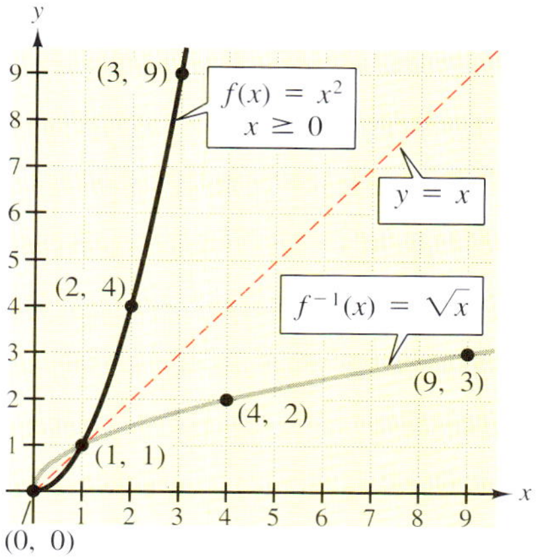

In our guidelines for finding the inverse of a function, we included an *algebraic* test for determining whether a function has an inverse. The reflective property of the graphs of inverse functions gives us a nice *geometric* test for determining whether a function has an inverse. We call this the **horizontal line test** for inverse functions.

Horizontal Line Test for Inverse Functions

A function f has an inverse function if and only if no *horizontal* line intersects the graph of f at more than one point.

We demonstrate the use of the horizontal line test for inverse functions in Example 9.

EXAMPLE 9 Applying the Horizontal Line Test for Inverse Functions

(a) The graph of the function $f(x) = x^3 - 1$ is shown in Figure 8.28. Because no horizontal line intersects the graph of f at more than one point, we conclude that f does possess an inverse function.

(b) The graph of the function $f(x) = x^2 - 1$ is shown in Figure 8.29. Because it is possible to find a horizontal line that intersects the graph of f at more than one point, we conclude that f does *not* possess an inverse function.

FIGURE 8.28

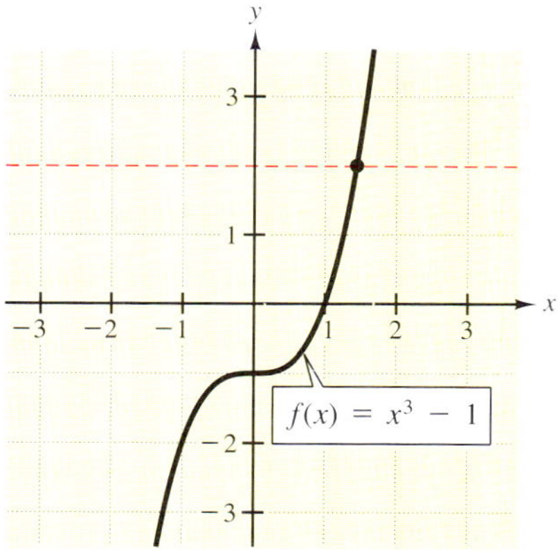

FIGURE 8.29

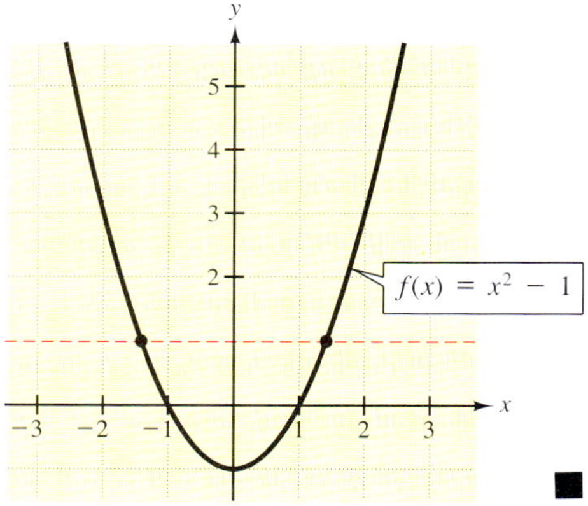

DISCUSSION PROBLEM ■ The Existence of an Inverse Function

Write a short paragraph describing why the following functions do or do not possess inverse functions.

(a) Let x represent the retail price of an item (in dollars), and let $f(x)$ represent the sales tax on the item. Assume that the sales tax is 6% of the retail price *and* that the sales tax is rounded to the nearest cent. Does this function possess an inverse? (*Hint:* Can you undo this function? For instance, if you know that the sales tax is \$0.12, can you determine *exactly* what the retail price is?)

(b) Let x represent the temperature in degrees Celsius, and $f(x)$ represent the temperature in degrees Fahrenheit. Does this function possess an inverse? (*Hint:* The formula for converting from degrees Celsius to degrees Fahrenheit is $F = \frac{9}{5}C + 32$.) ■

Warm-Up

The following warm-up exercises involve skills that were covered in earlier sections. You will use these skills in the exercise set for this section.

In Exercises 1–4, find the domain of the function.

1. $f(x) = x^3 - 2x$

2. $g(x) = \sqrt[3]{x}$

3. $h(x) = \sqrt{16 - x^2}$

4. $A(x) = \dfrac{3}{36 - x^2}$

In Exercises 5–10, use the given functions f and g to find the combinations $(f + g)(x)$, $(f - g)(x)$, $(fg)(x)$, $(f/g)(x)$, $(f \circ g)(x)$, and $(g \circ f)(x)$. Find the domain of each combination.

5. $f(x) = x^2 - 9$
$g(x) = 3x$

6. $f(x) = 2x - 3$
$g(x) = \frac{1}{2}x^2$

7. $f(x) = \sqrt[3]{x - 2}$
$g(x) = x^2$

8. $f(x) = \sqrt{x}$
$g(x) = x^2 - 4$

9. $f(x) = \dfrac{1}{x}$
$g(x) = \dfrac{1}{x - 2}$

10. $f(x) = x + 2$
$g(x) = \dfrac{1}{x + 2}$

8.4 EXERCISES

In Exercises 1–6, find the inverse of the function f informally. Verify that $f(f^{-1}(x))$ and $f^{-1}(f(x))$ are equal to the identity function.

1. $f(x) = 5x$

2. $f(x) = \frac{1}{3}x$

3. $f(x) = x + 10$

4. $f(x) = x - 5$

5. $f(x) = \sqrt[3]{x}$

6. $f(x) = x^5$

In Exercises 7–14, verify algebraically that the functions f and g are inverses of each other.

7. $f(x) = 10x$
$g(x) = \frac{1}{10}x$

8. $f(x) = 2x - 1$
$g(x) = \frac{1}{2}(x + 1)$

9. $f(x) = 2 - 3x$
$g(x) = \frac{1}{3}(2 - x)$

10. $f(x) = -\frac{1}{4}(x + 3)$
$g(x) = -4(x - 3)$

11. $f(x) = \sqrt[3]{x + 1}$
$g(x) = x^3 - 1$

12. $f(x) = x^5$
$g(x) = \sqrt[5]{x}$

13. $f(x) = \dfrac{1}{x}$
$g(x) = \dfrac{1}{x}$

14. $f(x) = \dfrac{1}{x - 3}$
$g(x) = 3 + \dfrac{1}{x}$

In Exercises 15–20, verify geometrically (as in Example 7) that the functions f and g are inverses of each other.

15. $f(x) = \frac{1}{3}x$
$g(x) = 3x$

16. $f(x) = \frac{1}{5}x - 1$
$g(x) = 5x + 5$

17. $f(x) = \sqrt{x + 1}$
$g(x) = x^2 - 1, \quad x \geq 0$

18. $f(x) = \sqrt{4 - x}$
$g(x) = 4 - x^2, \quad x \geq 0$

19. $f(x) = \frac{1}{8}x^3$
$g(x) = 2\sqrt[3]{x}$

20. $f(x) = \sqrt[3]{x + 2}$
$g(x) = x^3 - 2$

In Exercises 21–30, find the inverse of the given function.

21. $f(x) = 8x$

22. $f(x) = 7 - x$

23. $g(x) = 3 - 4x$

24. $g(x) = 6x + 1$

25. $h(x) = \sqrt{x}$

26. $h(x) = \sqrt{x + 5}$

27. $f(t) = t^3 - 1$

28. $h(t) = t^5$

29. $g(s) = \dfrac{5}{s}$

30. $f(s) = \dfrac{2}{3 - s}$

In Exercises 31–40, determine whether or not the given function has an inverse.

31. $f(x) = 9 - x^2$

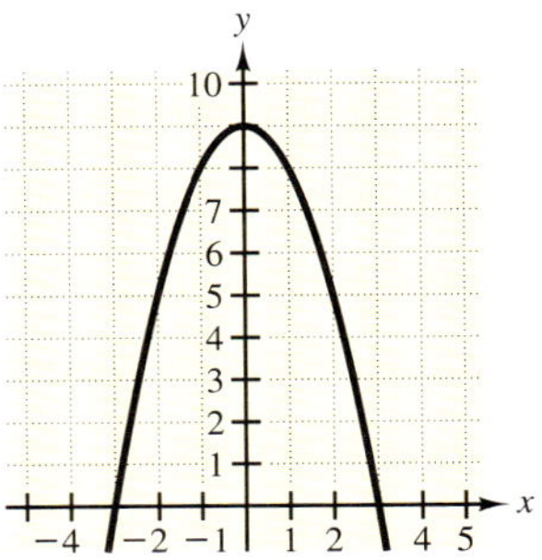

32. $f(x) = \frac{1}{5}x$

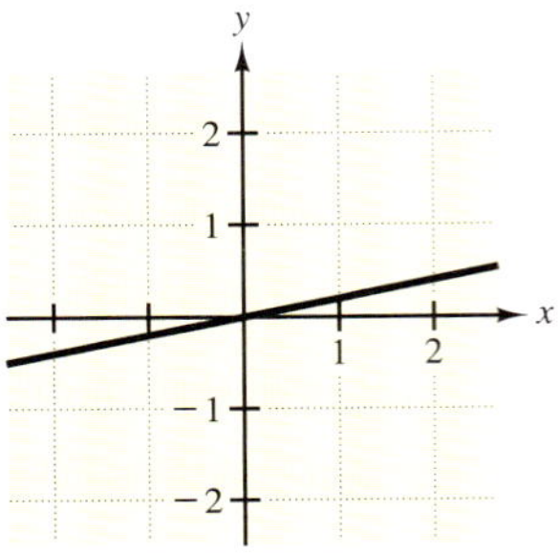

33. $f(x) = \frac{1}{4}x^3$

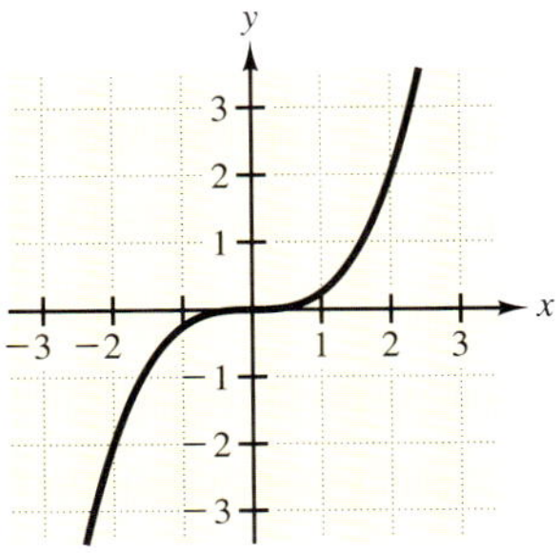

34. $f(x) = x^2 - 2$

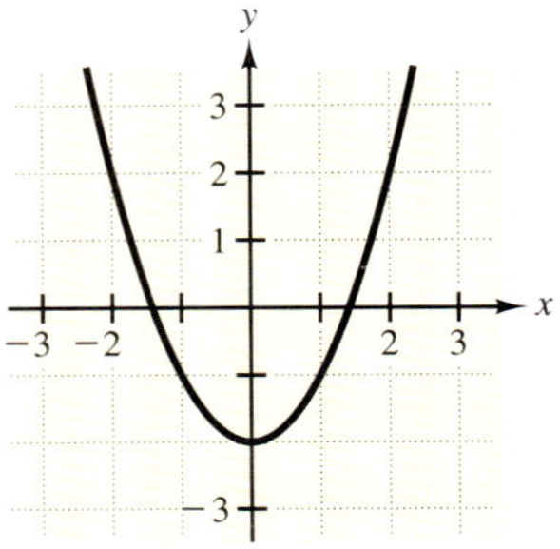

35. $g(x) = \sqrt{25 - x^2}$

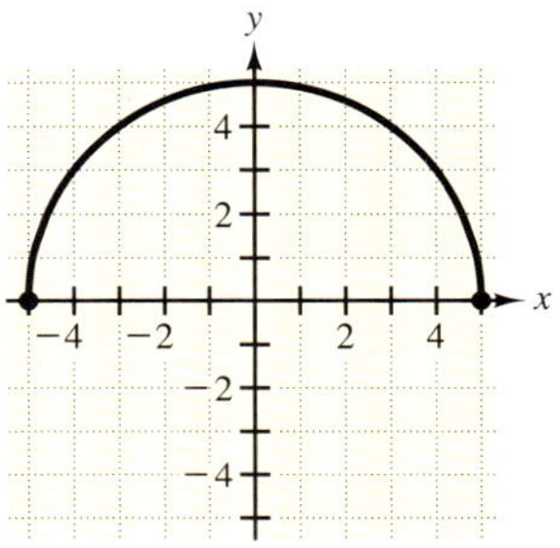

36. $g(x) = |x - 4|$

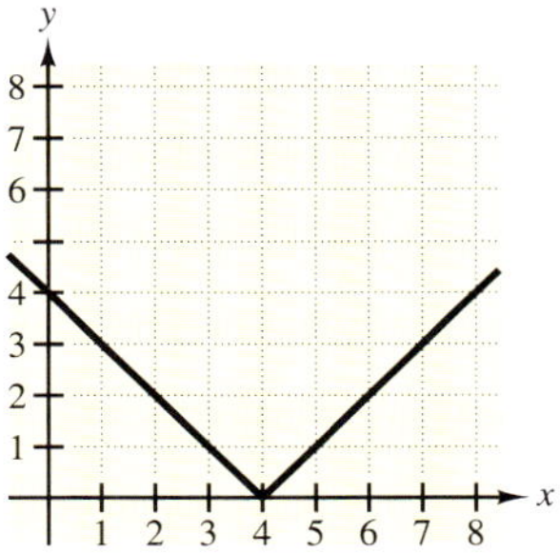

37. $f(t) = \sqrt[3]{5 - x}$

38. $h(t) = 4 - \sqrt[3]{x}$

39. $g(x) = x^4$

40. $f(x) = (x + 2)^5$

In Exercises 41–44, place a restriction on the domain of the given function so that the graph of the restricted function satisfies the horizontal line test for inverses. Find the inverse of the restricted function. (*Note:* There is more than one correct answer to each problem.)

41. $f(x) = x^4$

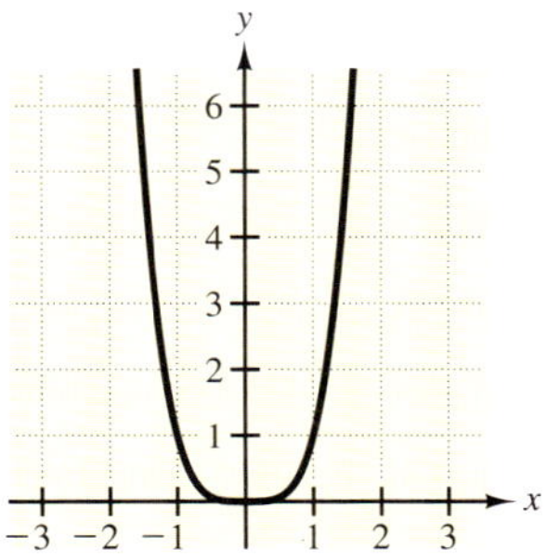

42. $f(x) = 9 - x^2$

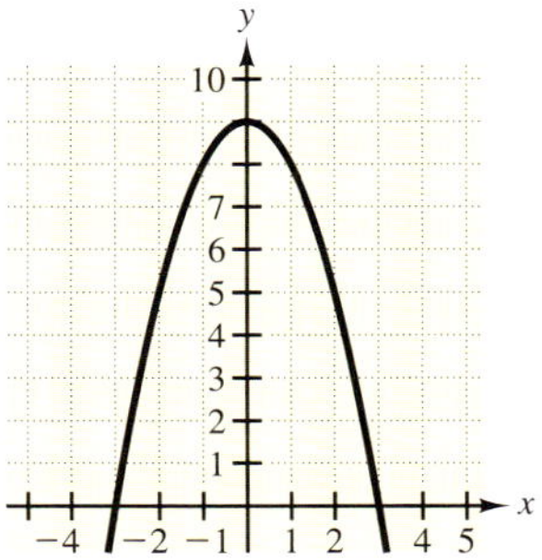

43. $f(x) = (x - 2)^2$

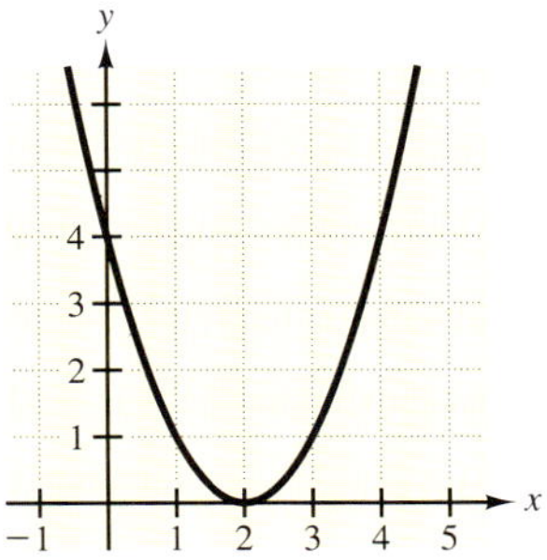

44. $f(x) = |x - 2|$

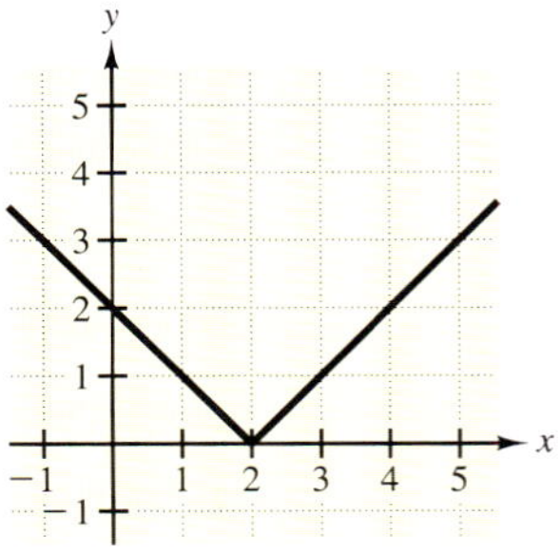

In Exercises 45–48, use the graph of f to complete the table and sketch the graph of f^{-1}.

45.

x	0	1	3	4
f^{-1}				

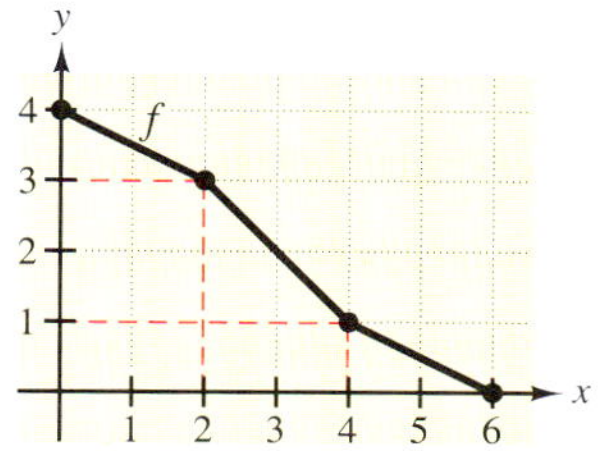

46.

x	−1	0	1	5
f^{-1}				

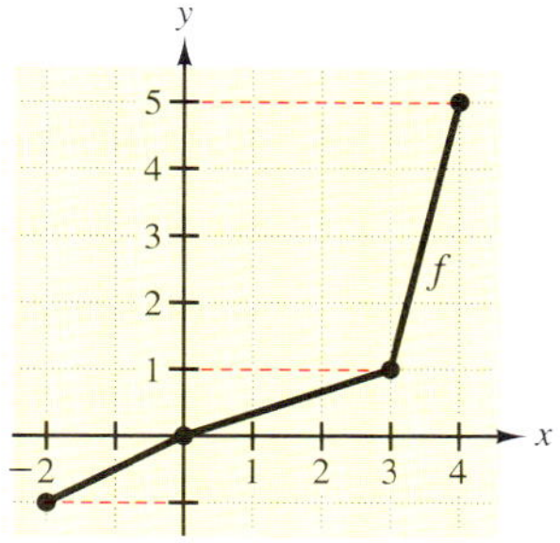

47.

x	−4	−2	2	3
f^{-1}				

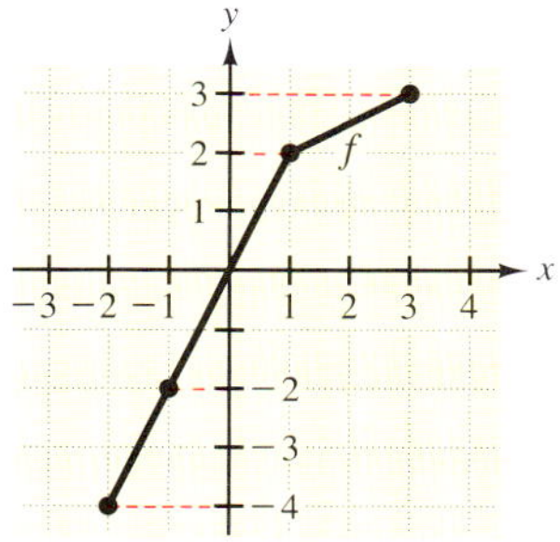

48.

x	−3	−2	0	6
f^{-1}				

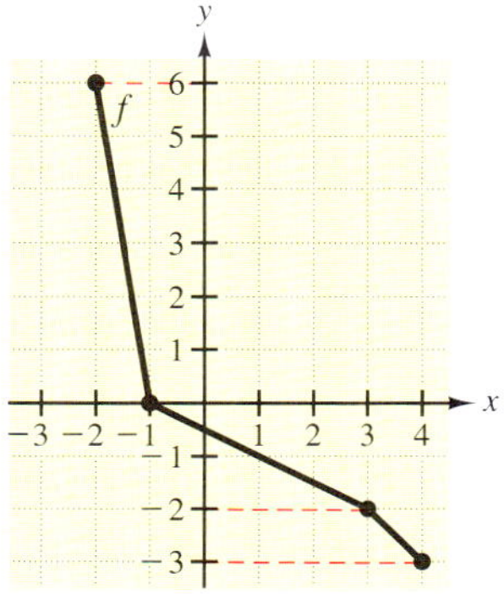

49. Consider the function $f(x) = 3 - 2x$.
(a) Find $f^{-1}(x)$.
(b) Find $(f^{-1})^{-1}(x)$.

50. Consider the functions $f(x) = 4x$ and $g(x) = x + 6$.
(a) Find $(f \circ g)(x)$.
(b) Find $(f \circ g)^{-1}(x)$.
(c) Find $f^{-1}(x)$ and $g^{-1}(x)$.
(d) Find $(g^{-1} \circ f^{-1})(x)$ and compare the result with that of part (b).

In Exercises 51 and 52, find the inverse of the function f. Use a graphing calculator to sketch the graphs of f and f^{-1} on the same screen using the given range settings.*

51. $f(x) = x^3 + 1$

RANGE
Xmin=−2
Xmax=4
Xscl=1
Ymin=−2
Ymax=4
Yscl=1
Xres=1

*The graphing calculator keystrokes given in this text correspond to the *TI-81* graphing calculator from *Texas Instruments*. However, you can also use other graphing calculators and computer graphing software for these examples and exercises.

52. $f(x) = \sqrt{x^2 - 4}, \quad x \geq 2$

RANGE
Xmin=0
Xmax=6
Xscl=1
Ymin=0
Ymax=6
Yscl=1
Xres=1

53. *Hourly Wage* Your wage is \$9.00 per hour plus \$0.65 for each unit produced per hour. Thus, your hourly wage y in terms of the number of units produced is given by $y = 9 + 0.65x$.

(a) Determine the inverse of the function.

(b) What does each variable represent in the inverse function?

(c) Determine the number of units produced when your hourly wage averages \$14.20.

SECTION 8.5 Variation and Mathematical Models

Direct Variation • Inverse Variation • Joint Variation

Direct Variation

One of the goals of applied mathematics is to find equations, which we call **mathematical models**, that describe real-world phenomena. In this section, we continue our study of methods for creating mathematical models by looking at models that are related to the concept of **variation**.

Direct Variation

The following statements are equivalent.

1. y **varies directly** as x.
2. y is **directly proportional** to x.
3. $y = kx$ for some constant k.

We call the number k the **constant of proportionality**.

In the mathematical model for direct variation, y is a *linear* function of x. That is,

$$y = kx.$$

To use this mathematical model in applications involving direct variation, we are usually given specific values of x and y, which then enables us to find the value for the constant k. This procedure is demonstrated in Examples 1 and 2.

EXAMPLE 1 ■ Direct Variation

Assume that the total revenue R (in dollars) obtained from selling x units of a given product is directly proportional to the number of units sold. When 10,000 units are sold, the total revenue is \$142,500.

(a) Find a mathematical model that relates the total revenue R to the number of units sold x.

(b) Find the total revenue obtained from selling 12,000 units.

Solution

(a) Since the total revenue is directly proportional to the number of units sold, we have the model

$$R = kx.$$

To find the value of the constant k, we use the fact that $R = 142{,}500$ when $x = 10{,}000$. Substituting these values into the given model produces

$$\underset{\overset{\downarrow}{R}}{142{,}500} = k(\underset{\overset{\downarrow}{x}}{10{,}000})$$

which implies that $k = 142{,}500/10{,}000 = 14.25$. Thus, the equation relating the total revenue to the total number of units sold is

$$R = 14.25x.$$

(b) When $x = 12{,}000$, the total revenue is

$$R = 14.25(12{,}000) = \$171{,}000.$$ ■

EXAMPLE 2 ■ Direct Variation

Hooke's Law for a spring states that the distance a spring is stretched (or compressed) is proportional to the force on the spring. A force of 20 pounds stretches the spring 5 inches.

(a) Find a mathematical model that relates the distance the spring is stretched to the force applied to the spring.

(b) How far will a force of 30 pounds stretch the spring?

Solution

(a) For this problem, we let d represent the distance (in inches) that the spring is stretched, and we let F represent the force (in pounds) that is applied to the spring. Since the distance d is proportional to the force F, we have the model

$$d = kF.$$

To find the value of the constant k, we use the fact that $d = 5$ when $F = 20$. Substituting these values into the given model produces

$$\overset{d}{\downarrow}\; \qquad \overset{F}{\downarrow}$$
$$5 = k(20)$$

which implies that $k = \frac{5}{20} = \frac{1}{4}$. Thus, the equation relating distance and force is

$$d = \frac{1}{4}F.$$

(b) When $F = 30$, the distance is

$$d = \frac{1}{4}(30) = 7.5 \text{ inches.}$$

(See Figure 8.30.)

FIGURE 8.30

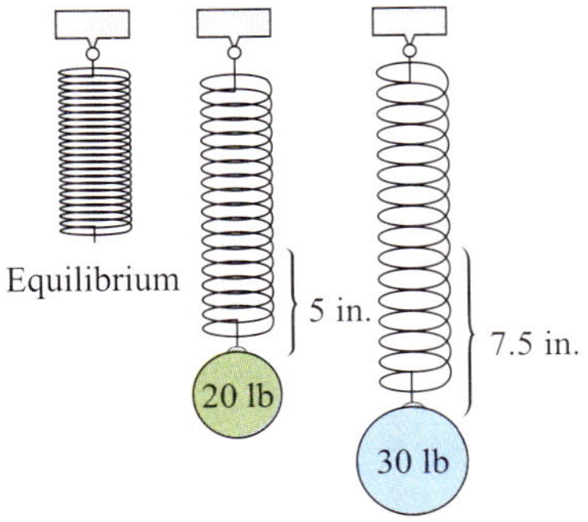

In Examples 1 and 2, the direct variations were such that an *increase* in one variable corresponded to an *increase* in the other variable. For instance, in the model $d = \frac{1}{4}F$, if the force F increases, then the distance d also increases. There are, however, other applications of direct variation in which an increase in one variable corresponds to a *decrease* in the other variable. For instance, in the model $y = -2x$, an increase in x will yield a decrease in y.

A second type of direct variation relates one variable to a *power* of another variable.

Direct Variation as *n*th Power

The following statements are equivalent.

1. y **varies directly as the *n*th power** of x.
2. y is **directly proportional to the *n*th power** of x.
3. $y = kx^n$ for some constant k.

EXAMPLE 3 ■ Direct Variation as a Power

The distance a ball rolls down an inclined plane is directly proportional to the square of the time it rolls. During the first second the ball rolls six feet.

(a) Find a mathematical model that relates the distance traveled to the time.

(b) How far will the ball roll during the first two seconds?

Solution

(a) Letting d be the distance (in feet) that the ball rolls and t be the time (in seconds), we obtain the model

$$d = kt^2.$$

Because $d = 6$ when $t = 1$, it follows that $k = 6$. Therefore, the equation relating distance to time is

$$d = 6t^2.$$

(b) When $t = 2$, the distance traveled is

$$d = 6(2^2) = 6(4) = 24 \text{ feet}.$$

(See Figure 8.31.)

FIGURE 8.31

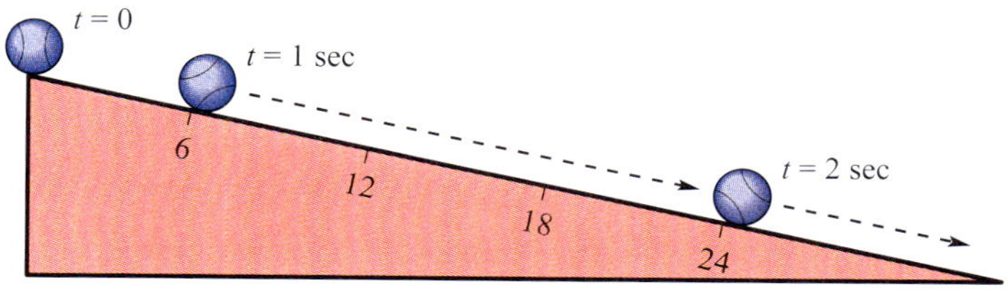

■

Inverse Variation

A third type of variation is called **inverse variation**. For this type of variation, we say that one of the variables is inversely proportional to the other variable.

Inverse Variation

The following statements are equivalent.

1. y **varies inversely** as x.
2. y is **inversely proportional** to x.
3. $y = \dfrac{k}{x}$ for some constant k.

If x and y are related by an equation of the form $y = k/x^n$, then we say that y varies inversely as the nth power of x (or y is inversely proportional to the nth power of x).

EXAMPLE 4 ■ Inverse Variation

The marketing department of a large company has found that the demand for one of its products varies inversely as the price of the product. (When the price is low, more people are willing to buy the product than when the price is high.) When the price of the product is \$7.50, the monthly demand is 50,000 units. Approximate the monthly demand if the price is reduced to \$6.00.

Solution

We let x represent the number of units that are sold each month (the demand), and let p represent the price per unit (in dollars). Because the demand is inversely proportional to the price, we obtain the model

$$x = \frac{k}{p}.$$

By substituting $x = 50{,}000$ when $p = 7.50$, we determine that $k = (7.5)(50{,}000) = 375{,}000$. Thus, the model is

$$x = \frac{375{,}000}{p}.$$

To find the demand that corresponds to a price of \$6.00, we substitute $p = 6$ into the equation and obtain a demand of

$$x = \frac{375{,}000}{6} = 62{,}500 \text{ units}.$$

Thus, if the price were lowered from \$7.50 per unit to \$6.00 per unit, we could expect the monthly demand to increase from 50,000 units to 62,500 units. ■

Some applications of variation involve problems with *both* direct and inverse variation in the same model. For instance, in the model

$$z = k\left(\frac{x}{y}\right)$$

we say that z is directly proportional to x *and* inversely proportional to y. We demonstrate this type of model in Example 5.

EXAMPLE 5 ■ Direct and Inverse Variation

A company determines that the demand for one of its products is directly proportional to the amount spent on advertising and inversely proportional to the price of the product. When \$40,000 is spent on advertising and the price per unit is \$20, the monthly demand is 10,000 units.

(a) If the amount of advertising was increased to \$50,000, how much could the price be increased to maintain a monthly demand of 10,000 units?

(b) If you were in charge of the advertising department, would you recommend this increased expense in advertising?

Solution

(a) We let x represent the number of units that are sold each month (the demand), a represent the amount spent on advertising (in dollars), and p represent the price per unit (in dollars). Because the demand is directly proportional to the advertising and inversely proportional to the price, we obtain the model

$$x = \frac{ka}{p}.$$

By substituting $x = 10{,}000$ when $a = 40{,}000$ and $p = 20$, we can determine that $k = (10{,}000)(20)/(40{,}000) = 5$. Thus, the model is

$$x = \frac{5a}{p}.$$

To find the price that corresponds to a demand of 10,000 and an advertising expense of \$50,000, we substitute $x = 10{,}000$ and $a = 50{,}000$ into the model and solve for p.

$$10{,}000 = \frac{5(50{,}000)}{p} \quad \Rightarrow \quad p = \frac{5(50{,}000)}{10{,}000} = \$25$$

(b) The total revenue for selling 10,000 units at \$20 is \$200,000, and for selling 10,000 units at \$25 is \$250,000. Thus, by increasing the advertising expense from \$40,000 to \$50,000, the company could increase its revenue by \$50,000. This implies that you should recommend the increased expense in advertising.

Amount of Advertising	*Price*	*Revenue*
\$40,000	\$20	$10{,}000 \times 20 = \$200{,}000$
\$50,000	\$25	$10{,}000 \times 25 = \$250{,}000$

■

Joint Variation

The model used in Example 5 involved both direct and inverse variation, and we used the word *and* to couple the two types of variation together. To describe two different *direct* variations in the same statement, we use the word *jointly*. For instance, the model

$$z = kxy$$

can be described by saying that z is *jointly* proportional to x and to y.

Joint Variation

The following statements are equivalent.

1. z **varies jointly** as x and y.
2. z is **jointly proportional** to x and to y.
3. $z = kxy$ for some constant k.

If x, y, and z are related by the equation $z = kx^n y^m$, then we say that z varies jointly as the nth power of x and the mth power of y.

EXAMPLE 6 Joint Variation

The *simple interest* for a certain savings account is jointly proportional to the time and the principal. After one quarter (three months), the interest for a principal of \$6,000 is \$120. How much interest would a principal of \$7,500 earn in five months?

Solution

To begin, we let I represent the interest earned (in dollars), P represent the principal (in dollars), and t represent the time (in years). Then, since the interest is jointly proportional to the time and the principal, we have the following model.

$$I = ktP$$

Because $I = 120$ when $P = 6{,}000$ and $t = \frac{1}{4}$, we find that $k = 120/\left(6{,}000 \cdot \frac{1}{4}\right) = 0.08$. Therefore, the model that relates the interest with the time and principal is

$$I = 0.08tP.$$

To find the interest earned for a principal of \$7,500 over a five-month period of time, we substitute $P = 7{,}500$ and $t = \frac{5}{12}$ into the model and obtain an interest of

$$I = 0.08\left(\frac{5}{12}\right)(7{,}500) = \$250.$$

DISCUSSION PROBLEM ■ You Be the Instructor

Suppose you are teaching an algebra class and are writing a test that covers this section. Write two test questions that you think give a fair representation of the material in this section. (Assume that your students can spend 10 minutes on each question.) ■

Warm-Up

The following warm-up exercises involve skills that were covered in earlier sections. You will use these skills in the exercise set for this section.

In Exercises 1–6, solve the given equation for k.

1. $4 = k(12)$ **2.** $250 = k(75)$ **3.** $46 = \dfrac{k}{0.02}$

4. $19 = \dfrac{k}{27}$ **5.** $360 = \dfrac{k(72)}{8}$ **6.** $700 = \dfrac{k(49)}{2^2}$

In Exercises 7–10, sketch the graph of $y = kx^2$ for the given value of k.

7. $k = \dfrac{1}{4}$ **8.** $k = 2$ **9.** $k = -4$ **10.** $k = -\dfrac{1}{2}$

8.5 EXERCISES

In Exercises 1–12, write a mathematical model for the given statement.

1. I varies directly as V.

2. C varies directly as r.

3. u is directly proportional to the square of v.

4. V varies directly as the cube of x.

5. p varies inversely as d.

6. S is inversely proportional to the square of v.

7. P is inversely proportional to the square root of $1 + r$.

8. A varies inversely as the fourth power of t.

9. A varies jointly as l and w.

10. V varies jointly as h and the square of r.

11. *Boyle's Law* If the temperature of a gas is not allowed to change, its absolute pressure P is inversely proportional to its volume V.

12. *Newton's Law of Universal Gravitation* The gravitational attraction F between two particles of masses m_1 and m_2 is proportional to the product of the masses and inversely proportional to the square of the distance r between the particles.

In Exercises 13–24, find the constant of proportionality and give the equation relating the variables.

13. s varies directly as t, and $s = 20$ when $t = 4$.

14. h is directly proportional to r, and $h = 28$ when $r = 12$.

15. F is directly proportional to the square of x, and $F = 500$ when $x = 40$.

16. v varies directly as the square root of s, and $v = 24$ when $s = 16$.

17. n varies inversely as m, and $n = 32$ when $m = 1.5$.

18. q is inversely proportional to p, and $q = \frac{3}{2}$ when $p = 50$.

19. F varies jointly as x and y, and $F = 500$ when $x = 15$ and $y = 8$.

20. V varies jointly as h and the square of b, and $V = 288$ when $h = 6$ and $b = 12$.

21. d varies directly as the square of x and inversely with r, and $d = 3{,}000$ when $x = 10$ and $r = 4$.

22. z is directly proportional to x and inversely proportional to the square root of y, and $z = 720$ when $x = 48$ and $y = 81$.

23. The total revenue R is directly proportional to the number of units x sold. When 500 units are sold, the revenue is \$3,875. Find the revenue when 635 units are sold.

24. What does the constant of proportionality measure in Exercise 23?

25. *Hooke's Law* A baby weighing $10\frac{1}{2}$ pounds compresses the spring of a baby scale 7 millimeters (see figure). Determine the weight of a baby that will compress the spring 12 millimeters.

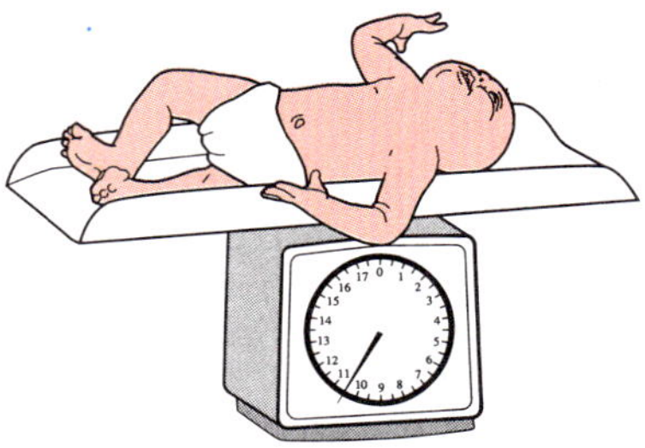

26. *Hooke's Law* A force of 50 pounds compresses the spring in a scale 1.5 inches. Find the distance the spring will be compressed when a 20-pound object is placed on the scale.

27. *Stopping Distance* The stopping distance d of an automobile is directly proportional to the square of its speed s. On a certain road surface a car requires 75 feet to stop when its speed is 30 miles per hour. Estimate the stopping distance if the brakes are applied when the car is traveling at 50 miles per hour under similar road conditions.

28. *Distance* Neglecting air resistance, the distance d that an object falls varies directly as the square of the time t it has been falling. If an object falls 64 feet in two seconds, determine the distance it will fall in six seconds.

29. *Power Generation* The power P generated by a wind turbine varies directly as the cube of the wind speed w. The turbine generates 750 watts of power in a 25-mile-per-hour wind. Find the power it generates in a 40-mile-per-hour wind.

30. *Velocity of a Stream* The diameter d of a particle moved by a stream is directly proportional to the square of the velocity v of the stream. A stream with a velocity of $\frac{1}{4}$ mile per hour can move coarse sand particles of about 0.02-inch diameter. What must the velocity be to carry particles with a diameter of 0.12 inch?

31. *Demand Function* A company has found that the daily demand x for its product is inversely proportional to the price p. When the price is \$5, the demand is 800 units. Approximate the demand if the price is increased to \$6.

32. *Weight of an Astronaut* The gravitational force F with which an object is attracted to the Earth is inversely proportional to the square of its distance r from the center of the Earth. If an astronaut weighs 190 pounds on the surface of the Earth ($r \approx 4{,}000$ miles), what will the astronaut weigh 1,000 miles above the Earth's surface?

33. *Amount of Illumination* The illumination I from a light source varies inversely as the square of the distance d from the light source (see figure). If you raise a study lamp from 18 inches to 36 inches over your study desk, the illumination will change by what factor?

34. *Frictional Force* The frictional force F between the tires and the road required to keep a car on a curved section of a highway is directly proportional to the square of the speed s of the car. If the speed of the car is doubled, the force will change by what factor?

35. *Load of a Beam* The load that can be safely supported by a horizontal beam varies jointly as the width of the beam and the square of its depth and inversely as the length of the beam. A beam of width 3 inches, depth 8 inches, and length 10 feet can safely support 2,000 pounds. Determine the safe load of a beam made from the same material if its depth is increased to 10 inches.

36. *Best Buy* The prices of 9-inch, 12-inch, and 15-inch diameter pizzas at a certain pizza shop are \$6.78, \$9.78, and \$12.18, respectively. One would expect that the price of a certain size pizza would be directly proportional to its surface area. Is that the case for this pizza shop? If not, which size pizza is the best buy?

In Exercises 37–40, use the given value of k to complete the table for the direct variation model $y = kx^2$. Plot the points on the rectangular coordinate system.

x	2	4	6	8	10
$y = kx^2$					

37. $k = 1$

38. $k = 2$

39. $k = \frac{1}{2}$

40. $k = \frac{1}{4}$

In Exercises 41–44, use the given value of k to complete the table for the inverse variation model $y = k/(x^2)$. Plot the points on the rectangular coordinate system.

x	2	4	6	8	10
$y = \frac{k}{x^2}$					

41. $k = 2$

42. $k = 5$

43. $k = 10$

44. $k = 20$

CHAPTER 8 SUMMARY

As you review and prepare for a test on this chapter, first try to obtain a global view of what was discussed. Then review the specific skills needed under each category.

Functions and Their Graphs

■ *Determine* whether a given set of ordered pairs or a given equation represents a function.	Section 8.1, Exercises 7–20
■ *Evaluate* a function at specified values of the independent variable.	Section 8.1, Exercises 25–40
■ *Find* the domain of a function.	Section 8.1, Exercises 45–60
■ *Create* an equation that represents a function, given a verbal description of the function.	Section 8.1, Exercises 61–68
■ *Sketch* the graph of a function and specify its domain and range.	Section 8.2, Exercises 1–20
■ *Sketch* the graph of an equation and use the vertical line test to see if it represents a function.	Section 8.2, Exercises 25–30
■ *Describe* transformations in the graphs of basic functions by using function notation.	Section 8.2, Exercises 31–50, Exercises 55–60

Combination and Inverse Functions

■ *Evaluate* combinations of functions.	Section 8.3, Exercises 1–24, Exercises 31–46
■ *Find* the inverse of a given function.	Section 8.4, Exercises 1–6, Exercises 21–30
■ *Determine* whether a given function has an inverse.	Section 8.4, Exercises 31–40

Variation and Mathematical Models

■ *Write* a mathematical model for different types of variation.	Section 8.5, Exercises 1–24
■ *Create and solve* mathematical models for variations described by verbal statements.	Section 8.5, Exercises 25–36

Chapter 8 Review Exercises

In Exercises 1–4, label the intercepts on the given graph and use the vertical line test to determine whether y is a function of x.

1. $9y^2 = 4x^3$

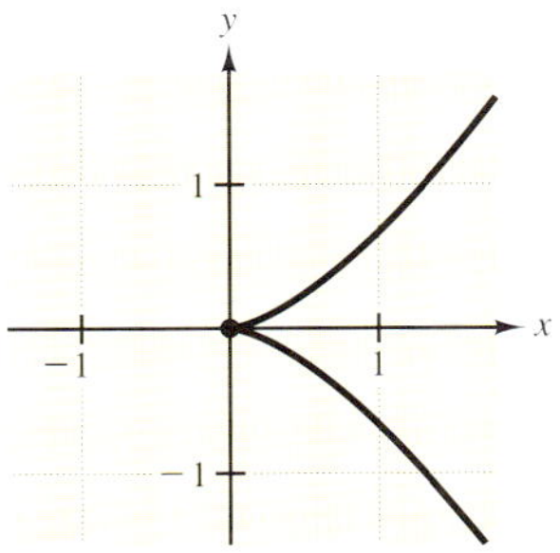

2. $y = 4x^3 - x^4$

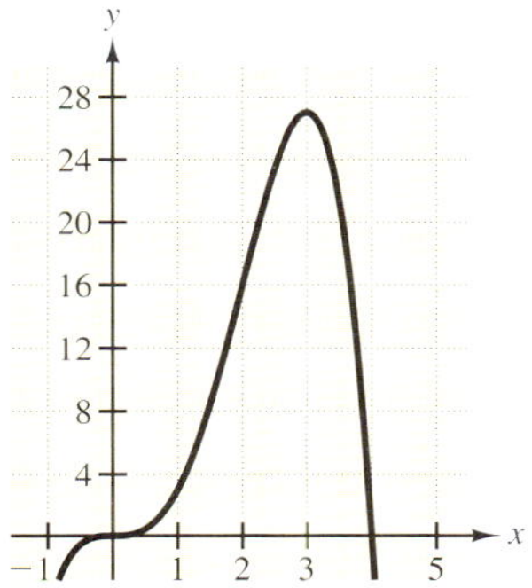

3. $y = x^2(x - 3)$

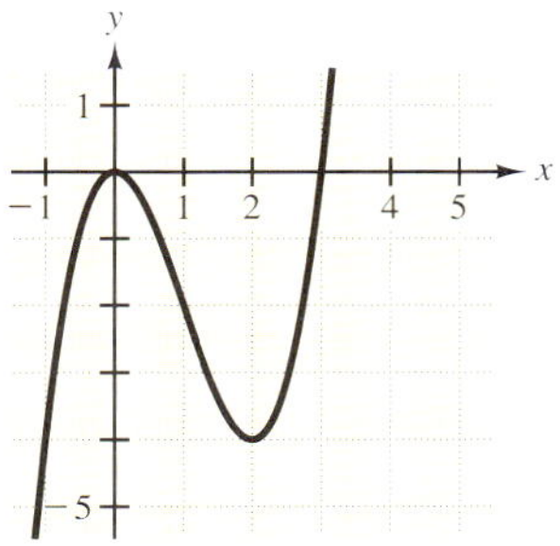

4. $x^3 + y^3 - 6xy = 0$

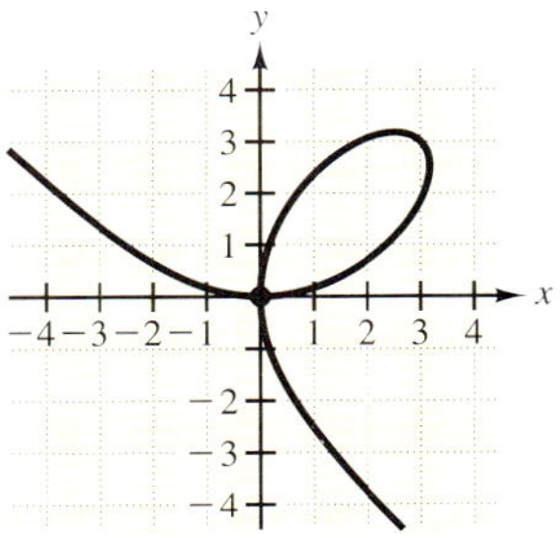

In Exercises 5–14, sketch the graph of the given equation. Use the vertical line test to determine whether y is a function of x.

5. $y = 6 - \frac{2}{3}x$

6. $y = \frac{5}{2}x - 4$

7. $2y - 4x + 3 = 0$

8. $3x + 2y + 12 = 0$

9. $y = 8 - 2|x|$

10. $y = \frac{1}{2}x(2 - x)$

11. $y^2 - x^2 = 1$

12. $x^2 + 4y^2 = 4$

13. $y = (x - 2)^2$

14. $y = 4 - (x - 3)^2$

In Exercises 15–24, evaluate the function at the specified value of the independent variable and simplify when possible.

Function	*Function Value*			
15. $f(x) = 4 - \frac{5}{2}x$	(a) $f(-10)$	(b) $f\left(\frac{2}{5}\right)$	(c) $f(t) + f(-4)$	(d) $f(x + h)$
16. $h(x) = x(x - 8)$	(a) $h(8)$	(b) $h(10)$	(c) $h(-3)$	(d) $h(t + 4)$
17. $f(t) = \sqrt{5 - t}$	(a) $f(-3)$	(b) $f(5)$	(c) $f\left(\frac{11}{3}\right)$	(d) $f(5z)$
18. $g(x) = \frac{\lvert x + 4\rvert}{4}$	(a) $g(0)$	(b) $g(-8)$	(c) $g(2) - g(-5)$	(d) $g(x - 2)$
19. $f(x) = \begin{cases} -3x, & \text{if } x \le 0 \\ 1 - x^2, & \text{if } x > 0 \end{cases}$	(a) $f(2)$	(b) $f\left(-\frac{2}{3}\right)$	(c) $f(1)$	(d) $f(4) - f(3)$
20. $h(x) = \begin{cases} x^3, & \text{if } x \le 1 \\ (x - 1)^2 + 1, & \text{if } x > 1 \end{cases}$	(a) $h(2)$	(b) $h\left(-\frac{1}{2}\right)$	(c) $h(0)$	(d) $h(4) - h(3)$

21. $f(x) = 3 - 2x$ (a) $\dfrac{f(x+2) - f(2)}{x}$ (b) $\dfrac{f(x-3) - f(3)}{x}$

22. $f(x) = x^3$ (a) $\dfrac{f(x+1) - f(1)}{x}$ (b) $\dfrac{f(x-5) - f(5)}{x}$

23. $g(x) = x^2$ (a) $g(x+h)$ (b) $\dfrac{g(x+h) - g(x)}{h}$

24. $f(x) = 8$ (a) $f(x+h)$ (b) $\dfrac{f(x+h) - f(x)}{h}$

In Exercises 25–30, find the domain of the given function.

25. $h(x) = 4x^2 - 7$

26. $g(s) = \dfrac{s+1}{(s-1)(s+5)}$

27. $g(x) = \dfrac{2x+5}{x^2 - 4x + 3}$

28. $f(x) = |x - 6| + 10$

29. $f(x) = \sqrt{5 - 2x}$

30. $G(x) = \sqrt{x^2 - 4}$

In Exercises 31–38, sketch the graph of the given function.

31. $g(x) = \frac{1}{8}x^2$

32. $h(x) = 9 - (x-2)^2$

33. $f(t) = \sqrt{\frac{t}{2}}$

34. $f(x) = |x + 1| - 2$

35. $g(x) = \frac{1}{4}x^3$

36. $h(x) = x(4 - x), \quad 0 \le x \le 4$

37. $f(x) = \begin{cases} 2 - (x-1)^2, & \text{if } x < 1 \\ 2 + (x-1)^2, & \text{if } x \ge 1 \end{cases}$

38. $f(x) = \begin{cases} 2x, & \text{if } x \le 0 \\ x^2 + 1, & \text{if } x > 0 \end{cases}$

In Exercises 39–44, evaluate the indicated combination of the functions f and g at the given values of the independent variable. (If not possible, state the reason.)

39. $f(x) = x^2, \quad g(x) = 4x - 5$ (a) $(f+g)(-5)$ (b) $(f-g)(0)$ (c) $(fg)(2)$ (d) $\left(\frac{f}{g}\right)(1)$

40. $f(x) = \frac{3}{4}x^3, \quad g(x) = x + 1$ (a) $(f+g)(-1)$ (b) $(f-g)(2)$ (c) $(fg)\left(\frac{1}{3}\right)$ (d) $\left(\frac{f}{g}\right)(2)$

41. $f(x) = \frac{2}{3}\sqrt{x}, \quad g(x) = -x^2$ (a) $(f+g)(1)$ (b) $(f-g)(9)$ (c) $(fg)\left(\frac{1}{4}\right)$ (d) $\left(\frac{f}{g}\right)(2)$

42. $f(x) = |x|, \quad g(x) = 3$ (a) $(f+g)(-2)$ (b) $(f-g)(3)$ (c) $(fg)(-10)$ (d) $\left(\frac{f}{g}\right)(-3)$

43. $f(x) = \frac{2}{x-1}, \quad g(x) = x$ (a) $(f+g)\left(\frac{1}{3}\right)$ (b) $(f-g)(3)$ (c) $(fg)(-1)$ (d) $\left(\frac{f}{g}\right)(5)$

44. $f(x) = \frac{1}{x}, \quad g(x) = \frac{1}{x-4}$ (a) $(f+g)(2)$ (b) $(f-g)(-2)$ (c) $(fg)\left(\frac{7}{2}\right)$ (d) $\left(\frac{f}{g}\right)(1)$

In Exercises 45–48, find the indicated composite function.

45. $f(x) = x + 2, \quad g(x) = x^2$ (a) $(f \circ g)(2)$ (b) $(g \circ f)(-1)$ (c) $(f \circ g)(x)$ (d) $(g \circ f)(x)$

46. $f(x) = \sqrt[3]{x}, \quad g(x) = x + 2$ (a) $(f \circ g)(6)$ (b) $(g \circ f)(64)$ (c) $(f \circ g)(x)$ (d) $(g \circ f)(x)$

47. $f(x) = \sqrt{x + 1}, \quad g(x) = x^2 - 1$ (a) $(f \circ g)(5)$ (b) $(g \circ f)(-1)$ (c) $(f \circ g)(x)$ (d) $(g \circ f)(x)$

48. $f(x) = \dfrac{1}{x - 5}, \quad g(x) = \dfrac{5x + 1}{x}$ (a) $(f \circ g)(1)$ (b) $(g \circ f)\left(\dfrac{1}{5}\right)$ (c) $(f \circ g)(x)$ (d) $(g \circ f)(x)$

In Exercises 49 and 50, find the domain of the composition (a) $f \circ g$, and (b) $g \circ f$.

49. $f(x) = \sqrt{x - 4}, \quad g(x) = 2x$

50. $f(x) = \dfrac{2}{x - 4}, \quad g(x) = x^2$

In Exercises 51–54, determine whether the given function has an inverse.

51. $f(x) = x^2 - 25$

52. $f(x) = \frac{1}{4}x^3$

53. $h(x) = 4\sqrt[3]{x}$

54. $g(x) = \sqrt{9 - x^2}$

In Exercises 55–60, find the inverse of the given function. (If not possible, state the reason.)

55. $f(x) = \frac{1}{4}x$

56. $f(x) = 2x - 3$

57. $h(x) = \sqrt{x}$

58. $g(x) = x^2 + 2, \quad x \geq 0$

59. $f(t) = |t + 3|$

60. $h(t) = t$

In Exercises 61–64, find the constant of proportionality and write an equation that relates the variables.

61. y varies directly as the cube root of x, and $y = 12$ when $x = 9$.

62. r varies inversely as s, and $r = 45$ when $s = \frac{3}{5}$.

63. T varies jointly as r and the square of s, and $T = 5{,}000$ when $r = 0.09$ and $s = 1{,}000$.

64. D is directly proportional to the cube of x and inversely proportional to y, and $D = 810$ when $x = 3$ and $y = 25$.

65. *Power Generation* The power generated by a wind turbine is given by the function $P = kw^3$ where P is the number of kilowatts produced at a wind speed of w miles per hour and k is the constant of proportionality.
(a) Find k if $P = 1{,}000$ when $w = 20$.
(b) Find the output for a wind speed of 25 miles per hour.

66. *Wire Length* A wire 100 inches long is to be cut into four pieces to form a rectangle whose shortest side has a length of x. Express the area A of the rectangle as a function of x.

67. *Hooke's Law* A force of 100 pounds stretches a spring four inches. Find the force required to stretch the same spring six inches.

68. *Stopping Distance* The stopping distance d of an automobile is directly proportional to the square of its speed s. How will the stopping distance be changed by doubling the speed of the car?

69. *Demand Function* A company has found that the daily demand x for its product varies inversely as the square root of the price p. When the price is \$25, the demand is approximately 1,000 units. Approximate the demand if the price is increased to \$28.

70. *Weight of an Astronaut* The gravitational force F with which an object is attracted to the Earth is inversely proportional to the square of its distance r from the center of the Earth. If an astronaut weighs 200 pounds on the surface of the Earth ($r \approx 4{,}000$ miles), what will the astronaut weigh 500 miles above the Earth's surface?

CHAPTER 8 TEST

Take this test as you would take a test in class. After you are done, check your work with the answers given in the back of the book.

1. Determine whether the equations represent y as a function of x.
(a) $3x^2 - y^2 = 9$ (b) $3x - y = 9$

2. Evaluate the function $g(x) = x/(x - 3)$ at the indicated values.
(a) $g(2)$ (b) $g\left(\frac{7}{2}\right)$

3. Evaluate the function $f(x) = 3x - 2$ at the indicated values.
(a) $f(t + 1)$ (b) $f(x + h) - f(x)$

4. Find the domain of each function.
(a) $h(t) = \sqrt{9 - t^2}$ (b) $f(x) = \dfrac{x + 1}{x - 4}$

5. A company allows each of its sales personnel \$150 per day for food and housing plus \$0.26 per mile for the use of the person's car while traveling. Write the daily cost C to the company for each salesperson as a function of the number of miles traveled, x. What is the domain of the function?

6. Sketch the graph of the function $g(x) = \sqrt{2 - x}$.

7. Use the graph of $f(x) = |x|$ to write an equation to represent the transformation of the given graph.

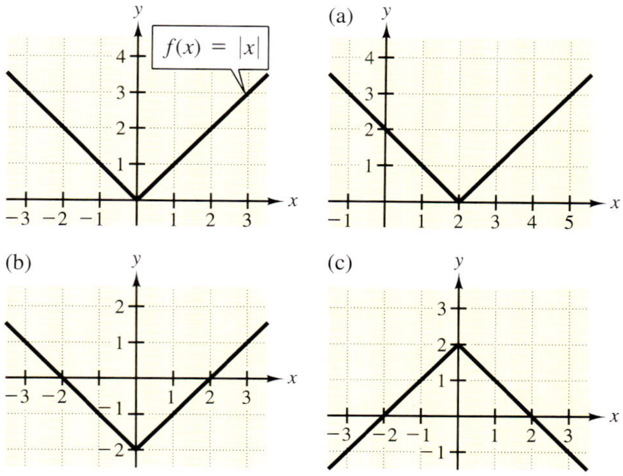

8. Use the vertical line test to determine whether the equation $y^2(4 - x) = x^3$ represents y as a function of x.

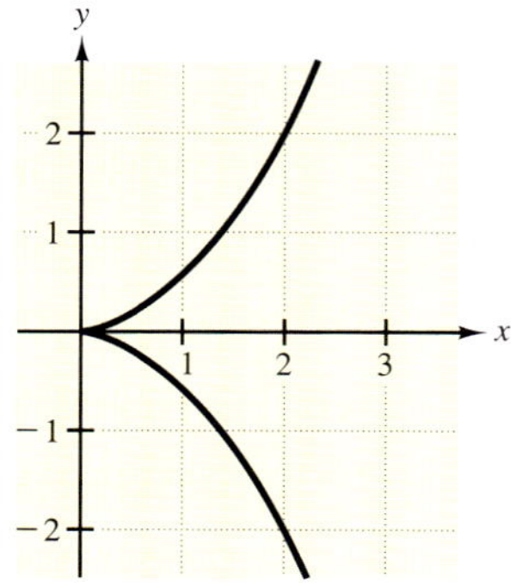

9. Given the functions $f(x) = \frac{1}{2}x$ and $g(x) = x^2 - 1$, find the following.
(a) $(f - g)(4)$ (b) $(fg)(-4)$
(c) $\left(\frac{f}{g}\right)(2)$ (d) $(f \circ g)(-5)$

10. Find the domain of the composite function $(f \circ g)(x)$ if $f(x) = \sqrt{25 - x}$ and $g(x) = x^2$.

11. Find the inverse of the function $f(x) = \frac{1}{2}x - 1$. Verify that $f(f^{-1}(x)) = x$.

12. Write a mathematical model for the statement "S varies directly as the square of x and inversely as y."

13. Find a mathematical model that relates u and v if v varies directly as the square root of u, and $v = \frac{3}{2}$ when $u = 36$.

14. Boyle's Law states that if the temperature of a gas is not allowed to change, its absolute pressure P is inversely proportional to its volume V. A large balloon is filled with 180 cubic meters of helium at atmospheric pressure (1 atm) at sea level. What is the volume of the helium if the balloon rises to an altitude where the atmospheric pressure is 0.75 atm? Assume that the temperature does not change.

CHAPTER NINE

Systems of Equations and Inequalities

Chapter Overview

This chapter surveys several topics of a branch of mathematics called linear algebra. The first section introduces two techniques for solving general systems of equations. Then, in Sections 9.2 through 9.5, we survey several different techniques for solving systems of linear equations. In the last section of the chapter, we discuss the graphs of linear inequalities.

SECTION 9.1

Introduction to Systems of Equations

Systems of Equations • Solving Systems of Equations by Substitution • Solving Systems of Equations by Graphing • Applications

Systems of Equations

Up to this point in the text, most problems have involved a single equation with one or two variables. Many problems in business and science involve a **system of equations** that consists of two or more equations, each involving two or more variables. Consider the following problem.

> A roofing contractor bought 30 bundles of shingles and four rolls of roofing paper for \$528. A second purchase (at the same prices) cost \$140 for eight bundles of shingles and one roll of roofing paper. Find the price per bundle of shingles and the price per roll of paper.

A verbal model for this problem consists of the following system.

Cost of 30 bundles	+	Cost of 4 rolls	=	528		First purchase
Cost of 8 bundles	+	Cost of 1 roll	=	140		Second purchase

If we let x be the price (in dollars) per bundle of shingles and y be the price (in dollars) per roll of paper, we obtain the following system of equations.

$$30x + 4y = 528 \qquad \text{Equation 1}$$
$$8x + y = 140 \qquad \text{Equation 2}$$

A **solution** of such a system is an ordered pair (x, y) of real numbers that satisfies *each* equation in the system. When we find the set of all solutions of the system of equations, we say that we are **solving the system of equations**.

EXAMPLE 1 ■ Checking Solutions of a System of Equations

Consider the following system of equations.

$$x + y = 6 \qquad \text{Equation 1}$$
$$2x - 5y = -2 \qquad \text{Equation 2}$$

Which of the given ordered pairs is a solution of this system of equations?

(a) $(3, 3)$ (b) $(4, 2)$

Solution

(a) To determine whether the ordered pair $(3, 3)$ is a solution of the given system of equations, we must substitute $x = 3$ and $y = 3$ into *each* of the given equations.

Substituting into Equation 1 produces

$3 + 3 = 6.$ Replace x by 3 and y by 3

Similarly, substituting into Equation 2 produces

$2(3) - 5(3) \neq -2.$ Replace x by 3 and y by 3

Since the ordered pair (3, 3) fails to check in *both* equations, we conclude that it is *not* a solution of the given system of equations.

(b) By substituting the coordinates of the ordered pair (4, 2) into the two given equations, we find that it is a solution of the first equation

$4 + 2 = 6$ Replace x by 4 and y by 2

and is also a solution of the second equation

$2(4) - 5(2) = -2.$ Replace x by 4 and y by 2

Therefore, we conclude that (4, 2) *is* a solution of the given system of equations. ■

Solving Systems of Equations by Substitution

One way to solve a system of two equations in two variables is to convert the system to *one* equation in *one* variable by an appropriate substitution. We illustrate this procedure in Example 2.

EXAMPLE 2 ■ The Method of Substitution: One-Solution Case

Solve the following system of equations.

$-x + y = 3$ Equation 1

$3x + y = -1$ Equation 2

Solution

We begin by solving for y in the first equation.

$y = x + 3$ Solve for y in Equation 1

Next, we substitute this expression for y into Equation 2.

$3x + y = -1$ Equation 2

$3x + (x + 3) = -1$ Replace y by $x + 3$

$4x = -4$

$x = -1$ Solve for x

At this point, we know that the x-coordinate of the solution is -1. To find the y-coordinate, we back-substitute the x-value into the revised Equation 1.

$y = x + 3$ Revised Equation 1

$y = -1 + 3$ Replace x by -1

$y = 2$ Solve for y

Thus, the solution is $(-1, 2)$. Check to see that it satisfies both original equations. ■

The term **back-substitute** implies that we work backwards. After finding a value for one of the variables, we substitute that value back into one of the equations in the original (or revised) system to find the value of the other variable.

When using the method of substitution, it does not matter which variable you choose to solve for first. Whether you solve for y first or x first, you will obtain the same solution. When making your choice, you should choose the variable that is easier to work with. For instance, in the system

$3x - 2y = 1$ Equation 1

$x + 4y = 3$ Equation 2

it is easier to solve for x first (in the second equation). On the other hand, in the system

$2x + y = 5$ Equation 1

$3x - 2y = 11$ Equation 2

it is easier to solve for y first (in the first equation).

We summarize the steps for using the method of substitution to solve a system of two equations involving two variables as follows.

Method of Substitution

To solve a system of two equations in two variables, use the following steps.

1. Solve one of the equations for one variable in terms of the other.
2. Substitute the expression found in Step 1 into the other equation to obtain an equation of one variable.
3. Solve the equation obtained in Step 2.
4. Back-substitute the solution from Step 3 into the expression obtained in Step 1 to find the value of the other variable.
5. Check the solution to see that it satisfies *each* of the original equations.

Both of the equations in Example 2 are linear. That is, the variables x and y occurred to the first power only. The method of substitution can also be used to solve a system of equations in which one or both of the equations is nonlinear. This possibility is illustrated in Examples 3 and 4.

EXAMPLE 3 ■ The Method of Substitution: Two-Solution Case

Solve the following system of equations.

$$x^2 + y^2 = 25 \qquad \text{Equation 1}$$
$$-x + y = -1 \qquad \text{Equation 2}$$

Solution

We begin by solving for y in the second equation.

$$y = x - 1 \qquad \text{Solve for } y \text{ in Equation 2}$$

Next, we substitute this expression for y into Equation 1.

$$\begin{aligned} x^2 + y^2 &= 25 && \text{Equation 1} \\ x^2 + (x - 1)^2 &= 25 && \text{Replace } y \text{ by } x - 1 \\ 2x^2 - 2x - 24 &= 0 \\ 2(x^2 - x - 12) &= 0 \\ 2(x - 4)(x + 3) &= 0 && \text{Factor} \\ x &= 4, -3 && \text{Solve for } x \end{aligned}$$

Finally, we back-substitute these values of x into the revised second equation to solve for y.

For $x = 4$: $\quad y = 4 - 1 = 3$

For $x = -3$: $\quad y = -3 - 1 = -4$

Thus, the given system of equations has two solutions: $(4, 3)$ and $(-3, -4)$. Try checking these solutions in each of the original equations. ■

EXAMPLE 4 ■ Method of Substitution: No-Solution Case

Solve the following system of equations.

$$x^2 - y = 0 \qquad \text{Equation 1}$$
$$x - y = 1 \qquad \text{Equation 2}$$

Solution

We begin by solving for y in the second equation.

$$y = x - 1 \qquad \text{Solve for } y \text{ in Equation 2}$$

Next, we substitute this expression for y into Equation 1.

$$\begin{aligned} x^2 - y &= 0 && \text{Equation 1} \\ x^2 - (x - 1) &= 0 && \text{Replace } y \text{ by } x - 1 \\ x^2 - x + 1 &= 0 \\ x &= \frac{1 \pm \sqrt{1 - 4}}{2} && \text{Use Quadratic Formula} \end{aligned}$$

Because the Quadratic Formula yields a negative number inside the square root, we conclude that the equation $x^2 - x + 1$ has no (real) solution. Hence, the given system has no (real) solution. ■

Solving Systems of Equations by Graphing

From Examples 2, 3, and 4, we can see that a system of two equations in two variables can have exactly one solution, more than one solution, or no solution. In practice we can gain insight about the location and number of solutions of a system of equations by sketching the graph of each equation in the same coordinate plane. The solutions of the system correspond to the **points of intersection** of the graphs. For instance, Figure 9.1 shows the graphs of the equations corresponding to the systems discussed in Examples 2, 3, and 4.

FIGURE 9.1

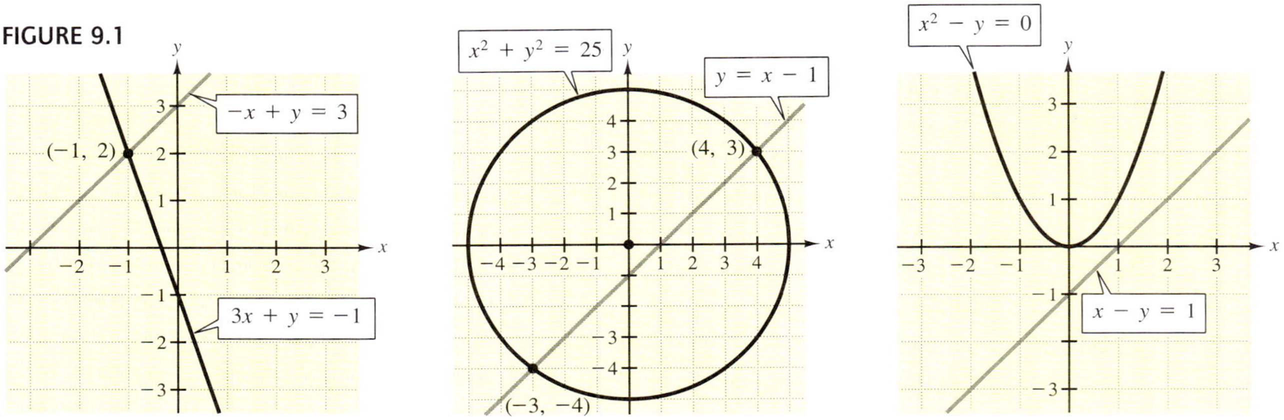

EXAMPLE 5 ■ The Graphical Method of Solving a System

Use the graphical method to solve the following system of equations.

$$2x + 3y = 7 \qquad \text{Equation 1}$$
$$2x - 5y = -1 \qquad \text{Equation 2}$$

Solution

Since both equations in the given system are linear, we know that they have graphs that are straight lines. To sketch these lines, we write each equation in slope-intercept form, as follows.

$$y = -\frac{2}{3}x + \frac{7}{3} \qquad \text{Equation 1}$$
$$y = \frac{2}{5}x + \frac{1}{5} \qquad \text{Equation 2}$$

The lines corresponding to these two equations are shown in Figure 9.2. From this figure, it appears that the two lines intersect in a single point, and that the coordinates of the point are approximately (2, 1). A check will show that this point is a solution of each of the original equations.

FIGURE 9.2

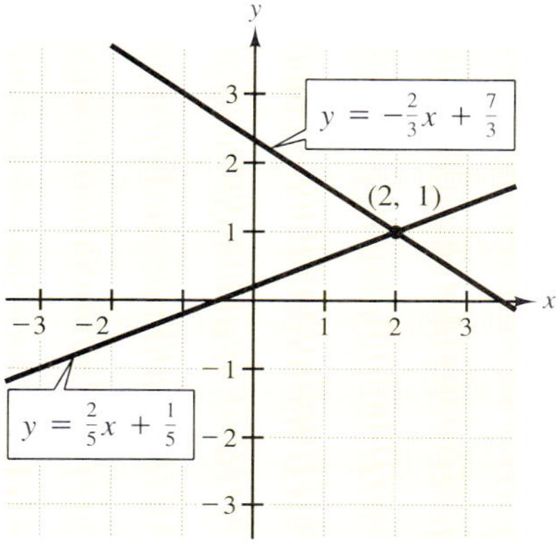

Applications

At the beginning of this section, we described an application involving the cost of construction materials. In Example 6 use the method of substitution to solve this problem.

EXAMPLE 6 ■ **An Application**

A roofing contractor bought 30 bundles of shingles and four rolls of roofing paper for \$528. A second purchase (at the same prices) cost \$140 for eight bundles of shingles and one roll of roofing paper. Find the price per bundle of shingles and the price per roll of paper.

Solution

As we did at the beginning of the section, we let x be the price (in dollars) per bundle of shingles and y be the price (in dollars) per roll of paper, and obtain the following system of equations.

$$30x + 4y = 528 \qquad \text{Equation 1}$$
$$8x + y = 140 \qquad \text{Equation 2}$$

Solving the second equation for y produces $y = 140 - 8x$, and substituting this expression into the first equation produces the following.

$$\begin{aligned} 30x + 4(140 - 8x) &= 528 \\ 30x + 560 - 32x &= 528 \\ -2x &= -32 \\ x &= 16 \end{aligned}$$

Back-substituting $x = 16$ into the revised second equation produces

$$y = 140 - 8(16) = 12.$$

Therefore, we conclude that the price of shingles is \$16 per bundle and the price of roofing paper is \$12 per roll. Check this solution in the original statement of the problem. ■

EXAMPLE 7 ■ An Application

A total of \$12,000 is invested in two funds paying 6% and 8% simple interest. If the interest for one year is \$880, how much of the \$12,000 was invested in each fund?

Solution

We let x represent the amount (in dollars) invested in the 6% fund and y represent the amount (in dollars) invested in the 8% fund. Since the total of the two invested amounts is \$12,000, we can write the following equation.

$$x + y = 12{,}000 \qquad \text{Equation 1}$$

To obtain a second equation, we use the fact that the interest for the two funds is \$880. This information can be represented by the following equation.

$$0.06x + 0.08y = 880 \qquad \text{Equation 2}$$

To solve this system, we solve the first equation for y to obtain $y = 12{,}000 - x$. Next, we substitute this expression into the second equation to obtain

$$\begin{aligned} 0.06x + 0.08(12{,}000 - x) &= 880 \\ 0.06x + 960 - 0.08x &= 880 \\ -0.02x &= -80 \\ x &= 4{,}000. \end{aligned}$$

Back-substituting this value for x into the revised first equation produces

$$y = 12{,}000 - 4{,}000 = 8{,}000.$$

Therefore, we conclude that \$4,000 was invested in the 6% fund and \$8,000 was invested in the 8% fund. ■

The total cost C of producing x units of a product typically has two components—the initial cost and the cost per unit. When enough units have been sold so that the total revenue R equals the total cost, we say that sales have reached the **break-even point**. We can find this break-even point by setting C equal to R and solving for x. In other words, the break-even point corresponds to the point of intersection of the cost and the revenue graphs.

EXAMPLE 8 ■ An Application: Break-Even Analysis

A small business invests \$14,000 in equipment to produce a product. Each unit of the product costs \$0.80 to produce and is sold for \$1.50. How many items must be sold before the business breaks even?

Solution

The total cost of producing x units is

$$C = \underbrace{0.80x}_{\text{Cost per unit}} + \underbrace{14{,}000}_{\text{Initial cost}}$$

and the revenue obtained by selling x units is

$$R = \underbrace{1.50x}_{\text{Price per unit}}.$$

Since the break-even point occurs when $R = C$, we have

$$1.5x = 0.8x + 14{,}000$$
$$0.7x = 14{,}000$$
$$x = 20{,}000.$$

Therefore, we conclude that the business must sell 20,000 units before it breaks even. Note in Figure 9.3 that sales less than the break-even point correspond to a loss for the business, while sales greater than the break-even point correspond to a profit for the business.

FIGURE 9.3

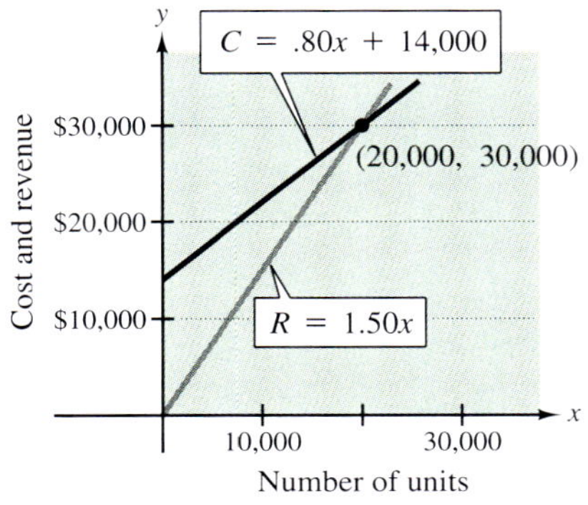

■

DISCUSSION PROBLEM ■ Creating a System of Equations with a Known Solution

Suppose you are tutoring another student and want to create several systems of equations that the student can use for practice. To begin, you want to find some equations that have relatively simple solutions. For instance, try to create a system of equations that has $(3, -2)$ as a solution. Write a short paragraph describing the process you used to create the system. ■

Warm-Up

The following warm-up exercises involve skills that were covered in earlier sections. You will use these skills in the exercise set for this section.

In Exercises 1–4, sketch the graph of the given equation.

1. $y = -\frac{1}{4}x + 8$

2. $y = 3(x - 2)$

3. $y - 2 = 2(x + 3)$

4. $y + 3 = \frac{1}{3}(x - 4)$

In Exercises 5 and 6, solve the given equation.

5. $2y - 3(4y - 2) = 5$

6. $8x + 6(3 - 2x) = 4$

In Exercises 7 and 8, find an equation of the line passing through the two given points.

7. $(-1, -2), (3, 6)$

8. $(1, 5), (6, 0)$

In Exercises 9 and 10, determine the slope of the given line.

9. $4x + 6y = 4$

10. $-9x + 5y = 10$

9.1 EXERCISES

In Exercises 1–4, determine whether the ordered pair is a solution of the system of equations.

1. $x + 2y = 9$
$-2x + 3y = 10$
(a) $(1, 4)$
(b) $(3, -1)$

2. $5x - 4y = 34$
$x - 2y = 8$
(a) $(0, 3)$
(b) $(6, -1)$

3. $-2x + 7y = 46$
$3x + y = 0$
(a) $(-3, 2)$
(b) $(-2, 6)$

4. $-5x - 2y = 23$
$x + 4y = -19$
(a) $(-3, -4)$
(b) $(3, 7)$

In Exercises 5–12, solve the given system by the method of substitution.

5. $x - 2y = 0$
$3x + 2y = 8$

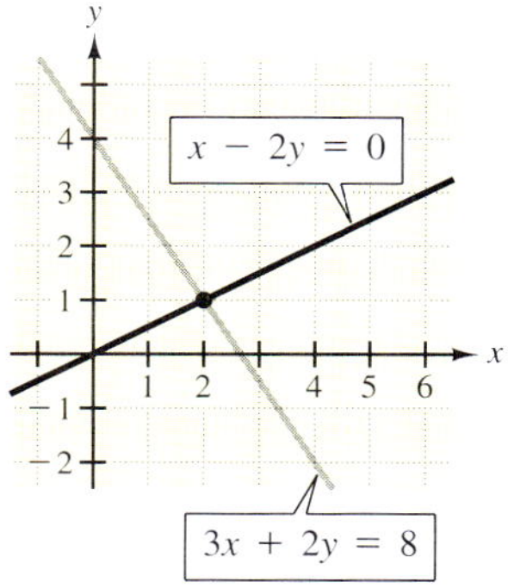

6. $x - y = 0$
$5x - 2y = 6$

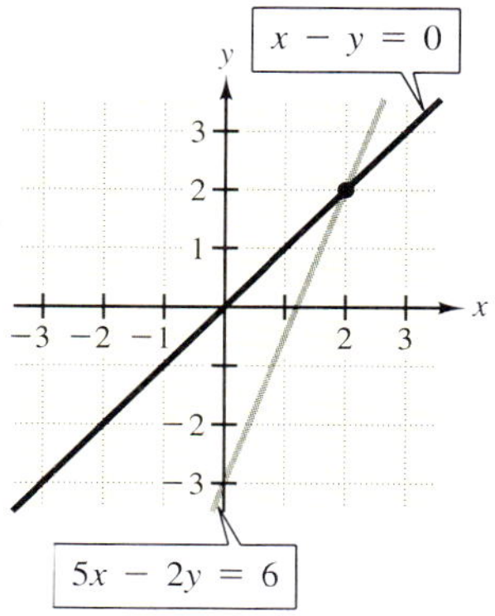

7. $x = 4$
$x - 2y = -2$

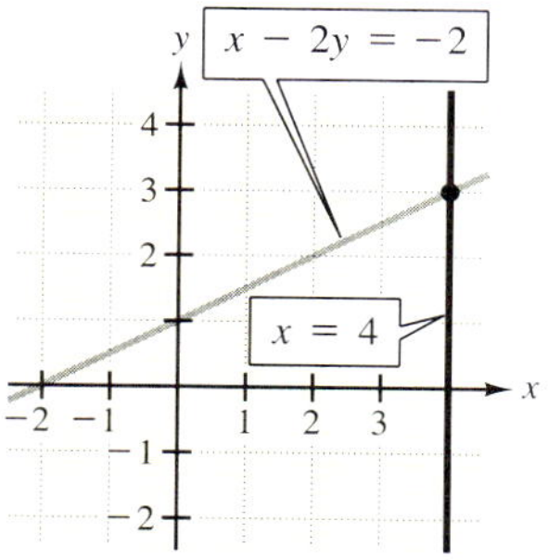

8. $y = 2$
$x - 6y = -6$

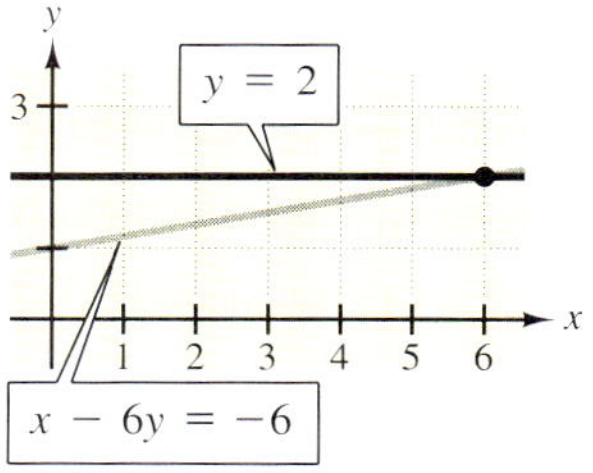

9. $7x + 8y = 24$
$x - 8y = 8$

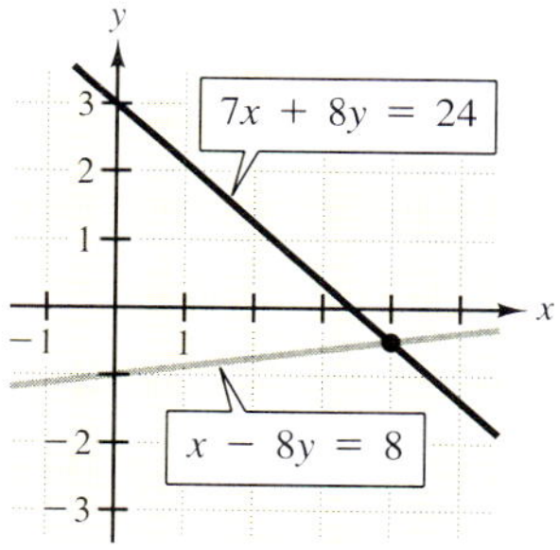

10. $x - 3y = -2$
$5x + 3y = 17$

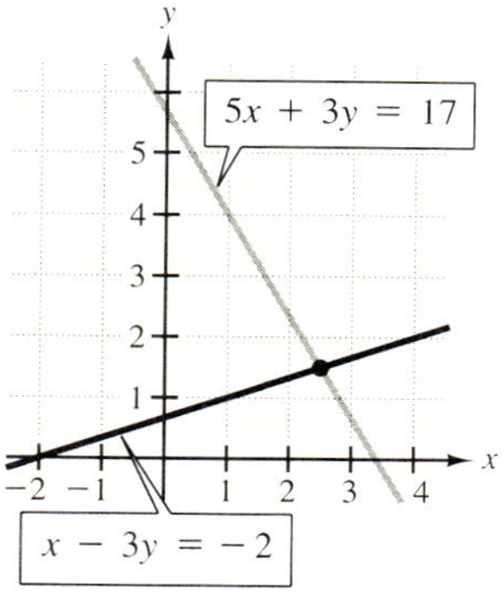

11. $x + y = 2$
$x^2 - y = 0$

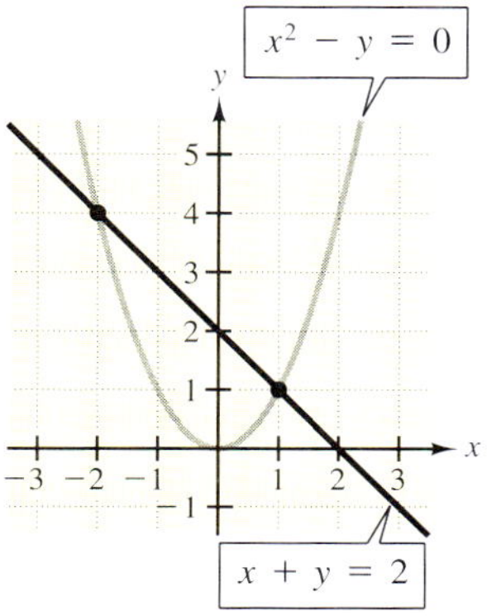

12. $2x + y = 10$
$x^2 + y^2 = 25$

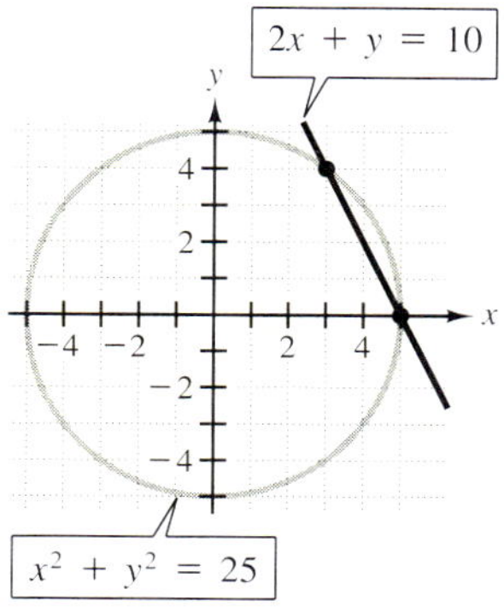

In Exercises 13–30, solve the given system by the method of substitution.

13. $x + y = 3$
$2x - y = 0$

14. $-x + y = 5$
$x - 4y = 0$

15. $x + y = 2$
$x - 4y = 12$

16. $x - 2y = -1$
$x - 5y = 2$

17. $x + 6y = 19$
$x - 7y = -7$

18. $x - 5y = -6$
$4x - 3y = 10$

19. $-13x + 16y = 10$
$5x + 16y = -26$

20. $2x + 5y = 29$
$5x + 2y = 13$

21. $4x - 14y = -15$
$18x - 12y = 9$

22. $5x - 24y = -12$
$17x - 24y = 36$

23. $3x + 2y = 90$
$xy = 300$

24. $x + 2y = 40$
$xy = 150$

25. $x^2 + y = 9$
$x - y = -3$

26. $x - y^2 = 0$
$x - y = 2$

27. $x^2 + y^2 = 25$
$2x - y = -5$

28. $x^2 + y^2 = 169$
$x + y = 7$

29. $y = \sqrt{4 - x}$
$x + 3y = 6$

30. $x^2 - y^2 = 16$
$3x - y = 12$

In Exercises 31–36, use the graphs of the equations to determine whether the system has any solutions. Find any solutions that exist.

31. $x + y = 4$
$x + y = -1$

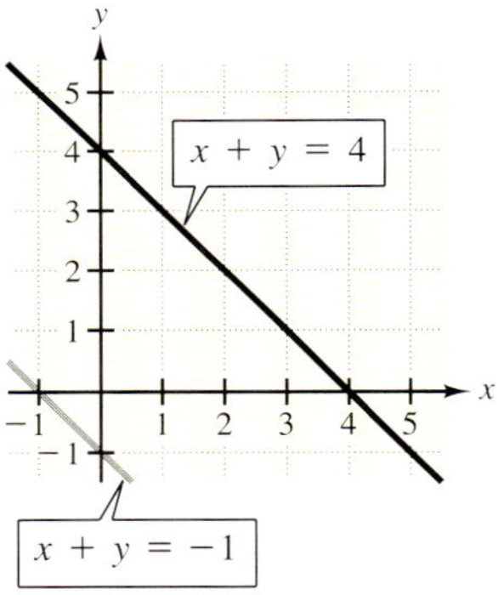

32. $-x + y = 5$
$x + 2y = 4$

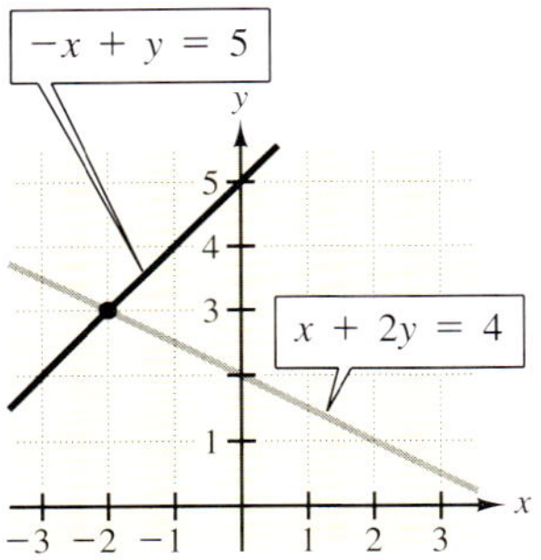

33. $5x - 3y = 4$
$2x + 3y = 3$

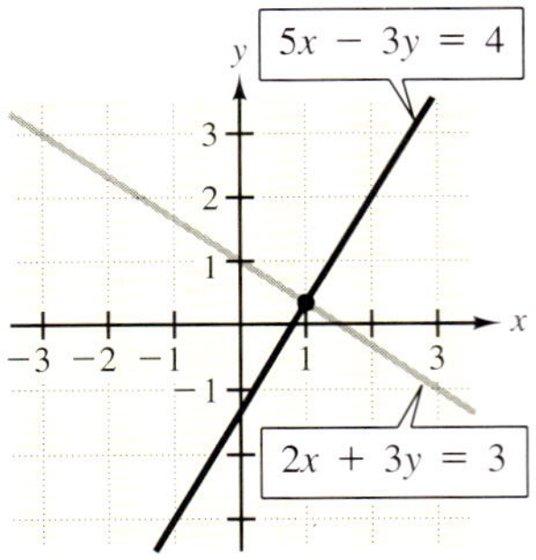

34. $2x - y = 4$
$-4x + 2y = -12$

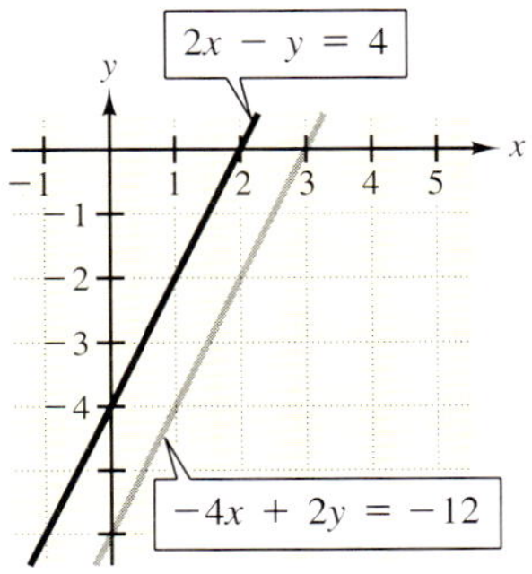

35. $x - 2y = 4$
$x^2 - y = 0$

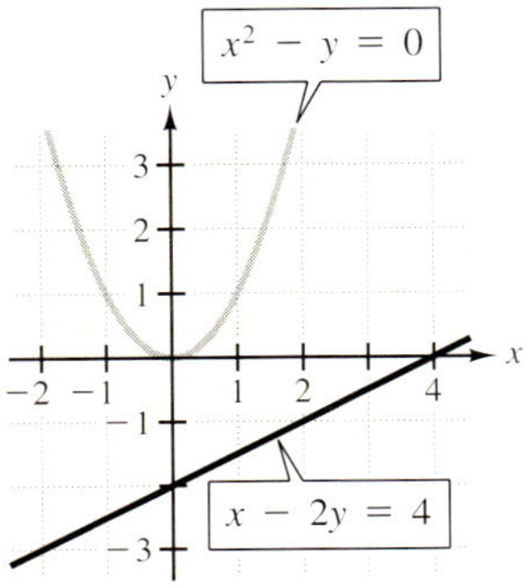

36. $y = \sqrt{x - 2}$
$x - 2y = 1$

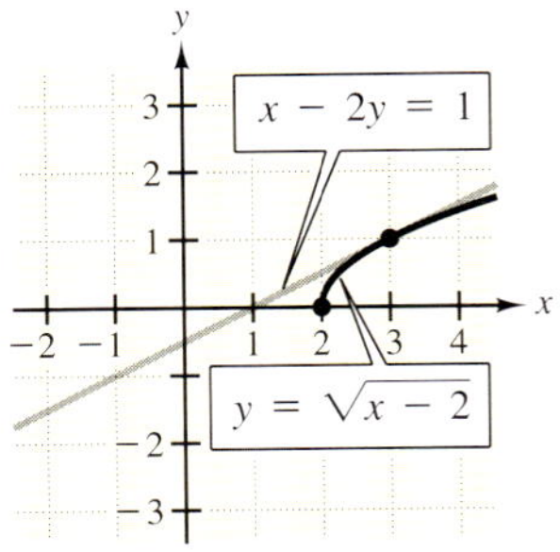

In Exercises 37–50, sketch the graphs of the given equations and find any solutions of the system of equations.

37. $-2x + y = 1$
$x - 3y = 2$

38. $5x - 6y = -30$
$5x + 4y = 20$

39. $2x - 5y = 20$
$4x - 5y = 40$

40. $5x + 3y = 24$
$x - 2y = 10$

41. $x - y = -3$
$2x - y = 6$

42. $x = 4$
$y = 3$

43. $-5x + 3y = 15$
$x + y = 1$

44. $9x + 4y = 8$
$7x - 4y = 8$

45. $\sqrt{x} - y = 0$
$x - 5y = -6$

46. $x^2 + y = 4$
$x + y = 6$

47. $x^2 - y^2 = 12$
$x - 2y = 0$

48. $x^2 + y^2 = 20$
$x + 3y = 10$

49. $16x^2 + 9y^2 = 144$
$4x + 3y = 12$

50. $y = x^2$
$y = -x^2 - 4x$

In Exercises 51 and 52, the graphs of the two equations appear parallel. Yet when each equation is written in slope-intercept form, we find that the slopes are not equal and thus, the graphs intersect. Find the point of intersection of the two lines.

51. $x - 100y = -200$
$3x - 275y = 198$

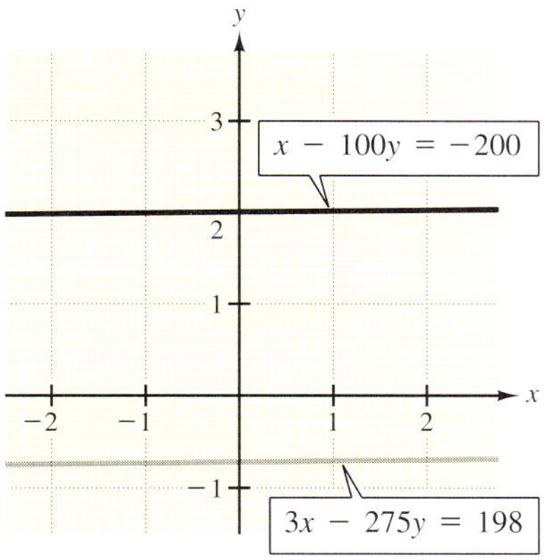

52. $35x - 33y = 0$
$12x - 11y = 92$

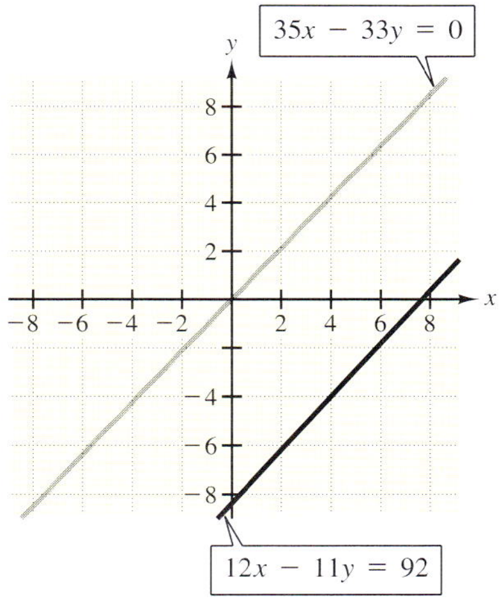

In Exercises 53–56, find two positive integers that satisfy the given requirements.

53. The sum of two numbers is 80 and their difference is 18.

54. The sum of a larger number and twice the smaller number is 61 and their difference is 7.

55. The sum of two numbers is 52 and the larger number is eight less than twice the smaller number.

56. The sum of two numbers is 160 and the larger number is three times the smaller number.

Dimensions of a Rectangle In Exercises 57–60, find the dimensions of the rectangle that meets the specified conditions.

	Perimeter	Relationship Between the Length and Width
57.	50 feet	Length is 5 feet greater than the width.
58.	320 inches	Width is 20 inches less than the length.
59.	68 yards	Width is seven-tenths the length.
60.	90 meters	Length is one and one-half times the width.

61. *Simple Interest* A combined total of \$20,000 is invested in two bonds that pay 8% and 9.5% simple interest. The annual interest is \$1,675. How much is invested in each bond?

62. *Simple Interest* A combined total of \$12,000 is invested in two bonds that pay 8.5% and 10% simple interest. The annual interest is \$1,140. How much is invested in each bond?

63. *Break-Even Analysis* A small business invests \$8,000 in equipment to produce a product. Each unit of the product costs \$1.20 to produce and is sold for \$2.00. How many items must be sold before the business breaks even?

64. *Break-Even Analysis* A business invests \$50,000 in equipment to produce a product. Each unit of the product costs \$19.25 to produce and is sold for \$35.95. How many items must be sold before the business breaks even?

65. *Number Problem* The sum of the digits of a given two-digit number is 9. If the digits are reversed, the number is decreased by 27. Find the number.

66. *Dimensions of a Corral* Suppose you have 250 feet of fencing to enclose two corrals of equal size (see figure). The combined area of the corrals is 2,400 square feet. Find the dimensions of each corral.

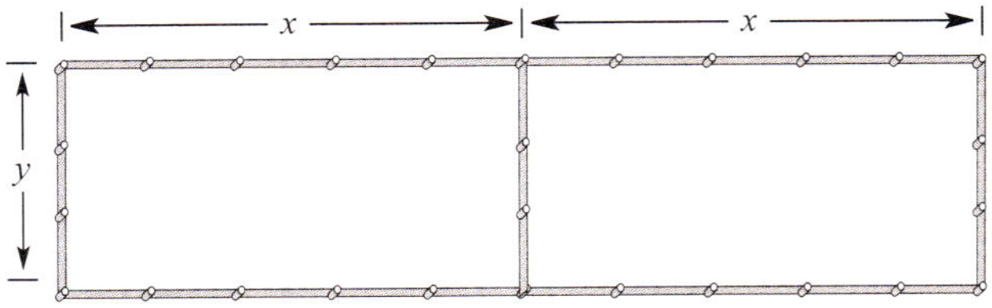

In Exercises 67–70, use the result of the discussion problem to create a system of equations having the given solution.

67. $(4, 5)$

68. $(-2, 6)$

69. $(-1, -2)$

70. $\left(\frac{1}{2}, 3\right)$

In Exercises 71 and 72, use a graphing calculator to sketch the graphs of the two equations on the same screen.* Use the TRACE key to approximate the points of intersection of the graphs. (Note that the up and down cursor control keys will move the cursor to the second graph.)

71. $y = x^3$
$y = x^3 - 3x^2 + 3x$

72. $y = \frac{1}{5}(24 - x)$
$y = \sqrt{64 - x^2}$

*The graphing calculator keystrokes given in this text correspond to the *TI-81* graphing calculator from *Texas Instruments*. However, you can also use other graphing calculators and computer graphing software for these examples and exercises.

SECTION 9.2

Systems of Linear Equations in Two Variables

The Method of Elimination • Graphical Interpretation of Solutions • Applications

The Method of Elimination

In Section 9.1 we discussed two methods of solving a system of equations (substitution and graphing). In this section we discuss a third method, called the **method of elimination**. The key step in the method of elimination is to obtain, for one of the variables, coefficients that differ only in sign, so that by *adding* the two equations this variable will be eliminated. This is true for the following system.

$$\begin{aligned} 2x + 7y &= 16 && \text{Equation 1} \\ -2x - 3y &= -8 && \text{Equation 2} \\ \hline 4y &= 8 && \text{Add equations} \end{aligned}$$

Note that by adding the two equations, we eliminated the variable x and obtained a single equation in y. Solving this equation for y produces $y = 2$, which we then back-substitute into one of the original equations to solve for x. We demonstrate the method of elimination in more detail in Example 1.

EXAMPLE 1 ■ The Method of Elimination

Solve the following system of linear equations.

$$\begin{aligned} 3x + 2y &= 4 && \text{Equation 1} \\ 5x - 2y &= 8 && \text{Equation 2} \end{aligned}$$

Solution

We begin by noting that the coefficients of y differ only in sign. Therefore, by adding the two equations, we can eliminate y.

$$\begin{aligned} 3x + 2y &= 4 && \text{Equation 1} \\ 5x - 2y &= 8 && \text{Equation 2} \\ \hline 8x \qquad &= 12 && \text{Add equations} \end{aligned}$$

Therefore, $x = \frac{3}{2}$. By back-substituting this value into the first equation, we can solve for y, as follows.

$$\begin{aligned} 3x + 2y &= 4 && \text{Equation 1} \\ 3\left(\frac{3}{2}\right) + 2y &= 4 && \text{Replace } x \text{ by } \tfrac{3}{2} \\ 2y &= -\frac{1}{2} \\ y &= -\frac{1}{4} && \text{Solve for } y \end{aligned}$$

Therefore, the solution is $\left(\frac{3}{2}, -\frac{1}{4}\right)$. Check this solution in the original system of linear equations. ■

Try using the method of substitution to solve the system given in Example 1. Which method do you think is easier? Many people find that the method of elimination is more efficient. We summarize the method of elimination as follows.

The Method of Elimination

To use the **method of elimination** to solve a system of two linear equations in x and y, use the following steps.

1. Obtain coefficients for x (or y) that differ only in sign by multiplying all terms of one or both equations by suitable constants.
2. Add the equations to eliminate one variable and solve the resulting equation.
3. Back-substitute the value obtained in Step 2 into either of the original equations and solve for the other variable.
4. Check your solution in both of the original equations.

To obtain coefficients (for one of the variables) that differ only in sign, we often need to multiply one or both of the equations by a suitable constant. This is demonstrated in the next two examples.

EXAMPLE 2 ■ The Method of Elimination

Solve the following system of linear equations.

$$4x - 5y = 13 \qquad \text{Equation 1}$$
$$3x - y = 7 \qquad \text{Equation 2}$$

Solution

For this system, we can obtain coefficients of y that differ only in sign by multiplying the second equation by -5.

$$4x - 5y = 13 \quad \Rightarrow \quad 4x - 5y = 13 \qquad \text{Equation 1}$$
$$3x - y = 7 \quad \Rightarrow \quad -15x + 5y = -35 \qquad \text{Multiply Equation 2 by } -5$$
$$-11x = -22 \qquad \text{Add equations}$$

Thus, we see that $x = 2$. By back-substituting this value of x into the second equation, we can solve for y.

$$3x - y = 7 \quad \text{Equation 2}$$
$$3(2) - y = 7 \quad \text{Replace } x \text{ by } 2$$
$$y = -1 \quad \text{Solve for } y$$

Therefore, the solution is $(2, -1)$. Check to see that it satisfies both original equations. ■

EXAMPLE 3 ■ **The Method of Elimination**

Solve the following system of linear equations.

$$5x + 3y = 9 \quad \text{Equation 1}$$
$$2x - 4y = 14 \quad \text{Equation 2}$$

Solution

We can obtain coefficients of y that differ only in sign by multiplying the first equation by 4 and the second equation by 3.

$$5x + 3y = 9 \Rightarrow 20x + 12y = 36 \quad \text{Multiply Equation 1 by 4}$$
$$2x - 4y = 14 \Rightarrow 6x - 12y = 42 \quad \text{Multiply Equation 2 by 3}$$
$$26x = 78 \quad \text{Add equations}$$

From this equation, we can see that $x = 3$. By back-substituting this value of x into the second equation, we can solve for y, as follows.

$$2x - 4y = 14 \quad \text{Equation 2}$$
$$2(3) - 4y = 14 \quad \text{Replace } x \text{ by } 3$$
$$-4y = 8$$
$$y = -2 \quad \text{Solve for } y$$

Therefore, the solution is $(3, -2)$. Check this solution in each of the original equations. ■

Graphical Interpretation of Solutions

As we observed in Section 9.1, it is possible for a *general* system of equations to have exactly one solution, two or more solutions, or no solution. We can strengthen this result somewhat for a system of *linear* equations. Specifically, if a system of linear equations has two different solutions, then it must have an infinite number of solutions! To see why this is true, consider the following graphical interpretations of a system of two linear equations in two variables. (Remember that the graph of a linear equation in two variables is a straight line.)

Graphical Interpretation of Solutions

For a system of two linear equations in two variables, the number of solutions is given by one of the following.

Number of Solutions	*Graphical Interpretation*
1. Exactly one solution	The two lines intersect at one point.
2. Infinitely many solutions	The two lines are identical.
3. No solution	The two lines are parallel.

These three possibilities are shown in Figure 9.4.

FIGURE 9.4

Consistent   Inconsistent

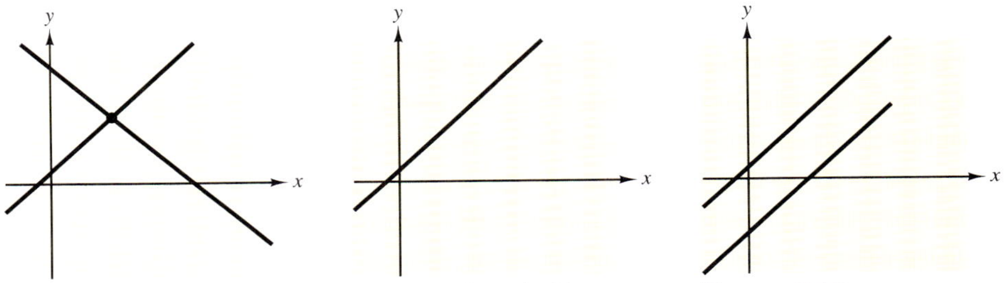

Two lines that intersect
Single point of intersection

Two lines that coincide
Infinitely many points of intersection

Two parallel lines
No point of intersection

A system of linear equations is called **consistent** if it has at least one solution, and it is called **inconsistent** if it has no solution. In the next example, note how we can use the method of elimination to determine that a system of linear equations has no solution.

EXAMPLE 4 ■ The Method of Elimination: No-Solution Case

Solve the following system of linear equations.

$$3x + 9y = 8 \qquad \text{Equation 1}$$
$$2x + 6y = 7 \qquad \text{Equation 2}$$

Solution

To obtain coefficients of x that differ only in sign, we multiply the first equation by 2 and the second equation by -3.

$$3x + 9y = 8 \Rightarrow 6x + 18y = 16 \qquad \text{Multiply Equation 1 by 2}$$
$$2x + 6y = 7 \Rightarrow -6x - 18y = -21 \qquad \text{Multiply Equation 2 by } -3$$
$$0 = -5 \qquad \text{False statement}$$

Since there are no values of x and y for which $0 = -5$, we conclude that the system is inconsistent and has no solution. The lines corresponding to the two equations given in this system are shown in Figure 9.5. Note that the two lines are parallel, and therefore have no point of intersection.

FIGURE 9.5

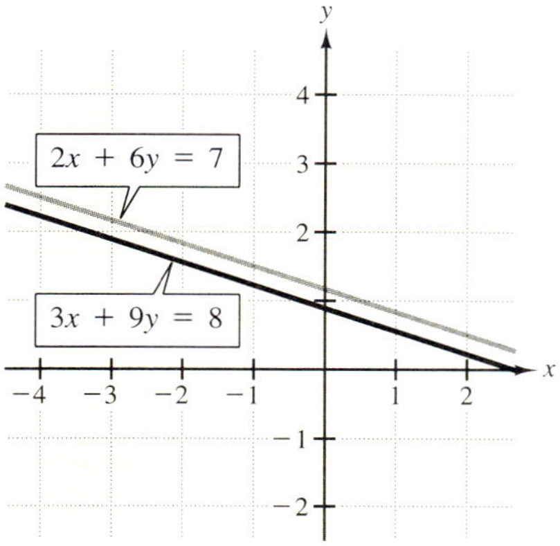

Example 5 shows how the method of elimination works with a system that has infinitely many solutions.

EXAMPLE 5 ■ The Method of Elimination: Many-Solutions Case

Solve the following system of linear equations.

$$-2x + 6y = 3 \qquad \text{Equation 1}$$
$$4x - 12y = -6 \qquad \text{Equation 2}$$

Solution

To obtain coefficients of x that differ only in sign, multiply the first equation by 2.

$$-2x + 6y = 3 \;\Rightarrow\; -4x + 12y = 6 \qquad \text{Multiply Equation 1 by 2}$$
$$4x - 12y = -6 \;\Rightarrow\; 4x - 12y = -6 \qquad \text{Equation 2}$$
$$0 = 0 \qquad \text{Add equations}$$

Since the two equations turned out to be equivalent, we conclude that the system has infinitely many solutions. The solution set consists of all points (x, y) lying on the line $-2x + 6y = 3$, as shown in Figure 9.6.

FIGURE 9.6

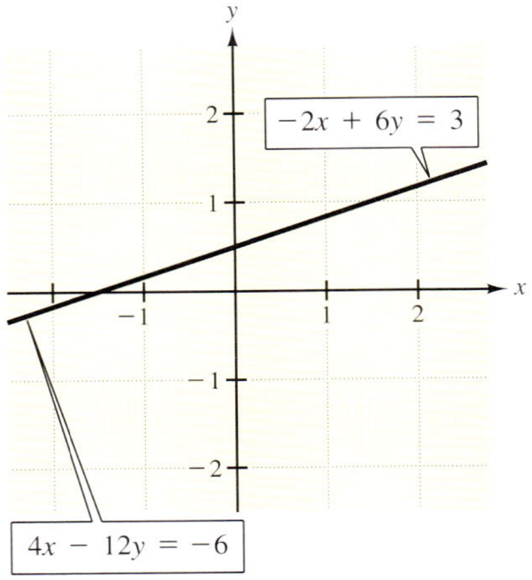

As our next example in this section, we show how the method of elimination works with a system of linear equations having decimal coefficients.

EXAMPLE 6 ■ Solving a Linear System Having Decimal Coefficients

Solve the following system of linear equations.

$$\begin{aligned} 0.02x - 0.05y &= -0.38 && \text{Equation 1} \\ 0.03x + 0.04y &= 1.04 && \text{Equation 2} \end{aligned}$$

Solution

Because the coefficients in this system have two decimal places, we begin by multiplying each equation by 100. (This produces a system in which the coefficients are all integers.)

$$\begin{aligned} 2x - 5y &= -38 && \text{Revised Equation 1} \\ 3x + 4y &= 104 && \text{Revised Equation 2} \end{aligned}$$

Now, to obtain coefficients of x that differ only in sign, we multiply the first equation by 3 and the second equation by -2.

$$\begin{aligned} 2x - 5y &= -38 \quad \Rightarrow & 6x - 15y &= -114 && \text{Multiply Equation 1 by 3} \\ \underline{3x + 4y} &\underline{= 104} \quad \Rightarrow & \underline{-6x - 8y} &\underline{= -208} && \text{Multiply Equation 2 by } -2 \\ & & -23y &= -322 && \text{Add equations} \end{aligned}$$

Thus, we find that

$$y = \frac{-322}{-23} = 14.$$

Back-substituting this value into Equation 2 produces the following.

$3x + 4y = 104$	Equation 2
$3x + 4(14) = 104$	Replace y by 14
$3x = 48$	
$x = 16$	Solve for x

Therefore, the solution is (16, 14). Check this solution in each of the original equations in the system. ■

Applications

We stated at the beginning of this chapter that systems of linear equations have many applications in science, business, health services, and government. The question that may come to mind is, How can I tell which application problems can be solved using a system of linear equations? The answer comes from the following considerations.

1. Does the problem involve more than one unknown quantity?
2. Are there two (or more) equations or conditions to be satisfied?

If one or both of these conditions occur, then the appropriate mathematical model for the problem may be a system of linear equations. Example 7 shows how to construct such a model.

EXAMPLE 7 ■ A Coin Mixture Problem

A cash register contains \$21.20 in dimes and quarters. If there are 119 coins in all, how many of each coin are in the cash register?

Solution

Note that this problem involves two unknowns: the number of dimes and the number of quarters.

Verbal model:

[Number of dimes] + [Number of quarters] = [119]

[Value of dimes] + [Value of quarters] = [\$21.20]

Labels:

x = Number of dimes
y = Number of quarters
$0.10x$ = Value of dimes (in dollars)
$0.25x$ = Value of quarters (in dollars)

System of equations:

$x + y = 119$	Equation 1
$0.10x + 0.25y = 21.20$	Equation 2

To solve this equation, we use the method of elimination. We begin by multiplying the second equation by 100 to eliminate the decimal points. Then, to obtain coefficients of x that differ only in sign, we multiply the first equation by -10.

$$\begin{array}{rcl} x + y = 119 & \Rightarrow & -10x - 10y = -1{,}190 \\ 0.10x + 0.25y = 21.20 & \Rightarrow & 10x + 25y = 2{,}120 \\ & & 15y = 930 \end{array}$$

Therefore, it follows that $y = 62$. By back-substituting this value of y into the first equation, we find that $x = 119 - 62 = 57$. Thus, we conclude that the cash register has 57 dimes and 62 quarters. ■

EXAMPLE 8 ■ A Mixture Problem

A company with two stores buys six large delivery vans and five small delivery vans. The first store receives four of the large vans and two of the small vans for a total cost of \$160,000. The second store receives two of the large vans and three of the small vans for a total cost of \$128,000. What is the cost of each type of van?

Solution

The two unknowns in this problem are the costs of the two types of vans.

Verbal model: $4\left(\text{Cost of large van}\right) + 2\left(\text{Cost of small van}\right) = \$160{,}000$

$2\left(\text{Cost of large van}\right) + 3\left(\text{Cost of small van}\right) = \$128{,}000$

Labels: x = Cost of large van (in dollars)
y = Cost of small van (in dollars)

System of equations:

$4x + 2y = 160{,}000$ Equation 1

$2x + 3y = 128{,}000$ Equation 2

To solve this system of linear equations, we use the method of elimination. To obtain coefficients of x that differ only in sign, we multiply the second equation by -2.

$$\begin{array}{rcl} 4x + 2y = 160{,}000 & \Rightarrow & 4x + 2y = 160{,}000 \\ 2x + 3y = 128{,}000 & \Rightarrow & -4x - 6y = -256{,}000 \\ & & -4y = -96{,}000 \end{array}$$

Thus, the cost of each small van is $y = \$24,000$. By back-substituting this value into Equation 1, we can find the cost of each large van.

$$\begin{aligned} 4x + 2y &= 160,000 && \text{Equation 1} \\ 4x + 2(24,000) &= 160,000 && \text{Replace } y \text{ by } 24,000 \\ 4x &= 112,000 \\ x &= 28,000 && \text{Solve for } x \end{aligned}$$

Thus, the cost of each large van is $x = \$28,000$. Check this solution in the original statement of the problem. ■

EXAMPLE 9 ■ **An Application Involving Two Speeds**

Suppose you take a motorboat trip on a river (18 miles upstream and 18 miles back downstream). You run the motor at the same speed going up and down the river, but because of the current of the river, the trip takes longer going upstream than it does downstream. You don't know the speed of the river's current, but you know that the trip upstream takes $1\frac{1}{2}$ hours and the trip downstream takes only 1 hour. From this information, can you determine the speed of the current?

Solution

One unknown in this problem is the speed of the current. The other unknown is the speed of the boat in still water. To set up a model, we use the fact that the effective speed of the boat (relative to the land) going upstream is $18/\left(1\frac{1}{2}\right) = 12$ miles per hour, and going downstream is $18/1 = 18$ miles per hour. In the following verbal model, note that the speed of the current fights against you going upstream but helps you going downstream.

Verbal model:

Boat speed (still water) − Speed of current = Upstream speed

Boat speed (still water) + Speed of current = Downstream speed

Labels:

x = Boat speed in still water (in miles per hour)
y = Current speed (in miles per hour)
Upstream speed = 12 miles per hour
Downstream speed = 18 miles per hour

System of equations:

$$\begin{aligned} x - y &= 12 && \text{Equation 1} \\ x + y &= 18 && \text{Equation 2} \end{aligned}$$

To solve this system of linear equations, we use the method of elimination.

$$\begin{aligned} x - y &= 12 && \text{Equation 1} \\ x + y &= 18 && \text{Equation 2} \\ \hline 2x \quad &= 30 && \text{Add equations} \end{aligned}$$

Thus, the speed of the boat in still water is $x = 15$ miles per hour. To find the speed of the current, we back-substitute this value into Equation 2.

$$x + y = 18 \quad \text{Equation 2}$$
$$15 + y = 18 \quad \text{Replace } x \text{ by } 15$$
$$y = 3 \quad \text{Solve for } y$$

Therefore, the speed of the current is 3 miles per hour. Check this solution in the original statement of the problem. ■

DISCUSSION PROBLEM ■ Creating Consistent and Inconsistent Systems

Consider the following system of linear equations.

$$x + y = 8$$
$$2x + 2y = k$$

(a) Find the value(s) of k for which the system has an infinite number of solutions.

(b) Find one value of k for which the system has no solutions.

(c) Can the system have a unique solution for some value of k? Why or why not? ■

Warm-Up

The following warm-up exercises involve skills that were covered in earlier sections. You will use these skills in the exercise set for this section.

In Exercises 1–4, simplify the given expression.

1. $(4x + 3y) - 3(5x + y)$

2. $(-15u + 4v) + 5(3u - 9v)$

3. $2x^2 + (2x - 3)^2 + 12x$

4. $y^2 - (y + 2)^2 + 4y$

In Exercises 5 and 6, solve the given equation.

5. $5x - (2x - 5) = 17$

6. $4t - 3(t + 1) = 8$

In Exercises 7–10, determine whether the lines represented by the pair of equations are parallel, perpendicular, or neither.

7. $3x - 4y = -10$
$4x + 3y = 11$

8. $6x - 18y = 8$
$-2x + 6y = 1$

9. $0.3x - 0.5y = 1$
$-1.2x + 2y = 2$

10. $2x + y = 2$
$3x + 4y = 1$

9.2 EXERCISES

In Exercises 1–10, solve the system of linear equations by elimination. Identify and label each line with the proper equation and label the point of intersection (if any).

1. $-x + 2y = 1$
$x - y = 2$

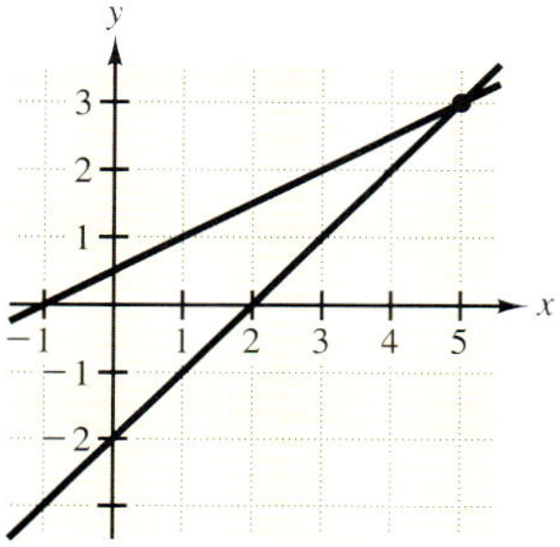

2. $3x + y = 3$
$2x - y = 7$

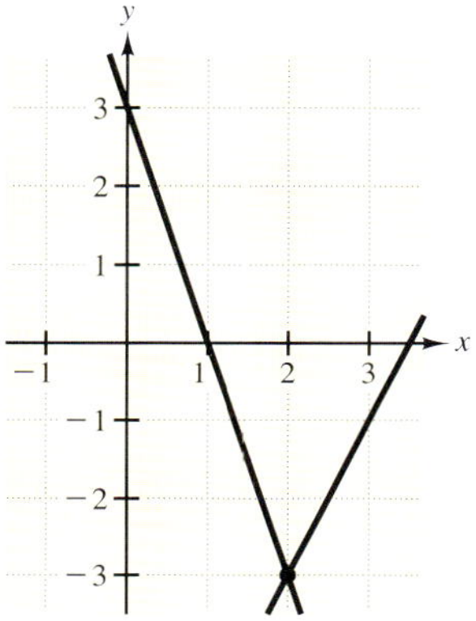

3. $x + y = 0$
$3x - 2y = 10$

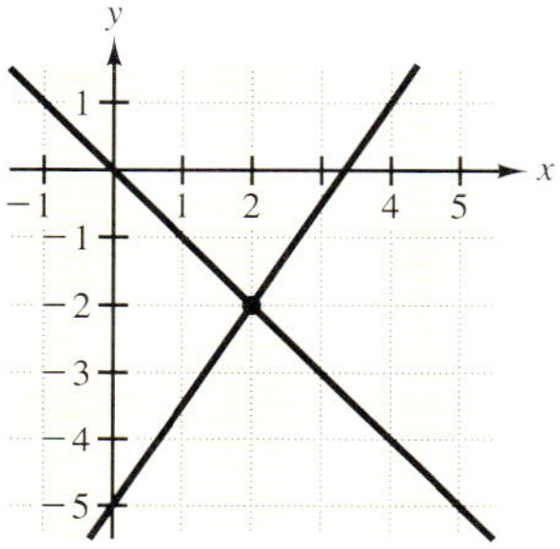

4. $-x + 2y = 2$
$3x + y = 15$

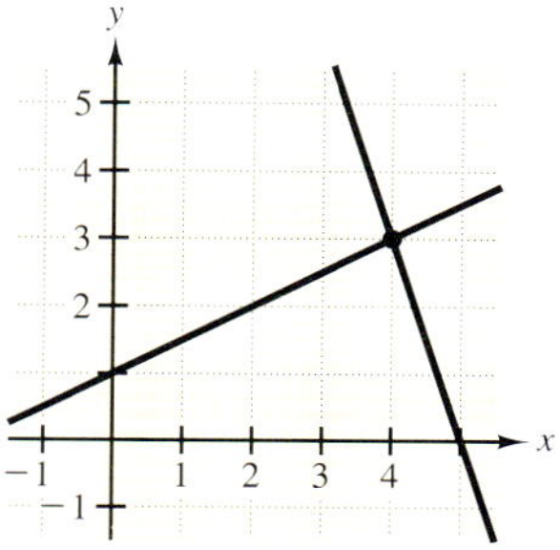

5. $x - y = 1$
$-3x + 3y = 8$

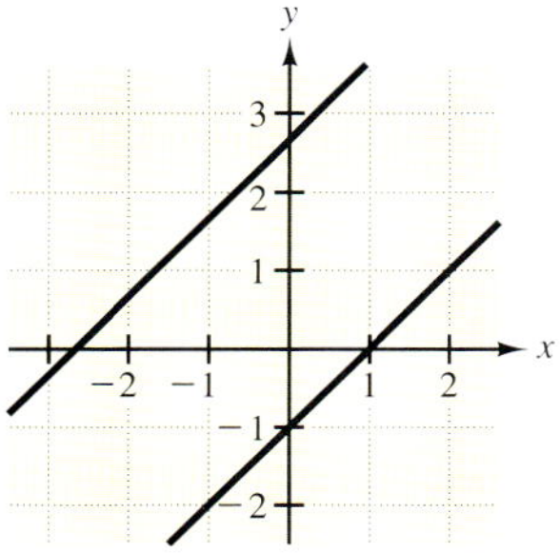

6. $3x + 4y = 2$
$0.6x + 0.8y = 1.6$

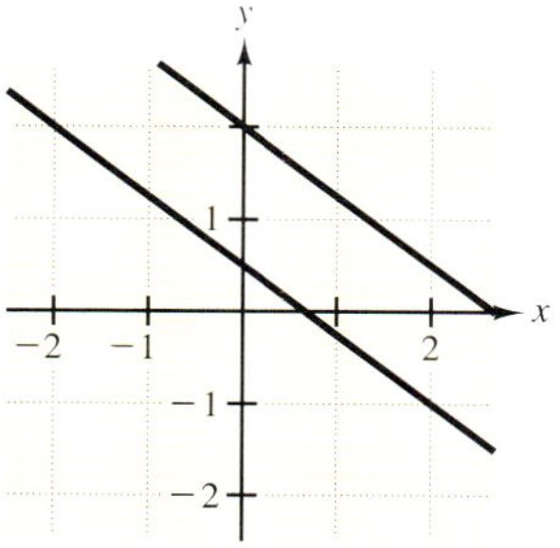

7. $x - 3y = 5$
$-2x + 6y = -10$

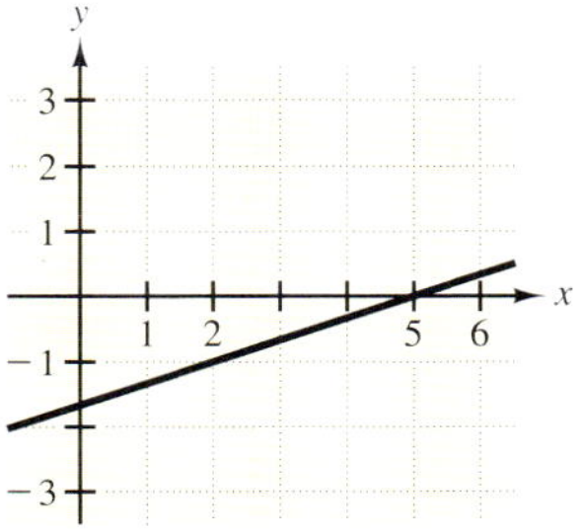

8. $x - 4y = 5$
$5x + 4y = 7$

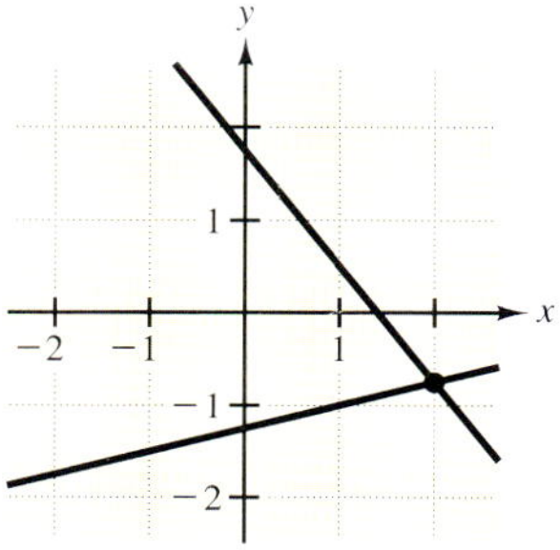

9. $2x - 8y = -11$
$5x + 3y = 7$

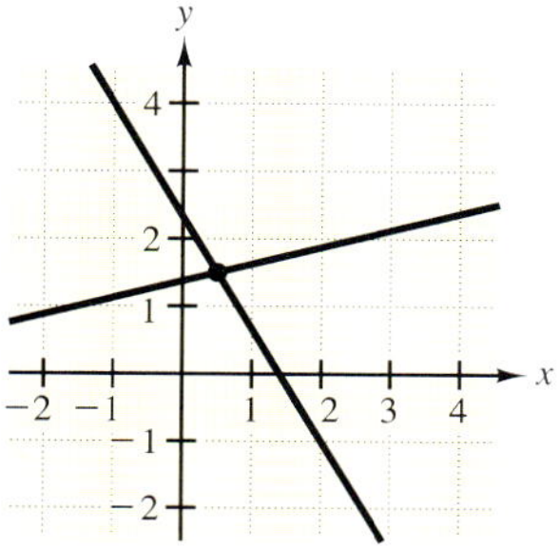

10. $3x + 4y = 0$
$9x - 5y = 17$

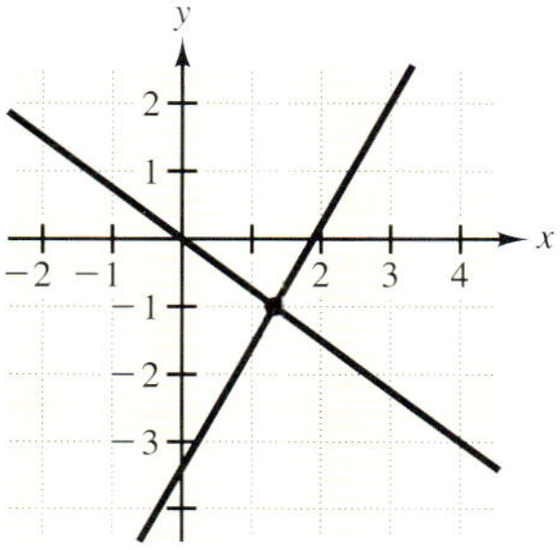

In Exercises 11–30, solve the system of linear equations by the method of elimination.

11. $3x - 2y = 5$
$x + 2y = 7$

12. $-x + 2y = 9$
$x + 3y = 16$

13. $4x + y = -3$
$-4x + 3y = 23$

14. $-3x + 5y = -23$
$2x - 5y = 22$

15. $x - 3y = 2$
$3x - 7y = 4$

16. $7r - s = -25$
$2r + 5s = 14$

17. $2u + 3v = 8$
$3u + 4v = 13$

18. $4x - 3y = 25$
$-3x + 8y = 10$

19. $12x - 5y = 2$
$-24x + 10y = 6$

20. $-2x + 3y = 9$
$6x - 9y = -27$

21. $\frac{2}{3}r - s = 0$
$10r + 4s = 19$

22. $x - y = -\frac{1}{2}$
$4x - 48y = -35$

23. $0.7u - v = -0.4$
$0.3u - 0.8v = 0.2$

24. $0.15x - 0.35y = -0.5$
$-0.12x + 0.25y = 0.1$

25. $5x + 7y = 25$
$x + 1.4y = 5$

26. $12b - 13m = 2$
$-6b + 6.5m = -2$

27. $2x = 25$
$4x - 10y = 0.52$

28. $6x - 6y = 25$
$3y = 11$

29. $2x + 3y = 0$
$3x + 5y = -1{,}000$

30. $0.4u + 1v = 800$
$0.7u + 2v = 1{,}850$

In Exercises 31–38, use the more convenient of the methods of substitution or elimination to solve the given system of equations.

31. $\frac{3}{2}x + 2y = 12$
$\frac{1}{4}x + y = 4$

32. $4x + y = -2$
$-6x + y = 18$

33. $y = 5x - 3$
$y = -2x + 11$

34. $3x + 2y = 5$
$y = 2x + 13$

35. $2x - y = 20$
$y = \sqrt{x} - 5$

36. $3x - 2y = -20$
$5x + 6y = 32$

37. $2x - 3y = -21$
$4y = 50$

38. $5x + 2y = 65$
$-4x + 3y = -6$

In Exercises 39 and 40, find a system of linear equations with integer coefficients that has the given point as a solution. (*Note:* There are many correct answers to each problem.)

39. $\left(3, -\frac{3}{2}\right)$

40. $(-8, 12)$

41. Determine the value of k so that the following system of linear equations is inconsistent.

$5x - 10y = 40$
$-2x + ky = 30$

42. Determine the value of k so that the following system is inconsistent.

$12x - 18y = 5$
$-18x + ky = 10$

In Exercises 43–48, write each equation of the given system of linear equations in slope-intercept form and use this form to determine whether the system has one solution, many solutions, or no solution. (It is not necessary to sketch the graphs of the equations.)

43. $4x - 5y = 3$
$-8x + 10y = -6$

44. $4x - 5y = 3$
$-8x + 10y = 14$

45. $-2x + 5y = 3$
$5x + 2y = 8$

46. $x + 10y = 12$
$-2x + 5y = 2$

47. $-10x + 15y = 25$
$2x - 3y = -24$

48. $4x - 5y = 28$
$-2x + 2.5y = -14$

Dimensions of a Rectangle In Exercises 49 and 50, find the dimensions of the rectangle that meets the specified conditions.

	Perimeter	Relationship Between the Length and Width
49.	220 meters	Length is 120% of the width.
50.	280 feet	Width is 75% of the length.

Coin Problem In Exercises 51–54, determine how many coins of each type will yield the given value.

	Coins	Value
51.	15 dimes and quarters	\$3.00
52.	32 dimes and quarters	\$6.95
53.	25 nickels and dimes	\$1.95
54.	50 nickels and quarters	\$7.50

55. *Gasoline Mixture* Twelve gallons of regular unleaded gasoline plus eight gallons of premium unleaded gasoline cost \$23.08. The price of premium unleaded is 11 cents more per gallon than the price of regular unleaded. Find the price per gallon for each grade of gasoline.

56. *Ticket Sales* Eight hundred tickets were sold for a theater production, and the receipts for the performance were \$8,600. The tickets for adults and students sold for \$12.50 and \$7.50, respectively. How many of each kind of ticket were sold?

57. *Nut Mixture* Ten pounds of mixed nuts sell for \$6.95 per pound. The mixture is obtained from two kinds of nuts, with one variety priced at \$5.65 per pound and the other at \$8.95 per pound. How many pounds of each variety of nuts were used in the mixture?

58. *Hay Mixture* How many tons of hay at \$125 per ton and \$75 per ton must be purchased to have 100 tons of hay with a value of \$90 per ton?

59. *Alcohol Mixture* How many liters of a 40% alcohol solution must be mixed with a 65% solution to obtain 20 liters of a 50% solution?

60. *Acid Mixture* Fifty gallons of 70% acid solution are obtained by mixing an 80% solution with a 50% solution. How many gallons of each solution must be used to obtain the desired mixture?

61. *Average Speed* A truck travels for four hours at an average speed of 42 miles per hour. How much longer must the truck travel at an average speed of 55 miles per hour so that the average speed for the total trip will be 50 miles per hour?

62. *Air Speed* An airplane flying into a headwind travels the 3,000-mile flying distance between two cities in 6 hours and 15 minutes. On the return flight, the distance is traveled in 5 hours. Find the speed of the plane in still air and the speed of the wind, assuming that both remain constant throughout the round trip.

63. *Rope Length* Suppose you must cut a rope that is 160 inches long into two pieces so that one piece is four times as long as the other. Find the length of each piece.

64. *Driving Distance* Two people share the driving on a trip of 300 miles. One person drives three times as far as the other. Find the distance that each person drives.

65. *Best-Fitting Line* The slope and y-intercept of the line $y = mx + b$ that best fits the three noncollinear points $(0, 0)$, $(1, 1)$, and $(2, 3)$ is given by the following system of linear equations.

$$3b + 5m = 7$$
$$3b + 3m = 4$$

(a) Solve the system and find the equation of the best-fitting line.
(b) Plot the three points and sketch the graph of the best-fitting line.

66. *Best-Fitting Line* The slope and y-intercept of the line $y = mx + b$ that best fits the three noncollinear points $(0, 3)$, $(1, 2)$, and $(2, 0)$ is given by the following system of linear equations.

$$3b + 5m = 2$$
$$3b + 3m = 5$$

(a) Solve the system and find the equation of the best-fitting line.
(b) Plot the three points and sketch the graph of the best-fitting line.

In Exercises 67 and 68, use a graphing calculator to sketch the graphs of the two equations on the same screen.* Use the TRACE key to approximate the points of intersection of the graphs.

67. $5x + 4y = 35$
$-x + 3y = 12$

68. $5x - 4y = 0$
$-3x + 8y = 14$

SECTION 9.3

Systems of Linear Equations in Three Variables

Row-Echelon Form • The Method of Elimination • Applications

Row-Echelon Form

The method of elimination introduced in Section 9.2 can be applied to a system of linear equations in more than two variables. In fact, this method easily adapts to computer use for solving systems of linear equations with dozens of variables.

When using the method of elimination to solve a system of linear equations, the goal is to rewrite the system in a form to which back-substitution can be applied. For instance, consider the following two systems of linear equations.

$$\begin{aligned} x - 2y + 2z &= 9 \\ -x + 3y &= -4 \\ 2x - 5y + z &= 10 \end{aligned} \qquad \begin{aligned} x - 2y + 2z &= 9 \\ y + 2z &= 5 \\ z &= 3 \end{aligned}$$

The system on the right is said to be in **row-echelon form**, which means that it has a "stair-step" pattern with leading coefficients of 1.

Which of these two systems do you think is easier to solve? After comparing the two systems, it should be clear that it is easier to solve the system on the right. We show how to use back-substitution to solve this system in Example 1.

EXAMPLE 1 ■ Using Back-Substitution

Solve the following system of linear equations.

$$x - 2y + 2z = 9 \qquad \text{Equation 1}$$
$$y + 2z = 5 \qquad \text{Equation 2}$$
$$z = 3 \qquad \text{Equation 3}$$

*The graphing calculator keystrokes given in this text correspond to the *TI-81* graphing calculator from *Texas Instruments*. However, you can also use other graphing calculators and computer graphing software for these examples and exercises.

Solution

From Equation 3, we know the value of z. To solve for y, we substitute $z = 3$ into Equation 2 to obtain

$$y + 2(3) = 5 \qquad \text{Substitute } z = 3$$
$$y = -1. \qquad \text{Solve for } y$$

Finally, we substitute $y = -1$ and $z = 3$ into Equation 1 to obtain

$$x - 2(-1) + 2(3) = 9 \qquad \text{Substitute } y = -1,\ z = 3$$
$$x = 1. \qquad \text{Solve for } x$$

Thus, the solution is $x = 1$, $y = -1$, and $z = 3$, which can also be written as the **ordered triple** $(1, -1, 3)$. ■

The Method of Elimination

Two systems of equations are called **equivalent** if they have the same solution set. To solve a system that is not in row-echelon form, we first convert it to an *equivalent* system that is in row-echelon form. To see how this is done, let's take another look at the method of elimination, as applied to a system of two linear equations.

EXAMPLE 2 ■ The Method of Elimination

Use the method of elimination to solve the following system of linear equations.

$$3x - 2y = -1 \qquad \text{Equation 1}$$
$$x - y = 0 \qquad \text{Equation 2}$$

Solution

1. $x - y = 0$
   $3x - 2y = -1$ — We can interchange two equations in the system.
2. $-3x + 3y = 0$ — We can multiply an equation by a nonzero number such as -3.
3. $-3x + 3y = 0$
   $3x - 2y = -1$
   $y = -1$ — We can add a multiple of one of the equations to another equation to obtain a new equation.
4. $x - y = 0$
   $y = -1$ — New system (in row-echelon form)

Now, using back-substitution on the row-echelon system, we find that the solution is $y = -1$ and $x = -1$. ■

We summarize the three basic types of operations that can be performed on a system of linear equations to produce an equivalent system as follows.

Operations that Lead to Equivalent Systems of Equations

Each of the following **row operations** on a system of linear equations produces an *equivalent* system of linear equations.

1. Interchange two equations.
2. Multiply one of the equations by a nonzero constant.
3. Add a multiple of one of the equations to another equation to replace the latter equation.

As shown in Example 2, rewriting a system of linear equations in row-echelon form usually involves a *chain* of equivalent systems, each of which is obtained by using one of the three basic row operations. This process is called **Gaussian elimination**, after the German mathematician Carl Friedrich Gauss (1777–1855). In Example 3 we show how to use Gaussian elimination to solve the system that was given at the beginning of the section.

EXAMPLE 3 ■ Using Elimination to Solve a System

Solve the following system of linear equations.

$$\begin{aligned} x - 2y + 2z &= 9 \\ -x + 3y &= -4 \\ 2x - 5y + z &= 10 \end{aligned}$$

Solution

Although there are several ways to begin, our goal is to develop a systematic procedure that can be applied to large systems. We work from the upper left corner, saving the x in the upper left position and eliminating the other x's from the first column.

$$\begin{aligned} x - 2y + 2z &= 9 \\ y + 2z &= 5 \\ 2x - 5y + z &= 10 \end{aligned}$$

Adding the first equation to the second equation produces a new second equation.

$$\begin{aligned} x - 2y + 2z &= 9 \\ y + 2z &= 5 \\ -y - 3z &= -8 \end{aligned}$$

Adding -2 times the first equation to the third equation produces a new third equation.

Now that all but the first x has been eliminated from the first column, we go to work on the second column. (We need to eliminate y from the third equation.)

$$\begin{aligned} x - 2y + 2z &= 9 \\ y + 2z &= 5 \\ -z &= -3 \end{aligned}$$

Adding the second equation to the third equation produces a new third equation.

Finally, we need a coefficient of 1 for z in the third equation.

$$\begin{aligned} x - 2y + 2z &= 9 \\ y + 2z &= 5 \\ z &= 3 \end{aligned}$$

Multiplying the third equation by -1 produces a new third equation.

This is the same system we solved in Example 1, and, as in that example, we conclude that the solution is

$$x = 1, \quad y = -1, \quad \text{and} \quad z = 3.$$

■

NOTE: The two systems of linear equations given in Examples 1 and 3 are equivalent to each other.

In Example 3 we can check the solution $x = 1$, $y = -1$, and $z = 3$ as follows.

$$\begin{aligned} \text{Equation 1:} &\quad (1) - 2(-1) + 2(3) = 9 \\ \text{Equation 2:} &\quad -(1) + 3(-1) = -4 \\ \text{Equation 3:} &\quad 2(1) - 5(-1) + (3) = 10 \end{aligned}$$

Check solution in each equation of original system.

EXAMPLE 4 ■ Using Elimination to Solve a System

Solve the following system of linear equations.

$$\begin{aligned} 4x + y - 3z &= 11 \\ 2x - 3y + 2z &= 9 \\ x + y + z &= -3 \end{aligned}$$

Solution

$$\begin{aligned} x + y + z &= -3 \\ 2x - 3y + 2z &= 9 \\ 4x + y - 3z &= 11 \end{aligned}$$

Interchange the first and third equations.

$$\begin{aligned} x + y + z &= -3 \\ -5y &= 15 \\ 4x + y - 3z &= 11 \end{aligned}$$

Adding -2 times the first equation to the second equation produces a new second equation.

$$\begin{aligned} x + y + z &= -3 \\ -5y \quad &= 15 \\ -3y - 7z &= 23 \end{aligned}$$

Adding -4 times the first equation to the third equation produces a new third equation.

$$\begin{aligned} x + y + z &= -3 \\ y \quad &= -3 \\ -3y - 7z &= 23 \end{aligned}$$

Multiplying the second equation by $-\frac{1}{5}$ produces a new second equation.

$$\begin{aligned} x + y + z &= -3 \\ y \quad &= -3 \\ -7z &= 14 \end{aligned}$$

Adding 3 times the second equation to the third equation produces a new third equation.

$$\begin{aligned} x + y + z &= -3 \\ y \quad &= -3 \\ z &= -2 \end{aligned}$$

Multiplying the third equation by $-\frac{1}{7}$ produces a new third equation.

Now that the system of equations is in row-echelon form, we can see that $z = -2$ and $y = -3$. Moreover, by back-substituting these values into Equation 1, we find that $x = 2$. Therefore, the solution is

$$x = 2, \qquad y = -3, \quad \text{and} \quad z = -2.$$ ■

We now look at an inconsistent system—one that has no solution. The key to recognizing an inconsistent system is that at some stage in the elimination process, we obtain an absurdity such as $0 = 7$. This is demonstrated in Example 5.

EXAMPLE 5 ■ An Inconsistent System

Solve the following system of linear equations.

$$\begin{aligned} x - 3y + z &= 1 \\ 2x - y - 2z &= 2 \\ x + 2y - 3z &= -1 \end{aligned}$$

Solution

$$\begin{aligned} x - 3y + z &= 1 \\ 5y - 4z &= 0 \\ x + 2y - 3z &= -1 \end{aligned}$$

Adding -2 times the first equation to the second equation produces a new second equation.

$$\begin{aligned} x - 3y + z &= 1 \\ 5y - 4z &= 0 \\ 5y - 4z &= -2 \end{aligned}$$

Adding -1 times the first equation to the third equation produces a new third equation.

$$\begin{aligned} x - 3y + z &= 1 \\ 5y - 4z &= 0 \\ 0 &= -2 \end{aligned}$$

Adding -1 times the second equation to the third equation produces a new third equation.

Because the third "equation" is impossible, we conclude that this system is inconsistent and therefore has no solution. Moreover, because this system is equivalent to the original system, we can conclude that the original system also has no solution. ■

As with a system of linear equations in two variables, the solution(s) to a system of linear equations in more than two variables must fall into one of three categories.

The Number of Solutions of a System of Linear Equations

For a system of linear equations, exactly one of the following is true.

1. There is exactly one solution.
2. There are infinitely many solutions.
3. There is no solution.

When a system of equations has no solution, we simply state that it is *inconsistent*. If a system has exactly one solution, we list the value of each variable. However, for systems that have infinitely many solutions, we encounter a certain awkwardness in listing the solutions. For example, we might give the solutions to a system in three variables as

$$(a, a + 1, 2a), \qquad \text{where } a \text{ is any real number.}$$

This means that for each real number a, we have a valid solution to the system. A few of the infinitely many possible solutions are found by letting $a = -1, 0, 1$, and 2 to obtain $(-1, 0, -2)$, $(0, 1, 0)$, $(1, 2, 2)$, and $(2, 3, 4)$, respectively. Now consider the solutions represented by

$$(b - 1, b, 2b - 2), \qquad \text{where } b \text{ is any real number.}$$

A few possible solutions are $(-1, 0, -2)$, $(0, 1, 0)$, $(1, 2, 2)$, and $(2, 3, 4)$, found by letting $b = 0, 1, 2$, and 3, respectively. Note that both descriptions result in the same collection of solutions. Thus, when comparing descriptions of an infinite solution set, keep in mind that there is more than one way to describe the set.

EXAMPLE 6 ■ A System with Infinitely Many Solutions

Solve the following system of linear equations.

$$
\begin{aligned}
x + y - 3z &= -1 && \text{Equation 1}\\
y - z &= 0 && \text{Equation 2}\\
-x + 2y &= 1 && \text{Equation 3}
\end{aligned}
$$

Solution

We begin by rewriting the system in row-echelon form as follows.

$$
\begin{aligned}
x + y - 3z &= -1\\
y - z &= 0\\
3y - 3z &= 0
\end{aligned}
$$

Adding the first equation to the third equation produces a new third equation.

$$
\begin{aligned}
x + y - 3z &= -1\\
y - z &= 0\\
0 &= 0
\end{aligned}
$$

Adding −3 times the second equation to the third equation produces a new third equation.

This means that Equation 3 depends on Equations 1 and 2 in the sense that it gives us no additional information about the variables. Thus, the original system is equivalent to the system

$$
\begin{aligned}
x + y - 3z &= -1\\
y - z &= 0.
\end{aligned}
$$

In this last equation, we solve for y in terms of z to obtain $y = z$. Back-substituting for y into the previous equation, we find x in terms of z, as follows.

$$
\begin{aligned}
x + z - 3z &= -1\\
x - 2z &= -1\\
x &= 2z - 1
\end{aligned}
$$

Finally, letting $z = a$, the solutions to the given system are all of the form

$$x = 2a - 1, \quad y = a, \quad z = a$$

where a is a real number. Thus, every ordered triple of the form

$$(2a - 1, a, a), \qquad a \text{ is a real number}$$

is a solution of the system. ■

MATH MATTERS

Circle Problem of Apollonius

Apollonius and Euclid are the two most famous mathematicians of the classical Greek period. Euclid lived about 300 B.C. and Apollonius lived about 100 years later. In one of the writings of Apollonius, he poses the following problem.

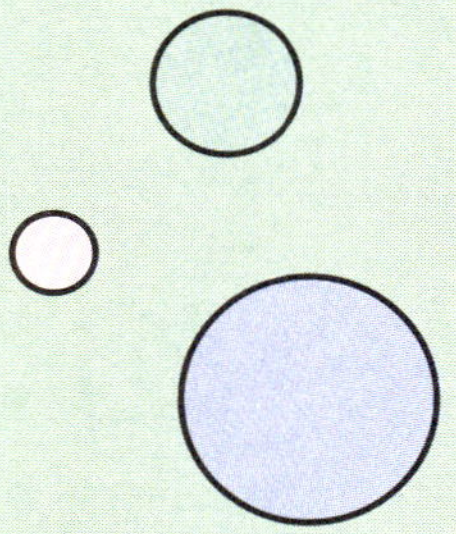

Suppose you are given three circles that have no points in common and no circle is inside either of the other two circles, as shown in the figure at right. Is it always possible to find a fourth circle that is tangent to each of the given circles? The answer is yes. In fact, it can be shown that the problem always has eight distinct solutions. Four are shown here. Can you find the other four? (The answers are given in the back of the text.)

Applications

We conclude this section with two applications involving systems of linear equations in three variables. In the first example, we show how to fit a parabola through three given points in the plane.

EXAMPLE 7 ■ **An Application: Curve Fitting**

Find a quadratic equation

$$y = ax^2 + bx + c$$

whose graph passes through the points $(-1, 3)$, $(1, 1)$, and $(2, 6)$.

Solution

Since the graph of $y = ax^2 + bx + c$ passes through the points $(-1, 3)$, $(1, 1)$, and $(2, 6)$, we have the following.

$$\begin{aligned}
&\text{When } x = -1,\ y = 3: && a(-1)^2 + b(-1) + c = 3\\
&\text{When } x = 1,\ y = 1: && a(1)^2 + b(1) + c = 1\\
&\text{When } x = 2,\ y = 6: && a(2)^2 + b(2) + c = 6
\end{aligned}$$

This produces the following system of linear equations in the variables a, b, and c.

$$\begin{aligned}
a - b + c &= 3\\
a + b + c &= 1\\
4a + 2b + c &= 6
\end{aligned}$$

The solution to this system turns out to be

$$a = 2, \qquad b = -1, \quad \text{and} \quad c = 0.$$

Thus, the equation of the parabola passing through the three given points is

$$y = 2x^2 - x$$

as shown in Figure 9.7.

FIGURE 9.7

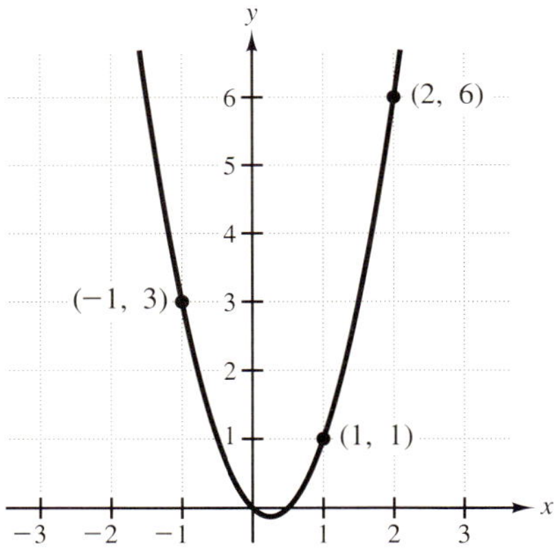

■

EXAMPLE 8 ■ An Application: Moving Object

The height at time t of an object that is moving in a (vertical) line with constant acceleration a is given by the **position equation**

$$s = \frac{1}{2}at^2 + v_0t + s_0.$$

The height s is measured in feet, the time t is measured in seconds, v_0 is the initial velocity (at time $t = 0$), and s_0 is the initial height. Find the values of a, v_0, and s_0, if $s = 164$ feet at 1 second, $s = 180$ feet at 2 seconds, and $s = 164$ feet at 3 seconds.

Solution

By substituting the three values of t and s into the position equation, we obtain three linear equations in a, v_0, and s_0.

$$\text{When } t = 1: \quad \frac{1}{2}a(1^2) + v_0(1) + s_0 = 164$$

$$\text{When } t = 2: \quad \frac{1}{2}a(2^2) + v_0(2) + s_0 = 180$$

$$\text{When } t = 3: \quad \frac{1}{2}a(3^2) + v_0(3) + s_0 = 164$$

By multiplying the first and third equations by 2, this system can be rewritten

$$\begin{aligned} a + 2v_0 + 2s_0 &= 328 \\ 2a + 2v_0 + s_0 &= 180 \\ 9a + 6v_0 + 2s_0 &= 328 \end{aligned}$$

and we apply elimination to obtain

$$\begin{aligned} a + 2v_0 + 2s_0 &= 328 \\ -2v_0 - 3s_0 &= -476 \\ 2s_0 &= 232. \end{aligned}$$

From the third equation, we find that $s_0 = 116$, so that back-substitution into the second equation yields

$$\begin{aligned} -2v_0 - 3(116) &= -476 \\ -2v_0 &= -128 \\ v_0 &= 64. \end{aligned}$$

Finally, by back-substituting $s_0 = 116$ and $v_0 = 64$ into the first equation, we have

$$\begin{aligned} a + 2(64) + 2(116) &= 328 \\ a &= -32. \end{aligned}$$

Thus, the position equation for this object is

$$s = -16t^2 + 64t + 116.$$

■

DISCUSSION PROBLEM ■ Computer Software

There are many computer software programs that will solve a system of linear equations. One is called MATRIXPAD and is available from D. C. Heath and Company. Use this software (or some other program) to solve the examples given in this section. For example, a screen from MATRIXPAD showing the solution to Example 8 is given below.

T-register	3	3 | 0		
Z-register	3	3 | 0		
Y-register	3	3 | 1		
	1	2	2	328
	2	2	1	180
	9	6	2	328
X-register	3	3 | 1		
	1	0	0	-32
	0	1	0	64
	0	0	1	116

Press <H> for Help   <Q> to Quit program

■

Warm-Up

The following warm-up exercises involve skills that were covered in earlier sections. You will use these skills in the exercise set for this section.

In Exercises 1–6, solve the system of linear equations.

1. $x + 3y = 17$
$2x + y = 14$

2. $4x + 2y = -2$
$3x - 2y = -5$

3. $-\frac{1}{2}x + \frac{3}{2}y = 2$
$2x - 6y = 1$

4. $\frac{2}{3}x + \frac{5}{6}y = 3$
$4x + 5y = 18$

5. $2x + 5y = -4$
$3x + 6y = -3$

6. $4x - 3y = -2$
$5x - 4y = -2$

In Exercises 7–10, simplify the given expression.

7. $(2x - 3y + 5z) + 3(x + y - 2z)$

8. $(x + 4y + 3z) - 3(2x - 10y + z)$

9. $-4(6x + 5y - 20z) + 3(8x + y - 15z)$

10. $5(2x - 3y + 4z) - 3(x - 5y - 2z)$

9.3 EXERCISES

In Exercises 1–4, use back-substitution to solve the system of linear equations.

1. $\begin{aligned} x - 2y + 4z &= 4 \\ 3y - z &= 2 \\ z &= -5 \end{aligned}$

2. $\begin{aligned} 5x + 4y - z &= 0 \\ 10y - 3z &= 11 \\ z &= 3 \end{aligned}$

3. $\begin{aligned} x - 2y + 4z &= 4 \\ y &= 3 \\ y + z &= 2 \end{aligned}$

4. $\begin{aligned} x &= 10 \\ 3x + 2y &= 2 \\ x + y + 2z &= 0 \end{aligned}$

In Exercises 5–26, use Gaussian elimination to solve the system of linear equations.

5. $\begin{aligned} x + y + z &= 6 \\ 2x - y + z &= 3 \\ 3x - z &= 0 \end{aligned}$

6. $\begin{aligned} x + y + z &= 2 \\ -x + 3y + 2z &= 8 \\ 4x + y &= 4 \end{aligned}$

7. $\begin{aligned} x + y + z &= -3 \\ 4x + y - 3z &= 11 \\ 2x - 3y + 2z &= 9 \end{aligned}$

8. $\begin{aligned} x - y + 2z &= -4 \\ 3x + y - 4z &= -6 \\ 2x + 3y - 4z &= 4 \end{aligned}$

9. $\begin{aligned} x + 2y + 6z &= 5 \\ -x + y - 2z &= 3 \\ x - 4y - 2z &= 1 \end{aligned}$

10. $\begin{aligned} x + 6y + 2z &= 9 \\ 3x - 2y + 3z &= -1 \\ 5x - 5y + 2z &= 7 \end{aligned}$

11. $\begin{aligned} 2x + 2z &= 2 \\ 5x + 3y &= 4 \\ 3y - 4z &= 4 \end{aligned}$

12. $\begin{aligned} 6y + 4z &= -12 \\ 3x + 3y &= 9 \\ 2x - 3z &= 10 \end{aligned}$

13. $\begin{aligned} 2x - 3y + 3z &= 5 \\ x - 3z &= 1 \\ 4x - 6y + 6z &= 10 \end{aligned}$

14. $\begin{aligned} 2x + y + 3z &= 1 \\ 2x + 6y + 8z &= 3 \\ 6x + 8y + 18z &= 5 \end{aligned}$

15. $\begin{aligned} 2x - 4y + z &= 0 \\ 3x + 2z &= -1 \\ -6x + 3y + 2z &= -10 \end{aligned}$

16. $\begin{aligned} 3x - y - 2z &= 5 \\ 2x + y + 3z &= 6 \\ 6x - y - 4z &= 9 \end{aligned}$

17. $\begin{aligned} y + z &= 5 \\ 2x + 4z &= 4 \\ 2x - 3y &= -14 \end{aligned}$

18. $\begin{aligned} 5x + 2y &= -8 \\ z &= 5 \\ 3x - y + z &= 9 \end{aligned}$

19. $\begin{aligned} 0.2x + 1.3y + 0.6z &= 0.1 \\ 0.1x + 0.3z &= 0.7 \\ 2x + 10y + 8z &= 8 \end{aligned}$

20. $\begin{aligned} 0.3x - 0.1y + 0.2z &= 0.35 \\ 2x + y - 2z &= -1 \\ 2x + 4y + 3z &= 10.5 \end{aligned}$

21. $\begin{aligned} x + 4y - 2z &= 2 \\ -3x + y + z &= -2 \\ 5x + 7y - 5z &= 6 \end{aligned}$

22. $\begin{aligned} 2x + 3y - z &= 1 \\ x + z &= 2 \\ x - 3y + 4z &= 5 \end{aligned}$

23. $\begin{aligned} x - 2y - z &= 3 \\ 2x + y - 3z &= 1 \\ x + 8y - 3z &= -7 \end{aligned}$

24. $\begin{aligned} 2x + z &= 1 \\ 5y - 3z &= 2 \\ 6x + 20y - 9z &= 11 \end{aligned}$

25. $\begin{aligned} 3x + y + z &= 2 \\ 4x + 2z &= 1 \\ 5x - y + 3z &= 0 \end{aligned}$

26. $\begin{aligned} 2x + 3z &= 4 \\ 5x + y + z &= 2 \\ 11x + 3y - 3z &= 0 \end{aligned}$

Curve Fitting In Exercises 27–30, find the equation of the parabola $y = ax^2 + bx + c$ that passes through the given points.

27. (0, −3), (1, −1), (2, 5)

28. (0, 0), (1, −1), (3, 3)

29. (−1, −3), (1, 1), (2, 0)

30. (−1, −1), (1, 1), (2, −4)

Curve Fitting In Exercises 31–34, find the equation of the circle $x^2 + y^2 + Dx + Ey + F = 0$ that passes through the given points.

31. (0, 0), (2, −2), (4, 0)

32. (0, 0), (0, 6), (−3, 3)

33. (3, −1), (−2, 4), (6, 8)

34. (0, 0), (0, 2), (3, 0)

In Exercises 35–38, find the position equation $s = \frac{1}{2}at^2 + v_0t + s_0$ for an object that has the indicated heights at the specified times.

35. $s = 128$ feet at $t = 1$ second
$s = 80$ feet at $t = 2$ seconds
$s = 0$ feet at $t = 3$ seconds

36. $s = 48$ feet at $t = 1$ second
$s = 64$ feet at $t = 2$ seconds
$s = 48$ feet at $t = 3$ seconds

37. $s = 32$ feet at $t = 1$ second
$s = 32$ feet at $t = 2$ seconds
$s = 0$ feet at $t = 3$ seconds

38. $s = 10$ feet at $t = 0$ second
$s = 54$ feet at $t = 1$ second
$s = 46$ feet at $t = 3$ seconds

39. Find a linear system of equations in three variables that has (4, −3, 2) as a solution. (There are many correct answers.)

40. *Rewriting a Fraction* The fraction $1/(x^3 - x)$ can be written as a sum of three fractions as follows.

$$\frac{1}{x^3 - x} = \frac{A}{x} + \frac{B}{x + 1} + \frac{C}{x - 1}$$

The numbers A, B, and C are the solutions of the system

$$\begin{aligned} A + B + C &= 0 \\ -B + C &= 0 \\ -A \qquad\quad &= 1. \end{aligned}$$

Solve the system and verify that the sum of the three resulting fractions is the original fraction.

SECTION 9.4

Matrices and Systems of Linear Equations

Matrices • Elementary Row Operations • Using Matrices to Solve a System of Linear Equations

Matrices

In this section we look at a streamlined technique for solving systems of linear equations. This technique involves the use of a rectangular array of real numbers called a **matrix**. (The plural of matrix is *matrices*.) Here is an example of a matrix.

$$\begin{bmatrix} 3 & -2 & 4 & 1 \\ 0 & 1 & -1 & 2 \\ 2 & 0 & -3 & 0 \end{bmatrix}$$

Note that this matrix has three rows and four columns, and we say that its **order** is 3×4, which we read as "3 by 4." Each number in the matrix is called an **entry** of the matrix. Example 1 gives some other examples of matrices.

EXAMPLE 1 ■ **Examples of Matrices**

The following matrices have the indicated orders.

(a) Order: 2×3

$$\begin{bmatrix} 1 & -2 & 4 \\ 0 & 1 & -2 \end{bmatrix}$$

(b) Order: 2×2

$$\begin{bmatrix} 0 & 0 \\ 0 & 0 \end{bmatrix}$$

(c) Order: 3×2

$$\begin{bmatrix} 1 & -3 \\ -2 & 0 \\ 4 & -2 \end{bmatrix}$$

A matrix with the same number of rows as columns is called a **square matrix**. For instance, the 2×2 matrix in part (b) is square. ■

A matrix derived from a system of linear equations (each written in standard form with the constant term on the right) is called the **augmented matrix** of the system. Moreover, the matrix derived from the coefficients of the system (but which does not include the constant terms) is called the **coefficient matrix** of the system. Here is an example.

System

$$\begin{aligned} x - 4y + 3z &= 5 \\ -x + 3y - z &= -3 \\ 2x - 4z &= 6 \end{aligned}$$

Coefficient Matrix

$$\begin{bmatrix} 1 & -4 & 3 \\ -1 & 3 & -1 \\ 2 & 0 & -4 \end{bmatrix}$$

Augmented Matrix

$$\left[\begin{array}{ccc:c} 1 & -4 & 3 & 5 \\ -1 & 3 & -1 & -3 \\ 2 & 0 & -4 & 6 \end{array}\right]$$

Note the use of 0 for the missing y-variable in the third equation, and also note the fourth column of constant terms in the augmented matrix.

When forming either the coefficient matrix or the augmented matrix of a system, you should begin by vertically aligning the variables in the equations.

Given System	*Line Up Variables*	*Form Augmented Matrix*
$\begin{aligned} x + 3y &= 9 \\ -y + 4z &= -2 \\ x - 5z &= 0 \end{aligned}$	$\begin{array}{rrrcr} x & +\ 3y & & = & 9 \\ & -y & +\ 4z & = & -2 \\ x & & -\ 5z & = & 0 \end{array}$	$\left[\begin{array}{rrr\|r} 1 & 3 & 0 & 9 \\ 0 & -1 & 4 & -2 \\ 1 & 0 & -5 & 0 \end{array}\right]$

EXAMPLE 2 ■ Forming Coefficient Matrices and Augmented Matrices

Form the coefficient matrix and augmented matrix for each of the following systems of linear equations.

(a) $\begin{aligned} -x + 5y &= 2 \\ 7x - 2y &= -6 \end{aligned}$ (b) $\begin{aligned} 3x + 2y - z &= 1 \\ x + 2z &= -3 \\ -2x - y &= 4 \end{aligned}$ (c) $\begin{aligned} x &= 3y - 1 \\ 2y - 5 &= 9x \end{aligned}$

Solution

(a)

System	*Coefficient Matrix*	*Augmented Matrix*
$\begin{aligned} -x + 5y &= 2 \\ 7x - 2y &= -6 \end{aligned}$	$\begin{bmatrix} -1 & 5 \\ 7 & -2 \end{bmatrix}$	$\left[\begin{array}{rr\|r} -1 & 5 & 2 \\ 7 & -2 & -6 \end{array}\right]$

(b) To begin, we rewrite the given system so that the variables line up and the constant terms are on the right.

System	*Coefficient Matrix*	*Augmented Matrix*
$\begin{array}{rrrcr} 3x & +\ 2y & -\ z & = & 1 \\ x & + & 2z & = & -3 \\ -2x & -\ y & & = & 4 \end{array}$	$\begin{bmatrix} 3 & 2 & -1 \\ 1 & 0 & 2 \\ -2 & -1 & 0 \end{bmatrix}$	$\left[\begin{array}{rrr\|r} 3 & 2 & -1 & 1 \\ 1 & 0 & 2 & -3 \\ -2 & -1 & 0 & 4 \end{array}\right]$

(c) To begin, we rewrite the given system so that the variables line up and the constant terms are on the right.

System	*Coefficient Matrix*	*Augmented Matrix*
$\begin{aligned} x - 3y &= -1 \\ -9x + 2y &= 5 \end{aligned}$	$\begin{bmatrix} 1 & -3 \\ -9 & 2 \end{bmatrix}$	$\left[\begin{array}{rr\|r} 1 & -3 & -1 \\ -9 & 2 & 5 \end{array}\right]$ ■

EXAMPLE 3 ■ Forming a Linear System from Its Augmented Matrix

Write a system of linear equations that is represented by each of the following augmented matrices.

(a) $\left[\begin{array}{rr\|r} 3 & -5 & 4 \\ -1 & 2 & 0 \end{array}\right]$ (b) $\left[\begin{array}{rr\|r} 1 & \frac{3}{2} & 2 \\ 0 & 1 & -3 \end{array}\right]$ (c) $\left[\begin{array}{rrr\|r} 2 & 0 & -8 & 1 \\ -1 & 1 & 1 & 2 \\ 5 & -1 & 7 & 3 \end{array}\right]$

Solution

(a) $$\begin{aligned} 3x - 5y &= 4 \\ -x + 2y &= 0 \end{aligned}$$

(b) $$\begin{aligned} x + \frac{3}{2}y &= 2 \\ y &= -3 \end{aligned}$$

(c) $$\begin{aligned} 2x \quad\quad - 8z &= 1 \\ -x + y + z &= 2 \\ 5x - y + 7z &= 3 \end{aligned}$$

■

Elementary Row Operations

In Section 9.3 we studied three operations that can be used on a system of linear equations to produce an equivalent system.

1. Interchange two equations.
2. Multiply an equation by a nonzero constant.
3. Add a multiple of an equation to another equation.

In matrix terminology these three operations correspond to **elementary row operations**. An elementary row operation on an augmented matrix of a given system of linear equations produces a new augmented matrix corresponding to a new (but equivalent) system of linear equations. Two matrices are said to be **row-equivalent** if one can be obtained from the other by a sequence of elementary row operations.

Elementary Row Operations

Any of the following **elementary row operations** performed on an augmented matrix will produce a matrix that is row-equivalent to the original matrix.

1. Interchange two rows.
2. Multiply a row by a nonzero constant.
3. Add a multiple of a row to another row.

Although elementary row operations are simple to perform, they involve a lot of arithmetic. Because it is easy to make a mistake, we suggest that you get in the habit of noting the elementary row operations performed in each step so that you can go back and check your work. People use different schemes to denote which elementary row operation has been performed. The scheme we use is to write an abbreviated version of the row operation to the left of the row that has been changed, as shown in Example 4.

EXAMPLE 4 ■ **Elementary Row Operations**

(a) Interchange the first and second rows.

Original Matrix

$$\begin{bmatrix} 0 & 1 & 3 & 4 \\ -1 & 2 & 0 & 3 \\ 2 & -3 & 4 & 1 \end{bmatrix}$$

New Row-Equivalent Matrix

$$\begin{matrix} R_2 \\ R_1 \\ \end{matrix} \begin{bmatrix} -1 & 2 & 0 & 3 \\ 0 & 1 & 3 & 4 \\ 2 & -3 & 4 & 1 \end{bmatrix}$$

(b) Multiply the first row by $\frac{1}{2}$.

Original Matrix

$$\begin{bmatrix} 2 & -4 & 6 & -2 \\ 1 & 3 & -3 & 0 \\ 5 & -2 & 1 & 2 \end{bmatrix}$$

New Row-Equivalent Matrix

$$\begin{matrix} \frac{1}{2}R_1 \rightarrow \\ \\ \end{matrix} \begin{bmatrix} 1 & -2 & 3 & -1 \\ 1 & 3 & -3 & 0 \\ 5 & -2 & 1 & 2 \end{bmatrix}$$

(c) Add -2 times the first row to the third row.

Original Matrix

$$\begin{bmatrix} 1 & 2 & -4 & 3 \\ 0 & 3 & -2 & -1 \\ 2 & 1 & 5 & -2 \end{bmatrix}$$

New Row-Equivalent Matrix

$$\begin{matrix} \\ \\ -2R_1 + R_3 \rightarrow \end{matrix} \begin{bmatrix} 1 & 2 & -4 & 3 \\ 0 & 3 & -2 & -1 \\ 0 & -3 & 13 & -8 \end{bmatrix}$$

Note that we write tne elementary row operation beside the row that we are *changing*. ■

In Section 9.3 we used Gaussian elimination with back-substitution to solve a system of linear equations. We now demonstrate the matrix version of Gaussian elimination. The two methods are essentially the same. The basic difference is that with matrices we do not need to keep writing the variables.

EXAMPLE 5 ■ **Using Elementary Row Operations to Solve a System**

Linear System

$$\begin{aligned} x - 2y + 2z &= 9 \\ -x + 3y &= -4 \\ 2x - 5y + z &= 10 \end{aligned}$$

Associated Augmented Matrix

$$\left[\begin{array}{ccc:c} 1 & -2 & 2 & 9 \\ -1 & 3 & 0 & -4 \\ 2 & -5 & 1 & 10 \end{array}\right]$$

Add the first equation to the second equation.

$$\begin{aligned} x - 2y + 2z &= 9 \\ y + 2z &= 5 \\ 2x - 5y + z &= 10 \end{aligned}$$

Add the first row to the second row ($R_1 + R_2$).

$$R_1 + R_2 \rightarrow \begin{bmatrix} 1 & -2 & 2 & \vdots & 9 \\ 0 & 1 & 2 & \vdots & 5 \\ 2 & -5 & 1 & \vdots & 10 \end{bmatrix}$$

Add -2 times the first equation to the third equation.

$$\begin{aligned} x - 2y + 2z &= 9 \\ y + 2z &= 5 \\ -y - 3z &= -8 \end{aligned}$$

Add -2 times the first row to the third row ($-2R_1 + R_3$).

$$-2R_1 + R_3 \rightarrow \begin{bmatrix} 1 & -2 & 2 & \vdots & 9 \\ 0 & 1 & 2 & \vdots & 5 \\ 0 & -1 & -3 & \vdots & -8 \end{bmatrix}$$

Add the second equation to the third equation.

$$\begin{aligned} x - 2y + 2z &= 9 \\ y + 2z &= 5 \\ -z &= -3 \end{aligned}$$

Add the second row to the third row ($R_2 + R_3$).

$$R_2 + R_3 \rightarrow \begin{bmatrix} 1 & -2 & 2 & \vdots & 9 \\ 0 & 1 & 2 & \vdots & 5 \\ 0 & 0 & -1 & \vdots & -3 \end{bmatrix}$$

Multiply the third equation by -1.

$$\begin{aligned} x - 2y + 2z &= 9 \\ y + 2z &= 5 \\ z &= 3 \end{aligned}$$

Multiply the third row by -1.

$$-R_3 \rightarrow \begin{bmatrix} 1 & -2 & 2 & \vdots & 9 \\ 0 & 1 & 2 & \vdots & 5 \\ 0 & 0 & 1 & \vdots & 3 \end{bmatrix}$$

At this point, we can use back-substitution, as we did in Example 1 in Section 9.3, to find that the solution is $x = 1$, $y = -1$, and $z = 3$. ■

The last matrix in Example 5 is said to be in **row-echelon form**. The term *echelon* refers to the stair-step pattern formed by the nonzero elements of the matrix. To be in this form, a matrix must have the following properties.

Definition of Row-Echelon Form of a Matrix

A matrix in **row-echelon form** has the following properties.

1. All rows consisting entirely of zeros occur at the bottom of the matrix.
2. For each row that does not consist entirely of zeros, the first nonzero entry is 1 (called a **leading 1**).
3. For two successive (nonzero) rows, the leading 1 in the higher row is farther to the left than the leading 1 in the lower row.

EXAMPLE 6 ■ Row-Echelon Form

The following matrices are in row-echelon form.

$$\text{(a)}\begin{bmatrix} 1 & 2 & -1 & 4 \\ 0 & 1 & 0 & 3 \\ 0 & 0 & 1 & -2 \end{bmatrix} \qquad \text{(b)}\begin{bmatrix} 0 & 1 & 0 & 3 \\ 0 & 0 & 1 & 3 \\ 0 & 0 & 0 & 0 \end{bmatrix}$$

$$\text{(c)}\begin{bmatrix} 1 & 2 & -3 & 4 \\ 0 & 1 & 1 & -1 \\ 0 & 0 & 1 & -3 \end{bmatrix} \qquad \text{(d)}\begin{bmatrix} 1 & 2 & -1 & 2 \\ 0 & 1 & 2 & -4 \\ 0 & 0 & 0 & 0 \end{bmatrix}$$

■

Using Matrices to Solve a System of Linear Equations

From the procedure used in Example 5, we give the following guidelines for using matrices to solve a system of linear equations.

Gaussian Elimination with Back-Substitution

To use matrices and Gaussian elimination to solve a system of linear equations, use the following steps.

1. Write the augmented matrix of the system of linear equations.
2. Use elementary row operations to rewrite the augmented matrix in row-echelon form.
3. Write the system of linear equations corresponding to the matrix in row-echelon form, and use back-substitution to find the solution.

Gaussian elimination with back-substitution works well for solving systems of linear equations with a computer. For this algorithm, the order in which the elementary row operations are performed is important. We operate from *left to right by columns*, using elementary row operations to obtain zeros in all entries directly below the leading 1's.

EXAMPLE 7 ■ Gaussian Elimination with Back-Substitution

Solve the following system.

$$\begin{aligned} 2x - 3y &= -2 \\ x + 2y &= 13 \end{aligned}$$

Solution

The augmented matrix for this system is as follows.

$$\left[\begin{array}{cc:c} 2 & -3 & -2 \\ 1 & 2 & 13 \end{array}\right]$$

We begin by obtaining a leading 1 in the upper left corner by interchanging the first and second rows, and then proceed to obtain zeros elsewhere in the first column.

$$\begin{matrix} R_2 \\ R_1 \end{matrix} \begin{bmatrix} 1 & 2 & \vdots & 13 \\ 2 & -3 & \vdots & -2 \end{bmatrix}$$ First column has leading 1 in upper left corner

$$-2R_1 + R_2 \rightarrow \begin{bmatrix} 1 & 2 & \vdots & 13 \\ 0 & -7 & \vdots & -28 \end{bmatrix}$$ First column has a zero under its leading 1

$$-\tfrac{1}{7}R_2 \rightarrow \begin{bmatrix} 1 & 2 & \vdots & 13 \\ 0 & 1 & \vdots & 4 \end{bmatrix}$$ Second column has leading 1 in second row

The system of linear equations that corresponds to the (row-echelon) matrix is as follows.

$$\begin{aligned} x + 2y &= 13 \\ y &= 4 \end{aligned}$$

Using back-substitution, we find that the solution of the system is $x = 5$ and $y = 4$. Check this solution in the original system of linear equations. ■

EXAMPLE 8 ■ **Gaussian Elimination with Back-Substitution**

Solve the following system.

$$\begin{aligned} 3x + 3y &= 9 \\ 2x - 3z &= 10 \\ 6y + 4z &= -12 \end{aligned}$$

Solution

$$\begin{bmatrix} 3 & 3 & 0 & \vdots & 9 \\ 2 & 0 & -3 & \vdots & 10 \\ 0 & 6 & 4 & \vdots & -12 \end{bmatrix}$$ Augmented matrix for given system of linear equations

$$\tfrac{1}{3}R_1 \rightarrow \begin{bmatrix} 1 & 1 & 0 & \vdots & 3 \\ 2 & 0 & -3 & \vdots & 10 \\ 0 & 6 & 4 & \vdots & -12 \end{bmatrix}$$ First column has leading 1 in upper left corner

$$-2R_1 + R_2 \rightarrow \begin{bmatrix} 1 & 1 & 0 & \vdots & 3 \\ 0 & -2 & -3 & \vdots & 4 \\ 0 & 6 & 4 & \vdots & -12 \end{bmatrix}$$ First column has zeros under its leading 1

$$-\tfrac{1}{2}R_2 \rightarrow \begin{bmatrix} 1 & 1 & 0 & \vdots & 3 \\ 0 & 1 & \frac{3}{2} & \vdots & -2 \\ 0 & 6 & 4 & \vdots & -12 \end{bmatrix}$$ Second column has leading 1 in second row

$$-6R_2 + R_3 \rightarrow \begin{bmatrix} 1 & 1 & 0 & \vdots & 3 \\ 0 & 1 & \frac{3}{2} & \vdots & -2 \\ 0 & 0 & -5 & \vdots & 0 \end{bmatrix}$$ Second column has zeros under its leading 1

$$-\tfrac{1}{5}R_3 \rightarrow \left[\begin{array}{ccc:c} 1 & 1 & 0 & 3 \\ 0 & 1 & \frac{3}{2} & -2 \\ 0 & 0 & 1 & 0 \end{array}\right]$$ Third column has leading 1 in third row

The system of linear equations that corresponds to this (row-echelon) matrix is as follows.

$$\begin{aligned} x + y \qquad &= 3 \\ y + \frac{3}{2}z &= -2 \\ z &= 0 \end{aligned}$$

Using back-substitution, we find that the solution is $x = 5$, $y = -2$, and $z = 0$. Check this solution in the original system of linear equations. ■

Each of the systems of linear equations that we have solved so far in this section has had only one solution. Remember from our earlier work in this chapter that it is possible for a system of linear equations to have no solution *or* an infinite number of solutions. We look at the use of matrices for solving such systems in Examples 9 and 10.

EXAMPLE 9 ■ A System with No Solution

Solve the following system of linear equations.

$$\begin{aligned} 6x - 10y &= -4 \\ 9x - 15y &= 5 \end{aligned}$$

Solution

$$\left[\begin{array}{cc:c} 6 & -10 & -4 \\ 9 & -15 & 5 \end{array}\right]$$ Augmented matrix for given system of equations

$$\tfrac{1}{6}R_1 \rightarrow \left[\begin{array}{cc:c} 1 & -\frac{5}{3} & -\frac{2}{3} \\ 9 & -15 & 5 \end{array}\right]$$ First column has leading 1 in upper left corner

$$-9R_1 + R_2 \rightarrow \left[\begin{array}{cc:c} 1 & -\frac{5}{3} & -\frac{2}{3} \\ 0 & 0 & 11 \end{array}\right]$$ First column has a zero under its leading 1

The system of linear equations that corresponds to the (row-echelon) matrix is

$$\begin{aligned} x - \frac{5}{3}y &= -\frac{2}{3} \\ 0 &= 11. \end{aligned}$$

Since the second equation in this system is not true for any values of x and y, we conclude that the original system of equations has no solution. ■

EXAMPLE 10 ■ A System with Infinitely Many Solutions

Solve the following system of linear equations.

$$\begin{aligned} 12x - 6y &= -3 \\ -8x + 4y &= 2 \end{aligned}$$

Solution

$$\begin{bmatrix} 12 & -6 & \vdots & -3 \\ -8 & 4 & \vdots & 2 \end{bmatrix}$$ Augmented matrix for given system of equations

$$\tfrac{1}{12}R_1 \rightarrow \begin{bmatrix} 1 & -\frac{1}{2} & \vdots & -\frac{1}{4} \\ -8 & 4 & \vdots & 2 \end{bmatrix}$$ First column has leading 1 in upper left corner

$$8R_1 + R_2 \rightarrow \begin{bmatrix} 1 & -\frac{1}{2} & \vdots & -\frac{1}{4} \\ 0 & 0 & \vdots & 0 \end{bmatrix}$$ First column has a zero under its leading 1

The system of linear equations that corresponds to the (row-echelon) matrix is

$$\begin{aligned} x - \frac{1}{2}y &= -\frac{1}{4} \\ 0 &= 0. \end{aligned}$$

Since the second equation in this system adds no information, we conclude that the system of equations has an infinite number of solutions, represented by all points (x, y) on the line $x - \frac{1}{2}y = -\frac{1}{4}$. Since this line can be written as $x = -\frac{1}{4} + \frac{1}{2}y$, we can write the solution set as

$$\left(-\frac{1}{4} + \frac{1}{2}a,\ a\right), \qquad \text{where } a \text{ is any real number.}$$ ■

Some calculators are capable of working with matrices. For example, to use a *TI-81* to write the matrix

$$A = \begin{bmatrix} 1 & -2 & 2 & \vdots & 9 \\ -1 & 3 & 0 & \vdots & -4 \\ 2 & -5 & 1 & \vdots & 10 \end{bmatrix}$$

in row-echelon form, we can use the following steps.*

1. Enter the matrix in the calculator by pressing [MATRIX], cursor to "Edit," press [ENTER], set the order of the matrix as 3×4, and enter the matrix entries by rows. Once you have entered the matrix, press [QUIT]. To check that the matrix has been entered correctly, press [A] and [ENTER].

*The graphing calculator keystrokes given in this text correspond to the *TI-81* graphing calculator from *Texas Instruments*. For other graphing calculators, the keystrokes may differ.

2. To write the matrix in row-echelon form, use the elementary row operations that appear in the "MATRIX" menu, as follows.

Row+([A],1,2) ENTER	Add row 1 to row 2
STO→ ENTER	Store the result as matrix A
*Row+(−2,[A],1,3) ENTER	Multiply row 1 by −2 and add to row 3
STO→ ENTER	Store the result as matrix A
Row+([A],2,3) ENTER	Add row 2 to row 3
STO→ ENTER	Store the result as matrix A
*Row(−1,[A],3) ENTER	Multiply row 3 by −1
STO→ ENTER	Store the result as matrix A

3. The result displayed on the screen should be as follows.

```
[1  -2  2  9]
[0   1  2  5]
[0   0  1  3]
```

You can now use back-substitution to solve this system, as was demonstrated in Example 5.

DISCUSSION PROBLEM ■ Too Much Information

Set up a system of linear equations that represents the following number problem. Can this problem be solved? Explain why or why not.

> The sum of three numbers is 9. The second number is twice the first number. The third number is half the sum of the first two numbers. The difference of the first and third numbers is 4. What are the numbers? ■

Warm-Up

The following warm-up exercises involve skills that were covered in earlier sections. You will use these skills in the exercise set for this section.

In Exercises 1–6, solve the system of equations.

1. $\begin{aligned} y &= x^2 \\ -3x + 2y &= 2 \end{aligned}$

2. $\begin{aligned} x - y^3 &= 0 \\ x - 2y^2 &= 0 \end{aligned}$

3. $\begin{aligned} x - 3y &= -7 \\ -2x + y &= -6 \end{aligned}$

4. $\begin{aligned} 2x + 3y &= 2 \\ 4x + 9y &= 5 \end{aligned}$

5. $\begin{aligned} x - y &= -1 \\ x + 2y - 2z &= 3 \\ 3x - y + 2z &= 3 \end{aligned}$

6. $\begin{aligned} 2x + y - 2z &= 1 \\ x - z &= 1 \\ 3x + 3y + z &= 12 \end{aligned}$

In Exercises 7–10, evaluate the given expression.

7. $3 - \frac{3}{2}(-4)$

8. $-2 + \frac{4}{5}(6)$

9. $5(-2) + 7\left(\frac{3}{4}\right)$

10. $-\frac{2}{3}(5) - \frac{1}{2}(-3)$

9.4 EXERCISES

In Exercises 1–4, determine the order of the matrix.

1. $\begin{bmatrix} 3 & -2 \\ -4 & 0 \\ 2 & -7 \\ -1 & -3 \end{bmatrix}$

2. $\begin{bmatrix} 4 & 0 & -5 \\ -1 & 8 & 9 \\ 0 & -3 & 4 \end{bmatrix}$

3. $\begin{bmatrix} 5 & -8 & 32 & -4 \\ 7 & 15 & 28 & -16 \\ -12 & 65 & -10 & 2 \end{bmatrix}$

4. $\begin{bmatrix} -2.3 & 4.5 & -25.1 & 7.5 & 1 \\ 11.8 & 82.3 & -19.6 & 4.0 & -2 \\ 0 & -9.3 & 24.0 & 8.7 & -5 \end{bmatrix}$

In Exercises 5–8, fill in the box by using elementary row operations to form a row-equivalent matrix.

5. $\begin{bmatrix} 1 & 4 & 3 \\ 2 & 10 & 5 \end{bmatrix}$

$\begin{bmatrix} 1 & 4 & 3 \\ 0 & \square & -1 \end{bmatrix}$

6. $\begin{bmatrix} 3 & 6 & 8 \\ 4 & -3 & 6 \end{bmatrix}$

$\begin{bmatrix} 1 & \square & \frac{8}{3} \\ 4 & -3 & 6 \end{bmatrix}$

7. $\begin{bmatrix} 1 & 1 & 4 & -1 \\ 3 & 8 & 10 & 3 \\ -2 & 1 & 12 & 6 \end{bmatrix}$

$\begin{bmatrix} 1 & 1 & 4 & -1 \\ 0 & 5 & \square & \square \\ 0 & 3 & \square & \square \end{bmatrix}$

$\begin{bmatrix} 1 & 1 & 4 & -1 \\ 0 & 1 & \square & \square \\ 0 & 3 & 20 & 4 \end{bmatrix}$

8. $\begin{bmatrix} 2 & 4 & 8 & 3 \\ 1 & -1 & -3 & 2 \\ 2 & 6 & 4 & 9 \end{bmatrix}$

$\begin{bmatrix} 1 & \square & \square & \square \\ 1 & -1 & -3 & 2 \\ 2 & 6 & 4 & 9 \end{bmatrix}$

$\begin{bmatrix} 1 & 2 & 4 & \frac{3}{2} \\ 0 & \square & -7 & \frac{1}{2} \\ 0 & 2 & \square & \square \end{bmatrix}$

In Exercises 9–16, convert the matrix to row-echelon form. (There is more than one correct answer.)

9. $\begin{bmatrix} 1 & 2 & 3 \\ 2 & -1 & -4 \end{bmatrix}$

10. $\begin{bmatrix} 1 & 3 & 6 \\ -4 & -9 & 3 \end{bmatrix}$

11. $\begin{bmatrix} 4 & 6 & 1 \\ -2 & 2 & 5 \end{bmatrix}$

12. $\begin{bmatrix} 3 & 2 & 6 \\ 2 & 3 & -3 \end{bmatrix}$

13. $\begin{bmatrix} 1 & 1 & 0 & 5 \\ -2 & -1 & 2 & -10 \\ 3 & 6 & 7 & 14 \end{bmatrix}$

14. $\begin{bmatrix} 1 & 2 & -1 & 3 \\ 3 & 7 & -5 & 14 \\ -2 & -1 & -3 & 8 \end{bmatrix}$

15. $\begin{bmatrix} 1 & -1 & -1 & 1 \\ 4 & -4 & 1 & 8 \\ -6 & 8 & 18 & 0 \end{bmatrix}$

16. $\begin{bmatrix} 1 & -3 & 0 & -7 \\ -3 & 10 & 1 & 23 \\ 4 & -10 & 2 & -24 \end{bmatrix}$

In Exercises 17–30, use matrices to solve the system of linear equations.

17. $\begin{aligned} x + 2y &= 0 \\ 3x + y &= 7 \end{aligned}$

18. $\begin{aligned} 2x + 6y &= 16 \\ 2x + 3y &= 7 \end{aligned}$

19. $\begin{aligned} 8x - 4y &= 7 \\ 5x + 2y &= 1 \end{aligned}$

20. $\begin{aligned} 2x - y &= -0.1 \\ 3x + 2y &= 1.6 \end{aligned}$

21. $\begin{aligned} -x + 2y &= 1.5 \\ 2x - 4y &= 3 \end{aligned}$

22. $\begin{aligned} x - 3y &= 5 \\ -2x + 6y &= -10 \end{aligned}$

23. $\begin{aligned} x \qquad - 3z &= -2 \\ 3x + y - 2z &= 5 \\ 2x + 2y + z &= 4 \end{aligned}$

24. $\begin{aligned} x - 2y - z &= 6 \\ y + 4z &= 5 \\ 4x + 2y + 3z &= 8 \end{aligned}$

25. $\begin{aligned} 2x + 4y \qquad &= 10 \\ 2x + 2y + 3z &= 3 \\ -3x + y + 2z &= -3 \end{aligned}$

26. $\begin{aligned} 2x - y + 3z &= 24 \\ 2y - z &= 14 \\ 7x - 5y \qquad &= 6 \end{aligned}$

27. $\begin{aligned} x + y - 5z &= 3 \\ x \qquad - 2z &= 1 \\ 2x - y - z &= 0 \end{aligned}$

28. $\begin{aligned} 2x \qquad + 3z &= 3 \\ 4x - 3y + 7z &= 5 \\ 8x - 9y + 15z &= 9 \end{aligned}$

29. $\begin{aligned} 2x \qquad + 4z &= 1 \\ x + y + 3z &= 0 \\ x + 3y + 5z &= 0 \end{aligned}$

30. $\begin{aligned} 3x + y - 2z &= 2 \\ 6x + 2y - 4z &= 1 \\ -3x - y + 2z &= 1 \end{aligned}$

31. *Simple Interest* A corporation borrowed \$1,500,000 to expand its product line. Some of the money was borrowed at 8%, some at 9%, and the remainder at 12%. The annual interest payment to the lenders was \$133,000. If the amount borrowed at 8% was four times the amount borrowed at 12%, how much was borrowed at each rate?

32. *Nut Mixture* A grocer wishes to mix three kinds of nuts costing \$3.50, \$4.50, and \$6.00 per pound to obtain 50 pounds of a mixture costing \$4.95 per pound. How many pounds of each variety should the grocer use if half the mixture is composed of the two cheapest varieties?

33. *Curve Fitting* Find the equation of the parabola

$$y = ax^2 + bx + c$$

that passes through the points $(1, 1)$, $(-3, 17)$, and $\left(2, -\frac{1}{2}\right)$.

34. *Curve Fitting* Find the equation of the circle

$$x^2 + y^2 + Dx + Ey + F = 0$$

that passes through the points $(1, 1)$, $(3, 3)$, and $(4, 2)$.

35. *Number Problem* The sum of three positive numbers is 33. The second number is three greater than the first, and the third is four times the first. Find the three numbers.

SECTION 9.5

Linear Systems and Determinants

The Determinant of a Matrix • Cramer's Rule for Solving a System of Linear Equations • Applications

The Determinant of a Matrix

In Section 9.4 we defined a **square** matrix as one that has the same number of rows and columns. Associated with each square matrix is a real number called its **determinant**.

Historically, the use of determinants arose from special number patterns that occur when solving a system of linear equations. For instance, the system

$$a_1x + b_1y = c_1$$
$$a_2x + b_2y = c_2$$

has a solution given by

$$x = \frac{c_1b_2 - c_2b_1}{a_1b_2 - a_2b_1} \quad \text{and} \quad y = \frac{a_1c_2 - a_2c_1}{a_1b_2 - a_2b_1}$$

provided $a_1b_2 - a_2b_1 \neq 0$. Note that the denominator of each fraction is the same, and we call this denominator the **determinant** of the coefficient matrix of the system.

Coefficient Matrix	*Determinant*
$A = \begin{bmatrix} a_1 & b_1 \\ a_2 & b_2 \end{bmatrix}$	$\det(A) = a_1b_2 - a_2b_1$

The determinant of the matrix A can also be denoted by vertical bars on both sides of the matrix, as indicated in the following definition.

Definition of the Determinant of a 2 × 2 Matrix

The **determinant** of the matrix

$$A = \begin{bmatrix} a_1 & b_1 \\ a_2 & b_2 \end{bmatrix}$$

is given by

$$\det(A) = \begin{vmatrix} a_1 & b_1 \\ a_2 & b_2 \end{vmatrix} = a_1b_2 - a_2b_1.$$

A convenient method for remembering the formula for the determinant of a 2×2 matrix is shown in the following diagram.

$$\det(A) = \begin{vmatrix} a_1 & b_1 \\ a_2 & b_2 \end{vmatrix} = a_1b_2 - a_2b_1$$

Note that the determinant is given by the difference of the products of the two diagonals of the matrix.

EXAMPLE 1 ■ The Determinant of a 2×2 Matrix

Find the determinant of each of the following matrices.

(a) $A = \begin{bmatrix} 2 & -3 \\ 1 & 4 \end{bmatrix}$ (b) $B = \begin{bmatrix} -1 & 2 \\ 2 & -4 \end{bmatrix}$ (c) $C = \begin{bmatrix} 1 & 3 \\ 2 & 5 \end{bmatrix}$

Solution

(a) $\det(A) = \begin{vmatrix} 2 & -3 \\ 1 & 4 \end{vmatrix} = 2(4) - 1(-3) = 8 + 3 = 11$

(b) $\det(B) = \begin{vmatrix} -1 & 2 \\ 2 & -4 \end{vmatrix} = (-1)(-4) - 2(2) = 4 - 4 = 0$

(c) $\det(C) = \begin{vmatrix} 1 & 3 \\ 2 & 5 \end{vmatrix} = 1(5) - 3(2) = -1$ ■

NOTE: Notice in Example 1 that the determinant of a matrix can be positive, zero, or negative.

The determinant of a 3×3 matrix is more difficult to evaluate than the determinant of a 2×2 matrix. To help with the evaluation, we use a technique called **expansion by minors**. This technique allows us to write the determinant of a 3×3 matrix in terms of three 2×2 determinants. The **minor** of an entry in a 3×3 matrix is the determinant of the 2×2 matrix that remains after deleting the row and column in which the entry occurs. Here are some examples.

Given Determinant	*Entry*	*Minor of Entry*	*Value of Minor*
$\begin{vmatrix} 1 & -1 & 3 \\ 0 & 2 & 5 \\ -2 & 4 & -7 \end{vmatrix}$	1	$\begin{vmatrix} 2 & 5 \\ 4 & -7 \end{vmatrix}$	$2(-7) - 4(5) = -34$
$\begin{vmatrix} 1 & -1 & 3 \\ 0 & 2 & 5 \\ -2 & 4 & -7 \end{vmatrix}$	-1	$\begin{vmatrix} 0 & 5 \\ -2 & -7 \end{vmatrix}$	$0(-7) - (-2)(5) = 10$
$\begin{vmatrix} 1 & -1 & 3 \\ 0 & 2 & 5 \\ -2 & 4 & -7 \end{vmatrix}$	5	$\begin{vmatrix} 1 & -1 \\ -2 & 4 \end{vmatrix}$	$1(4) - (-1)(-2) = 2$

To evaluate the determinant of a 3×3 matrix, we expand by minors along an *entire* row (or column) according to the following scheme.

Expanding by Minors

The determinant of the 3×3 matrix

$$A = \begin{bmatrix} a_1 & b_1 & c_1 \\ a_2 & b_2 & c_2 \\ a_3 & b_3 & c_3 \end{bmatrix}$$

is given by

$$\begin{vmatrix} a_1 & b_1 & c_1 \\ a_2 & b_2 & c_2 \\ a_3 & b_3 & c_3 \end{vmatrix} = (a_1)(\text{Minor of } a_1) - (b_1)(\text{Minor of } b_1) + (c_1)(\text{Minor of } c_1)$$

$$= a_1 \begin{vmatrix} b_2 & c_2 \\ b_3 & c_3 \end{vmatrix} - b_1 \begin{vmatrix} a_2 & c_2 \\ a_3 & c_3 \end{vmatrix} + c_1 \begin{vmatrix} a_2 & b_2 \\ a_3 & b_3 \end{vmatrix}.$$

This particular pattern is called **expanding by minors** along the first row. A similar pattern can be used to expand by minors along any row or column.

FIGURE 9.8
Sign pattern for 3×3 matrix

$$\begin{bmatrix} + & - & + \\ - & + & - \\ + & - & + \end{bmatrix}$$

The *signs* of the terms used in the expansion by minors follows the alternating pattern shown in Figure 9.8. For instance, the signs used to expand by minors along the second row are $-$, $+$, $-$, as follows.

$$\begin{vmatrix} a_1 & b_1 & c_1 \\ a_2 & b_2 & c_2 \\ a_3 & b_3 & c_3 \end{vmatrix} = -(a_2)(\text{Minor of } a_2) + (b_2)(\text{Minor of } b_2) - (c_2)(\text{Minor of } c_2)$$

EXAMPLE 2 ■ Finding the Determinant of a 3×3 Matrix

Find the determinant of the matrix $A = \begin{bmatrix} -1 & 1 & 2 \\ 0 & 2 & 3 \\ 3 & 4 & 2 \end{bmatrix}$.

Solution

By expanding by minors along the *first column,* we obtain the following.

$$\begin{aligned} \det(A) &= \begin{vmatrix} -1 & 1 & 2 \\ 0 & 2 & 3 \\ 3 & 4 & 2 \end{vmatrix} \\ &= (-1)\begin{vmatrix} 2 & 3 \\ 4 & 2 \end{vmatrix} - (0)\begin{vmatrix} 1 & 2 \\ 4 & 2 \end{vmatrix} + (3)\begin{vmatrix} 1 & 2 \\ 2 & 3 \end{vmatrix} \\ &= (-1)(4 - 12) - (0)(2 - 8) + (3)(3 - 4) \\ &= 8 - 0 - 3 \\ &= 5 \end{aligned}$$

■

Note in the expansion in Example 2 that a zero entry will always yield a zero term in the expansion by minors. Thus, when you are evaluating the determinant of a matrix, you should choose to expand along the row or column that has the most zero entries.

EXAMPLE 3 ■ **Finding the Determinant of a 3 × 3 Matrix**

Find the determinant of the following matrix.

$$A = \begin{bmatrix} 1 & 2 & 1 \\ 3 & 0 & 2 \\ 4 & 0 & -1 \end{bmatrix}$$

Solution

By expanding by minors along the *second column*, we obtain the following.

$$\begin{aligned} \det(A) &= \begin{vmatrix} 1 & 2 & 1 \\ 3 & 0 & 2 \\ 4 & 0 & -1 \end{vmatrix} \\ &= -(2)\begin{vmatrix} 3 & 2 \\ 4 & -1 \end{vmatrix} + (0)\begin{vmatrix} 1 & 1 \\ 4 & -1 \end{vmatrix} - (0)\begin{vmatrix} 1 & 1 \\ 3 & 2 \end{vmatrix} \\ &= -(2)(-3 - 8) + 0 - 0 \\ &= 22 \end{aligned}$$

■

NOTE: If you have a calculator that is capable of working with matrices, try using the calculator to find the determinant of a matrix. For instance, to find the determinant of the matrix in Example 3, enter the matrix as [A] in the calculator and then press MATRIX, cursor to "det," press ENTER, press [A], and press ENTER. The display should be 22, which is the value of the determinant of A.

Cramer's Rule for Solving a System of Linear Equations

So far in this chapter, we have discussed three methods for solving a system of linear equations: substitution, elimination with equations, and elimination with matrices. We now look at one more method, called **Cramer's Rule**, named after Gabriel Cramer (1704–1752). This rule uses determinants to write the solution of a system of linear equations. To see how Cramer's Rule works, let's take another look at the solution described at the beginning of this section. There, we pointed out that the system

$$\begin{aligned} a_1x + b_1y &= c_1 \\ a_2x + b_2y &= c_2 \end{aligned}$$

has a solution given by

$$x = \frac{c_1b_2 - c_2b_1}{a_1b_2 - a_2b_1} \quad \text{and} \quad y = \frac{a_1c_2 - a_2c_1}{a_1b_2 - a_2b_1}$$

provided $a_1b_2 - a_2b_1 \neq 0$. Each numerator and denominator in this solution can be expressed as a determinant, as follows.

$$x = \frac{c_1b_2 - c_2b_1}{a_1b_2 - a_2b_1} = \frac{\begin{vmatrix} c_1 & b_1 \\ c_2 & b_2 \end{vmatrix}}{\begin{vmatrix} a_1 & b_1 \\ a_2 & b_2 \end{vmatrix}}, \quad y = \frac{a_1c_2 - a_2c_1}{a_1b_2 - a_2b_1} = \frac{\begin{vmatrix} a_1 & c_1 \\ a_2 & c_2 \end{vmatrix}}{\begin{vmatrix} a_1 & b_1 \\ a_2 & b_2 \end{vmatrix}}$$

Relative to the original system, the denominator for x and y is simply the determinant of the *coefficient* matrix of the system. We denote this determinant by D. The numerators for x and y are denoted by D_x and D_y, respectively. They are formed by using the column of constants as replacements for the coefficients of x and y, as follows.

Coefficient Matrix	D	D_x	D_y
$\begin{bmatrix} a_1 & b_1 \\ a_2 & b_2 \end{bmatrix}$	$\begin{bmatrix} a_1 & b_1 \\ a_2 & b_2 \end{bmatrix}$	$\begin{bmatrix} c_1 & b_1 \\ c_2 & b_2 \end{bmatrix}$	$\begin{bmatrix} a_1 & c_1 \\ a_2 & c_2 \end{bmatrix}$

Cramer's Rule can be extended to a system of three linear equations in three variables, as indicated in the following.

Cramer's Rule

1. For the system of linear equations

$$a_1x + b_1y = c_1$$
$$a_2x + b_2y = c_2$$

the solution is given by

$$x = \frac{D_x}{D} = \frac{\begin{vmatrix} c_1 & b_1 \\ c_2 & b_2 \end{vmatrix}}{\begin{vmatrix} a_1 & b_1 \\ a_2 & b_2 \end{vmatrix}}, \quad y = \frac{D_y}{D} = \frac{\begin{vmatrix} a_1 & c_1 \\ a_2 & c_2 \end{vmatrix}}{\begin{vmatrix} a_1 & b_1 \\ a_2 & b_2 \end{vmatrix}}$$

provided $D \neq 0$.

2. For the system of linear equations

$$a_1x + b_1y + c_1z = d_1$$
$$a_2x + b_2y + c_2z = d_2$$
$$a_3x + b_3y + c_3z = d_3$$

the solution is given by

$$x = \frac{D_x}{D} = \frac{\begin{vmatrix} d_1 & b_1 & c_1 \\ d_2 & b_2 & c_2 \\ d_3 & b_3 & c_3 \end{vmatrix}}{\begin{vmatrix} a_1 & b_1 & c_1 \\ a_2 & b_2 & c_2 \\ a_3 & b_3 & c_3 \end{vmatrix}}, \quad y = \frac{D_y}{D} = \frac{\begin{vmatrix} a_1 & d_1 & c_1 \\ a_2 & d_2 & c_2 \\ a_3 & d_3 & c_3 \end{vmatrix}}{\begin{vmatrix} a_1 & b_1 & c_1 \\ a_2 & b_2 & c_2 \\ a_3 & b_3 & c_3 \end{vmatrix}}, \quad z = \frac{D_z}{D} = \frac{\begin{vmatrix} a_1 & b_1 & d_1 \\ a_2 & b_2 & d_2 \\ a_3 & b_3 & d_3 \end{vmatrix}}{\begin{vmatrix} a_1 & b_1 & c_1 \\ a_2 & b_2 & c_2 \\ a_3 & b_3 & c_3 \end{vmatrix}}$$

provided $D \neq 0$.

EXAMPLE 4 ■ **Using Cramer's Rule for a 2 × 2 System**

Use Cramer's Rule to solve the following system of linear equations.

$$\begin{aligned} 4x - 2y &= 10 \\ 3x - 5y &= 11 \end{aligned}$$

Solution

To begin, we find the determinant of the coefficient matrix.

$$D = \begin{vmatrix} 4 & -2 \\ 3 & -5 \end{vmatrix} = -20 - (-6) = -14$$

Because this determinant is not zero, we can apply Cramer's Rule to find the solution, as follows.

$$x = \frac{D_x}{D} = \frac{\begin{vmatrix} 10 & -2 \\ 11 & -5 \end{vmatrix}}{-14} = \frac{(-50) - (-22)}{-14} = \frac{-28}{-14} = 2$$

$$y = \frac{D_y}{D} = \frac{\begin{vmatrix} 4 & 10 \\ 3 & 11 \end{vmatrix}}{-14} = \frac{44 - 30}{-14} = \frac{14}{-14} = -1$$

Therefore, the solution is $(2, -1)$. Try checking this solution in the original system of equations. ■

EXAMPLE 5 ■ **Using Cramer's Rule for a 3 × 3 System**

Use Cramer's Rule to solve the following system of linear equations.

$$\begin{aligned} -x + 2y - 3z &= 1 \\ 2x + \phantom{2y - {}} z &= 0 \\ 3x - 4y + 4z &= 2 \end{aligned}$$

Solution

To begin, we find the determinant of the coefficient matrix.

$$\begin{aligned} D = \begin{vmatrix} -1 & 2 & -3 \\ 2 & 0 & 1 \\ 3 & -4 & 4 \end{vmatrix} &= -(2)\begin{vmatrix} 2 & -3 \\ -4 & 4 \end{vmatrix} + (0)\begin{vmatrix} -1 & -3 \\ 3 & 4 \end{vmatrix} - (1)\begin{vmatrix} -1 & 2 \\ 3 & -4 \end{vmatrix} \\ &= -2(-4) + 0 - (-2) \\ &= 10 \end{aligned}$$

Because this determinant is not zero, we can apply Cramer's Rule to find the solution, as follows.

$$x = \frac{D_x}{D} = \frac{\begin{vmatrix} 1 & 2 & -3 \\ 0 & 0 & 1 \\ 2 & -4 & 4 \end{vmatrix}}{10} = \frac{8}{10} = \frac{4}{5}$$

$$y = \frac{D_y}{D} = \frac{\begin{vmatrix} -1 & 1 & -3 \\ 2 & 0 & 1 \\ 3 & 2 & 4 \end{vmatrix}}{10} = \frac{-15}{10} = -\frac{3}{2}$$

$$z = \frac{D_z}{D} = \frac{\begin{vmatrix} -1 & 2 & 1 \\ 2 & 0 & 0 \\ 3 & -4 & 2 \end{vmatrix}}{10} = \frac{-16}{10} = -\frac{8}{5}$$

Therefore, the solution is $\left(\frac{4}{5}, -\frac{3}{2}, -\frac{8}{5}\right)$. Try checking this solution in the original system of equations. ■

When using Cramer's Rule, remember that the method does not apply if the determinant of the coefficient matrix is zero. For instance, the system

$$\begin{aligned} 6x - 10y &= -4 \\ 9x - 15y &= 5 \end{aligned}$$

has no solution (see Example 9 in Section 9.4), and the determinant of the coefficient matrix of this system is

$$D = \begin{vmatrix} 6 & -10 \\ 9 & -15 \end{vmatrix} = -90 - (-90) = 0.$$

Similarly, the system

$$\begin{aligned} 12x - 6y &= -3 \\ -8x + 4y &= 2 \end{aligned}$$

has infinitely many solutions (see Example 10 in Section 9.4), and the determinant of the coefficient matrix of this system is

$$D = \begin{vmatrix} 12 & -6 \\ -8 & 4 \end{vmatrix} = 48 - 48 = 0.$$

NOTE: Because of the many calculations involving determinants, Cramer's Rule is usually considered to be less efficient than using elimination to solve a system of linear equations. Moreover, Cramer's Rule is not as general as the elimination method because Cramer's Rule requires that the coefficient matrix of the system be square *and* that the system have exactly one solution.

Applications

We mentioned that using determinants with Cramer's Rule to solve a system of linear equations is not as efficient as the elimination method. *However*, determinants do have many other practical applications. We present three of these in this section. The first gives a formula for finding the area of a triangle whose vertices are given by three points in a rectangular coordinate system.

Area of a Triangle

The area of a triangle with vertices (x_1, y_1), (x_2, y_2), and (x_3, y_3) is given by

$$\text{Area} = \pm\frac{1}{2}\begin{vmatrix} x_1 & y_1 & 1 \\ x_2 & y_2 & 1 \\ x_3 & y_3 & 1 \end{vmatrix}$$

where the symbol $(\pm)$ indicates that the appropriate sign should be chosen to yield a positive area.

EXAMPLE 6 ■ Finding the Area of a Triangle

Find the area of the triangle whose vertices are (2, 0), (1, 3), and (3, 2), as shown in Figure 9.9.

Solution

We choose $(x_1, y_1) = (2, 0)$, $(x_2, y_2) = (1, 3)$, and $(x_3, y_3) = (3, 2)$. Then, to find the area of the triangle, we evaluate the determinant

$$\begin{aligned} \begin{vmatrix} x_1 & y_1 & 1 \\ x_2 & y_2 & 1 \\ x_3 & y_3 & 1 \end{vmatrix} &= \begin{vmatrix} 2 & 0 & 1 \\ 1 & 3 & 1 \\ 3 & 2 & 1 \end{vmatrix} \\ &= 2\begin{vmatrix} 3 & 1 \\ 2 & 1 \end{vmatrix} - 0\begin{vmatrix} 1 & 1 \\ 3 & 1 \end{vmatrix} + 1\begin{vmatrix} 1 & 3 \\ 3 & 2 \end{vmatrix} \\ &= 2(1) - 0 + 1(-7) \\ &= -5. \end{aligned}$$

Using this value, we conclude that the area of the triangle is

$$\text{Area} = -\frac{1}{2}\begin{vmatrix} 2 & 0 & 1 \\ 1 & 3 & 1 \\ 3 & 2 & 1 \end{vmatrix} = -\frac{1}{2}(-5) = \frac{5}{2}.$$

FIGURE 9.9

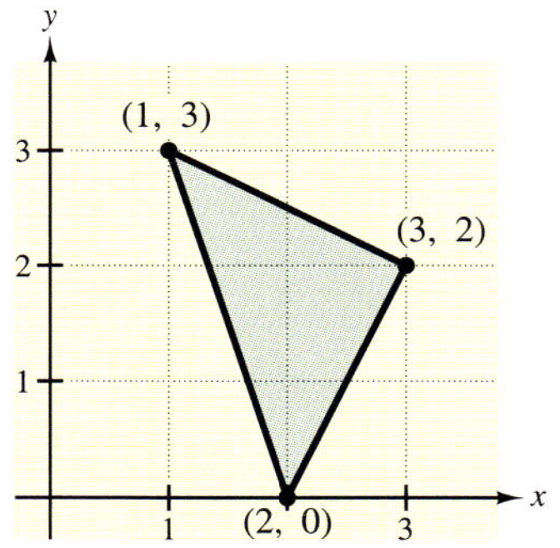

■

NOTE: To see the benefit of the "determinant formula for area," you should try finding the area of the triangle in Example 6 using the standard formula: area $= \frac{1}{2}$(base)(height).

Suppose the three points in Example 6 had been on the same line. What would have happened had we applied the area formula to three such points? The answer is that the determinant would have been zero. Consider, for instance, the three collinear points (1, 1), (3, 2), and (5, 3), as shown in Figure 9.10. The area of the "triangle" that has these three points as vertices is

$$\frac{1}{2}\begin{vmatrix} 1 & 1 & 1 \\ 3 & 2 & 1 \\ 5 & 3 & 1 \end{vmatrix} = \frac{1}{2}\left(1\begin{vmatrix} 2 & 1 \\ 3 & 1 \end{vmatrix} - 1\begin{vmatrix} 3 & 1 \\ 5 & 1 \end{vmatrix} + 1\begin{vmatrix} 3 & 2 \\ 5 & 3 \end{vmatrix}\right)$$

$$= \frac{1}{2}(-1 - (-2) + (-1)) = 0.$$

We generalize this result as follows.

FIGURE 9.10

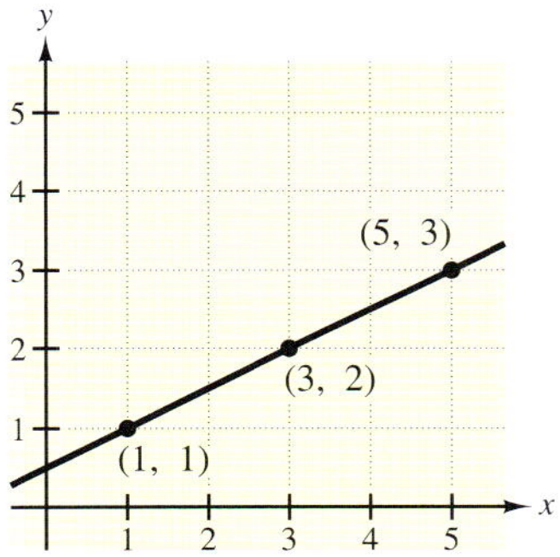

Test for Collinear Points

Three points (x_1, y_1), (x_2, y_2), and (x_3, y_3) are collinear (lie on the same line), if and only if

$$\begin{vmatrix} x_1 & y_1 & 1 \\ x_2 & y_2 & 1 \\ x_3 & y_3 & 1 \end{vmatrix} = 0.$$

EXAMPLE 7 ■ Testing for Collinear Points

Determine whether the points (−2, −2), (1, 1), and (7, 5) lie on the same line. (See Figure 9.11.)

FIGURE 9.11

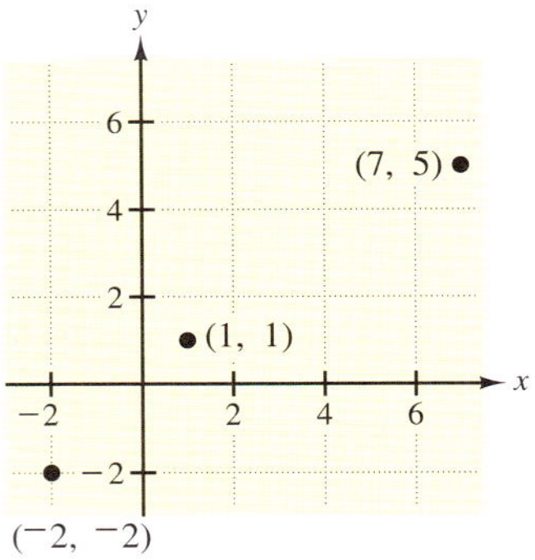

Solution

Letting $(x_1, y_1) = (-2, -2)$, $(x_2, y_2) = (1, 1)$, and $(x_3, y_3) = (7, 5)$, we have

$$\begin{aligned}\begin{vmatrix} x_1 & y_1 & 1 \\ x_2 & y_2 & 1 \\ x_3 & y_3 & 1 \end{vmatrix} &= \begin{vmatrix} -2 & -2 & 1 \\ 1 & 1 & 1 \\ 7 & 5 & 1 \end{vmatrix} \\ &= -2\begin{vmatrix} 1 & 1 \\ 5 & 1 \end{vmatrix} - (-2)\begin{vmatrix} 1 & 1 \\ 7 & 1 \end{vmatrix} + 1\begin{vmatrix} 1 & 1 \\ 7 & 5 \end{vmatrix} \\ &= -2(-4) - (-2)(-6) + 1(-2) \\ &= -6. \end{aligned}$$

Because the value of this determinant is not zero, we conclude that the three points do not lie on the same line. ■

The test for collinear points can be adapted to another use. That is, if we are given two points in a rectangular coordinate system, then we can find the equation of the line passing through the two points as follows.

Two-Point Form of the Equation of a Line

An equation of the line passing through the distinct points (x_1, y_1) and (x_2, y_2) is given by

$$\begin{vmatrix} x & y & 1 \\ x_1 & y_1 & 1 \\ x_2 & y_2 & 1 \end{vmatrix} = 0.$$

EXAMPLE 8 ■ Finding an Equation of the Line Passing Through Two Points

Find an equation of the line passing through the two points $(-2, 1)$ and $(3, -2)$.

Solution

Applying the determinant formula for the equation of the line passing through these two points produces

$$\begin{vmatrix} x & y & 1 \\ -2 & 1 & 1 \\ 3 & -2 & 1 \end{vmatrix} = 0.$$

To evaluate this determinant, we expand by cofactors along the first row to obtain the following.

$$x\begin{vmatrix} 1 & 1 \\ -2 & 1 \end{vmatrix} - y\begin{vmatrix} -2 & 1 \\ 3 & 1 \end{vmatrix} + 1\begin{vmatrix} -2 & 1 \\ 3 & -2 \end{vmatrix} = 3x + 5y + 1 = 0$$

Therefore, an equation of the line is

$$3x + 5y + 1 = 0.$$ ■

DISCUSSION PROBLEM ■ Determinant of a 3 × 3 Matrix

There is an alternative method that is commonly used for evaluating the determinant of a 3×3 matrix A. (This method *only* works for 3×3 matrices.) To apply this method, copy the first and second columns of A to form fourth and fifth columns. The determinant of A is then obtained by adding the products of three diagonals and subtracting the products of three diagonals, as shown in the following diagram.

$$A = \begin{bmatrix} 0 & 2 & 1 \\ 3 & -1 & 2 \\ 4 & -4 & 1 \end{bmatrix}$$

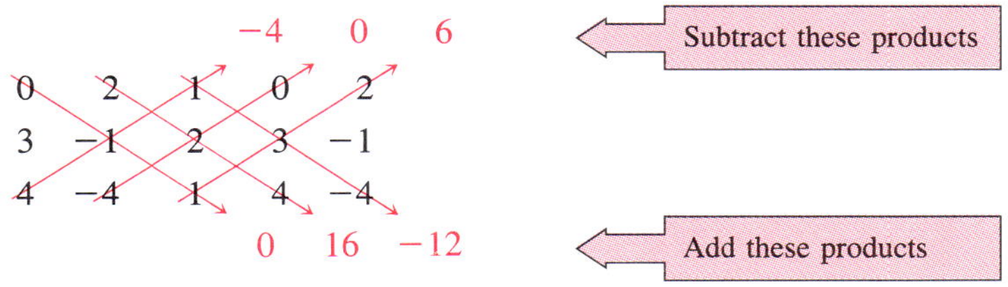

Thus, by adding the lower three products and subtracting the upper three products, we find the determinant of A to be

$$\det(A) = 0 + 16 - 12 - (-4) - 0 - 6 = 2.$$

Try using this technique to find the determinant of the matrices in Examples 2 and 3. Do you think this method is easier than expanding by cofactors? ■

Warm-Up

The following warm-up exercises involve skills that were covered in earlier sections. You will use these skills in the exercise set for this section.

In Exercises 1–10, evaluate the given quantity.

1. $2(25) - 5(8)$

2. $6(-8) - 4(2)$

3. $3(-2) - 7(4) + 8(-6)$

4. $10\left(\frac{3}{2}\right) - 6(-8) + (-3)(-30)$

5. $-4\left[-1(8) - 2\left(\frac{3}{4}\right) + (-4)\left(-\frac{3}{8}\right)\right]$

6. $\frac{5}{6}[10(-4) - (-1)(-6) + 4(5)]$

7. $\dfrac{2(-2) - 8(3)}{8(3/2) - 6(-1/3)}$

8. $\dfrac{5(3) - 4(-3)}{-6(2) - 5(-9)}$

9. $\dfrac{4(4) - 5(6) + 2(10)}{-5(1) - 10(3) + 8(5)}$

10. $\dfrac{-12(1/4) - 2(8) + 3(-1)}{3(2) - 4(15) + 10(6)}$

9.5 EXERCISES

In Exercises 1–10, find the determinant of the matrix.

1. $\begin{bmatrix} 2 & 1 \\ 3 & 4 \end{bmatrix}$

2. $\begin{bmatrix} -3 & 1 \\ 5 & 2 \end{bmatrix}$

3. $\begin{bmatrix} 5 & 2 \\ -6 & 3 \end{bmatrix}$

4. $\begin{bmatrix} 2 & -2 \\ 4 & 3 \end{bmatrix}$

5. $\begin{bmatrix} 5 & -4 \\ -10 & 8 \end{bmatrix}$

6. $\begin{bmatrix} 4 & -3 \\ 0 & 0 \end{bmatrix}$

7. $\begin{bmatrix} 2 & 6 \\ 0 & 3 \end{bmatrix}$

8. $\begin{bmatrix} -2 & 3 \\ 6 & -9 \end{bmatrix}$

9. $\begin{bmatrix} -7 & 6 \\ \frac{1}{2} & 3 \end{bmatrix}$

10. $\begin{bmatrix} \frac{2}{3} & \frac{5}{6} \\ 14 & -2 \end{bmatrix}$

In Exercises 11 and 12, evaluate the determinant of each matrix six different ways, by expanding cofactors along each row and column.

11. $\begin{bmatrix} 1 & 1 & 2 \\ 3 & 1 & 0 \\ -2 & 0 & 3 \end{bmatrix}$

12. $\begin{bmatrix} 2 & 1 & 3 \\ 1 & 4 & 4 \\ 1 & 0 & 2 \end{bmatrix}$

In Exercises 13–26, evaluate the determinant of the matrix. Expand by minors on the row or column that appears to make the computation easiest.

13. $\begin{bmatrix} 2 & 4 & 6 \\ 0 & 3 & 1 \\ 0 & 0 & -5 \end{bmatrix}$

14. $\begin{bmatrix} 2 & 3 & 1 \\ 0 & 5 & -2 \\ 0 & 0 & -2 \end{bmatrix}$

15. $\begin{bmatrix} -2 & 2 & 3 \\ 1 & -1 & 0 \\ 0 & 1 & 4 \end{bmatrix}$

16. $\begin{bmatrix} 3 & 2 & 2 \\ 2 & 2 & 2 \\ -4 & 4 & 3 \end{bmatrix}$

17. $\begin{bmatrix} 1 & 4 & -2 \\ 3 & 6 & -6 \\ -2 & 1 & 4 \end{bmatrix}$

18. $\begin{bmatrix} 2 & -1 & 0 \\ 4 & 2 & 1 \\ 4 & 2 & 1 \end{bmatrix}$

19. $\begin{bmatrix} -3 & 2 & 1 \\ 4 & 5 & 6 \\ 2 & -3 & 1 \end{bmatrix}$

20. $\begin{bmatrix} -3 & 4 & 2 \\ 6 & 3 & 1 \\ 4 & -7 & -8 \end{bmatrix}$

21. $\begin{bmatrix} 1 & 4 & -2 \\ 3 & 2 & 0 \\ -1 & 4 & 3 \end{bmatrix}$

22. $\begin{bmatrix} 6 & 8 & -7 \\ 0 & 0 & 0 \\ 4 & -6 & 22 \end{bmatrix}$

23. $\begin{bmatrix} 0.1 & 0.2 & 0.3 \\ -0.3 & 0.2 & 0.2 \\ 5 & 4 & 4 \end{bmatrix}$

24. $\begin{bmatrix} -0.4 & 0.4 & 0.3 \\ 0.2 & 0.2 & 0.2 \\ 0.3 & 0.2 & 0.2 \end{bmatrix}$

25. $\begin{bmatrix} x & y & 1 \\ 3 & 1 & 1 \\ -2 & 0 & 1 \end{bmatrix}$

26. $\begin{bmatrix} x & y & 1 \\ -2 & -2 & 1 \\ 1 & 5 & 1 \end{bmatrix}$

In Exercises 27–44, use Cramer's Rule to solve the system of equations. (If not possible, state the reason.)

27. $\begin{aligned} x + 2y &= 5 \\ -x + y &= 1 \end{aligned}$

28. $\begin{aligned} 2x - y &= -10 \\ 3x + 2y &= -1 \end{aligned}$

29. $\begin{aligned} 3x + 4y &= -2 \\ 5x + 3y &= 4 \end{aligned}$

30. $\begin{aligned} 18x + 12y &= 13 \\ 30x + 24y &= 23 \end{aligned}$

31. $\begin{aligned} 20x + 8y &= 11 \\ 12x - 24y &= 21 \end{aligned}$

32. $\begin{aligned} 13x - 6y &= 17 \\ 26x - 12y &= 8 \end{aligned}$

33. $-0.4x + 0.8y = 1.6$
$2x - 4y = 5$

34. $-0.4x + 0.8y = 1.6$
$0.2x + 0.3y = 2.2$

35. $3u + 6v = 5$
$6u + 14v = 11$

36. $3x_1 + 2x_2 = 1$
$2x_1 + 10x_2 = 6$

37. $4x - y + z = -5$
$2x + 2y + 3z = 10$
$5x - 2y + 6z = 1$

38. $4x - 2y + 3z = -2$
$2x + 2y + 5z = 16$
$8x - 5y - 2z = 4$

39. $3x + 4y + 4z = 11$
$4x - 4y + 6z = 11$
$6x - 6y = 3$

40. $14x_1 - 21x_2 - 7x_3 = 10$
$-4x_1 + 2x_2 - 2x_3 = 4$
$56x_1 - 21x_2 + 7x_3 = 5$

41. $3a + 3b + 4c = 1$
$3a + 5b + 9c = 2$
$5a + 9b + 17c = 4$

42. $2x + 3y + 5z = 4$
$3x + 5y + 9z = 7$
$5x + 9y + 17z = 13$

43. $5x - 3y + 2z = 2$
$2x + 2y - 3z = 3$
$x - 7y + 8z = -4$

44. $3x + 2y + 5z = 4$
$4x - 3y - 4z = 1$
$-8x + 2y + 3z = 0$

Area of a Triangle In Exercises 45–48, use a determinant to find the area of the triangle with the given vertices.

45. $(0, 0), (3, 1), (1, 5)$

46. $(-2, -3), (2, -3), (0, 4)$

47. $(-2, 1), (3, -1), (1, 6)$

48. $\left(0, \frac{1}{2}\right), \left(\frac{5}{2}, 0\right), (4, 3)$

Collinear Points In Exercises 49–52, use a determinant to determine whether the given points are collinear.

49. $(-1, 11), (0, 8), (2, 2)$

50. $(-1, -1), (1, 9), (2, 13)$

51. $\left(-2, \frac{1}{3}\right), (2, 1), \left(3, \frac{1}{5}\right)$

52. $\left(0, \frac{1}{2}\right), \left(1, \frac{7}{6}\right), \left(9, \frac{13}{2}\right)$

Equation of a Line In Exercises 53–56, use a determinant to find an equation of the line through the given points.

53. $(0, 0), (5, 3)$

54. $(-4, 3), (2, 1)$

55. $(10, 7), (-2, -7)$

56. $\left(-\frac{1}{2}, 3\right), \left(\frac{5}{2}, 1\right)$

Curve Fitting In Exercises 57–60, use Cramer's Rule to find the equation of the parabola $y = ax^2 + bx + c$ that passes through the given points.

57. $(0, 1), (1, -3), (-2, 21)$

58. $(2, 3), \left(-1, \frac{9}{2}\right), (-2, 9)$

59. $(1, -1), (-1, -5), \left(\frac{1}{2}, \frac{1}{4}\right)$

60. $(-2, 6), (1, 9), (3, 1)$

In Exercises 61 and 62, use the matrix capabilities (see the note in Example 3) of a graphing calculator to evaluate the determinant of the matrix.

61. $\begin{bmatrix} 5 & -3 & 2 \\ 7 & 5 & -7 \\ 0 & 6 & -1 \end{bmatrix}$

62. $\begin{bmatrix} -\frac{1}{2} & -1 & 6 \\ 8 & -\frac{1}{4} & -4 \\ 1 & 2 & 1 \end{bmatrix}$

In Exercises 63 and 64, use the matrix capabilities (see the note in Example 3) of a graphing calculator and Cramer's Rule to solve the system of equations.

63. $3x - 2y + 3z = 8$
$x + 3y + 6z = -3$
$x + 2y + 9z = -5$

64. $6x + 4y - 8z = -22$
$-2x + 2y + 3z = 13$
$-2x + 2y - z = 5$

65. (a) Use Cramer's Rule to solve the following system of linear equations.

$$kx + (1 - k)y = 1$$
$$(1 - k)x + ky = 3$$

(b) For what value(s) of k will the system in part (a) be inconsistent?

SECTION 9.6 Graphs of Linear Inequalities

Linear Inequalities in Two Variables • The Graph of a Linear Inequality in Two Variables

Linear Inequalities in Two Variables

A **linear inequality** in variables x and y is an inequality that can be written in one of the following forms (where a and b are not both zero).

$$ax + by < c \qquad ax + by \le c$$
$$ax + by > c \qquad ax + by \ge c$$

Here are some examples.

$$x - y > -3, \quad 4x - 3y < 7, \quad x \le 2, \quad \text{and} \quad y \ge -4$$

An ordered pair (x_1, y_1) is a **solution** of a linear inequality in x and y if the inequality is true when x_1 and y_1 are substituted for x and y, respectively.

EXAMPLE 1 ■ Verifying Solutions of Linear Inequalities

Determine whether the following points are solutions of $2x - 3y \ge -2$.

(a) (0, 0) (b) (2, 2) (c) (0, 1)

Solution

(a) To determine whether the point (0, 0) is a solution of the inequality, we substitute the coordinates of the point into the inequality as follows.

$2x - 3y \ge -2$ Given inequality

$2(0) - 3(0) \overset{?}{\ge} -2$ Replace x by 0 and y by 0

$0 \ge -2$ Inequality is satisfied

Since the inequality is satisfied, we conclude that the point (0, 0) is a solution.

(b) By substituting the coordinates of the point (2, 2) into the given inequality, we obtain $2(2) - 3(2) = -2$. Since -2 is greater than or equal to -2, we conclude that the point (2, 2) is also a solution of the inequality.

(c) By substituting the coordinates of the point (0, 1) into the given inequality, we obtain $2(0) - 3(1) = -3$. Since -3 is less than -2, we conclude that the point (0, 1) is *not* a solution of the inequality. ■

The Graph of a Linear Inequality in Two Variables

The **graph** of an inequality is the collection of all solution points of the inequality. To sketch the graph of a linear inequality such as $4x - 3y < 12$, we begin by sketching the graph of the *corresponding linear equation* $4x - 3y = 12$. This graph is made with a *dashed* line for the inequalities $<$ or $>$ and with a *solid* line for the inequalities $\leq$ or $\geq$. The graph of the equation (corresponding to a given linear inequality) separates the plane into two regions, called **half-planes**. In each half-plane, one of the following *must* be true.

1. All points in the half-plane are solutions of the inequality.
2. No point in the half-plane is a solution of the inequality.

Thus, we can determine whether the points in an entire half-plane satisfy the inequality by simply testing *one* point in the region. We summarize this graphing procedure as follows.

Sketching the Graph of a Linear Inequality in Two Variables

1. Replace the inequality sign by an equal sign, and sketch the graph of the resulting equation. (Use a dashed line for $<$ or $>$ and a solid line for $\leq$ or $\geq$.)
2. Test one point in each of the half-planes formed by the graph in Step 1. If the point satisfies the inequality, then shade the entire half-plane to denote that every point in the region satisfies the inequality.

EXAMPLE 2 ■ Sketching the Graph of a Linear Inequality

Sketch the graphs of the following linear inequalities.

(a) $x \geq -3$ (b) $y < 4$

Solution

(a) The graph of the corresponding equation $x = -3$ is a vertical line. The points that satisfy the inequality $x \geq -3$ are those lying to the right of (or on) this line, as shown in Figure 9.12.

(b) The graph of the corresponding equation $y = 4$ is a horizontal line. The points that satisfy the inequality $y < 4$ are those lying below this line, as shown in Figure 9.13.

FIGURE 9.12

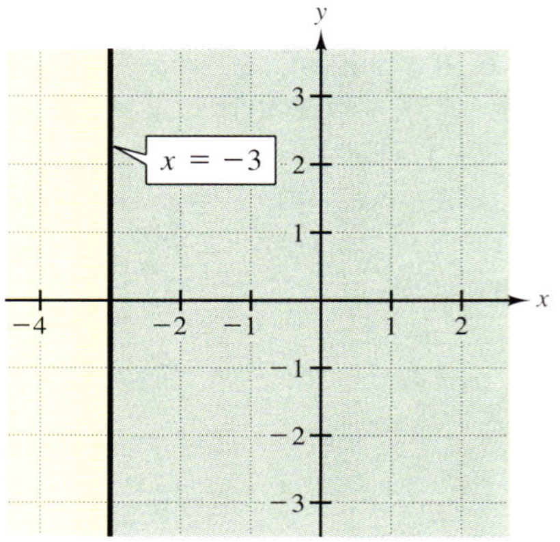

FIGURE 9.13

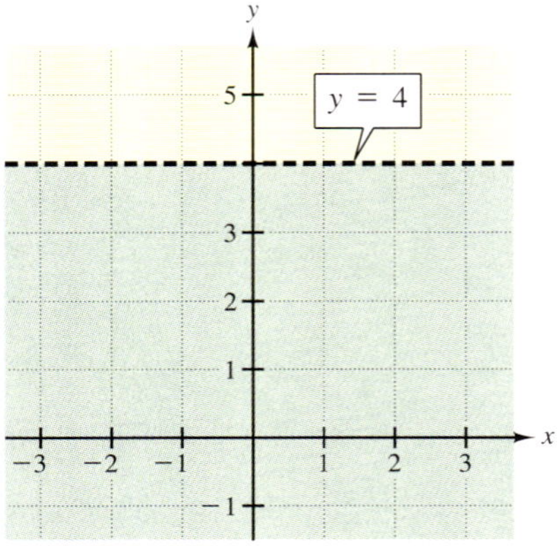

EXAMPLE 3 ■ Sketching the Graph of a Linear Inequality

Sketch the graph of the linear inequality

$$x + y > 3.$$

Solution

The graph of the corresponding equation $x + y = 3$ is a line, as shown in Figure 9.14. Since the origin $(0, 0)$ *does not* satisfy the inequality, the graph consists of the half-plane lying *above* the line. (Try checking a point above the line. Regardless of which point you choose, you will see that it satisfies the inequality.)

FIGURE 9.14

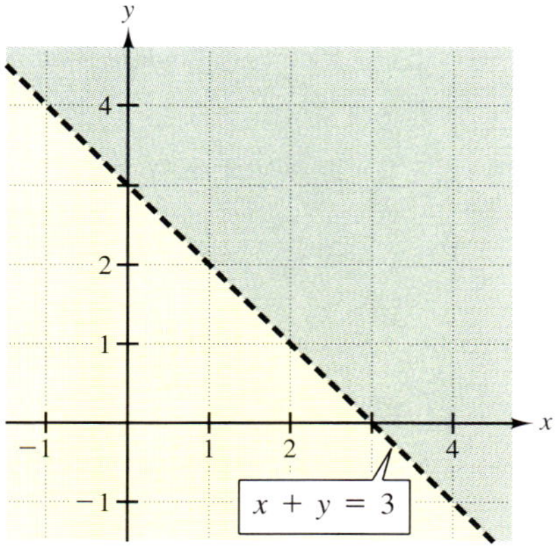

■

For a linear inequality in two variables, we can sometimes simplify the graphing procedure by writing the inequality in *slope-intercept* form. For instance, by writing $x + y > 3$ in the form

$$y > -x + 3$$

we can see that the solution points lie *above* the line $y = -x + 3$, as shown in Figure 9.14. Similarly, by writing the inequality $4x - 3y > 12$ in the form

$$y < \frac{4}{3}x - 4$$

we see that the solutions lie below the line $y = \frac{4}{3}x - 4$, as shown in Figure 9.15.

FIGURE 9.15

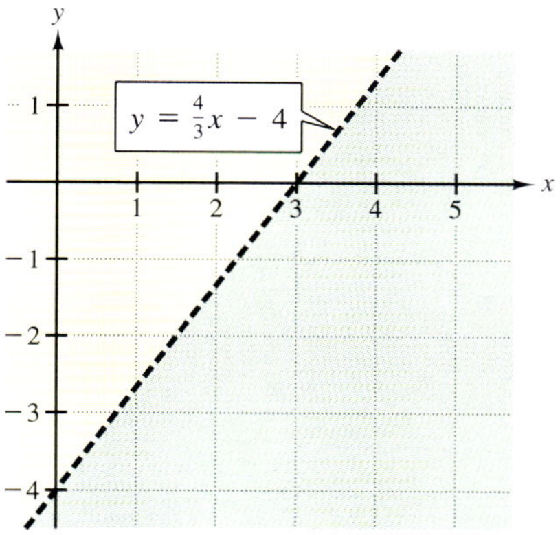

EXAMPLE 4 ■ **Sketching the Graph of a Linear Inequality**

Use the slope-intercept form of a linear equation as an aid in sketching the graph of the inequality

$$2x - 3y \le 15.$$

Solution

To begin, we rewrite the inequality in slope-intercept form.

$$2x - 3y \le 15 \quad \text{Given inequality}$$

$$-3y \le -2x + 15 \quad \text{Subtract } 2x \text{ from both sides}$$

$$y \ge \frac{2}{3}x - 5 \quad \text{Divide by } -3 \text{ and reverse inequality}$$

From this form, we can conclude that the solution is the half-plane lying *on* or *above* the line

$$y = \frac{2}{3}x - 5.$$

The graph is shown in Figure 9.16.

FIGURE 9.16

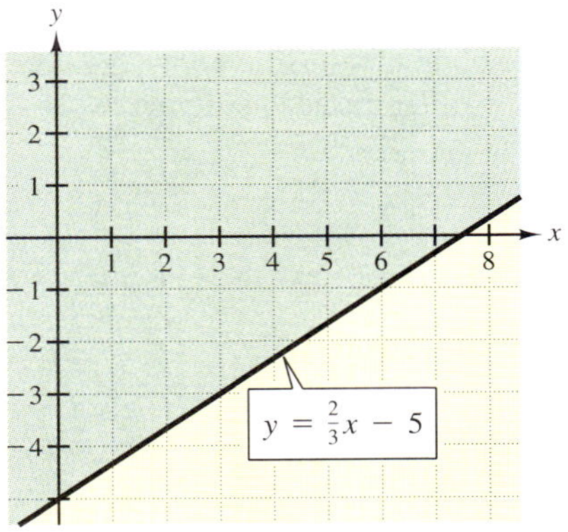

DISCUSSION PROBLEM ■ A Nonlinear Inequality

Write a short paragraph describing how you would sketch the graph of the nonlinear inequality

$$y < x^2 + 1.$$

Warm-Up

The following warm-up exercises involve skills that were covered in earlier sections. You will use these skills in the exercise set for this section.

In Exercises 1–6, identify the graph of the given equation. (Each graph is a line, a circle, an ellipse, or a hyperbola.)

1. $x^2 + y^2 = 25$

2. $y = (x - 2)^2 + 3$

3. $2x - 3y = 4$

4. $y = 8$

5. $\frac{x^2}{4} - y^2 = 1$

6. $\frac{x^2}{9} + \frac{y^2}{25} = 1$

In Exercises 7–10, sketch the graph of the given equation.

7. $x + 3y = 6$

8. $x = 4$

9. $x^2 + y^2 = 4$

10. $\frac{x^2}{4} + \frac{y^2}{9} = 1$

9.6 EXERCISES

In Exercises 1–6, match the inequality with its graph. [The graphs are labeled (a), (b), (c), (d), (e), and (f).]

1. $y \geq -2$

2. $x < -2$

3. $3x - 2y < 0$

4. $3x - 2y > 0$

5. $x + y < 4$

6. $x + y \leq 4$

(a)

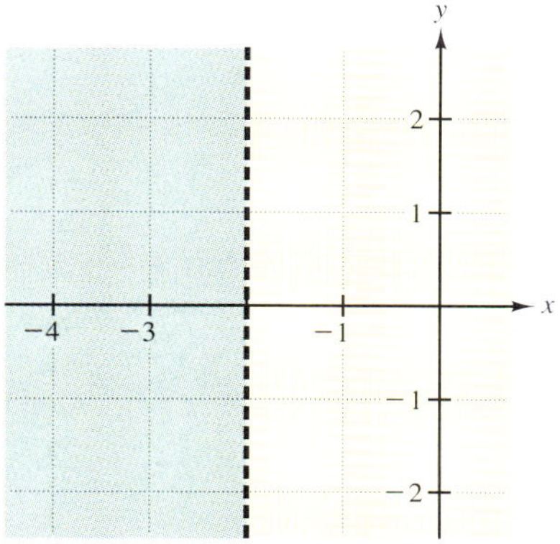

(b)

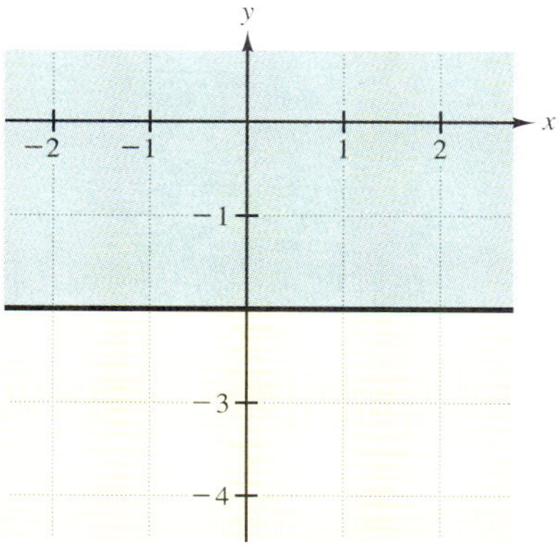

(c)

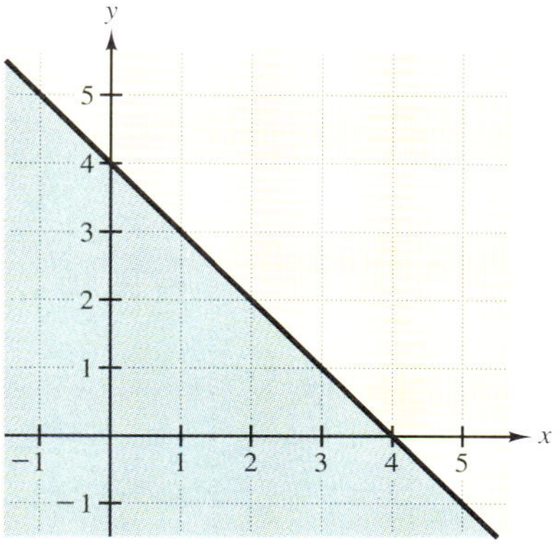

(d)

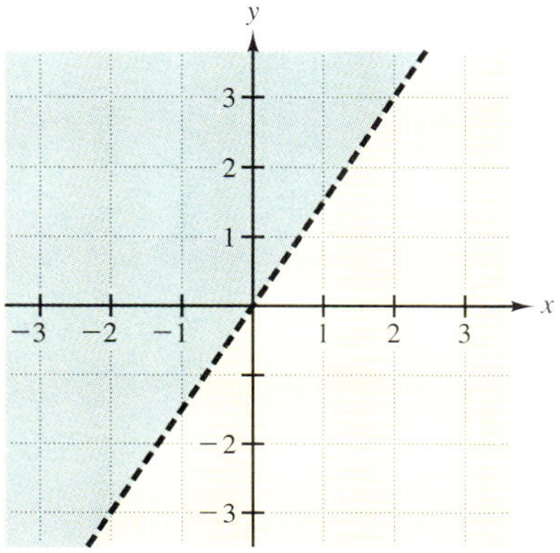

(e)

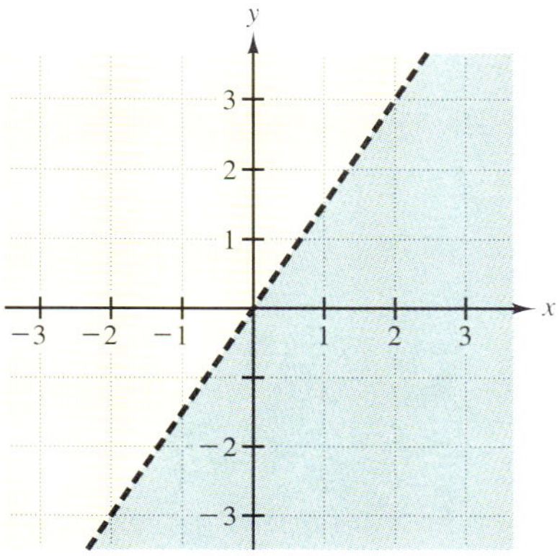

(f)

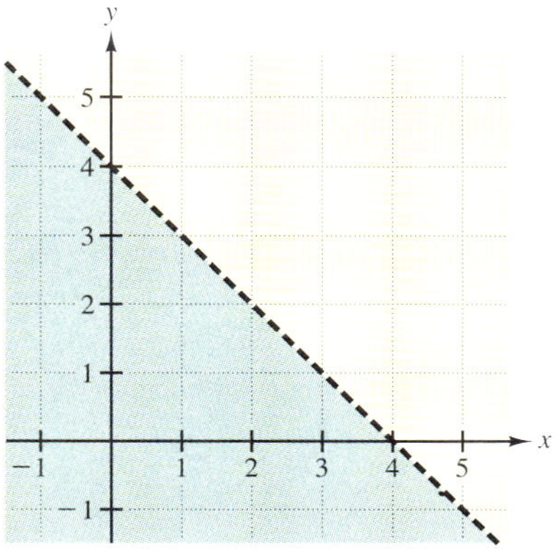

In Exercises 7–14, determine whether the points are solutions of the given inequality.

7. $x - 2y < 4$ (a) $(0, 0)$ (b) $(2, -1)$ (c) $(3, 4)$ (d) $(5, 1)$

8. $x + y < 3$ (a) $(0, 6)$ (b) $(4, 0)$ (c) $(0, -2)$ (d) $(1, 1)$

9. $3x + y \geq 10$ (a) $(1, 3)$ (b) $(-3, 1)$ (c) $(3, 1)$ (d) $(2, 15)$

10. $-3x + 5y \geq 6$ (a) $(2, 8)$ (b) $(-10, -3)$ (c) $(0, 0)$ (d) $(3, 3)$

11. $y > 0.2x - 1$	(a) $(0, 2)$	(b) $(6, 0)$	(c) $(4, -1)$	(d) $(-2, 7)$
12. $y < -3.5x + 7$	(a) $(1, 5)$	(b) $(5, -1)$	(c) $(-1, 4)$	(d) $(0, 4/3)$
13. $y \le 3 - \lvert x \rvert$	(a) $(-1, 4)$	(b) $(2, -2)$	(c) $(6, 0)$	(d) $(5, -2)$
14. $y \ge \lvert x - 3 \rvert$	(a) $(0, 0)$	(b) $(1, 2)$	(c) $(4, 10)$	(d) $(5, -1)$

In Exercises 15–30, sketch the graph of the given linear inequality.

15. $x \ge 2$

16. $x < -3$

17. $y < 5$

18. $y > 2$

19. $y > \frac{1}{2}x$

20. $y \le 2x$

21. $y \le x + 1$

22. $y > 4 - x$

23. $y - 1 > -\frac{1}{2}(x - 2)$

24. $y - 2 < -\frac{2}{3}(x - 1)$

25. $\frac{x}{3} + \frac{y}{4} \le 1$

26. $\frac{x}{2} + \frac{y}{6} \ge 1$

27. $x - 2y \ge 6$

28. $3x + 5y \le 15$

29. $0.2x + 0.3y < 2$

30. $x - 0.75y > 6$

In Exercises 31–36, write an inequality for the shaded region shown in the figure.

31.

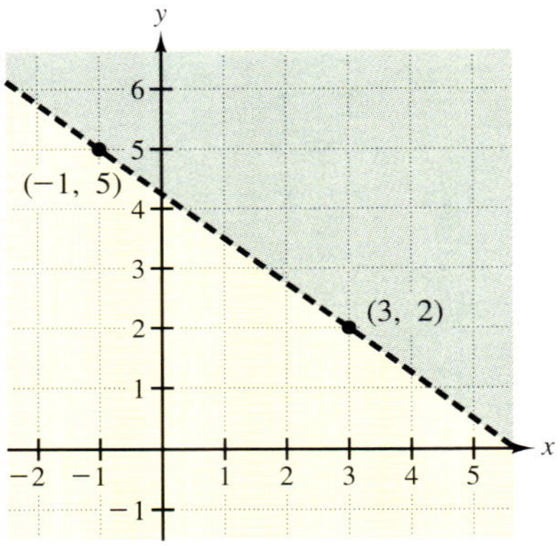

32.

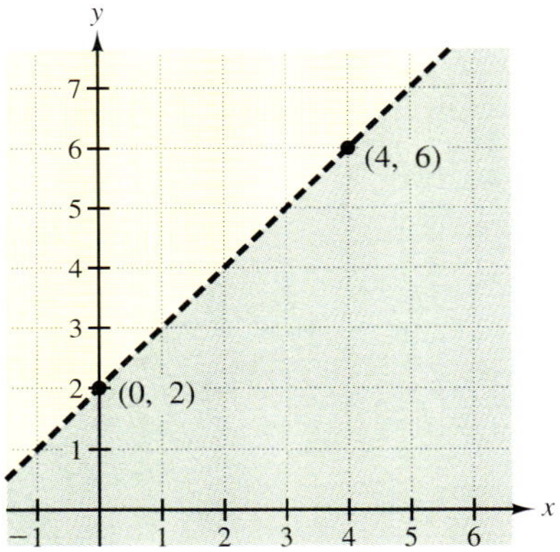

33.

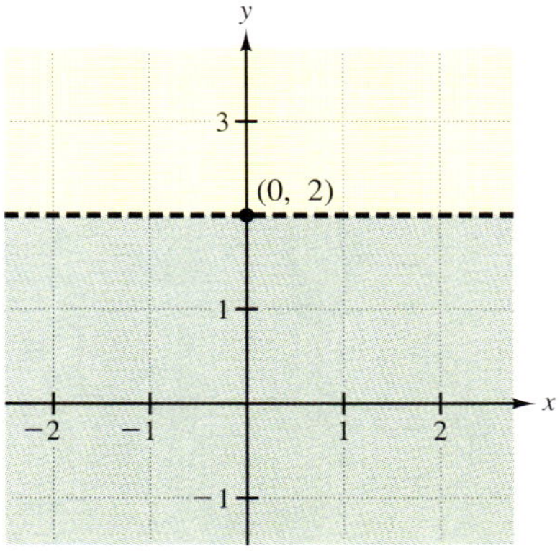

34.

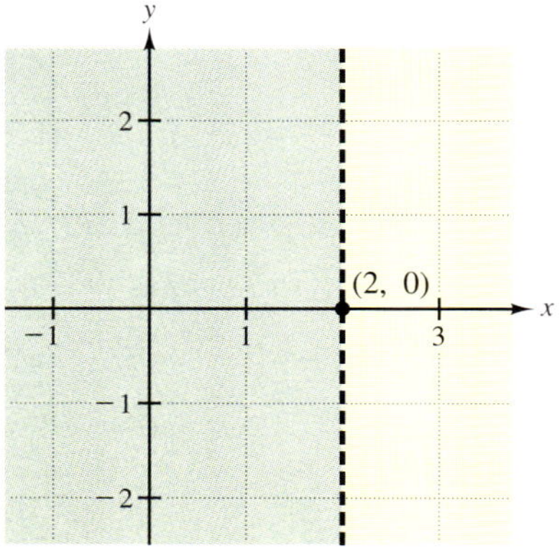

35.

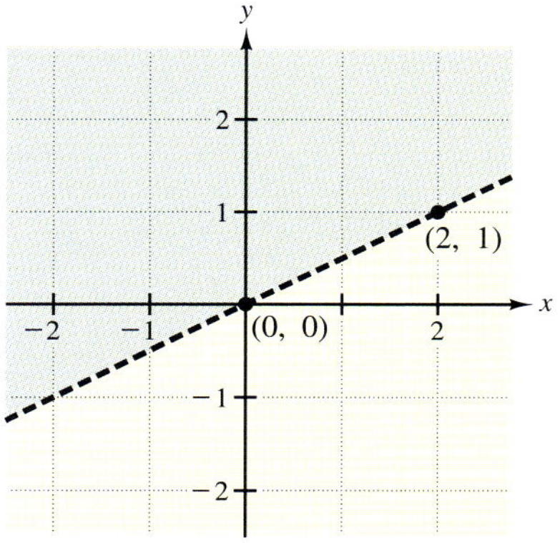

36.

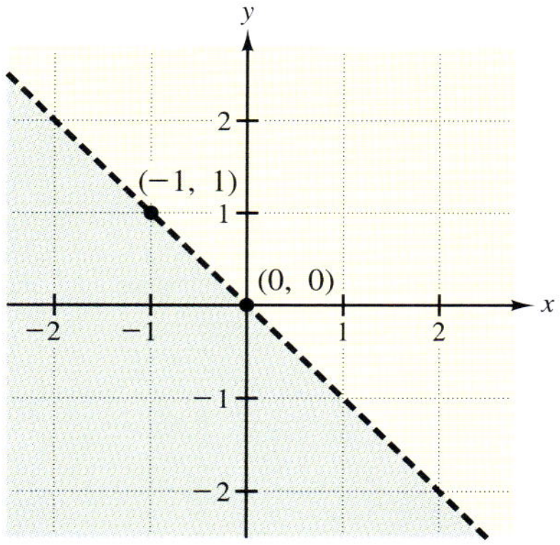

37. *Perimeter of a Rectangle* The perimeter of a rectangle of length x and width y cannot exceed 500 feet. Write a linear inequality for this constraint and sketch its graph.

38. *Storage Space* A warehouse for storing chairs and tables has 1,000 square feet of floor space. The space required for each chair and each table is 10 square feet and 15 square feet, respectively. Write a linear inequality for this space constraint if x is the number of chairs and y is the number of tables stored. Sketch a graph of the inequality.

CHAPTER 9 SUMMARY

As you review and prepare for a test on this chapter, first try to obtain a global view of what was discussed. Then review the specific skills needed under each category.

Solving Systems of Equations

- *Solve* a system of equations by the method of substitution. — Section 9.1, Exercises 5–30; Section 9.2, Exercises 31–38
- *Sketch and solve* a system of linear equations. — Section 9.1, Exercises 31–50
- *Create and solve* a system of linear equations from a verbal statement. — Section 9.1, Exercises 51–66; Section 9.2, Exercises 55–64; Section 9.3, Exercises 27–39
- *Solve* a system of linear equations by the method of elimination. — Section 9.2, Exercises 1–38
- *Solve* a system of linear equations using Gaussian elimination. — Section 9.3, Exercises 5–26

Using Matrices and Determinants to Solve Systems of Linear Equations

- *Convert* a matrix to row-echelon form by using elementary row operations. — Section 9.4, Exercises 5–16
- *Solve* a system of linear equations using matrices. — Section 9.4, Exercises 17–30
- *Evaluate* the determinant of a matrix. — Section 9.5, Exercises 1–26
- *Solve* a system of linear equations using Cramer's Rule. — Section 9.5, Exercises 27–44
- *Create and evaluate* determinants for finding areas of triangles, and equations of lines or parabolas. — Section 9.5, Exercises 45–60

Inequalities and Their Graphs

- *Determine* whether a specified point is a solution to a given inequality. — Section 9.6, Exercises 7–14
- *Sketch* a graph of the solution set of a linear inequality. — Section 9.6, Exercises 15–30

Chapter 9 Review Exercises

In Exercises 1–6, solve the system of linear equations by the method of substitution.

1. $\begin{aligned} 2x + 3y &= 1 \\ x + 4y &= -2 \end{aligned}$

2. $\begin{aligned} 3x - 7y &= 10 \\ -2x + y &= -14 \end{aligned}$

3. $\begin{aligned} -5x + 2y &= 4 \\ 10x - 4y &= 7 \end{aligned}$

4. $\begin{aligned} 5x + 2y &= 3 \\ 2x + 3y &= 10 \end{aligned}$

5. $\begin{aligned} 3x - 7y &= 5 \\ 5x - 9y &= -5 \end{aligned}$

6. $\begin{aligned} 24x - 4y &= 20 \\ 6x - y &= 5 \end{aligned}$

In Exercises 7–12, sketch the graphs of the given equations and find any solutions of the system of equations.

7. $\begin{aligned} x + y &= 2 \\ x - y &= 0 \end{aligned}$

8. $\begin{aligned} 2x &= 3(y - 1) \\ y &= x \end{aligned}$

9. $\begin{aligned} x - y &= 3 \\ -x + y &= 1 \end{aligned}$

10. $\begin{aligned} x + y &= -1 \\ 3x + 2y &= 0 \end{aligned}$

11. $\begin{aligned} 2x - y &= 0 \\ -x + y &= 4 \end{aligned}$

12. $\begin{aligned} x &= y + 3 \\ x &= y + 1 \end{aligned}$

In Exercises 13–18, solve the system of linear equations by the method of elimination.

13. $\begin{aligned} x + y &= 0 \\ 2x + y &= 0 \end{aligned}$

14. $\begin{aligned} 4x + y &= 1 \\ x - y &= 4 \end{aligned}$

15. $\begin{aligned} 2x - y &= 2 \\ 6x + 8y &= 39 \end{aligned}$

16. $\begin{aligned} 0.2x + 0.3y &= 0.14 \\ 0.4x + 0.5y &= 0.20 \end{aligned}$

17. $\begin{aligned} -x + y + 2z &= 1 \\ 2x + 3y + z &= -2 \\ 5x + 4y + 2z &= 4 \end{aligned}$

18. $\begin{aligned} 2x + 3y + z &= 10 \\ 2x - 3y - 3z &= 22 \\ 4x - 2y + 3z &= -2 \end{aligned}$

In Exercises 19–24, use matrices and elementary row operations to solve the system of linear equations.

19. $\begin{aligned} 5x + 4y &= 2 \\ -x + y &= -22 \end{aligned}$

20. $\begin{aligned} 2x - 5y &= 2 \\ 3x - 7y &= 1 \end{aligned}$

21. $\begin{aligned} 0.2x - 0.1y &= 0.07 \\ 0.4x - 0.5y &= -0.01 \end{aligned}$

22. $\begin{aligned} 2x + y &= 0.3 \\ 3x - y &= -1.3 \end{aligned}$

23. $\begin{aligned} x + 2y + 6z &= 4 \\ -3x + 2y - z &= -4 \\ 4x + 2z &= 16 \end{aligned}$

24. $\begin{aligned} 2x_1 + 3x_2 + 3x_3 &= 3 \\ 6x_1 + 6x_2 + 12x_3 &= 13 \\ 12x_1 + 9x_2 - x_3 &= 2 \end{aligned}$

In Exercises 25–30, solve the system of linear equations by using Cramer's Rule. (If not possible, state the reason.)

25. $\begin{aligned} 7x + 12y &= 63 \\ 2x + 3y &= 15 \end{aligned}$

26. $\begin{aligned} 12x + 42y &= -17 \\ 30x - 18y &= 19 \end{aligned}$

27. $\begin{aligned} 3x - 2y &= 16 \\ 12x - 8y &= -5 \end{aligned}$

28. $4x + 24y = 20$
$-3x + 12y = -5$

29. $-x + y + 2z = 1$
$2x + 3y + z = -2$
$5x + 4y + 2z = 4$

30. $2x_1 + x_2 + 2x_3 = 4$
$2x_1 + 2x_2 = 5$
$2x_1 - x_2 + 6x_3 = 2$

In Exercises 31–36, sketch the half-plane consisting of the solutions of the inequality.

31. $x - 2 \geq 0$

32. $y + 3 < 0$

33. $2x + y < 1$

34. $3x - 4y > 2$

35. $x \leq 4y - 2$

36. $(y - 3) \geq \frac{2}{3}(x - 5)$

In Exercises 37–40, use a determinant to find the equation of the line through the given points.

37. $(-4, 0), (4, 4)$

38. $(2, 5), (6, -1)$

39. $\left(-\frac{5}{2}, 3\right), \left(\frac{7}{2}, 1\right)$

40. $(-0.8, 0.2), (0.7, 3.2)$

In Exercises 41–44, use a determinant to find the area of the triangle with the given vertices.

41. $(1, 0), (5, 0), (5, 8)$

42. $(-4, 0), (4, 0), (0, 6)$

43. $(1, 2), (4, -5), (3, 2)$

44. $\left(\frac{3}{2}, 1\right), \left(4, -\frac{1}{2}\right), (4, 2)$

In Exercises 45 and 46, create a system of equations having the given solution. (Each problem has many correct answers.)

45. $\left(\frac{2}{3}, -4\right)$

46. $(-10, 12)$

47. *Break-Even Analysis* A small business invests \$25,000 in equipment to produce a product. Each unit of the product costs \$3.75 to produce and is sold for \$5.25. How many items must be sold before the business breaks even?

48. *Dimensions of a Rectangle* The perimeter of a rectangle is 480 meters, and its length is 150% of its width. Find the dimensions of the rectangle.

49. *Acid Mixture* One hundred gallons of a 60% acid solution are obtained by mixing a 75% solution with a 50% solution. How many gallons of each must be used to obtain the desired mixture?

50. *Rope Length* Suppose you must cut a rope that is 128 inches long into two pieces such that one piece is three times longer than the other. Find the length of each piece.

51. *Curve Fitting* Find an equation of the parabola $y = ax^2 + bx + c$ passing through the points $(-1, 2)$, $(0, 3)$, and $(1, 6)$.

52. *Number Problem* The sum of three positive numbers is 68. The second number is four greater than the first, and the third is twice the first. Find the three numbers.

CHAPTER 9 TEST

Take this test as you would take a test in class. After you are done, check your work with the answers given in the back of the book.

1. Determine which ordered pair is a solution of the system of equations.

$$\begin{aligned} 2x - 2y &= 1 \\ -x + 2y &= 0 \end{aligned}$$

(a) $(3, -4)$ (b) $\left(1, \frac{1}{2}\right)$

2. Solve the system by the method of substitution.

$$\begin{aligned} 5x - y &= 6 \\ 4x - 3y &= -4 \end{aligned}$$

3. Solve the nonlinear system by the method of substitution.

$$\begin{aligned} x + y &= 8 \\ xy &= 12 \end{aligned}$$

4. Solve the system of equations graphically.

$$\begin{aligned} x - 2y &= -1 \\ 2x + 3y &= 12 \end{aligned}$$

5. Solve the system of equations by the method of elimination.

$$\begin{aligned} 3x - 4y &= -14 \\ -3x + y &= 8 \end{aligned}$$

6. Solve the system of equations by the method of elimination.

$$\begin{aligned} 8x + 3y &= 3 \\ 4x - 6y &= -1 \end{aligned}$$

7. Use Gaussian elimination to solve the system of equations.

$$\begin{aligned} x + 2y - 4z &= 0 \\ 3x + y - 2z &= 5 \\ 3x - y + 2z &= 7 \end{aligned}$$

8. Use matrices to solve the system of equations.

$$\begin{aligned} x \qquad - 3z &= -10 \\ -2y + 2z &= 0 \\ x - 2y \qquad &= -7 \end{aligned}$$

9. Use Cramer's Rule to solve the system of equations.

$$\begin{aligned} 2x - 7y &= 7 \\ 3x + 7y &= 13 \end{aligned}$$

10. Evaluate the determinant of the matrix.

$$\begin{bmatrix} 3 & -2 & 0 \\ -1 & 5 & 3 \\ 2 & 7 & 1 \end{bmatrix}$$

11. Sketch the graph of the inequality $x + 2y \le 4$.

12. Find the value of a such that the system is inconsistent.

$$\begin{aligned} 5x - 8y &= 3 \\ 3x + ay &= 0 \end{aligned}$$

13. Find a system of linear equations with integer coefficients that has the solution $(5, -3)$. (The problem has many correct answers.)

14. Two people share the driving for a 200-mile trip. One person drives four times as far as the other. Write a system of linear equations that models the problem. Find the distance each person drives.

15. Find the equation of the parabola $y = ax^2 + bx + c$ that passes through the points $(0, 4)$, $(1, 3)$, and $(2, 6)$.

16. Use a determinant to find the area of the triangle with vertices $(0, 0)$, $(5, 4)$, and $(6, 0)$.

CHAPTER TEN

Exponential and Logarithmic Functions

Chapter Overview

So far we have dealt with algebraic functions, which include polynomial and rational functions. Functions that are not algebraic are called transcendental functions. Chapter 10 introduces two types of transcendental functions, called exponential functions and logarithmic functions. These functions are widely used in describing economic, biological, and physical phenomena such as compound interest, population growth, memory retention, earthquake intensity, and radioactive decay.

SECTION **10.1**

Exponential Functions

Exponential Functions • Graphs of Exponential Functions • The Natural Base e • Compound Interest

Exponential Functions

Whereas polynomial and rational functions have terms with variable bases and constant exponents, exponential functions have terms with *constant* bases and *variable* exponents. Here are some examples.

Polynomial or Rational Function	*Exponential Function*
x^2 (Constant exponent; Variable base)	2^x (Variable exponent; Constant base)
x^{-3} (Constant exponent; Variable base)	3^{-x} (Variable exponent; Constant base)

The definition of the **exponential function with base *a*** is as follows.

Definition of Exponential Function

The **exponential function *f* with base *a*** is denoted by

$$f(x) = a^x$$

where $a > 0$, $a \neq 1$, and x is any real number.

NOTE: We exclude the base $a = 1$ because $f(x) = 1^x = 1$ is a constant function, *not* an exponential function.

In Chapter 5 we learned to evaluate a^x for integer and rational values of x. For example, we know that

$$8^3 = 8 \cdot 8 \cdot 8 = 512 \quad \text{and} \quad 8^{2/3} = \left(\sqrt[3]{8}\right)^2 = 2^2 = 4.$$

However, to evaluate 8^x for any real number x, we need to interpret forms with *irrational* exponents, such as $8^{\sqrt{2}}$. For the purpose of this text, it is sufficient to think of

$$a^{\sqrt{2}}$$

where $\sqrt{2} \approx 1.414214$, as the number that has the successively closer approximations

$$a^{1.4},\ a^{1.41},\ a^{1.414},\ a^{1.4142},\ a^{1.41421},\ a^{1.414214},\ \ldots .$$

The properties of exponents that we discussed in Section 5.1 can be extended to cover exponential functions, as follows.

Properties of Exponential Functions

Let a be a positive real number, and let x and y be real numbers, variables, or algebraic expressions.

1. $a^x \cdot a^y = a^{x+y}$
2. $(a^x)^y = a^{xy}$
3. $\dfrac{a^x}{a^y} = a^{x-y}$
4. $a^{-x} = \dfrac{1}{a^x} = \left(\dfrac{1}{a}\right)^x$

To evaluate exponential functions with a calculator, you can use the exponential keys [y^x] (where y is the base and x is the exponent) or [∧]. The base is entered first, then the exponent. For example, to evaluate $3^{-1.3}$, we can use the following keystrokes.*

Keystrokes	*Display*	
3 [y^x] 1.3 [+/−] [=]	0.239741	Scientific
3 [∧] [(] [(−)] 1.3 [)] [ENTER]	0.239741	Graphing

EXAMPLE 1 ■ Evaluating Exponential Functions

Evaluate the following functions at the indicated values of x. Use a calculator only if it is more efficient or necessary.

Function	*Values*
(a) $f(x) = 2^x$	$x = 3,\ x = -4,\ x = \pi$
(b) $g(x) = 12^x$	$x = 3,\ x = -0.1,\ x = \frac{5}{7}$
(c) $h(x) = (1.085)^x$	$x = 0,\ x = -3$

Solution

Evaluation	*Comment*
(a) $f(3) = 2^3 = 8$	Calculator is not necessary
$f(-4) = 2^{-4} = \dfrac{1}{2^4} = \dfrac{1}{16}$	Calculator is not necessary
$f(\pi) = 2^\pi \approx 8.825$	Calculator is necessary
(b) $g(3) = 12^3 = 1{,}728$	Calculator is more efficient
$g(-0.1) = 12^{-0.1} \approx 0.7800$	Calculator is necessary
$g\left(\dfrac{5}{7}\right) = 12^{5/7} \approx 5.900$	Calculator is necessary
(c) $h(0) = (1.085)^0 = 1$	Calculator is not necessary
$h(-3) = (1.085)^{-3} \approx 0.7829$	Calculator is more efficient ■

*The graphing calculator keystrokes given in this text correspond to the *TI-81* graphing calculator from *Texas Instruments*. For other graphing calculators, the keystrokes may differ.

Graphs of Exponential Functions

The basic nature of the graph of an exponential function can be determined by plotting several points. (A computer or calculator with graphing capabilities would also help.)

EXAMPLE 2 ■ **Sketching the Graph of an Exponential Function ($y = a^x$)**

On the same coordinate plane, sketch the graphs of the following functions.

(a) $f(x) = 2^x$ (b) $g(x) = 4^x$

Determine the domain and range of each function.

Solution

Table 10.1 lists some values of each function, and Figure 10.1 shows their graphs. From the graphs, we can see that the domain and range of each function is as follows.

Domain: all real numbers

Range: all positive real numbers

TABLE 10.1

x	-2	-1	0	1	2	3
(a) 2^x	$\frac{1}{4}$	$\frac{1}{2}$	1	2	4	8
(b) 4^x	$\frac{1}{16}$	$\frac{1}{4}$	1	4	16	64

FIGURE 10.1

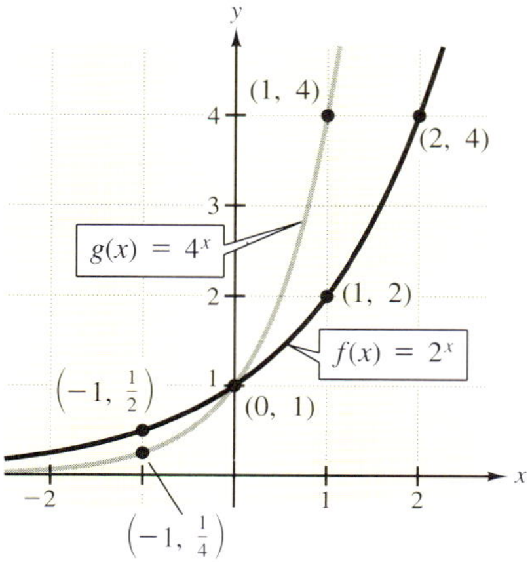

■

We know from our work with functions in Chapter 8 that the graph of $h(x) = f(-x) = 2^{-x}$ is a reflection of the graph of $f(x) = 2^x$ in the y-axis. This is reinforced in the following example.

EXAMPLE 3 ■ Sketching the Graph of an Exponential Function ($y = a^{-x}$)

On the same coordinate plane, sketch the graphs of the following functions.

(a) $y = 2^{-x}$ (b) $y = 4^{-x}$

Solution

Table 10.2 lists some values of each function, and Figure 10.2 shows their graphs.

TABLE 10.2

x	-3	-2	-1	0	1	2
(a) 2^{-x}	8	4	2	1	$\frac{1}{2}$	$\frac{1}{4}$
(b) 4^{-x}	64	16	4	1	$\frac{1}{4}$	$\frac{1}{16}$

FIGURE 10.2

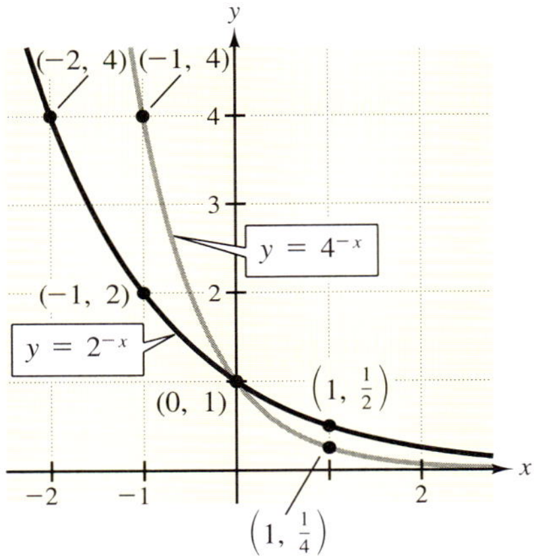

■

Examples 2 and 3 suggest that for $a > 1$, the graph of $y = a^x$ increases and the graph of $y = a^{-x}$ decreases. The graphs shown in Figure 10.3 are typical of the graphs of these two types of exponential functions. Note that each has a y-intercept at (0, 1).

FIGURE 10.3

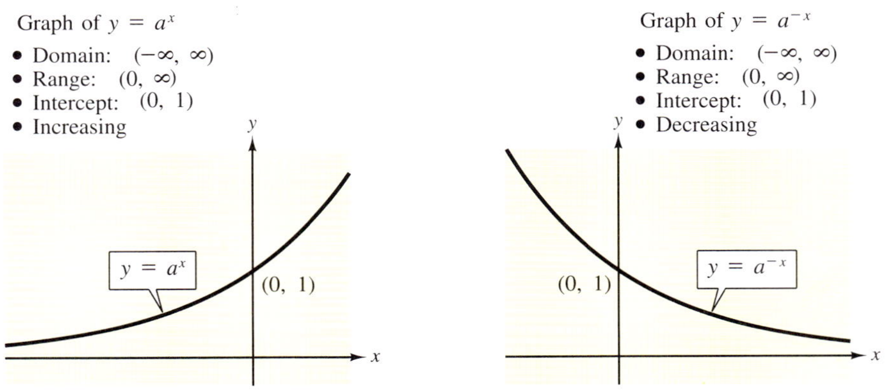

Characteristics of the Exponential Functions a^x and a^{-x} ($a > 1$)

EXAMPLE 4 ■ An Application: Radioactive Decay

Let y represent the mass of a particular radioactive element whose half-life is 25 years. The initial mass is 10 grams. After t years, the mass (in grams) is given by

$$y = 10\left(\frac{1}{2}\right)^{t/25}, \quad t \geq 0.$$

How much of the initial mass is present after 120 years?

Solution

When $t = 120$, the mass is given by

$$y = 10\left(\frac{1}{2}\right)^{120/25} = 10\left(\frac{1}{2}\right)^{4.8} \approx 0.359 \text{ grams.}$$

Thus, after 120 years, the mass has decayed from an initial amount of 10 grams to only 0.359 gram. Note in Figure 10.4 that the graph of the function shows the 25-year half-life. That is, after 25 years, the mass is 5 grams (half of the original). After another 25 years, the mass is 2.5 grams, and so on.

FIGURE 10.4

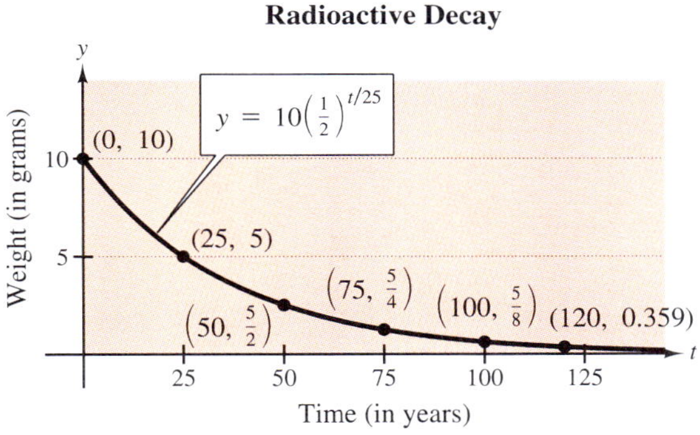

■

The Natural Base e

So far, we have used integers or rational numbers as bases of exponential functions. In many applications of exponential functions, the convenient choice for a base is the following irrational number, denoted by the letter e.

$$e \approx 2.71828 \ldots$$

This number is called the **natural base**.

The function $f(x) = e^x$ is called the **natural exponential function**. Its graph is shown in Figure 10.5. Be sure you understand that for this function, e is the constant number 2.71828 . . . , and x is a variable. To evaluate the natural exponential function, we need a calculator, preferably one having a natural exponential key $\boxed{e^x}$. Here are some examples of how to use such a calculator to evaluate the natural exponential function.

FIGURE 10.5

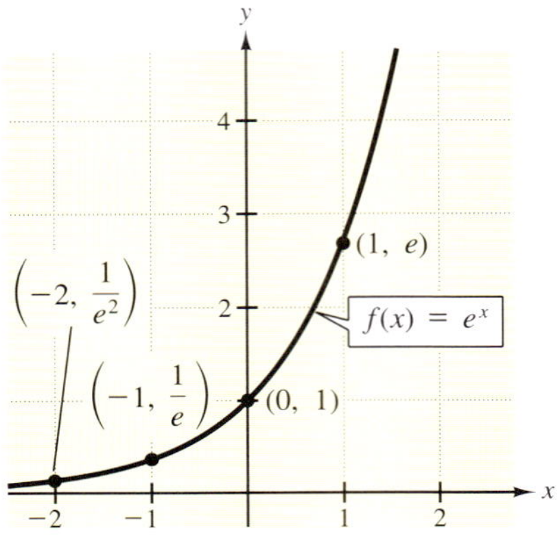

Scientific Calculator

Function Value	*Keystrokes*	*Calculator Display*
$f(2) = e^2$	2 [e^x]	7.3890561
$f(-3) = e^{-3}$	3 [+/−] [e^x]	0.049787
$f(0.32) = e^{0.32}$	.32 [e^x]	1.3771278

Graphing Calculator

Function Value	*Keystrokes*	*Calculator Display*
$f(2) = e^2$	[e^x] 2 [ENTER]	7.3890561
$f(-3) = e^{-3}$	[e^x] [(] [(−)] 3 [)] [ENTER]	0.049787
$f(0.32) = e^{0.32}$	[e^x] .32 [ENTER]	1.3771278

NOTE: Some calculators do not have a key labeled [e^x]. If your calculator doesn't have this key, but does have a key labeled [ln x], you will have to use the two-keystroke sequence [INV] [ln x] in place of [e^x].

Compound Interest

One of the most familiar uses of exponential functions involves **compound interest**. Suppose that a principal P is invested at an annual percentage rate r (in decimal form), compounded once a year. If the interest is added to the principal at the end of the year, then the balance is

$$A = P + Pr = P(1 + r).$$

This pattern of multiplying the previous principal by $(1 + r)$ is then repeated each successive year, as shown in Table 10.3.

TABLE 10.3

Time in Years	Balance After Each Compounding
0	$A = P$
1	$A = P(1 + r)$
2	$A = P(1 + r)(1 + r) = P(1 + r)^2$
3	$A = P(1 + r)^2(1 + r) = P(1 + r)^3$
$\vdots$	$\vdots$
t	$A = P(1 + r)^t$

To account for more frequent compounding of interest (such as quarterly or monthly compounding), we let n be the number of compoundings per year and let t be the number of years. Then the rate per compounding is r/n and the account balance after t years is

$$A = P\left(1 + \frac{r}{n}\right)^{nt}.$$

EXAMPLE 5 ■ Finding the Balance for Compound Interest

A sum of \$10,000 is invested at an annual percentage rate of 7.5%, compounded monthly. Find the balance in the account after 10 years.

Solution

Using the formula

$$A = P\left(1 + \frac{r}{n}\right)^{nt}$$

with $P = 10{,}000$, $r = 0.075$, $n = 12$ (for monthly compounding), and $t = 10$, we obtain the following balance.

$$A = 10{,}000\left(1 + \frac{0.075}{12}\right)^{12(10)} \approx \$21{,}120.65$$ ■

A second method used to compute interest is called **continuous compounding**. The formula for the balance for this type of compounding is given by

$$A = Pe^{rt}.$$

We summarize the formulas for both types of compounding as follows.

Formulas for Compound Interest

After t years, the balance A in an account with principal P and annual percentage rate r (in decimal form) is given by the following formulas.

1. For n compoundings per year: $A = P\left(1 + \frac{r}{n}\right)^{nt}$
2. For continuous compounding: $A = Pe^{rt}$

EXAMPLE 6 ■ Comparing Two Types of Compounding

A total of $15,000 is invested at an annual percentage rate of 8%. Find the balance after six years if the interest is compounded (a) quarterly and (b) continuously.

Solution

(a) Letting $P = 15{,}000$, $r = 0.08$, $n = 4$, and $t = 6$, we find the balance after six years at quarterly compounding to be

$$A = 15{,}000\left(1 + \frac{0.08}{4}\right)^{4(6)} = \$24{,}126.56.$$

(b) Letting $P = \$15{,}000$, $r = 0.08$, and $t = 6$, we find the balance after six years at continuous compounding to be

$$A = 15{,}000e^{0.08(6)} = \$24{,}241.12.$$

Note that the balance is greater with continuous compounding than with quarterly compounding. ■

NOTE: Example 6 illustrates the following general rule. For a given principal, interest rate, and time, the more often the interest is compounded per year, the greater the balance will be. Moreover, the balance obtained by continuous compounding is larger than the balance obtained by compounding n times per year.

DISCUSSION PROBLEM ■ A Calculator Experiment

The natural number e arises in a number of fascinating ways in mathematics. The following steps show one way that this number can be approximated with a calculator.

(a) Use a calculator to complete the following table for indicated values of x.

x	0	10	100	1,000	10,000
$\left(1 + \frac{1}{x}\right)^x$					

(b) Use the information given in the table to sketch the graph of the function

$$y = \left(1 + \frac{1}{x}\right)^x.$$

Does this graph appear to be approaching a horizontal asymptote?

(c) From parts (a) and (b), what conclusions can you draw about the value of

$$\left(1 + \frac{1}{x}\right)^x$$

as x gets larger and larger? ■

Warm-Up

The following warm-up exercises involve skills that were covered in earlier sections. You will use these skills in the exercise set for this section.

In Exercises 1–10, use the properties of exponents to simplify the given expression.

1. $x^5 \cdot x^3$

2. $(x + 10)^0, \quad x \neq -10$

3. $\dfrac{z^7}{z^4}, \quad z \neq 0$

4. $\dfrac{18(x-3)^5}{24(x-3)}, \quad x \neq 3$

5. $(x^2y)^4$

6. $\left(\dfrac{x}{2y}\right)^5, \quad y \neq 0$

7. $x^4y^{-3}, \quad y \neq 0$

8. $(a^2)^{-4}, \quad a \neq 0$

9. $(b^6)^{1/2}$

10. $(8x^3)^{1/3}$

10.1 EXERCISES

In Exercises 1–10, evaluate the function at the given values. Use a calculator only if it is more efficient or necessary.

1. $f(x) = 3^x$
(a) $x = -2$
(b) $x = 0$
(c) $x = 1$

2. $g(x) = 5^x$
(a) $x = -1$
(b) $x = 1$
(c) $x = 3$

3. $F(x) = 3^{-x}$
(a) $x = -2$
(b) $x = 0$
(c) $x = 1$

4. $G(x) = 5^{-x}$
(a) $x = -1$
(b) $x = 1$
(c) $x = \sqrt{3}$

5. $f(t) = 500\left(\dfrac{1}{2}\right)^t$
(a) $t = 0$
(b) $t = 1$
(c) $t = \pi$

6. $g(s) = 1{,}200\left(\dfrac{2}{3}\right)^s$
(a) $s = 0$
(b) $s = 2$
(c) $s = 4$

7. $f(x) = 1{,}000(1.05)^{2x}$
(a) $x = 0$
(b) $x = 5$
(c) $x = 10$

8. $P(t) = \dfrac{10{,}000}{(1.01)^{12t}}$
(a) $t = 2$
(b) $t = 10$
(c) $t = 20$

9. $f(x) = e^x$
(a) $x = -1$
(b) $x = 0$
(c) $x = \dfrac{1}{2}$

10. $A(t) = 200e^{0.1t}$
(a) $t = 10$
(b) $t = 20$
(c) $t = 40$

In Exercises 11–16, match the exponential function with its graph. [The graphs are labeled (a), (b), (c), (d), (e), and (f).]

11. $f(x) = 2^x$

12. $f(x) = -2^x$

13. $f(x) = 2^{-x}$

14. $f(x) = 2^x - 1$

15. $f(x) = 2^{x-1}$

16. $f(x) = 2^{x+1}$

(a)

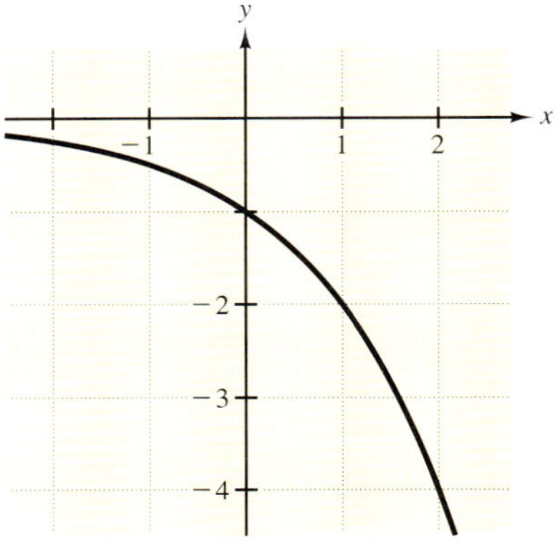

(b)

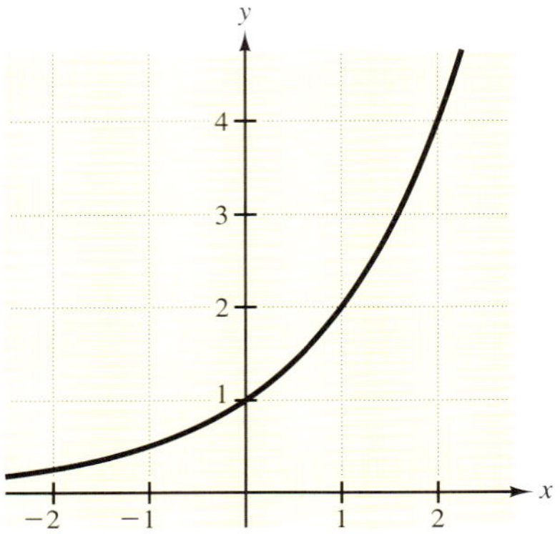

(c)

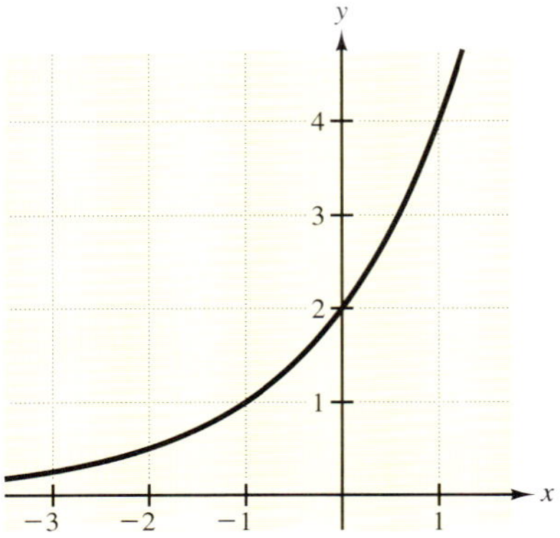

(d)

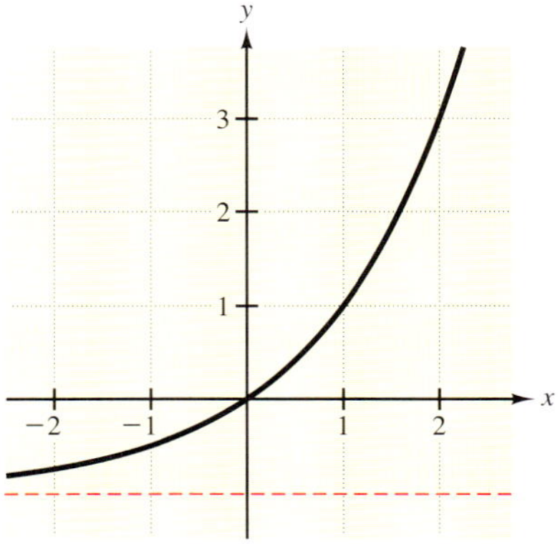

(e)

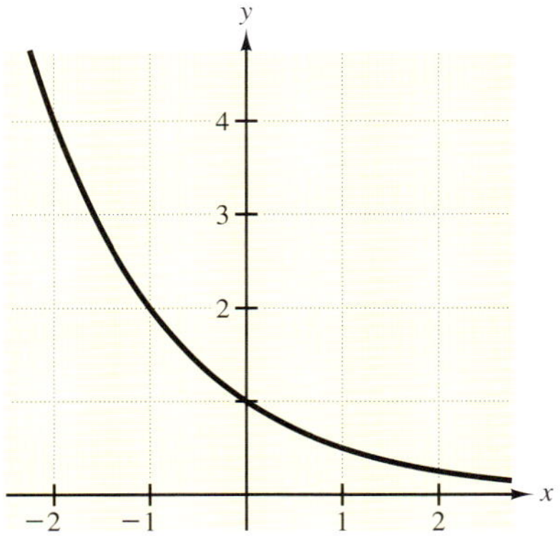

(f)

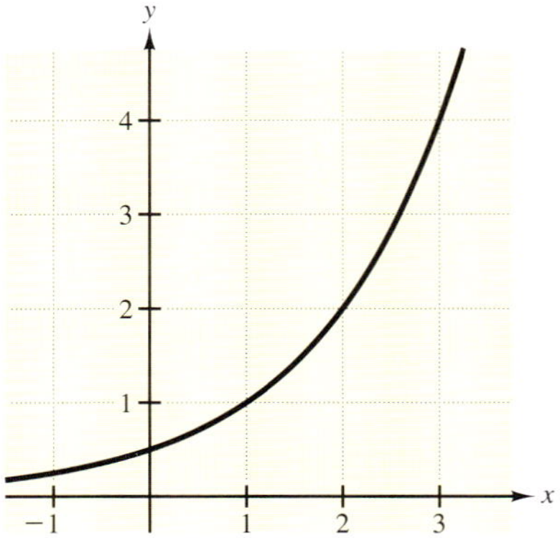

In Exercises 17–30, sketch the graph of the given function.

17. $f(x) = 3^x$

18. $f(x) = 3^{-x} = \left(\frac{1}{3}\right)^x$

19. $g(x) = 3^x - 2$

20. $g(x) = 3^x + 1$

21. $h(x) = \frac{1}{2}(3^x)$

22. $h(x) = \frac{1}{2}(3^{-x})$

23. $f(t) = 2^{-t^2}$

24. $f(t) = 2^{t^2}$

25. $h(x) = 2^{0.5x}$

26. $g(t) = 2^{-0.5t}$

27. $f(x) = e^{0.2x}$

28. $f(x) = e^{-0.2x}$

29. $P(t) = 100e^{-0.1t}$

30. $A(t) = 1{,}000e^{0.08t}$

31. *Population Growth* The population of the United States (in recent years) can be approximated by the exponential function

$$P(t) = 203(1.0118)^{t-1970}$$

where t is the year and P is the population in millions. Use the model to approximate the population for the years (a) 1995 and (b) 2000.

32. *Property Value* Suppose the value of a piece of property doubles every 15 years. If you buy the property for \$64,000, then its value t years after the date of the purchase should be

$$V(t) = 64{,}000(2)^{t/15}.$$

Use the model to approximate the value of the property (a) 5 years and (b) 20 years after it is purchased.

33. *Depreciation* After t years, the value of a car that cost \$16,000 is given by

$$V(t) = 16{,}000\left(\frac{3}{4}\right)^t.$$

Sketch a graph of the function and determine the value of the car two years after it was purchased.

34. *Depreciation* Suppose straight-line depreciation is used to determine the value of the car in Exercise 33. If the car depreciates \$3,000 per year, the model for its value after t years is given by

$$V(t) = 16{,}000 - 3{,}000t.$$

Sketch the graph of this line on the same coordinate axes as the graph in Exercise 33. If you were selling the car after having it for two years, which depreciation model would you prefer? If you sell it after four years, which model would be to your advantage?

35. *Inflation Rate* Suppose the annual rate of inflation averages 5% over the next 10 years. With this rate of inflation, the approximate cost C of goods or services during any year in that decade will be given by

$$C(t) = P(1.05)^t, \quad 0 \le t \le 10$$

where t is the time in years and P is the present cost. If the price of an oil change for your car is presently \$19.95, estimate the price 10 years from now.

36. *Price and Demand* The daily demand x for a certain product is given by

$$p = 25 - 0.4e^{0.02x}$$

where p is the price of the product. Find the price if the demand is (a) $x = 100$ units and (b) $x = 125$ units.

In Exercises 37–42, complete the following table to determine the balance A for P dollars invested at rate r for t years and compounded n times per year.

n	1	4	12	365	Continuous Compounding
A					

	Principal	*Rate*	*Time*
37.	$P = \$100$	$r = 8\%$	$t = 20$ years
38.	$P = \$2{,}000$	$r = 9\%$	$t = 10$ years
39.	$P = \$5{,}000$	$r = 10\%$	$t = 40$ years
40.	$P = \$1{,}500$	$r = 7\%$	$t = 2$ years
41.	$P = \$10{,}000$	$r = 9.5\%$	$t = 30$ years
42.	$P = \$400$	$r = 8\%$	$t = 50$ years

In Exercises 43–46, complete the following table to determine the principal P that will yield a balance of A dollars when invested at rate r for t years and compounded n times per year.

n	1	4	12	365	Continuous Compounding
P					

	Balance	*Rate*	*Time*
43.	$A = \$5{,}000$	$r = 7\%$	$t = 10$ years
44.	$A = \$100{,}000$	$r = 9\%$	$t = 20$ years
45.	$A = \$1{,}000{,}000$	$r = 10.5\%$	$t = 40$ years
46.	$A = \$2{,}500$	$r = 7.5\%$	$t = 2$ years

47. On the same set of coordinate axes, sketch the graph of the following functions.

(a) $f(x) = 2x$ (b) $f(x) = 2x^2$

(c) $f(x) = 2^x$ (d) $f(x) = 2^{-x}$

48. *Savings Plan* Suppose you decide to start saving pennies according to the following pattern. You save one penny the first day, two pennies the second day, four the third day, and so on. Each day you save twice the number of pennies as you did on the previous day. Which function of Exercise 47 models this problem and how many pennies do you save on the thirtieth day? (In the next chapter, you will learn how to find the total number saved.)

49. *Parachute Drop* A parachutist jumps from a plane and opens the parachute at a height of 2,000 feet (see figure). The height of the parachutist is then given by

$$h = 1{,}950 + 50e^{-1.6t} - 20t$$

where h is the height in feet and t is the time in seconds. (The time $t = 0$ corresponds to the time when the parachute is opened.) Find the height of the parachutist when $t = 0$, 25, 50, and 75. When do you think the parachutist will reach the ground?

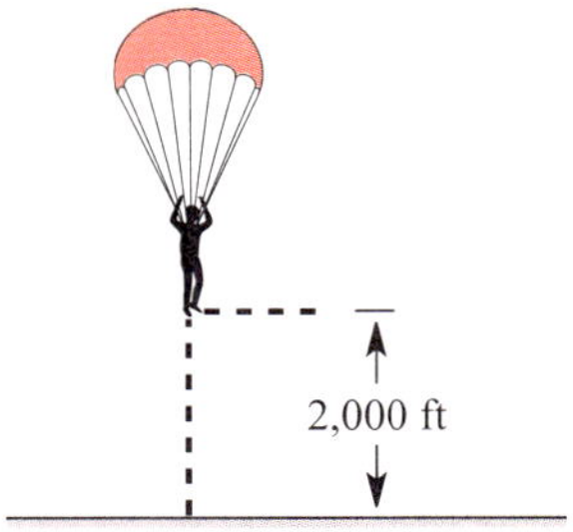

50. *Parachute Drop* A parachutist jumps from a plane and opens the parachute at a height of 3,000 feet. The height of the parachutist is then given by

$$h = 2{,}940 + 60e^{-1.7t} - 22t$$

where h is the height in feet and t is the time in seconds. (The time $t = 0$ corresponds to the time when the parachute is opened.) Find the height of the parachutist when $t = 0$, 50, and 100. When do you think the parachutist will reach the ground?

In Exercises 51–56, use a graphing calculator to sketch the graph of the function using the following settings.

RANGE
Xmin=−4
Xmax=4
Xscl=1
Ymin=0
Ymax=8
Yscl=1
Xres=1

51. $y = 5^{x/3}$

52. $y = 5^{(x-2)/3}$

53. $y = 5^{-x/3}$

54. $y = 5^{-x/3} + 2$

55. $y = \frac{3}{2}e^{x/3}$

56. $y = 6e^{-x^2/3}$

57. *Compound Interest* Compare the rate of increase of an investment of \$500 in two different accounts with respective interest rates of 6% and 8% compounded continuously. Perform the comparison by using a graphing calculator to graph the equations $y = 500e^{0.06x}$ and $y = 500e^{0.08x}$ on the same screen using the following settings.

RANGE
Xmin=0
Xmax=40
Xscl=5
Ymin=0
Ymax=6000
Yscl=500
Xres=1

SECTION 10.2

Logarithmic Functions

Logarithms and Logarithmic Functions • Graphs of Logarithmic Functions • The Natural Logarithmic Function • Change of Base

Logarithms and Logarithmic Functions

In Section 8.4 we discussed the concept of the inverse of a function. Moreover, we saw that if a function has the property that no horizontal line intersects the graph of a function more than once, then the function must have an inverse. By looking back at the graphs of the exponential functions introduced in Section 10.1, you will see that every function of the form $f(x) = a^x$ passes the "horizontal line test," and therefore must have an inverse. We call this inverse function the **logarithmic function with base *a***.

Definition of the Logarithmic Function

Let a and x be positive real numbers such that $a \neq 1$. The **logarithm of *x* with base *a*** is denoted by $\log_a x$ and is defined as follows.

$$y = \log_a x \quad \text{if and only if} \quad x = a^y$$

The function given by $f(x) = \log_a x$ is called the **logarithmic function with base *a***.

From this definition, we can see that the equations $y = \log_a x$ and $x = a^y$ are equivalent. The first equation is in *logarithmic* form and the second is in *exponential* form. From these equivalent equations it should also be clear that a *logarithm is an exponent*. For instance, since the exponent in the expression

Exponent

$$2^3 = 8$$

is 3, the value of the logarithm $\log_2 8$ is 3. That is,

$$\log_2 8 = 3.$$

Therefore, to evaluate the logarithmic expression $\log_a x$, we need to ask the question, To what power must a be raised to obtain x?

EXAMPLE 1 ■ Evaluating Logarithms

Evaluate the following logarithms.

(a) $\log_2 16$ (b) $\log_3 9$ (c) $\log_5 1$

Solution

In each case we need to answer the question, To what power must we raise the base to obtain the given number?

(a) The power to which 2 must be raised to obtain 16 is 4. That is,

$$2^4 = 16 \Longrightarrow \log_2 16 = 4.$$

(b) The power to which 3 must be raised to obtain 9 is 2. That is,

$$3^2 = 9 \Longrightarrow \log_3 9 = 2.$$

(c) The power to which 5 must be raised to obtain 1 is 0. That is,

$$5^0 = 1 \Longrightarrow \log_5 1 = 0.$$

■

EXAMPLE 2 ■ Evaluating Logarithms

Evaluate the following logarithms.

(a) $\log_4 2$ (b) $\log_{10} \dfrac{1}{10}$ (c) $\log_3 (-1)$

Solution

(a) The power to which 4 must be raised to obtain 2 is $\frac{1}{2}$. That is

$$4^{1/2} = 2 \Longrightarrow \log_4 2 = \frac{1}{2}.$$

(b) The power to which 10 must be raised to obtain $\frac{1}{10}$ is -1. That is,

$$10^{-1} = \frac{1}{10} \Longrightarrow \log_{10} \frac{1}{10} = -1.$$

(c) There is *no power* to which 3 can be raised to obtain -1. The reason for this is that for any real value of x, 3^x is a positive number. Therefore, $\log_3 (-1)$ is undefined. ■

The following properties of logarithms follow directly from the definition of the logarithmic function with base a.

Properties of Logarithms

Let a and x be positive real numbers such that $a \neq 1$. Then the following properties are true.

1. $\log_a 1 = 0$ because $a^0 = 1$.
2. $\log_a a = 1$ because $a^1 = a$.
3. $\log_a a^x = x$ because $a^x = a^x$.

The logarithmic function with base 10 is called the **common logarithmic function**. On most calculators, this function can be evaluated with the common logarithmic key [log], as illustrated in the next example.

EXAMPLE 3 ■ Evaluating Common Logarithms

Evaluate the following logarithms. Use a calculator only if necessary.

(a) $\log_{10} 100$ (b) $\log_{10} 0.01$ (c) $\log_{10} 5$

Solution

(a) Since $10^2 = 100$, it follows that

$$\log_{10} 100 = 2.$$

(b) Since $10^{-2} = \frac{1}{100} = 0.01$, it follows that

$$\log_{10} 0.01 = -2.$$

(c) There is no simple power to which 10 can be raised to obtain 5, so we use a calculator in this case.*

Keystrokes	*Display*	
5 [log]	0.69897	Scientific
[log] 5 [ENTER]	0.69897	Graphing

Therefore, rounded to three decimal places, we have

$$\log_{10} 5 \approx 0.699.$$ ■

*The graphing calculator keystrokes given in this text correspond to the *TI-81* graphing calculator from *Texas Instruments*. For other graphing calculators, the keystrokes may differ.

Graphs of Logarithmic Functions

To sketch the graph of $y = \log_a x$, we use the fact that the graphs of inverse functions are reflections of each other in the line $y = x$, as demonstrated in Example 4.

EXAMPLE 4 ■ **Graphs of Exponential and Logarithmic Functions**

On the same rectangular coordinate system, sketch the graphs of the following functions.

(a) $f(x) = 2^x$ (b) $g(x) = \log_2 x$

Solution

(a) For $f(x) = 2^x$, we make up the following table of values.

x	-2	-1	0	1	2	3
$f(x) = 2^x$	$\frac{1}{4}$	$\frac{1}{2}$	1	2	4	8

By plotting these points and connecting them with a smooth curve, we obtain the graph shown in Figure 10.6.

(b) Since $g(x) = \log_2 x$ is the inverse function of $f(x) = 2^x$, the graph of g is obtained by reflecting the graph of f in the line $y = x$, as shown in Figure 10.6.

FIGURE 10.6

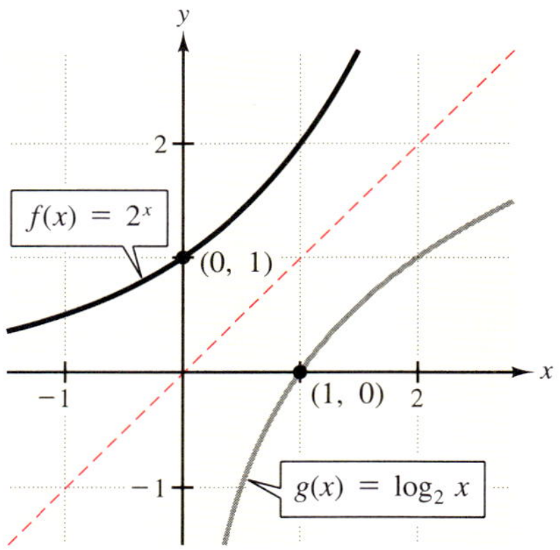

Inverse Functions ■

Notice from the graph of $g(x) = \log_2 x$ shown in Figure 10.6, that the domain of the function is the set of positive numbers and the range is the set of all real numbers.

We summarize the basic characteristics of the graph of a logarithmic function in Figure 10.7. In this figure note that the graph has one x-intercept, at (1, 0). Also note that the y-axis is a vertical asymptote of the graph.

FIGURE 10.7

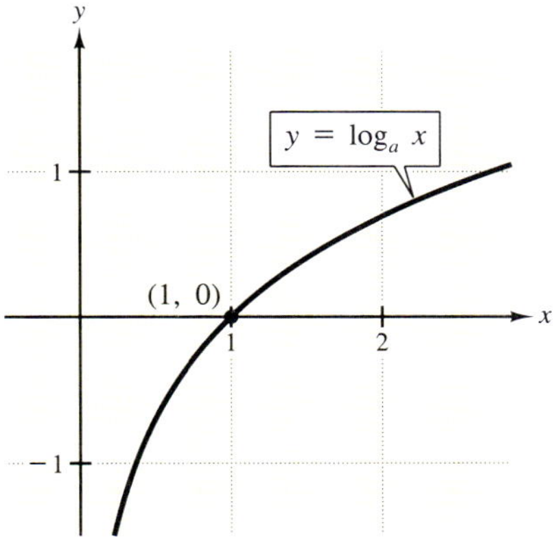

Graph of $y = \log_a x$, $a > 1$
- Domain: $(0, \infty)$
- Range: $(-\infty, \infty)$
- Intercept: $(1, 0)$
- Increasing

In Example 4 we used the inverse property of logarithmic functions to sketch the graph of $g(x) = \log_2 x$. You can also use a standard point-plotting approach, as demonstrated in Example 5.

EXAMPLE 5 ■ Sketching the Graph of a Logarithmic Function

Sketch the graph of the common logarithmic function

$$f(x) = \log_{10} x.$$

Solution

We begin by making up the following table of values. Note that some of the values can be obtained without a calculator, while others require a calculator. Using the points indicated by the table, we sketch the graph, as shown in Figure 10.8. Notice how slowly the graph rises for $x > 1$. In Figure 10.8 we would need to move out to $x =$ 1,000 before the graph would rise to $y = 3$.

	Without Calculator				With Calculator		
x	$\frac{1}{100}$	$\frac{1}{10}$	1	10	2	5	8
$\log_{10} x$	−2	−1	0	1	0.301	0.699	0.903

FIGURE 10.8

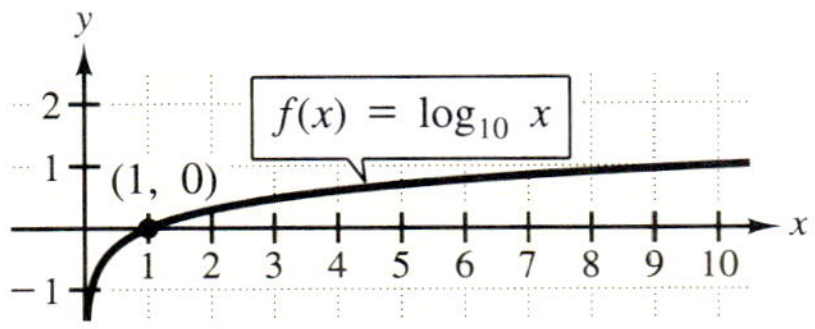

■

The Natural Logarithmic Function

As with exponential functions, the most widely used base for logarithmic functions is the number e. We call the logarithmic function with base e the **natural logarithmic function** and denote it by the special symbol $\ln x$, which is read as "el en of x."

The Natural Logarithmic Function

The function defined by

$$f(x) = \log_e x = \ln x$$

where $x > 0$, is called the **natural logarithmic function**.

The three properties of logarithms listed at the beginning of this section are also valid for natural logarithms.

Properties of Natural Logarithms

Let x be a positive real number. Then the following properties are true.

1. $\ln 1 = 0$ because $e^0 = 1$.
2. $\ln e = 1$ because $e^1 = e$.
3. $\ln e^x = x$ because $e^x = e^x$.

We use these properties to evaluate the natural logarithmic function in the following example.

EXAMPLE 6 ■ **Evaluating the Natural Logarithmic Function**

Using the property that $\ln e^x = x$, we obtain the following.

(a) $\ln e^2 = 2$

(b) $\ln \dfrac{1}{e} = \ln e^{-1} = -1$ ■

Using the points $(1/e, -1)$, $(1, 0)$, $(e, 1)$, and $(e^2, 2)$ we sketch the graph of the natural logarithmic function, as shown in Figure 10.9.

FIGURE 10.9

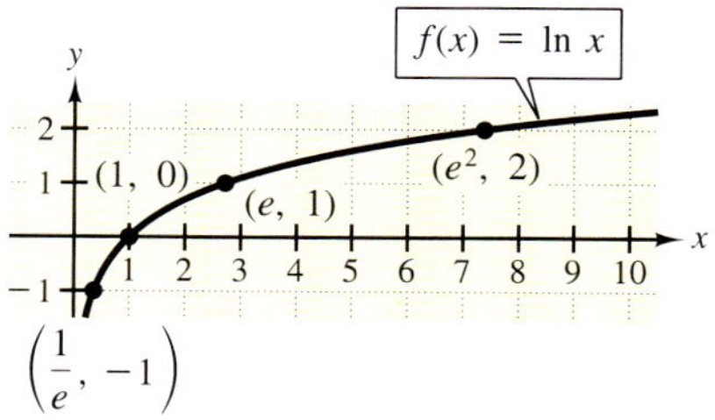

On most calculators, the natural logarithm key is denoted by [ln]. Here are some examples of using a calculator to evaluate the natural logarithmic function.

Scientific

Logarithm	*Keystrokes*	*Display*
ln 2	2 [ln]	0.6931472
ln 0.3	.3 [ln]	−1.2039728
ln (−1)	1 [+/−] [ln]	ERROR

Graphing

Logarithm	*Keystrokes*	*Display*
ln 2	[ln] 2 [ENTER]	0.6931472
ln 0.3	[ln] .3 [ENTER]	−1.2039728
ln (−1)	[ln] [(−)] 1 [ENTER]	ERROR

Be sure you see that ln (−1) gives an error. This occurs because the domain of $\ln x$ is the set of positive real numbers, which means that ln (−1) is undefined.

Change of Base

Although 10 and e are the most frequently used bases, we occasionally need to evaluate logarithms with other bases. In such cases the following **change of base formula** is useful.

Change of Base Formula

Let a, b, and x be positive real numbers such that $a \neq 1$ and $b \neq 1$. Then $\log_a x$ is given as follows.

$$\log_a x = \frac{\log_b x}{\log_b a} \quad \text{or} \quad \log_a x = \frac{\ln x}{\ln a}$$

The usefulness of this change of base formula is that we can use a calculator that has only the common logarithm key [log] and the natural logarithm key [ln] to evaluate logarithms to any base. This is demonstrated in Examples 7 and 8.

EXAMPLE 7 ■ Changing Base to Common Logarithms

Use *common* logarithms to evaluate the following.

$$\log_3 5$$

Solution

Using the change of base formula with $a = 3$, $b = 10$, and $x = 5$, we can convert to common logarithms by writing

$$\log_3 5 = \frac{\log_{10} 5}{\log_{10} 3}.$$

Now, using the following keystrokes

Keystrokes	*Display*	
5 [log] [÷] 3 [log] [=]	1.4649735	Scientific
[log] 5 [÷] [log] 3 [ENTER]	1.4649735	Graphing

we find that the value of $\log_3 5$ is

$$\log_3 5 \approx 1.465.$$

■

The evaluation in Example 7 used common logarithms, but we could just as easily have used natural logarithms, as demonstrated in Example 8.

EXAMPLE 8 ■ Changing Base to Natural Logarithms

Use *natural* logarithms to evaluate the following.

$$\log_6 2$$

Solution

Using the change of base formula with $a = 6$ and $x = 2$, we can convert to natural logarithms by writing

$$\log_6 2 = \frac{\ln 2}{\ln 6}.$$

Now, using the following keystrokes

Keystrokes	*Display*	
2 [ln] [÷] 6 [ln] [=]	0.3868528	Scientific
[ln] 2 [÷] [ln] 6 [ENTER]	0.3868528	Graphing

we find that the value of $\log_6 2$ is

$$\log_6 2 \approx 0.387.$$

(Try this problem with common logarithms to see that you obtain the same result.)

■

DISCUSSION PROBLEM ■ Identifying Graphs of Common Functions

Classify each of the following graphs as *linear*, *quadratic*, *exponential*, or *logarithmic*. In each case write a short paragraph to support your answer. Include statements regarding the domain, range, intercepts, and asymptotes of each.

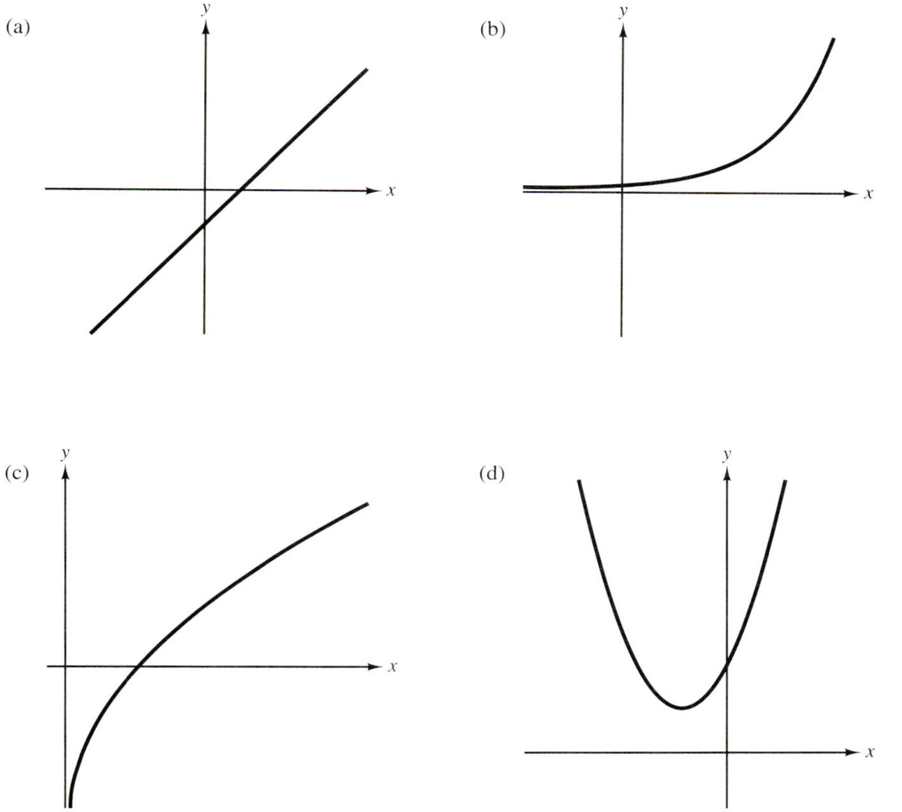

Warm-Up

The following warm-up exercises involve skills that were covered in earlier sections. You will use these skills in the exercise set for this section.

In Exercises 1–6, evaluate the given expression. Use a calculator only if it is necessary or more efficient.

1. $\left(\frac{3}{4}\right)^3$ **2.** $(3 \cdot 10)^2$ **3.** 2^{π}

4. $3^{\sqrt{2}}$ **5.** e^{-1} **6.** $e^{1/2}$

In Exercises 7–10, describe the relationship between the graphs of f and g.

7. $g(x) = f(x) - 4$ **8.** $g(x) = f(x - 4)$

9. $g(x) = -f(x)$ **10.** $g(x) = f(-x)$

10.2 EXERCISES

In Exercises 1–6, write the given logarithmic equation in exponential form.

1. $\log_5 25 = 2$

2. $\log_{10} 10{,}000 = 4$

3. $\log_4 \frac{1}{16} = -2$

4. $\log_3 \frac{1}{243} = -5$

5. $\log_{36} 6 = \frac{1}{2}$

6. $\log_{32} 4 = \frac{2}{5}$

In Exercises 7–12, write the given exponential equation in logarithmic form.

7. $7^2 = 49$

8. $5^4 = 625$

9. $3^{-2} = \frac{1}{9}$

10. $6^{-3} = \frac{1}{216}$

11. $8^{2/3} = 4$

12. $81^{3/4} = 27$

In Exercises 13–30, evaluate the given expression without the aid of a calculator. (If not possible, state the reason.)

13. $\log_2 8$

14. $\log_3 27$

15. $\log_2 \frac{1}{4}$

16. $\log_3 \frac{1}{9}$

17. $\log_4 1$

18. $\log_5 (-6)$

19. $\log_2 (-3)$

20. $\log_3 1$

21. $\log_{10} 10$

22. $\log_8 8$

23. $\log_{10} 1{,}000$

24. $\log_{10} \frac{1}{100}$

25. $\log_9 3$

26. $\log_{144} 12$

27. $\log_{16} 8$

28. $\log_{25} 125$

29. $\log_7 7^4$

30. $\log_2 0$

In Exercises 31–40, evaluate the given logarithm with the aid of a calculator. Round your answer to four decimal places.

31. $\log_{10} 31$

32. $\log_{10} 0.345$

33. $\log_{10} (\sqrt{2} + 4)$

34. $\log_{10} \frac{\sqrt{3}}{2}$

35. $\ln 25$

36. $\ln 6.57$

37. $\ln 0.75$

38. $\ln (\sqrt{3} - 1)$

39. $\ln \left(\frac{1 + \sqrt{5}}{3}\right)$

40. $\ln \left(1 + \frac{0.10}{12}\right)$

In Exercises 41 and 42, sketch the graph of f and g on the same set of coordinate axes. What can you conclude about the relationship between f and g?

41. $f(x) = \log_3 x, \quad g(x) = 3^x$

42. $f(x) = \log_4 x, \quad g(x) = 4^x$

In Exercises 43–48, use the graph of $y = \log_3 x$ (see Exercise 41) to match the given function to its graph. [The graphs are labeled (a), (b), (c), (d), (e), and (f).]

43. $f(x) = 4 + \log_3 x$

44. $f(x) = -2 + \log_3 x$

45. $f(x) = -\log_3 x$

46. $f(x) = \log_3 (-x)$

47. $f(x) = \log_3 (x - 4)$

48. $f(x) = \log_3 (x + 2)$

(a)

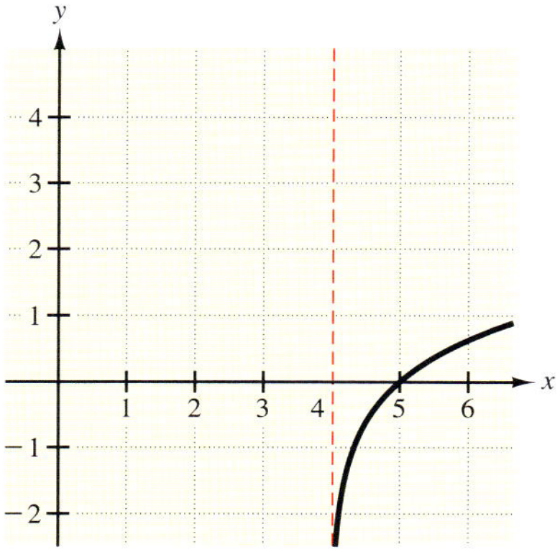

(b)

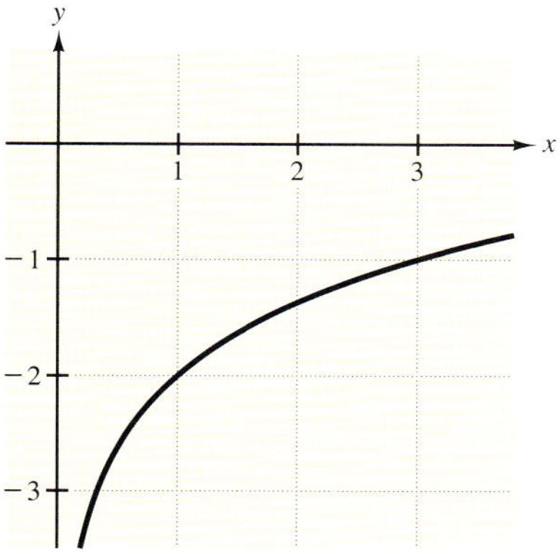

(c)

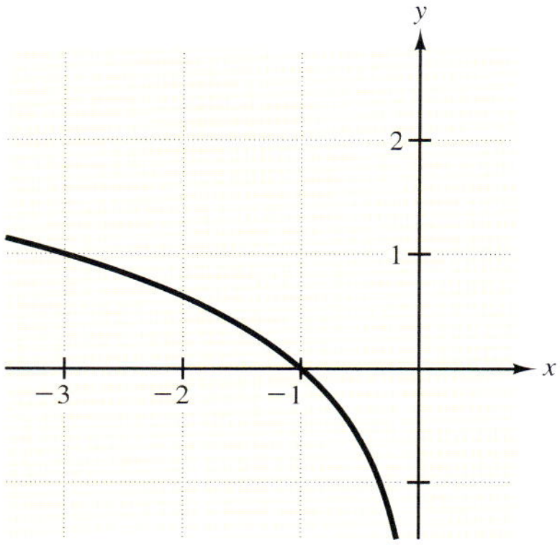

(d)

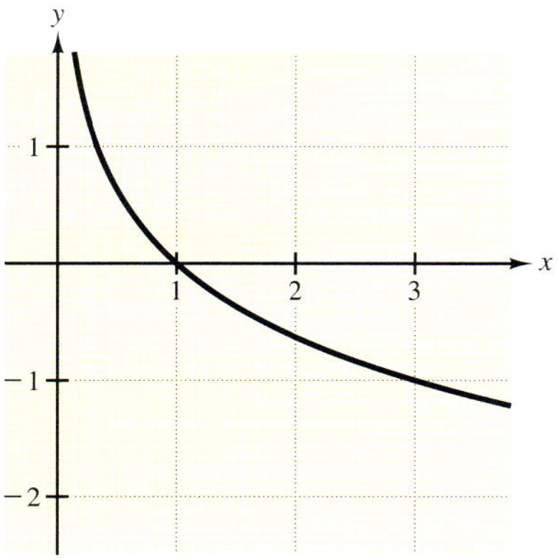

(e)

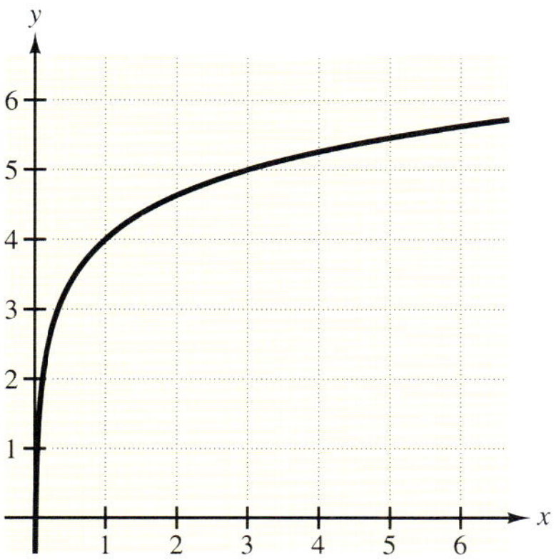

(f)

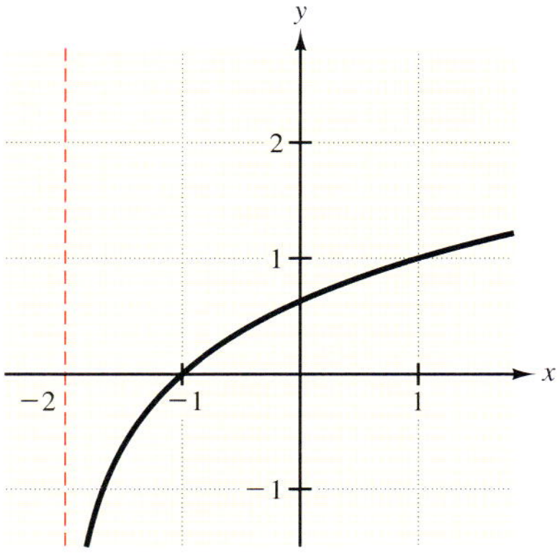

In Exercises 49–58, sketch the graph of the given function.

49. $f(x) = \log_5 x$

50. $g(x) = \log_8 x$

51. $g(t) = \log_2 t$

52. $h(s) = \log_3 s$

53. $f(x) = 3 + \log_2 x$

54. $f(x) = -2 + \log_3 x$

55. $f(x) = 3 \ln x$

56. $h(t) = 2 \ln t$

57. $f(x) = 1 + \ln x$

58. $h(x) = 2 + \ln x$

In Exercises 59–64, evaluate the given logarithm by means of the change-of-base formula. Use (a) the common logarithm key, and (b) the natural logarithm key on a calculator.

59. $\log_8 132$

60. $\log_5 510$

61. $\log_2 0.72$

62. $\log_{12} 0.6$

63. $\log_4 \sqrt{42}$

64. $\log_3 (1 + e^2)$

65. *Doubling Time* The time t in years for an investment to double in value when compounded continuously at the rate r is given by

$$t = \frac{\ln 2}{r}.$$

Complete the following table, which shows the "doubling time" for several annual percentage rates.

r	0.07	0.08	0.09
t			
r	0.10	0.11	0.12
t			

66. *Intensity of Sound* The relationship between the number of decibels B and the intensity of a sound I in watts per meter squared is given by

$$B = 10 \log_{10}\left(\frac{I}{10^{-16}}\right).$$

Determine the number of decibels of a sound with an intensity of 10^{-4} watts per meter squared.

In Exercises 67–72, answer the given question for the function $f(x) = \log_{10} x$. (Do not use a calculator.)

67. What is the domain of f?

68. Find the inverse function of f.

69. Find the interval in which $f(x)$ must lie, given that $1{,}000 \le x \le 10{,}000$.

70. Find the interval in which x must lie, given that $f(x)$ is negative.

71. By what amount will x increase, given that $f(x)$ is increased by one unit?

72. Find the ratio of a to b, given that $f(a) = 3 + f(b)$.

In Exercises 73–78, use a graphing calculator to sketch the graph of the function using the following settings.

RANGE
Xmin=0
Xmax=10
Xscl=1
Ymin=−3
Ymax=7
Yscl=1
Xres=1

73. $y = 5 \log_{10} x$

74. $y = 5 \log_{10} (x - 3)$

75. $y = -3 + 5 \log_{10} x$

76. $y = 5 \log_{10} (3x)$

77. $y = 2 \ln x$

78. $y = -2 \ln x$

SECTION 10.3

Properties of Logarithms

Properties of Logarithms • Rewriting Logarithmic Expressions • Applications

Properties of Logarithms

We know from the previous section that the logarithmic function with base a is the *inverse* of the exponential function with base a. Thus, it makes sense that each property of exponents should have a corresponding property of logarithms. For instance, the exponential property $a^0 = 1$ has the corresponding logarithmic property $\log_a 1 = 0$. In this section we will study the logarithmic properties that correspond to the following three exponential properties.

1. $a^n a^m = a^{n+m}$

2. $\dfrac{a^n}{a^m} = a^{n-m}$

3. $(a^n)^m = a^{nm}$

Properties of Logarithms

Let a be a positive real number such that $a \neq 1$, and let n be a real number. If u and v are real numbers, variables, or algebraic expressions such that $u > 0$ and $v > 0$, then the following properties are true.

Logarithm with Base a	*Natural Logarithm*
1. $\log_a (uv) = \log_a u + \log_a v$	$\ln (uv) = \ln u + \ln v$
2. $\log_a \dfrac{u}{v} = \log_a u - \log_a v$	$\ln \dfrac{u}{v} = \ln u - \ln v$
3. $\log_a u^n = n \log_a u$	$\ln u^n = n \ln u$

NOTE: There is no general property of logarithms that can be used to simplify $\log_a (u + v)$. Specifically, $\log_a (u + v)$ *does not equal* $\log_a u + \log_a v$.

EXAMPLE 1 ■ **Using Properties of Logarithms**

Use the fact that $\ln 2 \approx 0.693$, $\ln 3 \approx 1.099$, and $\ln 5 \approx 1.609$ to approximate the following.

(a) $\ln \dfrac{2}{3}$ (b) $\ln 10$ (c) $\ln 30$

Solution

(a) $\ln \frac{2}{3} = \ln 2 - \ln 3$ Property 2

$\approx 0.693 - 1.099$

$= -0.406$

(b) $\ln 10 = \ln (2 \cdot 5)$ Property 1

$= \ln 2 + \ln 5$

$\approx 0.693 + 1.609$

$= 2.302$

(c) $\ln 30 = \ln (2 \cdot 3 \cdot 5)$ Property 1

$= \ln 2 + \ln 3 + \ln 5$

$\approx 0.693 + 1.099 + 1.609$

$= 3.401$

Check these results on your calculator. ■

When using the properties of logarithms, we find that it helps to state the properties *verbally*. For instance, the verbal form of the property $\ln (uv) = \ln u + \ln v$ is, *The log of a product is the sum of the logs of the factors*. Similarly, the verbal form of the property $\ln (u/v) = \ln u - \ln v$ is, *The log of a quotient is the difference of the logs of the numerator and denominator*.

EXAMPLE 2 ■ **Using Properties of Logarithms**

Use the properties of logarithms to verify that

$$-\ln 2 = \ln \frac{1}{2}.$$

Solution

Using Property 3 of logarithms, we can write the following.

$$-\ln 2 = (-1)\ln 2 = \ln 2^{-1} = \ln \frac{1}{2}$$

Try checking this result on your calculator. ■

Rewriting Logarithmic Expressions

In Examples 1 and 2, we used the properties of logarithms to rewrite logarithmic expressions involving the log of a *constant*. A more common use of the properties is to rewrite the log of a *variable expression*. We illustrate this in the next several examples.

EXAMPLE 3 ■ Rewriting the Logarithm of a Product

Use the properties of logarithms to rewrite the following logarithmic expression.

$$\log_{10} 7x^3$$

Solution

$$\begin{aligned} \log_{10} 7x^3 &= \log_{10} 7 + \log_{10} x^3 && \text{Property 1} \\ &= \log_{10} 7 + 3 \log_{10} x && \text{Property 3} \end{aligned}$$ ■

When we rewrite a logarithmic expression as in Example 3, we say we are **expanding** the expression. The reverse procedure is demonstrated in Example 4, and is called **condensing** a logarithmic expression.

EXAMPLE 4 ■ Condensing a Logarithmic Expression

Use the properties of logarithms to condense the following logarithmic expression.

$$\ln x - \ln 3$$

Solution

$$\ln x - \ln 3 = \ln \frac{x}{3} \qquad \text{Property 2}$$ ■

Sometimes expanding or condensing logarithmic expressions involves several steps. In the next three examples, be sure that you can justify each step in the solution.

EXAMPLE 5 ■ Expanding a Logarithmic Expression

Expand the following logarithmic expression.

$$\log_2 3xy^2, \quad x > 0,\ y > 0$$

Solution

$$\begin{aligned} \log_2 3xy^2 &= \log_2 3 + \log_2 x + \log_2 y^2 && \text{Property 1} \\ &= \log_2 3 + \log_2 x + 2 \log_2 y && \text{Property 3} \end{aligned}$$ ■

EXAMPLE 6 ■ **Expanding a Logarithmic Expression**

Expand the following logarithmic expression.

$$\ln\sqrt{x^2 - 1}, \quad x > 1$$

Solution

$$\begin{aligned}
\ln\sqrt{x^2 - 1} &= \ln (x^2 - 1)^{1/2} && \text{Rewrite using fractional exponent} \\
&= \frac{1}{2}\ln (x^2 - 1) && \text{Property 3} \\
&= \frac{1}{2}\ln (x - 1)(x + 1) && \text{Factor} \\
&= \frac{1}{2}[\ln (x - 1) + \ln (x + 1)] && \text{Property 1} \\
&= \frac{1}{2}\ln (x - 1) + \frac{1}{2}\ln (x + 1) && \text{Distributive Property}
\end{aligned}$$

■

EXAMPLE 7 ■ **Condensing a Logarithmic Expression**

Use the property of logarithms to condense the following expression.

$$\ln 2 - 2 \ln x$$

Solution

$$\begin{aligned}
\ln 2 - 2 \ln x &= \ln 2 - \ln x^2, \quad x > 0 && \text{Property 3} \\
&= \ln \frac{2}{x^2}, \quad x > 0 && \text{Property 2}
\end{aligned}$$

■

There is a technical consideration that we should mention. When we expand or condense a logarithmic expression, it is possible to change the domain of the expression. For instance, the domain of the function

$$f(x) = 2 \ln x$$

is the set of positive real numbers, whereas the domain of

$$g(x) = \ln x^2$$

is the set of nonzero real numbers. Thus, when you expand or condense a logarithmic expression, you should check to see whether the rewriting changed the domain of the expression. In such cases we use the domain of the original function. Thus, we can write

$$f(x) = 2 \ln x = \ln x^2, \quad x > 0.$$

Applications

EXAMPLE 8 An Application: Human Memory Model

Students participating in a psychological experiment attended several lectures on a subject. Every month for a year after that, the students were tested to see how much of the material they remembered. The average scores for the group were given by the **human memory model**

$$f(t) = 80 - \ln (t + 1)^9, \quad 0 \le t \le 12$$

where t is the time in months. Find the average score for the group after two months and eight months.

Solution

To make the calculations easier, we begin by rewriting the model as

$$f(t) = 80 - 9 \ln (t + 1), \quad 0 \le t \le 12.$$

After two months, the average score will be

$$f(2) = 80 - 9 \ln 3 \approx 70.1 \qquad \text{Average score after two months}$$

and after eight months, the average score will be

$$f(8) = 80 - 9 \ln 9 \approx 60.2. \qquad \text{Average score after eight months}$$

The graph of the function is shown in Figure 10.10.

FIGURE 10.10

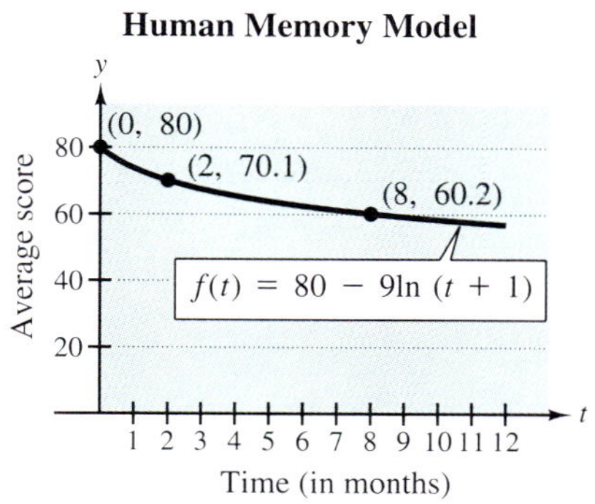

DISCUSSION PROBLEM Historical Use of Logarithms

Logarithms were invented in 1594 by John Napier, a Scottish mathematician. Using the reference books in your school's library, write a short paragraph describing the use of logarithms in calculating products, quotients, and powers of large numbers. For instance, how could a table of logarithms be used to approximate the number $(1.34)^{2.7}$? (Remember that electronic calculators and computers are 20th-century inventions.)

Warm-Up

The following warm-up exercises involve skills that were covered in earlier sections. You will use these skills in the exercise set for this section.

In Exercises 1–6, use the properties of exponents to simplify the given expression.

1. $(x^2 \cdot x^3)^4$ **2.** $4^{-2} \cdot x^2$ **3.** $\dfrac{15y^{-3}}{10y^2}$

4. $\left(\dfrac{3x^2}{2y}\right)^{-2}$ **5.** $\dfrac{3x^2y^3}{18x^{-1}y^2}$ **6.** $(x^2 + 1)^0$

In Exercises 7–10, rewrite the given equation in exponential form.

7. $\log_4 64 = 3$ **8.** $\log_3 \dfrac{1}{81} = -4$

9. $\ln e^{2x} = 2x$ **10.** $\ln 5 = 1.6094 \ldots$

10.3 EXERCISES

In Exercises 1–6, evaluate the given expression without using a calculator. (If not possible, state the reason.)

1. $\log_5 5^2$ **2.** $\log_6 2 + \log_6 3$ **3.** $\log_3 36 - \log_3 12$ **4.** $\log_{10} 1$

5. $\log_2 \dfrac{1}{8}$ **6.** $\log_3 (-3)$

In Exercises 7–20, use the properties of logarithms to expand the given expression as a sum, difference, or multiple of logarithms.

7. $\log_3 11x$ **8.** $\log_7 x^2$ **9.** $\log_2 \dfrac{z}{17}$ **10.** $\log_{10} \dfrac{7}{y}$

11. $\log_5 x^{-2}$ **12.** $\log_2 \sqrt{s}$ **13.** $\log_3 \sqrt[3]{x + 1}$ **14.** $\log_4 \dfrac{1}{\sqrt{t}}$

15. $\ln 3x^2y$ **16.** $\ln y(y - 1)^2$ **17.** $\log_2 \dfrac{x^2}{x - 3}$ **18.** $\log_5 \sqrt{\dfrac{x}{y}}$

19. $\ln \sqrt[3]{x(x + 5)}$ **20.** $\ln (3x(x - 5))^2$

In Exercises 21–34, use the properties of logarithms to condense the given expression as a logarithm of a single quantity.

21. $\log_2 3 + \log_2 x$

22. $\log_5 2x + \log_5 3y$

23. $\log_{10} 4 - \log_{10} x$

24. $\ln 10x - \ln z$

25. $4 \ln b$

26. $10 \log_4 z$

27. $-2 \log_5 2x$

28. $-5 \ln (x + 3)$

29. $\frac{1}{3} \ln (2x + 1)$

30. $-\frac{1}{2} \log_3 5y$

31. $2 \ln x + 3 \ln y - \ln z$

32. $4 \ln 3 - 2 \ln x - \ln y$

33. $4(\ln x + \ln y)$

34. $2[\ln x - \ln (x + 1)]$

In Exercises 35–40, determine whether the given equation is true or false.

35. $\log_2 8x = 3 + \log_2 x$

36. $\log_8 4 + \log_8 16 = 2$

37. $\log_3 (u + v) = \log_3 u + \log_3 v$

38. $\log_3 (u + v) = \log_3 u \cdot \log_3 v$

39. $\dfrac{\log_6 10}{\log_6 3} = \log_6 10 - \log_6 3$

40. $\ln e^{2-x} = 2 - x$

In Exercises 41–50, approximate the logarithm given that $\log_4 2 = 0.5000$ and $\log_4 3 \approx 0.7925$. (Do not use a calculator.)

41. $\log_4 4$

42. $\log_4 8$

43. $\log_4 6$

44. $\log_4 24$

45. $\log_4 \frac{3}{2}$

46. $\log_4 \frac{9}{2}$

47. $\log_4 \sqrt{2}$

48. $\log_4 \sqrt[3]{9}$

49. $\log_4 (3 \cdot 2^4)$

50. $\log_4 \sqrt{3 \cdot 2^5}$

In Exercises 51–56, simplify the given logarithmic expression.

51. $\log_4 \frac{4}{x}$

52. $\log_3 (3^2 \cdot 4)$

53. $\log_5 \sqrt{50}$

54. $\log_2 \sqrt{22}$

55. $\ln 3e^2$

56. $\ln \frac{6}{e^5}$

57. Approximate the natural logarithm of as many integers as possible between 1 and 20 using $\ln 2 \approx 0.6931$, $\ln 3 \approx 1.0986$, and $\ln 5 \approx 1.6094$. (Do not use a calculator.)

58. Use a calculator to demonstrate that

$$\frac{\ln x}{\ln y} \neq \ln \frac{x}{y} = \ln x - \ln y$$

by evaluating each of the three expressions when $x = 3$ and $y = 5$.

59. *Intensity of Sound* The relationship between the number of decibels B and the intensity of a sound I in watts per meter squared is given by

$$B = 10 \log_{10} \left(\frac{I}{10^{-16}}\right).$$

Use properties of logarithms to write the formula in simpler form, and determine the number of decibels of a sound with an intensity of 10^{-10} watts per meter squared.

60. *Human Memory Model* Students participating in a psychological experiment attended several lectures on a subject. Every month for a year after that, the students were tested to see how much of the material they remembered. The average scores for the group were given by the human memory model

$$f(t) = 80 - \log_{10} (t + 1)^{12}, \quad 0 \leq t \leq 12$$

where t is the time in months. Find the average score for the group after two months and eight months.

61. Use a graphing calculator to demonstrate that

$$\frac{\ln x}{\ln 2} \neq \ln x - \ln 2$$

by graphing the equations

$$y = \frac{\ln x}{\ln 2} \quad \text{and} \quad y = \ln x - \ln 2$$

on the same screen.* Use the following settings.

RANGE
Xmin=0
Xmax=10
Xscl=1
Ymin=−3
Ymax=4
Yscl=1
Xres=1

SECTION 10.4 Solving Exponential and Logarithmic Equations

Introduction • Solving Exponential Equations • Solving Logarithmic Equations • Applications

Introduction

So far in this chapter our study has focused on the definitions, graphs, and properties of exponential and logarithmic functions. In this section we will study procedures for *solving equations* that involve exponential or logarithmic expressions. As a simple example, consider the exponential equation

$$2^x = 16.$$

By rewriting this equation in the form

$$2^x = 2^4$$

we can see that the solution is $x = 4$. To solve this equation we use one of the following properties.

Properties of Exponential and Logarithmic Equations

Let a be a positive real number such that $a \neq 1$, and let x and y be real numbers. Then the following properties are true.

1. $a^x = a^y$ if and only if $x = y$.
2. $\log_a x = \log_a y$ if and only if $x = y$ $(x > 0, y > 0)$.

*The graphing calculator keystrokes given in this text correspond to the *TI-81* graphing calculator from *Texas Instruments*. However, you can also use other graphing calculators and computer graphing software for these examples and exercises.

EXAMPLE 1 ■ **Solving Exponential and Logarithmic Equations**

Solve the following equations.

(a) $4^{x+2} = 4^5$ (b) $\ln(2x - 3) = \ln 11$

Solution

(a) $4^{x+2} = 4^5$	Given equation
$x + 2 = 5$	Property 1
$x = 3$	Subtract 2 from both sides

Thus, the solution is 3. Check this solution in the original equation.

(b) $\ln(2x - 3) = \ln 11$	Given equation
$2x - 3 = 11$	Property 2
$2x = 14$	Add 3 to both sides
$x = 7$	Divide both sides by 2

Thus, the solution is 7. Check this solution in the original equation. ■

Solving Exponential Equations

In Example 1(a) we were able to solve the given equation because both sides of the equation were written in exponential form (with the same base). However, if only one side of the equation is written in exponential form, then it is more difficult to solve the equation. For example, how would you solve the following equation?

$$2^x = 7$$

To solve this equation, we must find the power to which 2 can be raised to obtain 7. To do this we use one of the following inverse properties of exponents and logarithms.

Inverse Properties of Exponents and Logarithms

	Base a	*Natural Base e*
1.	$\log_a(a^x) = x$	$\ln(e^x) = x$
2.	$a^{(\log_a x)} = x$	$e^{(\ln x)} = x$

Notice how we use the first property to solve the equation $2^x = 7$ in the next example.

EXAMPLE 2 ■ **Solving an Exponential Equation**

Solve the following equation.

$$2^x = 7$$

Solution

To isolate the x, we take the $\log_2$ of both sides of the equation, as follows.

$2^x = 7$	Given equation
$\log_2 (2^x) = \log_2 7$	Take $\log_2$ of both sides
$x = \log_2 7$	Inverse property

Thus, the solution of the equation is $x = \log_2 7 \approx 2.807$. Try checking this with your calculator by raising 2 to the $2^{2.807}$ power. The result should be $2^{2.807} \approx 7$. ■

EXAMPLE 3 ■ **Solving an Exponential Equation**

Solve the following equation.

$$2e^x = 10$$

Solution

$2e^x = 10$	Given equation
$e^x = 5$	Divide both sides by 2
$\ln (e^x) = \ln 5$	Take ln of both sides
$x = \ln 5$	Inverse property

Thus, the solution of the equation is $x = \ln 5 \approx 1.609$. Try checking this solution in the original equation. ■

EXAMPLE 4 ■ **Solving an Exponential Equation**

Solve the following equation.

$$5 + e^{x+1} = 20$$

Solution

$5 + e^{x+1} = 20$	Given equation
$e^{x+1} = 15$	Subtract 5 from both sides
$\ln (e^{x+1}) = \ln 15$	Take ln of both sides
$x + 1 = \ln 15$	Inverse property
$x = -1 + \ln 15$	Subtract 1 from both sides

Thus, the solution of the equation is $x = -1 + \ln 15 \approx 1.708$. Try checking this solution in the original equation. ■

Solving Logarithmic Equations

We know how to solve an exponential equation by *taking the logarithm of both sides*. To solve a logarithmic equation, we need to **exponentiate** both sides. For instance, to solve a logarithmic equation such as

$$\ln x = 2$$

we exponentiate both sides of the equation as follows.

$$\ln x = 2 \qquad \text{Given equation}$$
$$e^{(\ln x)} = e^2 \qquad \text{Exponentiate both sides}$$
$$x = e^2 \qquad \text{Inverse property}$$

This procedure is demonstrated in the next three examples.

We suggest the following guidelines for solving exponential or logarithmic equations.

Guidelines for Solving Exponential and Logarithmic Equations

1. To solve an exponential equation, first isolate the exponential expression, then take the logarithm of both sides of the equation and solve for the variable.
2. To solve a logarithmic equation, first isolate the logarithmic expression, then exponentiate both sides of the equation and solve for the variable.

EXAMPLE 5 ■ Solving a Logarithmic Equation

Solve the following equation.

$$2 \ln x = 5$$

Solution

We begin by isolating the logarithmic expression, as follows.

$$2 \ln x = 5 \qquad \text{Given equation}$$
$$\ln x = \frac{5}{2} \qquad \text{Divide both sides by 2}$$
$$e^{(\ln x)} = e^{5/2} \qquad \text{Exponentiate both sides}$$
$$x = e^{5/2} \qquad \text{Inverse property}$$

Thus, the solution is $x = e^{5/2} \approx 12.182$. Try checking this solution in the original equation. ■

EXAMPLE 6 ■ Solving a Logarithmic Equation

Solve the following equation.

$$3 \log_{10} x = 6$$

Solution

$3 \log_{10} x = 6$	Given equation
$\log_{10} x = 2$	Divide both sides by 3
$10^{(\log_{10} x)} = 10^2$	Exponentiate both sides
$x = 100$	Inverse property

Thus, the solution is $x = 100$. Try checking this solution in the original equation. ■

EXAMPLE 7 ■ Solving a Logarithmic Equation

Solve the following equation.

$$20 \ln 0.2x = 30$$

Solution

$20 \ln 0.2x = 30$	Given equation
$\ln 0.2x = 1.5$	Divide both sides by 20
$e^{(\ln 0.2x)} = e^{1.5}$	Exponentiate both sides
$0.2x = e^{1.5}$	Inverse property
$x = 5e^{1.5}$	Divide both sides by 0.2

Thus, the solution is $x = 5e^{1.5} \approx 22.408$. Try checking this solution in the original equation. ■

EXAMPLE 8 ■ Solving a Logarithmic Equation

Solve the following equation.

$$\log_{10} 2x - \log_{10} (x - 3) = 1$$

Solution

$\log_{10} 2x - \log_{10} (x - 3) = 1$	Given equation
$\log_{10} \dfrac{2x}{x - 3} = 1$	Condense the left side

$$10^{\log_{10}[(2x)/(x-3)]} = 10^1 \qquad \text{Exponentiate both sides}$$

$$\frac{2x}{x-3} = 10 \qquad \text{Inverse property}$$

$$2x = 10x - 30 \qquad \text{Multiply both sides by } x - 3$$

$$-8x = -30 \qquad \text{Subtract } 10x \text{ from both sides}$$

$$x = \frac{15}{4} \qquad \text{Solution}$$

Thus, the solution is $x = \frac{15}{4}$. Try checking this solution in the original equation. ■

Applications

In the following application, we use the formula for compound interest, $A = Pe^{rt}$.

EXAMPLE 9 ■ **An Application: Compound Interest**

A deposit of \$5,000 is placed in a savings account for two years. The interest for the account is compounded continuously. At the end of two years, the balance in the account is \$5,867.56. What is the annual percentage rate for this account?

Solution

Using the formula for continuously compounded interest, $A = Pe^{rt}$, we have the following solution.

Common formula: $A = Pe^{rt}$

Labels:
Principal $= P =$ \$5,000
Amount $= A =$ \$5,867.56
Time $= t = 2$ years
Annual percentage rate $= r$ (in decimal form)

Equation:

$$5{,}867.56 = 5{,}000e^{2r}$$

$$\frac{5{,}867.56}{5{,}000} = e^{2r}$$

$$1.1735 \approx e^{2r}$$

$$\ln 1.1735 \approx \ln (e^{2r})$$

$$0.16 \approx 2r$$

$$0.08 \approx r$$

Therefore, the annual percentage rate is 8%. Try checking this solution in the original statement of the problem. ■

DISCUSSION PROBLEM ■ Solving an Exponential Equation

The *polynomial* equation

$$x^2 - 3x + 2 = 0$$

can be solved by factoring the left side and setting each factor equal to zero, as follows.

$$(x - 2)(x - 1) = 0$$

$$x - 2 = 0 \Rightarrow x = 2$$

$$x - 1 = 0 \Rightarrow x = 1$$

Use this same procedure to solve the equation

$$e^{2x} - 3e^x + 2 = 0.$$

Warm-Up

The following warm-up exercises involve skills that were covered in earlier sections. You will use these skills in the exercise set for this section.

In Exercises 1–6, solve for x.

1. $10x - 3 = 22$
2. $3 - 2(x - 1) = 12$
3. $(x - 4)^2 - 36 = 0$
4. $x^2 - 5x + 6 = 0$
5. $-28 + x \ln 5 = 0$
6. $5xe^2 = 0$

In Exercises 7–10, simplify the given expression.

7. $\log_{10} 10^3$
8. $\log_5 5^6$
9. $\log_2 1$
10. $\ln e^4$

10.4 EXERCISES

In Exercises 1–16, solve the equation for x. (Do not use a calculator.)

1. $2^x = 2^5$
2. $5^x = 5^3$
3. $3^{x+4} = 3^{12}$
4. $10^{1-x} = 10^4$
5. $4^{x-1} = 16$
6. $3^{2x} = 81$
7. $5^x = \frac{1}{125}$
8. $4^{x+1} = \frac{1}{64}$
9. $\log_2 (x + 3) = \log_2 7$
10. $\log_5 2x = \log_5 36$
11. $\ln 5x = \ln 22$
12. $\ln (2x - 3) = \ln 17$
13. $\log_3 x = 4$
14. $\log_5 x = 3$
15. $\log_{10} 2x = 6$
16. $\log_2 (3x - 1) = 5$

In Exercises 17–20, use the inverse properties of the exponential and logarithmic functions to simplify the given expression.

17. $\ln e^{2x-1}$
18. $\log_3 3^{x^2}$
19. $10^{\log_{10} 2x}$
20. $e^{\ln (x+1)}$

In Exercises 21–40, solve the exponential equation. (Round your answer to two decimal places.)

21. $2^x = 45$
22. $5^x = 212$
23. $10^{2y} = 52$
24. $12^{x-1} = 1{,}500$

25. $\frac{1}{5}4^{x+2} = 300$ **26.** $3(2^{t+4}) = 350$ **27.** $4 + e^{2x} = 150$ **28.** $500 - e^{x/2} = 35$

29. $23 - 5e^{x+1} = 3$ **30.** $2e^x + 5 = 115$ **31.** $300e^{x/2} = 9{,}000$ **32.** $1{,}000^{0.12x} = 25{,}000$

33. $6{,}000e^{-2t} = 1{,}200$ **34.** $10{,}000e^{-0.1t} = 4{,}000$ **35.** $250(1.04)^x = 1{,}000$ **36.** $32(1.5)^x = 640$

37. $\frac{1{,}600}{(1.1)^x} = 200$ **38.** $\frac{5{,}000}{(1.05)^x} = 250$ **39.** $4(1 + e^{x/3}) = 84$ **40.** $50(3 - e^{2x}) = 125$

In Exercises 41–56, solve the logarithmic equation. (Round your answer to two decimal places.)

41. $\log_{10} x = 0$ **42.** $\ln x = 1$ **43.** $\log_{10} x = 3$ **44.** $\log_{10} x = -2$

45. $\log_{10} 4x = \frac{3}{2}$ **46.** $\log_{10} (x + 3) = \frac{5}{3}$ **47.** $\ln x = 2.1$ **48.** $\ln 2x = 3$

49. $\frac{2}{3}\ln (x + 1) = -1$ **50.** $8 \ln (3x - 2) = 1.5$ **51.** $2 \log_{10} (x + 5) = 15$ **52.** $-1 + 3 \log_{10} \frac{x}{2} = 8$

53. $\ln x^2 = 6$ **54.** $\ln \sqrt{x} = 6.5$ **55.** $\log_{10} x(x - 3) = 1$ **56.** $\log_{10} x(x + 1) = 0$

57. *Doubling Time* Solve the exponential equation

$$5{,}000 = 2{,}500e^{0.09t}$$

for t to determine the number of years for an investment of \$2,500 to double in value when compounded continuously at the rate of 9%.

58. *Doubling Time* Solve the exponential equation

$$10{,}000 = 5{,}000e^{10r}$$

for r to determine the interest rate required for an investment of \$5,000 to double in value when compounded continuously for 10 years.

59. *Intensity of Sound* The relationship between the number of decibels B and the intensity of a sound I in watts per meter squared is given by

$$B = 10 \log_{10} \left(\frac{I}{10^{-16}}\right).$$

Determine the intensity of a sound I if it registers 75 decibels on an intensity meter.

60. *Human Memory Model* The average score A for a group of students who took a test t months after the completion of a course is given by the memory model

$$A = 80 - \log_{10} (t + 1)^{12}.$$

Determine the time span after the completion of the course when the average score would fall to $A = 72$.

61. *Friction* In order to restrain an untrained horse, a person partially wraps the rope around a cylindrical post in the corral (see figure). If the horse is pulling on the rope with a force of 200 pounds, the force F in pounds required by the person is

$$F = 200e^{-0.2\pi\theta/180}$$

where θ is the angle of wrap in degrees. Find the smallest value of θ if F cannot exceed 80 pounds.

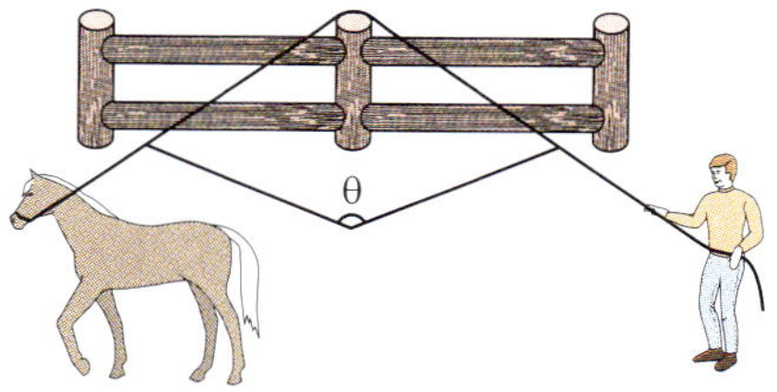

SECTION 10.5

Applications of Exponential and Logarithmic Functions

Compound Interest • Effective Yield • Growth and Decay • Intensity Models

Compound Interest

In Section 10.1 we introduced the following two formulas for compound interest. In these formulas A is the balance, P is the principal, r is the annual percentage rate (in decimal form), and t is the time in years.

n Compoundings per Year	*Continuous Compounding*
$A = P\left(1 + \frac{r}{n}\right)^{nt}$	$A = Pe^{rt}$

In Example 9 in the previous section, we demonstrated a procedure for finding the annual percentage rate for continuously compounded interest. In Example 1 we show how to find the annual percentage rate for interest that is compounded n times per year.

EXAMPLE 1 ■ Finding the Annual Percentage Rate

An investment of \$50,000 is made in an account that compounds interest quarterly. After four years, the balance in the account is \$71,381.07. What is the annual percentage rate for this account?

Solution

Common formula: $A = P\left(1 + \frac{r}{n}\right)^{nt}$

Labels:
Principal $= P =$ \$50,000
Amount $= A =$ \$71,381.07
Time $= t = 4$ years
Number of compoundings per year $= n = 4$
Annual percentage rate $= r$ (in decimal form)

Equation:

$$71{,}381.07 = 50{,}000\left(1 + \frac{r}{4}\right)^{(4)(4)}$$

$$\frac{71{,}381.07}{50{,}000} = \left(1 + \frac{r}{4}\right)^{16}$$

$$1.42762 \approx \left(1 + \frac{r}{4}\right)^{16}$$

$$
\begin{aligned}
(1.42762)^{1/16} &\approx 1 + \frac{r}{4} && \text{Raise both sides to } \tfrac{1}{16} \text{ power}\\
1.0225 &\approx 1 + \frac{r}{4}\\
0.0225 &\approx \frac{r}{4}\\
0.09 &\approx r
\end{aligned}
$$

Therefore, the annual percentage rate is 9%. Check this solution in the original statement of the problem. ■

NOTE: Notice in Example 1 that to get rid of the 16th power on the right side of the equation, we raised both sides of the equation to the $\frac{1}{16}$ power.

EXAMPLE 2 ■ Doubling Time or Continuous Compounding

An investment is made in a trust fund at an annual percentage rate of 8.75%, compounded continuously. How long will it take for the investment to double?

Solution

Common formula: $A = Pe^{rt}$

Labels:

Principal = P (in dollars)
Amount = $A = 2P$ (in dollars)
Time = t (in years)
Annual percentage rate = $r = 0.0875$

Equation:

$$
\begin{aligned}
2P &= Pe^{0.0875t}\\
2 &= e^{0.0875t}\\
\ln 2 &= \ln (e^{0.0875t})\\
\ln 2 &= 0.0875t\\
\frac{\ln 2}{0.0875} &= t\\
7.92 &\approx t
\end{aligned}
$$

Therefore, it will take approximately 7.92 years for the investment to double. Check this solution in the original statement of the problem. (Note that we do not have to know a dollar amount for the principal in order to answer this question.) ■

EXAMPLE 3 ■ Finding the Type of Compounding

Suppose you deposit \$1,000 in a savings account. At the end of one year your balance is \$1,077.63. If the bank tells you that the annual percentage rate for the account is 7.5%, how was the interest compounded?

Solution

If the interest had been compounded continuously at 7.5%, the balance would have been $A = 1{,}000e^{(0.075)(1)} = \$1{,}077.88$. Since the actual balance is slightly less than this, we use the formula for interest that is compounded n times per year.

Common formula: $A = P\left(1 + \frac{r}{n}\right)^{nt}$

Labels:
Principal $= P = \$1{,}000$
Amount $= A = \$1{,}077.63$
Time $= t = 1$ year
Annual percentage rate $= r = 0.075$
Number of compoundings per year $= n$

Equation:

$$1{,}077.63 = 1{,}000\left(1 + \frac{0.075}{n}\right)^n$$

$$1.07763 = \left(1 + \frac{0.075}{n}\right)^n$$

At this point, it is not clear what we should do to try to solve the equation for n. However, we know that the more common types of compoundings are annual ($n = 1$), quarterly ($n = 4$), monthly ($n = 12$), and daily ($n = 365$). By testing these four values of n, we find that the solution is $n = 12$.

n	1	4	12	365
$\left(1 + \frac{0.075}{n}\right)^n$	1.075	1.07714	1.07763	1.07788

Therefore, the interest was compounded monthly. ■

Effective Yield

In Example 3 notice that an investment of \$1,000 compounded monthly produced a balance of \$1,077.63 at the end of one year. Because \$77.63 of this amount is interest, we say that the **effective yield** for the investment is

$$\text{Effective yield} = \frac{\text{Year's interest}}{\text{Amount invested}} = \frac{77.63}{1{,}000} = 0.07763 = 7.763\%.$$

In other words, the effective yield for an investment collecting compound interest is the *simple interest rate* that would yield the same balance at the end of one year.

EXAMPLE 4 ■ Finding the Effective Yield

An investment is made in an account that pays 6.75% interest, compounded continuously. What is the effective yield for this investment?

Solution

Notice that we do not have to know the principal or the time that the money will be left in the account. Instead, we choose an arbitrary principal, such as \$1,000. Then, since effective yield is based on the balance at the end of *one* year, we use the following formula.

$$A = Pe^{rt} \qquad \text{Common formula}$$

$$A = 1{,}000e^{0.0675(1)} \qquad \text{Substitute known values}$$

$$A = 1{,}069.83$$

Now, since the account would earn \$69.83 in interest after one year for a principal of \$1,000, we conclude that the effective yield is

$$\text{Effective yield} = \frac{69.83}{1{,}000} = 0.06983 = 6.983\%.$$

■

MATH MATTERS

Animal Species

There are approximately six million different species of living organisms on earth—animals, plants, single-cell organisms, fungi, and so on. Of those that are animals, by far most are insects. What percent of the different species on earth are insects? (The answer is given in the back of the text.)

Type of Animal	*Example*	*Number of Species*
Insect	Ant, Butterfly, Honey Bee	950,000
Arachnid	Spider, Scorpion, Mite	110,000
Mollusk	Snail, Clam, Squid	45,000
Fish	Tuna, Salmon, Trout	30,000
Crustacean	Shrimp, Lobster, Crab	26,000
Bird	Robin, Ostrich, Eagle	8,650
Worm	Earthworm, Flatworm, Roundworm	6,700
Echinoderm	Starfish, Sea Urchin, Sea Cucumber	6,000
Reptile	Snake, Lizard, Alligator	5,180
Mammal	Human, Mouse, Whale	4,230
Amphibian	Frog, Toad, Salamander	3,000

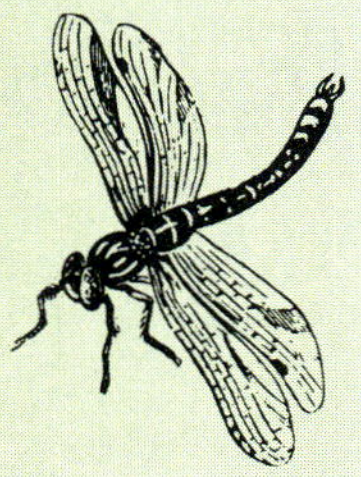

Growth and Decay

The balance in an account earning *continuously* compounded interest is one example of a quantity that increases over time according to the **exponential growth model**

$$y = Ce^{kt}.$$

Exponential Growth and Decay

The mathematical model for exponential growth or decay is given by

$$y = Ce^{kt}.$$

For this model, t is the time, C is the original size (or amount) of the quantity, and y is the size (or amount) after the time t. The number k is a constant that is determined by the rate of growth. If $k > 0$, the model represents **exponential growth**, and if $k < 0$, the model represents **exponential decay**.

One common application of exponential growth is in modeling the growth of a population, as shown in Example 5.

EXAMPLE 5 ■ Population Growth

Suppose you are doing research about a particular country and you discover that the country's population is two million in 1980 and three million in 1990. What would you predict the population of the country to be in the year 2000?

Solution

If we assumed a *linear growth model*, then we would simply predict the population in the year 2000 to be four million. (Try verifying this.) However, social scientists and demographers have discovered that exponential growth models are better than linear growth models for representing population growth. Thus, we use the exponential growth model

$$y = Ce^{kt}.$$

In this model we let $t = 0$ represent the year 1980. Then the given information about the population can be described by the following table.

t (years)	0	10	20
Ce^{kt} (million)	$Ce^{k(0)} = 2$	$Ce^{k(10)} = 3$	$Ce^{k(20)} = ?$

To find the population when $t = 20$, we must first find the values of C and k. From the table, we use the fact that $Ce^{k(0)} = Ce^0 = 2$ to conclude that $C = 2$. Then, using this value of C, we solve for k as follows.

$$Ce^{k(10)} = 3 \qquad \text{From table}$$
$$2e^{10k} = 3 \qquad \text{Substitute value of } C$$
$$e^{10k} = \frac{3}{2} \qquad \text{Divide both sides by 2}$$
$$\ln (e^{10k}) = \ln \frac{3}{2} \qquad \text{Take ln of both sides}$$
$$10k = \ln \frac{3}{2} \qquad \text{Inverse property}$$
$$k = \frac{1}{10} \ln \frac{3}{2} \qquad \text{Divide both sides by 10}$$
$$k \approx 0.0405$$

Finally, we can use this value of k to conclude that the population in the year 2000 is given by

$$2e^{0.0405(20)} \approx 2(2.25) = 4.5 \text{ million.}$$

Figure 10.11 graphically compares the exponential growth model with a linear growth model.

FIGURE 10.11

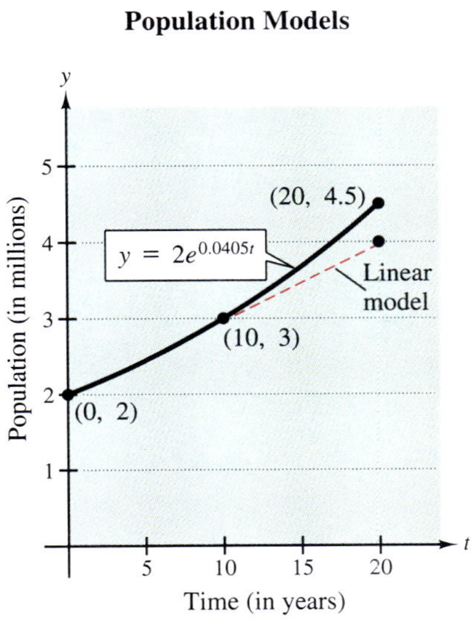

■

EXAMPLE 6 ■ Radioactive Decay

Radioactive iodine is a by-product of some types of nuclear reactors. Its **half-life** is 60 days. That is, after 60 days, a given amount of radioactive iodine will have decayed to half the original amount. Suppose a nuclear accident occurs and releases 20 grams of radioactive iodine. How long will it take for the radioactive iodine to decay to a level of 1 gram?

Solution

To solve this problem, we use the model for exponential decay.

$$y = Ce^{kt}$$

Next, we use the information given in the problem to set up the following table.

t (days)	0	60	?
Ce^{kt} (grams)	$Ce^{k(0)} = 20$	$Ce^{k(60)} = 10$	$Ce^{k(t)} = 1$

Since $Ce^{k(0)} = Ce^0 = 20$, we can conclude that $C = 20$. Then, using this value of C, we can solve for k as follows.

$$Ce^{k(60)} = 10 \qquad \text{From table}$$

$$20e^{60k} = 10 \qquad \text{Substitute value of } C$$

$$e^{60k} = \frac{1}{2} \qquad \text{Divide both sides by 20}$$

$$60k = \ln\frac{1}{2} \qquad \text{Take ln of both sides}$$

$$k = \frac{1}{60}\ln\frac{1}{2} \approx -0.01155 \qquad \text{Solve for } k$$

Finally, we can use this value of k to find the time when the amount is 1 gram, as follows.

$$Ce^{kt} = 1 \qquad \text{From table}$$

$$20e^{-0.01155t} = 1 \qquad \text{Substitute values of } C \text{ and } k$$

$$e^{-0.01155t} = \frac{1}{20} \qquad \text{Divide both sides by 20}$$

$$-0.01155t = \ln\frac{1}{20} \qquad \text{Take ln of both sides}$$

$$t = \frac{1}{-0.01155}\ln\frac{1}{20} \approx 259.4 \text{ days} \qquad \text{Solve for } t$$

Thus, 20 grams of radioactive iodine will have decayed to 1 gram after about 259.4 days. This solution is shown graphically in Figure 10.12.

FIGURE 10.12

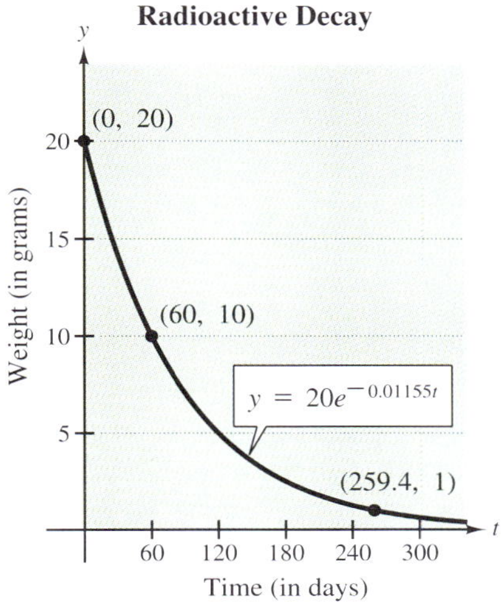

Intensity Models

On the **Richter scale**, the magnitude R of an earthquake can be measured by the **intensity model**

$$R = \log_{10} I$$

where I is the intensity of the shock wave. This intensity model is used in the following example.

EXAMPLE 7 ■ **Earthquake Intensity**

In 1906 San Francisco experienced an earthquake that measured 8.6 on the Richter scale. In 1989, another earthquake struck the San Francisco area. This earthquake measured 7.1 on the Richter scale. How many times greater in intensity was the 1906 earthquake than the 1989 earthquake?

Solution

The intensity for the 1906 earthquake is given as follows.

$$8.6 = \log_{10} I \qquad \text{Given equation}$$

$$10^{8.6} = 10^{(\log_{10} I)} \qquad \text{Exponentiate both sides}$$

$$10^{8.6} = I \qquad \text{Inverse property}$$

The intensity for the 1989 earthquake can be found in a similar way.

$$7.1 = \log_{10} I \qquad \text{Given equation}$$

$$10^{7.1} = 10^{(\log_{10} I)} \qquad \text{Exponentiate both sides}$$

$$10^{7.1} = I \qquad \text{Inverse property}$$

The ratio of the two intensities is

$$\frac{I \text{ for } 1906}{I \text{ for } 1989} = \frac{10^{8.6}}{10^{7.1}} = 10^{8.6-7.1} = 10^{1.5} \approx 31.6.$$

Therefore, we can conclude that the 1906 earthquake had an intensity that was about thirty-two times greater than the 1989 earthquake. ■

DISCUSSION PROBLEM ■ Creating Word Problems

In writing this text, we spent a lot of time looking for real-life applications. If you can think of any good applications that we can use in future editions of this text, we welcome your suggestions. Our mailing address is Roland Larson and Robert Hostetler, Department of Mathematics, The Pennsylvania State University, Erie, Pennsylvania, 16563–1500. If your application is used in a future edition, we will give you credit in the acknowledgment section in the front of the book. ■

Warm-Up

The following warm-up exercises involve skills that were covered in earlier sections. You will use these skills in the exercise set for this section.

In Exercises 1–10, solve the given equation. (Round your answer to two decimal places.)

1. $6 + 3(x - 5) = 27 - 5(x - 6)$ **2.** $5x^2 + 2x - 3 = 0$

3. $\sqrt{x - 5} = 6$ **4.** $\frac{4}{t} + \frac{3}{2t} = 1$ **5.** $\log_6 2x = 2$ **6.** $\log_2 (x - 5) = 6$

7. $e^{x/2} = 8$ **8.** $e^{x-5} = 550$ **9.** $300(1.01)^t = 1{,}200$ **10.** $100(1 - e^t) = 75$

10.5 EXERCISES

Annual Interest Rate In Exercises 1–6, find the annual interest rate given the principal P, the balance A, the time t, and the type of compounding.

	Principal	*Balance*	*Time*	*Compounding*
1.	\$500	\$1,004.83	10 years	Monthly
2.	\$3,000	\$21,628.70	20 years	Quarterly
3.	\$1,000	\$36,581.00	40 years	Daily
4.	\$200	\$314.85	5 years	Yearly
5.	\$750	\$8,267.38	30 years	Continuously
6.	\$2,000	\$4,234.00	10 years	Continuously

Doubling Time In Exercises 7–12, find the time for an investment to double in value given the principal P, the annual interest rate r, and the type of compounding.

	Principal	*Rate*	*Compounding*
7.	\$6,000	8%	Quarterly
8.	\$500	$5\frac{1}{4}$%	Monthly
9.	\$2,000	10.5%	Daily
10.	\$10,000	9.5%	Yearly
11.	\$1,500	7.5%	Continuously
12.	\$100	6%	Continuously

Type of Compounding In Exercises 13–16, determine the type of compounding given the principal P, the balance A, the time t, and the interest rate r. Solve the problem by trying the more common types of compounding.

	Principal	*Balance*	*Time*	*Rate*
13.	\$750	\$1,587.75	10 years	7.5%
14.	\$10,000	\$73,890.56	20 years	10%
15.	\$100	\$141.48	5 years	7%
16.	\$4,000	\$4,788.76	2 years	9%

Effective Yield In Exercises 17–22, find the effective yield given the annual interest rate and the type of compounding.

	Rate	*Compounding*
17.	8%	Continuously
18.	9.5%	Daily
19.	7%	Monthly
20.	8%	Yearly
21.	$5\frac{1}{4}$%	Daily
22.	9%	Quarterly

Principal In Exercises 23–26, determine the principal P that must be deposited in an account to obtain the balance A for the given interest rate r and type of compounding.

	Balance	*Rate*	*Time*	*Compounding*
23.	\$10,000	9%	20 years	Continuously
24.	\$5,000	8%	5 years	Continuously
25.	\$750	6%	3 years	Daily
26.	\$3,000	7%	10 years	Monthly

Balance After Monthly Deposits In Exercises 27–30, suppose you make monthly deposits of P dollars into a savings account at an annual interest rate r, compounded continuously. Find the balance A after t years given that

$$A = \frac{P(e^{rt} - 1)}{e^{r/12} - 1}.$$

	Principal	*Rate*	*Time*
27.	$P = \$30$	$r = 8\%$	$t = 10$ years
28.	$P = \$100$	$r = 9\%$	$t = 30$ years
29.	$P = \$50$	$r = 10\%$	$t = 40$ years
30.	$P = \$20$	$r = 7\%$	$t = 20$ years

Balance After Monthly Deposits In Exercises 31 and 32, suppose you make monthly deposits of \$30 into a savings account at an annual interest rate of 8%, compounded continuously (see figure).

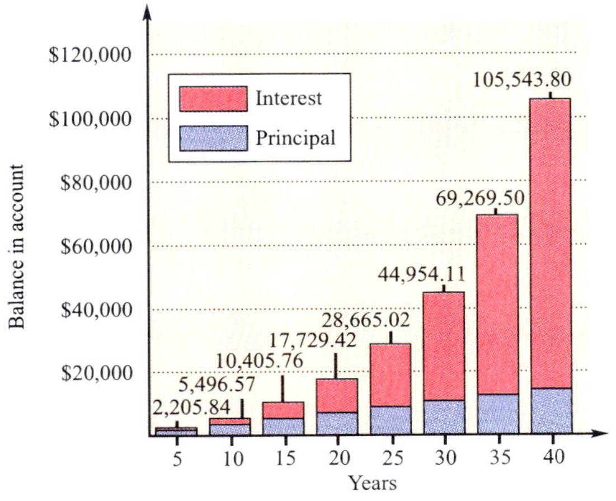

31. Find the total amount that has been deposited in the account in the first 20 years and the total interest earned.

32. Find the total amount that has been deposited in the account in 40 years and the total interest earned.

Exponential Growth and Decay In Exercises 33–36, find the constant k such that the exponential function $y = Ce^{kt}$ passes through the given points on the graph.

33.

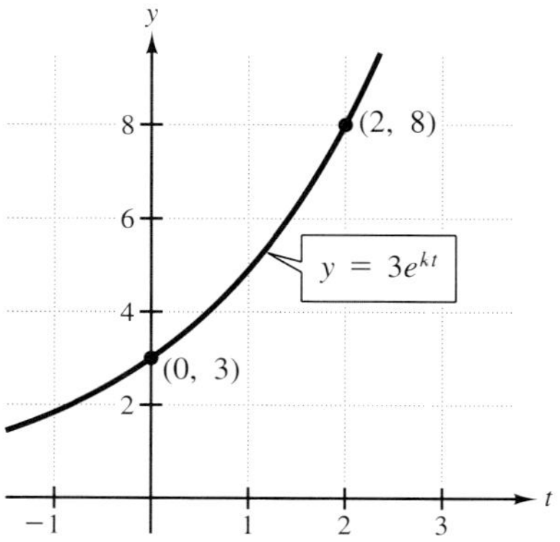

34.

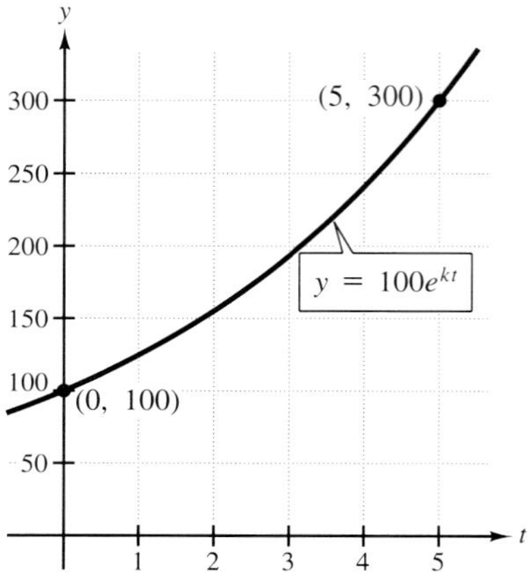

35.

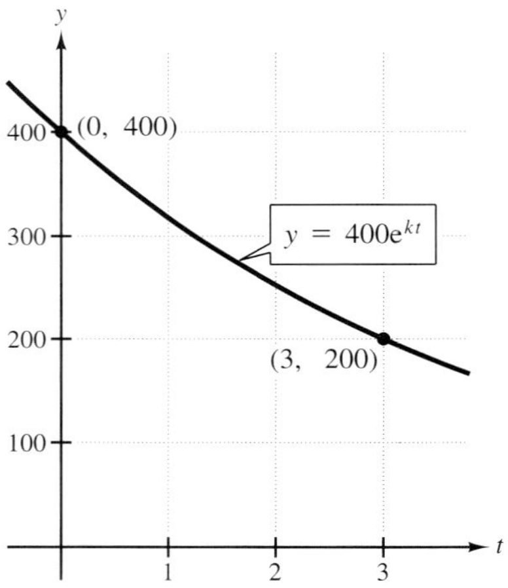

36.

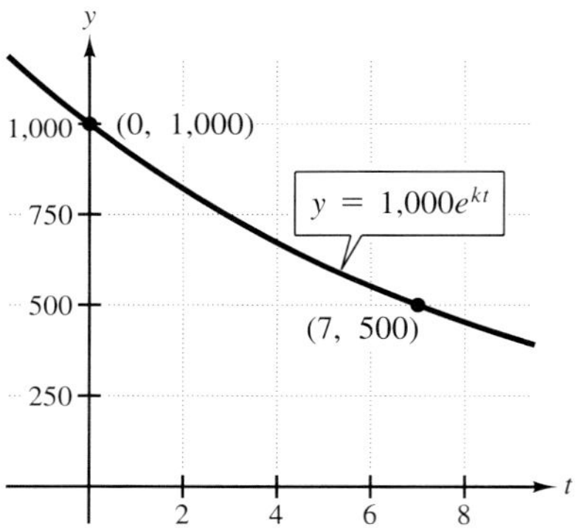

Population of a City In Exercises 37–40, the population of a city is given for the year 1985, and the predicted population for the year 2000. Find the constants C and k to obtain the exponential growth model $y = Ce^{kt}$ for the population growth. (Let $t = 0$ correspond to the year 1985.) Use the model to predict the population of the city in the year 2005.

City	*1985*	*2000*
37. Los Angeles	9.6 million	10.7 million
38. Houston	2.1 million	2.6 million
39. Dhaka, Bangladesh	3.3 million	6.5 million
40. Lagos, Nigeria	6.1 million	12.5 million

41. *Radioactive Decay* Radioactive radium (Ra^{226}) has a half-life of 1,620 years. If you start with 5 grams of the isotope, how much remains after 1,000 years?

42. *Radioactive Decay* The isotope Pu^{230} has a half-life of 24,360 years. If you start with 10 grams of the isotope, how much remains after 10,000 years?

43. *Carbon 14 Dating* C^{14} dating assumes that the carbon dioxide on earth today has the same radioactive content as it did centuries ago. If this is true, then the amount of C^{14} absorbed by a tree that grew several centuries ago should be the same as the amount of C^{14} absorbed by a tree growing today. A piece of ancient charcoal contains only 15% as much of the radioactive carbon as a piece of modern charcoal. How long ago did the tree burn to make the ancient charcoal if the half-life of C^{14} is 5,730 years?

44. *Depreciation* A new car that cost \$22,000 has a depreciated value of \$16,500 after one year. Find the value of the car when it is three years old by using the exponential model $y = Ce^{kt}$.

45. *Advertising Effect* The sales S (in thousands of units) of a product after spending x dollars in advertising is given by

$$S = 10(1 - e^{kx}).$$

(a) Find S as a function of x if 2,500 units are sold when \$500 is spent on advertising.
(b) How many units will be sold if advertising expenditures are raised to \$700?

46. *Population Growth* The population p of a certain species t years after it is introduced into a new habitat is given by

$$p(t) = \frac{5{,}000}{1 + 4e^{-t/6}}.$$

(a) Determine the population size that was introduced into the habitat.
(b) Determine the population size after nine years.
(c) After how many years will the population be 2,000?

47. *Earthquake Intensity* On March 27, 1964, Alaska experienced an earthquake that measured 8.4 on the Richter scale. On February 9, 1971, an earthquake in the San Fernando Valley, California, measured 6.6 on the Richter scale. How much greater in intensity was the Alaskan earthquake than the one in California?

48. *Earthquake Intensity* On March 10, 1933, Long Beach, California experienced an earthquake that measured 6.2 on the Richter scale. On December 3, 1988, an earthquake in Pasadena, California measured 5.0 on the Richter scale. How much greater in intensity was the Long Beach earthquake than the one in Pasadena?

Acidity Model In Exercises 49–52, use the acidity model $pH = -\log_{10} [H^+]$, where acidity (pH) is a measure of the hydrogen ion concentration $[H^+]$ (measured in moles of hydrogen per liter) of solution.

49. Find the pH of a solution that has a hydrogen ion concentration of 9.2×10^{-8}.

50. Compute the hydrogen ion concentration if the pH of a solution is 4.7.

51. A certain fruit has a pH of 2.5, and an antacid tablet has a pH of 9.5. The hydrogen ion concentration of the fruit is how many times the concentration of the tablet?

52. If the pH of a solution is decreased by one unit, the hydrogen ion concentration is increased by what factor?

53. *Sales Growth* Annual sales y of a product x years after it is introduced is approximated by

$$y = \frac{2,000}{1 + 4e^{-x/2}}.$$

(a) Use a graphing calculator to sketch the graph of the equation.* Use the following settings.

RANGE
Xmin=0
Xmax=10
Xscl=1
Ymin=0
Ymax=2000
Yscl=100
Xres=1

(b) Use the [TRACE] key to approximate annual sales when $x = 4$.

(c) Use the [TRACE] key to approximate the time when annual sales are $y = 1,100$ units.

(d) According to this model, the annual sales will approach what maximum level?

54. *Depreciation* After x years, the value y of a truck that cost \$32,000 is given by

$$y = 32,000(0.8)^x.$$

(a) Use a graphing calculator to sketch the graph of the equation. Use the following settings.

RANGE
Xmin=0
Xmax=8
Xscl=1
Ymin=0
Ymax=32000
Yscl=4000
Xres=1

(b) Use the [TRACE] key to approximate the value of the truck after one year.

(c) Use the [TRACE] key to approximate the time when the value of the truck will be \$16,000.

*The graphing calculator keystrokes given in this text correspond to the *TI-81* graphing calculator from *Texas Instruments*. However, you can also use other graphing calculators and computer graphing software for these examples and exercises.

CHAPTER 10 SUMMARY

As you review and prepare for a test on this chapter, first try to obtain a global view of what was discussed. Then review the specific skills needed under each category.

Exponential and Logarithmic Functions and Their Graphs

■ *Evaluate* an exponential expression on paper and with a calculator.	Section 10.1, Exercises 1–10
■ *Sketch* the graph of an exponential function.	Section 10.1, Exercises 17–30
■ *Create and/or evaluate* an exponential function from a verbal statement.	Section 10.1, Exercises 37–50
■ *Rewrite* a logarithmic equation in exponential form and conversely.	Section 10.2, Exercises 1–12
■ *Evaluate* a logarithmic expression on paper and with a calculator.	Section 10.2, Exercises 13–40
■ *Sketch* the graph of a logarithmic function.	Section 10.2, Exercises 41–58

Properties of Logarithms

■ *Rewrite* a logarithmic expression in expanded form using the properties of logarithms.	Section 10.3, Exercises 7–20
■ *Rewrite* a logarithmic expression in condensed form using the properties of logarithms.	Section 10.3, Exercises 21–34
■ *Approximate* the value of a given logarithmic expression given the values of specific logarithms.	Section 10.3, Exercises 41–50

Solving Exponential and Logarithmic Equations

■ *Solve* an exponential equation without a calculator.	Section 10.4, Exercises 1–16
■ *Solve* an exponential equation with a calculator.	Section 10.4, Exercises 21–40
■ *Solve* a logarithmic equation with a calculator.	Section 10.4, Exercises 41–56
■ *Create and/or solve* compound interest problems from verbal statements.	Section 10.5, Exercises 1–30
■ *Create and/or solve* exponential and logarithmic equations from verbal statements.	Section 10.5, Exercises 31–52

Chapter 10 Review Exercises

In Exercises 1–10, evaluate the function at the given values of the independent variable. Use a calculator only if it is more efficient or necessary.

1. $f(x) = 2^x$
   (a) $x = -3$ (b) $x = 1$ (c) $x = 2$

2. $g(x) = 2^{-x}$
   (a) $x = -2$ (b) $x = 0$ (c) $x = 2$

3. $g(t) = e^{-t/3}$
   (a) $t = -3$ (b) $t = \pi$ (c) $t = 6$

4. $h(s) = 1 - e^{0.2s}$
   (a) $s = 0$ (b) $s = 2$ (c) $s = \sqrt{10}$

5. $f(x) = \log_3 x$
   (a) $x = 1$ (b) $x = 27$ (c) $x = 0.5$

6. $g(x) = \log_{10} x$
   (a) $x = 0.01$ (b) $x = 0.1$ (c) $x = 30$

7. $f(x) = \ln x$
   (a) $x = e$ (b) $x = \frac{1}{3}$ (c) $x = 10$

8. $h(x) = \ln x$
   (a) $x = e^2$ (b) $x = \frac{5}{4}$ (c) $x = 1{,}200$

9. $g(x) = \ln e^{3x}$
   (a) $x = -2$ (b) $x = 0$ (c) $x = 7.5$

10. $f(x) = \log_2 \sqrt{x}$
    (a) $x = 4$ (b) $x = 64$ (c) $x = 5.2$

In Exercises 11–20, sketch the graph of the given function.

11. $y = 3^{x/2}$
12. $y = 3^{x/2} - 2$
13. $f(x) = 3^{-x/2}$
14. $f(x) = -3^{-x/2}$
15. $f(x) = 3^{-x^2}$
16. $g(t) = 3^{|t|}$
17. $f(x) = -2 + \log_3 x$
18. $f(x) = 2 + \log_3 x$
19. $g(x) = \ln 2x$
20. $f(x) = 3 + \ln x$

In Exercises 21–26, use the properties of logarithms to expand the expression as a sum, difference, or multiple of logarithms.

21. $\log_4 6x^4$
22. $\log_{10} 2x^{-3}$
23. $\log_5 \sqrt{x + 2}$
24. $\ln \sqrt[3]{\frac{x}{5}}$
25. $\ln \frac{x + 2}{x - 2}$
26. $\ln x(x - 3)^2$

In Exercises 27–32, use properties of logarithms to condense the given expression as a logarithm of a single quantity.

27. $\log_4 x - \log_4 10$
28. $5 \log_2 y$
29. $\log_8 16x + \log_8 2x^2$
30. $4(1 + \ln x + \ln x)$
31. $-2(\ln 2x - \ln 3)$
32. $-\frac{2}{3} \ln 3y$

In Exercises 33–38, determine whether the given equation is true or false.

33. $\log_4 \frac{16}{x} = 2 - \log_4 x$
34. $\log_2 4x = 2 \log_2 x$
35. $\frac{\ln 5x}{\ln 10x} = \ln \frac{1}{2}$
36. $\log_{10} 10^{2x} = 2x$
37. $e^{\ln t} = t$
38. $e^{2x} - 1 = (e^x + 1)(e^x - 1)$

In Exercises 39–44, approximate the logarithm given that $\log_5 2 \approx 0.43068$ and $\log_5 3 \approx 0.68260$. (Do not use a calculator.)

39. $\log_5 18$

40. $\log_5 \sqrt{6}$

41. $\log_5 \frac{1}{2}$

42. $\log_5 \frac{2}{3}$

43. $\log_5 (12)^{2/3}$

44. $\log_5 (5^2 \cdot 6)$

In Exercises 45–50, solve the equation without the aid of a calculator.

45. $2^x = 64$

46. $3^{x-2} = 81$

47. $4^{x-3} = \frac{1}{16}$

48. $\log_2 2x = \log_2 100$

49. $\log_3 x = 5$

50. $\log_5 (x - 10) = 2$

In Exercises 51–60, solve the given equation. (Round your answer to two decimal places.)

51. $3^x = 500$

52. $8^x = 1{,}000$

53. $2e^{x/2} = 45$

54. $100e^{-0.6x} = 20$

55. $\frac{500}{(1.05)^x} = 100$

56. $25(1 - e^t) = 12$

57. $\ln x = 7.25$

58. $\ln x = -0.5$

59. $\log_2 2x = -0.65$

60. $1 + \log_5 x = 4.8$

Balance in an Account In Exercises 61–64, complete the following table to determine the balance A for P dollars invested at rate r for t years, and compounded n times per year.

n	1	4	12	365	Continuous Compounding
A					

61. $P = \$500$ $r = 7\%$ $t = 30$ years

62. $P = \$100$ $r = 5\frac{1}{4}\%$ $t = 60$ years

63. $P = \$10{,}000$ $r = 10\%$ $t = 20$ years

64. $P = \$2{,}500$ $r = 8\%$ $t = 1$ year

Principal In Exercises 65 and 66, complete the following table to determine the principal P that will yield a balance of A dollars when invested at rate r for t years and compounded n times per year.

n	1	4	12	365	Continuous Compounding
P					

65. $A = \$50{,}000$ $r = 8\%$ $t = 40$ years

66. $A = \$1{,}000$ $r = 6\%$ $t = 1$ year

67. *Inflation Rate* If the annual rate of inflation averages 5% over the next 10 years, then the approximate cost C of goods or services during any year in that decade will be given by

$$C(t) = P(1.05)^t, \quad 0 \le t \le 10$$

where t is the time in years and P is the present cost. If the price of an oil change for your car is presently \$19.95, estimate the number of years until it will cost \$25.00.

68. *Doubling Time* Find the time for an investment of \$1,000 to double in value when invested at 8% compounded monthly.

69. *Product Demand* The daily demand x for a certain product is given by

$$p = 25 - 0.4e^{0.02x}$$

where p is the price of the product. Approximate the demand if the price is \$16.97.

70. *Sound Intensity* The relationship between the number of decibels B and the intensity of a sound I in watts per meter squared is given by

$$B = 10 \log_{10}\left(\frac{I}{10^{-16}}\right).$$

Determine the intensity of a sound in watts per meter squared if the decibel level is 125.

71. *Population Limit* The population p of a certain species t years after it is introduced into a new habitat is given by

$$p(t) = \frac{600}{1 + 2e^{-0.2t}}.$$

Complete the following table to determine the limiting size of the population in this habitat.

t	0	1	5	10	20	50	75
$p(t)$							

CHAPTER 10 TEST

Take this test as you would take a test in class. After you are done, check your work with the answers given in the back of the book.

1. Evaluate the exponential function $f(t) = 54\left(\frac{2}{3}\right)^t$ at the given values of t. (Use a calculator only if necessary.)
   (a) $f(0)$ (b) $f(2)$ (c) $f(-1)$ (d) $f\left(\frac{1}{2}\right)$
2. Sketch a graph of the function $f(x) = 2^{x/3}$.
3. Determine the amount in an account after 20 years if \$2,000 is invested at 7% compounded (a) quarterly and (b) continuously.
4. Write the logarithmic equation $\log_5 125 = 3$ in exponential form.
5. Write the exponential equation $4^{-2} = \frac{1}{16}$ in logarithmic form.
6. Evaluate $\log_8 2$ without the aid of a calculator.
7. Sketch the graphs of $f(x) = \log_5 x$ and $g(x) = 5^x$ on the same set of coordinate axes. What can you conclude about the relationship between f and g?
8. Use the properties of logarithms to expand $\log_4 \left(5x^2/\sqrt{y}\right)$ as a sum, difference, or multiple of logarithms.
9. Use the properties of logarithms to condense

   $$\ln x - 4 \ln y$$

   into a logarithm of a single quantity.
10. Simplify: $\log_5 (5^3 \cdot 6)$
11. Solve (without using a calculator): $\log_4 x = 3$
12. Solve: $10^{3y} = 832$
13. Solve: $400e^{0.08t} = 1{,}200$
14. Solve: $3 \ln (2x - 3) = 10$
15. Determine the principal that will yield \$100,000 when invested at 9% compounded quarterly for 25 years.
16. A principal of \$500 yields a balance of \$1,006.88 in 10 years when the interest is compounded continuously. What is the annual percentage rate?
17. A new car that cost \$18,000 has a depreciated value of \$14,000 after one year. Find the value of the car when it is three years old by using the exponential model $y = Ce^{kt}$.
18. The population of a certain species t years after it is introduced into a new habitat is given by

    $$p(t) = \frac{2{,}400}{1 + 3e^{-t/4}}.$$

    (a) Determine the population size that was introduced into the habitat.
    (b) Determine the population after four years.
    (c) After how many years will the population be 1,200?
19. Suppose the annual rate of inflation averages $4\frac{1}{2}$%. With this rate of inflation, the approximate cost C of goods or services during any year will be given by

    $$C(t) = P(1.045)^t$$

    where t is the time in years and P is the present cost.
    (a) If the price of a lawn tractor is presently \$8,500, estimate the price 10 years from now.
    (b) Estimate the time required for the price of goods and services to double at this annual rate of inflation.

CHAPTER ELEVEN

Additional Topics in Algebra

Chapter Overview

This chapter covers four basic topics: sequences, the Binomial Theorem, counting problems, and probability. Sequences are introduced in the first section. Then, in Sections 11.2 and 11.3, we look at two special types of sequences, called arithmetic sequences and geometric sequences. Section 11.4 discusses a very famous theorem in mathematics, called the Binomial Theorem. Sections 11.5 and 11.6 discuss counting problems and probability.

SECTION **11.1**

Sequences

Introduction • *Sequences* • *Factorial Notation* • *Sigma Notation*

Introduction

Suppose you were asked to choose between the following two contract offers for the next five years of employment.

Contract A	*Contract B*
\$20,000 the first year and a \$2,200 raise each year for the next four years	\$20,000 the first year and a 10% raise each year for the next four years

Which contract offers the largest total salary over the five-year period? Table 11.1 compares your salary for each contract.

TABLE 11.1

Year	Salary (Contract A)	Salary (Contract B)
1	\$ 20,000	\$ 20,000
2	22,200	22,000
3	24,400	24,200
4	26,600	26,620
5	28,800	29,282
Total	\$122,000	\$122,102

The salaries for Contract A represent the first five terms of what is called an **arithmetic sequence**, and the salaries for Contract B represent the first five terms of what is called a **geometric sequence**. We can see from this table that after five years the geometric sequence represents a better contract offer than the arithmetic sequence.

Sequences

A mathematical **sequence** is simply an ordered list of numbers. Each number in the list is called a **term** of the sequence. Sequences can have a finite number of terms or an infinite number of terms. For instance, the sequence of positive odd integers that are less than 15 is a *finite* sequence.

$1, 3, 5, 7, 9, 11, 13$ Finite sequence

Similarly, the sequence of positive odd integers is an infinite sequence.

$1, 3, 5, 7, 9, 11, 13, \ldots$ Infinite sequence

(Note that the three dots indicate that the sequence continues and has an infinite number of terms.)

Definition of a Sequence

An **infinite sequence** is an ordered list of real numbers.

$$a_1, a_2, a_3, a_4, a_5, \ldots, a_n, \ldots$$

Each number in the sequence is called a **term** of the sequence.

NOTE: On some occasions it is convenient to begin subscripting an infinite sequence with zero instead of one. In such cases the terms of the sequence are denoted by $a_0, a_1, a_2, a_3, a_4, a_5, \ldots, a_n, \ldots$.

EXAMPLE 1 ■ Finding the Terms of a Sequence

Write the first six terms of the infinite sequence whose nth term is given by

$$a_n = n^2 - 1.$$

(Assume the sequence begins with $n = 1$.)

Solution

The first six terms of the sequence are as follows.

$$a_1 = 1^2 - 1 = 0$$
$$a_2 = 2^2 - 1 = 3$$
$$a_3 = 3^2 - 1 = 8$$
$$a_4 = 4^2 - 1 = 15$$
$$a_5 = 5^2 - 1 = 24$$
$$a_6 = 6^2 - 1 = 35$$

Thus, if we wanted to represent the entire sequence, we could write the following.

$$0, 3, 8, 15, 24, 35, \ldots, n^2 - 1, \ldots$$ ■

EXAMPLE 2 ■ Finding the Terms of a Sequence

Write the first six terms of the infinite sequence whose nth term is given by

$$a_n = 3(2^n).$$

(Assume the sequence begins with $n = 0$.)

Solution

The first six terms of the sequence are as follows.

$$a_0 = 3(2^0) = 3 \cdot 1 = 3$$
$$a_1 = 3(2^1) = 3 \cdot 2 = 6$$
$$a_2 = 3(2^2) = 3 \cdot 4 = 12$$
$$a_3 = 3(2^3) = 3 \cdot 8 = 24$$
$$a_4 = 3(2^4) = 3 \cdot 16 = 48$$
$$a_5 = 3(2^5) = 3 \cdot 32 = 96$$

Thus, the entire sequence can be written as follows.

$$3, 6, 12, 24, 48, 96, \ldots, 3(2^n), \ldots$$

■

In the next example, notice how the formula for the nth term produces a sequence whose terms alternate in sign.

EXAMPLE 3 ■ **A Sequence Whose Terms Alternate in Sign**

Write the first six terms of the infinite sequence whose nth term is given by

$$a_n = \frac{(-1)^n}{2n - 1}.$$

(Assume the sequence begins with $n = 1$.)

Solution

The first six terms of the sequence are as follows.

$$a_1 = \frac{(-1)^1}{2(1) - 1} = -\frac{1}{1}$$
$$a_2 = \frac{(-1)^2}{2(2) - 1} = \frac{1}{3}$$
$$a_3 = \frac{(-1)^3}{2(3) - 1} = -\frac{1}{5}$$
$$a_4 = \frac{(-1)^4}{2(4) - 1} = \frac{1}{7}$$
$$a_5 = \frac{(-1)^5}{2(5) - 1} = -\frac{1}{9}$$
$$a_6 = \frac{(-1)^6}{2(6) - 1} = \frac{1}{11}$$

Thus, the entire sequence can be written as follows.

$$-1, \frac{1}{3}, -\frac{1}{5}, \frac{1}{7}, -\frac{1}{9}, \frac{1}{11}, \ldots, \frac{(-1)^n}{2n - 1}, \ldots$$

■

Factorial Notation

Some very important sequences in mathematics involve terms that are defined with special types of products, called **factorials**.

Definition of Factorial If n is a positive integer, then ***n* factorial** is defined by

$$n! = 1 \cdot 2 \cdot 3 \cdot 4 \cdots (n-1) \cdot n.$$

As a special case, we define zero factorial to be $0! = 1$.

The first several factorial values are as follows.

$0! = 1$ $\quad$ $1! = 1$

$2! = 1 \cdot 2 = 2$ $\quad$ $3! = 1 \cdot 2 \cdot 3 = 6$

$4! = 1 \cdot 2 \cdot 3 \cdot 4 = 24$ $\quad$ $5! = 1 \cdot 2 \cdot 3 \cdot 4 \cdot 5 = 120$

Many scientific calculators have a factorial key, denoted by [n!]. If your calculator has such a key, try using it to evaluate $n!$ for several values of n. You will see that the value of n does not have to be very large before the value of $n!$ is huge. For instance, $10! = 3{,}628{,}800$.

EXAMPLE 4 ■ Finding the Terms of a Sequence Involving Factorials

Write the first six terms of the sequence whose nth term is given by

$$a_n = \frac{1}{n!}.$$

(Assume the sequence begins with $n = 0$.)

Solution

The first six terms of the sequence are as follows.

$$a_0 = \frac{1}{0!} = \frac{1}{1} = 1$$

$$a_1 = \frac{1}{1!} = \frac{1}{1} = 1$$

$$a_2 = \frac{1}{2!} = \frac{1}{2}$$

$$a_3 = \frac{1}{3!} = \frac{1}{1 \cdot 2 \cdot 3} = \frac{1}{6}$$

$$a_4 = \frac{1}{4!} = \frac{1}{1 \cdot 2 \cdot 3 \cdot 4} = \frac{1}{24}$$

$$a_5 = \frac{1}{5!} = \frac{1}{1 \cdot 2 \cdot 3 \cdot 4 \cdot 5} = \frac{1}{120}$$

Thus, the entire sequence can be written as follows.

$$1, 1, \frac{1}{2}, \frac{1}{6}, \frac{1}{24}, \frac{1}{120}, \ldots, \frac{1}{n!}, \ldots$$ ■

Factorials follow the same conventions for order of operation as do exponents. For instance, when we write $2n!$, we mean $2(n!)$. Notice how these conventions are used in the next example.

EXAMPLE 5 ■ **Finding the Terms of a Sequence Involving Factorials**

Write the first six terms of the sequence whose nth term is given by

$$a_n = \frac{2n!}{(2n)!}.$$

(Assume the sequence begins with $n = 0$.)

Solution

The first six terms of the sequence are as follows.

$$a_0 = \frac{2(0!)}{0!} = \frac{2}{1} = 2$$

$$a_1 = \frac{2(1!)}{2!} = \frac{2}{2} = 1$$

$$a_2 = \frac{2(2!)}{4!} = \frac{4}{24} = \frac{1}{6}$$

$$a_3 = \frac{2(3!)}{6!} = \frac{12}{720} = \frac{1}{60}$$

$$a_4 = \frac{2(4!)}{8!} = \frac{48}{40{,}320} = \frac{1}{840}$$

$$a_5 = \frac{2(5!)}{10!} = \frac{240}{3{,}628{,}800} = \frac{1}{15{,}120}$$

Thus, the entire sequence can be written as follows.

$$2, 1, \frac{1}{6}, \frac{1}{60}, \frac{1}{840}, \frac{1}{15{,}120}, \ldots, \frac{2n!}{(2n)!}, \ldots$$ ■

In Example 5 we multiplied the numerators and denominators before reducing the fractions. When finding the terms of a sequence, reducing is often easier if you leave the numerator and denominator in factored form. For instance, notice the cancellation in the following fraction.

$$a_5 = \frac{2(5!)}{10!} = \frac{2 \cdot 1 \cdot 2 \cdot 3 \cdot 4 \cdot 5}{1 \cdot 2 \cdot 3 \cdot 4 \cdot 5 \cdot 6 \cdot 7 \cdot 8 \cdot 9 \cdot 10} = \frac{1}{3 \cdot 7 \cdot 8 \cdot 9 \cdot 10} = \frac{1}{15{,}120}$$

MATH MATTERS

Duo-Dice*

Ordinary playing dice are made in the shape of cubes. On the sides of each cube, the dice are imprinted with one, two, three, four, five, or six dots, as shown in the figure. On normal playing dice, these dots are usually recessed to prevent the paint from wearing off. These recessions, however, cause the dice to be unbalanced. The heaviest side is the side with only one recession, and the lightest side is the side with six recessions. In practice, this unbalancing is enough to create an increased probability that the lightest side (the side with six recessions) will turn up.

Ordinary six-sided dice

Twelve-sided duo-dice

To prevent the unbalancing caused by recessed dots, you can make the dice in the shape of dodecahedrons, as shown in the figure. On two opposing sides, one dot is recessed. On two other opposing sides, two dots are recessed, and so on. Dice made with this pattern are perfectly balanced (because each pair of opposing sides has the same number of receded dots). When two of these dice are used in any game that requires two six-sided dice, the probability of rolling a given total is the same as if you used two perfectly balanced six-sided dice.

*Duo-Dice are patented by Roland E. Larson (one of the authors of this text). (Patent Number: 4,465,279, August 14, 1984) If you would like a free pair of these dice, write to Roland E. Larson, Department of Mathematics, The Pennsylvania State University, Erie, Pennsylvania 16563–1500.

Sigma Notation

Many applications involve finding the sum of the first n terms of a sequence. A convenient shorthand notation for such a sum involves **sigma notation**. This name comes from the use of the uppercase Greek letter sigma, written as Σ.

Definition of Sigma Notation

The sum of the first n terms of the sequence whose nth term is a_n is

$$\sum_{i=1}^{n} a_i = a_1 + a_2 + a_3 + a_4 + \cdots + a_n$$

where i is called the **index of summation**, n is the **upper limit of summation**, and 1 is the **lower limit of summation**.

EXAMPLE 6 ■ Sigma Notation for Sums

Find the sum $\sum_{i=1}^{6} 2i$.

Solution

By letting i take on successive integer values from 1 to 6, we obtain the following.

$$\begin{aligned}\sum_{i=1}^{6} 2i &= 2(1) + 2(2) + 2(3) + 2(4) + 2(5) + 2(6) \\ &= 2 + 4 + 6 + 8 + 10 + 12 \\ &= 42\end{aligned}$$

■

In Example 6 the index of summation was i and the summation began with $i = 1$. Any letter can be used as the index of summation, and the summation can begin with any integer. For instance the index of summation in the next example is k and the summation begins with $k = 0$.

EXAMPLE 7 ■ Sigma Notation for Sums

Find the sum $\sum_{k=0}^{8} \frac{1}{k!}$.

Solution

By letting k take on successive integer values from 0 to 8, we obtain the following.

$$\begin{aligned}\sum_{k=0}^{8} \frac{1}{k!} &= \frac{1}{0!} + \frac{1}{1!} + \frac{1}{2!} + \frac{1}{3!} + \frac{1}{4!} + \frac{1}{5!} + \frac{1}{6!} + \frac{1}{7!} + \frac{1}{8!} \\ &= 1 + 1 + \frac{1}{2} + \frac{1}{6} + \frac{1}{24} + \frac{1}{120} + \frac{1}{720} + \frac{1}{5,040} + \frac{1}{40,320} \\ &\approx 2.71828\end{aligned}$$

Note that this sum is approximately $e = 2.71828 \ldots$. ■

EXAMPLE 8 ■ Sigma Notation for Sums

Find the following sums.

(a) $\sum_{i=1}^{4} 5$ (b) $\sum_{i=0}^{6} \frac{1}{2^i}$

Solution

(a) By letting i take on successive integer values from 1 to 4, we obtain the following.

$$\sum_{i=1}^{4} 5 = 5 + 5 + 5 + 5 = 20$$

Note that each of the terms of this sum is constant because $a_i = 5$ for all values of i.

(b) By letting i take on successive integer values from 0 to 6, we obtain the following.

$$\sum_{i=0}^{6} \frac{1}{2^i} = \frac{1}{2^0} + \frac{1}{2^1} + \frac{1}{2^2} + \frac{1}{2^3} + \frac{1}{2^4} + \frac{1}{2^5} + \frac{1}{2^6}$$

$$= \frac{1}{1} + \frac{1}{2} + \frac{1}{4} + \frac{1}{8} + \frac{1}{16} + \frac{1}{32} + \frac{1}{64}$$

$$= \frac{127}{64}$$

■

EXAMPLE 9 ■ Writing a Sum in Sigma Notation

Write the following sum using sigma notation.

$$\frac{2}{2} + \frac{2}{3} + \frac{2}{4} + \frac{2}{5} + \frac{2}{6}$$

Solution

To write this sum using sigma notation, we must find a pattern for the terms. After examining the terms, we see that each has a numerator of 2 and denominators that range over the integers from 2 to 6. Thus, one possible sigma notation for this sum is as follows.

$$\sum_{i=1}^{5} \frac{2}{i + 1} = \frac{2}{2} + \frac{2}{3} + \frac{2}{4} + \frac{2}{5} + \frac{2}{6}$$

■

EXAMPLE 10 ■ Writing a Sum in Sigma Notation

Write the following sum using sigma notation.

$$1 + 8 + 27 + 64 + 125$$

Solution

After examining the terms for this sum, we see that the terms are the cubes of the first five positive integers. That is, $1 = 1^3$, $8 = 2^3$, and so on. Thus, one possible sigma notation for this sum is as follows.

$$\sum_{i=1}^{5} i^3 = 1 + 8 + 27 + 64 + 125$$

■

DISCUSSION PROBLEM ■ Writing Sigma Notation in Different Ways

There is more than one way to write a given sum in sigma notation. For instance, the sum

$$1 + 2 + 3 + 4 + 5 + 6$$

could be written as

$$\sum_{i=1}^{6} i \quad \text{or} \quad \sum_{i=0}^{5} (i + 1).$$

Write each of the following sums in two ways and determine which is the simpler.

(a) $1 + 4 + 9 + 16 + 25 + 36$ (b) $3 + 5 + 7 + 9 + 11 + 13$ ■

Warm-Up

The following warm-up exercises involve skills that were covered in earlier sections. You will use these skills in the exercise set for this section.

In Exercises 1–10, perform the indicated operations and simplify.

1. $\frac{1}{3} \cdot \frac{3}{5} \cdot \frac{5}{7}$

2. $\frac{3 \cdot 6 \cdot 9 \cdot 12}{3^4}$

3. $\frac{1}{1 \cdot 2} + \frac{1}{2 \cdot 3} + \frac{1}{3 \cdot 4}$

4. $\frac{1}{2} + \frac{1}{3} + \frac{1}{4}$

5. $\frac{n(n-1)(n-2)}{(n-1)(n-2)}$

6. $\frac{n^2 - 1}{n + 1}$

7. $\frac{2n^2 + 7n - 15}{n^2 - 25}$

8. $\frac{n^2 + 3n + 2}{n^2 + 2n}$

9. $\frac{1}{n} + \frac{2}{n^2}$

10. $\frac{1}{n-1} + \frac{1}{n+1}$

11.1 EXERCISES

In Exercises 1–20, write the first five terms of the infinite sequence. (Assume the sequence begins with $n = 1$.)

1. $a_n = 2n$

2. $a_n = 3n$

3. $a_n = (-1)^n 2n$

4. $a_n = (-1)^{n+1} 3n$

5. $a_n = \left(\frac{1}{2}\right)^n$

6. $a_n = \left(\frac{1}{3}\right)^n$

7. $a_n = \left(-\frac{1}{2}\right)^n$

8. $a_n = \left(-\frac{1}{3}\right)^n$

9. $a_n = \dfrac{1}{n + 1}$

10. $a_n = \dfrac{3}{2n + 1}$

11. $a_n = \dfrac{2n}{3n + 2}$

12. $a_n = \dfrac{5n}{4n + 3}$

13. $a_n = \dfrac{(-1)^n}{n^2}$

14. $a_n = \dfrac{1}{\sqrt{n}}$

15. $a_n = \dfrac{2n}{n!}$

16. $a_n = \dfrac{n!}{(n - 1)!}$

17. $a_n = 5 - \dfrac{1}{2^n}$

18. $a_n = 7 + \dfrac{1}{3^n}$

19. $a_n = 2 + (-2)^n$

20. $a_n = \dfrac{1 + (-1)^n}{n^2}$

In Exercises 21–26, simplify the ratio of factorials.

21. $\dfrac{5!}{4!}$

22. $\dfrac{18!}{17!}$

23. $\dfrac{10!}{12!}$

24. $\dfrac{5!}{8!}$

25. $\dfrac{n!}{(n + 1)!}$

26. $\dfrac{(n + 2)!}{n!}$

In Exercises 27–40, find the specified sum.

27. $\sum_{k=1}^{6} 3k$

28. $\sum_{k=1}^{4} 5k$

29. $\sum_{i=0}^{4} (2i + 3)$

30. $\sum_{i=2}^{7} (4i - 1)$

31. $\sum_{j=1}^{5} \dfrac{(-1)^{j+1}}{j}$

32. $\sum_{j=0}^{3} \dfrac{1}{j^2 + 1}$

33. $\sum_{k=1}^{6} (-8)$

34. $\sum_{n=3}^{12} 10$

35. $\sum_{i=1}^{8} \left(\dfrac{1}{i} - \dfrac{1}{i + 1}\right)$

36. $\sum_{k=1}^{5} \left(\dfrac{2}{k} - \dfrac{2}{k + 2}\right)$

37. $\sum_{n=0}^{5} \left(-\dfrac{1}{3}\right)^n$

38. $\sum_{n=0}^{6} \left(\dfrac{3}{2}\right)^n$

39. $\sum_{k=1}^{6} \ln k$

40. $\sum_{k=2}^{4} \dfrac{k}{\ln k}$

In Exercises 41–54, write the given sum using sigma notation.

41. $1 + 2 + 3 + 4 + 5$

42. $8 + 9 + 10 + 11 + 12 + 13$

43. $2 + 4 + 6 + 8 + 10$

44. $24 + 30 + 36 + 42$

45. $\frac{1}{2(1)} + \frac{1}{2(2)} + \frac{1}{2(3)} + \frac{1}{2(4)} + \cdots + \frac{1}{2(10)}$

46. $\frac{3}{1+1} + \frac{3}{1+2} + \frac{3}{1+3} + \frac{3}{1+4} + \cdots + \frac{3}{1+50}$

47. $\frac{1}{1^2} + \frac{1}{2^2} + \frac{1}{3^2} + \frac{1}{4^2} + \cdots + \frac{1}{20^2}$

48. $\frac{1}{2^0} + \frac{1}{2^1} + \frac{1}{2^2} + \frac{1}{2^3} + \cdots + \frac{1}{2^{12}}$

49. $\frac{1}{3^0} - \frac{1}{3^1} + \frac{1}{3^2} - \frac{1}{3^3} + \cdots - \frac{1}{3^9}$

50. $\left(-\frac{2}{3}\right)^0 + \left(-\frac{2}{3}\right)^1 + \left(-\frac{2}{3}\right)^2 + \left(-\frac{2}{3}\right)^3 + \cdots + \left(-\frac{2}{3}\right)^{20}$

51. $\frac{1}{2} + \frac{2}{3} + \frac{3}{4} + \frac{4}{5} + \frac{5}{6} + \cdots + \frac{11}{12}$

52. $\frac{2}{4} + \frac{4}{7} + \frac{6}{10} + \frac{8}{13} + \frac{10}{16} + \cdots + \frac{20}{31}$

53. $1 + 1 + 2 + 6 + 24 + 120 + 720$

54. $1 + 1 + \frac{1}{2} + \frac{1}{6} + \frac{1}{24} + \frac{1}{120} + \frac{1}{720}$

Arithmetic Mean In Exercises 55–58, use the following definition of the *arithmetic mean* $\bar{x}$ of a set of n measurements $x_1, x_2, x_3, \ldots, x_n$ to find the mean (average) of the given measurements.

$$\bar{x} = \frac{x_1 + x_2 + x_3 + \cdots + x_n}{n}$$

55. 3, 7, 2, 1, 5

56. 84, 69, 66, 96

57. 0.5, 0.8, 1.1, 0.8, 0.7, 0.7, 1.0

58. −1.0, 4.2, 5.4, −3.2, 3.6

59. *Compound Interest* A deposit of \$500 is made in an account that earns 7% interest compounded monthly. The balance in the account after N months is given by

$$A_N = 500\left(1 + \frac{0.07}{12}\right)^N, \quad N = 1, 2, 3, \ldots .$$

(a) Compute the first eight terms of this sequence.
(b) Find the balance in this account after 20 years by computing A_{240}.

60. *Depreciation* At the end of each year the value of a car with an initial cost of \$16,000 is three-fourths what it was at the beginning of the year. Thus, after n years its value is given by

$$a_n = 16{,}000\left(\frac{3}{4}\right)^n, \quad n = 1, 2, 3, \ldots .$$

Find the value of the car five years after it was purchased by computing a_5.

SECTION 11.2 Arithmetic Sequences

Arithmetic Sequences • The Sum of an Arithmetic Sequence • Applications

Arithmetic Sequences

A sequence whose consecutive terms have a common difference is called an **arithmetic sequence**.

Definition of an Arithmetic Sequence

A sequence is called **arithmetic** if the differences between consecutive terms are the same. Thus, the sequence

$$a_1, a_2, a_3, a_4, \ldots, a_n, \ldots$$

is arithmetic if there is a number d such that

$$a_2 - a_1 = d, \qquad a_3 - a_2 = d, \qquad a_4 - a_3 = d$$

and so on. The number d is called the **common difference** of the arithmetic sequence.

EXAMPLE 1 ■ **Examples of Arithmetic Sequences**

(a) The sequence whose nth term is $3n + 2$ is arithmetic. For this sequence, the common difference between consecutive terms is 3.

$$\underbrace{5, 8}_{8 - 5 = 3}, 11, 14, \ldots, 3n + 2, \ldots$$

(b) The sequence whose nth term is $7 - 5n$ is arithmetic. For this sequence, the common difference between consecutive terms is -5.

$$\underbrace{2, -3}_{-3 - 2 = -5}, -8, -13, \ldots, 7 - 5n, \ldots$$

(c) The sequence whose nth term is $\frac{1}{4}(n + 3)$ is arithmetic. For this sequence, the common difference between consecutive terms is $\frac{1}{4}$.

$$\underbrace{1, \frac{5}{4}}_{\frac{5}{4} - 1 = \frac{1}{4}}, \frac{3}{2}, \frac{7}{4}, \ldots, \frac{n + 3}{4}, \ldots$$

■

In Example 1 notice that each of the arithmetic sequences has an nth term that is of the form $dn + c$, where the common difference of the sequence is d. We summarize this result as follows.

The nth Term of an Arithmetic Sequence

The nth term of an arithmetic sequence has the form

$$a_n = dn + c$$

where d is the common difference between the terms of the sequence, and c is a constant.

EXAMPLE 2 ■ Finding the nth Term of an Arithmetic Sequence

Find a formula for the nth term of the arithmetic sequence whose common difference is 2 and whose first term is 5.

Solution

We know that the formula for the nth term is of the form $a_n = dn + c$. Moreover, because the common difference is given to be $d = 2$, the formula must have the form

$$a_n = 2n + c.$$

Using the fact that the first term is

$$a_1 = 5 = 2(1) + c$$

it follows that $c = 3$. Thus, the formula for the nth term is

$$a_n = 2n + 3.$$

The sequence, therefore, has the following form.

$$5, 7, 9, 11, 13, \ldots, 2n + 3, \ldots$$ ■

EXAMPLE 3 ■ Finding the nth Term of an Arithmetic Sequence

Find a formula for the nth term of the arithmetic sequence whose common difference is 6 and whose *second* term is 11. What is the 19th term of this sequence?

Solution

We know that the formula for the nth term is of the form $a_n = dn + c$. Moreover, because the common difference is given to be $d = 6$, the formula must have the form

$$a_n = 6n + c.$$

Using the fact that the second term is

$$a_2 = 11 = 6(2) + c$$

it follows that $c = -1$. Thus, the formula for the nth term is

$$a_n = 6n - 1.$$

The sequence, therefore, has the following form.

$$5, 11, 17, 23, 29, \ldots, 6n - 1, \ldots$$

The 19th term of the sequence is

$$a_{19} = 6(19) - 1 = 113.$$

■

If we know the nth term of an arithmetic sequence *and* we know the common difference of the sequence, then we can find the $(n + 1)$th term by using the following **recursion formula**.

$$a_{n+1} = a_n + d$$

EXAMPLE 4 ■ **Using a Recursion Formula**

The 12th term of an arithmetic sequence is 52 and the common difference is 3. What is the 13th term of the sequence?

Solution

$$a_{13} = a_{12} + 3 = 52 + 3 = 55$$

■

The Sum of an Arithmetic Sequence

The sum of the first n terms of an arithmetic sequence is called the **nth partial sum** of the sequence. For instance, the fifth partial sum of the arithmetic sequence whose nth term is $3n + 4$ is

$$\sum_{i=1}^{5} (3i + 4) = 7 + 10 + 13 + 16 + 19 = 65.$$

There is a nice shortcut for finding the nth partial sum of an arithmetic sequence.

The nth Partial Sum of an Arithmetic Sequence

The nth partial sum of the arithmetic sequence whose nth term is a_n is given by

$$\sum_{i=1}^{n} a_i = a_1 + a_2 + a_3 + a_4 + \cdots + a_n = n\left(\frac{a_1 + a_n}{2}\right).$$

In other words, to find the sum of the first n terms of an arithmetic sequence, we find the average of the first term and nth term, and multiply by n.

EXAMPLE 5 ■ Finding the *n*th Partial Sum of an Arithmetic Sequence

Find the sum of the first 20 terms of the arithmetic sequence whose nth term is $4n + 1$.

Solution

The first term of this sequence is $a_1 = 4(1) + 1 = 5$ and the 20th term is $a_{20} = 4(20) + 1 = 81$. Therefore, the sum of the first 20 terms is given by

$$\sum_{i=1}^{n} a_i = n\left(\frac{a_1 + a_n}{2}\right)$$

$$\begin{aligned} \sum_{i=1}^{20} (4i + 1) &= 20\left(\frac{a_1 + a_{20}}{2}\right) \\ &= 20\left(\frac{5 + 81}{2}\right) \\ &= 20(43) \\ &= 860. \end{aligned}$$

■

EXAMPLE 6 ■ Finding the *n*th Partial Sum of an Arithmetic Sequence

Find the following sum.

$$7 + 10 + 13 + 16 + 19 + 22 + 25 + 28 + 31 + 34 + 37 + 40 + 43$$

Solution

One way to answer this question is to simply add all of the numbers. However, by recognizing that the numbers form an arithmetic sequence that has 13 terms, we can find the sum using the formula for the nth partial sum of an arithmetic sequence.

$$\text{Sum} = 13\left(\frac{7 + 43}{2}\right) = 13(25) = 325$$

Check this result on your calculator by actually adding the 13 terms. ■

EXAMPLE 7 ■ Finding the *n*th Partial Sum of an Arithmetic Sequence

Find the sum of the integers from 1 to 100.

Solution

Since the integers

$$1, 2, 3, 4, \ldots, 100$$

form an arithmetic sequence, we can find the sum as follows.

$$\begin{aligned} \sum_{i=1}^{100} i &= 100\left(\frac{1 + 100}{2}\right) \\ &= 100\left(\frac{101}{2}\right) \\ &= 5{,}050 \end{aligned}$$

■

Applications

The next two examples give you some idea of how arithmetic sequences can be used to solve problems in business and science.

EXAMPLE 8 ■ An Application: Total Sales

A small business sells \$100,000 worth of products during its first year. The owner of the business has a set goal of increasing annual sales by \$25,000 each year for nine years. Assuming that this goal is met, find the total sales during the first 10 years this business is in operation.

Solution

The annual sales during the first 10 years form the following arithmetic sequence.

\$100,000, \$125,000, \$150,000, \$175,000, \$200,000,
\$225,000, \$250,000, \$275,000, \$300,000, \$325,000

Using the formula for the *n*th partial sum of an arithmetic sequence, we find the total sales during the first 10 years as follows.

$$\text{Total sales} = 10\left(\frac{100{,}000 + 325{,}000}{2}\right) = 10(212{,}500) = \$2{,}125{,}000$$

From the bar graph shown in Figure 11.1, notice that the annual sales for this company follow a *linear growth* pattern. In other words, saying that a quantity increases arithmetically is the same as saying that it increases linearly.

FIGURE 11.1

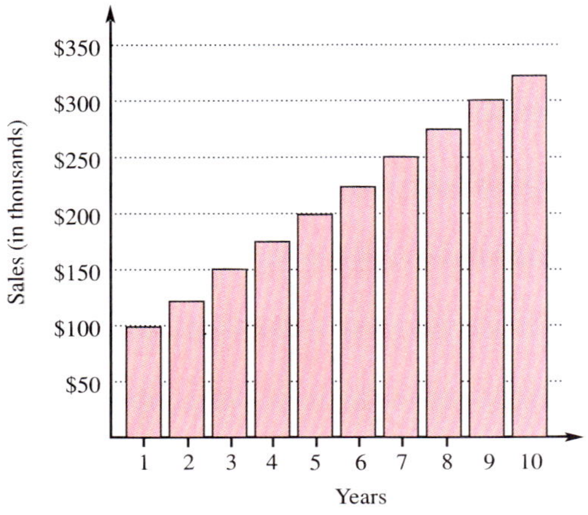

■

EXAMPLE 9 ■ An Application: Seating Capacity

An amphitheater has the seating arrangement shown in Figure 11.2. The theater has 40 rows, with 100 seats in the first row, 102 seats in the second row, 104 seats in the third row, and so on. What is the total seating capacity of the theater?

FIGURE 11.2

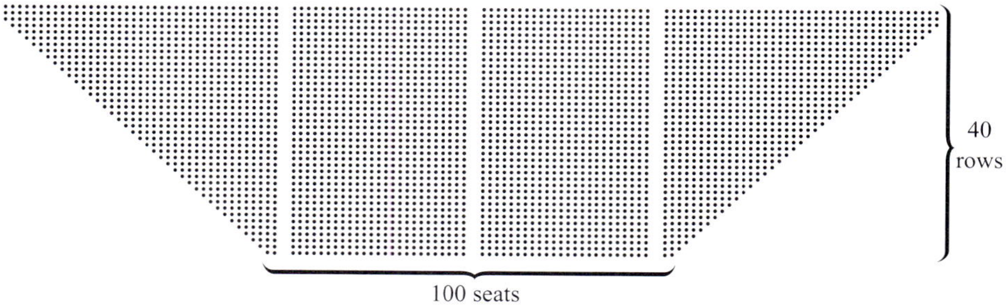

Solution

The total number of seats is given by the first 40 terms of the arithmetic sequence

100, 102, 104, 106, 108, ...

To find the number of seats in the 40th row, we must first find the 40th term of the sequence. Since the formula for the nth term of the sequence is $dn + c = 2n + c$ and the first term is $2(1) + c = 100$, it follows that $c = 98$. Thus, the 40th term of the sequence is

$$a_{40} = 2(40) + 98 = 178.$$

Now, using the formula for the nth partial sum of an arithmetic sequence, we find that the total number of seats is

$$\sum_{i=1}^{40} (2i + 98) = 40\left(\frac{100 + 178}{2}\right) = 40(139) = 5{,}560.$$ ■

DISCUSSION PROBLEM ■ An Experiment Involving an Arithmetic Sequence

The following sequence of perfect squares is *not* arithmetic.

1, 4, 9, 16, 25, 36, 49, 64, 81, . . .

However, we can form a related sequence that is arithmetic by finding the differences of consecutive terms as follows.

$$\begin{array}{ccccccccccccccccc} & & & 5 & & & & 9 & & & & 13 & & & & 17 & \\ 1 & & 4 & & 9 & & 16 & & 25 & & 36 & & 49 & & 64 & & 81 \\ & 3 & & & & 7 & & & & 11 & & & & 15 & & & \end{array}$$

Thus, the related arithmetic sequence is 3, 5, 7, 9, 11, 13, 15, 17.

Can you think of a way to find an arithmetic sequence that is related to the following sequence of perfect cubes?

1, 8, 27, 64, 125, 216, 343, 512, 729, . . . ■

Warm-Up

The following warm-up exercises involve skills that were covered in earlier sections. You will use these skills in the exercise set for this section.

In Exercises 1–4, find the sum.

1. $\sum_{i=1}^{6} \frac{1}{i}$

2. $\sum_{i=0}^{6} \frac{5}{i+1} - \frac{5}{i+3}$

3. $\sum_{n=1}^{4} 3^n$

4. $\sum_{n=1}^{5} (3n - 1)$

In Exercises 5–10, evaluate the given expression.

5. $15\left(\frac{5 + 33}{2}\right)$

6. $35\left(\frac{5 + 107}{2}\right)$

7. $11\left(\frac{(16/3) + (26/3)}{2}\right)$

8. $20\left(\frac{(7/4) + 16}{2}\right)$

9. $\frac{7}{2}[2(-3) + 6(4)]$

10. $\frac{9}{2}[2(4) + 8(3)]$

11.2 EXERCISES

In Exercises 1–8, find the common difference of the given arithmetic sequence.

1. $2, 5, 8, 11, \ldots$

2. $7, 9, 11, 13, \ldots$

3. $-1, 4, 9, 14, \ldots$

4. $-8, 0, 8, 16, \ldots$

5. $100, 94, 88, 82, \ldots$

6. $3{,}200, 2{,}800, 2{,}400, 2{,}000, \ldots$

7. $1, \frac{5}{3}, \frac{7}{3}, 3, \ldots$

8. $\frac{1}{2}, \frac{5}{4}, 2, \frac{11}{4}, \ldots$

In Exercises 9–16, determine whether the sequence is arithmetic. If it is, find the common difference.

9. $2, 4, 6, 8, \ldots$

10. $2, 6, 10, 14, \ldots$

11. $2, \frac{7}{2}, 5, \frac{13}{2}, \ldots$

12. $e, e^2, e^3, e^4, \ldots$

13. $32, 16, 0, -16, \ldots$

14. $32, 16, 8, 4, \ldots$

15. $\ln 4, \ln 8, \ln 12, \ln 16, \ldots$

16. $3.2, 4, 4.8, 5.6, \ldots$

In Exercises 17–24, write the first five terms of the arithmetic sequence. (Assume n begins with 1.)

17. $a_n = 3n + 4$

18. $a_n = 5n - 4$

19. $a_n = -2n + 8$

20. $a_n = -10n + 100$

21. $a_n = \frac{5}{2}n - 1$

22. $a_n = \frac{2}{3}n + 2$

23. $a_n = -\frac{1}{4}(n - 1) + 4$

24. $a_n = 4(n + 2) + 24$

In Exercises 25–32, find a formula for the nth term of the arithmetic sequence. (Assume n begins with 1.)

25. $a_1 = 3,\ d = \frac{1}{2}$

26. $a_1 = -1,\ d = 1.2$

27. $a_1 = 1{,}000,\ d = -25$

28. $a_1 = 64,\ d = -8$

29. $a_3 = 16,\ a_4 = 20$

30. $a_5 = 30,\ a_4 = 25$

31. $a_1 = 50,\ a_3 = 30$

32. $a_{10} = 32,\ a_{12} = 48$

In Exercises 33–40, write the first five terms of the arithmetic sequence.

33. $a_1 = 500,\ d = -50$

34. $a_1 = 10,\ d = 15$

35. $a_1 = 25,\ a_{k+1} = a_k + 3$

36. $a_1 = 250,\ a_{k+1} = a_k - 75$

37. $a_3 = 12,\ a_5 = 20$

38. $a_5 = 45,\ a_8 = 60$

39. $a_2 = 15,\ a_5 = 13.5$

40. $a_6 = 90,\ a_{10} = 80$

In Exercises 41–48, find the indicated sum.

41. $\sum_{n=1}^{20} n$

42. $\sum_{n=1}^{30} 4n$

43. $\sum_{n=1}^{50} (2n + 3)$

44. $\sum_{n=1}^{100} (4n - 1)$

45. $\sum_{n=1}^{40} (1{,}000 - 25n)$

46. $\sum_{n=1}^{20} (500 - 10n)$

47. $\sum_{n=1}^{500} \frac{n}{2}$

48. $\sum_{n=1}^{600} \frac{2n}{3}$

In Exercises 49–56, find the nth partial sum of the arithmetic sequence.

49. 5, 12, 19, 26, 33, . . . , $n = 12$

50. 2, 12, 22, 32, 42, . . . , $n = 20$

51. 200, 175, 150, 125, 100, . . . , $n = 8$

52. 800, 785, 770, 755, 740, . . . , $n = 25$

53. $-50, -38, -26, -14, -2$, . . . , $n = 50$

54. $-16, -8, 0, 8, 16$, . . . , $n = 30$

55. 1, 4.5, 8, 11.5, 15, . . . , $n = 12$

56. 2.2, 2.8, 3.4, 4.0, 4.6, . . . , $n = 12$

57. Find the sum of the first 75 positive integers.

58. Find the sum of the integers from 35 to 100 inclusive.

59. Find the sum of the first 50 even positive integers.

60. Find the sum of the first 100 positive odd integers.

61. *Salary Increases* In your new job you are told that your starting salary will be \$36,000 with an increase of \$2,000 at the end of each of the first five years. How much will you be paid through the end of your first six years of employment with the company?

62. *Would You Accept this Job?* Suppose you receive 25¢ on the first day of the month, 50¢ the second day, 75¢ the third day, and so on. Determine the total amount of money you will receive during a 30-day month.

63. *Auditorium Seating* Each row in a small auditorium has two more seats than the preceding row (see figure). Find the seating capacity of the auditorium if the front row seats 25 people and there are 15 rows of seats.

64. *Free-Falling Object* A free-falling object will fall 16 feet during the first second, 48 more feet during the second, 80 more during the third, and so on. What is the total distance the object will fall in eight seconds if this pattern continues?

65. *Baling Hay* In the first two trips baling hay around a large field, a farmer obtains 93 bales and 89 bales, respectively (see figure). Since each round gets shorter, the farmer estimates that the same pattern will continue. Estimate the total number of bales made if there are another six trips around the field.

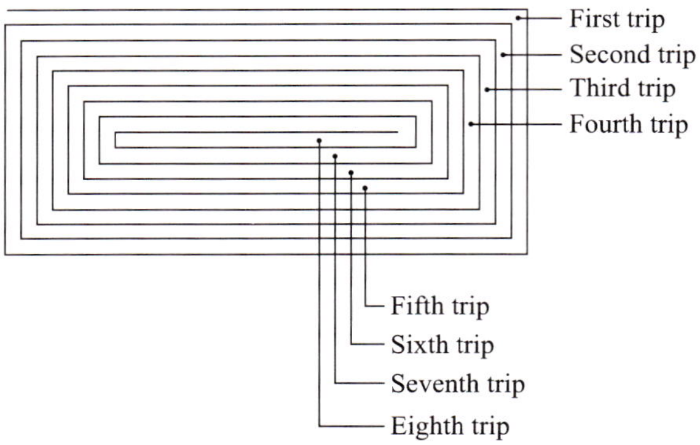

SECTION 11.3

Geometric Sequences

Geometric Sequences • Sum of a Geometric Sequence • Applications

Geometric Sequences

In Section 11.2 we saw that a sequence whose consecutive terms have a common *difference* is called an arithmetic sequence. In this section we will study another important type of sequence, called a **geometric sequence**. Consecutive terms of a geometric sequence have a common *ratio*, as indicated in the following definition.

Definition of a Geometric Sequence

A sequence is called **geometric** if the ratios of consecutive terms are the same. Thus, the sequence

$$a_1, a_2, a_3, a_4, \ldots, a_n, \ldots$$

is geometric if there is a number r, $r \neq 0$, such that

$$\frac{a_2}{a_1} = r, \quad \frac{a_3}{a_2} = r, \quad \frac{a_4}{a_3} = r$$

and so on. The number r is called the **common ratio** of the geometric sequence.

EXAMPLE 1 ■ Examples of Geometric Sequences

(a) The sequence whose nth term is 2^n is geometric. For this sequence, the common ratio between consecutive terms is 2.

$$\underbrace{2,\ 4,}_{\frac{4}{2}=2}\ 8,\ 16,\ \ldots,\ 2^n,\ \ldots$$

(b) The sequence whose nth term is $4(3^n)$ is geometric. For this sequence, the common ratio between consecutive terms is 3.

$$\underbrace{12,\ 36,}_{\frac{36}{12}=3}\ 108,\ 324,\ \ldots,\ 4(3^n),\ \ldots$$

(c) The sequence whose nth term is $\left(-\frac{1}{3}\right)^n$ is geometric. For this sequence, the common ratio between consecutive terms is $-\frac{1}{3}$.

$$\underbrace{-\frac{1}{3},\ \frac{1}{9},}_{\frac{1/9}{-1/3}=-\frac{1}{3}}\ -\frac{1}{27},\ \frac{1}{81},\ \ldots,\ \left(-\frac{1}{3}\right)^n,\ \ldots$$

■

In Example 1 notice that each of the geometric sequences has an nth term that is of the form a_1r^{n-1}, where the common ratio of the sequence is r. We summarize this result as follows.

The nth Term of a Geometric Sequence

The nth term of a geometric sequence has the form

$$a_n = a_1r^{n-1}$$

where r is the common ratio of consecutive terms of the sequence. Thus, every geometric sequence can be written in the following form.

$$\begin{array}{ccccccccc} a_1, & a_2, & a_3, & a_4, & a_5, & \ldots, & a_n, & \ldots \\ \downarrow & \downarrow & \downarrow & \downarrow & \downarrow & \ldots, & \downarrow & \ldots \\ a_1, & a_1r, & a_1r^2, & a_1r^3, & a_1r^4, & \ldots, & a_1r^{n-1}, & \ldots \end{array}$$

NOTE: If we know the nth term of a geometric sequence, then the $(n + 1)$th term can be found by multiplying by r. That is,

$$a_{n+1} = ra_n.$$

EXAMPLE 2 ■ Finding the nth Term of a Geometric Sequence

Find a formula for the nth term of the geometric sequence whose common ratio is 3 and whose first term is 1. What is the eighth term of this sequence?

Solution

We know that the formula for the nth term is of the form $a_n = a_1r^{n-1}$. Moreover, because the common ratio is given to be $r = 3$, the formula must have the form

$$a_n = a_1r^{n-1} = (1)(3^{n-1}) = 3^{n-1}.$$

The sequence, therefore, has the following form.

$$1, 3, 9, 27, 81, \ldots, 3^{n-1}, \ldots$$

The eighth term of the sequence is

$$a_8 = 3^{8-1} = 3^7 = 2{,}187.$$ ■

EXAMPLE 3 ■ **Finding the *n*th Term of a Geometric Sequence**

Find a formula for the nth term of the geometric sequence whose first two terms are 4 and 2. What is the 10th term of this sequence?

Solution

We know that the formula for the nth term is of the form $a_n = a_1r^{n-1}$. Moreover, because the common ratio is

$$\frac{a_2}{a_1} = \frac{2}{4} = \frac{1}{2}$$

the formula for the nth term must have the form

$$a_n = a_1r^{n-1} = 4\left(\frac{1}{2}\right)^{n-1}$$

The sequence, therefore, has the following form.

$$4, 2, 1, \frac{1}{2}, \frac{1}{4}, \ldots, 4\left(\frac{1}{2}\right)^{n-1}, \ldots$$

The 10th term of the sequence is

$$a_{10} = 4\left(\frac{1}{2}\right)^{10-1} = 2^2\left(\frac{1}{2^9}\right) = \frac{1}{2^7} = \frac{1}{128}.$$ ■

NOTE: Another way of solving Example 3 is simply to multiply 4 by $r = \frac{1}{2}$ several times to obtain the first 10 terms in the sequence.

$$\underset{\overset{\downarrow}{1}}{4},\ \underset{\overset{\downarrow}{2}}{2},\ \underset{\overset{\downarrow}{3}}{1},\ \underset{\overset{\downarrow}{4}}{\frac{1}{2}},\ \underset{\overset{\downarrow}{5}}{\frac{1}{4}},\ \underset{\overset{\downarrow}{6}}{\frac{1}{8}},\ \underset{\overset{\downarrow}{7}}{\frac{1}{16}},\ \underset{\overset{\downarrow}{8}}{\frac{1}{32}},\ \underset{\overset{\downarrow}{9}}{\frac{1}{64}},\ \underset{\overset{\downarrow}{10}}{\frac{1}{128}}$$

Sum of a Geometric Sequence

In Section 11.2 we found that there is a simple formula for finding the sum of the first n terms of an arithmetic sequence. There is also a formula for finding the sum of the first n terms of a geometric sequence.

The nth Partial Sum of a Geometric Sequence

The nth partial sum of the geometric sequence whose nth term is $a_n = a_1r^{n-1}$ is given by

$$\sum_{i=1}^{n} a_1r^{i-1} = a_1 + a_1r + a_1r^2 + a_1r^3 + \cdots + a_1r^{n-1} = a_1\left(\frac{r^n - 1}{r - 1}\right).$$

EXAMPLE 4 ■ Finding the nth Partial Sum of a Geometric Sequence

Find the following sum.

$$1 + 2 + 4 + 8 + 16 + 32 + 64 + 128$$

Solution

Because the terms of this sum form a geometric sequence whose common ratio is $r = 2$, we can use the formula for the nth partial sum of a geometric sequence. Moreover, since the first term of the sequence is $a_1 = 1$, it follows that the sum is given by

$$\begin{aligned}
\sum_{i=1}^{n} a_1r^{i-1} &= a_1\left(\frac{r^n - 1}{r - 1}\right) \\
\sum_{i=1}^{8} 2^{i-1} &= (1)\left(\frac{2^8 - 1}{2 - 1}\right) \\
&= \frac{256 - 1}{2 - 1} \\
&= 255.
\end{aligned}$$

■

EXAMPLE 5 ■ Finding the nth Partial Sum of a Geometric Sequence

Find the sum of the first five terms of the geometric sequence whose nth term is $a_n = \left(\frac{2}{3}\right)^n$.

Solution

For this geometric sequence, the first term is $a_1 = \frac{2}{3}$ and the common ratio is $r = \frac{2}{3}$. Therefore, the sum of the first five terms is as follows.

$$\sum_{i=1}^{n} a_1 r^{i-1} = a_1\left(\frac{r^n - 1}{r - 1}\right)$$

$$\sum_{i=1}^{5} \left(\frac{2}{3}\right)^i = \frac{2}{3}\left(\frac{(2/3)^5 - 1}{(2/3) - 1}\right)$$

$$= \frac{2}{3}\left(\frac{(32/243) - 1}{-1/3}\right)$$

$$= \frac{2}{3}\left(-\frac{211}{243}\right)(-3)$$

$$= \frac{422}{243}$$

$$\approx 1.737$$

■

EXAMPLE 6 ■ Finding the nth Partial Sum of a Geometric Sequence

Find the following sum.

$$\sum_{i=1}^{7} \frac{3^{i-1}}{2}$$

Solution

By writing out a few terms of the sum

$$\sum_{i=1}^{7} \frac{3^{i-1}}{2} = \sum_{i=1}^{7} \frac{1}{2}(3^{i-1}) = \frac{1}{2} + \frac{1}{2}(3) + \frac{1}{2}(3^2) + \cdots + \frac{1}{2}(3^6)$$

we see that $a_1 = \frac{1}{2}$ and $r = 3$. Thus, the sum is as follows.

$$\sum_{i=1}^{n} a_1 r^{i-1} = a_1\left(\frac{r^n - 1}{r - 1}\right)$$

$$\sum_{i=1}^{7} \frac{1}{2}(3^{i-1}) = \frac{1}{2}\left(\frac{3^7 - 1}{3 - 1}\right)$$

$$= \frac{1}{2}\left(\frac{2{,}187 - 1}{2}\right)$$

$$= \frac{2{,}186}{4}$$

$$= 546.5$$

■

Applications

We close this section by looking at two applications of geometric sequences.

EXAMPLE 7 ■ An Application: A Lifetime Salary

Suppose you accept a job that pays a salary of \$28,000 the first year. During the next 39 years, suppose you receive a 6% raise each year. What would your total salary be over the 40-year period?

Solution

Using a geometric sequence, your salary during the first year would be

$$a_1 = 28{,}000.$$

Then, with a 6% raise, your salary during the second and third years would be as follows.

$$a_2 = 28{,}000 + 28{,}000(0.6) = 28{,}000(1.06)^1$$

$$a_3 = 28{,}000(1.06) + 28{,}000(1.06)(0.06) = 28{,}000(1.06)^2$$

From this pattern, we can see that the common ratio of the geometric sequence is $r = 1.06$. Using the formula for the nth partial sum of a geometric sequence, we find that the total salary over the 40-year period is as follows.

$$\begin{aligned}\text{Total salary} &= a_1\left(\frac{r^n - 1}{r - 1}\right)\\ &= 28{,}000\left(\frac{(1.06)^{40} - 1}{1.06 - 1}\right)\\ &= 28{,}000\left(\frac{(1.06)^{40} - 1}{0.06}\right)\\ &\approx \$4{,}333{,}335\end{aligned}$$

■

EXAMPLE 8 ■ An Application: Increasing Annuity

Suppose you deposit \$100 in an account each month for two years. The account pays an annual percentage rate of 9%, compounded monthly. What is your balance at the end of two years? (This type of savings plan is called an **increasing annuity**.)

Solution

Using the formula for compound interest, we see that the first deposit would earn interest for the full 24 months. Thus, by the end of 24 months, the first deposit will have grown to a balance of

$$a_{24} = 100\left(1 + \frac{0.09}{12}\right)^{24} = 100(1.0075)^{24}.$$

Since the second deposit will have earned interest for only 23 months, its balance will be

$$a_{23} = 100\left(1 + \frac{0.09}{12}\right)^{23} = 100(1.0075)^{23}.$$

By continuing this procedure, we see that the total of the 24 deposits will be

$$\begin{aligned}\text{Total} &= a_1 + a_2 + \cdots + a_{24}\\ &= 100(1.0075)^1 + 100(1.0075)^2 + \cdots + 100(1.0075)^{24}\\ &= a_1\left(\frac{r^n - 1}{r - 1}\right)\\ &= 100(1.0075)\left(\frac{1.0075^{24} - 1}{1.0075 - 1}\right)\\ &= \$2{,}638.49.\end{aligned}$$

DISCUSSION PROBLEM An Infinite Summation

So far in this chapter we have been finding sums involving only a finite number of terms. There are some instances in mathematics in which we need to be able to add an infinite number of terms. For instance, the infinite geometric sequence

$$\frac{1}{2}, \frac{1}{4}, \frac{1}{8}, \frac{1}{16}, \frac{1}{32}, \ldots$$

can be added in *its entirety* to form the following **infinite series**.

$$\sum_{i=1}^{\infty} \frac{1}{2^i} = \frac{1}{2} + \frac{1}{4} + \frac{1}{8} + \frac{1}{16} + \frac{1}{32} + \cdots = 1$$

The following formula works for any infinite geometric sequence, provided that $-1 < r < 1$.

$$\sum_{i=1}^{\infty} a_1 r^{i-1} = a_1 + a_1 r + a_1 r^2 + a_1 r^3 + \cdots = \frac{a_1}{1 - r}$$

Use this formula to find the sum of the following infinite geometric series.

(a) $\displaystyle\sum_{i=1}^{\infty} \frac{1}{3^i}$ (b) $\displaystyle\sum_{i=1}^{\infty} \frac{3}{2^i}$

Warm-Up

The following warm-up exercises involve skills that were covered in earlier sections. You will use these skills in the exercise set for this section.

In Exercises 1–4, evaluate the quantity.

1. $\left(\frac{2}{3}\right)^4$ **2.** $\left(\frac{3}{5}\right)^3$ **3.** $\frac{2^{-3}}{3^2}$ **4.** $\frac{8^2}{4^3}$

In Exercises 5–10, simplify the expression.

5. $\frac{n!}{(n-1)!}$ **6.** $\frac{(n+1)!}{n!}$ **7.** $(3n)(4n^2)(5n^3)$

8. $n(2n)^4$ **9.** $\frac{(3n)^2}{18n}$ **10.** $\frac{18n}{10n^{-1}}$

11.3 EXERCISES

In Exercises 1–8, find the common ratio for the given geometric sequence.

1. $2, 6, 18, 54, \ldots$

2. $5, -10, 20, -40, \ldots$

3. $12, -6, 3, -\frac{3}{2}, \ldots$

4. $54, 18, 6, 2, \ldots$

5. $1, -\frac{3}{2}, \frac{9}{4}, -\frac{27}{8}, \ldots$

6. $9, 6, 4, \frac{8}{3}, \ldots$

7. $e, e^2, e^3, e^4, \ldots$

8. $1.1, (1.1)^2, (1.1)^3, (1.1)^4, \ldots$

In Exercises 9–16, determine whether the sequence is geometric. If it is, find the common ratio.

9. $10, 15, 20, 25, \ldots$

10. $10, 20, 40, 80, \ldots$

11. $64, 32, 16, 8, \ldots$

12. $64, 32, 0, -32, \ldots$

13. $1, -\frac{2}{3}, \frac{4}{9}, -\frac{8}{27}, \ldots$

14. $\frac{1}{3}, -\frac{2}{3}, \frac{4}{3}, -\frac{8}{3}, \ldots$

15. $10(1+0.02), 10(1+0.02)^2, 10(1+0.02)^3, 10(1+0.02)^4, \ldots$

16. $1, 0.2, 0.04, 0.008, \ldots$

In Exercises 17–24, write the first five terms of the geometric sequence.

17. $a_1 = 4,\ r = 2$

18. $a_1 = 3,\ r = 4$

19. $a_1 = 6,\ r = \frac{1}{3}$

20. $a_1 = 4,\ r = \frac{1}{2}$

21. $a_1 = 1,\ r = -\frac{1}{2}$

22. $a_1 = 32,\ r = -\frac{3}{4}$

23. $a_1 = 1{,}000,\ r = 1.01$

24. $a_1 = 4{,}000,\ r = \frac{1}{1.01}$

In Exercises 25–32, find the formula for the nth term of the geometric sequence. (Assume n begins with 1.)

25. $a_1 = 2,\ r = 3$

26. $a_1 = 5,\ r = 4$

27. $a_1 = 1,\ r = 2$

28. $a_1 = 1,\ r = -5$

29. $a_1 = 4,\ r = -\frac{1}{2}$

30. $a_1 = 9,\ r = \frac{2}{3}$

31. $a_1 = 8,\ a_2 = 2$

32. $a_1 = 18,\ a_2 = 8$

In Exercises 33–40, find the nth term of the geometric sequence.

33. $a_1 = 6,\ r = \frac{1}{2},\ n = 10$

34. $a_1 = 8,\ r = \frac{3}{4},\ n = 8$

35. $a_1 = 3,\ r = \sqrt{2},\ n = 10$

36. $a_1 = 500,\ r = 1.06,\ n = 40$

37. $a_1 = 4,\ a_2 = 3,\ n = 5$

38. $a_1 = 1,\ a_2 = 9,\ n = 7$

39. $a_1 = 1,\ a_3 = \frac{9}{4},\ n = 6$

40. $a_2 = 12,\ a_3 = 16,\ n = 5$

In Exercises 41–48, find the indicated sum.

41. $\sum_{i=1}^{10} 2^{i-1}$

42. $\sum_{i=1}^{6} 3^{i-1}$

43. $\sum_{i=1}^{12} 3\left(\frac{3}{2}\right)^{i-1}$

44. $\sum_{i=1}^{20} 12\left(\frac{2}{3}\right)^{i-1}$

45. $\sum_{i=1}^{15} 3\left(-\frac{1}{3}\right)^{i-1}$

46. $\sum_{i=1}^{8} 8\left(-\frac{1}{4}\right)^{i-1}$

47. $\sum_{i=1}^{20} 100(1.1)^{i-1}$

48. $\sum_{i=1}^{40} 50(1.07)^{i-1}$

In Exercises 49–56, find the nth partial sum of the geometric sequence.

49. $1, -3, 9, -27, 81, \ldots,\quad n = 10$

50. $3, -6, 12, -24, 48, \ldots,\quad n = 12$

51. $8, 4, 2, 1, \frac{1}{2}, \ldots,\quad n = 15$

52. $9, 6, 4, \frac{8}{3}, \frac{16}{9}, \ldots,\quad n = 10$

53. $1, \sqrt{2}, 2, 2\sqrt{2}, 4, \ldots,\quad n = 12$

54. $40, -10, \frac{5}{2}, -\frac{5}{8}, \frac{5}{32}, \ldots,\quad n = 10$

55. $30, 30(1.06), 30(1.06)^2, 30(1.06)^3, 30(1.06)^4, \ldots, n = 20$

56. $100, 100(1.08), 100(1.08)^2, 100(1.08)^3, 100(1.08)^4, \ldots,\quad n = 40$

57. *Depreciation* A company pays \$250,000 for a machine. During the next five years it will depreciate at the rate of 25% per year. (That is, at the end of each year the depreciated value is 75% of what it was at the beginning of the year.)
(a) Find a formula for the nth term of the geometric sequence that gives the value of the machine t full years after it was purchased.
(b) Find the depreciated value of the machine at the end of five full years.

58. *Population Increase* A city of 500,000 people is growing at the rate of 1% per year. (That is, at the end of each year the population is 1.01 times the population at the beginning of the year.)
(a) Find a formula for the nth term of the geometric sequence that gives the population t years from now.
(b) Estimate the population 20 years from now.

59. *Salary Increases* Suppose you accept a job that pays a salary of \$30,000 the first year. During the next 39 years, suppose you receive a 5% raise each year. What would your *total* salary be over the 40-year period?

60. *Salary Increases* Suppose you accept a job that pays a salary of \$30,000 the first year. During the next 39 years, suppose you receive a 5.5% raise each year. What would your *total* salary be over the 40-year period?

Increasing Annuity In Exercises 61–64, find the balance in an increasing annuity (see Example 8) when a principal of P dollars is invested at the beginning of each month for t years. Assume that the amount in the fund is compounded monthly at rate r.

61. $P = \$100$ $t = 10$ years $r = 9\%$

62. $P = \$50$ $t = 5$ years $r = 7\%$

63. $P = \$30$ $t = 40$ years $r = 8\%$

64. $P = \$200$ $t = 30$ years $r = 10\%$

65. *Would You Accept this Job?* Suppose you start work at a company that pays \$0.01 for the first day, \$0.02 for the second day, \$0.04 for the third day, and so on. If the daily wage keeps doubling, what would your total income be for working (a) 29 days and (b) 30 days?

66. *Would You Accept this Job?* Suppose you start work at a company that pays \$0.01 for the first day, \$0.03 for the second day, \$0.09 for the third day, and so on. If the daily wage keeps tripling, what would your total income be for working (a) 25 days and (b) 26 days?

67. *Area* The sides of a square are 12 inches in length. A new square is formed by connecting the midpoints of the sides of the original square, and two of the triangles are shaded (see figure). If this process is repeated five more times, what will the total area of the shaded region be?

68. *Area* The sides of a square are 12 inches in length. The square is divided into nine smaller squares and the center square is shaded (see figure). Each of the eight unshaded squares is then divided into nine smaller squares and the center square of each is shaded. If this process is repeated four more times, what will the total area of the shaded region be?

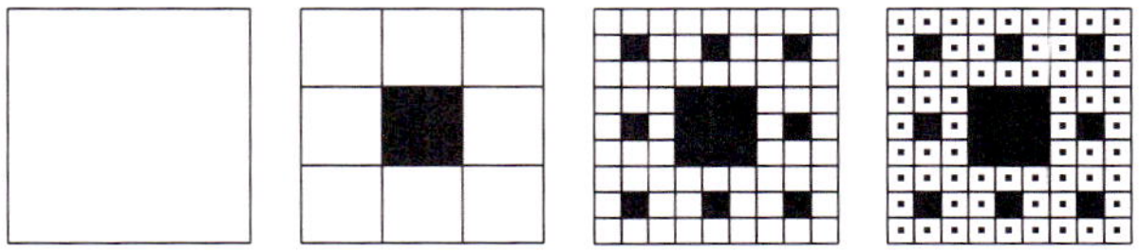

69. *Number of Ancestors* The number of direct ancestors a person has is given by

$$2 + 2^2 + 2^3 + 2^4 + \cdots + 2^n + \cdots.$$

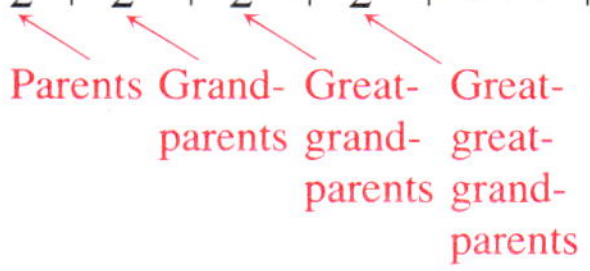

This formula is valid provided the person has no common ancestors. [A common ancestor is one to whom you are related in more than one way. For example, one of your great-grandmothers on your father's side might also be one of your great-grandmothers on your mother's side (see the following figure).] How many direct ancestors have you had who have lived since the year A.D. 1? Assume that the average time between generations was 30 years (resulting in 66 generations), so that the total is given by

$$2 + 2^2 + 2^3 + 2^4 + \cdots + 2^{66}.$$

Considering the total, is it reasonable to assume that you have had no common ancestors in the past 2,000 years?

Figure for 69

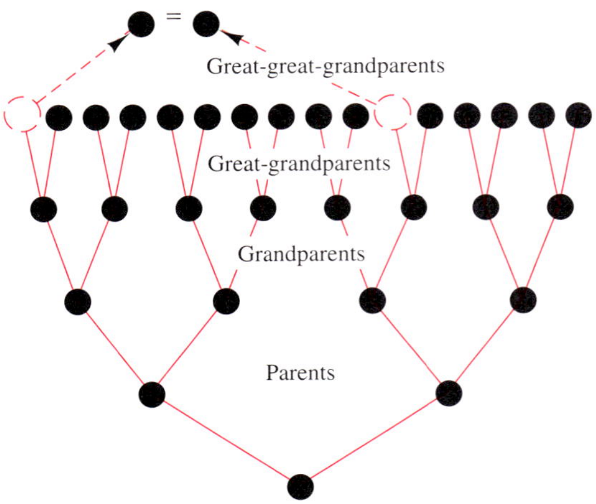

SECTION 11.4

The Binomial Theorem

Binomial Coefficients • Pascal's Triangle • Binomial Expansions

Binomial Coefficients

Recall that a **binomial** is a polynomial that has two terms. In this section we will look at a formula that gives a quick method of raising a binomial to a power. To begin, let's look at the expansion of $(x + y)^n$ for several values of n.

$$(x + y)^0 = 1$$
$$(x + y)^1 = x + y$$
$$(x + y)^2 = x^2 + 2xy + y^2$$
$$(x + y)^3 = x^3 + 3x^2y + 3xy^2 + y^3$$
$$(x + y)^4 = x^4 + 4x^3y + 6x^2y^2 + 4xy^3 + y^4$$
$$(x + y)^5 = x^5 + 5x^4y + 10x^3y^2 + 10x^2y^3 + 5xy^4 + y^5$$

There are several observations we can make about these expansions of $(x + y)^n$.

1. In each expansion, there are $n + 1$ terms.
2. In each expansion, x and y have symmetrical roles. The powers of x decrease by 1 in successive terms, whereas the powers of y increase by 1.
3. The sum of the powers of each term in a binomial expansion is n. For example, in the expansion of $(x + y)^5$, the sum of the powers of each term is 5, as follows.

$$(x + y)^5 = x^5 + 5\overbrace{x^4y^1}^{4+1=5} + 10\overbrace{x^3y^2}^{3+2=5} + 10x^2y^3 + 5xy^4 + y^5$$

4. The first term is x^n, the last term is y^n, and each of these terms has a coefficient of 1.
5. The coefficients increase and then decrease in a symmetric pattern. For $(x + y)^5$, the pattern is

$$1 \quad 5 \quad 10 \quad 10 \quad 5 \quad 1.$$

The most difficult part of a binomial expansion is finding the coefficients of the interior terms. To find these **binomial coefficients**, we use a well-known theorem called the **Binomial Theorem**.

The Binomial Theorem

In the expansion of $(x + y)^n$

$$(x + y)^n = x^n + nx^{n-1}y + \cdots + {}_nC_m x^{n-m}y^m + \cdots + nxy^{n-1} + y^n$$

the coefficient of $x^{n-m}y^m$ is given by

$${}_nC_m = \frac{n!}{(n - m)!m!}.$$

EXAMPLE 1 ■ Finding Binomial Coefficients

Find the following binomial coefficients.

(a) ${}_8C_2$ (b) ${}_{10}C_3$ (c) ${}_7C_0$

Solution

Note in parts (a) and (b) how we can factor the numerator factorial in order to cancel one of the denominator factorials.

(a) $${}_8C_2 = \frac{8!}{6!2!} = \frac{(8 \cdot 7) \cdot \cancel{6!}}{\cancel{6!} \cdot 2!} = \frac{8 \cdot 7}{2 \cdot 1} = 28$$

(b) $${}_{10}C_3 = \frac{10!}{7!3!} = \frac{(10 \cdot 9 \cdot 8) \cdot \cancel{7!}}{\cancel{7!} \cdot 3!} = \frac{10 \cdot 9 \cdot 8}{3 \cdot 2 \cdot 1} = 120$$

(c) $${}_7C_0 = \frac{7!}{7!0!} = 1$$

■

NOTE: When $m \neq 0$ or $m \neq n$, as in parts (a) and (b) above, there is a simple pattern for evaluating binomial coefficients.

$$_8C_2 = \frac{\overbrace{8 \cdot 7}^{\text{2 factors}}}{\underbrace{2 \cdot 1}_{\text{2 factorial}}} \quad \text{and} \quad _{10}C_3 = \frac{\overbrace{10 \cdot 9 \cdot 8}^{\text{3 factors}}}{\underbrace{3 \cdot 2 \cdot 1}_{\text{3 factorial}}}$$

In general, we have the following.

$$_nC_m = \frac{\overbrace{n(n-1)(n-2)\ldots}^{m \text{ factors}}}{m!}, \quad a < m < n$$

EXAMPLE 2 ■ **Finding Binomial Coefficients**

Find the following binomial coefficients.

(a) $_7C_3$ (b) $_7C_4$ (c) $_{12}C_1$ (d) $_{12}C_{11}$

Solution

(a) $_7C_3 = \dfrac{7 \cdot 6 \cdot 5}{3 \cdot 2 \cdot 1} = 35$

(b) $_7C_4 = \dfrac{7 \cdot 6 \cdot 5 \cdot 4}{4 \cdot 3 \cdot 2 \cdot 1} = 35$

(c) $_{12}C_1 = \dfrac{12}{1} = 12$

(d) $_{12}C_{11} = \dfrac{12!}{1!11!} = \dfrac{(12) \cdot \cancel{11!}}{1! \cdot \cancel{11!}} = \dfrac{12}{1} = 12$ ■

It is not a coincidence that the results in parts (a) and (b), and similarly in parts (c) and (d), are the same. In general, it is true that $_nC_m = {_nC_{n-m}}$. For instance,

$$_6C_0 = {_6C_6} = 1, \quad _6C_1 = {_6C_5} = 6, \quad \text{and} \quad _6C_2 = {_6C_4} = 15.$$

This shows the symmetric property of binomial coefficients that was identified earlier. Notice how this property is used in Example 3.

EXAMPLE 3 ■ **Finding a Binomial Coefficient**

Find the binomial coefficient $_{12}C_{10}$.

Solution

Rather than calculating the binomial coefficient

$$_{12}C_{10} = \frac{12 \cdot 11 \cdot 10 \cdot 9 \cdot 8 \cdot 7 \cdot 6 \cdot 5 \cdot 4 \cdot 3}{10 \cdot 9 \cdot 8 \cdot 7 \cdot 6 \cdot 5 \cdot 4 \cdot 3 \cdot 2 \cdot 1}$$

we can use the fact that ${}_{12}C_{10} = {}_{12}C_2$ as follows.

$$ {}_{12}C_{10} = {}_{12}C_2 = \frac{12 \cdot 11}{2 \cdot 1} = 66 $$

■

Pascal's Triangle

There is a convenient way to remember a pattern for binomial coefficients. By arranging the coefficients in a triangular pattern, we obtain the following array, which is called **Pascal's Triangle**. This triangle is named after the famous French mathematician Blaise Pascal (1623–1662).

$$
\begin{array}{ccccccccccccccc}
 & & & & & & & 1 & & & & & & & \\
 & & & & & & 1 & & 1 & & & & & & \\
 & & & & & 1 & & 2 & & 1 & & & & & \\
 & & & & 1 & & 3 & & 3 & & 1 & & & & \\
 & & & 1 & & 4 & & 6 & & 4 & & 1 & & & \\
 & & 1 & & 5 & & 10 & & 10 & & 5 & & 1 & & \\
 & 1 & & 6 & & 15 & & 20 & & 15 & & 6 & & 1 & \\
1 & & 7 & & 21 & & 35 & & 35 & & 21 & & 7 & & 1
\end{array}
$$

The first and last number in each row of Pascal's Triangle is 1. Every other number in each row is formed by adding the two numbers immediately above the number. For example, the two numbers above 35 are 15 and 20.

$$
\begin{array}{ccc}
15 & & 20 \\
 & 35 &
\end{array}
\qquad 15 + 20 = 35
$$

Pascal noticed that numbers in this triangle are precisely the same numbers that are the coefficients of binomial expansions, as follows.

$$
\begin{aligned}
(x + y)^0 &= 1 \\
(x + y)^1 &= 1x + 1y \\
(x + y)^2 &= 1x^2 + 2xy + 1y^2 \\
(x + y)^3 &= 1x^3 + 3x^2y + 3xy^2 + 1y^3 \\
(x + y)^4 &= 1x^4 + 4x^3y + 6x^2y^2 + 4xy^3 + 1y^4 \\
(x + y)^5 &= 1x^5 + 5x^4y + 10x^3y^2 + 10x^2y^3 + 5xy^4 + 1y^5 \\
(x + y)^6 &= 1x^6 + 6x^5y + 15x^4y^2 + 20x^3y^3 + 15x^2y^4 + 6xy^5 + 1y^6 \\
(x + y)^7 &= 1x^7 + 7x^6y + 21x^5y^2 + 35x^4y^3 + 35x^3y^4 + 21x^2y^5 + 7xy^6 + 1y^6
\end{aligned}
$$

Because the top row in Pascal's Triangle corresponds to the binomial expansion $(x + y)^0 = 1$, we call it the **zero row**. Similarly, the next row corresponds to the binomial expansion $(x + y)^1 = 1(x) + 1(y)$, and we call it the **first row**. In general, the ***n*th row** in Pascal's Triangle gives the coefficients of $(x + y)^n$.

EXAMPLE 4 ■ Using Pascal's Triangle

Use Pascal's Triangle to find the following binomial coefficients.

$$_8C_0, \quad _8C_1, \quad _8C_2, \quad _8C_3, \quad _8C_4, \quad _8C_5, \quad _8C_6, \quad _8C_7, \quad _8C_8$$

Solution

These nine binomial coefficients represent the eighth row of Pascal's Triangle. Thus, using the seventh row of the triangle, we can calculate the numbers in the eighth row, as follows.

1	7	21	35	35	21	7	1	
1	8	28	56	70	56	28	8	1
$_8C_0$	$_8C_1$	$_8C_2$	$_8C_3$	$_8C_4$	$_8C_5$	$_8C_6$	$_8C_7$	$_8C_8$

■

Binomial Expansions

As mentioned at the beginning of this section, when we write out the coefficients for a binomial that is raised to a power, we say that we are **expanding a binomial**. The formulas for binomial coefficients give us an easy way to expand binomials, as demonstrated in the next four examples.

EXAMPLE 5 ■ Expanding a Binomial

Write the expansion for the following expression.

$$(x + 1)^3$$

Solution

The binomial coefficients from the third row of Pascal's Triangle are 1, 3, 3, 1. Therefore, the expansion is as follows.

$$\begin{aligned}(x + 1)^3 &= (1)x^3 + (3)x^2(1) + (3)x(1^2) + (1)(1^3)\\ &= x^3 + 3x^2 + 3x + 1\end{aligned}$$

■

To expand binomials representing *differences*, rather than sums, we alternate signs. Here are two examples.

$$(x - 1)^3 = x^3 - 3x^2 + 3x - 1$$
$$(x - 1)^4 = x^4 - 4x^3 + 6x^2 - 4x + 1$$

EXAMPLE 6 ■ Expanding a Binomial

Write the expansion for the following expression.

$$(x - 2)^3$$

Solution

The binomial coefficients from the third row of Pascal's Triangle are 1, 3, 3, 1. Therefore, the expansion is as follows.

$$\begin{aligned}(x-2)^3 &= (1)x^3 - (3)x^2(2) + (3)x(2^2) - (1)(2^3)\\ &= x^3 - 6x^2 + 12x - 8\end{aligned}$$

EXAMPLE 7 ■ Expanding a Binomial

Write the expansion for the following expression.

$$(x+3)^4$$

Solution

The binomial coefficients from the fourth row of Pascal's Triangle are 1, 4, 6, 4, 1. Therefore, the expansion is as follows.

$$\begin{aligned}(x+3)^4 &= (1)x^4 + (4)x^3(3) + (6)x^2(3^2) + (4)x(3^3) + (1)(3^4)\\ &= x^4 + 12x^3 + 54x^2 + 108x + 81\end{aligned}$$

EXAMPLE 8 ■ Expanding a Binomial

Write the expansion for the following expression.

$$(x-2y)^4$$

Solution

The binomial coefficients from the fourth row of Pascal's Triangle are 1, 4, 6, 4, 1. Therefore, the expansion is as follows.

$$\begin{aligned}(x-2y)^4 &= (1)x^4 - (4)x^3(2y) + (6)x^2(2y)^2 - (4)x(2y)^3 + (1)(2y)^4\\ &= x^4 - 8x^3y + 24x^2y^2 - 32xy^3 + 16y^4\end{aligned}$$

DISCUSSION PROBLEM ■ The Rows of Pascal's Triangle

By adding the terms in each of the rows of Pascal's Triangle, we obtain the following.

Row 0: $1 = 1$

Row 1: $1 + 1 = 2$

Row 2: $1 + 2 + 1 = 4$

Row 3: $1 + 3 + 3 + 1 = 8$

Row 4: $1 + 4 + 6 + 4 + 1 = 16$

Can you find a pattern for this sequence? Use this pattern to find the sum of the terms in the 10th row of Pascal's Triangle. Then check your answer by actually adding the terms of the 10th row.

Warm-Up

The following warm-up exercises involve skills that were covered in earlier sections. You will use these skills in the exercise set for this section.

In Exercises 1–6, perform the indicated operations and simplify.

1. $-3x(x - 2)$ **2.** $(x + 3)(x - 3)$ **3.** $(x - 2)^2$

4. $(2x + 3)^2$ **5.** $x^2y(5xy^{-3})$ **6.** $(-2x^2y)^3$

In Exercises 7–10, evaluate the quantity.

7. $4!$ **8.** $\frac{8!}{7!}$ **9.** $\frac{12!}{9!}$ **10.** $\frac{10!}{2!8!}$

11.4 EXERCISES

In Exercises 1–10, evaluate ${}_nC_m$.

1. ${}_6C_4$ **2.** ${}_7C_3$

3. ${}_{10}C_5$ **4.** ${}_{12}C_9$

5. ${}_{20}C_{20}$ **6.** ${}_{15}C_0$

7. ${}_{50}C_{48}$ **8.** ${}_{200}C_1$

9. ${}_{25}C_4$ **10.** ${}_{18}C_5$

In Exercises 11–14, evaluate the given binomial coefficient ${}_nC_m$. Also, evaluate its symmetric coefficient ${}_nC_{n-m}$ to demonstrate that it is equal to the given binomial coefficient.

11. ${}_{15}C_3$ **12.** ${}_9C_4$

13. ${}_{25}C_5$ **14.** ${}_{30}C_3$

15. Find the ninth row of Pascal's Triangle.

16. Find the tenth row of Pascal's Triangle.

17. Use Pascal's Triangle to evaluate the coefficients.
(a) ${}_7C_3$ (b) ${}_9C_5$

18. Use Pascal's Triangle to evaluate the coefficients.
(a) ${}_6C_2$ (b) ${}_9C_3$

In Exercises 19–30, use the Binomial Theorem to expand the expression.

19. $(x + y)^4$ **20.** $(u + v)^6$

21. $(x + 1)^5$ **22.** $(x + 2)^5$

23. $(x + 3)^6$ **24.** $(x + 4)^4$

25. $(x - y)^5$ **26.** $(x - 5)^4$

27. $(u - 2v)^3$ **28.** $(2x + y)^5$

29. $(3a + 2b)^4$ **30.** $(4u - 3v)^3$

In Exercises 31–34, expand the expression using Pascal's Triangle to determine the coefficients.

31. $(x + y)^8$ **32.** $(r - s)^7$

33. $(x - 2)^6$ **34.** $(2x + 3)^5$

In Exercises 35–40, find the coefficient of the given term of the expression.

	Expression	*Term*
35.	$(x + 1)^{10}$	x^7
36.	$(x + 3)^{12}$	x^9
37.	$(x - y)^{15}$	x^4y^{11}
38.	$(x + y)^{10}$	x^7y^3
39.	$(2x + y)^{12}$	x^3y^9
40.	$(x - 3y)^{14}$	x^3y^{11}

Probability In Exercises 41–44, use the Binomial Theorem to expand the given expression. In the study of probability, it is sometimes necessary to use the expansion $(p + q)^n$, where $p + q = 1$.

41. $\left(\frac{1}{2} + \frac{1}{2}\right)^5$

42. $\left(\frac{2}{3} + \frac{1}{3}\right)^4$

43. $\left(\frac{1}{4} + \frac{3}{4}\right)^4$

44. $(0.4 + 0.6)^6$

In Exercises 45–48, use the Binomial Theorem to approximate the given quantity accurate to three decimal places. For example,

$$\begin{aligned}(1.02)^{10} &= (1 + 0.02)^{10}\\ &= 1 + 10(0.02) + 45(0.02)^2 + \cdots.\end{aligned}$$

45. $(1.02)^8$

46. $(2.005)^{10}$

47. $(2.99)^{12}$

48. $(1.98)^9$

SECTION 11.5 Counting Principles, Permutations, and Combinations

Simple Counting Problems • Counting Principles • Permutations • Combinations

Simple Counting Problems

In the last two sections of this chapter, we give a brief introduction to some of the basic counting principles and their application to probability. In the next section, you will see that much of probability has to do with counting the number of ways an event can occur. Examples 1, 2, and 3 describe some simple cases.

EXAMPLE 1 ■ **A Random Number Generator**

A random number generator (on a computer) selects an integer from 1 to 30. Find the number of ways the following events can occur.

(a) An even integer is selected.

(b) A number that is less than 12 is selected.

(c) A prime number is selected.

Solution

(a) Since half of the numbers between 1 and 30 are even, this event can occur in 15 different ways.

(b) The integers between 1 and 30 that are less than 12 are given by the following set.

$$\{1, 2, 3, 4, 5, 6, 7, 8, 9, 10, 11\}$$

Since this set has 11 members, we conclude that there are 11 different ways this event can happen.

(c) The prime numbers between 1 and 30 are as follows.

$$\{2, 3, 5, 7, 11, 13, 17, 19, 23, 29\}$$

Since this set has 10 members, we conclude that there are 10 different ways this event can happen. ■

EXAMPLE 2 ■ Selecting Pairs of Numbers at Random

Eight pieces of paper are numbered from 1 to 8 and placed in a box. One piece of paper is drawn from the box, its number is written down, and the piece of paper is replaced in the box. Then, a second piece of paper is drawn from the box, and its number is written down. Finally, the two numbers are added together. How many different ways can a total of 12 be obtained?

Solution

To solve this problem, we count the different ways that a total of 12 can be obtained using two numbers between 1 and 8.

$$\boxed{\text{First number}} + \boxed{\text{Second number}} = \boxed{12}$$

After considering the various possibilities, we see that this equation can be solved in the following five ways.

First Number	*Second Number*
4	8
5	7
6	6
7	5
8	4

Thus, a total of 12 can be obtained in five different ways. ■

Solving counting problems can be tricky. Often, seemingly minor changes in the statement of a problem can affect the answer. For instance, compare the counting problem in the next example with that given in Example 2.

EXAMPLE 3 ■ Selecting Pairs of Numbers at Random

Eight pieces of paper are numbered from 1 to 8 and placed in a box. Two pieces of paper are drawn from the box, and the numbers on each piece of paper are written down and totaled. How many different ways can a total of 12 be obtained?

Solution

To solve this problem, we count the different ways that a total of 12 can be obtained *using two different numbers* between 1 and 8.

First Number	*Second Number*
4	8
5	7
7	5
8	4

Thus, a total of 12 can be obtained in four different ways. ■

NOTE: The difference between counting problems in Examples 2 and 3 can be distinguished by saying that the random selection in Example 2 occurs **with replacement**, whereas the random selection in Example 3 occurs **without replacement**, which eliminates the possibility of choosing two 6's.

Counting Principles

In the first three examples, we looked at simple counting problems in which we can *list* each possible way that an event can occur. When it is possible, this is always the best way to solve a counting problem. However, some events can occur in so many different ways that it is not feasible to write out the entire list. In such cases we must rely on formulas and counting principles. The most important of these is called the **Fundamental Counting Principle**.

Fundamental Counting Principle

Let E_1 and E_2 be two events. The first event E_1 can occur in m_1 different ways. After E_1 has occurred, E_2 can occur in m_2 different ways. The number of ways that the two events can occur is

$$m_1 \cdot m_2.$$

NOTE: The Fundamental Counting Principle can be extended to three or more events. For instance, the number of ways that three events E_1, E_2, and E_3 can occur is $m_1 \cdot m_2 \cdot m_3$.

EXAMPLE 4 ■ Applying the Fundamental Counting Principle

How many different pairs of letters from the English alphabet are possible? (Disregard the difference between uppercase and lowercase letters.)

Solution

This experiment has two events. The first event is the choice of the first letter, and the second event is the choice of the second letter. Since the English alphabet contains 26 letters, it follows that each event can occur in 26 ways.

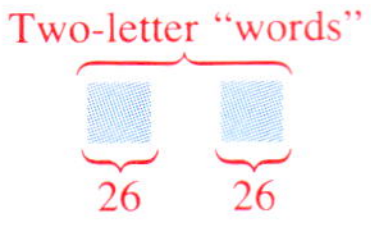

Thus, using the Fundamental Counting Principle, it follows that the number of two-letter words is

$$26 \cdot 26 = 676.$$ ■

EXAMPLE 5 ■ Applying the Fundamental Counting Principle

Telephone numbers in the United States have 10 digits. The first three make up the *area code*, and the next seven the *local telephone number*. How many different telephone numbers are possible within each area code? (Note that a local telephone number cannot begin with 0 or 1.)

Solution

Since the first digit cannot be 0 or 1, there are only 8 choices for the first digit. For each of the other six digits, there are 10 choices.

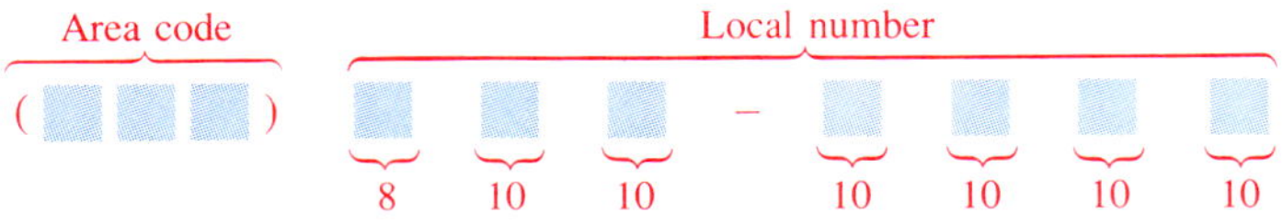

Thus, using the Fundamental Counting Principle, the number of local telephone numbers that are possible within each area code is

$$8 \cdot 10 \cdot 10 \cdot 10 \cdot 10 \cdot 10 \cdot 10 = 8{,}000{,}000.$$ ■

EXAMPLE 6 ■ Applying the Fundamental Counting Principle

Some states allow people to order special license plates for their cars. Suppose the state allows no more than six letters for the license plate. How many different six-letter license plates are possible?

Solution

There are 26 letters that can be used for each of the six letters on the plate.

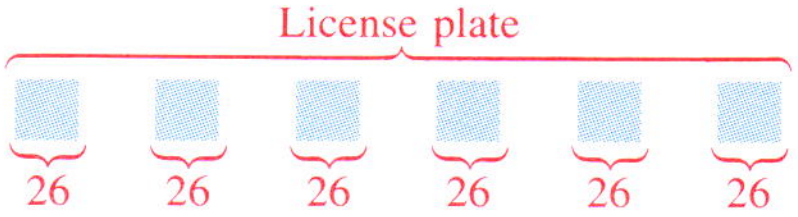

Thus, the number of different plates that are possible is

$$26 \cdot 26 \cdot 26 \cdot 26 \cdot 26 \cdot 26 = 26^6 = 308{,}915{,}776.$$

■

Permutations

One important application of the Fundamental Counting Principle is in determining the number of ways that n elements can be arranged (in order). We call an ordering of n elements a **permutation** of the elements.

Definition of Permutation

A **permutation** of n different elements is an ordering of the elements such that one element is first, one is second, one is third, and so on.

EXAMPLE 7 ■ **Listing Permutations**

Write the different permutations of the letters A, B, and C.

Solution

These three letters can be arranged in the following six different ways.

A, B, C,   B, A, C,   C, A, B
A, C, B,   B, C, A,   C, B, A

Thus, we see that these three letters have six different permutations. ■

In Example 7 we were able to list the different permutations of three letters. However, we could also have used the Fundamental Counting Principle. To do this, we could reason that there are three choices for the first letter, two choices for the second, and only one choice for the third, as follows.

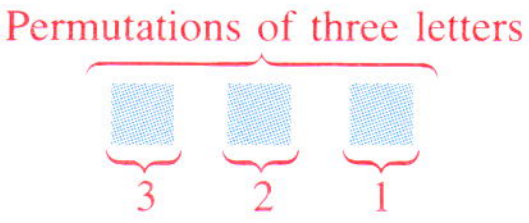

Thus, the number of permutations of three letters is

$$3 \cdot 2 \cdot 1 = 3! = 6.$$

EXAMPLE 8 ■ Finding the Number of Permutations of n Elements

How many different permutations are possible for the letters A, B, C, D, E, and F?

Solution

There are too many different permutations to list, so we use the following reasoning.

First position:	Any of the *six* letters
Second position:	Any of the remaining *five* letters
Third position:	Any of the remaining *four* letters
Fourth position:	Any of the remaining *three* letters
Fifth position:	Any of the remaining *two* letters
Sixth position:	The *one* remaining letter

Thus, the number of choices for the six positions is as follows.

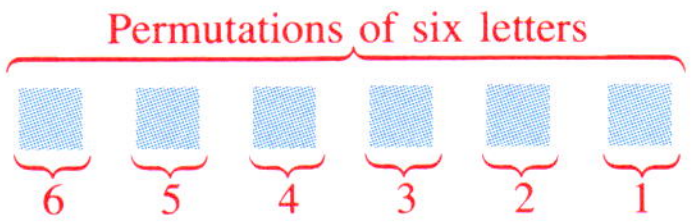

Using the Fundamental Counting Principle, we find the total number of permutations of the six letters is

$$6 \cdot 5 \cdot 4 \cdot 3 \cdot 2 \cdot 1 = 6! = 720.$$ ■

The results obtained in Examples 7 and 8 can be generalized to conclude that the number of permutations of n different elements is $n!$.

Number of Permutations of n Elements

The number of permutations of n elements is given by

$$n \cdot (n - 1) \cdot \cdot \cdot 4 \cdot 3 \cdot 2 \cdot 1 = n!.$$

In other words, there are $n!$ different ways that n elements can be ordered.

EXAMPLE 9 ■ Finding the Number of Permutations

Suppose you are a supervisor for 11 different employees. One of your responsibilities is to perform an annual evaluation for each employee, and then rank the 11 different performances. (In other words, one employee must be ranked as having the best performance, one must be ranked second, and so on.) How many different rankings are possible?

Solution

Since there are 11 different employees, you have 11 choices for first ranking. After choosing the first ranking, you can choose any of the remaining 10 for second ranking, and so on.

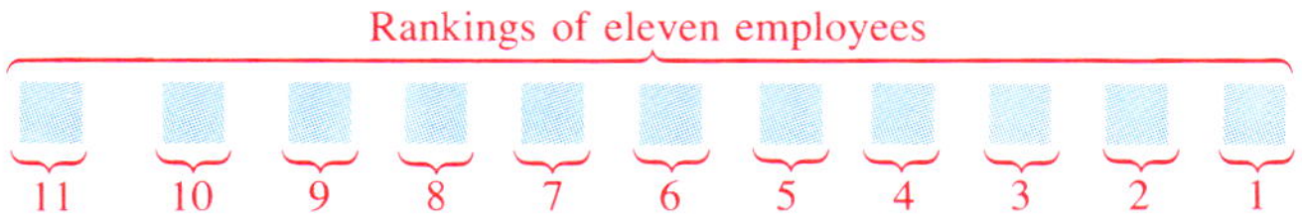

Thus, the number of different rankings is

$$11! = 39{,}916{,}800.$$ ■

Combinations

When counting the number of possible permutations of a set of elements, order is important. As a final topic in this section, we look at a method of selecting subsets of a larger set in which order is *not important*. We call such subsets **combinations of *n* elements taken *m* at a time**. For instance, the combinations

$$\{A, B, C\} \quad \text{and} \quad \{B, A, C\}$$

are equivalent because both sets contain the same three elements, and the order in which the elements are listed is *not important*. Hence, we would count only one of the two sets. A common example of how a combination occurs is a card game in which the player is free to reorder the cards after they have been dealt.

EXAMPLE 10 ■ Combination of *n* Elements Taken *m* at a Time

In how many different ways can three letters be chosen from the letters A, B, C, D, and E? (The order of the three letters is not important.)

Solution

The following subsets represent the different combinations of three letters that can be chosen from five letters.

{A, B, C}	{A, B, D}
{A, B, E}	{A, C, D}
{A, C, E}	{A, D, E}
{B, C, D}	{B, C, E}
{B, D, E}	{C, D, E}

From this list, we conclude that there are 10 different ways that three letters can be chosen from five letters. ■

The formula for the number of combinations of n elements taken m at a time is as follows.

Number of Combinations of *n* Elements Taken *m* at a Time

The number of combinations of n elements taken m at a time is

$$_{n}C_{m} = \frac{n!}{(n-m)!m!}.$$

Note that the formula for $_{n}C_{m}$ is the same one given for binomial coefficients. To see how this formula is used, let's solve the counting problem given in Example 10. In that problem we must find the number of combinations of 5 elements taken 3 at a time. Thus, $n = 5$, $m = 3$, and the number of combinations is

$$_{5}C_{3} = \frac{5!}{2!3!} = \frac{5 \cdot 4 \cdot 3}{3 \cdot 2 \cdot 1} = 10$$

which is the same answer we obtained in Example 10.

EXAMPLE 11 ■ Combinations of *n* Elements Taken *m* at a Time

A standard poker hand consists of five cards dealt from a deck of 52. How many different poker hands are possible? (After the cards are dealt, the player may reorder them, and therefore order is not important.)

Solution

We use the formula for the number of combinations of 52 elements taken five at a time, as follows.

$$_{52}C_{5} = \frac{52!}{47!5!} = \frac{52 \cdot 51 \cdot 50 \cdot 49 \cdot 48}{5 \cdot 4 \cdot 3 \cdot 2 \cdot 1} = 2{,}598{,}960 \text{ different hands}$$ ■

DISCUSSION PROBLEM ■ You Be the Instructor

Suppose you are teaching an algebra class and are writing a test for this chapter. Create two word problems that you think are appropriate for the test. One of the word problems should deal with permutations and the other should deal with combinations. (Assume that your students will have only five minutes to solve each problem.) ■

Warm-Up

The following warm-up exercises involve skills that were covered in earlier sections. You will use these skills in the exercise set for this section.

In Exercises 1–4, evaluate ${}_nC_m$.

1. ${}_5C_2$ **2.** ${}_{10}C_3$ **3.** ${}_9C_6$ **4.** ${}_{20}C_{20}$

In Exercises 5–10, evaluate the quantity.

5. $8 \cdot 2^3 \cdot 3^2$ **6.** $6^2 \cdot 5 \cdot 4$ **7.** $\dfrac{30!}{28!}$

8. $\dfrac{9!}{2!(4!)(3!)}$ **9.** $\dfrac{20!}{16!4!}$ **10.** $\dfrac{2 \cdot 4 \cdot 6 \cdot 8}{2^4}$

11.5 EXERCISES

Random Selection In Exercises 1–10, determine the number of ways the specified event can occur when one or more marbles are selected from a bowl containing 20 marbles numbered 1 through 20.

1. An odd number is obtained when one marble is drawn.

2. An even number is obtained when one marble is drawn.

3. A prime number is obtained when one marble is drawn.

4. A number greater than 12 is obtained when one marble is drawn.

5. A number divisible by 3 is obtained when one marble is drawn.

6. A number divisible by 6 is obtained when one marble is drawn.

7. Two marbles are drawn one after the other with the first being replaced before the second is drawn. The sum of the numbers is 8.

8. Two marbles are drawn one after the other with the first being replaced before the second is drawn. The sum of the numbers is 15.

9. Two marbles are drawn one after the other without replacement of the first one drawn. The sum of the numbers is 8.

10. Two marbles are drawn one after the other without replacement of the first one drawn. The sum of the numbers is 15.

11. *Identification Numbers* In a statistical study, each participant was given an identification label consisting of a letter of the alphabet followed by a single digit. How many distinct identification labels can be made in this way?

12. *Identification Numbers* How many identification labels (see Exercise 11) can be made by one letter of the alphabet followed by a two-digit number?

13. *License Plates* How many distinct automobile license plates can be formed by using a four-digit number followed by two letters?

14. *Three-Digit Numbers* How many three-digit numbers can be formed in each of the following situations?
(a) The hundreds digit cannot be 0.
(b) No repetition of digits is allowed.
(c) The number cannot be greater than 400.

15. *Staffing Choices* A small grocery store needs to open another checkout line. There are three people available who can run the cash register and two people available to bag groceries. How many different ways can the additional checkout line be staffed?

16. *Computer System* You are in the process of purchasing a new computer system. You must choose one of three monitors, one of two computers, and one of two keyboards. How many different configurations of the system are available to you?

17. *Toboggan Ride* Five people sit on a toboggan at the top of a hill. In how many ways can they be seated if only two of the five are willing to sit in the front?

18. *Task Assignment* Four people are assigned to four different tasks. In how many ways can the assignments be made if one of the four is not qualified for the first task?

19. *Permutations* List all the permutations of the letters X, Y, and Z.

20. *Permutations* List all the permutations of the letters A, B, C, and D.

21. *Seating Arrangement* In how many ways can five children be seated in a single row of chairs?

22. *Seating Arrangement* In how many ways can six people be seated in a six-passenger car?

23. *Time-Management Study* There are eight steps in accomplishing a certain task and these can be performed in any order. Management wants to test each possible order to determine which is least time-consuming. How many different orders will have to be tested?

24. *Time-Management Study* Repeat Exercise 23 if one step in accomplishing the task must be done first. (The other seven steps can be performed in any order.)

25. *Number of Subsets* List all the subsets with two elements that can be formed from the set of letters $\{A, B, C, D, E, F\}$.

26. *Number of Subsets* List all the subsets with three elements that can be formed from the set of letters $\{A, B, C, D, E, F\}$.

27. *Committee Selection* Three students are selected from a class of 20 to form a fund-raising committee. In how many ways can the committee be formed?

28. *Committee Selection* In how many ways can a committee of five be formed from a group of 30 people?

29. *Menu Selection* A group of four people go out to dinner at a restaurant. There are nine entrees on the menu and the four people decide that no two of them will order the same thing. How many ways can the four order from the nine entrees?

30. *Test Questions* A student is required to answer any nine questions from the 12 questions on the exam. In how many ways can the student select the questions?

31. *Number of Triangles* Eight points are located in the coordinate plane such that no three are collinear. How many different triangles can be formed having their vertices as three of the eight points?

32. *Relationships* The number of interpersonal relationships increases dramatically as the size of a group increases (see figure). Determine the number of different two-person relationships in a group having the following numbers.

(a) 3 (b) 4 (c) 6
(d) 8 (e) 10 (f) 12

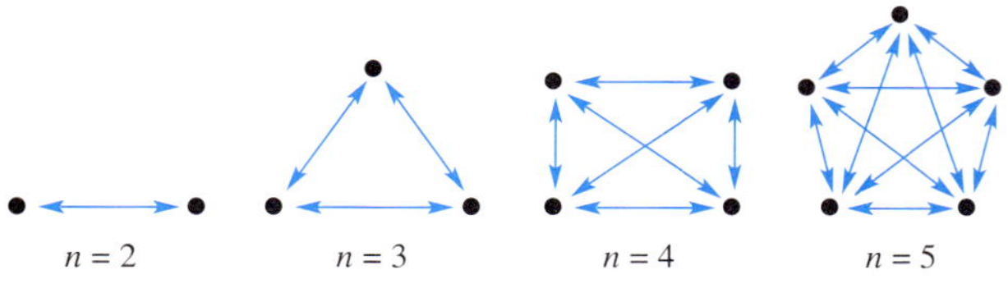

33. *Softball League* Six churches form a softball league. If each team must play every other team twice during the season, what is the total number of league games played?

34. *Group Selection* Four people are to be selected from four couples. In how many ways can this be done if (a) there are no restrictions, and (b) one person from each couple must be selected?

SECTION 11.6

Probability

Sample Space of an Experiment • The Probability of an Event

Sample Space of an Experiment

As a member of a complex society, you are used to living with varying amounts of uncertainty. For example, you may be questioning the likelihood of getting a good job after graduation, of winning a state lottery, or of having rain tomorrow.

In assigning measurements to uncertainties in everyday life, we often use ambiguous terminology, such as "fairly certain," "probable," or "highly unlikely." In mathematics we attempt to remove this ambiguity by assigning a number to the likelihood of the occurrence of an event. We call this measurement the **probability** that the event will occur. For example, if we toss a fair coin, we say that the probability that it will land heads up is one-half, or 0.5.

In the study of probability, we call any happening whose result is uncertain an **experiment**. The possible results of the experiment are called **outcomes**; the set of all possible outcomes of an experiment is called the **sample space** of the experiment; and any subcollection of a sample space is called an **event**. An event that consists of a single outcome is called a **simple event**. In this section we will deal only with sample spaces in which each outcome is equally likely, such as flipping a fair coin or tossing a fair die.

EXAMPLE 1 ■ Finding the Sample Space

An experiment consists of tossing a six-sided die.

(a) What is the sample space?

(b) Describe the event corresponding to a number greater than 2 turning up.

Solution

(a) The sample space consists of six outcomes, which we represent by the numbers 1 through 6. That is,

$$S = \{1, 2, 3, 4, 5, 6\}.$$

Note that each of the outcomes in the sample space is equally likely (assuming that the die is balanced).

(b) The *event* corresponding to a number greater than 2 turning up is the following subset of S.

$$E = \{3, 4, 5, 6\}$$ ■

To describe sample spaces in such a way that each outcome is equally likely, we must sometimes distinguish between various outcomes in ways that appear artificial. The next example illustrates such a situation.

EXAMPLE 2 ■ **Finding the Sample Space**

Find the sample spaces for the following.

(a) One coin is tossed. (b) Two coins are tossed. (c) Three coins are tossed.

Solution

(a) Since the coin will land either heads up (denoted by H) or tails up (denoted by T), the sample space is

$$S = \{H, T\}.$$

(b) Since either coin can land heads up or tails up, the possible outcomes are as follows.

HH = heads up on both coins
HT = heads up on first coin and tails up on second coin
TH = tails up on first coin and heads up on second coin
TT = tails up on both coins

Thus, the sample space is

$$S = \{HH, HT, TH, TT\}.$$

Note that we must distinguish between the two cases HT and TH, even though these two outcomes appear to be similar.

(c) Following the notation of part (b), the sample space is

$$S = \{HHH, HHT, HTH, HTT, THH, THT, TTH, TTT\}.$$ ■

The Probability of an Event

To calculate the probability of an event, we count the number of outcomes in the event and in the sample space. The *number of outcomes* in E is denoted by $n(E)$, and the number of outcomes in the sample space S is denoted by $n(S)$. Moreover, in this text, we assume that every sample space has a finite number of outcomes.

The Probability of an Event

If an event E has $n(E)$ equally likely outcomes and its sample space has $n(S)$ equally likely outcomes, then the **probability** of event E is

$$P(E) = \frac{n(E)}{n(S)}.$$

Because the number of outcomes in an event must be less than or equal to the number of outcomes in the sample space, we can see that the probability of an event must be a number between 0 and 1. That is, for any event E, it must be true that $0 \leq P(E) \leq 1$, as indicated in Figure 11.3.

FIGURE 11.3

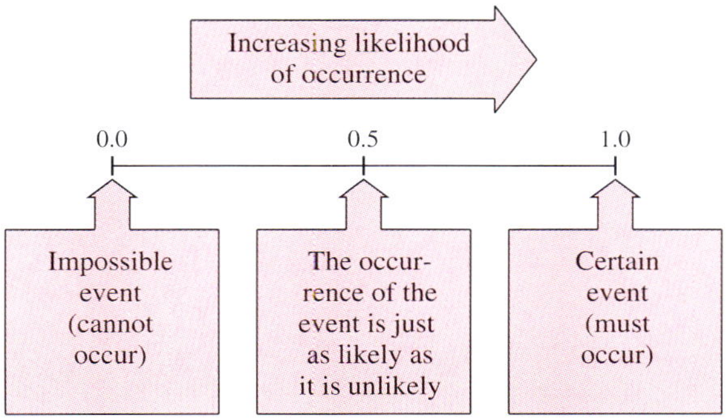

This, and other properties of the probability of an event are summarized as follows.

Properties of the Probability of an Event

Let E be an event that is a subset of a finite sample space S.

1. $0 \leq P(E) \leq 1$
2. If $P(E) = 0$, then the event E *cannot occur*, and we call E an **impossible event**.
3. If $P(E) = 1$, then the event E *must occur*, and we call E a **certain event**.

Historically, a great deal of interest in probability stemmed from its applications to gambling. Some simple cases are shown in the next three examples.

EXAMPLE 3 ■ Finding the Probability of an Event

Two coins are tossed. What is the probabilty that both land heads up?

Solution

Following the procedure in Example 2(b), we let

$$E = \{HH\} \qquad \text{and} \qquad S = \{HH, HT, TH, TT\}.$$

The probability of getting two heads is

$$P(E) = \frac{n(E)}{n(S)} = \frac{1}{4}.$$

■

EXAMPLE 4 ■ Finding the Probability of an Event

A card is drawn from a standard deck of playing cards. What is the probability that it is an ace?

Solution

Since there are 52 cards in a standard deck of playing cards and there are four aces (one in each suit), the probability of drawing an ace is

$$P(E) = \frac{n(E)}{n(S)} = \frac{4}{52} = \frac{1}{13}.$$

■

EXAMPLE 5 ■ Finding the Probability of an Event

Two six-sided dice are tossed, as shown in Figure 11.4. What is the probability that the total of the two dice is 7?

Solution

Since there are six possible outcomes on each die, we use the Fundamental Counting Principle to conclude that there are

$$6 \cdot 6 = 36 \text{ different outcomes}$$

when two dice are tossed. To find the probability of rolling a total of 7, we must first count the number of ways this can occur.

	Total of 7					
First Die	1	2	3	4	5	6
Second Die	6	5	4	3	2	1

Thus, a total of 7 can be rolled in six ways, which means that the probability of rolling a 7 is

$$P(E) = \frac{n(E)}{n(S)} = \frac{6}{36} = \frac{1}{6}.$$

FIGURE 11.4

■

EXAMPLE 6 ■ Comparing Probabilities

Twelve-sided dice can be constructed (in the shape of regular dodecahedrons) so that each of the numbers from 1 to 6 appears twice on each die, as shown in Figure 11.5. Show that these dice can be used in any game requiring ordinary six-sided dice without changing the probabilities of different outcomes.

FIGURE 11.5

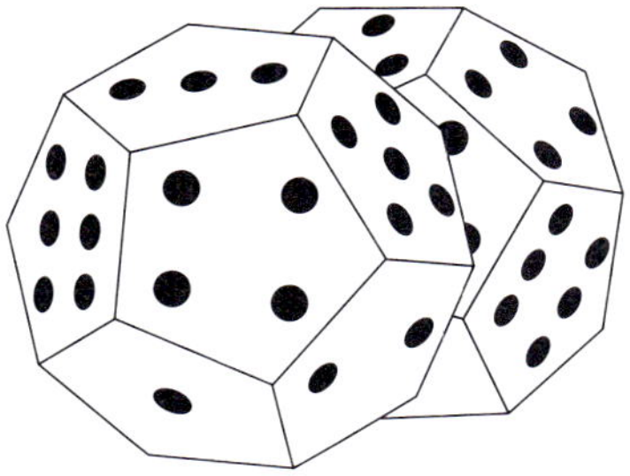

Solution

For an ordinary six-sided die, each of the numbers 1, 2, 3, 4, 5, and 6 occurs only once, so the probability of any particular number coming up is

$$P(E) = \frac{n(E)}{n(S)} = \frac{1}{6}.$$

For a twelve-sided die, each number occurs twice, so the probability of any particular number coming up is

$$P(E) = \frac{n(E)}{n(S)} = \frac{2}{12} = \frac{1}{6}.$$

Thus, the twelve-sided dice can be used in place of the six-sided dice without changing the probabilities. ■

We could have written out each sample space in Examples 3, 4, 5, and 6 and simply counted the outcomes in the desired events. For larger sample spaces, however, we must make more use of the counting principles discussed in the previous chapter.

EXAMPLE 7 ■ Finding the Probability of an Event

A state lottery is set up so that each player chooses six different numbers from 1 to 40. If these six numbers match the six numbers drawn by the lottery commission, the player wins (or shares) the top prize. What is the probability of winning the top prize in this game?

Solution

Since the order of the numbers is not important, we use our formula for the number of *combinations* of 40 elements taken six at a time to determine the size of the sample space.

$$n(S) = {}_{40}C_6 = \frac{40 \cdot 39 \cdot 38 \cdot 37 \cdot 36 \cdot 35}{6 \cdot 5 \cdot 4 \cdot 3 \cdot 2 \cdot 1} = 3{,}838{,}380$$

If a person buys only one ticket, the probability of winning the top prize is

$$P(E) = \frac{n(E)}{n(S)} = \frac{1}{3{,}838{,}380}.$$

■

DISCUSSION PROBLEM ■ A Computer Experiment

A well-known experiment in probability is called the "Birthday Problem." This problem concerns the probability that in a group of n people, at least two of the people will have the same birthday. The following BASIC program is a computer simulation of the birthday problem. Try running the program a few times with a computer that has the BASIC language. How many selections did the computer make each time before it found "two people with the same birthday"?

BASIC Program:

```
10 RANDOMIZE 10*VAL(MID$(TIME$,4,2)) + VAL(RIGHT$(TIME$,2))
20 DIM DAYS(365)
30 DONE = 0: SELECT = 0
40 WHILE (DONE = 0)
50   BIRTHDAY = INT(365*RND) + 1
60   IF DAYS(BIRTHDAY) = 1 THEN DONE = 1 ELSE DAYS(BIRTHDAY) = 1
70   IF DONE = 0 THEN PRINT BIRTHDAY;: SELECT = SELECT + 1
80 WEND
90 PRINT BIRTHDAY
100 PRINT "SELECTION NUMBER" ;SELECT+1; "IS A REPEAT."
110 END
```

■

Warm-Up

The following warm-up exercises involve skills that were covered in earlier sections. You will use these skills in the exercise set for this section.

In Exercises 1–10, evaluate the quantity.

1. $\frac{1}{6} + \frac{3}{4}$

2. $\frac{3}{8} + \frac{1}{4}$

3. $\frac{3}{5} + \frac{7}{10} - \frac{1}{15}$

4. $\frac{5}{9} + \frac{5}{12} - \frac{5}{16}$

5. $\frac{6 \cdot 5}{6!}$

6. $\frac{8 \cdot 7 \cdot 6}{8!}$

7. $\frac{{}_4C_2}{{}_8C_2}$

8. $\frac{{}_6C_3}{{}_{10}C_3}$

9. $\frac{7}{8} \cdot \frac{6}{7} \cdot \frac{5}{6}$

10. $1 - \left(\frac{2}{5}\right)^3$

11.6 EXERCISES

Sample Space In Exercises 1–6, determine the sample space for the given experiment.

1. A coin and a six-sided die are tossed.

2. A six-sided die is tossed twice and the sum is recorded.

3. One letter from the alphabet is chosen.

4. A taste tester has to rank three varieties of yogurt, A, B, and C, according to preference.

5. Two county supervisors are selected from five supervisors, A, B, C, D, and E, to study a recycling plan.

6. A salesperson makes a presentation about a product in three homes per day. In each home there may be a sale (denote by Y) or there may be no sale (denote by N).

Coin Tossing In Exercises 7–10, a coin is tossed three times. Find the probability of the specified event. (The sample space is given in Example 2.)

7. The event of getting two heads.

8. The event of getting a tail on the second toss.

9. The event of getting at least one head.

10. The event of getting no more than two heads.

Playing Cards In Exercises 11–14, a card is drawn from a standard deck of playing cards. Find the probability of the specified event.

11. The event of getting a red card.

12. The event of getting a queen.

13. The event of getting a face card.

14. The event of getting a black face card.

Tossing a Die In Exercises 15–18, a six-sided die is tossed. Find the probability of the specified event.

15. The event that the number that turns up is 5.

16. The event that the number that turns up is 7.

17. The event that the number that turns up is no more than 5.

18. The event that the number that turns up is at least 1.

19. *Multiple Choice Test* A student takes a multiple choice test in which there are five choices for each question. Find the probability that the question is answered correctly given the following conditions.
(a) The student has no idea of the answer and guesses at random.
(b) The student can eliminate two of the choices and guesses from the remaining choices.
(c) The student knows the answer.

20. *Multiple Choice Test* A student takes a multiple choice test in which there are four choices for each question. Find the probability that the question is answered correctly given the following conditions.
(a) The student has no idea of the answer and guesses at random.
(b) The student can eliminate two of the choices and guesses from the remaining choices.
(c) The student knows the answer.

21. *Random Selection* Twenty marbles numbered 1 through 20 are placed in a bag and one is selected. Find the probability that the number selected is the following.

(a) the number 12 (b) a prime number
(c) an odd number (d) a number less than 6

22. *Class Election* Three people are running for class president. From a small opinion poll, it is estimated that the probability that Candidate A will win is 0.5 and the probability that Candidate B will win is 0.3. What is the probability that the third candidate will win?

23. *Continuing Education* In a high school graduating class of 325 students, 255 are going to continue their education. What is the probability that a student selected at random from the class will not be furthering his or her education?

24. *Study Questions* An instructor gives the class a list of four study questions for the next exam. Two of the four study questions will be on the exam. Find the probability that a student who knows the material relating to three of the four questions will be able to answer both questions selected for the exam.

In Exercises 25–30, the sample spaces are larger and it will be necessary to use the counting principles discussed in Section 11.5.

25. *Game Show* On a game show you are given five digits to arrange in the proper order to give the price of a car. If you arrange them correctly, you win the car. Find the probability of winning if you know the correct position of only one digit and must guess at the digits in all the other positions.

26. *Committee Selection* A committee of three students is to be selected from a group of three girls and five boys. Find the probability that the committee is composed entirely of girls.

27. *Defective Units* A shipment of 10 food processors to a certain store contained two defective units. If you purchase two food processors as birthday gifts for friends, determine the probability that both are defective units.

28. *Book Selection* Four books are selected at random from a shelf containing six novels and four autobiographies. Find the probability that the four autobiographies are selected.

29. *Card Selection* Five cards are selected from a standard deck of 52 cards. Find the probability that you obtain the four aces.

30. *Card Selection* Five cards are selected from a standard deck of 52 cards. Find the probability of obtaining three of a kind and two of a kind.

31. If the probability of an event occurring is p, what is the probability that the event does not occur?

32. What is the sum of the probabilities of each element in a sample space?

CHAPTER 11 SUMMARY

As you review and prepare for a test on this chapter, first try to obtain a global view of what was discussed. Then review the specific skills needed under each category.

Sequences and Their Sums

- *Write* the first several terms of an infinite sequence, given an expression for a_n. — Section 11.1, Exercises 1–20
- *Evaluate* a sum expressed in sigma notation. — Section 11.1, Exercises 27–40
- *Write* a given sum in sigma notation form. — Section 11.1, Exercises 41–54
- *Determine* if a given sequence is arithmetic or geometric. — Section 11.2, Exercises 9–16; Section 11.3, Exercises 9–16
- *Write* the first several terms of an arithmetic sequence and of a geometric sequence. — Section 11.2, Exercises 17–24, Exercises 33–40; Section 11.3, Exercises 17–24
- *Create* a formula for the nth term of an arithmetic sequence and of a geometric sequence. — Section 11.2, Exercises 25–32; Section 11.3, Exercises 25–32
- *Find* the sum of an arithmetic sequence and of a geometric sequence. — Section 11.2, Exercises 41–56; Section 11.3, Exercises 41–56
- *Create and evaluate* the sum of an arithmetic sequence and of a geometric sequence from a verbal statement. — Section 11.2, Exercises 57–65; Section 11.3, Exercises 57–68

The Binomial Theorem

- *Evaluate* a binomial coefficient using the Binomial Theorem or Pascal's Triangle. — Section 11.4, Exercises 1–18
- *Expand* a binomial expression using the Binomial Theorem or Pascal's Triangle. — Section 11.4, Exercises 19–34
- *Evaluate* a specified coefficient in a binomial expansion. — Section 11.4, Exercises 35–40

Counting Principles and Probability

- *Calculate* the number of ways an event can occur using the Fundamental Counting Principle, permutations, or combinations. — Section 11.5, Exercises 1–34
- *Calculate* the sample space for a given experiment. — Section 11.6, Exercises 1–6
- *Find* the probability of the occurrence of a specified event. — Section 11.6, Exercises 7–32

Chapter 11 Review Exercises

In Exercises 1–4, use sigma notation to write the given sum.

1. $[5(1) - 3] + [5(2) - 3] + [5(3) - 3] + [5(4) - 3]$
2. $[90 - 10(1)] + [90 - 10(2)] + [90 - 10(3)] + [90 - 10(4)] + [90 - 10(5)]$
3. $\frac{1}{3(1)} + \frac{1}{3(2)} + \frac{1}{3(3)} + \frac{1}{3(4)} + \frac{1}{3(5)} + \frac{1}{3(6)}$
4. $\left(-\frac{1}{3}\right)^0 + \left(-\frac{1}{3}\right)^1 + \left(-\frac{1}{3}\right)^2 + \left(-\frac{1}{3}\right)^3 + \left(-\frac{1}{3}\right)^4$

In Exercises 5–8, simplify the ratio of factorials.

5. $\frac{20!}{18!}$
6. $\frac{50!}{53!}$
7. $\frac{n!}{(n-3)!}$
8. $\frac{(n-1)!}{(n+1)!}$

In Exercises 9–12, write the first five terms of the arithmetic sequence. (Assume the sequence begins with $n = 1$.)

9. $a_n = 132 - 5n$
10. $a_n = 2n + 3$
11. $a_n = \frac{3}{4}n + \frac{1}{2}$
12. $a_n = -\frac{3}{5}n + 1$

In Exercises 13–16, write the first five terms of the geometric sequence.

13. $a_1 = 10,\ r = 3$
14. $a_1 = 2,\ r = -5$
15. $a_1 = 100,\ r = -\frac{1}{2}$
16. $a_1 = 12,\ r = \frac{1}{6}$

In Exercises 17–26, find a formula for the nth term of the specified sequence. (Assume the sequence begins with $n = 1$.)

17. Arithmetic sequence: $a_1 = 10,\ d = 4$
18. Arithmetic sequence: $a_1 = 32,\ d = -2$
19. Arithmetic sequence: $a_1 = 1{,}000,\ a_2 = 950$
20. Arithmetic sequence: $a_1 = 12,\ a_2 = 20$
21. Geometric sequence: $a_1 = 1,\ r = -\frac{2}{3}$
22. Geometric sequence: $a_1 = 100,\ r = 1.07$
23. Geometric sequence: $a_1 = 24,\ a_2 = 48$
24. Geometric sequence: $a_1 = 16,\ a_2 = -4$
25. Geometric sequence: $a_1 = 12,\ a_4 = -\frac{3}{2}$
26. Geometric sequence: $a_2 = 1,\ a_3 = \frac{1}{3}$

In Exercises 27–40, find the specified sum.

27. $\sum_{k=1}^{4} 7$
28. $\sum_{k=1}^{4} \frac{(-1)^k}{k}$
29. $\sum_{n=1}^{4} \left(\frac{1}{n} - \frac{1}{n+1}\right)$
30. $\sum_{n=1}^{4} \left(\frac{1}{n} - \frac{1}{n+2}\right)$

31. $\sum_{k=1}^{12} (7k - 5)$

32. $\sum_{k=1}^{10} (100 - 10k)$

33. $\sum_{j=1}^{100} \frac{j}{4}$

34. $\sum_{j=1}^{50} \frac{3j}{2}$

35. $\sum_{n=1}^{12} 2^n$

36. $\sum_{n=1}^{12} (-2)^n$

37. $\sum_{k=1}^{8} 5\left(-\frac{3}{4}\right)^{k-1}$

38. $\sum_{k=1}^{10} 4\left(\frac{3}{2}\right)^{k-1}$

39. $\sum_{n=1}^{120} 500(1.01)^{n-1}$

40. $\sum_{n=1}^{40} 1{,}000(1.1)^{n-1}$

In Exercises 41–46, use the Binomial Theorem to expand the expression. Simplify your answer.

41. $(x + 1)^{10}$

42. $(u - v)^9$

43. $(y - 2)^6$

44. $(x + 3)^5$

45. $\left(\frac{1}{2} - x\right)^8$

46. $(3x - 2y)^4$

47. Find the sum of the first 50 positive integers that are multiples of 4.

48. Find the sum of the integers from 225 to 300.

49. *Auditorium Seating* Each row in a small auditorium has three more seats than the preceding row. Find the seating capacity of the auditorium if the front row seats 22 people and there are 12 rows of seats.

50. *Depreciation* A company pays $120,000 for a machine. During the next five years it will depreciate at the rate of 30% per year. (That is, at the end of each year the depreciated value is 70% of what it was at the beginning of the year.)

(a) Find a formula for the nth term of the geometric sequence that gives the value of the machine t full years after it was purchased.

(b) Find the depreciated value of the machine at the end of five full years.

51. *Population Increase* A city of 85,000 people is growing at the rate of 1.2% per year. (That is, at the end of each year the population is 1.012 times the population at the beginning of the year.)

(a) Find a formula for the nth term of the geometric sequence that gives the population t years from now.

(b) Estimate the population 50 years from now.

52. *Salary Increase* Suppose you accept a job that pays a salary of $32,000 the first year. During the next 39 years, suppose you receive a 5.5% raise each year. What would your total salary be over the 40-year period?

53. *Morse Code* In Morse code, all characters are transmitted using a sequence of dits or dahs. How many different characters can be formed by using a sequence of three dits and dahs? (These can be repeated. For example, dah-dit-dit represents the letter *d*.)

54. *Forming Line Segments* How many straight line segments can be formed by seven points such that no three are collinear?

55. *Committee Selection* Determine the number of ways a committee of five people can be formed from a group of 15 people.

56. *Program Listing* There are seven participants in a piano recital. In how many orders can their names be listed in the program?

57. *Rolling a Die* Find the probability of obtaining a number greater than 4 when a single six-sided die is rolled.

58. *Coin Tossing* Find the probability of obtaining at least one head when a coin is tossed four times.

59. *Book Selection* A child who does not know how to read carries a four-volume set of books to a bookshelf. Find the probability that the books are put on the shelf in the correct order.

60. *Rolling a Die* Are the chances of rolling a 3 with one six-sided die the same as the chances of rolling a total of 6 with two six-sided dice? If not, which has the greater probability of occurring?

CHAPTER 11 TEST

Take this test as you would take a test in class. After you are done, check your work with the answers given in the back of the book.

1. Write the first five terms of the infinite sequence $a_n = \left(-\frac{2}{3}\right)^{n-1}$. (Assume the sequence starts with $n = 1$.)

2. Evaluate: $\sum_{j=0}^{4} (3j + 1)$

3. Use sigma notation to write the sum

$$\frac{2}{3(1) + 1} + \frac{2}{3(2) + 1} + \cdots + \frac{2}{3(12) + 1}.$$

4. Write the first five terms of the arithmetic sequence with first term $a_1 = 12$ and common difference $d = 4$.

5. Find a formula for the nth term of the arithmetic sequence with first term $a_1 = 5{,}000$ and common difference $d = -100$.

6. Find the sum of the first 50 positive integers that are multiples of 3.

7. Find the common ratio of the geometric sequence

$$2, -3, \frac{9}{2}, -\frac{27}{4}, \frac{81}{8}, \ldots .$$

8. Find a formula for the nth term of the geometric sequence with first term $a_1 = 4$ and common ratio $r = \frac{1}{2}$.

9. Find the sum of the first eight terms of the geometric sequence for which $a_1 = 3$ and $a_2 = 6$.

10. Find the balance in an increasing annuity when a principal of \$50 is deposited at the beginning of each month for 25 years. Assume the amount in the fund is compounded monthly at 8%.

11. Evaluate: ${}_{20}C_3$

12. Use the Binomial Theorem to expand $(x - 2)^5$.

13. Find the coefficient of the term x^3y^5 in the expansion of $(x + y)^8$.

14. How many distinct license plates can be issued by using one letter followed by a three-digit number?

15. Four students are randomly selected from a class of 25 to answer questions from a reading assignment. In how many ways can four students be selected? (Assume that the order in which the students are selected is not important.)

16. A weather report indicates that the probability of snow tomorrow is 0.75. What is the probability that it will *not* snow?

17. A card is drawn from a standard deck of playing cards. Find the probability that it is a red face card.

18. Suppose two spark plugs require replacement in a four-cylinder engine. If the mechanic randomly removes two plugs, find the probability that they are the two defective plugs.

CUMULATIVE TEST: CHAPTERS 8–11

Take this test as you would take a test in class. After you are done, check your work with the answers given in the back of the book.

1. Determine whether the equation $x - y^3 = 0$ represents y as a function of x.

2. Find the domain of the function $f(x) = \sqrt{\dfrac{3}{x-2}}$.

3. Find $f(c + 3)$ if $f(x) = x^2 - 3x$.

4. Discuss the relationship between the graphs of $f(x)$ and $g(x) = f(x - 4) - 3$.

5. Given the functions $f(x) = \sqrt{x}$ and $g(x) = 2x - 1$, find the following.

(a) $(f - g)(4)$ (b) $(fg)(9)$

(c) $\left(\dfrac{f}{g}\right)(3)$ (d) $(f \circ g)(5)$

6. Find the inverse f^{-1} of the function $f(x) = 3x - 2$. Sketch the graphs of f and f^{-1} on the same set of coordinate axes.

7. The stopping distance d of a car is directly proportional to the square of its speed s. On a certain type of pavement, a car required 50 feet to stop when its speed was 25 miles per hour. Estimate the stopping distance when the speed of the car is 40 miles per hour.

8. Solve the system by the method of elimination.

$$\begin{aligned} 4x - 3y &= 8 \\ -2x + y &= -6 \end{aligned}$$

9. Use Cramer's Rule to find the y-coordinate of the solution to the following system of equations.

$$\begin{aligned} 2x - y \qquad &= 4 \\ 4y - 5z &= -23 \\ 3x + y - 2z &= -5 \end{aligned}$$

10. The sum of three positive numbers is 44. The second number is four greater than the first, and the third number is three times the magnitude of the first. Find the three numbers.

11. Graph the inequality $5x + 2y > 10$.

12. Sketch the graph of the functions $f(x) = e^x$ and $g(x) = \ln x$. What relationship exists between the functions f and g?

13. Use the properties of logarithms to condense $3(\log_2 x + \log_2 y) - \log_2 z$ into a logarithm of a single quantity.

14. Solve each of the following equations.

(a) $\log_x \left(\frac{1}{9}\right) = -2$ (b) $4 \ln x = 10$

(c) $500(1.08)^t = 2{,}000$ (d) $3(1 + e^{2x}) = 20$

15. Determine the effective yield of an 8% interest rate compounded continuously.

16. Determine the length of time for an investment of \$1,000 to quadruple in value if the investment earns 9% compounded continuously.

17. Find the sum of the first 20 terms of an arithmetic sequence if the first and second terms are $a_1 = 2$ and $a_2 = 6$.

18. Find the seventh term of a geometric sequence if $a_1 = 36$ and the common ratio is $r = -\frac{1}{3}$.

19. Find the coefficient of the term x^4y^3 in the expansion of $(x - y)^7$.

20. Five people are assigned to five different tasks in an assembly plant. How many ways can the assignments be made if only two of the five people are qualified to do the first task?

21. In how many ways can a committee of three be selected from a group of 10 people? (Assume that the order in which the people are selected is unimportant.)

22. Two people are running for political office. The latest polls indicate that one candidate is leading the other by eight percentage points. If one of the people who were polled is selected at random, what is the probability that the person indicated a preference for the candidate who is leading?

23. Find the probability of obtaining the sum of at least nine when two six-sided dice are tossed.

APPENDIX

Introduction to Logic

SECTION **A.1**

Introduction to Logic

Statements • *Truth Tables*

Statements

In everyday speech and in mathematics we make inferences that adhere to common **laws of logic**. These laws (or methods of reasoning) allow us to build an algebra of statements by using logical operations to form compound statements from simpler ones. One of the primary goals of logic is to determine the truth value (true or false) of a compound statement knowing the truth value of its simpler component statements. For instance, we will learn that the compound statement "The temperature is below freezing and it is snowing" is true only if both component statements are true.

Definition of a Statement

1. A **statement** is a sentence to which only one truth value (either true or false) can be meaningfully assigned.
2. An **open statement** is a sentence that contains one or more variables and becomes a statement when each variable is replaced by a specific item from a designated set.

NOTE: In the above definition, the word *statement* can be replaced by the word *proposition*.

EXAMPLE 1 ■ **Statements, Nonstatements, and Open Statements**

Statement	*Truth Value*
A square is a rectangle.	T
-3 is less than -5.	F
Nonstatement	*Truth Value*
Do your homework.	No truth value can be meaningfully assigned.
Did you call the police?	No truth value can be meaningfully assigned.
Open Statement	*Truth Value*
x is an irrational number.	We need a value of x.
She is a computer science major.	We need a specific person. ■

Symbolically, we represent statements by lowercase letters p, q, r, and so on. Statements can be changed or combined to form **compound statements** by means of the three logical operations **and**, **or**, and **not**, which we represent by $\wedge$ (and), $\vee$ (or), and $\sim$ (not). In logic we use the word *or* in the *inclusive* sense (meaning "and/or" in everyday language). That is, the statement "p or q" is true if p is true, q is true, or both p and q are true. The following list summarizes the terms and symbols used with these three operations of logic.

Operations of Logic

Operation	*Verbal Statement*	*Symbolic Form*	*Name of Operation*
$\sim$	not p	$\sim p$	**Negation**
$\wedge$	p and q	$p \wedge q$	**Conjunction**
$\vee$	p or q	$p \vee q$	**Disjunction**

Compound statements can be formed using more than one logical operation, as demonstrated in Example 2.

EXAMPLE 2 ■ Forming Negations and Compound Statements

The statements p and q are as follows.

p: The temperature is below freezing.
q: It is snowing.

Write the verbal form for each of the following.

(a) $p \wedge q$ (b) $\sim p$

(c) $\sim(p \vee q)$ (d) $\sim p \wedge \sim q$

Solution

(a) The temperature is below freezing and it is snowing.

(b) The temperature is not below freezing.

(c) It is not true that the temperature is below freezing or it is snowing.

(d) The temperature is not below freezing and it is not snowing. ■

EXAMPLE 3 ■ Forming Compound Statements

The statements p and q are as follows.

p: The temperature is below freezing.
q: It is snowing.

(a) Write the symbolic form for: *The temperature is not below freezing or it is not snowing.*

(b) Write the symbolic form for: *It is not true that the temperature is below freezing and it is snowing.*

Solution

(a) The symbolic form is: $\sim p \vee \sim q$

(b) The symbolic form is: $\sim(p \wedge q)$ ■

Truth Tables

To determine the truth value of a compound statement, we use charts called **truth tables**. The following tables represent the three basic logical operations.

TABLE 1 Negation

p	q	$\sim p$	$\sim q$
T	T	F	F
T	F	F	T
F	T	T	F
F	F	T	T

TABLE 2 Conjunction

p	q	$p \wedge q$
T	T	T
T	F	F
F	T	F
F	F	F

TABLE 3 Disjunction

p	q	$p \vee q$
T	T	T
T	F	T
F	T	T
F	F	F

For the sake of uniformity, all truth tables with two component statements will have T and F values for p and q assigned in the order shown in the first column of these three tables. Truth tables for several operations can be combined into one chart by using the same two first columns. For each operation a new column is added. Such an arrangement is especially useful with compound statements that involve more than one logical operation and for showing that two statements are logically equivalent.

Logical Equivalence Two compound statements are **logically equivalent** if they have identical truth tables. Symbolically, we denote the equivalence of the statements p and q by writing $p \equiv q$.

EXAMPLE 4 ■ Logical Equivalence

Use a truth table to show the logical equivalence of the statements $\sim p \wedge \sim q$ and $\sim(p \vee q)$.

Solution

TABLE 4

p	q	$\sim p$	$\sim q$	$\sim p \wedge \sim q$	$p \vee q$	$\sim(p \vee q)$
T	T	F	F	F	T	F
T	F	F	T	F	T	F
F	T	T	F	F	T	F
F	F	T	T	T	F	T

Identical (columns $\sim p \wedge \sim q$ and $\sim(p \vee q)$)

Since the fifth and seventh columns in Table 4 are identical, the two given statements are logically equivalent. ■

The equivalence established in Example 4 is one of two well-known rules in logic called **DeMorgan's Laws**. Verification of the second of DeMorgan's Laws is left as an exercise.

DeMorgan's Laws

1. $\sim(p \vee q) \equiv \sim p \wedge \sim q$
2. $\sim(p \wedge q) \equiv \sim p \vee \sim q$

Compound statements that are true, no matter what the truth values of component statements, are called **tautologies**. One simple example is the statement "*p* or not *p*," as shown in Table 5.

TABLE 5 $p \vee \sim p$ **is a tautology**

p	$\sim p$	$p \vee \sim p$
T	F	T
F	T	T

A.1 EXERCISES

In Exercises 1–12, classify the sentence as a statement, a nonstatement, or an open statement.

1. All dogs are brown.

2. Can I help you?

3. That figure is a circle.

4. Substitute 4 for x.

5. x is larger than 4.

6. 8 is larger than 4.

7. $x + y = 10$

8. $12 + 3 = 14$

9. Hockey is fun to watch.

10. One mile is greater than one kilometer.

11. It is more than one mile to the school.

12. Come to the party.

In Exercises 13–20, determine whether the open statement is true for the given values of x.

Open Statement	Values of x	
13. $x^2 - 5x + 6 = 0$	(a) $x = 2$	(b) $x = -2$
14. $x^2 - x - 6 = 0$	(a) $x = 2$	(b) $x = -2$
15. $x^2 \leq 4$	(a) $x = -2$	(b) $x = 0$
16. $\lvert x - 3 \rvert = 4$	(a) $x = -1$	(b) $x = 7$
17. $4 - \lvert x \rvert = 2$	(a) $x = 0$	(b) $x = 1$
18. $\sqrt{x^2} = x$	(a) $x = 3$	(b) $x = -3$
19. $\dfrac{x}{x} = 1$	(a) $x = -4$	(b) $x = 0$
20. $\sqrt[3]{x} = -2$	(a) $x = 8$	(b) $x = -8$

In Exercises 21–24, write the verbal form for each of the following.

(a) $\sim p$ (b) $\sim q$ (c) $p \wedge q$ (d) $p \vee q$

21. p: The sun is shining.
q: It is hot.

22. p: The car has a radio.
q: The car is red.

23. p: Lions are mammals.
q: Lions are carnivorous.

24. p: Twelve is less than fifteen.
q: Seven is a prime number.

In Exercises 25–28, write the verbal form for each of the following.

(a) $\sim p \wedge q$ (b) $\sim p \vee q$ (c) $p \wedge \sim q$ (d) $p \vee \sim q$

25. p: The sun is shining.
q: It is hot.

26. p: The car has a radio.
q: The car is red.

27. p: Lions are mammals.
q: Lions are carnivorous.

28. p: Twelve is less than fifteen.
q: Seven is a prime number.

In Exercises 29–32, write the symbolic form of the given compound statement. In each case let p represent the statement "It is four o'clock," and let q represent the statement "It is time to go home."

29. It is four o'clock and it is not time to go home.

30. It is not four o'clock or it is not time to go home.

31. It is not four o'clock or it is time to go home.

32. It is four o'clock and it is time to go home.

In Exercises 33–36, write the symbolic form of the given compound statement. In each case let p represent the statement "The dog has fleas," and let q represent the statement "The dog is scratching."

33. The dog does not have fleas or the dog is not scratching.

34. The dog has fleas and the dog is scratching.

35. The dog does not have fleas and the dog is scratching.

36. The dog has fleas or the dog is not scratching.

In Exercises 37–42, write the negation of the given statement.

37. The bus is not blue.

38. Frank is not six feet tall.

39. x is equal to 4.

40. x is not equal to 4.

41. The Earth is not flat.

42. The Earth is flat.

In Exercises 43–48, construct a truth table for the given compound statement.

43. $\sim p \wedge q$

44. $\sim p \vee q$

45. $\sim p \vee \sim q$

46. $\sim p \wedge \sim q$

47. $p \vee \sim q$

48. $p \wedge \sim q$

In Exercises 49–54, use a truth table to determine whether the given statements are logically equivalent.

49. $\sim p \wedge q, \quad p \vee \sim q$

50. $\sim(p \wedge \sim q), \quad \sim p \vee q$

51. $\sim(p \vee \sim q), \quad \sim p \wedge q$

52. $\sim(p \vee q), \quad \sim p \vee \sim q$

53. $p \wedge \sim q, \quad \sim(\sim p \vee q)$

54. $p \wedge \sim q, \quad \sim(\sim p \wedge q)$

In Exercises 55–58, determine whether the statements are logically equivalent.

55. (a) The house is red and it is not made of wood.
(b) The house is red or it is not made of wood.

56. (a) It is not true that the tree is not green.
(b) The tree is green.

57. (a) The statement that the house is white or blue is not true.
(b) The house is not white and it is not blue.

58. (a) I am not twenty-five years old and I am not applying for this job.
(b) The statement that I am twenty-five years old and applying for this job is not true.

In Exercises 59–62, use a truth table to determine whether the given statement is a tautology.

59. $\sim p \wedge p$

60. $\sim p \vee p$

61. $\sim(\sim p) \vee \sim p$

62. $\sim(\sim p) \wedge \sim p$

63. Use a truth table to verify the second of DeMorgan's Laws:

$$\sim(p \wedge q) \equiv \sim p \vee \sim q$$

SECTION A.2

Implications, Quantifiers, and Venn Diagrams

Implications • *Logical Quantifiers* • *Venn Diagrams*

Implications

A statement of the form "If p, then q," is called an **implication** (or a conditional statement) and is denoted by

$$p \rightarrow q.$$

We call p the **hypothesis** and q the **conclusion**. There are many different ways to express the implication $p \rightarrow q$, as shown in the following list.

Different Ways of Stating Implications

The implication $p \rightarrow q$ has the following equivalent verbal forms.

1. If p, then q.
2. p implies q.
3. p, only if q.
4. q follows from p.
5. q is necessary for p.
6. p is sufficient for q.

Normally, we think of the implication $p \rightarrow q$ as having a cause-and-effect relationship between the hypothesis p and the conclusion q. However, you should be careful not to confuse the truth value of the component statements with the truth value of the implication. The following truth table should help you keep this distinction in mind.

TABLE 6 **Implication**

p	q	$p \rightarrow q$
T	T	T
T	F	F
F	T	T
F	F	T

Note in Table 6 that the implication $p \rightarrow q$ is false only when p is true and q is false. This is like a promise. Suppose you promise a friend that "If the sun shines, I will take you fishing." The only way you can break your promise is for the sun to shine (p is true) and you do not take your friend fishing (q is false). If the sun doesn't shine (p is false), you have no obligation to go fishing, and hence, the promise cannot be broken.

EXAMPLE 1 ■ Finding Truth Values of Implications

Give the truth value of each implication.

(a) If 3 is odd, then 9 is odd.

(b) If 3 is odd, then 9 is even.

(c) If 3 is even, then 9 is odd.

(d) If 3 is even, then 9 is even.

Solution

	Hypothesis	*Conclusion*	*Implication*
(a)	T	T	T
(b)	T	F	F
(c)	F	T	T
(d)	F	F	T

■

The next example shows how to write an implication as a disjunction.

EXAMPLE 2 ■ Identifying Equivalent Statements

Use a truth table to show the logical equivalence of the following statements.

(a) If I get a raise, I will take my family on a vacation.

(b) I will not get a raise *or* I will take my family on a vacation.

Solution

We let p represent the statement "I will get a raise," and let q represent the statement "I will take my family on a vacation." Then, we can represent the statement in part (a) as $p \to q$ and the statement in part (b) as $\sim p \vee q$. The logical equivalence of these two statements is shown in the following truth table.

TABLE 7 $p \to q \equiv \sim p \vee q$

p	q	$\sim p$	$\sim p \vee q$	$p \to q$
T	T	F	T	T
T	F	F	F	F
F	T	T	T	T
F	F	T	T	T

Identical (columns $\sim p \vee q$ and $p \to q$)

Because the fourth and fifth columns of the truth table are identical, we can conclude that the two statements $p \to q$ and $\sim p \vee q$ are equivalent. ■

From Table 7 and the fact that $\sim(\sim p) \equiv p$, we can write the **negation of an implication**. That is, since $p \rightarrow q$ is equivalent to $\sim p \vee q$, it follows that the negation of $p \rightarrow q$ must be $\sim(\sim p \vee q)$, which by DeMorgan's Laws can be written as follows.

$$\sim(p \rightarrow q) \equiv p \wedge \sim q$$

For the implication $p \rightarrow q$, there are three important associated implications.

1. The **converse** of $p \rightarrow q$: $q \rightarrow p$
2. The **inverse** of $p \rightarrow q$: $\sim p \rightarrow \sim q$
3. The **contrapositive** of $p \rightarrow q$: $\sim q \rightarrow \sim p$

From Table 8 you can see that these four statements yield two pairs of logically equivalent implications.

TABLE 8

p	q	$\sim p$	$\sim q$	$p \rightarrow q$	$\sim q \rightarrow \sim p$	$q \rightarrow p$	$\sim p \rightarrow \sim q$
T	T	F	F	T	T	T	T
T	F	F	T	F	F	T	T
F	T	T	F	T	T	F	F
F	F	T	T	T	T	T	T

(Columns $p \rightarrow q$ and $\sim q \rightarrow \sim p$: Identical; columns $q \rightarrow p$ and $\sim p \rightarrow \sim q$: Identical)

NOTE: The connective "$\rightarrow$" is used to determine the truth values in the last three columns of Table 8.

EXAMPLE 3 ■ Writing the Converse, Inverse, and Contrapositive

Write the converse, inverse, and contrapositive for the implication "If I get a B on my test, then I will pass the course."

Solution

(a) *Converse:* If I pass the course, then I got a B on my test.

(b) *Inverse:* If I do not get a B on my test, then I will not pass the course.

(c) *Contrapositive:* If I do not pass the course, then I did not get a B on my test. ■

NOTE: In Example 3 be sure you see that neither the converse nor the inverse are logically equivalent to the original implication. To see this, consider that the original implication simply states that if you get a B on your test, then you will pass the course. The converse is not true because knowing that you passed the course does not imply that you got a B on the test. After all, you might have gotten an A on the test!

A **biconditional statement**, denoted by $p \leftrightarrow q$, is the conjunction of the implications $p \rightarrow q$ and $q \rightarrow p$. We often write a biconditional statement as "p if and only if q," or in shorter form as "p iff q." A biconditional statement is true when both components are true and when both components are false, as shown in the following truth table.

TABLE 9 Biconditional Statement: *p* if and only if *q*

p	q	$p \rightarrow q$	$q \rightarrow p$	$p \leftrightarrow q$	$(p \rightarrow q) \wedge (q \rightarrow p)$
T	T	T	T	T	T
T	F	F	T	F	F
F	T	T	F	F	F
F	F	T	T	T	T

The following list summarizes some of the laws of logic that we have discussed up to this point.

Laws of Logic

1. For every statement p, either p is true or p is false. — Law of Excluded Middle
2. $\sim(\sim p) \equiv p$ — Law of Double Negation
3. $\sim(p \vee q) \equiv \sim p \wedge \sim q$ — DeMorgan's Law
4. $\sim(p \wedge q) \equiv \sim p \vee \sim q$ — DeMorgan's Law
5. $p \rightarrow q \equiv \sim p \vee q$ — Law of Implication
6. $p \rightarrow q \equiv \sim q \rightarrow \sim p$ — Law of Contraposition

Logical Quantifiers

Logical quantifiers are words such as *some*, *all*, *every*, *each*, *one*, and *none*. Here are some examples of statements with quantifiers.

Some isosceles triangles are right triangles.

Every painting on display is for sale.

Not all corporations have male chief executive officers.

All squares are parallelograms.

Being able to recognize the negation of a statement involving a quantifier is one of the most important skills in logic. For instance, consider the statement "All dogs are brown." In order for this statement to be false, we do not have to show that *all* dogs are not brown, we must simply find at least one dog that is not brown. Thus, the negation of the statement is "Some dogs are brown."

Next we list some of the more common negations involving quantifiers.

Negating Statements with Quantifiers

	Statement	*Negation*
1.	All p are q.	Some p are not q.
2.	Some p are q.	No p is q.
3.	Some p are not q.	All p are q.
4.	No p is q.	Some p are q.

When using logical quantifiers, the word *all* can be replaced by the words *each* or *every*. For instance, the following are equivalent.

All p are q. Each p is q. Every p is q.

Similarly, the word *some* can be replaced by the words *at least one*. For instance, the following are equivalent.

Some p are q. At least one p is q.

EXAMPLE 4 ■ Negating Quantifying Statements

Write the negation of each of the following.

(a) All students study.

(b) Not all prime numbers are odd.

(c) At least one mammal can fly.

(d) Some bananas are not yellow.

Solution

(a) Some students do not study.

(b) All prime numbers are odd.

(c) No mammals can fly.

(d) All bananas are yellow. ■

Venn Diagrams

Venn diagrams are figures that are used to show relationships between two or more sets of objects. They can help us interpret quantifying statements. Study the following Venn diagrams in which the circle marked A represents people over six feet tall and the circle marked B represents the basketball players.

1. All basketball players are over six feet tall.

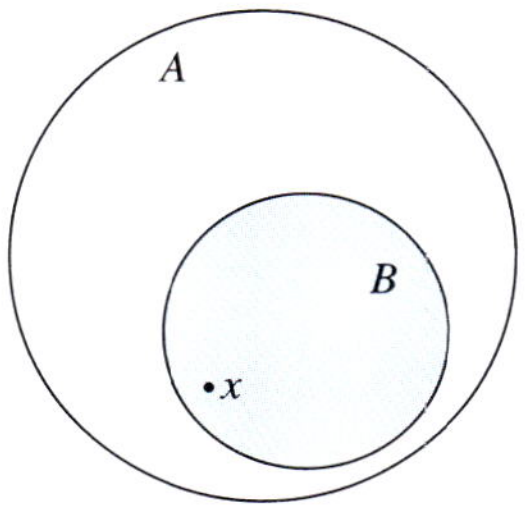

2. Some basketball players are over six feet tall.

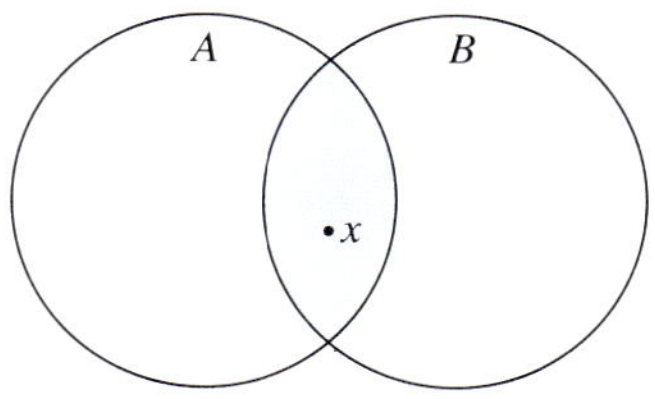

3. Some basketball players are not over six feet tall.

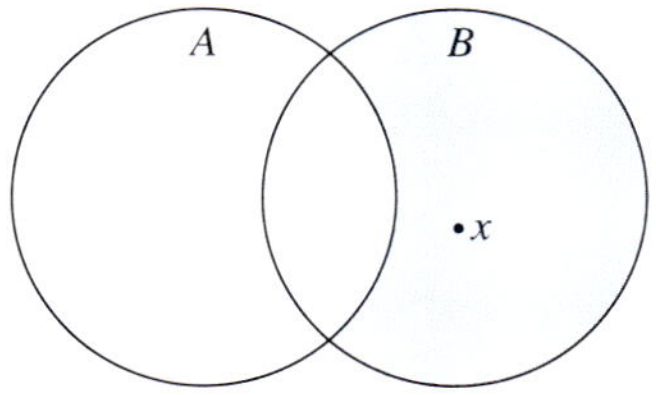

4. No basketball player is over six feet tall.

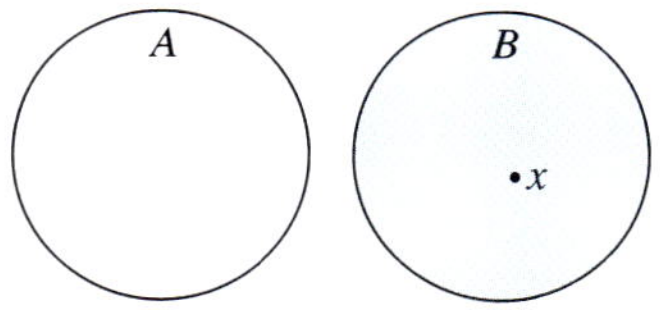

A.2 EXERCISES

In Exercises 1–4, write the verbal form for each of the following.

(a) $p \to q$ (b) $q \to p$ (c) $\sim q \to \sim p$ (d) $p \to \sim q$

1. p: The engine is running.
q: The engine is wasting gasoline.

2. p: The student is at school.
q: It is nine o'clock.

3. p: The integer is even.
q: It is divisible by two.

4. p: The person is generous.
q: The person is rich.

In Exercises 5–10, write the symbolic form of the compound statement. Let p represent the statement "The economy is expanding," and let q represent the statement "Interest rates are low."

5. If interest rates are low, then the economy is expanding.

6. If interest rates are not low, then the economy is not expanding.

7. An expanding economy implies low interest rates.

8. Low interest rates are sufficient for an expanding economy.

9. Low interest rates are necessary for an expanding economy.

10. The economy will expand only if interest rates are low.

In Exercises 11–20, give the truth value of the implication.

11. If 4 is even, then 12 is even.

12. If 4 is even, then 2 is odd.

13. If 4 is odd, then 3 is odd.

14. If 4 is odd, then 2 is odd.

15. If $2n$ is even, then $2n + 2$ is odd.

16. If $2n + 1$ is even, then $2n + 2$ is odd.

17. $3 + 11 > 16$ only if $2 + 3 = 5$.

18. $\frac{1}{6} < \frac{2}{3}$ is necessary for $\frac{1}{2} > 0$.

19. $x = -2$ follows from $2x + 3 = x + 1$.

20. If $2x = 224$, then $x = 10$.

In Exercises 21–26, write the converse, inverse, and contrapositive of the statement.

21. If the sky is clear, then you can see the eclipse.

22. If the person is nearsighted, then he is ineligible for the job.

23. If taxes are raised, then the deficit will increase.

24. If wages are raised, then the company's profits will decrease.

25. It is necessary to have a birth certificate to apply for the visa.

26. The number is divisible by three only if the sum of its digits is divisible by three.

In Exercises 27–40, write the negation of the statement.

27. Paul is a junior or senior.

28. Jack is a senior and he plays varsity basketball.

29. If the temperature increases, then the metal rod will expand.

30. If the test fails, then the project will be halted.

31. We will go to the ocean only if the weather forecast is good.

32. Completing the pass on this play is necessary if we are going to win the game.

33. Some students are in extracurricular activities.

34. Some odd integers are not prime numbers.

35. All contact sports are dangerous.

36. All members must pay their dues prior to June 1.

37. No child is allowed at the concert.

38. No contestant is over the age of twelve.

39. At least one of the $20 bills is counterfeit.

40. At least one unit is defective.

In Exercises 41–48, construct a truth table for the compound statement.

41. $\sim(p \to \sim q)$

42. $\sim q \to (p \to q)$

43. $\sim(q \to p) \land q$

44. $p \to (\sim p \lor q)$

45. $[(p \lor q) \land (\sim p)] \to q$

46. $[(p \to q) \land (\sim q)] \to p$

47. $(p \leftrightarrow \sim q) \to \sim p$

48. $(p \lor \sim q) \leftrightarrow (q \to \sim p)$

In Exercises 49–56, use a truth table to show the logical equivalence of the two statements.

49. $q \rightarrow p$ $\qquad$ $\sim p \rightarrow \sim q$

50. $\sim p \rightarrow q$ $\qquad$ $p \vee q$

51. $\sim(p \rightarrow q)$ $\qquad$ $p \wedge \sim q$

52. $(p \vee q) \rightarrow q$ $\qquad$ $p \rightarrow q$

53. $(p \rightarrow q) \vee \sim q$ $\qquad$ $p \vee \sim p$

54. $q \rightarrow (\sim p \vee q)$ $\qquad$ $q \vee \sim q$

55. $p \rightarrow (\sim p \wedge q)$ $\qquad$ $\sim p$

56. $\sim(p \wedge q) \rightarrow \sim q$ $\qquad$ $p \vee \sim q$

57. Select the statement that is logically equivalent to the statement "If a number is divisible by six, then it is divisible by two."
(a) If a number is divisible by two, then it is divisible by six.
(b) If a number is not divisible by six, then it is not divisible by two.
(c) If a number is not divisible by two, then it is not divisible by six.
(d) Some numbers are divisible by six and not divisible by two.

58. Select the statement that is logically equivalent to the statement "It is not true that Pam is a conservative and a Democrat."
(a) Pam is a conservative and a Democrat.
(b) Pam is not a conservative and not a Democrat.
(c) Pam is not a conservative or she is not a Democrat.
(d) If Pam is not a conservative, then she is a Democrat.

59. Select the statement that is *not* logically equivalent to the statement "Every citizen over the age of 18 has the right to vote."
(a) Some citizens over the age of 18 have the right to vote.
(b) Each citizen over the age of 18 has the right to vote.
(c) All citizens over the age of 18 have the right to vote.
(d) No citizen over the age of 18 can be restricted from voting.

60. Select the statement that is *not* logically equivalent to the statement "It is necessary to pay the registration fee to take the course."
(a) If you take the course, then you must pay the registration fee.
(b) If you do not pay the registration fee, then you cannot take the course.
(c) If you pay the registration fee, then you may take the course.
(d) You may take the course only if you pay the registration fee.

In Exercises 61–70, sketch a Venn diagram and shade the region that illustrates the given statement. Let *A* be a circle that represents people who are happy, and let *B* be a circle that represents college students.

61. All college students are happy.

62. All happy people are college students.

63. No college students are happy.

64. No happy people are college students.

65. Some college students are not happy.

66. Some happy people are not college students.

67. At least one college student is happy.

68. At least one happy person is not a college student.

69. Each college student is sad.

70. Each sad person is not a college student.

In Exercises 71–74, state whether the statement follows from the given Venn diagram. Assume that each area shown in the Venn diagram is non-empty. (*Note:* Use only the information given in the diagram. Do not be concerned with whether the statement is actually true or false.)

71. (a) All toads are green.
(b) Some toads are green.

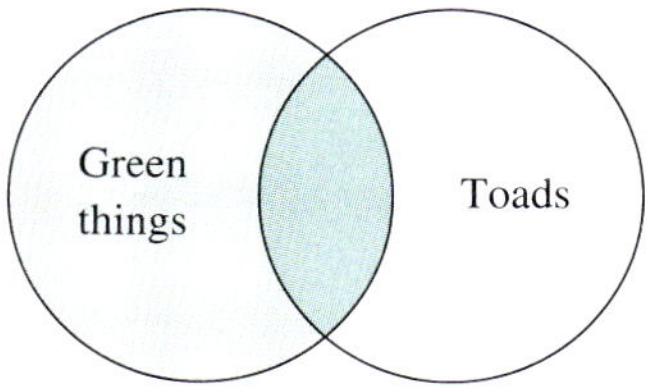

72. (a) All men are company presidents.
(b) Some company presidents are women.

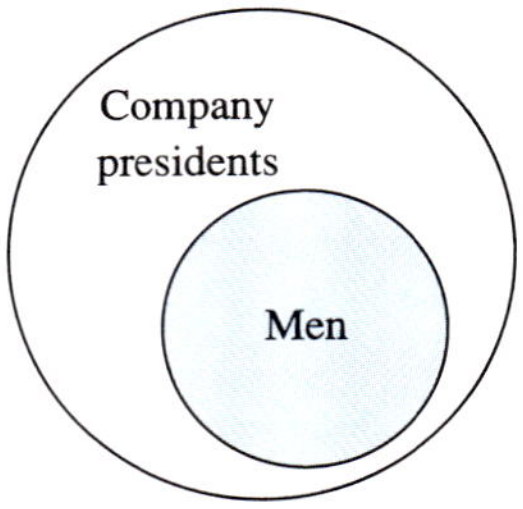

73. (a) All blue cars are old.
(b) Some blue cars are not old.

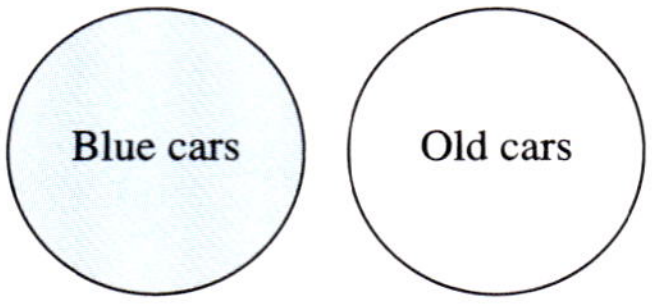

74. (a) No football players are over six feet tall.
(b) Every football player is over six feet tall.

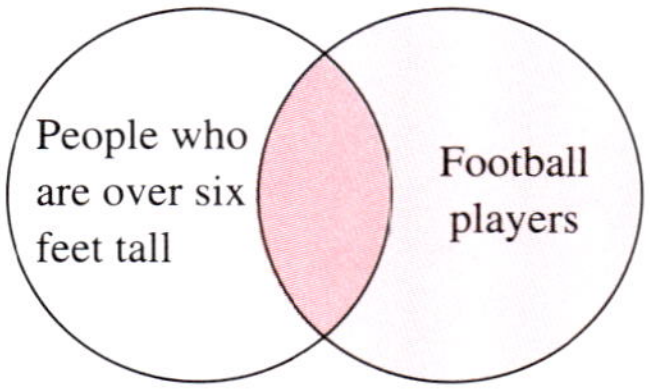

SECTION A.3 Logical Arguments

Arguments • Venn Diagrams and Arguments • Proofs

Arguments

An **argument** is a collection of statements, listed in order. The last statement is called the **conclusion** and the other statements are called the **premises**. An argument is **valid** if the conjunction of all the premises implies the conclusion. The most common type of argument takes the following form.

Premise #1: $p \to q$
Premise #2: p
Conclusion: q

This form of argument is called the **Law of Detachment** or ***Modus Ponens***. It is illustrated in the following example.

EXAMPLE 1 ■ A Valid Argument

Show that the following argument is valid.

Premise #1: If Sean is a freshman, then he is taking algebra.
Premise #2: Sean is a freshman.
Conclusion: Therefore, Sean is taking algebra.

Solution

We let p represent the statement "Sean is a freshman," and q represent the statement "Sean is taking algebra." Then the argument fits the Law of Detachment, which can be written as follows.

$$[(p \rightarrow q) \wedge p] \rightarrow q$$

The validity of this argument is shown in the following truth table.

TABLE 10 **Law of Detachment**

p	q	$p \rightarrow q$	$(p \rightarrow q) \wedge p$	$[(p \rightarrow q) \wedge p] \rightarrow q$
T	T	T	T	T
T	F	F	F	T
F	T	T	F	T
F	F	T	F	T

■

Keep in mind that the validity of an argument has nothing to do with the truthfulness of the premises or conclusion. For instance, the following argument is valid—the fact that it is fanciful does not alter its validity.

Premise #1: If I snap my fingers, elephants will stay out of my house.
Premise #2: I am snapping my fingers.
Conclusion: Therefore, elephants will stay out of my house.

We have discussed the most common form of logical argument. This and three other commonly used forms of valid arguments are summarized in the following list.

Four Types of Valid Arguments

Name	*Pattern*	
1. **Law of Detachment or *Modus Ponens***	Premise #1:	$p \rightarrow q$
	Premise #2:	p
	Conclusion:	q
2. **Law of Contraposition or *Modus Tollens***	Premise #1:	$p \rightarrow q$
	Premise #2:	$\sim q$
	Conclusion:	$\sim p$
3. **Law of Transitivity or Syllogism**	Premise #1:	$p \rightarrow q$
	Premise #2:	$q \rightarrow r$
	Conclusion:	$p \rightarrow r$
4. **Law of Disjunctive Syllogism**	Premise #1:	$p \vee q$
	Premise #2:	$\sim p$
	Conclusion:	q

EXAMPLE 2 ■ An Invalid Argument

Determine whether the following argument is valid.

Premise #1:	If John is elected, the income tax will be increased.
Premise #2:	The income tax was increased.
Conclusion:	Therefore, John was elected.

Solution

This argument has the following form.

Pattern		*Implication*
Premise #1:	$p \to q$	$[(p \to q) \wedge q] \to p$
Premise #2:	q	
Conclusion:	p	

This is not one of the four valid forms of arguments that we listed. We can construct a truth table to verify that the argument is invalid, as follows.

TABLE 11 An Invalid Argument

p	q	$p \to q$	$(p \to q) \wedge q$	$[(p \to q) \wedge q] \to p$
T	T	T	T	T
T	F	F	F	T
F	T	T	T	F
F	F	T	F	T

■

An invalid argument, like the one in Example 2, is called a **fallacy**. Other common fallacies are given in the following example.

EXAMPLE 3 ■ Common Fallacies

Each of the following arguments is invalid.

(a) *Arguing from the Converse:* If the football team wins the championship, then students will skip classes. The students skipped classes. Therefore, the football team won the championship.

(b) *Arguing from the Inverse:* If the football team wins the championship, then students will skip classes. The football team did not win the championship. Therefore, the students did not skip classes.

(c) *Arguing from False Authority:* Wheaties are best for you because Joe Montana eats them.

(d) *Arguing from an Example:* Beta Brand products are not reliable because my Beta Brand snowblower does not start in cold weather.

(e) *Arguing from Ambiguity:* If automobile carburetors are modified, the automobile will pollute. Brand X automobiles have modified carburetors. Therefore, Brand X automobiles pollute.

(f) *Arguing by False Association:* Joe was running through the alley when the fire alarm went off. Therefore, Joe started the fire. ■

EXAMPLE 4 ■ A Valid Argument

Determine whether the following argument is valid.

Premise #1:	You like strawberry pie or you like chocolate pie.
Premise #2:	You do not like strawberry pie.
Conclusion:	Therefore, you like chocolate pie.

Solution

This argument has the following form.

Premise #1:	$p \vee q$
Premise #2:	$\sim p$
Conclusion:	q

This argument is a disjunctive syllogism, which is one of the four common types of valid arguments. ■

In a valid argument, the conclusion drawn from the premise is called a **valid conclusion**.

EXAMPLE 5 ■ Making Valid Conclusions

Given the following two premises, which of the conclusions are valid?

Premise #1:	If you like boating, then you like swimming.
Premise #2:	If you like swimming, then you are a scholar.

(a) Conclusion: If you like boating, then you are a scholar.

(b) Conclusion: If you do not like boating, then you are not a scholar.

(c) Conclusion: If you are not a scholar, then you do not like boating.

Solution

(a) This conclusion is valid. It follows from the Law of Transitivity (or syllogism).

(b) This conclusion is invalid. The fallacy stems from arguing from the inverse.

(c) This conclusion is valid. It follows from the Law of Contraposition. ■

FIGURE A.1

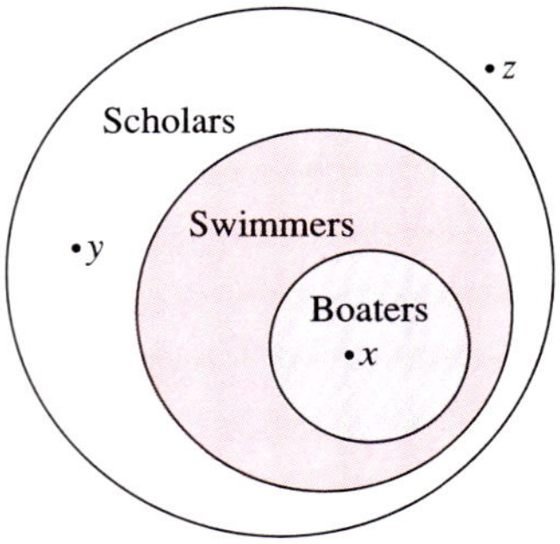

Venn Diagrams and Arguments

Venn diagrams can be used to informally test the validity of an argument. For instance, a Venn diagram for the premises in Example 5 is shown in Figure A.1. In this figure the validity of Conclusion (a) is seen by choosing a boater x in all three sets. Conclusion (b) is seen to be invalid by choosing a person y who is a scholar but does not like boating. Finally, person z indicates the validity of Conclusion (c).

Venn diagrams work well for testing arguments that involve quantifiers, as shown in the next two examples.

EXAMPLE 6 ■ Using a Venn Diagram to Show that an Argument Is Not Valid

Use a Venn diagram to test the validity of the following argument.

Premise #1:	Some plants are green.
Premise #2:	All lettuce is green.
Conclusion:	Therefore, lettuce is a plant.

Solution

FIGURE A.2

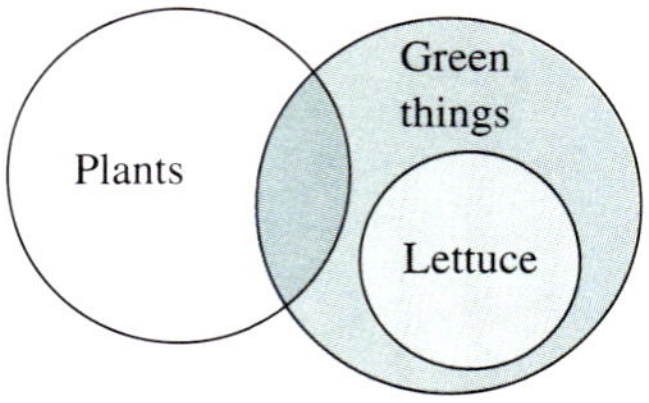

From the Venn diagram shown in Figure A.2, we can see that this is not a valid argument. Remember that even though the conclusion is true (lettuce is a plant), this does not imply that the argument is true. ■

NOTE: When you are using Venn diagrams, you must remember to draw the most general case. For example, in Figure A.2 the circle representing plants is not drawn entirely within the circle representing green things because we are told that only *some* plants are green.

EXAMPLE 7 ■ Using a Venn Diagram to Show that an Argument Is Valid

Use a Venn diagram to test the validity of the following argument.

Premise #1:	All good tennis players are physically fit.
Premise #2:	Some golfers are good tennis players.
Conclusion:	Therefore, some golfers are physically fit.

FIGURE A.3

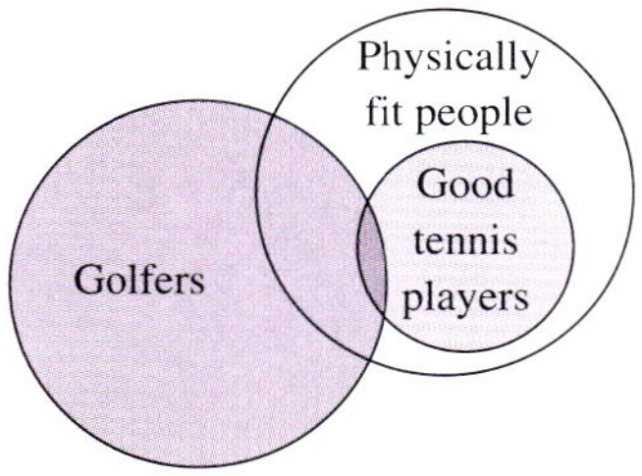

Solution

Since the set of golfers intersects the set of good tennis players, we see from Figure A.3 that the set of golfers must also intersect the set of physically fit people. Therefore, the argument is valid. ■

Proofs

What does the word *proof* mean to you? In mathematics we use the word *proof* to simply mean a valid argument. Many proofs involve more than two premises and a conclusion. For instance, the proof in Example 8 involves three premises and a conclusion.

EXAMPLE 8 ■ A Proof by Contraposition

Use the following three premises to prove that "It is not snowing today."

Premise #1:	If it is snowing today, Greg will go skiing.
Premise #2:	If Greg is skiing today, then he is not studying.
Premise #3:	Greg is studying today.

Solution

We let p represent the statement "It is snowing today," let q represent "Greg is skiing," and let r represent "Greg is studying today." Thus, the given premises have the following form.

Premise #1:	$p \to q$
Premise #2:	$q \to \sim r$
Premise #3:	r

By noting that $r \equiv \sim(\sim r)$, reordering the premises, and writing the contrapositives of the first and second premises, we can obtain the following valid argument.

Premise #3:	r
Contrapositive of Premise #2:	$r \to \sim q$
Contrapositive of Premise #1:	$\sim q \to \sim p$
Conclusion:	$\sim p$

Thus, we can conclude $\sim p$. That is, "It is not snowing today." ■

A.3 EXERCISES

In Exercises 1–4, use a truth table to show that the given argument is valid.

1. Premise #1: $p \to \sim q$
Premise #2: q
Conclusion: $\sim p$

2. Premise #1: $p \leftrightarrow q$
Premise #2: p
Conclusion: q

3. Premise #1: $p \vee q$
Premise #2: $\sim p$
Conclusion: q

4. Premise #1: $p \wedge q$
Premise #2: $\sim p$
Conclusion: q

In Exercises 5–8, use a truth table to show that the given argument is invalid.

5. Premise #1: $\sim p \to q$
Premise #2: p
Conclusion: $\sim q$

6. Premise #1: $p \to q$
Premise #2: $\sim p$
Conclusion: $\sim q$

7. Premise #1: $p \vee q$
Premise #2: q
Conclusion: p

8. Premise #1: $\sim(p \wedge q)$
Premise #2: q
Conclusion: p

In Exercises 9–22, determine whether the argument is valid or invalid.

9. Premise #1: If taxes are increased, then businesses will leave the state.
Premise #2: Taxes are increased.
Conclusion: Therefore, businesses will leave the state.

10. Premise #1: If a student does the homework, then a good grade is certain.
Premise #2: Liza does the homework.
Conclusion: Therefore, Liza will receive a good grade for the course.

11. Premise #1: If taxes are increased, then businesses will leave the state.
Premise #2: Businesses are leaving the state.
Conclusion: Therefore, taxes were increased.

12. Premise #1: If a student does the homework, then a good grade is certain.
Premise #2: Liza received a good grade for the course.
Conclusion: Therefore, Liza did her homework.

13. Premise #1: If the doors are kept locked, then the car will not be stolen.
Premise #2: The car was stolen.
Conclusion: Therefore, the car doors were unlocked.

14. Premise #1: If Jan passes the exam, she is eligible for the position.
Premise #2: Jan is not eligible for the position.
Conclusion: Therefore, Jan did not pass the exam.

15. Premise #1: All cars manufactured by the Ford Motor Company are reliable.
Premise #2: Lincolns are manufactured by Ford.
Conclusion: Therefore, Lincolns are reliable cars.

16. Premise #1: Some cars manufactured by the Ford Motor Company are reliable.
Premise #2: Lincolns are manufactured by Ford.
Conclusion: Therefore, Lincolns are reliable.

17. Premise #1: All federal income tax forms are subject to the Paperwork Reduction Act of 1980.
Premise #2: The 1040 Schedule A form is subject to the Paperwork Reduction Act of 1980.
Conclusion: Therefore, the 1040 Schedule A form is a federal income tax form.

18. Premise #1: All integers divisible by six are divisible by three.
Premise #2: Eighteen is divisible by six.
Conclusion: Therefore, eighteen is divisible by three.

19. Premise #1: Eric is at the store or the handball court.
Premise #2: He is not at the store.
Conclusion: Therefore, he must be at the handball court.

20. Premise #1: The book must be returned within two weeks or you pay a fine.
Premise #2: The book was not returned within two weeks.
Conclusion: Therefore, you must pay a fine.

21. Premise #1: It is not true that it is a diamond and it sparkles in the sunlight.
Premise #2: It does sparkle in the sunlight.
Conclusion: Therefore, it is a diamond.

22. Premise #1: Either I work tonight or I pass the mathematics test.
Premise #2: I'm going to work tonight.
Conclusion: Therefore, I will fail the mathematics test.

In Exercises 23–30, determine which conclusion is valid from the given premises.

23. Premise #1: If seven is a prime number, then seven does not divide evenly into twenty-one.
Premise #2: Seven divides evenly into twenty-one.
(a) Conclusion: Therefore, seven is a prime number.
(b) Conclusion: Therefore, seven is not a prime number.
(c) Conclusion: Therefore, twenty-one divided by seven is three.

24. Premise #1: If the fuel is shut off, then the fire will be extinguished.
Premise #2: The fire continues to burn.
(a) Conclusion: Therefore, the fuel was not shut off.
(b) Conclusion: Therefore, the fuel was shut off.
(c) Conclusion: Therefore, the fire becomes hotter.

25. Premise #1: It is necessary that interest rates be lowered for the economy to improve.
Premise #2: Interest rates were not lowered.
(a) Conclusion: Therefore, the economy will improve.
(b) Conclusion: Therefore, interest rates are irrelevant to the performance of the economy.
(c) Conclusion: Therefore, the economy will not improve.

26. Premise #1: It will snow only if the temperature is below 32° at some level of the atmosphere.
Premise #2: It is snowing.
(a) Conclusion: Therefore, the temperature is below 32° at ground level.
(b) Conclusion: Therefore, the temperature is above 32° at some level of the atmosphere.
(c) Conclusion: Therefore, the temperature is below 32° at some level of the atmosphere.

27. Premise #1: Smokestack emissions must be reduced or acid rain will continue as an environmental problem.
Premise #2: Smokestack emissions have not decreased.
(a) Conclusion: Therefore, the ozone layer will continue to be depleted.
(b) Conclusion: Therefore, acid rain will continue as an environmental problem.
(c) Conclusion: Therefore, stricter automobile emission standards must be enacted.

28. Premise #1: The library must upgrade its computer system or service will not improve.
Premise #2: Service at the library has improved.
(a) Conclusion: Therefore, the computer system was upgraded.
(b) Conclusion: Therefore, more personnel were hired for the library.
(c) Conclusion: Therefore, the computer system was not upgraded.

29. Premise #1: If Rodney studies, then he will make good grades.
Premise #2: If he makes good grades, then he will get a good job.
(a) Conclusion: Therefore, Rodney will get a good job.
(b) Conclusion: Therefore, if Rodney doesn't study, then he won't get a good job.
(c) Conclusion: Therefore, if Rodney doesn't get a good job, then he didn't study.

30. Premise #1: It is necessary to have a ticket and an ID card to get into the arena.
Premise #2: Janice entered the arena.
(a) Conclusion: Therefore, Janice does not have a ticket.
(b) Conclusion: Therefore, Janice has a ticket and an ID card.
(c) Conclusion: Therefore, Janice has an ID card.

In Exercises 31–34, use a Venn diagram to test the validity of the argument.

31. Premise #1: All numbers divisible by ten are divisible by five.
Premise #2: Fifty is divisible by ten.
Conclusion: Therefore, fifty is divisible by five.

32. Premise #1: All human beings require adequate rest.
Premise #2: All infants are human beings.
Conclusion: Therefore, all infants require adequate rest.

33. Premise #1: No person under the age of eighteen is eligible to vote.
Premise #2: Some college students are eligible to vote.
Conclusion: Therefore, some college students are under the age of eighteen.

34. Premise #1: Every amateur radio operator has a radio license.
Premise #2: Jackie has a radio license.
Conclusion: Therefore, Jackie is an amateur radio operator.

In Exercises 35–38, use the premises to prove the given conclusion.

35. Premise #1: If Sue drives to work, then she will stop at the grocery store.
Premise #2: If she stops at the grocery store, then she'll buy milk.
Premise #3: Sue drove to work today.
Conclusion: Therefore, Sue will get milk.

36. Premise #1: If Bill is patient, then he will succeed.
Premise #2: Bill will get bonus pay if he succeeds.
Premise #3: Bill did not get bonus pay.
Conclusion: Therefore, Bill is not patient.

37. Premise #1: If this is a good product, then we should buy it.
Premise #2: Either it was made by XYZ Corporation, or we will not buy it.
Premise #3: It is not made by XYZ Corporation.
Conclusion: Therefore, it is not a good product.

38. Premise #1: If the book is returned within two weeks, then there is no fine.
Premise #2: You pay a fine or you may not check out another book.
Premise #3: You are allowed to check out another book.
Conclusion: Therefore, the book was not returned within two weeks.

ANSWERS TO ODD-NUMBERED EXERCISES

Chapter 1

SECTION 1.1 *(page 10)*

1. **(a)** 1, 4, 6 **(b)** $-10, 0, 1, 4, 6$
(c) $-10, \frac{-2}{3}, \frac{-1}{4}, 0, \frac{5}{8}, 1, 4, 6$ **(d)** $-\sqrt{5}, \sqrt{3}, 2\pi$

3. **(a)**

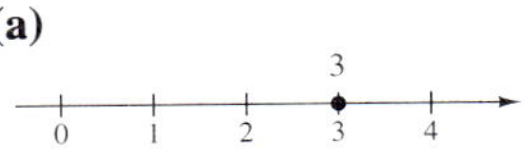

(b)

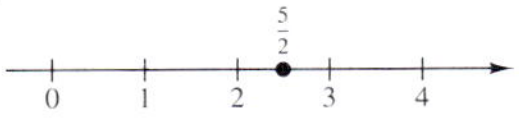

(c)

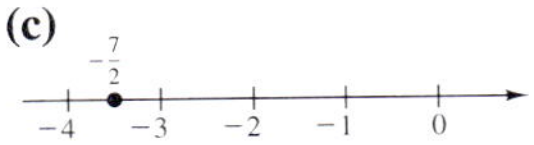

(d)

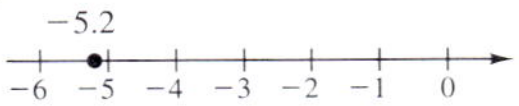

5. $-1 < 3$ **7.** $\frac{-9}{2} < -2$ **9.** $2 < 5$

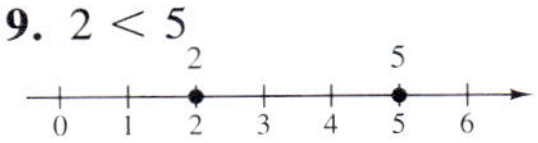

11. $-5 < 2$

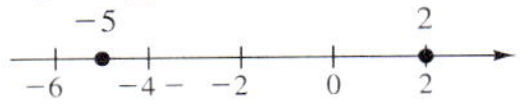

13. $-7 < -2$

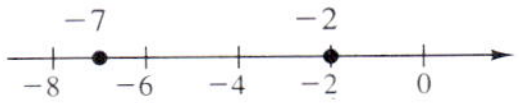

15. $\frac{1}{3} > \frac{1}{4}$

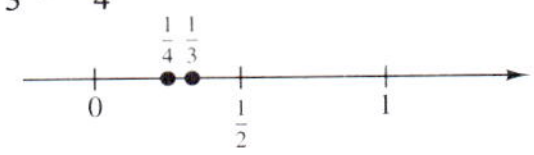

17. $-\frac{5}{8} < \frac{1}{2}$

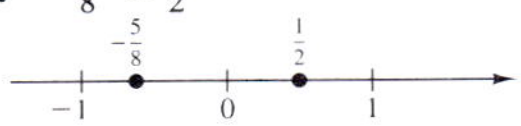

19. $-\frac{5}{3} < -\frac{3}{2}$ **21.** 6 **23.** 19 **25.** 50

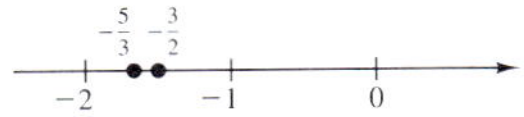

27. 8 **29.** 39 **31.** 10 **33.** 225 **35.** -16
37. -85 **39.** π **41.** $-34, 34$ **43.** 160, 160
45. $\frac{3}{11}, \frac{3}{11}$
47.

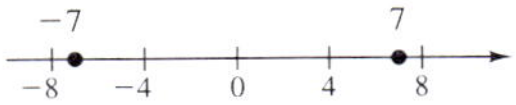

49.

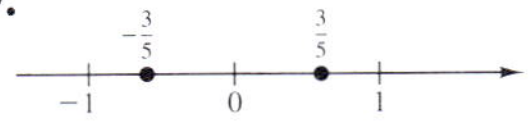

51. $x < 0$ **53.** $x \geq 0$ **55.** $u \geq 16$ **57.** $p < \$225$
59. True **61.** False

SECTION 1.2 *(page 22)*

Warm-Up *(page 21)*

1. $-\frac{5}{2} < \frac{5}{2}$

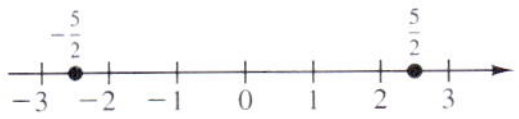

2. $6 > -\frac{1}{2}$

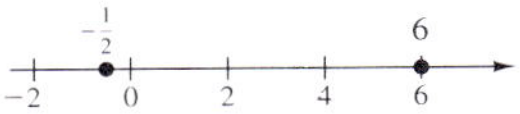

3. $-3 < -\frac{1}{3}$

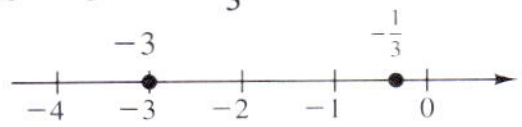

4. $-4.2 < -3.4$

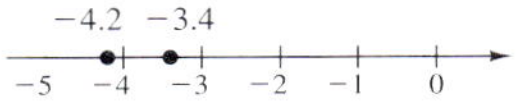

5. 25 **6.** 13 **7.** $-\frac{2}{3}$ **8.** -38
9. $y \leq 45$ **10.** $x \geq 15$

1. 45 **3.** 19 **5.** -19 **7.** -22 **9.** -21
11. -20 **13.** 0.7 **15.** $\frac{5}{4}$ **17.** $\frac{1}{2}$ **19.** $\frac{1}{10}$
21. $\frac{1}{24}$ **23.** $\frac{105}{8}$ **25.** 60 **27.** 45.95 **29.** -28
31. $5 + 5 + 5 + 5$ **33.** $(-4) + (-4) + (-4)$
35. $6\left(\frac{1}{4}\right)$ **37.** $4(-15)$ **39.** -30 **41.** 32.13
43. $\frac{1}{2}$ **45.** $-\frac{1}{3}$ **47.** 6 **49.** -2.5 **51.** 0.4
53. 2.71 **55.** 4.03 **57.** (4)(4)(4)
59. $\left(-\frac{4}{5}\right)\left(-\frac{4}{5}\right)\left(-\frac{4}{5}\right)\left(-\frac{4}{5}\right)\left(-\frac{4}{5}\right)\left(-\frac{4}{5}\right)$ **61.** $\left(\frac{5}{8}\right)^4$
63. $(-4)^6$ **65.** 16 **67.** -8 **69.** $\frac{64}{125}$ **71.** 36
73. 57 **75.** 135 **77.** $-\frac{17}{6}$ **79.** 72.2 **81.** 161
83. 14,425 **85.** 171.36 **87.** \$2,533.56
89. \$10,800 **91.** $\frac{17}{180}$ **93.** $5, 8, -5, 16$
95. 6.125 ft^3 **97.** False **99.** True

SECTION 1.3 *(page 32)*

Warm-Up *(page 32)*

1. 225 **2.** -240 **3.** -150 **4.** 120
5. $\frac{8}{5}$ **6.** $-\frac{1}{4}$ **7.** $\frac{6}{5}$ **8.** $\frac{3}{10}$ **9.** -27
10. $\frac{25}{64}$

1. Commutative Property of Addition
3. Commutative Property of Multiplication
5. Associative Property of Multiplication

7. Multiplicative Identity Property
9. Associative Property of Addition
11. Distributive Property **13.** Additive Identity Property
15. Additive Inverse Property
17. Multiplicative Inverse Property
19. Associative Property of Addition **21.** $(3 \cdot 6)y$
23. $-3(15)$ **25.** $5 \cdot 6 + 5 \cdot z$ **27.** $-x + 25$
29. $x + 8$ **31. (a)** -10 **(b)** $\frac{1}{10}$ **33. (a)** 16
(b) $-\frac{1}{16}$ **35. (a)** $-6z$ **(b)** $\frac{1}{6z}$ **37. (a)** $-x - 1$
(b) $\frac{1}{x+1}$ **39.** $x + (5 - 3)$ **41.** $(3 \cdot 4)5$
43. $(6 \cdot 2)y$ **45.** $(20 \cdot 2) + (20 \cdot 5)$
47. $5 \cdot 3x + 5 \cdot (-4)$
49. $x \cdot (-2) + 6 \cdot (-2)$
51. Given equation
Additive Property of Equality
Associative Property of Addition
Additive Inverse Property
Additive Identity Property
53. Given equation
Additive Property of Equality
Associative Property of Addition
Additive Inverse Property
Additive Identity Property
Multiplicative Property of Equality
Associative Property of Multiplication
Multiplicative Inverse Property
Multiplicative Identity Property
55. $3x + 15$
57. $-2x + 16$ **59.** 0 **61.** 28 **63.** 434
65. 62.82 **67.** $a(b + c) = ab + ac$ **69.** \$0.74
71.

$ac = bc, \quad c \neq 0$	Given equation
$\frac{1}{c}(ac) = \frac{1}{c}(bc)$	Multiplicative Property of Equality
$\frac{1}{c}(ca) = \frac{1}{c}(cb)$	Commutative Property of Multiplication
$\left(\frac{1}{c} \cdot c\right)a = \left(\frac{1}{c} \cdot c\right)b$	Associative Property of Multiplication
$1 \cdot a = 1 \cdot b$	Multiplicative Inverse Property
$a = b$	Multiplicative Identity Property

SECTION 1.4 *(page 42)*

Warm-Up *(page 41)*

1. 168 **2.** 20 **3.** $\frac{-21}{2}$ **4.** 7 **5.** $\frac{72}{5}$
6. $-\frac{19}{10}$ **7.** Multiplicative Inverse Property
8. Associative Property of Addition
9. Distributive Property
10. Additive Identity Property

1. $10x$, 5 **3.** $-3y^2$, $2y$, -8
5. $4x^2$, $-3y^2$, $-5x$, $2y$ **7.** x^2, $-2.5x$, $-\frac{1}{x}$ **9.** 5
11. $-\frac{3}{4}$ **13.** Commutative Property of Addition
15. Associative Property of Multiplication
17. Multiplicative Inverse Property
19. Distributive Property
21. Multiplicative Identity Property **23. (a)** $5x + 5 \cdot 6$
(b) $(x + 6)5$ **25. (a)** $(xy)6$ **(b)** $(6x)y$ **27. (a)** 0
(b) $-4 + 4$ **29.** $x \cdot x \cdot x \cdot x \cdot x \cdot x \cdot x$
31. $(-2x)(-2x)(-2x)$ **33.** $(5x)^4$ **35.** x^3y^3
37. x^{12} **39.** $27y^6$ **41.** $16x^2$ **43.** $6x^3y^4$
45. $-54u^5v^3$ **47.** $7x$ **49.** $8y$ **51.** $8x + 18y$
53. $\frac{11}{2}z^2 + \frac{3}{2}z + 10$ **55.** $4u^2v^2 + uv$
57. $8x^2 + 4x - 12$ **59.** $-18y^2 + 3y + 6$
61. $4(x - 2)$ **63.** $a(b + 2)$ **65.** $12x - 35$
67. $-3y^2 - 7y - 7$ **69.** $-2b^2 + 4b - 36$
71. $9x^3 - 5$ **73.** $2y^3 + y^2 + y$ **75. (a)** 3
(b) -10 **77. (a)** 7 **(b)** 7 **79. (a)** 0 **(b)** $\frac{3}{10}$
81. (a) 13 **(b)** -36 **83. (a)** 0 **(b)** Undefined
85. (a) 210 **(b)** 140
87. (a) $-7, -5, -3, -1, 1, 3$
(b) 2
(c) $\frac{3}{4}$ **89.** \$2,759.4009 million

91.
$$\begin{aligned} &= b_1h + \frac{1}{2}(b_2 - b_1)h \\ &= h\left[b_1 + \frac{1}{2}(b_2 - b_1)\right] \\ &= h\left[b_1 + \frac{1}{2}b_2 - \frac{1}{2}b_1\right] \\ &= h\left[\frac{1}{2}b_1 + \frac{1}{2}b_2\right] \\ &= \frac{1}{2}h(b_1 + b_2) \\ &= \frac{h}{2}(b_1 + b_2) \end{aligned}$$

93. 1,440 ft^2

SECTION 1.5 *(page 53)*

Warm-Up *(page 53)*

1. $-x + 10$ **2.** $11t - 5$ **3.** $43 - 6x$
4. $71 - 7x$ **5.** $y^2 - 12$ **6.** $u^2 - 16$
7. $2n - 3$ **8.** $16n + 6$ **9.** $\frac{11}{t}$ **10.** $\frac{13}{z}$

1. $8 + n$ **3.** $n + 5$ **5.** $15 - 3n$ **7.** $\frac{1}{3}n$
9. $0.30L$ **11.** $\frac{x}{6}$ **13.** $4c + 10$ **15.** $3x^2 - 4$
17. $|n - 5|$ **19.** $\frac{3 + 4x}{8}$
21. The sum of three times a number x and 2
23. Eight times the difference of a number x and 5
25. The ratio of y to 8
27. The sum of a number and 10, all divided by 3
29. A number x times the sum of x and 7 **31.** $0.25n$
33. $5m + 10n$ **35.** $d = 55t$ **37.** $t = \frac{100}{r}$
39. $0.45y$ **41.** $0.0125I$ **43.** $0.80L$
45. $8.25 + 0.60q$ **47.** s^2 **49.** $3.24l$ **51.** $4n + 2$
53. $4n + 4$ **55.** $6w^2$ ft^2
57. $-3, 2, 7, 12, 17, 22, 5, 5, 5, 5, 5$ **59.** a

CHAPTER 1 REVIEW EXERCISES *(page 57)*

1. $3 > -\frac{1}{8}$

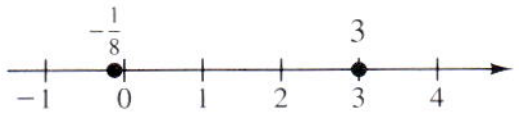

3. $-\frac{8}{5} < -\frac{2}{5}$

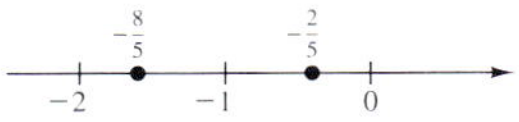

5. 7.2 **7.** -7.2 **9.** 11 **11.** 230 **13.** -38
15. $-4, 200$ **17.** 14 **19.** 0 **21.** $\frac{11}{21}$ **23.** $\frac{1}{6}$
25. $\frac{17}{8}$ **27.** $\frac{-1}{20}$ **29.** 2 **31.** $\frac{2}{5}$ **33.** 5.25
35. -216 **37.** $\frac{1}{12}$ **39.** 20
41. Additive Inverse Property **43.** Distributive Property
45. Associative Property of Addition
47. Commutative Property of Multiplication
49. Multiplicative Inverse Property **51.** $4s - 8t$
53. $-3y^2 + 10y$ **55.** $u - 3v$ **57.** $5x - 10$
59. $5x - y$ **61.** $20u$ **63.** $18b - 15a$ **65.** x^6
67. $-3x^3y^4$ **69.** $-64a^7$ **71.** (a) 0 (b) -3
73. (a) 15 (b) 3 **75.** 6,910 **77.** 55,578,932
79. \$644 **81.** 1 **83.** $200 - 3n$ **85.** $n^2 + 49$
87. The sum of two times a number y and 7
89. A number x decreased by five, all divided by four.
91. $0.18I$ **93.** $l(l - 5)$

CHAPTER 1 TEST *(page 60)*

1. $-\frac{5}{2} > |-3|$ **2.** 11.9 **3.** -57 **4.** $-\frac{1}{2}$
5. -18 **6.** $\frac{1}{6}$ **7.** $\frac{4}{27}$ **8.** -81 **9.** $-\frac{27}{125}$
10. 58 **11.** 15
12. Associative Property of Multiplication
13. Multiplicative Inverse Property **14.** $-5x$
15. $10x - 15$ **16.** $3x^2, -5x, 6$ **17.** $-8z^3$
18. $3x^4y^3$ **19.** $-2x^2 + 5x - 1$ **20.** a^2
21. (a) 4 (b) -12 **22.** $5n - 8$ **23.** $0.6l^2$
24. $4n + 2$ **25.** 16 ft **26.** 640 ft^3

Chapter 2

SECTION 2.1 *(page 70)*

Warm-Up *(page 69)*

1. $-2x$ **2.** $6x + 64$ **3.** $1 - 6x$
4. $-3x + 22$ **5.** $12y$ **6.** $0.02x + 100$
7. (a) -3 (b) 7 **8.** (a) 34 (b) -32
9. (a) 57 (b) 33 **10.** (a) $\frac{-13}{2}$ (b) -10

1. (a) No (b) Yes **3.** (a) Yes (b) No
5. (a) Yes (b) No
7. Given equation
Addition Property of Equality
Additive Inverse Property
Multiplication Property of Equality
Multiplicative Inverse Property
9. Given equation **11.** 4
Addition Property of Equality
Additive Inverse Property
Multiplication Property of Equality
Multiplicative Inverse Property
13. $-\frac{2}{3}$ **15.** 2 **17.** 3 **19.** $\frac{1}{3}$ **21.** -2
23. 0 **25.** Not possible as $7 \neq 0$. **27.** 11
29. -3 **31.** -2 **33.** -3 **35.** $\frac{6}{5}$ **37.** $\frac{25}{3}$
39. 50 **41.** $\frac{19}{10}$ **43.** -20 **45.** 0 **47.** $-\frac{10}{3}$
49. $-\frac{8}{31}$ **51.** 23 **53.** 12 **55.** 3.89 **57.** 30.28

59.

t	Width	Length	Area
1	300	300	90,000
1.5	240	360	86,400
2	200	400	80,000
3	150	450	67,500
4	120	480	57,600
5	100	500	50,000

61. 1.5 sec **63.** 6 hrs **65.** 1985 **67.** 1976

SECTION 2.2 *(page 84)*

Warm-Up *(page 83)*

1. -4 **2.** 27 **3.** 14
4. Not possible since $17 \neq 0$.
5. Not possible since $\frac{-19}{3} \neq 0$. **6.** -19 **7.** $\frac{8}{3}$
8. $\frac{940}{3}$ **9.** 0.8 **10.** $\frac{1}{10}$

1. A number + 30 = 82
$x + 30 = 82$

3. Perimeter = 3 × Length of side
$129 = 3x$

5. Total bill = Charge for parts + 35 × Number of hours worked
$380 = 275 + 35x$

7. 30, 0.30, $\frac{3}{10}$ **9.** 7.5%, 7.5, $\frac{3}{40}$
11. 12.5%, 12.5, 0.125 **13.** $41\frac{2}{3}$% **15.** 87.5
17. 69.36 **19.** 600 **21.** 350 **23.** 400
25. 12,000 **27.** 33% **29.** 175% **31.** $420
33. 15.625% **35.** 200 **37.** 475
39. 177.78%, 56.25% **41.** 2,004 **43.** 7,315
45. $\frac{3 \text{ inches}}{4 \text{ inches}}$ **47.** $\frac{1 \text{ pint}}{8 \text{ pints}}$ **49.** $\frac{1 \text{ milliliter}}{25 \text{ milliliters}}$ **51.** $\frac{1}{50}$
53. $\frac{85}{4}$ **55.** $0.0475 **57.** $0.0695 **59.** 32-oz bag
61. 12-oz tube **63.** 4 **65.** $\frac{15}{2}$ **67.** 16
69. 17.14 gal **71.** 20 pints **73.** $1,800
75. 2,667 units **77.** 112.5 minutes
79. Taxes 7.12% Wages 59.35%
Employee Benefits 10.98% Misc. 12.46%
Insurance 2.37% Supplies 2.67%
Utilities 1.19% Rent 3.86%

81. 3 **83.** $\frac{56}{11}$ **85.** 6.43 ft **87.** $\frac{4}{9}$

SECTION 2.3 *(page 99)*

Warm-Up *(page 99)*

1. $\frac{11}{4}$ **2.** 5 **3.** $\frac{5}{9}$ **4.** $\frac{10}{3}$ **5.** 1.875
6. 9,000 **7.** 150% **8.** $66\frac{2}{3}$ **9.** 80,000
10. 275

1. **(a)** 10.9 million **(b)** 8 **3.** $18.38, 40%
5. $152.00, 65% **7.** $22,250.00, 21%
9. $416.70, $191.70 **11.** $23.76, 48%
13. $111.00, 63% **15.** $61.75, $33.25
17. $1,145.00, 22% **19.** $54.15
21. Department store $223.96 **23.** 18.3%
25. 9 minutes, $2.06 **27.** Tax = $267
Total bill = $4,717
Amount financed = $3,717
29. 9% **31.** 30 15¢-stamps; 40 30¢-stamps **33.** 700
35. 500, $3.00; 400, $2.74; 300, $2.48; 200, $2.22; 100, $1.96; 0, $1.70 **(a)** Decreases **(b)** Decreases **(c)** Average of the prices of each component
37. Solution 1 = 50 gal
Solution 2 = 50 gal
39. Solution 1 = 8 qt
Solution 2 = 16 qt
41. $\frac{5}{6}$ gal **43.** 2,275 mi **45.** $\frac{100}{11}$ hrs
47. $\frac{2{,}000}{3}$ ft/sec **49.** 17.65 min **51.** 1,440 mi
53. 20 min **55.** 43.6 mph **57.** $\frac{1}{3}$, $\frac{1}{4}$, 1 hr 43 min
59. 18 **61.** 321, 322, 323 **63.** 84, 420
65. $R = \dfrac{E}{I}$ **67.** $r = \dfrac{A - P}{Pt}$ **69.** $b = \dfrac{2A - ah}{h}$
71. 14 **73.** 52 sq units **75.** $2,850
77. $3,571.43 **79.** $30,000 at 8.5%
$10,000 at 10%
81. 1985, $0.366/hr **83.** Bus drivers

SECTION 2.4 *(page 116)*

Warm-Up *(page 116)*

1. $-\frac{3}{4} > -5$ **2.** $-\frac{1}{5} > -\frac{1}{3}$ **3.** $\pi > -3$
4. $6 < \frac{13}{2}$ **5.** $\frac{7}{2}$ **6.** $\frac{48}{7}$ **7.** 40 **8.** 153
9. -12 **10.** 24

1. **(a)** Yes **(b)** No **(c)** Yes **(d)** No
3. **(a)** No **(b)** Yes **(c)** Yes **(d)** No **5.** (f)
6. (c) **7.** (a) **8.** (e) **9.** (d) **10.** (b)

11. $(-5, 3]$

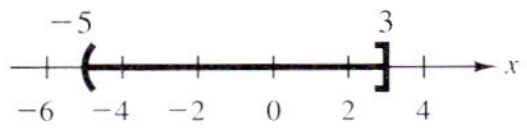

13. $\left[0, \frac{3}{2}\right)$

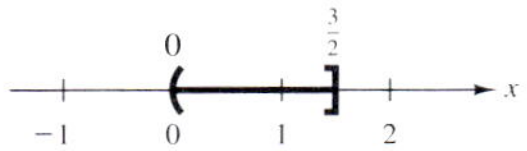

15. $\left(-\frac{15}{4}, -\frac{5}{2}\right)$

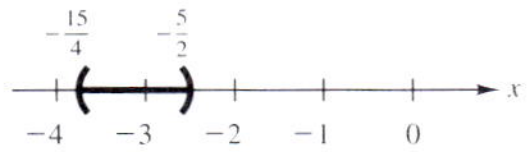

17. $x \le 2$

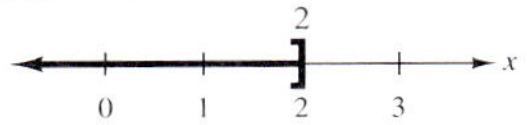

19. $x < \frac{11}{2}$

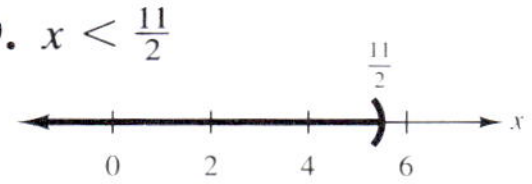

21. $x \le -4$

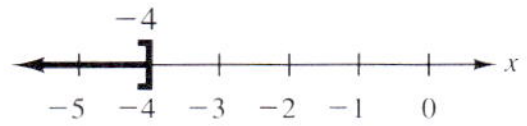

23. $x > 8$

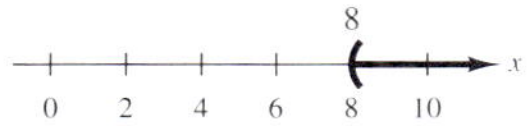

25. $x > 7$

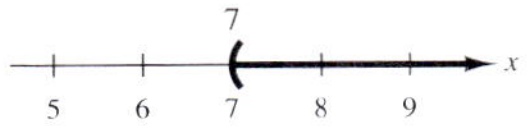

27. $x \ge 7$

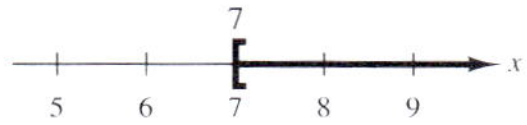

29. $x > -\frac{2}{3}$

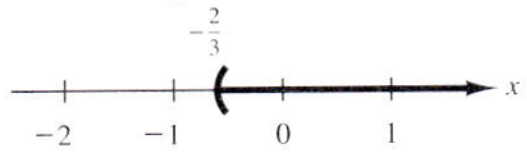

31. $x > \frac{9}{2}$

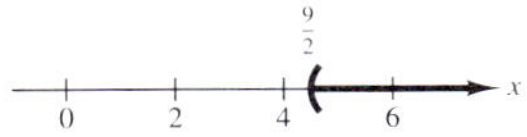

33. $y > 2$

35. $y \le -10$

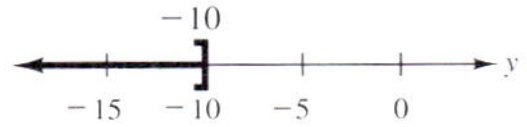

37. $x \ge -12$

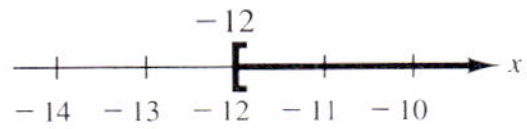

39. $\frac{5}{2} < x < 7$

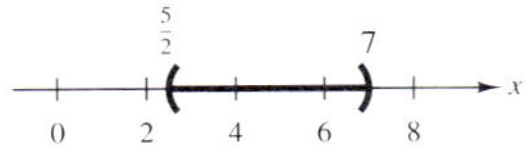

41. $-\frac{3}{2} < x < \frac{9}{2}$

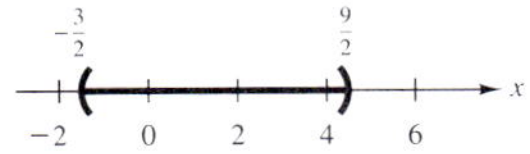

43. $1 < x < 10$

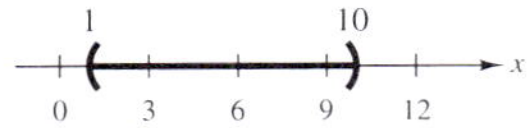

45. $-\frac{3}{5} < x < -\frac{1}{5}$ **47.** $x \ge 0$ **49.** $z \ge 2$

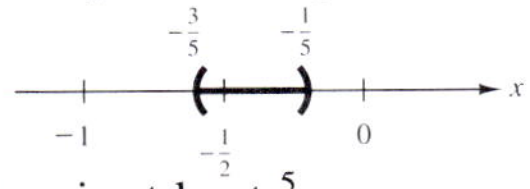

51. x is at least $\frac{5}{2}$.
53. z is more than 0, but not more than π. **55.** $2,600
57. Miami > New York **59.** Miles < 25,357
61. $x \ge 31$
63. $0 < t \le 6.38$ min, or if a portion of a minute is charged as a full minute, then $0 < t \le 6$.
65. $2 \le d \le 8$ **67.** $-8 < -t \le 5$ **69.** $3 \le n \le \frac{15}{2}$
71. $n > 21$ **73.** 1970–1974 **75.** C

SECTION 2.5 *(page 127)*

Warm-Up *(page 126)*

1. 15 **2.** 3.2 **3.** −72 **4.** −4 **5.** 18
6. −19.8 **7.** $-\frac{1}{2} < x < \frac{1}{2}$ **8.** $-\frac{1}{2} \le x \le \frac{3}{2}$
9. $-6 \le x \le 6$ **10.** $20 < x < 30$

1. 45, −45 **3.** 0 **5.** 21, 11 **7.** 11, −14
9. $\frac{16}{3}$, 16 **11.** No solution **13.** $-\frac{11}{5}, \frac{17}{5}$
15. $-\frac{39}{2}, \frac{15}{2}$ **17.** 18.75, −6.25 **19.** −3, 7
21. 11, 13
23. **(a)** Yes **(b)** No **(c)** No **(d)** Yes
25. **(a)** No **(b)** Yes **(c)** Yes **(d)** No
27. **(a)** Yes **(b)** No **(c)** No **(d)** Yes
29. **(a)** No **(b)** Yes **(c)** Yes **(d)** No

31.

33.

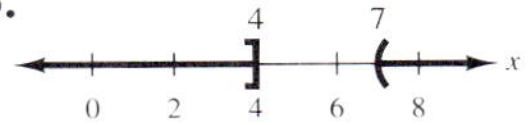

35. $-4 < y < 4$

37. $-2 \le y \le 6$

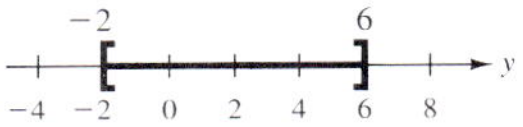

39. $-6 < y < 2$

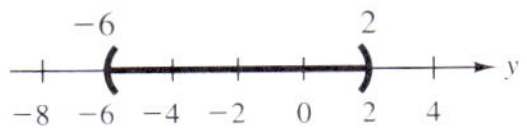

41. $y \ge 4$ or $y \le -4$

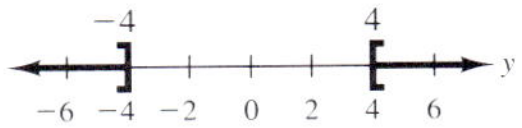

43. $y > 6$ or $y < -2$

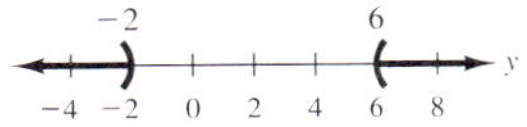

45. $y \ge 2$ or $y \le -6$

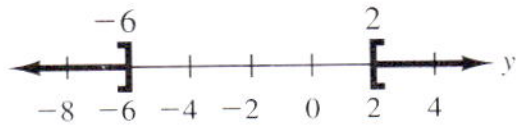

47. $-7 < x < 7$

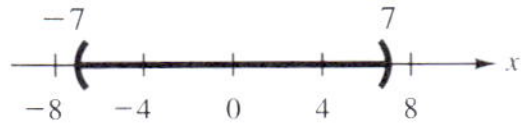

49. $-9 \le y \le 9$

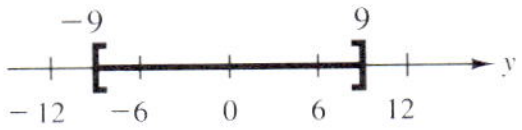

51. $x > \frac{2}{3}$ or $x < -2$

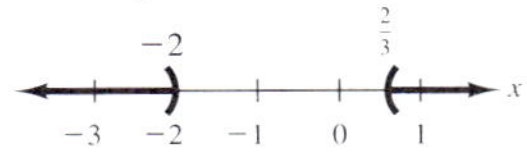

53. $x > 7$ or $x < 3$

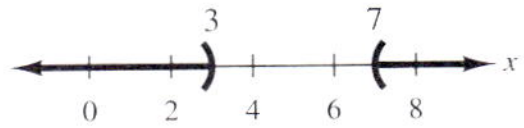

55. $-5 < x < 35$ **57.** $-82 \le x \le 78$
59. $t \ge \frac{5}{2}$ or $t \le \frac{-15}{2}$ **61.** $s > 23$ or $s < -17$
63. $z > 110$ or $z < -50$
65.

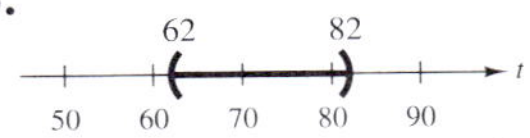

67. $|x| \le 2$
69. $|x - 10| < 3$ **71.** $|x - 19| < 3$ **73.** $|x| < 3$
75. $|x - 5| > 6$

CHAPTER 2 REVIEW EXERCISES *(page 130)*

1. **(a)** No **(b)** Yes **3.** **(a)** Yes **(b)** No **5.** 2
7. 14 **9.** −8.2 **11.** $\frac{16}{3}$ **13.** 2 **15.** $\frac{20}{17}$
17. 3.38 **19.** 33.32 **21.** 87, 0.87, $\frac{87}{100}$ **23.** 65
25. 3,000 **27.** 125%
29. 10.2%, 11.6%, 9.9%, 13.2% **31.** 3.5%
33. \$350.93 **35.** \$194.25 **37.** 15.5%
39. 25 dimes, 35 quarters **41.** 6.35 hrs
43. 2,800 mi **45.** $\frac{18}{7} \approx 2.57$ hrs **47.** \$340
49. \$210,526.32
51. \$20,000 at 8.5% \$30,000 at 10% **53.** 8 in. × 6 in. **55.** $\frac{4 \text{ ft}}{3 \text{ ft}}$
57. $\frac{3 \text{ min}}{20 \text{ min}}$ **59.** $\frac{7}{2}$ **61.** $\frac{25}{2}$ **63.** \$1,856.25
65. 487.5 mi **67.** 3 **69.** $x > 3$

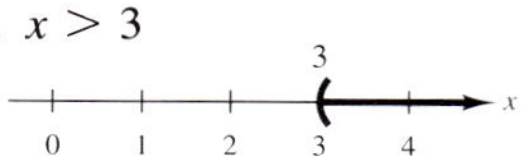

71. $y > -\frac{70}{3}$

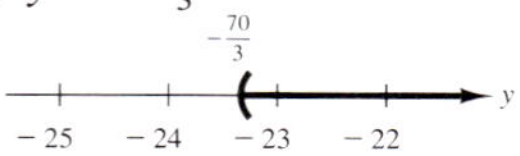

73. $-20 < x \le 20$

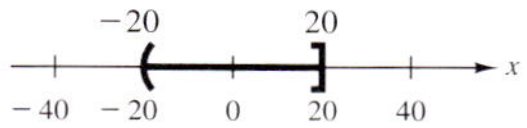

75. $-16 < x < -1$

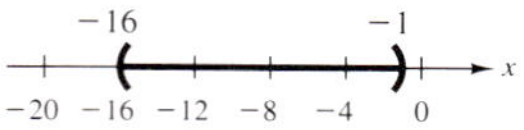

77. $-4 < x < 11$

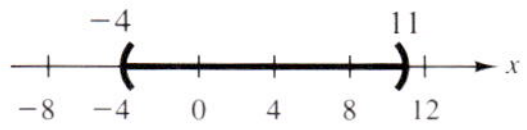

79. $x > 7$ or $x < 1$

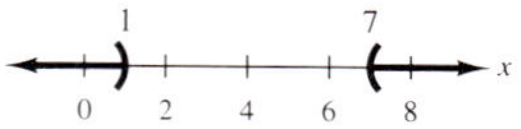

81. $b > 5$ or $b < -9$

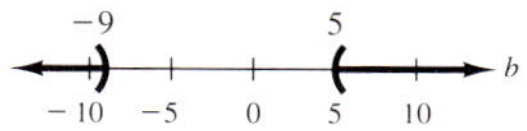

83. $z \le 10$ **85.** $7 \le y < 14$

CHAPTER 2 TEST *(page 134)*

1. **(a)** No **(b)** Yes **2.** 4 **3.** 4 **4.** 24
5. $\frac{19}{2}$ **6.** 14.83 **7.** 25 **8.** 864 **9.** 150%
10. \$8,000 **11.** 15-oz pkg **12.** \$1,466.67
13. $2\frac{1}{2}$ hrs **14.** 12 dimes, 18 quarters
15. $33\frac{1}{3}\,L$ of 10%; $66\frac{2}{3}\,L$ of 40% **16.** 40 min
17. 13, 52 **18.** \$2,000
19. $x > 2$

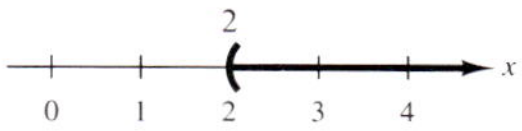

20. $-7 < x \le 1$

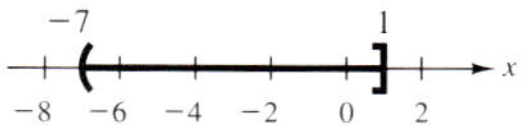

21. $1 \le x \le 5$

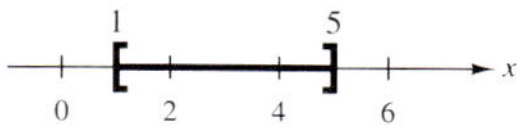

22. $x > -3$ or $x < -5$

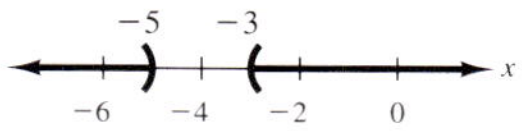

23. $t \ge 8$

Chapter 3

SECTION 3.1 *(page 143)*

Warm-Up *(page 142)*

1. $23y - 69$ **2.** $12 - 30z$ **3.** $-10 + 4x$
4. $-25x + 20$ **5.** $32x, -10y, -7$
6. $-3s, 4t, 1$ **7.** $-4y$ **8.** $\frac{7}{24}x + 8$
9. $-30x^2 + 23x + 3$
10. $4x^3 + 12x^2y + 4xy^2 + y^3$

1. $10x - 4$, 1, 10 **3.** $-3y^4 + 5$, 4, −3
5. $4t^5 - t^2 + 6t + 3$, 5, 4 **7.** −4, 0, −4
9. $-16t^2 + v_0t$, 2, −16 **11.** Binomial
13. Trinomial **15.** Monomial
17. Division by x not allowed
19. Exponent in y^{-3} is not nonnegative. **21.** $3x^3$
23. $8x^2 - 9$ **25.** $7x^2 + 3$ **27.** 13
29. $-2y^4 - 5y + 4$ **31.** $6x^2 - 7x + 8$
33. $-2x^2 + 15$ **35.** $2x^2 - 3x$ **37.** $2p^2 - 2p - 5$
39. 4 **41.** $x^2 - 3x + 2$ **43.** $-2y^3$ **45.** $q^2 + 15$
47. $7x^3 + 2x$ **49.** $4x^3 + x + 4$ **51.** $x^2 - 2x + 5$
53. $-4x^3 - 2x + 13$ **55.** $2x^2 + 9x - 11$
57. $2x^3 - 2x + 3$ **59.** $7x^3 + 22x^2 + 4$
61. $29s + 8$ **63.** $3t^2 + 29$ **65.** $3v^2 + 78v + 27$
67. **(a)** 16 **(b)** 0 **(c)** −16 **(d)** 16
69. **(a)** −27 **(b)** −16 **(c)** 0 **(d)** $\frac{9}{16}$
71. **(a)** 32 **(b)** 7 **73.** **(a)** 50 **(b)** 146 **(c)** 114 **(d)** 50 **75.** Dropped, 100 ft
77. Thrown downward, 50 ft **79.** 224 ft, 216 ft, 176 ft
81. \$15,000 **83.** $[(2x - 3)x + 10]x - 12$
85. $[(x - 4)x + 2]x + 10$ **87.** $[(4x - 3)x - 2]x + 25$
89. $36x$
91. **(a)** $y = 31.43 - 0.55t + 0.013t^2$ **(b)** Decreasing
1975 29.005 gal
1985 26.105 gal

SECTION 3.2 *(page 155)*

Warm-Up *(page 154)*

1. $(-5)^3$ **2.** y^4 **3.** $\left(\frac{2}{3}\right)\left(\frac{2}{3}\right)\left(\frac{2}{3}\right)\left(\frac{2}{3}\right)$
4. $(2.1)(2.1)(2.1)(2.1)(2.1)$ **5.** $(2x)(2x)(2x)$
6. $(-y)(-y)$ **7.** $7x - 15$ **8.** $3y + 36$
9. $-4u$ **10.** $-5v - 18$

1. $t^{3+4} = t^7$ **3.** $(-5)^5x^5 = -3{,}125x^5$ **5.** $\dfrac{y^4}{5^4} = \dfrac{y^4}{625}$
7. $x^{6-4} = x^2$ **9.** $3^{7-3}x^{5-3} = 3^4x^2 = 81x^2$
11. **(a)** $-3x^8$ **(b)** $9x^8$ **13.** **(a)** $-125z^3$ **(b)** $25z^2$
15. **(a)** $625z^4$ **(b)** $-3125z^8$ **17.** **(a)** $\dfrac{4xy}{3}$ **(b)** $\dfrac{3}{4xy}$
19. **(a)** $\dfrac{8(x+3)^2}{3}$ **(b)** $\dfrac{x^2+6x+9}{18}$ **21.** $16a^3$
23. $20x^3$ **25.** $10y - 2y^2$ **27.** $8x^3 - 12x^2 + 20x$
29. $-10x - 6x^3 + 14x^4$ **31.** $75x^3 + 30x^2$
33. $48y^2 + 32y - 3$ **35.** $6x^2 + 7xy + 2y^2$
37. $-a^2 + 19a$ **39.** $2st - 6t^2$
41. $x^4 - 2x^3 - 3x^2 + 8x - 4$
43. $2u^3 + 13u^2 + 11u - 20$ **45.** $x^3 - 5x^2 + 10x - 6$
47. $7x^3 + 7x^2 - 33x + 27$ **49.** $t^4 - t^2 + 4t - 4$
51. $x^5 - 3x^3 - 4x$ **53.** $-6x^3 + x^2 - 1$
55. $2u^3 + u^2 - 7u - 6$ **57.** $x^2 - 4$ **59.** $4 - 49y^2$
61. $4x^2 - \frac{1}{16}$ **63.** $0.04t^2 - 0.25$
65. $x^2 + 10x + 25$ **67.** $25x^2 - 20x + 4$
69. $4x^2 - 28xy + 49y^2$
71. $x^2 + 2xy + y^2 + 4x + 4y + 4$
73. $u^2 - 2uv + v^2 + 6u - 6v + 9$ **75.** $12t$
77. $x^3 + 9x^2 + 27x + 27$ **79.** $u^3 + 3u^2v + 3uv^2 + v^3$
81. $8x^2 + 26x$ **83.** **(a)** $5w$ **(b)** $\frac{3}{2}w^2$
85. $1{,}000 + 2{,}000r + 1{,}000r^2$
87. $A = (x+a)(x+b) = x^2 + ax + bx + ab$ (FOIL)
89. **(a)** $x^2 - 1$ **(b)** $x^3 - 1$ **(c)** $x^4 - 1$; $x^5 - 1$

MATH MATTERS: THE ROLLING COIN PROBLEM *(page 161)*

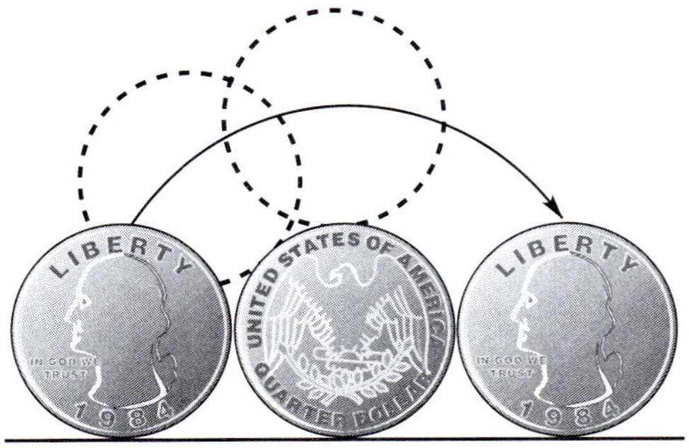

SECTION 3.3 *(page 166)*

Warm-Up *(page 166)*

1. $24x^2 - 36x$ **2.** $28y - 21y^2$
3. $-60x + 42x^2$ **4.** $-2y^2 - 2y$ **5.** $2t$
6. $2z^2 + 3z - 35$ **7.** $121 - x^2$
8. $36r^2 - 25s^2$ **9.** $x^3 - 8$ **10.** $x^3 + 1$

1. 6 **3.** $3x$ **5.** $6z^2$ **7.** $14b^2$ **9.** $21(x+8)^2$
11. $8(z-1)$ **13.** $2(2u+5)$ **15.** $6(4x^2-3)$
17. $x(2x+1)$ **19.** $7u(3u-2)$ **21.** $3y(x^2y-5)$
23. $4(7x^2+4x-2)$ **25.** $x^2(14x^2+21x+9)$
27. $(y-3)(2y+5)$ **29.** $(a+6)a(1-a)$
31. $(10+b)(c+7)(-c-6)$ **33.** $-(x-10)$
35. $-7(2x-1)$ **37.** $-(x^2-4x-5)$ **39.** $(6x+5)$
41. $(10y-3)$ **43.** $(2x+y)$ **45.** $(x+25)(x+1)$
47. $(y-6)(y+2)$ **49.** $(x+2)(x^2+1)$
51. $(a-4)(a^2+2)$ **53.** $(z+3)(z^3-2)$
55. $(x+8)(x-8)$ **57.** $(4y+3)(4y-3)$
59. $(15+3y)(15-3y)$ **61.** $\left(u+\frac{1}{4}\right)\left(u-\frac{1}{4}\right)$
63. $[9+(z+5)][9-(z+5)]$
65. $(x-2)(x^2+2x+4)$ **67.** $(y+4)(y^2-4y+16)$
69. $(2t-3)(4t^2+6t+9)$
71. $(3u+1)(9u^2-3u+1)$ **73.** $2(2-5x)(2+5x)$
75. $(y-3)(y+3)(y^2+9)$
77. $2(x-3)(x^2+3x+9)$ **79.** $6{,}399$
81. $(2x^n-5)(2x^n+5)$

83. $x^2(3x+4) - (3x+4) = (x-1)(x+1)(3x+4)$
$3x(x^2-1) + 4(x^2-1) = (x-1)(x+1)(3x+4)$

85. $p = 800 - 0.25x$ **87.** Width $= 45 - l$
89. $P(1+rt)$ **91.** $\pi(R-r)(R+r)$

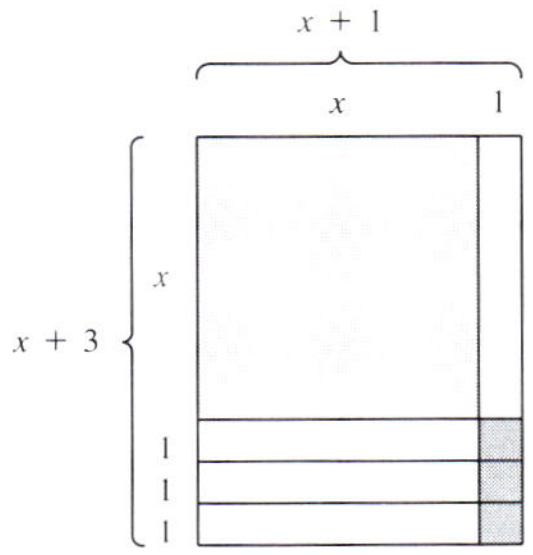

SECTION 3.4 *(page 179)*

Warm-Up *(page 178)*

1. $x^2 + 2x + 1$ **2.** $x^2 - 10x + 25$
3. $4 - 4y + y^2$ **4.** $6 + y - 2y^2$
5. $10z^2 - 29z - 21$ **6.** $2t^3 - 5t^2 - 12t$
7. $3x(x - 7)$ **8.** $-(x - 3)(x^2 + 1)$
9. $(2t - 13)(2t + 13)$
10. $(y - 4)(y^2 + 4y + 16)$

1. $(x + 2)^2$ **3.** $(a - 6)^2$ **5.** $(5y - 1)^2$
7. $(3b + 2)^2$ **9.** $(u + 4v)^2$ **11.** ± 18 **13.** ± 12
15. 16 **17.** 9 **19.** 14 **21.** $x + 1$ **23.** $y - 5$
25. $z - 2$ **27.** $(x + 3)(x + 1)$ **29.** $(x - 3)(x - 2)$
31. $(y + 10)(y - 3)$ **33.** $(t - 7)(t + 3)$
35. $(x - 8)(x - 12)$ **37.** $(x + 12)(x + 18)$
39. $(x - 7y)(x + 5y)$ **41.** $\pm 9, \pm 11, \pm 19$
43. $\pm 4, \pm 20$ **45.** $\pm 12, \pm 36$ **47.** $8, -16$
49. $2, -18$ **51.** $5x + 3$ **53.** $5a - 3$ **55.** $2y - 9$
57. $(3x + 1)(x + 1)$ **59.** $(2x - 3)(x - 3)$
61. $(3b - 1)(2b + 7)$ **63.** $(4x - 3y)(6x + y)$
65. Not factorable **67.** $-(2x - 3)(x + 2)$
69. $-(4x + 1)(15x - 1)$ **71.** $(x + 2)(3x + 4)$
73. $(2x - 1)(3x + 2)$ **75.** $(3x - 1)(5x - 2)$
77. $(3x^3)(x - 4)$ **79.** $2t(5t - 9)(t + 2)$
81. $(3x - 2)(7x - 2)$ **83.** $(3 - z)(9 + z)$
85. $2(3x - 1)(9x^2 + 3x + 1)$ **87.** 2,704
89. $(x + 3)(x + 1)$

SECTION 3.5 *(page 189)*

Warm-Up *(page 189)*

1. $\frac{22}{5}$ **2.** -14 **3.** 5 **4.** $\frac{4}{3}$
5. $2v(2v + 9)$ **6.** $(4x - 5)(4x + 5)$
7. $(17 - x)(-5 + x)$ **8.** $(x + 9)(x - 3)$
9. $-(4x - 3)(2x - 1)$ **10.** $(x - 12)(6x - 1)$

1. 0, 8 **3.** $3, -10$ **5.** $2, -4$ **7.** $-\frac{5}{2}, -\frac{1}{3}$
9. $0, \frac{3}{2}, \frac{-25}{2}$ **11.** ± 5 **13.** ± 4 **15.** $6, -2$
17. $0, \frac{-5}{3}$ **19.** $2, -10$ **21.** $5, -2$ **23.** 4
25. -8 **27.** $\frac{3}{2}$ **29.** $7, -\frac{1}{2}$ **31.** $\frac{5}{4}, -5$
33. $-2, 5$ **35.** $-12, 6$ **37.** $-\frac{7}{2}, 5$ **39.** $-7, 0$
41. $-6, 5$ **43.** 0, 7, 12 **45.** $0, -\frac{1}{3}, \frac{1}{2}$
47. $\pm 4, 25$ **49.** $\pm 3, -2$ **51.** 15 ft $\times$ 22 ft
53. $b = 8, h = 12$ **55.** 12, 13 **57.** 20 sec
59. **(a)** $V = \text{length} \cdot \text{width} \cdot \text{height}$
$V = x \cdot x \cdot 5 = 5x^2$
(b)

x	2	4	6	8
V	20	80	180	320

(c) 12 square inches **61.** $x = 0$ and $\dfrac{-b}{a}$
63. **(a)** $-8, 0, 10, 22, 36, 52$
(b) P increases by $2(x + 1)$.
(c) 9
65. **(a)** $x = \frac{-20}{3}, -2$
(b) $x = \frac{-5}{4}, -\frac{1}{2}$

CHAPTER 3 REVIEW EXERCISES *(page 193)*

1. $-4(x - 2) = -4x + 8$ **3.** $(x + 3)^2 = x^2 + 6x + 9$
5. $-216z^3$ **7.** $8u^2v^2$ **9.** $6x - x^2$
11. $-2t^2 + 5t + 5$ **13.** $-9x^3 + 9x - 4$
15. $6y^2 - 2y + 15$ **17.** $-8x^4 - 32x^3$
19. $15x^2 - 11x - 12$ **21.** $4x^3 - 5x + 6$
23. $u^2 - 8u + 7$ **25.** $16x^2 - 56x + 49$
27. $25u^2 - 64$ **29.** $u^2 - v^2 - 6u + 9$
31. $3x^2(2 + 5x)$ **33.** $-14(x + 5)(5x + 23)$
35. $(3a - 10)(3a + 10)$ **37.** $(u - 3)(u + 15)$
39. $(x - 20)^2$ **41.** $4(s + 5)^2$
43. $4a(1 - 4a)(1 + 4a)$ **45.** $4(2x - 3)(2x - 1)$
47. $(2x + 1)(4x^2 - 2x + 1)$ **49.** $(2u - 7)^2$
51. Not factorable **53.** Not factorable **55.** ± 6
57. $\pm 1, \pm 7, \pm 13, \pm 29$ **59.** $12, -8$ **61.** 0, 3
63. $10, -10$ **65.** 10, 15 **67.** $-\frac{4}{3}, 2$ **69.** $-4, 9$
71. $10p^3 - 20p^4 + 10p^5$ **73.** **(a)** $4l - 10$
(b) $l^2 - 5l$ **75.** 3 sec **77.** 24×18 in.

CHAPTER 3 TEST *(page 195)*

1. Degree: 3; Leading coefficient: -5.2 **2.** $6a^2 - 3a$
3. $-2y^2 - 2y$ **4.** $8x^2 - 4x + 10$ **5.** $11t + 7$
6. $\dfrac{y^3}{8}$ **7.** $2x^2 + 7xy - 15y^2$ **8.** $3x^2 - 6x + 3$
9. $6s^3 - 17s^2 + 26s - 21$ **10.** $16x^2 - 24x + 9$
11. $16 - a^2 - 2ab - b^2$ **12.** $6y(3y - 2)$
13. $\left(v - \frac{4}{3}\right)\left(v + \frac{4}{3}\right)$ **14.** $(x + 2)(x - 2)(x - 3)$
15. $(3u - 1)^2$ **16.** $2(x - 5)(3x + 2)$ **17.** $1, -5$
18. $3, -\frac{4}{3}$ **19.** $x^2 + 26x$ **20.** 6×9 cm

Chapter 4

SECTION 4.1 *(page 203)*

Warm-Up *(page 203)*

1. (-20) **2.** (70) **3.** (-21) **4.** (-27)
5. $\frac{15}{26}$ **6.** $\frac{11}{23}$ **7.** $\frac{5}{8}$ **8.** $\frac{7}{12}$
9. $20a(x^2 + 3x + 4)$ **10.** $y^2(x - 8)(2x + 7)$

1. $(-\infty, 8) \cup (8, \infty)$ **3.** $(-\infty, -4) \cup (-4, \infty)$
5. $(-\infty, \infty)$ **7.** $(-\infty, -4) \cup (-4, 4) \cup (4, \infty)$
9. $(-\infty, -1) \cup (-1, 5) \cup (5, \infty)$
11. $\{1, 2, 3, 4, \ldots\}$ **13.** $(x + 3)$ **15.** $(3)(x + 16)^2$
17. $(x)(x - 2)$ **19.** $(x + 2)$ **21.** $\dfrac{x}{5}$ **23.** $6y$
25. $\dfrac{6x}{5y^3}$ **27.** x **29.** $\frac{1}{2}$ **31.** $-\frac{1}{3}$ **33.** $\dfrac{3y^2}{y^2 + 1}$
35. $\dfrac{y - 8}{15}$ **37.** $\dfrac{1}{a + 3}$ **39.** $\dfrac{x}{x - 7}$ **41.** $\dfrac{y(y + 2)}{y + 6}$
43. $\dfrac{-1}{2x + 3}$ **45.** $\dfrac{5x + 4}{5x + 2}$ **47.** $\dfrac{3xy + 5}{y^2}$
49. $\dfrac{3(m - 2n)}{m + 2n}$ **51.** $\frac{3}{2} \neq 9$ **53.** $\frac{16}{21} \neq \frac{3}{4}$
55. $x^n - 2$
57.

-1	0	1	2	Undefined	4	5
-1	0	1	2	3	4	5

59. $x = 90, 630$

61. **(a)** $\dfrac{2{,}500 + 9.25x}{x}$
(b) $x = \{1, 2, 3, 4, \ldots\}$
(c) \$34.25

63. **(a)** Van $= 45(t + 3)$
Car $= 60t$
(b) $\dfrac{4t}{3(t + 3)}$

65. $\dfrac{107.1 + 12.64t + 0.54t^2}{31.6 + 0.51t - 0.14t^2}$

SECTION 4.2 *(page 212)*

Warm-Up *(page 212)*

1. $\frac{3}{10}$ **2.** $\frac{1}{20}$ **3.** $\frac{5}{56}$ **4.** $-\frac{15}{112}$ **5.** 5
6. $-\frac{4}{35}$ **7.** $(x - 1)(x - 2)$
8. $(x - 1)(2x + 7)$ **9.** $(x + 1)(11x - 5)$
10. $(2x - 7)^2$

1. x^2 **3.** $(u + 1)$ **5.** $(-1)(2 + x)$ **7.** $\dfrac{3x}{2}$
9. $\dfrac{s^3}{6}$ **11.** $24u^2$ **13.** 24 **15.** $\dfrac{2uv(u + v)}{3(3u + v)}$
17. -1 **19.** $-\dfrac{x + 8}{x^2}$ **21.** $\dfrac{2(r + 2)}{11r}$ **23.** $2t + 5$
25. $\dfrac{xy(x + 2y)}{(x - 2y)}$ **27.** $\dfrac{(x - y)^2}{x + y}$ **29.** $\dfrac{(u + v)^2}{(u - v)^2}$
31. $\dfrac{(x - 1)(2x + 1)}{(3x - 2)(x + 2)}$ **33.** $\dfrac{3y^2}{2ux^2}$ **35.** $\dfrac{3}{2(a + b)}$
37. $x^4y(x + 2y)$ **39.** $\dfrac{3x}{10}$ **41.** $-\dfrac{5x}{2}$
43. $\dfrac{(x + 3)(4x + 1)}{(3x - 1)(x - 1)}$ **45.** $\dfrac{(x + 2)(x + 3)}{x}$
47. $\dfrac{(x + 1)(2x - 5)}{x}$ **49.** $\dfrac{x^4}{x^{2n} + 2x^n + 1}$
51. **(a)** 0
(b) Undefined
(c) $\frac{3}{2}$
(d) $\frac{1}{24}$

53. **(a)** $\frac{1}{20}$ min
(b) $\dfrac{x}{20}$ min
(c) $\frac{7}{4}$ min

55. $\dfrac{2w^2 + 3w}{6}$

MATH MATTERS: ROMAN NUMERALS *(page 217)*

(a) 1776
(b) 1492
(c) 1620

SECTION 4.3 *(page 224)*

Warm-Up *(page 224)*

1. $\frac{7}{9}$ **2.** $\frac{1}{2}$ **3.** $\frac{3}{16}$ **4.** $-\frac{3}{5}$ **5.** $\frac{37}{36}$ **6.** $\frac{37}{48}$
7. $-\frac{146}{225}$ **8.** $\frac{181}{21}$ **9.** 26 **10.** $\frac{5}{64}$

1. $-\frac{2}{9}$ **3.** $\dfrac{2 + y}{2}$ **5.** $-\dfrac{3}{a}$ **7.** $\dfrac{x + 6}{3}$ **9.** $-\frac{4}{3}$
11. $\dfrac{1}{x - 3}$ **13.** 1 **15.** $20x^3$ **17.** $15x^2(x + 5)$
19. $6x(x + 2)(x - 2)$ **21.** $\dfrac{2n^2(n + 8)}{6n^2(n - 4)}, \dfrac{10(n - 4)}{6n^2(n - 4)}$
23. $\dfrac{2(x + 3)}{x^2(x + 3)(x - 3)}, \dfrac{5x(x - 3)}{x^2(x + 3)(x - 3)}$

25. $\dfrac{(x-8)(x-5)}{(x+5)(x-5)^2}, \dfrac{9x(x+5)}{(x+5)(x-5)^2}$ **27.** $\dfrac{25-12x}{20x}$

29. $\dfrac{7(a+2)}{a^2}$ **31.** 0 **33.** $\dfrac{3(x+2)}{x-8}$

35. $\dfrac{5(5x+22)}{x+4}$ **37.** $\dfrac{1}{2x(x-3)}$ **39.** $\dfrac{x^2-7x-15}{(x+3)(x-2)}$

41. $\dfrac{x-2}{x(x+1)}$ **43.** $\dfrac{5(x+1)}{(x+5)(x-5)}$ **45.** $\dfrac{4}{x^2(x^2+1)}$

47. $\dfrac{x^2+3x+9}{x(x^2-9)}$ **49.** $\dfrac{4x}{(x-4)^2}$

53. $\dfrac{x}{x-1}$ **55.** $\dfrac{y-x}{xy}$ **51.** $\dfrac{2(4x^2+5x-3)}{x^2(x+3)}$

57. **(a)** $\dfrac{5u-2v}{(u-v)^2}$ **(b)** $\dfrac{1}{(x+1)(x-1)(x+2)}$

59. $\dfrac{x}{2(3x+1)}$ **61.** $-4x-1$ **63.** $\frac{3}{4}$

65. $\dfrac{y+1}{y-3}$ **67.** $y-x$ **69.** $\dfrac{5t}{12}$ **71.** $\dfrac{5x}{24}$

73. $\dfrac{x}{4}, \dfrac{x}{3}, \dfrac{5x}{12}$ **75.** $\dfrac{R_1R_2}{R_1+R_2}$

SECTION 4.4 *(page 236)*

Warm-Up *(page 236)*

1. $\frac{1}{4}$ **2.** $\frac{29}{90}$ **3.** $-\dfrac{4x}{3}$ **4.** $\dfrac{t^2}{9}$ **5.** $\dfrac{2u}{9v^2}$

6. $-\dfrac{r^3}{7}$ **7.** $-15x^4+12x^3$ **8.** $70y^3-20y$

9. $2x^3+7x^2-3x-4$ **10.** $8x^2+10x+3$

1. $3z+5$ **3.** $\frac{5}{2}z^2+z-3$ **5.** $7x^2-2x$

7. $-10z^2-6$ **9.** $4z^2+\frac{3}{2}z-1$ **11.** $m^3+2m-\dfrac{7}{m}$

13. $\frac{5}{2}x-4+\frac{7}{2}y$ **15.** $4(x+7)$ **17.** $x-5$

19. $x+10$ **21.** $x+7$ **23.** $y+3$ **25.** $6t-5$

27. $4x-1$ **29.** $x^2-5x+25$ **31.** x^2+4

33. $2+\dfrac{5}{x+2}$ **35.** $x-4+\dfrac{32}{x+4}$

37. $5x-8+\dfrac{19}{x+2}$ **39.** $4x-\dfrac{25}{3}+\dfrac{35}{3(3x+2)}$

41. $\dfrac{6}{5}z+\dfrac{41}{25}+\dfrac{41}{25(5z-1)}$ **43.** $2x^2+x+4+\dfrac{6}{x-3}$

45. $x^5+x^4+x^3+x^2+x+1$ **47.** $x+2$

49. $x^{2n}+x^n+4$ **51.** $(x-7)(x-8)$

53. $(2a-5)(a+9)$ **55.** $(15x+10)\left(x-\dfrac{4}{5}\right)$

57. $(2t^2+5t-6)(t+5)$ **59.** $(x^3-3x+1)(x-1)$

61. $x^2-x+4-\dfrac{17}{x+4}$ **63.** $x^3-2x^2-4x-7-\dfrac{4}{x-2}$

65. $5x^2-25x+125-\dfrac{613}{x+5}$

67. $\dfrac{1}{10}x+\dfrac{41}{50}+\dfrac{291}{250(x-0.2)}$ **69.** -8 **71.** $2x$

73. $7uv$

75. Invalid. Only common factors can be cancelled.

77. Valid **79.** $x^3-5x^2-5x-10$ **81.** x^2-3

83. **(a)**

36	$\frac{51}{8}$	-12	16	148

(b)

36	$\frac{51}{8}$	-12	16	148

(c) The value of the polynomial at $x=r$ is equal to the remainder when the polynomial is divided by $x-r(66)$.

SECTION 4.5 *(page 247)*

Warm-Up *(page 246)*

1. $\frac{16}{3}$ **2.** -19 **3.** 3 **4.** $\frac{5}{2}$ **5.** $\frac{2}{3}, -8$
6. $16, -\frac{3}{2}$ **7.** $0, \frac{21}{2}$ **8.** 5 **9.** $6, -7$
10. 0, 8

1. (d) **3.** (b), (c) **5.** $\frac{3}{2}$ **7.** $\frac{1}{4}$ **9.** $-\frac{8}{3}$
11. 10 **13.** $\frac{7}{4}$ **15.** $\frac{1}{3}$ **17.** 61 **19.** $\frac{18}{5}$ **21.** 3
23. 3 **25.** $-\frac{11}{5}$ **27.** $\frac{4}{3}$ **29.** $-\frac{4}{15}$
31. No solution **33.** No solution **35.** 3 **37.** ± 6
39. ± 4 **41.** $8, -9$ **43.** 3, 13 **45.** 0, 2
47. $8, -3$ **49.** 2, 3 **51.** 8 **53.** 40 mph
55. 15 **57.** 85% **59.** 3 hrs; $\frac{15}{8}$ min; $\frac{5}{3}$ hrs
61. 15 hrs; $22\frac{1}{2}$ hrs **63.** $11\frac{1}{4}$ hrs **65.** 1986

CHAPTER 4 REVIEW EXERCISES *(page 251)*

1. $(-\infty, 8) \cup (8, \infty)$ **3.** $(-\infty, 1) \cup (1, 6) \cup (6, \infty)$

5. $\dfrac{2x^3}{5}$ **7.** $\dfrac{b-3}{6(b-4)}$ **9.** -9 **11.** $\dfrac{y}{8x}$

13. $12z(z-6)$ **15.** $-\frac{1}{4}$ **17.** $3x^2$ **19.** $\dfrac{125y}{x}$

21. $\dfrac{x(x-1)}{x-7}$ **23.** $-\frac{7}{9}$ **25.** $-\frac{13}{48}$

27. $\dfrac{4x + 3}{(x + 5)(x - 12)}$ **29.** $\dfrac{5x^3 - 5x^2 - 31x + 13}{(x + 2)(x - 3)}$
31. $\dfrac{x + 24}{x(x^2 + 4)}$ **33.** $\dfrac{6(x - 9)}{(x + 3)^2(x - 3)}$ **35.** $\dfrac{6(x + 5)}{x(x + 7)}$
37. $\dfrac{3t^2}{5t - 2}$ **39.** $-\dfrac{a^2 - a - 16}{(4a^2 + 16a + 1)(a - 4)}$
41. $2x^2 - \frac{1}{2}$ **43.** $2x^2 + \dfrac{4}{3}x - \dfrac{8}{9} + \dfrac{10}{9(3x - 1)}$
45. $x^2 - 2$ **47.** $x^2 + 5x - 7$
49. $x^3 + 3x^2 + 6x + 18 + \dfrac{29}{x - 3}$ **51.** -40 **53.** $\frac{3}{2}$
55. 5 **57.** 6, -4 **59.** 2, -6 **61.** 56 mph
63. 4 **65.** $6\frac{2}{3}$ min

CHAPTER 4 TEST *(page 253)*

1. $(-\infty, -5) \cup (-5, 5) \cup (5, \infty)$ **2.** $x + 2$ **3.** $-\frac{1}{3}$
4. $\dfrac{2a + 3}{5}$ **5.** $\dfrac{5z}{3}$ **6.** $\dfrac{4}{y + 4}$ **7.** $\dfrac{(2x + 3)^2}{x + 1}$
8. $\dfrac{14y^6}{15}$ **9.** $\dfrac{x^3}{4}$ **10.** $x^3(x + 3)^2(x - 3)$
11. $-\dfrac{2x^2 - 2x - 1}{x + 1}$ **12.** $\dfrac{5x^2 - 15x - 2}{x^2 - x - 6}$
13. $\dfrac{5x^3 - x^2 - 7x - 5}{x^2(x^2 + 2x + 1)}$ **14.** $-(3x + 1)$
15. $t^2 + 3 - \dfrac{6t - 6}{t^2 - 2}$ **16.** $5x^2 - 19x + 98 - \dfrac{500}{x + 5}$
17. $2x^3 + 6x^2 + 3x + 9 + \dfrac{20}{x - 3}$ **18.** 22
19. $-1, -\frac{15}{2}$ **20.** No solution **21.** $6\frac{2}{3}$ hrs, 10 hrs

CUMULATIVE TEST: CHAPTERS 1–4 *(page 254)*

1. $-\frac{10}{27}$ **2.** $8a^5b^7$ **3.** $2x^3y - 11xy$ **4.** $3n - 8$
5. $\frac{3}{2}$ **6.** $-\frac{3}{2}$ **7.** 13% **8.** \$1,408.75 **9.** $\frac{14}{3}$
10. $\frac{13}{2}$ **11.** $x \geq 103$
12. $x \leq -1$ or $x \geq 5$ **13.** $t^2 - 9t$

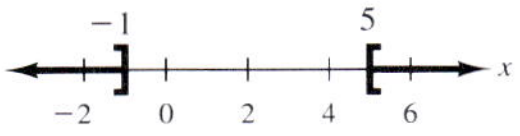

14. $x^2 - 2xy + y^2 + 4x - 4y + 4$
15. $(y - 3)^2(y + 3)$ **16.** $-4, \frac{5}{3}$
17. $\dfrac{x(x + 2)(x + 4)}{9(x - 4)}$ **18.** $\dfrac{3x + 5}{x(x + 3)}$ **19.** $x + y$
20. $x^2 - 3x + 9$ **21.** 2

Chapter 5

SECTION 5.1 *(page 262)*

Warm-Up *(page 262)*

1. a^{10} **2.** x^5y^4 **3.** $9y^3$ **4.** $4rs^2$
5. $-t^6$ **6.** $16z^2$ **7.** $36x^5y^8$ **8.** $-128u^5v^7$
9. $\dfrac{16a^8}{81b^4}$ **10.** $\dfrac{9x^2}{16y^6}$

1. $\frac{1}{25}$ **3.** $-\frac{1}{1{,}000}$ **5.** $-\frac{1}{243}$ **7.** 64 **9.** -32
11. $\frac{3}{2}$ **13.** 1 **15.** 1 **17.** 729 **19.** 100,000
21. $\frac{1}{64}$ **23.** $\frac{1}{16}$ **25.** $\frac{3}{16}$ **27.** $\frac{64}{121}$ **29.** $\frac{16}{15}$
31. y^2 **33.** x^6 **35.** x^6 **37.** a **39.** t^2
41. $-\dfrac{12}{xy^3}$ **43.** $\dfrac{y^4}{9x^4}$ **45.** $\dfrac{3x^5}{y^4}$ **47.** $\dfrac{81v^8}{u^6}$
49. $\dfrac{v^2}{uv^2 + 1}$ **51.** $\dfrac{ab}{b - a}$ **53.** $\dfrac{x^{12}y^8}{16}$
55. 3.6×10^6 **57.** 6.843×10^7 **59.** 3.81×10^{-8}
61. 60,000,000 **63.** 0.0000001359 **65.** 0.0093
67. 6.8×10^5 **69.** 4×10^3 **71.** 9×10^{15}
73. 4.70×10^{11} **75.** 3.46×10^{10} **77.** 9.3×10^7
79. 1.6×10^{-5} years **81.** 3.3×10^5

SECTION 5.2 *(page 272)*

Warm-Up *(page 272)*

1. 169 **2.** -169 **3.** $\frac{8}{27}$ **4.** $\frac{3}{10}$ **5.** 48
6. $\frac{27}{8}$ **7.** $\dfrac{1}{x^3}$ **8.** $-\dfrac{15y^4}{x^2}$ **9.** $\dfrac{9y^2}{4x^2}$ **10.** $\dfrac{v^4}{u^5}$

1. 7 **3.** 4.2 **5.** Square root, $\sqrt{2{,}025}$ **7.** 9
9. $\frac{3}{4}$ **11.** 8 **13.** Not a real number **15.** $-\frac{2}{3}$
17. 0.4 **19.** 13 **21.** 5 **23.** $-\frac{1}{4}$ **25.** 3
27. Not a real number **29.** -0.3 **31.** Irrational
33. Rational **35.** $16^{1/2} = 4$ **37.** $27^{2/3} = 9$
39. $\sqrt[4]{256^3} = 64$ **41.** 5 **43.** 8 **45.** $\frac{1}{4}$ **47.** $\frac{4}{9}$
49. $\frac{3}{11}$ **51.** 8.5440 **53.** 1.0420 **55.** 0.0038
57. 4.3004 **59.** 66.7213 **61.** $|t|$ **63.** y^3
65. x^3 **67.** $\dfrac{9y^{3/2}}{x^{2/3}}$ **69.** $\dfrac{3y^2}{4z^{4/3}}$ **71.** $x^{1/4}$ **73.** $c^{1/2}$
75. $\sqrt[8]{y}$ **77.** $2x^{3/2} - 3x^{1/2}$ **79.** $1 + 5y$

81. (a) 3.4 (b) 142 (c) 22 (d) 1,632
83. 13 in. × 13 in. × 13 in. **85.** 0.026 in.

MATH MATTERS: STANDARD PREFIXES USED WITH THE METRIC SYSTEM (page 277)

1,000,000,000 nanoseconds in a second
1,000 millimeters in a meter
1,000 watts in a kilowatt

SECTION 5.3 *(page 281)*

Warm-Up *(page 281)*

1. 100 **2.** -10 **3.** $-\frac{3}{5}$ **4.** $\frac{2}{5}$ **5.** $x^{11/6}$
6. $a^{1/4}$ **7.** $16x^4$ **8.** $4{,}096x^2y^8$ **9.** $\dfrac{16}{x^{1/4}y^{1/4}}$
10. $\dfrac{8y^2}{z^{5/2}}$

1. $(0, \infty)$ **3.** $(-\infty, \infty)$ **5.** $(0, \infty)$ **7.** $\sqrt{30}$
9. $\sqrt[3]{110}$ **11.** $\sqrt{\dfrac{15}{31}}$ **13.** $\sqrt[5]{\dfrac{152}{3}}$ **15.** $3\sqrt{35}$
17. $3\sqrt[4]{11}$ **19.** $\frac{1}{3}\sqrt{35}$ **21.** $\dfrac{10}{\sqrt[3]{11}}$ **23.** $2\sqrt{5}$
25. $3\sqrt{3}$ **27.** $2\sqrt[3]{3}$ **29.** $10\sqrt[4]{3}$ **31.** $\frac{1}{2}\sqrt{15}$
33. $\frac{1}{4}\sqrt[3]{35}$ **35.** $\frac{1}{3}\sqrt[5]{15}$ **37.** 0.02 **39.** $20\sqrt[3]{300}$
41. $3x^2\sqrt{x}$ **43.** $xy\sqrt[3]{x}$ **45.** $2xy\sqrt[5]{y}$ **47.** $\frac{1}{5}\sqrt{13}$
49. $\dfrac{2}{y}\sqrt[5]{x^2}$ **51.** $\dfrac{3a\sqrt[3]{2a}}{b^3}$ **53.** $\dfrac{\sqrt{3}}{3}$ **55.** $4\sqrt{3}$
57. $\dfrac{\sqrt[4]{20}}{2}$ **59.** $\frac{3}{2}\sqrt[3]{2}$ **61.** $\dfrac{\sqrt{y}}{y}$ **63.** $\dfrac{\sqrt[3]{18xy^2}}{3y}$
65. $\dfrac{a^2\sqrt[3]{a^2b}}{b}$ **67.** $2\sqrt{2}$ **69.** $-5\sqrt[4]{7} - 11\sqrt[4]{3}$
71. $24\sqrt{2} - 6$ **73.** $30\sqrt[3]{2}$ **75.** Not possible
77. $13\sqrt{y}$ **79.** $(10 - z)\sqrt[3]{z}$ **81.** $\dfrac{2\sqrt{5}}{5}$ **83.** $\dfrac{9\sqrt{5}}{5}$
85. 89.44 cycles/sec **87.** $>$ **89.** $>$ **91.** 1

SECTION 5.4 *(page 288)*

Warm-Up *(page 287)*

1. $34 - 5x$ **2.** $18x - 28$ **3.** $t^2 + \frac{2}{3}t + \frac{1}{9}$
4. $4x^2 - 20x + 25$ **5.** $25x^2 - 16$
6. $2 - \frac{23}{2}x - 3x^2$ **7.** $2xy\sqrt[3]{2x}$ **8.** $6xy\sqrt{2y}$
9. $(x - 4)\sqrt{3}$ **10.** $2\sqrt{7x}$

1. 4 **3.** $2\sqrt{5} - \sqrt{15}$ **5.** $2\sqrt{10} + 8\sqrt{2}$ **7.** -1
9. 4 **11.** $8\sqrt{5} + 24$
13. $\sqrt{15} + 3\sqrt{3} - 5\sqrt{5} - 15$ **15.** $y + 4\sqrt{y}$
17. $9x - 25$ **19.** $2x + 20\sqrt{2x} + 100$
21. $45x - 17\sqrt{x} - 6$ **23.** $x - y$ **25.** $2 - 7\sqrt[3]{4}$
27. $\sqrt[3]{4x^2} + 10\sqrt[3]{2x} + 25$
29. $2y - 10\sqrt[3]{2y} + 10\sqrt[3]{4y^2} - 100$
31. $(\sqrt{3} + 3\sqrt{3})$ **33.** $(4 - 3x)$ **35.** $(2u + \sqrt{2u})$
37. $\dfrac{1 - 2\sqrt{x}}{3}$ **39.** $\dfrac{-1 + \sqrt{3y}}{4}$ **41.** $2 - \sqrt{5}, -1$
43. $\sqrt{11} + \sqrt{3}, 8$ **45.** $\sqrt{x} + 3, x - 9$
47. $\sqrt{2u} + \sqrt{3}, 2u - 3$ **49.** $\dfrac{\sqrt{22} + 2}{3}$
51. $4(3 - \sqrt{7})$ **53.** $\dfrac{4\sqrt{7} + 11}{3}$ **55.** $\dfrac{(\sqrt{15} + \sqrt{3})x}{4}$
57. $\dfrac{(\sqrt{5} + 1)t^{3/2}}{2}$ **59.** $\dfrac{(\sqrt{x} - 5)(2\sqrt{x} + 1)}{4x - 1}$
61. $192\sqrt{2}$

SECTION 5.5 *(page 296)*

Warm-Up *(page 296)*

1. 2 **2.** $\frac{9}{2}$ **3.** $\frac{27}{5}$ **4.** $\frac{1}{6}$ **5.** $2, -25$
6. $12, -114$ **7.** $x - 9$ **8.** $\sqrt{5u}$
9. $4t + 12\sqrt{t} + 9$ **10.** $25\sqrt{2x}$

1. 400 **3.** 49 **5.** No solution **7.** 525 **9.** 90
11. $\frac{44}{3}$ **13.** $\frac{14}{25}$ **15.** $-\frac{2}{3}$ **17.** 4 **19.** No solution
21. 7 **23.** -15 **25.** 8 **27.** 1, 3 **29.** $\frac{1}{2}$
31. 10 **33.** $4\sqrt{2} \approx 5.66$ **35.** $\dfrac{5\sqrt{3}}{2} \approx 4.33$
37. $8\sqrt{6} \approx 19.60$ ft **39.** $40\sqrt{2} \approx 56.57$ ft/sec
41. 56.25 ft **43.** 64 ft **45.** 1.82 ft **47.** 500 units
49. $h = \dfrac{\sqrt{S^2 - r^4\pi^2}}{\pi r}$

SECTION 5.6 *(page 306)*

Warm-Up *(page 305)*

1. $4\sqrt{3}$ **2.** $6\sqrt{3}$ **3.** $23\sqrt{2}$ **4.** 150
5. $\frac{4\sqrt{10}}{5}$ **6.** $\frac{5(\sqrt{3}+1)}{4}$ **7.** $0, \frac{35}{2}$
8. $-4, \frac{5}{4}$ **9.** $-2, 9$ **10.** $-1, \frac{3}{2}$

1. $2i$ **3.** $-12i$ **5.** $\frac{2}{5}i$ **7.** $0.3i$ **9.** $2\sqrt{2}i$
11. $\sqrt{7}i$ **13.** $10i$ **15.** $3\sqrt{2}i$ **17.** -4
19. $-3\sqrt{6}$ **21.** $-2\sqrt{3}-3$ **23.** $5\sqrt{2}-4\sqrt{5}$
25. $a=3, b=-4$ **27.** $a=-4, b=-2\sqrt{2}$
29. $a=2, b=-2$ **31.** $10+4i$ **33.** $-14, -40i$
35. 13 **37.** $-14+20i$ **39.** $\frac{13}{6}+\frac{3}{2}i$
41. $-3+49i$ **43.** -36 **45.** $-36i$ **47.** $27i$
49. $-65-10i$ **51.** $20-12i$ **53.** $-40-5i$
55. $-7-24i$ **57.** $-24+2\sqrt{3}i$ **59.** $2+11i$
61. $2-i, 5$ **63.** $5+\sqrt{6}i, 31$ **65.** $-2+8i, 68$
67. $-10i, 100$ **69.** $1-\sqrt{3}i, 4$ **71.** $2+2i$
73. $-\frac{24}{53}+\frac{84}{53}i$ **75.** $-\frac{6}{5}+\frac{2}{5}i$ **77.** $-10i$
79. $\frac{8}{5}-\frac{1}{5}i$ **81.** $1-\frac{6}{5}i$ **83.** $-\frac{53}{25}+\frac{29}{25}i$ **85.** $2a$
87. $2bi$

89. $(-1+2i)^2+2(-1+2i)+5$
$= 1-4i+4i^2-2+4i+5$
$= (-4+4)i+(1-4-2+5)$
$= 0$

CHAPTER 5 REVIEW EXERCISES *(page 309)*

1. $\frac{1}{72}$ **3.** $\frac{125}{8}$ **5.** 36 million **7.** $\frac{x^6}{y^8}$ **9.** $\frac{1}{t^3}$
11. $\frac{27}{y^3}$ **13.** 1.2 **15.** $\frac{5}{6}$ **17.** 12 **19.** 10.63
21. 81 **23.** 0.04 **25.** $x^{7/12}$ **27.** $\frac{3}{x^{1/4}y^{2/5}}$
29. $6\sqrt{10}$ **31.** $0.5x^2\sqrt{y}$ **33.** $2ab\sqrt[3]{6b}$ **35.** $\frac{\sqrt{30}}{6}$
37. $\frac{\sqrt{3x}}{2x}$ **39.** $\frac{\sqrt[3]{4x^2}}{x}$ **41.** $-24\sqrt{10}$
43. $7\sqrt[4]{y+3}$ **45.** $12\sqrt{5}+41$
47. $6\sqrt{2}+7\sqrt{6}-4\sqrt{3}-14$ **49.** $\frac{(\sqrt{x}+10)^2}{x-100}$
51. 225 **53.** 105 **55.** $-3, -5$ **57.** $4\sqrt{3}i$
59. $\frac{3}{4}-\sqrt{3}i$ **61.** $8-3i$ **63.** -90 **65.** 25
67. $59+74i$ **69.** $\frac{9}{17}+\frac{2}{17}i$ **71.** $40\sqrt{29}\approx 215.4$ ft
73. 1.37 ft

CHAPTER 5 TEST *(page 311)*

1. $\frac{3}{8}$ **2.** $\frac{3}{5t}$ **3.** $\frac{y^2}{xy^2+1}$ **4.** 3.2×10^{-5}
5. 30,400,000 **6.** 3.0×10^{-5} **7.** $\frac{1}{9}$ **8.** $x^{1/3}$
9. $\sqrt[3]{30}$ **10.** $\frac{4}{3}\sqrt{2}$ **11.** $2\sqrt[3]{3}$ **12.** $\frac{\sqrt{6}}{2}$
13. $\frac{\sqrt[3]{2}}{2}$ **14.** $-10\sqrt{3x}$ **15.** $5\sqrt{3x}+3\sqrt{5}$
16. $16-8\sqrt{2x}+2x$ **17.** $7\sqrt{3}(4y+3)$
18. No solution **19.** 9 **20.** $2-2i$ **21.** $13+13i$
22. $-8+4i$ **23.** $-\frac{8}{41}+\frac{10}{41}i$ **24.** $3\sqrt{15}\approx 11.62$ ft

Chapter 6

SECTION 6.1 *(page 321)*

Warm-Up *(page 321)*

1. $(4x+11)(4x-11)$ **2.** $(3t-4)^2$
3. $(x-3)(5x+2)$ **4.** $(x-10)(x-4)$
5. $y(2y-1)(2y+3)$ **6.** $4x(x^2-3x+4)$
7. $-\frac{6}{5}$ **8.** $\frac{7}{2}$ **9.** $-9, \frac{15}{2}$ **10.** $\frac{8}{5}, 12$

1. $0, 3$ **3.** $9, 12$ **5.** $\pm\frac{5}{2}$ **7.** 6 **9.** -30
11. $1, 6$ **13.** $\frac{1}{2}, -\frac{5}{6}$ **15.** $\frac{5}{3}, 6$ **17.** ± 3
19. ± 8 **21.** ± 8 **23.** $\pm\frac{4}{5}$ **25.** $\pm\frac{15}{2}$
27. $9, -17$ **29.** $2.5, 3.5$ **31.** $2\pm\sqrt{7}$
33. $\frac{-1\pm 5\sqrt{2}}{2}$ **35.** $\frac{3\pm 7\sqrt{2}}{4}$ **37.** $\pm 2i$
39. $3\pm 5i$ **41.** $-\frac{4}{3}\pm 4i$ **43.** $-6\pm\frac{11}{3}i$
45. $\frac{2}{3}\pm\frac{1}{3}i$ **47.** $1\pm 3\sqrt{3}i$ **49.** $-\frac{7}{3}\pm\frac{\sqrt{38}}{3}i$
51. $\pm 4, \pm 2$ **53.** $\frac{1}{27}, -27$ **55.** $\pm 2\sqrt{2}$ **57.** ± 10
59. $-5, 15$ **61.** $5\pm 10i$ **63.** $\pm\sqrt{15}, \pm\sqrt{5}i$
65. $\pm 2, \pm 2i$ **67.** $(x-5)(x+2)=x^2-3x-10=0$
69. $[x-(1+\sqrt{2})][x-(1-\sqrt{2})]=x^2-2x-1=0$
71. 9 sec **73.** 1980

SECTION 6.2 *(page 329)*

Warm-Up *(page 329)*

1. $x^2 + 6x + 10$ **2.** $x^2 + 20x + 150$ **3.** $u^2 - u + \frac{1}{4}$ **4.** $v^2 - 24v + 44$ **5.** $\pm\frac{2}{7}$ **6.** $\pm\frac{3}{4}i$ **7.** $4 \pm 6i$ **8.** $-2 \pm 2\sqrt{5}$ **9.** $\frac{2}{3} \pm \frac{\sqrt{5}}{3}$ **10.** $\frac{3}{8}$, 12

1. 16 **3.** 100 **5.** $\frac{25}{4}$ **7.** $\frac{9}{25}$ **9.** $\frac{9}{625}$ **11.** 0, 25 **13.** 1, 7 **15.** 4, −6 **17.** −3, −4 **19.** −3, 6

21. $2 + \sqrt{7} \approx 4.65$, $2 - \sqrt{7} \approx -0.65$ **23.** $-2 + \sqrt{7} \approx 0.65$, $-2 - \sqrt{7} \approx -4.65$

25. $2 + \sqrt{3} \approx 3.73$, $2 - \sqrt{3} \approx 0.27$ **27.** $-1 + \sqrt{2}i \approx -1 + 1.41i$, $-1 - \sqrt{2}i \approx -1 - 1.41i$

29. $5 + 3\sqrt{3} \approx 10.20$, $5 - 3\sqrt{3} \approx -0.20$ **31.** $-10 + 3\sqrt{10} \approx -0.51$, $-10 - 3\sqrt{10} \approx -19.49$

33. $\frac{3 + 2\sqrt{7}}{3} \approx 2.10$, $\frac{3 - 2\sqrt{7}}{3} \approx -1.43$ **35.** $\frac{-5 + \sqrt{13}}{2} \approx -0.70$, $\frac{-5 - \sqrt{13}}{2} \approx -4.30$

37. $\frac{-3 + \sqrt{17}}{2} \approx 0.56$, $\frac{-3 - \sqrt{17}}{2} \approx -3.56$ **39.** $\frac{1}{2} + \frac{\sqrt{3}}{2}i \approx 0.5 + 0.87i$, $\frac{1}{2} - \frac{\sqrt{3}}{2}i \approx 0.5 - 0.87i$

41. $\frac{-4 + \sqrt{10}}{2} \approx -0.42$, $\frac{-4 - \sqrt{10}}{2} \approx -3.58$ **43.** $\frac{-9 + \sqrt{21}}{6} \approx -0.74$, $\frac{-9 - \sqrt{21}}{6} \approx -2.26$

45. $\frac{-1 + \sqrt{10}}{2} \approx 1.08$, $\frac{-1 - \sqrt{10}}{2} \approx -2.08$ **47.** $\frac{-5 + \sqrt{17}}{2} \approx -0.44$, $\frac{-5 - \sqrt{17}}{2} \approx -4.56$

49. $\frac{7 + \sqrt{57}}{2} \approx 7.27$, $\frac{7 - \sqrt{57}}{2} \approx -0.27$ **51.** $1 \pm \sqrt{3}$ **53.** $4 \pm 2\sqrt{2}$

55. **(a)** $x^2 + 8x$ **(b)** $x^2 + 8x + 16$ **(c)** $(x + 4)^2$ **57.** 42 or 58; both are correct

SECTION 6.3 *(page 339)*

Warm-Up *(page 339)*

1. 3, 7 **2.** 2, −16 **3.** 3, $-\frac{10}{3}$ **4.** $\frac{3}{5}$, 3 **5.** $\frac{-3 \pm \sqrt{17}}{2}$ **6.** $-1 \pm \frac{\sqrt{3}}{3}i$ **7.** 3 **8.** $\sqrt{11}i$ **9.** $4\sqrt{6}$ **10.** $3\sqrt{17}$

1. $2x^2 + 2x - 7 = 0$ **3.** $-x^2 + 10x - 5 = 0$ **5.** 4, 7 **7.** −2, −4 **9.** $\frac{2}{3}$ **11.** $-\frac{3}{2}$ **13.** $-\frac{1}{2}, \frac{2}{3}$ **15.** Two distinct irrational solutions **17.** Two complex solutions **19.** One rational solution **21.** $1 \pm \sqrt{5}$ **23.** $-2 \pm \sqrt{3}$ **25.** $-3 \pm 2\sqrt{3}$ **27.** $5 \pm \sqrt{2}$ **29.** $-\frac{3}{2} \pm \frac{\sqrt{3}}{2}i$ **31.** $\frac{1 \pm \sqrt{3}}{2}$ **33.** $\frac{-2 \pm \sqrt{10}}{2}$ **35.** $\frac{-1 \pm \sqrt{5}}{3}$ **37.** $\frac{1 \pm \sqrt{5}}{5}$ **39.** $\frac{-1 \pm \sqrt{10}}{5}$ **41.** ±13 **43.** 0, −15 **45.** $\frac{9}{5}, \frac{21}{5}$ **47.** 8, 16 **49.** $-4 \pm 3i$ **51.** 0.372, 3.228 **53.** 0.200, 99.800 **55.** $\frac{5 \pm \sqrt{185}}{8}$ **57.** $\frac{3 + \sqrt{17}}{2}$ **59.** **(a)** 2.5 sec **(b)** $\frac{5 + 5\sqrt{3}}{4} \approx 3.415$ sec **61.** 1968

SECTION 6.4 *(page 348)*

Warm-Up *(page 347)*

1. −7 **2.** 0, $\frac{2}{15}$ **3.** $\frac{5}{2}$, −1 **4.** 10, −12 **5.** 1, −17 **6.** 1 **7.** $-\frac{3}{2} \pm \frac{\sqrt{11}}{2}i$ **8.** $\frac{-3 \pm \sqrt{29}}{2}$ **9.** 6, −1 **10.** 3, −1

1. 15, 16 **3.** 14, 16 **5.** 12, 13 **7.** 12 in., 108 in.^{2} **9.** 10 ft, 70 ft **11.** 24 in., 64 in. **13.** 12 km, 180 km^2 **15.** 120 m^2, 440 m **17.** 100 × 125 ft **19.** Base: 24 in.; height: 16 in. **21.** 8% **23.** 18 doz, \$1.20/doz **25.** 48 **27.** 15.86 mi **29.** 400 mi/hr **31.** 9.1 hrs, 11.1 hrs **33.** 3 sec **35.** 9.5 sec **37.** 4.7 sec

SECTION 6.5 *(page 360)*

Warm-Up *(page 360)*

1. $x \le 5$

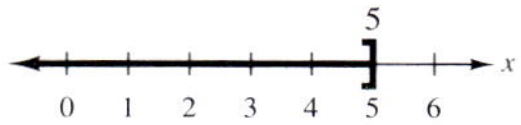

2. $x \ge -\frac{5}{3}$

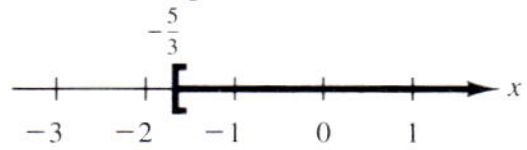

3. $x < \frac{3}{2}$

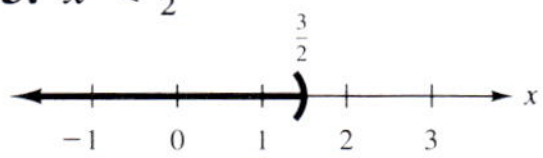

4. $x < 5$

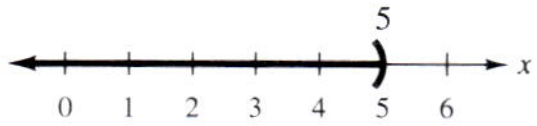

5. $1 < x < 5$

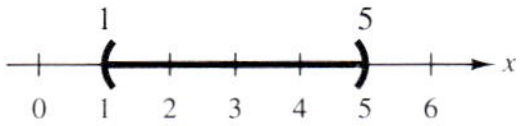

6. $x < 2,\ x > 8$

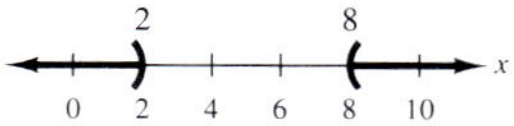

7. Negative **8.** Zero **9.** Positive
10. Negative

1. 3

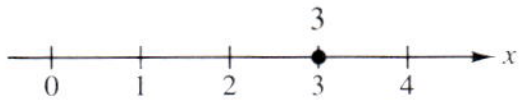

3. $0,\ \frac{5}{2}$

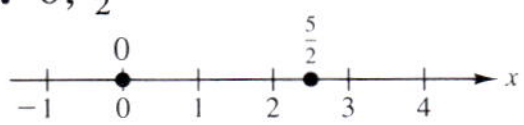

5. 1, 3

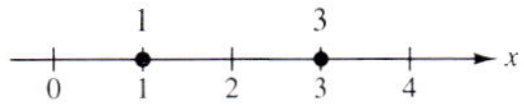

7. Negative: $x < 4$
Positive: $x > 4$

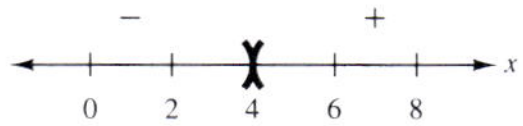

9. Negative: $0 < x < 4$
Positive: $x < 0$ or $x > 4$

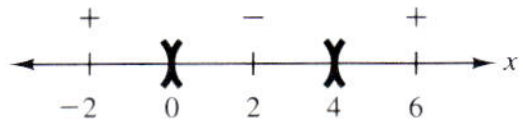

11. Negative: $-1 < x < 5$
Positive: $x < -1$ or $x > 5$

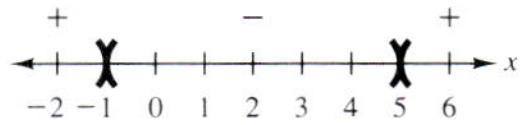

13. $x \ge -3$

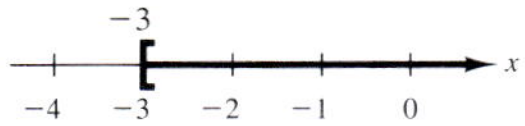

15. $0 < x < 2$

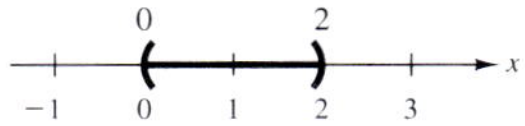

17. $x < 0$ or $x > 2$

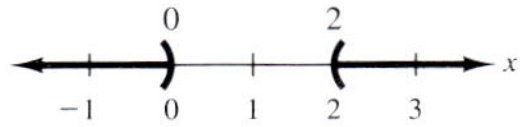

19. $x \le -2$ or $x \ge 2$

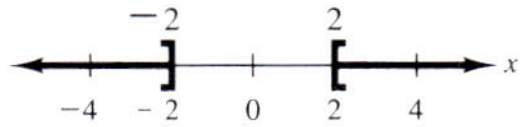

21. $-5 \le x \le 2$

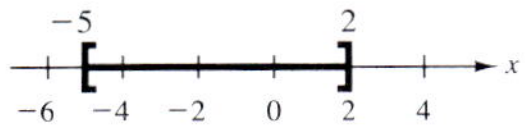

23. $u < -\frac{1}{2}$ or $u > 4$

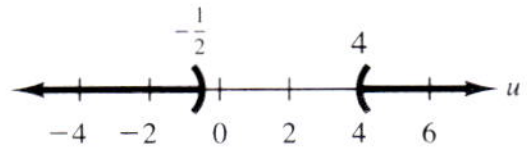

25. No solution

27. $-\infty < x < \infty$

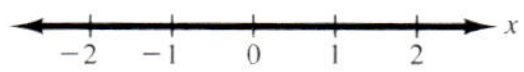

29. $x < 2 - \sqrt{2}$ or
$x > 2 + \sqrt{2}$

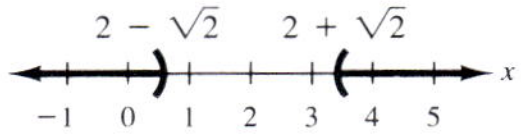

31. No solution

33. $x < 5 - \sqrt{6}$ or
$x > 5 + \sqrt{6}$

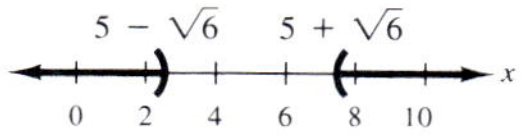

35. $-2 < x < 0$ or $x > 2$

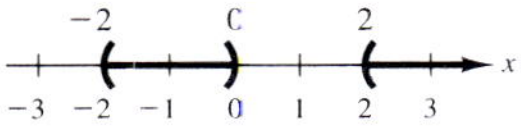

37. 3

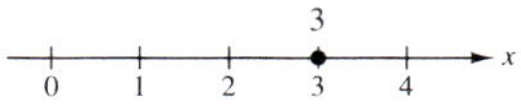

39. $0,\ -5$

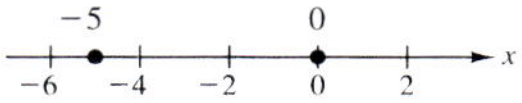

41. $x > 3$

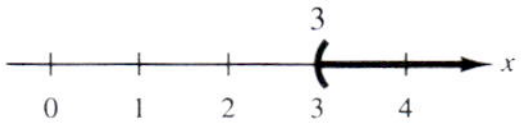

43. $x < 3$

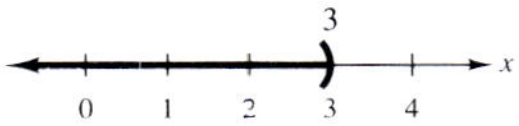

45. $0 \le x < 4$

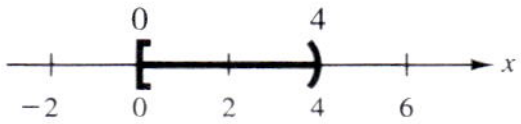

47. $y \le 3$ or $y > \frac{11}{2}$

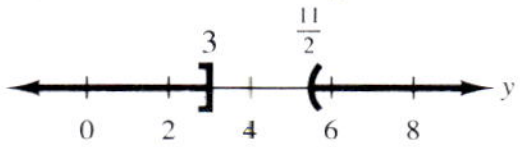

49. $4 < x < 7$

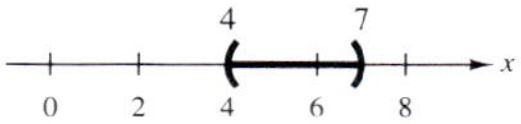

51. $x < 3$ or $x \ge 6$

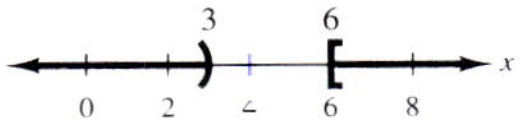

53. $3 < t < 5$ **55.** $12 < l < 20$

CHAPTER 6 REVIEW EXERCISES *(page 363)*

1. $0,\ -12$ **3.** $-10,\ \frac{8}{3}$ **5.** $\pm\frac{1}{2}$ **7.** $-\frac{5}{2}$
9. $-9,\ 10$ **11.** ± 100 **13.** $\pm\frac{3}{2}$ **15.** $-4,\ 36$
17. $3 \pm 2\sqrt{3}$ **19.** $\frac{3}{2} \pm \frac{\sqrt{3}}{2}i$ **21.** $\frac{-5 \pm \sqrt{19}}{2}$
23. $5,\ -6$ **25.** $3,\ -\frac{7}{2}$ **27.** $\frac{10}{3} \pm \frac{5\sqrt{2}}{3}i$
29. $3 \pm 5\sqrt{10}$ **31.** $-7,\ 12$ **33.** $-2,\ 20$
35. $3 \pm i$ **37.** $-\frac{1}{2},\ -1$ **39.** $\frac{3 \pm \sqrt{17}}{2}$

41. $x < 3$

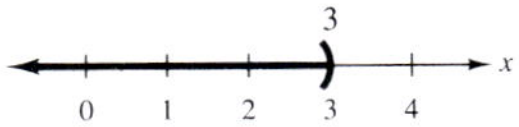

43. $0 < x < 7$

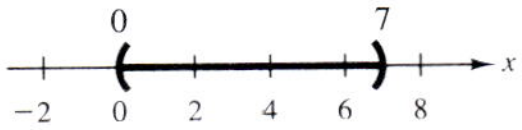

45. $x \le -2$ or $x \ge 6$

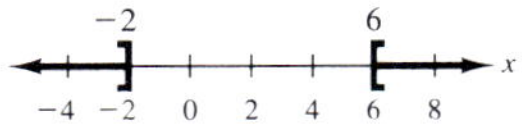

47. $-4 < x < \frac{5}{2}$

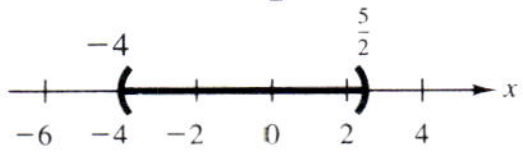

49. $x \le 0$ or $x > \frac{7}{2}$

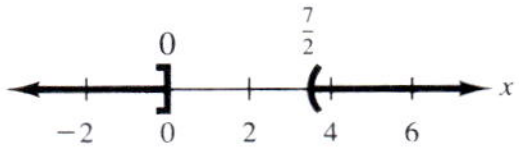

51. 11, 12

53. $\frac{5\sqrt{2}}{2} \approx 3.54$ sec

55. (a) $P = 2l + 2w = 48$
$l + w = 24$
$w = 24 - l$

(b) $A = lw$
$w = 24 - l$
$A = l(24 - l)$

(c)

l	A
2	44
4	80
6	108
8	128
10	140
12	144
14	140
16	128
18	108

57. 15 **59.** $\frac{9 + 3\sqrt{17}}{2} \approx 10.68$ hrs, $\frac{15 + 3\sqrt{17}}{2} \approx 13.68$ hrs

CHAPTER 6 TEST *(page 365)*

1. $-5, 10$ **2.** $-\frac{3}{8}, 3$ **3.** 1.7, 2.3 **4.** $-3 \pm 9i$

5. $C = \frac{9}{4}$ **6.** $\frac{3 \pm \sqrt{3}}{2}$

7. -56; two complex solutions **8.** $\frac{4 \pm \sqrt{7}}{3}$

9. $\frac{7 + \sqrt{61}}{2}$

10. $0 < x < 3$

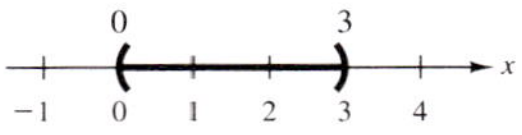

11. $x \le -2$ or $x \ge 6$

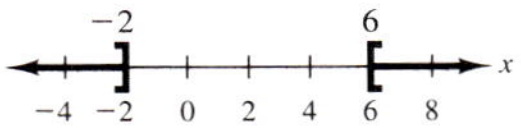

12. $2 < x < \frac{11}{4}$

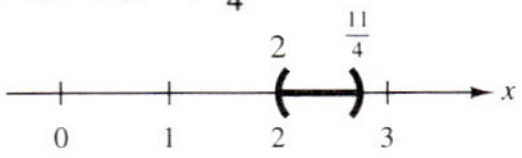

13. $(x + 4)(x - 5) = x^2 - x - 20 = 0$ **14.** 14, 15

15. 12×20 ft **16.** 50 mi/hr **17.** $\frac{\sqrt{10}}{2} \approx 1.58$ sec

Chapter 7

SECTION 7.1 *(page 377)*

Warm-Up *(page 377)*

1. $-6 < 2$

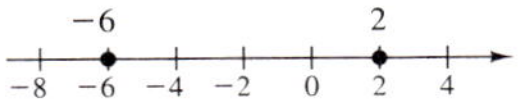

2. $\frac{11}{3} > 2$

3. $-3 > -7$

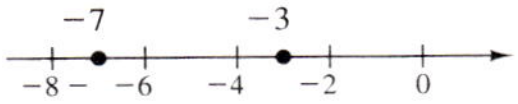

4. $-5 < 0$

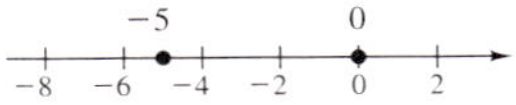

5. $\frac{13}{3} > -\frac{3}{2}$

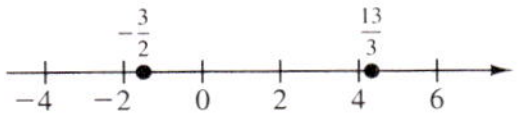

6. $\frac{3}{5} < \frac{13}{16}$

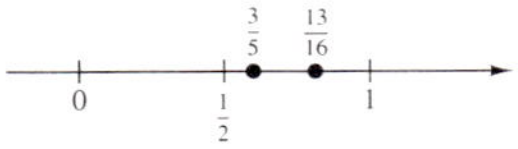

7. 3 **8.** 27 **9.** $\frac{33}{4}$ **10.** 3

1.

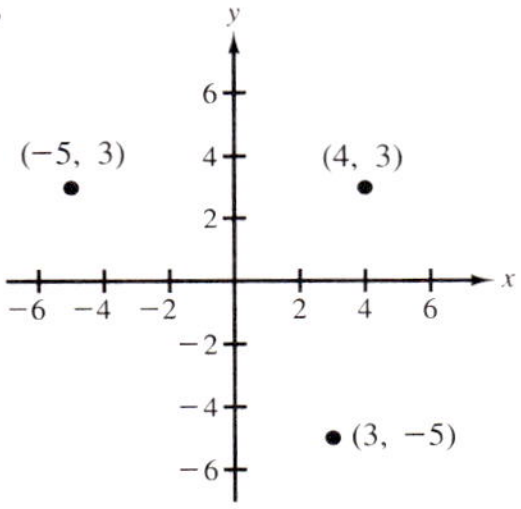

3.

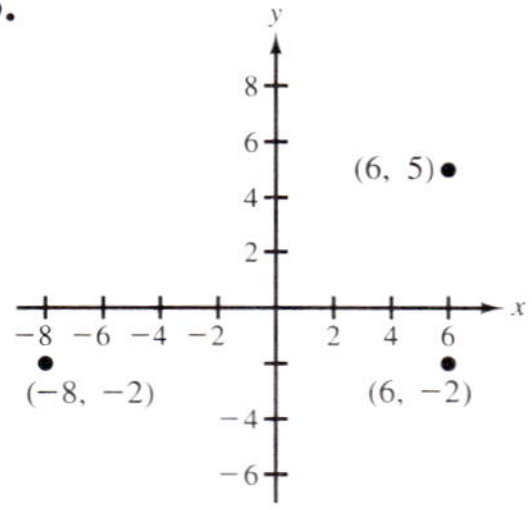

5.

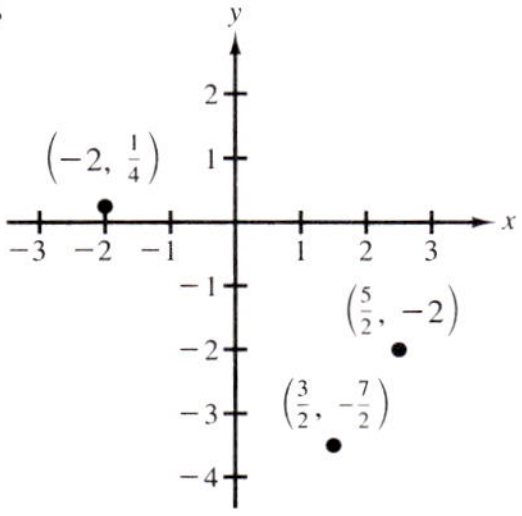

7. (a) Quadrant IV
(b) Quadrant II

9. A: $(4, -2)$
B: $(-3, -2.5)$
C: $(3, 0.5)$

11.

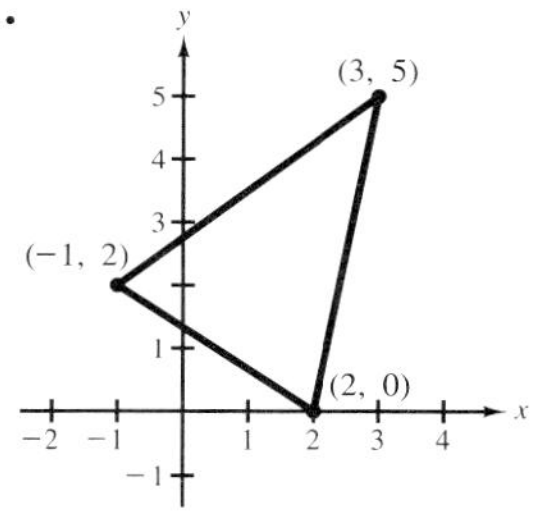

13.

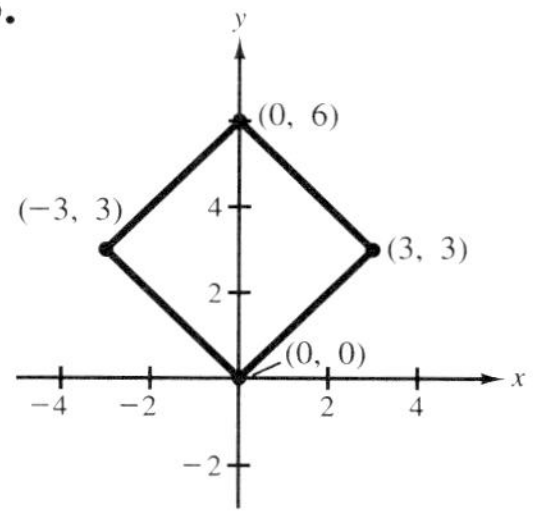

15.

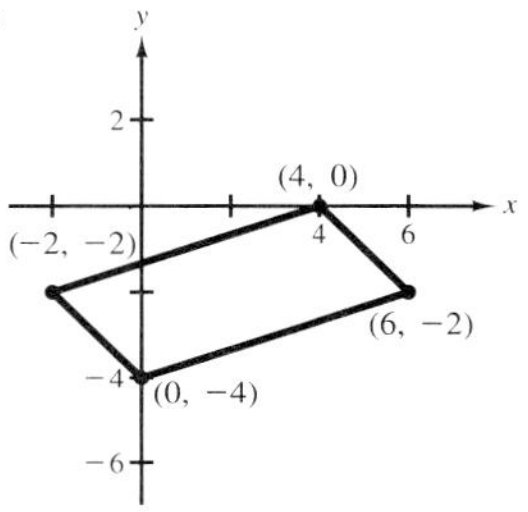

17. Quadrant II, Quadrant III

19. Quadrant I, Quadrant III

21. $(3, -4)$

23. $-10, -10$

25.

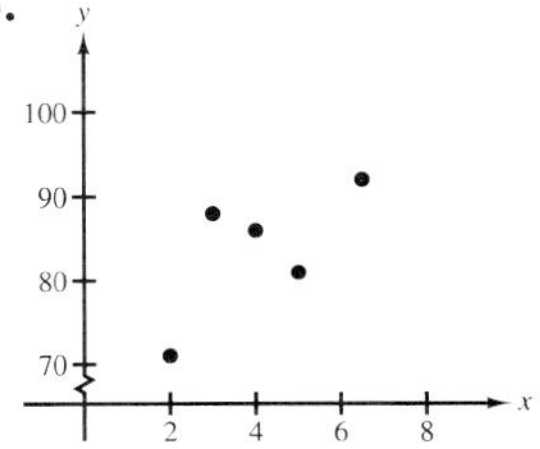

27.

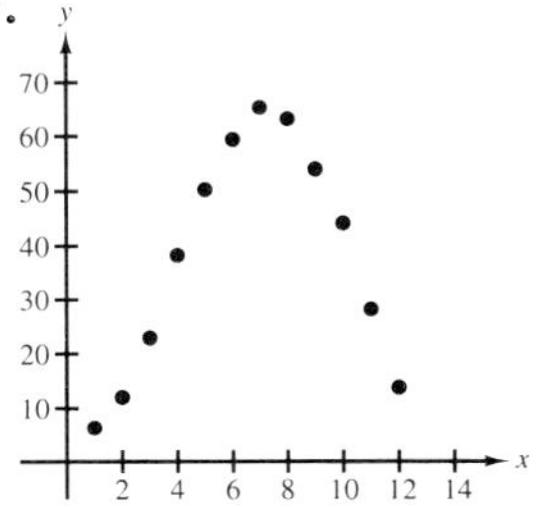

29. **(a)** Yes **(b)** No **(c)** No **(d)** Yes

31. **(a)** No **(b)** Yes **(c)** Yes **(d)** No

33. **(a)** Yes **(b)** Yes **(c)** Yes **(d)** No

35.

x	-2	0	2	4	6
$y = 5x - 1$	-11	-1	9	19	29

37.

x	-4	$\frac{2}{5}$	4	8	12
$y = -\frac{5}{2}x + 4$	14	3	-6	-16	-26

39.

x	$y = 28x + 3{,}000$
100	5,800
150	7,200
200	8,600
250	10,000
300	11,400

41. 7

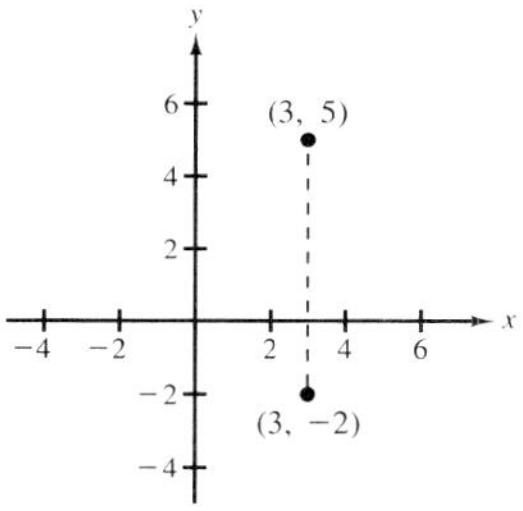

43. 7

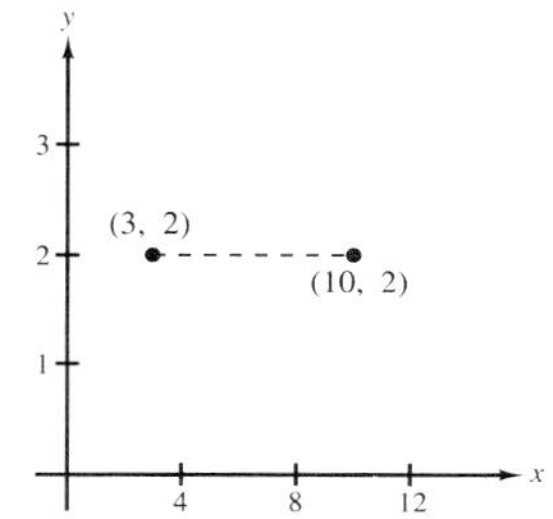

45. 11

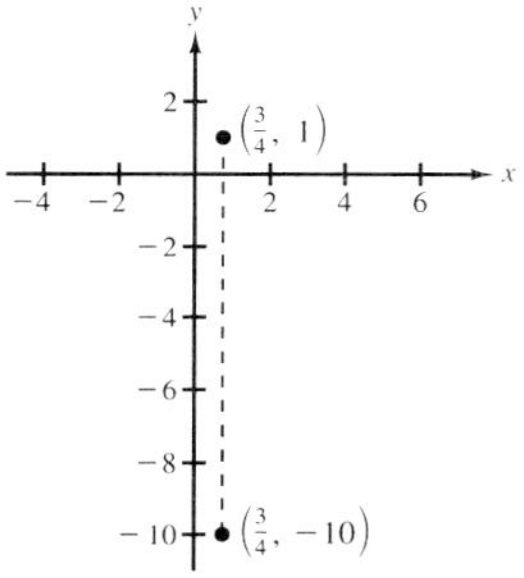

47. **(a)** $(x, y) = (10, 2)$ **(b)** 6, 8 **(c)** 10 **(d)** 10

49. **(a)** $(x, y) = (4, -4)$ **(b)** 8, 7 **(c)** $\sqrt{113}$ **(d)** $\sqrt{113}$

51. 5 **53.** $8\sqrt{2}$ **55.** $2\sqrt{13}$ **57.** $\dfrac{\sqrt{113}}{3}$

59. Collinear **61.** Not collinear

63. $3 + \sqrt{26} + \sqrt{29} \approx 13.48$

65. $(1, 4)$

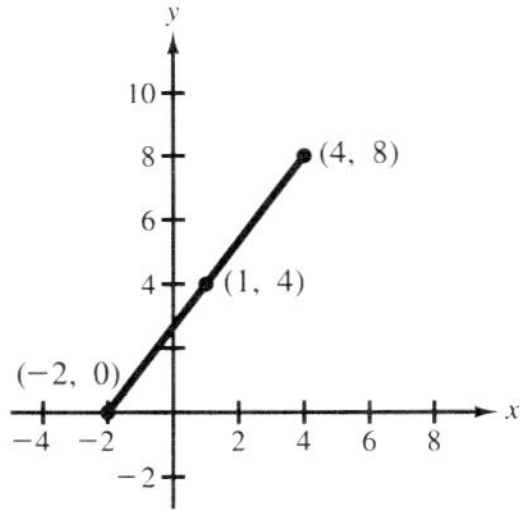

67. $\left(\frac{7}{2}, \frac{9}{2}\right)$

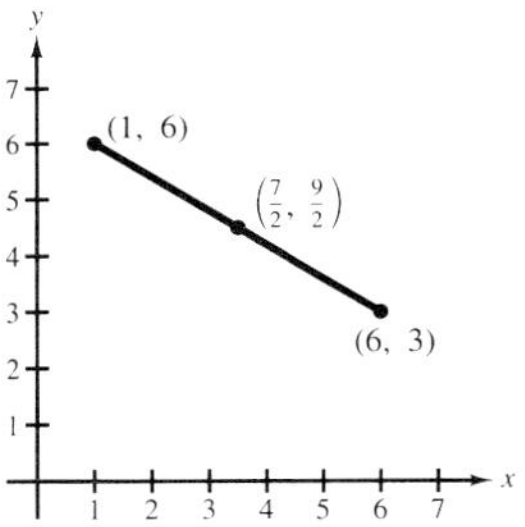

69. Reflection through the y-axis

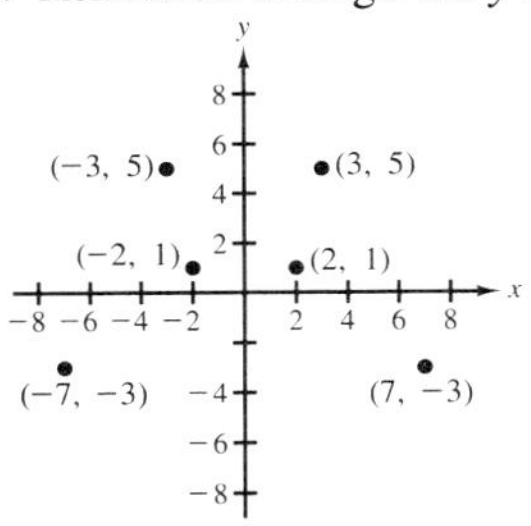

71. 18 feet

MATH MATTERS: FINDING THE SHORTEST PATH (page 373)

At first, it might seem that the answer is for the spider to walk straight down the wall, across the floor, and up the other wall. The total length of this path is 42 feet. There is, however, a shorter path. To see this, imagine that the sides of the room are unfolded as shown in the following figure. By imposing a rectangular coordinate system on the unfolded walls, we see that the spider starts at the point $(0, -1)$ and wants to travel to the point $(24, 31)$. The distance between these two points is $\sqrt{24^2 + 32^2} = 40$. Thus, the shortest distance is not the 42-foot path, but the diagonal path, running across five of the six surfaces of the room, as shown in the accompanying figure.

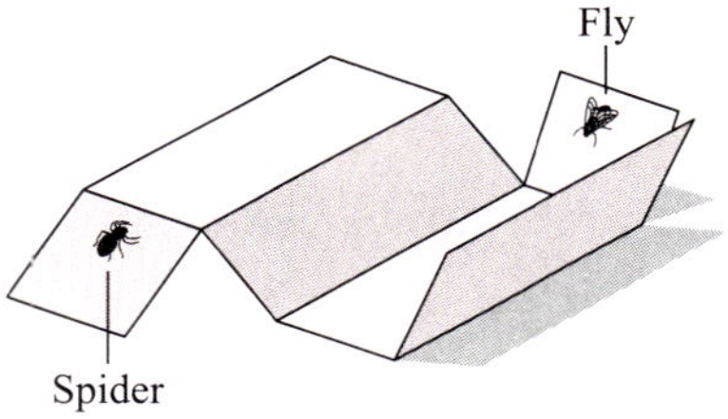

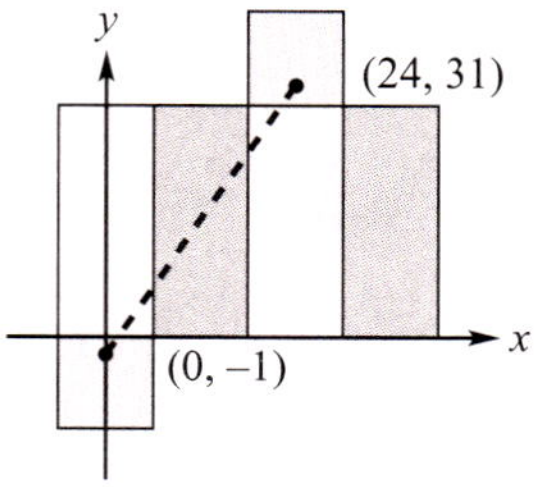

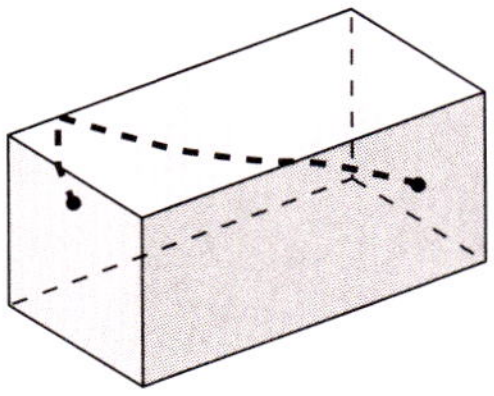

SECTION 7.2 (page 391)

Warm-Up (page 391)

1.

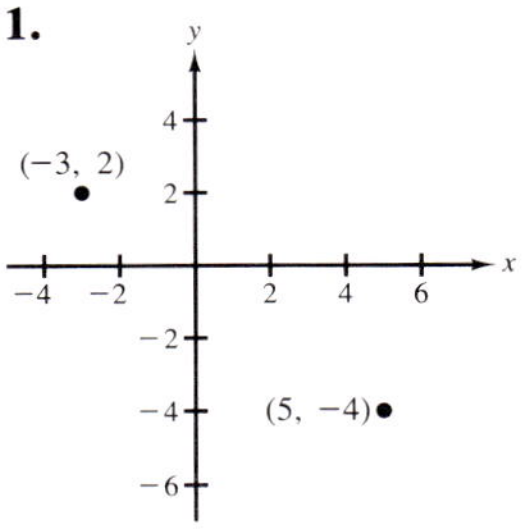

2.

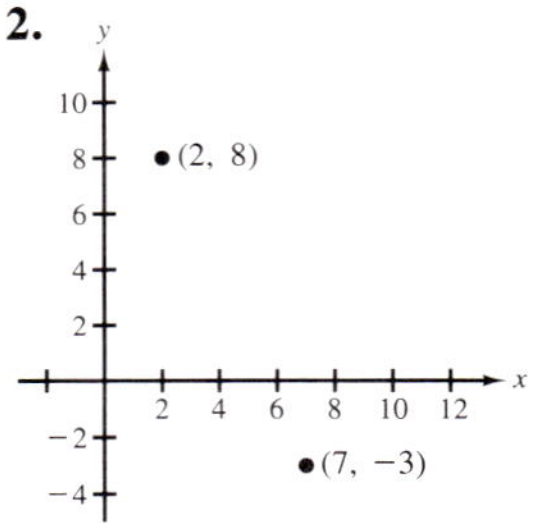

3.

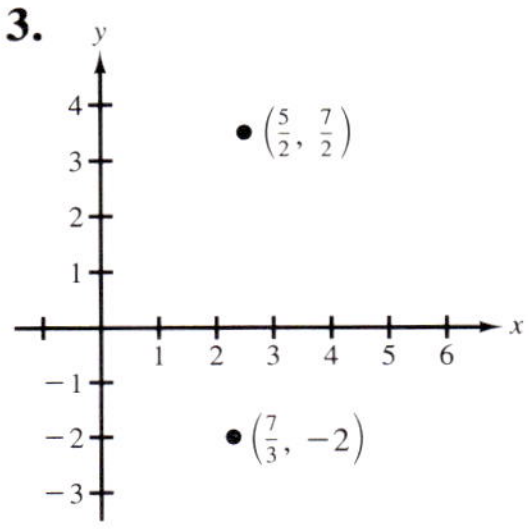

4.

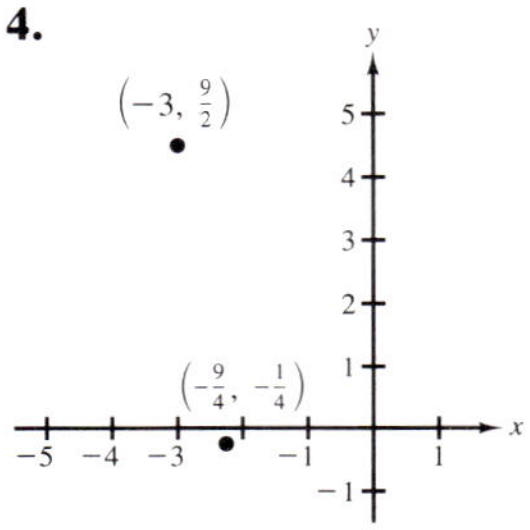

5. $\frac{48}{5}$ **6.** 6 **7.** 13 **8.** $-\frac{11}{3}$
9. $\frac{130}{19} \approx 6.84$ **10.** 7

1.

x	-4	-2	0	2	4
y	11	7	3	-1	-5
(x, y)	$(-4, 11)$	$(-2, 7)$	$(0, 3)$	$(2, -1)$	$(4, -5)$

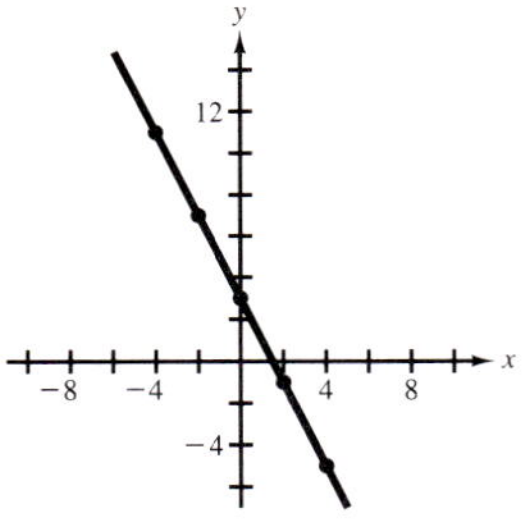

3.

x	± 2	-1	0	2	± 3
y	0	3	4	0	-5
(x, y)	$(\pm 2, 0)$	$(-1, 3)$	$(0, 4)$	$(2, 0)$	$(\pm 3, -5)$

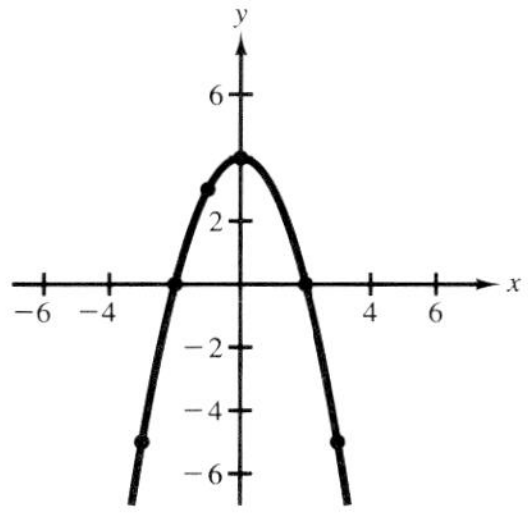

5.

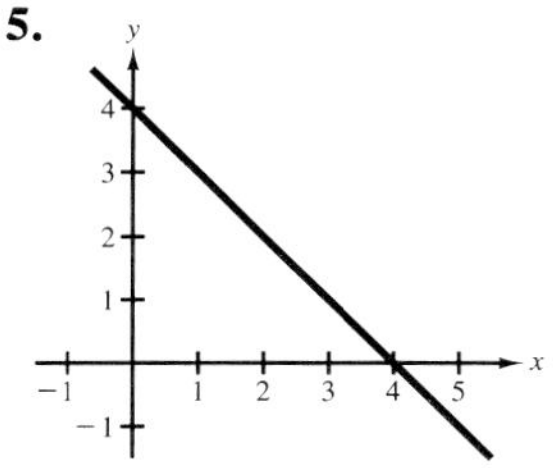

7.

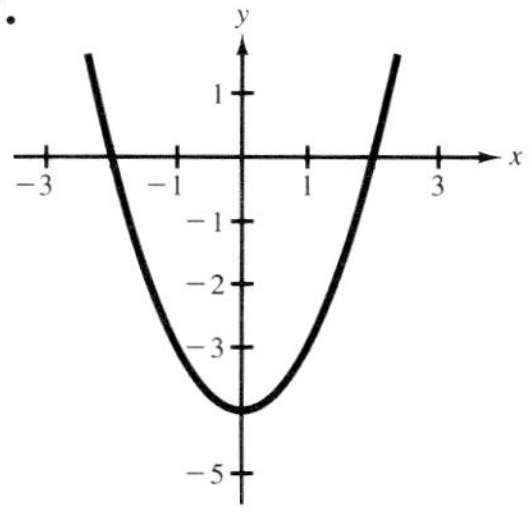

9.

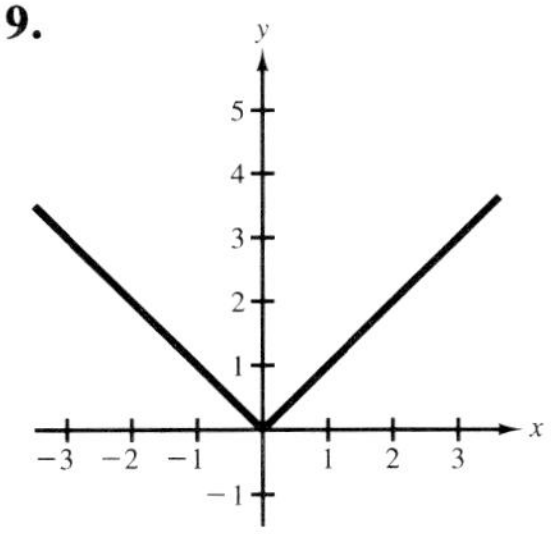

11. (10, 0), (0, 5) **13.** (−20, 0), (0, 15)
15. (±5, 0), (0, −25) **17.** (0, 3)
19. $\left(\frac{2}{3}, 0\right)$, (−1, 0), (0, −2)

21.

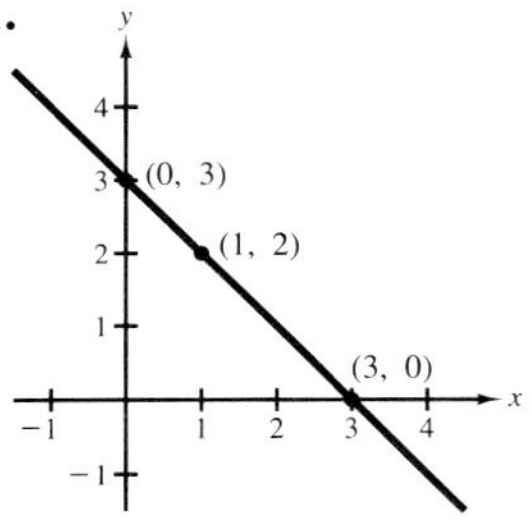

23.

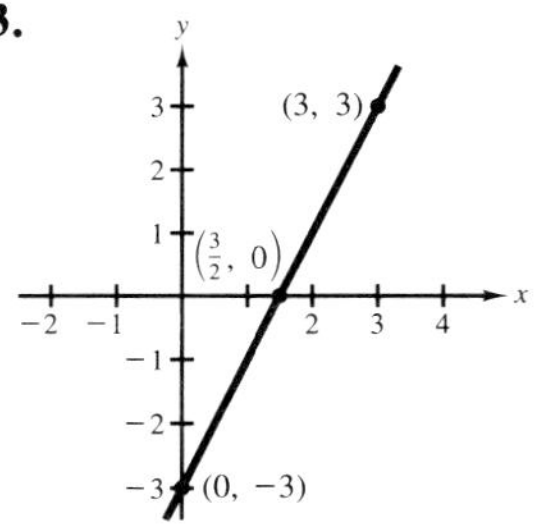

25.

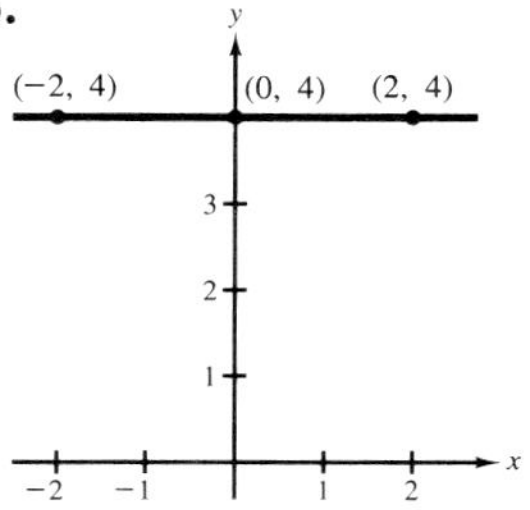

27.

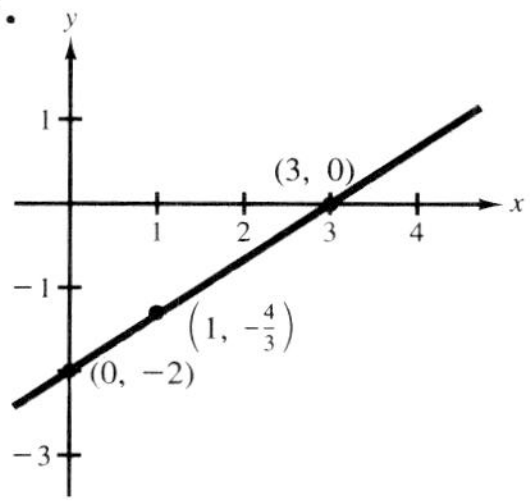

29.

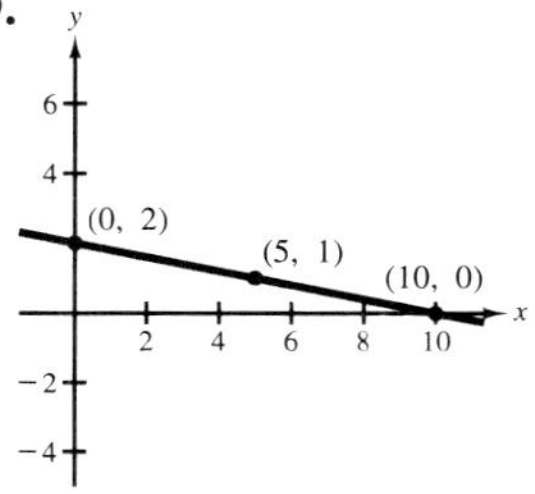

31.

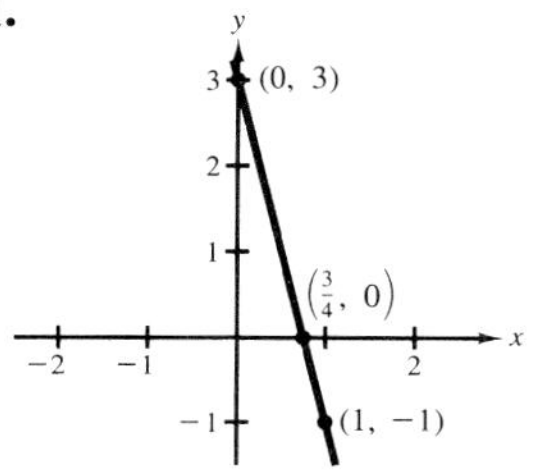

33.

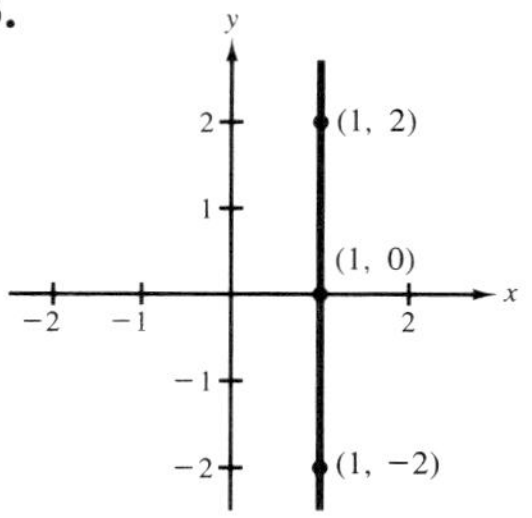

35.

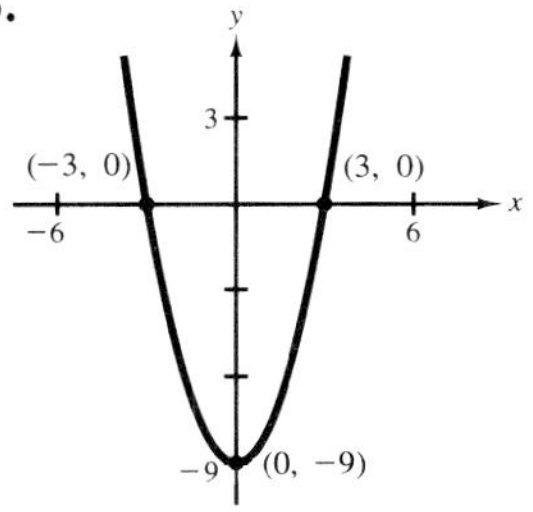

37.

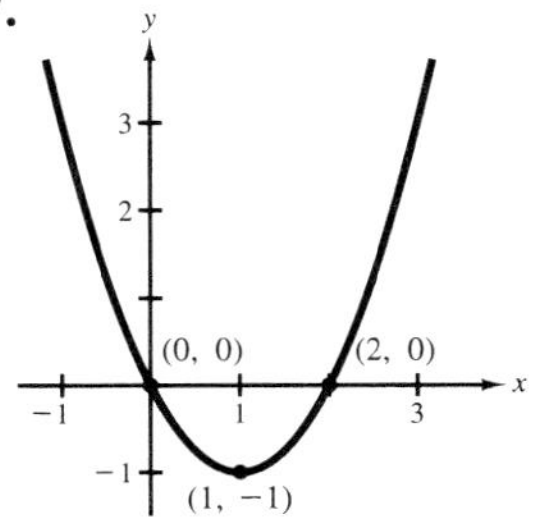

39.

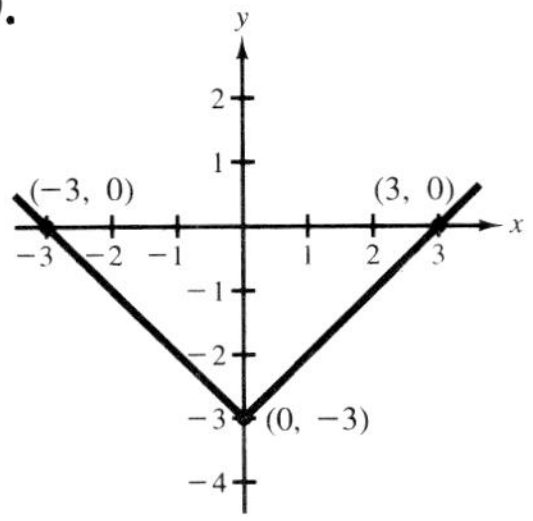

41.

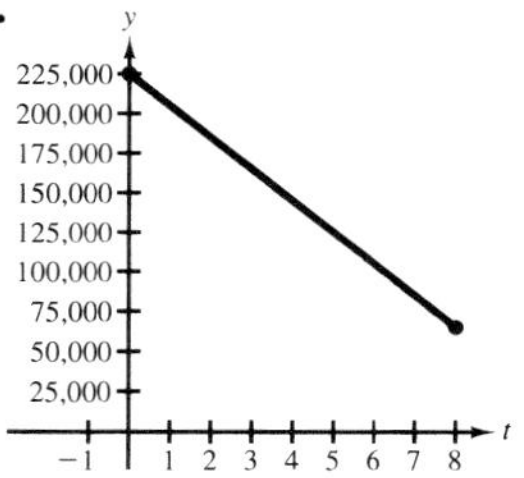

43. **(a)** $y = x^2$

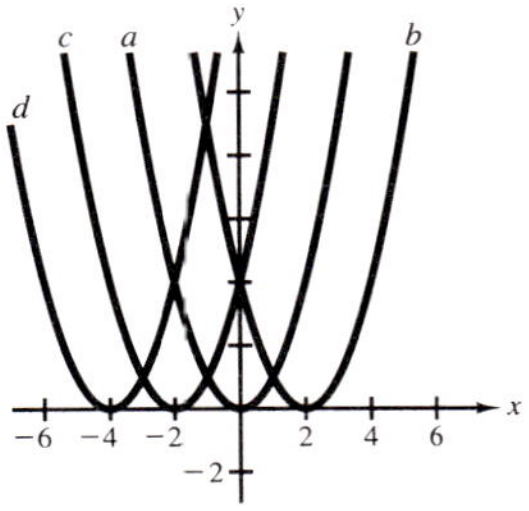

(b) Translation of a 2 units to right
(c) Translation of a 2 units to left
(d) Translation of a 4 units to left

SECTION 7.3 *(page 400)*

Warm-Up *(page 400)*

1. $y = -3x + 4$ **2.** $y = x$ **3.** $y = \frac{2}{3}(1 - x)$
4. $y = \frac{2}{5}(2x + 1)$ **5.** $y = \frac{1}{4}(5 - 3x)$
6. $y = \frac{2}{3}(x + 3)$ **7.** $y = 4 - x^2$
8. $y = -\frac{2}{3}(x^2 + 1)$ **9.** $y = \frac{1}{3}(4 - x^2)$
10. $y = \frac{1}{2}(x^2 - 9)$

1.
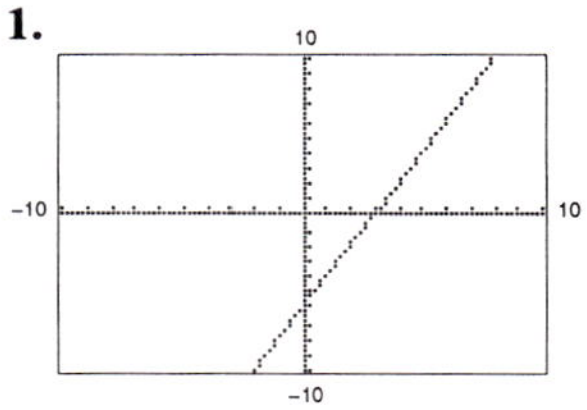

3.
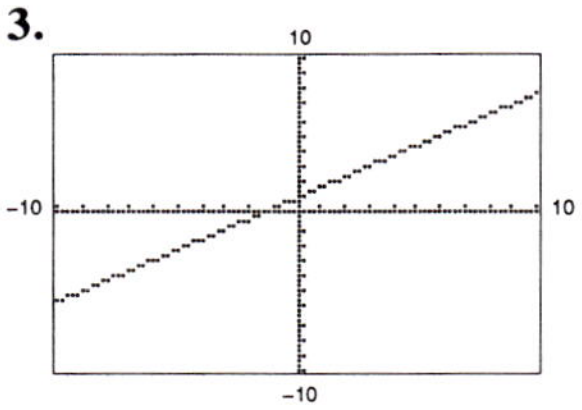

5.
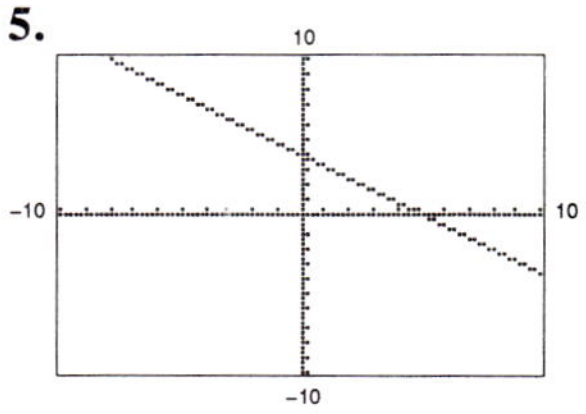

7.
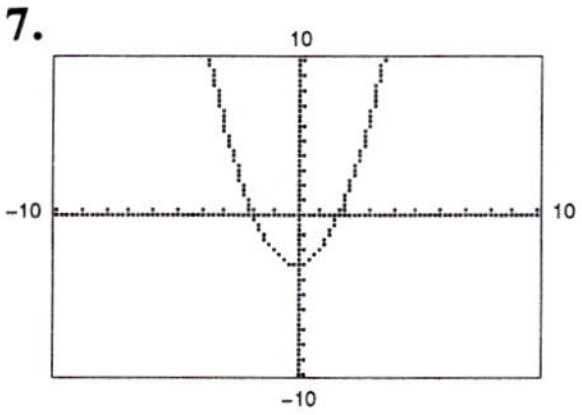

9.
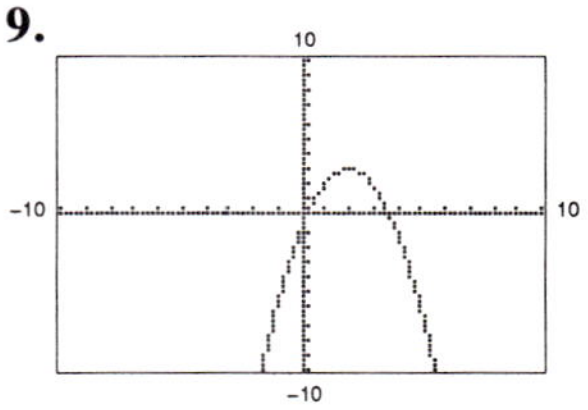

11.
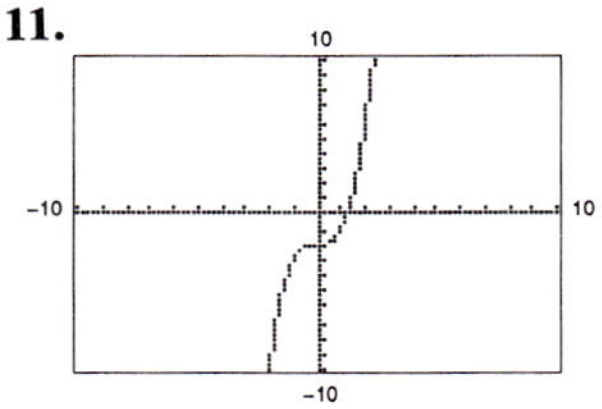

13.
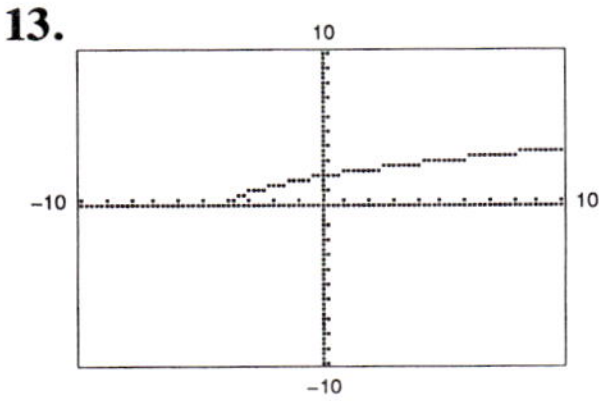

15.
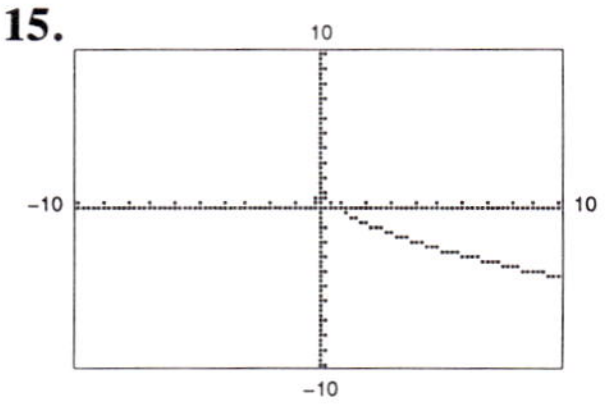

17.
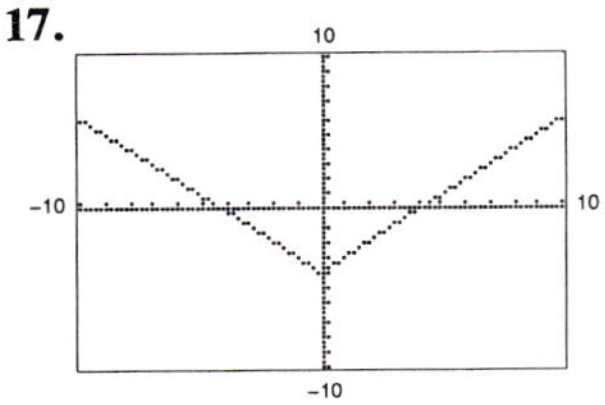

19.
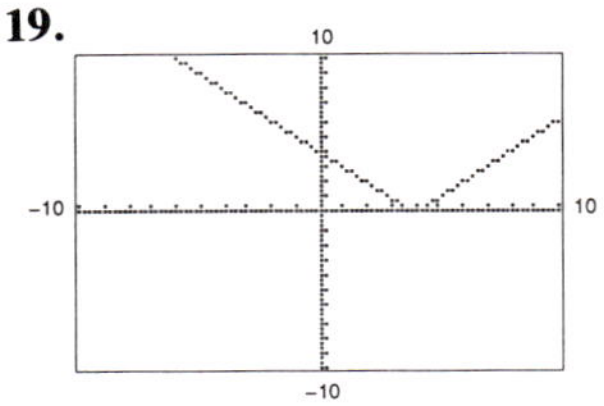

21. **(a)**
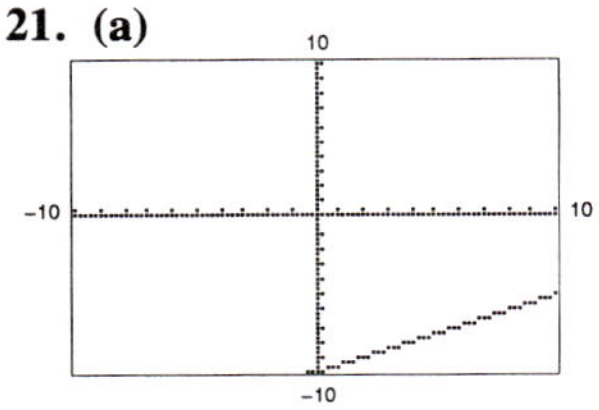

(b)
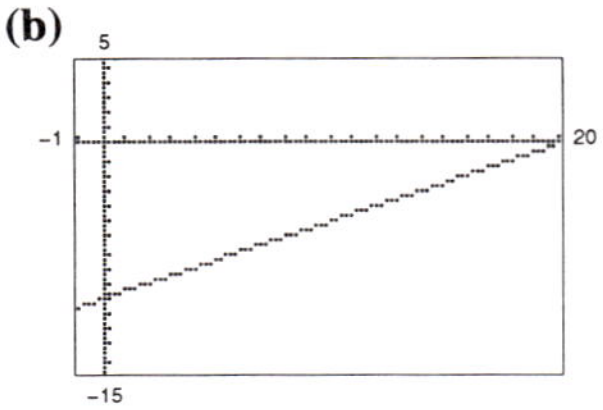

23. **(a)**
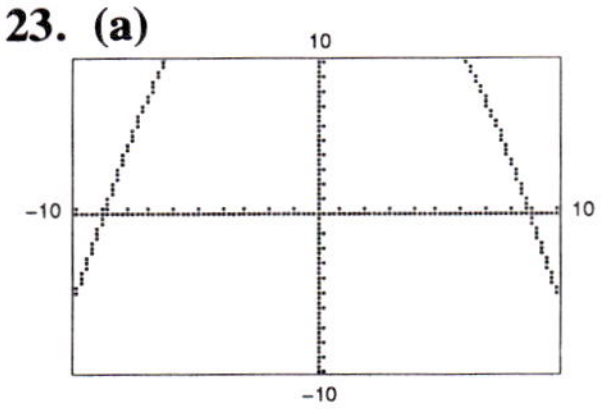

(b)
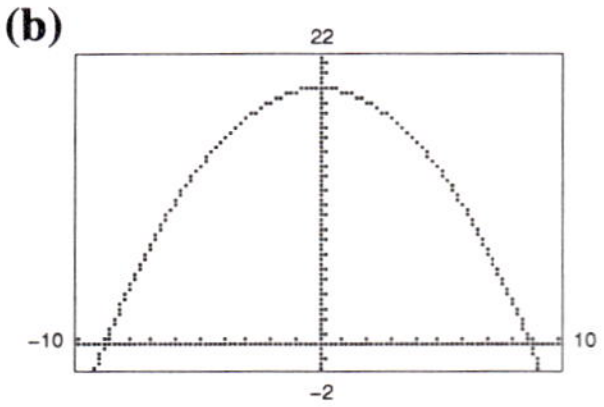

25. **(a)**
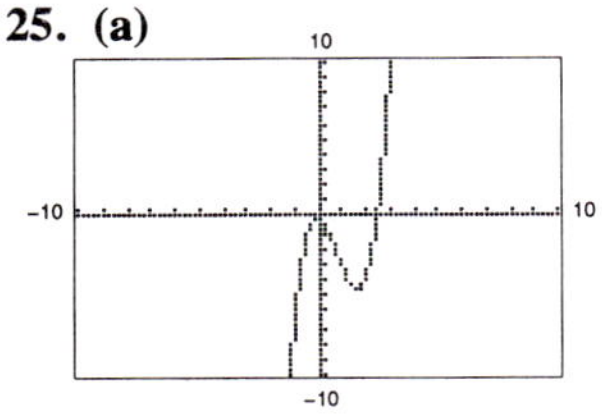

(b)
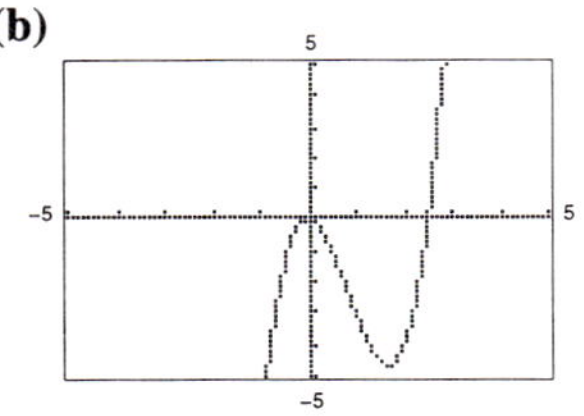

27.

RANGE
Xmin = −2
Xmax = 8
Xscl = 1
Ymin = −1
Ymax = 17
Yscl = 1
Xres = 1

29.

RANGE
Xmin = −2
Xmax = 10
Xscl = 1
Ymin = −5
Ymax = 5
Yscl = 1
Xres = 1

31.

RANGE
Xmin = −5
Xmax = 5
Xscl = 1
Ymin = −3
Ymax = 6
Yscl = 1
Xres = 1

33.

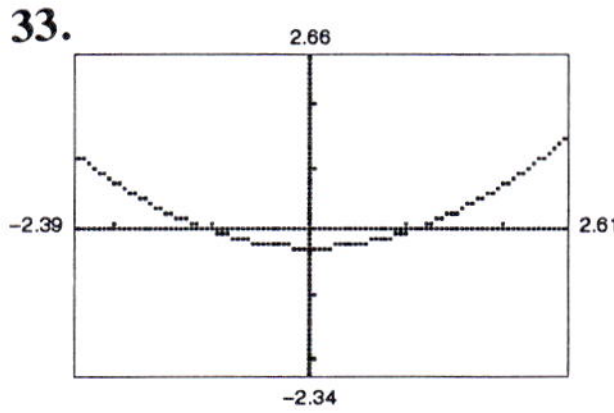

35.

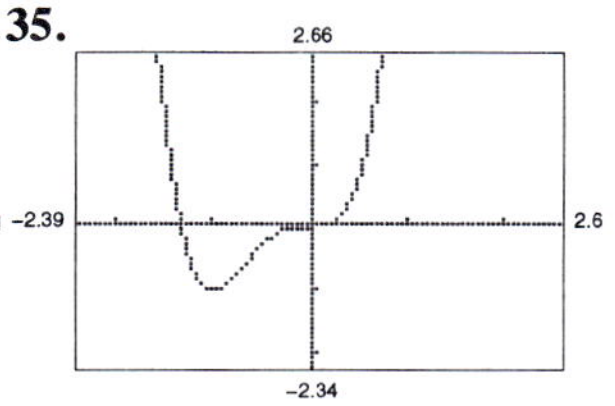

37.

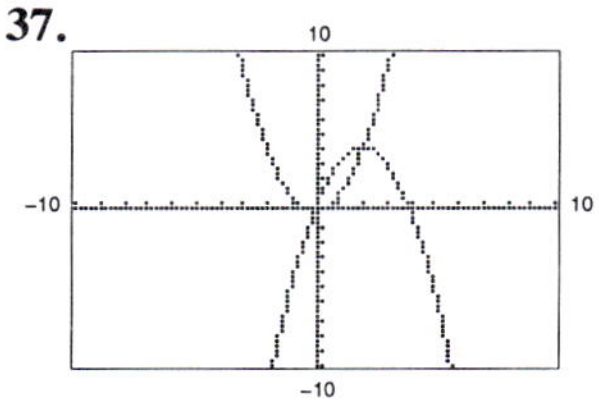

39.

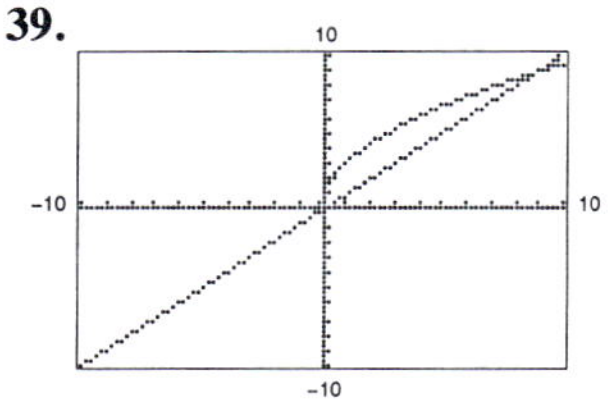

41. 2 **43.** 0 **45.** −2.5, 3 **47.** −0.75, 1, 2.75

49.

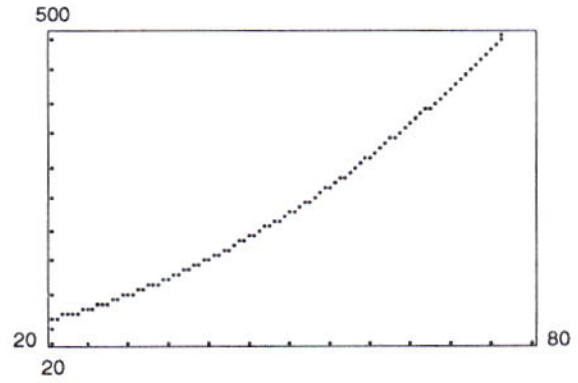

51.

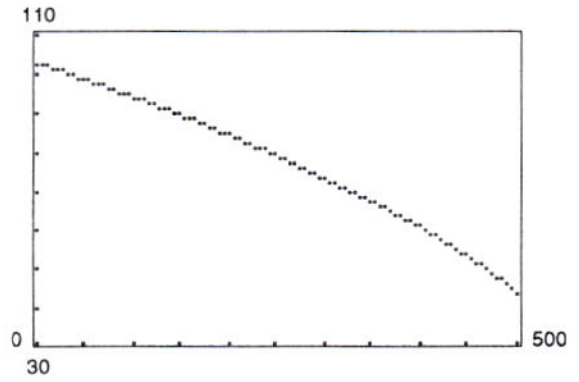

SECTION 7.4 *(page 415)*

Warm-Up *(page 415)*

1. $\frac{1}{2}$ **2.** $\frac{1}{2}$ **3.** $-\frac{29}{2}$ **4.** $-\frac{1}{2}$ **5.** $\frac{16}{5}$
6. $-\frac{7}{3}$ **7.** $y = \frac{5}{8}(x - 1)$ **8.** $y = -\frac{9}{2}x$
9. $y = 2x - 1$ **10.** $y = \frac{1}{4}(3x + 20)$

1. $\frac{2}{3}$ **3.** −2 **5.** Undefined
7. **(a)** L_3 **(b)** L_2 **(c)** L_1

9. $\frac{5}{7}$, rises

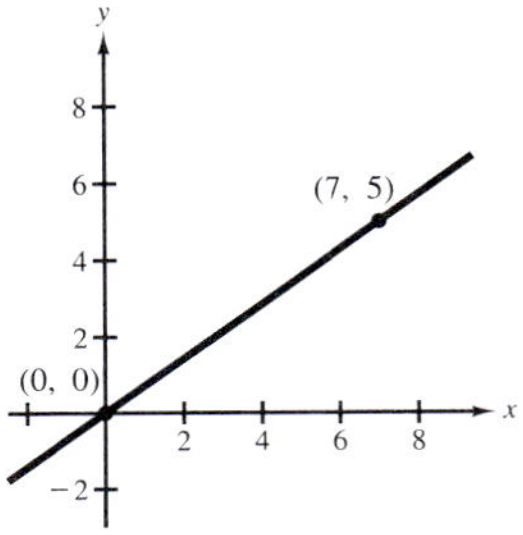

11. $\frac{1}{2}$, rises

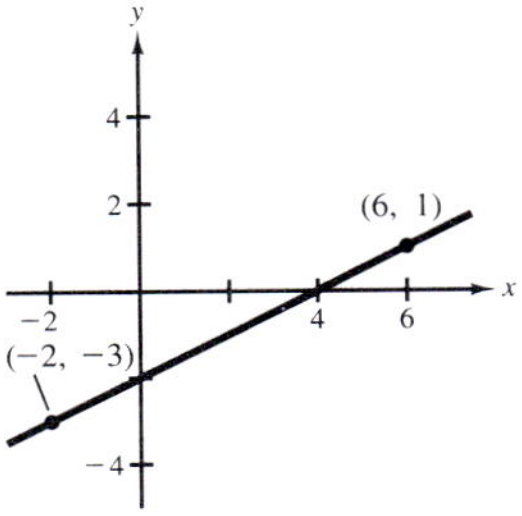

13. $-\frac{3}{2}$, falls

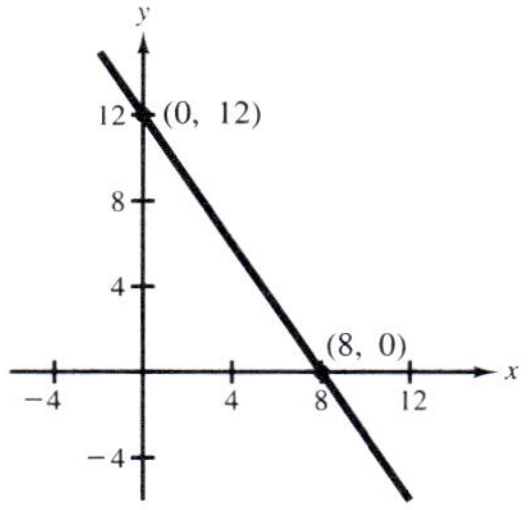

15. Undefined, vertical

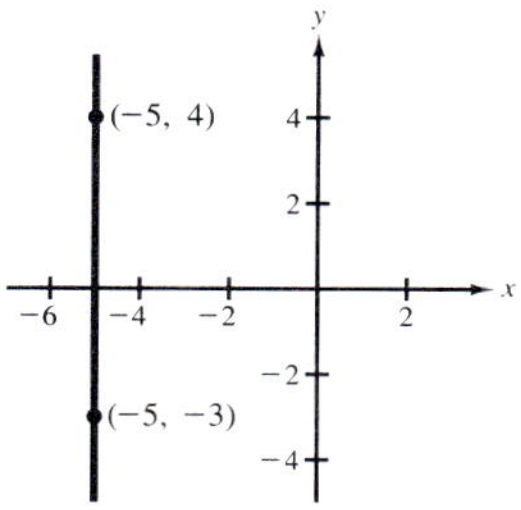

17. 0, horizontal

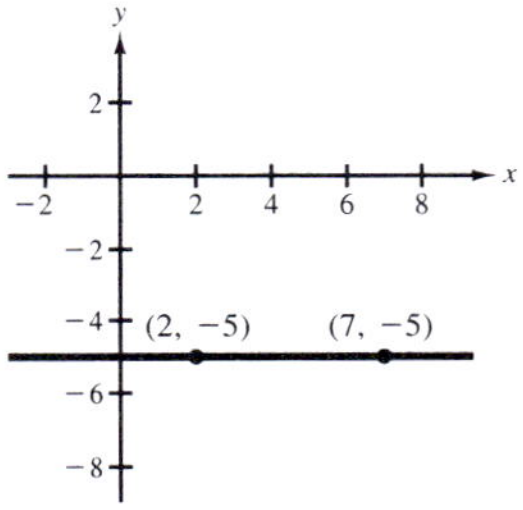

19. $-\frac{18}{17}$, falls

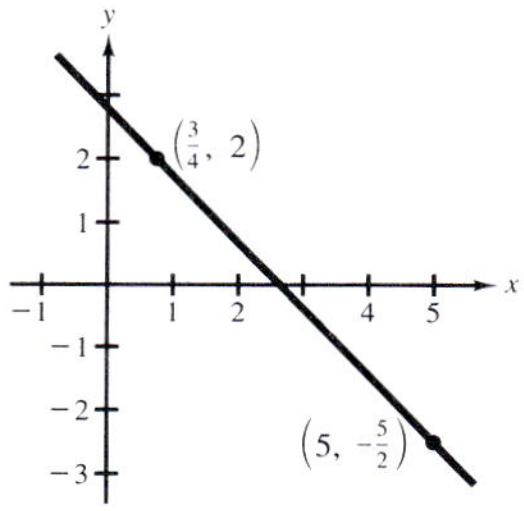

21. $\frac{-7}{8.4}$, falls

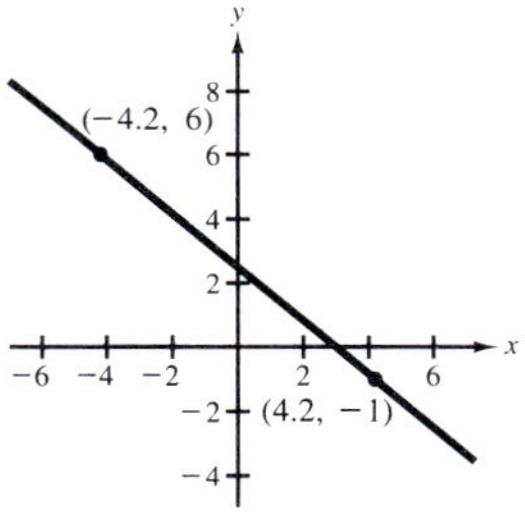

23. $\frac{29}{9}$, rises

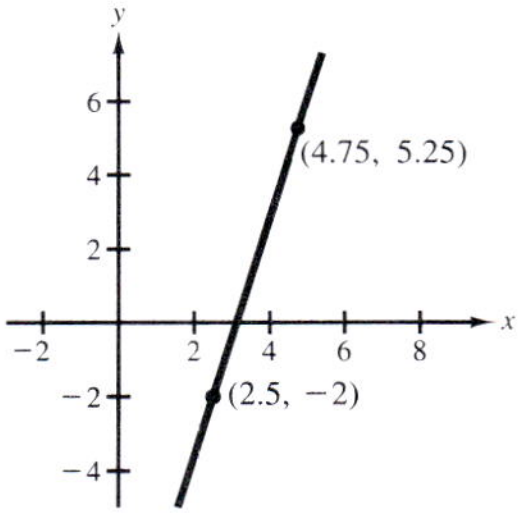

25. $\frac{5}{3}$, rises

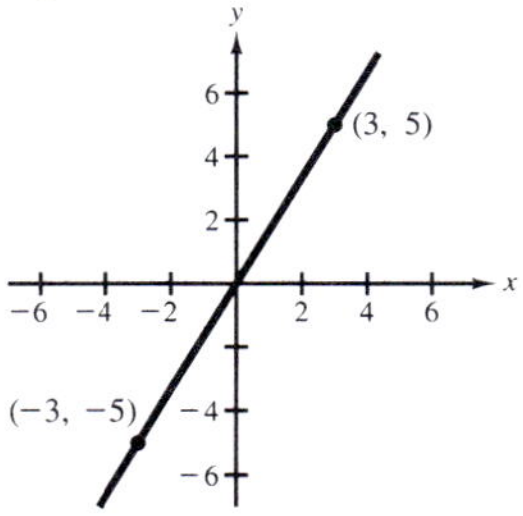

27. $x = 1$ **29.** $y = -15$ **31.** (0, 2), (1, 2)
33. (4, −1), (5, 2) **35.** (1, 2), (2, 1)
37. (−2, 4), (1, 8) **39.** (4, 0), (4, 1)

41.

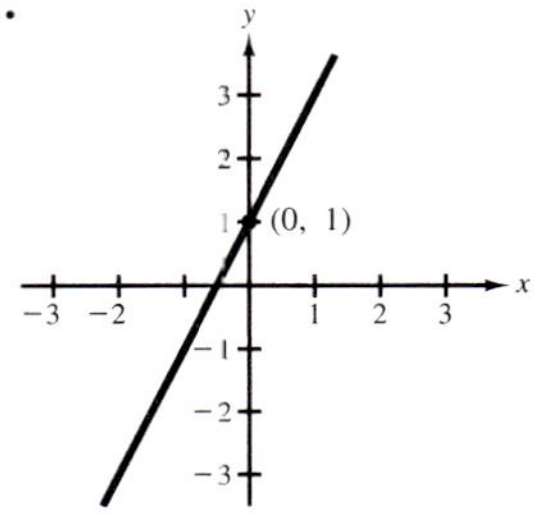

43.

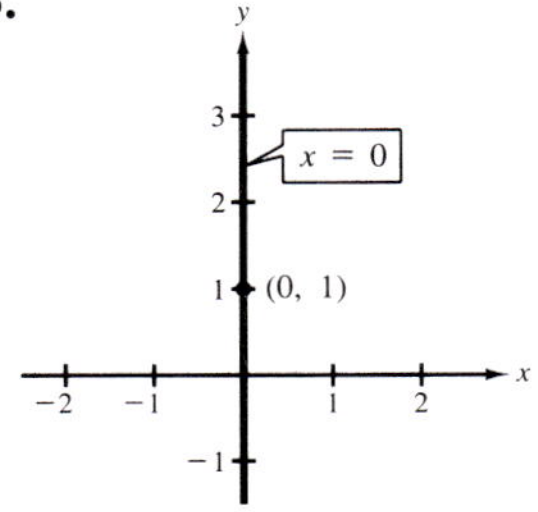

45.

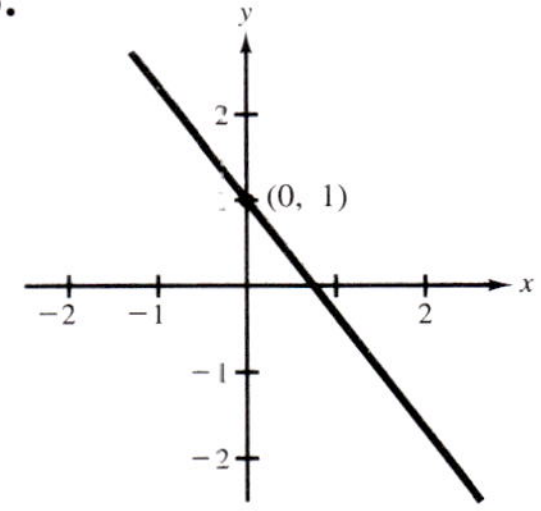

47.

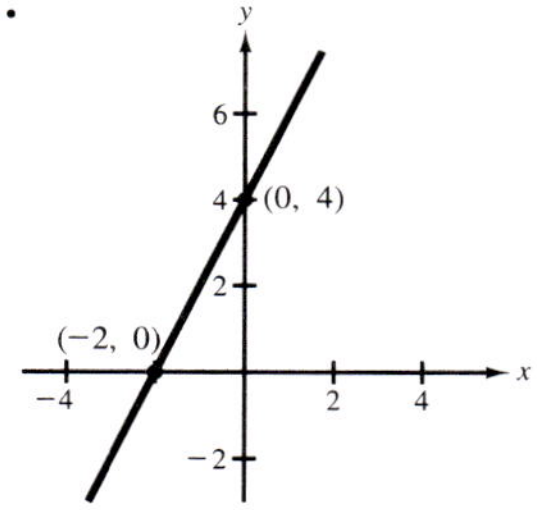

49.

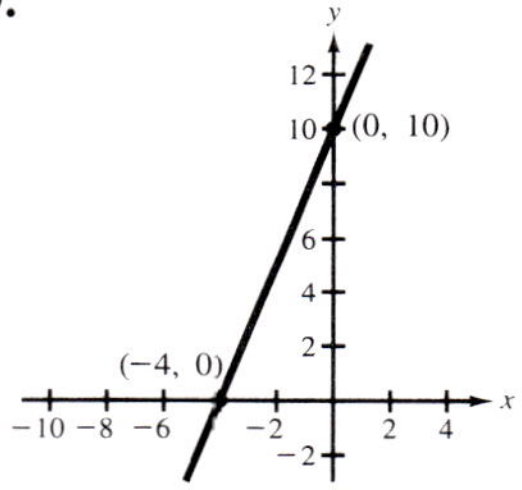

51. $m = 3$, $(0, -2)$

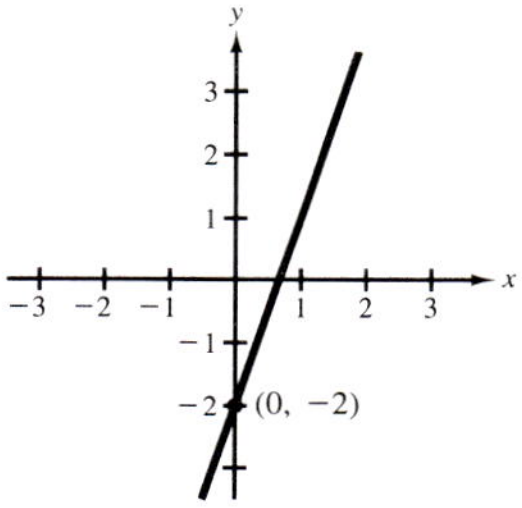

53. $m = -1$, $(0, 0)$

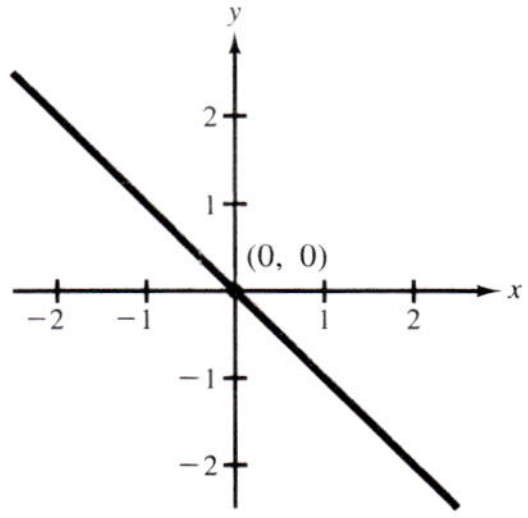

55. $m = -\frac{3}{2}$, $(0, 1)$

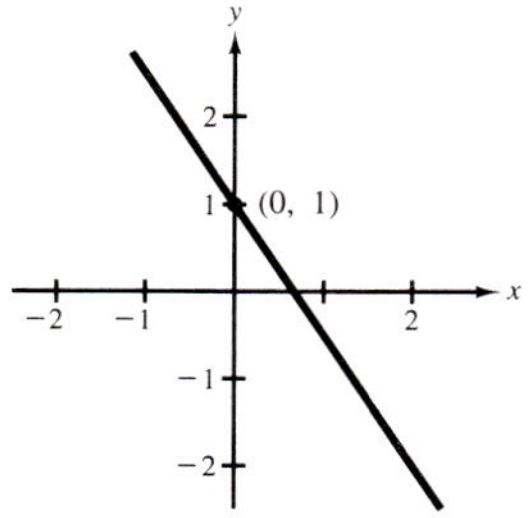

57. $m = \frac{1}{4}$, $\left(0, \frac{1}{2}\right)$

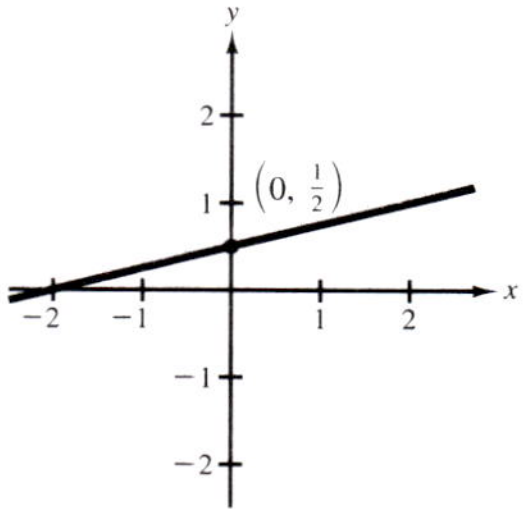

59. $m = 0$, $(0, 2)$

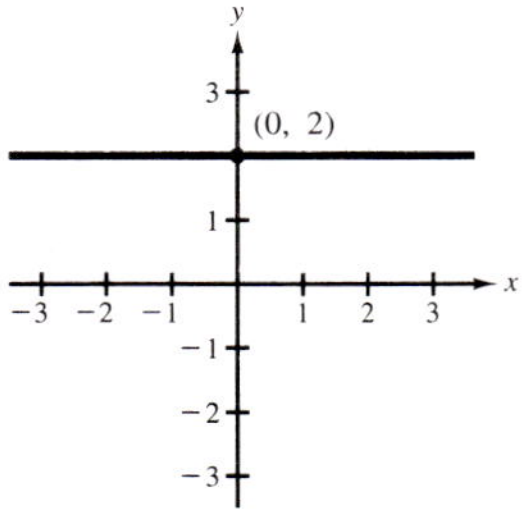

61. Perpendicular **63.** Parallel

65. Parallel **67.** Perpendicular

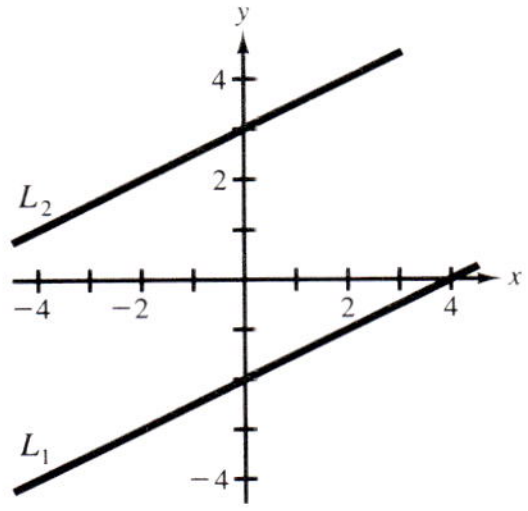

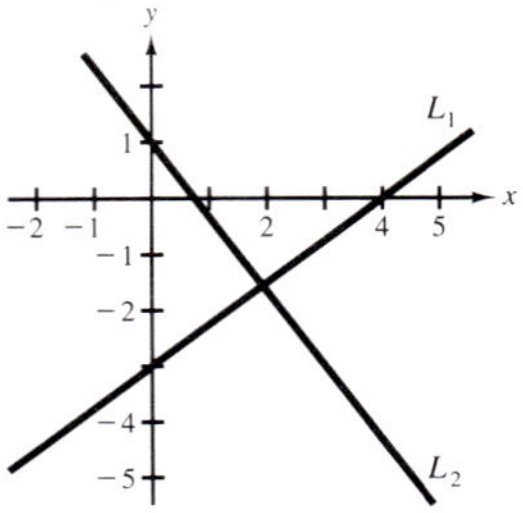

69. **(a)** 1988 **(b)** 1985 **71.** $166{,}666\frac{2}{3}$ ft ≈ 3.16 mi

SECTION 7.5 *(page 429)*

Warm-Up *(page 429)*

1. $-\frac{1}{4}$ **2.** 1 **3.** Undefined **4.** $-\frac{9}{10}$

5.

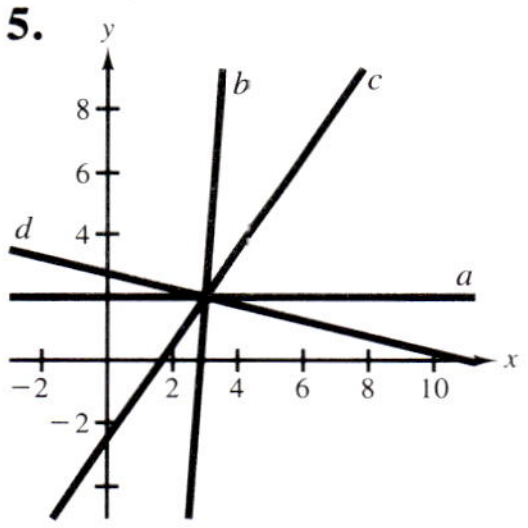

6.

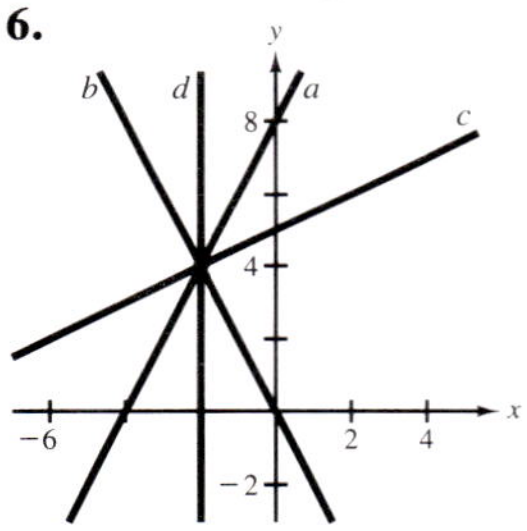

7. $y = -\frac{3}{5}x + \frac{21}{5}$ **8.** $y = \frac{4}{3}x - 9$

9. $y = -4x + 18$ **10.** $y = x + 9$

1. $m = 6$, $(-4, 3)$ **3.** $m = \frac{3}{4}$, $\left(-2, -\frac{5}{8}\right)$

5. $m = \frac{2}{3}$, $(0, -2)$ **7.** $m = \frac{3}{2}$, $(0, 0)$

9. $m = \frac{5}{2}$, $(0, 12)$

11. $y = -\frac{1}{2}x$

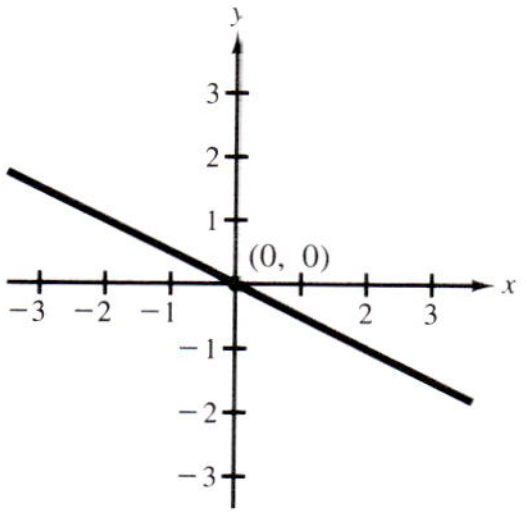

13. $y + 4 = 3x$

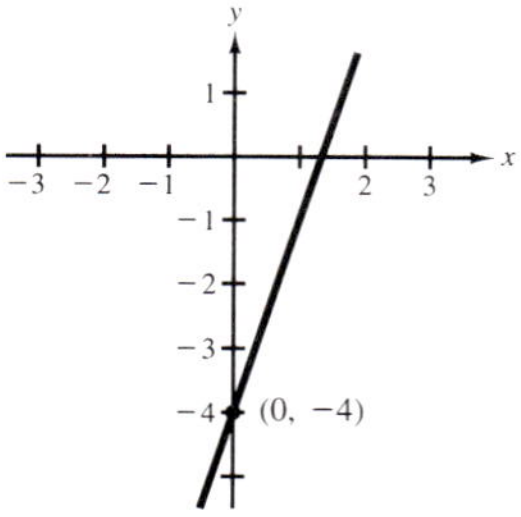

15. $y - \frac{7}{2} = -4(x + 2)$

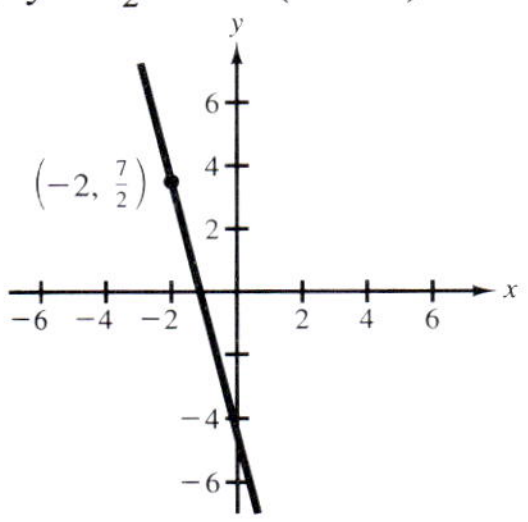

17. $3x - 2y = 0$

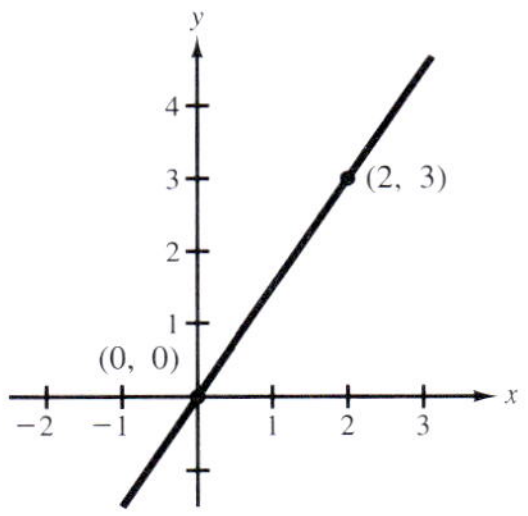

19. $3x + 7y - 15 = 0$

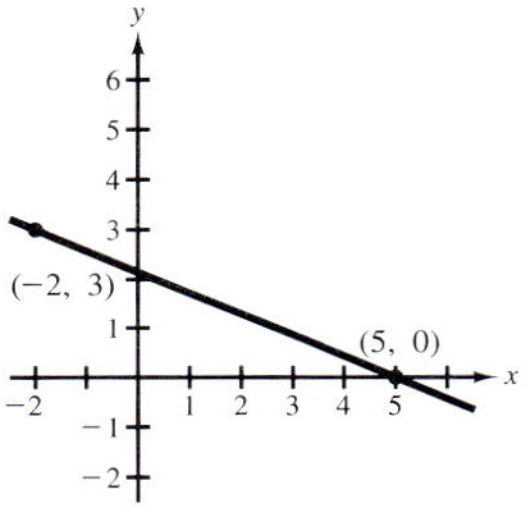

21. $x - 2y + 6 = 0$

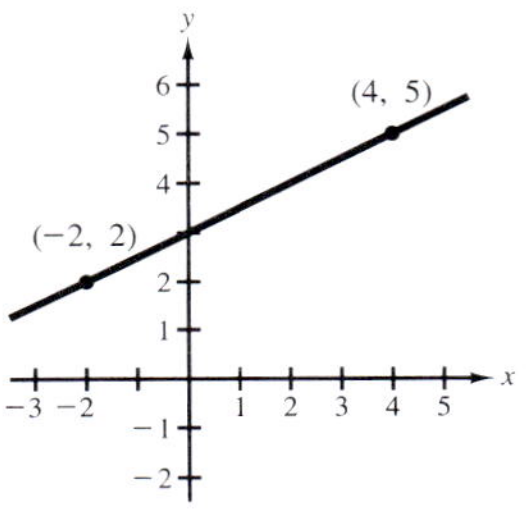

23. $y - 3 = 0$

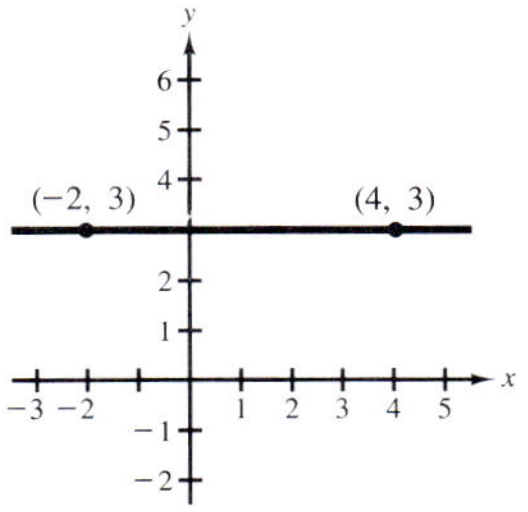

25. $14x + 6y - 39 = 0$

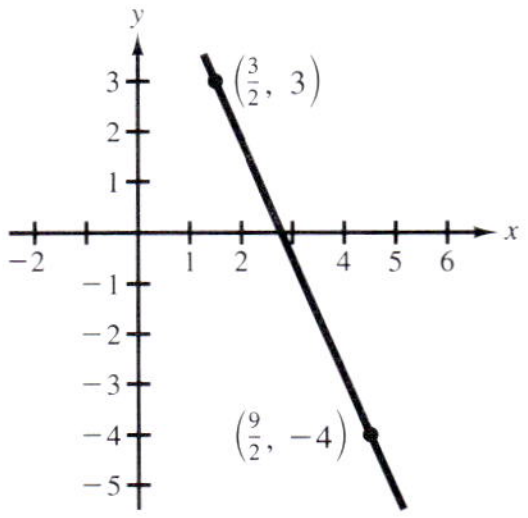

27. $x - 2y - 12 = 0$ **29.** $2x + y - 4 = 0$

31. $2x + 3y - 10 = 0$ **33.** $x - 2 = 0$

35. $8x - 6y + 9 = 0$ **37.** $2x + 5y = 0$

39. $5x + 34y - 67 = 0$ **41.** $2x - 15 = 0$

43. $4x + 5y - 11 = 0$

45. **(a)** $3x - y - 5 = 0$
(b) $x + 3y - 5 = 0$

47. **(a)** $5x + 4y + 9 = 0$
(b) $4x - 5y + 40 = 0$

49. **(a)** $y - 2 = 0$
(b) $x + 1 = 0$

51. $\frac{x}{3} + \frac{y}{2} = 1$ **53.** $\frac{-6x}{5} - \frac{3y}{7} = 1$

55.

x	0	50	100	500	1,000
C	5,000	6,000	7,000	15,000	25,000

57. $S = 1{,}500 + 0.03M$ **59.** $S = 0.70L$

61. $y = 72 + 5t$

63. **(a)** $P = 1{,}200 - 15x$
(b) 42
(c) 48

65. **(a)** and **(b)**

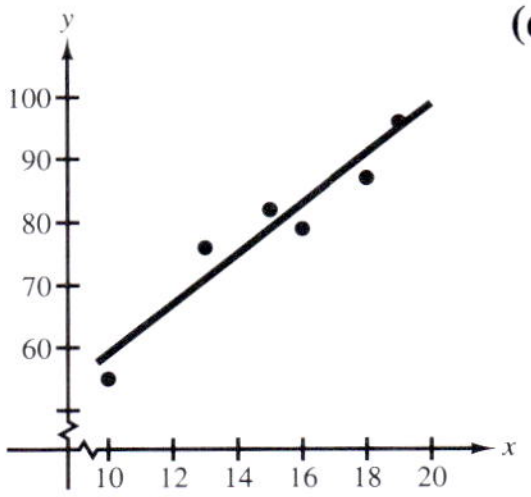

(c) $y = 4x + 19$
(d) 87

67.

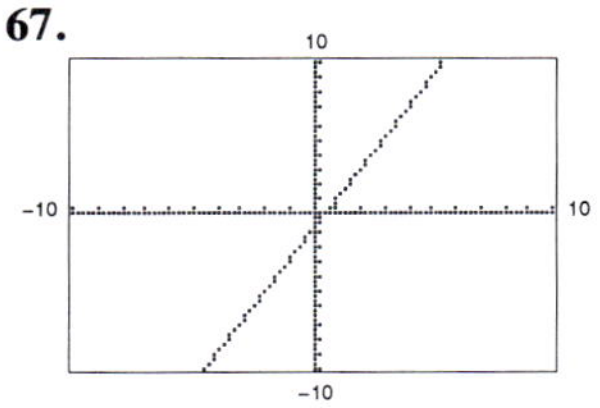

69.

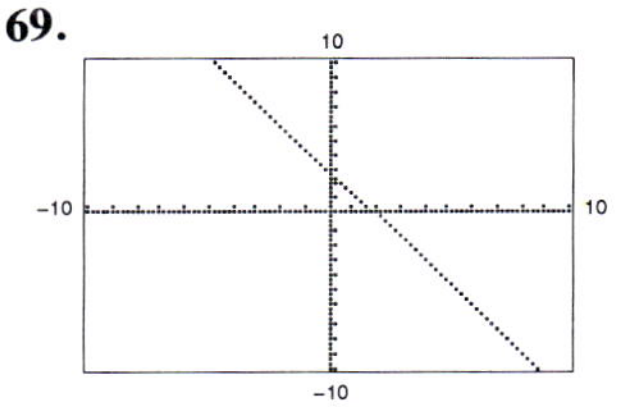

71. Perpendicular

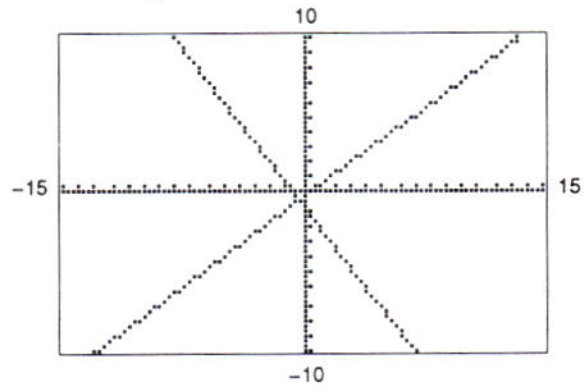

SECTION 7.6 *(page 443)*

Warm-Up *(page 443)*

1.

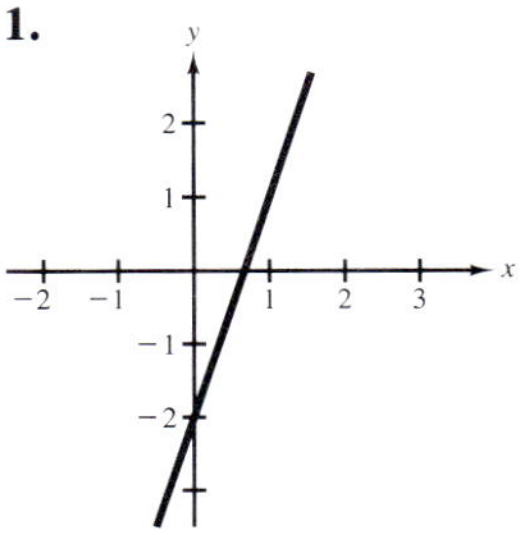

2.

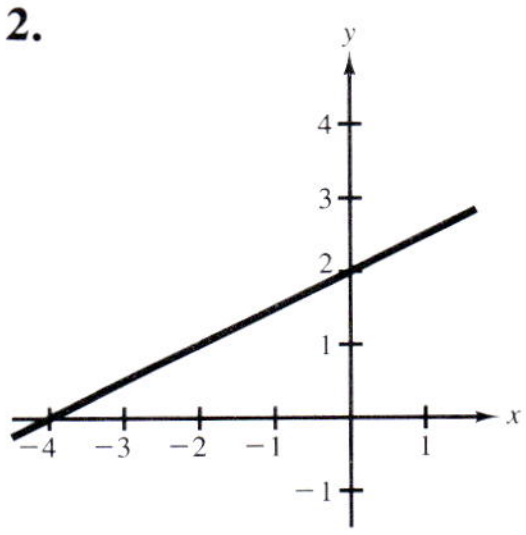

3.
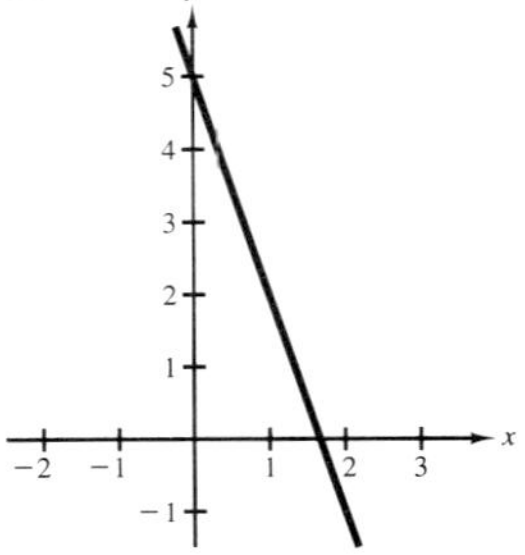

4.
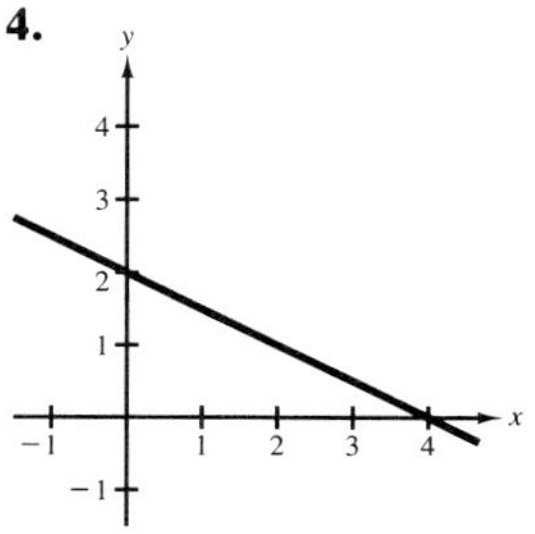

5. 2 **6.** 20 **7.** 0, −8 **8.** $\dfrac{-1 \pm \sqrt{41}}{4}$

9. $2x^2 + 20x + 20$ **10.** $-3x^2 + 12x - 19$

1. (e) **2.** (f) **3.** (b) **4.** (c) **5.** (d) **6.** (a)
7. Upward, (0, 2) **9.** Downward, (10, 4)
11. Upward, (0, −6) **13.** (±5, 0) **15.** (0, 0), (9, 0)
17. $\left(\frac{3}{2}, 0\right)$ **19.** No x-intercepts
21. $y = (x - 2)^2 + 3$, (2, 3)
23. $y = -(x - 1)^2 - 6$, (1, −6)
25. $y = 2\left(x + \frac{3}{2}\right)^2 - \frac{5}{2}$, $\left(-\frac{3}{2}, -\frac{5}{2}\right)$

27.
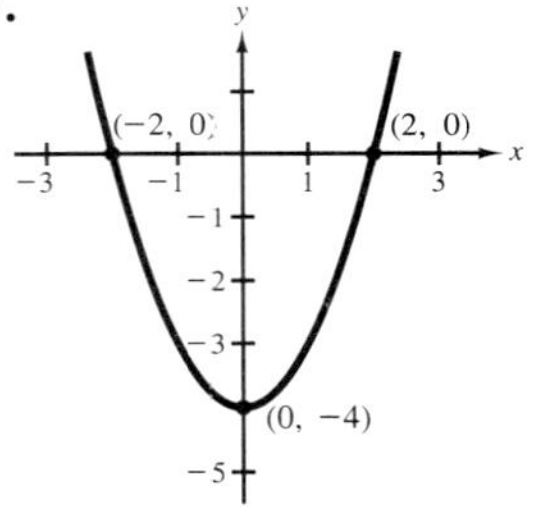

29.
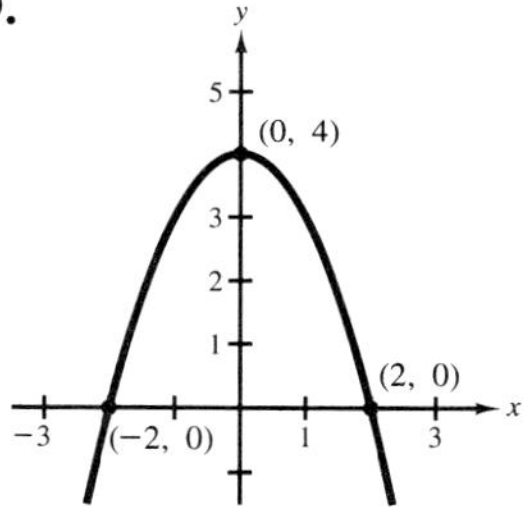

31.
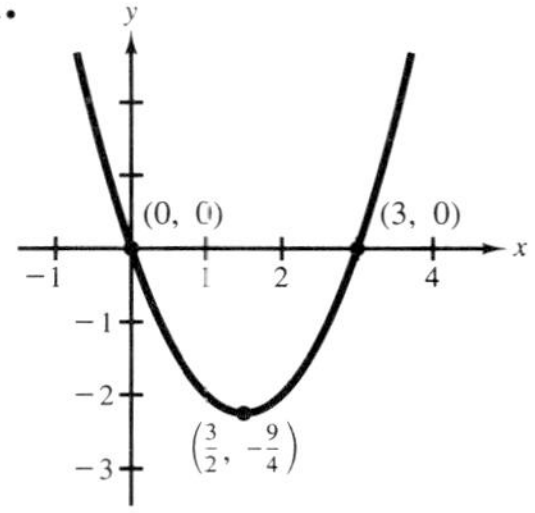

33.
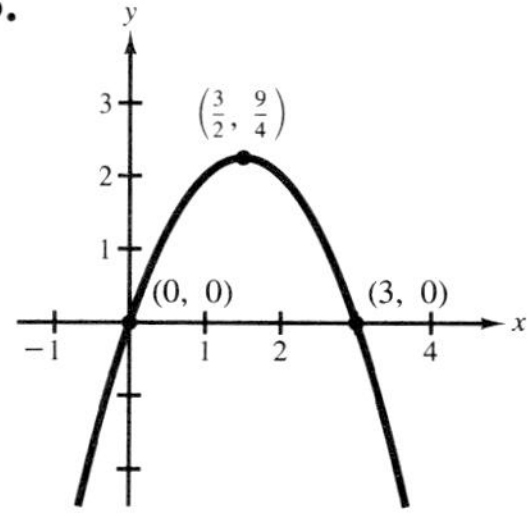

35.
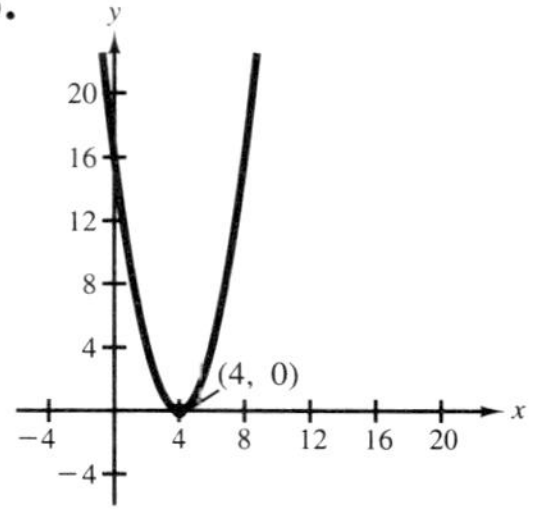

37.
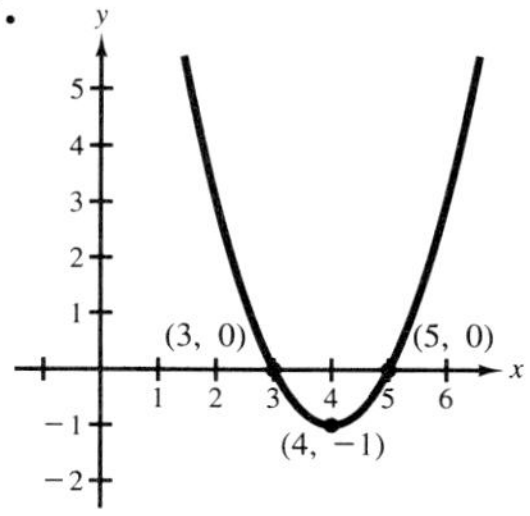

39.
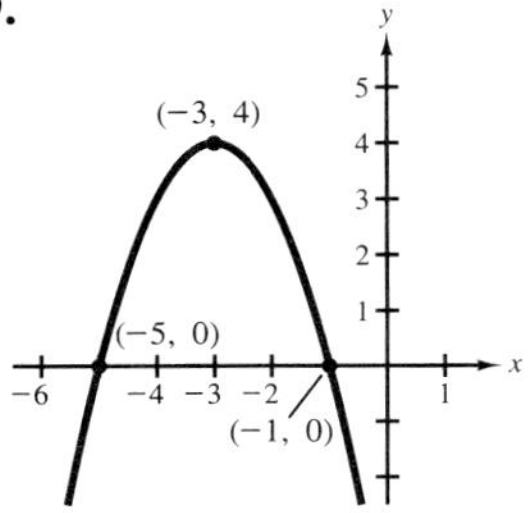

41.
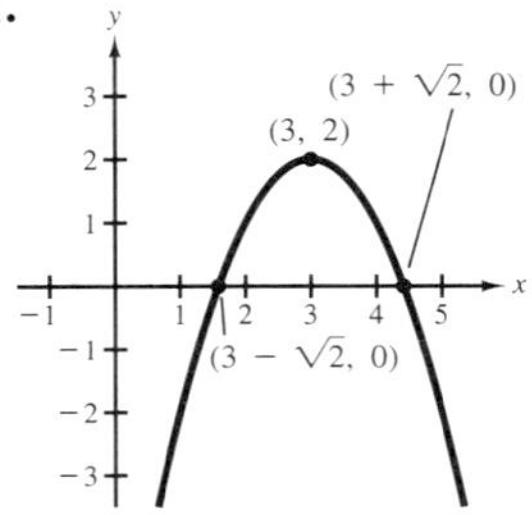

43.
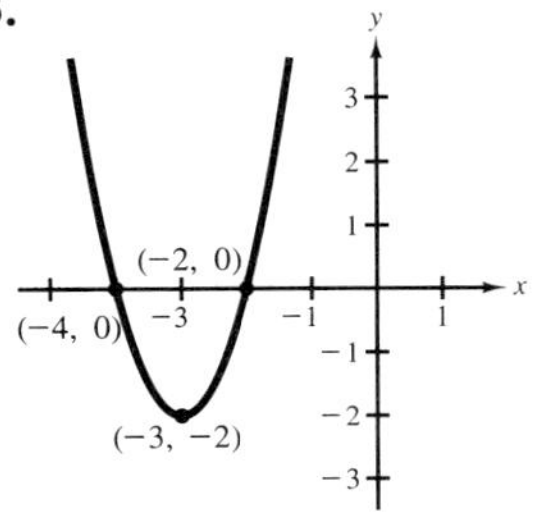

45.
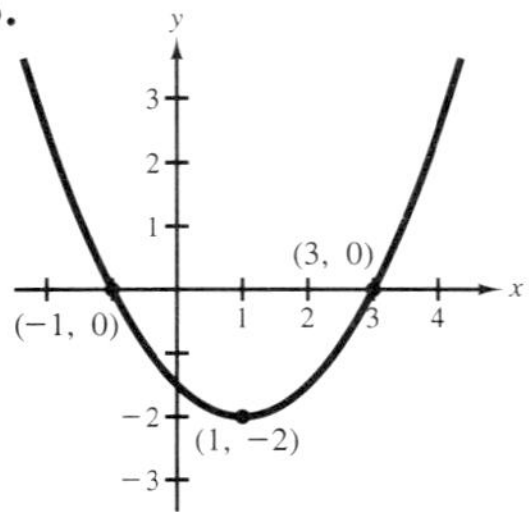

47.
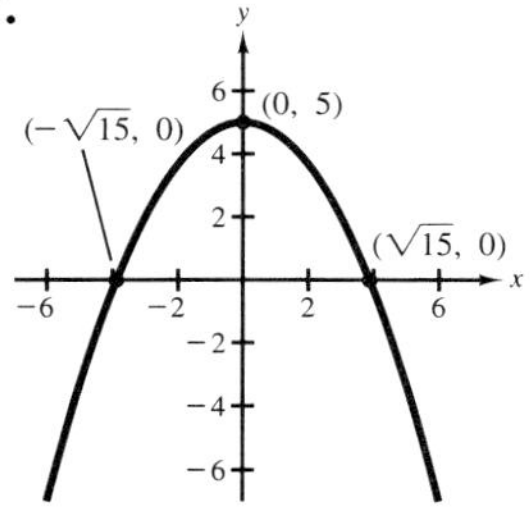

49.
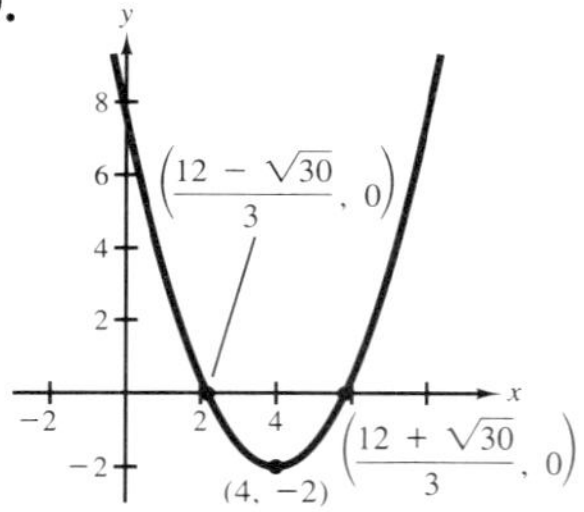

51.
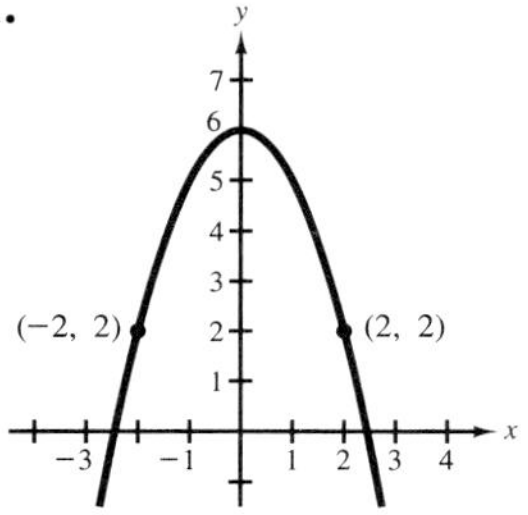

53. $\left(3 \pm \sqrt{2}, 3\right)$
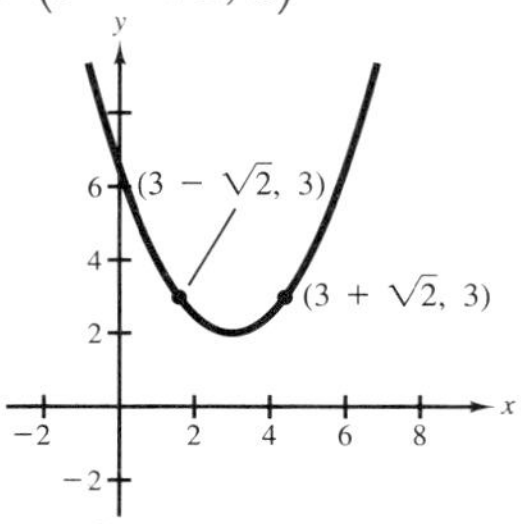

55. $y = x^2 - 4x + 5$ **57.** $y = -x^2 - 6x - 5$
59. $y = x^2 - 4x$ **61.** $y = \frac{1}{2}x^2 - 3x + \frac{13}{2}$
63. $y = -4x^2 - 8x + 1$ **65.** $y = \frac{1}{25}x^2 - \frac{2}{5}x + 3$
67. (a)
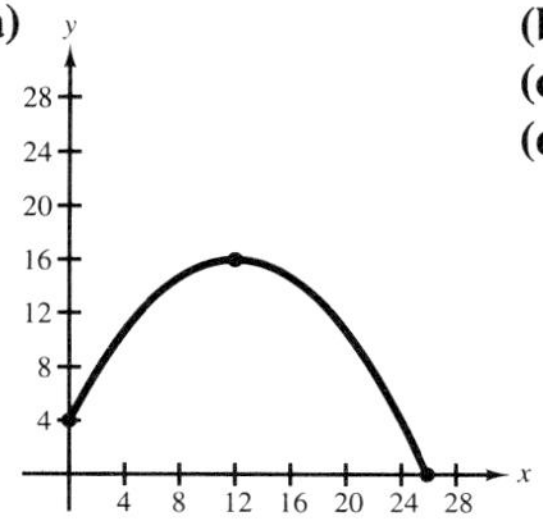

(b) 4 ft
(c) 16 ft
(d) $12 + 8\sqrt{3} \approx 25.9$ ft

69. **(a)** $P = R - C$

$= (100 + x)[90 - x(0.15)] - (100 + x)60$

$= 9{,}000 + 75x - 0.15x^2 - 6{,}000 - 60x$

$= 3{,}000 + 15x - \frac{3}{20}x^2$

(b) Vertex: (50, 3,375), 150 units

(c) No, Yes

71. **(a)** 1,000 ft

(b) 400 ft

(c) 16, 64, 144, 256, 400

73. $x \approx 50.421053$

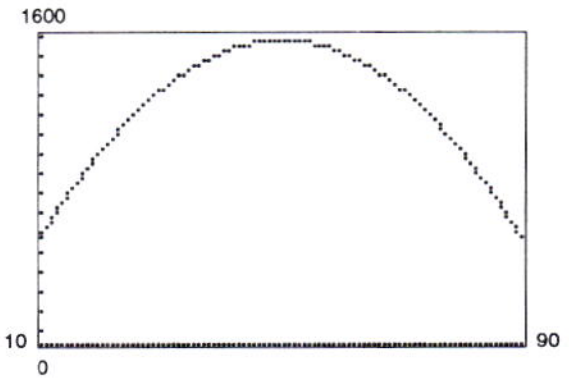

SECTION 7.7 *(page 460)*

Warm-Up *(page 459)*

1. $y = \frac{1}{3}(9 - 2x)$ **2.** $y = \pm\sqrt{36 - x^2}$

3. $y = \pm\sqrt{\dfrac{6 - 2x^2}{3}}$ **4.** $y = \pm\sqrt{2x^2 - 3}$

5. $15\sqrt{2}$ **6.** $15\sqrt{6}$

7.

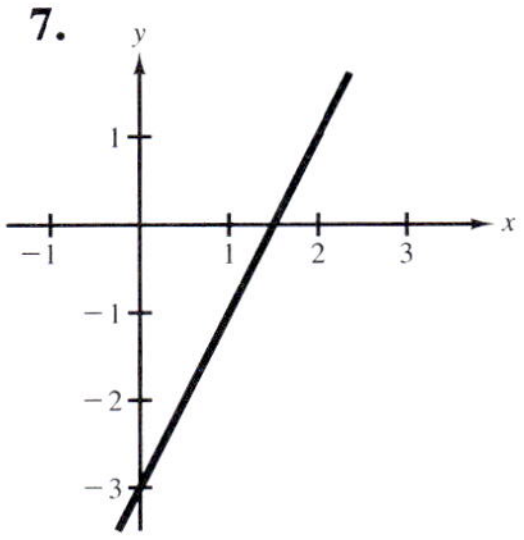

8.

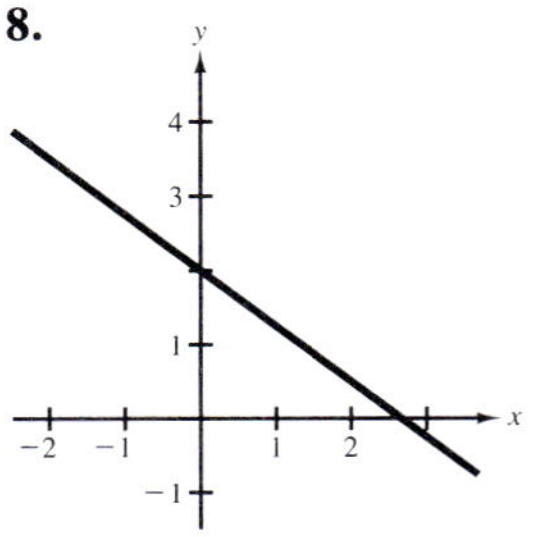

9.

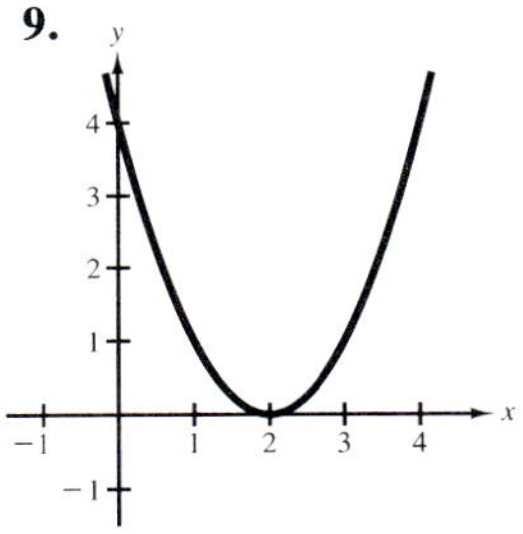

10.

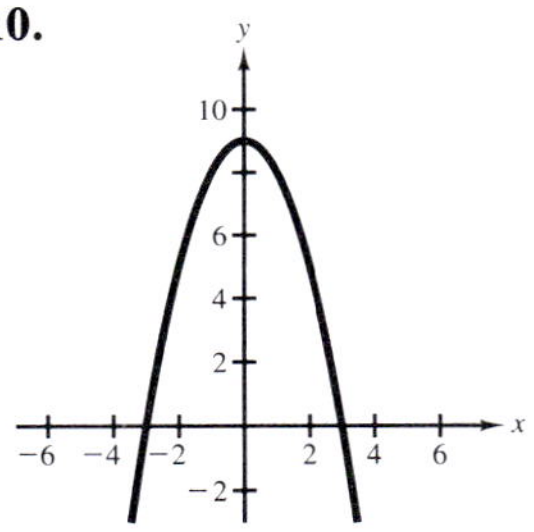

1. (c) **2.** (f) **3.** (e) **4.** (b) **5.** (a) **6.** (d)

7. $x^2 + y^2 = 25$ **9.** $x^2 + y^2 = \frac{4}{9}$

11. $x^2 + y^2 = 64$ **13.** $x^2 + y^2 = 29$

15. $r = 4$

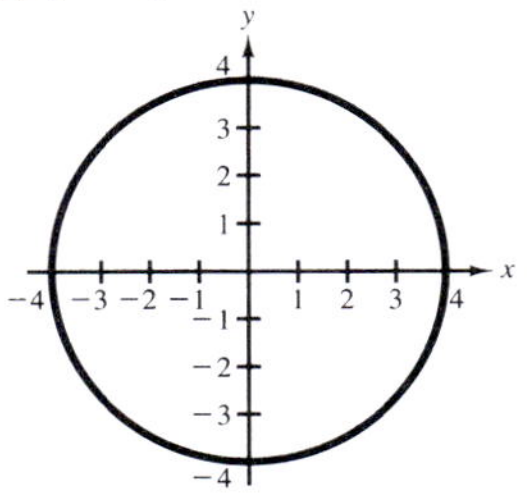

17. $r = \frac{1}{2}$

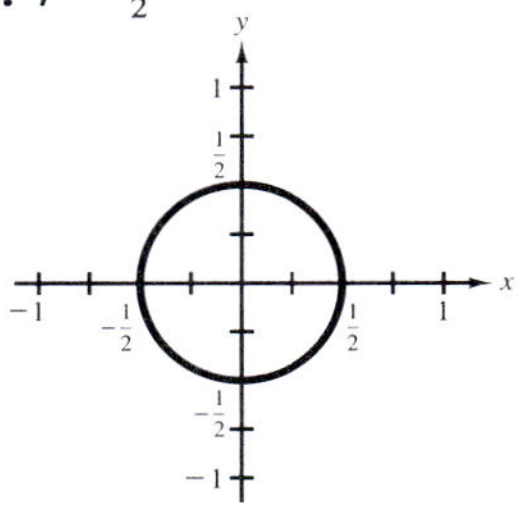

19. $r = \frac{12}{5}$

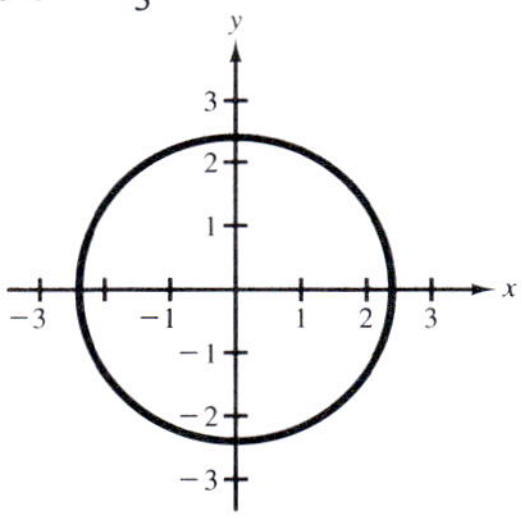

21. $\dfrac{x^2}{16} + \dfrac{y^2}{9} = 1$ **23.** $\dfrac{x^2}{9} + \dfrac{y^2}{16} = 1$

25. $\dfrac{x^2}{100} + \dfrac{y^2}{36} = 1$

27. Vertices: $(\pm 4, 0)$
Co-vertices: $(0, \pm 2)$

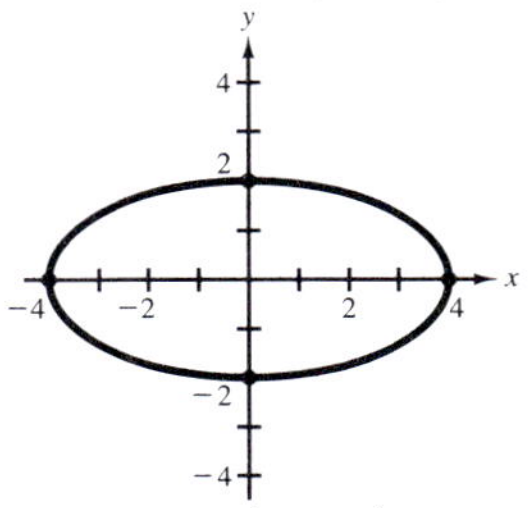

29. Vertices: $(0, \pm 4)$
Co-vertices: $(\pm 2, 0)$

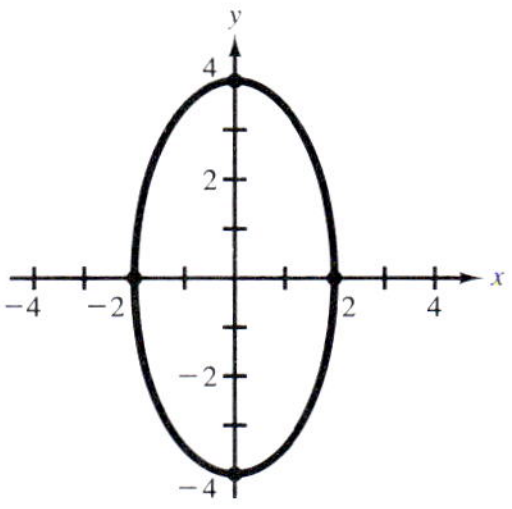

31. Vertices: $\left(\pm\frac{5}{3}, 0\right)$
Co-vertices: $\left(0, \pm\frac{4}{3}\right)$

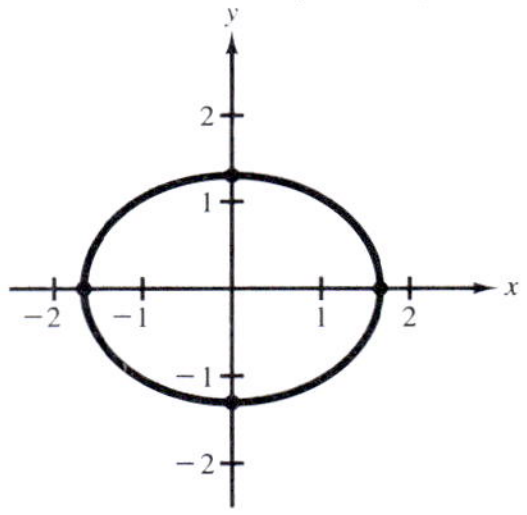

33. Vertices: $(0, \pm 2)$
Co-vertices: $(\pm 1, 0)$

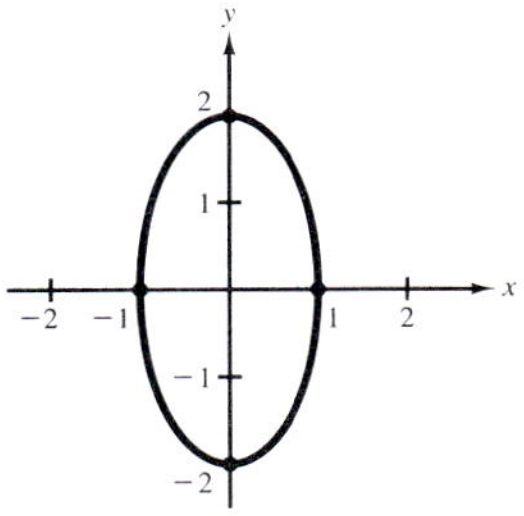

35. Vertices: $(\pm 3, 0)$
Asymptotes: $y = \pm x$

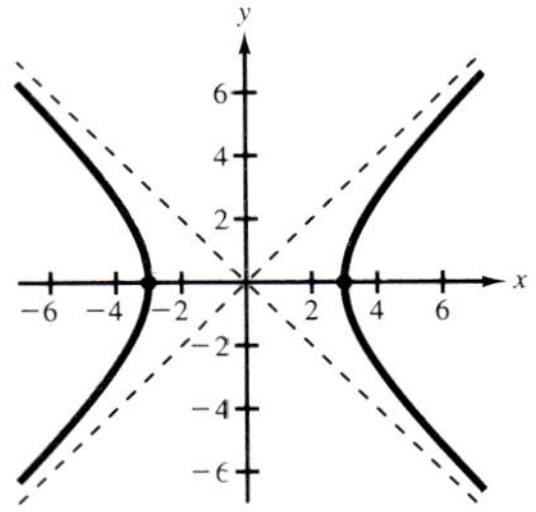

37. Vertices: $(0, \pm 3)$
Asymptotes: $y = \pm x$

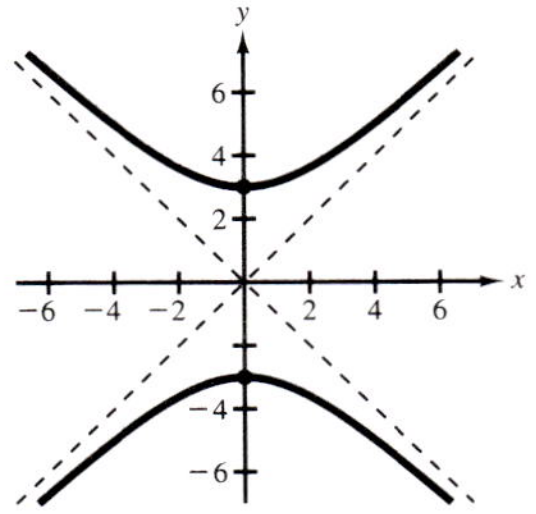

39. Vertices: $(\pm 3, 0)$
Asymptotes: $y = \pm \frac{5}{3}x$

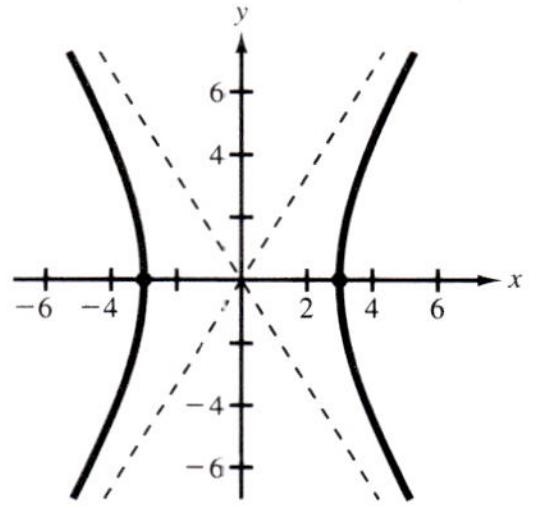

41. Vertices: $(0, \pm 3)$
Asymptotes: $y = \pm \frac{3}{5}x$

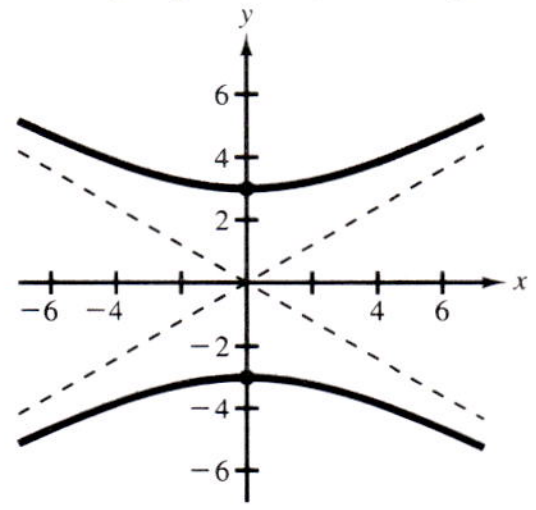

43. Vertices: $(\pm 4, 0)$
Asymptotes: $y = \pm \frac{1}{2}x$

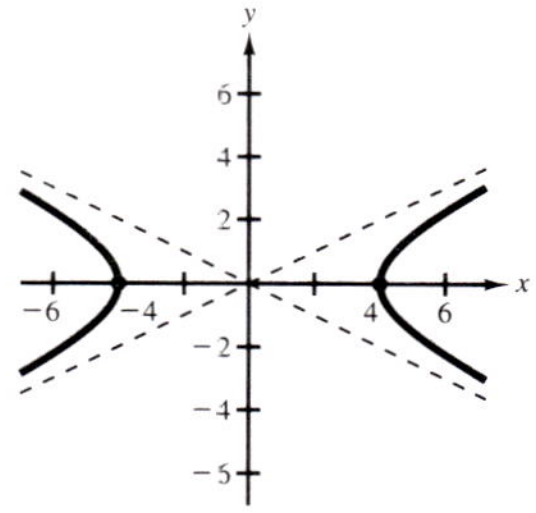

45. $\dfrac{x^2}{16} - \dfrac{y^2}{64} = 1$ **47.** $\dfrac{y^2}{16} - \dfrac{x^2}{64} = 1$

49. $\dfrac{x^2}{81} - \dfrac{y^2}{36} = 1$ **51.** Parabola **53.** Ellipse

55. Hyperbola **57.** Circle **59.** Line

61.

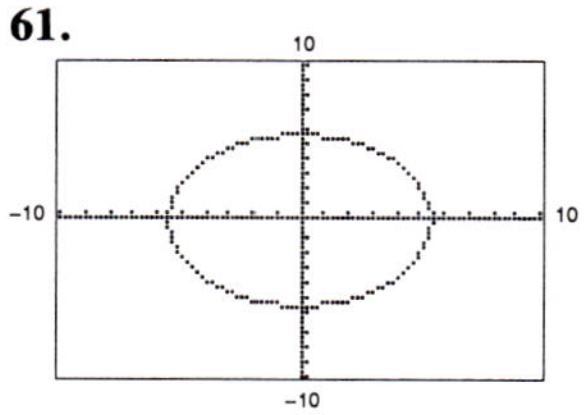

63.

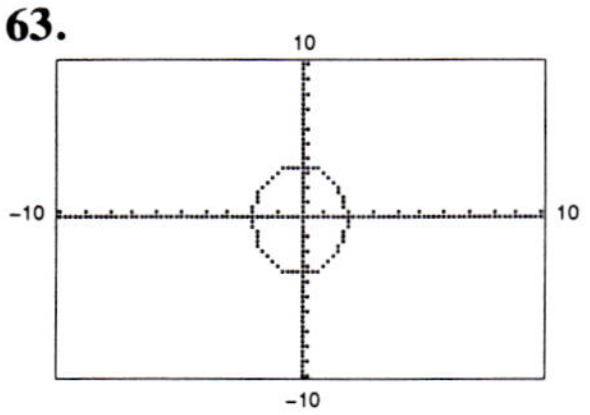

65.

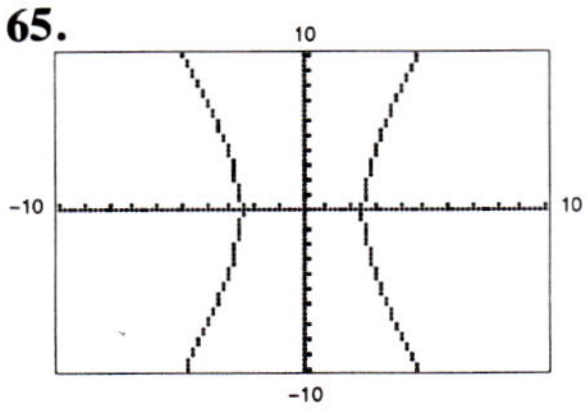

67. $x^2 + y^2 = 4{,}500^2$

69. Circle: $x^2 + y^2 = 25^2$
Width: $2y = 2\sqrt{625 - x^2}$
Area: $(2x)(2y) = 4x\sqrt{625 - x^2}$

71. $\sqrt{304} \approx 17.4$ ft **73.**

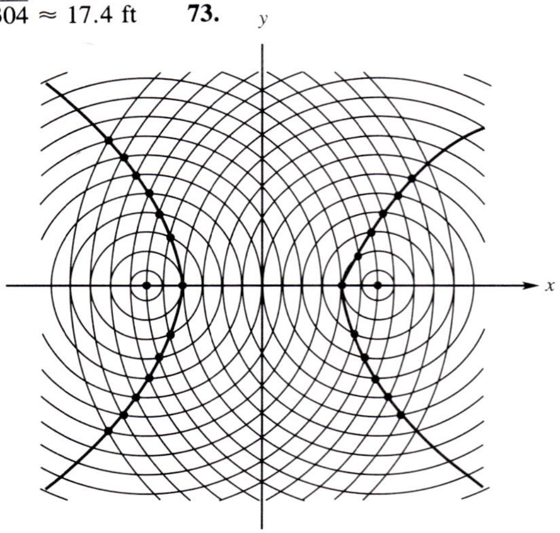

CHAPTER 7 REVIEW EXERCISES *(page 465)*

1.

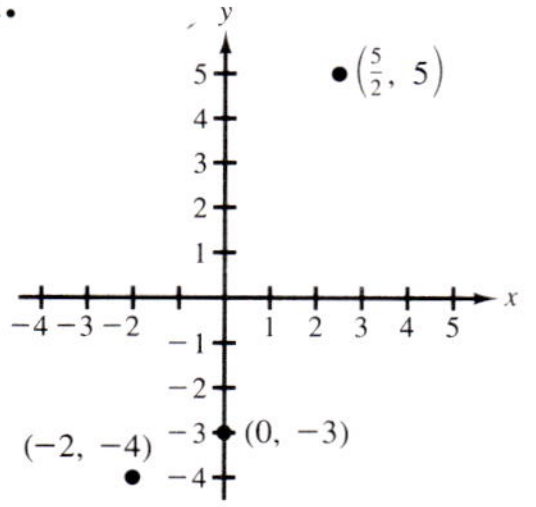

3. 5

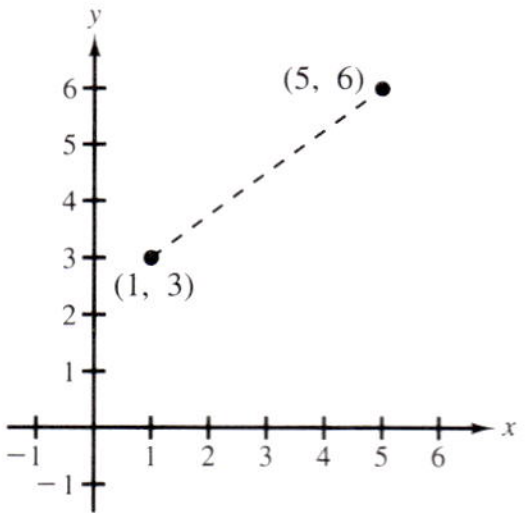

5. $3\sqrt{5}$

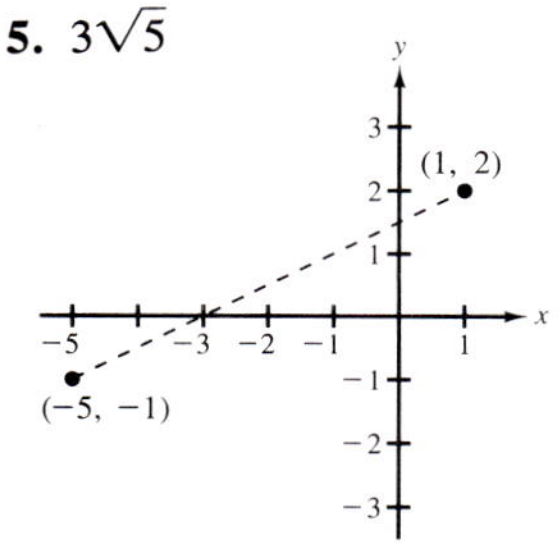

7.

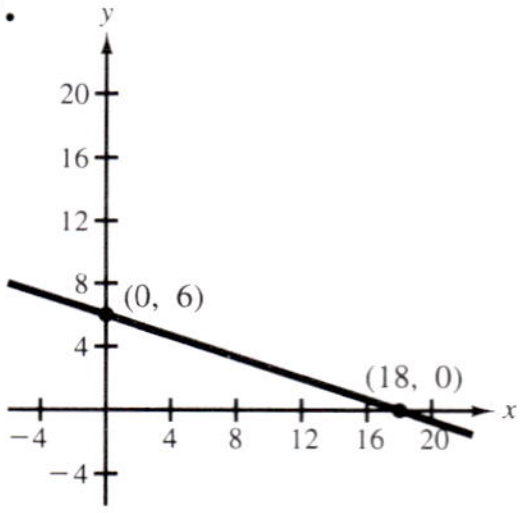

9.

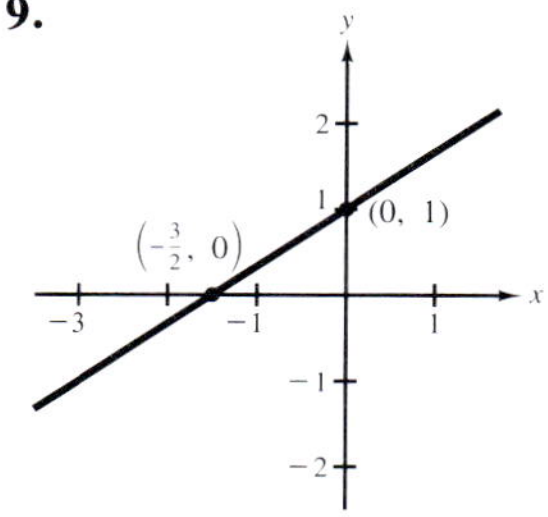

11.

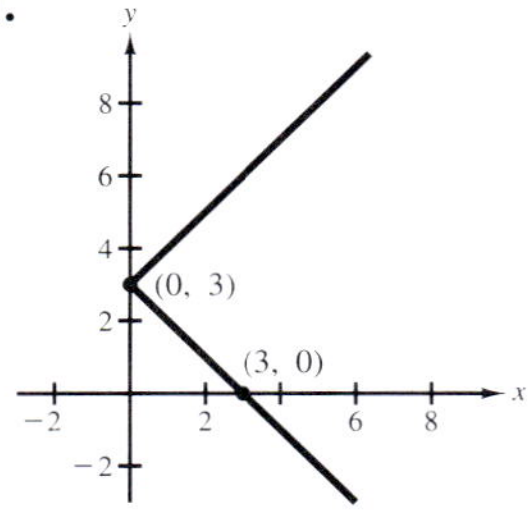

13. $\frac{2}{7}$ **15.** 0 **17.** $\frac{3}{2}$ **19.** $\frac{1}{4}$

21.

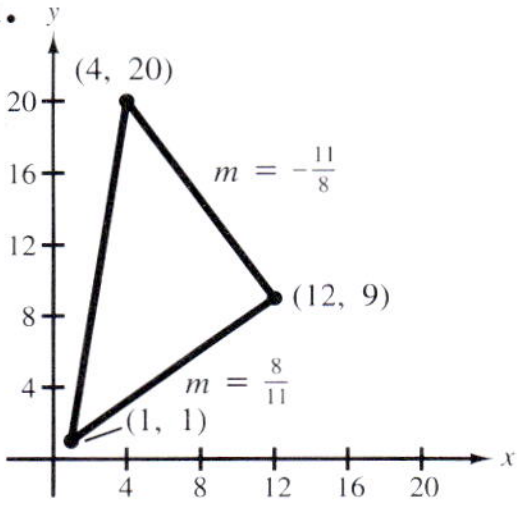

23. $(1, -1), (0, 2)$ **25.** $(7, 6), (11, 11)$

27. $2x - y - 6 = 0$ **29.** $4x + y = 0$

31. $2x + 3y - 17 = 0$ **33.** $x - 7 = 0$

35. $x + 2y + 6 = 0$ **37.** $y - 10 = 0$

39. $9x - 24y - 8 = 0$ **41.** **(a)** $x + 2y - 9 = 0$ **(b)** $2x - y + 7 = 0$

43. **(a)** $8x - 6y + 15 = 0$ **(b)** $24x + 32y - 105 = 0$

45. Parabola

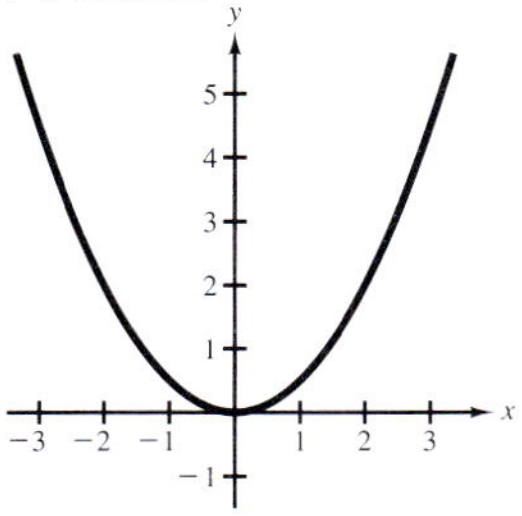

47. Hyperbola

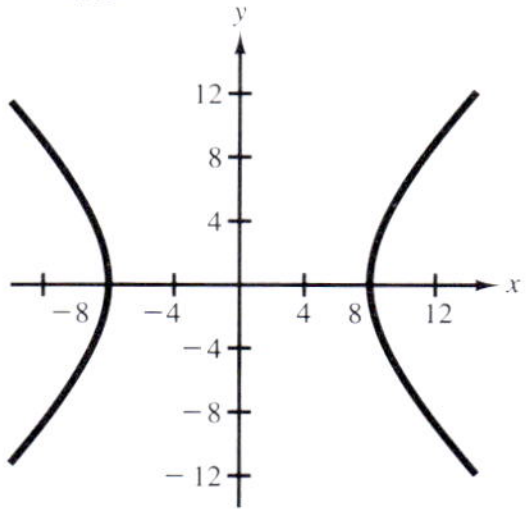

49. Parabola

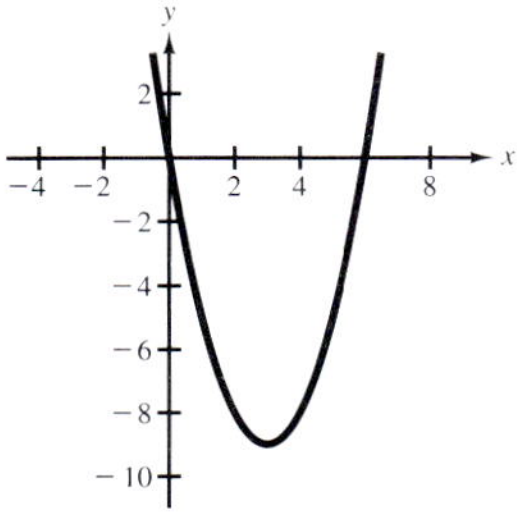

51. Ellipse

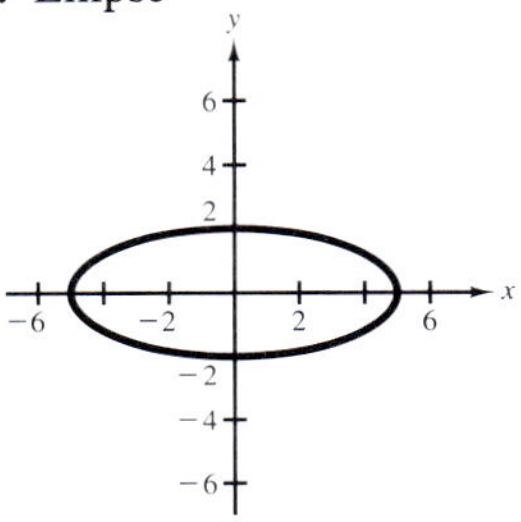

53. Circle

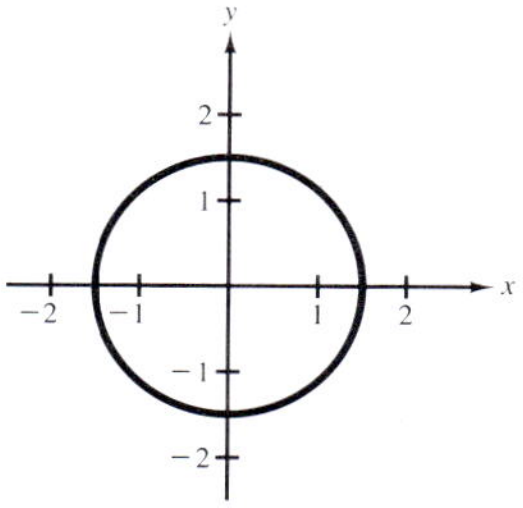

55. $y = \frac{1}{16}x^2 - \frac{5}{8}x + \frac{25}{16}$ **57.** $\dfrac{x^2}{4} + \dfrac{y^2}{25} = 1$

59. $x^2 + y^2 = 400$ **61.** $5.63\sqrt{2} \approx 7.96$

63. $C = 8.55x + 25{,}000$, $P = 4.05x - 25{,}000$ **65.** $x^2 + y^2 = 5{,}000^2$

CHAPTER 7 TEST *(page 468)*

1. Quadrant II, Quadrant IV **2.** 5

3. y-intercept: $(0, -3)$; x-intercepts: $(3, 0), (-1, 0)$

4.

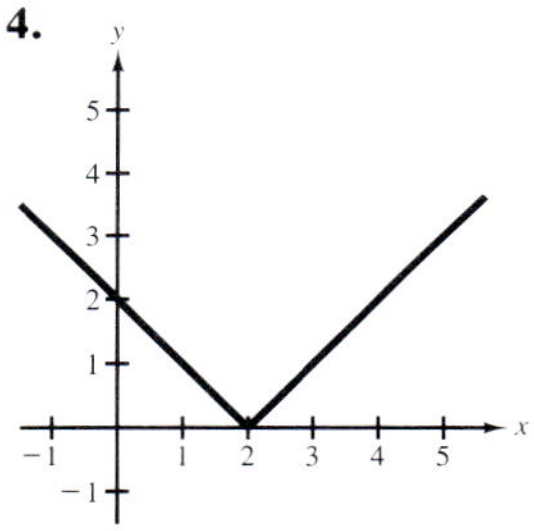

5. **(a)** $-\frac{2}{3}$ **(b)** Undefined

6.

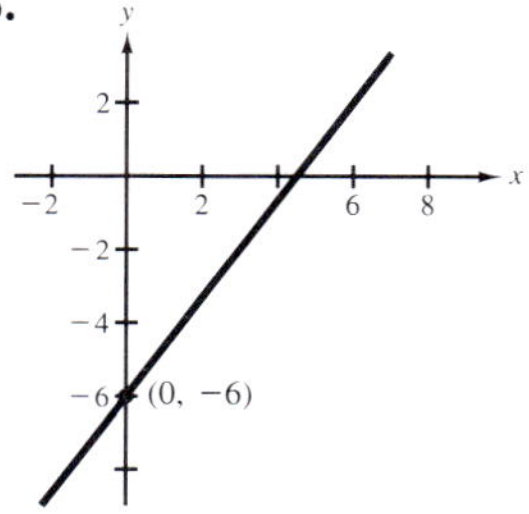

7.

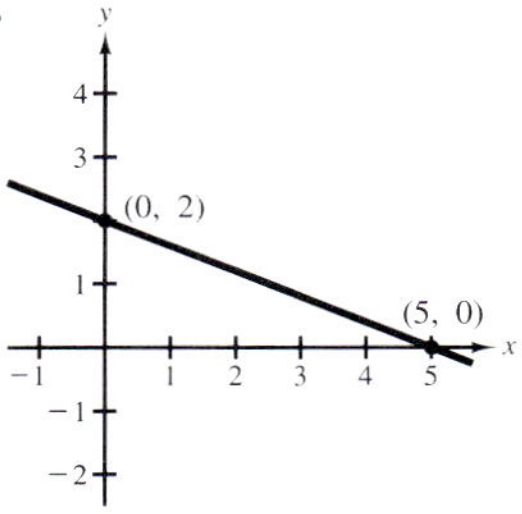

8. $y = -\frac{3}{2}x + 4$ **9.** $x - 2y - 55 = 0$ **10.** $\frac{3}{5}$

11. Vertex: $(2, 8)$; Intercepts: $(0, 0), (4, 0)$

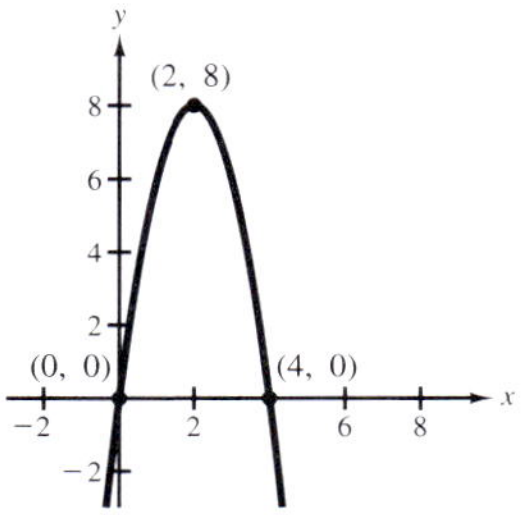

12. $y = \frac{2}{3}x^2 - 4x + 4$ **13.** $x^2 + y^2 = 25$

14. $\frac{x^2}{9} + \frac{y^2}{100} = 1$ **15.** $\frac{x^2}{9} - \frac{y^2}{9/4} = 1$

16. (a)

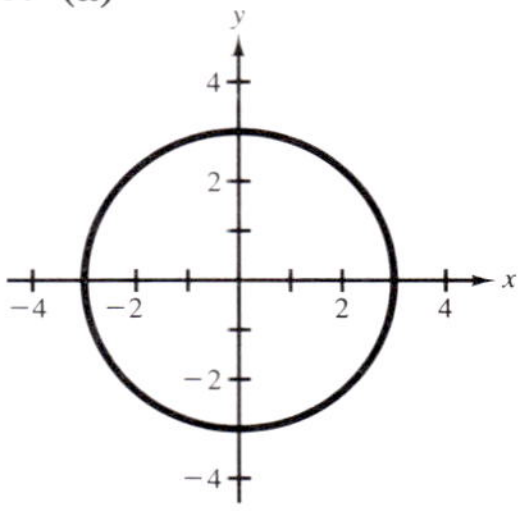

(b)

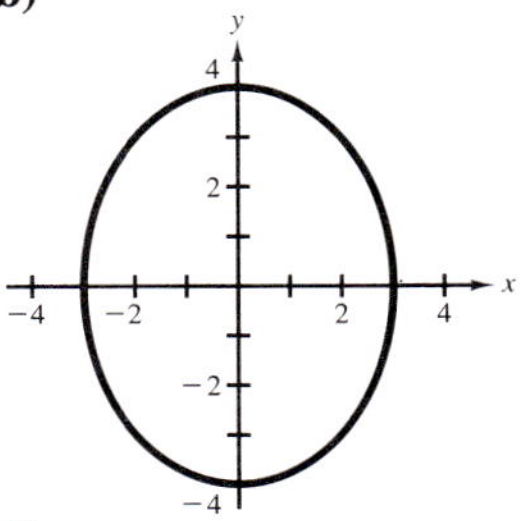

(c)

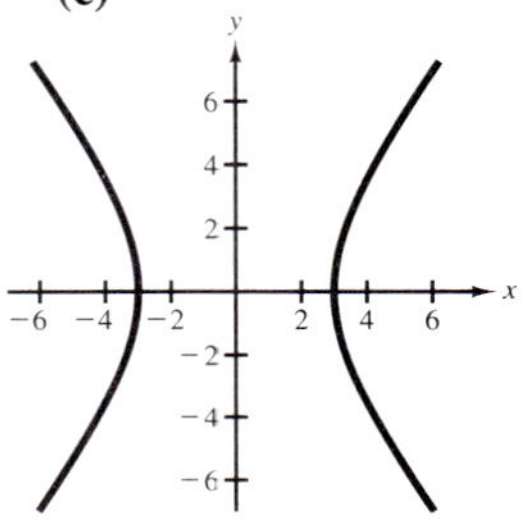

(d)

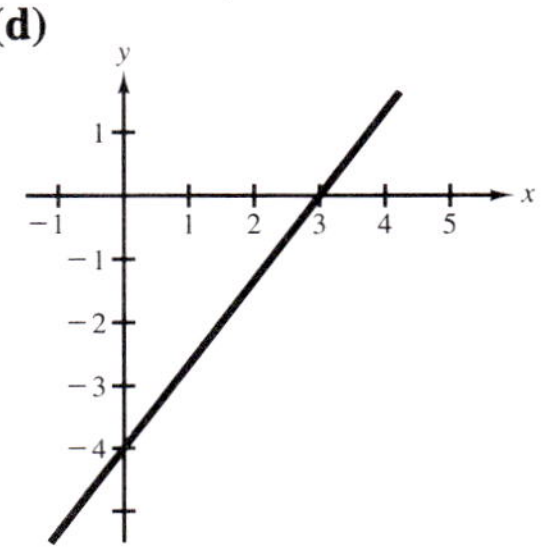

17. Approximate intercepts: $(8.5, 0)$, $(-2, 0)$
Vertex: $\left(3\frac{1}{3}, 8\frac{1}{3}\right)$

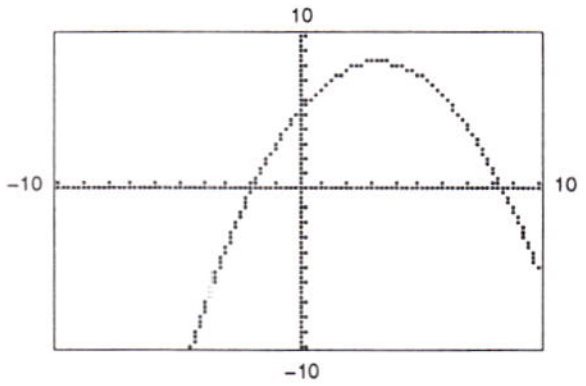

18. $V = 16{,}000 - 2{,}250t$, $2\frac{2}{3}$ yrs **19.** 120

CUMULATIVE TEST: CHAPTERS 5–7 *(page 469)*

1. $-\frac{2y^6}{3x^4}$ **2.** $t^{1/2}$ **3.** 1.6×10^7 **4.** $35\sqrt{5x}$

5. $2x - 6\sqrt{2x} + 9$ **6.** $\sqrt{10} + 2$ **7.** $-4 + 3\sqrt{2}i$

8. $5 \pm 5\sqrt{2}i$ **9.** $\frac{-3 \pm \sqrt{3}}{3}$ **10.** 6

11. $-1 \pm \sqrt{5}$ **12.** 15 hrs, 17 hrs

13. (a) 10
(b) $3x - 4y + 12 = 0$

14. (a)

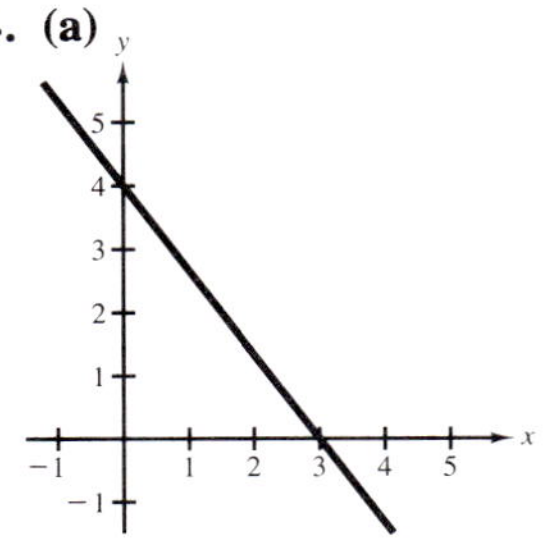

(b)

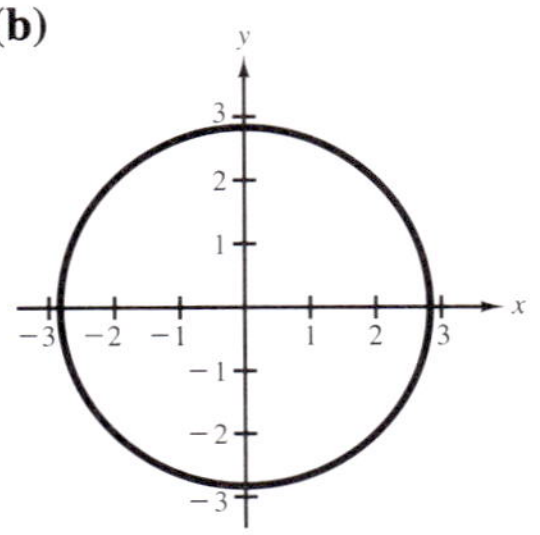

(c)

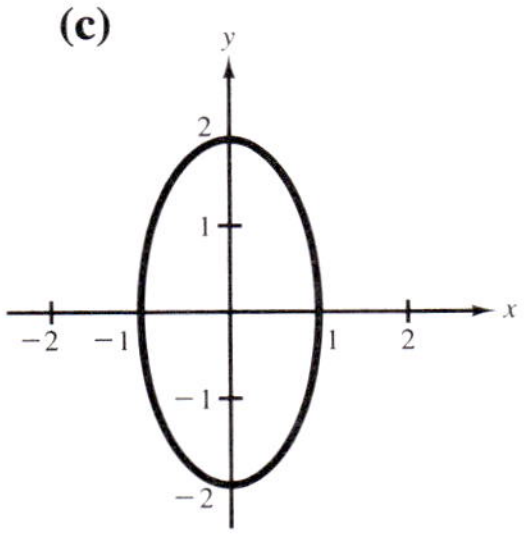

(d)

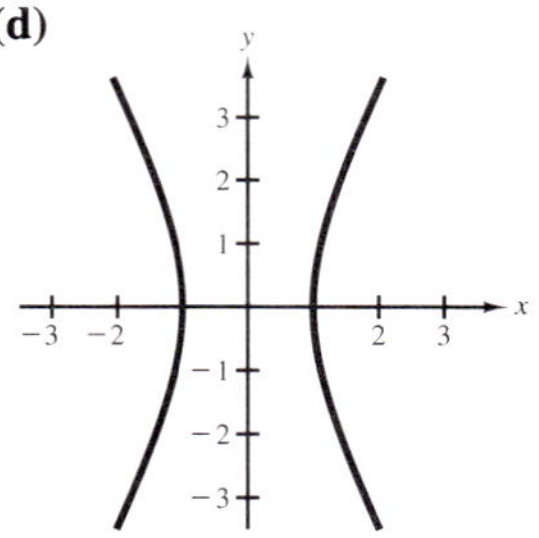

15.

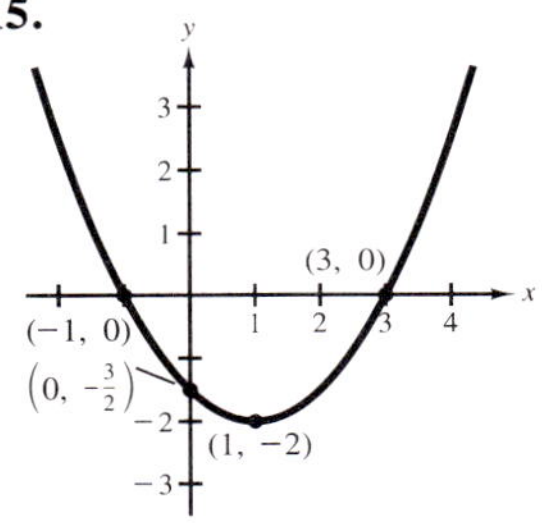

16. $\frac{x^2}{400} + \frac{y^2}{10} = 1$

17. $P = 250{,}000 + 25{,}000t$, \$500,000 **18.** 11 ft

Chapter 8

SECTION 8.1 *(page 479)*

Warm-Up *(page 479)*

1. $2h$ **2.** $3t$ **3.** $2x^2 - 16x + 27$
4. $-3x^2 - 12x - 2$ **5.** $-8x - 10$
6. $-2x^2 + 14x$

7. (a) -4
(b) 8
(c) 8
(d) 71

8. (a) 0
(b) 9
(c) 0
(d) 8.75

9. (a) 0
(b) Undefined
(c) 8
(d) -6

10. (a) 2
(b) -2
(c) $\frac{5}{2}$
(d) $\frac{101}{10}$

1. $(3, 150), (2, 100), (8, 400), (6, 300), \left(\frac{1}{2}, 25\right)$

3. (1986, New York), (1987, Minnesota), (1988, Los Angeles), (1989, Oakland), (1990, Cincinnati)

5. **(a)** 80.6 **(b)** 89.3 **(c)** 74.8 **(d)** 63.9

7. **(a)** Function **(b)** Not a function **(c)** Function **(d)** Not a function

9. Function **11.** Function **13.** Function

15. Not a function **17.** Function **19.** Not a function

21. **(a)** 2 **(b)** -2 **(c)** k **(d)** $k + 1$

23. **(a)** 3, 3 **(b)** $-4, -4$ **(c)** s, s **(d)** $s - 2, s - 2$

25. **(a)** 29 **(b)** 11 **(c)** $12a - 2$ **(d)** $12a + 5$

27. **(a)** 8 **(b)** $\frac{2}{9}$ **(c)** $2y^2$ **(d)** 26

29. **(a)** 2 **(b)** 3 **(c)** $\sqrt{\frac{31}{3}}$ **(d)** $\sqrt{5(z + 1)}$

31. **(a)** 0 **(b)** $-\frac{3}{2}$ **(c)** $-\frac{5}{2}$ **(d)** $\frac{3(x + 4)}{x - 1}$

33. **(a)** 2 **(b)** -2 **(c)** 10 **(d)** -8

35. **(a)** 0 **(b)** $\frac{7}{4}$ **(c)** 3 **(d)** 0

37. **(a)** 2 **(b)** $\frac{2(x - 6)}{x}$

39. **(a)** $1 - (x + h)^2$ **(b)** $-2x - h$

41. Domain $= \{0, 2, 4, 6\}$
Range $= \{0, 1, 8, 27\}$

43. Domain: All real numbers r such that $r \geq 0$
Range: All real numbers C such that $C \geq 0$

45. All real values x such that $x \neq 3$

47. All real values of x

49. All real values t such that $t \neq 0$ and $t \neq -2$

51. All real values x such that $x \neq 1$ and $x \neq 2$

53. All real values x such that $x \geq \frac{1}{2}$

55. All real values x such that $-5 \leq x \leq 5$

57. All real values x **59.** All real values t

61. $P(x) = 4x$ **63.** $V(x) = x^3$

65. $V(x) = x(24 - 2x)^2 = 4x(12 - x)^2$ **67.** $A(x) = (32 - x)^2$

69. **(a)** Correct **(b)** Not correct

SECTION 8.2 *(page 494)*

Warm-Up *(page 493)*

1.
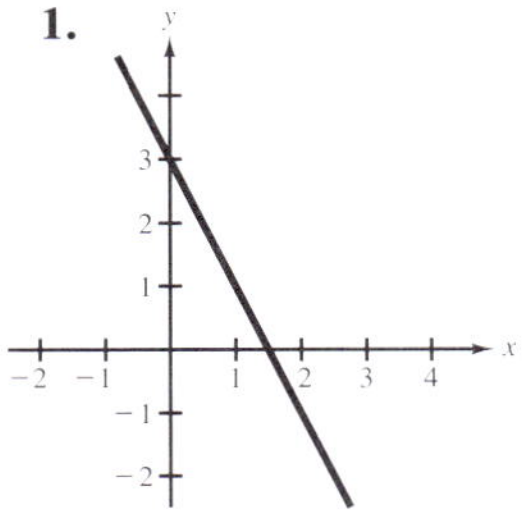

2.
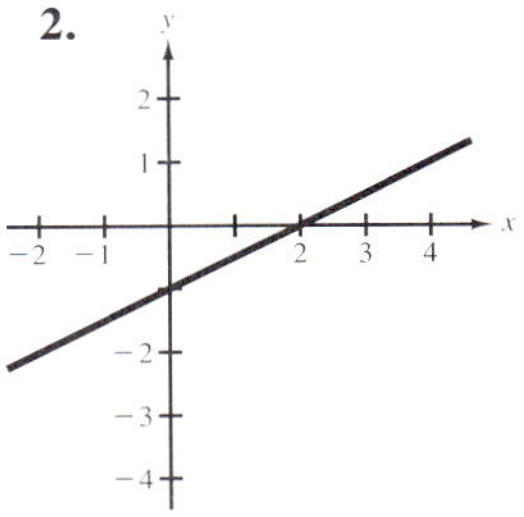

3.
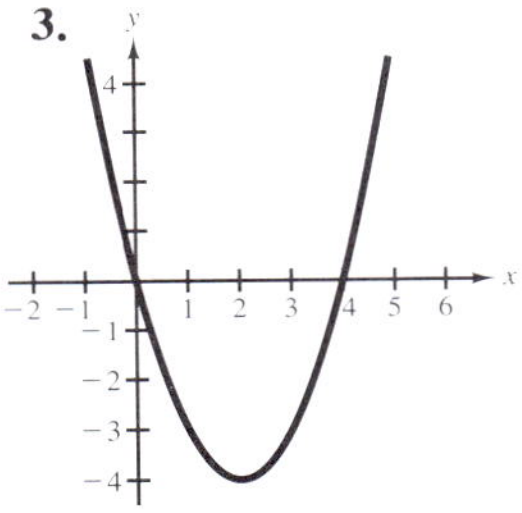

4.
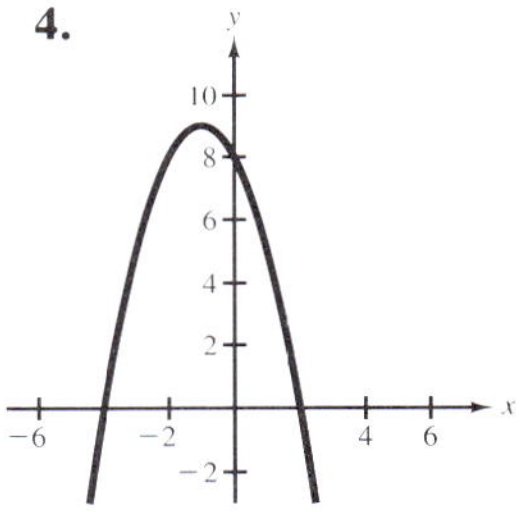

5.
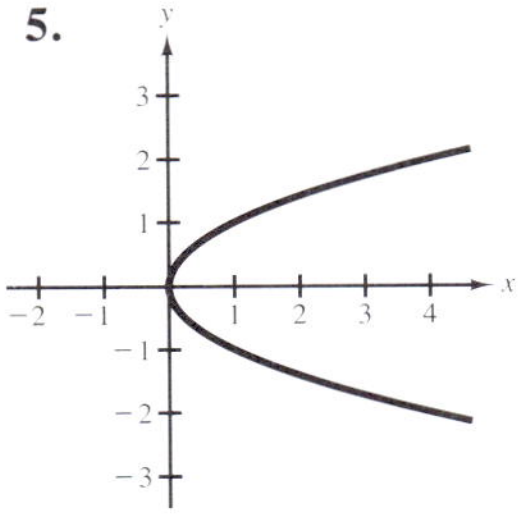

6.
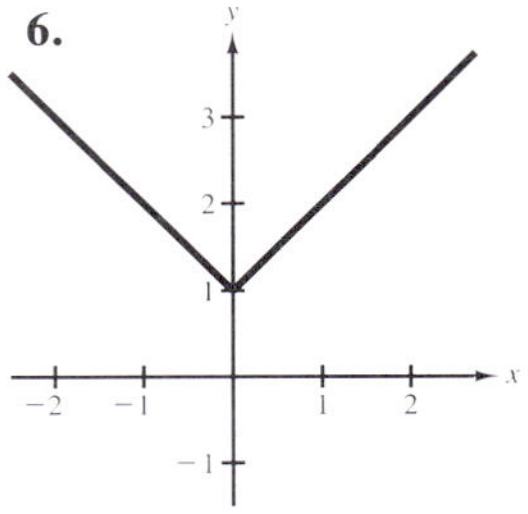

7. **(a)** 12 **(b)** $\frac{3}{16}$

8. **(a)** -7 **(b)** $-2x$

9. **(a)** $\frac{1}{3}$ **(b)** $\frac{c - 6}{c + 4}$

10. **(a)** $2\sqrt{3}$ **(b)** $\sqrt{t - 1}$

1. Domain: $-\infty < x < \infty$
Range: $-\infty < y < \infty$

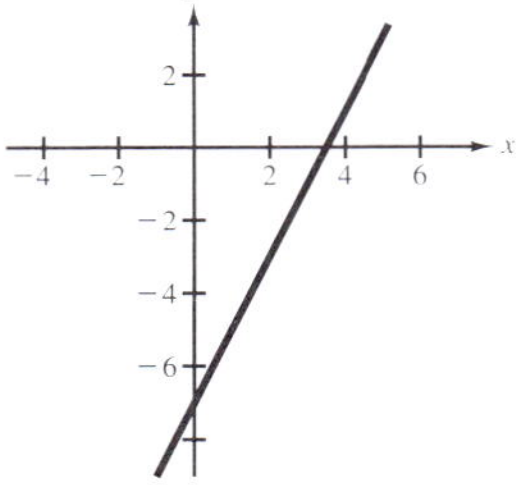

3. Domain: $-\infty < x < \infty$
Range: $0 \leq y < \infty$

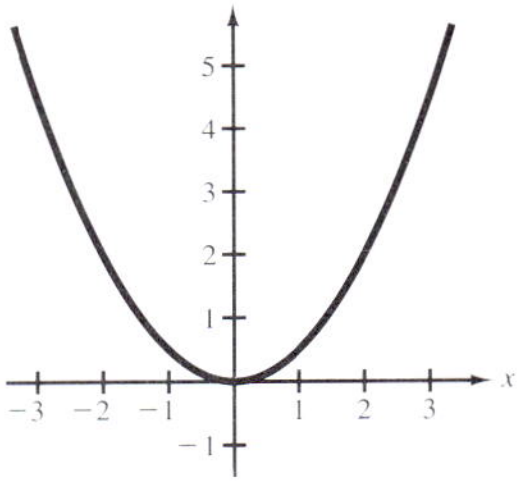

5. Domain: $-\infty < x < \infty$
Range: $-1 \le y < \infty$

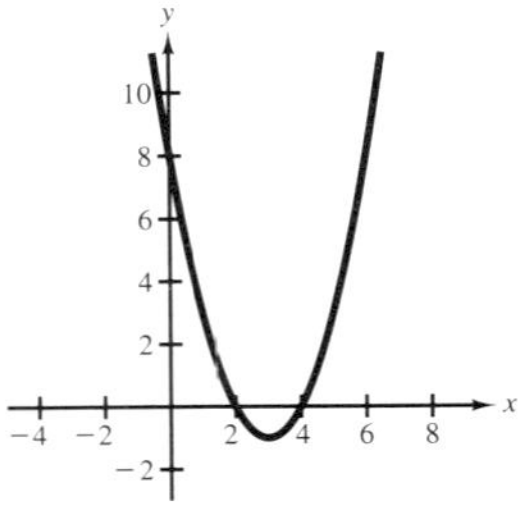

7. Domain: $2 \le t < \infty$
Range: $0 \le y < \infty$

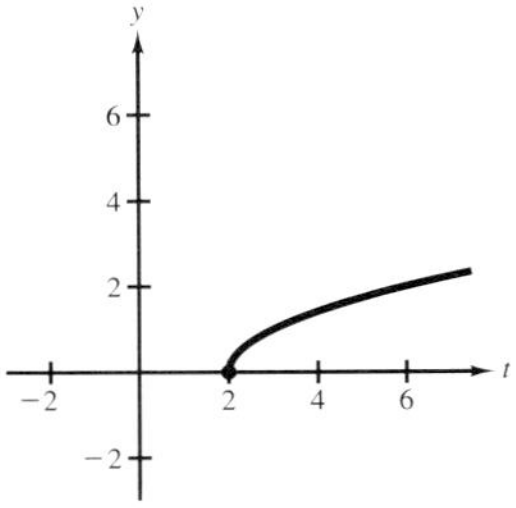

9. Domain: $-\infty < s < \infty$
Range: $-\infty < y < \infty$

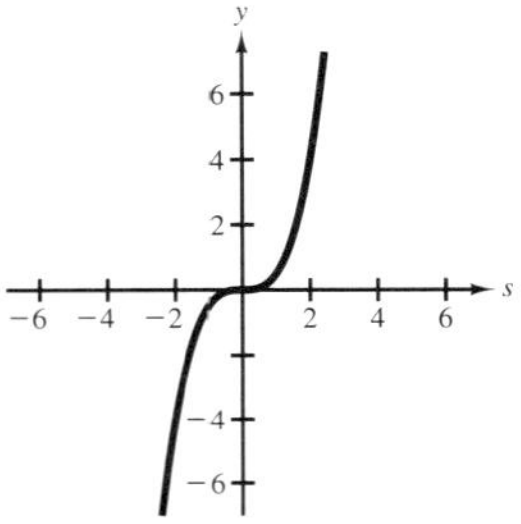

11. Domain: $-\infty < x < \infty$
Range: $0 \le y < \infty$

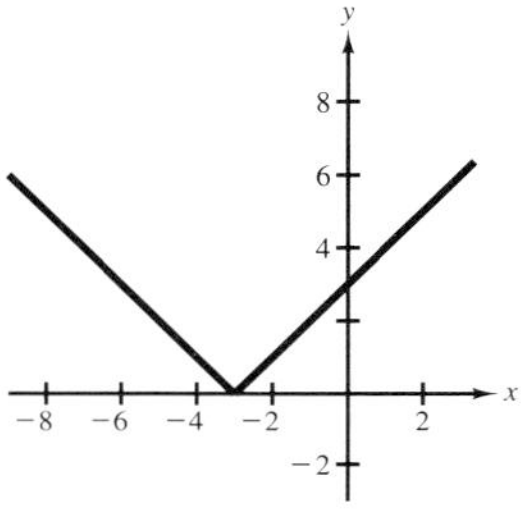

13. Domain: $0 \le x \le 2$
Range: $0 \le y \le 6$

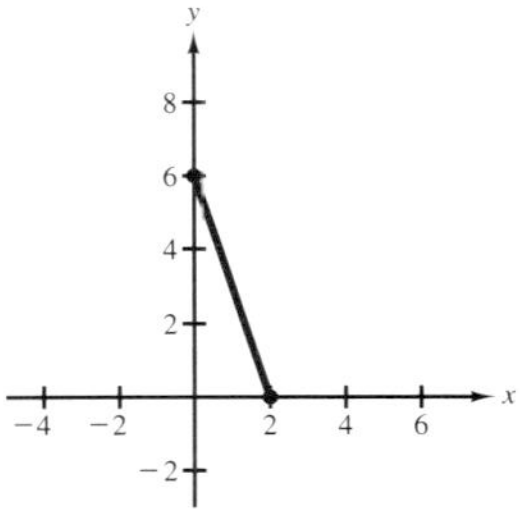

15. Domain: $-2 \le x \le 2$
Range: $-8 \le y \le 8$

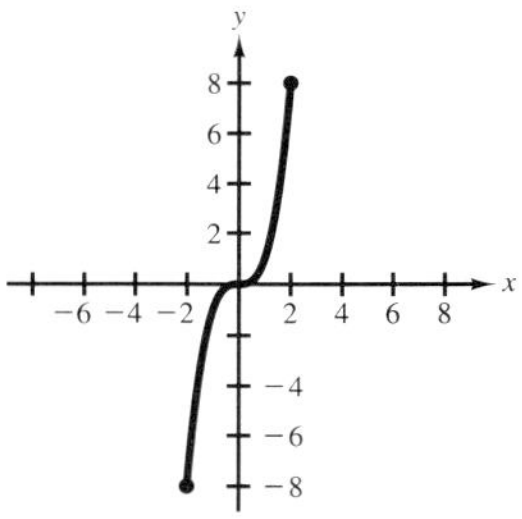

17. Domain: $-\infty < x < \infty$
Range: $-\infty < y \le 6$

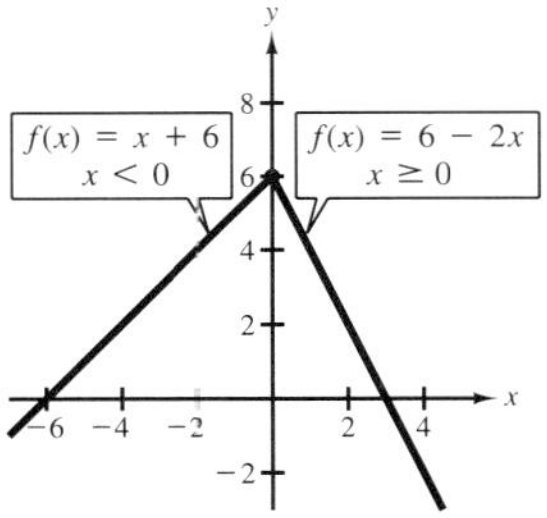

19. Domain: $-\infty < x < \infty$
Range: $-\infty < y < \infty$

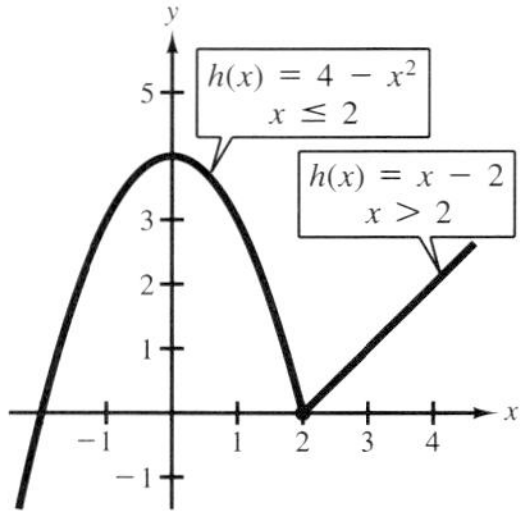

21. Function **23.** Not a function

25. Function

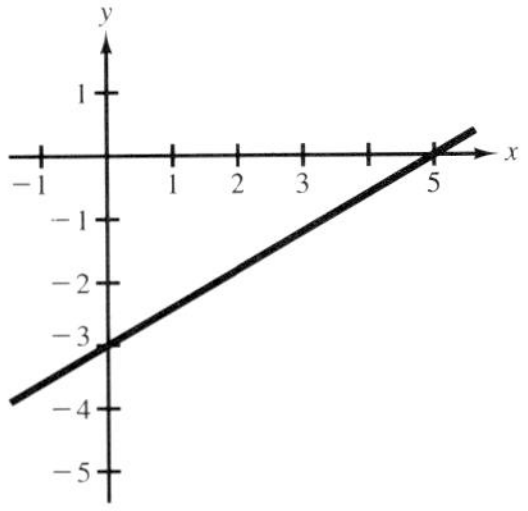

27. Not a function

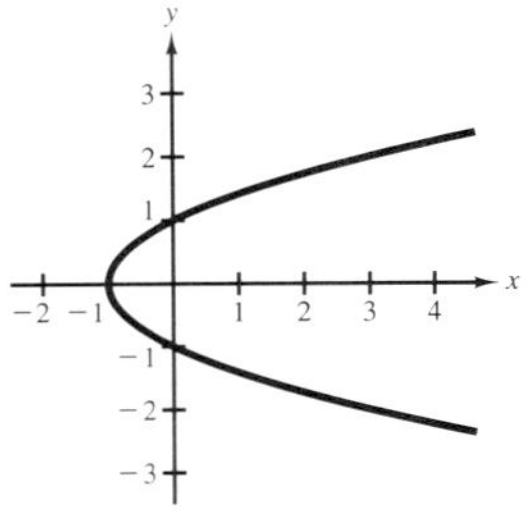

29. Function

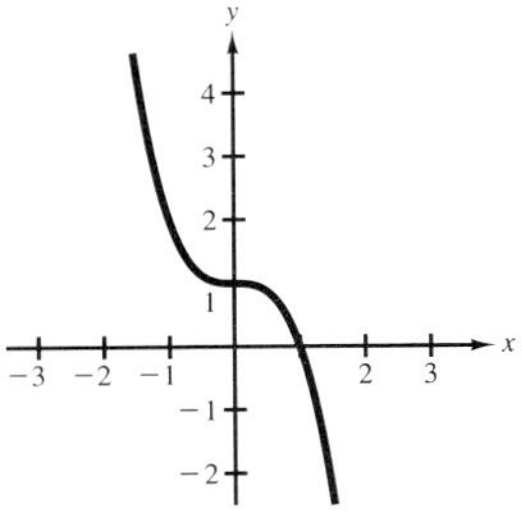

31. Vertical translation 2 units upward

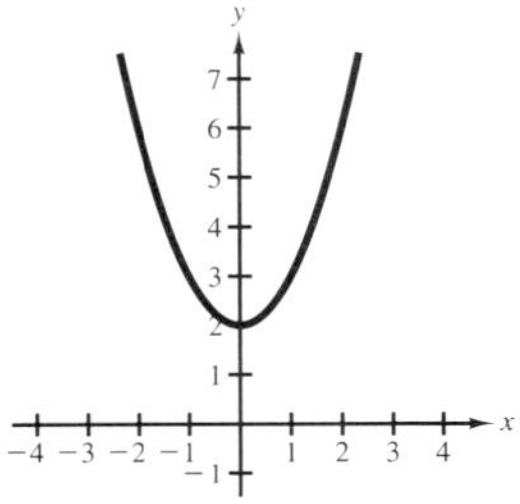

33. Horizontal translation 2 units to the left

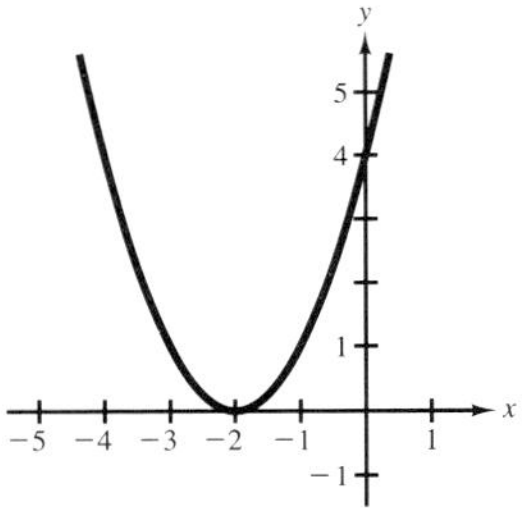

35. Reflection in the x-axis

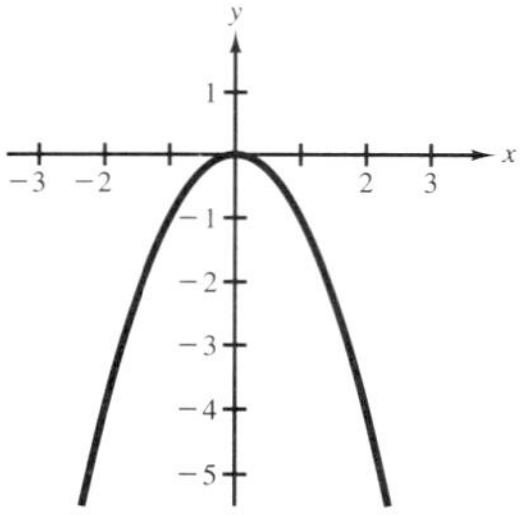

37. Vertical translation 3 units upward

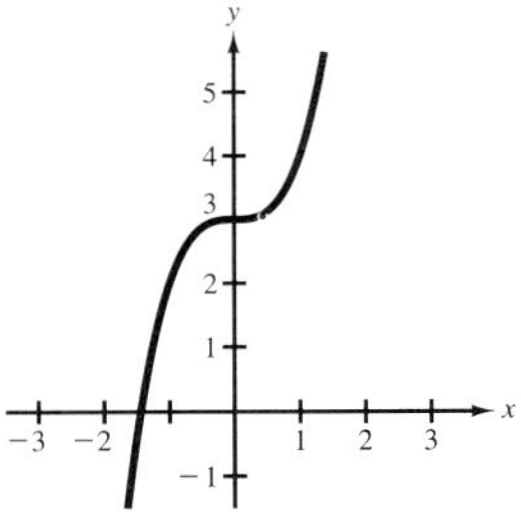

39. Horizontal translation 3 units to the right

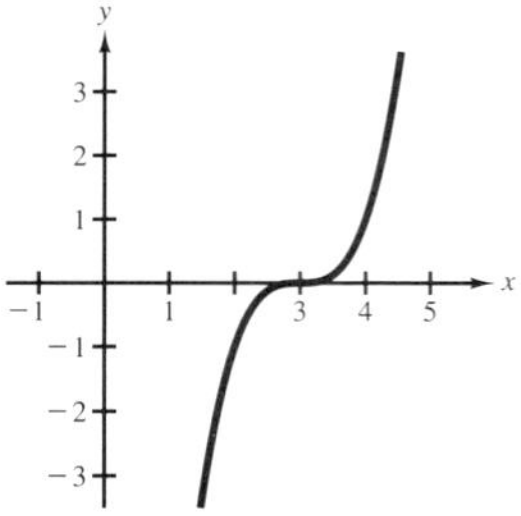

41. Reflection in the y-axis

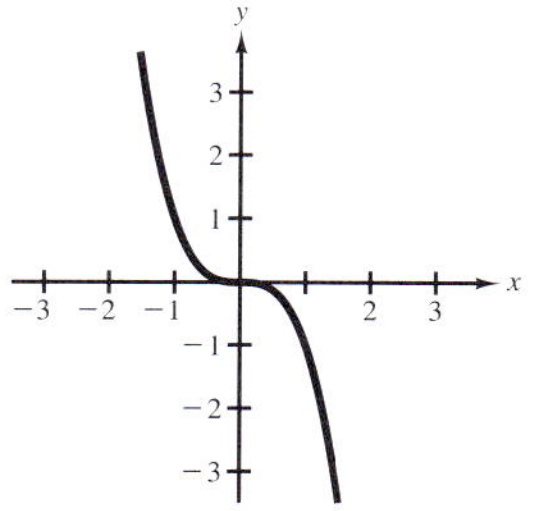

43. Reflection in the x-axis followed by a horizontal translation 1 unit to the right followed by a vertical translation 2 units upward

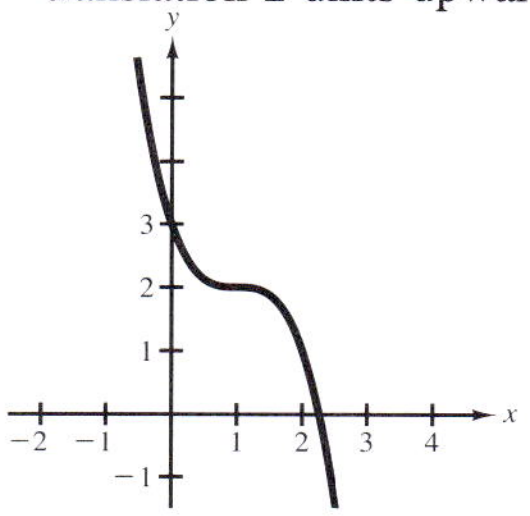

45. $y = -\sqrt{x}$ **47.** $y = \sqrt{x + 2}$ **49.** $y = \sqrt{-x}$

51. $f(x) = \sqrt{4 - x^2}$ $g(x) = -\sqrt{4 - x^2}$

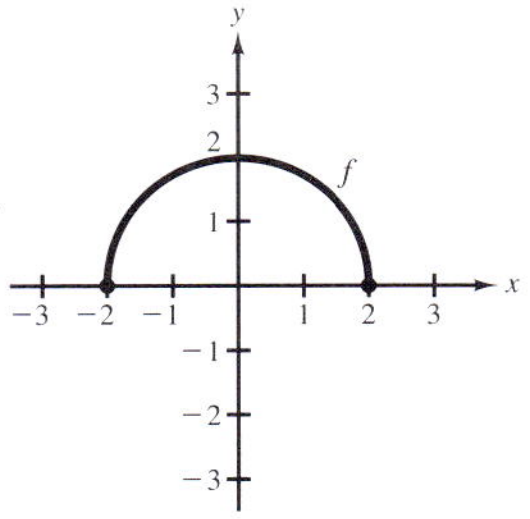

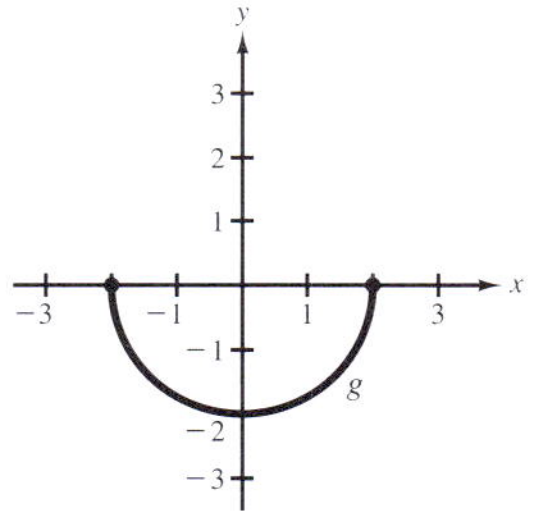

53. $f(x) = \frac{1}{2}\sqrt{4 - x^2}$ $g(x) = -\frac{1}{2}\sqrt{4 - x^2}$

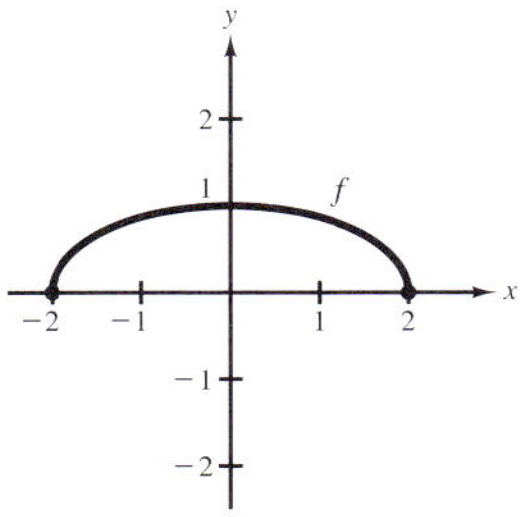

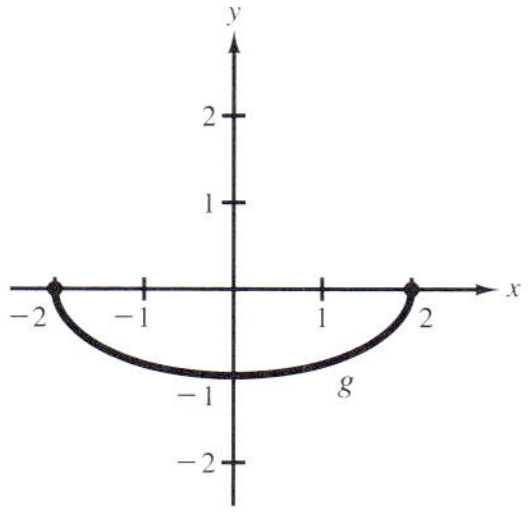

55. Horizontal translation 5 units to the right

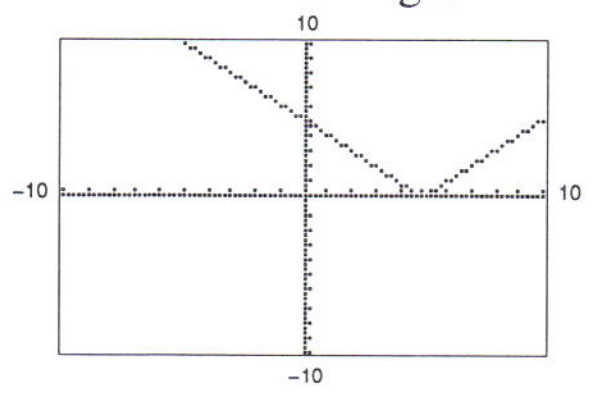

57. Vertical translation 5 units downward

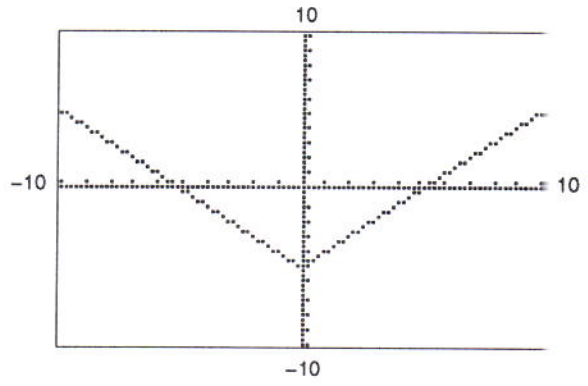

59. Reflection in the x-axis

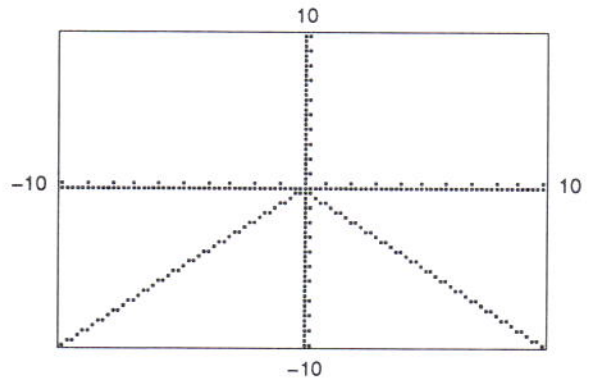

SECTION 8.3 *(page 501)*

Warm-Up *(page 501)*

1. All real values x
2. All real values x such that $x \leq 10$
3. All real values x such that $x \neq 5$
4. All real values x such that $x = 5$ and $x \neq -5$
5.

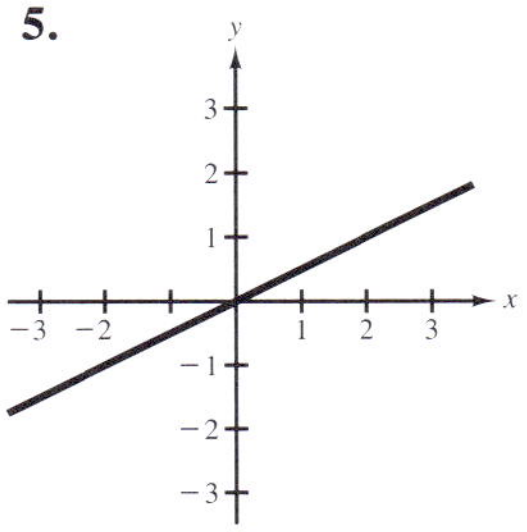

6.

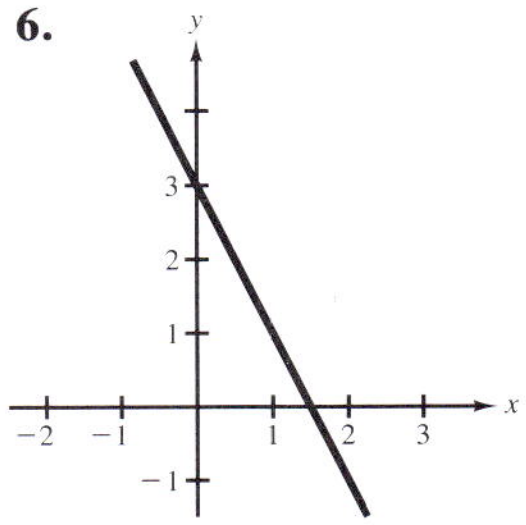

7.

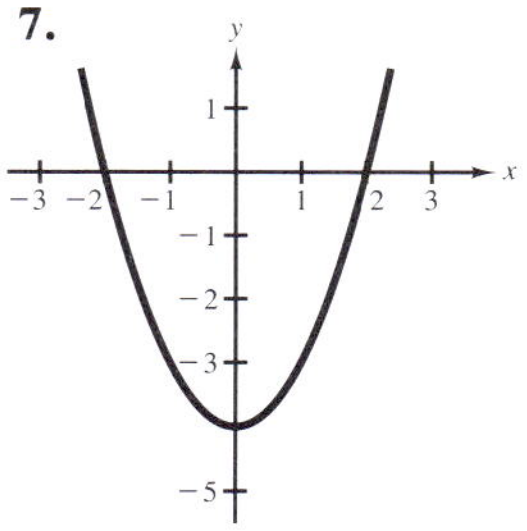

8.

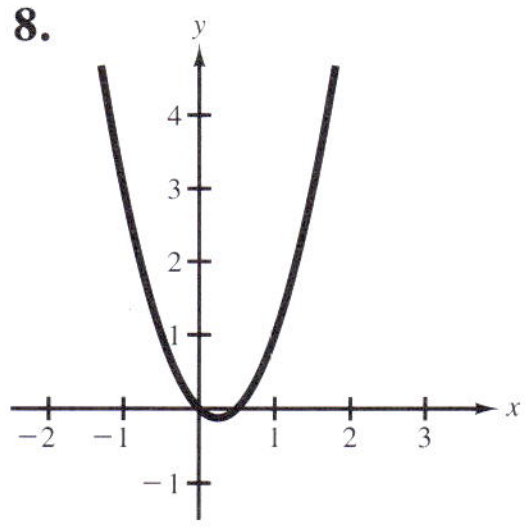

9.

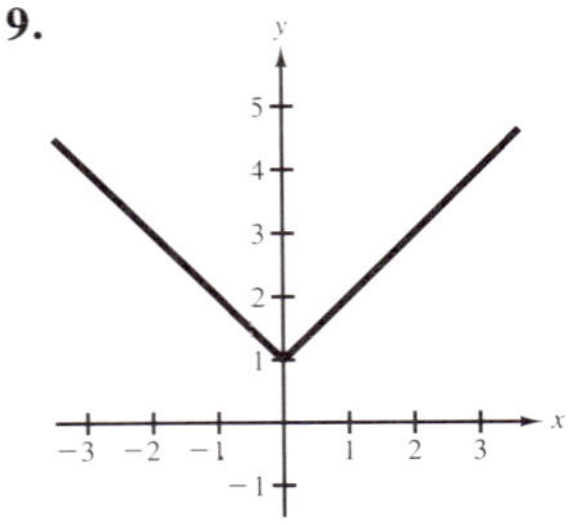

10.

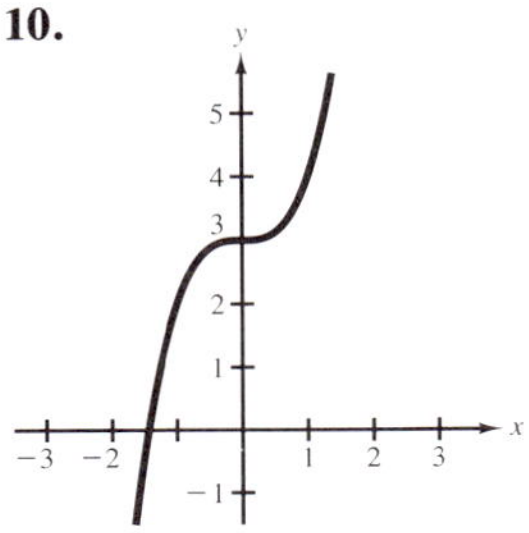

1. $(f+g)(x) = x^2 + 2x, \quad -\infty < x < \infty$
$(f-g)(x) = 2x - x^2, \quad -\infty < x < \infty$
$(fg)(x) = 2x^3, \quad -\infty < x < \infty$
$\left(\frac{f}{g}\right)(x) = \frac{2}{x}, \quad -\infty < x < 0,\ 0 < x < \infty$

3. $(f+g)(x) = x^2 + 4x - 12, \quad -\infty < x < \infty$
$(f-g)(x) = -x^2 + 4x + 6, \quad -\infty < x < \infty$
$(fg)(x) = 4x^3 - 3x^2 - 36x + 27, \quad -\infty < x < \infty$
$\left(\frac{f}{g}\right)(x) = \frac{4x-3}{x^2-9}, \quad -\infty < x < -3,\ -3 < x < 3,\ 3 < x < \infty$

5. $(f+g)(x) = \frac{1}{x} + 5x, \quad -\infty < x < 0,\ 0 < x < \infty$
$(f-g)(x) = \frac{1}{x} - 5x, \quad -\infty < x < 0,\ 0 < x < \infty$
$(fg)(x) = 5, \quad -\infty < x < \infty$
$\left(\frac{f}{g}\right)(x) = \frac{1}{5x^2}, \quad -\infty < x < 0,\ 0 < x < \infty$

7. $(f+g)(x) = \sqrt{x+4} + \sqrt{4-x}, \quad -4 \le x \le 4$
$(f-g)(x) = \sqrt{x+4} - \sqrt{4-x}, \quad -4 \le x \le 4$
$(fg)(x) = \sqrt{16-x^2}, \quad -4 \le x \le 4$
$\left(\frac{f}{g}\right) = \frac{\sqrt{x+4}}{\sqrt{4-x}}, \quad -4 \le x < 4$

9. **(a)** 11 **(b)** 0

11. **(a)** 19 **(b)** -79

13. 10 $\frac{2}{5}$

15. $\frac{1}{3}$ Undefined (division by zero)

17. **(a)** 13 **(b)** 16 **(c)** $x^2 - 3$ **(d)** $(x-3)^2$

19. **(a)** 0 **(b)** 3 **(c)** $3|x-1|$ **(d)** $3|x-3|$

21. **(a)** 3 **(b)** 8 **(c)** $\sqrt{x+5}$ **(d)** $\sqrt{x}+5$

23. **(a)** $\frac{1}{4}$ **(b)** $\frac{1}{3}$ **(c)** $\frac{1}{\sqrt{x-3}}$ **(d)** $\frac{1}{\sqrt{x}-3}$

25. $f \circ g$: All real values x
$g \circ f$: All real values x

27. $f \circ g$: Real values x such that $x \ge 2$
$g \circ f$: Real values x such that $x \ge 0$

29. $f \circ g$: All real values x
$g \circ f$: Real values x such that $x \ne -9$

31. $t^2 - 3t$ **33.** $t^2 - 2t + 3$ **35.** 5 **37.** 10

39. $5y^2 - 15y + 3$ **41.** -2 **43.** 0 **45.** -1

47.

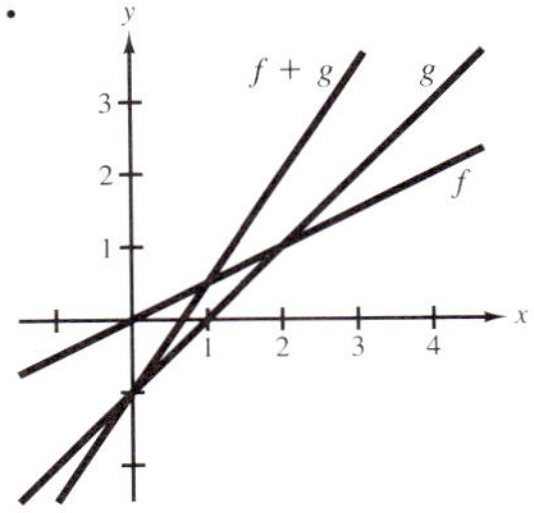

49.

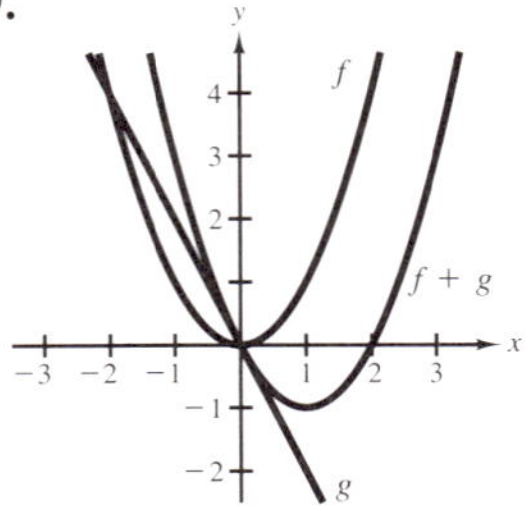

51. $T(x) = \frac{3}{4}x + \frac{1}{15}x^2$

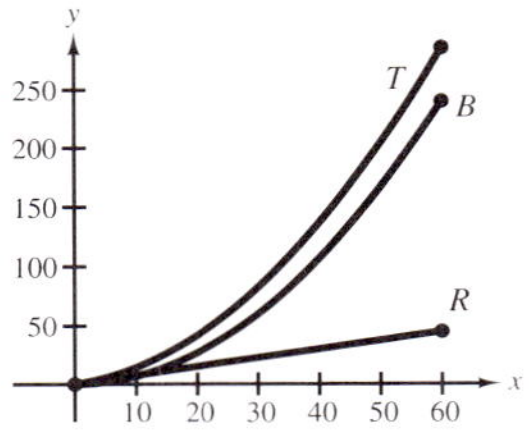

53.

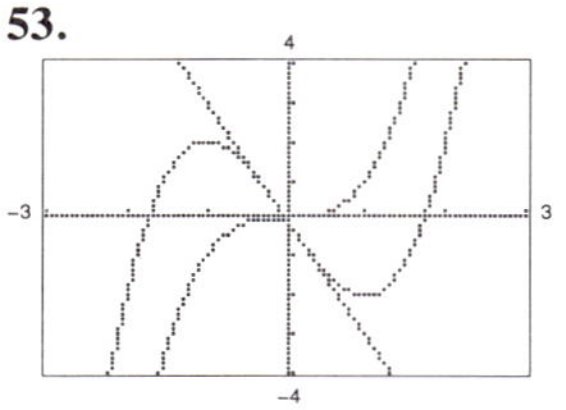

$f(x)$ is most significant in determining the magnitude of the sum.

SECTION 8.4 *(page 513)*

Warm-Up *(page 512)*

1. All real values x **2.** All real values x

3. Real values x such that $-4 \le x \le 4$

4. Real values x such that $x \ne -6$ and $x \ne 6$

5. $(f+g)(x) = x^2 + 3x - 9, \quad -\infty < x < \infty$
$(f-g)(x) = x^2 - 3x - 9, \quad -\infty < x < \infty$
$(fg)(x) = 3x^3 - 27x, \quad -\infty < x < \infty$
$\left(\frac{f}{g}\right)(x) = \frac{x^2-9}{3x}, \quad -\infty < x < 0,\ 0 < x < \infty$
$(f \circ g)(x) = 9x^2 - 9, \quad -\infty < x < \infty$
$(g \circ f)(x) = 3(x^2 - 9), \quad -\infty < x < \infty$

6. $(f+g)(x) = \frac{1}{2}x^2 + 2x - 3, \quad -\infty < x < \infty$

$(f-g)(x) = -\frac{1}{2}x^2 + 2x - 3, \quad -\infty < x < \infty$

$(fg)(x) = x^3 - \frac{3}{2}x^2, \quad -\infty < x < \infty$

$\left(\frac{f}{g}\right)(x) = \frac{4x-6}{x^2}, \quad -\infty < x < 0,\ 0 < x < \infty$

$(f \circ g)(x) = x^2 - 3, \quad -\infty < x < \infty$

$(g \circ f)(x) = \frac{1}{2}(2x-3)^2, \quad -\infty < x < \infty$

7. $(f+g)(x) = \sqrt[3]{x-2} + x^2, \quad -\infty \le x < \infty$

$(f-g)(x) = \sqrt[3]{x-2} - x^2, \quad -\infty \le x < \infty$

$(fg)(x) = x^2\sqrt[3]{x-2}, \quad -\infty \le x < \infty$

$\left(\frac{f}{g}\right)(x) = \frac{\sqrt[3]{x-2}}{x^2}, \quad -\infty < x < 0,\ 0 < x < \infty$

$(f \circ g)(x) = \sqrt[3]{(x^2-2)}, \quad -\infty < x < \infty$

$(g \circ f)(x) = \sqrt[3]{(x-2)^2}, \quad -\infty < x < \infty$

8. $(f+g)(x) = \sqrt{x} + x^2 - 4, \quad \infty \le x < \infty$

$(f-g)(x) = \sqrt{x} - x^2 + 4, \quad \infty \le x < \infty$

$(fg)(x) = \sqrt{x}(x^2-4), \quad 0 \le x < \infty$

$\left(\frac{f}{g}\right)(x) = \frac{\sqrt{x}}{x^2-4}, \quad 0 \le x < 2,\ 2 < x < \infty$

$(f \circ g)(x) = \sqrt{x^2-4}, \quad -\infty < x \le -2,\ 2 \le x < \infty$

$(g \circ f)(x) = x - 4, \quad -\infty < x < \infty$

9. $(f+g)(x) = \frac{2(x-1)}{x(x-2)}, \quad -\infty < x < 0,\ 0 < x < 2,\ 2 < x < \infty$

$(f-g)(x) = \frac{2}{x(2-x)}, \quad -\infty < x < 0,\ 0 < x < 2,\ 2 < x < \infty$

$(fg)(x) = \frac{1}{x(x-2)}, \quad -\infty < x < 0,\ 0 < x < 2,\ 2 < x < \infty$

$\left(\frac{f}{g}\right)(x) = \frac{x-2}{x}, \quad -\infty < x < 0,\ 0 < x < \infty$

$(f \circ g)(x) = x - 2, \quad -\infty < x < \infty$

$(g \circ f)(x) = \frac{x}{1-2x}, \quad -\infty < x < \frac{1}{2},\ \frac{1}{2} < x < \infty$

10. $(f+g)(x) = \frac{x^2+4x+5}{x+2}, \quad -\infty < x < -2,\ -2 < x < \infty$

$(f-g)(x) = \frac{x^2+4x+3}{x+2}, \quad -\infty < x < -2,\ -2 < x < \infty$

$(fg)(x) = 1, \quad -\infty < x < \infty$

$\left(\frac{f}{g}\right)(x) = (x+2)^2, \quad -\infty < x < \infty$

$(f \circ g)(x) = \frac{2x+5}{x+2}, \quad -\infty < x < -2,\ -2 < x < \infty$

$(g \circ f)(x) = \frac{1}{x+4}, \quad -\infty < x < -4,\ -4 < x < \infty$

1. $f^{-1}(x) = \frac{1}{5}x$ **3.** $f^{-1}(x) = x - 10$ **5.** $f^{-1}(x) = x^3$

7. $f(g(x)) = f\left(\frac{1}{10}x\right) = 10\left(\frac{1}{10}x\right) = x$

$g(f(x)) = g(10x) = \frac{1}{10}(10x) = x$

9. $f(g(x)) = f\left(\frac{1}{3}[2-x]\right) = 2 - 3\left(\frac{1}{3}[2-x]\right) = 2 - (2-x) = x$

$g(f(x)) = g(2-3x) = \frac{1}{3}(2-[2-3x]) = \frac{1}{3}(3x) = x$

11. $f(g(x)) = f(x^3-1) = \sqrt[3]{(x^3-1)+1} = \sqrt[3]{x^3} = x$

$g(f(x)) = g\left(\sqrt[3]{x+1}\right) = \left(\sqrt[3]{x+1}\right)^3 - 1 = x + 1 - 1 = x$

13. $f(g(x)) = f\left(\frac{1}{x}\right) = \frac{1}{(1/x)} = x$

$g(f(x)) = g\left(\frac{1}{x}\right) = \frac{1}{(1/x)} = x$

15.

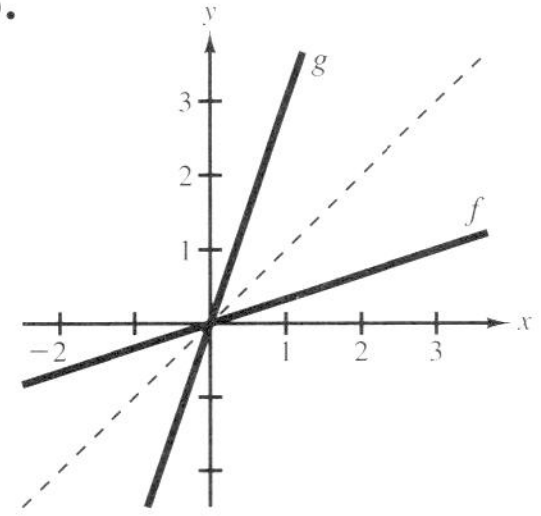

17.

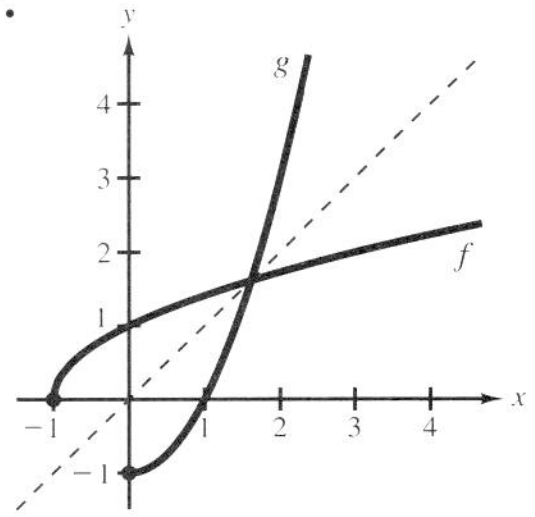

19.

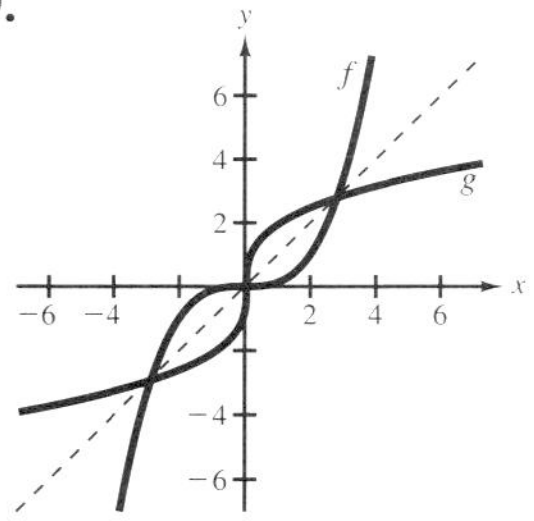

21. $f^{-1}(x) = \frac{1}{8}x$ **23.** $g^{-1}(x) = \frac{1}{4}(3-x)$

25. $h^{-1}(x) = x^2, \quad x \ge 0$ **27.** $f^{-1}(t) = \sqrt[3]{t+1}$

29. $g^{-1}(s) = \frac{5}{s},\ s \ne 0$ **31.** No **33.** Yes **35.** Yes

37. Yes **39.** No **41.** $x \ge 0, \quad f^{-1}(x) = \sqrt[4]{x}$

43. $x \ge 2, \quad f^{-1}(x) = \sqrt{x} + 2$

45.

x	0	1	3	4
f^{-1}	6	4	2	0

47.

x	-4	-2	2	3
f^{-1}	-2	-1	1	3

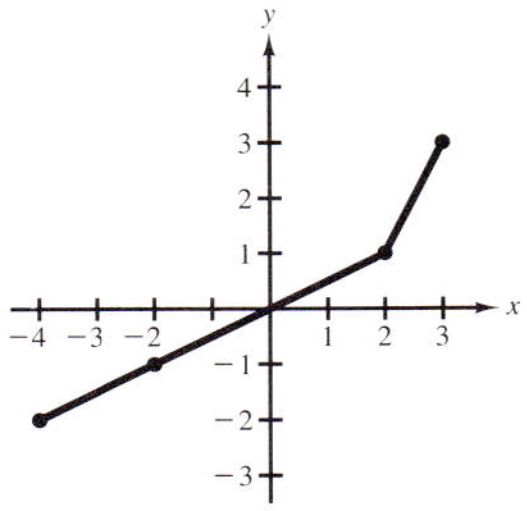

49. **(a)** $f^{-1}(x) = \frac{1}{2}(3 - x)$
(b) $(f^{-1})^{-1}(x) = 3 - 2x$

51. $f^{-1}(x) = \sqrt[3]{x - 1}$

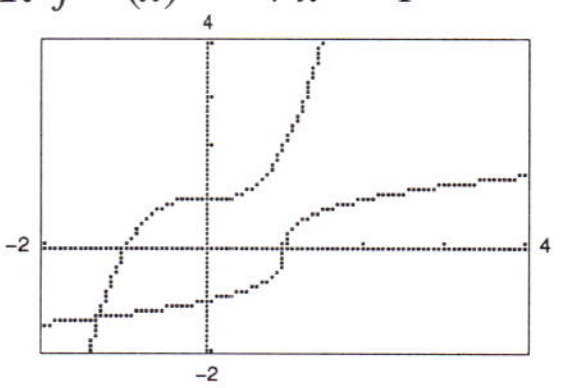

53. **(a)** $y = \frac{20}{13}(x - 9)$
(b) x: hourly wage
y: number of units produced
(c) 8

MATH MATTERS: ANNUAL SALARY VERSUS WAGE (page 508)

(a) \$17,500 a year
(b) \$12.50 an hour

SECTION 8.5 *(page 523)*

Warm-Up *(page 523)*

1. $\frac{1}{3}$ **2.** $\frac{10}{3}$ **3.** 0.92 **4.** 513 **5.** 40
6. $\frac{400}{7}$

7.

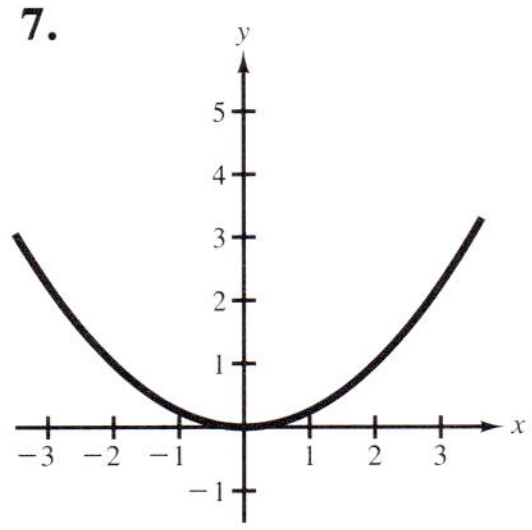

8.

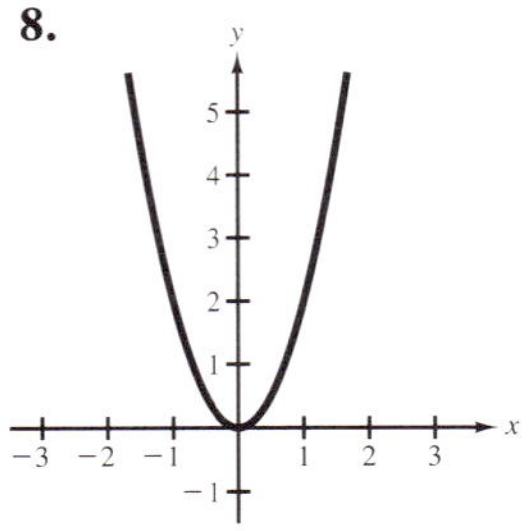

9.

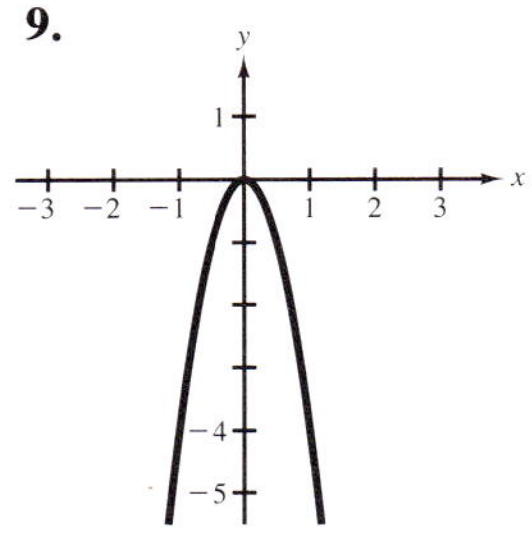

10.

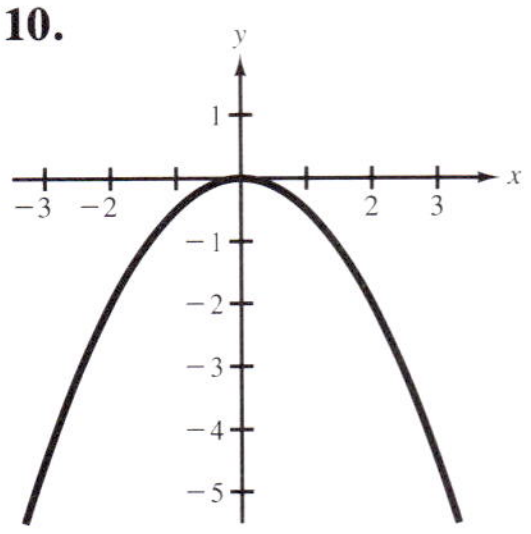

1. $I = kV$ **3.** $u = kv^2$ **5.** $p = \dfrac{k}{d}$

7. $P = \dfrac{k}{\sqrt{1 + r}}$ **9.** $A = klw$ **11.** $P = \dfrac{k}{V}$

13. $s = 5t$ **15.** $F = \frac{5}{16}x^2$ **17.** $n = \dfrac{48}{m}$

19. $F = \frac{25}{6}xy$ **21.** $d = \dfrac{120x^2}{r}$ **23.** \$4,921.25

25. 18 lbs **27.** $208\frac{1}{3}$ ft **29.** 3,072 watts
31. 667 units **33.** $\frac{1}{4}$ **35.** 3,125 lbs

37.

x	2	4	6	8	10
$y = kx^2$	4	16	36	64	100

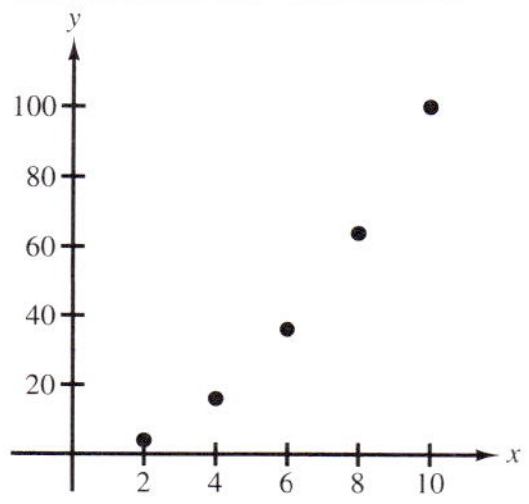

39.

x	2	4	6	8	10
$y = kx^2$	2	8	18	32	50

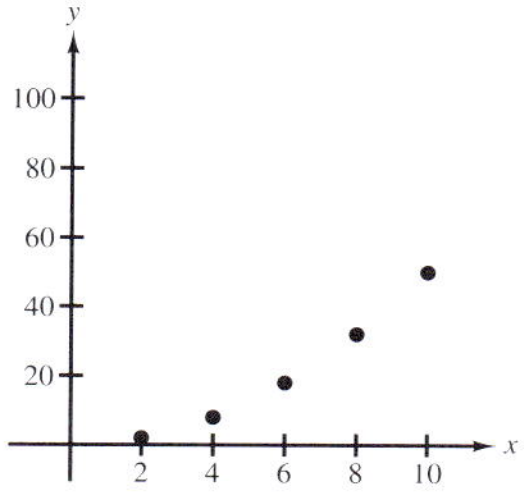

41.

x	2	4	6	8	10
$y = \frac{k}{x^2}$	$\frac{1}{2}$	$\frac{1}{8}$	$\frac{1}{18}$	$\frac{1}{32}$	$\frac{1}{50}$

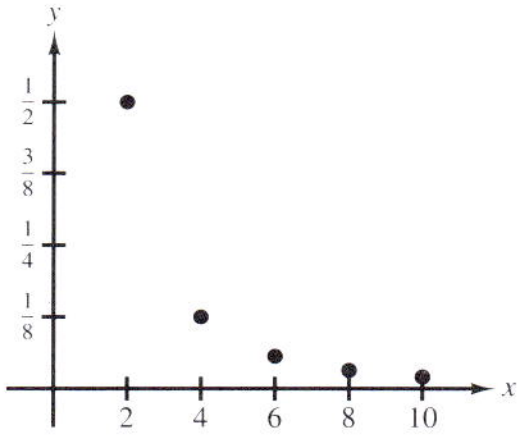

43.

x	2	4	6	8	10
$y = \frac{k}{x^2}$	$\frac{5}{2}$	$\frac{5}{8}$	$\frac{5}{18}$	$\frac{5}{32}$	$\frac{1}{10}$

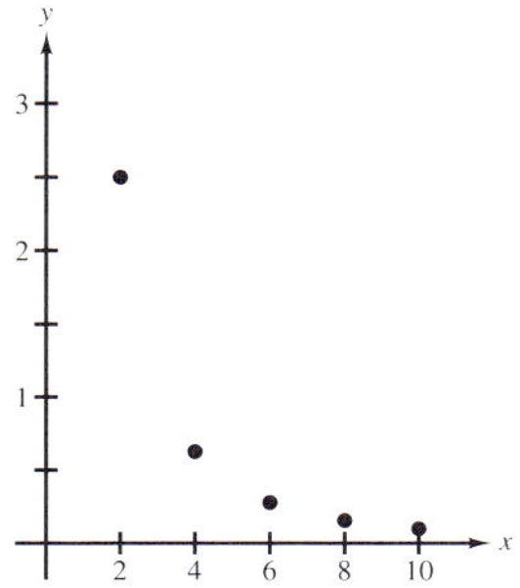

CHAPTER 8 REVIEW EXERCISES *(page 527)*

1. Not a function **3.** Function

5. Function

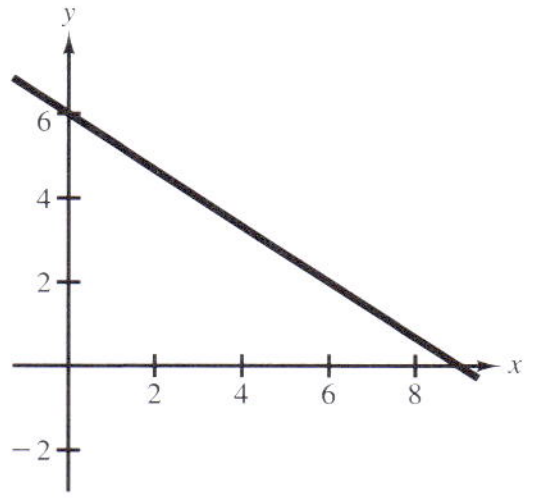

7. Function

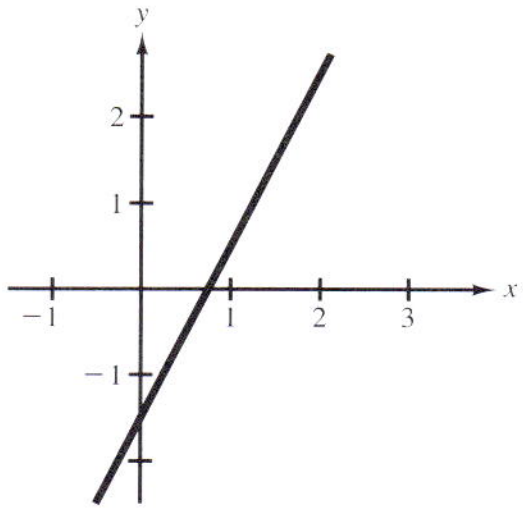

9. Function

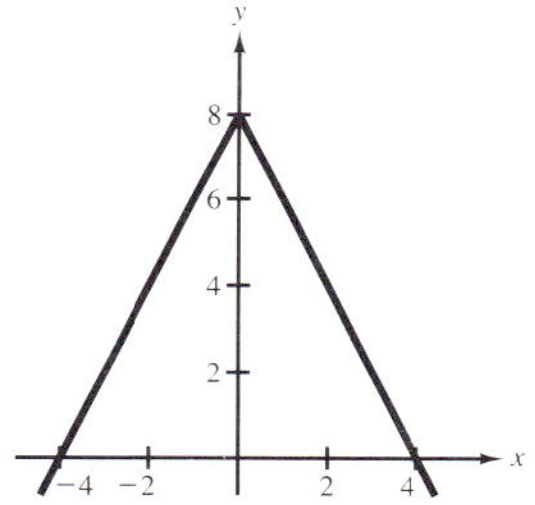

11. Not a function

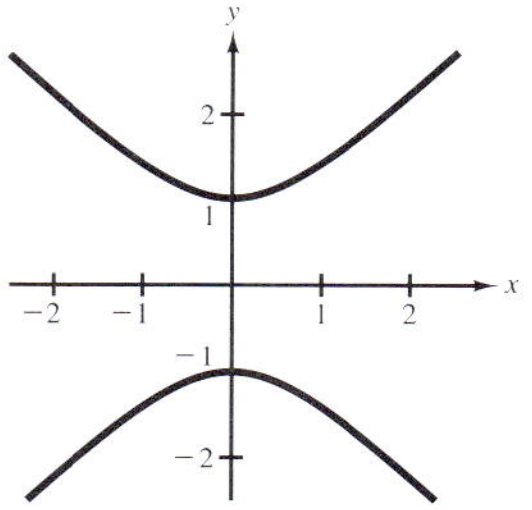

13. Function

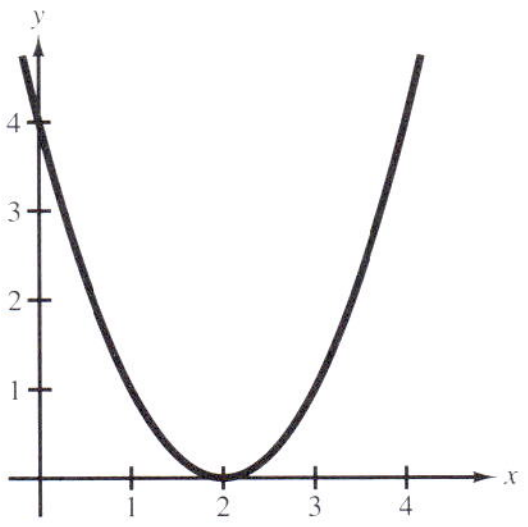

15. **(a)** 29
(b) 3
(c) $\dfrac{36 - 5t}{2}$
(d) $4 - \frac{5}{2}(x + h)$

17. **(a)** $2\sqrt{2}$
(b) 0
(c) $\dfrac{2\sqrt{3}}{3}$
(d) $\sqrt{5 - 5z}$

19. **(a)** -3
(b) 2
(c) 0
(d) -7

21. **(a)** -2
(b) $\dfrac{2(6 - x)}{x}$

23. **(a)** $(x + h)^2$
(b) $2x + h$

25. All real values x

27. All real values x such that $x \neq 1$ and $x \neq 3$

29. All real values x such that $x \leq \frac{5}{2}$

31.

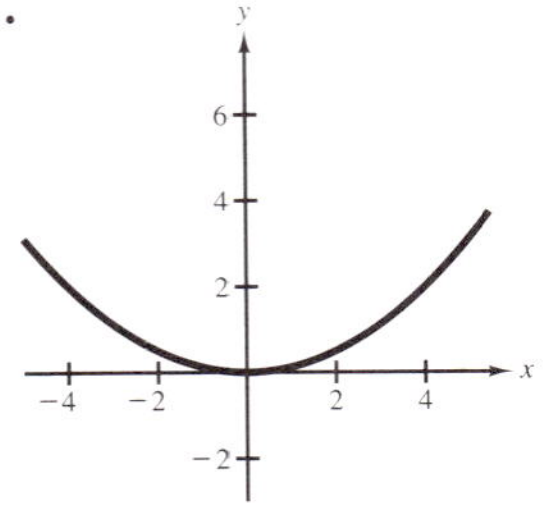

33.

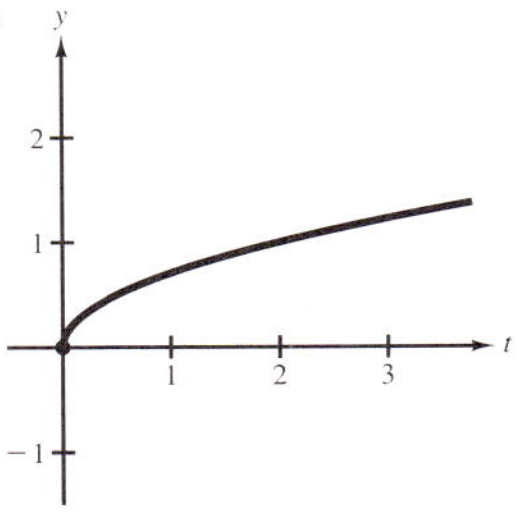

35.

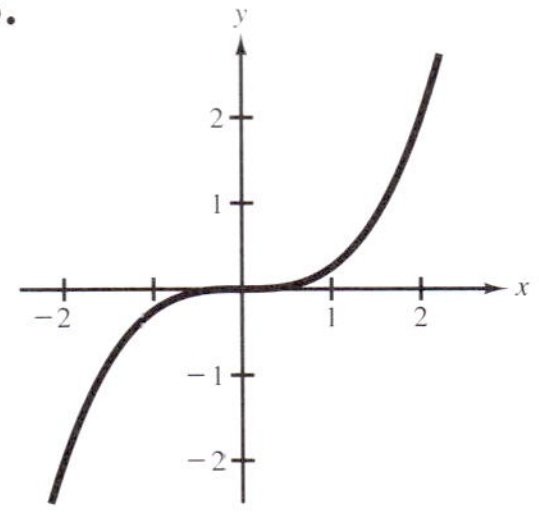

37.

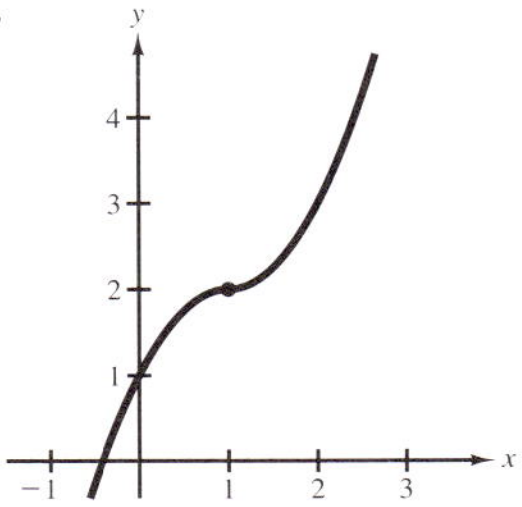

39. **(a)** 0 **(b)** 5 **(c)** 12 **(d)** -1

41. **(a)** $-\frac{1}{3}$ **(b)** 83 **(c)** $-\frac{1}{48}$ **(d)** $-\frac{\sqrt{2}}{6}$

43. **(a)** $-\frac{8}{3}$ **(b)** -2 **(c)** 1 **(d)** $\frac{1}{10}$

45. **(a)** 6 **(b)** 1 **(c)** $x^2 + 2$ **(d)** $(x + 2)^2$

47. **(a)** 5 **(b)** -1 **(c)** $|x|$ **(d)** x

49. $f \circ g$: $2 \leq x < \infty$
$g \circ f$: $4 \leq x < \infty$

51. No **53.** Yes **55.** $f^{-1}(x) = 4x$

57. $h^{-1}(x) = x^2, \quad x \geq 0$

59. $f^{-1}(t)$ does not exist since f is not one to one.

61. $y = 4\sqrt[3]{3x}$ **63.** $T = \frac{1}{18}rs^2$

65. **(a)** $k = \frac{1}{8}$ **(b)** 1,953.125 kilowatts **67.** 150 lbs

69. 945 units

CHAPTER 8 TEST *(page 530)*

1. **(a)** Not a function **(b)** Function **2.** **(a)** -2 **(b)** 7

3. **(a)** $3t + 1$ **(b)** $3h$ **4.** **(a)** All real values t such that $-3 \leq t \leq 3$ **(b)** All real values x such that $x \neq 4$

5. $C = 150 + 0.26x, \; x \geq 0$

6.

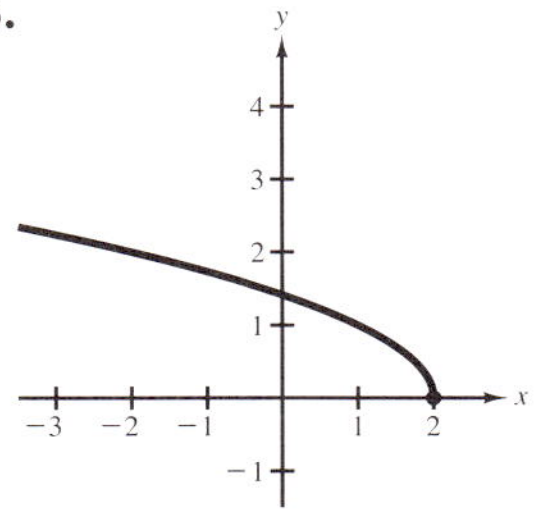

7. **(a)** $y = |x - 2|$ **(b)** $y = |x| - 2$ **(c)** $y = 2 - |x|$ **8.** Not a function **9.** **(a)** -13 **(b)** -30 **(c)** $\frac{1}{3}$ **(d)** 12

10. All real values x such that $-5 \leq x \leq 5$

11. $f^{-1}(x) = 2(x + 1)$ **12.** $S = \frac{kx^2}{y}$ **13.** $v = \frac{1}{4}\sqrt{u}$

14. 240 cubic meters

Chapter 9

SECTION 9.1 *(page 540)*

Warm-Up *(page 540)*

1.

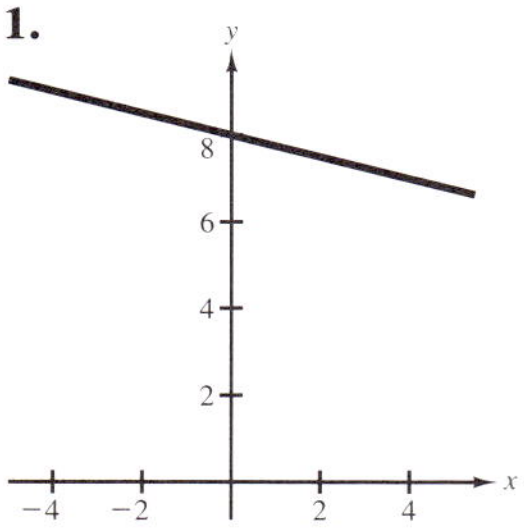

2.

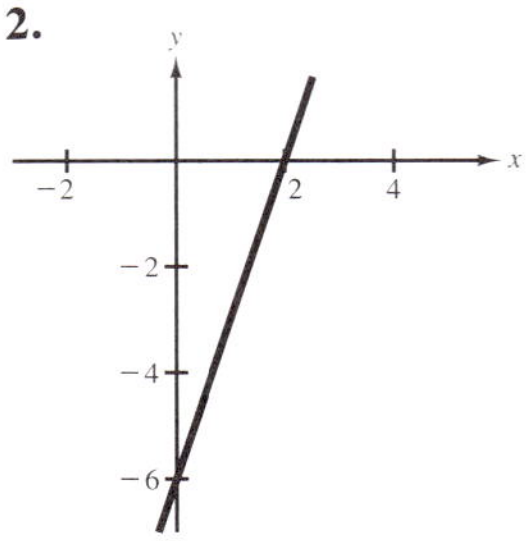

3.

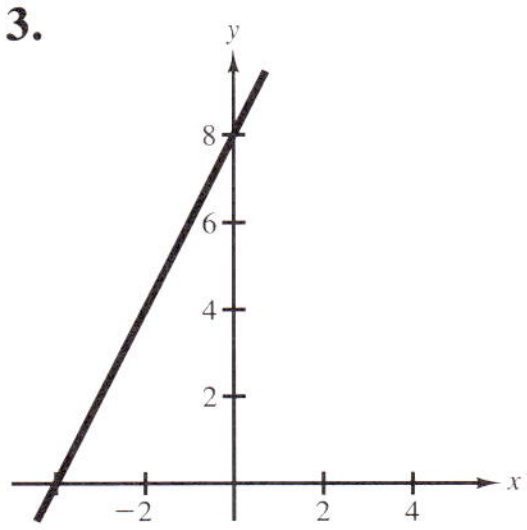

4.

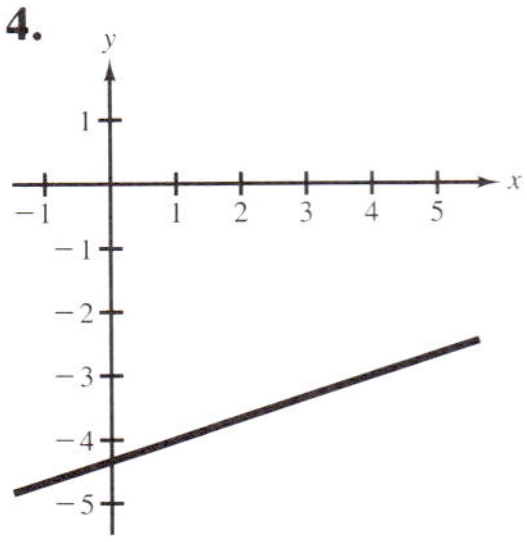

5. $\frac{3}{5}$ **6.** $\frac{7}{2}$ **7.** $2x - y = 0$

8. $x + y - 6 = 0$ **9.** $-\frac{2}{3}$ **10.** $\frac{9}{5}$

1. **(a)** Yes **(b)** No **3.** **(a)** No **(b)** Yes
5. (2, 1) **7.** (4, 3) **9.** $\left(4, -\frac{1}{2}\right)$
11. (−2, 4), (1, 1) **13.** (1, 2) **15.** (4, −2)
17. (7, 2) **19.** (−2, −1) **21.** $\left(\frac{3}{2}, \frac{3}{2}\right)$
23. (10, 30), (20, 15) **25.** (2, 5), (−3, 0)
27. (0, 5), (−4, −3) **29.** (0, 2), (3, 1)
31. No solutions **33.** $\left(1, \frac{1}{3}\right)$ **35.** No solutions

37. (−1, −1)

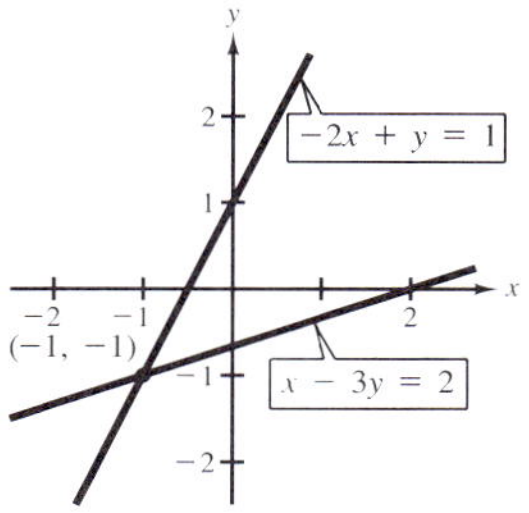

39. (10, 0)

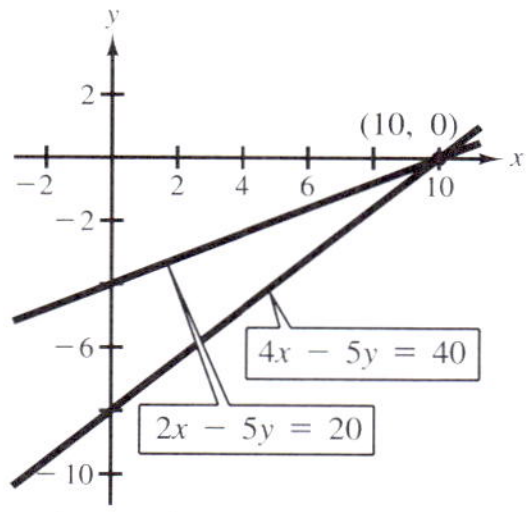

41. (9, 12)

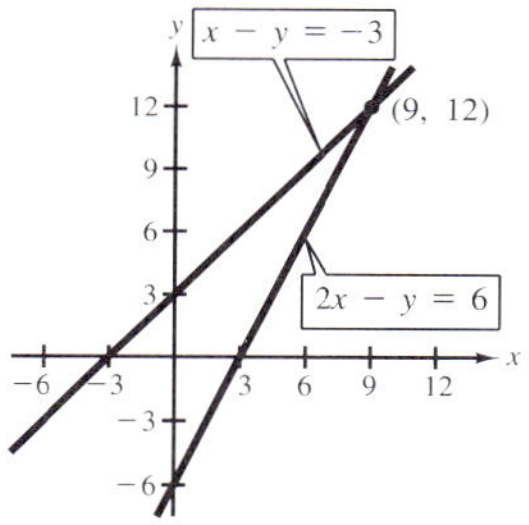

43. $\left(-\frac{3}{2}, \frac{5}{2}\right)$

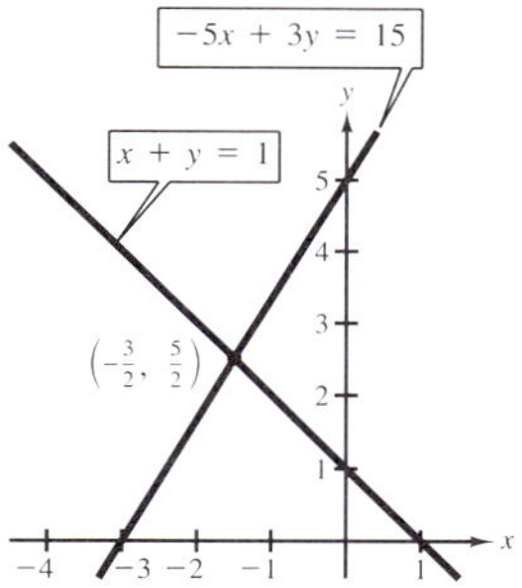

45. (4, 2), (9, 3)

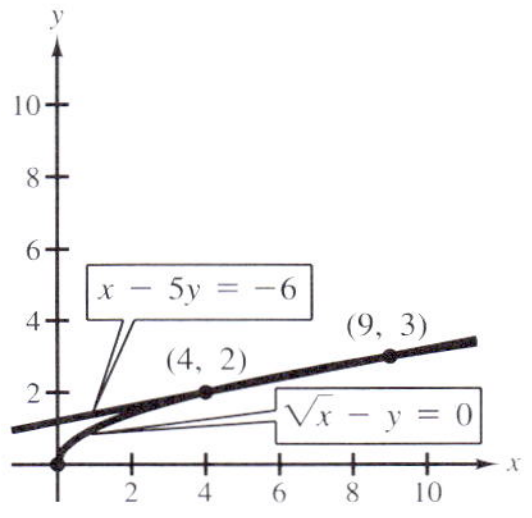

47. (4, 2), (−4, −2)

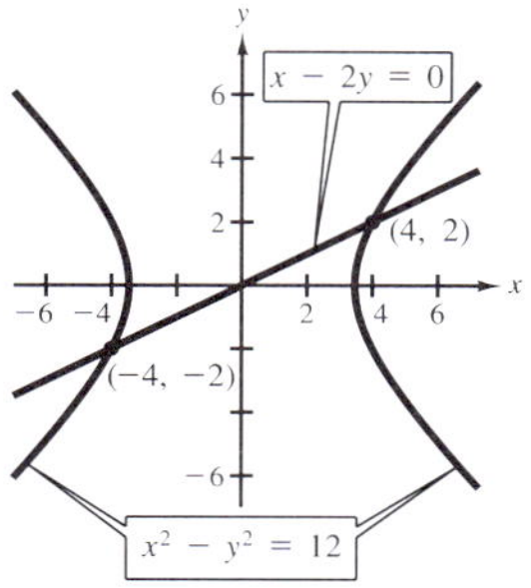

49. (0, 4), (3, 0)

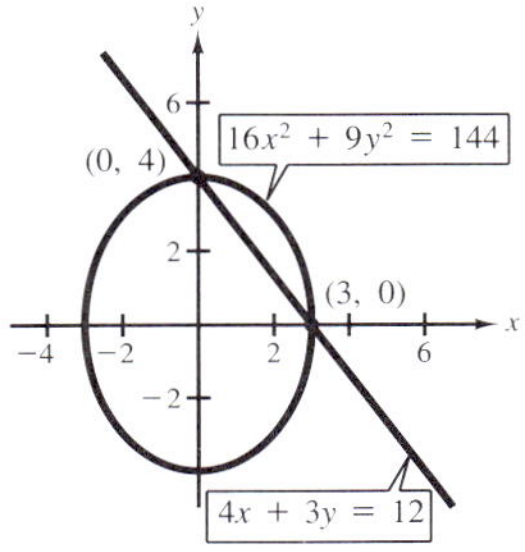

51. $\left(2{,}992, \frac{798}{25}\right)$ **53.** 31, 49 **55.** 20, 32
57. 10 × 15 ft **59.** 14 × 20 yd
61. \$15,000 at 8%
\$5,000 at 9.5% **63.** 10,000 units **65.** 63

67. $\begin{aligned} x - 2y &= -6 \\ -x + y &= 1 \end{aligned}$ **69.** $\begin{aligned} 3x - y &= -1 \\ -2x + y &= 0 \end{aligned}$

71. (0, 0), (1, 1)

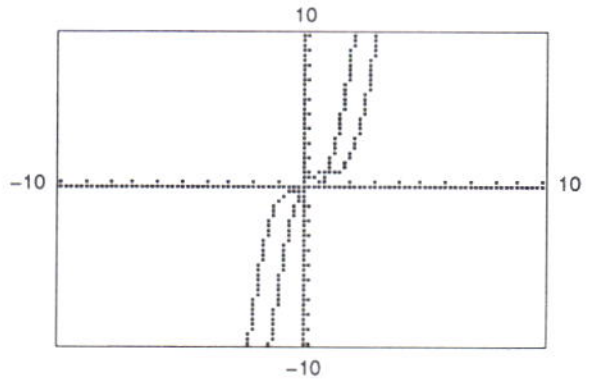

SECTION 9.2 *(page 555)*

Warm-Up *(page 554)*

1. $-11x$ **2.** $-41v$ **3.** $6x^2 + 9$ **4.** −4
5. 4 **6.** 11 **7.** Perpendicular **8.** Parallel
9. Parallel **10.** Neither

1. (5, 3)

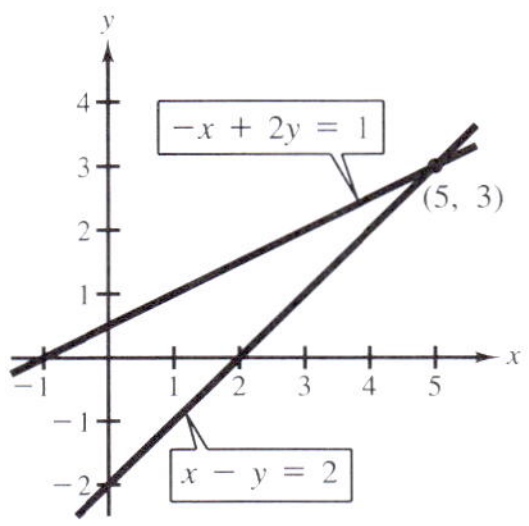

3. (2, −2)

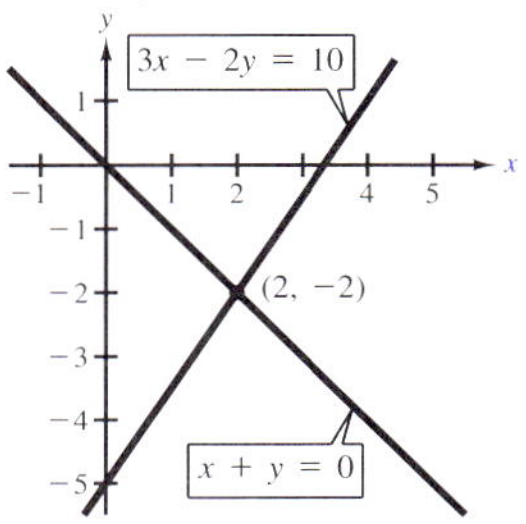

5. No solution

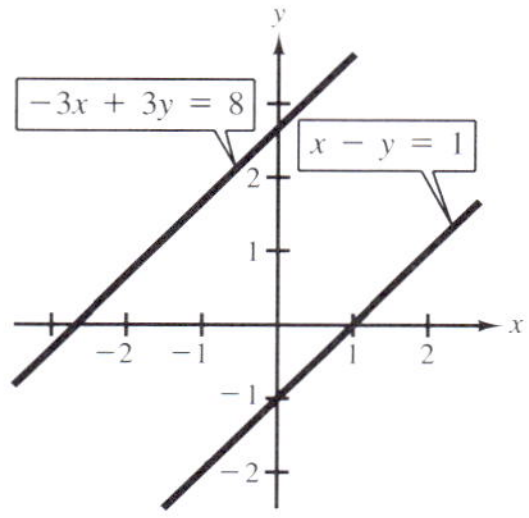

7. Infinite number of solutions

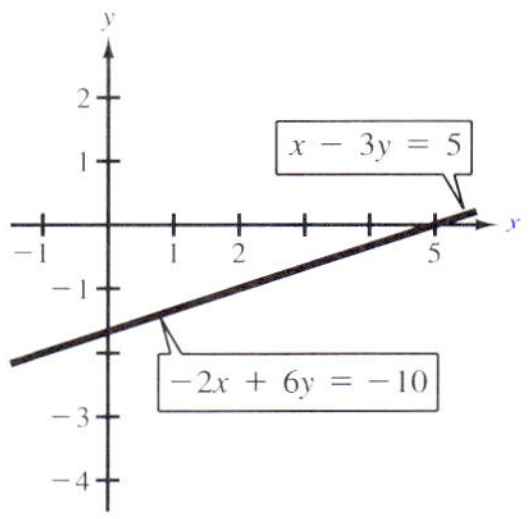

9. $\left(\frac{1}{2}, \frac{3}{2}\right)$

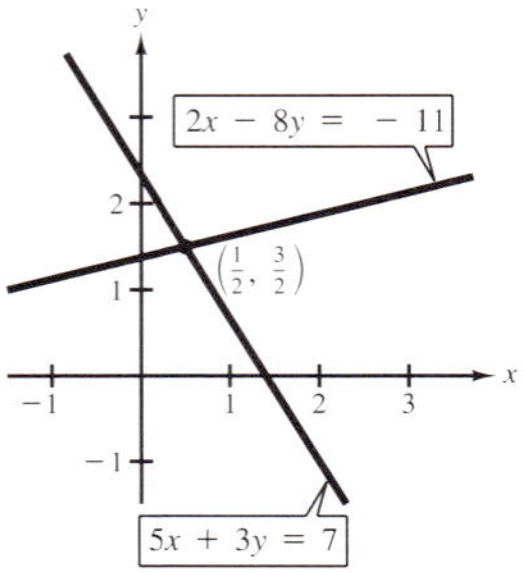

11. (3, 2) **13.** (−2, 5) **15.** (−1, −1)
17. (7, −2) **19.** Inconsistent **21.** $\left(\frac{3}{2}, 1\right)$
23. (−2, −1) **25.** Infinite number of solutions
27. $\left(\frac{25}{2}, \frac{1,237}{250}\right)$ **29.** (3,000, −2,000) **31.** (4, 3)
33. (2, 7) **35.** (9, −2) **37.** $\left(\frac{33}{4}, \frac{25}{2}\right)$

39. $x + 2y = 0$
$4x + 2y = 9$

41. $k = 4$ **43.** Many solutions **45.** One solution
47. No solution **49.** 50 × 60 m
51. 5 dimes, 10 quarters **53.** 11 nickels, 14 dimes
55. Regular unleaded: $1.11
Premium unleaded: $1.22
57. $5.65 variety: 6.1 lb
$8.95 variety: 3.9 lb
59. 40% solution: 12 liters
65% solution: 8 liters
61. 6.4 hr
63. 32 in., 128 in.
65. **(a)** $y = \frac{3}{2}x - \frac{1}{6}$
(b)

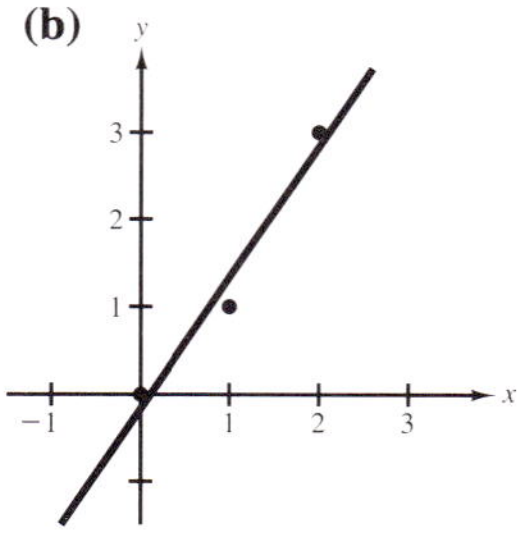

67. (3, 5)

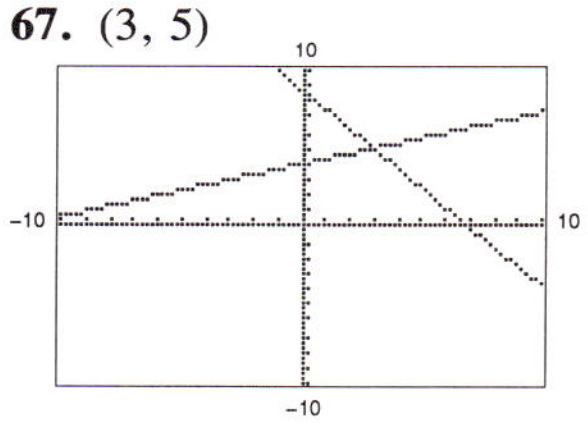

SECTION 9.3 *(page 569)*

Warm-Up *(page 568)*

1. (5, 4) **2.** (−1, 1) **3.** Inconsistent
4. Many solutions **5.** (3, −2) **6.** (−2, −2)
7. $5x - z$ **8.** $34y - 5x$ **9.** $35z - 17y$
10. $7x + 26z$

1. (22, −1, −5) **3.** (14, 3, −1) **5.** (1, 2, 3)
7. (2, −3, −2) **9.** No solution **11.** (−4, 8, 5)
13. $(3a + 1, 3a - 1, a)$ **15.** (1, 0, −2)
17. (−4, 2, 3) **19.** (1, −1, 2)
21. $\left(\frac{6}{13}a + \frac{10}{13}, \frac{5}{13}a + \frac{4}{13}, a\right)$, where a is any real number
23. $\left(\frac{7}{5}a + 1, \frac{1}{5}a - 1, a\right)$ where a is any real number
25. $\left(\frac{1}{4} - \frac{1}{2}a, \frac{1}{2}a + \frac{5}{4}, a\right)$, where a is any real number
27. $y = 2x^2 - 3$ **29.** $y = 2x - x^2$
31. $x^2 + y^2 - 4x = 0$ **33.** $x^2 + y^2 - 6x - 8y = 0$
35. $s = -16t^2 + 144$ **37.** $s = -16t^2 + 48t$
39. $x + 2y - z = -4$
$y + 2z = 1$
$3x + y + 3z = 15$

MATH MATTERS: CIRCLE PROBLEM OF APOLLONIUS *(page 565)*

1.

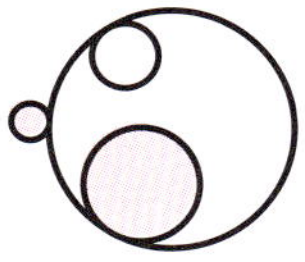

2.

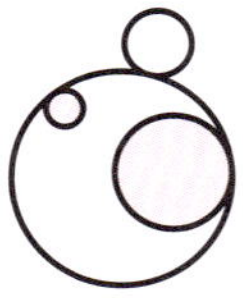

3.

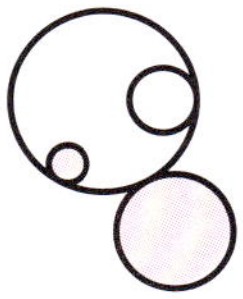

4.

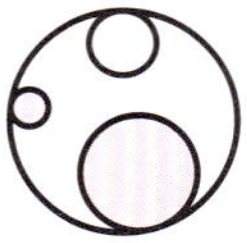

SECTION 9.4 *(page 581)*

Warm-Up *(page 580)*

1. (2, 4), $\left(-\frac{1}{2}, \frac{1}{4}\right)$ **2.** (0, 0), (8, 2) **3.** (5, 4)
4. $\left(\frac{1}{2}, \frac{1}{3}\right)$ **5.** (1, 2, 1) **6.** (4, −1, 3) **7.** 9
8. $\frac{14}{5}$ **9.** $-\frac{19}{4}$ **10.** $-\frac{11}{6}$

1. 4 × 2 **3.** 3 × 4 **5.** $\begin{bmatrix} 1 & 4 & 3 \\ 0 & 2 & -1 \end{bmatrix}$

7. $\begin{bmatrix} 1 & 1 & 4 & -1 \\ 0 & 5 & -2 & 6 \\ 0 & 3 & 20 & 4 \end{bmatrix}$ **9.** $\begin{bmatrix} 1 & 2 & 3 \\ 0 & 1 & 2 \end{bmatrix}$

$\begin{bmatrix} 1 & 1 & 4 & -1 \\ 0 & 1 & -\frac{2}{5} & \frac{6}{5} \\ 0 & 3 & 20 & 4 \end{bmatrix}$

11. $\begin{bmatrix} 1 & \frac{3}{2} & \frac{1}{4} \\ 0 & 1 & \frac{11}{10} \end{bmatrix}$ **13.** $\begin{bmatrix} 1 & 1 & 0 & 5 \\ 0 & 1 & 2 & 0 \\ 0 & 0 & 1 & -1 \end{bmatrix}$

15. $\begin{bmatrix} 1 & -1 & -1 & 1 \\ 0 & 1 & 6 & 3 \\ 0 & 0 & 1 & \frac{4}{5} \end{bmatrix}$ **17.** $\left(\frac{14}{5}, -\frac{7}{5}\right)$ **19.** $\left(\frac{1}{2}, -\frac{3}{4}\right)$

21. Inconsistent **23.** $(4, -3, 2)$ **25.** $(1, 2, -1)$
27. $(2a + 1, 3a + 2, a)$ **29.** Inconsistent
31. 8%: \$800,000 **33.** $y = \frac{1}{2}x^2 - 3x + \frac{7}{2}$
9%: \$500,000
12%: \$200,000
35. 5, 8, 20

SECTION 9.5 *(page 594)*

Warm-Up *(page 593)*

1. 10 **2.** -56 **3.** -82 **4.** 153 **5.** 32
6. $-\frac{65}{3}$ **7.** -2 **8.** $\frac{9}{11}$ **9.** $\frac{6}{5}$ **10.** $-\frac{11}{3}$

1. 5 **3.** 27 **5.** 0 **7.** 6 **9.** -24 **11.** -2
13. -30 **15.** 3 **17.** 0 **19.** -75 **21.** -58
23. -0.22 **25.** $x - 5y + 2$ **27.** $(1, 2)$
29. $(2, -2)$ **31.** $\left(\frac{3}{4}, -\frac{1}{2}\right)$
33. Determinant of the coefficient matrix is zero.
35. $\left(\frac{2}{3}, \frac{1}{2}\right)$ **37.** $(-1, 3, 2)$ **39.** $\left(1, \frac{1}{2}, \frac{3}{2}\right)$
41. $(1, -2, 1)$
43. Determinant of the coefficient matrix is zero. **45.** 7
47. $\frac{31}{2}$ **49.** Collinear **51.** Not collinear
53. $3x - 5y = 0$ **55.** $7x - 6y - 28 = 0$
57. $y = 2x^2 - 6x + 1$ **59.** $y = -3x^2 + 2x$
61. 248 **63.** $\left(\frac{51}{16}, -\frac{7}{16}, -\frac{13}{16}\right)$
65. **(a)** $\left(\frac{4k - 3}{2k - 1}, \frac{4k - 1}{2k - 1}\right)$
(b) $\frac{1}{2}$

SECTION 9.6 *(page 601)*

Warm-Up *(page 600)*

1. Circle **2.** Parabola **3.** Line **4.** Line
5. Hyperbola **6.** Ellipse

7.
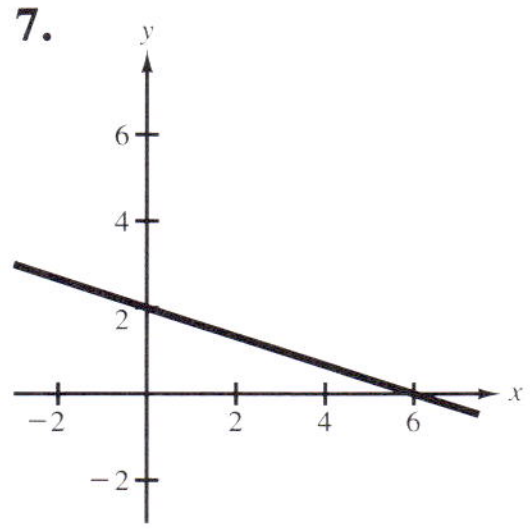

8.
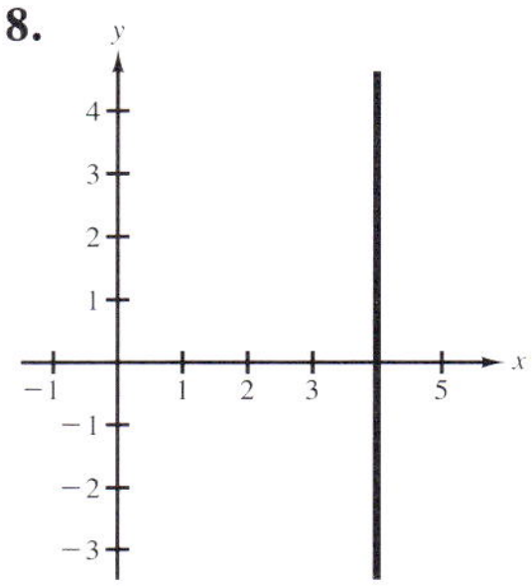

9.
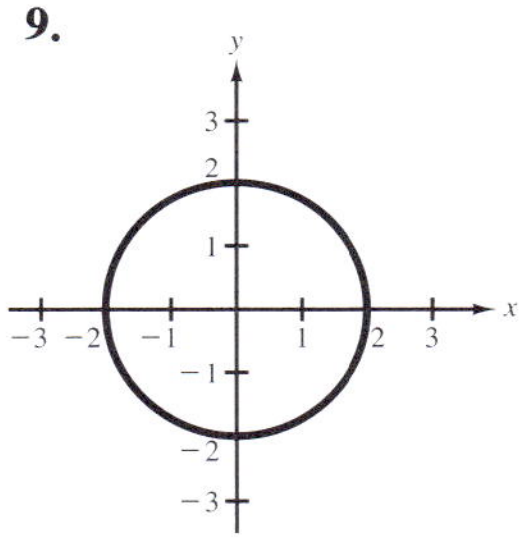

10.
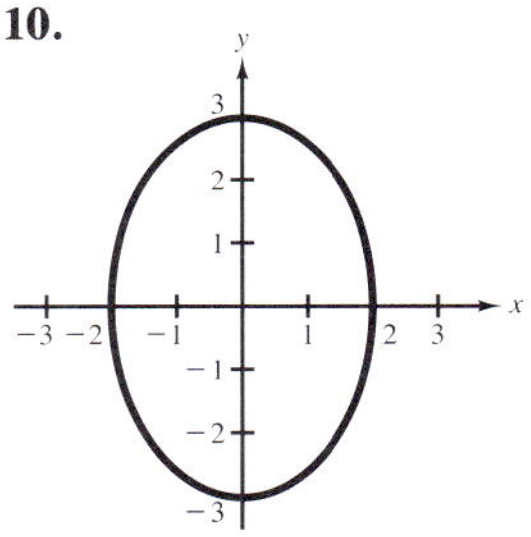

1. (b) **2.** (a) **3.** (d) **4.** (e) **5.** (f) **6.** (c)
7. **(a)** Yes **(b)** No **(c)** Yes **(d)** Yes
9. **(a)** No **(b)** No **(c)** Yes **(d)** Yes
11. **(a)** Yes **(b)** No **(c)** No **(d)** Yes
13. **(a)** No **(b)** Yes **(c)** No **(d)** Yes

15.
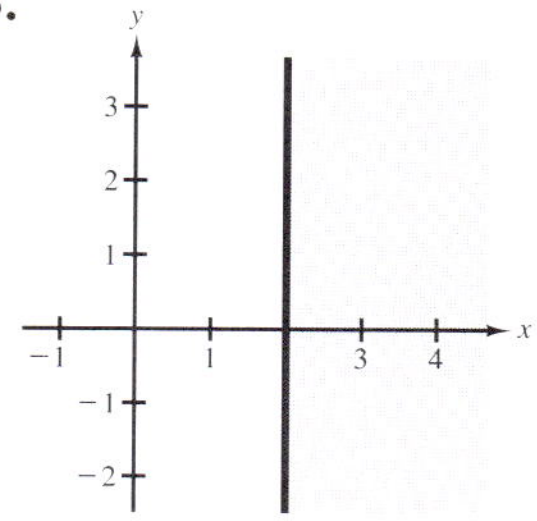

17.
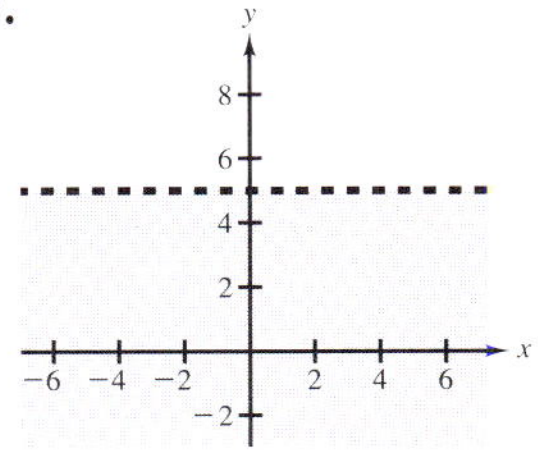

19.

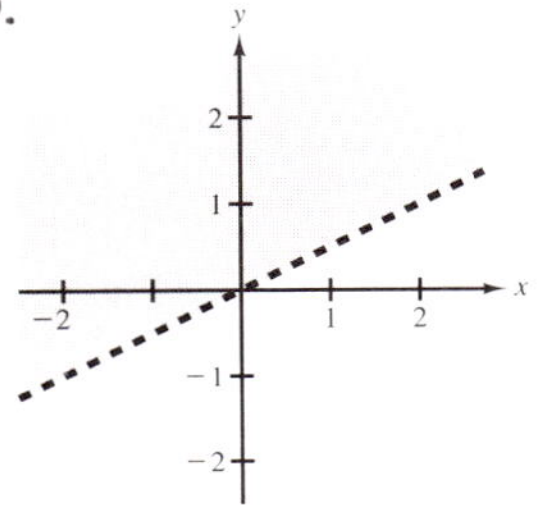

21.

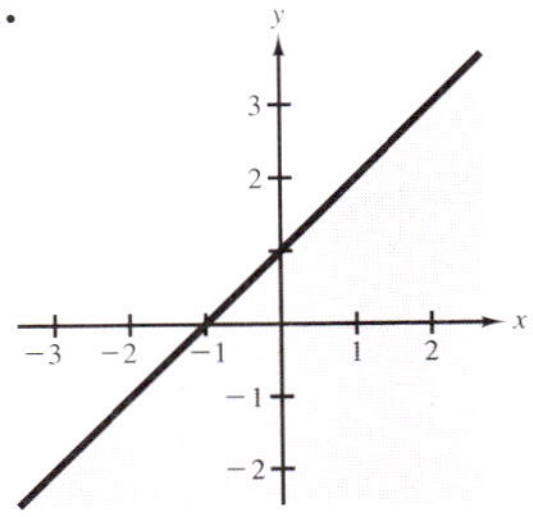

23.

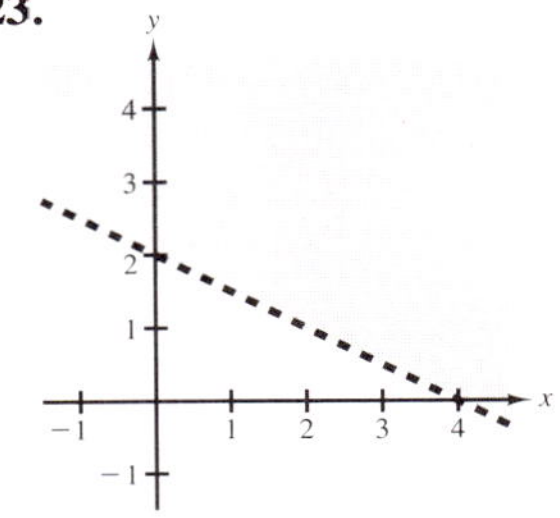

25.

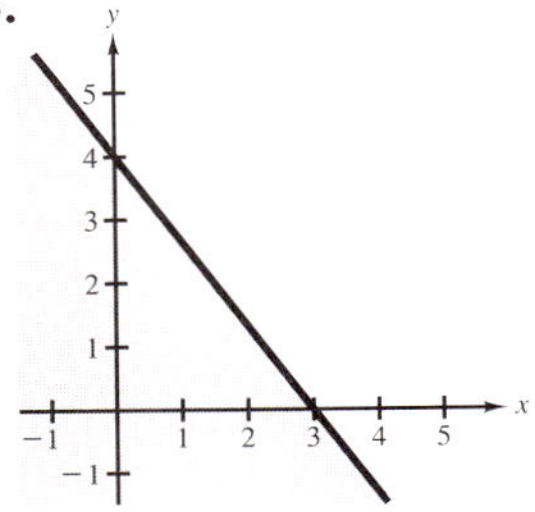

27.

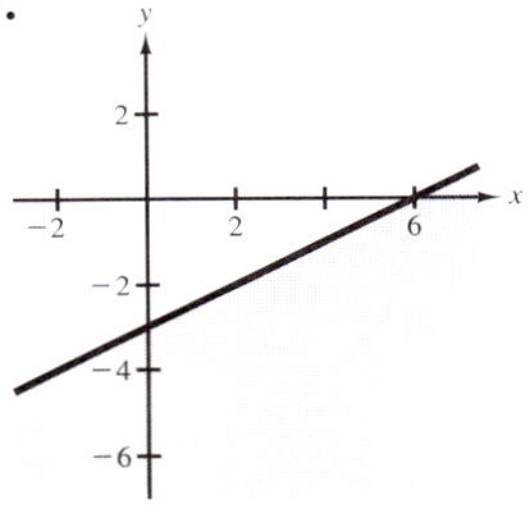

29.

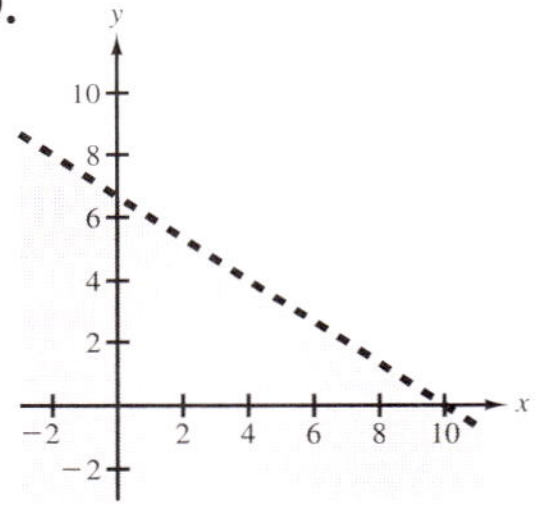

31. $3x + 4y - 17 > 0$ **33.** $y < 2$ **35.** $x - 2y < 0$

37. $2x + 2y \le 500$

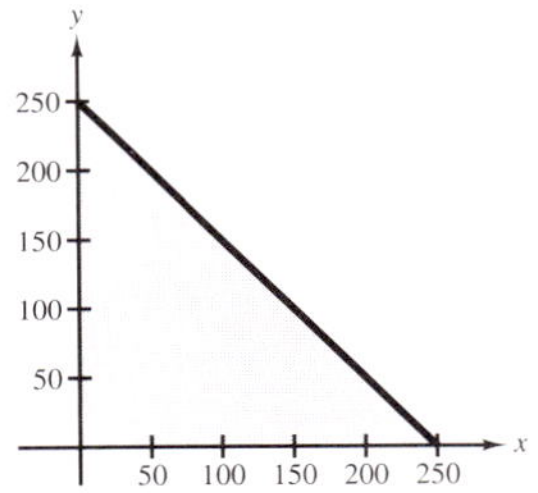

CHAPTER 9 REVIEW EXERCISES *(page 605)*

1. $(2, -1)$ **3.** Inconsistent **5.** $(-10, -5)$

7. $(1, 1)$ **9.** Inconsistent

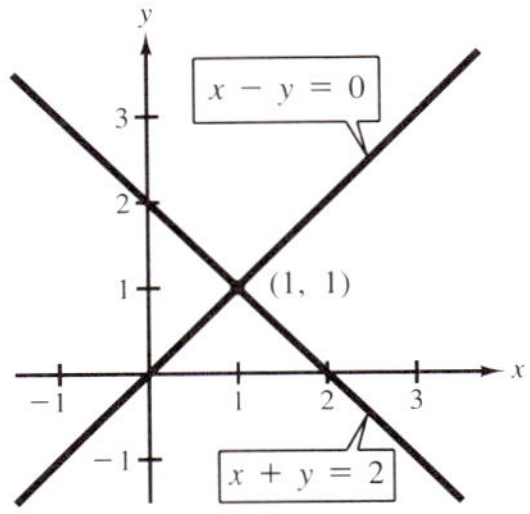

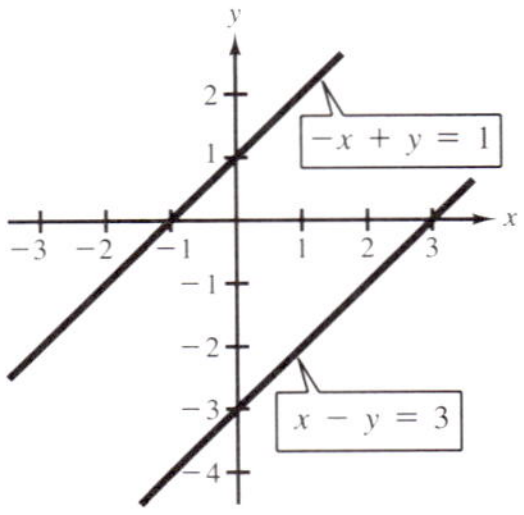

11. $(4, 8)$

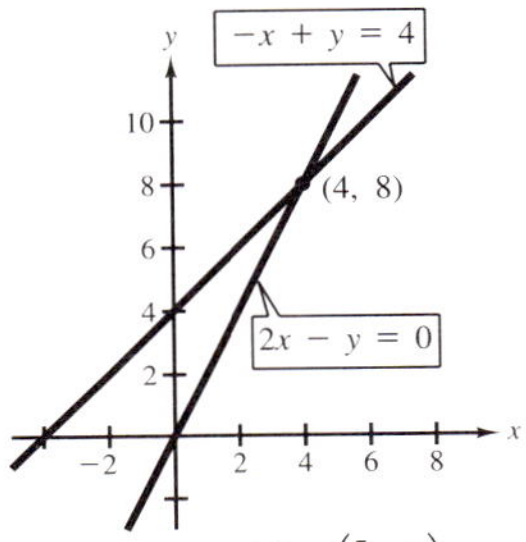

13. $(0, 0)$ **15.** $\left(\frac{5}{2}, 3\right)$ **17.** $(2, -3, 3)$

19. $(10, -12)$ **21.** $\left(\frac{3}{5}, \frac{1}{2}\right)$ **23.** $\left(\frac{24}{5}, \frac{22}{5}, -\frac{8}{5}\right)$

25. $(-3, 7)$ **27.** Inconsistent **29.** $(2, -3, 3)$

31.

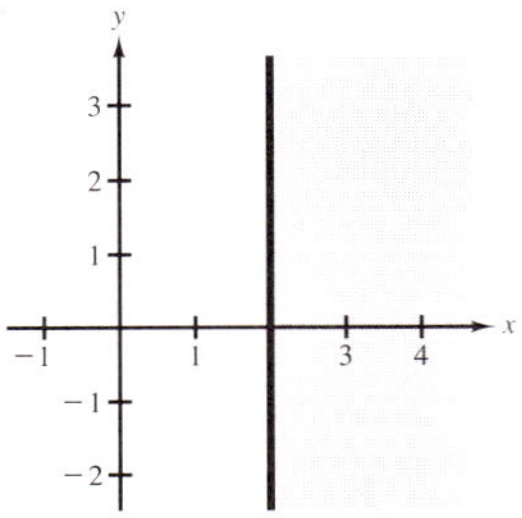

33.

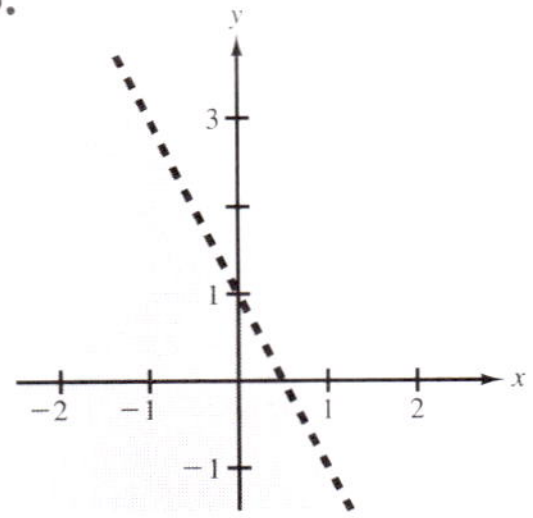

35.

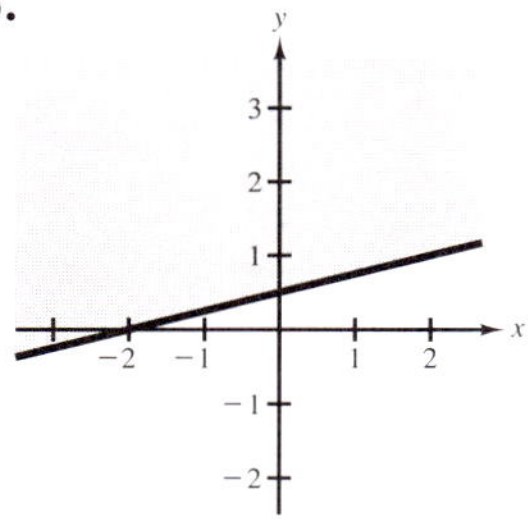

37. $x - 2y + 4 = 0$ **39.** $2x + 6y - 13 = 0$
41. 16

43. 7 **45.** $3x + y = -2$
$6x + y = 0$ **47.** 16,667 units

49. 75% solution: 40 gal
50% solution: 60 gal **51.** $y = x^2 + 2x + 3$

CHAPTER 9 TEST *(page 607)*

1. (b) **2.** (2, 4) **3.** (2, 6), (6, 2) **4.** (3, 2)
5. $(-2, 2)$ **6.** $\left(\frac{1}{4}, \frac{1}{3}\right)$ **7.** $(2, -1, 0)$
8. $(-1, 3, 3)$ **9.** $\left(4, \frac{1}{7}\right)$ **10.** -62
11.

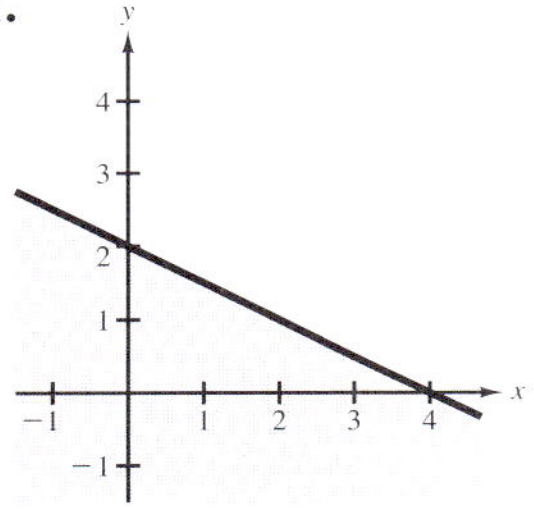

12. $-\frac{24}{5}$ **13.** $x + 2y = -1$
$x + y = 2$ **14.** $x + y = 200$
$4x - y = 0$
40 mi, 60 mi

15. $y = 2x^2 - 3x + 4$ **16.** 12

Chapter 10

SECTION 10.1 *(page 617)*

Warm-Up *(page 617)*

1. x^8 **2.** 1 **3.** z^3 **4.** $\frac{3}{4}(x - 3)^4$ **5.** x^8y^4
6. $\dfrac{x^5}{32y^5}$ **7.** $\dfrac{x^4}{y^3}$ **8.** $\dfrac{1}{a^8}$ **9.** b^3 **10.** $2x$

1. **(a)** $\frac{1}{9}$ **(b)** 1 **(c)** 3
3. **(a)** 9 **(b)** 1 **(c)** $\frac{1}{3}$
5. **(a)** 500 **(b)** 250 **(c)** 56.657
7. **(a)** 1,000 **(b)** 1,628.895 **(c)** 2,653.298
9. **(a)** 0.368 **(b)** 1 **(c)** 1.649
11. (b) **12.** (a)
13. (e) **14.** (d) **15.** (f) **16.** (c)

17.

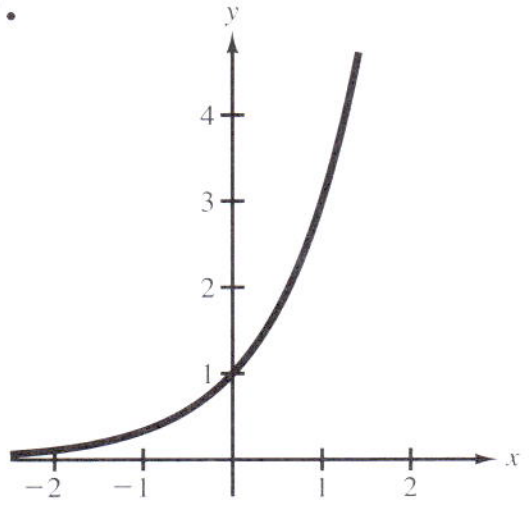

19.

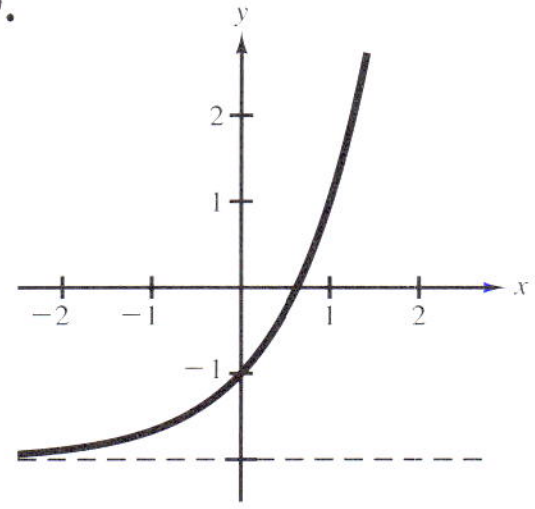

21.

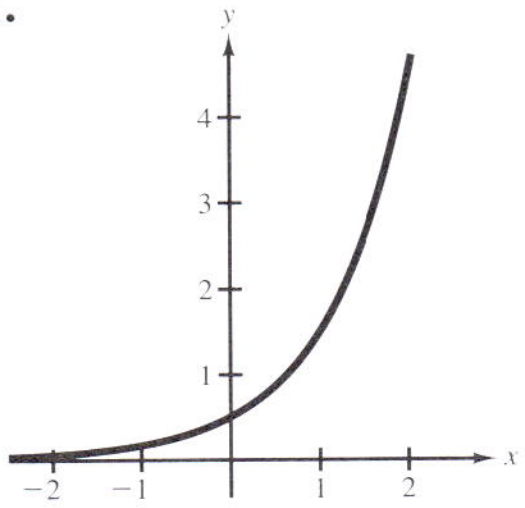

23.

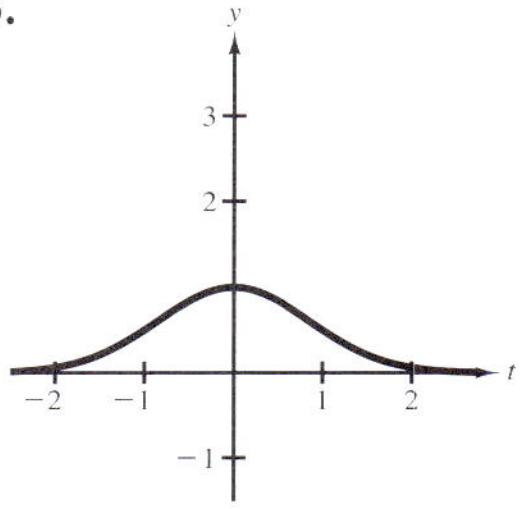

25.

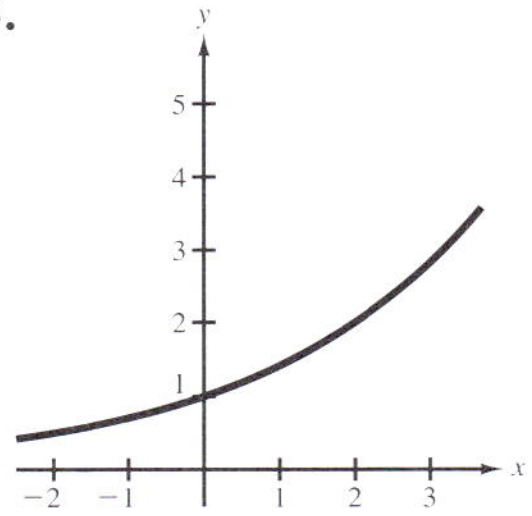

27.

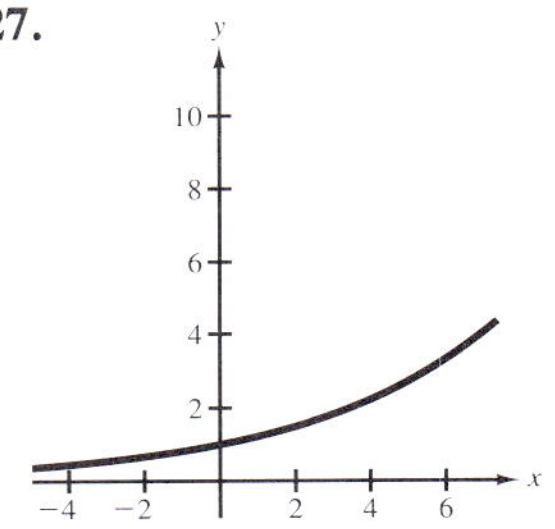

29.

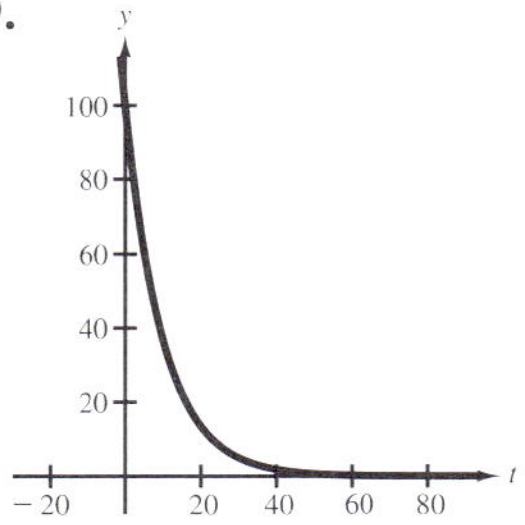

31. **(a)** 272.184 million
(b) 288.627 million

33. \$9,000

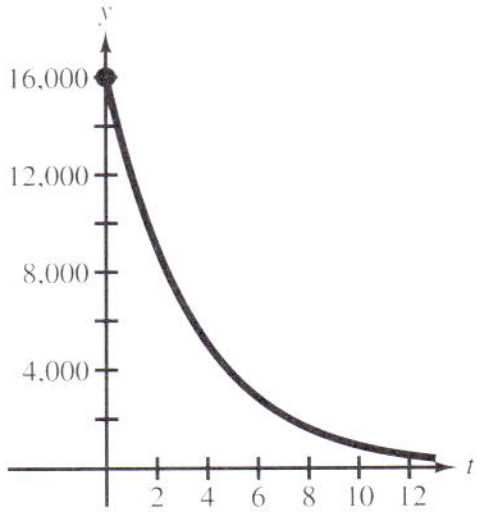

35. \$32.50

37.

n	A
1	\$466.10
4	487.54
12	492.68
365	495.22
Continuous Compounding	495.30

39.

n	A
1	\$226,296.28
4	259,889.34
12	268,503.32
365	272,841.22
Continuous Compounding	272,990.75

41.

n	A
1	\$152,203.13
4	167,212.90
12	170,948.62
365	172,813.72
Continuous Compounding	172,877.82

43.

n	P
1	\$2,541.75
4	2,498.00
12	2,487.98
365	2,483.09
Continuous Compounding	2,482.93

45.

n	P
1	\$18,429.30
4	15,830.43
12	15,272.04
365	15,004.64
Continuous Compounding	14,995.58

47.

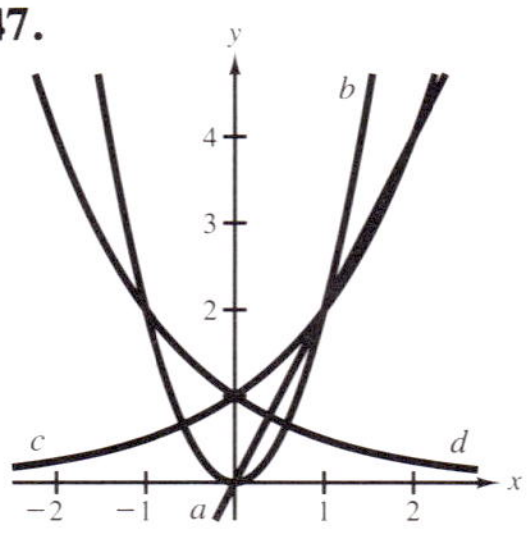

49.

t	h
0	2,000 ft
25	1,450 ft
50	950 ft
75	450 ft

Ground level: 97.5 sec

51.

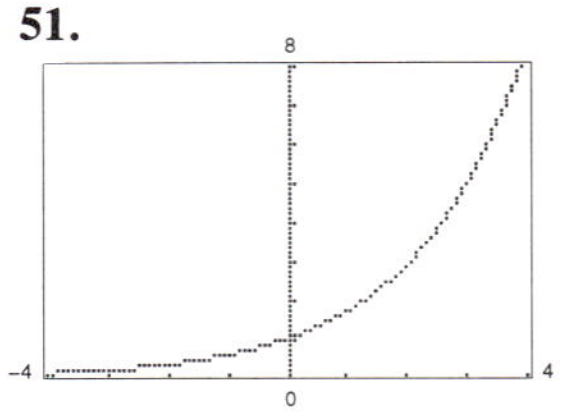

53.

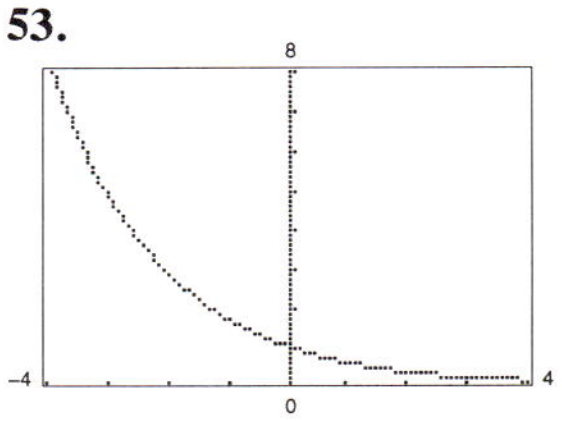

55.

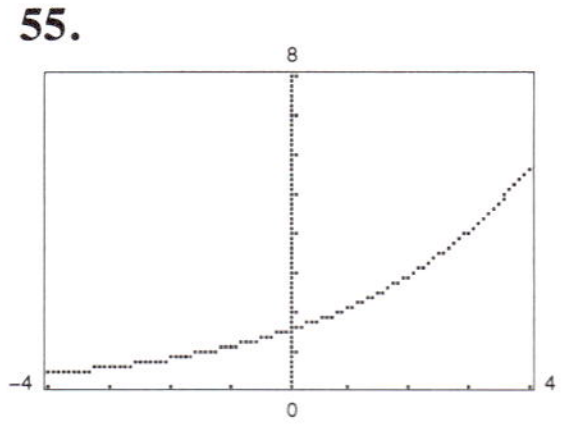

57.

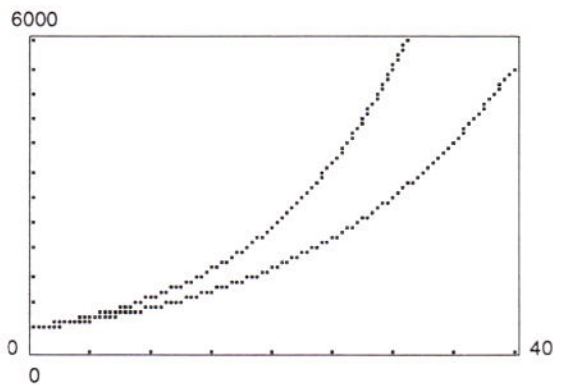

SECTION 10.2 *(page 630)*

Warm-Up *(page 629)*

1. $\frac{27}{64}$ **2.** 900 **3.** 8.825 **4.** 4.729
5. 0.368 **6.** 1.649
7. Vertical translation 4 units downward
8. Horizontal translation 4 units to the right
9. Reflection in the x-axis
10. Reflection in the y-axis

1. $5^2 = 25$ **3.** $4^{-2} = \frac{1}{16}$ **5.** $36^{1/2} = 6$
7. $\log_7 49 = 2$ **9.** $\log_3 \frac{1}{9} = -2$ **11.** $\log_8 4 = \frac{2}{3}$
13. 3 **15.** -2 **17.** 0
19. There is no power to which 2 can be raised to obtain -3.
21. 1 **23.** 3 **25.** $\frac{1}{2}$ **27.** $\frac{3}{4}$ **29.** 4
31. 1.4914 **33.** 0.7335 **35.** 3.2189
37. -0.2877 **39.** 0.0757
41. f is the inverse of g.

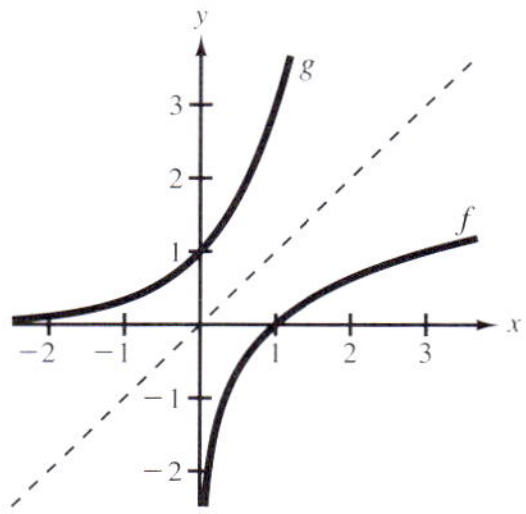

43. (e) **44.** (b) **45.** (d) **46.** (c) **47.** (a)
48. (f)
49.

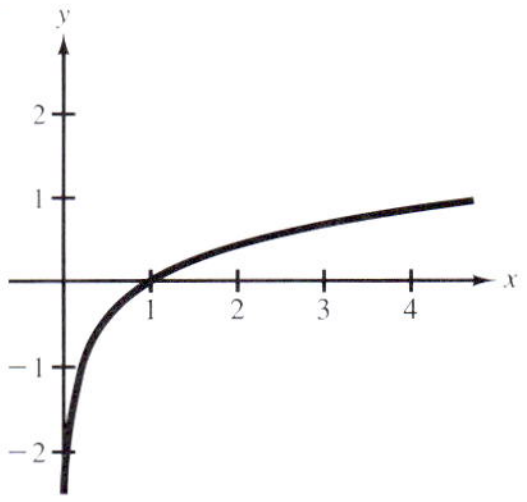

51.

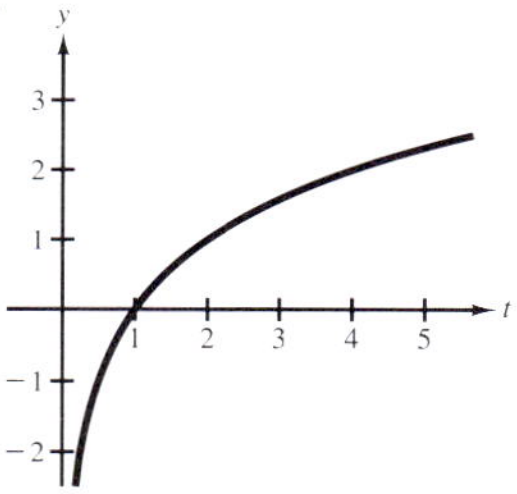

53.

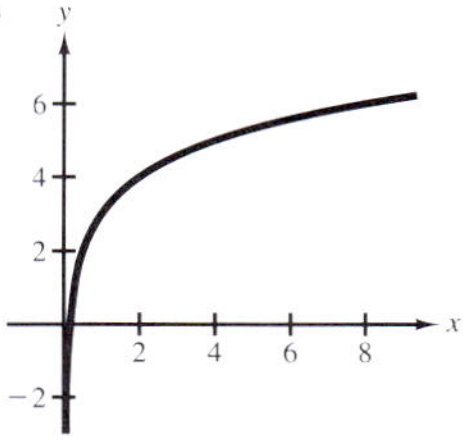

55.

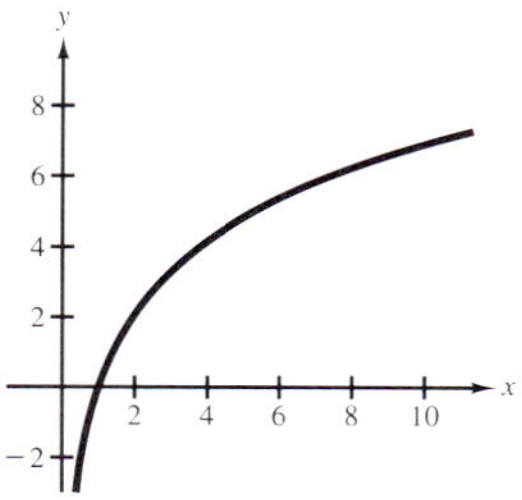

57.

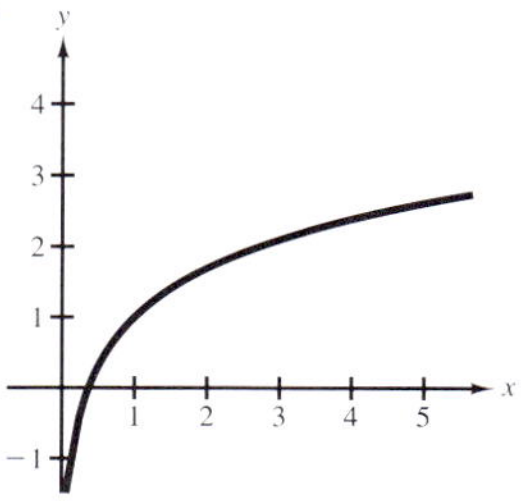

59. 2.3481 **61.** -0.4739 **63.** 1.3481
65.

r	0.07	0.08	0.09	0.10	0.11	0.12
t	9.90	8.66	7.70	6.93	6.30	5.78

67. $0 < x < \infty$ **69.** $3 \le f(x) \le 4$
71. A factor of 10
73.

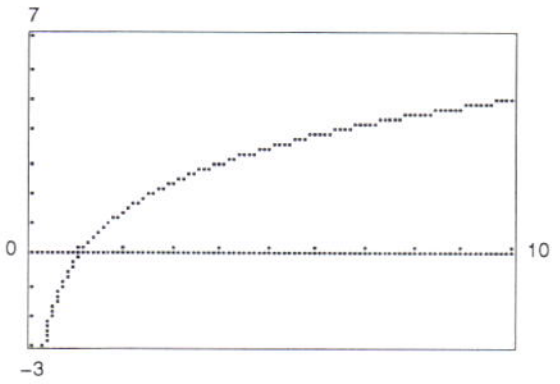

75.

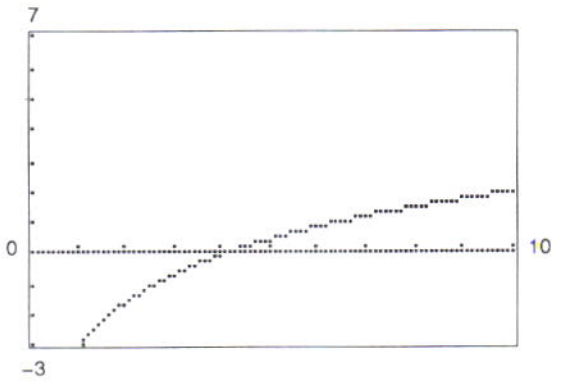

77.

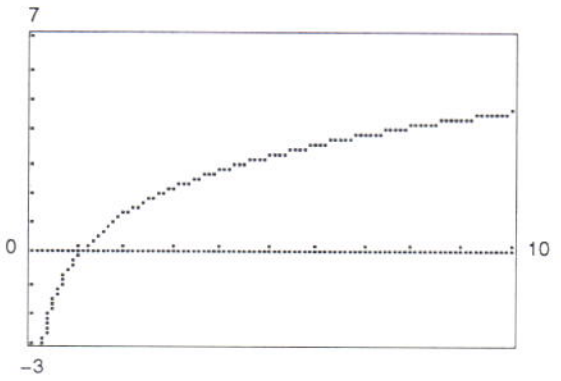

SECTION 10.3 *(page 638)*

Warm-Up *(page 638)*

1. x^{20} **2.** $\dfrac{x^2}{16}$ **3.** $\dfrac{3}{2y^5}$ **4** $\dfrac{4y^2}{9x^4}$ **5.** $\frac{1}{6}x^3y$
6. 1 **7.** $4^3 = 64$ **8.** $3^{-4} = \frac{1}{81}$
9. $e^{2x} = e^{2x}$ **10.** $e^{1.6094\ldots} = 5$

1. 2 **3.** 1 **5.** -3 **7.** $\log_3 11 + \log_3 x$
9. $\log_2 z - \log_2 17$ **11.** $-2 \log_5 x$
13. $\frac{1}{3} \log_3 (x + 1)$ **15.** $\ln 3 + 2 \ln x + \ln y$
17. $2 \log_2 x - \log_2 (x - 3)$ **19.** $\frac{1}{3}[\ln x + \ln(x + 5)]$
21. $\log_2 3x$ **23.** $\log_{10} \dfrac{4}{x}$ **25.** $\ln b^4$
27. $\log_5 (2x)^{-2}$ **29.** $\ln \sqrt[3]{2x + 1}$ **31.** $\ln \dfrac{x^2y^3}{z}$
33. $\ln(xy)^4$ **35.** True **37.** False **39.** False
41. 1 **43.** 1.2925 **45.** 0.2925 **47.** 0.2500
49. 2.7925 **51.** $1 - \log_4 x$ **53.** $1 + \frac{1}{2} \log_5 2$
55. $2 + \ln 3$

57. ln 4 ≈ 1.3862
ln 6 ≈ 1.7917
ln 8 ≈ 2.0793
ln 9 ≈ 2.1972
ln 10 ≈ 2.3025
ln 12 ≈ 2.4848
ln 15 ≈ 2.7080
ln 16 ≈ 2.7724
ln 18 ≈ 2.8903
ln 20 ≈ 2.9956

59. $10[\log_{10} I + 16]$, 60 decibels

61.

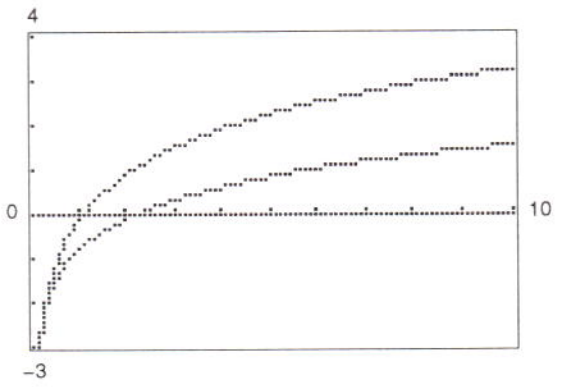

SECTION 10.4 *(page 646)*

Warm-Up *(page 646)*

1. $\frac{5}{2}$ **2.** $-\frac{7}{2}$ **3.** 10, −2 **4.** 2, 3 **5.** $\frac{28}{\ln 5}$
6. 0 **7.** 3 **8.** 6 **9.** 0 **10.** 4

1. 5 **3.** 8 **5.** 3 **7.** −3 **9.** 4 **11.** $\frac{22}{5}$
13. 81 **15.** 500,000 **17.** $2x - 1$ **19.** $2x$
21. 5.49 **23.** 0.86 **25.** 3.28 **27.** 2.49
29. 0.39 **31.** 6.80 **33.** 0.80 **35.** 35.35
37. 21.82 **39.** 8.99 **41.** 1 **43.** 1,000
45. 7.91 **47.** 8.17 **49.** −0.78 **51.** ≈ 31,622,772
53. 20.09 **55.** −2, 5 **57.** 7.70 yrs **59.** $10^{-8.5}$
61. 262.5°

SECTION 10.5 *(page 656)*

Warm-Up *(page 656)*

1. $\frac{33}{4}$ **2.** $\frac{3}{5}$, −1 **3.** 41 **4.** $\frac{11}{2}$ **5.** 18
6. 69 **7.** 4.16 **8.** 11.31 **9.** 139.32
10. −1.39

1. 7% **3.** 9% **5.** 8% **7.** 8.75 yrs
9. 6.60 yrs **11.** 9.24 yrs **13.** Continuous
15. Quarterly **17.** 8.33% **19.** 7.23% **21.** 5.39%
23. $1,652.99 **25.** $626.46 **27.** $5,496.57
29. $320,250.81 **31.** Total deposits: $7,200.00
Total interest: $10,529.42

33. $k = \frac{1}{2}\ln\frac{8}{3} \approx 0.4904$ **35.** $k = -\frac{\ln 2}{3} \approx -0.2310$

37. $y = 9.6e^{0.0072t}$; 11.1 million
39. $y = 3.3e^{0.0452t}$; 8.1 million **41.** 3.2595 gr
43. 15,683 yrs **45.** (a) $S = 10(1 - e^{-0.0575x})$
(b) 3,314 units
47. 63 times as great **49.** 7.04 **51.** 10^7 times
53. (a)

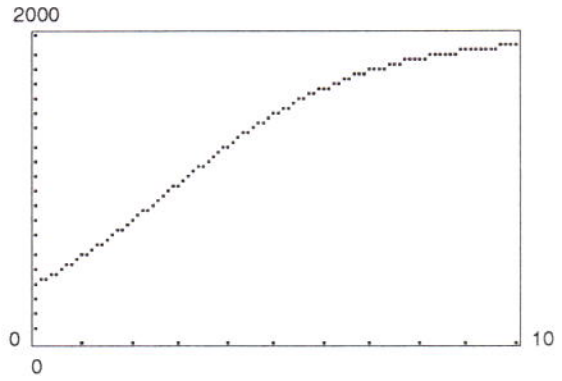

(b) 1,298 units
(c) 3.2 yrs
(d) 2,000 units

MATH MATTERS: ANIMAL SPECIES *(page 651)*

79.5%

CHAPTER 10 REVIEW EXERCISES *(page 662)*

1. (a) $\frac{1}{8}$ (b) 2 (c) 4
3. (a) 2.718 (b) 0.351 (c) 0.135
5. (a) 0 (b) 3 (c) −0.631
7. (a) 1 (b) −1.099 (c) 2.303
9. (a) −6 (b) 0 (c) 22.5

11.

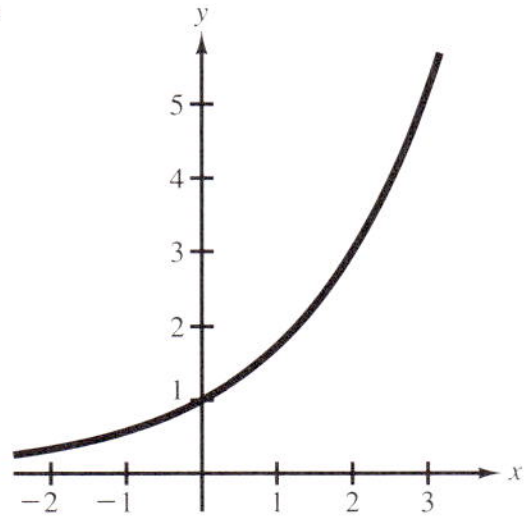

13.

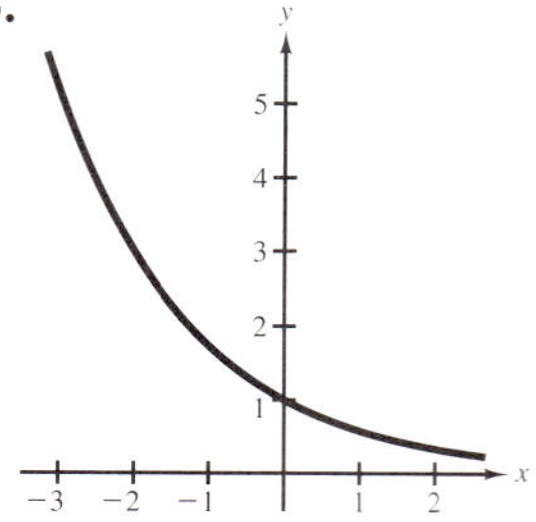

15.

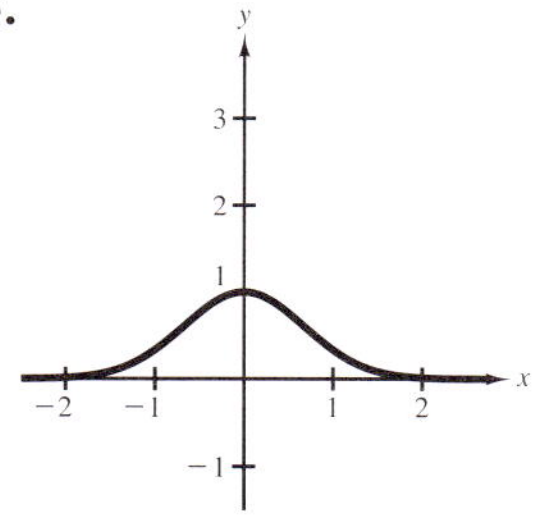

17.

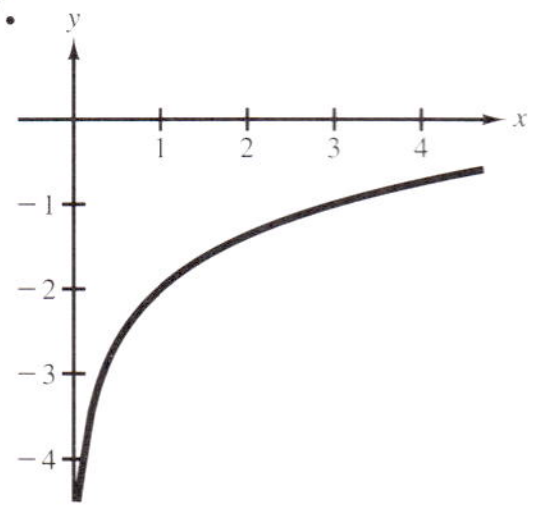

19.

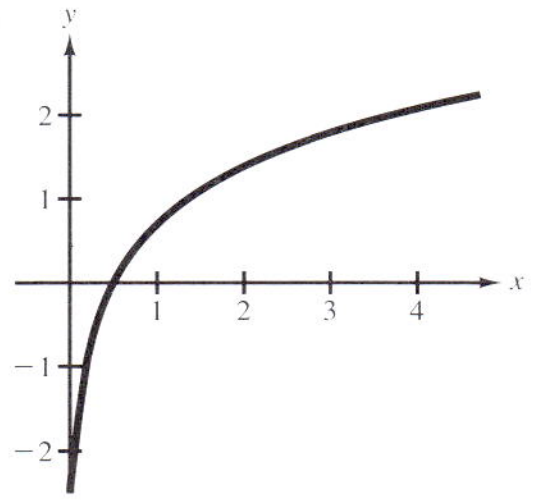

21. $\log_4 6 + 4\log_4 x$ **23.** $\frac{1}{2}\log_5 (x + 2)$

25. $\ln(x + 2) - \ln(x - 2)$ **27.** $\log_4 \dfrac{x}{10}$

29. $\log_8 32x^3$ **31.** $\ln\left(\dfrac{3}{2x}\right)^2$ **33.** True **35.** False

37. True **39.** 1.79588 **41.** −0.43068

43. 1.02931 **45.** 6 **47.** 1 **49.** 243 **51.** 5.66

53. 6.23 **55.** 32.99 **57.** 1,408.10 **59.** 0.32

61.

n	A
1	\$3,806.13
4	4,009.59
12	4,058.25
365	4,082.26
Continuous Compounding	4,083.08

63.

n	A
1	\$67,275.00
4	72,095.68
12	73,280.74
365	73,870.32
Continuous Compounding	73,890.56

65.

n	P
1	\$2,301.55
4	2,103.50
12	2,059.87
365	2,038.83
Continuous Compounding	2,038.11

67. 4.6 yrs **69.** 150 units

71.

t	0	1	5	10	20	50	75
$p(t)$	200	227	346	472	579	600	600

CHAPTER 10 TEST *(page 665)*

1. **(a)** 54
(b) 24
(c) 81
(d) $18\sqrt{6} \approx 44.09$

2.

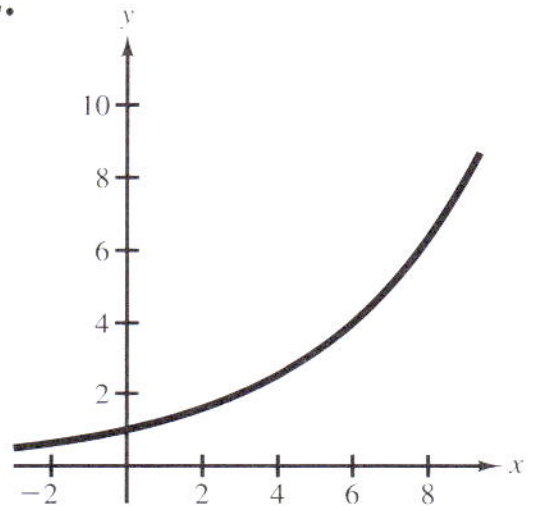

3. **(a)** \$8,012.78
(b) \$8,110.40

4. $5^3 = 125$ **5.** $\log_4 \frac{1}{16} = -2$

6. $\frac{1}{3}$

7. $g = f^{-1}$

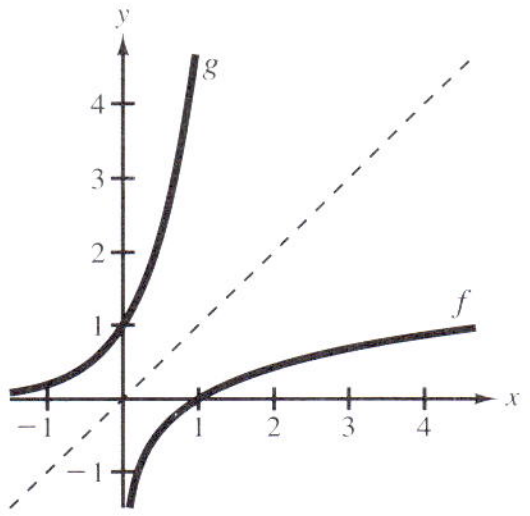

8. $\log_4 5 + 2\log_4 x - \frac{1}{2}\log_4 y$ **9.** $\ln \dfrac{x}{y^4}$

10. $3 + \log_5 6$ **11.** 64 **12.** 0.9734 **13.** 13.73

14. 15.52 **15.** \$10,806.08 **16.** 7%

17. \$8,469.14 **18.** **(a)** 600
(b) 1,141
(c) 4.4 yrs

19. **(a)** \$13,200.24
(b) 15.7 yrs

Chapter 11

SECTION 11.1 *(page 676)*

Warm-Up *(page 675)*

1. $\frac{1}{7}$ **2.** 24 **3.** $\frac{3}{4}$ **4.** $\frac{13}{12}$ **5.** n
6. $n - 1$ **7.** $\dfrac{2n-3}{n-5}$ **8.** $\dfrac{n+1}{n}$ **9.** $\dfrac{n+2}{n^2}$
10. $\dfrac{2n}{n^2-1}$

1. 2, 4, 6, 8, 10 **3.** $-2, 4, -6, 8, -10$
5. $\frac{1}{2}, \frac{1}{4}, \frac{1}{8}, \frac{1}{16}, \frac{1}{32}$ **7.** $-\frac{1}{2}, \frac{1}{4}, -\frac{1}{8}, \frac{1}{16}, -\frac{1}{32}$
9. $\frac{1}{2}, \frac{1}{3}, \frac{1}{4}, \frac{1}{5}, \frac{1}{6}$ **11.** $\frac{2}{5}, \frac{4}{8}, \frac{6}{11}, \frac{8}{14}, \frac{10}{17}$
13. $-1, \frac{1}{4}, -\frac{1}{9}, \frac{1}{16}, -\frac{1}{25}$ **15.** $2, 2, \frac{4}{3}, \frac{2}{3}, \frac{4}{15}$
17. $\frac{9}{2}, \frac{19}{4}, \frac{39}{8}, \frac{79}{16}, \frac{159}{32}$ **19.** 0, 4, 0, 4, 0 **21.** 5
23. $\frac{1}{132}$ **25.** $\dfrac{1}{n+1}$ **27.** 63 **29.** 35 **31.** $\frac{47}{60}$
33. -48 **35.** $\frac{8}{9}$ **37.** $\frac{182}{243}$ **39.** 6.5793 **41.** $\sum_{k=1}^{5} k$
43. $\sum_{j=1}^{5} 2j$ **45.** $\sum_{n=1}^{10} \dfrac{1}{2n}$ **47.** $\sum_{n=1}^{20} \dfrac{1}{n^2}$
49. $\sum_{k=0}^{9} \dfrac{(-1)^k}{3^k}$ **51.** $\sum_{n=1}^{11} \dfrac{n}{n+1}$ **53.** $\sum_{j=0}^{6} j!$ **55.** $\frac{18}{5}$
57. 0.8 **59. (a)** \$502.92
\$505.85
\$508.80
\$511.77
\$514.75
\$517.76
\$520.78
\$523.82
(b) \$2,019.37

SECTION 11.2 *(page 684)*

Warm-Up *(page 684)*

1. $\frac{49}{20}$ **2.** $\frac{455}{72}$ **3.** 120 **4.** 40 **5.** 285
6. 1,960 **7.** 77 **8.** $\frac{355}{2}$ **9.** 63 **10.** 144

1. 3 **3.** 5 **5.** -6 **7.** $\frac{2}{3}$ **9.** Arithmetic, 2
11. Arithmetic, $\frac{3}{2}$ **13.** Arithmetic, -16
15. Not arithmetic **17.** 7, 10, 13, 16, 19
19. 6, 4, 2, 0, -2 **21.** $\frac{3}{2}, 4, \frac{13}{2}, 9, \frac{23}{2}$
23. $4, \frac{15}{4}, \frac{7}{2}, \frac{13}{4}, 3$ **25.** $\frac{1}{2}n + \frac{5}{2}$ **27.** $-25n + 1{,}025$
29. $4n + 4$ **31.** $-10n + 60$
33. 500, 450, 400, 350, 300 **35.** 25, 28, 31, 34, 37
37. 4, 8, 12, 16, 20 **39.** 15.5, 15, 14.5, 14, 13.5
41. 210 **43.** 2,700 **45.** 19,500 **47.** 62,625
49. 522 **51.** 900 **53.** 12,200 **55.** 243
57. 2,850 **59.** 2,550 **61.** \$246,000 **63.** 585
65. 632 bales

SECTION 11.3 *(page 693)*

Warm-Up *(page 693)*

1. $\frac{16}{81}$ **2.** $\frac{27}{125}$ **3.** $\frac{1}{72}$ **4.** 1 **5.** n
6. $n + 1$ **7.** $60n^6$ **8.** $16n^5$ **9.** $\dfrac{n}{2}$
10. $\frac{9}{5}n^2$

1. 3 **3.** $-\frac{1}{2}$ **5.** $-\frac{3}{2}$ **7.** e **9.** Not geometric
11. Geometric, $\frac{1}{2}$ **13.** Geometric, $-\frac{2}{3}$
15. Geometric, 1.02 **17.** 4, 8, 16, 32, 64
19. $6, 2, \frac{2}{3}, \frac{2}{9}, \frac{2}{27}$ **21.** $1, -\frac{1}{2}, \frac{1}{4}, -\frac{1}{8}, \frac{1}{16}$
23. 1,000, 1,010, 1,020.1, 1,030.3, 1,040.6
25. $2(3)^{n-1}$ **27.** 2^{n-1} **29.** $4\left(-\frac{1}{2}\right)^{n-1}$ **31.** $8\left(\frac{1}{4}\right)^{n-1}$
33. $\frac{3}{256}$ **35.** $48\sqrt{2}$ **37.** $\frac{81}{64}$ **39.** $\frac{243}{32}$ or $-\frac{243}{32}$
41. 1,023 **43.** 772.478 **45.** 2.250 **47.** 5,727.500
49. $-14{,}762$ **51.** 16.000
53. $63\left(1 + \sqrt{2}\right) \approx 152.095$ **55.** 1,103.568
57. (a) $250{,}000(0.75)^t$ **59.** \$3,623,993.23
(b) \$59,326.17
61. \$19,496.56 **63.** \$105,428.44
65. (a) \$5,368,709.11 **67.** 70.875 in.^{2}
(b) \$10,737,418.23
69. 1.476×10^{20}, No

SECTION 11.4 *(page 702)*

Warm-Up *(page 702)*

1. $-3x^2 + 6x$ **2.** $x^2 - 9$ **3.** $x^2 - 4x + 4$
4. $4x^2 + 12x + 9$ **5.** $\dfrac{5x^3}{y^2}$ **6.** $-8x^6y^3$
7. 24 **8.** 8 **9.** 1,320 **10.** 45

1. 15 **3.** 252 **5.** 1 **7.** 1,225 **9.** 12,650
11. 455 **13.** 53,130
15. 1, 9, 36, 84, 126, 126, 84, 36, 9, 1
17. **(a)** 35 **(b)** 126 **19.** $x^4 + 4x^3y + 6x^2y^2 + 4xy^3 + y^4$
21. $x^5 + 5x^4 + 10x^3 + 10x^2 + 5x + 1$
23. $x^6 + 18x^5 + 135x^4 + 540x^3 + 1{,}215x^2 + 1{,}458x + 729$
25. $x^5 - 5x^4y + 10x^3y^2 - 10x^2y^3 + 5xy^4 - y^5$
27. $u^3 - 6u^2v + 12uv^2 - 8v^3$
29. $81a^4 + 216a^3b + 216a^2b^2 + 96ab^3 + 16b^4$
31. $x^8 + 8x^7y + 28x^6y^2 + 56x^5y^3 + 70x^4y^4 + 56x^3y^5 + 28x^2y^6 + 8xy^7 + y^8$
33. $x^6 - 12x^5 + 60x^4 - 160x^3 + 240x^2 - 192x + 64$
35. 120 **37.** $-1{,}365$ **39.** 1,760
41. $\frac{1}{32} + \frac{5}{32} + \frac{10}{32} + \frac{10}{32} + \frac{5}{32} + \frac{1}{32}$
43. $\frac{1}{256} + \frac{12}{256} + \frac{54}{256} + \frac{108}{256} + \frac{81}{256}$ **45.** 1.172
47. 510,568.785

SECTION 11.5 *(page 711)*

Warm-Up *(page 711)*

1. 10 **2.** 120 **3.** 84 **4.** 1 **5.** 576
6. 720 **7.** 870 **8.** 1,260 **9.** 4,845
10. 24

1. 10 **3.** 8 **5.** 6 **7.** 7 **9.** 6 **11.** 260
13. 6,760,000 **15.** 6 **17.** 48
19. *xyz, xzy, yxz, yzx, zxy, zyx* **21.** 120 **23.** 40,320
25. {*A*, *B*}, {*A*, *C*}, {*A*, *D*}, {*A*, *E*}, {*A*, *F*}, {*B*, *C*}, {*B*, *D*}, {*B*, *E*}, {*B*, *F*}, {*C*, *D*}, {*C*, *E*}, {*C*, *F*}, {*D*, *E*}, {*D*, *F*}, {*E*, *F*}
27. 1,140 **29.** 126 **31.** 56 **33.** 30

SECTION 11.6 *(page 719)*

Warm-Up *(page 719)*

1. $\frac{11}{12}$ **2.** $\frac{5}{8}$ **3.** $\frac{37}{30}$ **4.** $\frac{95}{144}$ **5.** $\frac{1}{24}$
6. $\frac{1}{120}$ **7.** $\frac{3}{14}$ **8.** $\frac{1}{6}$ **9.** $\frac{5}{8}$ **10.** $\frac{117}{125}$

1. {(*h*, 1), (*h*, 2), (*h*, 3), (*h*, 4), (*h*, 5), (*h*, 6), (*t*, 1), (*t*, 2), (*t*, 3), (*t*, 4), (*t*, 5), (*t*, 6)}
3. {*A*, *B*, *C*, *D*, *E*, . . . , *X*, *Y*, *Z*}
5. {*AB*, *AC*, *AD*, *AE*, *BC*, *BD*, *BE*, *CD*, *CE*, *DE*}
7. $\frac{3}{8}$ **9.** $\frac{7}{8}$ **11.** $\frac{1}{2}$ **13.** $\frac{3}{13}$ **15.** $\frac{1}{6}$ **17.** $\frac{5}{6}$
19. **(a)** $\frac{1}{5}$ **(b)** $\frac{1}{3}$ **(c)** 1 **21.** **(a)** $\frac{1}{20}$ **(b)** $\frac{2}{5}$ **(c)** $\frac{1}{2}$ **(d)** $\frac{1}{4}$ **23.** $\frac{14}{65}$ **25.** $\frac{1}{24}$
27. $\frac{1}{45}$ **29.** $\frac{1}{54{,}145}$ **31.** $1 - p$

CHAPTER 11 REVIEW EXERCISES *(page 722)*

1. $\sum_{k=1}^{4} (5k - 3)$ **3.** $\sum_{n=1}^{6} \frac{1}{3n}$ **5.** 380
7. $n(n - 1)(n - 2)$ **9.** 127, 122, 117, 112, 107
11. $\frac{5}{4}, 2, \frac{11}{4}, \frac{7}{2}, \frac{17}{4}$ **13.** 10, 30, 90, 270, 810
15. $100, -50, 25, -\frac{25}{2}, \frac{25}{4}$ **17.** $6 + 4n$
19. $1{,}050 - 50n$ **21.** $\left(-\frac{2}{3}\right)^{n-1}$ **23.** $24(2)^{n-1}$
25. $12\left(-\frac{1}{2}\right)^{n-1}$ **27.** 28 **29.** $\frac{4}{5}$ **31.** 486
33. $\frac{2{,}525}{2}$ **35.** 8,190 **37.** 2.5711
39. 115,019.3447
41. $x^{10} + 10x^9 + 45x^8 + 120x^7 + 210x^6 + 252x^5 + 210x^4 + 120x^3 + 45x^2 + 10x + 1$
43. $y^6 - 12y^5 + 60y^4 - 160y^3 + 240y^2 - 192y + 64$
45. $x^8 - 4x^7 + 7x^6 - 7x^5 + \frac{35}{8}x^4 - \frac{7}{4}x^3 + \frac{7}{16}x^2 - \frac{1}{16}x + \frac{1}{256}$ **47.** 5,100 **49.** 462
51. **(a)** $85{,}000(1.012)^t$ **(b)** 154,328
53. 8 **55.** 3,003 **57.** $\frac{1}{3}$ **59.** $\frac{1}{24}$

CHAPTER 11 TEST *(page 724)*

1. $1, -\frac{2}{3}, \frac{4}{9}, -\frac{8}{27}, \frac{16}{81}$ **2.** 35 **3.** $\sum_{k=1}^{12} \frac{2}{3k + 1}$
4. 12, 16, 20, 24, 28 **5.** $5100 - 100n$ **6.** 3,825
7. $-\frac{3}{2}$ **8.** $4\left(\frac{1}{2}\right)^{n-1}$ **9.** 765 **10.** \$47,868.33
11. 1,140 **12.** $x^5 - 10x^4 + 40x^3 - 80x^2 + 80x - 32$
13. 56 **14.** 26,000 **15.** 12,650 **16.** 0.25
17. $\frac{3}{26}$ **18.** $\frac{1}{6}$

CUMULATIVE TEST: CHAPTERS 8–11 *(page 725)*

1. *y* is a function of *x*.
2. All real values *x* such that $x > 2$ **3.** $c(c + 3)$
4. *g* is a horizontal translation 4 units to the right and a vertical translation 3 units downward.

5. **(a)** -5
(b) 51
(c) $\dfrac{\sqrt{3}}{5}$
(d) 3

6. $f^{-1}(x) = \frac{1}{3}(x + 2)$

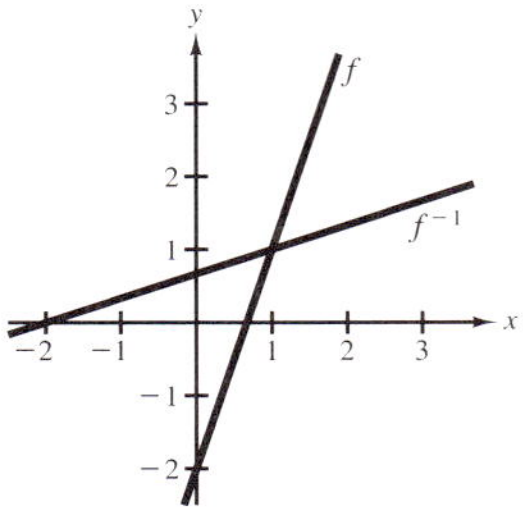

7. 128 ft **8.** (5, 4) **9.** (1, −2, 3)
10. 8, 12, 24
11.

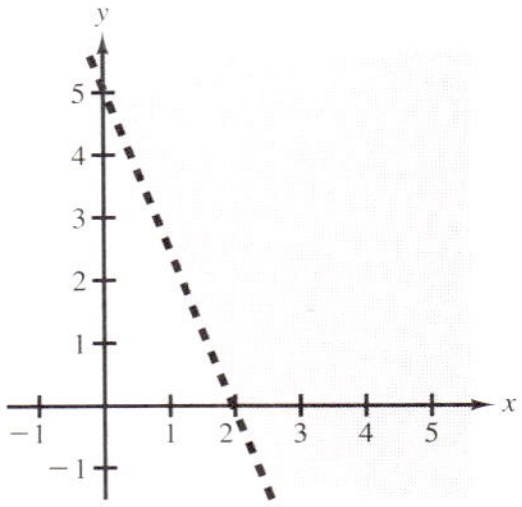

12. $g = f^{-1}$

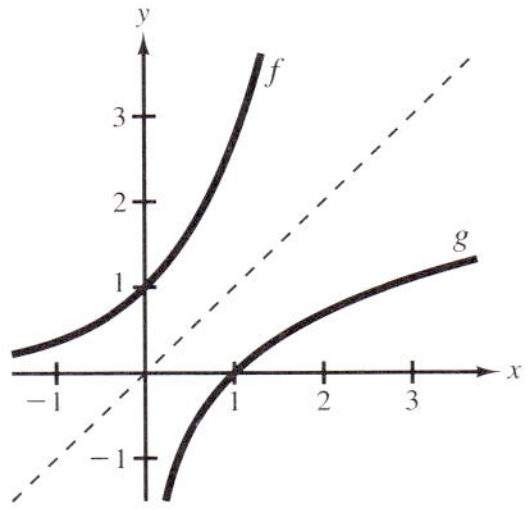

13. $\log_2 \dfrac{(xy)^3}{z}$ **14.** **(a)** 3
(b) 12.1825
(c) 18.0129
(d) 0.8673

15. 8.33% **16.** 15.4 yrs **17.** 800 **18.** $\frac{4}{81}$
19. −35 **20.** 48 **21.** 120 **22.** 0.54 **23.** $\frac{5}{18}$

Appendix Introduction to Logic

SECTION A.1 *(page A5)*

1. Statement **3.** Open statement
5. Open statement **7.** Open statement
9. Nonstatement **11.** Open statement
13. **(a)** True **(b)** False **15.** **(a)** True **(b)** True
17. **(a)** False **(b)** False **19.** **(a)** True **(b)** False
21. **(a)** The sun is not shining.
(b) It is not hot.
(c) The sun is shining and it is hot.
(d) The sun is shining or it is hot.
23. **(a)** Lions are not mammals.
(b) Lions are not carnivorous.
(c) Lions are mammals and lions are carnivorous.
(d) Lions are mammals or lions are carnivorous.
25. **(a)** The sun is not shining and it is hot.
(b) The sun is not shining or it is hot.
(c) The sun is shining and it is not hot.
(d) The sun is shining or it is not hot.
27. **(a)** Lions are not mammals and lions are carnivorous.
(b) Lions are not mammals or lions are carnivorous.
(c) Lions are mammals and lions are not carnivorous.
(d) Lions are mammals or lions are not carnivorous.
29. $p \wedge \sim q$ **31.** $\sim p \vee q$ **33.** $\sim p \vee \sim q$
35. $\sim p \wedge q$ **37.** The bus is blue.
39. x is not equal to 4. **41.** The Earth is flat.
43.

p	q	$\sim p$	$\sim p \wedge q$
T	T	F	F
T	F	F	F
F	T	T	T
F	F	T	F

45.

p	q	$\sim p$	$\sim q$	$\sim p \vee \sim q$
T	T	F	F	F
T	F	F	T	T
F	T	T	F	T
F	F	T	T	T

47.

p	q	$\sim q$	$p \vee \sim q$
T	T	F	T
T	F	T	T
F	T	F	F
F	F	T	T

49. Not logically equivalent **51.** Logically equivalent
53. Logically equivalent **55.** Not logically equivalent
57. Logically equivalent **59.** Not a tautology
61. A tautology
63.

Identical (columns $\sim(p \wedge q)$ and $\sim p \vee \sim q$)

p	q	$\sim p$	$\sim q$	$p \wedge q$	$\sim(p \wedge q)$	$\sim p \vee \sim q$
T	T	F	F	T	F	F
T	F	F	T	F	T	T
F	T	T	F	F	T	T
F	F	T	T	F	T	T

SECTION A.2 *(page A13)*

1. **(a)** If the engine is running, then the engine is wasting gasoline.
(b) If the engine is wasting gasoline, then the engine is running.
(c) If the engine is not wasting gasoline, then the engine is not running.
(d) If the engine is running, then the engine is not wasting gasoline.

3. **(a)** If the integer is even, then it is divisible by two.
(b) If it is divisible by two, then the integer is even.
(c) If it is not divisible by two, then the integer is not even.
(d) If the integer is even, then it is not divisible by two.

5. $q \rightarrow p$ **7.** $p \rightarrow q$ **9.** $p \rightarrow q$ **11.** True

13. True **15.** False **17.** True **19.** True

21. Converse: If you can see the eclipse, then the sky is clear.
Inverse: If the sky is not clear, then you cannot see the eclipse.
Contrapositive: If you cannot see the eclipse, then the sky is not clear.

23. Converse: If the deficit increases, then taxes were raised.
Inverse: If taxes are not raised, then the deficit will not increase.
Contrapositive: If the deficit does not increase, then taxes were not raised.

25. Converse: It is necessary to apply for the visa to have a birth certificate.
Inverse: It is not necessary to have a birth certificate to not apply for the visa.
Contrapositive: It is not necessary to apply for the visa to not have a birth certificate.

27. Paul is not a junior and not a senior.

29. The temperature will increase and the metal rod will not expand.

31. We will go to the ocean and the weather forecast is not good.

33. No students are in extracurricular activities.

35. Some contact sports are not dangerous.

37. Some children are allowed at the concert.

39. None of the $20 bills is counterfeit.

41.

p	q	$\sim q$	$p \rightarrow q$	$p \rightarrow \sim q$	$\sim(p \rightarrow \sim q)$
T	T	F	T	F	T
T	F	T	F	T	F
F	T	F	T	T	F
F	F	T	T	T	F

43.

p	q	$p \rightarrow q$	$q \rightarrow p$	$\sim(q \rightarrow p)$	$\sim(q \rightarrow p) \wedge q$
T	T	T	T	F	F
T	F	F	T	F	F
F	T	T	F	T	T
F	F	T	T	F	F

45.

p	q	$\sim p$	$p \vee q$	$(p \vee q) \wedge (\sim p)$	$[(p \vee q) \wedge (\sim p)] \rightarrow q$
T	T	F	T	F	T
T	F	F	T	F	T
F	T	T	T	T	T
F	F	T	F	F	T

47.

p	q	$\sim p$	$\sim q$	$p \to \sim q$	$\sim q \to p$	$p \leftrightarrow \sim q$	$(p \leftrightarrow \sim q) \to \sim p$
T	T	F	F	F	T	F	T
T	F	F	T	T	T	T	F
F	T	T	F	T	T	T	T
F	F	T	T	T	F	F	T

49.

p	q	$\sim p$	$\sim q$	$q \to p$	$\sim p \to \sim q$
T	T	F	F	T	T
T	F	F	T	T	T
F	T	T	F	F	F
F	F	T	T	T	T

Identical

51.

p	q	$\sim q$	$p \to q$	$\sim(p \to q)$	$p \wedge \sim q$
T	T	F	T	F	F
T	F	T	F	T	T
F	T	F	T	F	F
F	F	T	T	F	F

Identical

53.

p	q	$\sim q$	$p \to q$	$(p \to q) \vee \sim q$	$p \vee \sim p$
T	T	F	T	T	T
T	F	T	F	T	T
F	T	F	T	T	T
F	F	T	T	T	T

Identical

55.

p	q	$\sim p$	$\sim p \wedge q$	$p \to (\sim p \wedge q)$
T	T	F	F	F
T	F	F	F	F
F	T	T	T	T
F	F	T	F	T

Identical

57. (c) **59.** (a)

61.

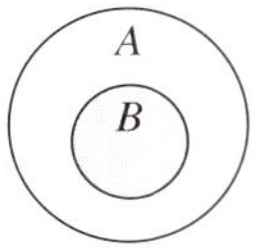

63.

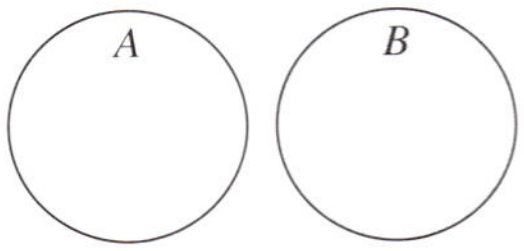

65.

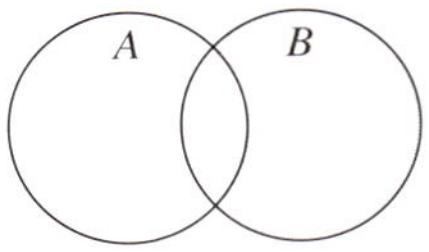

67.

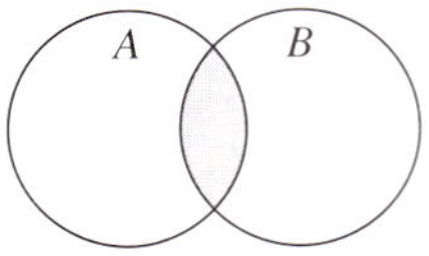

69.

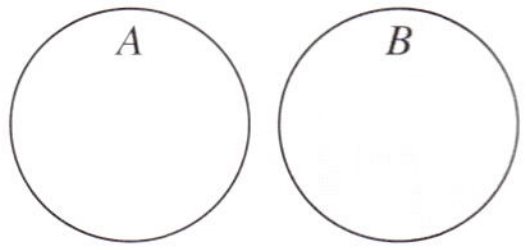

71. **(a)** Statement does not follow
(b) Statement follows

73. **(a)** Statement does not follow
(b) Statement does not follow

SECTION A.3 *(page A22)*

1.

p	q	$\sim p$	$\sim q$	$p \to \sim q$	$(p \to \sim q) \wedge q$	$[(p \to \sim q) \wedge q] \to \sim p$
T	T	F	F	F	F	T
T	F	F	T	T	F	T
F	T	T	F	T	T	T
F	F	T	T	T	F	T

3.

p	q	$\sim p$	$p \vee q$	$(p \vee q) \wedge \sim p$	$[(p \vee q) \wedge \sim p] \to q$
T	T	F	T	F	T
T	F	F	T	F	T
F	T	T	T	T	T
F	F	T	F	F	T

5.

p	q	$\sim p$	$\sim q$	$\sim p \to q$	$(\sim p \to q) \wedge p$	$[(\sim p \to q) \wedge p] \to \sim q$
T	T	F	F	T	T	F
T	F	F	T	T	T	T
F	T	T	F	T	F	T
F	F	T	T	F	F	T

7.

p	q	$p \vee q$	$(p \vee q) \wedge q$	$[(p \vee q) \wedge q] \to p$
T	T	T	T	T
T	F	T	F	T
F	T	T	T	F
F	F	F	F	T

9. Valid **11.** Invalid **13.** Valid **15.** Valid
17. Invalid **19.** Valid **21.** Invalid **23.** (b)
25. (c) **27.** (b) **29.** (c)
31. Valid

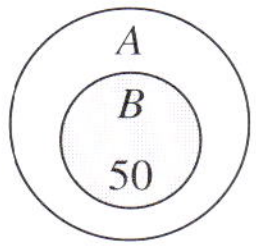

A = All numbers divisible by five
B = All numbers divisible by ten

33. Invalid

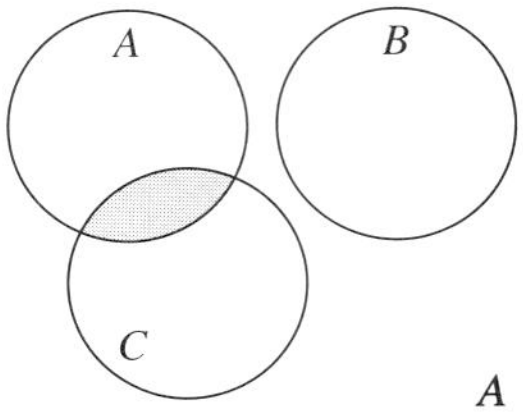

A = People eligible to vote
B = People under the age of 18
C = College students

35. Let p represent the statement "Sue drives to work," let q represent "She will stop at the grocery store," and let r represent "She'll buy milk."

First write:

Premise #1: $p \rightarrow q$
Premise #2: $q \rightarrow r$
Premise #3: p

Reorder the premises:

Premise #3: p
Premise #1: $p \rightarrow q$
Premise #2: $q \rightarrow r$
Conclusion: r

Then we can conclude r. That is, "Sue will get milk."

37. Let p represent "This is a good product," let q represent "We will buy it," and let r represent "The product was made by XYZ Corporation."

First write:

Premise #1: $p \rightarrow q$
Premise #2: $r \vee \sim q$
Premise #3: $\sim r$

Note that $p \rightarrow q \equiv \sim q \rightarrow \sim p$, and reorder the premises:

Premise #2: $r \vee \sim q$
Premise #3: $\sim r$
(Conclusion from Premise #2, Premise #3: $\sim q$)
Premise #1: $\sim q \rightarrow \sim p$
Conclusion: $\sim p$

Then we can conclude $\sim p$. That is, "It is not a good product."

INDEX OF APPLICATIONS

Biology and Life Science Applications

Business Applications

Chemistry and Physics Applications

Construction Problems

Consumer Applications

Geometric Applications

Interest Rate Applications

Miscellaneous Applications

Time and Distance Applications

U.S. Demographics Applications

INDEX

E

F

G

H

I

J

L

M

N

O

Graphing Linear Equations

A **linear equation in two variables** is an equation of first degree in both variables.

The **graph of an equation** is the set of all points in the rectangular coordinate system whose coordinates are solutions of the equation.

The **graph of a linear equation in two variables** is a straight line.

To find the ***x*-intercepts** (where the graph intersects the x-axis), let y be zero and solve the equation for x.

To find the ***y*-intercepts** (where the graph intersects the y-axis), let x be zero and solve the equation for y.

Slope, $m = \dfrac{y_2 - y_1}{x_2 - x_1} = \dfrac{\text{Change in } y}{\text{Change in } x}$.

Slope-intercept form of the equation of a line:
$y = mx + b$, m is the slope, and $(0, b)$ is the y-intercept.

Point-slope form of the equation of a line:
$y - y_1 = m(x - x_1)$, m is the slope, and (x_1, y_1) is a point on the line.

Vertical line
Undefined slope

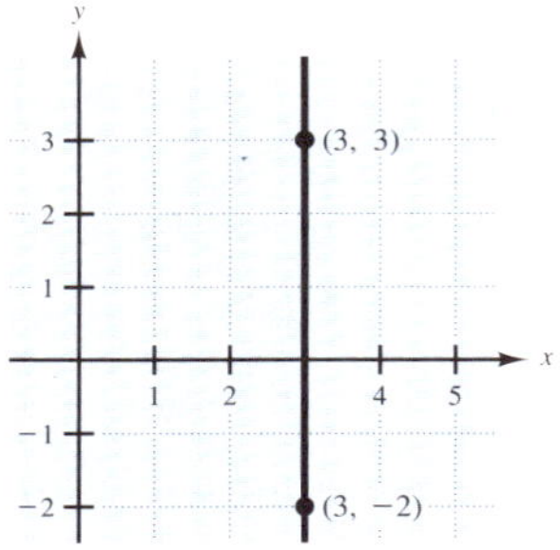

Line falls
Negative slope

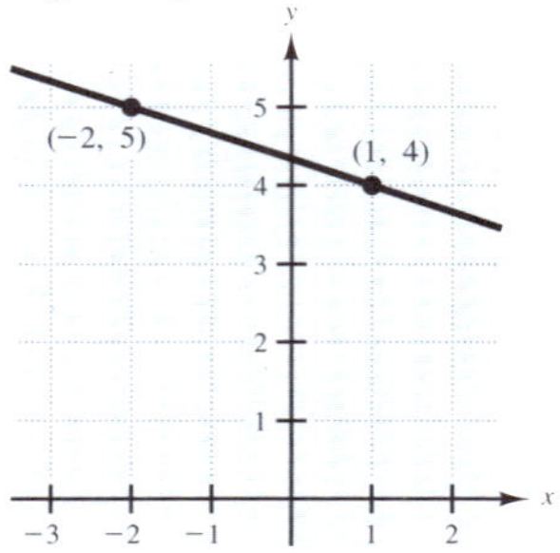

Horizontal line
Zero slope

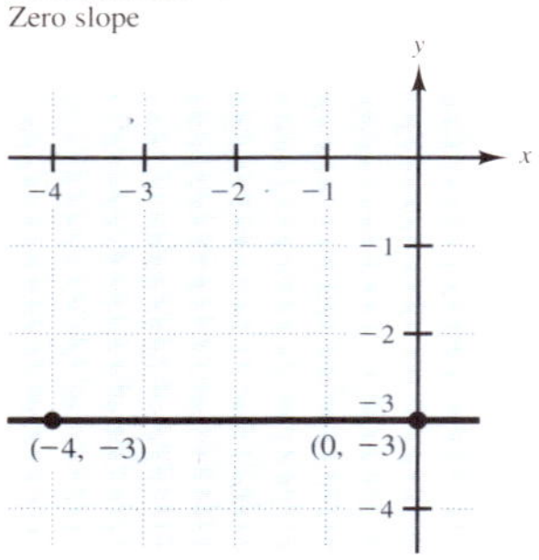

Line rises
Positive slope

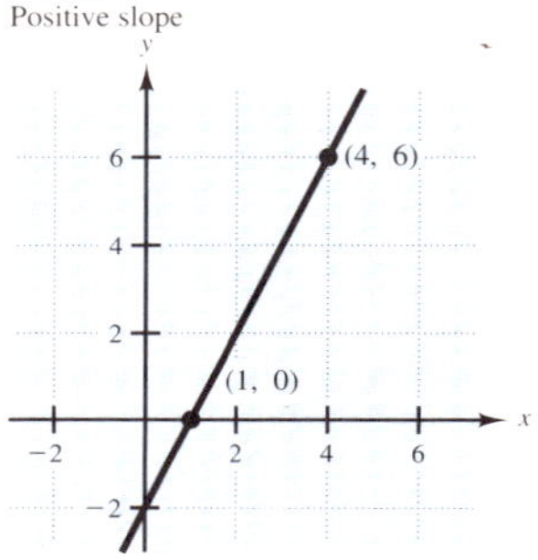

Graphs of Systems of Linear Equations

The **solution** of a system of linear equations is an ordered pair (a, b) that satisfies each of the equations. A system of linear equations may have no solution, exactly one solution, or infinitely many solutions.

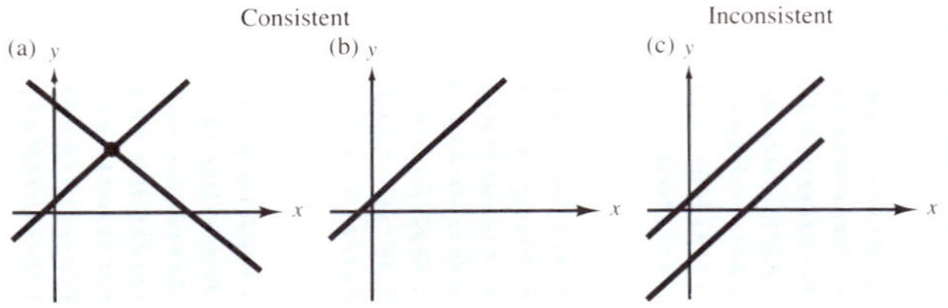

(a) Two lines that intersect at a single point
(b) Two lines that coincide with infinitely many points of intersection
(c) Two parallel lines with no point of intersection

Graphs of Quadratic Equations

Standard form of a quadratic equation:
$ax^2 + bx + c = 0$, a, b, and c are real numbers with $a \neq 0$. A quadratic equation can be solved by factoring, extracting square roots, completing the square, or using the Quadratic Formula.

Extracting square roots: If $u^2 = d$, where $d > 0$, then $u = \pm\sqrt{d}$.

Quadratic Formula: $x = \dfrac{-b \pm \sqrt{b^2 - 4ac}}{2a}$

Discriminant: $b^2 - 4ac$
If $b^2 - 4ac > 0$, then the equation has two real number solutions.
If $b^2 - 4ac = 0$, then the equation has one (repeated) real number solution.
If $b^2 - 4ac < 0$, then the equation has no real number solutions.

The graph of $y = ax^2 + bx + c$, $a \neq 0$, is called a **parabola** which opens up if $a > 0$ and opens down if $a < 0$.

Parabola Opens Up

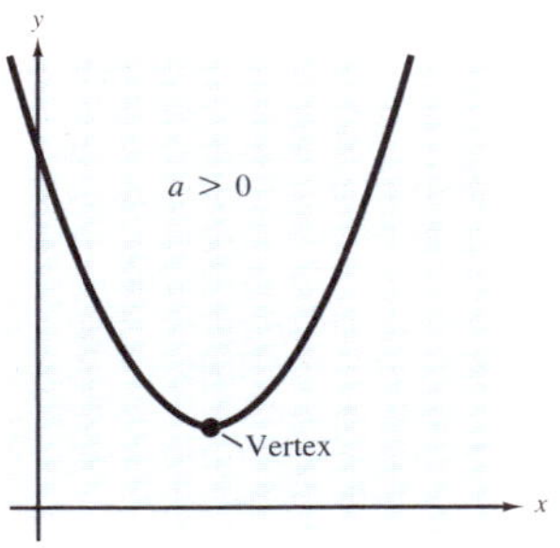

Parabola Opens Down

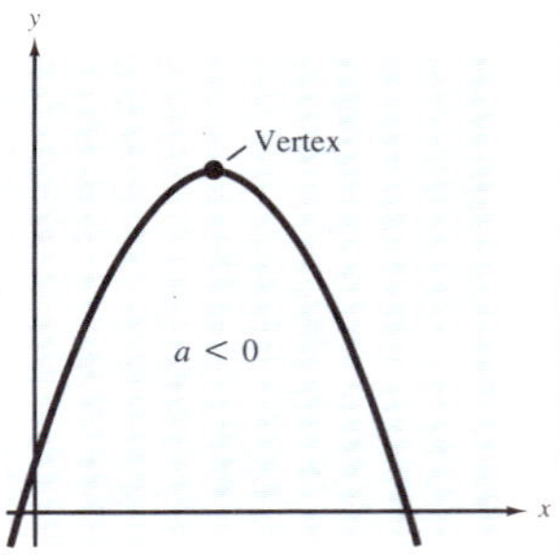